Springer Proceedings in Energy

The series Springer Proceedings in Energy covers a broad range of multidisciplinary subjects in those research fields closely related to present and future forms of energy as a resource for human societies. Typically based on material presented at conferences, workshops and similar scientific meetings, volumes published in this series will constitute comprehensive state-of-the-art references on energy-related science and technology studies. The subjects of these conferences will fall typically within these broad categories:

- Energy Efficiency
- Fossil Fuels
- Nuclear Energy
- Policy, Economics, Management & Transport
- Renewable and Green Energy
- Systems, Storage and Harvesting

Materials for Energy eBooks Volumes in the Springer Proceedings in Energy will be available online in the world's most extensive eBook collection, as part of the Springer Energy eBook Collection. Proposals for new volumes should include the following:

- name, place and date of the scientific event
- a link to the committees (local organization, international advisors etc.)
- description of the scientific aims and scope of the meeting
- list of invited/plenary speakers
- an estimate of the proceedings book details (number of pages/articles, requested number of bulk copies, submission deadline).

Please send your proposals to Dr. Maria Bellantone, Senior Publishing Editor, Springer (maria.bellantone@springer.com).

More information about this series at http://www.springer.com/series/13370

Dennis Johansson
Hans Bagge
Åsa Wahlström
Editors

Cold Climate HVAC 2018

Sustainable Buildings in Cold Climates

Cold Climate HVAC 2018
The 9th International Cold Climate Conference
Sustainable new and renovated buildings in cold climates
Kiruna – Sweden 12-15, March 2018

Springer

Editors
Dennis Johansson
Division of Building Services
Lund University
Lund, Sweden

Åsa Wahlström
Division of Building Services
Lund University
Lund, Sweden

Hans Bagge
Division of Building Physics
Lund University
Lund, Sweden

ISSN 2352-2534 ISSN 2352-2542 (electronic)
Springer Proceedings in Energy
ISBN 978-3-030-00661-7 ISBN 978-3-030-00662-4 (eBook)
https://doi.org/10.1007/978-3-030-00662-4

Library of Congress Control Number: 2018955160

This Springer imprint is published by the registered company Springer Nature Switzerland AG
The registered company address is: Gewerbestrasse 11, 6330 Cham, Switzerland

Preface

The 9th international Cold Climate HVAC 2018 conference was held in the city of Kiruna, Sweden, from March 12 to 15. The conference series was initiated by the Scandinavian Federation of Heating, Ventilation, and Sanitary Engineering Associations (SCANVAC) in 1994 and has been held every third year in a northern circumpolar community. This was the first time the conference was held in Sweden and was the most northern Cold Climate HVAC conference ever. The municipality of Kiruna is facing an extensive transformation due to an expanding iron ore mine which is undermining the existing city. The result of this is that one-third of the city needs to be relocated which leads to new real estate developments in an arctic climate. The conference theme was sustainable new and renovated buildings in cold climates. The conference was organized and run by the Department of Building Services, in cooperation with the Department of Building Physics, at Lund University.

Program Committee:

Dennis Johansson, Chair, Building Services, Lund University, Sweden
Hans Bagge, Co-Chair, Building Physics, Lund University Sweden
Åsa Wahlström, Senior Advisor, Building Services, Lund University, Sweden

Over 4 days, approximately 150 delegates from 19 countries from the northern part of Europe, Asia and North America met, made connections and listened to almost 100 presentations on scientific papers that examined northern sustainable buildings and their HVAC systems. The program included interesting and informative booths and exhibitions from nine companies and partners. Conference keynote speakers gave lectures on topics ranging from the transformation of Kiruna city and the structural and social issues in the movement of its buildings, the construction and background of Icehotel Jukkasjärvi and their new year-round Icehotel 365. They also presented the technical, social, and cultural challenges of the design of buildings and communities in remote northern areas. The program also included study visits looking at the local architecture of Kiruna, the new city hall under construction, the Icehotel Jukkasjärvi, and the iron ore mine and its impressive HVAC system. Pre- and post-conference tours were made to Esrange Space Centre and the Abisko National Park and mountain area.

The numerous connections made during the conference and tour between researchers from academy, practitioners, and developers from industry help in bringing improvements to the future sustainability of both new and renovated northern buildings.

A total of 112 full papers were submitted to the conference and 91 papers were accepted for publication. Each paper was categorized under one topic and in this book, the articles are presented by topic. The topics are:

- Energy and power efficiency and low energy buildings
- Renovation of buildings
- Efficient HVAC components
- Heat pump and geothermal systems
- District and city energy systems
- Buildings in operation
- Building simulation
- Transdisciplinary connections and social aspects
- Indoor environment and health
- Moisture safety and hygrothermal aspects
- Codes, regulations, standards, and policies
- Other aspects of buildings in cold climates

The papers were single-blind reviewed by the scientific committee and each paper had at least two reviewers. The scientific committee members, listed below, consisted of researchers and experts from different parts of the world who had expertise in one or more of the conference topics.

The Scientific Committee members:

Angela Sasic, Chalmers University of Technology, Sweden
Berit Time, SINTEF, Norway
Bert Phillips, UNIES Ltd., Canada
Bill Semple, University of Calgary, Canada
Björn Marteinsson, University of Iceland, Iceland
Boualem Ouazia, National Research Council Canada, Canada
Carey Simonson, University of Saskatchewan, Canada
Carl-Erik Hagentoft, Chalmers University of Technology, Sweden
D. Y. Zheldakov, Research Institute of building physics of RAACS, Moscow, Russia
Eckhard Groll, Purdue University, USA
Erich Binder, WorleyParsons Ltd., Canada
Guangyu Cao, Norwegian University of Science and Technology, Norway
Hans Martin Mathisen, Norwegian University of Science and Technology, Norway
Henrik Davidsson, Lund University, Sweden
Hiroshi Yoshino, Tohoku University, Japan
Hua Ge, Concordia University, Canada

Ivo Martinac, Royal Institute of Technology, Sweden
Jan Sundell, Tsinghua University, China
Jan-Olof Dalenbäck, Chalmers University of Technology, Sweden
Jarek Kurnitski, Tallin University of Technology, Estonia
Jeffrey D. Spitler, Oklahoma State University, USA
Jesper Arfvidsson, Lund University, Sweden
Juha Vinha, Tampere University of Technology, Finland
Jutta Schade, Luleå Technical University, Sweden
Kristina Mjörnell, RISE, Sweden
Kwang Woo Kim, Seoul National University, South Korea
Lars-Erik Harderup, Lund University, Sweden
Leif Gustafsson, Växjö Linnaeus University, Sweden
Lin Duanmu, Dalian University of Technology, China
Linda Hägerhed, University of Borås, Sweden
Marianna Brodach, Moscow Architectural Institute, Russia
Martin Kotol, Technical University of Denmark, Greenland
Mika Vuolle, EQUA, Sweden
Olli Seppanen, FINVAC, Finland
Paula Femenias, Chalmers University of Technology, Sweden
Pawel Wargocki, Technical University of Denmark, Denmark
Petra Vladykova Bednarova, Swegon, Sweden
Robert Cerný, Czech Technical University in Prague, Czech Republic
Saqib Javed, Lund University, Sweden
Shuichi Hokoi, Kyoto University, Japan
Stefan Olander, Lund University, Sweden
Sven Werner, Halmstad University, Sweden
Svend Svendsen, Technical University of Denmark, Denmark
Targo Kalames, Tallinn University of Technology, Estonia
Thomas Lützkendorf, Karlsruhe Institute of Technology, Germany
Thomas Olofsson, Luleå Technical University, Sweden
Thoms Olofsson, Umeå University, Sweden
Toke Rammer Nielsen, Technical University of Denmark, Denmark
Ulla Janson, Skanska AB, Sweden
V. G. Gagarin, Research Institute of building physics of RAACS, Russia
Vojislav Novakovic, Norwegian University of Science and Technology, Norway
Xudong Yang, Tsinghua University, China
Åke Blomsterberg, Lund University, Sweden

Lund, Sweden
May 2018

Dennis Johansson
Hans Bagge
Åsa Wahlström

Contents

Part I
Energy and Power Efficiency and Low Energy Buildings

Passive House Construction Above the Arctic Circle

Stefan Dehlin, Catrin Heincke and Peter Koskinen

Abstract In Tuolluvaara, just outside Kiruna in northern Sweden, NCC has worked with Kiruna Municipality and Tekniska Verken i Kiruna AB to build the passive house project "Sjunde Huset", a full-scale demonstration of a low-energy semi-detached dwelling for a sub-arctic climate. Based on NCC's Cube concept for energy-efficient houses, the semis are built to FEBY12's passive house criteria. The building serves as a test bed for the design, material choices, technical solutions and construction processes associated with energy-efficient construction in a sub-arctic climate. Features include a system for mechanical ventilation with heat recovery (MVHR). The building project has been prompted by the major social changes taking place in Kiruna and Gällivare and the newly tightened EU directives on energy consumption. This article presents the building and its energy-efficient solutions, along with measurements of the buildings heat loss factor (HLF), the ventilation system's efficiency, and the specific energy consumption. We can report that Sjunde Huset meets FEBY12's passive house requirements and that it is perfectly possible to build low-energy homes in a sub-arctic climate, with the potential to reproduce it at a lower production cost.

Keywords Energy-efficiency · Near-zero energy · Passive house
Sub-arctic

1 Background

The EU directive aimed at increasing the proportion of near-zero energy buildings (Directive 2010/31/EU of the European Parliament and of the Council of 19 May 2010 on the energy performance of buildings) [1] states that such buildings must

S. Dehlin (✉)
NCC Infrastructure, Vallgatan 3, 170 80 Solna, Sweden
e-mail: stefan.dehlin@ncc.se

C. Heincke · P. Koskinen
NCC Division Building Sweden, Vallgatan 3, 170 80 Solna, Sweden

D. Johansson et al. (eds.), *Cold Climate HVAC 2018*,
Springer Proceedings in Energy, https://doi.org/10.1007/978-3-030-00662-4_1

have an extremely high level of energy performance. Another requirement is that much of the small amount of energy that is supplied to these buildings should be from renewable sources, including renewable energy on-site or from the local vicinity.

Finding economically viable total solutions for whole areas and urban districts will be an important factor in fulfilling increasingly tough energy performance criteria. At the same time it is important not to ignore other aspects of sustainability such as making the areas attractive to live in and spend time in.

The major social changes that are taking place in Kiruna and Gällivare and the newly tightened EU directives on energy consumption provide a unique opportunity to get things right from the beginning and create a new society that is sustainable and attractive. Though, the conditions for building in a sub-arctic climate are challenging.

This article presents a full-scale demonstration of a low-energy house, "Sjunde Huset" (Seventh House), for a sub-arctic climate, one that serves as a test bed for the design, material choices, technical solutions and construction processes associated with energy-efficient construction in a sub-arctic climate. It showcases the building and its energy-efficient solutions, along with the results of measurements to ascertain the buildings heat loss factor, ventilation system's efficiency and specific energy consumption. A more detailed description of the building Sjunde Huset, its technology, metrics, experiences from the production phase and feedback from the tenants can be found in the report Dehlin et al. [2], published by the Swedish Energy Agency.

2 Sjunde Huset

2.1 Sjunde Huset as a Passive House

Sjunde Huset (Seventh House), as shown in Fig. 1, is a passive house that took NCC's "Cube concept" for cost and energy-efficient houses as the basis for its planning and construction. The Swedish definition for passive houses was formulated by Forum for energy efficient buildings (FEBY) based on the German definition [3]. However adjusted to generally used standards in Sweden slightly influence energy calculation results [4]. NCC puts the development and production costs at around 70% more than for a regular Cube house. One goal was for the building to be prefabricated as elements that could be assembled on-site, since conventional construction methods are not applicable. BASTA-approved materials have been used where possible.

The building is located in Tuolluvaara, which will be part of the new Kiruna, in northern Sweden. The climate is classed as sub-arctic, with long, cold winters and short, mild summers. By November 2014, Sjunde Huset was completed and ready to be lived in.

Fig. 1 Sjunde Huset

2.2 *Energy Efficiency Measures*

2.2.1 Building Envelope

To create an energy-efficient house, heat losses through the building envelope have to be minimized, which in practice means a building design that is compact and a building envelope that is well-insulated and airtight, with few thermal bridges. A highly efficient ventilation system with a high level of heat recovery is also required to minimize heat losses and keep the need for heating very low.

The timber-framed building comprises two semi-detached apartments, each with a floor space of 140 m^2. Apartments 1 (left) and 2 (right) have the same basic features: First floor containing the kitchen, living room that can be sectioned off to form an extra bedroom, laundry room, and toilet and shower. Second floor containing one large and two smaller bedrooms, toilet and shower, a family room that could be sectioned off to create another bedroom, and a walk-in closet.

The design of the house and its location on the plot has been adapted to the sub-arctic climate in order to reduce heat losses through the building envelope. The building has a compact shape that avoids angles in the facade to minimize the potential for thermal bridges. To reduce the ability for cold air to enter the apartments, porches have been built onto the entrances, forming airlocks between inside and out. This space is not heated and has walls with a "low standard of insulation". The foundation slab is insulated from the building so that it is not a thermal bridge, and it is designed to minimize the risk of frost heave. There is also a storeroom and a carport outside the "warm shell". The balcony is a freestanding structure to avoid thermal bridges.

The windows and doors have low thermal transmittance (U-value), 0.65 and 0.7 W/m^2 K. The outer walls have additional rigid polyiso foam (PIR) insulation, which is an efficient, high-performance (0.023 W/m^2 K) insulation material. The roof is insulated with 1000 mm loose-fill wool insulation. The foundation slab is insulated with 400 mm graphite foam. The average thermal transmittance for the building envelope, U_m (insulation value), is less than 0.16 W/m^2 K, whereas the Swedish building standards require a value of 0.40 W/m^2 K.

Ensuring that the insulated structure is airtight is important for various reasons. Air permeability reduces the infiltration of cold air, which causes drafts and increases the need for heating, and the exfiltration of air out of the building. It is important to minimize exfiltration in order to avoid the damp problems that can occur when warm, moist indoor air cools down inside the wall and forms condensation.

2.2.2 Green Roof and Stormwater Management

The building has a gently sloping roof (<6°) covered with moss/sedum/grass. There are numerous benefits to this type of roof: the sedum roof has an insulating effect (the soil layer is thermally dense), absorbs air pollution and reduces stormwater run-off. Sjunde Huset is testing two different types of sedum roof (sedum and sedum-herb) to see which kind is most effective in managing stormwater—which roof is best at absorbing rain, meltwater and water pollution—test carried out by LTU, see e.g. [5]. Alongside the sedum roof, which reduces the risk of large flows, there are also swales and soakaways in the ground. Stormwater is thus managed locally within the plot.

2.2.3 Solar Panels

Each apartment is equipped with 15 m^2 solar panels. In climates with low sun and large amounts of snow, solar panels provide valuable additional energy even during the winter. To optimize their use in the winter, the solar panels have been placed on the facades rather than on the roof. In this position, reflected light from the snow can also drive up the production of electricity. During the summer, with its long period of midnight sun, electricity production will be substantial, of around 500 kWh/year from the solar panels, which have an estimated total production of around 900 kWh/year.

2.2.4 District Heating

The homes are heated via a local heating system that in turn is heated by the municipal district heating system. The heating system is fueled by burnable waste from households and industry and leftover heat from the LKAB mining factory.

The heat is distributed within the homes by an airborne heating system. The district heating is fed directly into the ventilation unit's "heating" battery and into the washing machine, tumble dryer and dishwasher. The district heating network is laid together with the water and sewage pipes. The district heating was also used to thaw the ground before construction work began and to heat the site offices. This brought down electricity use in the production phase.

2.2.5 Ventilation System

The homes each have their own ventilation system (UNI3 from Flexit), with heat recovery and they are placed in the walk-in closet on the upper floor. The unit has a rotating heat exchanger and a heating battery that is connected to the district heating network. The air is distributed to the bedrooms and family room through valves placed at ceiling height. The extract air is placed in the kitchen, bathroom and laundry. In addition to the task of supplying the homes with new clean air, the ventilation system also has to meet the heating needs in the building. Since the heating is provided through the ventilation system, there are no radiators in the rooms.

The airtight building envelope ensures the optimal operation of the ventilation system, which recovers the heat from the extract air, reduces heat leakage and removes cold drafts. The MVHR unit has a high-performance heat exchanger—83% at a design winter outdoor temperature (DVUT) of −30 °C according to the supplier's specification. District heating supplies the heating battery, as well as the washing machine, tumble dryer and dishwasher.

The ventilation unit has three operational flows: high and low speed plus a booster setting that only runs for 30 min. The default for the ventilation unit is high speed, which has been designed and calibrated as 55 l/s [0.4 l/s, m^2 A_{temp} (A_{temp}—interior area that is heated)]. A site visit in November 2016 revealed that the right-hand apartment 2 had the flow set at low speed, probably ever since it was handed over. Low speed equates to a flow of around 40 l/s. The supply air temperature is controlled by the desired temperature in the room, which is measured in the extract air.

The ventilation unit has a built-in defrosting function. This function needs to be activated manually, but this rarely happens, since problems of frosting are rare in units with rotating heat exchangers. The risk is greater in Sjunde Huset, due to the sub-arctic climate. However, the defrosting function appears not to have been activated for any of the building's units.

2.2.6 White Goods and Lighting

Sjunde Huset has been fitted with low-energy white goods and lighting in order to further reduce the demand for energy. The white goods are connected to the district heating, which is expected to cut electricity demand by 80%. Since Kiruna has a long period of darkness with polar nights, the homes have been equipped with various types of facade and outdoor lighting. All use LEDs, which is expected to bring a saving of up to 80% for the same light strength, compared with regular filament bulbs.

2.2.7 Co-location of District Heating with Water and Sewage Pipes

The residential area in which Sjunde Huset is located has special infrastructure for water, wastewater and district heating. A low-temperature district heating network has been laid together with the water and sewage pipes, which are warmed by the heat given off. Placing all the pipes together makes it possible to bury them at a depth of just 70 cm. Normally in Kiruna, they would have to be dug down 3 m to remain frost-free. Since Kiruna is built on bedrock, this dramatically reduces the cost of running the pipes. Further studies into co-location of pipes are set to be conducted and reported via the cross-disciplinary project Attract [5].

2.2.8 Showers

Instead of a conventional shower system, Sjunde Huset is fitted with energy-smart water-recycling showers from Orbital. These showers use technology developed by NASA to clean and re-use the water: Water that runs into the drain is collected, filtered and pumped back up to the shower head, allowing the same water to be used again and again. The integral water treatment unit filters out particles, bacteria and viruses. The water is only flushed away once the shower is over. With this system, a ten-minute shower uses only five liters of water. The water loses about two degrees in temperature from the shower head to the drain, so it needs to be reheated by an electric heater that is part of the shower system. Electric heaters in showers meet the energy requirement, but may require special dispensation within the Swedish building regulations in order not to affect the passive house certification.

2.2.9 Display

The apartments in the Sjunde Huset building are equipped with displays in the hall plus temperature sensors in all the rooms. This makes it easy for residents to keep an eye on their use of hot water and energy. The hall also has a central control switch that turns off all the lights in the house in one go.

3 Measurements and Analysis

The aim of building Sjunde Huset has been to show that despite its sub-arctic location, it is possible to build a low-energy house that still meets the Swedish standards for passive houses. Sjunde Huset is built to FEBY12's passive house criteria for the conditions of Climate Zone I [1]. The criteria include requirements concerning specific energy consumption and the buildings heat loss factor. Since the standard for specific energy consumption is high, the ventilation needs to perform extremely well.

The heat loss factor is a measure of the heat that leaks out from the building when it is coldest outdoors, and the criteria for compliance depend on where in the country the building is located and what type of heating system exists. For Sjunde Huset, the heat loss figure must be no higher than 19.8 W/m^2 A_{temp}. According to FEBY12, the building's specific energy consumption $Es_{upplied}$ must not exceed 63 kWh/m^2 A_{temp}/year, since the building is classed as not electrically heated. Air leakage q50 through the building envelope must be no more than 0.30 l/s per m^2 of enclosed area at a pressure difference of 50 Pa.

3.1 Measurements Conditions and Deviations

The measurement values for temperatures and air humidity in the ventilation units and in various rooms used for the technical analysis and later also for the production of the heat loss factor figures are average values on a 15-min basis. This is because the minute by minute data produced files that were too large to handle.

The system efficiency has been calculated on an annual basis (January 1, 2016 to February 20, 2017), but also for the period October 2016 to February 2017, which is part of the heating season. This is to see how performance differs between the heating season and on an annual basis, expecting the efficiency to be slightly higher.

There is a meter to measure the heat supplied to the heating battery in the unit, which has been monitored and compared with the calculated energy consumption in the battery. Sensors from the manufacturer Regin have been installed in the ventilation unit to log temperatures at five points. The five sensors measure the temperature of the outdoor air, supply air (before the heating battery), supply air (after the heating battery), extract air and in the exhaust air. In addition to temperatures, the sensors also measure relative humidity at these points. Key limitations and deviations in the input data and conditions that affect the measurements:

- Collection of measurement data during the time periods January 1 to April 30, 2016 and August 20, 2016 to February 20, 2017 due to router problems: problems transferring measurement data to the database and overwritten measurement data.
- Purchase of incorrect sensors for the ventilation units' exhaust air that can only measure temperatures in the range 0–50 °C. All minus temperatures have been recorded as 0 °C. When calculating system efficiency, the temperature of the supply air has been used instead.
- The substation that Tekniska Verken, the district heating supplier in Kiruna, owns has long had a factory-controlled night reduction that was not known. This means that at night the heating battery for the ventilation has not been able to

deliver the necessary quantity of heating, which in turn has led to a substantial drop in the indoor temperature. The night reduction function was removed November 22, 2016.
- The tenants have not had the apartments as their permanent residence, which means that they have been used considerably less than was planned and expected.

3.2 Calculation Conditions Heating Battery Energy Consumption

Energy consumption in the battery has been calculated using measured temperatures and produced with the help of the following formula:

$$Q_{Battery} = \sum_{i=1}^{i=n} \rho \cdot C_p \cdot q\left(T_{supply\,air,i} - T_{hr,i}\right) \tag{1}$$

$Q_{Battery}$	Battery energy consumption [kWh]
ρ	Air density, 1.2 kg/m^3
C_p	Specific thermal capacity, 1 kJ/kg K
q	Air flow [m^3/h]
$T_{supply\ air,I}$	Supply air at each 15-min value
$T_{hr,i}$	Temperature after heat recovery at each 15-min value.

3.3 Calculation Conditions for the Heat Loss Factor

The following formula has been used to calculate the heat loss factor (HLF) using measurements from the building:

$$\mathrm{HLF_{WDT}} = \mathrm{H_T} \cdot (21 - \mathrm{WDT})/\mathrm{A_{temp}} \tag{2}$$

H_T is the building's heat loss coefficient [W/K] and WDT is the design winter outdoor temperature for the location in question. To determine the heat loss factor, WDT is chosen for a time constant of max 12 days, which equates to −24.3 °C. If you choose a time constant of just one day, the WDT instead becomes −30.3 °C. The thermal loss coefficient has been calculated for a time constant of both 1 day and 12 days to see how the two differ. H_T is produced from the formula:

$$\mathrm{H_T} = \mathrm{U_m} \cdot \mathrm{A_{encl}} + \rho \cdot \mathrm{C_p} \cdot \mathrm{q_{leak}} + \rho \cdot \mathrm{C_p} \cdot \mathrm{d} \cdot \mathrm{q_{vent}} \cdot (1 - \eta) \tag{3}$$

U_m	The building envelope's average U-value
A_{encl}	The building envelope's enclosed area, measured internally
$\rho \cdot C_p \cdot q_{leak}$	Thermal energy losses due to air leakage q_{leak} [m^3/s], air density ρ [kg/m^3], and thermal capacity C_p [kJ/kg K]
$\rho \cdot C_p \cdot q_{vent}\,(1-\eta) \cdot d$	Thermal energy losses due to ventilation with regard to the system's efficiency, η, density, ρ, thermal capacity, C_p, relative operating time, *d*.

The calculation of air leakage, q_{leak}, for an MVHR system takes account of the building's location and balancing of the ventilation, in line with EN ISO 13789:2008:

$$q_{leak} = q_{50} \cdot A_{encl} \cdot e \Big/ \left(1 + f/e\left(\left(q_{supl} - q_{extr}\right)/\, q_{50} \cdot A_{encl}\right)^2\right) \tag{4}$$

$q_{supl} - q_{extr}$	Air surplus between supply air and extract air [l/s]
q50	Leak flow at 50 Pa pressure difference between inside and out [m^3/s]
e and f	Tabulated wind protection coefficients.

Since the flow of the ventilation is balanced, $q_{supl} - q_{extr}$ becomes 0, such that the simplified equation becomes:

$$q_{leak} = q_{50} \cdot A_{encl} \cdot e \tag{5}$$

3.4 Calculation Conditions Specific Energy Consumption

The requirement concerning specific energy consumption, $E_{supplied}$, includes purchased energy for heating, hot water and the building's installations (fans and pumps).

$$E_{supplied} = E_{heating} + E_{domestic\ electricity} + E_{hot\ water} \tag{6}$$

The "NASA" showers with their recycled hot water contain a small electric battery, which technically means this is a "non-pure system that is heated with both electricity and district heating". However, since the electric batteries are "extremely limited" in scope, the building is instead classed as having a "pure heating system", with a FEBY12 requirement of 63 kWh/m^2 A_{temp}/year that must not be exceeded.

3.5 Measurement Results

Below are the results from the calculations made using the above conditions. For a more detailed description and more graphs, see report Dehlin et al. [2].

3.5.1 Ventilation System Efficiency

The average system efficiency on an annual basis comes out at 81% for the unit in apartment 1 and 79% for the unit in apartment 2. Relating the system efficiency to the outdoor temperature reveals that the heat exchanger is most stable when it is cold outside. When the weather is warmer, the results show a broader spread: System efficiency approaches 100% when the temperature outside is the same as the desired temperature inside. The efficiency approaches 0% when the outdoor air temperature is higher than the supply air temperature, which gives a low efficiency figure in the calculations, sometimes even negative.

Figure 2 shows the graph for system efficiency in both units during the "heating period" from October 1, 2016 until February 20, 2017 (plotted as daily averages for clarity). The figure for average efficiency during this period is 81% for the unit in apartment 1 and 80% for apartment 2.

At the start of the period, the system efficiency for the two units differs somewhat. The efficiency of the unit in apartment 1 is higher than the efficiency of the unit in apartment 2. One explanation may be that the intake duct for apartment 1 faces south, so the effect of the sun is greater here than for apartment 2 (north-facing).

3.5.2 Battery Energy Consumption

See Table 1: In apartment 1, the battery used 4482 kWh between October 1, 2016 and February 20, 2017 and in apartment 2 the corresponding figure is 4016 kWh. These figures are taken from the heat supply meter and relate to the period October 1, 2016 and February 20, 2017, not the whole heating season. Since the apartments are identical, the measured energy demand should be the same, but in fact they differ by almost 500 kWh. A logical explanation for the lower energy use in apartment 2 is that the air flow was lower for a time; see section "Measurements conditions and deviations".

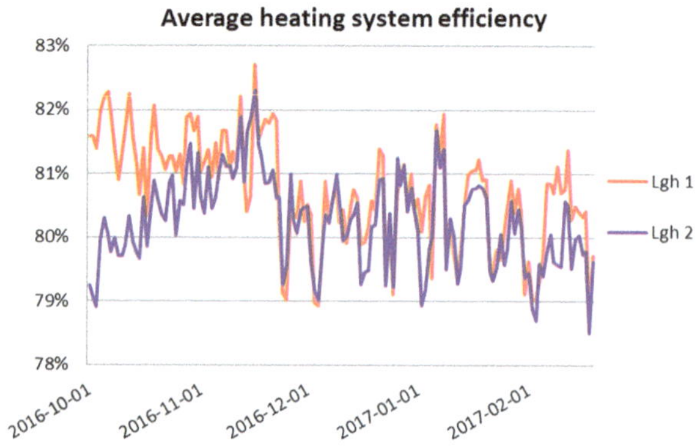

Fig. 2 Average heating system efficiency per day during the heating season

Table 1 Measured and estimated energy in heat exchanger and battery

	Apartment 1 (kWh)	Apartment 2 (kWh)
Measured energy battery	4482	4016
Estimated energy battery	3909	4026
Estimated recovered energy exchanger	5030	4454
Estimated energy demand without exchanger	8939	8480

In apartment 1, the heat exchanger recovered 56% of the energy during the stated period, while in apartment 2 the heat exchanger recovered 53% during same period.

3.5.3 Heat Loss Factor

Table 2 presents the results of the heat loss factor for both WDTmin (−24.3 °C) and WDTmax (−30.3 °C). As clearly shown in the table, the heat loss factor is well below the requirement set by FEBY12 of 19.8 W/m^2 A_{temp}. The heat loss factor, calculated at an early stage and for a time constant of 12 days, equating to WDTmax, also works out at 16.3 W/m^2, which means that there is a good match between the measured value and the estimated value.

3.5.4 Specific Energy Consumption

Measured values for calculating specific energy consumption for the period July 15, 2015 to July 15, 2016 were produced by Daniel Risberg at LTU. However, these values are not used in their pure form. They are adapted to deviations relating to low use, low indoor temperature and high outdoor temperature (compared with a normal year).

Estimated specific energy consumption, $E_{supplied}$, is 56.7 kWh/m^2/year, which is well within FEBY12's passive house requirement of 63 kWh/m^2 A_{temp}/year.

3.5.5 Air Permeability

Air permeability tests were performed in September 2014. The left and right apartments were tested separately. The test was carried out in line with standard EN-13829 *Thermal performance of buildings*.

Table 2 Heat loss figures for Sjunde Huset

	HLF_{WDTmin}	HLF_{WDTmax}
Apartment 1	14.20	16.10
Apartment 2	14.30	16.20
Whole building	14.20	16.10

The calculated air permeability of the total enclosed area ranged between 0.23 and 0.27 l/s m^2, and the project's agreed air permeability requirement was 0.31 l/s m^2. Both the apartments meet the air permeability requirement of 0.31 l/s m^2 within a 10% margin of error, irrespective of whether the party wall separating the apartments is included in the area calculation.

4 Conclusions

The full-scale demonstration building, constructed to FEBY12's passive house criteria for the conditions of Climate Zone I, has definitely helped to increase our knowledge of low-energy houses and innovative new technical solutions.

This article presents the building and its energy-efficient solutions, along with the results of measurements of the ventilation system's efficiency, the heat loss factor and specific energy consumption. Measurements have been affected by certain issues, e.g. sporadic occupancy, purchase of the wrong sensors and various operational problems. We can still report that Sjunde Huset meets FEBY12's passive house requirements and that it is perfectly possible to build low-energy homes in a sub-arctic climate.

We can also draw a number of important and well-founded conclusions about the heat recovery system, based on measurements and feedback from the tenants: The heat recovery system requires particularly intensive upkeep, with regular calibration of air flows, maintenance—such as duct cleaning and filter replacement—and switching between summer and winter settings in order to deliver good comfort and a pleasant indoor environment. The heat recovery ventilation system (supply/extract air units in the ceiling, extract air units in the wetrooms and kitchen) leads to a certain degree of uneven temperature distribution in the house—colder on the lower floor than on the upper floor—along with cold floors and a dry indoor climate.

The starting point for the development of Sjunde Huset was NCC's Cube concept house, which was upgraded to a passive house. Having created a passive house concept for Kiruna's conditions, there is now the potential to reproduce it at a lower production cost. Future houses will also cost less to live in due to their low energy consumption. A functional and sustainable passive house concept should now be an attractive alternative for newbuild housing in Kiruna.

References

1. Sweden's Center for Near-Zero Energy Houses Homepage, http://www.nollhus.se/feby-12. Last accessed 18 Aug 2017 (in Swedish)
2. S. Dehlin, C. Heincke, J. Redman Hedensjö, P. Koskinen, Sjunde huset – ett lågenergiparhus i Kiruna. Projekt – Energimyndigheten (2017) (in Swedish)

3. Passive House Institute, Dr Wolfgang Feist. (2007). Passive House Planning Package 2007
4. A. Zalejska-Jonsson, Low-energy residential buildings evaluation from investor and tenant perspectives. Licentiate Thesis, KTH, Sweden (2011)
5. Attract Homepage, https://www.ltu.se/proj/Attract-Hallbar-samhallsplanering. Last accessed 14 Aug 2017 (in Swedish)

In Situ Thermal Resistance Testing of an Energy Efficient Building Envelope in the Canadian Arctic

Carsen Banister, Michael Swinton, Travis Moore and Dennis Krys

Abstract A structural insulated panel (SIP) demonstration house was commissioned in Iqaluit, Nunavut, Canada, along with instrumentation to assess thermal performance. Instrumentation panels were developed to perform in situ measurements. Laboratory testing was completed to confirm the accuracy of the instrumentation panels and measure the thermal resistance of the insulation panels at various mean temperatures. In situ testing confirmed the variability in R-value with temperature seen in laboratory testing. The thermal resistance decreases with decreasing temperature, which is of concern in this particularly cold climate. Regardless, the envelope thermal performance was consistently superior to a 2″ × 6″ (38 mm 89 mm) wood stud construction with batt insulation. The in situ R-value of the SIP varied from approximately RSI 4.4 (R25) to RSI 5.3 (R30) between October, 2016 and April, 2017. The cumulative apparent R-value was also calculated according to the method defined in ISO 9869, resulting in values of RSI 5.07 (R28.8), 4.84 (R27.5), and 4.97 (R28.2) for the wall, floor, and ceiling modules, respectively.

Keywords Structural insulated panel · Thermal resistance · Cold climate

1 Introduction

The far north of Canada presents tough challenges in terms of housing and building construction. Remoteness causes increased cost of supplies and labour [8]. The extremely cold climate causes increased heating costs and durability issues leading to health problems [3]. This intent of this project is to investigate the performance of a structural insulated panel (SIP) system that can be manufactured in remote communities. The housing technology studied in this work seeks to relieve several of the adverse conditions experienced in Canada's remote and northern communities.

C. Banister (✉) · M. Swinton · T. Moore · D. Krys
National Research Council Canada, Ottawa, ON K1K 4R7, Canada
e-mail: carsen.banister@nrc-cnrc.gc.ca

D. Johansson et al. (eds.), *Cold Climate HVAC 2018*,
Springer Proceedings in Energy, https://doi.org/10.1007/978-3-030-00662-4_2

A demonstration house was commissioned in Iqaluit, Nunavut, Canada by Qikiqtaaluk Corporation, the Inuit birthright corporation of the region. Qikiqtaaluk Corporation is interested in housing solutions that are durable, rapidly constructed, energy efficient, and can be manufactured in the territory. Iqaluit is a city of 7740 which lies on Baffin Island in the Canadian Arctic at 64°N 59°W. The average annual Celsius heating degree days for Iqaluit over the 24 years preceding the year 2017 was 9693 [1], approximately double the heating degree days of Ottawa, Ontario, Canada, which is located at 45°N 76°W.

The purpose of the house was to demonstrate the technique of constructing with structural insulated panel systems in a cold climate such as Nunavut and to assess its energy performance in the Iqaluit climate.

The interior dimensions of the demonstration house are approximately 4.5 m by 5.4 m and the building is partitioned into three spaces: a bedroom, a washroom, and a utility space. The construction consists of a 15 cm thick panel system which is highly insulating, in comparison to buildings further south, at an R-value of RSI 4.53 (R25.7) at −20 °C exterior temperature, +20 °C interior temperature. The key to this being a high performance building is that it balances the aspects of low air leakage, modularity, prefabrication, and minimal thermal bridging to give high overall performance of the integrated system.

2 Methods and Materials

A custom instrumentation module was developed, built, laboratory tested, and installed in the field to measure the thermal resistance of the building envelope in situ at the Iqaluit site. The instrumentation module consists of a 1 m squared calibrated expanded polystyrene (XPS) sheet, a heat flux transducer (HFT), and several redundant thermocouples on either side of the instrumentation module. R-value can be calculated via two different techniques: one using the HFT and the temperature difference across the building envelope (Eq. 1) and a second using three temperature readings across the envelope and module along with the known R-value of the instrumentation module (Eq. 2).

$$R_{wall} = \frac{\Delta T_{\text{wall}}}{q} \tag{1}$$

$$R_{wall} = R_{\text{reference}} \cdot \frac{\Delta T_{\text{wall}}}{\Delta T_{\text{reference}}} \tag{2}$$

where R_{wall} is the thermal resistance of the wall assembly, ΔT_{wall} is the temperature difference across the wall assembly, q is the heat flux across both the wall assembly and the instrumentation module, $R_{\text{reference}}$ is the thermal resistance of the instrumentation module, and $\Delta T_{\text{reference}}$ is the temperature difference across the instrumentation module.

The R-values of the four instrumentation modules built were measured according to ASTM C518 in March 2016, enabling calibration of the measurement technique. The four individual RSI values measured were 3.26, 3.25, 3.32, and 3.32 $K \cdot m^2/W$ for the 1 inch (25 mm) nominal specimens. Since conditions are not steady-state for the in situ measurements, heat flows and driving forces must be summed up over a sufficiently long period, as per Eq. 3 later in this paper.

The evaluation consisted of first testing a sample wall panel's thermal resistance in the laboratory at various mean temperatures using the ASTM C1363 Guarded Hot Box (GHB) test method and the ASTM C518 Heat Flow Meter test method. A full scale square 2.44 m (8′ × 8′) wall sample was received from the manufacturer for testing in the GHB and is shown in Fig. 1, along with heat flow instrumentation installed. A close-up photo of one of the instrumentation modules installed on the wall sample tested in the GHB is shown in Fig. 2. A 0.61 m × 0.61 m specimen was used for the Heat Flow Meter test method. Both specimens consisted of a polyisocyanurate core encased in a fiberglass shell.

After the laboratory testing, the instrumentation modules were installed on-site in Iqaluit on the demonstration house in April 2016. The completed demonstration house in Iqaluit is shown in Fig. 3, along with one of the installed instrumentation modules shown in Fig. 4. The house is supported on the permafrost by a multipoint foundation frame system. The oil tank at the side the right of the house is not used; heating is provided by wall-mounted electric infrared heating panels. The door shown is the only point of entry/exit and there are a total of three windows installed in the house, all of which are quad pane with an additional fifth, unsealed protective pane on the outboard side.

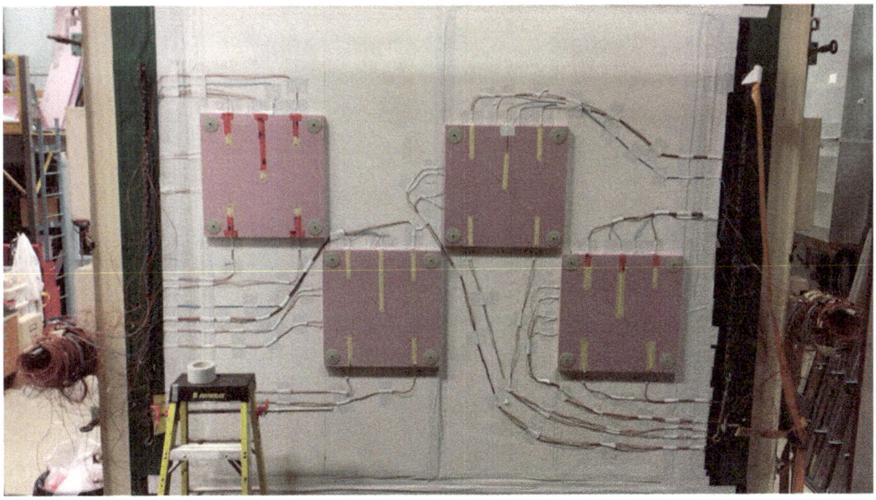

Fig. 1 Four instrumentation modules mounted on the 6″ wall sample

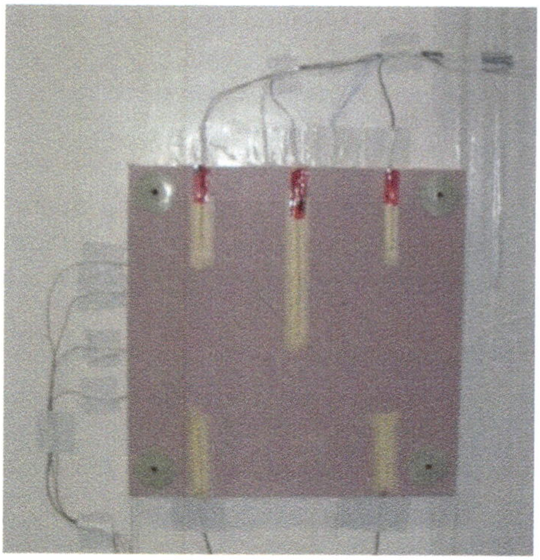

Fig. 2 An individual instrumentation module mounted on the wall sample tested in the GHB

Fig. 3 Qikiqtaaluk demonstration house in Iqaluit

The airtightness of the house was measured, since uncontrolled ventilation is a major source of energy usage for the building stock in Canada. In addition, excessive air leakage, whether due to lack of an air barrier system or defects, can cause durability and/or comfort issues.

Fig. 4 Floor instrumentation module installed on site

3 Results

3.1 Airtightness Testing

Two air leakage tests were conducted on the house on June 1, 2016, which found the airflow rate at 50 Pa of pressure difference to be an average of 425 L/min (15 CFM). The individual test results were very consistent, reading 396 and 453 L/min (14 and 16 CFM) at 50 Pa. The reported air leakage rates of 0.27 and 0.32 air changes per hour (ACH) at 50 Pa is notably lower than the maximum prescribed by the R2000 program of 1.5 ACH @ 50 Pa [4], meaning that the demonstration house is very airtight. It is also lower than the target of less than 0.6 ACH50 set by the Passive House Institute and compares very favourably to a conscientiously constructed wood stud frame test house of similar size built in Grenoble, France, which had a measured 0.67 ACH50 and had no windows [6]. This highlights the very high airtightness of the demonstration house built and tested in this work.

Ventilation. Although a highly airtight building is desirable because it limits uncontrolled ventilation, the shelter would require a ventilation system under regular occupancy. The recommended ventilation rate based on ASHRAE 62.2 was reported as 354 L/min (12.5 CFM) [5]. It should be noted that 50 Pa of pressure difference will not steadily occur for this building, hence the need for a ventilation system if regularly occupied. Further, since occupancy could at times be higher than assumed by ASHRAE 62.2, demand-controlled ventilation based on carbon dioxide and relative humidity may be a suitable strategy. This would also have the added benefit of reducing or pausing ventilation during times of low or no occupancy, which could have pronounced energy savings due to the extremely cold climate. This is an area of future work for this housing technology.

3.2 Laboratory Testing of Thermal Performance

The sample installed and tested in the GHB allowed comparison between the R-value measurements of the GHB and the instrumentation panels measuring in situ R-value. Since the instrumentation panels each have both a heat flux transducer and thermocouples across the instrumentation panel, the R-value can be calculated via two different methods for each panel (denoted as HFT and TC, respectively). The results of these comparisons between each calculation method and the GHB result are shown in Fig. 5.

The comparison between panels, calculation methods, and the GHB result are all very consistent. Maximum deviation between all results is R1.3, or about 5%, which is similar in variation to the GHB uncertainty of 6%.

Effect of Mean Temperature on Thermal Resistance. In addition to the above comparisons, the performance of the panelized wall was measured for two cold side temperatures in the GHB and for five different average temperatures in the heat flux meter apparatus in accordance with ASTM C518.

This enabled analysis of the impact of mean temperature on R-value on the building panel and these results are presented in Fig. 6. An approximate correlation was derived based on the lab results and the expression is reported in the figure. The relationship was used to predict the R-value of the field results in Iqaluit based on exterior temperature. This provided a ready means to check the expected field results, based on the more reliable laboratory test data.

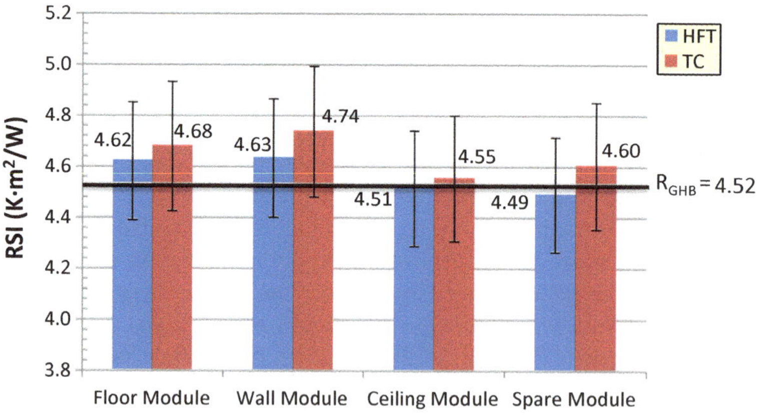

Fig. 5 Results of the instrument panels compared to Guarded Hot Box test

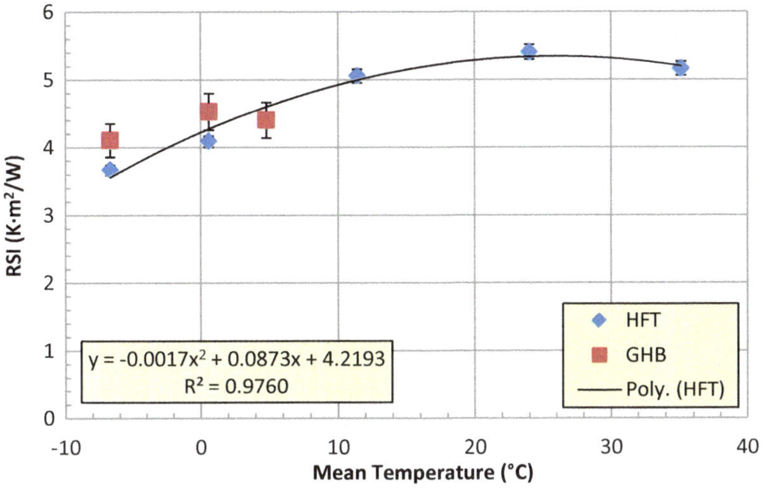

Fig. 6 Correlation of the lab test results of R-value and cold side temperature

The results from both ASTM C1363 and ASTM C518 consistently show decreasing thermal resistance with decreasing mean temperature below a value of 25 °C, a phenomenon that is known to occur in certain polyisocyanurate foams due to condensation of the blowing agents at decreasing mean temperatures. The thermal resistance (RSI) of the building panels evaluated ranged from 5.41 m^2 K/W at a mean temperature of ~24 °C, to 4.10 m^2 K/W at a mean temperature of ~−7 °C.

Comparison to Wood Stud Construction. The results of the two GHB tests are shown in Fig. 7 in comparison to two previous laboratory tests of 2″ × 6″ wood stud wall assemblies with batt insulation undertaken at the National Research Council Canada. This comparison shows the overall higher performance of the SIP wall, regardless of the decrease of its R-value with decreasing exterior temperature.

The comparison readily shows that the SIP construction performs notably better than the wood stud construction, by over 30% at −20 °C exterior temperature and by about 18% at −35 °C exterior temperature. This suggests an overall annual thermal transmittance of the SIP being about 20–25% less than the wood stud construction. Not included here are the energy savings resulting from reduced uncontrolled air leakage in the case of the SIP compared to wood stud construction, which are expected to be significant due to the extremely cold climate.

3.3 *In Situ Testing of Thermal Performance*

Heat Fluxes. The measured heat fluxes through the floor, wall, and ceiling are shown in Fig. 8. It should be noted that the heat fluxes of wall and roof are negative at points in time during the summer period, indicating a direction reversal of heat

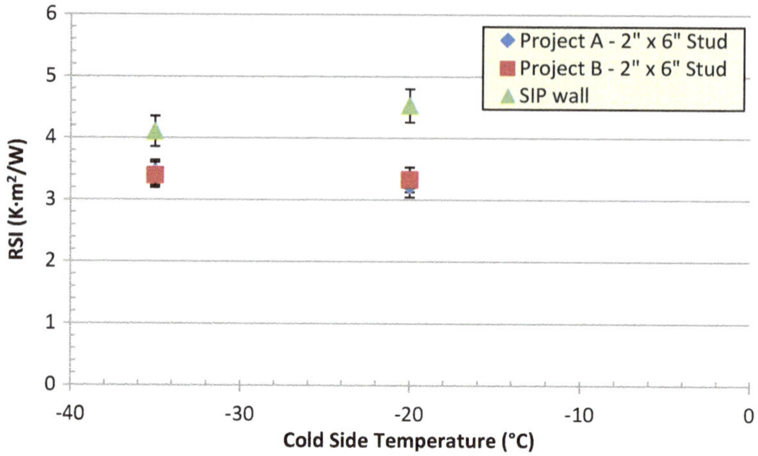

Fig. 7 R-value comparison of SIP wall and previously measured 2″ × 6″ wood stud walls

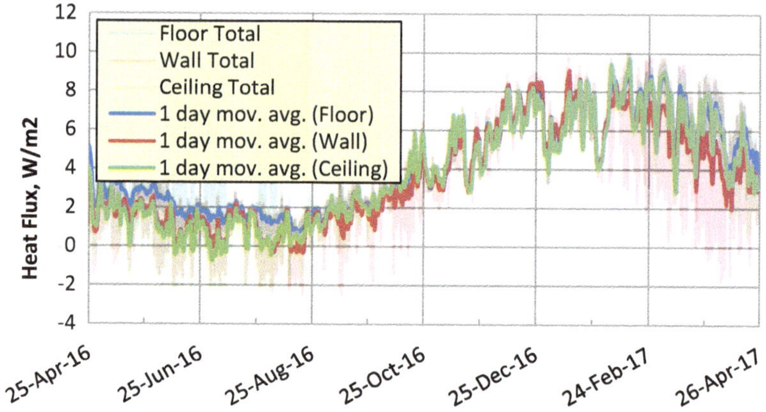

Fig. 8 Monitored heat flux through the floor, wall, and ceiling

flow. This is due to solar gains on the walls and roof that raised the respective exterior surface temperatures above interior surface temperatures, resulting in heat gain by the building. The heat flux of the floor stays positive throughout summer, meaning that the floor continually loses heat throughout the year.

Total Heat Loss. The heat fluxes of each component were summed over the entire monitoring period with the resulting total heat loss per square meter shown for each component in Table 1. Assuming a cost of electricity of 60¢ per kWh (Canadian dollars) [7], the unit heating cost of these components ranges from a low of $22.14/m^2 per year for the southwest wall to $25.86/m^2 per year for the floor.

Table 1 Total heat flux through the floor, wall and ceiling for April 25, 2016 to April 24, 2017

	Floor	Wall	Ceiling
Total heat loss per surface area since start (kWh/m^2)	43.1 ± 2.2	36.9 ± 1.8	38.7 ± 1.9
Cost of heat per surface area @ \$0.60/kWh (\$/m^2)	\$25.86 ± 1.32	\$22.14 ± 1.08	\$23.22 ± 1.14
Heat loss compared to floor	–	86%	90%
Savings relative to floor	–	14%	10%

The total heat loss of the wall and roof can be compared to the heat loss through the floor as shown in the two last rows of Table 1. Although there are small differences in the apparent R-value of these components, as discussed in the next section, it is presumed at this point that the savings in heat loss for the wall and the ceiling compared to the floor are mostly due to solar gains on the exterior component surfaces, since the exposed floor receives no direct solar gains (only reflected radiation).

Cumulative R-Value. The in situ measurements enabled the monitoring of the heat flux across the building envelope at multiple locations. Measurement of the temperatures across the envelope enables the calculation of cumulative thermal resistance, $R_{cumulative}$. This is performed according to Eq. 3, in accordance with [2].

$$R_{culumative} = \frac{\sum (T_{in} - T_{out})}{\sum Q_{out}} \quad (3)$$

where T_{in} is the interior surface temperature, T_{out} is the exterior surface temperature, and Q_{out} is heat flux positive in the outward direction. Since the measurements are in situ and the conditions are ever-changing, the calculation must be performed using summations over a suitably long period of time in order to dampen out and eliminate the transient heat transfer effects.

The cumulative R-value approaches a steady value over time, since it is a ratio of the summation of temperature difference driving force divided by the summation of heat transfer. The final values for cumulative apparent R-value at the end of the full year of monitoring for the wall, floor, and ceiling are presented in Table 2.

Table 2 Cumulative apparent R-values of SIP components over full year of monitoring

Component	Cumulative apparent RSI ($K \cdot m^2/W$) from April 25, 2016 to April 24, 2017	Cumulative apparent R-value ($hour \cdot ft^2 \cdot F/BTU$) from April 25, 2016 to April 24, 2017
Wall	5.07 ± 0.26	28.8 ± 1.5
Floor	4.84 ± 0.25	27.5 ± 1.4
Ceiling	4.97 ± 0.25	28.2 ± 1.4
Average	4.93 ± 0.25	28.0 ± 1.4

In the above table, it can be seen that the 3 measured components had very close cumulative apparent R-values. The floor is the most reliable reading, since it is minimally affected by solar gains, which prevent accurate calculations during some hours of the year for the wall and ceiling. The maximum difference between each of the component values are within the uncertainty of the method based on the accuracy of the heat flux meters and thermocouples.

The average cumulative apparent RSI value of 5.07 (R28.0) for all components for an annual average floor outdoor surface temperature of −6.5 °C for the monitoring period in Iqaluit is in agreement with the Guarded Hot Box lab result for the full scale specimen of RSI 4.51 (R25.6) at −20 °C (see Fig. 6).

4 Summary and Conclusions

A demonstration house constructed out of structural insulated panels was commissioned in Iqaluit by Qikiqtaaluk Corporation. Instrumentation panels to measure in situ thermal performance were developed and installed by the National Research Council Canada and a full year of monitoring was completed. The insulation panels were laboratory tested, along with the instrumentation panels, prior to deployment in the field. The laboratory tests demonstrated that the instrumentation panels functioned accurately and consistently. In addition, the typical behavior of polyisocyanurate foams' decreasing thermal resistance with decreasing temperature was demonstrated.

Monitoring has been conducted for over a year at the time of writing, with many temperatures and heat fluxes being measured throughout the house. Cumulative apparent R-values were consistent between the wall, floor and ceiling, with an average of about R28 between all the components. These field measurements are also consistent with the performance expected based on lab test results. It was seen that the R-value performance drops to a minimum of approximately R24 as the outdoor temperature drops to −30 °C or below; however, the performance is restored as temperature regains. Ongoing monitoring is planned in order to evaluate long-term performance.

References

1. Environment and Climate Change Canada, Canada Weather Stats. Retrieved August 25, 2017, from https://www.weatherstats.ca/ (2017)
2. International Standards Organization, ISO 9869-1:2014, Thermal insulation—building elements—in-situ measurements of thermal resistance and thermal transmittance (2014)
3. K. Minich, H. Saudny, C. Lennie, M. Wood, L. Williamson-Bathory, Z. Cao, G.M. Egeland, Inuit housing and homelessness: results from the International Polar Year Inuit Health Survey 2007–2008. Int. J. Circumpolar Health **70**(5), 520–531 (2011). https://doi.org/10.3402/ijch.v70i5.17858

4. Natural Resources Canada, 2012 R-2000 Standard (2012)
5. Nunavut Housing Corporation, *Building Air Leakage Test—QC Panel House* (2016)
6. A. Piot, M. Woloszyn, J. Brau, C. Abele, Experimental wooden frame house for the validation of whole building heat and moisture transfer numerical models. Energy Build. **43**(6), 1322–1328 (2011). https://doi.org/10.1016/J.ENBUILD.2011.01.008
7. Qulliq Energy Corporation, Customer types. Retrieved August 24, 2017 from http://www.qec.nu.ca/customer-care/accounts-and-billing/customer-types (2017)
8. S. Roszler, *Building Skills: A Construction Trades Training Facility for the Eastern Canadian Arctic*. Massachusetts Institute of Technology (2005)

Energy Consumption of an Energy Efficient Building Envelope in the Canadian Arctic

Carsen Banister, Michael Swinton, Travis Moore, Dennis Krys and Iain Macdonald

Abstract A demonstration house was built and commissioned in Iqaluit, Nunavut, Canada. The purpose of this work is to evaluate the energy consumption of the high performance building, while considering the unique social, economic, and logistical challenges for such a remote location. At 4.5 m by 5.4 m internal dimensions, the building has approximately 24 m^2 of floor area and is a 15 cm thick structural insulated panel (SIP) system with an R-value of 24 (RSI 4.23) at a panel mean temperature of 0 °C. A full year of monitoring has been conducted thus far, between April, 2016 and April, 2017. The cold climate required heating during all but a few hours of the year, with the outdoor ambient temperature ranging from −39 °C to +21 °C and a total of 9540 °C heating degree days for calendar year 2016. Daily heating energy consumption ranged from a peak of 30.4 kWh in the winter down to a minimum of 0.9 kWh for a small number of days during the summer when outdoor ambient temperatures neared 20 °C. The total heating electricity consumed for the period of April 25, 2016 to April 25, 2017, including electronics and lighting, was 4945 kWh. Based on the floor area, the building had an energy use intensity of 206 kWh per m^2.

Keywords Structural insulated panel · Energy performance · Cold climate

1 Introduction

A research house (shown in Fig. 1) was built and commissioned by Qikiqtaaluk Corporation and instrumented by National Research Council Canada (NRC) in Iqaluit, Nunavut, Canada, a city of 7740 people which lies on Baffin Island in the Canadian Arctic at 64°N 59°W. Qikiqtaaluk Corporation is interested in housing solutions that are durable, rapidly constructed, energy efficient, and can be manufactured in the territory. The house is constructed of an innovative structural

C. Banister (✉) · M. Swinton · T. Moore · D. Krys · I. Macdonald
National Research Council Canada, Ottawa, ON K1K 4R7, Canada
e-mail: carsen.banister@gmail.com

D. Johansson et al. (eds.), *Cold Climate HVAC 2018*,
Springer Proceedings in Energy, https://doi.org/10.1007/978-3-030-00662-4_3

insulated panel (SIP) design which comprises the floor, walls, and roof/ceiling of the building. The primary purpose of this investigation was to assess the energy performance of the building envelope over the course of at least a full elapsed year. Monitoring began on April 25, 2016 and is ongoing at the time of writing. A counterpart publication [1] describes the in situ R-value measurements of the building envelope over the course of the year and may provide additional background information for the work presented here.

Iqaluit is situated in a harsh northern climate, with approximately double the heating degree days of Ottawa, Ontario, Canada, which is located at 45°N 76°W. At the north end of Baffin Island is Pond Inlet at 73°N 78°W with a staggering average 12,000 HDD [2], about 30% greater than Iqaluit. This is of relevance because the energy performance in other cold climate locations is of interest.

There is little to no information on building heating energy usage in Nunavut, either in literature or in governmental reports. In fact, the Survey of Household Energy Use by Natural Resources Canada includes very detailed statistical data for all regions in Canada *except* all three Canadian territories (Yukon, Northwest Territories, and Nunavut) [3]. Thus, the results cannot readily be compared to like results in locations with similar conditions. In addition, the lack of existing information on thermal performance of buildings in the far north indicates a need for continued study and dissemination of information.

This house is taken to be energy efficient, as its air leakage is minimal at about 0.30 air changes per hour (ACH), it has minimal thermal bridging, and it can be prefabricated in northern remote communities.

Fig. 1 Qikiqtaaluk demonstration house in Iqaluit

2 Methods and Materials

2.1 Instrumentation

Heat Transfer Instrumentation. Instrumentation for measuring the heat transfer across the building envelope was installed on the wall, floor, and ceiling of the house. The primary technique used included designing and building custom instrumentation panels which consisted of a calibrated 1 in. specimen of EPS insulation, several thermocouples, and a heat flux transducer. The approach enables calculation of building envelope R-value by two methods. Details of the instrumentation for heat transfer, along with results and conclusions, are presented in the counterpart paper [1].

Heat Source Instrumentation. The demonstration building is fitted with three infrared (IR) heating panels, each mounted to a wall of one of the three rooms in the interior space. Fuel oil heating is common practice in the locale; however, it was not used for this project for at least two reasons. First, a gravity fed heater supplied by an outdoor tank cannot be metered accurately. Second, a main goal of this work is to develop systems and methods to offset the use of fossil fuels in remote and/or Arctic communities. The electricity consumption of the components within the house was measured using power meters seen in Fig. 2.

2.2 Modelling of House Heating Consumption

A model was created in ESP-r to define the test building in Iqaluit as shown in Fig. 3. The building was modelled as an "Energy Efficient Building" with an assumed constant air leakage of 0.1 air changes per hour.

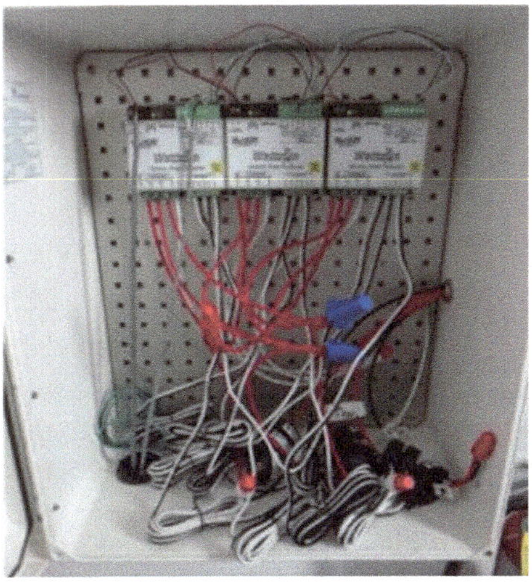

Fig. 2 Installed power meters to measure electricity consumption

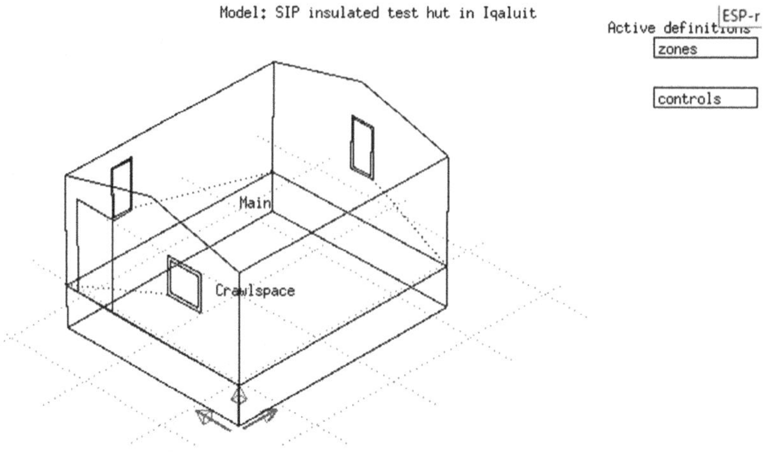

Fig. 3 ESP-r building energy model screenshot

Model Description. The model was defined in the information provided in Tables 1, 2, 3 and 4: window dimensions by location (Table 1); building dimensions (Table 2); building construction materials characterization (Table 3), and glazing configuration (Table 4).

Weather Data. The overall weather data used for the modelling process were based on data recorded by Environment and Climate Change Canada from the Iqaluit Airport weather station [4]. Solar radiation data were not directly available for this location, as pyranometers had not been installed. Sky cover information was available, which was used in conjunction with models by Risk Sciences International, Inc. to estimate what the solar radiation component values might have been. In some periods of time on the order of a few hours, it was necessary to perform linear interpolation to provide missing data. This was seen to be an

Table 1 Window dimensions by location

Dimension type	Orientation	West wall	South wall	North wall
Height from floor (mm)		1113	1029	895
Rough opening	Width	870	559	540
	Height	870	1178	1162
Glazing area	Width	702	464	464
	Height	702	1076	1073

Table 2 Building dimensions

Component	Orientation	Size (m)
Roof	–	0.6
Main zone	Width	4.5
	Length	5.4
	Height	2.5

Table 3 Building constructions

Component	Material	Thickness (mm)	R (BTU/h °F ft^2)	RSI (m^2 K/W)
Wall	Insulation	157	27.4	4.8
	Siding	13		
Floor	Insulation	158	27.4	4.8
Roof	Insulation	158	27.4	4.8

Table 4 Triple glazing window construction

Layer	Thickness (mm)
Clear glass	6
Argon fill	12

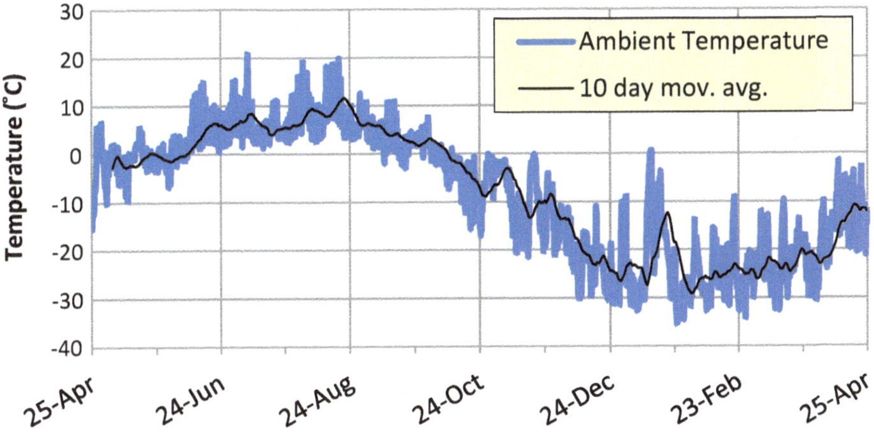

Fig. 4 Iqaluit ambient dry bulb temperature weather data used for modelling

appropriate approach due to the short length of time for each occurrence of missing data. The ambient dry bulb temperature data used for modelling are shown in Fig. 4.

3 Results

3.1 House Electrical Consumption

The total daily electricity consumption of the Qikiqtaaluk Demonstration House is plotted in Fig. 5. The total daily consumption includes the electricity consumption for the three IR panel heaters (one in each room), lights and power outlets. The energy consumed and monitored from the outside lights has been removed from

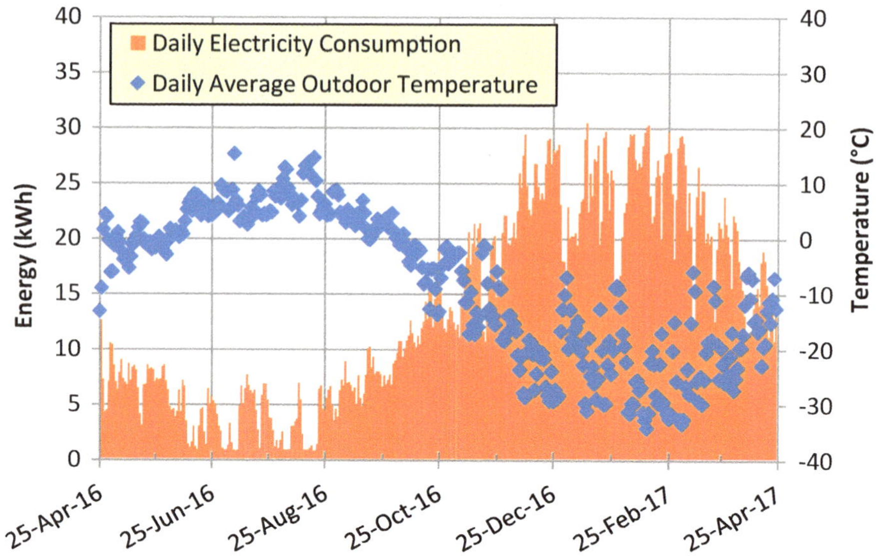

Fig. 5 Daily electrical consumption and corresponding daily average outdoor temperature

this data. Thus the values in Fig. 5 represent the energy consumed inside the house since the start of monitoring on April 25, 2016 until April 24, 2017. This energy usage represents the amount of heat required to maintain the indoor air temperature. Any energy consumed by lights or equipment inside the building contributes to heating the space, thus this energy would have to be added by either the heating system or lights and equipment, regardless.

However, it is important to note that the building was unoccupied during the monitoring period, since an active use for the space has not yet been identified. The internal heat gains due to the occupants and their activity would reduce the energy required to be delivered by the heating system. In an opposite way, the house would require ventilation air if occupied, which would increase the heating energy consumption. Future plans to continue to monitor building performance, add a ventilation system, and add occupancy of the building would enable an enhanced ability to predict the actual energy consumption of the building in use.

From the data provided in Fig. 5 it can be seen that when this building is subjected to the Iqaluit climate, it requires heating for nearly the entire year. Indeed, there are very few days when heating is not required.

Daily average outdoor temperatures were calculated from the data measured at the Iqaluit Airport are also shown to highlight the dependence of total consumption on outdoor temperature. The relationship between daily total consumption and daily average outdoor temperature is shown explicitly in Fig. 6. The maximum daily consumption is about 30 kWh per day at a daily average outdoor temperature of just under −30 °C. The minimum daily consumption at the bottom right of the graph when there is no heating load is about 0.88 kWh per day, which is what is

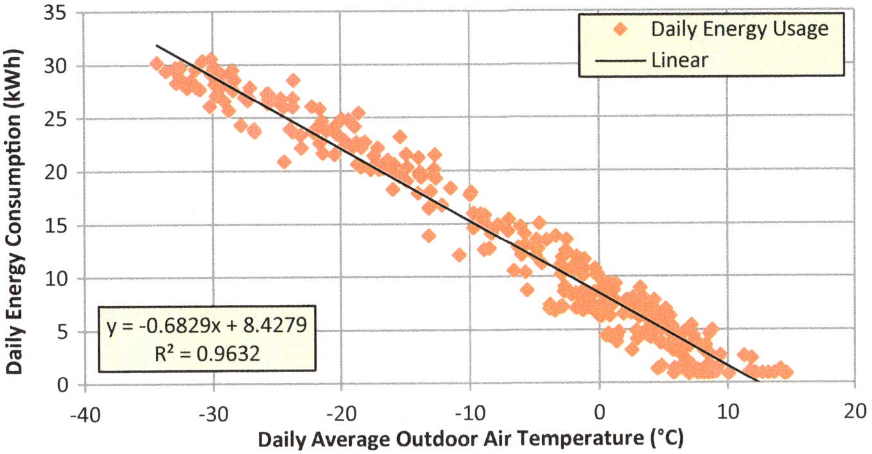

Fig. 6 Relationship between daily energy consumption and daily average outdoor temperature

consumed by the data acquisition system (including computer, modem, and other devices). This represents about 37 Watts of continuously connected power.

The total electrical energy consumed inside the house for the period of April 25, 2016 to April 24, 2017 was 4945 kWh. At an assumed rate of CAD$0.60 per kWh [5], the cost of heating for this period is CAD$2967. It was necessary to use electrical energy to heat the house rather than oil, since the gravity-fed oil heater could not be accurately metered. Ideally, renewable resources such as solar and wind energy could be used to reduce cost, increase self-sufficiency, and reduce environmental impact. An alternative energy source, such as small wind power would be less expensive per unit of heating energy delivered [6], therefore reducing this cost.

Based on the floor area of 24.3 m^2, the energy use intensity (EUI) of the building for the monitoring period in Iqaluit was 203 kWh/m^2/year or 0.73 GJ/m^2/year. As anticipated, this compares favourably to typical construction in Canada. The 2011 Survey of Household Energy Use by Natural Resources Canada was consulted for comparison [3], which is a very detailed analysis of residential buildings and their energy use. Values for EUI are available for many regions of Canada, but does not include the three territories. For buildings under 56 m^2, the average EUI in Québec was estimated at 0.98 GJ/m^2/year and for Ontario was 0.83 GJ/m^2/year.

It must be noted that these figures are not adjusted based on climate. Although the provincial figures are averages, it is reasonable to estimate that the climatic load in these two provinces is on the order of half that of Iqaluit, supported by this being true for Ottawa in comparison to Iqaluit, as noted earlier in this report.

3.2 Modelling Results

The total modelled heat loss for Iqaluit was 4801 kWh, whereas the actual measured consumption was 4935 kWh. This represents a 3% under-prediction of the energy consumption by the model. This is quite good agreement between the total annual energy consumption between the modelled and measured test house. However, the agreement should be accompanied with a reasonable amount of skepticism, as it is possible the errors of some effects are counterbalancing each other. Potential factors contributing to the uncertainty of the model are presented later in this section. Despite these possibilities, it is evident that the building energy model is representative of the overall performance of the house.

Figure 7 shows the relationship between the modelled energy consumption and the outdoor temperature. The annual model results are presented in comparison to the measured results in Fig. 8.

As can be seen in Figs. 7 and 8, the model underestimates the energy consumption of the prefabricated house on milder days by a small degree (i.e. <5%). It then overestimates the energy consumption during colder days, i.e., those below approximately −25 °C daily average outdoor air temperature. As a result, the slope of the linear trendline for the model is steeper than that of the measured data.

This indicates that for climates with milder annual average weather, the model might be expected to under predict energy consumption slightly more, and for colder annual average weather, the model might be expected to over predict energy consumption slightly. Despite this, the already good accuracy enables a reasonable prediction of energy usage performance via parametric studies.

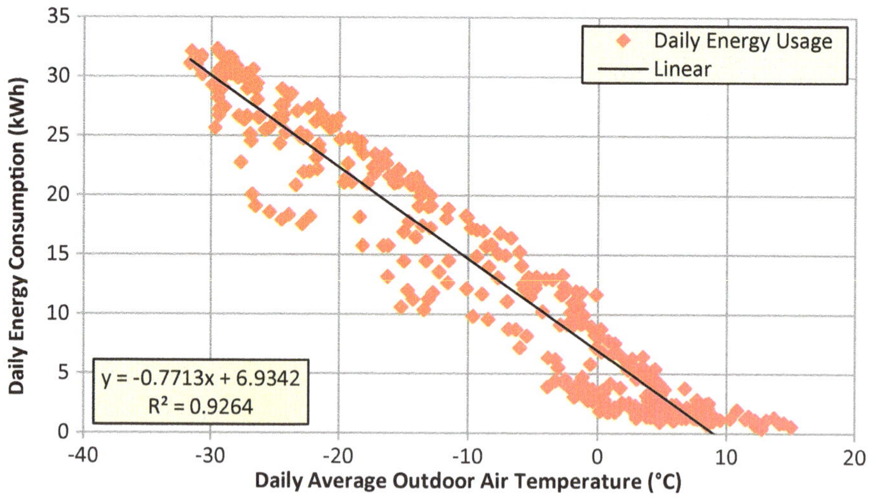

Fig. 7 Modelled daily energy consumption versus daily average outdoor air temperature

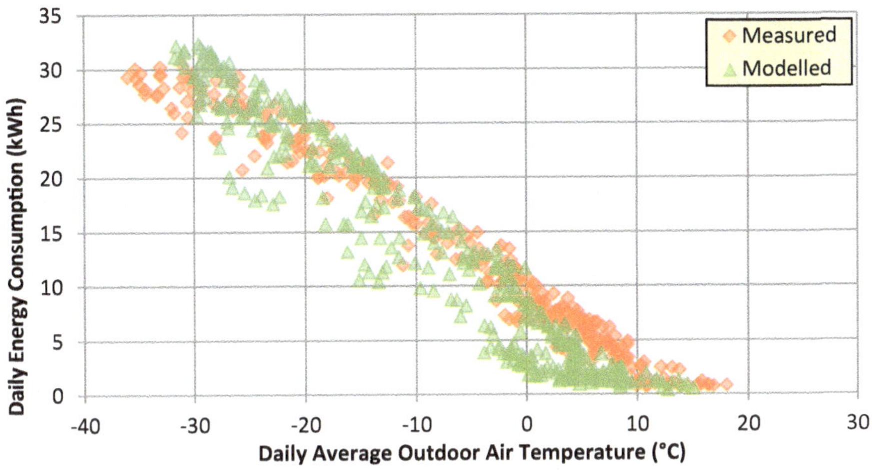

Fig. 8 Comparison of modelled and measured daily energy consumption data

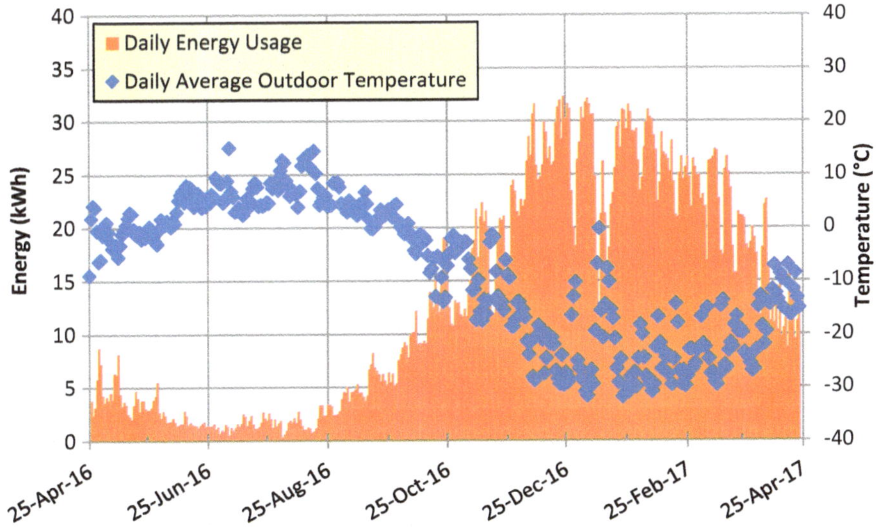

Fig. 9 Modelled daily electricity consumption and average outdoor air temperature

The same modelling results data are shown in Fig. 9, but presented chronologically over the monitoring period.

Differences Between Measured and Modelled. The model results were found to agree with the measured data to a reasonably high accuracy (i.e. <5% deviation). As with all modelling, there is always deviation from actual results due to many factors, in this case, such as:

- Uncertainty in values for solar radiation;
- Exactness of orientation;
- Deviations between actual material properties and modelled properties;
- Underlying assumptions of or in model governing equations; and
- Other unknown factors to be discovered;

Despite these limitations, there is opportunity to adjust the model to provide more accurate results given some additional information in respect to climate loads, building orientation and variations in material properties. Accurate solar radiation measurements on site would help correct some potential uncertainty, as high differences between modelled and measured were seen on cold days having significant solar radiation.

Given that it is impossible to model a building to perfection, the deviation between the model and measured values for a given location can be noted and used to adjust for results in a different location if reasonable understanding of the reasons for deviation is thought to be known and this knowledge can be applied to the transformation.

4 Summary and Conclusions

A demonstration house was built in Iqaluit and instrumented to monitor the energy performance of the structural insulated panel technology from which it is built. To date, over a year of elapsed monitoring has been conducted, and is still ongoing at the time of writing. The energy consumption results of the house have been presented, along with a comparison to the results from a building energy model.

The total energy consumed by the house from April 25, 2016 to April 24, 2017 was 4935 kWh, which is an energy use intensity of 203 kWh/m^2/year or 0.73 GJ/m^2/year. Lack of data for residential energy consumption in such climates makes comparison to other technologies difficult. In addition, it is important to note that the house was unoccupied during the period monitored thus far, therefore energy consumption is expected to be higher during occupancy due to the need for mechanical, heat recovery ventilation.

The building configuration was replicated in ESP-r to develop a model for future work and was benchmarked against the measured annual energy consumption. The model underestimated energy consumption slightly at mild outdoor temperatures and overestimated at cold outdoor temperatures, but was within 3% of measured consumption overall, demonstrating excellent agreement. The benchmarked model will be used to support further development of this residential building solution in harsh cold climates.

Building physics, energy systems, and mechanical systems modelling can be used in the future to advance the design and functionality of this building solution. Some potential investigations or advancements include: various locations; optimization, e.g., window placement, adding a vestibule, adding insulation; ventilation strategy;

occupancy scenarios; humidity loads, e.g., from cooking; predicting thermal comfort; interactions with energy system(s); and alternate heating system(s).

References

1. C.J. Banister, M. Swinton, T. Moore, D. Krys, In-situ thermal resistance testing of a high performance building envelope in the Canadian Arctic, in *9th International Cold Climate HVAC Conference*, Springer, Kiruna, SE (2018)
2. Environment and Climate Change Canada. Canada Weather Stats [Internet]. 2017 [cited 2017 Aug 25]. Available from: https://www.weatherstats.ca/
3. Natural Resources Canada. Survey of Household Energy Use (2011)
4. Environment and Climate Change Canada. Historical Data, Iqaluit A [Internet]. 2016 [cited 2017 Aug 28]. Available from: http://climate.weather.gc.ca/historical_data/search_historic_data_e.html
5. Qulliq Energy Corporation. Customer Types [Internet]. 2017 [cited 2017 Aug 24]. Available from: http://www.qec.nu.ca/customer-care/accounts-and-billing/customer-types
6. C. Budischak, D. Sewell, H. Thomson, L. Mach, D.E. Veron, W. Kempton, Cost-minimized combinations of wind power, solar power and electrochemical storage, powering the grid up to 99.9% of the time. J. Power Sour. **225**, 60–74 (2013)

Increasing Buildings Automation Systems Efficiency with Real-Time Simulation Trough Improved Machine Self-learning Algorithms

Andris Krūmiņš, Nikolajs Bogdanovs, Romualds Beļinskis, Kristīne Mežale and Miks Garjāns

Abstract Energy savings with building's automation is related to the functionality of the building's automation system, equipment automation level, control algorithms and user-selected operating parameters. It is difficult and time-consuming process to customize each of these algorithms to a level, which would ensure the rational use of energy resources. To increase the adjusting quality that controls building's engineering system it is necessary to implement new advanced methods, technologies and control algorithms within building's automation system. This work presents the method of creating self-learning system, which is capable of detecting negative influences of different control algorithms. This self-learning system also provides error detection in engineering systems of the building. The system was designed in Matlab and Simulink programming environments. The article also describes the architecture of designing a complex wireless network system with real time (online) connection, including automatics and building's engineering systems. This complex system is also capable of collecting data to perform an effective intellectual analysis.

Keywords Self-learning system · Matlab and simulink · Wireless sensor network

1 Introduction

There are problems with energy efficiency of buildings in Latvia. There is a necessity for algorithms and systems improving the situation. In order to achieve this goal and improve building's energy efficiency it is necessary to develop self-learning algorithms based on the climate of Baltic region and building's load in particular. Individually created algorithms using obtained data will be developed for each type of the building.

The main part of the task is to analyze operating conditions and energy efficiency of heating units. Determined necessary parameters of air supply and program

A. Krūmiņš · N. Bogdanovs (✉) · R. Beļinskis · K. Mežale · M. Garjāns
BMS Department, Lafivents Ltd., Bauskas street 58, Riga, Latvia
e-mail: nikolajs@lafivents.lv

D. Johansson et al. (eds.), *Cold Climate HVAC 2018*,
Springer Proceedings in Energy, https://doi.org/10.1007/978-3-030-00662-4_4

calculated physical energy allows choosing correct configuration for operating parameters of heating units. Coincidences from parameter settings of air supply on heating units configuration and control system, should be investigated further. Necessary air supply can be adjusted to insure comfortable conditions. Adjustments related to comfort zone, comparing to temperature adjustments with deadbands, requires control system of higher-level, but it gives the opportunity to calculate optimal air supply parameters, which requires minimum energy consumption. Developed heating energy calculation program can be supplemented with optimal air supply parameter calculations. This option will allow calculating potential energy savings, which can be achieved by control system of higher-level. Results can be used as key factors for better control system utility. To reduce consumption of energy resources it is necessary to create a wireless sensor system with self-learning algorithms, which will assist in energy efficiency and reveal weak points of systems exploitation. The goal of this work is to create a sensor network that determines energy efficiency in buildings according to the EN 15603 standard. The European EN 15603 standard sets the general structure of the assessment of general energy consumption in buildings, and also a technique, which should be used when determining the nominal power parameters.

One of main difficulties in creating a sensor data collection system is designing inexpensive and easily customized system, which will transfer data to the remote server (Cloud). The server should contain a database and quick assessment method for energy consumption of the building and create long-term forecasts of energy consumption. By using collected data from experimental objects it is necessary to create a database on different groups of building types and their operational parameters. The developed sensor system will help adjusting the system of existing buildings by providing deeper energy consumption analysis and comparison, assessing energy class and providing reliable data for possible buildings or its engineering system upgrades.

The research of building's automation systems with different control algorithms is relevant for industry and academics. Analyses of control algorithm and energy consumption have been comprehensively disclosed in author's various papers. They have made a great effort to develop different control algorithms.

Chai [1] has described installation and temperature control algorithm depending on external temperature changes. The author describes the goal of energy control in office environment, by adjusting thermostat and minimizing maximum load. Developed models and simulations are accordingly implemented into Matlab to evaluate maximum load savings.

Lynggaard's [2] promotion work is based on the combination of established theories. This work describes artificial intelligence of automated system based on analysis of building's sensor database. The researcher analyzed methodology offering iteratively-based experimental research strategies.

Sandeep [3] describes analytical model of automated system which is knowledge-based system "KBS". Based on this database the system can decide how to effectively control the building.

Badlani's [4] work describes smart home systems that decrease energy consumption by studying human behavior pattern.

The main contributions of this paper are Building Management System (BMS) with self-learning algorithms based on WSN of a heterogeneous network specified for providing energy consumption analysis and estimating energy efficiency upgrading actions.

2 Overall System Architecture

Multi-store buildings, which initially were not planned to integrate building's control systems raises a complex problem condition for modernization. Installation of wired sensor system in such building becomes impractical or even impossible. Therefore, effective wireless sensor systems are necessary to be able to provide measurement delivery considering specifics of multi-store buildings.

Research direction, described in this work, leads to the creation of a wireless sensor data transmission network to ensure data collection from sensors and determine the loss of energy resources of the building. The specific task is to create the wireless sensor network in which the initial data flow concentrates in a special concentrator with further data transmission via high-speed channels to main network server. Such method is economically more effective comparing to systems without the concentrator. Although, despite the message being small in amount, there is a chance of electromagnetic incompatibility effect for channels. To reduce the noise, lines are separated in points of concentration, which do not cross the frequency of low channels and their throughput is defined by the sensor group. Data from developed sensor network is automatically transferred to Matlab and SQL (Structured Query Language) Database. The program written in Matlab has analysis algorithm for calculating the necessary heating power and work conditions of Heating systems. Calculation program for energy consumption is the first step for in developing a good climate control system for building areas. The program graphically analyzes the processes and gives information about the necessity to adjust. Power calculations of heating units show the necessary power requirements for different regimes. In this test bed of prototype architecture (see Fig. 1) the Sensor node senses the data from the sensor and the data received at the end base station (Raspberry Pi).

Base station with MQTT (Message Queuing Telemetry Transport) protocol sends all data to the Web cloud to Matlab. We managed to integrate it into building's old automatic equipment by using Matlab and OPC (Open Platform Communications) standard. We used Matlab machine learning toolbox and weather forecast. OPC is the interoperability standard for secure and reliable exchange of data in industrial automation area. By analyzing next day's weather forecast and the database with data from wireless temperature sensors, a self-learning algorithm was created that allows adjusting automatic equipment of the building.

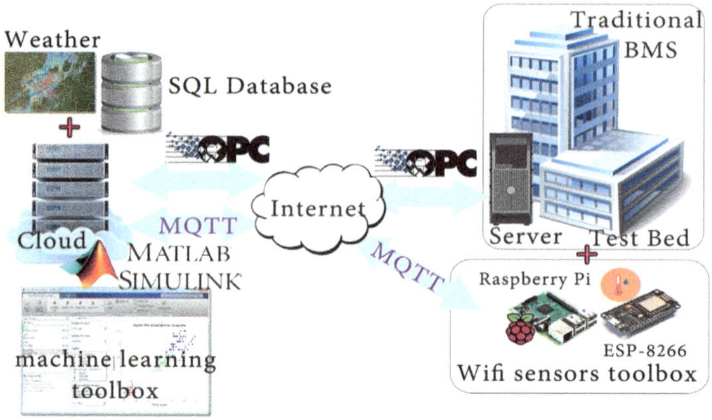

Fig. 1 Test bed prototype

Using PID control for Building Management System (BMS) is not effective method to control the heating system. Thus, a considerable change in set point occurs. Integral terms cause an overshooting error during the rise [5].

For the purpose of optimizing the control algorithm, a modular heat flow numerical model of a heating system integrated with BMS was developed by discrediting time using Finite Difference Method in self-learning model. It was developed in Matlab and Simulink (see Fig. 2). To calculate the supply temperature we use the following parameters in this algorithm:

- Temperature increase/decrease gradient.
- Supply temperature compared to the weather forecasted.
- Inertia time constant of valve.
- Limit value for heating circuit power.
- Comparison of consumption energy with supply temperature.
- Adjustment of control algorithm based on the analysis of data archive.

3 Wireless Sensor Network Design

The Wireless Sensor Network (WSN) was implemented by using a topology in beacon mode where sensors collect data and deliver it to the base station which is the concentrator of the network. Offered WSN design is shown in Fig. 3. This system works in 25 m range from the base station and it is suitable for monitoring [6–8]. The IEEE 802.11 devices are capable of transmitting data for long distances by passing them through intermediate devices reaching longer transmission range for the data. The main feature of ESP-8266 technology is the fact that at small energy consumption it supports not only simple network topologies, but also self-organized and self-recovering mesh topology with the retransferring and routing of messages [9].

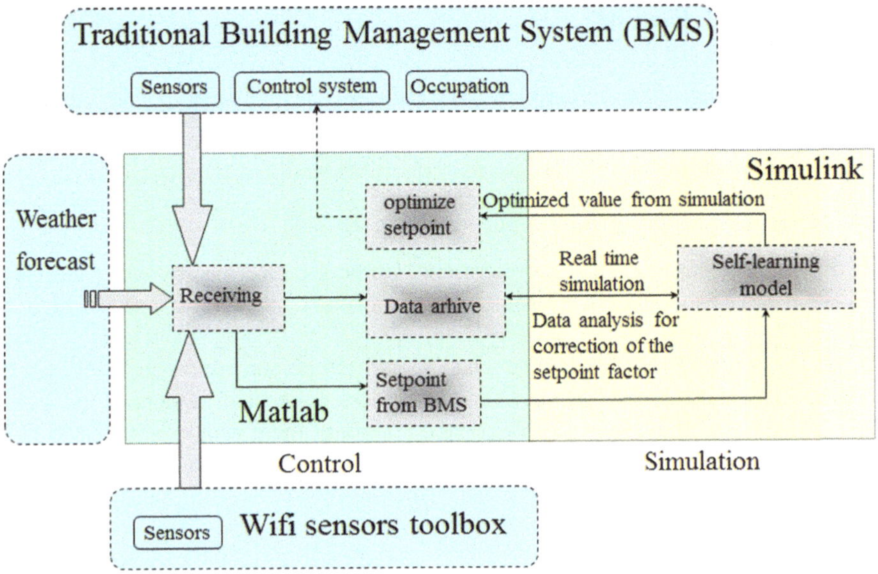

Fig. 2 Self-learning implementation scheme

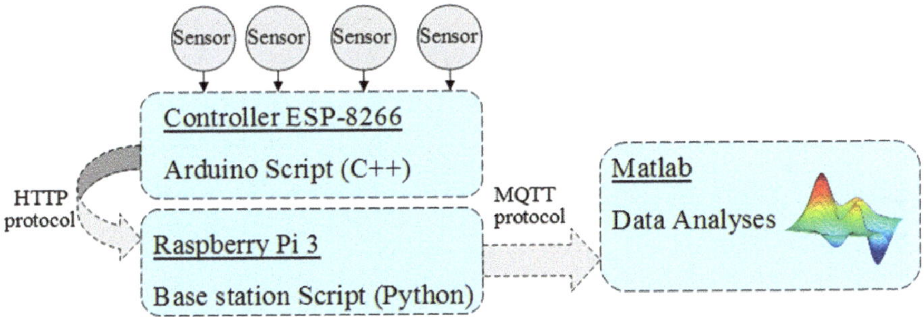

Fig. 3 WSN toolbox software overview

We designed a system for collecting and analyzing the data from sensors. This system includes the base station (Raspberry Pi), the router (ESP-8266) and sensors (see Fig. 3).

The dynamic structure of the network allows eliminating accidents quickly by laying out new paths of data transfer bypassing the faulty section. 16-bit addressing allows using more than 65 thousand nodes [10].

4 Test Bed Results

At the beginning of our experiment we analyzed the regularity of thermal energy consumption depending on outside temperature over the 5 year period. The results show that the approximation of stored experimental data leads to a conclusion that this dependence is a linear function, see Fig. 4.

During next stage we carried out following actions:

1. Dividing heating unit into northern and southern part. Thus, by using the valve we had a chance to control southern and northern heating units separately (see Fig. 5) [11, 12].
2. To optimize the heating unit we had to add temperature sensors. We did it by choosing wireless ESP controllers and sensors connected to them. This system allows establishing sensors quickly and integrates them with existing BMS [13].
3. By using the Matlab machine learning toolbox and weather forecast a self-learning algorithm was created, which allows controlling pumps and heating unit valves more effectively and correctly [13–16].

As a result of our actions we managed to optimize the heating unit and decrease the consumption of thermal energy by ≈38% as you can see in Fig. 6.

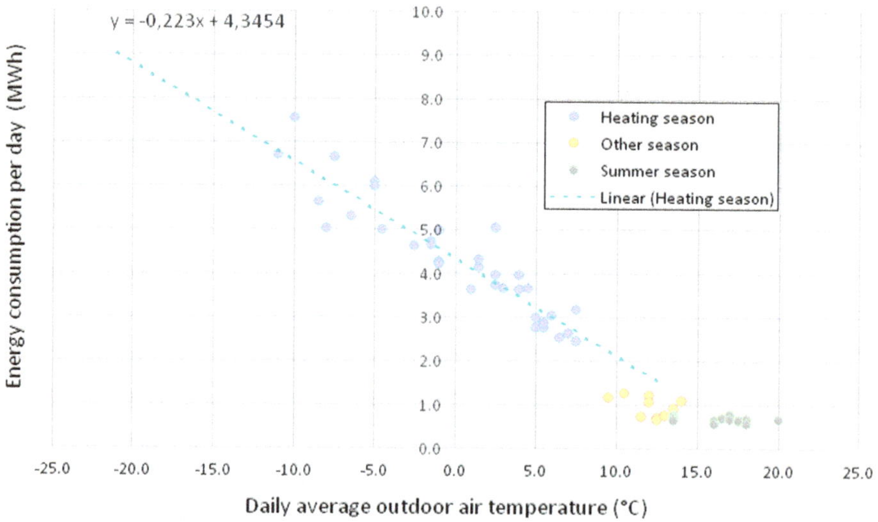

Fig. 4 Thermal energy consumption depending on outdoor temperature over the 5 year period

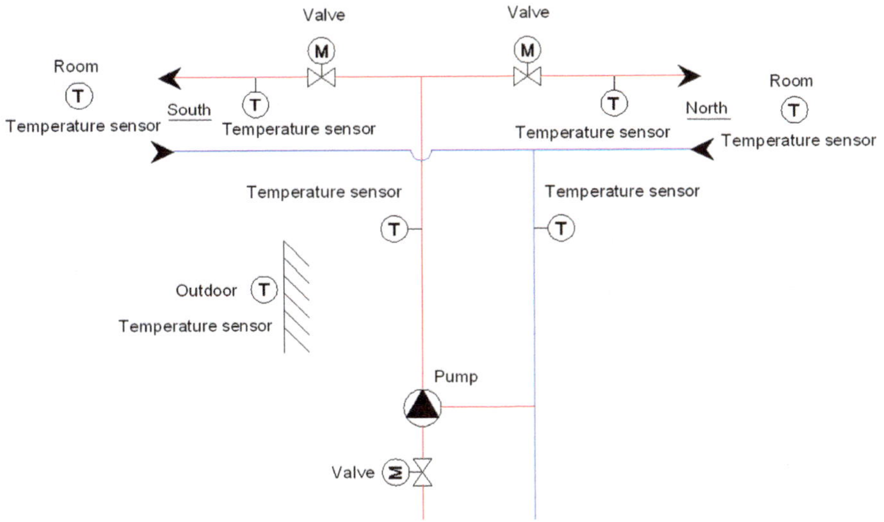

Fig. 5 Additional temperature sensors for heating center

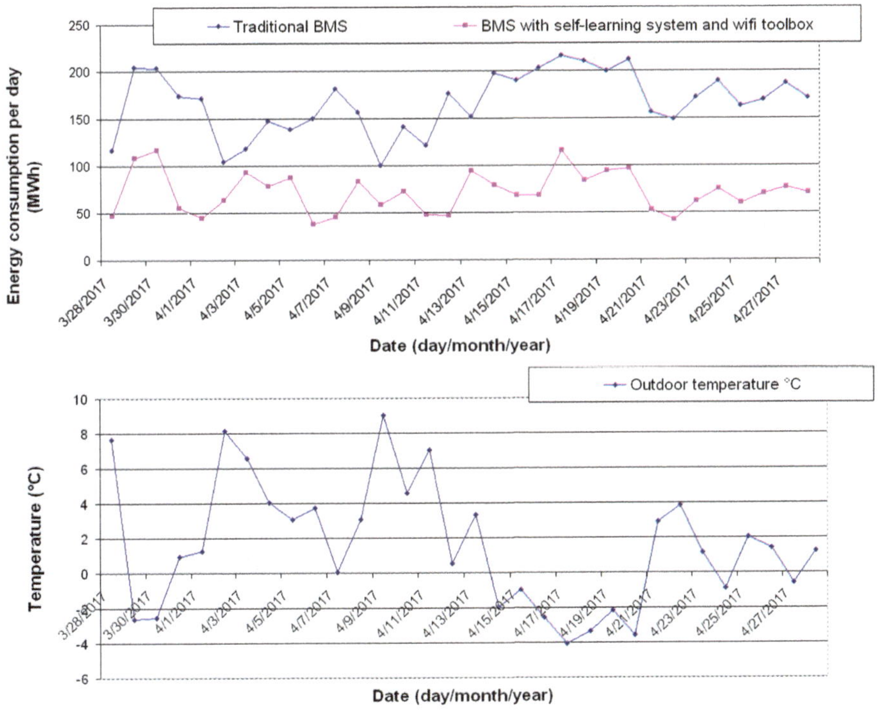

Fig. 6 Heat energy consumption

5 Conclusion

The proposed self-learning system in this paper is designed with wireless sensor network and Raspberry Pi as the base station, and also ESP-8266 as the sensor node. This type of wireless sensor and self-learning algorithm control system improves the effectiveness and the efficiency of consumed heating resources. Described Self-learning system was developed and successfully tested. The main and major advantage of this system lies in the integration of the gateway node of wireless sensor network into Matlab and Simulink and traditional Building's Management System.

Research results will be applied on energy-efficiency exploitation of engineering networks in existing buildings, which will help to increase energy-efficiency. Based on this research, recommendations will be made for the building owners about the possibilities for exploiting existing engineering systems and constructing buildings more efficiently.

Future's plan is to address all above-mentioned problems included in this research by providing following solutions:

- The problem of ensuring the wireless network throughput and delays could be solved by the method of queuing theory of closed systems. Developed optimal assistant path for data transmission in wireless sensors system.
- When using the number of nodes (wireless controllers) >300, network's goodput falls to unacceptable level due to increasing traffic service. In future the program—the simulator, could solve this problem. Simulators demonstrate how the performance was influenced by configuration of wireless sensors systems.
- To create an algorithm in Matlab and Simulink optimizing the work of ventilation based on specifics Baltic climate.

Acknowledgements The research received funding from the ERAF Post-doctoral Research Support Program project Nr. 1.1.1.2/16/I/001 Research application "Wireless sensor networks for a building's energy efficiency evaluation and data exchange with the building's management systems" Nr. 1.1.1.2./VIAA/1/16/228.

References

1. B. Chai, W. Tushar, N. U. Hassan, C. Yuen, Z. Yang, Managing energy consumption in buildings through offline and online control of HVAC systems, in *2016 IEEE Region 10 Conference (TENCON)—Proceedings of the International Conference*, ISSN 2159-3450, pp. 3368–3373 (2016)
2. P. Lynggaard, Artificial intelligence and internet of things in a "smart home" context: a distributed system architecture. Ph.D. Thesis, Press/Media, 268 p. (2014)
3. A. Badlani, S. Bhanot, Smart home system design based on artificial neural networks. ISSN: 2078-0958 (Print); 2078-0966 (Online), in *Proceedings of the World Congress on Engineering and Computer Science 2011*, vol. I WCECS, pp. 106–111. October 19–21, San Francisco, USA (2011)

4. K. Sandeep, A.Q. Mohammed, Application of AI in home automation. Int. J. Eng. Technol. **4** (6), pp. 803–807 (0975-4024) (2012)
5. R. Gorzenski, A. Gorka, E. Szczechowiak, Guidelines for the control algorithm of heating installation in LARS low-energy system. Technical Transactions Civil Engineering, R.109, z. -B, pp. 139–148 (2012)
6. N. Patil, K. Anand, Warad: IOT and Raspberry PI based environmental monitoring application. Int. J. Electr. Electron. Comput. Syst. (IJEECS) **4**(4), 36–41 (2016) ISSN (Online): 2347-2820
7. V. Raghavan, H. Shahnasser, Embedded wireless sensor network for environment monitoring. J. Adv. Comput. Netw. **3**(1), 13–17 (2015)
8. A. Gaddam, M. Al-Hrooby, W.F. Esmael, Designing a wireless sensors network for monitoring and predicting droughts, in *Proceedings of the 8th International Conference on Sensing Technology*, September 2–4, Liverpool, UK, pp. 210–215 (2014)
9. T. Nouidui, M. Wetter, Tool coupling for the design and operation of building energy and control systems based on the functional mock-up interface standard, in *Proceedings of the 10th International Modelica Conference*, March 10–12, Lund, Sweden, pp. 311–320 (2014)
10. S. Ferdoush, X. Li, Wireless sensor network system design using Raspberry Pi and Arduino for environmental monitoring applications. Procedia Comput. Sci. **34**, 103–110 (2014) (The 9th International Conference on Future Networks and Communications (FNC-2014))
11. H. Hadwan, P. Reddy, Smart home control by using Raspberry Pi & Arduino UNO. Int. J. Adv. Res. Comput. Commun. Eng. **5**(4), 283–288 ISSN (Online) 2278-1021 (2016)
12. J.L.G. Ortega, L. Han, N. Whittacker, N. Bowring, A machine-learning based approach to model user occupancy and activity patterns for energy saving in buildings, in *Science and Information Conference 2015*, July 28–30, London, UK, pp. 474–482 (2015)
13. A. Capozzoli, F. Lauro, I. Khan, Fault detection analysis using data mining techniques for a cluster of smart office buildings. Expert Syst. Appl. **42**, 4324–4338 (2015)
14. C. Fan, F. Xiao, C. Yan, A framework for knowledge discovery in massive building automation data and its application in building diagnostics. Autom. Constr. **50**, 81–90 (2015)
15. A. Zanella, N. Bui, A. Castellani, L. Vangelista, M. Zorzi, Internet of things for smart cities. Internet Things J IEEE **1**, 22–32 (2014)
16. S. Treado, Y. Chen, Saving building energy through advanced control strategies. Energies **6**, 4769–4785 (2013)

Analysis of Various Ventilation Solutions for Residential and Non-residential Buildings in Latvia and Estonia

Jurgis Zemitis, Anatolijs Borodinecs and Targo Kalamees

Abstract As the newly built and renovated buildings consume less energy for heating needs, due to better and thicker insulation, the relative energy consumption for ventilation increases. This leads to necessity for increased effectiveness of ventilation systems, but such systems are more expensive in installation therefore the most economically feasible solution must be found in each case. A specific attention should be paid to such unclassified buildings as dormitories and barracks where occupancy profile and density differs from residential buildings which are already widely analyzed. This paper presents study results of cost analysis for different ventilation strategies for case study multi-story apartment building in Latvia and Estonia. The compared ventilation strategies include natural ventilation through windows, natural ventilation by having inlet valves with natural exhaust, hybrid ventilation with inlet devices in walls and mechanical exhaust, decentralized mechanical ventilation with room based heat recovery, decentralized mechanical ventilation with apartment based heat recovery and building based centralized ventilation system. For each of these system types installation costs are estimated, based on necessary equipment and actual market prices. Afterwards annual running and maintenance costs are calculated and obtained data compared to select the optimal solution. The results show that the most cost effective system in longer time period is centralized ventilation system which serves whole staircase. Although the simpler solutions like natural or hybrid ventilation systems with air inlets through walls and mechanical exhausts are initially cheaper the energy costs to heat up the incoming air are high and therefore cost inefficient in longer time period.

Keywords Ventilation · Efficiency · Cost analysis

J. Zemitis (✉) · A. Borodinecs
Institute of Heat, Gas and Water Technology, Riga Technical University, Kipsalas street 6A, Riga, Latvia
e-mail: jurgis.zemitis@rtu.lv

T. Kalamees
Tallinn University of Technology, Ehitajate tee 5, Tallinn 19086, Estonia

D. Johansson et al. (eds.), *Cold Climate HVAC 2018*,
Springer Proceedings in Energy, https://doi.org/10.1007/978-3-030-00662-4_5

1 Introduction

One of the most important engineering systems from all is the ventilation. It has always been present in the buildings but the exact solutions have changed during the years. The ventilation system has evolved starting from simple natural ventilation with supply through construction cracks or windows to fully mechanical ducted supply/exhaust ventilation system with various in between solutions. This makes the choosing of the optimal ventilation solution difficult in each specific situation as each type ensures different comfort and control level but also varies in installation and running costs. In existing researches [1] it is stated that a VHR systems can give substantial final energy reduction, but the primary energy benefit depends strongly on the type of heat supply system. However for renovated buildings or some lower priced newly built buildings it could either not be possible or feasible to install such and simpler solutions could be chosen. This leads to necessity to carefully choose the appropriate ventilation system type to maximize energy savings while providing good indoor air quality. Investigation [2] have shown that in renovated apartment buildings with natural passive stack ventilation, the indoor air quality is quite bad and has high CO_2 concentration and relative humidity level. Similar data on IAQ problems in multi apartment buildings shows research [3] done in Estonia. Some authors [4] have already performed analysis on how ventilation rates vary depending on chosen system. Others [5, 6] have analyzed the system influence on IAQ.

Although the ventilation solutions for living building sector have been more widely analyzed it is important to provide the solutions also for such unclassified buildings as military barracks, shooting ranges or armories. As there is a strong potential to significantly reduce energy consumption also for them. For example shooting ranges have a need for large amounts of ventilation air and therefore consume a lot of energy however according to some studies [7] state that in case when conventional ventilation methods are used it is not possible to obtain suitably low concentrations of harmful substances in indoor shooting ranges and only a low turbulence displacement ventilation system provides the correct conditions for rapid conveyance of harmful substances to the extraction outlets in the area of the bullet trap. This means that a special attention must be paid in choosing the exact solution.

The paper presents the findings of calculation results of performed cost analysis of different ventilation solutions to determine the most feasible solution for longer running time periods.

2 Methods

A calculation to determine the installation costs of various ventilation systems and their annual running costs is performed in this paper. The comparison is made for typical apartment building located in Latvia and Estonia. The method to achieve this is based on following steps: Determining the necessary ventilation air volumes for each country; Designing various ventilation system types for an apartment; Estimating construction costs of ventilation systems; Performing life cycle costs analysis for each ventilation system including installation costs, annual maintenance and necessary energy for heating and electricity.

2.1 *Description of Case Building*

All further calculations are based on a chosen case study building. The building type is a multi-apartment building with 5 stories and 30 flats for each staircase section. The ceiling height is 2.5 m. In building two types of apartments are present. 25 apartments in each staircase are one bedroom (living area 35 m^2) while five apartments have two bedrooms (living area 47 m^2). The ventilation volume is calculated for each type of flat and the total ventilation volume necessary for one staircase section.

To determine the design ventilation rate a calculation of both supply and exhaust was performed. Afterwards the largest value was applied as a design value for each flat type. In Estonia multi-apartment building exhaust airflow can be lower if required supply airflow is guaranteed. Knowing the ventilation air volume for the each flat the air for whole staircase section can be calculated. It involves multiplying the calculated ventilation air volume of one apartment with the number of apartments located in one buildings section (staircase) and with the number of stories, in this case five. This is necessary as the centralized AHU unit could serve each staircase separately therefore dividing building into smaller sections.

For all ventilation design cases the general rule was to supply the air in the bedrooms and living rooms while there are two spate exhaust systems—in bathroom and kitchen. The entrance area will be ventilated with the transfer air (Table 1).

Table 1 Calculated ventilation air volumes for Latvian and Estonian case study building

Room Nr.	Room type	Latvia		Estonia	
		Supply air m^3/h	Exhaust air m^3/h	Supply air m^3/h	Exhaust air m^3/h
1	Bedroom	55 (90 for 2-room apartment)	–	50	–
2	Kitchen	–	90	–	22 (30 for 2-room apartment)
3	Bathroom	–	50	–	54
4	Storage	–	–	–	10
Total for an apartment (m^3/h)		55 (90)	140	50	86
Total for whole staircase section (m^3/h)		$6 \cdot 5 \cdot 140 = 4200$		$(5 \cdot 86 + 119) \cdot 5 = 2745$	

2.2 Design Examples of Ventilation Systems

To prepare the cost analysis between the most common ventilation system types they all were fully designed for an apartment. The analyzed ventilation types included: natural ventilation by openable windows and natural exhaust (see Fig. 1), natural ventilation by having inlet valves and natural exhaust (see Fig. 1), hybrid type ventilation by having inlet devices in walls and mechanical exhaust (see Fig. 2), decentralized mechanical supply and exhaust with heat recovery (room based heat recovery) (see Fig. 2), decentralized mechanical supply and exhaust with heat recovery (apartment based heat recovery) (see Fig. 3), centralized mechanical supply and exhaust with heat recovery (see Fig. 3). The shown air volumes in all figures are for case of Latvia as it had the higher necessary ventilation volumes according to regulations.

In Fig. 1 two natural ventilation solutions are showed. They are still very common for existing, non-renovated buildings in Latvia and Estonia. The air supply through openable windows or building cracks is the supply type that was foreseen in Soviet time buildings while air supply through special air inlets is usually designed for renovation projects of such buildings.

For each ventilation type, a specification is provided to show the necessary elements and prices of them. The specification does not include fittings, mounting elements or work price. To compensate this, final cost of the system is assumed to be higher by 10–30% depending on the predicted extra elements. These values are later used to compare the overall installation and running costs between various solutions (Tables 2 and 3).

The left figure above shows the most commonly used ventilation system in newly built objects in Latvia and Estonia as it is relatively cheap and provides

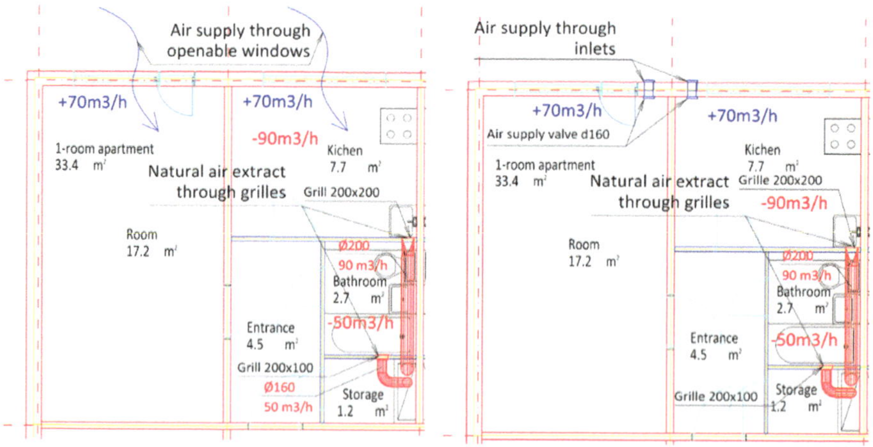

Fig. 1 Natural ventilation system through openable windows and exhaust (left); Natural ventilation system through air inlets and exhaust (right)

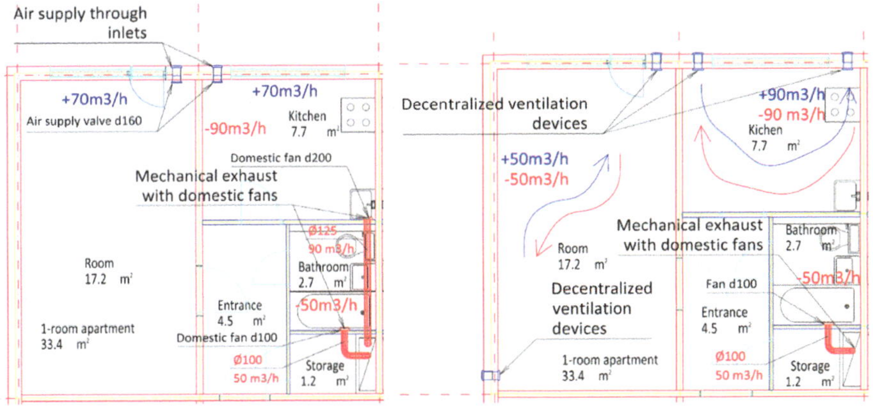

Fig. 2 Hybrid type ventilation system with supply through air inlets and mechanical exhaust (left); Decentralized ventilation system with room based mechanical supply and exhaust with room based heat recovery (right)

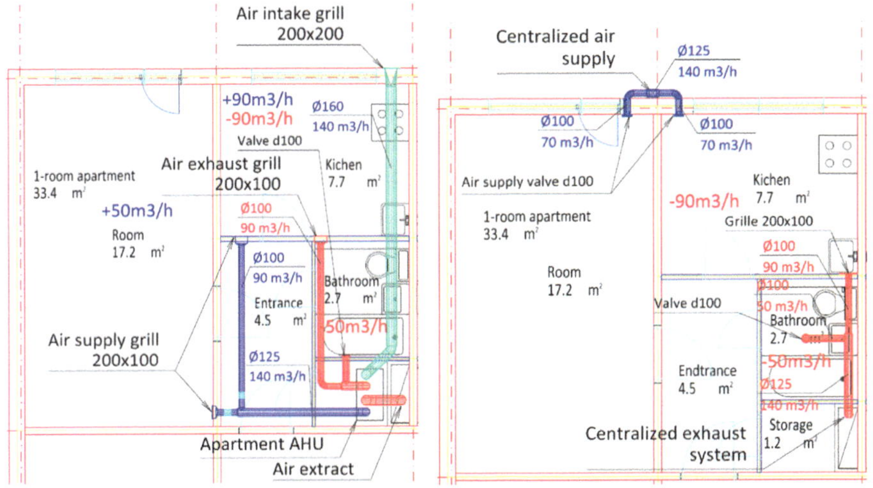

Fig. 3 Decentralized ventilation system with apartment based mechanical supply and exhaust with apartment based heat recovery (left); Centralized ventilation system with building based mechanical supply and exhaust with heat recovery (right)

reliable ventilation volume with mechanical extract ventilators. In more advanced cases the extract duct is equipped with roof ventilator and a special extract device that evens the ventilation volume by compensating the natural stack effect (Tables 4 and 5).

In Fig. 3 fully mechanical ventilation systems with controlled supply/exhaust ventilation volumes are shown. Such type of system grants the highest precision in setting the necessary ventilation volume as well as provides the possibility to set up

Table 2 Specification of ventilation system through openable windows and natural exhaust

Name	Size (mm)	Units	Quantity	Average price in EU for one unit (EUR)
Extract grille	200 × 100	Pcs.	1	17.00
Extract grille	200 × 200	Pcs.	1	20.00
Duct	Ø200	m	2	1.90
Duct	Ø160	m	3	1.50
Total cost with 30% added				60.00

Table 3 Specification of ventilation system through air inlets and natural exhaust

Name	Size (mm)	Units	Quantity	Average price in EU for one unit (EUR)
Supply vents	Ø160	Pcs.	2	120.00
Extract grille	200 × 100	Pcs.	1	17.00
Extract grille	200 × 200	Pcs.	1	20.00
Duct	Ø200	m	2	1.90
Duct	Ø160	m	3	1.50
Total cost with 30% added				370.00

Table 4 Specification of hybrid type ventilation system with supply through air inlets and mechanical exhaust

Name	Size (mm)	Units	Quantity	Average price in EU for one unit (EUR)
Supply vents	Ø160	Pcs.	2	120.00
Domestic type extract fan	Ø100	Pcs.	1	70.00
Domestic type extract fan	Ø200	Pcs.	1	90.00
Duct	Ø125	m	2	1.20
Duct	Ø100	m	3	0.90
Total cost with 30% added				525.00

Table 5 Specification of decentralized ventilation system with room based mechanical supply and exhaust with room based heat recovery

Name	Size	Units	Quantity	Average price in EU for one unit (EUR)
Paired decentralized ventilation devices (50 and 90 m^3/h)	–	Pcs.	4	485.00
Control unit	–	Pcs.	1	320
Domestic type extract fan	Ø100 mm	Pcs.	1	70.00
Duct	Ø100 mm	m	2	0.90
Total cost with 10% added				2560.00

daily ventilation schedule which reduces the energy consumption. The mechanical ventilation system can either be located in each apartment or made centralized in AHU located in attic (Tables 6 and 7).

Table 6 Specification of decentralized ventilation system with apartment based mechanical supply and exhaust with apartment based heat recovery

Name	Size	Units	Quantity	Average price in EU for one unit (EUR)
AHU (140 m^3/h)	–	Pcs.	1	1800.00
Air intake grill	200 × 200 mm	Pcs.	1	40.00
Extract air roof hood	Ø160 mm	Pcs.	1	200.00
Air supply grill	200 × 100 mm	Pcs.	2	17.00
Air exhaust grill	200 × 100 mm	Pcs.	1	17.00
Air exhaust valve	Ø100	Pcs.	1	5.00
Duct	Ø160 mm	m	10	1.50
Duct	Ø125 mm	m	4	1.20
Duct	Ø100 mm	m	7	0.90
Silencers	Ø100 mm/ L = 1000 mm	Pcs.	4	75.00
Total cost with 20% added				2900.00

Table 7 Specification of centralized ventilation system with building based mechanical supply and exhaust with heat recovery

Name	Size	Units	Quantity	Average price in EU for one unit (EUR)
For whole building				
Centralized AHU (4200 m^3/h)	–	Pcs.	1	9500.00
Air intake grill	Ø560 mm	Pcs.	1	100.00
Extract air roof hood	Ø560 mm	Pcs.	1	700.00
Silencers	Ø560 mm/ L = 1000 mm	Pcs.	4	250.00
Duct	Ø560 mm	m	20	17.30
For apartment				
Air supply valve	Ø100 mm	Pcs.	2	5.00
Air exhaust grill	200 × 100 mm	Pcs.	1	17.00
Air exhaust valve	Ø100 mm	Pcs.	1	5.00
Duct	Ø160 mm	m	3	1.50
Duct	Ø125 mm	m	4	1.20
Total cost with 30% added[a]				560.00

[a]The total cost includes the cost of all units located in the apartment and 1/30 of whole price for whole building units, as there are thirty apartments that would be served by the AHU

3 Results

The results are based on the comparison of calculation results of economic analysis for designed ventilation systems as cost analysis has a major importance in choosing the appropriate ventilation system. To compare cost efficiency of different ventilation systems the following factors was taken into account: installation costs, maintenance costs, heating costs to heat up the ventilation air during heating period, electricity consumption for powering the ventilation system, assuming that the ventilation system works continuously through whole year. The cost comparison is done for one staircase of previously described case study building and with following assumptions (Table 8).

Table 8 Comparison of installation and running costs for Latvian/Estonian case study (upper numbers are for Latvia, lower for Estonia)

Type of ventilation system	Installation costs[a] (EUR)	Maintenance costs[b] (EUR)	Heating costs[c] (EUR)	Powering costs[d] (EUR)	Total Annual costs (EUR)
Natural by opening windows and natural exhaust	1800	– –	8005 8110	– –	8005 8110
Natural by having inlet valves and natural exhaust	11,100	300 450	8005 8110	– –	8305 8560
Hybrid by having inlet devices in walls and mechanical exhaust	15,750	450 600	8005 8110	1155 1025	9610 9465
Decentralized mechanical supply and exhaust with heat recovery (room based system)	76,800	750 750	1600 (eff. 0.80) 1620 (eff. 0.8)	535 580 (0.16 W/ m^3/h)	2885 2950
Decentralized mechanical supply and exhaust with heat recovery (apartment based system)	87,000	1050 1500	1200 (eff. 0.85) 2030 (eff. 0.75)	3455 (SFP 1.0) 3210 (SFP 1.6)	5705 6740
Centralized mechanical supply and exhaust with heat recovery	16,760	1000 1000	1600 (eff. 0.80) 2030 (eff. 0.75)	2070 (SFP 1.2) 4010 (SFP 2.0)	4670 7040

[a]Cost of installing all necessary equipment

[b]Annual maintenance cost for all necessary equipment for whole staircase section

[c]Cost of heating supply air for one heating season assuming the Heating degree days for base indoor temperature of +21.0 °C for Latvia 4263, for Estonia 5656 (8) and assuming that the heating occurs by district heating system with following costs—for Latvia 55.55 EUR/MWh; for Estonia 65 EUR/MWh

[d]Annual cost of all energy necessary to power the ventilation devices for whole staircase apartment assuming that they are powered by electricity with the cost of 0.169 EUR/kWh for Latvia; 0.15 EUR/kWh for Estonia

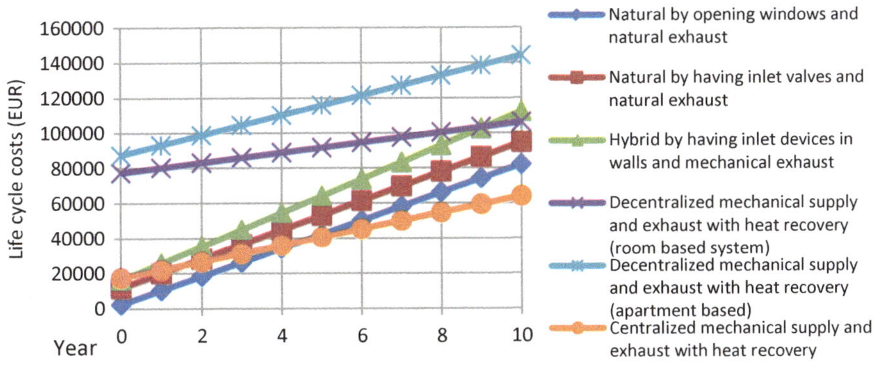

Fig. 4 Life cycle cost comparison of various ventilation solutions for 10 year period for Latvia

Figure 4 represents the life cycle cost analysis for 10 year period of different ventilation scenarios in case of Latvian ventilation volumes and climate.

4 Discussion

The results show that different ventilation solutions can very noticeably vary in installation and running costs. In general it can be concluded that the systems can be divided in three price ranges—the simplest one is just having natural exhaust channels and supply through windows (~1800 EUR), the middle price range includes more advanced systems such as natural system with inlet valves, hybrid system with mechanical exhaust and centralized ventilation system (~15,000 EUR), while the most pricy systems are either room based or apartment based decentralized ventilation systems (80–90 thousand EUR). However these costs must be looked at in combination with annual running and maintenance costs. For this three different levels also stand out—the most cost efficient is the room based mechanical system, the second are the centralized and apartment based full mechanical ventilation systems as they have high heat recovery, and lastly the most inefficient systems are the ones without heat recovery.

Figure 4 shows that in general the most economically feasible choice would be to use centralized, building based ventilation system with one AHU for each staircase. In this case the costs of installation and maintenance would be reduced by spreading it evenly through all the occupants. Also the maintenance would be lower as each flat only needs to take care of one unit. The second most economical system is just having natural ventilation through openable windows. Although in such case no heat is recovered but there is no need for electricity or maintenance.

However it must be noted that to make the final decision when choosing ventilation type not only economical factor must be accounted for as the most important task of the ventilation system is to provide good indoor climate even if it means larger investments during construction phase. This means that additional factors like automation level, human comfort, noise generation from ventilation system or outside, possibility to filter incoming air, etc. must be introduced to account for this.

Also the existing experience of Estonia shows that room based ventilation is not recommended as it generates high noise, often not enough space in external wall is present, the wind flows through the device and the actual efficiency is smaller than specified in technical data. This has led to trend to avoid such solution.

5 Conclusions

The results for case study building showed that the design ventilation air volume between Latvia and Estonia for the specific multi apartment building can vary about 1.5 times—4200 m^3/h in case of Latvia, while for Estonia 2745 m^3/h.

Installation cost for various ventilation system types where calculated and the results showed that the cost can vary from 1800 EUR for simple natural exhaust system up to 87,000 EUR if each flat has a separate full supply/exhaust ventilation system with AHU. The more common solutions as hybrid or centralized mechanical ventilation systems would cost around 15,000 EUR per staircase section.

The ventilation running costs that include maintenance, heating and electricity can vary from 3000 EUR in case of room based decentralized ventilation system up to 9500 EUR in case of hybrid type ventilation system. The running cost between Latvia and Estonia are quite similar as the necessary ventilation volume for Estonia is lower but the price for heating energy is higher. According to cost simulation for 10 year running period for the analyzed ventilation system the results showed that the most cost-effective systems is the centralized full mechanical supply/exhaust system.

Acknowledgements This study was supported by European Regional Development Fund project Nr.1.1.1.1/16/A/048 "Nearly zero energy solutions for unclassified buildings".

This research was supported by the Estonian Centre of Excellence in Zero Energy and Resource Efficient Smart Buildings and Districts, ZEBE, grant TK146 funded by the European Regional Development Fund, and by the Estonian Research Council with Institutional research funding grant IUT1-15.

This work has been supported by the European Regional Development Fund within the Activity 1.1.1.2 "Post-doctoral Research Aid" of the Specific Aid Objective 1.1.1 "To increase the research and innovative capacity of scientific institutions of Latvia and the ability to attract external financing, investing in human resources and infrastructure" of the Operational Programme "Growth and Employment" (No. 1.1.1.2/VIAA/1/16/033).

References

1. A. Dodoo, L. Gustavsson, R. Sathre, Primary energy implications of ventilation heat recovery in residential buildings. Energy Build. **43**, 1566–1572 (2011)
2. I. Dimdina, A. Lešinskis, E. Krumiņš, V. Krumiņš, L. Šnidere, V. Zagorskis, Indoor air quality and energy efficiency in multi-apartment buildings before and after renovation, in *12th International Conference on Air Distribution in Rooms „Roomvent 2011": Book of Abstracts*, pp. 30–30. Norway, Trondheim (2011)

3. T.-A. Koiv, H. Voll, A. Mikola, K. Kuusk, M. Maivel, Indoor climate and energy consumption in residential buildings in Estonian climatic conditions. WSEAS Trans. Environ. Dev. **6**, 247–256 (2010)
4. R.R.H. Santos, M.S.V. Leal, Energy vs. ventilation rate in buildings: a comprehensive scenario-based assessment in the European context. Energy Build. **54**, 111–121 (2012)
5. T.B. David, M.S. Waring, Impact of natural versus mechanical ventilation on simulated indoor air quality and energy consumption in offices in fourteen U.S. cities. Build. Environ. **104**, 320–336 (2016)
6. M.F. Silva, S. Maas, H.A. Souza, A.P. Gomes, Post-occupancy evaluation of residential buildings in Luxembourg with centralized and decentralized ventilation systems, focusing on indoor air quality (IAQ). Assessment by questionnaires and physical measurements. Energy Build. **148**, 119–127 (2017)
7. G. Mirbach, Ventilation in indoor shooting ranges, in *Proceedings of the Workshop on Indoor Shooting Ranges*, pp. 105–112, Rome, Italy (2005)
8. T. Kalamees, J. Kurnitski, Estonian test reference year for energy calculations. Estonian Acad. Sci. Eng. **1**(12), 40–58 (2006)

Performances of Gas-Water Direct-Contact Heat Transfer

Feng Li, Lin Duanmu, Lin Fu and Xiling Zhao

Abstract Compared with indirect-contact heat exchanger, gas-water direct-contact heat exchanger has superiority of decreasing metal heat-exchange surface, small temperature difference and volume, less investment and good antiseptic effect. This paper studies the droplets movement characteristics and performance of gas-water heat and mass transfer. The gas-water direct-contact heat transfer differential equation has been set up to obtain temperature distribution of flue gas and water. The main factors are also analyzed. In this paper, a direct-contact heat transfer model is established for a project, and the results of theoretical calculation and engineering operation are compared. The model has been verified.

Keywords Direct-contact · Flue-gas condensation · Motion characteristics
Heat and mass exchange · Heat transfer model · Test verification

1 Introduction

Nowadays the increasing attention has been paid to energy conservation and environmental protection. As a clean energy, the natural-gas consumption has increased rapidly. As a heating source, gas-fired boilers have been widely promoted in northern cities. However, due to the high exhaust temperature, the condensing heat in the flue gas is difficult to be recovered. Scholars have studied the flue-gas waste heat recovery of the gas-fired boiler [1–5]. It has been proposed to recover the condensing heat of flue gas by the absorption heat pump [6], which has solved the

F. Li (✉) · L. Duanmu
School of Civil Engineering, Faculty of Infrastructure Engineering,
Dalian University of Technology, Liaoning, People's Republic of China
e-mail: lifengfenggg@126.com

L. Fu · X. Zhao
Department of Building Science, School of Architecture,
Tsinghua University, Beijing, People's Republic of China

D. Johansson et al. (eds.), *Cold Climate HVAC 2018*,
Springer Proceedings in Energy, https://doi.org/10.1007/978-3-030-00662-4_6

problem that the return-water temperature is higher, and the condensing heat of the flue gas can not be recovered.

The atomized low-temperature water directly contacts the flue gas. The flue gas is cooled below dew point and water vapor in the flue gas releases condensation heat, which has achieved the purpose of recovering waste heat and condensed water. Direct contact heat exchangers are generally combined with absorption heat pumps. The flue gas is cooled by low-temperature water produced from the absorption heat pump in the direct contact heat exchanger, the heated low-temperature water enters the absorption heat pump to be cooled and then is pumped to the direct contact heat exchanger. An automatic alkali adding device is arranged in the direct contact heat exchanger for the low-temperature water, and the key parts of equipment are protected against corrosion, which has solved the corrosion problem of equipment.

Theoretical analysis of direct contact heat transfer is carried out in the literature [7–9], and the analytical solution for the flue-gas water heat transfer efficiency is obtained. However, it is assumed that the particle size of droplets and physical properties of flue gas are constant during the heat transfer, which simplifies the calculation and also affects the calculation accuracy. The above factors are considered in this paper, a micro model of heat and mass exchange between flue gas and water droplets is established, and the numerical solution of heat transfer between flue gas and water droplets is obtained.

2 Analysis of Motion Characteristics of Water Droplets

To study the heat and mass exchange characteristics of flue gas and water, the motion characteristics of water droplets should be firstly studied and the velocity distribution of water droplets can be obtained. The relative motion of water droplets and flue gas is not all downstream or upstream. The velocity of water droplets relative to flue gas can be decomposed into vertical and horizontal directions (Fig. 1).

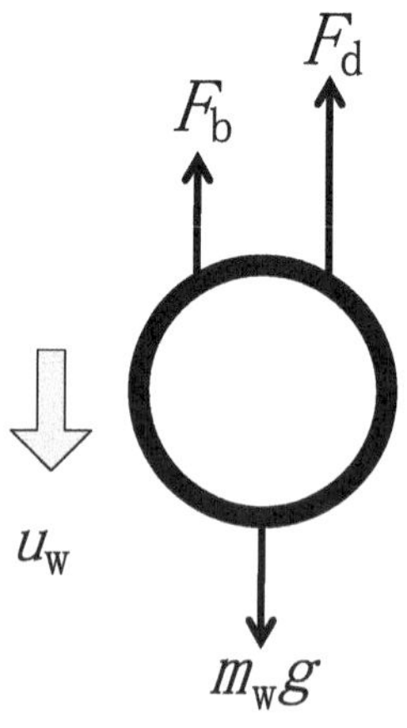

Fig. 1 The force analysis of water droplet in the vertical direction

For example, in the vertical direction, water droplets move in the initial velocity of u0 relative to the flue gas. Water droplets are affected by three forces, such as downward gravity (m_wg), upward buoyancy (F_b) and upward frictional resistance (F_d). Buoyancy is negligible compared to gravity, and the force analysis of water drops is as follows:

$$m_w \frac{du_{w,v}}{d\tau} = m_w g - F_d \tag{1}$$

where, m_w is weight of a single drop, kg; $u_{w,v}$ is droplet velocity in the vertical direction, m/s; τ is time, s; and g is gravitational acceleration (9.8 m/s^2).

The calculation formula for gravity:

$$m_w g = \frac{1}{6} \pi d_w^3 \rho_w g \tag{2}$$

where, d_w is droplet size, m; ρ_w is droplet density, kg/m^3.

The calculation formula for resistance [11]:

$$F_d = \frac{1}{4} \pi d_w^2 C_d \times \frac{1}{2} \rho_f u_{w,v}^2 \tag{3}$$

where, C_d is resistance coefficient.

At low Reynolds number (Re < 1000), optimum resistance coefficient expression [10, 11] is as follows:

$$C_d = 24(1 + 0.197\,\mathrm{Re}^{0.63} + 2.6 \times 10^{-4}\,\mathrm{Re}^{1.38})/\mathrm{Re} \tag{4}$$

$$\mathrm{Re} = d_w u_{w,v}/\mu \tag{5}$$

where, μ is coefficient of dynamic viscosity, Pa.s.

Therefore, it is as below:

$$\frac{du_{w,v}}{d\tau} = g - \frac{F_d}{m_w} = g - \frac{3}{4} C_d \rho_f u_{w,v}^2 / (\rho_w d_w) \tag{6}$$

Assume that:

$$f(u_{w,v}) = g - \frac{3}{4} C_d \rho_f u_{w,v}^2 / (\rho_w d_w) \tag{7}$$

That is:

$$\frac{du_{w,v}}{f(u_{w,v})} = d\tau \tag{8}$$

Since the calculation formula in C_d contains u_w, C_d cannot be considered as a constant. Therefore, it is difficult to obtain analytic solutions by integration, and numerical analysis would be used.

In order to calculate the velocity distribution of water droplets, the motion time of water droplets is divided into several motion ranges at $d\tau$ intervals. The initial velocity of water droplet is $u_{w,v}(0)$, and the exit velocity of droplets at the nth motion interval is $u_{w,v}(n)$. When $d\tau$ is small enough, higher calculation accuracy can be guaranteed. Iterative method is used. According to formula (1)–(8) and initial velocity $u_{w,v}(0)$, the $u_{w,v}(1)_1$ can be calculated, in which subscript denotes iteration times. Assume that $u_{w,v}(0)_2 = (u_0 + u_w(1)_1)/2$, so the $u_{w,v}(1)_2$ can be calculated. It is by analogy until "$u_{w,v}(1)_{n+1} - u_{w,v}(1)_n$" is within the acceptable accuracy range. Taking "$u_{w,v}(1) = u_{w,v}(1)_{n+1}$" as the initial value of the relative velocity for the next period of time, so that the velocity distribution over the entire time period can be calculated.

In the calculation, as the droplet velocity decreases, "$f(u_w) = 0$" will appear. At this point, the vertical relative velocity of the water droplet will no longer change, and this velocity is defined as the settling velocity relative to the flue gas. Figure 2 illustrates the settling velocity of water droplet with different sizes in flue gas (50 °C). With the increasing of droplet size, the settling velocity increases approximately linearly.

The horizontal velocity distribution is similar to the above method. In the horizontal direction, it is only affected by resistance, so the force analysis of water drops is as follows:

$$\frac{du_{w,h}}{d\tau} = -\frac{3}{4}C_d\rho_f u_{w,h}^2/\rho_w \tag{9}$$

where, $u_{w,h}$ is droplet velocity in the horizontal direction, m/s The numerical method is also used to solve the problem.

A gas boiler flue-gas waste heat recovery project in Beijing is taken for example. The water droplets are ejected vertically at 15 m/s. the flue gas with 2 m/s flow across the water droplets. The height of direct contact heat exchanger is 1.5 m. The physical parameters of water droplets and flue gas change little in the heat transfer, which are calculated as the constant values. The flue-gas density is 1.09 kg/m^3, the

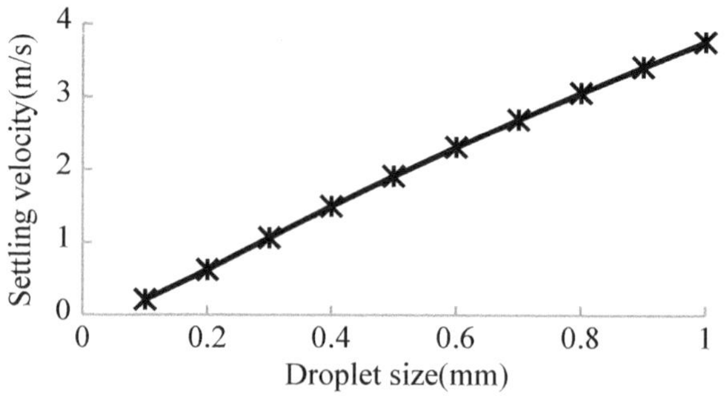

Fig. 2 The settling velocity of water droplet with different sizes in flue gas (50 °C)

droplet density is 1000 kg/m^3, and the dynamic viscosity coefficient of flue gas is 1.86×10^{-5} Pa.s. In order to ensure the calculation accuracy, $d\tau$ is taken as 0.01 s. The trajectories and relative velocities of water droplets with different sizes are shown in Figs. 3 and 4. As can be seen from the figures, the larger the droplet size, the slower the vertical velocity decreases and the shorter in the horizontal direction water droplets moves, which is more easier to avoid the phenomenon that water droplets are taken away by flue gas.

In the above calculations, all values of Re are less than 1000, so the applicable conditions of C_d formula are satisfied.

3 Analysis of Heat and Mass Transfer Between Water Droplets and Flue Gas

The heat and mass transfer process of water droplets with flue gas containing water vapor is a complex physical phenomenon. In order to facilitate the analysis, the following assumptions are made:

(1) The droplet is spherically symmetric and the temperature is uniform distribution, so the lumped parameter method can be used;
(2) The radiative heat transfer between droplets and flue gas is ignored;

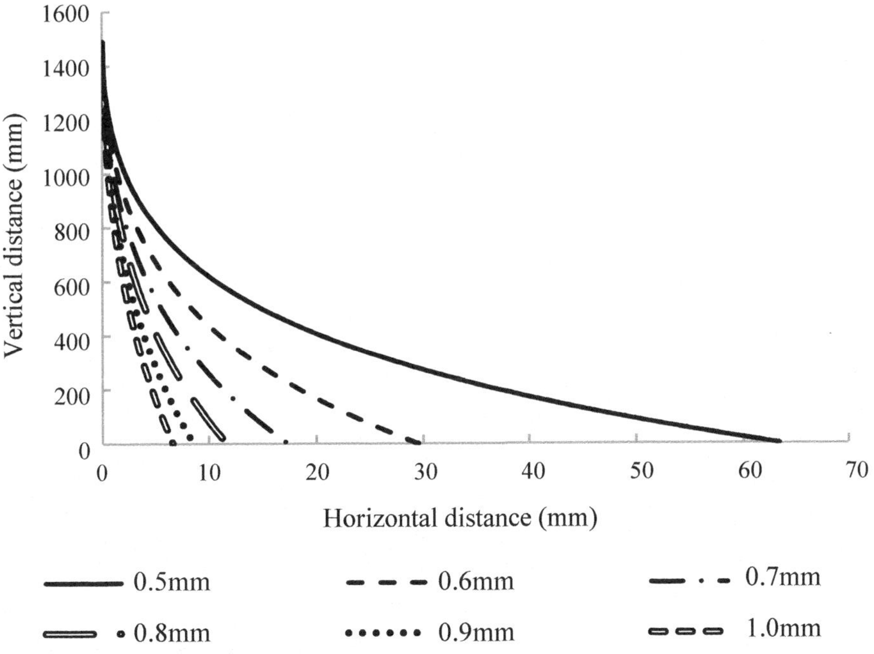

Fig. 3 Trajectories of droplets with different sizes

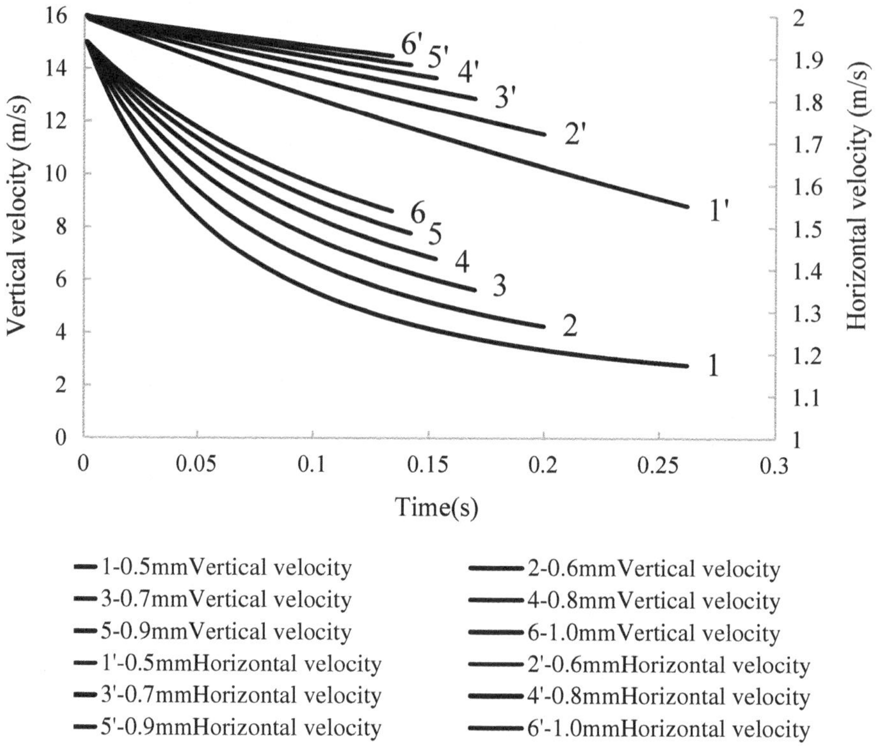

Fig. 4 Relative velocity of droplets with different sizes

(3) The flue gas and water vapor are ideal gases;
(4) The influence of mass transfer on heat transfer is ignored [7];
(5) The Lewis criterion is established [7].

In the flue gas waste heat recovery system of gas boiler, the heat and mass transfer characteristics of direct contact heat transfer between water droplets and flue gas are different from those of conventional spray chambers:

(1) The main component of natural gas is CH_4, and flue gas is rich in water vapor. In the mass transfer process, the water vapor pressure in the flue gas is generally higher than the saturated vapor pressure in the boundary layer of the water droplet. Therefore, the direction of mass transfer is generally opposite to the direction of the water air mass transfer in the spray chamber.
(2) In the gas-water heat and mass transfer process, when the flue-gas temperature is below the dew point temperature, condensed water vapor in the flue gas includes not only the vapor mass transfer, but also the condensation of water vapor caused by the temperature drop of the flue gas, which is different from the conventional surface heat transfer and water-air mass transfer.

According to Fick mass diffusion law, the mass diffusion equation of spherically symmetric model in the radial direction is as follow:

$$\dot{m} = -D\rho_{\mathrm{f}}\frac{dY}{dr} + Y\dot{m} \tag{10}$$

where, $\dot{m}$ is mass diffusion rate per unit area, kg/m^2/s; D is mass diffusion coefficient, m^2/s; ρ_{f} is flue-gas density, kg/m^3; Y is water vapor mass ratio; r is radial coordinate, which is 0 at the center of the droplet and expressed as subscript s on the surface of the droplet, m.

Y can be calculated by the following formula [12]:

$$Y = \frac{P_{\mathrm{w}}M_{\mathrm{w}}}{P_{\mathrm{w}}M_{\mathrm{w}} + (P - P_{\mathrm{w}})M_{\mathrm{f}}} \tag{11}$$

where, P_{w} is water vapor pressure, Pa; M_{w} is molecular weight of water (18); P is flue-gas pressure, Pa; M_{f} is molecular weight of flue gas (29).

According to the literature [12], the mass transfer formula per unit time through the water droplet surface is as follows:

$$m_{\mathrm{z}} = 2\pi d_{\mathrm{w}}\frac{\lambda_{\alpha}}{c_{\mathrm{p},\alpha}}\ln(1 + B_{\mathrm{M}}) \tag{12}$$

where, m_{z} is total mass diffusion rate of water droplet surface, kg/s; λ_{α} is thermal conductivity of flue gas, W/m/K; $c_{p,\alpha}$ is specific heat capacity of flue gas, J/kg/K.

Defining mass transfer number B_{M}:

$$B_{\mathrm{M}} = \frac{Y_{\mathrm{s}} - Y_{\infty}}{1 - Y_{\mathrm{s}}} \tag{13}$$

where, subscript s represents the parameters of water droplets surface, and subscript ∞ is the parameter of flue gas.

The heat transfered by flue gas to droplets is:

$$Q_{\mathrm{h}} = \pi d_{\mathrm{w}}^2 h(T_{\infty} - T_{\mathrm{s}}) \tag{14}$$

where, h is the convective heat transfer coefficient of droplets and flue gas, which can be calculated by following formula [13, 14]:

$$Nu = \frac{hd_{\mathrm{w}}}{\lambda_{\alpha}} = 2 + 0.6\,\mathrm{Re}^{1/2}\mathrm{Pr}^{1/3} \tag{15}$$

Therefore, it can be deduced:

$$Q_{\mathrm{h}} = \pi d_{\mathrm{w}} \lambda_{\alpha} (T_{\infty} - T_{\mathrm{s}})(2 + 0.6\,\mathrm{Re}^{1/2}\,\mathrm{Pr}^{1/3}) \tag{16}$$

The condensation heat of water vapor is:

$$Q_{\mathrm{c}} = m_{\mathrm{c}} L \tag{17}$$

where, m_{c} is condensate water flow, kg/s; L is latent heat of water vapor, kJ/kg.

In the process of heat and mass transfer, the mass and temperature of water droplets are increased, so that heating capacity of droplets is as follows:

$$Q = Q_{\mathrm{h}} + Q_{\mathrm{c}} = \pi d_{\mathrm{w}} \lambda_{\alpha} (T_{\infty} - T_{\mathrm{s}})(2 + 0.6\,\mathrm{Re}^{1/2}\mathrm{Pr}^{1/3}) + m_{\mathrm{c}} L \tag{18}$$

The change rate of droplet temperature is as follows:

$$\frac{dT_{\mathrm{s}}}{d\tau} = \frac{Q}{c_{\mathrm{p}} m_{\mathrm{w}}} \tag{19}$$

where, c_{p} is specific heat of water at constant pressure, kJ/(kg · °C).

According to the conservation of energy, the amount of heat obtained by water droplets and released by the flue gas should be equal. The heat released by flue gas includes sensible heat and condensation heat of the water vapor. The condensation heat of water vapor is Q_{c}, and the sensible heat Q_{h} is as follows:

$$Q_{\mathrm{h}} = \pi d_{\mathrm{w}} \lambda_{\alpha} (T_{\infty} - T_{\mathrm{s}})(2 + 0.6\,\mathrm{Re}^{1/2}\mathrm{Pr}^{1/3}) = \frac{c_{\mathrm{p},\alpha} m_{\mathrm{f}} dt_{\infty}}{d\tau} \tag{20}$$

$$m_{\mathrm{f}} = \frac{m_{\mathrm{w}}}{k} \tag{21}$$

where, m_{f} is flue-gas mass flow, kg/s; k is water-gas ratio, which is mass flow ratio of water to flue gas.

The relationship between m_{c} and m_{z} should be paid attention to. When "$T_{\infty} - dt_{\infty}$" is above the dew point temperature of flue gas, the mass transfer can be considered as diffusive mass transfer, and m_{c} is m_{z}. When "$T_{\infty} - dt_{\infty}$" is below the dew point temperature of flue gas, mass transfer not only includes diffusive mass transfer, but may also mass transfer due to the condensation of water vapor. So the amount of water vapor remaining in the flue gas (residual water vapor) should be calculated according to the accumulated diffusive mass transfer. Then, according to the flue-gas temperature of "$T_{\infty} - dt_{\infty}$", the amount of saturated water vapor in the flue gas is determined. When the amount of residual water vapor is greater than saturated water vapor, the condensation of water vapor will occur, and the amount of condensed water vapor will be added to the calculation.

In the process of heat and mass transfer, various physical properties and droplets sizes are changing, which are correlated with the nonlinear flue-gas temperature or droplets temperature. If they are regarded as variables, it is difficult to get a theoretical solution. And if they are regarded as fixed values, the calculation results will be inaccurate. In order to ensure the accuracy of calculation, numerical analysis method is used.

According to motion characteristics of water droplets, movement time of water droplets is divided into several sections at $\Delta\tau$ intervals. The infinitesimal equations are established for each section. For a section, as initial parameters of water droplets and flue gas are known, the heat transfer and mass transfer in the section can be calculated, and the parameters of calculated water droplets and flue gas are taken as the initial conditions of next section. Water droplets and flue gas can flow downstream, upstream, and across.

It is assumed that size of the water droplet is 0.5 mm, initial flue-gas temperature is 70 °C, inlet temperature of low temperature is 20 °C, water-gas ratio is 5, heat exchanger height is 1.5 m, and water droplets fall perpendicular to flue gas at a speed of 15 m/s. Based on these, the variation of exhaust temperature after heat exchanger with the change of parameters is studied.

(1) Droplet size

Droplet size ranges from 0.3 to 1.0 mm. As shown in Fig. 5, the flue-gas temperature decreases with decreasing droplet size. This is because in the case of constant mass flow of water, reducing droplet size is equivalent to increasing the contact area. Moreover, the droplet size decreases, causing water droplets to slow down more quickly. The contact time between water droplets and flue gas increases when the heat exchanger height is constant. When droplet size decreases to a certain extent (0.3 mm), the exhaust temperature can even be reduced to the water inlet temperature of 20 °C. At this point, it is meaningless to reduce the droplet size to recover the waste heat.

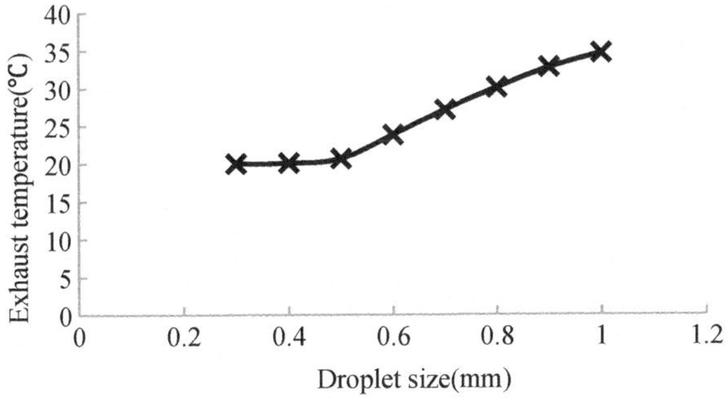

Fig. 5 The relationship between droplet size and exhaust gas temperature

(2) Water-gas ratio

Water-gas ratio ranges from 1 to 10. As shown in Fig. 6, the exhaust temperature decreases with the increase of water-gas ratio, that is because increasing water-gas ratio is equivalent to increasing heat exchange contact area. When water-gas ratio increases to 8, exhaust temperature is reduced to 20 °C, and increasing water-gas ratio will not improve heat transfer effect.

(3) Heat exchanger height

The height of the heat exchanger ranges from 0.5 to 2.0 m. As shown in Fig. 7, increasing the height of the heat exchanger equals to increasing the contact time between the flue gas and water droplets. When the height of the heat exchanger is increased to 2 m, the exhaust gas temperature is about 20 °C.

(4) Initial water droplet velocity

Initial water droplet velocity ranges from 5 to 30 m/s. As shown in Fig. 8, the exhaust temperature increases with the initial velocity of water droplets. Initial water droplet velocity will increase convective heat transfer coefficient of flue gas and water, but will decrease the contact time.

The higher the initial water droplet velocity is, the worse the heat transfer effect is, and the initial kinetic energy of droplet is increased, which can increase the power consumption of water pump. Therefore, the initial water droplet velocity

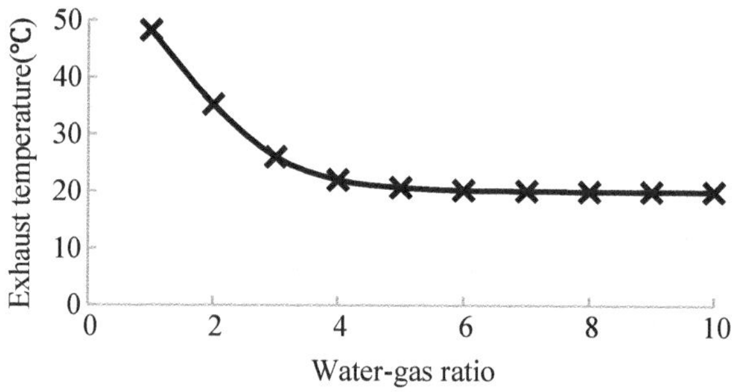

Fig. 6 The relationship between water-gas ratio and exhaust temperature

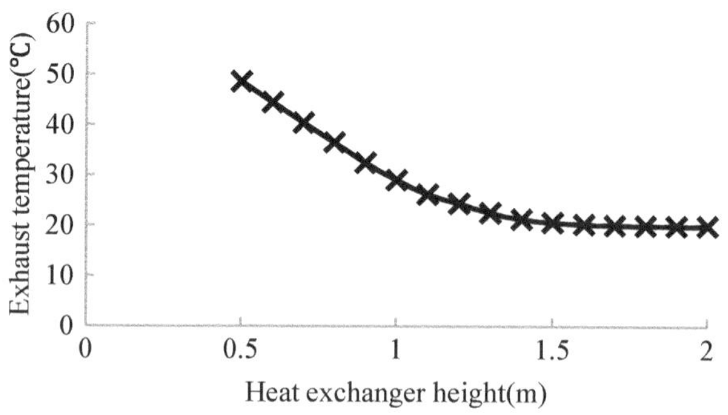

Fig. 7 The relationship between height of heat exchanger and exhaust temperature

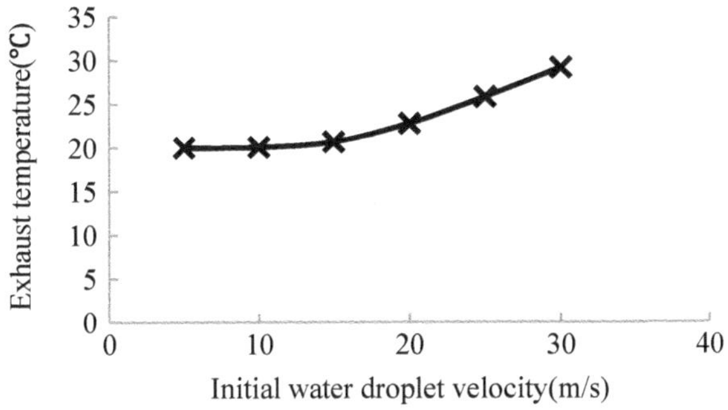

Fig. 8 The relationship between initial water droplet velocity and exhaust temperature

should not be too high, but increasing initial water droplet velocity can prevent low-size droplets from being removed by high velocity flue gas.

(5) Initial flue-gas temperature

Initial flue-gas temperature ranges from 50 to 100 °C. As shown in Fig. 9, exhaust temperature varies little with increasing initial flue-gas temperature. When initial flue-gas temperature increases from 50 to 100 °C, the exhaust temperature increases less than 1 °C. This is because increasing flue-gas temperature does not reduce the contact time and area between flue gas and water droplets.

(6) Initial water temperature

Initial water temperature ranges from 10 to 40 °C. As shown in Fig. 10, exhaust temperature increases approximately linearly with initial water temperature, and initial water temperature is the limiting temperature that flue gas can be cooled to.

The contact area and contact time of flue gas and water droplets have greater influence on heat transfer effect. Therefore, the size, water-gas ratio and height of heat exchanger are the main factors affecting the heat transfer. The initial water temperature is the limiting factor affecting the exhaust temperature, which determines the cooling limit of the exhaust temperature.

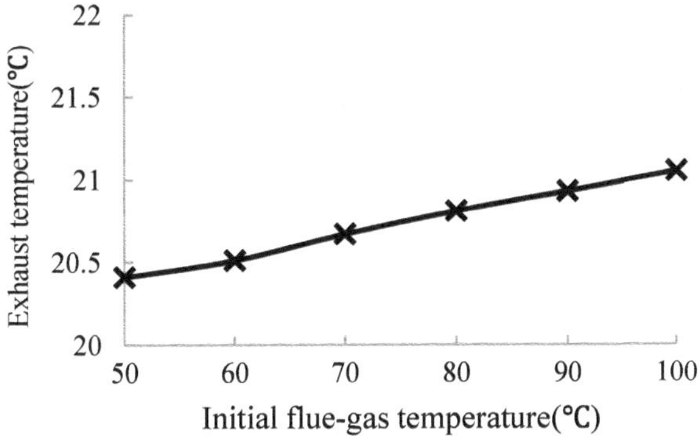

Fig. 9 The relationship between initial flue-gas temperature and exhaust temperature

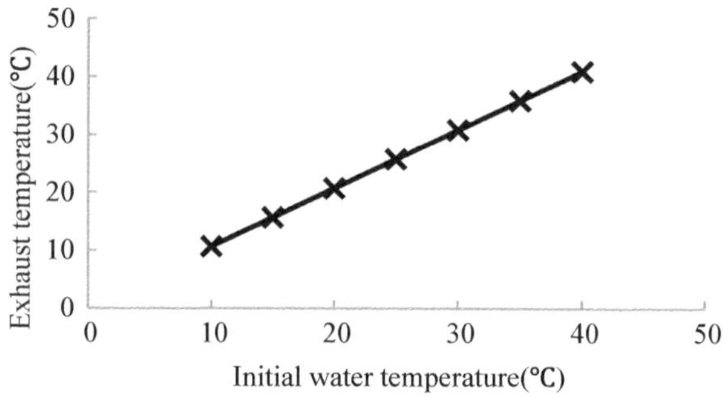

Fig. 10 The relationship between initial water temperature and exhaust temperature

4 Experimental Analysis of Direct Contact Heat Transfer

There are three 29 MW hot-water boilers and a 14 MW hot-water boiler in a gas-fired boiler room in Beijing. The temperature of flue gas entering the chimney is generally 60–80 °C.

In this project, an absorption heat pump and direct contact heat exchangers are added, which is used to recover flue-gas waste heat of a 29 MW boiler. The absorption heat pump generates low-temperature water by using gas for driving to recover flue-gas waste heat and heats the return water of heat supply network. In heat exchangers, flue gas exchanges heat with low-temperature water to release sensible heat and latent heat, and is discharged to the atmosphere through the chimney. The low-temperature water is pumped into absorption heat pump after heating in direct contact heat exchangers.

There are three direct contact heat exchangers, in which the size of a single heat exchanger (length × width × height) is 2 m × 2.45 m × 1.5 m. Water droplets can be regarded as free fall motion with initial velocity 15 m/s. According to the heat exchanger height 1.5 m, it can be calculated that time of water droplets from the outlet of the nozzle to the surface of the water is 0.28 s. The flue gas moves along the length direction of the heat exchanger with 2 m/s, and the time is 3 s through the heat exchanger.

The mesh is divided according to the movement time, and the time interval is 0.02 s. In the vertical direction, it can be divided into 14 sections, where the outlet temperature of water in the first section is equal to the inlet temperature of water in the second section, and so on. It can be divided into 150 sections in the length direction, where the outlet temperature of flue gas in the first section is equal to the inlet temperature of flue gas in the second section, and so on. The mesh is shown in Fig. 11.

After the flue gas passes through the left heat exchanger, the flue-gas temperature of each micro segment is mixed as the initial temperature of the middle heat exchanger, and so on. The water drop passes along the right heat exchanger and flows into the pool, and mixing temperature is used as the initial temperature of the middle heat exchanger, and so on.

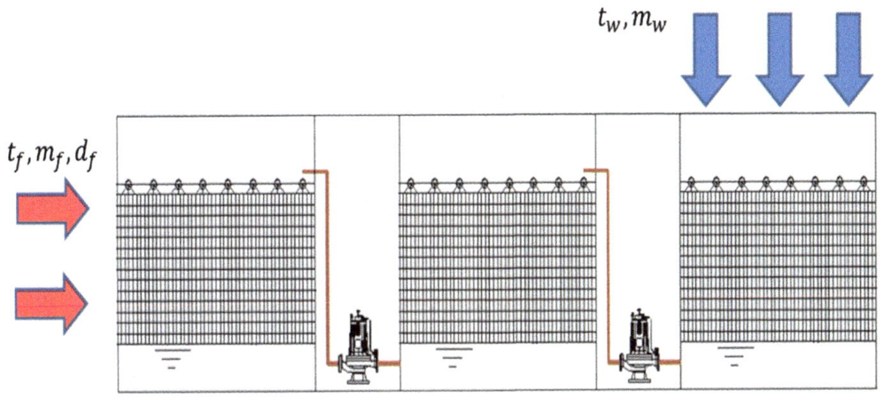

Fig. 11 The model of direct contact heat exchanger

Figures 12 and 13 compares simulation results with experimental results.

By comparison, the error of water side and flue-gas side is within 2 °C, and the average error is 0.98 and 0.95 °C, respectively. The main reasons for the errors are that:

1. The sizes of water droplets are in normal distribution, not a constant;
2. The water droplets do not drop completely vertically, but fall at a certain angle;
3. In the process of water droping, collision is inevitable.

All the above problems are simplified in this model, and further research is needed.

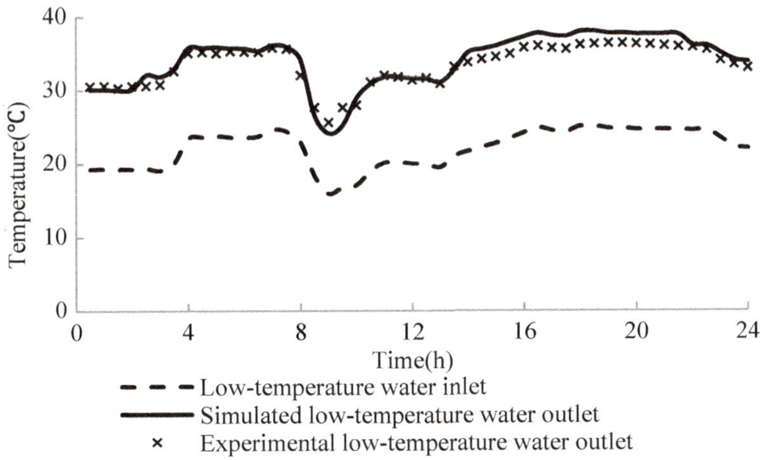

Fig. 12 The comparison between simulation and experimental of water outlet

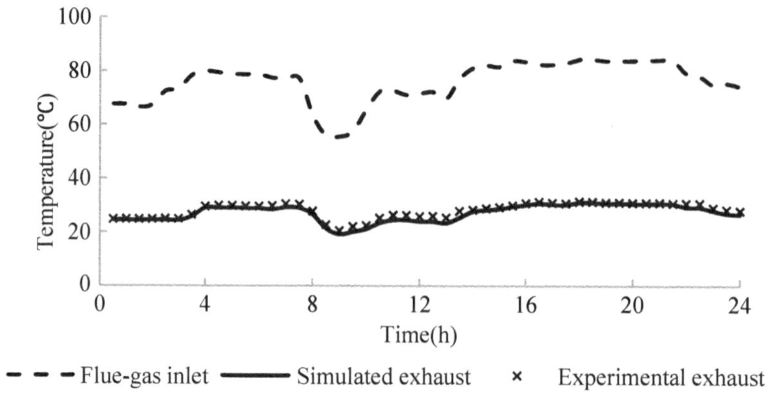

Fig. 13 The comparison between simulation and experimental of flue-gas outlet

5 Conclusion and Discussion

In this paper, the direct contact heat transfer of flue gas and water has been analyzed and validated by experiments. The following conclusions are obtained:

(1) The motion characteristics of water droplets are analyzed theoretically, and the numerical solutions of velocity distribution of water droplets are obtained. The larger the droplet size, the higher the settling velocity.
(2) Heat-exchange characteristics of flue gas and water droplets have been analyzed, and the main factors have been studied. The results show that the water droplet size, water-gas ratio and heat exchanger height are the main factors influencing heat transfer.
(3) The direct contact heat exchangers of flue-gas waste heat recovery project in a gas-fired boiler were measured and the accuracy of theoretical calculation results has been verified.

References

1. D.F. Che, Y.H. Liu, C.Y. Gao, Evaluation of retrofitting a conventional natural gas fired boiler into a condensing boiler. Energy Convers. Manag. **45**, 3251–3266 (2004)
2. R.M. Lazzarin, Condensing boilers in buildings and plants refurbishment. Energy Build. **47**, 61–67 (2002)
3. T.K. Zhelev, K.A. Semkov, Cleaner flue gas and energy recovery through pinch analysis. J. Clean. Prod. **12**, 165–170 (2004)
4. J. Kuck, Efficiency of vapor pump equipped condensing boilers. Appl. Therm. Eng. **16**(3), 233–244 (1996)
5. M. Lianbo, Application and analysis of heat pump system applied to the flue gas heat recovery for gas boiler. HV&AC **44**(12), 49–54 (2014)

6. F. Lin, T. Guansan, S. Jun et al., Combining absorption heat pump with gas boiler for exhaust condensing heat recovery in district heating system. Acta Energiae Solaris Sinica **24**(5), 620–624 (2003)
7. L. Hua, Z. Xian, F. Lin, Heat transfer performance of direct-contact flue-gas condensation heat exchanger. HV&AC **44**(9), 97–100+15 (2014)
8. L. Hua, Z. Xian, F. Lin, Thermal calculation method for direct-contact flue-gas-water condensation heat exchangers. HV&AC **45**(2), 111–114 (2015)
9. L. Hua, Z. Xian, F. Lin, Experimental research on influence factors of direct-contact flue-gas-water condensation heat exchange. HV&AC **45**(7), 90–95 (2015)
10. J. Xin, Z. Zhou, H. Xin, et al., Validation analysis of different mass transfer formula in single droplet evaporation model. CIESC J. **63**(6), 1704–1708 (2012) (http://www.cnki.net/KCMS/detail/detail.aspx?QueryID=1&CurRec=1&filename=HGSZ201206009&dbname=CJFD2012&dbcode=CJFQ&pr=&urlid=&yx=&uid=WEEvREcwSlJHSldRa1FhdkJkcGp4QmVpMjRmOGZMNE8zekVSY-lrMkdNbz0=$9A4hF_YAuvQ5obgVAqNKPCYcEjKensW4ggI8Fm4gTkoUKaID8j8gFw!!&v=MjcwMTdZUjhlWDFMdXhZUzdEaDFUM3FUcldNMUZyQ1VSTDJlWitSdkZ5amxWTHJCTFNyWWRMRzRIOVBNcVk5RmI=)
11. H. Xin, Z. Zhou, B Chen et al., Comparative analysis of single droplet evaporation models in cryogen spray cooling. J. Eng. Thermophys. **30**(4), 653–656 (2009) (http://www.cnki.net/KCMS/detail/detail.aspx?QueryID=7&CurRec=1&filename=GCRB200904032&dbname=CJFD2009&dbcode=CJFQ&pr=&urlid=&yx=&uid=WEEvREcwSlJHSldRa1FhdkJkcGp4QmVpMjRmOGZMNE8zekVSYVlrMkdNbz0=$9A4hF_YAuvQ5obgVAqNKPCYcEjKensW4ggI8Fm4gTkoUKaID8j8gFw!!&v=MzExMDBlWDFMdXhZUzdEaDFUM3FUcldNMUZyQ1VSTDJlWitSdkZ5amxWYnpLSWk3WmJMRzRIdGpNcTQ5R1pvUjg=)
12. N. Liu, X. Zhang, Theoretical analysis of droplet evaporation and numerical calculation of its unsteady-state period. Refrigeration **26**(3), 45–48 (2007)
13. Y. Zhang, *Building Environment Mass Transfer* (China Architecture and Building Press, Beijing, 2006)
14. W. Ranz, W. Marshall, Chem. Eng. Prog. **48**, 141 (1952)

Impacts of Energy Efficient Constant Output Heating on the Moisture Conditions of Unoccupied Summer Cottages in Finland

Juha Vinha, Jarkko Piironen and Mihkel Kiviste

Abstract Finland has around half a million summer cottages. An increasing demand exist for using these cottages also in other seasons. The influence of constant output heating is studied for energy saving purposes while avoiding moisture damage in the cottages. The proposed type of heating requires only a fraction of amount of electrical energy in comparison with conventional heating. Field measurements, which are presented in current paper, aim at indicating the actual moisture conditions in unoccupied Finnish summer cottages. Field measurements in seven non-insulated massive log walled cottages in Tampere region (Finland) have been performed during 2007–2009. Based on the values of indoor and outdoor RH (at one-hour interval) the monthly average differences between indoor and outdoor water vapour content were determined. Prevailing conditions of moisture deficit were found in summer cottage case buildings during the measurement period. These results oppose to the usual conditions of moisture excess in Finnish residential buildings. Despite the difference in moisture conditions the envelope structures in summer cottages and in residential buildings are often similar.

Keywords Field measurements · Unoccupied summer cottage

1 Introduction

Research has drawn out the advantages of development and application of energy-efficient heating, ventilation, and air conditioning (HVACs) systems [1–6]. Finland has around half a million summer cottages. There exists a trend of constructing new cottages as well as renovating the building structures and replacing HVAC systems in existing ones. The objective of replacing heating systems in

J. Vinha (✉) · J. Piironen · M. Kiviste
Tampere University of Technology, Korkeakoulunkatu 5, 33101 Tampere, Finland
e-mail: juha.vinha@tut.fi

M. Kiviste
e-mail: mihkel.kiviste@tut.fi

D. Johansson et al. (eds.), *Cold Climate HVAC 2018*,
Springer Proceedings in Energy, https://doi.org/10.1007/978-3-030-00662-4_7

existing cottages is to avoid moisture damage with minimal costs, while the occupants are away [1, 7, 8].

The purpose of typical electric heating is to maintain certain thermal conditions in summer cottages. Maintaining constantly high indoor temperatures is very energy consuming. Current annual electricity consumption of free-time residences is clearly exceeding 500 GWh and due to the increasing rate of construction of new summer cottages it is expected reach 1000 GWh in the future, which is 10% of heating and domestic electricity consumption. Therefore, there lies considerable energy saving potential.

Studied constant output heating require only 5–15 W of electricity per floor square meter, which is rather marginal compared to conventional electric heating [1].

In heated and occupied residential buildings in Nordic climate the indoor air usually consists more water vapour than outdoor air. Therefore, there exists constant conditions of moisture excess. That is caused by residential moisture production of inhabitants and heated indoor air, which has naturally higher water vapour absorption capacity in comparison with outdoor air.

The moisture conditions in summer cottages are expected to be different since there exists less moisture production from inhabitants (used only occasionally) and also the indoor temperature is kept on lower level. However, the actual moisture conditions and their temporal change in summer cottages was not exactly known. Field measurements, which are presented in current paper, aim at indicating the actual moisture conditions in unoccupied Finnish summer cottages.

The study is a part of research project "Eco-efficient irregular heating (EREL)" carried out in 2007–2009. This project studied free-time residences based on continuous energy and eco-efficiency in order to pursue finding and developing optional technologies, services and practices, which enable to increase the eco-efficiency of summer cottages.

2 Materials and Methods

2.1 Case Study Buildings

Seven uninsulated massive log summer cottages in Tampere region, Finland (Fig. 1) were selected for the study. The field measurements consisted of temperature, relative humidity as well as air-tightness measurements of the log buildings. In five case buildings the temperature and relative humidity were measured for two consecutive years: 2007–2009. In two case buildings the measurements were performed for one year: in 2007–2008 and 2008–2009, respectively.

The general information of the case study buildings is presented in Table 1. All the case buildings have natural ventilation through ventilation holes in external walls. The occupants have possibility to control the ventilation by opening or closing the ventilation holes. The information about ventilation openings (open, partially open or closed) is presented in Table 1. The ceiling and floor structures in cottage no. 5 have been retrofitted during moving.

Fig. 1 Front and side view of two Finnish summer cottages

2.2 *Temperature and Relative Humidity Measurements*

Temperature and relative humidity dataloggers (Comark Diligence EV N2003) with battery power were applied for the measurements. The measuring range for relative humidity was 0…97% in a precondition that the humidity is not condensed. The accuracy for temperature is ±0.5 °C and ±3% for relative humidity. Dataloggers were programmed for one hour measuring interval of temperature and relative humidity measurements. Due to limitations in memory capacity, the measuring data were collected in two stages: in spring 2008 and in spring 2009 (at the end of measurements). Four, five or six loggers were placed to each case building. One logger was recording the outdoor conditions while the rest were recording indoor conditions. The outdoor dataloggers were placed under the protection from rain and direct sunlight i.e. to the terrace or under the shelter. Same outdoor logger was placed for case buildings 4 and 5, since they were located at the same plot. The indoor loggers were placed to inner surface of external wall and window and so called cold surface, where conditions were expected to change more slowly compared to indoor air. If possible, windows in different directions were included to account also sun radiation. Figure 2 shows the location of dataloggers in case building 5.

Table 1 General information of the case study buildings

No. of case building	1	2	3	4	5	6	7
Measurement period	07–08	07–09	07–09	07–09	07–09	07–09	08–09
Location	Tampere	Keuruu	Orivesi	Lavia	Lavia	Keuruu	Pälkäne
Year of construction	1989	2000	2007	1985[a]	2002	1993	1988
Indoor area [m^2]	53	24	75	17	66	27	28
Indoor volume [m^3]	135	68	182	40	181	78	69
Envelope weighed U-value [W/($m^2 \cdot$ K)]	Not calculated	0.66	0.45	0.60	0.51	0.60	0.63
Airtightness: air change rate, n_{50} [h^{-1}]	24.0	26.9	15.7	30.0	5.1	29.7	6.3
Ventilation holes	Opened	Partially opened	Partially opened	Partially opened	Partially opened	Closed	Closed
Heating in winter 07–08	No	No	No	No	Basic heating	Constant 220 W	Not included
Heating in winter 08–09	Not included	Constant 210 W	Constant 780 W[b]	Constant 110 W[b]	Constant 800 W	Constant 220 W	Constant 140 W
Main direction of windows	West	East	West	East	East	East	East

Notes [a]Retrofits in 2000, [b]part of the time unheated

3 Results and Discussion

The average daily conditions of different case buildings are presented in Fig. 3. Figure 3 shows the indoor air condition at the centre of each cottage (logger 1 in Fig. 2). Indoor and outdoor conditions of unheated case buildings from 1.12.07 to 23.2.08 are presented in left and constant output heated case buildings from 1.12.08 to 23.2.09 are presented in right column in Fig. 3. Upper figures show measured relative humidities, middle show measured temperatures and lower show water vapour concentrations, which were calculated from the measured temperature and relative humidity values. Since the indoor air conditions are measured in consecutive years, to enable better comparison, the outdoor air conditions in Fig. 3 are taken from the nearest weather station of Finnish Meteorological Institute (in Tampere, abbreviation TRE in Fig. 3).

The following findings were made from the measurement results from unheated summer cottages (Fig. 3 left).

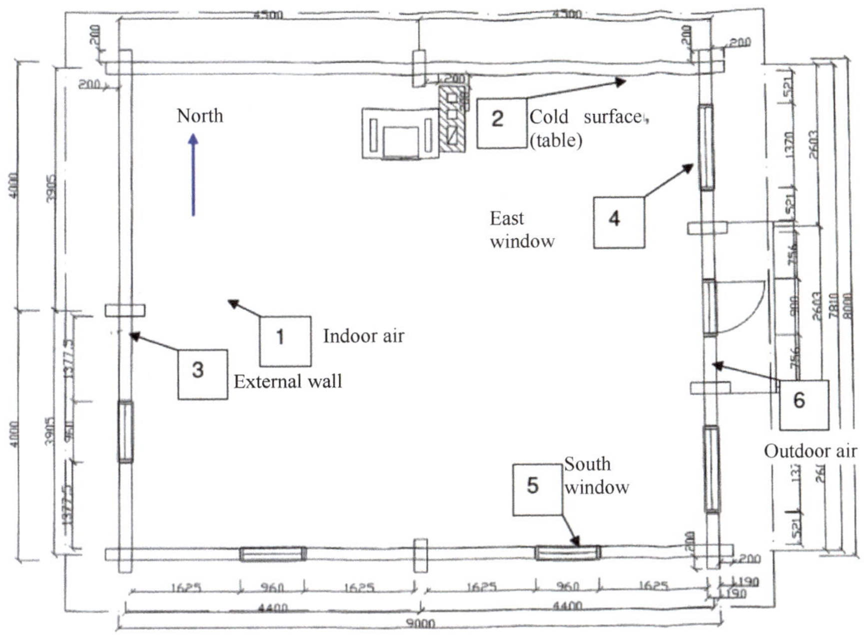

Fig. 2 Location of dataloggers in case building 5

- The relative humidity of indoor air stays mainly under 80%, but could be occasionally 10–15% higher on window surface making it the most susceptible place for condensation. However, usually condensation does not occur in unoccupied premises.
- Indoor temperature and humidity conditions mainly follow outdoor conditions. Indoor temperature and water vapour content follow outdoor changes with some hours delay. The surface of the materials also influences indoor air conditions.
- The thermal conductivity of envelope and ventilation have a major impact on the indoor temperature and water vapour content.
- The indoor water vapour content is lower than that of outdoor water vapour content in most times. The opposite happens in spring and during the residence. The moisture absorption capacity of log surfaces in indoor air also affect the situation.
- Winter visits dry the cottage. Drying is more significant the more the visits occur and last.

The following findings were made in constant output heated summer cottages (Fig. 3 right):

- The temperature increase (heating) +3 °C decreases about 10% relative humidity. Heating has lower impact on window surfaces than for indoor air conditions.

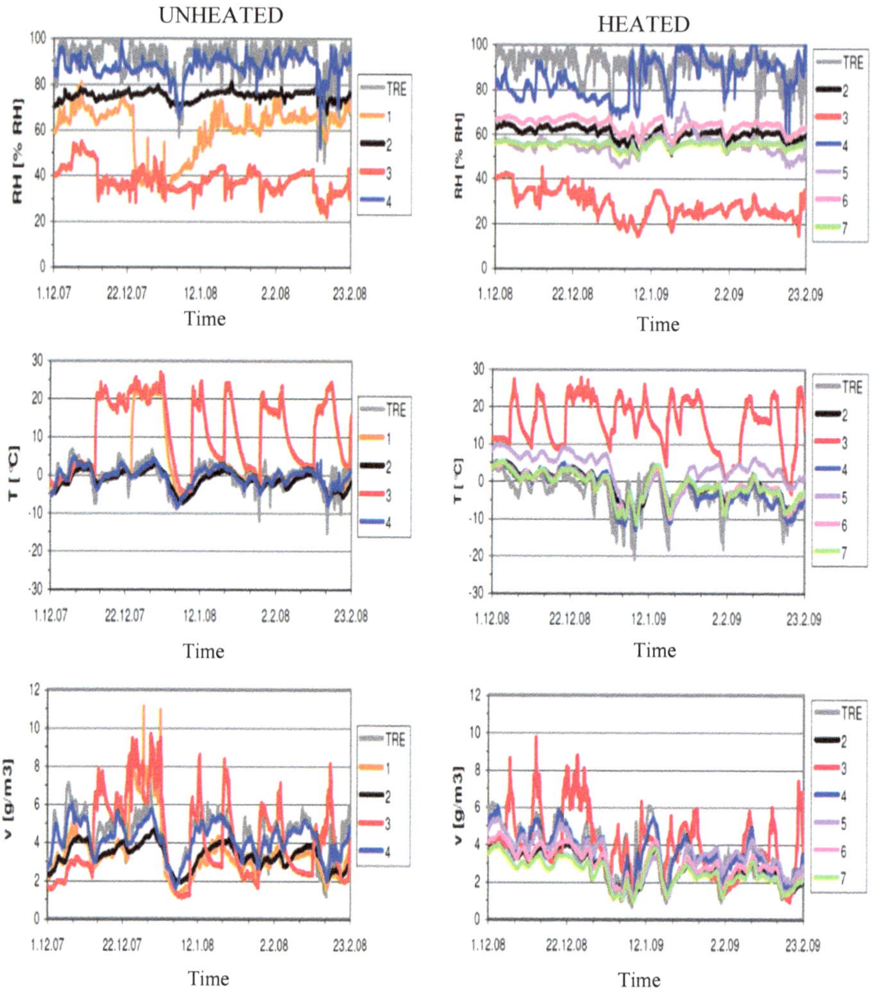

Fig. 3 Daily average indoor and outdoor conditions of unheated case buildings from 1.12.07 to 23.2.08 are presented in left and constant output heated case buildings from 1.12.08 to 23.2.09 are presented in right. Upper figures show relative humidities, middle show temperatures and lower show water vapour concentrations

- Changing conventional heating to constant output heating does not affect indoor conditions significantly. Constant output heating could improve the conditions in autumn and spring.
- Similarly, outdoor conditions affect indoor conditions. However, indoor temperature stays a few degrees higher.
- Heating does not affect water vapour content, thus the difference between indoor and outdoor water vapour content is the same as in unheated summer cottages.

- Heating is not necessary in airtight cottage with nearly continuous residence. Heating could decrease the indoor relative humidity to aversely low levels.
- Decreasing ventilation could save heating power. The effect is greater in more airtight envelopes (Table 1). Closing ventilation in an airtight unoccupied cottages (cottages 6 and 7 in Table 1) did not have adverse effect on the relative humidity or water vapour content even in spring (upper and lower figure in Fig. 3).

Based on indoor and outdoor RH logger measurements the monthly average differences between indoor and outdoor water vapour content were found (Fig. 4).

Figure 4 shows that indoor moisture deficit related to outdoor exist in case buildings during most times a year excluding spring. This situation is the opposite to the residential buildings in continuous use. The greatest indoor moisture deficit could be found in late summer and early autumn. Indoor moisture deficit decreases during winter and could turn to moisture excess in spring. In addition, occupant visits also produce moisture. It can be noticed from Fig. 4 that cottage 3 has higher indoor moisture related to outdoors in both studied winters, which was related to frequent winter use of the cottage.

It should be taken into account field measurements were performed only in uninsulated massive log cottages. The results might not be valid in e.g. timber-frame or inside insulated log cottages, because the water absorption capacity of inner surface of log wall could differ significantly. Conventional heating could be changed to constant output heating only, if the risks for possible freezing of home appliances and water systems are eliminated.

Additional research is needed in timber-frame summer cottages with different insulation materials in order to account the whole Finnish summer cottage building stock. Also, additional field measurements in roof, floor and foundations (with and without crawl spaces) need to be performed for more reliable understanding of the performance of constant output heating. Additional research is necessary for the importance of ventilation, temporary breaks of heating and intermittent heating for possible heating optimization without increasing mould risks.

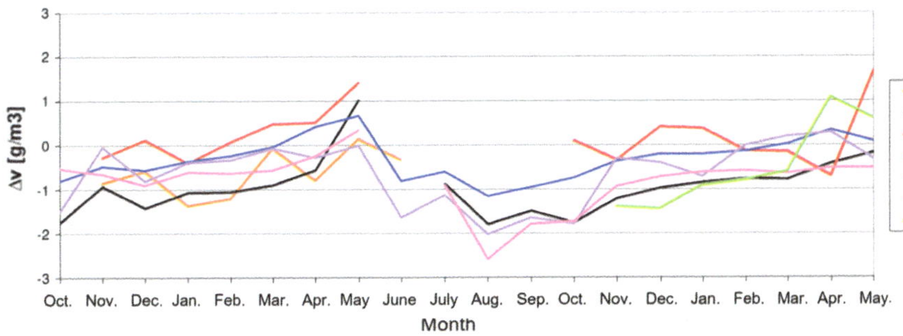

Fig. 4 The monthly average differences between the indoor and outdoor water vapour content of studied case buildings during 2007–2009

4 Conclusions

Current study presents the results of field measurements of indoor air conditions from seven unoccupied non-insulated massive log-walled cottages in Finland (Tampere region) during 2007–2009.

In general, changing conventional heating to constant output heating does not affect indoor air conditions significantly. Constant output heating could improve the conditions in autumn and spring. A temperature increase of 3 °C was found to decrease relative humidity about 10%.

The monthly average moisture content of indoor air was found to be lower than the average moisture content of outdoor air indicating an average moisture deficit of indoor air in unoccupied Finnish summer cottages. The indoor moisture deficit related to outdoor exist in case buildings during most times a year excluding spring in some cases. These results oppose to the usual moisture excess of full-time residential buildings in Finnish climate.

References

1. J. Piironen, J. Vinha, M. Kiviste, Modelling hygrothermal performance of roof and floor structures with an energy-efficient constant output heating. Energy Procedia **132**, 694–699 (2017)
2. J. Burgett, A. Chini, P. Oppenheim, Specifying residential retrofit packages for 30% reductions in energy consumption in hot–humid climate zones. Energ. Effi. **6**(3), 523–543 (2013)
3. E.H.C.P. Mathews, D.C. Botha, A. Arndt, Malan.: HVAC control strategies to enhance comfort and minimise energy usage. Energy Build. **33**(8), 853–863 (2001)
4. L. Pérez-Lombard, J. Ortiz, J.F. Coronel, I.R. Maestre, A review of HVAC systems requirements in building energy regulations. Energy Build. **43**, 255–268 (2011)
5. R.J. Meyers, E.D. Williams, H.S. Matthews, Scoping the potential of monitoring and control technologies to reduce energy use in homes. Energy Build. **42**, 563–569 (2010)
6. M.M. Ardehali, T.F. Smith, Evaluation of variable volume and temperature HVAC system for commercial and residential buildings. Energy Convers. Manag. **37**(9), 1469–1479 (1996)
7. A. Rytkönen, A. Kirkkari, *Eco-Efficiency of Holiday Homes*. The Finnish Environment 6/2010. Ministry of the Environment, Department of the Built Environment, Helsinki, 112 p. (in Finnish, 2010)
8. J. Piironen, J. Vinha, *Constant Output Heating and its Effect on the Hygrothermal Performance of Log Summer Houses*. Research report no. 150. Tampere University of Technology, Tampere. 79 p. + app. 16 p. (in Finnish, 2010)

Setback Efficiency of Limited-Power Heating Systems in Cold Climate

Tuule Mall Kull, Raimo Simson and Jarek Kurnitski

Abstract The main aim of this work is to analyse the energy saving potential and peak power impact resulting from the temperature setback approach. This paper analyses low energy buildings incorporating high-efficiency heating systems with limited power, as additional power for district heating and heat pump systems will need costly investments. The setback efficiency is estimated for different types of heating systems. Underfloor heating is compared to radiators both for limited and excess power. Based on estimated time constants, suitable heat-up time is calculated to minimise the time when temperatures stay below setpoint during occupancy. The energy saving potential of night-time and weekend setback periods in an office are analysed. It is found that the energy saving potential of setback is low under given constraints. Therefore, for modern buildings the cost-optimality has to be assessed separately for specific cases.

Keywords Temperature setback efficiency · Low-energy buildings
Limited power · Cold climate

1 Introduction

The method of periodically decreasing the set temperature of heating systems in buildings when the rooms are vacant, often called intermittent heating or the setback approach is a widely used method for energy saving. In several studies [1–3] energy saving potentials of up to 20% were identified. In single cases, the observed reductions are much higher [4] or much lower [5]. In the mentioned studies, mostly moderately insulated buildings are considered with simple setback control mechanisms based on pre-defined set-temperature schedules. However, such an approach

T. M. Kull (✉) · R. Simson · J. Kurnitski
Tallinn University of Technology, Ehitajate tee 5, 19086 Tallinn, Estonia
e-mail: tuule.kull@ttu.ee

J. Kurnitski
School of Engineering, Aalto University, Rakentajanaukio 4A, 02150 Espoo, Finland

D. Johansson et al. (eds.), *Cold Climate HVAC 2018*,
Springer Proceedings in Energy, https://doi.org/10.1007/978-3-030-00662-4_8

generates discomfort during the times when people arrive and the temperature has not achieved its set value. In recent years, most intermittent heating control systems for low energy buildings include advanced control methods to solve this problem [6]. However, these are not simple to apply. Applying setback temperatures requires over-dimensioned heating systems to enable fast heat up times [7]. However, a typical advantage of modern low energy buildings is the utilization of low peak-power heating systems which reduces the building's investment cost.

Our assumption is that only very low energy savings can be achieved by temperature setback in modern well insulated buildings and therefore the required investment in over-sized heating systems is not profitable. Therefore, the efficiency of intermittent heating in modern and old buildings is compared in this work.

2 Approach

2.1 The Building Description

Envelope. The room model used for the simulations is a 13.3 m^2 office with a 3-m^2 window facing north. We have previously described this model with all construction specifications for modern and old buildings as well as heavy and light construction types in [8]. The room has one external wall and an external floor over outdoor air, therefore, its heating demand is larger than for an average office building. This case is defined as 'standard' office. For comparison, a similar office room with less insulation is defined as 'old' office. The third configuration is referred to as 'modern' as it has standard constructions but its floor is adiabatic and the window is south-facing. The total heat loss coefficient (without ventilation system) is 7, 9 or 18 W/K for modern, standard, and old buildings respectively.

Ventilation and internal gains. We have redefined the internal loads and ventilation control according to the Estonian norms for office simulations [9], meaning that the ventilation airflow is 2 l/s/m^2 during the occupancy hours (7 a.m. to 6 p.m. at workdays) and 1 h before and after this timeframe. The usage profile is depicted in Fig. 1. These usage factors are multiplied with 5.8 W/m^2 for occupant heat gain, and 9.5 W/m^2 for heat gains from lighting and electrical appliances. During the weekends and holidays the building is not in use. The supply air temperature of the ventilation is 18 °C. For the modern offices, 80% heat recovery from the exhaust air is assumed, whereas the old building has no heat recovery at all. The infiltration in modern buildings is 1.3 l/s; in the old buildings, it is included in the ventilation.

Heating systems. Two types of limited power heating systems are simulated: ideal heaters and electric floor heaters. Ideal heaters represent radiators (Rad) supplied by a district heating system, while electric floor heating represents underfloor heating (UFH) supplied by a heat pump. Using the electric/ideal systems replaces here the function of raising the heating curve to achieve maximum output power of the systems during the heat-up. The nominal power of the modern systems is 273 W,

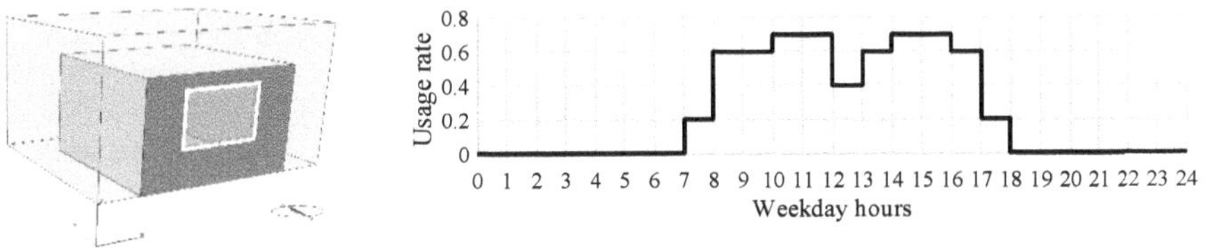

Fig. 1 The isometric view of the simulated zone (on the left, screenshot from software IDA-ICE) and the usage factor profile

of standard systems 437 W, and for old systems 1656 W. For the comparison with a gas boiler supplied radiator-based heating system, which can be easily over dimensioned without distinct cost increase, ideal heater cases with 684 W in standard light and 1367 W in heavy offices are simulated and defined as 'over-dim-Rad'. These are dimensioned according [7] to weekend setbacks.

Simulation. The building is simulated in IDA-ICE 4.7.1 software [10] using Estonian TRY [11] as climate data. The heating season is assumed to be from 1st of October to 30th of April.

2.2 *The Control Description*

Reference case. The performance of the setback control is evaluated by comparing the required heating energy demand against the demand of a reference case, where a constant temperature of 21 °C is maintained by a proportional-integral (PI) controller in the same room. Here, the performance is defined as the heating demand for both space heating and supply air heating by the air-handling unit.

Control algorithm. The setback control also keeps the air temperature during the occupancy hours at 21 °C with PI control, only it is reduced to 18 °C during unoccupied hours. However, to ensure comfort conditions when occupancy starts the required heat-up time, until comfort temperature is reached, is calculated.

If that calculated heat-up time is longer than the actual time left to the start of occupancy hours, the set temperature is changed to 21 °C overriding the initial PI control. If the temperature rises faster than estimated, the set temperature is turned back to 18 °C again.

Heat-up time calculation. The heat-up time is the time the system needs to heat the room up to 21 °C again from setback. It is calculated every 5 min. For that, the one-time-constant model for the building is used. From the heat balance equation of

$$C\frac{d\theta_{in}}{dt} = H(\theta_{out} - \theta_{in}) + \Phi \tag{1}$$

Table 1 Calculated input parameters used for preheat-control in intermittent heating

Energy-efficiency level	Str. mass	Heat emitter	Abbrev.	τ_n (h)	τ_{wnd} (h)	Φ (W)	H (W/K)	C_{20mm} (kJ/K)
Standard	Heavy	UFH	S_H_UFH	50	225	437	9	1677
Standard	Heavy	Rad	S_H_Rad	50	225	437	9	1677
Standard	Light	UFH	S_L_UFH	50	150	437	9	1561
Standard	Light	Rad	S_L_Rad	50	150	437	9	1561
Old	Heavy	Rad	O_H_Rad	25	125	1656	18	1677
Old	Light	Rad	O_L_Rad	25	75	1656	18	1561
Modern	Heavy	UFH	M_H_UFH	50	300	273	7	1677
Modern	Heavy	Rad	M_H_Rad	50	300	273	7	1677
Standard	Heavy	Over-dim Rad	S_H_O_Rad	50	225	1367	9	1677
Standard	Light	Over-dim Rad	S_L_O_Rad	50	150	684	9	1561

for the indoor temperature θ_{in}, the solution for the heat-up time is derived:

$$t = -\tau \cdot \ln\left(\frac{\Phi/\mathrm{H} - \theta_{set} + \theta_{out}}{\Phi/\mathrm{H} - \theta_{in} + \theta_{out}}\right). \tag{2}$$

Φ is the heating power in watts, H is the heat loss coefficient (W/K), θ_{out} is the exterior air temperature, θ_{set} is 21 °C and θ_{in} is the current indoor air temperature. τ is the time constant in seconds, however in this work it is always converted to hours. It is calculated as $\tau = \mathrm{C}/\mathrm{H}$. C represents the heat capacity of the air and structures (J/K). For the calculation of time constant for night setback, the surface layers up to 20 mm depth are included into the heat capacity calculation. The active layer depth of 100 mm is used for weekend setback [12]. The time constants are quantized; they are rounded to the closest 25 h. This is done to use approximate values, as the exact values are not known in real cases. The used values are shown in Table 1. The 100 mm heat capacity values are approximately four times higher than the 20 mm values shown in the table (7449 kJ/K for heavy and 5002 kJ/K for light).

3 Results and Discussion

3.1 Energy Performance

The simulated energy demands for all observed cases are shown in Table 2. It can be seen that the energy consumption in the air-handling unit is almost equal for all the modern and standard cases; it is zero for old buildings, as the supply air is not heated. The total reduction of energy demand resulting from the intermittent heating

Table 2 Energy need results for constant temperature and setback control cases

	Space heating [kWh/(m^2 a)]		Air handling unit [kWh/(m^2 a)]		Total [kWh/(m^2 a)]	
	21 °C	Setback	21 °C	Setback	21 °C	Setback
S_H_UFH	52	47	15	17	68	64
S_H_Rad	48	44	16	17	65	62
S_L_UFH	53	47	15	17	68	64
S_L_Rad	48	43	16	17	64	60
O_H_Rad	207	195	0	0	**207**	**195**
O_L_Rad	206	194	0	0	**206**	**194**
M_H_UFH	17	15	14	15	*31*	*30*
M_H_Rad	15	13	15	16	*30*	*29*
S_H_O_Rad	49	44	16	17	65	61
S_L_O_Rad	48	42	16	17	64	60

operation is shown in Fig. 2. All observed cases result in heating demands reduced by approximately 4–7% when setback control is compared against the constant temperature reference cases. However, the absolute reduction differs significantly between construction types, as the net heat demand between the evaluated cases ranges from 29 to 195 kWh/(m^2 a). While for the old buildings the reduction is about 12 kWh/(m^2 a), then for the south-oriented low energy buildings the reduction is only 1 kWh/(m^2 a). For heavy construction, setback efficiency is in all cases marginally less than for the corresponding light construction case.

3.2 Temperature Performance

Weekly fluctuations. The resulting air temperatures in the observed office during a two-week period in winter are depicted for all simulated cases in Fig. 3. In Fig. 3a, we can see that for the well-insulated room, air temperatures do not decrease to 18 °C (lower horizontal interrupted line), staying even for weekend setback above 19 °C. Moreover, the graph shows that the PI-control cannot hold a constant temperature during the day (set temperature level of 21 °C shown in upper horizontal dashed line) and the room overheats for the floor-heating case.

Figure 3b illustrates the known observation that a room with higher heat capacity cools down slower. In case of a light building structure, the temperature can decrease to 18 °C even for the floor heating case the Fig. 3b shows. Still, the temperature change is slow and the set temperatures difficult to maintain.

Compared to these, changes in temperature for radiator cases in Fig. 3c are faster. PI control with the radiators maintains the temperature setpoint well. However, during the heat-up fluctuations occur.

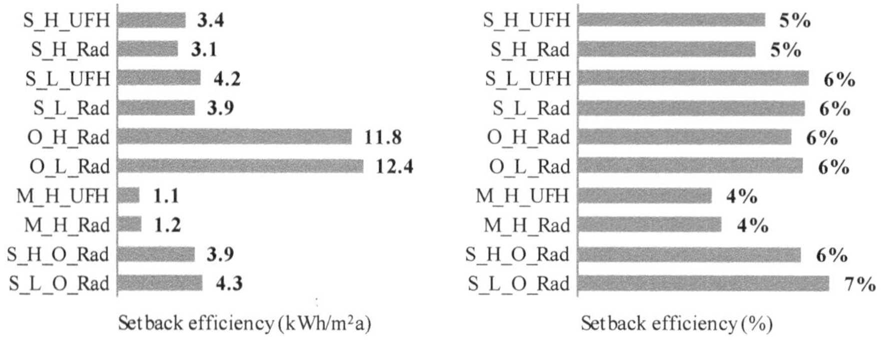

Fig. 2 Energy performance of the intermittent heating, absolute difference from the reference cases on the left and relative on the right. The abbreviations in the labels are explained in Table 1

In Fig. 3d, we can see that the over-dimensioned radiators allow for stronger room temperature reductions than in the corresponding cases with regular heating dimensioning in Fig. 3c. However, Table 2 shows that the resulting reduction of energy demand is not higher than 1 kWh/m^2 a.

Figure 3e shows that in old buildings the temperature drops to 18 °C almost immediately and, due to the high available heating power, raises up to 21 °C fast as well. Therefore, in the old building case, the setback potential is fully exploited.

Heat-up performance. In all the temperature graphs (Fig. 3), we can see that for the heavy building cases the temperatures fluctuate before reaching the set temperature. This is because the actual temperature increase is significantly faster than modelled and the system lets the room cool down again until it calculates that heat-up should be started again. In Fig. 4. we can see that the fluctuation after weekend setback (on Monday) is more significant than after night-setback (Friday). While on Friday, the setpoint (21 °C) is reached by the start of occupancy in most of the evaluated office rooms, it is not the case on Monday. However, the temperatures are above 20 °C when occupancy begins in all observed scenarios. On Friday, as an exceptional case, S_L_UFH cools down fast but does not manage to heat up on time.

Heating season. The overall temperature performance during the heating season is depicted in duration graphs in Fig. 5. Figure 5a illustrates the difference between the two heat emitters. We can see that the floor heating is not keeping the given set temperatures, as the graphs are smooth, whereas the plateaus in the radiator graphs show that to some extent set temperatures are maintained. In Fig. 5b, we can see that there is a very clear difference between the different insulation levels. While the plateaus are very clear in the old house case, the modern south-oriented building has very small energy losses and it has significantly higher temperatures. The standard cases can be found between these two extremes. Figure 5c compares heavy and light structures in old and modern buildings. The slower cool-down of the heavy buildings results in higher temperatures in the duration graph.

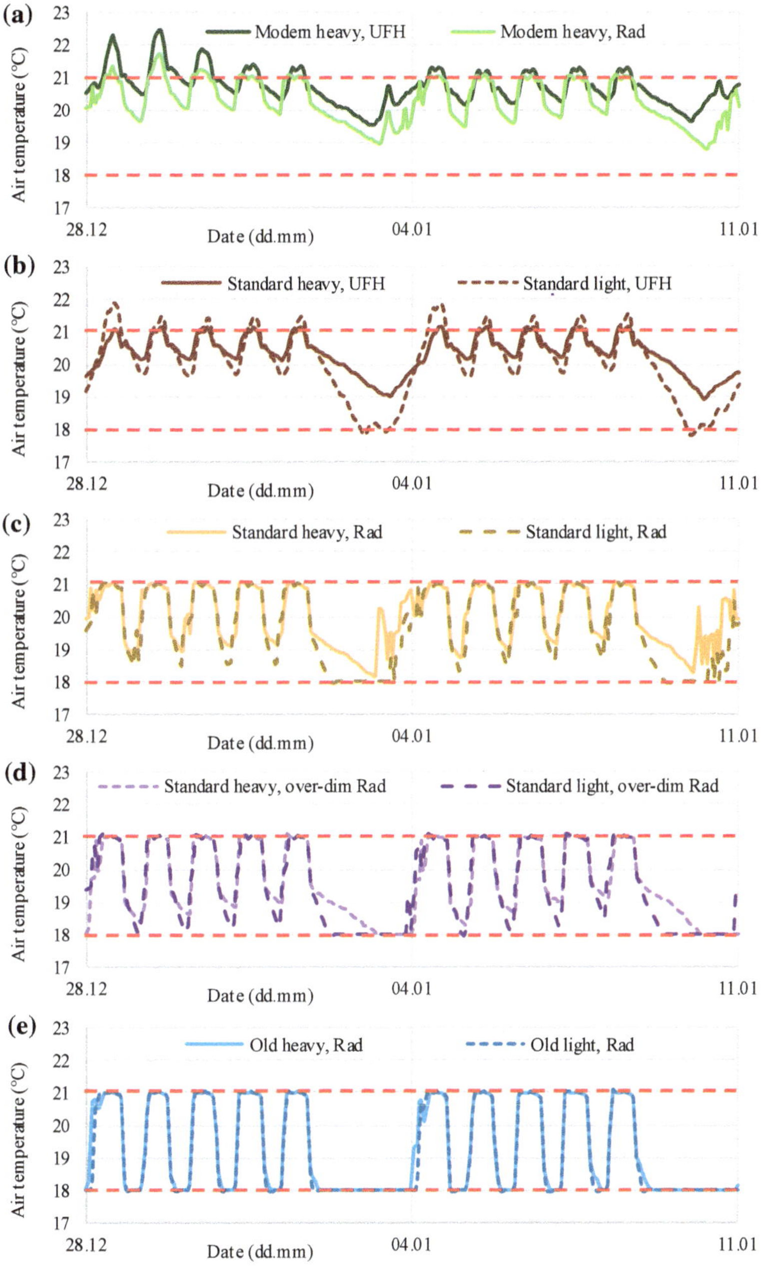

Fig. 3 Air temperature fluctuations during two winter weeks for all simulated cases

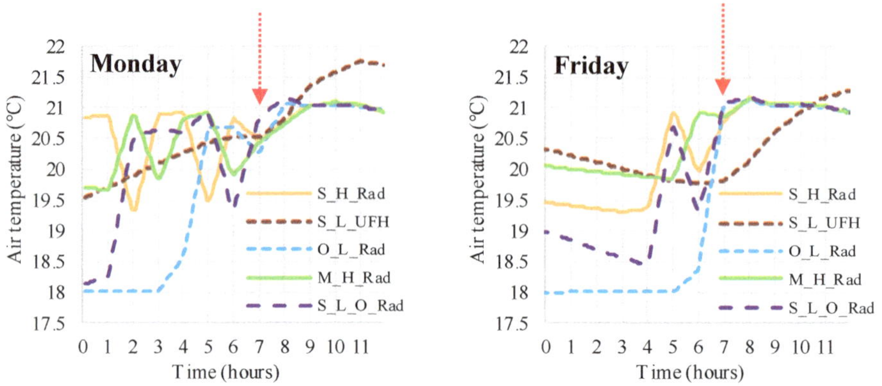

Fig. 4 Temperature performance during heat-up times in the first 12 h on Monday, January 4 (left) and Friday, January 8 (right). Occupancy start at 7 a.m. and is marked with arrows

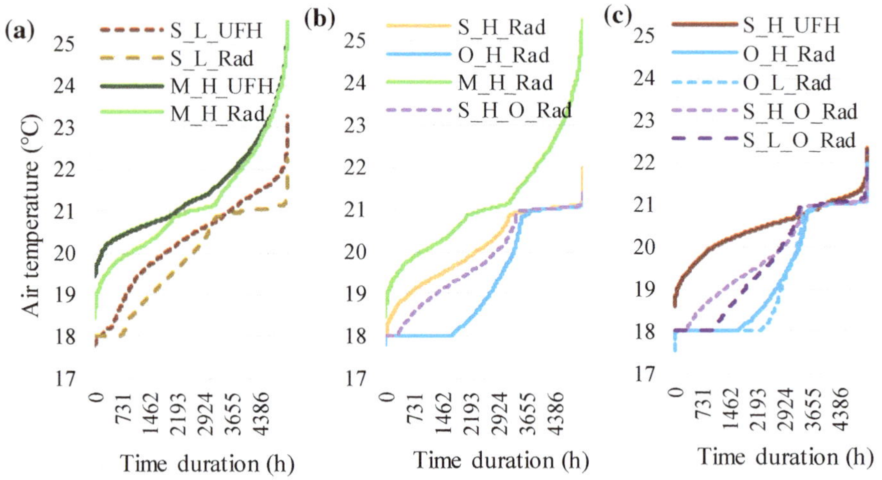

Fig. 5 Temperature duration graphs over the heating season

4 Conclusion

This research shows that the absolute setback efficiency in modern office buildings is significantly lower than in old buildings. This is mainly because the air temperature does not drop as fast and it stays above allowed minimal limit during the nighttime setbacks. For floor heating cases, this applies even for weekend setbacks. It has been shown that buildings with lower thermal mass and faster reacting heating system have higher energy conservation potential when applying intermittent heating operation. The setback control for old office buildings is always profitable. However, for modern and especially modern buildings with slow reacting heating system, the benefits of setback control are low, especially in absolute numbers. Before applying, the saving needs to be weighed against potential discomfort and additional cost for every specific case.

References

1. B. Xu, S. Zhou, W. Hu, An intermittent heating strategy by predicting warm-up time for office buildings in Beijing. Energy Build (2017)
2. J.W. Moon, S.H. Han, Thermostat strategies impact on energy consumption in residential buildings. Energy Build. **43**(2–3), 338–346 (2011)
3. W. Wang, J. Zhang, W. Jiang, B. Liu, Energy performance comparison of heating and air-conditioning systems for multi-family residential buildings. HVAC&R Res. **17**(3), 309–322 (2011)
4. W. Guo, D.W. Nutter, Setback and setup temperature analysis for a classic double-corridor classroom building. Energy Build. **42**(2), 189–197 (2010)
5. Z. Wang, B. Lin, Y. Zhu, Modeling and measurement study on an intermittent heating system of a residence in Cambridgeshire. Build. Environ. **92**, 380–386 (2015)
6. Y. Wang, J. Kuckelkorn, Y. Liu, A state of art review on methodologies for control strategies in low energy buildings in the period from 2006 to 2016. Energy Build. **147**, 27–40 (2017)
7. Energy performance of buildings—Method for calculation of the design heat load—Part 1: Space heating load. FprEN 12831-1, European Standard (2016)
8. T.M. Kull, R. Simson, M. Thalfeldt, J. Kurnitski, Influence of time constants on low energy buildings' heating control, pp. 11–14 (2017)
9. Estonian Regulation No 58: Methodology for calculating the energy performance of buildings. Ministry of Economic Affairs and Communications (2015)
10. "IDA Indoor Climate and Energy 4.7.1." Equa Simulations AB, 2016
11. T. Kalamees, J. Kurnitski, Estonian test reference year for energy calculations. Proc. Est. Acad. Sci. Eng. **12**(1), 40–58 (2006)
12. Energy performance of buildings—Calculation of energy use for space heating and cooling, vol. 2008. ISO 13790, International Standard (2008)

Integration of Building Integrated Photovoltaic/Thermal (BIPV/T) System with Heat Recovery Ventilators for Improved Performance Under Extreme Cold Climates

Riccardo Toffanin, Hua Ge and Andreas Athienitis

Abstract The reliable operation of Heat Recovery Ventilator (HRV) is critical for maintaining a healthy indoor environment to remove contaminants and moisture, however, it remains a challenge in the Northern Canada due to the frequent frosting under the extreme cold conditions. The heat generated by a building-integrated photovoltaic/thermal (BIPV/T) system can be used to pre-heat the incoming fresh air in HRV in order to reduce its defrost cycle, therefore, improving the reliability of HRV to provide adequate ventilation required. In this case, the BIPV/T needs to be designed for higher air temperature rise, which may not be optimum for the thermal energy and PV power generation. Therefore, system integration and optimization for coupling BIVP/T with HRVs is required. Depending on the level of thermal energy available and the outlet air temperature from the BIPV/T system, a control strategy needs to be developed to optimize the operation of HRVs. This paper presents the analysis of four different BIPV/T configurations and their integration with HRVs for a 120 m^2 house located in Iqaluit, NU, Canada through modelling. Results show that the outlet air of a BIPV/T façade installation can be 14.8 °C higher than outdoor air on a clear sky winter day and that the defrost cycle can be reduced by 13%, up to 619 h annually.

Keywords Building integrated photovoltaics/thermal (BIPV/T)
Heat recovery ventilator (HRV) · System integration · Defrost cycle
Extreme cold climate · Canadian northern region

R. Toffanin
KTH, Royal Institute of Technology, Stockholm, Sweden
e-mail: toffanin@kth.se

H. Ge (✉) · A. Athienitis
Concordia University, Montréal, Canada
e-mail: hua.ge@concordia.ca

D. Johansson et al. (eds.), *Cold Climate HVAC 2018*,
Springer Proceedings in Energy, https://doi.org/10.1007/978-3-030-00662-4_9

1 Introduction

Extreme weather (temperatures below −40 °C, snow storms, hail, etc.), high dependence on fossil fuels and high electricity cost (10 times higher than the Canadian average) constitute not only obstacles but also opportunities for Northern Canada housing, creating a push towards the increase of building efficiency and the exploitation of renewable source of energy [1]. Among the different sustainable housing practices, the implementation of BIPV/T systems is promising because it introduces an envelope concept that simultaneously generates electricity and recovers thermal energy, transforming the building skin into an active system [1]. The heat recovered from the photovoltaic (PV) panels can be used for pre-heating the supply inlet air of a HRV, a device that not only reduces the energy required for heating but also offers health benefits for the building occupants. Thus, the coupling of the two systems can help prevent frost formation, a major problem for HRVs in cold climates. The formation of frost in HRV reduces the efficiency of the system and can provoke damage or system failure, if it is not controlled [2]. As a consequence of the coupling, the reliability of the system, which is critical to maintain a healthy indoor environment through the removal of moisture and contaminants, can be improved. The aim of this paper is to study different integration scenarios of BIPV/T systems with HRV. Four different configurations are analysed for a 120 m^2 house located in Iqaluit, NU, Canada through a MATLAB model and compared in order to determine the reduction of defrost cycle and the energy generation of each scenario.

1.1 Building Integrated Photovoltaics/Thermal (BIPV/T)

A BIPV/T system is a photovoltaic system that is incorporated in the envelope of a building and produces both electricity and thermal energy recovering heat from the PV panels [3]. The system is not only a simple addition to the building but also an active part of its envelope, changing the various loads (heating, cooling and lighting), that can be coupled with other energy systems such as a heat recovery ventilator or a heat pump [4]. The PV panels form the exterior layer/cladding of the walls or roof and as such also act as a rainscreen managing various environmental loads.

1.2 Heat Recovery Ventilator (HRV)

Air-to-air heat exchangers used as HRVs are devices that exchange sensible heat between two air-steams [5, 6]. Their main use in residential or commercial building is the pre-heating/cooling of supply air. Major advantages of HRVs include not only energy saving and reduction of fossil fuel consumption but also removal of indoor air contaminants and prevention of respiratory disorders [2]. However, problems in their operation can occur if implemented in cold climates. At low outdoor air temperatures, frost formation within the exchanger core starts due to the condensation of water-vapour in the exhaust air [7]. As a consequence, the efficiency of the system is reduced and, if no measures are implemented to stop the phenomenon, failure and/or damage can occur [2]. Different techniques are implemented to solve this problem; such as defrost methods like recirculation of exhaust air or supply fan shutoff, or frost control methods like the pre-heating of cold supply air before its entering the exchanger core. The last strategy can significantly affect the system efficiency and be very expensive due to the use of fossil fuel, but, at the same time, it offers a good possibility of integration with BIPV/T [5–8].

1.3 Integration of BIPV/T with HRV

The integration and optimization of these two technologies is a recent topic of research, especially for cold climates. An interesting study was conducted by Natural Resources Canada—CanmetENERGY. In this paper, Maayan Tardif et al. [9] studied the performance of different heat management strategies in an all-electric Canadian home located in Montréal, QC, in which one of the scenarios used the pre-heated air from a roof BIPV/T system as supply inlet air of a HRV. This case consisted of a 12.75 m^2 installation on 40° slope south-facing roof. The annual electric energy consumption for heating and domestic hot water was 12338 kWh for the base case with electric furnace (no BIPV/T and HRV). During the heating season and whenever solar radiation was above 50 W/m^2, pre-heated air was fed to the HRV; on the other hand, when solar radiation was not available and during the non-heating season, outdoor air was used as inlet of the HRV. Results showed a promising maximum increase in pre-heated air temperature of 21 °C, however, the reduction of heating and domestic hot water energy consumption was only 1%, as a result of the small amount of energy recovered due to the low availability of solar radiation during the heating season.

2 Methodology

Different configurations of the BIPV/T system in terms of tilt angle of the surface and air speed in the cavity behind the PV panels are considered. The first scenario (Scenario Façade 90) is a south-facing façade installation of PV panels (90° from the horizontal). The second (Scenario Roof 10), third (Scenario Roof 20), and fourth (Scenario Roof 30) scenarios model a south-facing roof installation of the system with roof tilt angle among the most common roof inclination for Northern

Canada (namely 10°, 20°, and 30°) [1]. For every single configuration, a one-year simulation with hourly weather data at different air speeds (from 0.5 to 2.5 m/s) is performed using a code implemented in MATLAB. After the sensitivity analysis, the air speed that maximizes the integration with the heat recovery ventilator system is chosen and the main results are presented and compared. In the last part of the paper, the main findings are summarized and conclusions are drawn.

2.1 Location

The selected location for the simulation is Iqaluit (63.75°N, 68.55°W) in the Nunavut territory. It was chosen for the simulation because it represents an example of the challenges related to housing in extreme cold climates and for the availability of reliable weather data for a Typical Meteorological Year (TMY) [10]. Table 1 shows the typical climatic parameters for the location, while daily average, maximum and minimum temperatures over 30 years in °C are presented in Table 2 [11]. The outdoor temperature and the Global Horizontal Radiation (GHR) for an average day[1] and a clear sky day[2] of summer and winter are shown in Fig. 1 (left) and Fig. 1 (right), respectively.

2.2 House Model

The building considered is a 12 m by 10 m house with attic and a floor height of 3.43 m representing a low-energy house built in this region (see Fig. 2) [12]. The external envelope is an R-45 (8.0 m^2 K/W) Structural Insulated Panel (SIP) wall, a pre-fabricated high-performance envelope system, whereas the ceiling envelope is R-70 (12.4 m^2 K/W). The indoor conditions are 22 °C and 30% of relative humidity and the fresh air intake necessary to satisfy Indoor Air Quality (IAQ) is 0.035 m^3/s as defined by ANSI/ASHRAE Standard 62.2-2016 [5, 13].

As a benchmark for the energy consumption of the house, the E/2 Northern Sustainable House (NSH) is considered, which is a 127.8 m^2 one-storey household with three bedrooms, designed and built by the Nunavut Housing Corporation under the NSH initiative of Canada Mortgage and Housing Corporation (CMHC), due to its representative role for the energy-efficient and cultural appropriate

[1]An average day is estimated by calculating the mean of the weather parameters of the days of the seasonal typical week indicated in the TMY data. For Iqaluit, the typical summer week is July 13 to July 19, and the typical winter week is January 20 to January 26 [10].

[2]January 29 and June 16 of a TMY were chosen as representative clear sky days, due to their low value of daily average sky cover, a value of 1.63 and 0.58, respectively [10].

Table 1 Typical climatic parameters for Iqaluit [11]

Heating degree days (HDD 18)	10117.4 days
Cooling degree days (CDD 18)	0 days
Annual beam (direct normal) radiation	1421.5 kWh/m^2
Annual diffuse radiation	426.0 kWh/m^2
Annual days with daily GHR below 200 Wh/m^2	60 days

Table 2 Monthly temperature of Iqaluit [11]

Temperature	Jan	Feb	Mar	Apr	May	Jun	Jul	Aug	Sep	Oct	Nov	Dec
Average (°C)	−26.6	−28	−23.7	−14.8	−4.4	3.6	7.7	6.8	2.2	−4.9	−12.8	−22.7
Maximum (°C)	−22.5	−23.8	−18.8	−9.9	−0.9	6.8	11.6	10.3	4.7	−2	−8.9	−18.5
Minimum (°C)	−30.6	−32.2	−28.6	−19.6	−7.8	0.3	3.7	3.3	−0.4	−7.7	−16.7	−26.9

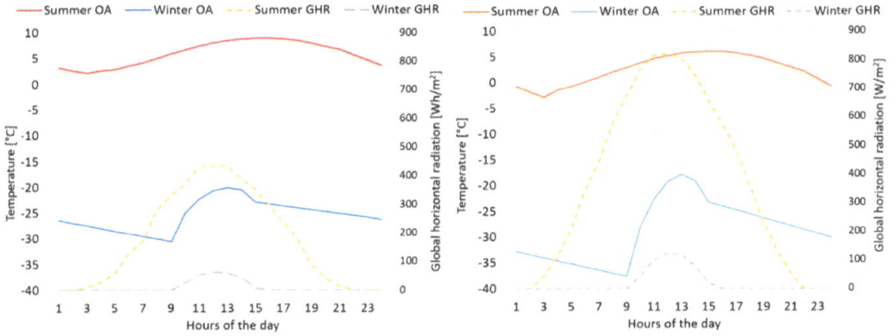

Fig. 1 Representative weather data for Iqaluit [10, 11]

housing models for Northern Canada [14]. The household has an annual consumption of 9180 kWh of electricity and 20,556 kWh of heating oil. The main energy-performance related details of the house are presented in [14].

2.3 *BIPV/T Model*

The system is an air-based open-loop BIPV/T, which uses Canadian Solar poly-Si modules with efficiency of 16.16% and nominal power of 310 W at standard conditions [15]. It is mechanically ventilated by a fan in order to provide a constant

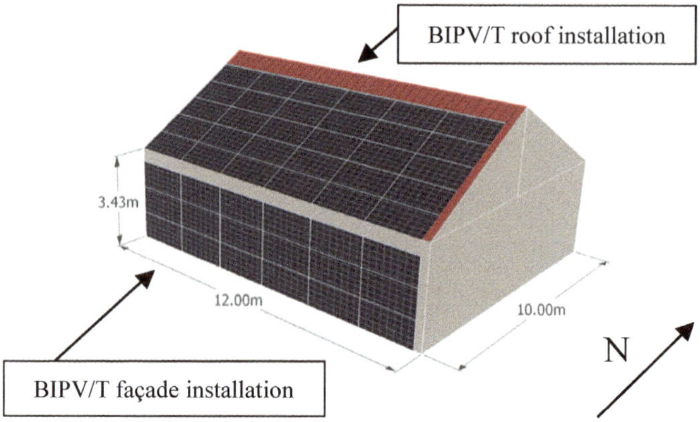

Fig. 2 3D model of the house

air flow as modelled in Yang [16]. Its consumption varies from approximately 5 kWh/year at 0.5 m/s to approximately 500 kWh/year at 2.5 m/s with a 20% efficiency [3]. The reduction of collected energy due to snow cover on a roof tilted at an angle below 60° is neglected.

Performance parameters. The annual electricity production of the BIPV/T system is calculated using Eq. 1:

$$E_{PV} = \sum_{i=1}^{8760} S_i \alpha_w \eta_{PV_i} A_w - E_{fan_i} \tag{1}$$

where S_i is the hourly solar radiation received on the surface in kWh/m^2, α_w is the absorptivity of the BIPV cladding, η_{PV_i} is the hourly PV module efficiency, A_w is the surface area in m^2, and E_{fan_i} is the hourly electricity fan consumption in kWh. The heat gained by the air during its passage through the cavity is:

$$Q_{air} = \sum_{i=1}^{8760} \dot{m}_i c_{air} \left(T_{outlet,a_i} - T_{o,i} \right) \tag{2}$$

where $\dot{m}_i$ is the hourly mass flow rate in kg/s, c_{air} is the specific heat of air in kJ/kg K, T_{outlet,a_i} is the hourly temperature of air at the outlet of the cavity in °C calculated with the model presented in Kayello et al. [12], and T_{o_i} is the hourly outdoor temperature in °C. The electrical and thermal efficiency of the system are [17]:

$$\eta_{el} = \frac{E_{PV}}{SA_w} \tag{3}$$

$$\eta_{thermal} = \frac{Q_{air}}{SA_w} \tag{4}$$

where S is the annual solar irradiance on the surface in kWh/m^2. The combined efficiency is as follows [17]:

$$\eta_{combined} = \frac{E_{PV} + Q_{air}}{SA_w} \tag{5}$$

The electricity savings are calculated with the following equation:

$$E_{savings} = \frac{E_{PV}}{E_{E/2NSH}} 100 \tag{6}$$

$$HO_{savings} = \frac{Q_{air}}{Q_{E/2NSH}} 100 \tag{7}$$

where $E_{E/2NSH}$ and $Q_{E/2NSH}$ are the electricity and heating oil consumption in kWh of the E/2 NSH house, respectively.

2.4 HRV

The heat recovery ventilator uses outdoor air either directly or pre-heated by the BIPV/T system depending on the temperature. The operating strategy is summarized in Table 3. When Pre-heated Air (PA) temperature is lower than Outdoor Air (OA) temperature, HRV draws outdoor air directly. When the PA temperature is higher than OA temperature but lower than 25 °C, the air from the BIPV/T system is utilized in the HRV. When the PA temperature is higher than 25 °C and there is heating demand from the house, BIPV/T outlet air can be supplied directly to the house for Direct Space Heating (DSH) while indoor air can be exhausted directly without heat recovery. When the PA temperature is higher than 25 °C while there is no heating demand from the house, the thermal energy generated by BIPV/T can be used for other applications such as pre-heating domestic hot water or exhausted while the OA will be used in HRV.

Table 3 HRV control strategy

$T_{outlet,a} < T_o \Rightarrow$	OA to HRV and PA to other purposes
$T_{outlet,a} \geq T_o \Rightarrow$	$T_{outlet,a} \leq 25$ °C $\Rightarrow$ PA to HRV
	$T_{outlet,a} > 25$ °C during heating season $\Rightarrow$ PA to DSH
	$T_{outlet,a} > 25$ °C during non-heating season $\Rightarrow$ OA to HRV

Note that the air entering the HRV is a portion of the outlet air of the BIPV/T system, as the HRV provides solely the flow rate required for an adequate IAQ, and, thus, the surface area of the PV modules required for HRV purposes is smaller and defined as:

$$A_{HRV} = h_{PV} \times \frac{\dot{V}_{IAQ}}{l \times V_{air}} \quad (8)$$

where h_{PV} is the height of the PV installation in m, $\dot{V}_{IAQ}$ is the IAQ air flow rate in m^3/s, l is the width of the air cavity in m, and V_{air} is the air speed in the cavity in m/s.

An important parameter to be considered is the frosting limit, which is defined as the lowest inlet supply air temperature that does not lead to frost formation. A typical value is −5 °C for relative humidity levels of typical indoor air in sensible heat exchanger cores as experimentally verified in [2, 5, 7, 18], hence in cold climate it is necessary to implement a defrost mechanism in order to avoid the frost formation on the surfaces of the heat-exchanger core [19].

Performance parameters. The following parameters are used to evaluate the integration of the BIPV/T with HRV:

- frost risk (FR) time t_{FR}: number of hours when OA temperature is lower or equal to the frosting limit;
- frost control (FC) time t_{FC}: number of hours when PA temperature is higher than the frosting limit while the OA temperature is below or equal to the frosting limit;
- pre-heated air time t_{PA}: number of hours when PA is used as supply inlet air of the HRV (PA mode);
- outdoor air time t_{OA}: number of hours when OA is used as supply inlet air of the HRV (OA mode);
- direct space heating time t_{DSH}: number of hours when PA is used for direct space heating (DSH mode).

3 Simulation

Input data are divided into fixed parameters, which are constant in every scenario, and variable parameters, which represent the key variables to be used for identifying the best configuration of the system and are changed in different cases. The variable parameters include the tilt angle of the surface β_c and the air speed in the cavity V_{air}. An overview of the scenarios is shown in Table 4.

Table 4 Overview of the scenarios

Scenario	Tilt angle	Number of PV modules	Surface area (m^2)	Area sufficient for HRV (m^2)	Air speed in BIPV/T cavity
Façade 90	90°	18	34.54	From 8.25 to 1.65	From 0.5 m/s (0.15 m^3/s) to 2.5 m/s (0.73 m^3/s)
Roof 10	10°	30	57.56	From 13.75 to 2.75	
Roof 20	20°	30	57.56	From 13.75 to 2.75	
Roof 30	30°	30	57.56	From 13.75 to 2.75	

3.1 Sensitivity Analysis

The electricity and thermal energy generation[3] per installed m^2 are presented in Fig. 3 (left) and Fig. 3 (right), respectively. It can be seen that Scenario Façade 90 has the best energy production performance, followed by Scenario Roof 30, Roof 20, and Roof 10, due to better exploitation of low radiation period for closeness to the optimal tilt angle of the location (the latitude). The four scenarios have a similar trend for electricity reaching a peak between 0.75 and 1.25 m/s and then decreasing of approximately 3.5% as the fan consumption rises with the increase of velocity, while for thermal energy there is an upward trend with an increase approximately 200% due to higher air mass flow rate.

As for the integration with HRV, the frost control time hours fall while the speed of the air in the cavity rises as shown in Fig. 4 (left). Scenario Façade 90 reaches 619 h at 0.5 m/s and then is the second best until the air speed is just below 1.25 m/s when it is overcome by Scenario Roof 20. This is because the air is pre-heated in the façade installation for a shorter distance compared to the roof scenarios, thereby an increasing air speed gains lower thermal energy. The performance of Scenario Façade 90 is worse compared to the remaining scenarios, as it declines significantly faster than the others (37.80%), while the three roof scenarios have an approximately constant downward trend (decrease between 27.76 and 30.83%). Higher roof inclination has greater number of frost control hours.

The maximum difference in temperature between OA and PA for an average summer and winter day are shown in Fig. 4 (centre) and Fig. 4 (right), respectively. It can be seen that in the two seasons and for every scenario the maximum temperature difference decreases as the air speed rises. This is due to a higher air mass flow rate of air passing through the cavity. The façade installation performs better in

[3]In this paper, electricity generation includes also the fan electricity consumption, thus, it is a net electricity production. However, in general electricity generation does not consider the consumption of the fan.

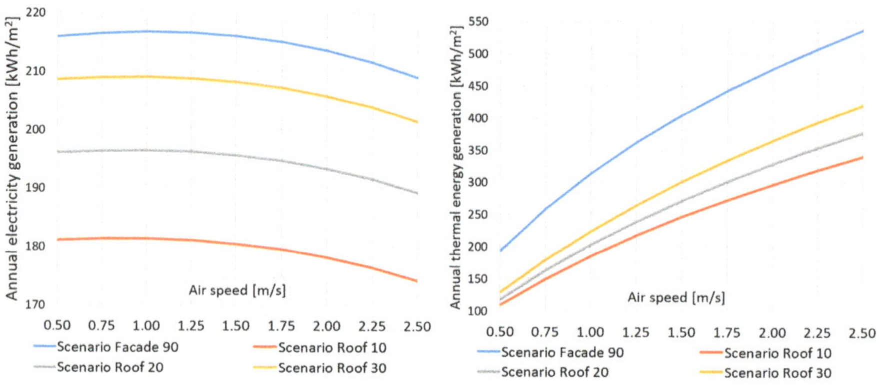

Fig. 3 Sensitivity analysis: annual electricity (left) and thermal energy (right) generation per installed m^2

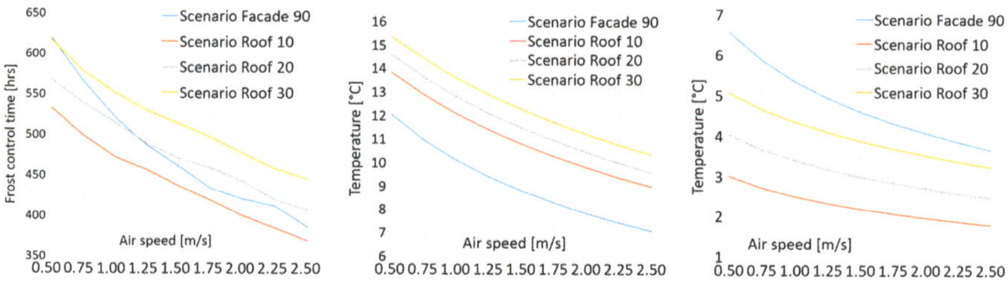

Fig. 4 Sensitivity analysis: frost control time (left) and maximum temperature difference between OA and PA for an average summer (centre) and winter (right) day

winter when the sun is low on the horizon with a maximum increase of 6.58 °C whereas the 30° slope roof configuration has the best performance in summer with a 15.34 °C temperature rise.

4 Results

The results of the simulation with an air speed of 0.5 m/s, which maximizes the frost control time and, therefore, the integration with HRV, are summarized in Table 5.

4.1 General Results

It can be seen that, while the electricity production increases from 7454.49 kWh (81.20% of the electricity requirement of the household) in Scenario Façade 90 to

Table 5 Main results of the simulation

Scenario	Facade 90	Roof 10	Roof 20	Roof 30
Annual electricity production (kWh)	7454.49	10422.56	11289.25	12002.86
Annual electricity production (kWh/m^2)	215.83	181.06	196.11	208.51
Electricity savings (%)	81.20	113.54	122.98	130.75
Annual thermal energy production (kWh)	6647.72	6214.23	6806.17	7388.97
Annual thermal energy production (kWh/m^2)	192.47	107.95	118.23	128.36
Heating oil savings (%)	32.34	30.23	33.11	35.95
Electrical efficiency (%)	16.08	16.11	16.05	15.99
Thermal efficiency (%)	14.34	9.61	9.68	9.84
Combined efficiency (%)	30.42	25.72	25.72	25.83
Maximum BIPV/T cladding temperature (°C)	51.05	50.14	52.83	55.04
Frost risk time (h)	4763	4763	4763	4763
Frost control time (h)	619	532	568	616
Pre-heated air time (h)	8568	8543	8492	8427
Outdoor air time (h)	0	0	0	0
DSH time (h)	192	217	268	333
FC time relative to FR time (%)	13.00	11.17	11.93	12.93

12002.86 kWh (130.75% of the building's need) in Scenario Roof 30, the heat generation decreases from 6647.72 kWh in the façade installation to 6214.23 kWh in the 10° tilt angle roof configuration but it increases from Scenario Roof 20 (6806.17 kWh) to Scenario Roof 30 (7388.97 kWh). This different patterns are because façade and roof scenarios differ in both surface areas, air path length, and amount of solar radiation. When the energy production is analysed per installed m^2, both the electricity and thermal output follow the same trend: Façade 90 is the highest followed by Roof 30, Roof 20 and Roof 10. The thermal output is 192.47 kWh/m^2 for Façade 90, 128.36 kWh/m^2 for Roof 30, 118.23 kWh/m^2 for Roof 20 and 107.95 kWh/m^2 for Roof 10. The electricity output for is 215.83 kWh/m^2 for Façade 90, 208.51 kWh/m^2 for Roof 30, 196.11 kWh/m^2 for Roof 20 and 181.06 kWh/m^2 for Roof 10.

Thermal efficiency is 5% higher in Scenario Façade 90 than in the roof configurations (between 9.61 and 9.84%) whereas electrical efficiency remains approximately constant among the different scenarios (approximately 16.05%). As the variation in thermal efficiency is higher than the variation of the electrical efficiency, combined efficiency follows the same trend of thermal efficiency.

4.2 *Integration with HRV*

Figure 5 shows the frost control time and frost control time relative to the frost risk time. Outdoor temperature is below −5 °C for 4763 h, which corresponds to the

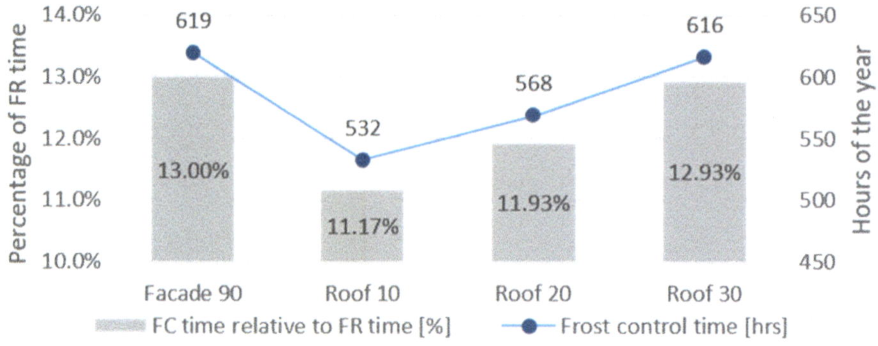

Fig. 5 Frost control time and frost control time relative to frost risk time

54.37% of the year. The coupling of BIPV/T with HRV reduces the formation of frost in all four scenarios. The maximum reduction of frost condition is achieved by Scenario Façade 90 with a decrease of 13.00%, followed by Scenario Roof 30, Scenario Roof 20, and Scenario Roof 10 in this order. As a consequence, the HRV system avoids entering the defrost cycle for 619 h with Scenario Façade 90, for 532 h with Scenario Roof 10, for 568 h with Scenario Roof 20, and 616 h with Scenario Roof 30 compared to a system without BIPV/T air preheating.

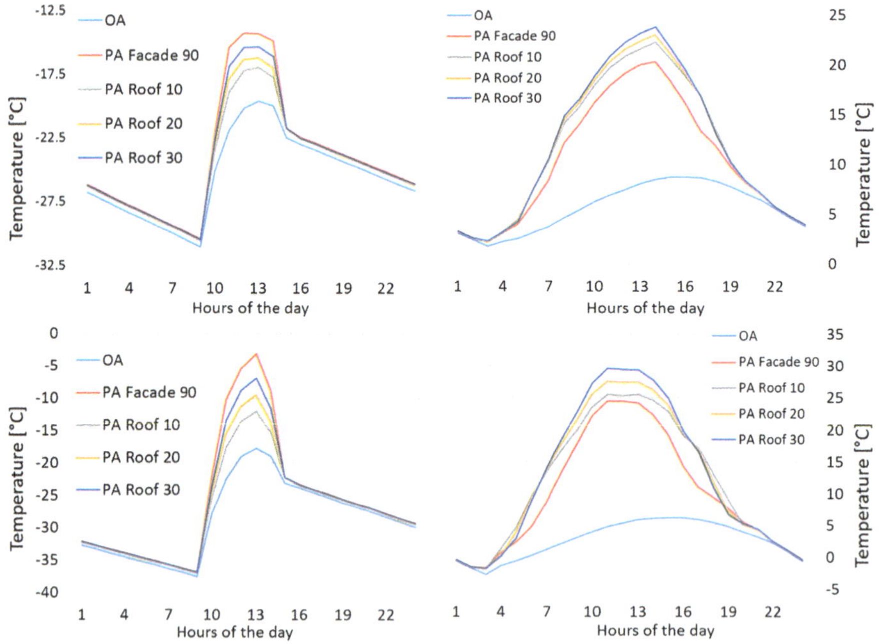

Fig. 6 PA and OA temperatures of a typical winter (top left) and summer (top right) day, and of a winter clear sky (bottom left) and summer clear sky (bottom right) day

Figure 6 (top) shows the daily trend of outdoor temperature and air outlet temperature for an average summer day and an average winter day. In summer, Scenario Roof 30 presents the best performance with a maximum increase in temperature of 15.34 °C, followed by Scenario Roof 20 and Roof 10 with 14.59 °C and 13.81 °C, respectively, while Scenario Façade 90 increases the temperature of 12.01 °C. However, in winter, Scenario Façade 90 produces an increase of 6.57 °C, which is a temperature rise higher than the one achieved in Scenario Roof 10 (3.01 °C), in Scenario Roof 20 (4.04 °C), and in Scenario 4 (5.07 °C). On a clear sky day, the same trend of performance is seen but the increase in temperature is higher as shown in Fig. 6 (bottom). In summer, PA is 24.84 °C hotter than OA for Scenario Roof 30, 22.91 °C for Scenario Roof 20, 20.76 °C for Scenario Roof 10, and 19.76 °C for Scenario Façade 90. On the other hand, in winter the façade installation produces air 14.81 °C hotter than the ambient, while the increase in temperature of PA is 5.74 °C for Scenario Roof 10, 8.34 °C for Scenario Roof 20, and 10.91 °C for Scenario Roof 30.

5 Conclusion

Four different configurations of a BIPV/T system with HRV have been examined for a 120 m^2 house in Iqaluit, NU, Canada. Annual simulations were performed to analyse the configurations of the system with different tilt angles of the surface and air speeds in the cavity. The sensitivity analysis revealed that the cavity air speed that maximizes the electricity production does not optimize the integration with HRV. Indeed, an air speed of 0.5 m/s achieves the highest frost control time whereas the scenarios reach the peak of the electricity production with velocity between 0.75 and 1.25 m/s. For this reason, the 0.5 m/s air speed was chosen for the further analysis of the systems. Results show that, even if Scenario Façade 90 has the smallest PV installation area, the façade installation is able to generate more energy during the winter months and, therefore, to optimize the integration with HRV in the period when it is more needed, achieving a maximum temperature increase between PA and OA of 6.57 °C and of 14.81 °C during an average and clear sky winter day, respectively. As a consequence, the pre-heated air from the façade installation of the PV modules reaches a temperature above the frosting limit for 619 h, therefore, reducing the need of defrost cycle by 13% and it achieves a heating oil saving of 32.34% compared to a system without BIPV/T pre-heating. This paper shows that the integration of BIPV/T with HRV is useful to improve the performance of a heat recovery ventilator in cold climates reducing the time needed for defrost cycles, especially if the PV modules are installed on the south-facing façade of a building. Future work should focus on the analysis of the energy consumption of HRV and investigate the energy benefit provided by the integration with BIPV/T.

References

1. Y. Chen, *Design and Evaluation of Facade-Integrated Solar Technologies Suitable for High-Latitude Applications* (Department of Building, Civil and Environmental Engineering, Concordia University, Montréal, Canada, 2012)
2. C. Beattie, P. Fazio, R. Zmeureanu, J. Rao, A preliminary study of the performance of sensible and latent heat exchanger cores at the frosting limit for use in Arctic housing. Energy Procedia **78**, 2596–2601 (2015)
3. L. Candanedo Ibarra, *Modelling and Evaluation of the Performance of Building-Integrated Open Loop Air-based Photovoltaic/Thermal Systems* (Department of Building, Civil and Environmental Engineering, Concordia University, Montréal, Canada, 2010)
4. T. Yang, A. Athienitis, A review of research and developments of building-integrated photovoltaic/thermal (BIPV/T) systems. Renew. Sustain. Energy Rev. **66**, 886–912 (2016)
5. C. Beattie, *Experimental Study of Air-to-Air Heat/Energy Exchangers for Use in Arctic Housing* (Department of Building, Civil and Environmental Engineering, Concordia University, Montréal, Canada, 2017)
6. E. Phillips, R. Chant, D. Fisher, V. Bradley, Comparison of freezing control strategies for residential air-to-air heat recovery ventilators. ASHRAE Trans. **95**(2), 484–490 (1989)
7. J. Kragh, J. Rose, S. Svendsen, Mechanical ventilation with heat recovery in cold climates, in *7th Symposium on Building Physics in Nordic Countries*, Reykjavik, vol. 1 (2005)
8. M. Rafati Nasr, M. Fauchoux, R. Besant, C. Simonson, A review of frosting air-to-air energy exchangers. Renew. Sustain. Energy Rev. **30**, 538–554 (2014)
9. J. Maayan Tardif, J. Tamasauskas, V. Delisle, M. Kegel, Performance of air based BIPV/T heat management strategies in a Canadian home. Procedia Environ. Sci. **38**, 140–147 (2017)
10. Government of Canada, Engineering Climate Datasets. Accessed 2017. Available at: HYPERLINK http://climate.weather.gc.ca
11. Government of Canada: Canadian Climate Normals 1971-2000 Station Data: Iqaluit A, in: Government of Canada. Accessed 1 June 2017. Available at: HYPERLINK http://climate.weather.gc.ca
12. A. Kayello, H. Ge, A. Athienitis, Attic ventilation in Northern Canadian climates, in *15th Canadian Conference on Building Science and Technology*, Vancouver, British Columbia, Canada, pp. 1–16 (2017)
13. ASHRAE, *Ventilation and Acceptable Indoor Air Quality in Residential Buildings* (Atlanta, 2016)
14. Canada Mortgage and Housing Corporation (CMHC), *Arviat E/2 Northern Sustainable House Energy Consumption Performance Assessment* (Ottawa, 2015)
15. Canadian Solar Inc.: Max Power CS6X-310 | 315 | 320P. (Accessed September 2017)
16. T. Yang, *A Numerical and Experimental Investigation of Enhanced Open-Loop Air-Based Building-integrated Photovoltaic/Thermal (BIPV/T) Systems* (Department of Building, Civil and Environmental Engineering, Concordia University, Montréal, Canada, 2015)
17. A. Athienitis, J. Bambara, B. O'Neill, J. Faille, A prototype photovoltaic/thermal system integrated with transpired collector. Sol. Energy **85**(2011), 139–153 (2010)
18. W. Fisk, R. Chant, K. Archer, D. Hekmet, F. Offerman III, S. Pedersen, Onset of freezing in residential air-to-air heat exchangers. ASHRAE Trans. **91**(Part 1B), 145–158 (1985)
19. Natural Resources Canada, *Heat Recovery Ventilators* (Office of Energy Efficiency, Natural Resources Canada, Ottawa, Canada, 2012)

Demand Controlled Ventilation in Residential Buildings

Huijuan Chen and Caroline Markusson

Abstract The energy use for ventilation (heating and fan electricity) accounts for a large part of the energy use in residential buildings. For residential buildings, in many cases the building is occupied only part of the day, and further the pollution and moisture load generated by household activities varies during a day. Using demand controlled ventilation (DCV) has a great energy saving potential both regarding fan and heating energy. However, it is important how the ventilation is controlled in order to ensure an adequate indoor air quality, thermal comfort and avoid damages on the building. In this study different control strategies, control parameters, number of sensors and placing of sensors, number of zones are tested by modeling a single family house. Conclusions from the study are that the size of the energy saving depends on control strategy and system design and it is important to design and choose appropriate control strategy to obtain a good indoor environment.

Keywords Demand controlled ventilation · DCV · Control strategies
Energy saving

1 Introduction

The role of a ventilation system is to provide an acceptable indoor air quality (IAQ) and thermal comfort for occupants in buildings. The energy use for ventilation, both for heating and fan electricity, accounts for a large part of energy use in residential buildings, varying form 15 to 45%, depending on the type of ventilation system and building [1–3]. According to Swedish building regulation [4], the

H. Chen · C. Markusson (✉)
RISE, Research Institutes of Sweden, Göteborg, Sweden
e-mail: Caroline.markusson@ri.se

H. Chen
e-mail: Huijuan.chen@ri.se

D. Johansson et al. (eds.), *Cold Climate HVAC 2018*,
Springer Proceedings in Energy, https://doi.org/10.1007/978-3-030-00662-4_10

ventilation flow rate required is 0.35 l/s per m^2; it is allowed to be reduced to a minimum flow rate of 0.1 l/s per m^2 if no occupants are at home.

For residential buildings, in many cases the building is occupied only for part of the day, and further the pollution and moisture load generated by household activities varies during a day [5, 6]. Use of demand controlled ventilation (DCV) has a great energy saving potential both regarding fan and heating energy. The ventilation flow rate can be controlled by the demand of IAQ and temperature. However, it is important how the ventilation is controlled in order to ensure adequate indoor air quality, thermal comfort and avoid damages on the building. Further the size of the energy saving depends to a great extent on control strategy and ventilation system design.

The performances of DCV have been investigated previously for office buildings [7], schools [8] and residential buildings [9, 10] etc. For office and schools, demand is commonly measured by CO_2 or presence. While for residential buildings, the demand is more complex. Choice of control strategy, control parameters, e.g., temperature (T), relative humidity (RH), carbon dioxide (CO_2), difference of the CO_2 concentration between indoor and outdoor (ΔCO_2), difference of the moisture content between indoor and outdoor (Δx) or combination of several variables, as well as the their respective threshold values are important to demand controlled ventilation.

Nielsen and Drivsholm [11] tested demand controlled ventilation for a single family house where all sensors and controls are located in the air handing unit. In their study, ΔCO_2 and Δx between the exhaust air and outdoor are used as control parameters and are set with different limits in the range of 100–200 ppm and 1–2 g/kg, respectively. It is found that with the setting of the maximum value of 150 ppm for ΔCO_2 and 2 g/kg for Δx, the highest saving on heat (23%) and fan energy (35%) is achieved among all cases tested in their study. However, sensor placement and control strategy are also important aspects of DCV. Most of the previous studies have been focused on demand controlled ventilation where the demand is measured in the exhaust air and the flow rate is switched between a high and low flow rate. Other strategies, such as individual room control, can be more energy efficient and needs to be evaluated.

In this study different control strategies, control parameters, number of sensors and placement of sensors, number of zones are tested by modeling a low energy single family house with a ventilation system with heat recovery.

Further, saving on space heating is modelled for additional three cases; (1) for a cold climate—Kiruna, (2) for a conventional house with an overall heat transfer coefficient typical for Swedish single family houses and finally (3) for a single family house with only exhaust ventilation. All simulations are based on the same model (the low energy house model), this by moving the house to another climate zone (Kiruna) or reducing the insulation layer and increasing thermal bridges or by changing ventilation system design to an exhaust ventilation system.

The low energy house is modelled in IDA ICE with the aim to test and evaluate demand control ventilation with different control strategies and system designs, including control parameters, number of sensors and placing of sensors (see below

for examples). The study discuss how the size of the energy saving depends on control strategy and system design and it is important to design and choose appropriate control strategy to obtain a good indoor environment.
Control parameters evaluated in project are:

- temperature
- carbon dioxide
- relative humidity
- difference in absolute humidity indoor and outdoor.

Sensor placings evaluated in the project are:

- centrally sensor placing, one or more sensors are placed centrally, e.g. in exhaust duct
- multiple sensor placing, one or more sensors are placed in each zone of the building.

Different divisions/zoning of the house evaluated in the project are:

- every room is a zone
- every floor is a zone
- the building is a single zone.

2 The Reference House

In the project a reference house is used for the model. The house, situated at RISEs premises in Borås (Fig. 1), Sweden, is a low energy single family house (average heat transfer coefficient, $U_{average} = 0.16\ W/m^2\ K$) with a total floor area of 155 m^2 and two floor levels. The ventilation system is a balanced mechanical ventilation system with heat recovery, the design supply flow rate is 60 l/s and the design exhaust flow rate is 66 l/s. The efficiency of the heat exchanger is, according to manufacturer data, 82%.

3 Method

3.1 The Model

A model of the single family house is created in IDA ICE, and the construction and the configuration corresponds to the reference house situated in Borås (see Fig. 1). Result from the model for the heating demand is compared with measurements from the reference house and show good agreement [12].

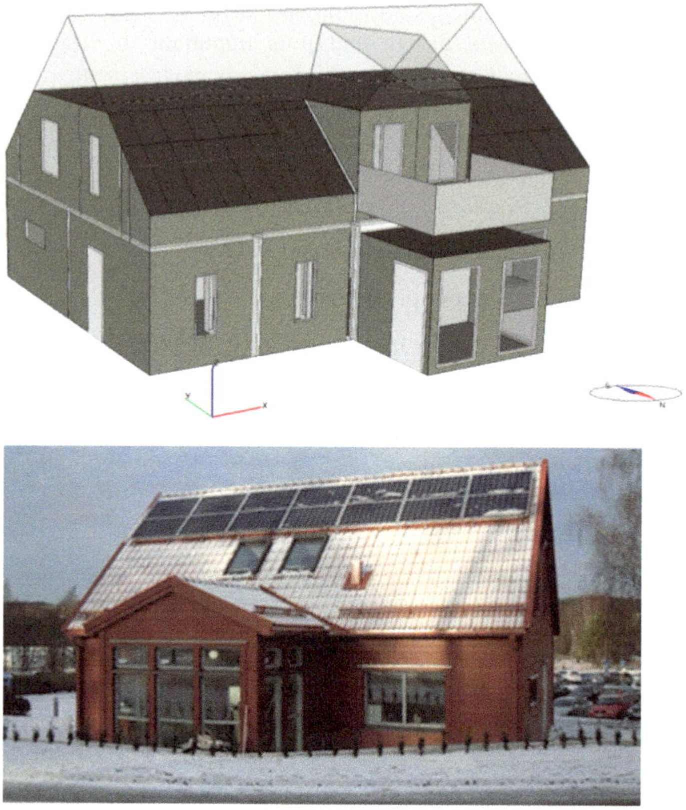

Fig. 1 The IDA model (above) and the reference house (below)

The program of IDA Indoor Climate and Energy (ICE) 4.6 is used to test and evaluate the DCV system with different control strategies. IDA is a dynamic simulation tool for studying indoor climate in the zones as well as energy use in the zones and the entire building for general-purpose. It models buildings, systems and its controllers, and it provides user interface to define, build up and simulate different cases which makes it possible to simulate a wide range of system designs and configurations [13].

Input data to the model climate and load profiles are important. The climate data used in the model is for Gothenburg-Landvetter, which is fairly representative for Swedish conditions with a yearly average temperature of 8 °C. The weather data files are available in the IDA program and they are derived from integrated surface hourly weather data originally archived at the national climate data center [13].

The load profiles used in the IDA model include heat emission, CO_2 generation and water vapor generation. The internal heat load (i.e. heat generated by lighting and other equipment) of 30 kWh/m^2 is used, suggested by SVEBY [14] as a standard value for residential building energy simulation. The internal heat load is distributed to the rooms with specific schedules and unevenly distributed over the year i.e., higher household electricity use in the winter and lower in the summer.

The generation rate of CO_2 and moisture by occupants depends on metabolism varying for each activity, which is modelled to be linearly varying with metabolism in this study. According to EN 15251 [15], the generation rate is 11.875 l/h/met for CO_2 and 34.375 g/h/met for moisture for an adult. In this simulation, 0.8 met is considered for sleeping, 1.0 met is for resting and 1.5 met is for cooking. The ambient air CO_2 level is assumed to be 400 ppm.

The total daily moisture production designed in the IDA modelling is about 4 kg/day for weekdays and 5 kg/day for weekends, which is within the range reported by Mattson [16]. The moisture production is considered due to household activities of showering, preparing food, washing and drying cloths, dishwashing, as well as the occupants themselves.

Further, three additional cases have been modeled to obtain result for colder climate and for typical Swedish single family house. The model of the reference house has been moved or modified to obtain result for a:

1. cold climate, this by moving the modelled reference house to Kiruna which is situated in a cold climate.
2. conventional Swedish single family house with balanced ventilation and heat recovery, this by changing the average heat transfer coefficient to 0.28 $W/m^2 \cdot K$ by decreasing insulation layer and increasing thermal bridges.
3. conventional Swedish single family house with exhaust ventilation, this by changing the average heat transfer coefficient to 0.28 $W/m^2 \cdot K$ and changing ventilation system design to an exhaust ventilation system.

3.2 *Modelled Control Strategies and Sensor Placements*

Four different ventilation control strategies are designed and implemented in the IDA model including multiple-zone control, two-zone control, one-zone control and exhaust duct control, see Table 1. Four parameters are chosen as control parameters: CO_2, RH, T, and Δx between indoor and outdoor. A short description of those control strategies and threshold values for control parameters are presented in Table 1.

The reference case is the existing ventilation system with constant air volume with supply flow rate of 60 l/s. In the case of demand control ventilation the maximum ventilation flow rate is 60 l/s and the minimum flow rate corresponds to 0.1 $l/s/m^2$.

The demand ventilation flow rate is determined by proportional integral (PI) control. The total supply flow rate is balanced with exhaust flow rate, by feeding back the signal from supply to exhaust if there is a demand for increasing the supply flow rate, and vise versa if there is a demand for increasing the exhaust flow rate.

Table 1 Description of the modelled cases using different control strategies, sensor placement and zoning

Strategy	Sensor placement	Threshold	Description
Multiple-zone control	Living rooms, kitchen, bed rooms, laundry room and bathroom	CO_2: 700–1000 ppm RH: 20–75% T: 25 °C Δx: 2.0–2.5 g/kg	Every room is a zone. Each zone has its own control signal
Two-zone control	See above	See above	Every floor is a zone. The control signal is the maximum signal among the rooms connected to each floor level
One-zone control	See above	See above	The building is single zone. The control signal is the maximum signal among all rooms inside the house
Exhaust duct control	Exhaust duct	CO_2: 400–600 ppm RH: 20–75% T: 25 °C Δx: 2.0–2.5 g/kg	The building is a single zone. The control signal comes from the exhaust duct

The maximum CO_2 concentration is set to be 1000 ppm which is a typical value considered giving an acceptable IAQ. A threshold of 2.5 g/kg in absolute humidity between indoor air and outdoor air is used for ensuring adequate removal of the moisture produced in the house, as recommend by Public health agency of Sweden [17].

Notice, Table 1, that threshold for exhaust control is set differently compared to the other cases in order to give adequate IAQ. With the setting of high- and low-limits for CO_2 of 400–600 ppm, a good indoor environment could be maintained based on IDA simulation results.

4 Results

4.1 Energy Saving

Simulation results from different ventilation control strategies in Table 1 are compared with the reference case (constant air volume system), and results are summarized in Table 2. The multiple-zone control strategy gives the highest ventilation flow rate reduction of 57% and space heating of 25%, while the exhaust duct control yields the least saving potential, i.e., 32% for flow rate and 14% for space heating. The space heating includes the total heating demand of the house.

The flow rate reduction and fan electric energy saving is much higher in the winter time than in the summer, as the ventilation fan is almost always on full speed in the summer months due to overheating.

Table 2 Simulation results in terms of ventilation flow rate and space heating energy for the low-energy single family house (Gothenburg)

Case	The average ventilation flow rate during the year (l/s)	Reduction of ventilation flow rate (%)	Space heating demand (kWh/ m^2 year)	Reduction of space heating in kWh/m^2 and in %	Fan electrical energy (kWh)	Reduction of the fan electrical energy (%)
Reference case	60	/	16	/	1090	/
Multiple-zone control	26	57	12	4 and 25	84	92
Two-zone control	29	52	12	4 and 23	126	88
One-zone control	38	37	13	3 and 17	284	74
Exhaust duct control	41	32	13	3 and 14	350	68

Table 3 Simulation results of space heating energy for another climate and a conventional single family house

	Low energy house (Kiruna)		Conventional house with balanced ventilation (Gothenburg)		Conventional house with exhaust ventilation (Gothenburg)	
Case	Space heating demand	Absolut reduction of space heating	Space heating demand	Absolut reduction of space heating	Space heating demand	Absolut reduction of space heating
kWh/m^2 year						
Reference case	43	/	30	/	68	/
Multiple-zone control	36	7	26	4	42	26
One-zone control	39	4	28	2	51	17

In Table 3 space heating demand is presented for the additional three cases;

1. reference house in a cold climate (Kiruna)
2. conventional single family house ($U_{avarage} = 0.28\ W/m^2 \cdot K$) with balanced ventilation and heat recovery
3. conventional house ($U_{avarage} = 0.28\ W/m^2 \cdot K$) with exhaust ventilation.

As the ventilation flow rate is nearly the same for all cases, only the space heating demand is presented in Table 3.

The saving in heating demand in kWh/m^2 will of course be much higher when the reference house is placed in a cold climate of Kiruna compared to when the reference house is placed in Gothenburg, Tables 2 and 3. The saving potential for multiple zone control is 7 and 4 kWh/m^2 for the reference house placed in Kiruna and Gothenburg respectively. For the single-zone control the saving is 4 and 2 kWh/m^2 for the reference house placed in Kiruna and Gothenburg respectively. The house area of 155 m^2 will give a total saving per year, for heating, for the multiple-zone control of 1085 and 620 kWh for the reference house placed in Kiruna and Gothenburg respectively.

If comparing the reference house placed in Gothenburg with a conventional house ($U_{avearge}$ = 0.28 W/m^2 · K) with balanced ventilation and heat recovery also placed in Gothenburg the result for the saving is similar for the two cases, this since both cases has the same ventilation system design and performance.

If comparing the reference house and conventional house ($U_{avearge}$ = 0.28 W/m^2 · K) with balanced ventilation and heat recovery with a conventional house with only exhaust ventilation the difference is significant. For a conventional house with only exhaust air the saving potential for heating is for multiple zone control 26 kWh/m^2 and for single-zone control 17 kWh/m^2. When considering a house area of 155 m^2 the saving is 4030 and 2635 kWh per year for multiple-zone and single-zone control respectively, see Table 3.

4.2 Control Parameters

Results from the simulation shows that during the winter months the CO_2 and Δx are the most important control parameters for living spaces (living room, kitchen, bed room etc.). Simulations from multiple, two-zone and one-zone control show that Δx is even more important than CO_2, and suggest that Δx could work as single control parameter for living spaces during winter months for those control strategies. However, this depends to a great extent to choice of moisture load, which is in data in the model, and need to be investigated further with field measurements in real conditions. If only CO_2 is used, the limit for Δx will be exceeded for example during preparation and cooking of food.

For the bathroom during wintertime the Δx is always the only needed control parameter. If instead the relative humidity is preferred as a control signal one has to be attentive regarding threshold/limits used for the control and to be aware that the often used limits for RH (e.g. 75%) will not work properly.

During summertime temperature alone is the most (and only) important control parameter due to high temperature indoor. This is especially true for this house, since the house is a low energy house. Due to the high temperatures in the living spaces during summertime the supply flow rate controlled by the temperature is already at maximum at all times and thus the control signal from dx or RH in the bathroom cannot change flow rate. In Table 4 the parameters which are important for each season and living area are summarized.

Table 4 Summary of parameters are important for each season and living area

	Living spaces (living room, kitchen, bed room)		Bathroom	
	Summer	Winter	Summer	Winter
CO_2		x		
Temperature	x		x[a]	x
Δx		x	x[a]	x
RH		Possibly		Possibly

[a]Will be override by temperature signal from living space due to overtemperatures

4.3 IAQ and Demand Control

For all results presented in Table 2, the IAQ and thermal comfort as well as the humidity level in all rooms including bathrooms is controlled and within the limits (Table 1) at all time.

Regarding IAQ and thermal comfort all the three cases where the sensors/control signal are placed in each room (multiple-zone control, two-zone control, one-zone control) will give more or less the same result.

Since exhaust duct control is more easily implemented in reality both regarding cost and technical installation results from this case will be presented more thoroughly below in Figs. 2 and 3. Figure 2 shows the CO_2-level in the exhaust duct and the control signals from exhaust duct. Figure 3 shows the IAQ and ventilation flow rate for kitchen, master bedroom and bathroom. According to Fig. 3a, c, d, IAQ in the living room and kitchen, master bedroom and bathroom is maintained at an acceptable level: the maximum CO_2 concentration and RH is about 820 ppm and 40%, respectively, which are below the maximum limits. The common supply and extract air is continuously adjusted by the control signal from the exhaust duct and divided into the rooms, Fig. 3b, d, f.

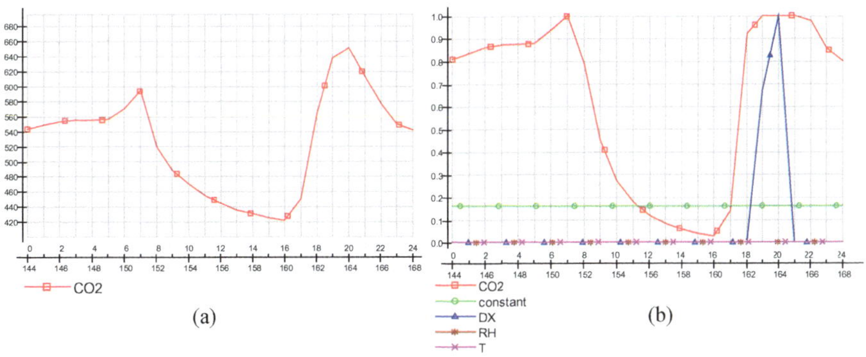

Fig. 2 Exhaust duct control for a winter day (2015-01-07): **a** CO_2 concentration in the exhaust duct and **b** control signals

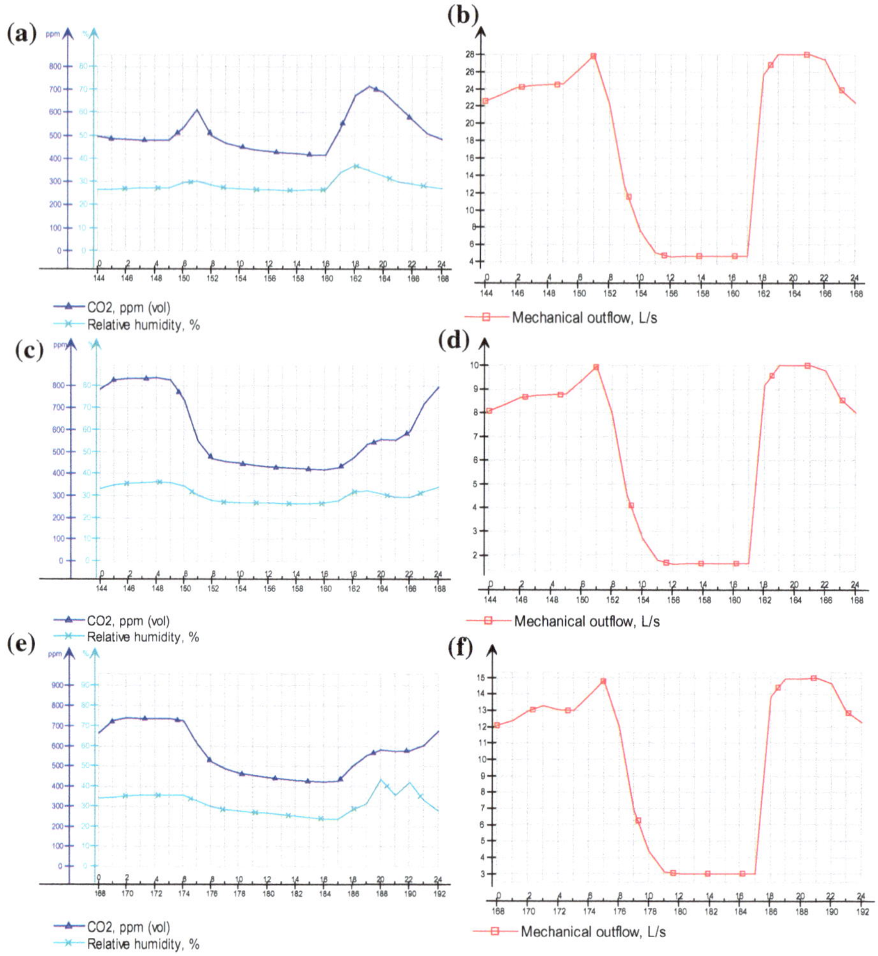

Fig. 3 IAQ and ventilation flow rate for several selected rooms for a winter day (2015-01-07). Living room and kitchen: **a** IAQ and **b** ventilation flow rate; master bedroom: **c** IAQ and **d** ventilation flow rate; bathroom: **e** IAQ and **f** ventilation flow rate

Comparing the CO_2-levels in Figs. 2a and 3c one can see that the CO_2 concentration in the exhaust duct (Fig. 2a) is lower than that in the rooms (Fig. 3a, c, e) when the house is occupied, this is due to that the air in the exhaust duct is diluted as all exhaust air from all the rooms are going through the exhaust duct. During the midnight, the CO_2 concentration in exhaust air is about 550 ppm while in the bedroom is about 800 ppm. This means that the control signal from the exhaust duct seldom reaches levels above the set point which results in an insufficient ventilation flow rate.

In a larger houses the total air flow rate will be higher (as the flow rate is based on square meter) resulting in an even larger discrepancy between the actual room levels and the level in the exhaust duct. If exhaust duct control is to be installed and used in buildings this has to be considered and a method for calculating correct set point for the exhaust duct control is needed.

Thus when placing control sensors in the exhaust duct the flow rate is reduced mainly during unoccupied hours, while when many zones can be individually controlled the air flow will be reduced continuously depending on demand for each zone yielding a higher energy saving.

5 Conclusions

Different strategies of demand controlled ventilation have been tested for a single family house by performing IDA simulations. The program IDA has been shown to be a useful tool for testing demand control strategies and detecting problems. It is concluded that there is an energy saving potential in using demand control ventilation without impairing IAQ in residential buildings. However, it is important to design and choose an appropriate control strategy including control parameters and sensor placing to obtain a good indoor environment, thermal comfort and avoiding damages on the building. Further, the energy saving depends to a great extent on how many zones the building is divided into, also depends on the climate zone and ventilation system design. The multiple zone control will always yield a higher saving potential, however this system is much more complex both in terms of number of mechanical devices as well as control strategy, and thus will be more expensive and require more maintenance compared to single zone control systems. For residential houses with balanced ventilation and heat recovery, the saving potential in kWh is small for both multiple and single zone control and the investment needed for demand control ventilation can be hard to justify. In residential houses without heat recovery the saving potential is much higher and demand control ventilation can be an option.

References

1. Energiläget 2015, Energimyndigheten (2015)
2. Energieffektiva småhus- En marknadsöversikt för dig som ska bygga nytt, Energimyndigheten (2011)
3. Att ventilera hus- Fakta och bakgrund om ventilation av hus och lägenheter, Joakim Achim Friedrich, Hållbar utveckling Väst, Göteborg (2011)
4. BBR Swedish Building Regulations: [Online] http://www.boverket.se
5. S. Geving, J. Holme, J.A. Jenssen, Indoor air humidity in Norwegian houses, in *Proceedings of 8th Symposium on Building Physics in the Nordic Countries*, pp. 801–807 (2008)

6. T. Kalamees, J. Kurnitski, J. Vinha, Indoor humidity loads in Finnish detached houses, in *Proceedings of the 7th Symposium on Building Physics in the Nordic Countries*, pp. 13–15, Iceland (2005)
7. Y. Fan, K. Ito, Integrated building energy computational fluid dynamics simulation for estimating the energy-saving effect of energy recovery ventilator with CO_2 demand-controlled ventilation system in office space. Indoor Built Environ. **23**(6), 785–803 (2014)
8. B.J. Wachenfeldt, M. Mysen, P.G. Schild, Air flow rates and energy saving potential in schools with demand-controlled displacement ventilation. Energy Build. **39**(10), 1073–1079 (2007)
9. V. Pavlovas, Demand controlled ventilation: a case study for existing Swedish multifamily buildings. Energy Build. **36**(10), 1029–1034 (2004)
10. A. Hesaraki, S. Holmberg, Demand-controlled ventilation in new residential buildings: consequences on indoor air quality and energy savings. Indoor Built Environ. **24**(2), 162–173 (2015)
11. T.R. Nielsen, C. Drivsholm, Energy efficient demand controlled ventilation in single family houses. Energy Build. **42**(11), 1995–1998 (2010)
12. P. Ylmén, J. Persson, Monitoring of pilot sites. Grant agreement: ENER/FP7/285173/NEED4B (2017)
13. Equa Simulation Technology, AB, IDA Simulation Environment. Version 2.11, Reference manual, Sundbyberg, Sweden, www.equa.se (1999)
14. Sveby. Brukarindata för energiberäkningar i bostäder, projekt rapport (2009)
15. CEN. Criteria for the indoor environment, including thermal, indoor air quality, light and noise. Brussels (2005)
16. B. Mattson, *Extract of raw data from the measurements in BETSI-study* (Boverket, Karlskrona, 2010)
17. FoHMFS 2014:18, Folkhälsomyndighetens allmänna råd om ventilation

Is It Possible to Build Near Zero Energy Single Family Buildings in Very Cold Arctic Climate?

Svein Ruud

Abstract According to the Energy Performance of Buildings Directive recast of 2010 all new buildings in European Union shall be nearly zero energy buildings from 2021 and onwards. However only very few of the countries in the Europan Union have to deal with very cold arctic climatic conditions, i.e. Sweden and Finland. Further, the required energy performance of a nearly zero energy building should be calculated on the basis of a methodology, which may be differentiated at national and regional level. This means that the definition may be quite different from member state to member state. This paper presents how different solutions regarding design of building envelopes combined with different heating and ventilation systems in single-family buildings can meet the energy requirements for nearly zero energy buildings. The main focus is on how the proposed Swedish building regulations for nearly zero energy buildings affects the possibility to build single-family buildings in the very north arctic part of Sweden. The main conclusions from this study is that the building envelopes in most cases need to be improved compared today existing standard and that direct electric heating will not be an option. Further, also with improved building envelopes solutions with extract air heat pumps will have difficulties in meeting the tougher energy requirements, especially in single-storey buildings. However geothermal heat pumps will meet the requirements even with existing building envelopes and even with only exhaust ventilation without heat recovery. Also air-water heat pumps combined with ventilation heat recovery can meet the requirement.

Keywords Nearly zero energy · Energy performance · Arctic climate
Heat pump · Mechanical ventilation

S. Ruud (✉)
Built Environment, Energy and Circular Economy,
RISE Research Institutes of Sweden, Borås, Sweden
e-mail: svein.ruud@ri.se

D. Johansson et al. (eds.), *Cold Climate HVAC 2018*,
Springer Proceedings in Energy, https://doi.org/10.1007/978-3-030-00662-4_11

1 Background

According to the Energy Performance of Buildings Directive recast of 2010 [1] all new buildings in the European Union shall be nearly zero energy buildings from 2021 and onwards. However only very few of the countries in the Europan Union have to deal with very cold arctic climatic conditions, i.e. Sweden and Finland. Further, the required energy performance of a nearly zero energy building should be calculated on the basis of a methodology, which may be differentiated at national and regional level. This means that the definition may be quite different from member state to member state. After years of investigations Boverket—the National Board of Housing, Building and Planning—finally presented a Swedish definition of nearly zero energy building in the summer of 2017 [2]. However a final determination of primary energy factors to be used in 2021 remains to be decided. Thus this study has used the values that were proposed by Boverket in the spring of 2017.

2 Purpose

The purpose of this paper is to presents how different solutions regarding design of building envelopes combined with different heating and ventilation systems in single-family buildings can meet the proposed energy requirements for nearly zero energy buildings in Sweden. The main focus is on how the proposed Swedish building regulations for nearly zero energy buildings affects the possibility to build single-family buildings, especially in the very north arctic part of Sweden.

3 Method

Different solutions regarding design of building envelopes have been combined with different heating and ventilation systems. The energy use for the different combinations has then been calculated using the energy calculation program TMF Energi [3]. Based on these results the energy performance has been calculated and compared with the proposed Swedish requirements for nearly zero energy buildings.

3.1 Description of the Studied Single Family Houses

The purpose of the study was to see how the Swedish definition of a near zero energy buildings, the proposed factors and maximum primary energy number for 2021 affects the possibility to design different building envelopes in combination with different heating and ventilation systems. Values for different possible building designs from very compact, extremely well insulated and air-tight envelopes to less compact and less insulated and air-tight envelopes have been developed. These different building designs have been combined with several heating and ventilation systems.

The building physic properties of the houses. The first building type to be studied is very compact single-storey buildings of different sizes and with two levels of thermal insulation and air-tightness. Table 1 shows values for an improved envelope performance compared to today standard, but still with reasonable additional costs. Table 2 shows values for an extreme building envelope where the additional costs may not be justified. With the same building system the U_m-value increases somewhat as A_{temp} decreases and the ratio A_{om}/A_{temp} is also increased. This is the reason why Boverket allows higher primary energy numbers when A_{temp} is below 130 m^2.

The second building type to be studied is a single-storey single-family house with an angular extension and with three different levels of thermal insulation and air-tightness; today standard, improved standard and extreme standard (Table 3).

The third building type to be studied is a 1½ storey single-family house with an angular roof extension and with three different levels of building standard. This is today the most common single family building type (Table 4).

Table 1 Very compact, well insulated and air-tight single-storey house

Tempered floor area (A_{temp}) (m^2)	90	130	160
Number of rooms and kitchen	3	4	5
Building inside envelope area (A_{om}) (m^2)	283.3	386.1	463.3
Average thermal transmittance (U_m) (W/m^2 K)	0.212	0.205	0.202
Average air leakage rate at (q_{50}) (l/s m^2)	0.30	0.30	0.30

Table 2 Very compact, extremely well insulated and air-tight single-storey house

Tempered floor area (A_{temp}) (m^2)	90	130	160
Number of rooms and kitchen	3	4	5
Building inside envelope area (A_{om}) (m^2)	283.3	386.1	463.3
Average thermal transmittance (U_m) (W/m^2 K)	0.168	0.163	0.160
Average air leakage rate at (q_{50}) (l/s m^2)	0.15	0.15	0.15

Table 3 Single-storey single-family house with an angular extension

Tempered floor area (A_{temp}) (m^2)	160	160	160
Number of rooms and kitchen	5	5	5
Building inside envelope area (A_{om}) (m^2)	500	500	500
Average thermal transmittance (U_m) (W/m^2 K)	0.24	0.20	0.16
Average air leakage rate at (q_{50}) (l/s m^2)	0.60	0.30	0.15

Table 4 1½-storey single family house with an angular roof extension

Tempered floor area (A_{temp}) (m^2)	160	160	160
Number of rooms and kitchen	5	5	5
Building inside envelope area (A_{om}) (m^2)	400	400	400
Average thermal transmittance (U_m) (W/m^2 K)	0.26	0.22	0.18
Average air leakage rate at (q_{50}) (l/s m^2)	0.60	0.30	0.15

Table 5 Very compact 2-storey single family house

Tempered floor area (A_{temp}) (m^2)	160	160	160
Number of rooms and kitchen	5	5	5
Building inside envelope area (A_{om}) (m^2)	350	350	350
Average thermal transmittance (U_m) (W/m^2 K)	0.27	0.23	0.19
Average air leakage rate at (q_{50}) (l/s m^2)	0.60	0.30	0.15

The fourth building type to be studied is a very compact 2 storey single-family house and with three different levels of building standard (Table 5).

The heating and ventilation systems studied. Nine different heating and ventilation systems have been studied. Some of them are the most commonly used today and other are expected to be used more as the energy requirements are getting tougher.

For all studied systems the today best available technologies are assumed to be used, i.e. very energy efficient pumps and fans, inverter controlled compressors, low stand-by power, very well insulated ducts, pipes and hot water storage, feedback space heating control, very efficient ventilation heat recovery and demand controlled ventilation (when applicable). The main parts of the systems are placed in the insulated part of the building envelope. The different heating and ventilation systems studied are described in Table 6 together with the abbreviations used in the results section.

3.2 *Description of the Calculation Program Used*

All energy calculations has been made with a for the purpose modified version of the energy calculation program TMF Energi [3]. It is a calculation tool developed by RISE, Research Institutes of Sweden (former SP) [4], for TMF, the national trade and employers' association of the wood processing and furniture industry in Sweden, for calculation of specific energy use in single family houses according to the energy requirement in the Swedish building regulations. It has been used for more than ten years and has been updated regularly along with changes in the building regulations. It is used by a majority of the manufacturers of detached single-family houses.

Table 6 Description of the heating and ventilation systems studied

System	Abbreviation
Geothermal heat pump+exhaust air	GHP+E
Geothermal heat pump+exhaust and supply air with heat recovery	GHP+ESH
Air-water heat pump+exhaust air	AWHP+E
Air-water heat pump+exhaust and supply air with heat recovery	AWHP+ESH
Electricity+thermal solar+exhaust and supply air with heat recovery	EL+TS+ESH
Exhaust air heat pump+peak electric heating	EHP+EL[a]
Exhaust air heat pump+peak district heating	EHP+DH
District heating+exhaust and supply air with heat recovery	DH+ESH
Biofuel heating+thermal solar+exhaust and supply air with heat recovery	BF+TS+ESH

[a]A more simple exhaust air heat pump that keeps the exhaust air above 0 °C

3.3 Description of the Swedish Definition of Near Zero Energy Buildings

The definition of nearly zero energy buildings in Sweden is given by Boverket in its latest mandatory provisions and general recommendations, BBR 25 [2]. The energy performance is calculated as a primary energy number (EP_{pet}) according to the following formula:

$$EP_{\text{pet}} = \frac{\sum_{i=1}^{6}\left(\frac{E_{\text{uppv,i}}}{F_{\text{geo}}} + E_{\text{kyl,i}} + E_{\text{tvv,i}} + E_{\text{f,i}}\right) \times PE_{\text{i}}}{A_{\text{temp}}}$$

where

$E_{uppv,i}$	is the space heating energy for energy carrier i
$E_{kyl,i}$	is the space comfort cooling energy for energy carrier i
$E_{tvv,i}$	is the domestic hot water heating energy for energy carrier i
$E_{f,i}$	is the building property energy for energy carrier i
F_{geo}	is a geographic adjustment factor
PE_i	is the primary energy factor for energy carrier i
A_{temp}	is the “floor area” intended to be heated to more than 10 °C.

Each municipality has been allocated a geographical adjustment factor to compensate for their different climate conditions. The geographical adjustment factors are in the range from 0.8 in the very south to 1.9 in the very north. This means that the same maximum primary energy number can apply for all single-family houses regardless of in which municipality it is built. In this study 0.8 applies to Malmö, 1.0 applies to Stockholm and 1.9 applies to Kiruna.

In BBR 25 the primary energy factor for electricity is set to 1.6 and for all other energy carries it is set to 1.0. In the spring of 2017 Boverket proposed that in 2021 the primary energy factor for electricity shall set to 2.5 and for all other energy carries it shall remain at 1.0. But as the primary energy factor for electricity in Europe is expected to be about 2.0 in 2021 many respondents had objections about this. Hence a final determination of primary energy factors to be used in 2021 remains to be decided. However if the primary energy factor for electricity is decreased it is likely that the primary energy factors for other energy carriers is also decreased as well as the maximum allowed primary energy number. The ratio between electricity and other energy carriers as well as the required energy performance will then in practice remain the same. In this study the primary energy factors proposed by Boverket for 2021 has been used as well as the proposed maximum primary energy number. For single-family houses the proposed maximum primary energy number is 90 kWh/m^2 a if the heated floor area is above 130 m^2, linearly increased to 110 kWh/m^2 a as the heated floor area is decreased to 90 m^2 and 110 kWh/m^2 a if the heated floor area is below 90 m^2,

In addition to the requirements above the building regulations also have requirements on the maximum installed electric power for heating, ventilation and hot water production. These requirements also remains to be decided for 2021, but it is likely that they for single-family houses will remain almost the same as in BBR 25 and as the energy requirements gets tougher the requirements on maximum installed electric power will in most cases not imply a limitation.

A third requirement in the building regulation is on the average thermal transmittance of the building envelope. The maximum allowed average thermal transmittance is proposed to decrease from 40 to 30 W/m^2 K in 2021. However, this is also not a limitation as in almost all cases a much better building envelope is required to meet the proposed maximum primary energy number.

3.4 Determination of Normal Use During a Normal Year

A determination of normal use of different building types during a normal year is given by Boverket in its mandatory provisions and general recommendations, BEN 2 [5]. It is a new regulation establishing how to calculate energy use during normal use of a building during a normal year and how to normalize the measured energy consumption during any year to a normal usage and to a normal year. The main rule of the Swedish building regulations is that the energy performance shall be verified by measurements within two years of completion, but it is also possible to verify by calculations on a completed building. For a single-family house BEN 2 states the following standardized values for normal use:

Use of household electricity	30 kWh/m^2 a
Use of domestic hot water	18–20 kWh/m^2 a
Indoor temperature (minimum)	21 °C

Furthermore each Swedish municipality has been assigned a specific standardized climatic condition for a normal year to be used both for calculation of and normalization of measured energy use.

4 Results

The results of the calculations are presented in Tables 7, 8, 9, 10, 11, 12, 13, 14, 15, 16, 17, 18, 19, 20 and 21. Calculated primary energy numbers that meets the proposed required level for near zero energy buildings are marked in bold green. Calculated values that do not meet the required level are marked in bold red and values on the verge of meeting the requirements are marked in bold orange. As the primary energy number is a rather abstract value without any direct correlation with the real energy consumption of the houses, the bought energy including household electricity is also presented in each table. The total bought energy is presented as well as divided into electricity, district heating and biofuel where applicable.

Table 7 A_{temp} **90** m^2 (U_m 0.212 W/K m^2, q_{50} 0.30 l/s m^2), $EP_{pet,\ max}$ **110** kWh/m^2 a

	GHP +E	GHP +ESH	AWHP +E	AWHP +ESH	EL+TS +ESH	EHP +EL	EHP +DH	DH +ESH	BF+TS +ESH
EP_{pet}	**73.8**	**72.0**	**100.0**	**92.4**	**146.2**	**119.3**	**112.8**	**93.4**	**96.1**
BE_{total}	6706	6401	8813	8102	13,180	9819	13,192	15,624	14,626
$BE_{electr.}$	6706	6401	8813	8102	13,180	9819	7132	3125	3317
$BE_{distr.}$	–	–	–	–	–	–	6059	12,499	–
$BE_{biof.}$	–	–	–	–	–	–	–	–	11,309

Table 8 A_{temp} **130** m^2 (U_m 0.205 W/K m^2, q_{50} 0.30 l/s m^2), $EP_{pet,\ max}$ **90** kWh/m^2 a

	GHP +E	GHP +ESH	AWHP +E	AWHP +ESH	EL+TS +ESH	EHP +EL	EHP +DH	DH +ESH	BF+TS +ESH
EP_{pet}	**69.3**	**67.2**	**85.2**	**85.3**	**135.7**	**112.0**	**104.4**	**86.7**	**88.3**
BE_{total}	9250	8784	11,320	11,004	17,717	13,486	19,340	20,965	19,710
$BE_{electr.}$	9250	8784	11,320	11,004	17,717	13,486	8767	4498	4690
$BE_{distr.}$	–	–	–	–	–	–	10,573	16,467	–
$BE_{biof.}$	–	–	–	–	–	–	–	–	15,021

Table 9 A_{temp} **160** m^2 (U_m 0.202 W/K m^2, q_{50} 0.30 l/s m^2), $EP_{pet,\ max}$ **90** kWh/m^2 a

	GHP +E	GHP +ESH	AWHP +E	AWHP +ESH	EL+TS +ESH	EHP +EL	EHP +DH	DH +ESH	BF+TS +ESH
EP_{pet}	**67.8**	**65.1**	**82.5**	**83.2**	**130.9**	**108.9**	**103.1**	**83.6**	**84.8**
BE_{total}	11,200	10,566	13,610	13,267	21,063	16,235	25,040	24,905	23,460
$BE_{electr.}$	11,200	10,566	13,610	13,267	21,063	16,235	9690	5527	5719
$BE_{distr.}$	–	–	–	–	–	–	15,350	19,378	–
$BE_{biof.}$	–	–	–	–	–	–	–	–	17,741

Table 10 A_{temp} **90** m^2 (U_m 0.168 W/K m^2, q_{50} 0.15 l/s m^2), $EP_{pet, max}$ **110** kWh/m^2 a

	GHP +E	GHP +ESH	AWHP +E	AWHP +ESH	EL+TS +ESH	EHP +EL	EHP +DH	DH +ESH	BF+TS +ESH
EP_{pet}	**66.5**	**62.6**	**87.5**	**80.3**	**115.5**	**94.9**	**98.5**	**75.7**	**76.9**
BE_{total}	6228	5801	7972	7289	10,711	8276	11,124	12,666	11,788
$BE_{electr.}$	6228	5801	7972	7289	10,711	8276	6920	3104	3296
$BE_{distr.}$	–	–	–	–	–	–	4204	9562	–
$BE_{biof.}$	–	–	–	–	–	–	–	–	8492

Table 11 A_{temp} **130** m^2 (U_m 0.163 W/K m^2, q_{50} 0.15 l/s m^2), $EP_{pet, max}$ **90** kWh/m^2 a

	GHP +E	GHP +ESH	AWHP +E	AWHP +ESH	EL+TS +ESH	EHP +EL	EHP +DH	DH +ESH	BF+TS +ESH
EP_{pet}	**62.6**	**58.7**	**82.3**	**74.1**	**107.9**	**90.2**	**91.6**	**70.6**	**70.8**
BE_{total}	8624	8006	10,993	9928	14,482	11,497	16,533	17,107	15,966
$BE_{electr.}$	8624	8006	10,993	9928	14,482	11,497	8585	4470	4662
$BE_{distr.}$	–	–	–	–	–	–	7947	12,637	–
$BE_{biof.}$	–	–	–	–	–	–	–	–	11,304

Table 12 A_{temp} **160** m^2 (U_m 0.160 W/K m^2, q_{50} 0.15 l/s m^2), $EP_{pet, max}$ **90** kWh/m^2 a

	GHP +E	GHP +ESH	AWHP +E	AWHP +ESH	EL+TS +ESH	EHP +EL	EHP +DH	DH +ESH	BF+TS +ESH
EP_{pet}	**60.8**	**56.8**	**81.1**	**71.3**	**103.8**	**87.6**	**90.6**	**68.1**	**67.7**
BE_{total}	10,404	9628	13,355	11,872	17,195	13,842	21,557	20,302	18,969
$BE_{electr.}$	10,404	9628	13,355	11,872	17,195	13,842	9538	5494	5686
$BE_{distr.}$	–	–	–	–	–	–	12,019	14,808	–
$BE_{biof.}$	–	–	–	–	–	–	–	–	13,282

Table 13 A_{temp} 160 m^2 (U_m **0.24** W/K m^2, q_{50} **0.60** l/s m^2), $EP_{pet, max}$ **90** kWh/m^2 a

	GHP +E	GHP +ESH	AWHP +E	AWHP +ESH	EL+TS +ESH	EHP +EL	EHP +DH	DH +ESH	BF+TS +ESH
EP_{pet}	**77.5**	**77.5**	**100.1**	**89.7**	**175.5**	**142.4**	**122.0**	**109.1**	**113.4**
BE_{total}	12,403	12,070	15,707	14,105	27,436	20,025	30,393	32,455	30,931
$BE_{electr.}$	12,403	12,070	15,707	14,105	27,436	20,025	9881	5582	5773
$BE_{distr.}$	–	–	–	–	–	–	20,511	26,873	–
$BE_{biof.}$	–	–	–	–	–	–	–	–	25,158

Table 14 A_{temp} 160 m^2 (U_m **0.20** W/K m^2, q_{50} **0.30** l/s m^2), $EP_{pet, max}$ **90** kWh/m^2 a

	GHP +E	GHP +ESH	AWHP +E	AWHP +ESH	EL+TS +ESH	EHP +EL	EHP +DH	DH +ESH	BF+TS +ESH
EP_{pet}	**68.7**	**67.5**	**86.0**	**84.1**	**138.9**	**116.8**	**107.3**	**88.2**	**89.9**
BE_{total}	11,386	10,848	14,027	13,428	22,214	17,119	26,222	26,271	24,804
$BE_{electr.}$	11,386	10,848	14,027	13,428	22,214	17,119	9736	5537	5729
$BE_{distr.}$	–	–	–	–	–	–	16,486	20,734	–
$BE_{biof.}$	–	–	–	–	–	–	–	–	19,075

Table 15 A_{temp} 160 m² (U_m **0.16** W/K m², q_{50} 0.**15** l/s m²), $EP_{pet,\ max}$ **90** kWh/m² a

	GHP +E	GHP +ESH	AWHP +E	AWHP +ESH	EL+TS +ESH	EHP +EL	EHP +DH	DH +ESH	BF+TS +ESH
EP_{pet}	**62.8**	**58.8**	**80.3**	**75.3**	**110.5**	**93.7**	**94.3**	**71.9**	**71.9**
BE_{total}	10,633	9859	13,337	12,339	18,146	14,531	22,597	21,436	20,069
$BE_{electr.}$	10,633	9859	13,337	12,339	18,146	14,531	9587	5502	5694
$BE_{distr.}$	–	–	–	–	–	–	13,010	15,935	–
$BE_{biof.}$	–	–	–	–	–	–	–	–	14,374

Table 16 A_{temp} 160 m² (U_m **0.26** W/K m², q_{50} **0.60** l/s m²), $EP_{pet,\ max}$ **90** kWh/m² a

	GHP +E	GHP +ESH	AWHP +E	AWHP +ESH	EL+TS +ESH	EHP +EL	EHP +DH	DH +ESH	BF+TS +ESH
EP_{pet}	**83.4**	**81.0**	**89.7**[a]	**84.8**	**151.8**	**121.4**	**110.7**	**95.6**	**98.2**
BE_{total}	13,359	12,647	14,459	13,512	24,050	17,625	27,177	28,447	26,954
$BE_{electr.}$	13,359	12,647	14,459	13,512	24,050	17,625	9776	5555	5747
$BE_{distr.}$	–	–	–	–	–	–	17,401	22,892	–
$BE_{biof.}$	–	–	–	–	–	–	–	–	21,207

[a]The maximum electric power for heating and domestic hot water may be too high

Table 17 A_{temp} 160 m² (U_m **0.22** W/K m², q_{50} **0.30** l/s m²), $EP_{pet,\ max}$ **90** kWh/m² a

	GHP +E	GHP +ESH	AWHP +E	AWHP +ESH	EL+TS +ESH	EHP +EL	EHP +DH	DH +ESH	BF+TS +ESH
EP_{pet}	**75.2**	**70.3**	**84.7**	**78.7**	**123.1**	**102.1**	**99.5**	**79.2**	**79.8**
BE_{total}	12,312	11,408	13,852	12,731	19,948	15,464	24,033	23,580	22,160
$BE_{electr.}$	12,312	11,408	13,852	12,731	19,948	15,464	9653	5517	5710
$BE_{distr.}$	–	–	–	–	–	–	14,380	18,063	–
$BE_{biof.}$	–	–	–	–	–	–	–	–	16,450

Table 18 A_{temp} 160 m² (U_m **0.18** W/K m², q_{50} **0.15** l/s m²), $EP_{pet,\ max}$ **90** kWh/m² a

	GHP +E	GHP +ESH	AWHP +E	AWHP +ESH	EL+TS +ESH	EHP +EL	EHP +DH	DH +ESH	BF+TS +ESH
EP_{pet}	**68.4**	**62.0**	**79.7**	**69.2**	**100.8**	**85.2**	**89.3**	**66.3**	**65.8**
BE_{total}	11,528	10,443	13,182	11,624	16,760	13,572	21,186	19,783	18,468
$BE_{electr.}$	11,528	10,443	13,182	11,624	16,760	13,572	9522	5490	5683
$BE_{distr.}$	–	–	–	–	–	–	11,664	14,293	–
$BE_{biof.}$	–	–	–	–	–	–	–	–	12,785

Table 19 A_{temp} 160 m² (U_m **0.27** W/K m², q_{50} **0.60** l/s m²), $EP_{pet,\ max}$ **90** kWh/m² a

	GHP +E	GHP +ESH	AWHP +E	AWHP +ESH	EL+TS +ESH	EHP +EL	EHP +DH	DH +ESH	BF+TS +ESH
EP_{pet}	**78.1**	**74.5**	**84.0**[a]	**81.9**	**138.5**	**109.6**	**104.2**	**88.0**	**89.7**
BE_{total}	12,699	11,952	13,776	13,105	22,145	16,300	25,351	26,189	24,723
$BE_{electr.}$	12,699	11,952	13,776	13,105	22,145	16,300	9708	5539	5731
$BE_{distr.}$	–	–	–	–	–	–	15,643	20,649	–
$BE_{biof.}$	–	–	–	–	–	–	–	–	18,992

Table 20 A_{temp} 160 m^2 (U_m **0.23** W/K m^2, q_{50} **0.30** l/s m^2), $EP_{pet,\ max}$ **90** kWh/m^2 a

	GHP +E	GHP +ESH	AWHP +E	AWHP +ESH	EL+TS +ESH	EHP +EL	EHP +DH	DH +ESH	BF+TS +ESH
EP_{pet}	**70.5**	**65.7**	**80.4**	**73.3**	**113.7**	**93.8**	**94.7**	**73.8**	**73.9**
BE_{total}	11,830	10,926	13,341	12,105	18,605	14,526	22,691	21,983	20,600
$BE_{electr.}$	11,830	10,926	13,341	12,105	18,605	14,526	9595	5506	5699
$BE_{distr.}$	–	–	–	–	–	–	13,096	16,477	–
$BE_{biof.}$	–	–	–	–	–	–	–	–	14,901

Table 21 A_{temp} 160 m^2 (U_m **0.19** W/K m^2, q_{50} **0.20** l/s m^2), $EP_{pet,\ max}$ **90** kWh/m^2 a

	GHP +E	GHP +ESH	AWHP +E	AWHP +ESH	EL+TS +ESH	EHP +EL	EHP +DH	DH +ESH	BF+TS +ESH
EP_{pet}	**64.9**	**58.6**	**75.9**	**65.5**	**94.5**	**79.8**	**85.8**	**62.7**	**61.8**
BE_{total}	11,173	10,109	12,748	11,190	15,860	12,966	20,227	18,707	17,433
$BE_{electr.}$	11,173	10,109	12,748	11,190	15,860	12,966	9474	5482	5675
$BE_{distr.}$	–	–	–	–	–	–	10,753	13,225	–
$BE_{biof.}$	–	–	–	–	–	–	–	–	11,758

4.1 *Single Storey Very Compact Single Family Houses*

Well-insulated and air-tight houses of different sizes. For these houses the heating and ventilation systems direct electricity (EL+TS+ESH) and exhaust air heat pump with peak electric heating (EHP+EL) fail to meet the suggested requirement level for near zero energy buildings. Exhaust air heat pump with peak district heating (EHP+DH) is on the verge to meet the requirement for the 90 m^2 house but does not meet it for the larger houses. Biofuel heating with thermal solar, supply and exhaust ventilation with heat recovery (BF+TS+ESH) are on the verge to meet the requirement for the 130 m^2 house and meet the requirement for the larger and the smaller house. The systems with geothermal heat pumps and meets the requirement for all house sizes, even without ventilation heat recovery.

Extremely well-insulated and air-tight houses of different sizes. Also for these houses the heating and ventilation system direct electricity (EL+TS+ESH) fail to meet the suggested requirement. Exhaust air heat pump with peak electric heating (EHP+EL) are on the verge to meet the requirement for the 130 m^2 house and meet the requirement for the larger and the smaller house. Exhaust air heat pump with peak district heating (EHP+DH) meet the requirement for the 90 m^2 house and are on the verge to meet the requirement for the larger houses. All other heating and ventilation systems meet the requirement.

4.2 Less Compact Single-Storey Single-Family House with an Angular Extension and with Different Insulation and Air-Tightness Standard

For the house with today normal insulation and air-tightness standard it is only the geothermal heat pump that meets the requirement. The air-water heat pump combined with supply and exhaust ventilation and heat recovery (AWHP+ESH) is on the verge to meet the requirement for that house. The direct electric system (EL+TS+ESH) and the systems with exhaust air heat pumps (EHP+EL, EHP+DH) do not meet the requirement even with the extreme building envelope. The systems with district heating (DH+ESH) and biofuel (BF+TS+ESH) are on the verge to meet the requirement with the improved building envelope and meet the requirement with the extreme envelope.

4.3 1½-Storey Single-Family House with an Angular Roof Extension and with Different Insulation and Air-Tightness Standard

For the house with today normal insulation and air-tightness standard it is only the systems with geothermal heat pump (GHS+E, GHS+ESH) and the air-water heat pump combined with supply and exhaust ventilation and heat recovery (AWHP +ESH) that meets the requirement. The system with air-water heat pump and ventilation heat recovery (AW+ESH) are on the verge to meet the requirement. The direct electric system (EL+TS+ESH) do not meet the requirement even with the extreme building envelope. Systems with exhaust air heat pumps (EHP+EL, EHP +DH) meet or are at the verge of the requirement with the extreme building envelope. The systems with district heating (DH+ESH) and biofuel (BF+TS+ESH) meet the requirement with the improved building envelope.

4.4 2-Storey Very Compact Single-Family House with Different Insulation and Air-Tightness Standard

The systems with district heating (DH+ESH) and biofuel (BF+TS+ESH) are on the verge to meet the requirement with the standard building envelope and meet the requirement with the improved envelope. The systems with district heating (DH+ESH) and biofuel (BF+TS+ESH) are on the verge to meet the requirement with the standard building envelope. The direct electric system (EL +TS+ESH) do not meet the requirement even with the extreme building envelope. Systems with exhaust air heat pumps (EHP+EL, EHP+DH) meet the requirement with the extreme building envelope.

5 Conclusions

The main conclusions from this study is that the building envelopes in most cases need to be improved compared today existing standard and that direct electric heating will not be an option. Further, also with improved building envelopes solutions with extract air heat pumps will have difficulties in meeting the tougher energy requirements in the very north arctic part of Sweden, especially in single-storey buildings. However geothermal heat pumps will meet the requirements even with today existing building envelope standard and even with only exhaust ventilation without heat recovery. But solutions with no preheating of the outdoor supply air are from a thermal comfort point of view not recommended in a cold arctic climate. Air-water heat pumps may in many cases also meet the Swedish requirement for near zero energy buildings if combined with ventilation heat recovery. The challenge for that solution in a very cold arctic climate is to not exceed the maximum allowed electric power for heating and domestic hot water when the heat pump is shut off.

References

1. Directive 2010/31/EU of the European Parliament and of the Council of 19 May 2010 on the energy performance of buildings (recast)
2. Boverket's building regulations—mandatory provisions and general recommendations, BFS 2011:6 with amendments up to BFS 2017:5 BBR 25 (in Swedish)
3. TMF Energi is a calculation tool developed by RISE (formerly SP) for TMF, the national trade and employers' association of the wood processing and furniture industry in Sweden, for calculation of energy use in single family buildings according to the energy requirement in the Swedish building regulations
4. The RISE institutes Innventia, SP Technical Research Institute of Sweden and Swedish ICT have in 2017 merged into RISE Research Institutes of Sweden AB in order to become a stronger research and innovation partner for businesses and society
5. Boverket's regulations and general advice on determination of the energy consumption of the building in normal use during a normal year, BFS 2016:12 with amendments up to BFS 2017:6 BEN 2 (in Swedish)

The Energy Performance of Green Roof in Sub-arctic Climate

Jutta Schade and Farshid Shadram

Abstract Green roofs are complex technology systems, adopting a vegetation layer on the outermost surface of the building shell. A proper design implement environmental and energy benefits. Green roof are aimed to reduce roof temperature and thus the summer solar gains, without worsening the winter energy performance. Most studies evaluating green roof performance have been conducted in warmer climates. There are very limited studies of green roofs in cold climate. Some research has investigated the thermal effect of the snow layer on green roof. But no study has so far evaluated the energy performance of green roof in sub-arctic climate. This study evaluates the heat flow and thermal effect on a green roof situated on a passive house building in the sub-arctic town Kiruna, Sweden for a period from 25th of October—4th of January. The ongoing measurements of temperature and heat flux is done on an extensive green roof and compared to the same roof covered solely by a roofing felt layer. The fluctuation in temperature was consistently higher for the roof with the roofing felt layer than for the green roof. But the surface temperature of both roofs was getting more and more align as the roofs are covered by snow during November and December. However during December month the green roof had a higher heat flux out of the building compared to the black roof.

Keywords Green roof · Building energy

1 Introduction

Green roofs are considered to be an effective contribution to the resolution to several environmental problems at the building and urban levels. They are considered to be environmental benefiting the building and its surrounding by

J. Schade (✉) · F. Shadram
Department of Civil, Environmental and Natural Resources Engineering, Luleå University of Technology, 97187 Luleå, Sweden
e-mail: Jutta.Schade@ltu.se

D. Johansson et al. (eds.), *Cold Climate HVAC 2018*,
Springer Proceedings in Energy, https://doi.org/10.1007/978-3-030-00662-4_12

improving storm water management [1, 2] water run-off quality [3] urban air quality [4] reducing of the urban heat island effect [5] and preventing noise pollution [6]. Furthermore, are green roof known to improve building energy efficiency by enhancing the heat transfer trough roof [2, 7]. Green roofs can reduce the heat flux through a building envelope since the growing media acts as an isolation layer, the plants shade the roof and are believed to provide transpiration cooling [8, 9]. Supplementary the higher thermal mass of a green roof system improve the thermal performance of the roof [10].

The main function of green roofs in terms of energy efficiency is to prevent the solar radiation which heats building's interior spaces. Green roofs are able to reflect 27% of solar radiation and absorb 60% of the radiation through photosynthesis and transmit the remainder as much as 13% to the growing medium [11]. An experimental study conducted in Japan showed that green roofs can decrease the surface temperature of the roof around 30–60 °C [11] As Saadatian [12] summarized green roofs field studies were mostly been limited to cooling effects and to measure surface temperatures of roofs.

There are very limited studies of green roofs in cold climate. One of the few is the study conducted by Getter et al. [13] in the Midwestern U.S climate, with hot humid summers and cold snowy winters. Their result showed that the green roof impact on the heat flux through the roof was greatest during the summer by a heat flux reduction of 167% where the reduction in the winter had an average of 13%. The temperature difference of the green roof and gravel roof showed that the gravel roof was up to 20 °C warmer during the summer. However, during spring and autumn the green and gravel roofs showed similar responses, as well as during winter when the roofs were cover with snow. Getter et al. [13] concluded that heat transfer and thermal differences between green and gravel roofs appear to be primarily influenced by solar radiation, ambient outside temperature and volumetric moisture content of the growing medium.

The green roof studied in this paper, is situated in the north of Sweden in Kiruna on Sweden's northernmost passive house. It was built to demonstrate and evaluate the environmental benefits, as reducing of storm water flows and enhancing the water quality and increasing building energy efficiency in a subarctic climate. The climate characterized by long, usually very cold winters and short cold to mild summers. Kiruna is situated above the polar circle, which include polar night during the winter and midnight sun during the summer. This special condition of no sunlight for approximately 28 days during the winter and 24 h of sunlight for approximately 50 days during the summer in combination with a rather cold climate makes this study unique. The investigations are still ongoing regarding hydraulic and hydrological performance, storm water contamination, heat flux and surface temperature. The measurements started in October 2016.

The object of this study was to quantify thermal properties of an extensive green roof installed on a good isolated roof (U-value 0.035 WK/m^2K) against a conventional black roof with the same good isolation in a subarctic climate in Kiruna Sweden for a period of 2 months and 10 days (25th of October—4th of January).

2 Method

2.1 *Roof Construction*

The extensive green roof with a total area of 160 m^2, was installed on a passive house named “sjunde huset” in the north of Sweden in Kiruna. The roof is divided in two equal sized parts of 40 m^2, the left side is a moss-sedum roof. The other half is a sedum, wild flowers and grass roof with 30 mm vegetation, 40 mm roof soil, 40 mm grodan TT 100/40 see Fig. 1. As the schematic picture of the roof section shows, the roof is isolated with 1000 mm loose fill mineral wool and has including the studs a U-value from 0.035 W/m^2 K. The roof is tilt by 1,26/12. The green roof was installed October 2014.

2.2 *Environmental Monitoring*

The environmental monitoring was done on the right side of the roof, i.e. the sedum, wild flower and grass roof. To get a reference area with a conventional black roof, an area of approximately 1 m^2 has been removed see Fig. 1.

The temperature and heat flux were recorded using tow CR800 datalogger (Campbell Scientific, Inc. Logan, UT) one for recording the data on the roof and one for recording the data inside the building directly from the sealing. The temperature on the surface of the green roof and on the black roof (removed part) was measured with a thermocouple (Campbell Scientific Type T107 Temperature Probe). Thermocouple accuracy ranged from ±0.4 °C, ±0.9 °C for temperatures in the range of −24 to 48 °C and −35 to 50 °C, respectively. Inside the building directly against the sealing, on the opposite side of the measured spots, tow heat flux sensors pairs (HFP01SC-05 heat Flux sensor) has been installed and one thermocouple (Campbell Scientific Type T107 Temperature Probe) in-between the two pairs. The heat flux sensor has an accuracy of ±2% (Fig. 2).

Fig. 1 Left: picture of the green roof in Kiruna with sedum on the left and sedum wild flowers and grass on the right. Right: schematic picture of the roof sections

Fig. 2 Left: installation of the heat flux sensor pairs and the thermocouple. Right: removed part as reference area

The collection of the data started on the 25th of October 2015. The datalogger was programmed using the PC200Wdatalogger support software (Campbell Scientific, Inc. Logan, UT) to sample the temperature and heat flux every 15 min. The data were averaged and reported at hourly intervals. However due to some short circuit on the roof equipment there is a lack of date under the period of January–March 2016. Therefore only the data from the autumn 25th of October 2016 to 4th of January 2017 will be analyzed in this paper. The monitoring of the roof is still ongoing.

Climate data as outside temperature, sun and snow were collected from the closest weather station 1.5 km south of study side (Kiruna airport).

3 Results and Analysis

The measured data for the heat flux of the green roof and black roof are presented in combination with the monthly measured solar global radiation from the weather station, Figs. 3, 5, 7 and 9. Negative and positive heat flux represents heat entering and leaving the building, respectively. Further the measured temperatures from the roof surfaces and the inside sealing of the building both for the green and black roof are presented together with the monthly outside air temperature, Figs. 4, 6, 8 and 10.

During October a higher fluctuation of the heat flux for the green roof can be seen in comparison to the black roof. Further the surface temperature variations of the black roof are higher than for the green roof surface. During this period the temperature was around 0 °C and relatively high solar radiation during day time, which may lead to thawing and refreezing on the green roof surface and therefore may be an explanation for this behavior.

Considering the measured snow data, the first snow arrived at the 3rd of November and covered the roof with 3 cm of snow, on the 16th of November the

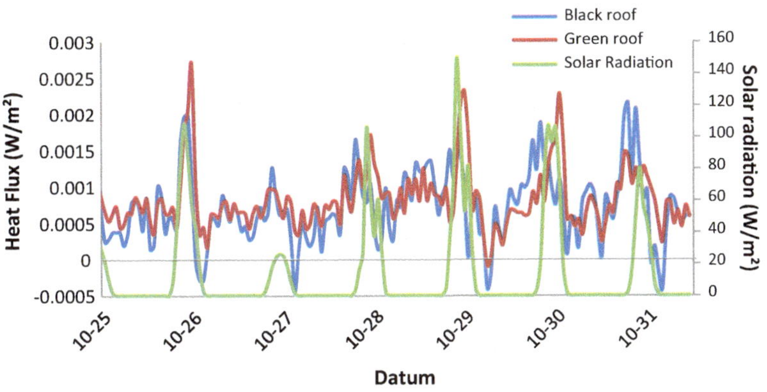

Fig. 3 Heat flux and solar radiation data for the green roof and the black roof over the time period 25th–31th of October 2016

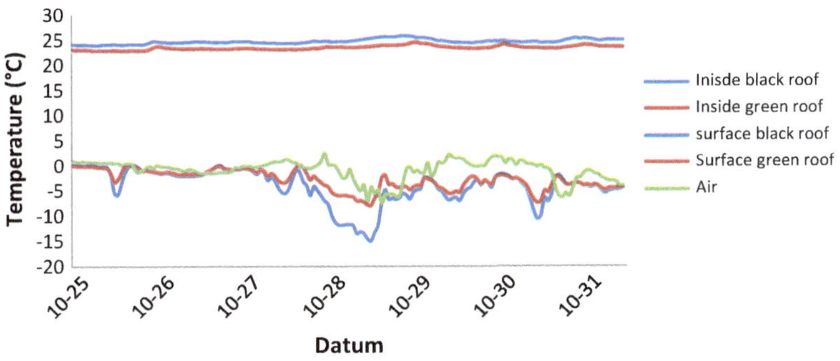

Fig. 4 Temperature data for the green roof, the black roof and outside air over the time period 25th–31st of October 2016

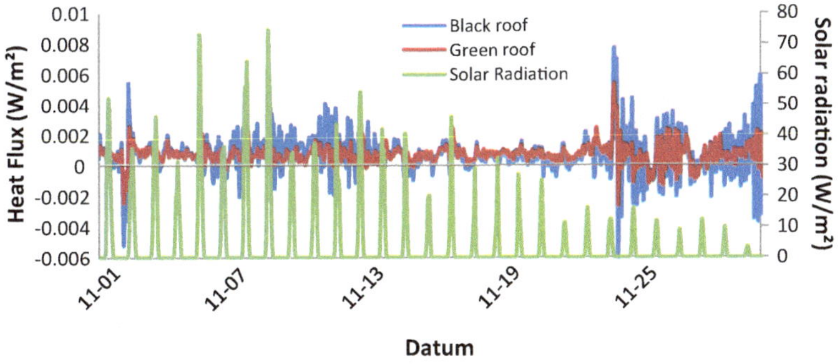

Fig. 5 Heat flux and solar radiation data for the green roof and the black roof over the time period 1st–30th of November 2016

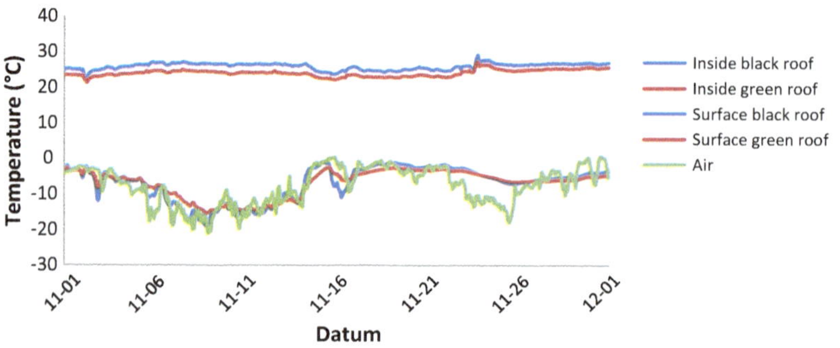

Fig. 6 Temperature data for the green roof, the black roof and outside air over the time period 1st–30th of November 2016

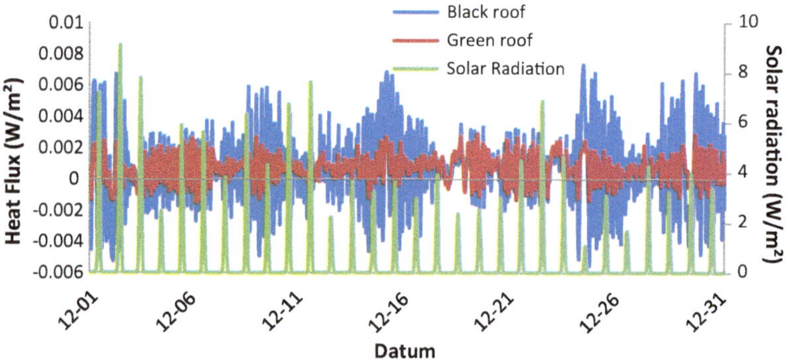

Fig. 7 Heat flux and solar radiation data for the green roof and the black roof over the time period 1st–31st of December 2016

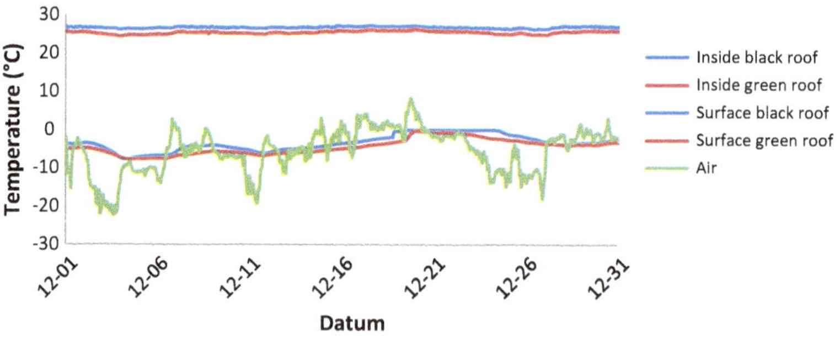

Fig. 8 Temperature data for the green roof, the black roof and the outside temperature over the time period 1st–31st of December

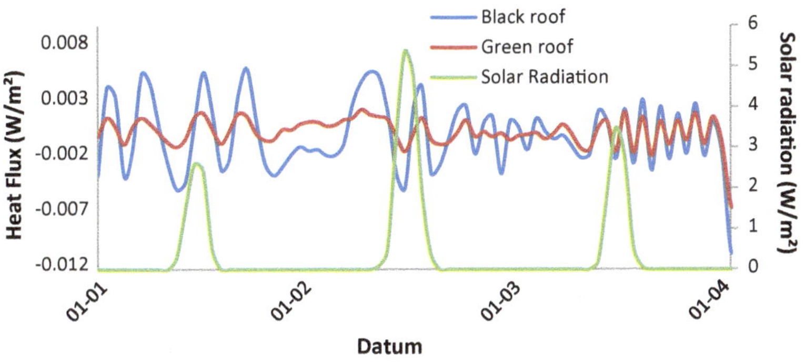

Fig. 9 Heat flux and solar radiation data for the green roof and the black roof over the time period of 1st–4th of January 2017

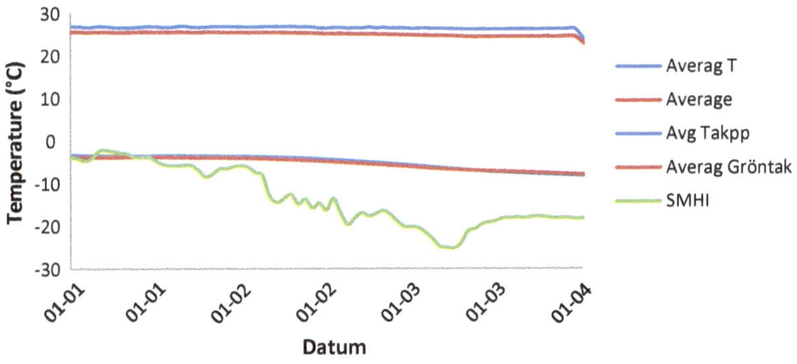

Fig. 10 Temperature data for the green roof, the black roof and the outside temperature over the time period of 1st–4th of January 2017

layer of snow increased up to 5 cm. From the 17th of November the roof was constantly covered by snow layer at between 10 and 53 cm until the 4th of January. This leads to the fact that during this period relatively small heat flux variations can be observed. Furthermore, the surface temperature of the black and green roof is getting more and more align as the roofs are covered by snow. However, at some occasions higher variations of the heat flux can be observed, which explained by very low outdoor temperatures during those occasions.

Similar effects of outdoor temperature can be seen during December. During periods with low outdoor temperature also high heat flux variation can be observed. And this trend continues as it can be seen in Figs. 9 and 10. The surface temperature of the black and green roof is very similar under the increased snow layer.

Table 1 shows the monthly cumulative heat flux of the green and black roof. Interestingly it should be mentioned that during December month when the roofs were covered with snow and the surfaces temperature were similar for both roof types, the green roof had a higher heat flux out of the building (0.53 W/m^2 for the

Table 1 Monthly cumulative heat flux (W/m^2) of a green roof and black roof

Month	Cumulative heat flux (W/m^2)		Seasonal average reduction (%)
	Green roof	Black roof	
October (25th–30th)	0.13	0.11	−14
November	0.51	0.55	6.6
December	0.53	0.43	−22.3
January (1–4th)	0.01	−0.02	177

Note the heat flux sensor was placed such that a negative and positive signifies heat entering and leaving the building respectively

green roof compared to 0.43 W/m^2 for the black roof). This observation could be not found in other studies where snow was covering the roof during the winter time. However, the results for the remaining month agree with previous studies [13, 14].

While analyzing the data it should be considered that possibilities were limited where to install the sensors on the inside of the roof since most space of the roof inside was covered with an installation layer. Therefore, the two measured areas; green roof and black roof are close to each other, which imply that both areas possibly affect each other. Further the black roof area is relatively small and gets affected by the surrounding of the green roof.

The heat flux sensors were installed pairwise for the green and black roof area, to minimize measurement errors. Furthermore this was done to get an indication of the potential effect of heat flows closer to the edge of green roof to black roof. As the pairwise heat flux sensors show barely any difference in the measured heat flow, the affecting seems to be low.

4 Discussion and Conclusion

The results are demonstrating how roof temperature and heat flux are influenced by an extensive green roof in Kiruna during the season 25th of October 2016 until 4th of January 2017.

The period from November to January showed quite similar responses for the green roof and the black roof in contemplation of heat transfer and surface temperature after the roofs were covered with snow.

This study agrees with other researchers result that heat transfer and thermal difference between green roof and black roof appear to be mainly influenced by solar radiation, outside temperature, wind velocity and probably the volumetric moisture content of the growing medium.

Research teams as Getter et al. [13] claim that the volumetric moisture content of the growing medium influences the heat transfer and thermal difference this data are monitored in this study as well and need to be further evaluated and analyzed.

The monthly cumulative heat flux data for the green and black roof, show similarities to the result of the studies [13, 14] conducted in climate zones with longer snow periods during the winter time. However, contradictory results were observed for December, consequently this needs further research in terms of data collection and analysis in order to explain this behavior.

To be able to conclude if green roofs in subarctic climates can lower energy consumption a longer time period need to be analyzed. Especially the spring and summer time with snow melting and higher solar radiation in combination with night frost will be of interest.

Acknowledgements We greatly acknowledge the financial support of the Swedish Research Council Formas grant 2014-854.

References

1. J. Mentens, D. Raes, M. Hermy, Green roofs as a tool for solving the rainwater runoff problem in the urbanized 21st century? Landscape Urban Plann. **77**, 217–226 (2006)
2. R. Fioretti, A. Palla, L.G. Lanza, P. Principi, Green roof energy and water related performance in the Mediterranean climate. Build. Environ. **45**, 1890–1904 (2010)
3. J.C. Berndtsson, L. Bengtsson, K. Jinno, Runoff water quality from intensive and extensive vegetated roofs. Ecol. Eng. **35**, 369–380 (2009)
4. J. Yang, Q. Yu, P. Gong, Quantifying air pollution removal by green roofs in Chicago. Atmos. Environ. **42**, 7266–7273 (2008)
5. D. Banting, H. Doshi, J. Li, P. Missios, *Report on the Environmental Benefits and Costs of Green Roof Technology for the City of Toronto* (2005)
6. T. Van Renterghem, D. Botteldooren, In-situ measurements of sound propagating over extensive green roofs. Build. Environ. **46**, 729–738 (2011)
7. M. Santamouris, C. Pavlou, P. Doukas, G. Mihalakakou, A. Synnefa, A. Hatzibiros et al., Investigating and analysing the energy and environmental performance of an experimental green roof system installed in a nursery school building in Athens, Greece. Energy **32**, 1781–1788 (2007)
8. T.G. Theodosiou, Summer period analysis of the performance of a planted roof as a passive cooling technique. Energy Build. **35**, 909–917 (2003)
9. Q. Meng, W. Hu, Roof cooling effect with humid porous medium. Energy Build. **37**, 1–9 (2005)
10. S. Ouldboukhitine, R. Belarbi, R. Djedjig, Characterization of green roof components: measurements of thermal and hydrological properties. Build. Environ. **56**, 78–85 (2012)
11. N.H. Wong, D.K.W. Cheong, H. Yan, J. Soh, C.L. Ong, A. Sia, The effects of rooftop garden on energy consumption of a commercial building in Singapore. Energy Build. **35**, 353–364 (2003)
12. O. Saadatian, K. Sopian, E. Salleh, C.H. Lim, S. Riffat, E. Saadatian et al., A review of energy aspects of green roofs. Renew. Sustain. Energy Rev. **23**, 155–168 (2013)
13. K.L. Getter, D.B. Rowe, J.A. Andresen, I.S. Wichman, Seasonal heat flux properties of an extensive green roof in a Midwestern U.S. climate. Energy Build. **43**, 3548–3557 (2013)
14. G. Spolek, Performance monitoring of three ecoroofs in Portland, Oregon. Urban Ecosyst. **11**, 349–359 (2009)

Performance Evaluation of a Passive House in Sub-arctic Climate

Shimantika Bhattacharjee, Sofia Lidelöw and Jutta Schade

Abstract As the operational energy use in buildings contributes highly to the total energy used and greenhouse gases emitted in the cold climate regions of Europe, buildings which are more energy-efficient and less carbon-intensive during operation are key to meet sustainability objectives in these regions. Yet, research shows that the practice of passive or low-energy buildings in the sub-arctic climate of northern Sweden is comparatively less than in the southern region. Moreover, previous studies did not explicitly examine the performance of low energy buildings in sub-arctic climate in relation to established building energy efficiency standards. Consequently, knowledge regarding the energy performance of low-energy buildings in such climate is limited. Therefore, the aim is to evaluate the performance, in terms of indoor temperature and energy use for heating, domestic hot water and electricity of a new-built passive house titled "Sjunde Huset" in the sub-arctic town of Kiruna. It is Sweden's northernmost house designed to fulfil the Swedish passive-house criteria of a maximum heat loss factor of 17 W/m^2 and a maximum annual energy use of 63 kWh/m^2. The implemented passive design strategies include a highly insulated, compact and airtight building envelope with a vestibule, mechanical ventilation with heat recovery and renewable energy production through photovoltaic solar cells. The house is connected to district heating and is equipped with energy-efficient appliances to allow low occupant energy use. Ongoing performance evaluation is based on building simulation and measurements of energy and temperature in different zones of the building. Energy performance deviations between occupied and non-occupied zones are explored through internal heat gain evaluations. The indoor temperature is also evaluated to assess the temperature variations throughout the year. The ongoing research further evaluate a comparative simulated and measured energy analysis of heating, hot water and electricity based on both the international passive house standard and the Swedish passive house criteria "Feby 12".

S. Bhattacharjee (✉) · S. Lidelöw · J. Schade
Department of Civil, Environmental and Natural Resources Engineering, Luleå University of Technology, 97187 Luleå, Sweden
e-mail: shimantika.bhattacharjee@ltu.se

D. Johansson et al. (eds.), *Cold Climate HVAC 2018*,
Springer Proceedings in Energy, https://doi.org/10.1007/978-3-030-00662-4_13

Keywords Passive house · Sub-arctic climate · Performance evaluation
Energy efficiency · Indoor temperature

1 Introduction

1.1 Background and Problem Identification

Technological advancements, infrastructure growth and urbanization have globally accelerated the energy used by human society. Building energy use is responsible for a significant amount of the total energy used worldwide. In Europe, different initiatives were taken into consideration for attaining the progressive sustainability objectives in the building sector. For example, the European Union (EU) set up a target for the year 2020 to improve the energy efficiency by 20% (from 1990 levels), reduce the GHG emissions by 20% and increase the share for renewable energy sources by 20%.

The operational energy use in buildings contributes highly to the total energy used and the greenhouse gases emitted in cold-climate regions. Buildings that are more energy-efficient and less carbon-intensive during operation are therefore key to meet sustainability objectives in these regions. In southern Sweden, the number of buildings that are in compliance with the Swedish passive house criteria "Feby" is steadily increasing. However, realizing such energy-efficient houses in the sub-arctic climate zone of northern Sweden is more challenging due to the cold, long and snowy winters and the mild, short summers. Moreover, there exists a lack of systematic and documented measurements regarding the monitoring and analysis of the energy-efficiency solutions, particularly solutions that have been developed and tested for this climate zone. Previous studies did not explicitly examine the performance of low-energy buildings in sub-arctic climate in relation to established building energy-efficiency standards. However, a few studies were found relevant for this particular study. Risberg et al. [1] performed an analysis of different heating systems in a single-family house situated in Kiruna and found that floor heating is the most efficient system in terms of achieving the desired indoor temperature and a comfortable indoor climate. Another study by Mukhopadhyaya et al. [2] regarding the application and thermal performance of vacuum insulation panel (VIP) in buildings situated in the northern Canadian city of Yukon, which is also in the sub-arctic climate zone showed that VIP, if appropriately designed and constructed, can be an attractive option for retrofitted building envelopes in such climate. Vladykova et al. [3] highlighted the challenges and potential technical barriers of building energy-efficient buildings in Greenland, which is under the arctic climate region. Yet, even though the arctic climate is more severe in terms of temperature, snow, wind and storms than the sub-arctic climate, Vladykova's et al. study gives some valuable insights. For example, it was found that a ventilation system with heat recovery needs to adjust to larger temperature differences and frequent condensation problems from air and snow can occur which may eventually reduce the heat exchanger efficiency [3].

Sweden's northernmost passive house, titled "Sjunde Huset", was built in the town of Kiruna in 2014 with the aim to develop, demonstrate and evaluate a passive house concept for the sub-arctic climate region. Kiruna has a sub-arctic climate, with monthly temperatures normally averaging below 0 °C during October to April and just above 10 °C during July to August. Being situated well above the Arctic Circle, it also experiences a pronounced seasonal cycle of solar radiation with only around 15% of the annual hours of sunshine hours occurring during the five coldest months between October and February. Consequently, the house receives small heat gains from solar radiation during the period when the need for heating is greatest. The energy used for space heating the existing buildings in the sub-arctic part of northern Sweden is comparatively higher than the Swedish average, which raises prospect for reducing the region's overall energy use through constructing more energy-efficient houses such as passive houses. In "Sjunde Huset" passive strategies were selected and evaluated based on their performance effectiveness in cold climate. According to a recently published report by NCC [4], which presents the construction process, the technical installations and the first measurement results from "Sjunde huset", the energy use for heating and hot water was estimated using building energy simulations to 54 kWh/m^2, year, which would fulfil the Swedish passive-house requirement of 63 kWh/m^2, year.

As part of an on-going performance evaluation of "Sjunde Huset", this paper presents and contrasts design-stage simulations and measured data for one year in terms of energy use and indoor temperature of the building. Since there is no established standards for energy-efficient buildings in sub-arctic climate this study can serve as an inspiration for future passive or low-energy house practices in northern Sweden. Building energy efficient houses in sub-arctic climate is challenging due to the extreme climate conditions. Yet, a few demonstration projects have been performed in northern most Sweden but characteristic of most of these is the lack of systematic and documented measurements and evaluations of the implemented energy-efficiency solutions during operation. The on-going performance evaluation of "Sjunde Huset" seeks to contribute to fill this gap.

1.2 Aims and Objectives

The aim is to make a simulated and measured comparative analysis in terms of indoor temperature and operational energy use between an occupied and non-occupied zone of the passive house "Sjunde Huset". More specifically the objectives are to:

- Evaluate certain discrepancies between the simulated (design-stage) and measured (operational-stage) energy performance.
- Evaluate the indoor temperature to assess the variations throughout the year.
- Evaluate the simulated and measured performance based on both the international passive house standard and the Swedish passive house criteria.

2 Method

The passive house titled "Sjunde huset" has a heated floor area of 280 m^2, which is equally distributed between two apartments. One of these apartments (hereafter referred to as "apartment 2") was selected as a case for this particular study, which is divided into three phases.

First, dynamic simulations were performed in IDA ICE advanced version 4.7 (Equa, Sweden) to replicate the building's energy use as it was estimated by the contractor in the design stage. Furthermore, measured energy use for one year of operation was collected. Discrepancies between the design stage and operational stage energy use in occupied and non-occupied zones were analyzed.

Second, measured indoor temperatures in different rooms of apartment 2 were collected for the studied year. Annual measured data for space heating that enabled a comparison between the measured and simulated heating energy use was available only for apartment 2, which was a motivation for selecting apartment 2 as the study object here. The third phase includes the comparison between the simulated and measured energy use of "Sjunde huset" with both the Swedish passive house criteria (Feby) and the international passive house standards.

2.1 Description of the Passive House "Sjunde Huset"

"Sjunde Huset" is a two-storey house which contains two separated apartments with a heated floor area of 140 m^2 each (see Figs. 1 and 2). The house is developed from NCC's conceptualized house to an energy-efficient passive house to fulfill the Swedish passive house criteria "Feby". As such, it has a highly insulated, airtight and compact building envelope (see Table 1) with vestibules that reduce air infiltration through the entrance doors. The house is also equipped with mechanical exhaust and supply air ventilation system with heat recovery to minimize the ventilation heat losses and energy-efficient lighting and appliances. Space and hot-water heating is provided through district heating which is converted to air heating via a heating coil. The "Nasa Shower" being equipped with water recycling potential is estimated to reduce the energy for each shower by 80% and water use by 90%. Each apartment is equipped with around 15 m^2 of solar panels for electricity production. The solar panels are placed vertically on the façade to make use of the reflection from snow and to minimize snow accumulation. The roof is green with sedum plants, herbs and grasses. The appliances such as washing machine and dishwasher is connected to the district heating. Energy efficiency measures along with innovative techniques has been implemented in this case study building to reduce the heating, domestic hot water use and electricity use.

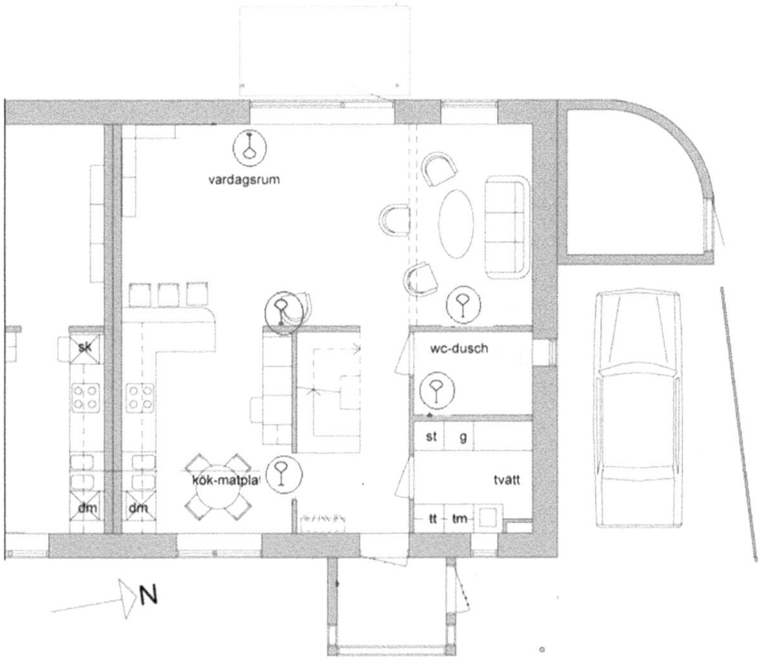

Fig. 1 Ground floor plan of the passive house "Sjunde Huset". The red circles represent the sensors placed inside the house for measuring indoor temperature

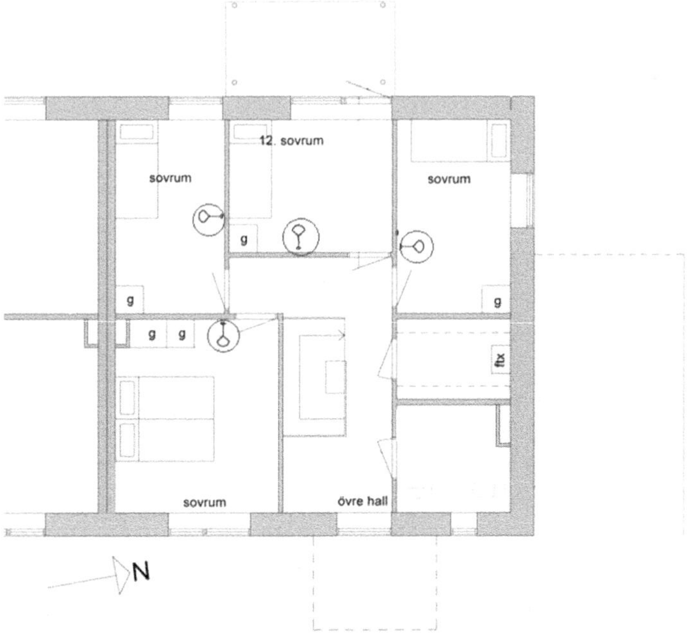

Fig. 2 First floor plan. The red circles represent the sensors placed inside the house for measuring indoor temperature

2.2 Input Data

See Tables 1 and 2.

Table 1 Thermal characteristics of the building envelope

Building parts	Construction	U-value/(W $(m^2/(\cdot K))$
External wall	50 mm façade panel, 6 mm Minerit board, 120 mm Mineral Wool/studs cc 600, 140 mm PIR, 45 mm Mineral Wool/studs cc 450, 13 mm Plasterboard	0.079
Roof	1000 mm loose-fill Mineral Wool insulation including studs	0.0349
Intermediate floor	Double gypsum on 95 mm frame, 30 mm light insulation	0.6187
External floor	100 mm concrete and 400 mm cellular plastic	0.0756
Window	Triple pane glazing, g-value was considered as 0.37	0.65
Door	Wood	0.7
Overall heat loss coefficient	–	0.16

Source NCC [4]

Table 2 Input data for "Sjunde Huset" used for design stage simulations

Infiltration rate/(L/s · m^2) at ± 50 Pa pressure difference	0.05
Building envelope area (m^2)	527
Heating (°C)	21
Cooling (°C)	25
HVAC template	CAV

Source NCC [4]

2.3 Measurement Data

Energy and temperature data for the entire year of 2016 was analyzed. Energy data for heating energy use in apartment 2 was collected from the district-heating meter and supplied by Tekniska Verken i Kiruna AB. Apartment 2 was essentially non-occupied throughout the analyzed year; hence, the measured values represent the space heating use as no domestic hot water was used during that period. This facilitated the comparison between the simulated non-occupied heating energy use and measured non-occupied heating energy use.

3 Result and Analysis

3.1 Simulated and Measured Energy Use

According to the aggregated results shown in Table 3, the simulated heating energy use is higher than the measured heating energy use for the analyzed year. The simulated and measured energy use varies since the measured energy use was obtained through static metered data, while, on the contrary, the simulated energy use was obtained through dynamic simulations. According to the simulations, the heating energy use is higher for the non-occupied situation than for the occupied situation. This is mainly due to the lack of internal heat gains from occupants and their use of electric equipment and hot water. Hence, more heat must be supplied by the heating system to maintain the set-point indoor temperature, especially during the coldest months when the solar gains are minimal.

Energy use for DHW (domestic hot water) can be estimated to 20 kWh/m^2 annually according to SVEBY [5]. However, "Sjunde huset" is equipped with "NASA" technology showers that recycle used water and provide an estimated energy saving of 80% [4]. That is why DHW use was estimated to 10 kWh/m^2 in the design stage (see Table 3). When the apartment is finally occupied, measurement data will be gathered to analyze the actual energy savings of using this technology.

The monthly variations in the simulated energy use for heating, hot water and building electricity for the non-occupied scenario are shown in Fig. 3. This scenario corresponds best to the real case since apartment 2 was essentially unoccupied throughout the measurement period presented in this paper. According to the simulation results shown in Fig. 3, the heating energy use is decreasing towards the warmer summer months and starts to increase gradually towards the colder winter months. The simulated heating energy use show a maximum value of 1373 kWh in January and a minimum value of 2.3 kWh in June.

Table 3 Simulated and measured energy use in apartment 2 of the passive house "Sjunde Huset"

	Energy use/(kWh/m^2. year)		
	Simulated occupied	Simulated non-occupied	Measured non-occupied
Heating	40	45	32
Domestic hot water	10	0	–
Kitchen ventilation losses	1.7	0	–
Ventilation losses through windows, doors and other openings	4	0	–
Auxiliary energy for fans and pumps	4.9	4.9	–
Hot water circulation system losses	0.6	0.6	–
Total energy use	61	51	32

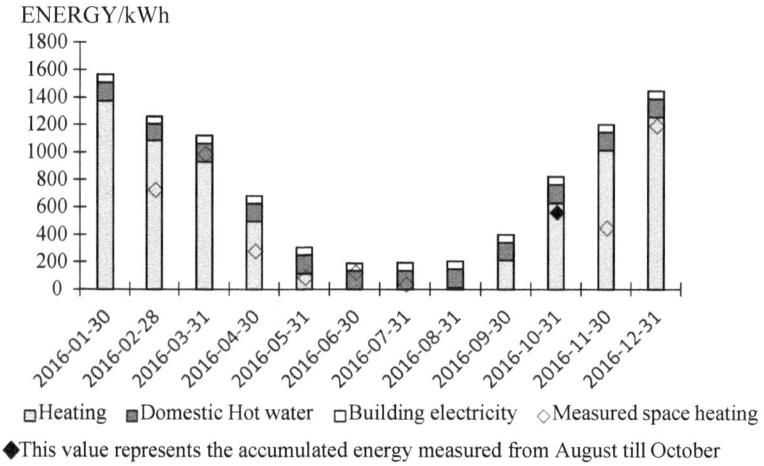

Fig. 3 Simulated unoccupied and measured unoccupied energy use for heating, domestic hot water and building electricity in "Sjunde Huset". Source of measured energy use: Tekniska Verken i Kiruna AB, 2016

As can be seen from Fig. 3, the measured heating energy use has a similar trend over the year as the simulated heating energy use. Due to problems with the metering, the measured value shown for October represents the accumulated heating energy use during August to October. This could explain the slightly higher measured value for October than for November.

3.2 *Indoor Temperatures*

Temperature sensors were placed in different locations of apartment 2 in order to evaluate the indoor temperature throughout the year, see Figs. 1 and 2. Table 4 shows the evaluated temperature sensors and their respective placement.

Table 4 List of sensors and their respective placement in apartment 2

Sensors	Place	Height from the floor level/meter
GT10	Living room (middle)	1.59
GT11	Living room (northeastern facade)	1.60
GT12	Near the staircase	1.57
GT14	Washroom(ground floor)	1.58
GT15	Bedroom 1	1.59
GT17	Bedroom 2	1.57
GT18	Bedroom 3	1.58
GT19	Bedroom 4	1.57

Table 5 shows the yearly average, minimum and maximum indoor temperature measured through each of the sensors during 2016. The sensors exhibited, to a varying extent, some erratic temperature data during the period of January until June. The erratic values were omitted before the calculations.

The average indoor temperatures ranged between 19 and 21 °C over the measured periods; this range is relatively low as 21 °C, is generally considered as optimum temperature indoor. The minimum indoor temperatures obtained from the sensors are only 15–16 °C, which is due to the fact that the heating system did not work properly especially during October to November and, being an unoccupied zone, it took until November before anybody realized the problem. The maximum indoor temperatures of 33–34 °C were recorded in May and July for the sensors GT17 and GT19, respectively. This is a consequence of the fact that the zone was unoccupied.

Occupants opening windows or doors could have contributed to reduce the indoor temperature as the maximum outdoor temperature recorded was as 24 °C in May and 25 °C in July [6]. In this context, it should be noted that this passive house

Table 5 Yearly indoor temperature obtained from measured data for different sensors during 2016 Yearly indoor temperature (°C)

Sensors	Average	Maximum	Minimum	Available data
GT10	19.3	27.2	15.1	June-Dec
GT11	19.2	26.6	15.0	Feb-Dec
GT12	20.0	27.3	15.0	June-Dec
GT14	19.0	25.9	15.0	Feb-Dec
GT15	20.0	29.2	15.0	Feb-Dec
GT17	19.6	34.1	15.0	May-Dec
GT18	20.0	28.0	15.1	June-Dec
GT19	21.1	33.4	15.1	June-Dec

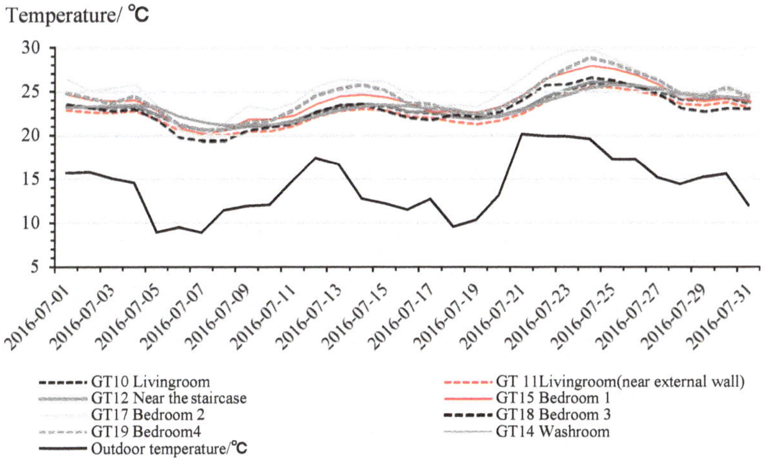

Fig. 4 Measured daily average indoor and outdoor temperature during July 2016

is not equipped with any active cooling system since it is situated in the sub-arctic climate with its short and cold to mild summers.

The above-mentioned section briefly described the indoor temperature variations throughout the year. Two specific months from summer and winter were chosen for deeper analysis of the temperature fluctuations (see Figs. 4 and 5). More specifically, the aim of this part was to analyze whether the outdoor temperature could have influenced the indoor temperatures measured in "Sjunde huset". Outdoor temperature data for the year 2016 was collected from the nearest weather station which is situated 2.5 km away from Sjunde Huset [6].

Figure 4 shows the temperature variations indoor and outdoor during a typical summer month July. The X-axis show the dates and the Y-axis show the temperature range. The graph is showing similar trends for the indoor and outdoor temperature. The living room, which is situated on the ground floor level, has indoor temperatures ranging from 20 to 26 °C. The bedrooms, which are situated at the 2.8 m level have slightly higher indoor temperature than the zones situated on the ground level. The daily average temperature peaked at 30 °C in bedroom 2 during the third week of July although the recorded daily average outdoor temperature for that particular day was 20 °C.

It seems that opening windows or ventilating the space could have helped to reduce the indoor temperature as the outdoor temperatures were mild and did not exceed 20 °C during July.

Figure 5 shows the indoor and outdoor temperature variations for the month of December. It can be seen from the graph that, being situated in the subarctic climate zone, the outdoor temperature is considerably lower than the indoor temperature as it is ranging from 0.9 to −19 °C.

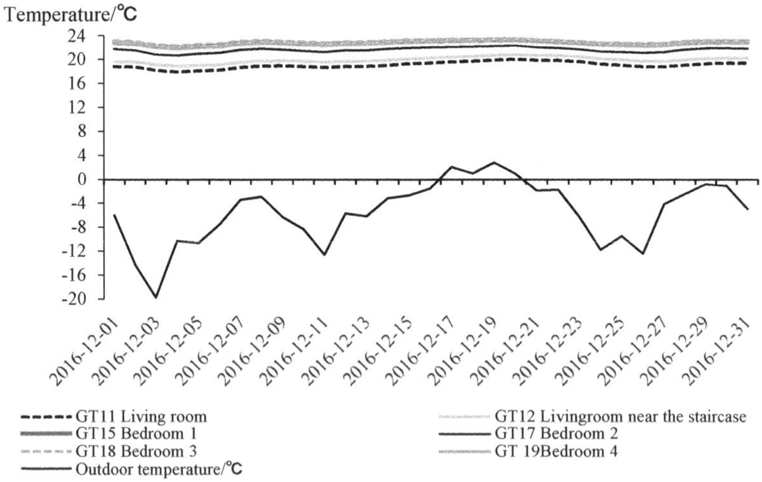

Fig. 5 Measured daily average temperature during December 2016

There were some issues regarding the battery of the heat exchanger, which was not working properly. The heating system was adjusted during November and therefore the indoor temperature in December show an optimum range between approximately 18 and 23 °C. However, zones situated on the first floor level had slightly higher temperature indoor compared to the temperature in the corresponding zones situated on the ground floor.

3.3 *Comparison Between the International Passive House Standard and Swedish Passive House Criteria (Feby)*

Table 6 shows different criteria of the international passive house standard [8] and the Swedish passive house criteria (Feby) [7]. According to the international passive house criteria the energy use for space heating should not exceed 15 kWh/m^2. It is comparatively rigid compared to the Swedish standards for subarctic climate. The scenario is also true for peak heat load standards. The subarctic climate is more severe in terms of extremely lower outdoor temperature together with heavy snow and polar nights. During winter there prevails almost negligible amount of solar irradiation so it cannot contribute to internal gains when the need for heating is greatest. Therefore, it requires more energy for space heating to maintain an optimum temperature indoor. Therefore, it can be seen that the peak heat load is comparatively higher in the sub-arctic climate zone.

Both the simulated and the measured heating energy use (Table 3) shows that "Sjunde huset" fulfils the Swedish passive house criteria of 63 kWh/m^2, year for a non- electrically heated residential house with a heated floor area smaller than 400 m^2 (Table 6) [7]. However, the comparatively stricter international passive house criteria for heating energy use is not met [8].

Table 6 Comparison between selected criteria of the international passive house standard and Swedish passive house criteria (Feby)

Input data	International passive house standard	Swedish passive house criteria (Feby) for sub-arctic climate
Air leakage rate	$n_{50} \leq 0.6$ 1/h	Air leakage through the building envelope at ±50 pa pressure difference q_{50} 0.3 l/(s.m^2)
Heating energy use	≤ 15 kWh/(m^2. a) (excluding Domestic hot water)	58 kWh/m^2 for non-electrically heated home[a] and 29 kWh/m^2 for electrically heated houses[b] (including domestic hot water)
Peak heat load	≤ 10 W/(m^2. a)	17 W/m^{2b}

[a]+5 kWh/m^2 for buildings with a heated floor area below 400 m^2
[b]+2 kWh/m^2 for buildings with a heated floor area below 400 m^2

The air leakage rate of the house was measured to 0.25 l/(s.m^2), which is higher than the targeted value which was used also for the design-stage simulation (Table 1). Still, the measured air leakage rate complies with both Swedish and the international passive house criteria.

4 Discussion and Conclusion

The simulated (design-stage) and measured (operational-stage) amount of energy used for heating the studied passive house "Sjunde huset" in Kiruna during one year of operation were found to be in the same range. Yet, the simulated heating energy use was somewhat higher than the actual, measured one. This result is likely linked to the fact that the house remained essentially unoccupied throughout the measured year, which meant that no domestic hot water was used and a low indoor temperature remained unnoticed. Consequently, the heating system was not properly adjusted. The simulated and measured heating energy use showed similar variations over the year, with considerably higher values during the winter than during the summer months. Still, the house needed active supply of heating even at times during the summer months due to the subarctic climate; for example, the outdoor temperature in July was ranging between 9 and 20 °C. The simulations showed that the heating energy use can be expected to decrease as residents eventually move into the apartment and contribute with body heat as well as heat gains from their use of electric equipment and hot water. However, in case the heating system yet again would be malfunctioning and the indoor temperatures drop to uncomfortably low levels occupants could be expected to start using extra heating devices, which would lead to increase the heating energy use. Measurements are still on-going, not just regarding the heating energy use but also regarding other factors affecting the total building energy use, and future work are targeting more detailed evaluations as the house becomes permanently and fully occupied.

The evaluation of the measured indoor temperature turned out to be difficult since the heating system was not working properly throughout the measured period. This problem was not detected until November as the building was unoccupied. However, the system was then adjusted, and, after that, the indoor temperatures remained stable even when outdoor temperatures reached down to −20 °C. This shows the importance of a properly functioning and well-maintained heating system, especially in cold climates where indoor temperatures otherwise can turn uncomfortably low. Furthermore, overheating, with indoor temperatures of up to 33 °C, was noticed in nearly all evaluated rooms during the studied summer month (July), while the outdoor temperature remained moderate (around 20 °C). This showed clearly that the solar radiation had a big impact on the indoor temperature in "Sjunde huset". Consequently, the effect of solar radiation on indoor temperature levels needs to be further evaluated.

Both the simulated and the measured heating energy use of "Sjunde huset" fulfils the Swedish passive house criteria, which was the target for the demonstration. However, the simulated values are about two times higher than the international passive house criteria for heating energy use. Hence, even though this house has a very well- insulated thermal envelope that fulfils the criteria regarding maximum air leakages, the international passive house criteria seem hard to reach in the sub-arctic climate of Kiruna.

Acknowledgements We greatly acknowledge the financial support of Vinnova (Sweden's Innovation Agency), HLRC (Hjalmar Lundbohm Research Center) and the partner companies through the project Attract (Attractive and Sustainable Cities in Cold Climates) and of the European Union Interreg North program through the project EEBAK (Energi Effektiva Byggnader i Arktiska Kommuner).

References

1. D. Risberg, M. Vesterlund, L. Westerlund, J. Dahl, CFD simulation and evaluation of different heating systems installed in low energy building located in sub-arctic climate. Build. Environ. **89**, 160–169 (2015)
2. P. Mukhopadhyaya, D. MacLean, J. Korn, D. Van Reenen, S. Molleti, Building application and thermal performance of vacuum insulation panels (VIPs) in Canadian subarctic climate. Energy Build. **85**, 672–680 (2014)
3. P. Vladyková, C. Rode, T.R. Nielsen, S. Pedersen, Passive houses for arctic: measures and alternatives, in *13th International Passive House Conference*, Germany 2009
4. NCC Report. Sjunde huset–ett lågenergiparhus i Kiruna, Slutrapport 38548-1, Energymyndigheten, May 2017
5. SVEBY: Branschstandard för energi i byggnader. Brukarindata för energiberäkningar i bo- städer. Projektrapport, April, 2009
6. Swedish Meteorological and Hydrological Institute (SMHI) Homepage, https://opendata-download-metobs.smhi.se/explore/?parameter=0. Last accessed 14 Sep 2017
7. FEBY (Forum for energy-efficient buildings) Homepage, http://www.nollhus.se/. Last accessed 14 Sep 2017
8. Passive House Institute Homepage (Passivhaus Institute), www.passiv.de, http://www.passiv.de/downloads/03_building_criteria_en.pdf. Last accessed 15 Oct 2015

The Study of Energy-Saving Window Technology Adaption for Green Buildings in the Severe Cold Region of Northern China

Liang Yu and Xiaotong Wang

Abstract This paper studies the technical adaptability of energy-saving windows in green buildings based on thermal comfort and thermal defect for the severe cold climate region of northern China. Firstly, this paper makes statistics and analysis on the energy-saving window technology of green buildings in severe cold region. Secondly, this paper selects a representative public green building to tests its thermal comfort and thermodynamic disfigurement. The results show that the link between walls and window frames become weak links in technology. The large area of glass maintenance structure does not apply to the south side of the underground although it can save the lighting energy.

Keywords Severe cold region · Energy-saving window · Green building

1 The Present Development Situation of Green Building in Severe Cold Region of China

Nowadays green buildings have been greatly developed in China. By the end of September 2016, China's green buildings had got 5396 certifications nationwide including 4515 Chinese Green Building Stars Logo certifications and 895 LEED certifications. The severe cold region located in the northeast of China (Liaoning, Jilin, Heilongjiang Province) possesses 192 green building certifications including 173 Chinese Green Building Star Logo Certification projects and 19 LEED certification projects [1, 2], as shown in the Figs. 1 and 2.

L. Yu (✉) · X. Wang
Shenyang Jianzhu University, No. 9, Hunnan East Road, Hunnan New District, Shenyang, Liaoning, People's Republic of China
e-mail: sy_yuliang@163.com; 923303247@qq.com

D. Johansson et al. (eds.), *Cold Climate HVAC 2018*,
Springer Proceedings in Energy, https://doi.org/10.1007/978-3-030-00662-4_14

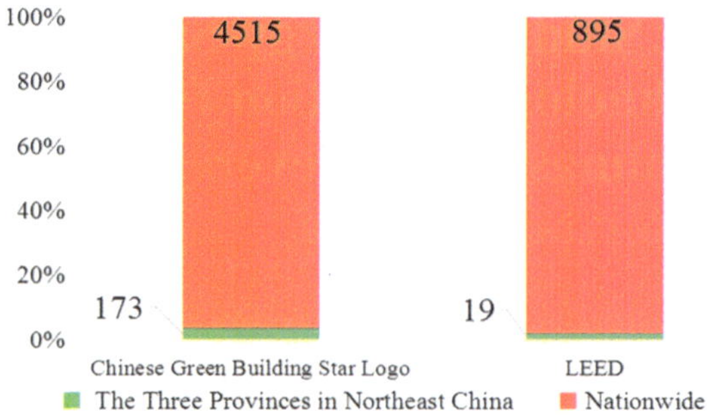

Fig. 1 The proportion of the northeast of China's certifications in national wide

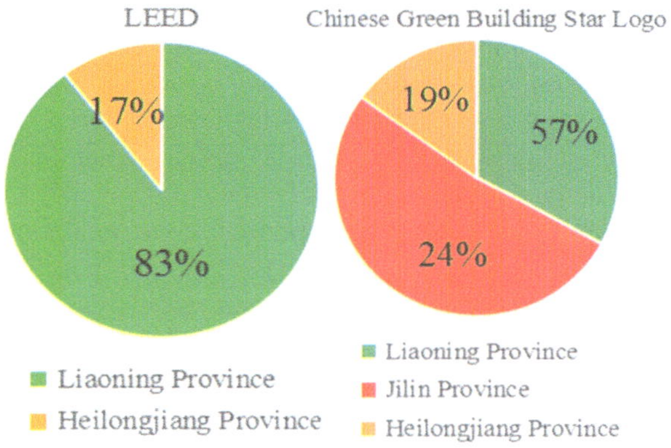

Fig. 2 The proportion of the northeast of China about every certification

2 Analysis of the Study on the Energy-Saving Windows of the Green Buildings in the Northeast of China

The performance of the building envelope plays an important role in building energy consumption. Outside window, which is the weakest link in the building envelope, can account for more than 50% of the total heat consumption. And it can have a great impact on building energy efficiency [3]. In this part, the technology of energy-saving window of green buildings in the northeast of China (Liaoning, Jilin, Heilongjiang Province) is considered. The frequency of usage and characteristics of different kinds of glass and window frames are summarized as follow (shown in Figs. 3 and 4).

Figure 3 shows that double layer low-E glass is popular with the green buildings in the northeast of China. It uses a vacuum deposition technique that deposits a layer of low radiation coating on the glass surface. Compared with ordinary glass, it has low infrared emission rate and long wave radiation rate and high infrared reflectance. From Fig. 4 we can see that broken bridge aluminum alloy frame makes a big part of the pie. It uses heat insulation material and aluminum alloy frame to form good heat insulator. With good thermal insulation effect, it has been

Fig. 3 Different kinds of glasses used in green buildings in the northeast of China

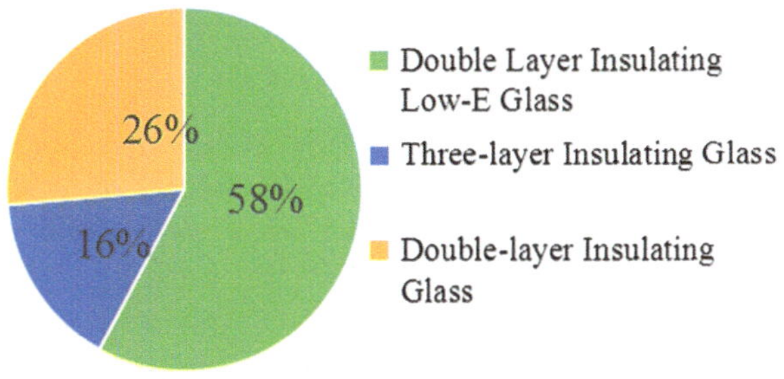

Fig. 4 Different kinds of window frame used in green buildings in the northeast of China

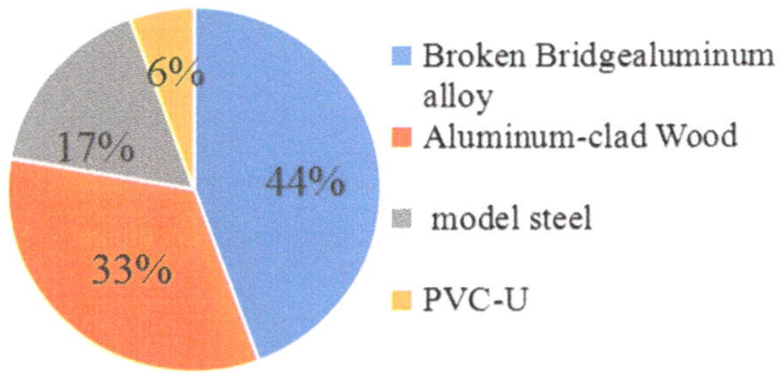

used in China since the 1990s and its excellent energy-saving effect has made it still popular with users.

3 Adaptability Analysis of Energy-Saving Window

The purpose of this part is to study the adaptability of excellent thermal performance window technology in practical use. A public green building in severe cold region has been chosen. The adaptability of its energy-saving windows will be studied from thermal comfort and thermal defect.

3.1 Preparatory Work

The building named Sino-German Energy Conservation Demonstration Center (Fig. 5) is located in Shenyang Jianzhu University in Liaoning Province. The total height is 12.0 m with two floors and one floor underground. The underground floor is 569.5 m^2, the first floor is 573.5 m^2, and the second floor is 457.7 m^2. In 2015, the building received a three-star green building design identification certification from Chinese Green Building Stars Logo. The window of the building is a single frame three-glass aluminum window (with argon gas, warm edge, tempered glass), and its main performance parameters are shown in Table 1. The 5 temperature and

Fig. 5 Architectural appearance

Table 1 External window performance parameter table

Size		Heat transfer coefficient W/(m^2 · K)	Shading coefficient	Opening area ratio (%)
Width (mm)	Height (mm)			
4200	2900	0.9	0.5	35.47
1200	2900	0.9	0.5	82.76
6525	2900	0.9	0.5	30.47
7450	2900	0.9	0.5	17.21

Fig. 6 Automatic recorder

humidity automatic recorders are used to test the thermal comfort (Fig. 6). The model is RR002. The range is −10 to 50 °C and the precision is ±0.5 °C. One infrared thermograph is used to test the thermal defect (Fig. 7). Its model is Fluke TiR1. The range is—20 to +100 °C, and the accuracy is 2 °C.

Fig. 7 Infrared thermograph

3.2 *Thermal Comfort Analysis of Indoor Environment*

The test was carried out in accordance with the relevant requirements of JGJ/T 177-2009 [4]. Here, it is stipulated that the temperature and humidity detection time should be no less than 6 h continuously, and the data recording time interval should not exceed 30 min. The test points shall be 700–1800 mm away from the ground and shall not be directly affected by solar radiation and heat and moisture sources. According to the above regulations, combined with the actual situation of the building, we will make the following arrangements for the test. Our test data was from March 1, 2017 to March 31, 2017, when was the last heating month. The test condition is set as the air conditioning shutdown, the whole day low temperature hot water surface radiant heating. The data collection interval is 10 min. The test point layout is shown in Figs. 8, 9 and 10. They are located 1200 mm away from the ground.

According to JGJ/T 177-2009 [4], this paper deals with the temperature and humidity data of 5 test points and draws the results as below (Figs. 11 and 12).

According to the requirements of GB/T 50378-2014[5], the test results should meet the requirements of GB 50736-2012 [6], as shown in Table 2. In winter, when the relative humidity is more than 60%, it will cause the human body's thermal discomfort. When the indoor temperature is 18 degrees, the person who is dressed properly and is quiet will not feel cold. On this basis, indoor comfort in winter was divided into two levels. The thermal comfort of the level 1 is better than that of the level 2.

From Table 2, the winter room temperature shall not be less than 18 °C, relative humidity shall not be less than 30%. The results in Fig. 11 show that the four test points reach the standard value except for point 1. According to the results in Fig. 12, the relative humidity of 5 test points is lower than the standard value. The main reasons are as below.

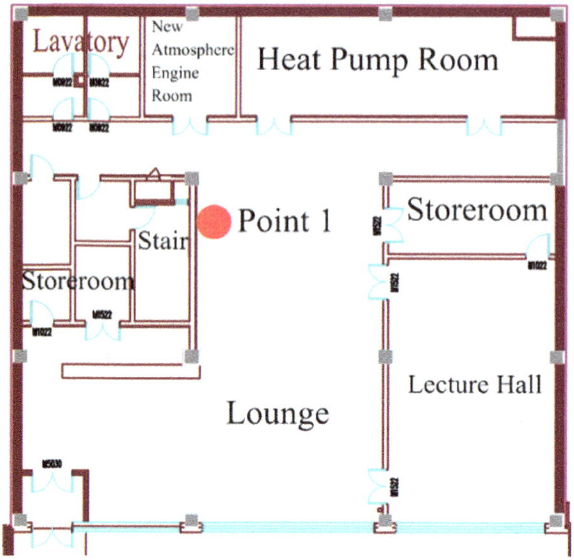

Fig. 8 The distribution of underground

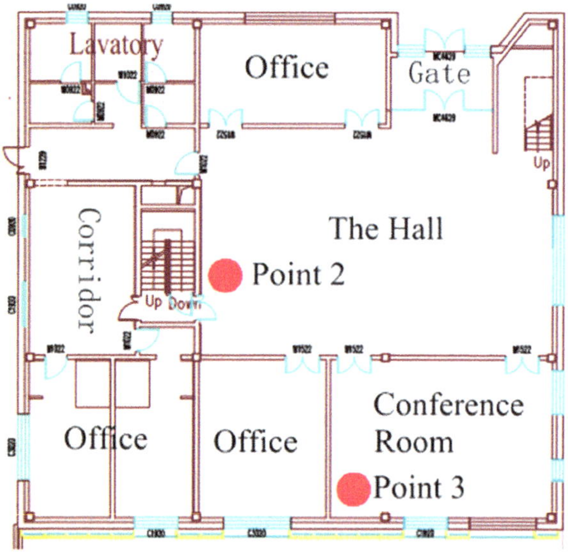

Fig. 9 The distribution of points on the first floor

Test point 1 is set on the underground floor. Its south side facade has a large transparent enclosure structure (Figs. 13 and 14). This increases the quality of natural lighting in the interior, but this also makes the side significantly heat loss than the other orientation structures. Finally, the indoor temperature of the floor is lower than others.

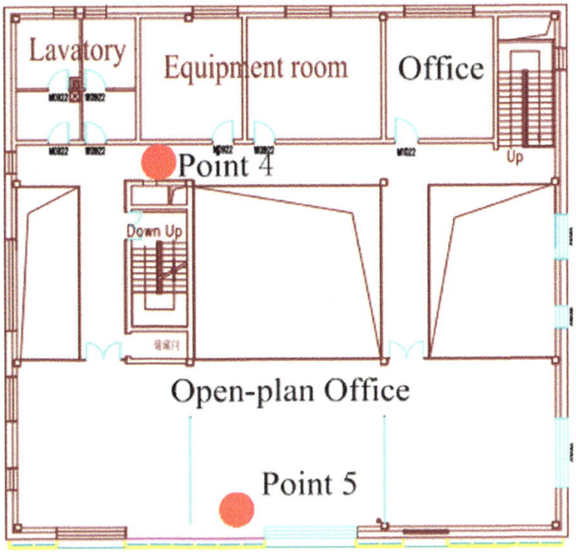

Fig. 10 The distribution of points on the second floor

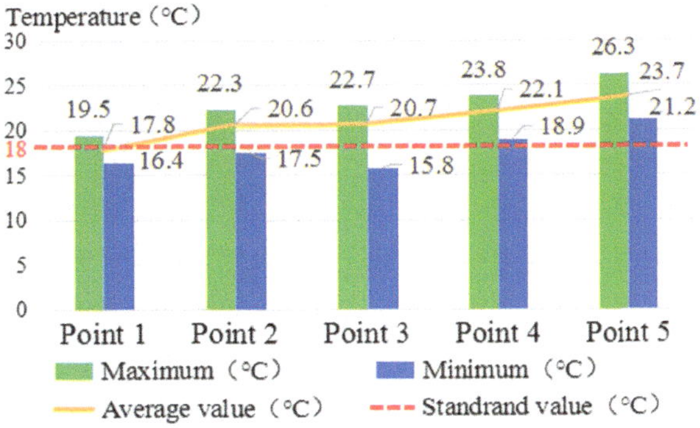

Fig. 11 Indoor temperature

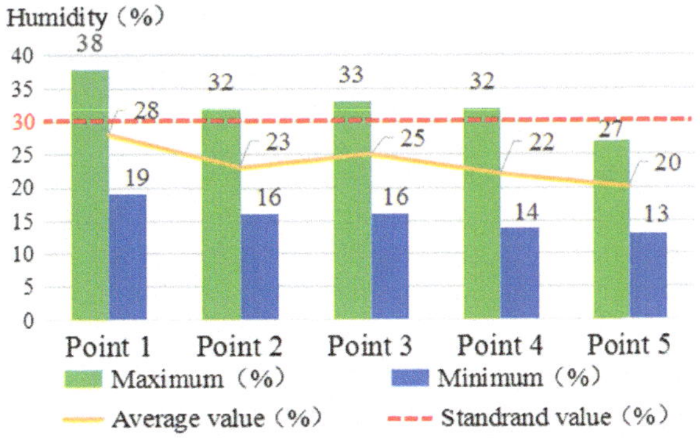

Fig. 12 Indoor humidity

Table 2 Indoor computation parameter

Parameter	Thermal comfort level	Temperature (°C)	Relative humidity (%)
Winter	I	22–24	30–60
	II	18–21	≤ 60

Fig. 13 The underground south side 1

Fig. 14 The underground south side 2

In addition, the window of the project has a seal level of 8 and there is less open window behavior in winter. The limited humidity of the office building that lacked the source of the scattered wet source was heated under the heating of the geothermal coil, so the indoor environment became drier and drier.

As a result of these reasons, none of the 5 measuring points can meet the requirements of thermal comfort.

3.3 Thermal Defects Test

In this part, Fluke TiR1 type infrared thermograph is used to detect the defects of energy-saving windows of the building. Firstly, we made infrared imaging of the four facades of the building. Secondly, we selected a window that had a large defect area in the image for a single image. Finally, a detailed analysis was made using the SmartView software, and the results were calculated by the formula 1 in the JGJ/T 132-2009 [7].

$$\beta = \Psi\left|\frac{T_1 - T_2}{T_1 - T_0}\right| \times 100\% \quad (1)$$

In the formula:

Ψ—The ratio of area of surface defect to the area of body area.

β—The increased energy consumption caused by the thermal defect.

T_1—The average temperature of the subject area (excluding the defect area).

T_2—The average temperature of the defect area.

T_0—The environment temperature.

On the north facade, the window of the east side of the second floor is selected as the representative. In the image generated by the infrared thermograph (Fig. 15), we can clearly see that the window frame is deep red, indicating that there is a thermal defect. With the help of SmartView3.14 software, further quantitative analysis of the defect is carried out.

The temperature distribution of the whole window can be clearly seen from the above image. By counting the number of pixels of each pixel, we can get the number of pixels of different temperatures. According to JGJ/T 132-2009 [7], the result of the calculation of relative area and energy consumption is shown as below (Table 3).

Next, we use the same method steps to analyze the energy saving windows in three directions: East, South and West.

In the eastern panorama, the window in the first floor is selected as the representative, and the pixel temperature distribution histogram is obtained by SmartView3.14 software as below (Fig. 16).

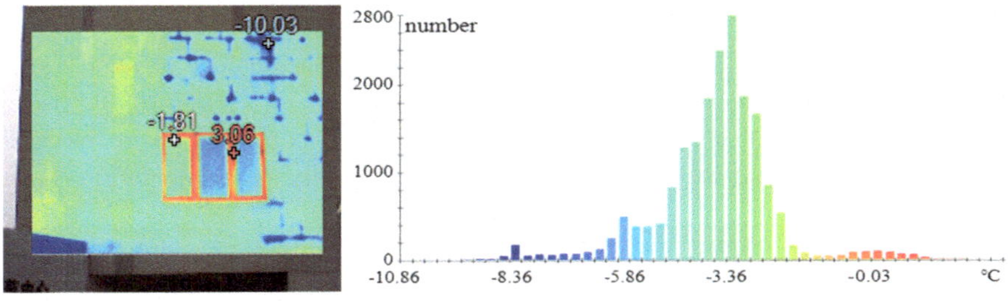

Fig. 15 Infrared thermal image of the north side window and its pixel histogram

Table 3 Image processing results

Environment (°C)	Body (°C)	Defects (°C)	Body (m^2)	Defects (m^2)	The relative area (%)	Energy consumption increase ratio (%)
22.00	−3.85	−1.14	5.56	1.76	31.71	3.32

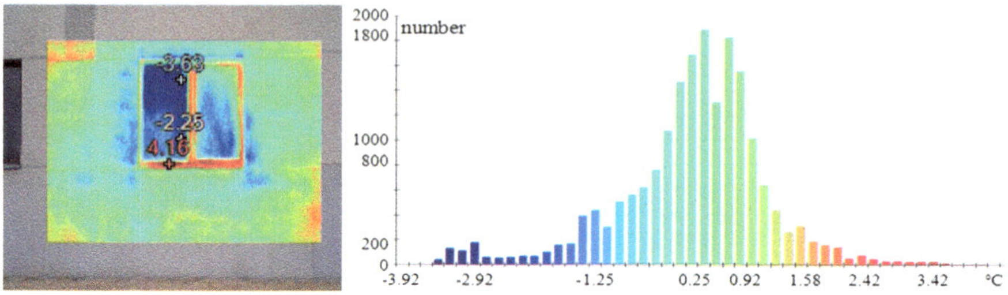

Fig. 16 Infrared thermal image of the east side window and its pixel histogram

Table 4 Image processing results

Environment (°C)	Body (°C)	Defects (°C)	Body (m^2)	Defects (m^2)	The relative area (%)	Energy consumption increase ratio (%)
22.00	−2.81	1.95	3.71	1.16	31.71	6.09

According to the statistics of the temperature pixels above, the processing results are shown as below (Table 4).

On the south facade, the window on the west side of the first floor was selected as the representative. The pixel temperature distribution histogram is shown as below (Fig. 17).

According to the statistics of the temperature pixels above, the processing results are calculated according to the provisions of JGJ-T132-2009 [7] as below (Table 5).

In the western panorama, the window in the ground floor is selected as the representative. With the help of SmartView3.14, the pixel temperature distribution histogram is shown as follows (Fig. 18).

According to the statistics of the temperature pixels above, the results of the treatment are calculated according to the provisions of JGJ/T 132-2009 [7], as shown below (Table 6).

According to the regulations [7], the tested outer surface defect area and subject area ratio should be less than 20%, and the single block defect area should be less than 0.5 m^2. The tested inner surface's ratio of energy consumption increased by the defect area should be less than 5%. The results of the four orientations are as follows.

As it is shown in Table 7, there are thermal defects in the four orientations of the window, and all are in the window frame and the glass joint. According to the regulations [7] above, four windows are not in line with the energy-saving standard.

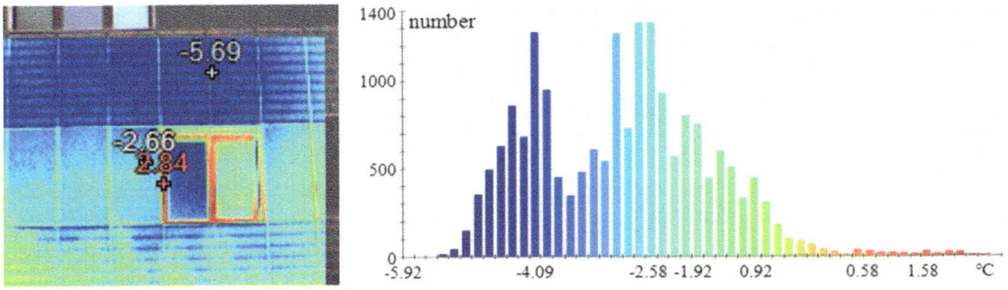

Fig. 17 Infrared thermal image of the south side window and its pixel histogram

Table 5 Image processing results

Environment (°C)	Body (°C)	Defects (°C)	Body (m^2)	Defects (m^2)	The relative area (%)	Energy consumption increase ratio (%)
22.00	−2.54	0.92	3.71	1.18	31.71	4.47

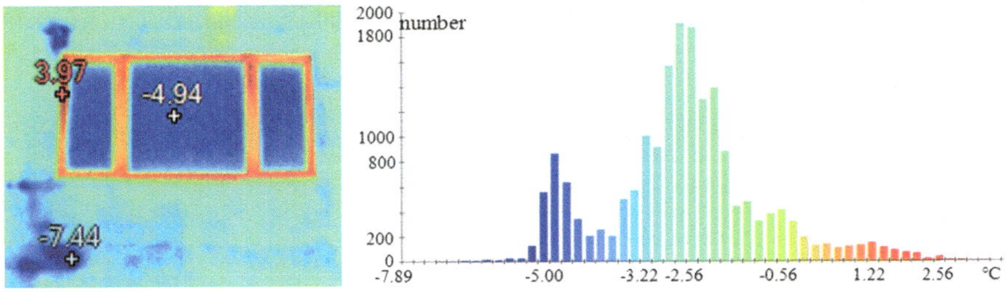

Fig. 18 Infrared thermal image of the west side window and its pixel histogram

Table 6 Image processing results

Environment (°C)	Body (°C)	Defects (°C)	Body (m^2)	Defects (m^2)	The relative area (%)	Energy consumption increase ratio (%)
22.00	−4.10	−0.20	5.56	1.18	31.71	4.74

Table 7 Summary of processing results

Orientation	North	East	South	West
The relative area (%)	31.71	31.71	31.71	31.71
Energy consumption increase ratio (%)	3.32	6.09	4.47	4.74

4 Conclusion

The energy-saving window should not only be displayed in the superior parameter value, but also in the actual operation of the building to play a real energy saving role that can reflect its significance. According to the analysis of these results above, this kind of energy-saving windows used in this building have thermal defects. And the indoor environment cannot meet the requirement of thermal comfort. We think that this kind of energy-saving window doesn't have the adaptability to the building named Sino-German Energy Conservation Demonstration Center.

According to the above analysis, the following suggestions are proposed for the adaptability of this kind of energy-saving window to the building.

For thermal comfort, this building is in the severe cold region and it is not suitable for large areas of transparent enclosure. Although the heat transfer coefficient of the windows is very low, the design causes huge heat loss in winter. The balance between lighting and heat preservation should be found when making design.

For thermal defect, the link between walls and window frames lead to a loss of heat in winter. The energy-saving window should be maintained regularly after it is put into use. Effective remedial measures should be taken for small cracks in the connection between windows and walls.

Acknowledgements This study was supported by the China National Key R&D Program (Grant No. 2016YFC0700100).

References

1. Green Building Map. http://www.gbmap.org/
2. Chinese Green Building Evaluation Label. http://www.cngb.org.cn/index.action?sid=40288840
3. X. Fu, *Energy Saving Technology in Hot Summer and Cold Winter Area* (China Construction Industry Press, Beijing, 2002), pp. 124–158
4. *JGJ/T 177-2009 Standard for Energy Test of Public Building.* (Beijing, Ministry of Housing and Urban-rural Development of the People's Republic of China, 2009) p. 4
5. *GB/T 50378-2014 Assessment Standard for Green Buildings.* (Beijing, Ministry of Housing and Urban-rural Development of the People's Republic of China, 2015) p. 103
6. *GB 50736-2012 Design Code for Heating Ventilation and Air Conditioning of Civil Buildings.* (Beijing, Ministry of Housing and Urban-rural Development of the People's Republic of China, 2012) p. 4
7. *JGJ/T 132-2009 Standard for Energy Efficiency Test of Residential Buildings.* (Beijing, Ministry of Housing and Urban-rural Development of the People's Republic of China, 2009) pp. 9–12

Part II
Renovation of Buildings

Tenants' Priority of Renovation Measures and Their Willingness to Pay Higher Rent to Implement These

Kristina Mjörnell and Carolina Hiller

Abstract In 2013 a web-based questionnaire was sent out to residents living in apartments in multifamily housing areas with the aim to get their opinions on what renovation measures they prioritized and what they would be willing to pay in terms of percentage rent increase. On a neighbourhood level, the result showed that higher indoor standard followed by higher standard on the exterior façade and windows were on the top of the list followed by better light environment, parking spaces and waste rooms. On apartment level, the most important measures were renovation of kitchen and bathroom followed by a reduced noise from neighbours and increased thermal comfort. The willingness or ability to pay for these measures was however quite low. More than one third of the tenants could not accept any rent increase while nearly half of the residents could accept a rent increase of 1–10% and very few could accept rent increases above 10%. Considering the cost for implementing the desired measures, the rent increase would probably be considerably higher than 10%. This is why it is of crucial importance to have a close dialogue with residents at an early stage in the renovation process in order to find the most cost-efficient package of renovation measures that responds to the technical, environmental and social needs of the buildings and their residents.

Keywords Renovation measures · Rent increase · Thermal comfort Indoor environment · Tenants

K. Mjörnell (✉) · C. Hiller
RISE Research Institutes of Sweden, Stockholm, Sweden
e-mail: kristina.mjornell@ri.se

C. Hiller
e-mail: carolina.hiller@ri.se

K. Mjörnell · C. Hiller
Building Physics, LTH, Lund University, Lund, Sweden

D. Johansson et al. (eds.), *Cold Climate HVAC 2018*,
Springer Proceedings in Energy, https://doi.org/10.1007/978-3-030-00662-4_15

1 Introduction

An important part of the multifamily building stock built in Sweden during the record years 1961–1975 is still in need for extensive renovation measures. Common for many of these buildings are that neglected maintenance has led to technical shortcomings such as; high energy use and low thermal comfort due to bad insulation, unsatisfactory air tightness and leaky windows, inefficient heating systems and insufficient ventilation, moisture damage due to leaking building envelope and leaking pipes. Complaints about cold floors, drafts from windows and doors, smells from own and neighbours' cooking as well as disturbing sounds from neighbours are more prominent in multi-family houses built in 1961–1985 [1]. However, the people living in these buildings are not willing to or do not even have the ability to pay higher rents that are entailed by extensive renovations. In some cases, the building owner implements renovation measures considered unnecessary among residents, causing considerable rent increases, forcing some people to leave their homes [2]. A study of renovation needs and average incomes in Swedish multi-family housing areas show that extensive renovation needs often coincide with low income of the tenants [3].

Former studies show that tenants place a higher value on visible features rather than on those that have high effect on energy efficiency [4]. Improvements in flats, relative to those in the building as a whole, have been stressed as most important [5]. A case study, involving interviews with tenants in a Swedish renovated multifamily housing area, shows that several participants are initially worried about how much the renovation would increase their rent. The results from interviews show a higher acceptability of a rent increase for changes that one thought of as beneficial for oneself (for example, increased security), than changes that was perceived as of less/ no visible value (i.e. energy saving equipment), or even disadvantageous to oneself (colder indoor temperatures and less ability to control the temperatures than before as well as damages and disturbance during renovation) [6].

However, there have not been many surveys about the residents' requests for renovation and their ability to pay for the increased rents after renovation. The authors therefore formulated four questions with multiple-choice answers concerning the tenants' preferences of renovation of their living environment and apartments as well as how much higher rent they would be willing to pay after renovation.

2 Method of Investigation

2.1 Questionnaire

In December 2013 a web based questionnaire was sent out to residents living in apartments (approximately 80% rented and 20% owned) in multifamily housing areas, covering a number of issues on their living conditions as well as on other topics of current interest. The questions were formulated by the authors of this paper but the survey was practically handled by the company Sweden Research, which meant that they identified a representative sample, distributed and collected the web based questionnaires. Sweden Research conducts studies and analyses on issues concerning social sustainability and urban development for authorities, municipalities, and other publicly financed organisations. Since 2009, they have conducted four surveys among residents living in the Swedish multifamily housing areas mainly built during the record years in 163 out of 290 Swedish municipalities. The selected residential areas are attributed to a lower socio-economic status of residents than in surrounding areas.

The selection of sample was made by distributing an invitation to a survey to several hundred thousand residents. The first about 1000 respondents that qualify to participate based on a number of background parameters and criteria were included in the survey. One such criterion was that the respondent lived in a multifamily housing area not located in the city centre. The language used in the web based questionnaire was Swedish.

2.2 *Respondents*

Over a thousand (1021) respondents living in seventy-seven multifamily housing areas replied to the questions considering the perception of their housing area, the labour market and what types of housing are most desirable. One section of the questionnaire included questions with the aim to get the residents opinions on what renovation measures they prioritized and what they were willing to pay in terms of percentage rent increase if these measures were carried out. The data presented in this paper was based on the response on these questions and has been analysed by the authors. The respondents were tenants the age of 14 or older, living in apartments in multifamily buildings that were mainly built during the record years 1961–1973.

Among the respondents, 53% were women and 47% men, and 72.5% had parents that are both born in Sweden. Almost a third had completed 3-year high school and, in addition to that, 42.8% had finished university (at least 2 years). Many of the respondents had full-time jobs (35.7%), were retired (24.5%) or were students (16.5%), whereas 7.4% stated that they were unemployed. One third worked in the private sector (33.2%), almost one sixth within the municipal or local governmental sector (15.3%) and 5.2% in the governmental sector. The average income was 19 336 SEK/month, which is approximately 2000 Euros/month.

The respondents represented a small sample of people living in multifamily houses in Sweden. In 2013 more than 2 378 000 people, all ages, lived in rental apartments in Sweden. The average income of the persons that participated in this survey was below the national average of 21 858 SEK/month (year 2013). This figure included people living in all types of dwellings [7]. The unemployment rate was however lower than the national average at the time, which was 8% [8].

2.3 Questions Asked

Four questions were formulated by the authors regarding what renovation measures were the most prioritized by tenants and what were their willingness to pay for an increased rent for these measures. Two of the questions had multiple-choice answers, where a maximum of three alternatives could be chosen. The respondents were also given the opportunity to give a free text answer. The questions are declared in full text in the results Sect. 3 together with the answers (original questions were in Swedish).

3 Results

3.1 Responses to Questions

The Tables 1, 2, 3 and 4 in this chapter show in percentage how many of the respondents chose the different alternatives for each question.

3.2 Results from Free Text Answers

The questionnaire gave the possibility for the tenants to give a free text answer. As seen in the tables above 8.2% respectively 6.4% of the respondents took the opportunity to express their own opinions about their wishes and needs for a possible refurbishment of the housing area as well as a possible renovation of their apartments. Many of these tenants expressed that they were satisfied with their housing conditions. The more critical free text answers could be divided into the following categories:

- Safety and security: Invest in safety, install intercom instead of key, safety doors, safer at night,
- Apartment: New wallpaper, floors, change of floorplan, install kitchen fan with light, better heating, reduce draught, better ventilation, noise insulation to avoid noise from neighbours, better indoor quality, glazed-in balconies, refurbish balconies, prohibit smoking inside, eliminate moisture, radon and mould.
- Common facilities: Storage rooms, bike and motorcycle parking, garage, carport, laundry rooms, elevators, weight training facilities, pool, build new top floor with apartments, waste handling, room for bulky waste, total renovation of facades.
- Outdoors: Better and nicer roads and path ways, nicer gardens, better outdoor environment.
- Social: Tenant involvement, local garden and city culture.
- Services in the neighbourhood: Lively centre, better shops, better services.

Table 1 Responses to question Q1

Multiple-choice alternatives	%
Better green spaces	15.9
Better internal standard of the houses, such as bathroom, kitchen, changing pipes	**37.6**
Better external standards on the houses, such as new facade and windows	**27.2**
Better parking places	21.2
Better bike and moped parking	8.6
More common areas, such as barbecue areas, playgrounds, skate parks	15.2
Better storage facilities (basements/garbage rooms)	13.2
Better room to store and sort waste	18.9
Better access to public transport	12.5
Better common rooms for gathering	7.1
Better lighting environment, outdoor lighting	**25.6**
Better bus, train and subway station	6.4
Otherwise, namely	8.2

In case of possible refurbishment in your area, what do you think would be the most important changes? Mark the most important things to you (maximum three). The alternatives that many respondents marked most important are bolded

Table 2 Responses to question Q2

Multiple choice alternative	%
Nothing, 0%	**33.9**
A little higher, 1–10%	**49.7**
Somewhat higher, 11–20%	7.7
Pretty much higher, 21–30%	0.8
Much higher, more than 30%	0.3
Do not know	7.6

If your house were undergoing refurbishment and the measures you indicated in question 1 were taken care of, how much higher rent would you accept to pay before you would start looking for a new apartment? The alternatives that many respondents marked are bolded

4 Discussion

The findings of the questions covered by this paper were that more than one third of the tenants could not accept any rent increase while almost half could accept a rent increase of 1–10% and very few could accept rent increases above 10%. These findings correspond well with the predictions made in [9]. Our experience is however, that tenants are very sensitive even to minor rent increases, wherefore a finer span would have been even more interesting. Similar to findings in [5],

Table 3 Responses to question Q3

Multiple-choice alternatives	%
Renovate kitchen	**29.1**
Renovate bathroom	**23.5**
Better locks on doors	6.8
Change of sewage pipes	12.0
Better air indoors	9.9
Better heat/indoor temperature	**21.1**
Reduce the draught	13.1
Better daylight in the apartment	2.2
Better lighting in stairs and entrances	2.4
Reduce noise from neighbours	**24.0**
Reduce noise from the outside	11.4
Less smell from stairwells or other apartments	7.6
Clean stairwells	5.6
New laundry rooms	9.0
New facade on the houses	8.5
New windows	14.5
Better accessibility	2.1
Otherwise, namely	6.4
I see no need for renovation	17.0

In case of possible renovation in your apartment, what do you think would be the most important measures? Mark the most important things to you (maximum three). The alternatives that many respondents marked most important are bolded

Table 4 Responses to question Q4

Multiple choice alternative	%
Nothing, 0%	**33.1**
A little higher, 1–10%	**46.5**
Somewhat higher, 11–20%	10.0
Pretty much higher, 21–30%	1.6
Much higher, more than 30%	0.4
Do not know	8.4

If your apartment was renovated and measures you indicated in question 3 were taken care of, how much higher rent would you accept before you would start to look for a new apartment? The alternatives that many respondents marked are bolded

measures in the respondents' own apartments were more preferred compared to measures in common areas, where renovation of kitchens and bathrooms were the highest priority together with noise reduction from neighbouring apartments. Next on the priority list was a thermal comfort aspect, namely better heat/indoor temperature. However, there were also a number of people who see no need for renovation of the apartments. Whether these tenants were among those not willing to pay any rent increase is not known to the authors. Many of the aspects that people thought were in need of renovation in this survey confirm findings of other studies such as the comprehensive national investigation of the Swedish housing stock [1]. This regards for example noise disturbance and thermal comfort problems. Kitchen and bathrooms are clearly part of the visual appearance of a dwelling which usually is of great importance to residents [4].

In the present study, other important issues were renovation measures related to the building itself. This was followed by a number of issues related to everyday situations such as better lighting environment, parking and waste facilities. Further measures that were thereafter desired and affecting the housing area were related to creation and social activities. Free text answers were only given by a small proportion of the respondents and several of these were satisfied with their housing situation. Others expressed a variety of desired measures, where issues related to safety and indoor environment, common facilities and outdoor environment were frequently stated. Although quite a large group of the tenants were willing to pay up to 10% more rent after renovation, the renovation measures the tenants expressed a need for or required are quite extensive and usually implies a rent increase of typically up to 50% [9]. To implement the necessary renovation measures, new financing models are required rather than just raising the rent. [10]. Nevertheless, this discrepancy between the costs for desired renovation measures and the willingness/ability to pay a higher rent for these measures means that it is of crucial importance to have a close dialogue with residents at an early stage in the renovation process. A successful tenant dialogue will increase the possibility to find the most cost-efficient package of renovation measures that responds to the technical, environmental and social needs of the buildings and their residents [11]. The respondents represent residents living in rental multifamily houses and people who generally have a lower capacity for large living costs compared to the average person in Sweden, considering income levels. One should note that the survey in this study addressed individuals and not households, hence the households' total income levels are not known. But generally, low income households often live in rental housing in Sweden [12] and most of the respondents in the survey lived in in houses from the "Record years", which are housing areas where the incomes tend to be even lower than in other housing areas with rental apartments [13].

5 Conclusions

The main conclusions of this study can be summarised as:

- The willingness or ability to pay for renovation measures is quite low. Several tenants would however accept a higher rent if desired renovation measures are implemented, but maximum 10% increase.

- Renovation measures related to the residents' own apartments are preferred compared to common areas. These measures include renovation of kitchens and bathrooms as well as noise reduction from neighbouring apartments.
- Regarding the common areas, the tenants would preferably see renovation measures related to everyday situations such as better lighting environment, parking and waste facilities.
- The costs for the desired renovation measures do not correspond to what the tenants are willing or able to pay in rent increase.

On large, the conclusions of this study confirm results of previous studies. The findings imply that there will be a need for other approaches than just to increase the rent in order to meet the need of extensive renovation of the Swedish multifamily housing stock. This is especially true for houses built in the period of 1961–1975 within the "Record years", where generally the economic situation for the low income households is strained. This could mean that new financial models are needed together with establishing a sincere dialogue with the tenants in order to find feasible solutions.

References

1. BETSI, Enkätundersökning om boendes upplevda inomhusmiljö och ohälsa—resultat från projektet BETSI, in *Questionnaire Survey on Residents Experienced Indoor Environment and Health—Results från the BETSI Project.* ISBN pdf: 978-91-86342-45-6, www.boverket.se (2009)
2. S. Westin, in *"... but Where Will you then Reside?" Renovation from a Tenant Perspective* (Institute for Housing and Urban Research, Uppsala University, Uppsala, Sweden, 2017)
3. T. Johansson, M. Mangold, T. Olofsson, Development of an energy atlas for renovation of the multifamily building stock in Sweden. Appl. Energy **203**, 723–736 (2017). ISSN 0306-2619, E-ISSN 1872-9118
4. Y. Phillips, Landlords versus tenants: information asymmetry and mismatched preferences for home energy efficiency. Energy Policy **45**, 112–121 (2012)
5. L.J. Vale, Public housing redevelopment: seven kinds of success. Housing Policy Debate **7**, 491–534 (1996)
6. E. Pedersen, A. Blomsterberg, J. Eng. Architect. **4**(1), 17–32 (2016). ISSN: 2334-2986 (Print), 2334-2994 (Online)
7. Statistics Sweden, in *Statistikdatabasen* (2017a), http://www.statistikdatabasen.scb.se. Retrieved from 23 Aug 2017
8. Statistics Sweden, in *Arbetslösheten tog fart under finanskrisen* (2017b), http://www.scb.se. Retrieved from 23 Aug 2017
9. S. Bergenstråhle, P. Palmstierna, *Var tredje kan tvingas flytta—En rapport om effekterna av hyreshöjningar i samband med standardhöjande åtgärder i Gothenburg* (2017). https://www.hyresgastforeningen.se/pressmeddelanden/1751221/var-tredje-hyresgast-kan-bli-tvungen-att-flytta. Retrieved from 30 Aug 2017
10. H. Lind, *Ekonomiska aspekter på renoveringar av bostäder* (KTH, Stockholm, 2014)
11. J. Stenberg, in *Medskapande renovering, Co-creation in renovation*, pp 77–88. Chapter in anthology about social sustainability with focus on renovation of housing. Social hållbarhet med fokus på bostadsrenovering—en antologi, Sustainable Integrated Renovation, Report

2015:4. ISBN 978-91-88001-80-1. http://www.renoveringscentrum.lth.se/fileadmin/renoveringscentrum/SIRen/Publikationer/Social_haallbarhet_vid_bostadsrenovering_-_en_antologi.pdf. Retrieved from 30 Aug 2017
12. Statistics Sweden, in *Låginkomsttagare bor ofta i hyresrätt* (2017c), http://www.scb.se Retrieved from 23 Aug 2017
13. Fastighetsägarna, in *Segregation bland hyresgäster i Stockholms län—En beskrivning av hyresgästers inkomster* (Fastighetsägarna, Stockholm, 2013)

Upgrading of a Typical Norwegian Existing Wooden House According to the EnerPHit Standard

Bozena Dorota Hrynyszyn and Laurina Cornelia Felius

Abstract The building sector has a key role to play in implementing the EU energy efficiency objectives. Around 40% of the energy consumption and a third of CO_2 emissions comes from buildings. With the adoption of Nearly Zero Energy Buildings throughout the EU from 2020 onwards, these figures will be reduced in a perceptible and sustainable way [1]. To achieve such a significant reduction before 2030 with the current low new built rate, a comprehensive effort on upgrading existing buildings is necessary. Thus, we should aim for more optimized refurbishment solutions, in addition to building new and more energy efficient buildings. Upgrading existing buildings to higher energy standards is usually far more difficult than obtaining the same standards in new buildings. In many cases, upgrading to the nZEB-level [2] is unlikely to be cost effective. Thus, the Passive House Institute has published the international EnerPHit Standard for retrofit of existing buildings [3]. This article compares the Norwegian energy standards with the international EnerPHit Standard for retrofitting. The article also analyses the upgrade potential of a typical Norwegian wooden house from 1960 to 70 by following the EnerPHit Standard using the "step-by-step" method. Two different programs are used for energy simulation: The Passive House Planning Package and SIMIEN. Furthermore, the article briefly discusses how building automation can be used as a next step to increase energy efficiency. Upgrading of a typical Norwegian wooden house from 1960 to 70 is not free from challenges, but it is possible to achieve the EnerPHit Standard following the "step-by-step" method.

Keywords Effective energy use · Retrofit · Building simulation
Building automation

B. D. Hrynyszyn (✉) · L. C. Felius
Department of Civil and Environmental Engineering,
Norwegian University of Science and Technology (NTNU), Trondheim, Norway
e-mail: bozena.d.hrynyszyn@ntnu.no

D. Johansson et al. (eds.), *Cold Climate HVAC 2018*,
Springer Proceedings in Energy, https://doi.org/10.1007/978-3-030-00662-4_16

1 The International Standard for Retrofit EnerPHit Versus the Norwegian Standards for Residential Buildings

1.1 Introduction

Upgrading existing buildings to higher energy standards is usually far more difficult than obtaining the same standards in new buildings. In many cases, upgrading to the nZEB-level [2] is unlikely to be cost effective.

Thus, the Passive Institute in Darmstadt has published a new standard for modernization of existing buildings, also called the "step-by-step" method [3]. The international EnerPHit Standard is an addition to the previously formulated international standards for new buildings in the classes: Passive House Classic, Passive House Plus and Passive House Premium [4]. Formulation of the EnerPHit Standard was preceded by the research project EuroPHit, supported by the Intelligent Energy Europe Program of the European Union and coordinated by the Passive House Institute in Darmstadt.

In Norway, in addition to the minimum energy requirements defined in the Norwegian Building Code, there are two standards for higher energy classes for residential [5] and commercial [6] buildings: low energy and passive house. These standards are however not coordinated with the cited international definition of the Passive House and they have lower ambitions than the international standards. There are no guidelines or standards for refurbishment of existing buildings in Norway, and the standards with higher energy classes are still not defined for other building typologies.

The standard NS-EN 15232: 2012 "Energy performance of buildings—Impact of Building Automation, Controls and Building Management" [7] is used in Norway, but there are only few reference projects that take into account energy savings as a result of a higher degree of automation in combination with the optimizing of building envelope. These kind of measures are also not included in the international standards, even though the benefits of it may be particularly relevant in countries with challenges related to difficult local climate conditions.

1.2 The International Energy Standard for Retrofit, EnerPHit Standard

The EnerPHit Standard proposed by the Passive House Institute is a guideline for reasonable thermal upgrading of existing buildings. The standard is versatile and applicable to different building typologies in different climatic zones [8].

The standard proposes two different methods to achieve the refurbishment criteria:

1. EnerPHit criteria for energy retrofit with passive house components, which is recommended to use for buildings with one or more obstacles concerning energy-relevant upgrades. This can be related to for instance the building location, geometry and the existing building technology.
2. EnerPHit criteria for energy retrofit with the energy demand method, which is recommended for buildings with favorable conditions.

Both methods take into account the climatic conditions of the building. The second method has been used for achieving the EnerPHit criteria for the case study in this article. These main energy-relevant criteria are related to maximum heating demand, 30 kWh/m^{2}a in the cold climate, and maximum cooling and dehumification demand, corresponding with Passive House requirements [3].

1.3 Norwegian Energy Standards

The requirements for energy efficiency in the Norwegian Building Code, TEK 17 [9], are stated as a maximum netto energy demand, in kWh/m^2 BRA per year, for each building typology. An alternative solution to fulfilling the energy efficiency requirements for residential buildings is satisfying the energy characteristics for different components. The actual values can deviate from these requirements, as long as the building's heat loss factor does not increase and as long as the minimum requirements for energy characteristics are met.

The standard for passive house and low-energy classification for residential buildings in Norway, NS 3700:2013 [5], has criteria for the maximum netto energy demand for heating, as opposed to TEK 17, that gives requirements for the total netto energy demand. This is a fixed value for houses larger than 250 m^2, but for smaller houses, the value depends on the heated floor area and on the average outdoor temperature. In addition, the calculated amount of electrical and fossil energy should be less than the total energy demand minus 50% of the demand for domestic hot water. There are requirements regarding U-values of windows and doors and of system parameters, but for the opaque building envelope there are no requirements, only typical values. These standards only consider new residential buildings and commercial buildings, and do not take the specific situation in the case of existing buildings into account.

2 Case Study: A Typical Wooden House

2.1 Introduction to the Case Study

A 60-70-year-old wooden house is usually still in good technical condition, and therefore often becomes an object of modernization. As a result, modernization of existing houses to adjust them to the present-day energy standards becomes an

Fig. 1 A typical Norwegian wooden house from the 60s and 70s [11]

important construction task. However, building components have different life durations, and in many cases, they do not need to be replaced simultaneously which can be challenging. The "step-by-step" method of modernization of existing houses can therefore be a good solution for these refurbishment projects.

The research object (see Fig. 1), is a characteristic detached wooden house with a heated floor area (BRA) of ca. 90 m^2. This one floor house with a suspended floor above a ventilated crawl space and a ventilated, cold loft represents one of the smaller typical houses from the 60s and 70s. It has a simple economic floor plan, but the building form cannot be considered compact. The house is built with standard components from this period. The thickness of insulation for the building envelope is 100 mm in the floor, 150 mm in the external walls, and 150 mm in the loft. The construction method of the house is the traditional, most common wooden construction in Norway, timber framing [10], and was also common in the rest of Scandinavia in the 60s and 70s. Wood or brick was used as external cladding.

Four theoretical locations with different local climate in Scandinavia have been chosen—all classified as cold climates according to the Passive House Institute in order to compare the results of simulation. The locations are Bergen, Oslo and Trondheim in Norway and Gothenburg in Sweden.

2.2 Method

Two simulation tools have been used to simulate the energy consumption for the case study. The Passive House Planning Package (PHPP) [12] was used to evaluate the refurbishment potential of the house following the EnerPHit Standard, preliminary in the energy class Classic. PHPP, developed by the Passive House Institute, offers the possibility to calculate versions of the building in parallel, which is particularly useful when evaluating the "step-by-step" method. SIMIEN [13] was used to simulate the building according to Norwegian standards. It uses a dynamic calculation method that is described in the Norwegian standards and is an approved tool in Norway for simulating the energy demand of a building.

3 Upgrading the Case Study Following the EnerPHit Standard and the Norwegian Standards

3.1 Results of Simulation According to the EnerPHit Standard

The software tool PHPP was used to simulate the energy demand for heating of the house as well as to evaluate the upgrade potential towards the EnerPHit Standard. The following steps were applied:

1. The existing windows and external doors are replaced with new components according to the Passive House Standard, and a new ventilation system with heating recovery and a heat pump are installed.
2. The existing insulation in the loft (150 mm) is replaced with 300 mm inflated cellulose insulation, still maintaining the cold loft principle.
3. For the floor, 150 mm is added to the existing insulation (100 mm). The new insulation fills the remaining free space in the existing floor construction. The existing floor may also be replaced with a slab-on-grade, which can provide even better safety against moisture damage.
4. For the external walls, 200 mm is added to the existing insulation (150 mm). The new insulation may be added as a double layer, i.e. 100 mm on the outside and 100 mm on the inside of the existing wall or only on the inside or outside of the existing construction.

The new construction maintains the general principle for heat and moisture transport. In addition to a better thermal comfort and a significantly lowered energy demand for heating, the new construction is better protected against any moisture damage due to the fact that the new construction is designed as a nearly thermal-bridge-free construction and the new, replaced and continuous vapor barrier layer ensures a high air tightness from the inside. The air change rate at 50 Pa of the new building envelope is less than 0.6 1/h. Assuming that the general rules for the placement and installation of the new moisture barriers are taken care of, the new construction will provide a good protection against moisture.

A new balanced passive house ventilation with heat recovery of ca. 80% will be installed already during step 1 of renovation to ensure a proper ventilation according to the increased airtightness of the renovated envelope.

The climate in Gothenburg, Sweden is similar to the climate in Bergen, Norway according to the standard climate data in PHPP for these locations, and the results of simulations are nearly identical when the same components are used. Figure 2 illustrates the preliminary results of the simulations of the analyzed house for location in Gothenburg and in Bergen. The simulations show how much the heating demand of the building varies during the "step by step" renovation (see Figs. 2, 3 and 4).

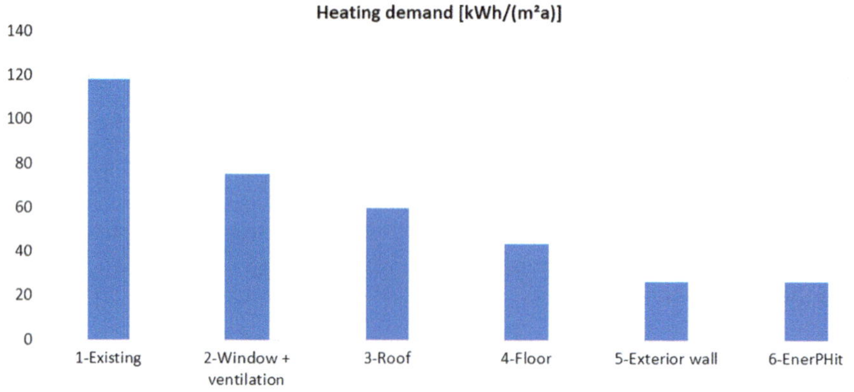

Fig. 2 Summary of the modernization steps in the PHPP, Gothenburg/Bergen

Specific building characteristics with reference to the treated floor area					Criteria	Alternative criteria	Fullfilled?[2]
	Treated floor area	m²	91,2				
Space heating	Heating demand	kWh/(m²a)	27	≤	30	-	yes
	Heating load	W/m²	14	≤	-	-	
Space cooling	Cooling & dehum. demand	kWh/(m²a)	-	≤	-	-	-
	Cooling load	W/m²	-	≤	-	-	
	Frequency of overheating (> 25 °C)	%	7	≤	10		yes
	Frequency of excessively high humidity (> 12 g/kg)	%	0	≤	20		yes
Airtightness	Pressurization test result n_{50}	1/h	0,3	≤	1,0		yes
Non-renewable Primary Energy (PE)	PE demand	kWh/(m²a)	127	≤	-		-
Primary Energy Renewable (PER)	PER demand	kWh/(m²a)	65	≤	77	77	yes
	Generation of renewable energy (in relation to pro-jected building footprint area)	kWh/(m²a)	0	≥	-	-	

[2] Empty field: Data missing; '-': No requirement

I confirm that the values given herein have been determined following the PHPP methodology and based on the characteristic values of the building. The PHPP calculations are attached to this verification.	EnerPHit Classic?	yes

Fig. 3 Verification in the PHPP according to the energy demand method, Gothenburg

Specific building characteristics with reference to the treated floor area					Criteria	Alternative criteria	Fullfilled?[2]
	Treated floor area	m²	91,2				
Space heating	Heating demand	kWh/(m²a)	27	≤	30	-	yes
	Heating load	W/m²	11	≤	-	-	
Space cooling	Cooling & dehum. demand	kWh/(m²a)	-	≤	-	-	-
	Cooling load	W/m²	-	≤	-	-	
	Frequency of overheating (> 25 °C)	%	2	≤	10		yes
	Frequency of excessively high humidity (> 12 g/kg)	%	0	≤	20		yes
Airtightness	Pressurization test result n_{50}	1/h	0,3	≤	1,0		yes
Non-renewable Primary Energy (PE)	PE demand	kWh/(m²a)	127	≤	-		-
Primary Energy Renewable (PER)	PER demand	kWh/(m²a)	59	≤	75	75	yes
	Generation of renewable energy (in relation to pro-jected building footprint area)	kWh/(m²a)	0	≥	-	-	

[2] Empty field: Data missing; '-': No requirement

I confirm that the values given herein have been determined following the PHPP methodology and based on the characteristic values of the building. The PHPP calculations are attached to this verification.	EnerPHit Classic?	yes

Fig. 4 Verification in the PHPP according to the energy demand method, Bergen

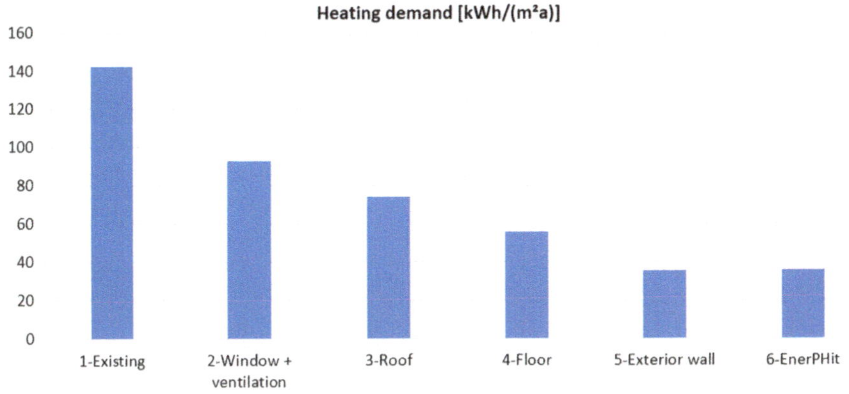

Fig. 5 Summary of the modernization steps in the PHPP worksheet "Variants", Oslo

The simulation shows that it is possible to reduce the energy demand for heating of the house from ca. 118 kWh/m^2 per year to ca. 27 kWh/m^2 per year, i.e. under the maximum 30 kWh/m^2 per year required by the EnerPHit Standard.

The preliminary results also show that it is impossible to successfully upgrade the house if it were located in Oslo or Trondheim using the same components (see Fig. 5).

However, it is relatively easy to achieve the EnerPHit Standard in Oslo using for instance 150 mm insulation extra in the roof (450 mm in total). In Trondheim it is also possible by adding 300 mm additional insulation, for instance by adding an extra 150 mm in the roof and 150 mm in the floor. In this case, other measures may also be considered that improve efficient energy management, for instance using building automation (read more in Chap. 4).

3.2 Results of Simulation According to the Norwegian Standards

The software tool SIMIEN was also used to simulate the energy demand of the house as well as to evaluate the upgrade potential of the house towards Norwegian standards.

Figure 6 shows the results from five different scenarios in the Oslo climate: the existing building, and four levels of refurbishment according to the Norwegian standards. It shows the simulated netto energy demand, both the total and the specific demand for heating, and the maximum demand according to the standards. Since TEK 17 was recently published with stricter energy efficiency requirements, and there has not been an update for NS 3700:2013, the low-energy 2 scenario results in a higher energy demand than the TEK 17 scenario. The simulated netto

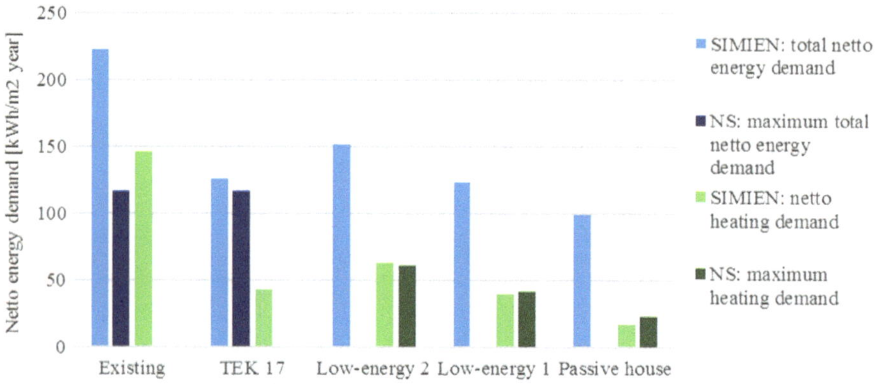

Fig. 6 Simulation results in SIMIEN from different scenarios in the Oslo climate

energy demand in the TEK 17 scenario is higher than the maximum value. However, the results showed that the criteria for the alternative solution, satisfying energy characteristics for components, are met.

4 Effective Energy Use in Residential Buildings with the Use of Building Automation

In general, building automatic control systems (BACS) can be said to function at several levels: following a fixed user-defined set point (D), or following a schedule or pattern depending on time (C), presence (B) or demand (A). The standard NS-EN 15232: 2012 [7] gives an overview of efficiency factors of the building automation control system (BACS) in different types of buildings and for different levels of automation. These factors for residential buildings are given in Table 1, and can be used to estimate the energy savings potential of implementing building automation.

The case study uses electricity for heating and lighting, and has a balanced ventilation system after refurbishment. This means that the building control system can be kept reasonably simple. The number of possible functions involved is however still considerable. The house classifies as D, which means that the BACS system is non-efficient and in need for retrofitting. The energy savings potential is 8–16% for electrical energy and 10-29% for thermal energy, depending on the BACS level after upgrading (see Table 1 and Fig. 7). The actual savings depend on

Table 1 BACS efficiency factors for residential buildings [7]

Category	Electrical energy				Thermal energy			
	D	C	B	A	D	C	B	A
Housing	1.08	1	0.93	0.92	1.10	1	0.88	0.81

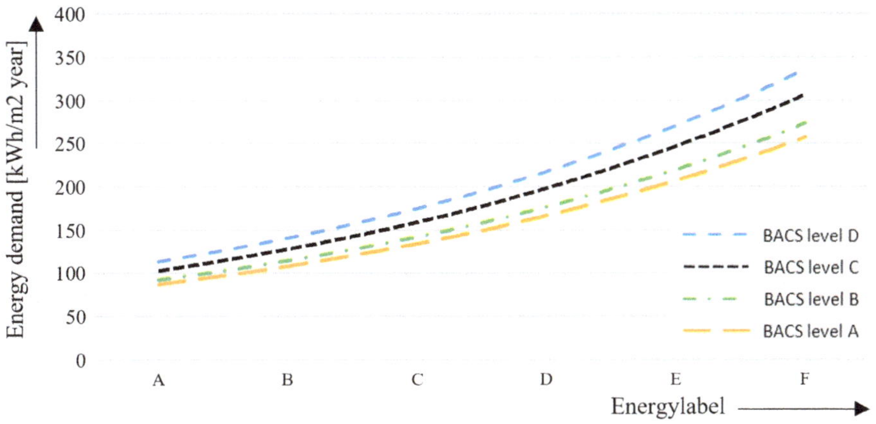

Fig. 7 Energy savings potential for implementing building automation

the choice of HVAC systems and energy sources, and on the complexity of the BACS. Figure 7 also shows that implementing building automation alone is not enough to reach Passive House ambitions. Retrofitting of the building envelope is needed first, after which building automation can be implemented to further increase energy efficiency.

5 Discussion

The preliminary results of the calculations in PHPP according to the energy demand method from EnerPHit Standard show that the "step-by-step" method can be successfully used for retrofitting existing wooden houses in Scandinavia. Of course, the method can also be used to simulate a comprehensive retrofitting that is carried out during one phase only. The parallel simulation options in PPHP can be used to compare both complete refurbishments as well as compare the benefits of individual components.

The preliminary results show that with some reasonable measures it is possible to reduce the existing building energy demand for heating by almost 80%. The "step-by-step" method is especially relevant for upgrading the chosen house located in Bergen, Norway as well as in Gothenburg, Sweden. For other locations, such as Oslo and Trondheim in Norway, more measures need to be implemented to reach the level from the EnerPHit Standard, but it is still reasonable. The simulations indicate that for buildings with more favorable conditions, for instance more compact buildings, even better results can be expected.

The Norwegian standards does not take into account limitations related to refurbishment existing building. The preliminary results of the simulations in SIMIEN show that it can be more challenging to meet the requirements for some typical building components (see Table 1), compared to the preliminary results of calculations in PHPP according to the international EnerPHit Standard (read more in Chap. 3). The preliminary results of calculations in PHPP according to the EnerPHit Standard demonstrate that the standard can be successfully used in Scandinavian.

For more challenging climates, building automation could be used as an energy saving measure when it is difficult to reach the EnerPHit criteria with thermal upgrades only. The savings potential of building automation alone is however not enough to reach the Passive House ambitions. Therefore, retrofitting of the building envelope is needed first. Building automation can then be implemented to further reduce the energy demand by increasing energy efficiency of the HVAC systems. The savings potential of building automation is further limited by factors such as feasibility and cost-efficiency. Most energy savings can be achieved when HVAC systems are demand-controlled, but for residential units with poor existing HVAC structures this could result in high costs. It may be more realistic to upgrade to a time or presence controlled system.

5.1 Limitations and Further Research

- The existing house's energy demand for heating is currently covered completely by electricity. In the preliminary calculation in PHPP, a heat pump is implemented during step 1, which will reduce the amount of the delivered energy, (see Fig. 8).
- The use of Photovoltaic panels may be considered to cover the remaining electricity demand, but it is not taken into account in this article. Other heating alternatives are not taken account in this article, but heating systems based on non-renewable energy sources or heating systems that increase the amount of the delivered energy, like district heating, are not recommended.
- The benefits of building automation are not taken into account in the simulations, neither in PHPP nor in SIMIEN.

	Select the active variant here >>>>>>>	6-External wall	Existing	Windows + ventilation	Heat pump	Roof	Floor	External wall
	Units	6	1	2	3	4	5	6
Heating demand	kWh/(m²a)	26,7	118,2	75,3	74,4	59,7	47,0	26,7
Heating load	W/m²	11,3	41,5	23,3	23,2	19,4	16,8	11,3
Cooling & dehum. demand	kWh/(m²a)							
Cooling load	W/m²							
Frequency of overheating (> 25 °C)	%	1,5	0,1	0,2	0,2	0,2	1,0	1,5
PER demand	kWh/(m²a)	58,9	209,5	154,3	97,9	85,5	75,1	58,9
EnerPHit Classic?	yes / no	yes	no	no	no	no	no	yes

Fig. 8 Summary of the modernization steps in the PHPP, Bergen

- Standard climate conditions for these four analyzed locations according to PHPP are taken into account in the preliminary calculations. It is recommended to use specific local climate data for specific individual locations in practice.
- The architectural measure to make the building form more compact, by for example adding an extra floor, is not taken account in the calculations.

6 Conclusion

Upgrading of a typical Norwegian wooden house to the EnerPHit Standard is not free from challenges. These challenges can be related to the existing building location, geometry, construction, and building technology. Despite the limitations, it is possible to upgrade this house to the EnerPHit Standard using the "step-by-step" method. By using modern technology and components as well as renewable energy sources, it is possible to achieve not only a more energy efficient building, but also an environmentally friendly, user friendly and cost-effective living space, even in the challenging cold climate.

References

1. IEA Homepage, in *Energy Climate and Change; World Energy Outlook Special Report* (2015), https://www.iea.org/publications/freepublications/publication/WEO2015SpecialReportonEnergyandClimateChange.pdf. Last accessed Jan 2017
2. PHI Homepage, *PassREg Final report Passive House Regions with Renewable Energies*. (Passive House Institute Dr. Wolfgang Feist, 2015), http://www.passREg.eu. Last accessed Aug 2017
3. PHI Homepage, *Criteria for the Passive House, EnerPHit and PHI Low Energy Building Standard*, https://passipedia.org/_media/picopen/phi_building_criteria_draft.pdf. Last accessed Nov 2016
4. Zeno Bastian: Passive House Institute Introduces New an Updated Building Certification Criteria, Editor, Passive House Institute, Germany 7. Passivhus Norden Copenhagen (2015)
5. Standard Norge, *NS 3700:2013 Kriterier for passivhus og lavenergibygninger-Boligbygninger* (Standard Norge, Lysaker, 2013)
6. Standard Norge, *NS 3701:2012 Kriterier for passivhus og lavenergibygninger-Yrkesbygninger* (Standard Norge, Lysaker, 2012)
7. Standard Norge, *NS-EN 15232:2012 Bygningens energiyttelse- Innvirkning ved bruk av bygningsautomation og bygningsadministrasjon* (Standard Norge, Lysaker, 2012)
8. PHI Homepage, *Step by Step Retrofits with Passive House Components*. In: Editor, Zeno Bastian (2016), www.europhit.eu. Last accessed Jan 2017
9. DiBK Homepage, https://dibk.no/byggereglene/byggteknisk-forskrift-tek17/14/14-2/. Last accessed 28 Oct 2017
10. Knut Ivar Edvardsen, Trond Ramstad, Håndbok 53 Trehus. SINTEF Byggforsk, 2010
11. Nasjonalmuseet, Den Norske Stats Husbank Type 131, http://samling.nasjonalmuseet.no/en/object/NMK.2006.hus070.001. Last accessed 12 July 2017
12. Passive House Institute, Passive House Planning Package (PHPP, Version 9)
13. Programbyggerne ANS, SIMIEN (Version 6.005), http://www.programbyggerne.no/. Last accessed 28 Aug 2018

Adding Glazing as an Energy Saving Renovation Measure in Cold Climates

Birgitta Nordquist, Kimmo Hilliaho and Petter Wallentén

Abstract Adding exterior insulation as an energy saving renovation measure is not always possible for cultural heritage reasons. This study explores the energy saving that can be made if glazing instead is added to a heavy structure as a brick building creating a double skin façade. Extensive measurements have been made in a full-scale building. The building has then been modelled in IDA-ICE and simulations, validated by the measurements, have been made for both an outdoor climate representing the southern part of Sweden, Malmö and for one climate representing the northern part of Finland; Sodankylä. The annual heating energy savings have been calculated for various design combinations; different U-values and for both mechanical exhaust respectively mechanical supply and exhaust ventilation systems. The results show that an energy saving is achieved in the order of 8–38% depending on the design of the building; the glazing and the ventilation system. As both a southern and northern climate of Scandinavia are studied the results indicates how this type of renovation measure would perform in general in this part of the world.

Keywords Energy saving · Double skin-façade · Renovation · Brick building Simulations

B. Nordquist (✉)
Building Services, Building and Environmental Technology,
Lund Institute of Technology, Lund University, Lund, Sweden
e-mail: birgitta.nordquist@hvac.lth.se

K. Hilliaho
Service Life Engineering of Structures Group, Structural Engineering Unit,
Department of Civil Engineering, Tampere University of Technology,
Tampere, Finland

P. Wallentén
Building Physics, Building and Environmental Technology,
Lund Institute of Technology, Lund University, Lund, Sweden

D. Johansson et al. (eds.), *Cold Climate HVAC 2018*,
Springer Proceedings in Energy, https://doi.org/10.1007/978-3-030-00662-4_17

1 Introduction

If the environmental challenges are to be met in terms of reducing the total energy use in the building sector, renovation measures of the existing building stock must be included, since the majority of houses in e.g. 30 years are already built. Knowledge about different technical measures that can be applied is therefore important to retrieve.

One possible renovation measure is to add glazing to an existing building. Studies including temperate and hot climates are reported [1–3]. However as the energy demands for lower energy use are getting stronger it becomes increasingly more relevant to also learn even more about the energy performance of the construction in colder climates. This includes attaining knowledge on how large energy saving that is possible to achieve. This knowledge can contribute to the decision making process when choosing between different energy saving measures to apply in existing buildings.

When undertaking several renovation measures on a 1930-ties brick building, situated in a cold climate, adding exterior insulation to the brick façade was first intended. For cultural heritage reasons this was not allowed and the idea of adding glazing to the facades came up. A mechanically supply and exhaust ventilation system with a rotary heat exchanger was also installed. Situated in an area with sustainable renovation goals the building was extensively equipped with temperature, relative humidity and air flow sensors in several points of the building and ventilation system to allow study of the performance to be possible.

Double skin facades are extensively studied for double glazing. However adding glazing on heavy structures as brick is not that widely studied especially lacking field-measurements and simulations [4]. This paper will contribute with knowledge about the possible energy savings for full scale buildings to this identified gap by studying this for the 1930-ties brick building. The aim is to attain values of the possible energy saving that can be achieved with this type of renovation measure in cold climates.

As a colder climate is in focus calculations should be made for both southern and northern parts of Scandinavia with the aim to cover the borders and thereby most outdoor climates of Scandinavia.

This article focuses on the heating period and the possible energy savings for this period of the year. The conditions for the thermal indoor climate and performance during the summer period should also be addressed but is excluded here. It is also limited to the saving in terms of kWh and the cost and monetary consideration is not included.

2 Method

2.1 Description of the Building and the Installation Systems

The building envelope consisted of 1 ½ bricks. 8 mm single glazing with a U-value of 5.8 W/m^2 K and double glazing on the top with 2.6 W/m^2 K had been added 0.75 m outside the brick walls on the south, east and west facades. Adding glazing on the north façade was not possible for space reasons and is usually not done due to lack of insolation. The building and the added glazing on the south and east façade can be seen in Fig. 1.

Different installation systems were installed to address the heating and cooling needs during the year.

During the heating season the building was heated with a hydronic radiator system supplied by district heating. The intention was also to pre-heat the outdoor air by letting it pass through the cavity between the old brick façade and the added glazing before passing through the ventilation unit. When the temperature was above 20 °C in the cavity the air path was changed so that the outdoor air entered the ventilation unit directly without being pre-heated in the cavity. The outdoor air was then passed through an air intake on the north façade.

During the summer season the outdoor air was always passed directly to the ventilation unit. To reduce the possible over-heating in the cavity, several cooling options were installed including a separate ventilation system. A concrete duct had been placed in the ground. Outdoor air entered the duct via a vertical concrete duct

Fig. 1 The studied building located in Sege Park in Malmö, with the added glazing visible on the south and east façade. The inlet pipe to the added concrete duct for the cooling of the cavity can be seen on the left part of the picture

situated among trees. A supply fan was located in the vertical duct. The air was intended to be cooled in the ground and then supplied to the bottom of the cavity. A corresponding exhaust fan removed the same amount of air from the cavity via high placed ventilation ducts. It is worth mentioning that the cooled air, with an increased relative humidity due to cooling, is not intended to be supplied to the building. The building was equipped with another ventilation unit previously mentioned.

Ventilation windows, which were opened by motors when the cavity temperature exceeded 23 °C were placed at the top of the cavity.

The air supply rate to the building is 50 l/s fulfilling the Swedish building regulation of 0.35 l/s m^2 floor area for dwellings and the exhaust rate is 53 l/s. The summer air flow through the cooling system of the cavity is designed to be 150 l/s for both supply and exhaust sides.

2.2 *Energy Calculation*

An extensive set of sensors for temperature, relative humidity, air flows etc. were installed in the building and measurements were performed over in total 2 years. The existing (as designed) building was modelled in IDA-ICE. The IDA-ICE model was validated for winter, spring/autumn respectively summer conditions by these measurements. A detailed description of the model and the measurements are made in [5].

Two locations in Scandinavia have been studied, Malmö in the southern part of Sweden and Sodankylä in the northern part of Finland. The latitude is 55° 36′N and the longitude for Malmö is 12° 59′E or 55.605 and 13.0038 respectively. Sodankylä has a latitude of 67° 24′ 57.6″N and a longitude of 26° 35′ 20.3″E or 67.415999 and 26.588973 respectively. A climate file for the year of 2014 was used for Malmö. The specific year of 2014 was used to make it possible to validate the model with the full-scale measurements which comprised of values attained during 2014. The Sodankylä weather file in the simulations was the test reference year for heating and cooling of buildings in Finland (version 2012). The Finnish test reference year, 2012, is based on the weather logs at Vantaa, Jyväskylä and Sodankylä weather stations from 1980 to 2009. They consist of weather data for twelve months, which have weather conditions close to the long-term climatological average [6].

Full year simulations were made and a large amount of parameters were attained. In this paper the energy use for heating has been retrieved and summarized.

3 Results

3.1 *Energy Saving of Renovated Building*

The simulated energy need for heating during the heating season in Malmö are presented in Table 1 for a number of different designs; before renovation and after with various characteristics.

Table 1 Need of heating energy and energy saving in percentage of energy need before renovation in Malmö, southern part of Sweden for two types of ventilation systems; exhaust respectively mechanical supply and exhaust system

Case	Annual heating energy need (kWh)	Energy saving in whole percentage (%)
Before energy renovation measures (exhaust ventilation)	20,558	–
Exhaust ventilation		
Only glazing, no pre-heating of outdoor air via cavity	18,709	9
Glazing + pre-heating of outdoor air via cavity below 20 °C	17,850	13
Glazing + pre-heating of outdoor air via cavity below 30 °C	17,262	16
Glazing + pre-heating of outdoor air via cavity below 50 °C	17,124	17
Supply and exhaust ventilation with heat exchanger	16,384	20
Only glazing, no pre-heating of outdoor air via cavity	14,672	29
Glazing + pre-heating of outdoor air via cavity below 20 °C	14,686	29
Glazing + pre-heating of outdoor air via cavity below 30 °C	14,355	30
Glazing + pre-heating of outdoor air via cavity below 50 °C	14,332	30

The same design values as in the existing building were applied in the simulations. The set point of changing the path of the outdoor air between pre-heating in the cavity respectively directly to the ventilation unit was in the existing building 20 °C. The measurements showed that the air temperature in the cavity can increase above 20 °C during the winter depending on among all the insolation [7]. This means that there may exist a heating need also when the outdoor air is not passed via the cavity. The effect of a set point of 30 and 50 °C were therefore also studied.

The simulated energy need for heating during the heating season in Sodankylä are presented in Table 2.

The energy saving in terms of percentages was larger for Malmö. The reductions in terms of kWh was however larger for Sodankylä.

The design of the glazing were also changed for the Malmö location to study the effect of double respectively triple glazing as well. This is presented in Table 3.

The energy saving of the heating energy need was between 8 and 38% depending on the design. Both the U-value and the g-value and the combination of them have influence on the energy need.

Table 2 Need of heating energy and energy saving in percentage of energy need before renovation in Sodankylä, northern part of Finland for two types of ventilation systems; exhaust respectively mechanical supply and exhaust system

Case	Annual heating energy need (kWh)	Energy saving in whole percentage (%)
Before energy renovation measures (exhaust ventilation)	49,167	–
Exhaust ventilation	45,054	8
Only glazing, no pre-heating of outdoor air via cavity	43,467	12
Glazing + pre-heating of outdoor air via cavity below 20 °C	43,467	12
Glazing + pre-heating of outdoor air via cavity below 30 °C	43,147	12
Glazing + pre-heating of outdoor air via cavity below 50 °C	43,080	12
Supply and exhaust ventilation with heat exchanger	40,706	17
Only glazing, no pre-heating of outdoor air via cavity	37,859	23
Glazing + pre-heating of outdoor air via cavity below 20 °C	37,460	24
Glazing + pre-heating of outdoor air via cavity below 30 °C	37,422	24
Glazing + pre-heating of outdoor air via cavity below 50 °C	37,409	24

4 Discussion

In the existing building the outdoor air is not passed through the cavity when the cavity temperature exceeds 20 °C. The increased set point from 20 to 30 °C resulted in somewhat increased energy savings. However increasing further to 50 ° C was marginal. There will also emerge disadvantages with increasing the supply air temperature for air quality reasons which also must be considered.

One specific year was used as a climate file for the Malmö location to allow validation with the full-scale measurements. The year was compared to a normal year for the period of 1999–2010 in [4]. The weather data match each other pretty closely although the radiation level for the monitored year was higher than for the long-term—average especially for the summer months. Higher radiation levels could imply overestimated energy saving during the winter period and increased over-heating during the summer time. This paper focuses on the heating period and during the months of October-April the deviation is small except for April. So except for April the used whether file should not imply a significant difference from normal conditions for the Malmö location. The assessment is therefore made that the studied data represent normal conditions.

Table 3 Need of heating energy and energy saving in percentage of energy need before renovation in Malmö, southern part of Sweden with improved glazing for two types of ventilation systems; exhaust respectively mechanical supply and exhaust system

Case	Annual heating energy need (kWh)	Energy saving in whole percentage (%)
Before energy renovation measures (exhaust ventilation)	20,558	–
Exhaust ventilation		
Only Single glazing, U-value = 5.7 W/m^2 K, g = 0.82	18,709	9
Single glazing + pre-heating of outdoor air	17,850	13
Only Double glazing, U-value = 2.6 W/m^2 K, g = 0.73	17,310	16
Double glazing + pre-heating of outdoor air	16,125	22
Only Triple glazing, U-value = 1.7 W/m^2 K, g = 0.63	16,642	19
Triple glazing + pre-heating of outdoor air via cavity	15,364	25
Supply and exhaust ventilation with heat exchanger	16,384	20
Only Single glazing	14,672	29
Single glazing + pre-heating of outdoor air	14,686	29
Only Double glazing U-value = 2.6 W/m^2 K, g = 0.73	13,367	35
Double glazing + pre-heating of outdoor air	13,376	35
Only Triple glazing, U-value = 1.7 W/m^2 K, g = 0.63	12,766	38
Triple glazing + pre-heating of outdoor air via cavity	12,728	38
Only Triple glazing, U-value = 0.7 W/m^2 K, g = 0.24	13,313	35
Triple glazing + pre-heating of outdoor air via cavity	13,429	35

U-value is the heat transfer coefficient for the glazing, the g-value is the solar transmittance of the glazing

The cost and profitability has not been studied. This depends on a number of parameters such as: energy price, political decisions as subsidies for energy saving renovations, future energy demands and also the condition of the building and cultural heritage aspects. These parameters will vary depending on the conditions and prerequisites that are valid for each object. The level of deterioration of a brick façade and needed measures to repair and subsequent costs for this will affect the economic considerations when choosing between different renovation measures. If the brick façade has deteriorated and something needs to be done to repair it, the studied solution might be an alternative which will stop further deterioration and protect the façade in the future.

The intention of this study was to contribute with levels of possible energy savings for this type of renovation. This can constitute a part of the input to the profitability calculation the designer makes.

5 Conclusions

The conclusion can be drawn that the calculated energy saving is in the span of 8–38% depending on the chosen type of glazing. For the existing glazing design the energy reduction is 9–30% compared to before the renovation in the southern part of Sweden (Malmö) and 8–24% in the northern part of Finland (Sodankylä). The saving is about 3000–6000 kWh in southern Sweden and about 6000–12,000 kWh in northern Finland for the existing glazing design.

As both a southern and northern outdoor climate of Scandinavia are studied the results indicates how this type of renovation measure would perform in general in this cold part of the world. The simulation results indicates that energy savings are possible to make which makes this construction worth considering when exploring different options.

Acknowledgements *The Swedish Energy Agency* and Tampere University are acknowledged for financing the study. The study is also a result of a co-operation between the Tampere University in Finland and Lund University in Sweden. The city of Malmö is gratefully acknowledged for allowing the analysis to be made, for supplying the measurement data, and for undertaking and financing the renovation including the measurement equipment.

References

1. P. Wong, *Natural Ventilation in Double-Skin Façade Design for Office Building on Hot and Humid Climate* (University of New South Wales, Australia, 2008)
2. A. Ghaffarianhoseini, U. Berardi, J. Tookey, D. Hin Wa Li, S. Kariminia, Exploring the advantages and challenges of double-skin facades (DSFs). Renew. Sustain. Energy Rev. **60**, 1052–1065 (2016). Elsevier
3. Y. Wang, Y. Chen, J. Zhou, Dynamic modeling of ventilated double skin façade in hot summer and cold winter zone in China. Building and Environment **106**, 365–377 (2016). Elsevier
4. K. Hilliaho, Energy Saving Potential and Interior Temperatures of Glazed Spaces—Evaluation Through Measurements and Simulations, PhD thesis. Tampere University of Technology, Tampere (2017)
5. K. Hilliaho, B. Nordquist, P. Wallentén, Energy saving and indoor climate effects of an added glazed façade to a brick wall building: Case study. J. Build. Eng. **7**, 246–262 (2016)
6. K. Jylhä, T. Kalamees, H. Tietäväinen, K. Ruosteenoja, J. Jokisalo, R. Hyvönen, S. Ilomets, S. Saku, A. Hutila, Test reference year 2012 for building energy demand and impacts of climate change, Reports 2011:6, Finnish meteorological Institute, Helsinki, 2011, 112 p. (in Finnish)
7. G. Avgenakis, Inglasning av tegelfasad, Energianalys samt fältstudie på ett flerbostadshus och på Byggnad 22. Report TVIT-5059, in Swedish, Avd. för Installationsteknik, Lund University, (2017)

The Most Cost-Effective Energy Solution in Renovating a Multi-family House

Elaheh Jalilzadehazhari and Krushna Mahapatra

Abstract The Swedish government aims to reduce total energy demand per heated building area by 50% until 2050. A large number of residential buildings, built within the so-called "Million homes program" in Sweden, need major renovations, which offers an opportunity to implement energy efficiency measures and thereby, reduce total energy demand. The best way to encourage the implementation of a major renovation is to demonstrate a practical method which reduces energy demand and provides economic benefits. Hence, this study aims to determine the most cost-effective energy solution in renovating a multi-family residential building. Multiple energy renovation measures were simulated on a case study to reduce the space heating and domestic hot water by 50%. The case study building was built within the "Million homes program" and is located in Växjö, Swedish climate zone 3. Design Builder software was used for analysing the pre-renovation energy performance of the building. The renovation measures comprised different insulation thicknesses of external walls, attic and ground floors, windows with different U-values, a mechanical ventilation with heat recovery system, and solar system for supporting space heating and domestic hot water. Later, a multi-objective optimization was accomplished for analysing every possible combination of renovation measures. The most cost-effective energy solution was obtained by calculating the net present value in a lifetime of 30 and 50 years and discount rate of 1, 3 and 5%. Comparing the implications of two different lifetimes on net present value with implications of three different discount rates on net present value shows that lifetime has more influence on net present value. Furthermore, the results show the capability of the multi-objective optimization method in analysing multiple renovation solution.

Keywords Net present value · Energy renovation measures · Multi-objective optimization

E. Jalilzadehazhari (✉) · K. Mahapatra
Linnaeus University, Växjö 35195, Sweden
e-mail: elaheh.Jalilzadehazhari@lnu.se

D. Johansson et al. (eds.), *Cold Climate HVAC 2018*,
Springer Proceedings in Energy, https://doi.org/10.1007/978-3-030-00662-4_18

1 Introduction

The latest recast of the EU Energy Performance of Buildings Directive (EPBD) asks all member states to implement changes in building regulations to reduce the total energy demand [1]. Although EPBD's requirements target mostly new buildings, the directive asked all EU members to determine cost-effective energy solutions in renovating existing buildings [1], because the existing older buildings within the EU region have significant energy savings potential [2]. Furthermore, the annex to the EPBD has mentioned that the minimum comfort thresholds, defined at national level, should be ensured in calculating the energy demand [1]. In Sweden, actions have been taken by the government to reduce the total energy demand per heated building area by 20 and 50% until 2020 and 2050 respectively, compared to the energy demand in 1995 [3]. The total energy demand of the existing residential buildings in Sweden was about 147 TWh in 2013, which corresponds to 40% of the total energy demand in this country at the same year [4]. A large share of the existing residential buildings in Sweden were constructed in 1960s and 1970s, known as "Million homes program" [5]. According to Swedish energy agency [6], over half of the total energy consumption in Swedish residential buildings is for space heating and domestic hot water. Major renovation of these buildings can provide a significant opportunity to implement energy efficiency measures and thereby, reduce the energy need for space heating and domestic hot water. At this point, Mahapatra and Olsson [7] and Dodoo et al. [8] discussed that renovating buildings to passive house standard can significantly reduce energy demand for space heating and domestic hot water. Janson [9] listed energy renovation measures, which can be implemented in renovating Swedish residential buildings to reduce energy need for space heating and domestic hot water. The list comprised energy renovation measures such as improved airtightness of buildings' envelopes, energy efficient windows and doors, efficient water taps, improved thermal insulation, installation of energy efficient mechanical ventilation system, energy-efficient home appliances [9], and other renewable-based energy systems such as heat pumps and solar energy systems [10].

But, according to Niemelä et al. [2], the best way to encourage the implementation of a major renovation is to demonstrate a practical method which reduces energy demand and also provides economic benefits. Bonakdar et al. [11], Bonakdar et al. [12] and Gustafsson et al. [13] discussed about energy and cost effectiveness of implementing energy renovation measures in Swedish residential buildings. Although these studies presented useful results, but they have mainly considered a few renovation measures in Sweden, without analysing all possible combinations of the measures [12–14]. For instance, Gustafsson et al. [13] analysed five different renovation measures including insulation of external walls (with U-value of 0.6 and 0.26 W/m^2 K), roof (with U-value of 0.6 and 0.15 W/m^2 K), energy efficient windows (with U-value of 2.58 and 1.2 W/m^2 K), doors (with U-value of 2.72 and 1.38 W/m^2 K), and four different energy-efficient mechanical ventilation systems. However, only three renovation solutions were defined based on *combination* of the renovation measures. The effectiveness of each solution was later analysed with respect to life cycle cost, primary energy consumption, payback time and carbon dioxide emission. Such analyses are limited to a small number of solutions (based on combination of measures), hence limits the feasibility of finding *the most* cost-effective energy renovation solution [2, 14]. Bonakdar et al. [11] studied the contribution of Swedish climate zones to the cost-optimal renovation strategy. Four renovation measures were defined and analysed in a multi-family house building, including insulation of external walls, basement walls, attic floor and the installation of energy-efficient windows. However, they analysed the effectiveness of an individual renovation strategy on cost but not the effectiveness of combination of measures, where there may exist the most cost-effective energy renovation solution.

In line with previous studies, Chantrelle et al. [15] attempt to develop a tool by using an optimization method, which was used to specify a cost-effective energy solution in renovating an educational building in France. The renovation measures included renovation on external walls, roof, ground floor, intermediate floor, internal partition wall and windows. Furthermore, they analysed the performance of the renovation solutions in improving thermal comfort. The presented results indicate that optimization can be used as an aid in specifying a cost-effective energy renovation solution. Because, it can analyse a large number of renovation solutions based on combination of measures, that increases the likelihood of finding the most cost-effective energy renovation solution. Niemelä et al. [2] used a multi-objective optimization method to specify the most cost-effective solution in renovating an educational building to near zero energy building in Finland. The renovation measures comprised the renovation of existed ventilation system, installation of windows with smaller U-value, using additional insulation layer for external walls and roof, installation a ground source heat pump system and PV-panels for electricity production. The results show that multi-objective optimization is a beneficial method in analysing multiple solutions and specifying the most cost-effective energy renovation solution. However, no study, to the best of authors knowledge, have used a multi-objective optimization as a method in renovating a multi-family building in Sweden.

Considering the abovementioned limitations, this study aims to find the most cost-effective energy renovation solution in renovating a multi-family residential building, constructed during the "Million homes program". Multiple energy renovation measures, suggested by Janson [9] and Kjellsson [10], were implemented to reduce the energy need for space heating and domestic hot water by 50%, compared to 1995.[1] Later, a multi-objective optimization was accomplished for analysing every possible combination of renovation measures. The most cost-effective energy solution was obtained by calculating the net present value in a lifetime of 30 and 50 years and discount rate of 1, 3 and 5%.

---

[1]Between 1990 and 2000, the energy demand for space heating and domestic hot water in multi-family house buildings was about 130 (kWh/m^2) [16].

2 Case Study

The multi-family building was located in Växjö municipality in Swedish climate zone 3. The building was constructed between 1966 and 1968. It has three residential-floors above the ground with four separate apartments in each floor. The total heated area of each floor is about 375 m^2, with ventilated volume of 937.5 m^3. The forth floor, which includes some storage unit for the apartment, had a wooden pitched roof and considered as an unheated area.

2.1 Energy Performance Simulations

The energy performance of the building was studied using Design Builder software, version 5.0.3.007 [17]. Design Builder is a graphical user interface for EnergyPlus. EnergyPlus is one of the frequently used tools in analysing the energy performance of a building [18]. The energy performance evaluation was started by constructing a 3D model and providing data regarding material specifications of envelopes including windows, doors, external and internal walls, ground floor and attic floor. Furthermore, building's location, geometry, occupancy schedule, heating and ventilation system were specified for executing simulations. The building was connected to the district heating system to cover the demand for space heating and domestic hot water. Table 1 presents a summary of the simulation layout of the building, applied in Design Builder software.

The internal gain from occupants was assumed to be 60 W/ person, which corresponds to activity level of seated at rest [19]. Furthermore, the heat gain from electrical lighting was assumed to be about 82% of the energy used by lighting system.[2] The thermal resistance of clothing follows ISO7730 Standard and was set to be 0.5 (clo) in summer and 1 (clo) in winter [20]. The building was assumed to be occupied 24 h during weekends and 16 h (between 16:00 pm to 08:00 am) during working days. Furthermore, the density of occupancy was assumed to be 0.03 (person/m^2). The simulated initial energy demand before implementing energy renovation measures for space heating, domestic hot water and electricity for ventilation was about 95 (kWh/m^2), 24.8 (kWh/m^2) and 5 (kWh/m^2) respectively.

2.2 Renovation Measures

In this study, the renovation measures were comprised of improving the U-value of the building's envelopes, upgrading the ventilation system and installing solar

[2]In Design Builder database, it has been stated that visible fraction of a fluorescent lighting system is about 18%, while the radiant and convictive fractions are 42 and 40% respectively.

Table 1 Simulation layout of the studied building in initial status (The presented information in Table 1 was an assumption regarding initial status of the building and was provided by Växjöbostäder)

General properties	Values
U-value of Windows	2.90 (W/m^2 K)
U-value of Doors	3.00 (W/m^2 K)
U-value of the attic floor	0.51 (W/m^2 K)
U-value of the east and west façade	0.367 (W/m^2 K)
U-value of the north and south façade	0.279 (W/m^2 K)
U-value of the ground floor slab	2.70 (W/m^2 K)
Efficiency of fan for ventilation	0.5
Air tightness	0.8 (l/m^2 s) at differential pressure of ± 50 Pa
Electrical lighting	Fluorescent 9.9 (W/m^2) power

heating system for supporting space heating and domestic hot water. Table 2 lists the renovation measures analysed in this study. The U-values of the measures were equal or lower than the requirements of the Swedish building code for new buildings (BBR 2015) [21]. The heating system was kept unchanged, because this type of biomass-based district heating systems has low environmental impact and about 80% of the multi-family buildings in Sweden are connected to such a system [13]. The U-value of external walls, attic floor and ground floor were improved by replacing the old insulation layer with a cellulose insulation layer having a conductivity of 0.04 (W/m^2 K). Because old insulation layers can be degraded after their lifetime, which has negative impact on conductivity of insulation layer [22]. Cellulose was selected due to its small environmental impact compared to Rockwool insulation material [23]. The pitched roof, which had initially no insulation layer, was not considered in defining renovation measures. Because the forth floor (area under the pitched roof) is indeed an unheated area. Furthermore, adding some insulation layer in pitched roof may have negative impact by accumulating moisture in the roof [24].

The existing ventilation system was upgraded by using a heat recovery system with an efficiency of 76% [25]. A solar heating system, comprised of 18 vacuum solar collectors, was installed on the roof with a 50° slope of surface and facing south.

Combination of the above energy renovation measures resulted in 625 renovation solutions. There is a need to clarify that specifying the most cost-effective energy renovation solution may not necessarily include all abovementioned renovation measures. However, evaluating different renovation measures helps to monitor the degree of their effect in reducing energy consumption and increases the likelihood of finding the most cost-effective energy renovation solution.

Since simulating and analysing so many solutions is infeasible, multi-objective optimization was performed to evaluate the cost and energy performance of the 625 renovation solutions.

Table 2 Renovation measures

Building envelop component	Simulated alternatives	BBR 2015 requirement U-value (W/m^2 K)	U-value (W/ m^2 K)	Efficiency
Windows	–	1.3	–	
	Type 1		0.8	
	Type 2		0.9	
	Type 3		1	
	Type 4		1.1	
	Type 5		1.2	
External walls	–	0.18	–	
	Type 1		0.09	
	Type 2		0.1	
	Type 3		0.12	
	Type 4		0.14	
	Type 5		0.18	
Attic floor	–	(No requirements have been found in BBR 2015 considering the maximum U-value for attic floor)	–	
	Type 1		0.08	
	Type 2		0.09	
	Type 3		0.1	
	Type 4		0.12	
	Type 5		0.13	
Ground floor	–	0.15		
	Type 1		0.08	
	Type 2		0.09	
	Type 3		0.1	
	Type 4		0.12	
	Type 5		0.15	
Heat recovery	–		–	76%
Solar collectors	–		–	1119 (kWh/ year)

2.3 *Multi-objective Optimization*

A multi-objective optimization involves in the minimization of optimization functions by interacting optimization variables [26]. In this study, the multi-objective optimization was accomplished by using the non-dominated sorting genetic algorithm-II (NSGA-II) in Design Builder software. The first step in performing the multi-objective optimization was to specify optimization variables. All considered renovation measures in Table 2 were defined as optimization variables. Later, the thermal discomfort hours (all clo) $\leq$ 200 [h] was defined as constraints function [27]. This means that optimization process searched for renovation solutions in which the total thermal discomfort hours are less than 200 h. This decision was made to fulfil the EPBD's requirements in terms of ensuring the minimum thermal comfort threshold in renovating the building. Furthermore, the execution time required for performing an optimization was recorded. The processor of the computer, used in running simulations and optimization was an AMD FX-7600P Radeon (4 cores, 2700 MHz) with 8 GB installed physical memory (RAM).

The optimization process provided information regarding energy need for space heating, domestic hot water and electricity for running the ventilation system of all 625 renovation solutions. The total energy demand of the renovation solutions was later calculated as the sum of energy need for space heating, domestic hot water and electricity for running ventilation system.

Furthermore, the optimization process uses provided information in Design Builder regarding investment cost of each renovation measure and calculates the total investment cost for all 625 renovation solutions. The investment cost in this study, was based on material cost for each renovation measure. The investment cost for five types of windows was defined separately as price per unit window (SEK/m^2) [28]. Design builder calculates automatically the total investment cost for five types of windows based on the unit price of the windows multiplied by the total windows area. The investment cost of insulation layer was defined as price per volume cellulose (SEK/m^3) [29]. Design Builder quantifies automatically the required volume of cellulose for five types of external walls, attic and ground floors and calculates their respective investment cost. All results obtained from optimization processes were saved in excel files. A constant expense for installing heat recovery for ventilation [30–32] and solar heating system [33] was added to the investment cost of each of the 625 renovation solutions.

To find the most cost-effective energy renovation solution, the net present value of the 625 renovation solutions was calculated. Net present value shows the difference between total energy saving cost during a building's lifetime and investment and maintenance cost [34]. A higher net present value corresponds to the most profitable and cost-effective energy renovation solution. The net present value of the solutions with 50% reduced energy demand were calculated using Eq. 1 in Microsoft Excel.

$$NPV = \sum_{t=0}^{n} D'_t * \frac{1}{(1+r)^t} - (I_0 + U) \quad (1)$$

$$D'_t = (E_0 - E_t) * \alpha(1+\beta)^t \quad (2)$$

Table 3 Investment, maintenance and life time of the renovation measures

Renovation measures	Investment cost	Maintenance cost	Lifetime
Windows [28]			
Type 1	4770 (SEK/m^2)	–	30
Type 2	4430 (SEK/m^2)	–	30
Type 3	4000 (SEK/m^2)	–	30
Type 4	3310 (SEK/m^2)	–	30
Type 5	2870 (SEK/m^2)	–	30
Insulation layer [38]	500 (SEK/m^3)	–	30
Heat recovery system [30–32]	140 (SEK/m^2)	15 (SEK/ m^2)	15
Solar heating system [33, 35]	6950 (SEK/m^2)	200 (SEK/ year)	50

where

NPV is the net present value during lifespan of n year;

D'_t is energy cost saved annually;
E_0 is the initial annual energy demand before renovations
E_t is the annual energy demand after implementing renovations
α is energy price per kwh/m^2 in 2016
β is inflation in energy price (%)
t: lifespan of n years
r: discount rate
I_0 is the investment cost
U is the lifetime maintenance cost

In calculating the net present value, a discount rate of 1, 3 and 5% with a lifetime of 30 and 50 years were considered [11, 13]. This decision was made to analyse the effect of the discount rate and lifetime of profitability of the 625 renovation solutions. The energy price for heating and electricity was 0.74 (SEK/ kWh) and 1.38 (SEK/ kWh) in 2016 respectively, including 1% of inflation rate [13]. The investment cost for heat recovery system, solar heating, windows and insulation layer were taken from [30–32],[3] [28, 33, 35, 36], respectively. At the end of the lifetime, the renovation measures were refurbished. Accordingly, the investment cost of refurbishment was also included in calculating the net present value. The maintenance cost for the ventilation system was assumed to be 1% of the investment cost [13]. The maintenance cost for the solar heating system was assumed to be 200 (SEK/year) [37]. Table 3 shows the applied investment, maintenance and life time of the renovation measures.

[3]The investment cost for heat recovery system can vary greatly between different references. Therefore, the considered value in this study was taken to fit within the range.

3 Results

The total required time for performing the optimization was about 60 h, which can be considered as a long execution time. Thus, performing an optimization may require a computer with more powerful processor. Furthermore, the initial results showed that energy need for space heating and domestic hot water of all 625 renovation solutions were smaller than 75 (kWh/m^2) (50% of heating demand in 1995). The total thermal discomfort hours among all renovation solutions were less than 200 h. Total energy consumption of the renovation solutions was varied between 24.1 and 35.6 (kWh/m^2), which corresponds to 80.7 and 71.5% reduction in total energy consumption respectively. The minimum energy demand of 24.1 kWh/m^2 represents a renovation solution with a mechanical ventilation heat recovery system, solar heating system, window with U-value of 0.8 (W/m^2 K), external wall with U-value of 0.09 (W/m^2 K), attic floor with U-value of 0.08 (W/m^2 K) and floor with U-value of 0.08 (W/m^2 K). The investment cost of this renovation solution was about 1860000 SEK, which corresponds to 1650 SEK/m^2. The maximum energy demand of 35.6 kWh/m^2 corresponds to the previously mentioned mechanical ventilation heat recovery system and solar heating system, but with installation of windows with U-value of 1.2 (W/m^2 K), external wall with U-value of 0.18 (W/m^2 K), attic floor with U-value of 0.13(W/m^2 K) and floor with U-value of 0.15 (W/m^2 K). The investment cost of this renovation solution was about 1520000 SEK, which corresponds to 1350 SEK/m^2.

Figure 1 shows the net present value of 625 solutions with a lifetime of 30 years and three different discount rates. Analysing the results show that, by increasing the discount rate from 1% to 3% and 5%, the net present value of the 625 renovation solutions decreased significantly. When the discount rate is 1%, the highest net present value was about 453 (SEK/m^2 of floor area), which represents the most cost-effective energy renovation solution. This renovation solution corresponds to the combination of the mechanical ventilation heat recovery system, solar heating system, windows with U-value of 1.2 (W/m^2 K), external wall with U-value of 0.1 (W/m^2 K), attic floor with U-value of 0.12 (W/m^2 K) and floor with U-value of 0.08 (W/m^2 K). The total energy demand of this solution was 31.3 (kWh/m^2). However, by increasing the discount rate from 1 to 3 and 5%, the highest net present value fell about 507.8 and 847 (SEK/m^2 of floor area) respectively. With a discount rate of 3 and 5%, the net present value of all renovation solution was negative. Accordingly, none of the renovation solutions can be considered as cost-effective solutions.

Figure 2 shows the net present value of 625 renovation solution with a lifetime of 50 years and three different discount rates. As was found earlier, increasing the discount rate from 1 to 3 and 5% decreases the net present values significantly. With a discount rate of 1%, all 625 renovation solutions had a positive net present value. The highest net present value, representing the most cost-effective energy renovation solution, was about 1869.5 (SEK/m^2 of floor area), which corresponds to the renovation solution with the mechanical ventilation heat recovery system, solar

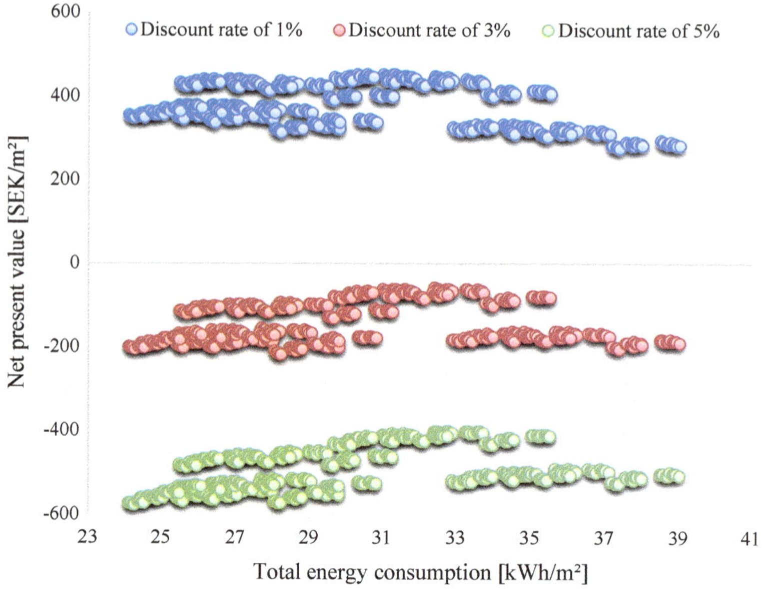

Fig. 1 Net present value of renovation solutions in a lifetime of 30 years together with their respective total energy demand

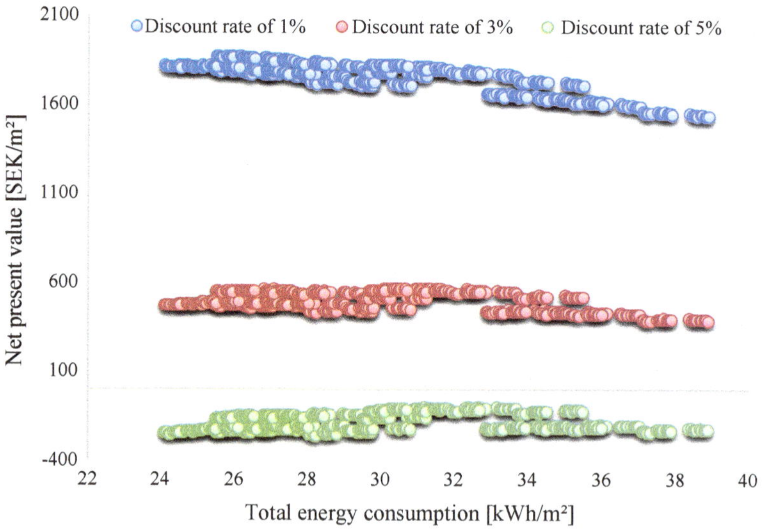

Fig. 2 Net present value of renovation solutions in a lifetime of 50 years together with their respective total energy demand

heating system, windows with U-value of 1 (W/m^2 K), external wall with U-value of 0.09 (W/m^2 K), attic floor with U-value of 0.08 (W/m^2 K) and floor with U-value of 0.08 (W/m^2 K). The total energy demand of this solution was 25.5 (kWh/m^2). With a discount rate of 3%, highest net present value was changed to 558.8 (SEK/m^2 of floor area), which corresponds to 1310.7 (SEK/ m^2 of floor area) fall in highest net present value. Furthermore, the most cost-effective energy renovation solution was changed to the renovation solution with the mechanical ventilation heat recovery system, solar heating system, windows with U-value of 1.2 (W/m^2 K), external wall with U-value of 0.1 (W/m^2 K), attic floor with U-value of 0.12 (W/m^2 K) and floor with U-value of 0.08 (W/m^2 K). The total energy demand of this solution was 31.3 (kWh/m^2). By increasing the discount rate from 1 to 5%, the net present value of all renovation solutions was negative. Accordingly, none of the renovation solutions can be considered as cost-effective solution.

4 Conclusion

The aim of this study was to find the most cost-effective energy renovation solution in renovating a residential building in Swedish climate zone 3, Växjö. For this purpose, the initial energy performance of the building was analysed using Design Builder software. Multiple energy renovation measures were simulated to reduce the energy need for space heating and domestic hot water by 50%, compared to 1995. The renovation measures were comprised of improving the U-value of the building's envelopes, upgrading the ventilation system and installing solar heating system for supporting space heating and domestic hot water. Later, a multi-objective optimization was accomplished for analysing every possible combination between renovation measures. The most cost-effective energy solution was obtained by calculating the net present value in a lifetime of 30 and 50 years and discount rate of 1, 3 and 5%.

As point of energy consumption, analysing the results show that a reduction around 80.7% in total energy consumption can be achieved by implementing energy renovation measures. Considering cost evaluations, net present value of the renovation solutions was sensitive both for changes in discount rate and lifetime period. However, comparing the implications of two different lifetimes on net present value with implications of three different discount rates on net present value shows that lifetime has more contribution on net present value. For instance, when discount rate was 1%, the highest net present value with a lifetime of 50 years was about four times greater than the highest net present value with a lifetime of 30 years.

When discount rate was changed from 1 to 3%, the fall in the highest net present value with a lifetime of 50 years was 2.6% greater than the fall in the highest net present value with a lifetime of 30 years. However, with a discount rate of 3%, the net present value of all renovation solutions, with a lifetime of 50 years, was still positive, while the net present value of all renovation solutions, with a lifetime of

30 years, was negative. Accordingly, implementing renovation solutions with a discount rate of 3% can only be cost-effective when the lifetime is 50 years. When discount rate was changed from 1% to 5%, the fall in the highest net present value with a lifetime of 50 years was 2.3% greater than the fall in the highest net present value with a lifetime of 30 years. With a discount rate of 5%, none of renovation solutions could be considered as cost-effective solution either with a lifetime of 30 or 50 years.

Furthermore, the results show the capability of the multi-objective optimization method (in this study NSGA-II algorithm) in analysing multiple renovation solution. However, one of the main concerns in performing an optimization using NSGA-II is the long execution time. Accordingly, performing an optimization may require a computer with more powerful processor.

The presented results in this study can be applied for buildings, which have similar characteristics and located in Swedish climate zone 3. Furthermore, the presented method in this study can be used for analysing the cost-effectiveness of other energy renovation measures and their combinations in Swedish residential buildings. This process may help to specify suitable renovation solutions, which fit for other climate zones. The future work of this study will include the analysis of required monetary instruments for implementing energy renovation measures, such as provided subsides by banks and studying how it affects the rental rates.

References

1. E. Recast, Directive 2010/31/EU of the European parliament and of the council of 19 May 2010 on the energy performance of buildings (recast). Official J. Eur. Union **18**(06), 2010 (2010)
2. T. Niemelä, R. Kosonen, J. Jokisalo, Cost-optimal energy performance renovation measures of educational buildings in cold climate. Appl. Energy **183**, 1005–1020 (2016)
3. National audit Office, *EU and Sweden's climate and energy Climate targets*, (In Swedish: EU: s och Sveriges klimat-och energimål) RiR 2013:19 Sweden 9 (2013)
4. Swedish Environmental Protection Agency, *The environment control* (Title is Swedish: Styr med sikte på miljömålen) (2015), pp. 1–140
5. L. Itard, F. Meijer, E. Vrins, H. Hoiting, Building renovation and modernisation in Europe: state of the art review. Final Report ERABUILD, Delft OTB Research Institute for Housing, Urban and Mobility Studies, Delft University of Technology 1–232 (2008)
6. Swedish energy agency, *Energy Situation* (Title in Swedish: Energiläget) (Bromma, Sweden) (2017), pp. 1–86
7. K. Mahapatra, S. Olsson, Energy performance of two multi-story wood-frame passive houses in Sweden. Buildings **5**(4), 1207–1220 (2015)
8. A. Dodoo, L. Gustavsson, R. Sathre, Life cycle primary energy implication of retrofitting a wood-framed apartment building to passive house standard. Resour. Conserv. Recycl. **54**(12), 1152–1160 (2010)
9. U. Janson, *Passive houses in Sweden-experiences from design and construction, energy and building design* (Lund University, Lund, 2008)
10. E. Kjellsson, *Solar collectors combined with ground-source heat pumps in dwellings-analyses of system performance* (Byggnadsfysik LTH, Lunds Tekniska Högskola, 2009)

11. F. Bonakdar, A. Sasic Kalagasidis, K. Mahapatra, The implications of climate zones on the cost-optimal level and cost-effectiveness of building envelope energy renovation and space heat demand reduction. Buildings **7**(2), 39 (2017)
12. F. Bonakdar, L. Gustavsson, A. Dodoo, *Implications of Energy Efficiency Renovation Measures for a Swedish Residential Building on Cost, Primary Energy Use and Carbon Dioxide Emission*, (ECEEE 2013, Belambra Les Criques, France, June 3–8, 2013), pp. 1287–1296
13. M. Gustafsson, M.S. Gustafsson, J.A. Myhren, C. Bales, S. Holmberg, Techno-economic analysis of energy renovation measures for a district heated multi-family house. Appl. Energy **177**, 108–116 (2016)
14. M. Gustafsson, C. Dipasquale, S. Poppi, A. Bellini, R. Fedrizzi, C. Bales, F. Ochs, M. Sié, S. Holmberg, Economic and environmental analysis of energy renovation packages for European office buildings. Energy Build. **148**, 155–165 (2017)
15. F.P. Chantrelle, H. Lahmidi, W. Keilholz, M.E. Mankibi, P. Michel, Development of a multicriteria tool for optimizing the renovation of buildings. Appl. Energy **88**(4), 1386–1394 (2011)
16. U. Janson, *Passive Houses in Sweden. From Design to Evaluation of Four Demonstration Projects* (Lund, Sweden) (2010), p. 390
17. Design Builder, 2017. www.designbuilder.co.uk
18. A.-T. Nguyen, S. Reiter, P. Rigo, A review on simulation-based optimization methods applied to building performance analysis. Appl. Energy **113**, 1043–1058 (2014)
19. A. Standard, *Standard 55-2010: Thermal Environmental Conditions for Human Occupancy* (ASHRAE, Atlanta, USA, 2010)
20. ISO7730-Standard, 7730, *Ergonomics of the Thermal Environment—Analytical Determination and Interpretation of Thermal Comfort using Calculation of the PMV and PPD Indices and Local Thermal Comfort Criteria*, International Organization for Standardization: Geneva, Switzerland (2005)
21. National Board of Housing Building and Planning, *Building Regulation* (In Swedish: Boverkets byggregler, BFS 2011:6 ändrad t.o.m. BFS 2015:3Avsnitt 9 Energihushållning), National board of housing, building and planning [In Swedish: Boverket] (2015)
22. F. Tittarelli, F. Stazi, G. Politi, C. di Perna, P. Munafò, *Degradation of Glass Mineral Wool Insulation after 25 Years in Masonry Cavity Walls* (2013)
23. U.Y.A. Tettey, A. Dodoo, L. Gustavsson, Effects of different insulation materials on primary energy and CO_2 emission of a multi-storey residential building. Energy Build. **82**, 369–377 (2014)
24. A. Kaufmann, H.M. Künzel, J. Radoń, Preventing moisture problems in retrofitted pitched roofs. ACTA Scientiarum Polonorum, Architectura **5**(1), 69–79 (2006)
25. Swedish Energy Agency, *FTX System for a House with 130 m^2 of Living Space* (In Swedish: FTX-aggregat hus med 130 m^2 boyta) (2017). http://www.energimyndigheten.se/tester/tester-a-o/ftx-aggregat/ftx-aggregat-hus-med-130-m-boyta/. Last accessed 07 July 2017
26. S. Koziel, X.-S. Yang, *Computational Optimization, Methods and Algorithms* (Springer, Berlin, 2011)
27. P. Hoes, J. Hensen, M. Loomans, B. De Vries, D. Bourgeois, User behavior in whole building simulation. Energy Build. **41**(3), 295–302 (2009)
28. Elitfonster, 2016. www.elitfonster.se
29. L. Gustavsson, A. Joelsson, Energy conservation and conversion of electrical heating systems in detached houses. Energy Build. **39**(6), 717–726 (2007)
30. S. Ruud, *Economic Heating Systems for Low Energy Buildings–Calculation, Comparison and Evaluation of Different System Solutions* (Borås, Sweden, SP Technical Research Institute of Sweden, 2010)
31. Å. Wahlström, Å. Blomsterberg, D. Olsson, *Heat Recovery System for Existing Multi-Family Houses* (In Swedish: Värmeåtervinningssystem för befintliga flerbostadshus). Pre study for technology procurement [In Swedish: Förstudie inför teknikupphandling] (2009)
32. NIBE, 2017. www.nibe.se. Last accessed 27 June 2017

33. Solra, 2017. www.solra.se
34. K. Blok, *The Effectiveness of Policy Instruments for Energy-Efficiency Improvement in Firms: The Dutch Experience* (Springer, Berlin, 2004)
35. Byggshop, 2017. www.byggshop.se
36. M. Wetter, J. Wright, A comparison of deterministic and probabilistic optimization algorithms for non smooth simulation-based optimization. Build. Environ. **39**(8), 989–999 (2004)
37. L.R. Bernardo, H. Davidsson, E. Andersson, Retrofitted solar domestic hot water systems for swedish single-family houses—evaluation of a prototype and life-cycle cost analysis. Energies **9**(11), 953 (2016)
38. isolerproffs, 2017. http://www.isolerproffs.se. Last accessed 27 June 2017

Operational Characterisation of Neighbourhood Heat Energy After Large-Scale Building Retrofit

Paul Beagon, Fiona Boland and James O'Donnell

Abstract To achieve housing retrofit targets, traditional house-by-house approaches must scale. Neighbourhood retrofit also facilitates community participation. This paper aims to quantitatively characterise the heat energy demand of similar homes in a post-retrofit neighbourhood. The method employs the Modelica AixLib library, dedicated to building performance simulation. A modern semi-detached house is modelled as thermal network. The passive thermal network is calibrated against an equivalent EnergyPlus model. The developed Modelica model then generates time series heat energy demand to meet occupant comfort. This model separates heating for internal space and domestic hot water. Simulation results are gathered for a range of house occupancy profiles, with varying heating schedules and occupant quantities. The calibration results compare the time series of internal house temperature produced by the EnergyPlus and Modelica simulations. Modelica simulations of two heating schedules generate distinct annual demand curves against occupant quantity. As expected in a modern house, domestic hot water accounts for a relatively high proportion of heat energy. Over a year it ranges between 20 and 45% depending on occupant profile. Overall conclusions are threefold. Firstly, occupant profiles of a modern semi-detached house increase annual heat energy demand by 77%, and the coincidence of daily peak demand persists across occupant profiles. Furthermore, percentages of domestic hot water demand start from 20 or 24% and plateau at 39 or 45% depending on space heating schedule. A statistical distribution of energy demand by neighbourhood homes is possible. Its curve plot is not perfectly normal, skewing to larger energy demands.

P. Beagon (✉) · J. O'Donnell
School of Mechanical and Materials Engineering, University College Dublin, Belfield, 4, Dublin, Ireland
e-mail: paul.beagon@ucdconnect.ie

P. Beagon · J. O'Donnell
Energy Institute, University College Dublin, Belfield, 4, Dublin, Ireland

F. Boland
Royal College of Surgeons in Ireland, Lower Mercer Street, 2, Dublin, Ireland

D. Johansson et al. (eds.), *Cold Climate HVAC 2018*,
Springer Proceedings in Energy, https://doi.org/10.1007/978-3-030-00662-4_19

Keywords Building retrofit · Building simulation · Modelica · AixLib library Neighbourhood scale · Statistical distribution

1 Introduction

National and international policies aim to reduce greenhouse gas emissions and reduce reliance on fossil fuels. An enduring energy demand and cause of greenhouse gas (GHG) is domestic heating, accounting for 57% of British heat use [1]. Domestic appliances and lighting form the remaining energy, but are excluded from this research.

Large-scale retrofit is a potential solution, albeit complex and challenging to deliver [2, 3]. In recent energy analysis of a semi-detached house in an oceanic climate, domestic super insulation outperforms expansion of local renewable energy [4]. Policy documents concur that reducing energy demand of existing buildings by retrofit, precedes any integration of energy storage or local energy generation [5].

After an energy retrofit, the energy demand of the similar homes will reduce both on average and in variation. Improved statistical parameters facilitate the aggregation of homes as distributed energy resources. Such aggregation of a retrofitted neighbourhood into a distributed energy system is proposed by Koch [6].

The modelling and simulation of building archetypes is one method to estimate the statistical parameters of a retrofitted neighbourhood or national building stock [7]. Building archetypes already exist for European countries thanks to a large research effort [8]. In parallel, Annex 60 of IEA-EBC (International Energy Agency Energy in Buildings and Communities Programme) delivered Modelica for building performance simulation [9]. One Modelica library is AixLib, that is capable of modelling heat demand of a city district [10].

The overall aim of this paper is to quantitatively characterise the heat energy demand of structurally similar housing in a post-retrofit neighbourhood. Responding to identified research gaps [6], statistical distributions of heat demand per home are calculated. The main objectives of this study are to employ time domain simulation that responds to standard weather files to (i) quantify annual heat demand per standard house, split by space and water heating, (ii) compute the annual heat demand per house as this varies with occupant schedule and occupant quantity, and (iii) illustrate the coincidence shared by heat energy time series caused by different occupancy profiles. It is this coincidence, or simultaneity, that drives energy demand peaks and sizing of energy supply infrastructure.

This research formulates and tests a modern semi-detached home as a thermal network or "RC" model. The RC name derives from the resistor and capacitance elements in the electrical analogy of a thermal model. RC models are proposed due to their aggregation potential into neighbourhood or city district scale [11]. The heat energy demand is split into space heating and domestic hot water (DHW) heating.

First, the model is tested with a weather profile for correlation with the popular building simulator EnergyPlus [12, 13]. Second, the space heating and domestic water heating (DHW) demand are computed for different occupant profiles. The argument that DHW demand will persist or even grow are strong [14]. As a result, the proportion of DHW of total heat demand is quantified. Third, a distribution of housing gas demand of an example neighbourhood is presented. Fourth, the coincidence of peak heat energy demand by different occupant profiles is illustrated. This last point affects utilities and energy network operators. The retrofit of an entire neighbourhood affects both home occupants and energy suppliers, especially where heating fuel switches from gas to electricity.

2 Methodology

The process to produce data and statistical distributions of household heat energy requires multiple steps (Fig. 1). Two initial steps develop and calibrate a thermal network model. Subsequently the process amends model parameters, executes simulations and analyses energy data. Electrical energy demand by plug loads and appliances is excluded.

The calibration step references a building model configured in the well-known EnergyPlus software. The model is available from other research [15]. EnergyPlus was released in 2001 after five years of US federal funded development. It is accepted as a very popular and commonly used building energy simulator [16]. The reference EnergyPlus model resembles the archetype of a modern or retrofitted semi-detached house. Modelica can model and simulate the same house as a RC

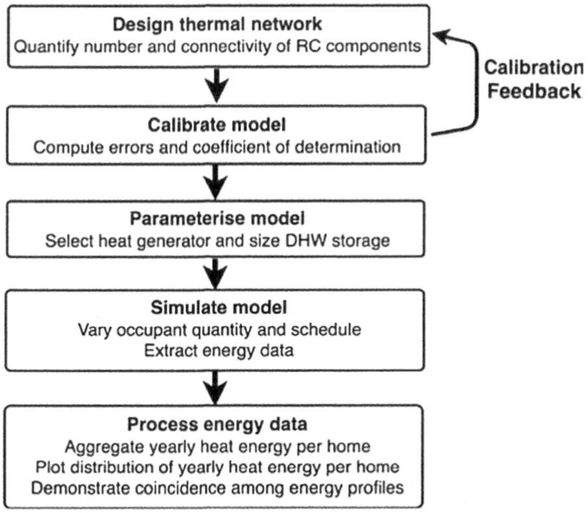

Fig. 1 Method process, develop and calibrate a thermal network model, parameterise its heating components, simulate occupancy profiles and process energy results

network. Iterative comparisons of simulation results guide the RC network design, shown as a feedback loop between the initial two steps (Fig. 1).

Subsequently, model parameters of heat generator type and storage tank size are configured. Simulations are now possible over different occupancy profiles which combine occupant quantity and space heating schedule.

2.1 RC Model of the Fabric

As mentioned in the Introduction, the model and simulation rely on components and examples from AixLib Modelica library [10]. Buildings as thermal networks are configurable in AixLib. The RC model is developed from existing low order thermal network model for dynamic simulation [11]. That model separates equivalent air temperatures of wall and windows for each building orientation. Thereby, the effect of solar radiation is more accurately calculated. Radiated heat transfers across thermal network resistors, for example from windows to internal walls, impeded by resistor $R_{WinIntRad}$. Such resistor values are based on a heat transfer coefficient (α_{rad}); assumed constant in simulated temperature range at 5 W/(m^{2}K). A convective heat transfer coefficient (α_{conv}) of 2.7 W/(m^{2}K) is assumed. The α_{conv} coefficient determines resistances, for example, to convective heat transfer at windows R_{WinCon}. Resistances and capacitances are specific to the building construction; external wall (ext), internal walls and objects (int), and floor plate (floor). U-values and geometry of external, internal and floor structures, determine the building specific RC values. Retrofit of building improves (decreases) the U values with associated increase in resistances.

In order to reduce the over-responsiveness of simulated internal temperature, lumped capacitances are split [17]. External wall capacitances are split into C_{ext1} or C_{ext2}, and floor capacitances C_{floor1} and C_{floor2}; interconnected in Fig. 2. Internal capacitance C_{int1} and C_{int2} account for internal furnishings and the relatively small internal wall capacitance.

2.2 Calibration of the Building Fabric

Initially the modern semi-detached house is modelled in EnergyPlus which simulates building state variables such as internal temperature. EnergyPlus allows removal of all heating equipment and internal gains; reducing the model to a building fabric. Similarly, the Modelica RC is disconnected from heating and internal loads are zeroed.

Identical external loads of solar radiation and outside dry bulb temperature must be applied to each model of the building fabric. IWEC (International Weather for Energy Calculations) provides standardised weather data [18].

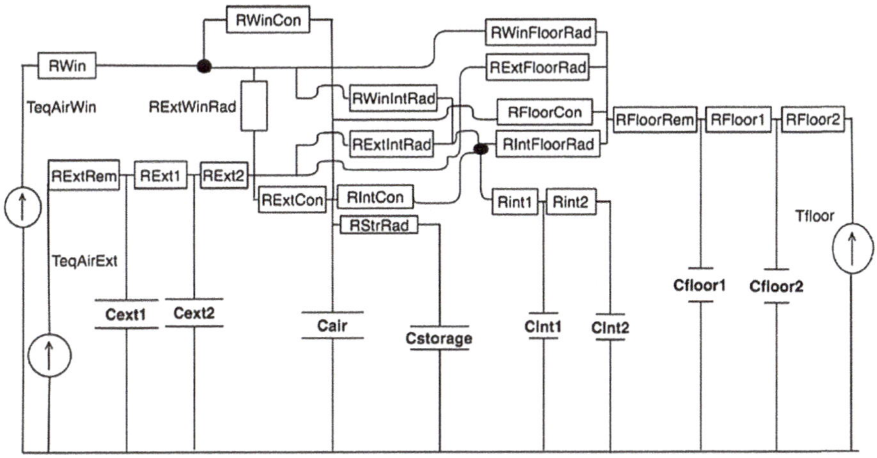

Fig. 2 Thermal network of semi-detached building extended from [11]. The external wall, building internals and floor are configured as 3R-2C sub-models

Fig. 3 Internal temperature of empty home fabric simulated by EnergyPlus and Modelica RC models

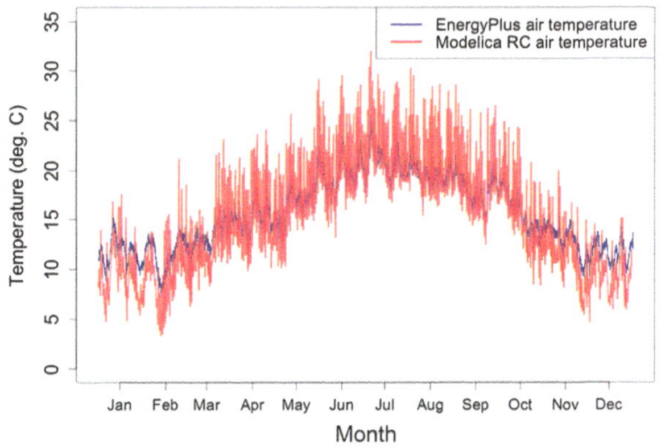

An iterative process leads to the thermal network in Fig. 2. Every iteration compares the two time series of simulated internal temperatures. Standard error metrics of mean absolute error (MAE) and root mean square error (RMSE) are calculated, along with the coefficient of determination (R^2). Time series data points are available for almost every hour of the 12 months (Fig. 3). The sample size is $n = 8736$. As a trade-off between model simplicity and accuracy, the selected RC model computes inside temperature comparisons of MAE = 1.39, RMSE = 1.92 and $R^2 = 0.9215$.

2.3 Simulation of Different Schedules and Occupancies

Realistic domestic energy evaluation comprises two demands; space heating and domestic hot water (DHW). Any future retrofit of a home will increase the proportion of energy demand by DHW. Therefore, the first simulated heat generator is a water based heat exchanger to meet both demands. This component can be considered as a 20 kW boiler of constant 85% efficiency. By its location outside the house thermal envelope, for example a garage, the remaining 15% of boiler energy is lost to outside air.

A water heater is connected to both radiators and a hot water storage tank. The storage tank is sized at 80 litres, with each occupant consuming 40 litres per day. Storage size influences the total heat transferred to a potentially empty home. Energy losses from unneeded air heating highlight the importance of accurate controls. Two heating schedules are applied: economical and comfort. The combination of heating schedule and occupant quantity is termed in the methodology as occupant profile.

The economical schedule rations space heating during working hours when the internal temperature is allowed to drop from 20 °C (Table 1). The minimum of 10 °C internal temperature rarely occurs due to building thermal mass and DHW storage. The comfort schedule maintains comfortable temperatures of 20 °C during all waking hours. Both schedules minimise the heating of DHW during the morning while meeting occupant demand.

3 Results

The results are converted to primary energy. The popular metric of energy use intensity (EUI) expresses demand in kWh/m^2/year. The division by floor area (m^2) normalises the performance metric across different sized buildings. Many energy performance certificates (EPC) adopt the same metric. One EPC, the building energy rating (BER) for Irish homes, allocates modelled EUI into discrete bands.

Table 1 Air temperature setpoints for both economical and comfort schedules

Time	Economical setpoint (°C)	Time	Comfort setpoint (°C)
00:00–07:00	10	00:00–08:00	10
07:00–08:00	20	08:00–23:00	20
08:00–17:30	10	23:00–24:00	10
17:30–19:00	20	–	–
19:00–21:00	10	–	–
21:00–23:00	20	–	–
23:00–24:00	10	–	–

$$Total\ heating\ energy = Total\ space\ heating\ energy + Total\ DHW\ heating\ energy \quad (1)$$

$$Total\ gas\ secondary\ EUI = (Total\ heating\ energy/0.85)/floorspace \quad (2)$$

$$Gas\ primary\ EUI = Gas\ secondary\ EUI * 1.1 \quad (3)$$

The pre-requisite computation of gas secondary EUI assumes a boiler efficiency of 85%. An 85% boiler efficiency is the EnergyPlus default and approximately average in [19]. Assuming gas heating fuel, multiplication by 1.1 converts secondary gas EUI to primary (or "source") EUI [20]. In future studies, electrical heating would incorporate a larger primary energy factor; calculated for Ireland in 2017 as 2.08 [21].

3.1 Gas Consumption for Heating Under Economical and Comfort Schedules

The total gas energy results, varied by occupant quantity, are presented for economical (Table 2) and comfort schedules (Table 3). Both results exclude space heating during three summer months; June, July and August. A model of the same house in EnergyPlus corroborates the results; simulating approximately 8000 kWh total energy per year [22]. Scaling down from total energy to heating energy, it lies between the gas used by three or four occupants in Table 2.

The aforementioned BER initially increments bands by 25–225 kWh/m^2/year. Subsequent BER bands enlarge to cover the breadth of low performance buildings. Table 2 displays a range of gas primary EUI from 73.2 to 105.8 kWh/m^2/year. Range magnitude is 32.6 kWh/m^2/year. Table 3 displays a range of gas primary

Table 2 Gas energy metrics of simulated economical schedule, without summer space heating

Number of occupants	Total gas energy consumed (kWh)	DHW proportion of total gas energy	Gas secondary EUI (kWh/m^2/year)	Gas primary EUI (kWh/m^2/year)
1	5922	0.24	66.5	73.2
2	6609	0.33	74.3	81.7
3	7219	0.38	81.1	89.2
4	7756	0.40	87.1	95.8
5	8012	0.41	90.0	99.0
6	8194	0.42	92.1	101.3
7	8429	0.44	94.7	104.2
8	8566	0.45	96.2	105.8

Table 3 Gas energy metrics of simulated comfort schedule, without summer space heating

Number of occupants	Total gas energy consumed (kWh)	DHW proportion of total gas energy	Gas secondary EUI (kWh/m^2/year)	Gas primary EUI (kWh/m^2/year)
1	7688	0.20	86.4	95.0
2	8319	0.27	93.5	102.9
3	8856	0.32	99.5	109.5
4	9451	0.34	106.2	116.8
5	9876	0.35	111.0	122.1
6	10,079	0.37	113.2	124.5
7	10,299	0.38	115.7	127.3
8	10,429	0.39	117.2	128.9

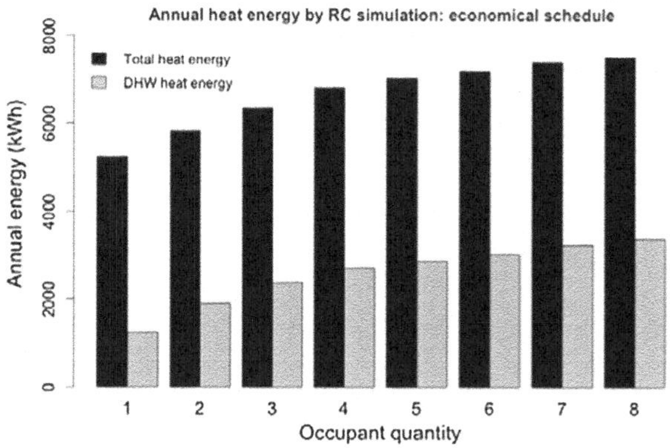

Fig. 4 Total and DHW secondary heat energy against occupant quantity under economical schedule, without summer space heating

EUI from 95.0 to 128.9 kWh/m^2/year. Range magnitude is 33.9 kWh/m^2/year, slightly more than under the economical schedule. Under both heating schedules, increasing occupancy may re-allocate similar homes across at least two BER bands. In practice, BER bands are selected after additional accounting for pump and lighting energy.

The occupants consume the aforementioned gas energy to provide the heating services; heating space and DHW. Heating energy is calculated before application of heat generator efficiency and independent of fuel. The increases in total heating energy and its proportion of DHW energy are plotted in Figs. 4 and 5. The figures exclude summer space heating during the summer months of June, July and August. The figures illustrate plateauing of both heat energy measurements.

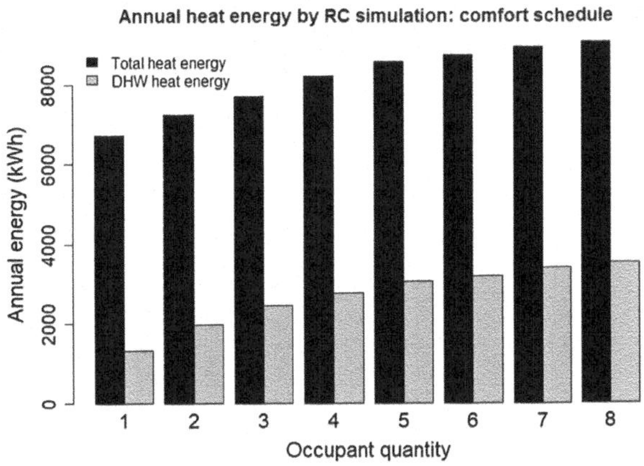

Fig. 5 Total and DHW secondary heat energy against occupant quantity under comfort schedule, without summer space heating

3.2 *Combination of Gas Use and Occupant Quantity Distributions*

A social housing retrofit project took place in Ireland during 2014 with available dataset size n = 188 [23]. Household interviews report occupant quantities ranging from one to eight persons. The median and mean averages equal 3.00 and 2.89 respectively.

The distribution of home gas demand by similar buildings in a neighbourhood can be considered as a combination of two distributions. The first is building demand distributed by occupant quantity; second a probability distribution of occupant quantity across the homes of a neighbourhood. The homes are of similar construction.

Two distributions of neighbourhood gas demand appear in Fig. 6. They distinguish the home gas demands by heating schedules (Table 1). As expected in the literature of neighbourhood scale retrofit [6], the better performing schedule displays lower variance. The mean (μ) and standard deviation (σ) of annual gas consumption (kWh) by each schedule are: economical $\mu = 7032$ and $\sigma = 737$, comfort $\mu = 8755$ and $\sigma = 754$.

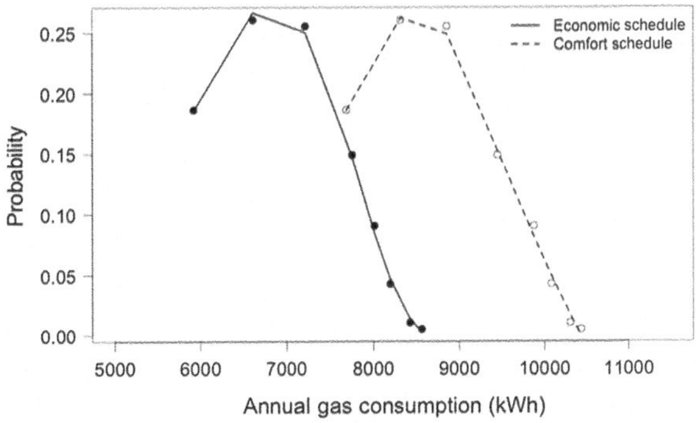

Fig. 6 Distributions of simulated house gas consumption in a neighbourhood, economical and comfort occupancy schedules

Table 4 Peak gas daily demand, simulated for occupant quantities under both economical and comfort schedules

Occupant quantity	Economical schedule peak day number	Economical schedule peak daily gas (kWh)	Comfort schedule peak day	Comfort schedule peak daily gas (kWh)
1	42	33.12	42	50.71
2	42	37.12	42	53.35
3	348	35.48	39	54.32
4	4	38.90	42	58.32
5	4	37.82	42	57.42
6	42	38.42	4	55.05
7	42	38.97	42	57.41
8	42	39.22	42	58.38

3.3 *Peak Gas Demand on the Utility Network*

Energy demands made by homes impact the energy distribution systems [1, 24]. This paper examines gas supply, but electrification of heat would affect the electricity grid. Peak gas energy demand and associated day of occurrence appear in Table 4. Strong coincidence of peak demand is displayed by the buildings. All but two households reach peak demand on two identical days: 4 and 42. The first day occurs in January and the other in February. The coincidence of time series gas demand of two different occupant profiles with summer space heating is demonstrated in Fig. 7.

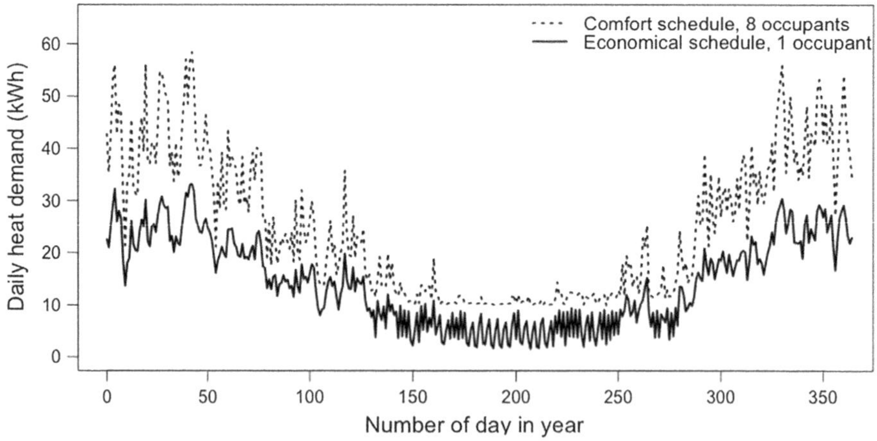

Fig. 7 Simulated daily heating gas consumption for two occupant profiles with summer space heating: an economical schedule for one occupant and a comfort schedule for eight occupants

Despite their mutual difference in annual demand, the daily demands caused by different occupancy profiles continue to coincide temporally. Most homes peak their daily gas consumption on identical days (Fig. 7). Small peaks recur in the economical schedule during the summer months (days 150 to 250). They reflect daily variations in heat increments to DHW storage.

4 Conclusion and Future Work

A thermal network (RC) model of a semi-detached house was implemented in Modelica AixLib library. By reducing computational cost, RC models are better suited for neighbourhood scale simulations of heat energy demand. With the IWEC v1 sample days of Dublin weather, the Modelica simulated internal temperature responds faster than the equivalent EnergyPlus simulation. Nevertheless, an annual time series MAE (mean absolute error) of the passive fabric is relatively small at 1.39°C. The slower thermal dynamics of EnergyPlus reduce annual heat demand by 12% when simulating the same house. Putting this difference in perspective, Waltz [25] views 10% simulation error as acceptable. Further corroboration if energy use results appears in Sect. 3.1. An archetype of a semi-detached house modelled in EnergyPlus, locates annual heat energy in a distribution curve of annual heat energy produced by Modelica simulation.

As expected, domestic hot water increases its proportion of household heating demand with increasing building performance and occupant quantities. The proportion can exceed 40%, and care is needed to control DHW heat losses throughout the day.

A combination of per home heat demands (Tables 2 and 3), with a real distribution of occupant quantities produces two overlapping distributions of annual heat demand. Each distribution curve represents one of two distinct heating schedules applied to similar houses of a neighbourhood (Fig. 6). By initial inspection of the distribution curves, asymmetry and skewness to higher heat demands prevent a classic "bell curve" associated with normal distributions. Nevertheless, statistical distributions could be applied to retrofit planning at neighbourhood scale. Increasing the sample size would reduce variation and increase prediction accuracy of heat energy savings.

Future work includes increasing sample size and producing statistical distributions of heat energy demand for the most common housing archetypes. Other research [6] has already proposed that energy use of similar homes in a neighbourhood is best expressed as a statistical distribution function. The distribution function promises more accurate prediction of neighbourhood energy demand compared to individual homes. Ultimately, the prediction of energy savings by home retrofit at neighbourhood scale, will be more accurate.

Time series of home gas demands illustrate coincident peaks, regardless of occupant profile. This means that percentage reductions in annual energy demand due to retrofit may not repeat in reductions to peak demand. Energy utilities must size networks to fulfil aggregated peak demand. Such network sizing is very important if gas heating switches to electricity.

Acknowledgements Paul Beagon gratefully acknowledges that his research is supported in part by a research grant from Science Foundation Ireland (SFI) under the SFI Strategic Partnership Programme Grant Number SFI/15/SPP/E3125.

Paul thanks Moritz Lauster of RWTH Aachen University and E. ON Energy Research Center, for his kind and useful advice on AixLib library.

References

1. N. Good, L. Zhang, A. Navarro-Espinosa, P. Mancarella, High resolution modelling of multi-energy domestic demand profiles. Appl. Energy **137**, 193–210 (2015)
2. R. Gupta, M. Gregg, S. Passmore, G. Stevens, Intent and outcomes from the retrofit for the future programme: key lessons. Build. Res. Inf. **43**, 435–451 (2015)
3. P. Jones, S. Lannon, J. Patterson, Retrofitting existing housing: how far, how much? Build. Res. Inf. **41**, 532–550 (2013)
4. P. Moran, J. Goggins, M. Hajdukiewicz, Super-insulate or use renewable technology? Life cycle cost, energy and global warming potential analysis of nearly zero energy buildings (NZEB) in a temperate oceanic climate. Energy Build. **139**, 590–607 (2017)
5. B.V. Mathiesen, D. Drysdale, H. Lund, S. Paardekooper, I. Ridjan, D. Connolly, J.S. Jensen, *Future Green Buildings* (Aalborg University, Department of Development and Planning, Aalborg, 2016)
6. A. Koch, S. Girard, K. McKoen, Towards a neighbourhood scale for low- or zero-carbon building projects. Build. Res. Inf. **40**, 527–537 (2012)
7. O. Neu, B. Sherlock, S. Oxizidis, D. Flynn, D. Finn, Developing building archetypes for electrical load shifting assessment: analysis of Irish residential stock, in *CIBSE ASHRAE Technical Symposium: Moving to a New World of Building Systems Performance*, Dublin, Ireland (2014), pp. 1–19
8. M. Badurek, M. Hanratty, B. Sheldrick, D. Stewart, *Building Typology Brochure Ireland* (2014)
9. IEA Annex 60: Summary—IEA EBC Annex 60, http://www.iea-annex60.org/about.html
10. D. Müller, M. Lauster, A. Constantin, M. Fuchs, P. Remmen, AIXLIB—An open-source Modelica library within the IEA-EBC Annex 60 framework, in *BauSIM* (2016), pp. 3–6
11. M. Lauster, J. Teichmann, M. Fuchs, R. Streblow, D. Mueller, Low order thermal network models for dynamic simulations of buildings on city district scale. Build. Environ. **73**, 223–231 (2014)
12. B.D.B. Crawley, L.K. Lawrie, EnergyPlus: energy simulation program. ASHRAE J. **42**, 49–56 (2000)
13. D.B. Crawley, L.K. Lawrie, F.C. Winkelmann, W.F. Buhl, Y.J. Huang, C.O. Pedersen, R.K. Strand, R.J. Liesen, D.E. Fisher, M.J. Witte, J. Glazer, EnergyPlus: creating a new-generation building energy simulation program. Energy Build. **33**, 319–331 (2001)
14. D. Connolly, H. Lund, B.V. Mathiesen, S. Werner, B. Möller, U. Persson, T. Boermans, D. Trier, P.A. Østergaard, S. Nielsen, Heat roadmap Europe: combining district heating with heat savings to decarbonise the EU energy system. Energy Policy **65**, 475–489 (2014)
15. J. Egan, D. Finn, P.H. Deogene Soares, V.A. Rocha Baumann, R. Aghamolaei, P. Beagon, O. Neu, F. Pallonetto, J. O'Donnell, Definition of a useful minimal-set of accurately-specified input data for building energy performance simulation. Energy Build. **165**, 172–183 (2018)

16. A. Garnier, J. Eynard, M. Caussanel, S. Grieu, Predictive control of multizone heating, ventilation and air-conditioning systems in non-residential buildings. Appl. Soft Comput. **37**, 847–862 (2015)
17. A. Tindale, Third-order lumped-parameter simulation method. Build. Serv. Eng. Res. Technol. **14**, 87–97 (1993)
18. NREL: Weather Data by Location EnergyPlus—Europe WMO Region 6, Ireland, https://energyplus.net/weather-location/europe_wmo_region_6/IRL//IRL_Dublin.039690_IWEC
19. L. Peeters, J. Van der Veken, H. Hens, L. Helsen, W. D'haeseleer, Control of heating systems in residential buildings: current practice. Energy Build. **40**, 1446–1455 (2008)
20. SEAI: What are the electricity factors used in the latest version of DEAP?, http://www.seai.ie/Your_Building/BER/BER_FAQ/FAQ_DEAP/Results/What_are_the_electricity_factors_used_in_the_latest_version_of_DEAP.html
21. SEAI: Derivation of Primary Energy and CO_2 Factors for Electricity in DEAP (2016)
22. J. Egan, *The Definition and Application of a Useful Minimum Set of Accurately-Defined Data for Building Energy Performance Simulation* (2015)
23. B. Coyne, S. Lyons, D. McCoy, *The Effects of Home Energy Efficiency Upgrades on Social Housing Tenants: Evidence from Ireland* (2016)
24. J. Salom, A.J. Marszal, J. Widén, J. Candanedo, K.B. Lindberg, Analysis of load match and grid interaction indicators in net zero energy buildings with simulated and monitored data. Appl. Energy **136**, 119–131 (2014)
25. J.P. Waltz, *Computerized Building Energy Simulation Handbook* (Marcel Dekker, New York, 2000)

Potentials and Challenges for Integrating PV in Roof Renovation of Multi-residential Houses—A Questionnaire Survey

Liane Thuvander, Paula Femenías, Johan Gren and Peter Kovacs

Abstract A questionnaire survey investigates the conditions for future roof renovations, driving forces and incentives as well as barriers to install roof mounted or roof integrated photovoltaics (PV) among Swedish owners and managers of multi-residential buildings. Respondents were identified through a database hosted by the Swedish Energy Agency holding information about all projects that received subsidies for installation of PV between 2009 and 2016. The final sample comprised 77 organizations and the response rate was 36%. The questionnaire covers general information about the responding companies' property portfolios; roof renovations in general; routines, motives, and driving forces for installation of roof PV; and planned roof renovations. Results show that the main cause for conducting roof renovations is end of life-time and improvement of energy efficiency. The initiative to install PV is mostly taken by the board of a Housing Association, the management team, or the board of a company. Standard PV modules mounted onto the roof is the predominant choice. Better profitability is needed to encourage more PV installations, for example, through higher subsidy levels and long-term security of the value of produced electricity. None of respondents asks for more appealing design of PV products, better internal organization, or improved knowledge about operation and management of the PV plant.

Keywords Photovoltaics · PV · Questionnaire survey · Roof · Renovation

---

L. Thuvander (✉) · P. Femenías · J. Gren
Chalmers University of Technology, Sven-Hultins Gata 6,
41767 Göteborg, Sweden
e-mail: liane.thuvander@chalmers.se

P. Kovacs
RISE, Research Institutes of Sweden, Brinellgatan 4, 50115 Borås, Sweden

D. Johansson et al. (eds.), *Cold Climate HVAC 2018*,
Springer Proceedings in Energy, https://doi.org/10.1007/978-3-030-00662-4_20

1 Introduction

There is a need for cost efficient renovation solutions resulting in energy efficiency and climate benefits and which are applicable to a larger part of the existing housing stock. Installation of photovoltaic systems (PV) is one way forward. Large parts of the existing housing stock constructed during the 1960s and 1970s have flat roofs or roofs with small inclinations in need of renovation which are also favorable for installation of PV systems. The price of PV has fallen dramatically over the past decade [1] and for less advanced roof installations of PV there are well functioning systems and experienced suppliers. However, little is known about the actual extent of roof renovations in the existing multi-residential building stock. The same goes for Housing Companies' decisions about potentials and limitations related to roof renovations in general and the integration of PV in particular.

Previous Swedish studies have investigated experiences from installing PV among broader stakeholder groups [2], for example suppliers, installers and property managers in general and including different types of buildings (medium-sized facilities) such as multi-residential buildings, schools, offices, and sports arenas [2] or explored more narrow stakeholder groups [3], such Housing Associations that manages residents' owned condominiums in multi-residential buildings. Also, the focus varied from process description for installing PV including procurement and operation [3, 4] to mapping mounting systems for PV with regards to maintenance and moisture damages of building components [2].

This paper presents an overview of larger roof renovations with focus on the integration of PV. Based on 28 responses from a questionnaire survey, the driving forces, incentives and barriers to install roof mounted or roof integrated PV in multi-residential buildings are investigated as perceived by managers and owners of multi-residential buildings.

2 Method

A questionnaire survey was carried out studying Swedish housing owners and managers of rental and resident-owned condominiums in multi-residential buildings that have installed PV with subsidies from the Swedish Energy Agency. The questionnaire focused on PV installations in connection to roof renovations.

A database with information about all projects in Sweden that have received subsidies was used to identify relevant respondents (extract from 18 January 2017). Out of 6222 approved projects, 195 were related to multi-residential buildings and 133 individual stakeholders could be identified. After removing non-housing companies such as construction companies or energy providers 101 organizations remained. Thereafter, contact details such as e-mail or phone numbers were gathered manually through searches on the Internet. The final sample consisted of 77 potential respondents.

In a first step, a request to participate in the questionnaire survey was sent by e-mail to the managers of the property development divisions at the Housing Companies, to chair members of the board in the Housing Associations, or other relevant persons in the Housing Companies or the Housing Associations. After a positive response, the questionnaire was sent out with the choice to either fill in an interactive pdf-file or a web-based version of the questionnaire. Two subsequent reminders were sent out. Four respondents declined to participate. Data has been stored using the web-based survey tool SurveyMonkey®. The study was carried out May–July 2017.

The questionnaire was divided into six sections with in total 26 questions (q) covering general information about the companies' property portfolio, size and type of tenancy; implemented roof renovations (6 q); implemented installation of PV roofs—routines, motives, and driving forces (6 q); planned roof renovations (5 q); as well as general reflections and background information (1 q each).

3 Results

3.1 Responses and Respondents

In total, 28 respondents filled in the questionnaire which corresponds to a response rate of 36%, see Table 1. There are equal number of responses from housing companies (public and private) that offers rental housing (RH) and resident owned housing (ROH) Associations. The response rate for public owned rental apartments is much lower than the total response rate and, in contrary, the response for resident own housing is much higher, almost 50%.

The property portfolios differ considerably between the respondents, up to a factor 100 distinguishes the largest from the smallest (Fig. 1). Most of the smaller organizations are ROH, but some smaller privately-owned RH are also represented. The average size of the 124,629 apartments represented through the respondents' organizations is around 77 m^2. Around two-thirds of the responding organizations have a policy regarding PV installation and/or using renewable energy sources.

Table 1 Response rate and type of tenure

Type of tenure	Number of organizations	Number of responses	Response rate (%)
Rental apartments—public owned	33	10	30
Rental apartments—privately owned	11	3	27
Resident owned housing	27	13	48
Other	6	2	33
Total	77	28	36

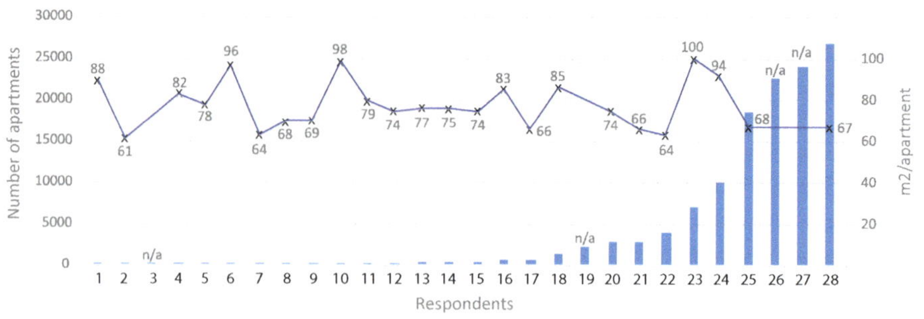

Fig. 1 Number of apartment per responding organization and average apartment size [N = 28]

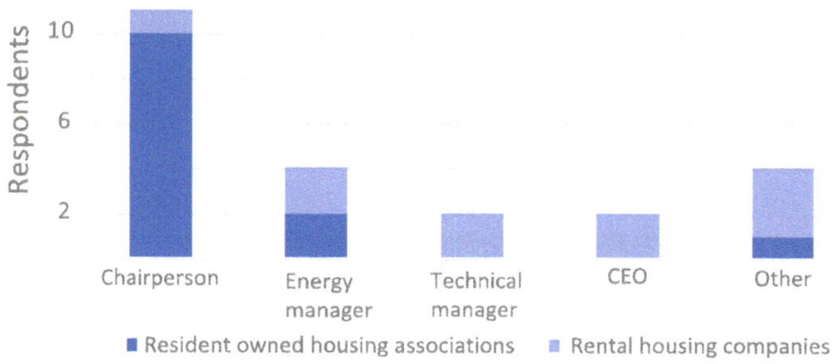

Fig. 2 Position of the respondents [N = 23]

The positions of the respondents in their respective companies/associations are diverse: chairman, energy manager, technical manager, CEO, and other, Fig. 2. In the group called others, the following positions are included: assistant project manager, suppleant of the board, owner, technical consultant, or property manager.

3.2 *Roof Renovations in General*

The main reason for carrying out a roof renovation, i.e. replacing the roof or conducting a major roof renovation is, not surprisingly, end of technical life-time of the roof. To some extent, also improvement of energy efficiency is a driver (Fig. 3).

Roofs renovated the past five years are mostly in the categories pitched roofs and stock constructed between 1961 and 1980. The number of roof renovations in stocks from the period 1941–60 is much higher than what would be expected considering the relative size of the stock in our study. Flat roofs and tarred roofs and especially the combination of these two categories are perceived as in need of renovation more often than other roof types. When asking for a roof type or

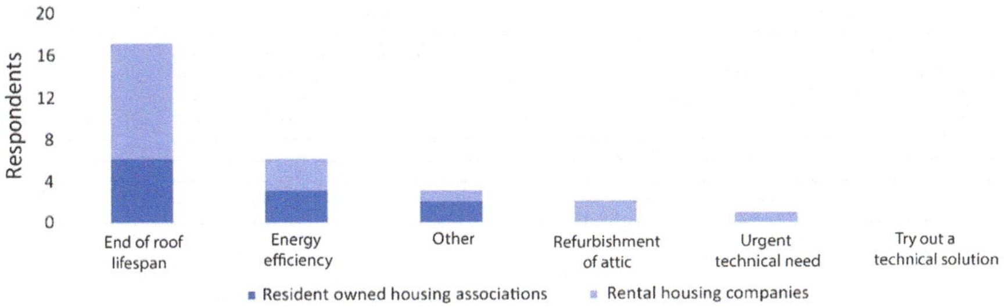

Fig. 3 Causes for roof renovations [N = 23]

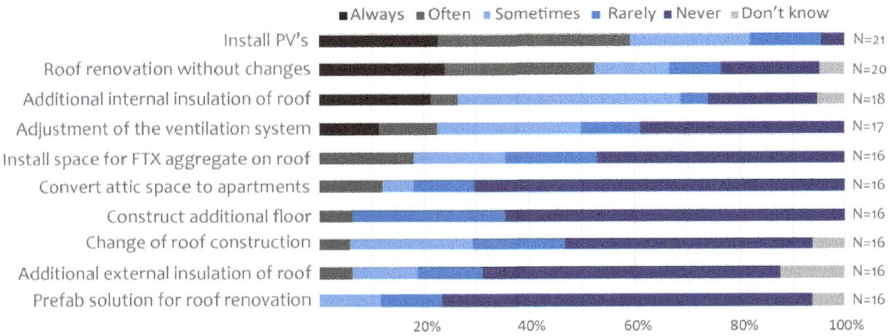

Fig. 4 Measures carried out in conjunction with roof renovations [N = 24]

construction period which has been renovated more often than others flat roofs are mentioned several times.

The most common measures carried out in conjunction with a roof renovation are: replacing the roof without alterations or to install solar panels, Fig. 4. More than half of the respondents 'always' or 'often' install PV or carry out roof renovations without changes.

Among the responding organizations, it is uncommon to renovate roofs and install PV in the same go, most do either or. Reasons for renovating roofs without installing PV are mostly due to the roofs facing an unfavorable direction, having an inappropriate shape, and low profitability/high investment costs of PV, Fig. 5. Other reasons are unclear tax regulations and uncertainties around the future value of the produced electricity. The question about motives for not installing PV has a low response rate, notably for ROHs. Probably this is a consequence of our population (projects with subsidies for PV installation) and that the ROHs are small and only have renovated one roof, i.e. with PV.

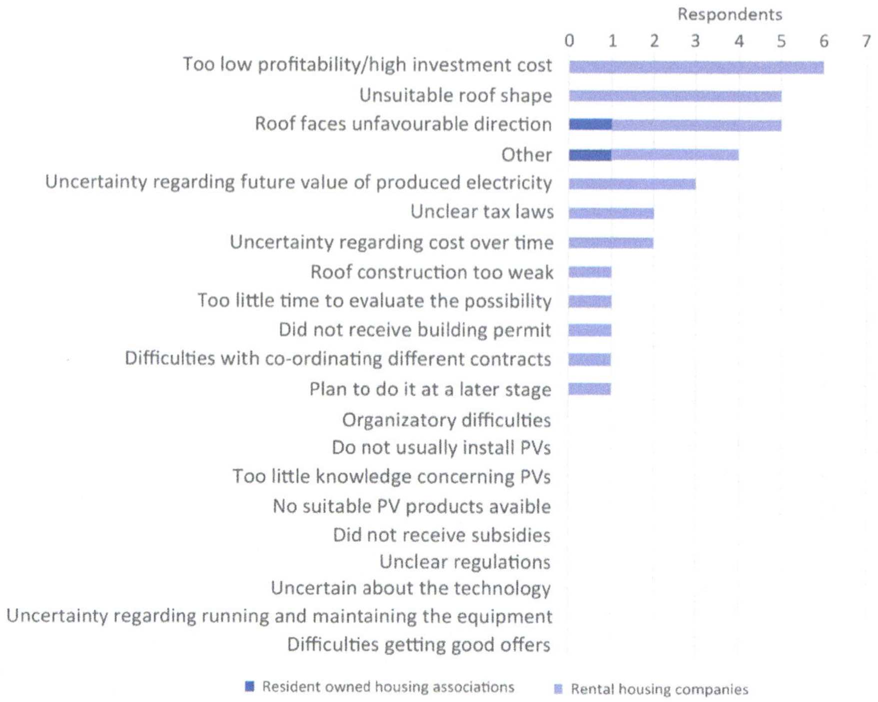

Fig. 5 Reasons for not installing PV in a roof renovation. Up to 5 options could be selected by the respondents (N = 11)

3.3 Implemented PV Installations on Roofs: Routines, Motives and Driving Forces

Standard PV modules mounted onto the roof is the predominant choice, selected by 10 respondents. Another 3 respondents state that they installed roof-integrated standard PV-modules and one states that pre-fabricated roof modules with integrated PV were selected. None of the responding organizations mounted specially designed solutions. The initiative to install PV is most often taken by the Board of the Housing Association of the ROH or in the case of RH organizations, the management team of the business board of the company/association, Fig. 6.

The motives for installing PV vary widely. The most common motives are energy saving, the idea that it is an effective environmental measure, reaching self-sufficiency/contributing to being self-sufficient with electricity; to access subsidies; and marketing of an environmental profile, Fig. 7.

Regarding lessons learnt from carrying out a roof renovation with PV, out of nineteen respondents, twelve state that their organizations are satisfied with how the PV project turned out and would not change anything in a new project. A potential bias related to this is that the respondents want to illustrate how successful their

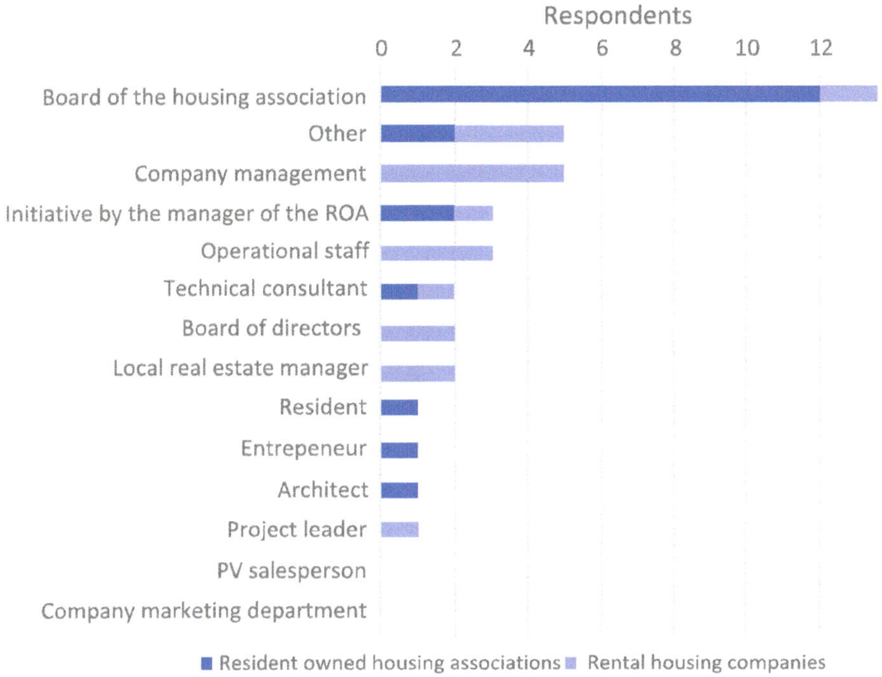

Fig. 6 Stakeholders that initiated roof renovation with PV. Others are, for example: politicians, technical manager, researcher, owner of construction company with experience of installing PV. Up to 5 options could be selected by the respondents (N = 24)

organization was and, in case of the ROH, possibly the main workload was delegated to an external project leader and, thus, the respondent is not acquainted with all the details of the project. Among the comments from those who suggested changes they would like to make slight changes to the procurement process and clarify roles; have a turnkey contract instead of contracts with two separate contractors; or add details in the procurement process, contract, follow-up; renovate the roof at the same time and not install hybrid PV; mount a net between the PV modules and the roof to avoid birds to build nests; and install larger PV plants.

3.4 *Future Roof Renovations*

Regarding plans for roof renovations the up-coming five years, many respondents plan either to install PV without renovating the roofs or to install PV and renovate the roof at the same time (8 respondents for each option, out of 20). Eleven respondents identified the following types of roofs to be renovated: both pitched roofs (4) and flat roofs (3); mostly in buildings constructed 1961–80 (6 responses), and buildings constructed 1941–60 (3 responses) and 1981–2000.

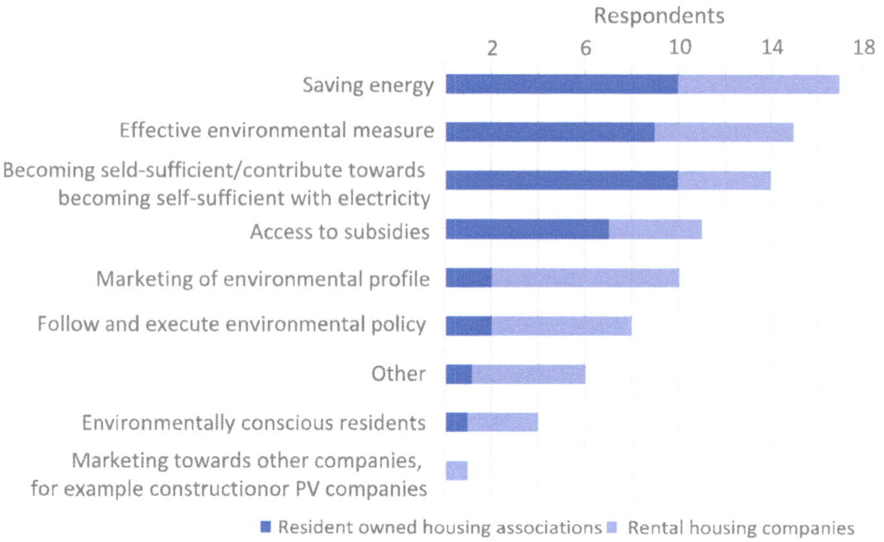

Fig. 7 Motives for installation of roof PV. Others are: owner directive, synergies, supplement to existing heating system, feels good to take a step towards a cleaner world, economy in general, and hundred percent renewable energy. Up to 5 options could be selected by the respondents [N = 24]

On the question of what is needed to encourage more PV installations, better profitability was mentioned most often, Fig. 8. Other important factors relate to economy, such as higher subsidy rates and more security regarding the future value of produced electricity. None of respondents see a need for more appealing/more beautiful design of PV products, better organization within the own company, or better knowledge about operation and maintenance of the PV plant.

Pre-fabricated roof elements with integrated PV are of no interest to 13 of the respondents (10 ROH, 3 RH). At the same time 8 respondents are positive to this kind of solution as a renovation and energy measure and 3 respondents stated other reasons. The mentioned reasons for selecting “other” indicated that the respondents did not understand the questions properly/do not know what is meant by a pre-fabricated roof element with integrated PV.

The respondents were also asked to rank different methods to handle surplus electricity generated from the PV in order to counteract a negative economic result. The two most favorably ways are to introduce a common subscription for electricity for the whole building so that the solar electricity can be used in the apartments and to introduce batteries to store surplus electricity generated during day-time to enable using it in the evening and night.

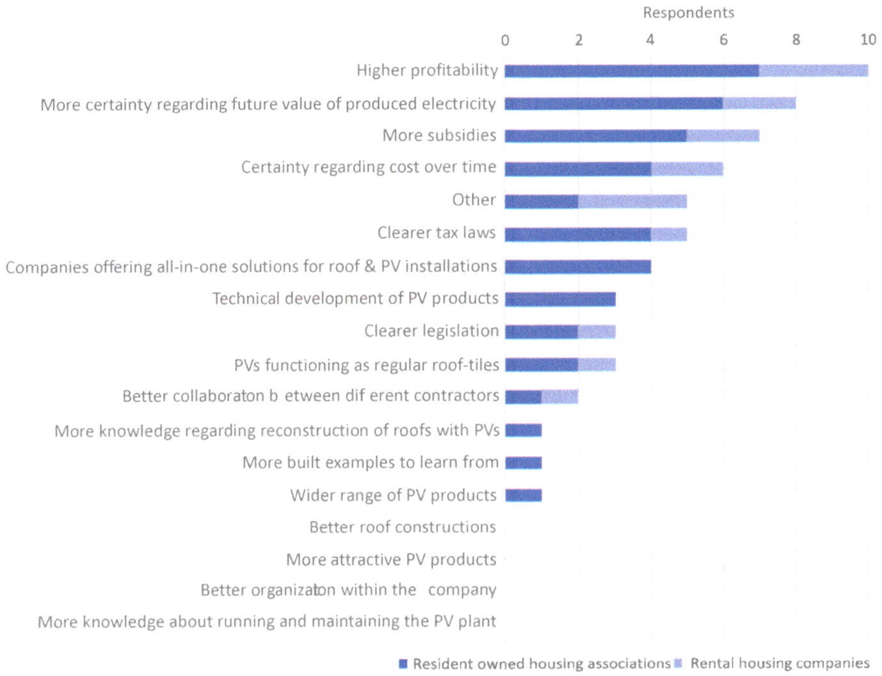

Fig. 8 Motivations to increase PV installations. Up to 5 options could be selected by the respondents [N = 20]

4 Discussion and Conclusion

Regarding roof renovations in general, it is interesting to mention that only a few of the responding organizations conduct a roof renovation with installation of PVs. Instead they do either a renovation or install PV. At the same time, the main reason for conducting a roof renovation is the technical end-of-life time of the roof (Fig. 3). This could be a good opportunity to integrate PV when carrying out a roof renovation but the challenge is to stimulate the PV thinking among mainstream housing managers. Our results differ from the study of Sommerfeldt et al. [3] in which motives for installing PV are among others the combination of roof renovations with installation of PV in order to utilize scaffolding efficiently and by that decrease the installation costs for PV. Other motives for installing PV according to Sommerfeldt et al. [3] are the decrease of operations costs by self-produced electricity, energy efficiency, contribution to sustainability, and marketing of environmental profile. All those are also important motives in the study of Olsson et al. [2] and in our survey.

Economic aspects have been mentioned as an important factor for not installing PV (Fig. 5) or to expand future installation of PV (Fig. 8). The availability of subsidies is possibly a crucial barrier and subsidies will still have a major importance in the near future in order to create a mass market for PV installations. With respect to the small amount of installations, there are possibly too few examples to learn from and there is a lack of well-documented knowledge. For the ROH associations, their national organisation could play an important role to convince, collect, share, and communicate knowledge about roof PV.

The managers of ROH associations are usually non-professionals in construction and their decisions often based on inspirations from other ROH associations or the city scape. The social impact between individual is important and every new PV installation increases the probability for further PV installations in the neighborhood [5]. Also, according to Palm [5], the contact between relatives and friends is more important rather than neighbors and the contact between the individuals can act as a complement to professional consulting.

The addition of an extra floor/attic on top of the building with new apartments is not considered as an option among most of the respondents. The most common measure is to replace the roof without alterations of the existing form and size. Prefabricated roof elements (see Fig. 4) still seem to be rare in roof renovations. Possibly because there are too few available (more less none?) solutions, a lack of knowledge, i.e. this solution is not known among the housing owners and managers, or that the existing are not applicable because of the nature of the planned roof renovations. A conclusion could be that future development of prefabricated roof elements with integrated PV, thus, should focus on roof constructions without addition of an additional floor. These should provide joints adaptable to the different existing roof constructions. Even if the interest in prefabricated solutions is low today, there is a certain curiosity and willingness to apply it in future roof renovations.

Interesting to mention is that none of respondents see a need for more appealing design of PV products, better organization within the own company, or better knowledge about operation and management of the PV plant. These results contradict findings from a process-simulation workshop, where the participants identified the those aspects as important for a more extensive implementation of roof PV in renovations [4].

Our questionnaire was directed to all Swedish housing owners and managers of rental housing and resident-owned housing in multi-residential buildings that have installed PV with subsidies. This gives a certain bias as the respondents in general have a positive attitude towards PV and the barriers are possibly perceived differently from those who have not installed PV. However, we can take advantage from the knowledge the respondents have from carrying renovations with PVs and identify barriers and potentials in relation to the installation of PV and not only reasons for opting out PV. One can speculate that few housing owners of multi-residential buildings will install PV without subsidies. By that, our study probably covers most of the installed PV in this type of buildings. Furthermore, the responses include several smaller ROH organizations, sometimes only owning one property and consequently not conducting many other roof renovations. Finally, our research provides first result regarding PV integration in roof renovations, motivations as well as barriers. For a better understanding of the roof renovations and the potentials for integration of PV in multi-residential buildings, future research needs to focus on the renovation process in general and reasoning behind decision-making involving not only housing managers but also other stakeholders.

Acknowledgements The project is a collaboration project between RISE (former SP), Chalmers University of Technology, Dalarna University, Skanska, Wästbygg, the Swedish Construction Federation, Solkompaniet, White architects, Vätterhem and Riksbyggen. This research was conducted with financial support from E2B2/the Swedish Energy Agency and SBUF. We also want to thank all the respondents of the survey.

References

1. J. Lindahl, *National Survey Report of PV Power Applications in Sweden, in IEA International Energy Agency* (Swedish Energy Agency, Stockholm, 2015)
2. D. Olsson, C. Heincke, P. Wennerhag, *Erfarenheter från medelstora solcellsinstallationer på byggnader. En intervju- och enkätstudie med fokus på montage* (Energiforsk, Stockholm, 2015)
3. N. Sommerfeldt, H. Muyingo, T. af Klintberg, *Solceller ur flera persepktiv. Handbok för beslutsfattare. KTH, skolan för industriell teknik och management* (institutionen för energiteknik, Stockholm, 2016)
4. P. Kovács, L. Thuvander, P. Femenías, O.P. Hemlin, D. Larsson, *Nya utsikter för solceller vid takrenovering* (Bygg and Teknik, Stockholm, 2017), pp. 12–17
5. A. Palm, *Residential Solar Photovoltaics Deployment: Barriers and Drivers in Space* (Lund University, IIIEE, 2017)

Improving the Indoor Climate and Energy Saving in Renovated Apartment Buildings in Estonia

Anti Hamburg and Targo Kalamees

Abstract Energy saving is one of the driving forces in renovation of buildings. Ideally, energy savings should cover the cost of renovation. For purposes of cost efficiency, energy use before and after renovation should be known as accurately as possible. If the energy saving target is too ambitious, energy use after renovation could increase notably and, vice versa, if the target is too low, renovation may not be feasible. In this study we analyze how well the energy saving targets are achieved in renovated apartment buildings in Estonia. The analysis is based on measurements and simulations of indoor climate and energy use in 20 comprehensively renovated apartment buildings. A professional designer and consultant have made an energy audit and design solution before the renovation. Our task was to check the energy audit and compare target and real energy use. We found out that in most cases energy auditors have not assessed existing structures and ventilation correctly, and that basic energy audits should be more detailed in order to assess the existing buildings' energy consumption. Energy saving targets after renovation were also overoptimistic. Based on our research the Estonian energy renovation grant scheme was upgraded.

Keywords Energy saving · Energy audit · Renovation · Apartment buildings

1 Introduction

Energy use in buildings is the largest segment of energy use. Although the requirements for energy use of new buildings have been tightened since the energy crisis in 1970s, the energy use of existing buildings is still high [1] compared to what we expect from today's new buildings and from future near-zero energy buildings. Because the replacement rate of the existing building stock is only some percentages per year, the renovation and improvement of energy performance of

A. Hamburg (✉) · T. Kalamees
Tallinn University of Technology, Ehitajate tee 5, 19086 Tallinn, Estonia
e-mail: anti@tktk.ee

D. Johansson et al. (eds.), *Cold Climate HVAC 2018*,
Springer Proceedings in Energy, https://doi.org/10.1007/978-3-030-00662-4_21

existing building stock plays an important role in reaching national energy efficiency targets. Depending on the Member State, only 0.4–1.2% of the building stock is renovated each year [2]. Baek [3] showed that lack of awareness, information, and regulatory system as well economic reasons are the major barriers to improving the energy performance of existing residential buildings. Kuusk [4] showed that the apartment associations' investment capability is not sufficient to achieve the energy efficiency level of new buildings or low-energy buildings and subsides will increase investments of apartment associations into energy efficiency improvements.

Many studies have shown that investments in energy performance and comprehensive renovation of existing apartment buildings would be economically viable in longer terms [5–10]. In reality the cost effectiveness depends on how accurately energy saving targets are achieved. Branco et al. [11] showed after a 3-year experimental study that the real annual energy use was 268.3 kWh/m^2 instead of initially predicted 44.4 kWh/m^2 because the theoretical value does not take into account real conditions. Cali et al. [12] evaluated refurbished German dwellings and showed the average energy performance gap variation between 41 and 117% during different years. Majcen [13] analyzed Dutch social housing stock, renovated between 2010 and 2013, and showed that the energy performance gap is lower in more efficient buildings.

In Estonia the majority of the multistore apartment buildings were built during the period from 1960 to 1990, employing similar construction solutions. The priority of this dwelling programme was to build as quickly as possible and energy efficiency was not considered important during that period. Systematic renovation of residential buildings started in 2000s when also the energy performance regulation entered into force. During the period between 2010 and 2014 a total of 663 apartment buildings were renovated under the renovation grant scheme and supported by Ministry of Economic Affairs and Communications fund Kredex. To receive finance support, 3 levels for thermal energy saving were established (30, 40 and 50%). The government grant was 15, 25, or 35% from total cost, respectively. The total investment of apartment associations and the grant scheme amounted to 151 million euros, of which 38 million was in form of grants. In this study we analyze how the indoor climate and energy saving improved in these renovated apartment buildings.

2 Methods

In our case study we had analyzed 20 apartment building with different building types (CE: Prefabricated concrete element, AAC: lightweight concrete, Brick) and renovation solutions (Table 1). The heating system was renovated in all buildings. Renovation of ventilation system varied from system cleaning to installing fresh

Table 1 Studied renovated buildings

Code	Building type	Constr. year	Heating area (m^2)	Floors	Renovation works and additional insulation			
					Ex. wall	Roof	Windows	Vent
15.1	Brick	1970	3163	5	10 cm			
15.2	CE	1973	1718	4	15–20 cm		Partly	FAI
15.3	CE	1969	2959	5		23 cm	–	
15.4	ACC	1984	1737	3	15 cm	20 cm	Partly	FAI
15.5	CE	1976	3075	5				
25.1	ACC	1975	777	2	15 cm	25 cm		FAI
25.2	Brick	1982	2623	5	15 cm	25 cm		FAI
25.3	CE	1988	3519	5	15 cm	20 cm	Partly	FAI
25.4	Brick	1975	550	2	15 cm	25 cm	Partly	
25.5	Brick	1971	1903	2	10–15 cm	30 cm	Partly	FAI
35.1	Brick	1978	1064	3	12 cm	13 cm	Partly	SERU
35.2	ACC	1979	1285	3	15 cm	13 cm	Partly	SERU
35.3	Brick/ACC	1982	1527	4	15 cm	25 cm	Partly	EXHP
35.4	Brick/ACC	1979	1041	3	15 cm	23 cm	Partly	EXHP
35.5	Brick/ACC	1979	1162	3	10 cm	23 cm	Partly	EXHP
35.6	Brick	1991	1151	5	15 cm	8 cm	Partly	EXHP
35.7	Brick/ACC	1972	1026	3	5–15 cm	23 cm	Partly	SERU
35.8	CE	1970	5030	9	15 cm	23 cm	Partly	EXHP
35.9	ACC	1981	940	2	15–20 cm	20 cm	Partly	SERU
35.10	Brick	1971	561	2	15 cm	10 cm		AHU

air inlets (FAI) or installation of a completely new ventilation system (SERU: supply-exhaust room units, EXHP: exhaust ventilation with heat pump heat recovery, AHU: central air handling unit).

Energy consumption data before and after reconstruction was collected by building managers. Energy balance contains use of electricity (including household electricity), heating and domestic hot water (DHW). We checked all energy audits made by consultants using the same method. In addition to the original energy balance, we separated energy for production and circulation of DHW. Because apartment buildings with district heating have only one heating meter that measures energy for room heating and DHW we calculated energy for DHW based on water use (DHW is 45% from whole water usage) and temperature rise (50 °C). This calculation based on measured values [16]. The circulation heat loss is calculated based on difference of theoretical and measured energy use for DHW during summer months.

We measured indoor temperature, ventilation airflow and CO_2 concentration in all buildings in order to compare thermal energy use with the indoor climate situation. The table (see Table 1) shows building codes involved in government financial support. 15 means 15% financial support and 30% heating energy saving, 25 means 25% support and 40% heating energy saving and 35 means 35% support and 50% heating energy saving.

3 Results

Our check showed that there are various methodological errors in existing energy audits. That's why we decided to make new calculations for all existing buildings based on measured energy data by using the same methodology. Figure 1 (left) shows the adjusted energy consumption in the pre-renovation situation. The graph shows energy consumption for heating, DHW, DHW circulation and electrical lighting and equipment. A comparison of adjusted energy consumption with auditor's values shows that the adjusted values are of the same magnitude or lower. Higher energy consumption in the pre-renovation situation (calculated by the auditor) also allows improved energy savings (Fig. 2). The main reasons why the energy consumption calculated by the auditor differed from our adjusted values are:

- Energy use that depends on outdoor temperature (heating, ventilation) must be based on the reference year. However, in some energy audits also the consumption of DHW was based on the reference year, although it does not depend on outdoor temperature.
- In some energy audits, the electricity consumption on heating, DHW, lighting and appliances was wrongly allocated. In some cases, the consumptions was calculated twice.
- Some audits had no electricity consumption at all and the auditor had for analysis used the estimated amount of energy for electricity.

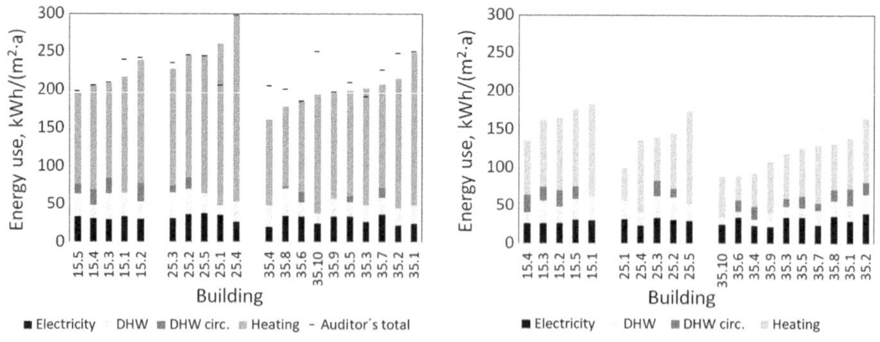

Fig. 1 Energy use before (left) and after (right)

Fig. 2 Target and measured energy save compared with financial support target level (line)

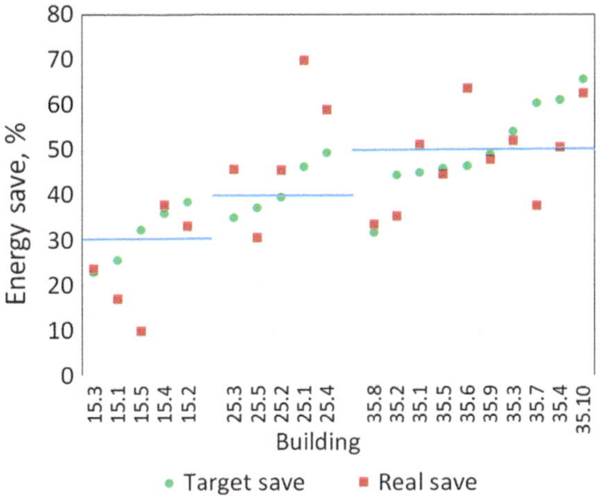

When comparing the energy consumption before and after renovation (Fig. 1), we see that average energy saving in buildings is 37%. Energy consumption after renovation from total energy was in first group (energy saving target 30%, financial support 15%) in average 22%, in the second group it was 44% (energy saving target 40%, financial support 25%), and in the last group it was 40% (energy saving target 50%, financial support 35%). When we compare total energy consumption after renovation the best results were shown by comprehensively renovated buildings with 35% financial support. In these buildings average delivered energy consumption per heated area is 119 kWh/(m^2 a). After renovation, in the buildings with 15% financial support it is in average 165 kWh/(m^2 a) which is more than 25% bigger than in comprehensively renovated buildings.

The renovation grant scheme for buildings depends directly on energy savings achieved (heating, DHW). When comparing energy saving for room heating, DHW and DHW circulation, only half of the buildings fulfil the support criterion (Fig. 2). The reason why many buildings fail to meet the criterion is due to the differences in the calculation of energy saving. Energy use for 9 buildings (1.3; 1.5; 1.2; 2.2; 2.4; 3.8; 3.3; 3.7 and 3.10) was calculated correctly as required for the grant.

For the rest of the buildings, savings are calculated only on the heat energy used for space heating. For buildings where thermal heat consumption was calculated solely on the basis of thermal energy required for heating and ventilation air heating, the achievable energy savings (round dots) is below the level of thermal energy required under the financing requirements of the grant (Fig. 2).

In Fig. 2, the round dots indicate the energy saving projected by the auditor and square dots indicate real savings. One of the reasons why the target savings were not achieved is related to errors in the calculation of thermal energy use before the renovation.

Another reason why the target and real energy savings vary may be due to the difference between the calculated and actual temperature and different ventilation

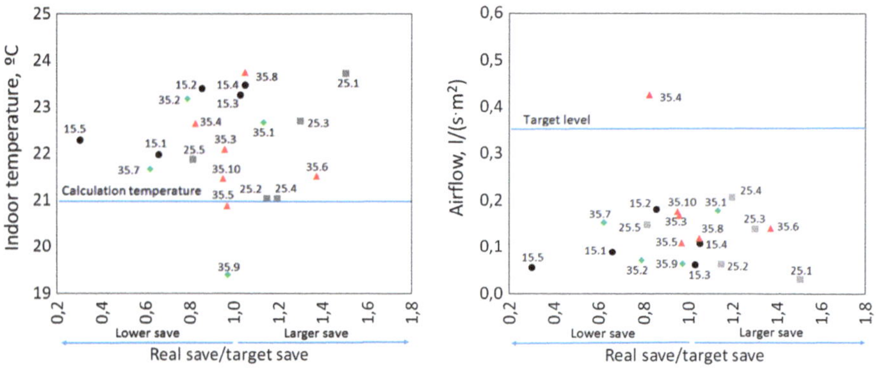

Fig. 3 The comparison of indoor temperature (left) and ventilation airflow with target energy saving for space heating (right)

airflow. When analyzing indoor air temperature and ventilated airflow after renovation, we can see that they do not correlate with achievable energy savings (Fig. 3). Most buildings have higher temperatures than in the calculations and airflow is in average twice lower than the required level [0.35 l/(s·m^2)]. Only some buildings (35.2, 35.3, and 35.4) where indoor air temperature is near 22 degrees and airflow per heated area is 0.2 l/s·m^2 can reach the target energy saving with energy efficiency by the fifth energy saving criterion. When we compare the target and real energy savings of various buildings with air temperature and airflow, then in buildings 15.1, 15.5, 25.5, and 35.7 there is no explicit correlation between the measured values. Therefore, we can say that the calculation of the thermal energy savings made by the auditor of these buildings was too optimistic. Looking at the energy savings achieved and comparing them with the measured airflow and indoor temperatures, we can say that in buildings 25.2, 25.4, 35.3, 35.5, 35.6, 35.7, 35.9, and 35.10 the thermal energy savings were achieved at the expense of indoor climate quality. If the airflow of these buildings is at the required level, achieving energy efficiency would be difficult (Figs. 2 and 3 right). In buildings 35.3 and 35.5 the achievement of energy efficiency may be related to the low efficiency of the exhaust air heat pump and in buildings 35.7 and 35.9 with the low efficiency of space-based ventilation equipment.

4 Discussion

Half of the studied buildings achieved the target thermal energy savings. In several buildings, the real energy savings are higher than calculated. This is due to the lower ventilation airflow in buildings. This result is distressful, because energy savings cannot be achieved at the expense of worse indoor climate. The airflow was at the required level only in one building. As a result of our study, we can say that it is not possible to ensure proper airflow with natural ventilation. Of ventilation equipment, also room-based ventilation equipment proved problematic (noise, draft, efficiency, etc.). Therefore we no longer recommend to use these units for renovation of residential buildings in cold climate. That has been shown also by Simson [17].

We found a number of calculation errors in energy audits. Most of the errors were related to the reduction of heat energy use to the reference year and wrong allocation of electricity use for heating, DHW, lighting and appliances. In some cases, the energy auditor had also taken twice into account some energy use. This shows that there is a need for a common method of energy auditing. Better control would help to avoid such mistakes. In the future, there should be trained consultants who could check the most common errors.

There is no requirement to separate hot water circulation from domestic hot water supply in Estonian energy efficiency regulation. Heat losses from hot water circulation was a problem in houses that had a local electric boiler but after renovation are using district heating (35.2; 35.3; 35.4). In those buildings domestic hot water circulation losses after renovation were about 10 kWh/($m^2 \cdot a$) and DHW and DHW circulation was after renovation that much bigger. This shows us that we also need a calculation method for hot water circulation.

In four apartment buildings (15.1, 15.5, 25.5, and 35.7) where measured indoor temperature was comparable to calculated temperature and real airflow was more than half lower than required, it was clear that energy saving calculations made by auditors contain mistakes. It is likely that auditors showed better energy saving target in order to secure financial support. This problem showed us that thermal energy saving is not a very good base point for financial support and one possibility is to show only target heating energy consumption after renovation which is also connected with Estonian energy labeling calculations.

The second possibility why auditors' energy saving targets were too high may have been that the existing energy auditing form for calculating heat losses is too simplified. Current form enables to take into account thermal conductivity heat losses through envelopes and envelop junctions. Comparing renovated buildings' energy consumption we can also analyse other parameters which should be differently taken into account [15]. This requires updating the energy auditing methodology.

A comparison of thermal energy efficiency levels between different renovation packages shows that there is almost statistical significance ($p = 0.07$) between buildings with minor renovation (target level 30 and 15% financial support) and others. This shows that minor renovation do not guarantee energy savings and it would not be feasible for the state to support it. The importance to comprehensive renovation was showed also by Kuusk [4] and Majcen [13].

5 Conclusion

The energy saving target was achieved only in 40% of buildings with minor renovations (heat saving target: 30%), 40% in buildings with average renovations (heat saving target: 40%) and 50% in building with comprehensive energy renovation (heat saving target: 50%), all together in 11 buildings.

In the course of the study we found mistakes in calculated energy consumption by auditors. There were problems in analyzing existing energy consumption data. In the future it is important to improve control to avoid such mistakes. For this we should in addition to supplementary training of auditors we also need to train consultants to detect possible mistakes in audits. This requires updating the existing energy auditing form and methodology. The majority of studied buildings had problems with ventilation, indicating that energy saving comes partly at the expense of quality of indoor climate. Therefore, it is important to find better ventilation systems that guarantee required airflow since analysed systems mostly did not enable it.

From knowledge collected from this research it is important to ensure that in the future the renovation grant scheme is no longer linked to the energy saving target but with the final energy use that is also linked with Estonian energy label calculations.

Based on our research the Estonian energy renovation grant scheme was upgraded.

Acknowledgements This article was supported by the Estonian Centre of Excellence in Zero Energy and Resource Efficient Smart Buildings and Districts, ZEBE, grant TK146, funded by the European Regional Development Fund, and by the Estonian Research Council, with Institutional research funding grant IUT1−15. Authors would also like to Fund Kredex for cooperation and financial support for our research work.

References

1. T. Csoknyai, S. Hrabovszky-Horváth, Z. Georgiev, M. Jovanovic-Popovic, B. Stankovic, O. Villatoro et al., Building stock characteristics and energy performance of residential buildings in Eastern-European countries. Energy Build. **132**, 39–52 (2016)
2. 2016/0381 (COD), *Amending Directive of the European Parliament and of the Council Amending Directive 2010/31/EU on the Energy Performance of Buildings* (Brussels, European Commission, 2016). Accessed 30 Nov 2016
3. C. Baek, S. Park, Policy measures to overcome barriers to energy renovation of existing buildings. Renew. Sustain. Energy Rev. **16**, 3939–3947 (2012)
4. K. Kuusk, T. Kalamees, Retrofit cost-effectiveness: Estonian apartment buildings. Build. Res. Inf. **44**(8), 920–934 (2016)
5. J. Kurnitski, A. Saari, T. Kalamees, M. Vuolle, J. Niemelä, T. Tark, Cost optimal and nearly zero energy performance requirements for buildings in Estonia. Est. J. Eng. **19**(3), 183–202 (2013)

6. E. Arumägi, T. Kalamees, Analysis of energy economic renovation for historic wooden apartment buildings in cold climates. Appl. Energy **115**, 540–548 (2014)
7. T. Niemelä, R. Kosonen, J. Jokisalo, Cost-effectiveness of energy performance renovation measures in Finnish brick apartment buildings. Energy Build. **137**, 60–75 (2017)
8. M. Ferreira, M. Almeida, A. Rodrigues, Cost-optimal energy efficiency levels are the first step in achieving cost effective renovation in residential buildings with a nearly-zero energy target. Energy Build. **133**, 724–737 (2016). https://doi.org/10.1016/j.enbuild.2016.10.017
9. A. Saari, T. Kalamees, J. Jokisalo, R. Michelsson, K. Alanne, J. Kurnitski, Financial viability of energy-efficiency measures in a new detached house design in Finland. Appl. Energy **92**, 76–83 (2012)
10. F. Bonakdar, A. Dodoo, L. Gustavsson, Cost-optimum analysis of building fabric renovation in a Swedish multi-story residential building. Energy Build. **84**, 662–673 (2014)
11. J. Rose, K. Kuusk, K.E. Thomsen, T. Kalamees, O.C. Mørck, The economic challenges of deep energy renovation: differences, similarities, and possible solutions in Northern Europe—Estonia and Denmark. ASHRAE Transactions **122**(1), 58–68 (2016)
12. G. Branco, B. Lachal, P. Gallinelli, W. Weber, Predicted versus observed heat consumption of a low energy multifamily complex in Switzerland based on long-term experimental data. Energy Build. **36**, 543–555 (2004)
13. D. Calì, T. Osterhage, R. Streblow, D. Müller, Energy performance gap in refurbished German dwellings: lesson learned from a field test. Energy Build. **127**, 1146–1158 (2016)
14. D. Majcen, L. Itard, H. Visscher, Actual heating energy savings in thermally renovated Dutch dwellings. Energy Policy **97**, 82–92 (2016)
15. N. Kalinic, M. Krarti, Evaluation of measurement and verification procedures for retrofit savings using calibrated energy building models, in *ASME 2010 4th International Conference on Energy Sustainability*, vol. 2. (ASME, 2010), pp. 363–373
16. A. Toode, T.A. Kõiv, Investigation of the domestic hot water consumption in apartment buildings. Proc. Estonian Acad. Sci. Eng. **11**(3), 207–214 (2005)
17. R. Simson, A. Mikola, T.-A. Koiv, The impact of air pressure conditions on the performance of single room ventilation units in renovated multi-storey apartment buildings. Int. J. Vent. **10** (11), 227–236 (2014)

Building Refurbishment from a Life Cycle Perspective—An Environmental Return on Investment Approach

Helena Nydahl, Staffan Andersson, Anders P. Åstrand and Thomas Olofsson

Abstract This study applies an environmental return on investment approach to evaluate building refurbishment from a life cycle perspective. The used methodology focuses on the changes introduced by refurbishment, i.e. added embodied environmental impact and changed operational environmental impact, from a life cycle perspective with the technical service life of the refurbishment measure as a time limit. The methodology is applied to a case study in Umeå, located 455 km south of the Arctic Circle, with a unique set of data on reduction in operational energy. The result show the environmental impact, energy (Joule) and GWP (CO_2-eq), in terms of environmental return on investment of the case study refurbishment measures. The case study shows that the methodology is a useable approach to compare refurbishment measures from a life cycle perspective. It is possible to use the methodology as a tool at an early stage in planning of sustainable building refurbishment from a life cycle perspective. For a widespread use of a tool based on an environmental return on investment approach, further research on guidelines for sustainable environmental return on investment values is required.

Keywords Building refurbishment · Life cycle assessment (LCA)
Technical service life

1 Introduction

Refurbishment of buildings represents 17% of the primary energy savings potential in the EU until 2050 [1]. Building refurbishment from a life cycle perspective is required to meet the nearly zero-energy policies [2] and continue cutting emissions. Anand et al. suggest that further Life Cycle Assessment (LCA) studies on building refurbishment is needed and that more case studies are required to build references

H. Nydahl (✉) · S. Andersson · A. P. Åstrand · T. Olofsson
Umeå University, 90187 Umeå, Sweden
e-mail: helena.nydahl@umu.se

D. Johansson et al. (eds.), *Cold Climate HVAC 2018*,
Springer Proceedings in Energy, https://doi.org/10.1007/978-3-030-00662-4_22

[3]. Vilches et al. claims in a recent review study that there are only 13 European studies published on refurbishment LCA [1] out of these, 4 studies are made on multi-family buildings.

There are challenges and research opportunities associated with LCA. Building service life is a parameter containing many variables making it difficult to estimate. In LCA studies the building service life varies from 50 to 150 years [1, 4], making it problematic to compare LCA studies.

The challenges associated with LCA have a negative effect on the usage of the tool. The Swedish National Board of housing, building and planning believe that a simplified LCA tools is crucial to stimulate and increase the use of LCA in the Swedish building sector [5].

An environmental return on investment approach could provide a simplified tool for refurbishment LCA. Similar to the approach presented by Bhandari et al. in an energy return on energy invested evaluation of photovoltaic (PV) technologies [6].

The scope of this study is to apply an environmental return on investment approach as a simplified tool for planning of sustainable refurbishment based on a life cycle perspective and the technical service life of the refurbishment measures.

Vesterberg et al. has during a time period of five years (2010–2015) studied the refurbishment of multi-family buildings in the "Sustainable Ålidhem" project in Umeå [7]. The study has resulted in a unique set of data on the real reduction in operational energy for individual refurbishment measures. The methodology of this study will be applied to the Ålidhem case study to assess the refurbishment measures made.

2 Methodology

The methodology focuses on the changes introduced by refurbishment, i.e. added embodied environmental impact and changed operational environmental impact, from a life cycle perspective with the technical service life of the refurbishment measure as a time limit.

System boundaries of the refurbishment measures follows the EN 15978:2011 standard [8] and includes as follow;

- **Production stage** including raw material supply, transport and manufacturing. Impact of this stage is found in Environmental Product Declarations (EPD).
- **Construction stage** including transport and construction installation process.
- **Use stage** including environmental impact linked to use, maintenance, repair, operational use of energy and water.
- **End-of-life stage** including environmental impact from deconstruction, reuse, transport and disposal.

The EN 15978:2011 standard does not credit benefits and loads beyond the system boundary, e.g. combined heat- and power production from waste material.

The total environmental impact of the refurbishment measure, according to EN 15978:2011, is referred to as the Added Embodied Environmental Impact (AEEI) in this study. The refurbishment measure affects the operational energy use resulting in a change in operational environmental impact. Environmental impact includes energy, measured in Joule, and Global Warming Potential (GWP), measured in CO_2-eq, in this study. Sustainability is assessed trough these units.

The refurbishment measures are made from an efficiency aspect, to reduce operational energy during the use stage of the building life cycle, thus reducing environmental impact during the use stage. Figure 1 illustrates the effect of refurbishment as an immediate increase in environmental impact because of AEEI from the refurbishment measure. As a result of the refurbishment measure there is a reduction in environmental impact over time, illustrated in the figure. At the environmental payback, the AEEI is balanced by the reduction in operational environmental impact. The refurbishment measure is advantageous in those cases when the technical service life is longer than the payback time.

The Environmental Payback Time (EPBT) is calculated according to Eq. 1 [6]. AEEI is divided by the change (Δ) in operational environmental impact (OEI) per year.

$$EPBT_{Energy/GWP} = \frac{AEEI_{Energy/GWP}}{\Delta OEI_{Energy/GWP}} \quad (1)$$

Besides EPBT another important key number is Environmental Return on Investment (EROI), calculated according to Eq. 2 [6]. EROI accounts for the technical service life (TSL) of the refurbishment measure.

$$EROI_{Energy/GWP} = \frac{TSL}{EPBT_{Energy/GWP}} \quad (2)$$

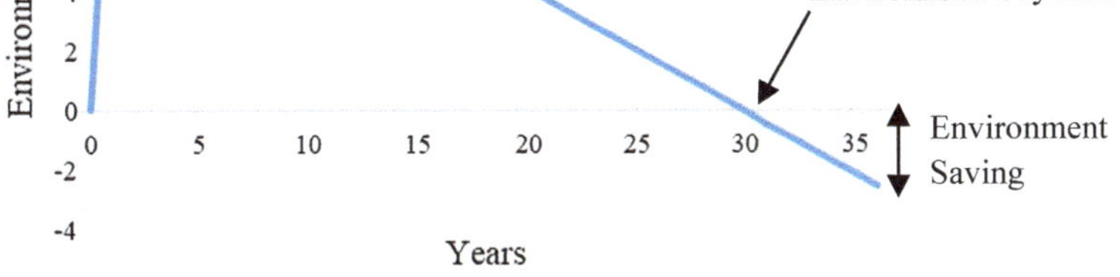

Fig. 1 The environmental payback of a refurbishment measure. AEEI and annual reduction in operational environmental impact due to refurbishment is displayed

In a previous work, Hall et al. analyzed the minimum needed return on energy invested for a sustainable society [9]. In their study, the downstream energy use associated with refining, transporting and using oil and ethanol in the transport sector was studied. The result of this study was a desirable $EROI_{Energy}$ of 3, due to downstream energy use and needed infrastructure.

A refurbishment measure with an EROI greater than one results of course in an environmental saving and a positive environmental impact. If today's energy system and current climate development is considered sustainable, an EROI of one would be sufficient. But taken into account the global need of a transition to renewable energy sources, reduction in greenhouse gas emissions and protection of finite natural resources, an EROI value greater than one would be needed for both energy and GWP. In this study, we have used a minimum EROI of 3, the same as Hall et al. [9], for both energy and GWP, as a reasonable demand for building refurbishment.

3 Case Study Data

The methodology presented in this study has been applied and tested on the "Sustainable Ålidhem" refurbishment project conducted by the municipal housing company in Umeå, AB Bostaden. The project included 405 apartments distributed on 21 buildings connected to the district heating in Umeå for space heating and domestic hot water and to the national grid for electricity. During the project, measuring equipment was installed in a reference building, where no refurbishment measures were made, and a Pilot building that was refurbished. The refurbishment measures studied are;

- Installation of an energy recovery ventilation system with a weight of 691 kg.
- Installation of 3-glass windows with a U-value of 1.1 $WK^{-1}m^{-2}$. The total installed area of windows where 114 m^2.
- Roof insulation of 50 cm loose glass wool and a total weight of 3 ton.
- Additional glass wool wall insulation. 2.8 cm on short sides and 9.0 cm on long sides summing up to a total weight of 1.6 ton.
- PV was mounted on the roof. 70 m^2 of CIGS PV and 55 m^2 Multi-Si PV.

The annual reduction in operational environmental impact as a result of a refurbishment measure on the pilot building is shown in Table 1. The annual reduction in operational energy has been measured by measuring equipment [7] and annual reduction in operational GWP has been calculated from the reduction in combined heat- and power production (CHP). The district heating in Umeå is based on low impact bio-based fueled CHP with an emission factor of 0.011 ton CO_2-eq/GJ [10].

As a refurbishment measure PV were mounted on the roof of the pilot building. Two different types of PV where installed, the CIGS PV and the Multi-Si PV. Table 2 shows the amount of energy produced by the different types of PV and the GWP savings as a result of the renewable energy produced.

Table 1 Annual reduction in operational environmental impact due to reduced heating demand

Refurbishment measure	Annual reduction in operational energy (GJ/year)	Annual reduction in operational GWP (ton CO_2-eq/year)
Energy recovery ventilation system	263	2.89
Windows	37.8[a]	0.42
Roof insulation	76.0[a]	0.84
Wall insulation	23.4[a]	0.26

[a][7]

Table 2 Annual reduction in operational environmental impact due to installation of PV, replaces Nordic electricity mix

Refurbishment measure	Annual reduction in operational energy (GJ/year)	Annual reduction in operational GWP (ton CO_2-eq/year)
CIGS PV	18.8	0.69
Multi-Si PV	18.1	0.66

Table 3 Added embodied energy and GWP per weight or area unit to refurbishment measure

Refurbishment measure	Added embodied energy (GJ/m^2)	Added embodied energy (GJ/ton)	Added embodied GWP (ton CO_2-eq/m^2)	Added embodied GWP (ton CO_2-eq/ton)
Energy recovery ventilation system[a]		33.9		1.90
Windows[b]	0.13		0.11	
Roof insulation[c]		19.5		3.71
Wall insulation[c]		19.5		3.71
CIGS PV[d]	1.52		0.08	
Multi-Si PV[d]	2.23		0.12	

[a]Wheel Air Handling Unit [13], [b]Aluminium-Clad Wood Window [14]
[c]Glass wool insulation [12], [d]PV system UCTE [15]

The GWP savings are calculated assuming that the electricity produced by the PV replaces the Nordic electricity mix [11] and that the net energy production of the PV does not decrease over time. The emission factor (0.036 ton CO_2-eq/GJ) of the Nordic electricity mix is a mean value over the period 2005–2009 with the allocation method *Consideration of gross import/export* [11]. According to Martinsson et al. this is the allocation option that can be considered most similar to reality [11].

Added embodied energy and GWP per refurbishment measure is displayed in Table 3. The environmental impact from the construction stage is not included.

Table 4 Technical service life of refurbishment measures

Refurbishment measure	Technical service life (years)
Energy recovery ventilation system	20
Windows	35
Roof insulation	50
Wall insulation	50
CIGS PV	30
Multi-Si PV	30

The data in Table 3 has been calculated based on information gathered from various scientific reports, referenced in the tables. The only exception is the added embodied GWP for the roof and wall insulation. The amount of electricity and fossil energy used in production of glass wool is found in the report by Tettey et al. [12]. The added embodied GWP is calculated from the environmental data of the Nordic electricity mix, given by Martinsson et al. [11] and oil, fossil gas and coal data was taken from Engström et al. [10].

To calculate the environmental return on investment the technical service life of the refurbishment measure is needed. Table 4 shows the technical service life of refurbishment measures relevant to the case study. Technical service life of the refurbishment measures is collected from the references presented in Table 3, except the technical service life of the roof and wall insulation, collected from a study by Sohn et al. [16].

4 Result and Discussion

The methodology described in this study has been applied to the Sustainable Ålidhem refurbishment project case study. The results presented in this section is linked to the Pilot building and the conditions of that particular building. The AEEI of the refurbishment measures is plotted in Fig. 2. The environmental impact from the construction stage is not included in the result. According to Liljenström et al. the environmental impact of the construction stage accounts for about 16% of total AEEI [17]. If the construction stage would be included, the AEEI would increase with 19% and the EROI would decrease by 16%. This simplification does not affect the internal relations between the results since it applies to all refurbishment measures, but it does affect the relation to a minimum sustainable EROI.

The AEEI of the windows shows the lowest energy impact, but has the highest GWP impact. This shows that the manufacturing process of aluminium-clad wood windows is fossil charged.

Environmental return on investment is plotted in Fig. 3. The figure shows the relationship between technical service life and environmental payback time.

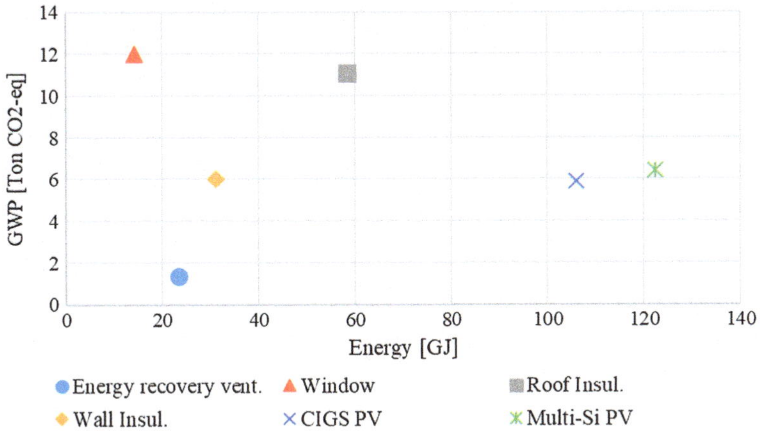

Fig. 2 Added embodied environmental impact of the refurbishment measures

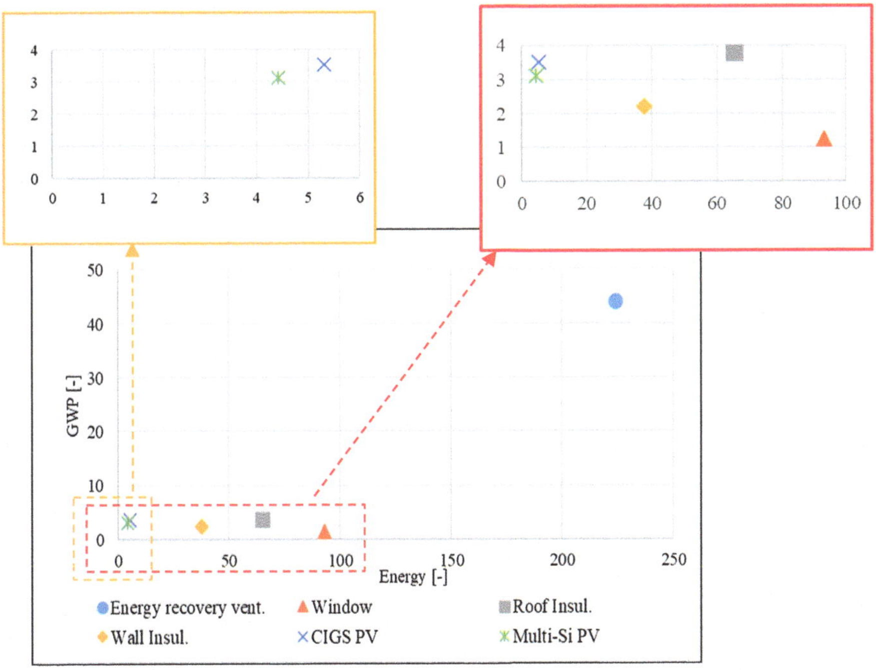

Fig. 3 Environmental return on investment of the refurbishment measures

Figure 3 shows that the choice to change from exhaust air ventilation to an energy recovery ventilation system in the Pilot building was successful with a high EROI. The window and wall insulation has the lowest $EROI_{GWP}$. As said in the method, the minimum EROI for a sustainable refurbishment measure is set to 3 in this study. With this reasoning, it would not be sustainable to invest in windows and wall insulation. If the construction stage would be included in the calculations, the $EROI_{GWP}$ of the window and wall insulation would be even lower.

Roof insulation and PV has an $EROI_{GWP}$ of somewhere between 3 and 4. A reduction of 16%, by including construction stage, would move them to the verge of being sustainable investments, with a minimum sustainable $EROI_{GWP}$ of 3. The PV has a low $EROI_{Energy}$ as well. If construction stage would be included the $EROI_{Energy}$ of the Multi-Si PV would get lower than 3.

Depending on the priorities of the building owner, information of this kind at the planning stage of the refurbishment of the Pilot building could have influenced the decision making to not invest in PV, because of its low EROI for both energy and GWP. Priorities of the building owner might also have influenced the investment in window and wall insulation because of their low $EROI_{GWP}$.

The results presented in this study is linked to the Pilot building, the results are dependent on the source for heating and electricity, geographical location and original construction design of the building.

The emission factor (ton CO_2-eq/GJ) of the energy mix is of importance in the evaluation of refurbishment measures. This is illustrated in Fig. 4. The results for the pilot building is compared to a different case of emission factor for heating and electricity, marked with red marks in Fig. 4.

The Pilot building is heated by the Umeå district heating, based on bio fuels with low emission factor (0.011 ton CO_2-eq/GJ) [10]. In Fig. 4 this is compared to a Swedish district heating based on fossil fuels with a higher emission factor (0.027 ton CO_2-eq/GJ) [10].

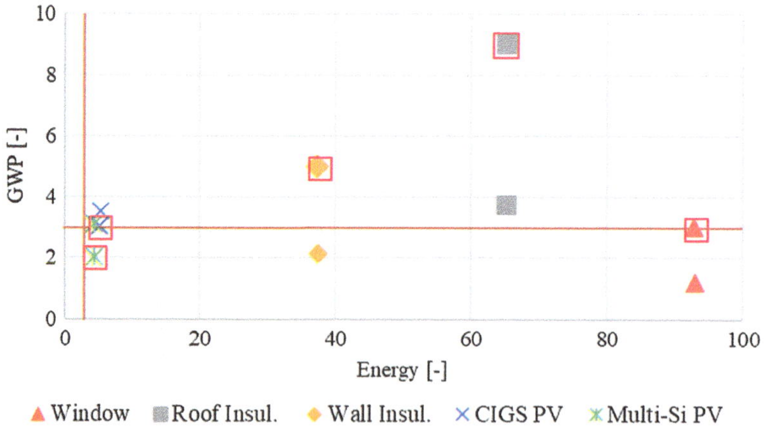

Fig. 4 Illustrating the methodology dependence on emission factor for heating and electricity by comparing DH fuel-composition and Nordic electricity mix allocation. Red marks indicate a different case of heating and electricity source than the Pilot building and the orange lines indicate $EROI_{GWP}$ and $EROI_{Energy}$ of 3

For the electricity, two different kinds of allocation methods, with different emission factors, are compared to each other. The electricity used in the Pilot building is believed to be the Nordic electricity mix with the allocation method *Consideration of gross import/export* with a high emission factor (0.036 ton CO_2-eq/GJ) [11]. In Fig. 4 this is compared to a lower climate impact Nordic electricity mix with the allocation method *without regard to distribution, import or export* (0.029 ton CO_2-eq/GJ) [11].

The results shown in Fig. 4 clearly illustrates that the EROI results are dependent on the source for heating and electricity. An increased use of fossil fuels for heating affects the $EROI_{GWP}$ of the window and insulation, resulting in a $EROI_{GWP}$ higher than 3 for the wall insulation. The change in Nordic electricity mix allocation results in a lower EROI for both energy and GWP for the PV. The PV indicators plotted in Fig. 4 are close to the minimum sustainable indicator lines for $EROI_{GWP}$ and $EROI_{Energy}$ of 3, before and after change in allocation method for the Nordic electricity mix. But the result is also dependent on the AEEI of the refurbishment measures, a less fossil charged production would result in a higher $EROI_{GWP}$ and thus make the questionable refurbishment measures sustainable.

The methodology, an EROI approach, including AEEI and changed operational environmental impact based on a life cycle perspective and technical service life, can be implemented into excel and become a simplified tool for planning of sustainable refurbishment from a life cycle perspective. This would make the methodology usable in the building sector. Hopefully this kind of tool would stimulate and increase the use of LCA in the building sector.

A excel tool could be developed by including the possibility to choose from different sources for heating and electricity and enable options for geographical placement of the building being assessed. Annual reduction in operational energy could be calculated using simulation tools, such as IDA ICE, when data is not available. Further studies should be conducted to develop a toolbox for planning of sustainable refurbishment from a life cycle perspective.

In accordance with the EN 15978:2011 standard [8], the methodology does not credit benefits and loads beyond the system boundary. If the methodology would credit benefits and loads beyond the system boundary, it would affect the AEEI and the EROI. Assessing benefits and loads beyond the system boundary is complex and paved with many allocation choices. In a simplified tool for planning of sustainable refurbishment, it would not be suitable to include benefits and loads beyond the system boundary.

A key element for the future use of a tool based on this methodology, is the minimum EROI values. In this study, the minimum value for sustainable EROI for both energy and GWP has been set to 3. However, for future applications, guidelines for sustainable EROI values for both energy and GWP are needed.

5 Conclusion

The case study show that the methodology is an accessible approach to comparing refurbishment measures based on a life cycle perspective and the technical service life. There is a possibility to implement the methodology into excel as a tool to be used at an early stage in planning of sustainable building refurbishment from a life cycle perspective. Further development of the tool would include different sources for heating and electricity, possibility to choose geographical location and enable calculation of the annual reduction in operational energy through a simulation tool. For the widespread use of a tool based on an EROI approach, the development of guidelines on sustainable EROI values is required.

Acknowledgements We gratefully acknowledge the financial support for this project from AB Bostaden and the Industrial Doctoral School at Umeå University.

References

1. A. Vilches, A. Garcia-Martinez, B. Sanchez-Montañes, Life cycle assessment (LCA) of building refurbishment: a literature review. Energy Build. **135**, 286–301 (2017)
2. J. Buzek, D. López Garrido, Directive 2010/31/EU of the European parliament and of the council of 19 on the energy performance of buildings. Official J. Eur. Union **3**, 13–35 (2010)
3. C.K. Anand, B. Amor, Recent developments, future challenges and new research directions in LCA of buildings: a critical review. Renew. Sustain. Energy Rev. **67**, 408–416 (2017)
4. J.D. Silvestre, A. Silva, J. De Brito, Uncertainty modelling of service life and environmental performance to reduce risk in building design decisions. J. Civ. Eng. Manage. **21**, 308–322 (2015)
5. Boverket (ed.), *Byggnaders klimatpåverkan utifrån ett livscykelperspektiv* (The Swedish National Board of Housing, Building and Planning, Sweden, 2015), p. 95
6. K.P. Bhandari, J.M. Collier, R.J. Ellingson, D.S. Apul, Energy payback time (EPBT) and energy return on energy invested (EROI) of solar photovoltaic systems: A systematic review and meta-analysis. Renew. Sustain. Energy Rev. **47**, 133–141 (2015)
7. J. Vesterberg, S. Andersson, R. Söderström, *Validering lärande och utveckling av hållbarhetsmål i Hållbara Ålidhem* (Slutrapport, Karlskrona, 2017)
8. ISO, *Sustainability of Construction Works—Assessment of Environmental Performance of Buildings—Calculation Method* (Swedish standards institute, 2011), p. 72
9. C. Hall, S. Balogh, D. Murphy, What is the minimum EROI that a sustainable society must have? Energies **2**, 25–47 (2009)
10. R. Engström, J. Gode, U. Axelsson, *Vägledning till metodval vid beräkning av påverkan från förändrad energianvändning på de svenska miljömålen. Framtagen med stöd av Miljömålsrådet, Energimyndigheten och Naturvårdsverket.* (Stockholm, 2009), pp. 76
11. F. Martinsson, J. Gode, J. Arnell, J. Höglund, *Emissionsfaktor för nordisk elmix* (IVL Swedish Environmental Research Institute, Stockholm, 2012), p. 35
12. U.Y.A. Tettey, A. Dodoo, L. Gustavsson, Effects of different insulation materials on primary energy and CO_2 emission of a multi-storey residential building. Energy Build. **82**, 369–377 (2014)
13. M. Nyman, C.J. Simonson, Life-cycle assessment (LCA) of air-handling units with and without air-to-air energy exchangers. ASHRAE Trans. **110**, 399–409 (2004)

14. K. Switala-Elmhurst, in *Life Cycle Assessment of Residential Windows: Analyzing the Environmental Impact of Window Restoration Versus Window Replacement*, ed. by P. Udo-Inyang, B. Flamm, M. Henry, S. Neretina, S. Serrano, B. Van Aken, (ProQuest Dissertations Publishing, Michigan, 2014)
15. M.J. de Wild-Scholten, Energy payback time and carbon footprint of commercial photovoltaic systems. Sol. Energy Mater. Sol. Cells **119**, 296–305 (2013)
16. J.L. Sohn, P.P. Kalbar, G.T. Banta, M. Birkved, Life-cycle based dynamic assessment of mineral wool insulation in a Danish residential building application. J. Clean. Prod. **142**, 3243–3253 (2017)
17. C. Liljenström, T. Malmqvist, M. Erlandsson, J. Fredén, I. Adolfsson, G. Larsson, *Byggproduktionens miljöpåverkan i förhållande till driften* (KTH Arkitektur och samhällsbyggnad, Sweden, 2014), p. 39

Optimizing the Life Cycle Costs of Building Components with Regard to Energy Renovation

A. Farahani and J. Dalenbäck

Abstract Considering the high share of residential buildings in the total energy use in Sweden, having the ambitious national energy and climate goals in mind, the real estate sector and its issues have been under a lot of attention during the past few decades. The Swedish real estate sector has often been identified with its ambitious public housing program during the record years (1960–1974). This was at the time the largest housing program per capita in the world where more than a million apartments were built in a nation with a population of 8 million. These apartments once being the pride of a nation, are facing a lot of problems today, ranging from vacancy and unacceptable physical condition to very poor energy performance. These buildings at the verge of their service/economic life are in need of extensive maintenance and renovation measures. Considering the technological development today, the problem with maintenance and renovation remains to be the financial constraints. These are what makes planning for maintenance and renovation complicated and cost inefficient. Although there are tools that can help property managers with maintenance and renovation planning, they all fail to address the complexity of the decision-making process in a multi-objective criteria under financial and time constraints. In this study, the focus is on the life cycle economy of the building components subject to energy performance improvements during renovation. A systematic approach has been proposed that can be used to budget and plan renovation with regard to energy efficiency under budget constraints. This approach utilizes a modified condition/deterioration model of the method Schroeder to simulate the maintenance effect on the condition state of building components in order to obtain the cost-optimal maintenance regime under given restrictions. This methodology can be used to compare the cost effectiveness of different energy-renovation scenarios and determine the optimal renovation plan for a single or a combination of buildings with regard to owners' objectives and existing constraints. The results from this study illustrates how prioritizing action plans can affect the life cycle costs of building components.

A. Farahani (✉) · J. Dalenbäck
Chalmers University of Technology, 412 96 Gothenburg, Sweden
e-mail: Abolfazl.farahani@chalmers.se

D. Johansson et al. (eds.), *Cold Climate HVAC 2018*,
Springer Proceedings in Energy, https://doi.org/10.1007/978-3-030-00662-4_23

Keywords Energy renovation · Energy efficiency · Maintenance
Life cycle cost

1 Introduction

The residential building stock in Europe is rather old. After the second world war, together with an outstanding rapid change of economic growth, the residential stock was largely developed so as today buildings built before 1970 comprise around half of the total residential stock in most EU countries, [1]. Considering the introduction of thermal regulations during 1970s, the majority of these buildings were built with already poor energy performance. Today, these buildings at the verge of their service lives, are in need of extensive maintenance and renovation measures to not only improve the energy performance but also to comply with current standards and regulations.

Considering the high costs of extensive renovation measures, [2], the economy of implementing energy efficiency measures (EEM) becomes of greater importance. Having in mind the high investment costs of EEMs, an alternative is to combine the investment costs of EEMs with anyway costs in order to save on the fixed costs. Current renovation/replacement planning follows an opportunistic approach, where in the case of failure a decision is made to whether implement an EEM or not. Or on the other hand, maintenance has been neglected in favor of deep renovation which also creates opportunity for energy efficiency improvements.

The aftermath of energy efficiency investment has extensively been studied [3–6] while there are very few studies available treating the effect of timing in the life cycle costs (LCC) of respective building components, [7]. This is partly due to the lack of information as well as uncertainty in estimating the service life of different building components under different maintenance regimes, [8]. The questions regarding the timing of energy efficiency investments is: if it is better to implement EEMs prematurely or by the end of a component's service life? Or simply what is the best time to implement an EEM. In this study, the systematic approach to maintenance and renovation planning proposed by [9] was used to address the aforementioned questions. It is important to mention that in the following analysis the profitability of the energy efficiency investment itself was not of interest but the LCC of the respective building component.

2 Method

In this study façade rendering was chosen as an example to demonstrate the given methodology. Four scenarios were discussed to illustrate the effects of energy efficiency timing and maintenance planning on the life cycle costs of façade rendering. The chosen EEM was the façade insulation with an estimated marginal cost of 40% of the reinstatement cost, [10]. Table 1 shows the characteristics of the chosen component, namely the façade rendering and the assumptions made in this study regarding the costs of energy, EEM investment cost and so on.

As mentioned earlier there is in general lack of knowledge regarding the service life of building components and the corresponding maintenance effect. To overcome this issue in this study the method proposed by Farahani [9] was used to simulate the estimated service life of façade rendering under different maintenance regimes. Thereafter the LCC analysis has been carried out for each simulated scenario to create a cost profile of different maintenance regimes.

For the simulation of the component's condition behavior in Farahani's [9] approach, Schroeder method [12] was utilized for the estimation of condition behavior and the deterioration model was modified by non-repeating alteration of the condition state to predict the effect of maintenance on the condition state and consequently the estimated service life of respective building components.

For the first scenario in this study the maintenance interval of 15 years (a common interval) was used to calculate the LCC of the façade rendering. The reinstatement cost was applied at year 0, whereas the implementation costs of energy efficiency measure was applied at the end of the service life for each scenario. 20 years of age was considered as the longest maintenance interval in order to keep the acceptable condition level for the façade rendering (technical limit). For the second scenario, the LCC of an opportunistic premature implementation of energy efficiency measure was calculated for a hypothetic implementation year. The result then was compared to the first scenario taking into account the energy savings made during the time between the premature implementation of the energy efficiency measure and the service life of the façade rendering calculated in the first scenario. In the third scenario, the proper maintenance regime which resulted in the same service life as the one used in scenario 2 were found and the LCC were calculated to study the difference between the opportunistic premature implementation and the planned premature implementation of the energy efficiency measure. Finally, for the forth scenario, all possible maintenance scenarios were tested

Table 1 Costs, assumptions and energy use data for the façade rendering, [10, 11]

Component	Façade rendering
Reinstatement cost (per reinstatement activity)	694 Kr/m^2
Maintenance cost (per maintenance activity)	147 Kr/m^2
Marginal cost of implementing EEM	277 Kr/m^2
Building energy use	140 kWh/m^2. year
Energy saving potential	25%
Energy price (average)	0.8 Kr/kWh
Growth rate in energy price	1%/year
Inflation	2%
Growth rate in construction and services costs	1%/year
Discount rate	4%
Initial service life	35 years

against each other to find the cost-optimal year at which the respective energy efficiency measure should be implemented.

It is important to mention that all the scenarios were only simulated for the maintenance regimes which kept the façade rendering on proper working condition. Therefore, retroactive maintenance regimes were excluded from the results of the simulated scenarios.

3 Results and Discussion

For the first scenario, the estimated service life was simulated for a common 15 years maintenance interval. Figure (see Fig. 1) shows the condition behavior of the façade rendering under considered maintenance regime. The condition state in the following figures represents three characteristics of building components, namely: functionality; safety and aesthetic. Condition state at 100% represents the new component whereas 20% identifies the minimum accepted quality therefore the end of service life of the respective component. Below 20% are the condition states at which there is high risk of failure. As it can be seen, the estimated service life in this case reached 60 years of age at point of which the part is replaced and the EEM is implemented. These two actions together are called energy renovation in the figure and is used in the rest of this study.

In this scenario, the life cycle costs (net present value) are calculated as below:

$$LCC_{sc.1} = C_{rein,t_0} + \sum_{i=1}^{n} \left(C_{m,i} \cdot (1 + r_c)^{t_i} / (1 + r_d)^{t_i} \right) + C_{eren,t_{esl_1}} \cdot \left((1 + r_c)^{t_{esl_1}} / (1 + r_d)^{t_{esl_1}} \right). \quad (1)$$

where, $LCC_{sc.1}$ is the net present value of the life cycle costs for the first scenario (Kr/m^2); C_{rein,t_0} is the reinstatement cost (Kr/m^2); $C_{m,i}$ is the maintenance cost in

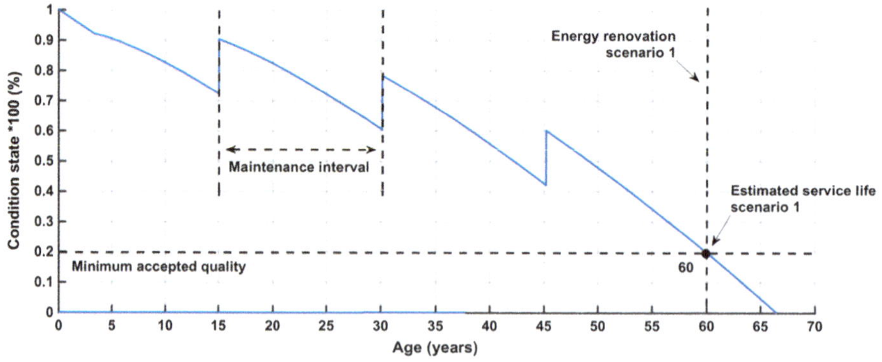

Fig. 1 Condition state of the façade rendering under the maintenance regime with 15 years interval and an energy renovation at 60 years of age

different iterations (Kr/m^2); $C_{eren,t_{esl}}$ is the costs of energy renovation (Kr/m^2); esl_1 is the estimated service life in the first scenario (years); r_c is the rate of growth in construction and services costs on top of inflation; r_d is the discount rate and t is the time (years); i is the maintenance counter; n is the number of maintenance measures carried out.

Using the values given in Table 1, the $LCC_{sc.1}$ was calculated at 1140 Kr/m^2. It is important to note that, since the difference between the 4 scenarios were of interest in this study, the total cost of energy use was not included in the calculation of the LCC in scenario 1. In other scenarios however, the cost of saved energy during the time between the energy renovations in two scenarios were added.

In the second scenario, it was assumed that due to other component's failure (for example, windows had to be replaced) an opportunity arose for renovation of the façade rendering earlier than the end of its expected service life, in this case 9 years earlier at 51 years of age, (see Fig. 2).

In this case, considering the earlier energy renovation time than in the first scenario, the LCC include the saved energy costs between the year 51 and year 60 (assuming that the same energy saving measure is to be implemented in year 60 in scenario 1) in addition to the salvage value of the façade rendering by the year 60.

The life cycle cost is calculated as below:

$$\begin{aligned} LCC_{sc.2} = & C_{rein,t_0} + \sum_{i=1}^{n} \left(C_{m,i} \cdot (1+r_c)^{t_i} / (1+r_d)^{t_i} \right) \\ & + C_{eren,t_{pren}} \cdot \left((1+r_c)^{t_{pren}} / (1+r_d)^{t_{pren}} \right) \\ & + C_{eren,t_{esl_1}} \cdot \left((t_{esl_1} - t_{pren}) / t_{esl_1} \right) \left((1+r_c)^{t_{esl_1}} / (1+r_d)^{t_{esl_1}} \right) \\ & - \sum_{j=t_{pren}}^{t_{esl_1}} r_s (E \cdot C_e) \left((1+r_e)^{t_j} / (1+r_d)^{t_j} \right) \end{aligned} \qquad (2)$$

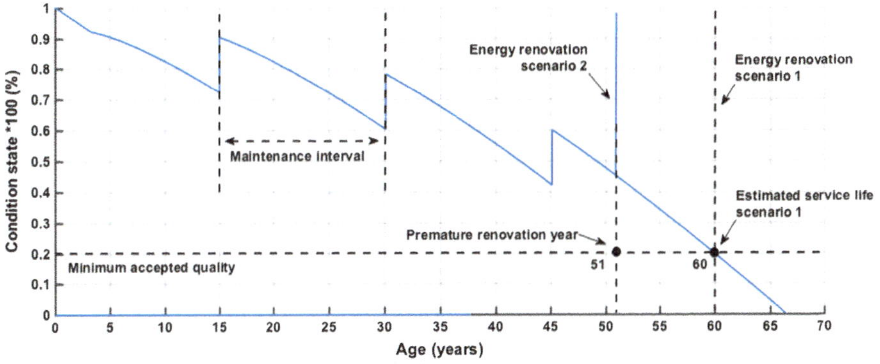

Fig. 2 Condition state of the façade rendering under the maintenance regime with 15 years interval and a premature energy renovation at 51 years of age

where, t_{pren} is the time of premature energy renovation; E is the energy use (kWh/m^2, year); C_e is the cost of energy (Kr/kWh); r_s is the energy saving ratio (%); r_e is the growth rate in energy price. The salvage value in this study is calculated using a linear value depreciation.

For this scenario, the LCC was calculated at 1155 and 15 Kr/m^2 higher than the LCC of the first scenario showing that even considering the savings made through implementation of the EEM earlier in time did not justify the premature replacement of the façade rendering. If the energy savings between the two scenarios were not taken into account, which normally is the case, the offset would increase to 60 Kr/m^2 at the LCC of 1200 Kr/m^2. This is equal to 22% of the energy efficiency implementation costs.

As it was shown, opportunities would be financially justified only if the effects of the decision made were evaluated in a life cycle perspective. Therefore, in this case, the decision to carry out energy renovation at earlier time would only be financially justified if the savings made through combining the renovation of the two parts (windows and façade rendering) resulted in savings greater than the difference between the two scenarios, 15 and 60 Kr/m^2 with and without energy savings respectively.

In the third scenario, the aim was to study the effect of a different maintenance regime which resulted in the estimated service life of 51 years in an ideal case, on the LCC of the façade rendering. Using the proposed method in [9] a maintenance regime with an interval of 17.5 years was found to be the ideal scenario which resulted in 51 years of estimated service life, see Fig. 3.

For this scenario, the life cycle costs were calculated using the equation below, Eq. 3.

$$\begin{aligned} LCC_{sc.3} = {} & C_{rein,t_0} + \sum_{i=1}^{n} \left(C_{m,i} \cdot (1+r_c)^{t_i} / (1+r_d)^{t_i} \right) \\ & + C_{eren,t_{esl_3}} \cdot \left((1+r_c)^{t_{esl_3}} / (1+r_d)^{t_{esl_3}} \right) \\ & + C_{eren,t_{esl_1}} \cdot \left(\left(t_{esl_1} - t_{t_{esl_3}} \right) / t_{esl_3} \right) \left((1+r_c)^{t_{esl_1}} / (1+r_d)^{t_{esl_1}} \right) \\ & - \sum_{j=t_{esl_3}}^{t_{esl_1}} r_s (E \cdot C_e) \left((1+r_e)^{t_j} / (1+r_d)^{t_j} \right) \end{aligned} \tag{3}$$

The only difference between Eqs. 2 and 3 is the value depreciation base as the estimated service life in the third scenario was calculated at 51 years comparing to the second scenario which was based on the estimated value for the first scenario, namely 60 years. Using Eq. 3, the LCC in the third scenario was calculated at 1122 and 33 Kr/m^2 lower than the LCC of the second scenario and even 18 Kr/m^2 lower than the original plan (the first scenario).

The difference in the calculated LCC highlights the importance of proper maintenance planning in the life cycle economy of a given building component. Since the assumption regarding the opportunity given in the second scenario was

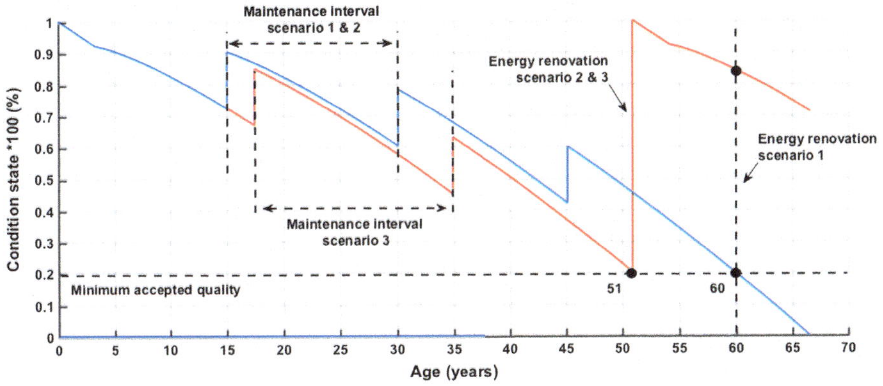

Fig. 3 Condition state of the façade rendering under the maintenance regimes in two different scenarios with 15 and 17.5 years of intervals

based on an unexpected failure of the other component, namely windows, it is very difficult for the property manager to foresee the failure and plan maintenance for the other component accordingly. However, considering the nature of building deterioration, the time to failure can still be estimated and using a proper maintenance regime the LCC can be optimized. It should be noted that, this study neglects the maintenance regime applied to the other component (windows) entirely, otherwise as shown in [9], two building components which share certain fixed costs can be managed and undergo an optimized maintenance and renovation strategy in order to minimize costs while sustaining the acceptable working condition.

It was assumed that the first scenario in this study represents a proper maintenance regime for the façade rendering with no regard to energy efficiency. In the first scenario, the LCC was calculated for the case where an energy efficiency measure was to be implemented in connection with the replacement of the façade rendering. In the fourth scenario, the aim was to evaluate the LCC of earlier energy renovations using different maintenance intervals in order to find the optimum plan for the given building component.

Longer maintenance intervals result in shorter estimated service lives, thus earlier renovation years. In order to examine the effects of earlier energy renovations on the LCC of the façade rendering, the estimated service lives for maintenance intervals longer than 15 years (the first scenario) were needed. Using the same method [6], these values were simulated and the resulting LCCs were calculated against the LCC of the first scenario, see Fig. 4. In order to see the effects of the EEM on the LCC of the façade rendering, all the LCCs were calculated for two cases. First, the LCC including the energy savings made through implementation of the EEM, the blue line in the figure. Second, the LCC excluding the energy savings, the orange line in the figure. In another word, the orange line represents a practice where property managers do not include opportunity value (energy savings between the two alternatives) in the calculation of the LCC. And the blue line represents a complete LCC analysis.

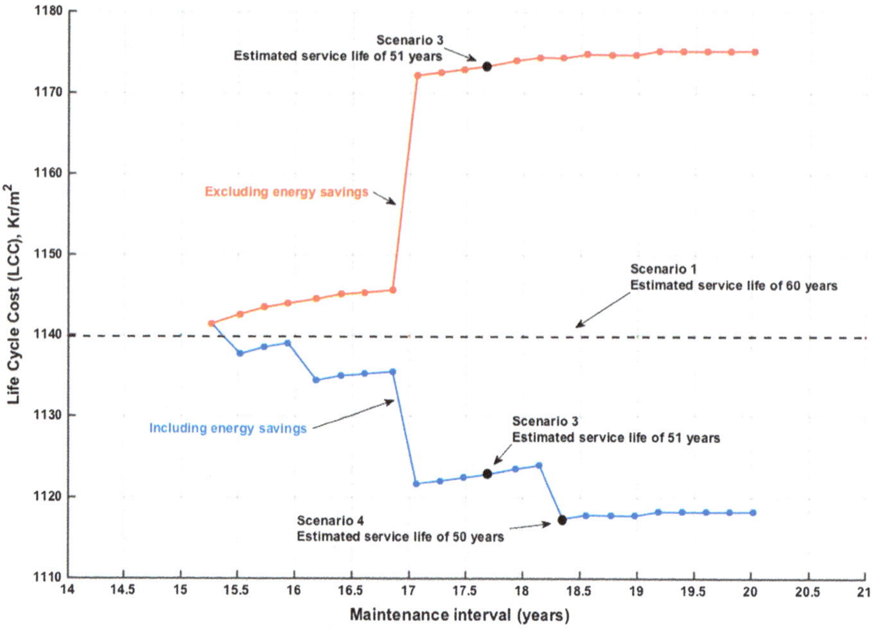

Fig. 4 The calculated LCC values (with and without energy savings) for different maintenance intervals in comparison to the LCC of the first scenario

In Fig. 4, except for the first point on both the orange and the blue lines, taking on the life cycle assessment approach illustrates the positive effect of energy efficiency measure (EEM) on the LCC of the façade rendering. As it is shown the given maintenance regime in scenario 1 was the most cost-effective plan comparing to the rest of the simulated alternatives (the orange line). Yet taking into account the energy savings can potentially lower the LCC by 23 Kr/m^2 with a maintenance interval of 18.3 years comparing to the first scenario or 83 Kr/m^2 comparing to the second scenario without energy savings. That is equivalent to 8 and 30% of the energy efficiency investment costs, respectively.

The results showed that, in general, earlier implementation of the EEM noticeably lowered the LCC of the façade rendering. Considering that buildings comprise of many different components, when assessed together, the property manager can take advantage of the presented results and choose the appropriate renovation year accordingly. It should however be noted that, the magnitude of the results strongly depends on the assumptions made and the calculated amount of energy saved. Nonetheless, large or small, whatever the magnitude of the savings, this approach can be used to optimize planning and estimate renovation time at minimized life cycle costs specially under budget constraints.

4 Conclusion

The results given in this study showed the importance of life cycle assessment in the cost analysis of energy renovation measures (EEM). What is often missed in economic assessments of energy renovation measures is the cost effects of proper maintenance planning in the economic evaluation of presented opportunities. Today the focus is on the aftermath of energy renovation as buildings are old and in need of immediate attention. This study showed that considering the uncertain nature of building deterioration, a systematic planning can still help property managers to optimize costs and create opportunities rather than waiting for them to show up.

As discussed earlier, a proper maintenance regime with regard to energy efficiency could save up to 30% of the investment cost required for the implementation of the respective EEM. Considering that combining a number of building components (with sharing fixed costs) results in lower investment required, using the proposed approach for more than one component can potentially result in greater savings than what was calculated in this study.

In general, the results indicate that earlier implementation of EEMs make economic sense, however the magnitude of the results can vary depending on the input values for the discount rate, energy price growth rate, savings made, energy use, etc.

The simulated LCC profile can help property managers to more reliably adjust maintenance and renovation plans based on the available budget and/or technical constraints.

References

1. European Commission, https://ec.europa.eu/energy/en/eu-buildings-factsheets
2. I. Artola, K. Rademaekers, R. Williams, J. Yearwood, *Boosting Building Renovation: What Potential and Value for Europe?* http://www.europarl.europa.eu/RegData/etudes/STUD/2016/587326/IPOL_STU(2016)587326_EN.pdf, (2016)
3. J. Morrissey, R.E. Horne, Life cycle cost implications of energy efficiency measures in new residential buildings. Energy Build. **43**, 915–924 (2011). https://doi.org/10.1016/j.enbuild.2010.12.013
4. L.F. Cabeza, L. Rincon, V. Vilarino, G. Perez, A. Castell, Life cycle assessment (LCA) and life cycle energy analysis (LCEA) of buildings and the building sector: a review. Renew. Sustain. Energy Rev. **29**, 394–416 (2014). https://doi.org/10.1016/j.rser.2013.08.037
5. M. Jakob, Marginal costs and co-benefits of energy efficiency investments. Energy Policy. **34**, 172–187 (2006). https://doi.org/10.1016/j.enpol.2004.08.039
6. S. Gustafsson, B.G. Karlsson, Life-cycle cost minimization considering retrofits in multi-family residences. Energy Build. **14**, 9–17 (1989). https://doi.org/10.1016/0378-7788(89)90024-8
7. A.H. Sanstad, C. Blumstein, S.E. Stoft, How high are option values in energy-efficiency investments? Energy Policy. **23**, 739–743 (1995). https://doi.org/10.1016/0301-4215(95)00065-Q
8. ISO15686-8, *Part 8: Reference Service Life and Service-Life Estimation* (2011)

9. A. Farahani, *A systematic approach to maintenance and renovation planning under budget constraints* (Constr. Manag, Econ, 2017)
10. K. Westerbjörk, *Halvera mera 1 +2 analysis*, Stockholm (2015)
11. INCIT: UNDERHÅLLSKOSTNADER - REPAB FAKTA., Mölndal (2017)
12. J. Schröder, *Zustandsbewertung grosser Gebäudebestände. Schweizer Ing. und Archit* (1989)

The Challenge of Energy Efficiency in Kiruna's Heritage Buildings

Andrea Luciani, Sofia Lidelöw, Shimantika Bhattacharjee and Tomas Örn

Abstract The town of Kiruna, founded in 1900 in the northernmost part of Sweden, is nowadays in the middle of an impressive urban transformation: due to the impacts of mining activities a large part of the city center has to be moved or rebuilt. Among the buildings to be moved and kept in use are some of the so-called 'Bläckhorn' timber houses, designed by Gustaf Wickman in the early 20th century as residential units for the workers of the mining company LKAB and part of the original core of Kiruna. This has raised several questions on the sustainability of renovating historic buildings in a sub-arctic climate. In order to explore the challenge of increasing the energy efficiency of the Bläckhorn houses, data on their constructional and historical features as well as their thermal and energy performance have been collected. The paper addresses the following issues. Historic buildings are often blamed for their poor energy efficiency without considering their usually high constructional quality. What do we know about the real performances of these buildings? Energy retrofits in non-monumental and inhabited historic buildings are often guided by practical and operational needs rather than by their heritage significance. Can a value-based approach affect the improvement of energy efficiency? In a subarctic climate, even simple interventions can help to save a considerable amount of energy in historic buildings. To which extent the energy performances of the Bläckhorn houses could be increased without affecting their heritage values?

Keywords Energy efficiency · Historic buildings · Cultural heritage values

A. Luciani (✉) · S. Lidelöw · S. Bhattacharjee · T. Örn
Department of Civil, Environmental and Natural Resources Engineering, Luleå University of Technology, 97187 Luleå, Sweden
e-mail: andrea.luciani@ltu.se

D. Johansson et al. (eds.), *Cold Climate HVAC 2018*,
Springer Proceedings in Energy, https://doi.org/10.1007/978-3-030-00662-4_24

1 Introduction

In 2004, the city administration of Kiruna announced the need to move a significant part of the town in order to allow Luossavaara-Kirunavaara AB (LKAB) to continue mining the rich iron ore deposits in the area. The subsidence caused by mining activities was expected to affect the original core of Kiruna, founded in 1900 in the northernmost part of Sweden as LKAB's company town. Many of the buildings affected are officially protected for their recognized cultural significance. An agreement between LKAB and the local authority of Kiruna will allow some of them to be moved, preserved and kept in use. This has raised a rather controversial discussion about the urban transformation of Kiruna and the future of its built heritage [23, 28].

Among the buildings to be moved are the so-called Bläckhorn timber houses, built in the early 20th century as residential units for the workers of the mining company. They have been chosen as a case study within which the challenge of increasing the energy efficiency of heritage buildings is explored.

Existing buildings, in particular the oldest ones, are often blamed for their poor energy performances. Insulation standards in the years of construction of the Bläckhorn houses (1900s) were low, which has a major impact on their heating energy use, especially in Kiruna's cold climate and considering that these houses were not designed for our contemporary thermal comfort levels.

A study by Johansson et al. [13] has shown that Kiruna would be able to meet the Swedish national energy reduction target by 2050 thanks just to the transformation of the existing building stock, but only by replacing all the buildings affected by subsidence with new ones with passive-house standard. On the other hand, the heritage buildings to be moved have been exempted from meeting the present-day building requirements for energy efficiency.

The aim of this study is thus to understand to which extent the energy performances of the Bläckhorn houses could be increased without affecting their heritage values. Data on their constructional and historical features, on their thermal behavior and energy use are here presented and discussed.

2 The Bläckhorn Houses as Part of Kiruna's Built Heritage

The "Bläckhorn" houses were named after their shape, which resembles an inkwell. They were designed as multifamily residential units for the workers of the LKAB by the architect Gustaf Wickman and they embody many of those typical construction features which have been described as "Kiruna style" [2, 7].

The houses adopted a six-rooms layout, with two double-room apartments in the bottom floor and two single-room apartments in the attic. Four different variants of Bläckhorn houses were identified among the around 50 surviving buildings [7]: one variant with brick masonry (1901), a timber one with two entrances (1901–1904) and two more with four entrances (1904–7 and 1907–9).

This study focuses on the houses identified as B52 and B53, which belong to the first timber variant and were among the first Bläckhorn houses to be built. They were also among the first ones to be moved to their new location in the summer 2017 (Fig. 1).

The B52 and B53 houses are part of a designated area of national interest [24]. The cultural values of these buildings have been officially described as follows: "[They] are characteristic buildings for Kiruna, and the oldest of its kind with many followers. The buildings are of high architectural and building-technical quality and relatively in good conditions, which contributes to very high cultural values." […] "The cultural values of the buildings are primarily in the general historical context, with the link to Kiruna's oldest epoch, its genesis and build-up to become a regional and national economy, thriving to be a coherent symbiosis between industry and society." (Joseph [14]—translated from Swedish).

Over the years, many changes have occurred. Already before the 1940s, the entrance porches were replaced by closed vestibules and some originally unheated spaces were transformed to heated living spaces on both floors. Furthermore, the inner layout changed and the original four small and modest apartments were merged to obtain two larger ones. In the following years, windows and doors were replaced, while the external timber paneling has been subjected to maintenance interventions. B52 was moved already in the 1960s and a heated basement in concrete was built in the new location.

Also the heating systems have been updated; the original fireplaces were substituted by radiators and nowadays the energy for space heating is exclusively supplied by Kiruna's district heating system.

Despite that both buildings were included in the research project, the analyses presented in this paper will focus on house B52 because, compared to the measurement data of B53, those collected for this building are of better quality (more continuous in time and more representative for the situation).

3 Contextualization: Heritage Values and Energy Efficiency

Energy retrofits can have a huge impact on heritage buildings. For this reason, both European and Swedish legislation generally allow exemption from building energy requirements if cultural values are affected. The EU directive on the energy performance of buildings exempts those with cultural, architectural or aesthetical values (Directive 2010/31/EU:153/19). Similarly, the Swedish Planning and Building Act [26] includes the possibility to exempt buildings with particular historical, cultural or artistic value. The same is true for the Swedish building code [4], which specifically recommends that windows with a "highly significant cultural value" should not be replaced [4, Sect. 9:91–9:92]. Pracchi [22] highlighted that

Fig. 1 The Bläckhorn house B52 in June 2017, during the moving (© J. Ylitalo)

this legislative framework can lead to an overuse of exemptions to avoid problems with heritage buildings.

Dialogue between different stakeholders is not facilitated by the general perception of the conservation field as rigid and monolithic, without acknowledging the deep change that has occurred in the last decades. Traditional key-concepts such as reversibility, authenticity and safeguarding of original forms and materials, which were codified in the Venice charter [9], have been questioned [17, 19]. Similarly, the idea of the objects being truthful and, hence, authentic is under discussion [20]. Simplified, contemporary conservation theory does not necessarily prioritize the original materials and forms, but rather look at the values associated with the object, and is therefore more open to recognize conflicting perspectives and accept change [1]. A sound application of a value-based approach can thus help to enlarge the space of action for energy retrofits of heritage buildings and overcome a precautionary use of exemptions.

Kiruna's Bläckhorn houses exemplify a typical situation where giving up on improving the energy performance of a heritage building can seem the best option but risks being a lost opportunity. In its building permit application regarding the moving of buildings B52 and B53, the owner referred to the possibility to exempt the buildings from new technical demands [15, 16]. Swedish legislation prescribed that, when a building is moved and re-built in a new place, it has to meet the requirements for new buildings. In 2007 exemptions were introduced to avoid the risk of what the Government referred to as "cultural destruction" [25:8], which could occur when large numbers of historical buildings are to be moved, as in the case of Kiruna. The word "moving" was added to the paragraph in the Swedish Planning and Building Act that describes how alterations of a building must consider its cultural and architectural values [26, Sect. 8:17].

The city of Kiruna has agreed to exempt the moved Bläckhorn houses from meeting the requirement for new buildings due to their high cultural-historical value. Yet, in the building permission only accessibility requirements are referred to, while energy demands are not explicitly quoted [15, 16].

4 Methods

In order to obtain a deep knowledge of the analyzed building, a wide range of methods rooted in different fields (building conservation, construction history, civil and energy engineering…) were combined. Legnér and Luciani [18] proposed a similar analytical approach, involving the use of bibliographical, historical and archive sources together with instrumental analysis, to understand the thermal behavior of a heritage building.

Quantitative information (measured and calculated energy and temperature data) and qualitative information (cultural value assessment) were discussed and analysed in a multidisciplinary framework in order to individuate appropriate measures for the energy retrofit of the studied building.

The method developed and used, as well as the multidisciplinary composition of the research group, reflects that contained in the recently approved European "Guidelines for improving the energy performance of historic buildings" (EN 16883:2017) [8] which proposes "a procedure for selecting measures to improve energy performance, based on an investigation, analysis and documentation of the building including its heritage significance".

The collection of quantitative data went on from December 2014 to September 2016. The total power and energy supplied for space heating and domestic hot water was measured using a Saber energy meter (KYAB, Sweden) connected to the district-heating sub-station in both the houses. The Saber energy meter separates the energy usage for heating tap water from the space heating energy usage with a ±2–10% accuracy by means of an integrated algorithm method [29, 30]. Indoor and outdoor temperatures were measured using factory-calibrated sensors (range −40 to +80 °C, accuracy ±0.1 °C). The indoor sensor was placed in a living room on an outer wall, away from heat sources and direct sunlight. The outdoor sensor was placed on a façade facing south but in a location that minimized the level of incident sunlight and snowfall. Sampling of the amount of energy and power supplied by the district-heating system as well as indoor and outdoor temperatures was conducted once every minute. The sampled values were automatically converted and registered as hourly averages.

To complement these measurements, a thermographical survey of the building was performed in February 2017 using a FLIR T620bx thermal imaging camera (thermal sensitivity: < 0.04 °C, resolution: 640 × 480 pixels).

5 Results and Analysis

5.1 Temperatures and Heating Energy Use

Figure 2 represents the heating energy use in one year from February 2015 to February 2016. Despite the gaps in the data, the trend of the cumulative energy use line is still clearly identifiable. The daily averages of heating power used are higher in the coldest months, but in the subarctic climate of Kiruna some heating power is continuously required even during summer.

A total heating energy use of 80,099 kWh was measured in the surveyed year. Considering a heated floor area of 340 m^2, it gives a yearly heating energy use of about 200 kWh/m^2/year, which can be considered a typical performance given the climate zone and the age of the building. According to the energy declaration of the B52 house, its heating energy use is considered somewhat higher than the statistical average of two-family buildings of similar age, which is reasonable given the northern location. However, compared to today's standards the house has a low energy performance. For example, the Swedish building code prescribes a maximum energy use of about 115 kWh/m^2/year, including energy used for space heating, hot tap water and building operation, for a corresponding new multi-family building.

The heating power signature of the house (Fig. 3) shows a strong correlation between the heating energy use and the difference between indoor and outdoor temperature. The use of the latter parameter was preferred to the simple outdoor temperature because of the surveyed fluctuations in indoor temperature (see Fig. 4). This result indicates that the driving factors for the heating energy use are transmission and ventilation losses through the building envelope and, as a consequence,

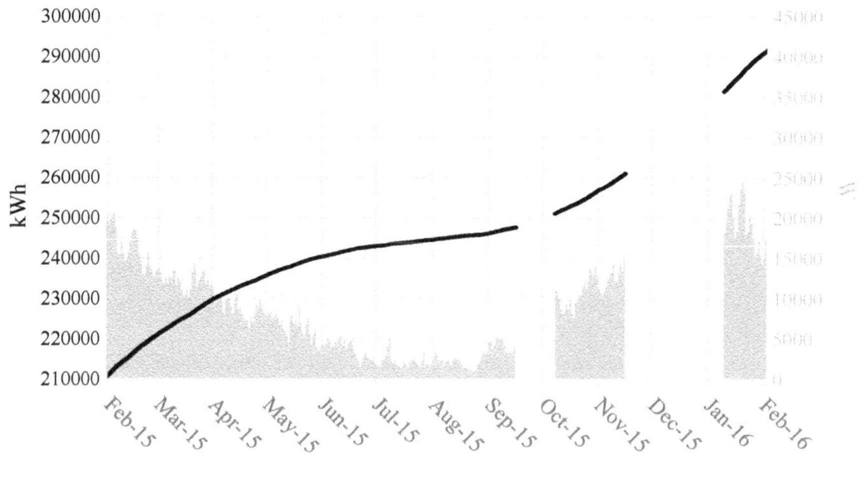

Fig. 2 Heating energy use of house B52. The grey bars represent the daily average use of heating power (W) and the black line shows the cumulative heating energy (kWh) used over the year

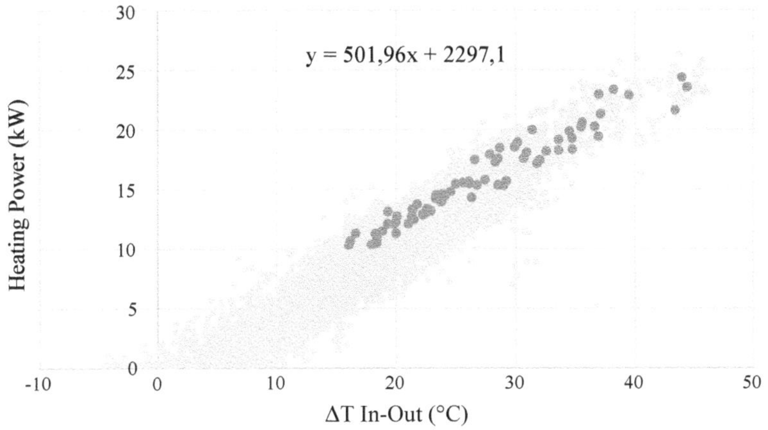

Fig. 3 Heating power signature of B52. The hourly values of heating power are plotted against the corresponding difference between indoor and outdoor T. Dots in darker grey represent the daily average values in the period 21/10-21/2. The equation is derived from these latter values

that improvements in the thermal building envelope are needed to increase the energy efficiency of the building.

The measured indoor temperatures were often quite high, especially in the winter months (Fig. 4). It seems unlikely that this overheating was intentional or even needed to compensate for cold wall surfaces, since the indoor temperature sensor was placed on an outdoor wall and the thermal imaging analysis did not detect excessively cold inner wall surfaces.

Figure 4 shows also the daily variation of the overall heat loss coefficient (W/K), which was calculated as the 24 h average of the values plotted in Fig. 3. It presents higher and more constant values in the coldest and darkest months, since in this period of the year the influence of solar gains is very limited. Nordström et al. [21] showed that the energy signature is most robust when the difference between indoor and outdoor temperatures is large; hence, it is advantageous to estimate it under cold conditions. Sjögren et al. [27] also pointed out that the energy signature is sensitive to heat contributions from the sun. For these two reasons, data from the four months around the winter solstice (21 October–21 February), when the solar radiation in Kiruna yields a minimal contribution to the heating of the house, were used for the heating power signature (darker dots in Fig. 3). It was also preferred to use daily averages in order to compensate night/day cycles. The overall heat loss coefficient we obtain based on these premises is 502 W/K. To verify the validity of this value, which is based on measured data, and to better understand how different elements of the building contribute to its overall thermal performance, thermal transmittances of the thermal envelope were calculated.

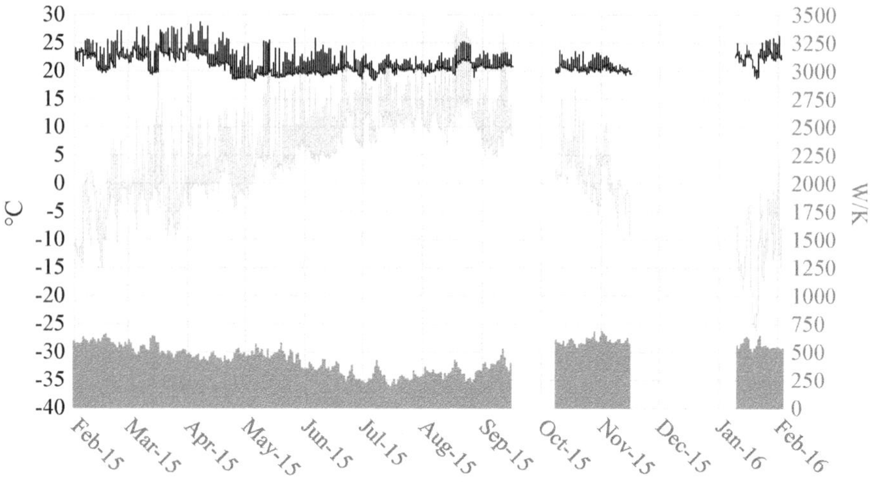

Fig. 4 Indoor (black line) and outdoor (light grey line) temperatures compared with the daily variation of the overall heat loss coefficient (dark grey bars) in house B52

5.2 *Constructional Features and Performance Evaluation*

The first step for the calculation of the heat transfers through the building envelope of house B52, was to define the current composition of its structural elements. This was a challenging task, since the Bläckhorn houses have been deeply and continuously modified over time and destructive analyses were not allowed. Beside direct visual inspection, where possible, the sizes and composition of the structural elements of the building envelope were estimated on the basis of literature, archive sources, drawings, photos and other documentation. Besides sources directly referring to the analyzed building, documentation and constructional descriptions of other Bläckhorn houses were used as well as general knowledge and literature on Swedish timber-house construction typologies from the same construction years [5].

The main source of information was a construction contract preserved in the LKAB archives that describes the works performed on the Bläckhorn houses from 1903 on. Yet, it was not always possible to use it as a reliable source, as the construction process of some elements changed over time. As an example, the contract describes outer walls composed by prefabricated 4-inch standing plank covered with impregnated sheathing paper and 25 mm panelling on both sides. This technique, though, was not yet in use when house B52 was built and historical sources rather suggest the use of thicker standing-logs assembled following the "Resvirke" technique [7:161–62]. The picture in Fig. 5 shows a Bläckhorn house with exposed logged structure without panelling. A 1943 survey of the house shows 22 cm-thick outer walls and panelled façades, a thickness compatible with a logged structure to which a timber panelling was later added.

Fig. 5 Detail of a picture by Borg Mesch (© Kiruna Bildsamlingen) representing house B54 (ca 1901-02). Timber logs are disposed vertically on the bottom floor, horizontally in the attic

Old pictures and drawings, together with the visual inspection of the existing elements, also suggest that a large part of the windows of the Bläckhorn houses were changed in the 70 s and that most of these are still in place.

The result of this part of the work was the definition of a plausible composition of the building elements of the B52 house that, with minor modifications, can be representative also for other variants of the Bläckhorn houses. Once the sizes and composition of the structural elements of the building envelope were established, transmission heat transfers were calculated for the total surface of the building envelope in contact with heated air.

The thermal transmittance (U-value) of the outer walls and the attic was calculated in accordance with the simplified calculation method described in ISO 6946 [12]. The thermal transmittance of elements in contact with the ground was calculated in accordance with ISO 13370 [10]. The average thermal transmittance of the building envelope was calculated as the transmission heat transfer between interior and exterior environments divided by the thermal envelope area according to ISO 13789 [11]. The thermal conductivities or resistances of the constituent materials/products and the thermal transmittance of windows, doors, and linear thermal bridges were drawn from tabulated, standard values. Point-type thermal bridges were neglected in the current work. The results of these calculations are summarised in Table 1.

The Swedish National Board of Housing, Building and Planning [6] conducted a survey on energy and thermal performance of 1400 residential buildings chosen as

Table 1 Thermal transmittances of structural elements in contact with heated air for the current construction of the house B52

	Thermal transmittance		Description
	(W/m^2 K)	(W/K)	
Attic	0.39	40	Timber frame structure with 170 mm sawdust fill insulation
Outer wall	0.48	110	220 mm wood structure with no insulation in most walls
Basement	0.93	205	120 mm concrete floor; 250 mm lightweight-concrete masonry wall
Windows	2.1	58	Triple pane, wood frame
Doors	1.4	7.6	Wooden with 50 mm foam fill insulation
Linear thermal bridges	0.062	27	
	Average: 0.77	Total: 448	

being "statistically representative" of the existing building stock. According to the results presented, multifamily buildings built before 1960 generally have U-values of 0.58 ± 0.07 W/m^2 K for exterior walls, 0.36 ± 0.07 W/m^2 K for the attic and 2.2 ± 0.07 W/m^2 K for the windows [6]. In comparison with the results presented by Boverket, house B52 represents a quite typical Swedish multi-family building from those construction years. Nevertheless, we should bear in mind that the changes in the construction over time have probably increased the insulation level of the house (e.g. the replacement of windows and doors or the likely addition of insulation on the attic). At the same time, a large part of the transmission heat losses at the time of our analysis originated from the heated basement (Fig. 6), which was built in the 1960s, when the B52 house was moved for the first time.

In comparison with the overall heat loss coefficient derived from the measured data (502 W/K; Fig. 3), the calculated total transmission heat losses (448 W/K; Table 1) is lower. This was expected since the measured heat loss include also ventilation heat losses caused by air leakages through the thermal envelope. The contribution of the chimney could also be important. Not so much for an air draught, since it is no longer in use and has been sealed, but because its brick structure likely acts as a huge thermal bridge (see Fig. 6), which was not considered in the transmission heat loss calculation.

6 A Value-Based Retrofitting of a Bläckhorn House

Proposals for an energy retrofit of the Bläckhorn house B52 were designed on the basis of a cultural value assessments of the impact they could have on the character-defining elements of the building. The proposed measures are intended to

Fig. 6 Infrared image of house B52 from outside. The poor thermal performances of many components of the building envelope are clear. Outdoor temperature: −13 °C

merge traditional conservation considerations (such as minimal intervention, compatibility with the existing materials and reversibility) with a life-cycle perspective in a pragmatic and cost-effective approach. For this reason, the proposed improvements are based on inexpensive technologies that are easy to access in the market and to install, especially considering that the houses are to be emptied for the moving. As an example, the main outer walls are left untouched in their exterior paneling and in their core of timber logs. The exterior paneling is considered crucial to keep a connection of the building with its historical context, considering also its movement to a new location and despite the maintenance cycles that have surely changed its aspect and materials over time. The timber log structure represents an irreplaceable material document of the technological evolution of the local construction techniques and of the passage from traditional to standardized methods in the early history of Kiruna architectures. The choice was thus an internal insulation, with the replacement of just the existing and relatively recent fiberboards with new wood-fiber insulation boards, which were considered compatible with the timber log structure.

Berggren and Wall [3] have shown that an external insulation would be preferred to reduce the risk of mould growth in a cold climate, especially considering the expected consequences of climate change in Sweden. However, the risk of mould growth should be smaller in the subarctic climate, with relatively low temperatures and the choice of a breathable, vapour-permeable material for insulation will ensure that the wall will maintain its ability to dehydrate moisture also after refurbishment. Moreover, the internal insulation could be quite easily removed if problems would appear in the future due to changed climatic conditions.

The existing windows are neither original nor of particular cultural value. Nevertheless it is proposed to keep them and to increase the U-value of the envelope by the addition of an inner secondary glazing, to be installed in continuity with the inner insulation. This solution would contribute to lower the thermal bridges and at the same time would let the existing windows complete their life cycle before being replaced. Beside saving some embodied energy in the present, this will give more freedom of choice and flexibility in the future. When the substitution of the existing elements will be needed, the rather good thermal performance already guaranteed by the secondary glazing could allow decision-makers to prioritize other requirements (e.g. the aesthetic quality or the visual compatibility with the other historic components).

The proposed improvements of the structural elements are presented in Table 2, including the reduction (in %) of the thermal transmittance and of the heating energy use of the building that each individual measure would result in. The heating energy savings were estimated using heating-degree days (Kiruna, base temperature 17 °C) for the monitoring period.

From these figures, it results that by only changing the ground construction from the heated basement to a contemporary, insulated crawl-space basement (as in the new location of the building), the heating energy use decreases by an estimated 31%. This means that the move of the house has already lead to a considerably increased energy efficiency. Even excluding the contribution of the change of the

Table 2 Thermal transmittances of structural elements in contact with heated air after the proposed refurbishment of the house B52

	Thermal transmittance		Description	Reduction of thermal transmittance (%)	Estimated savings of heating energy use (%)
	(W/m^2K)	(W/K)			
Attic	0.17	17	Removal of sawdust fill and addition of 300 mm cellulose loose-fill insulation	5	4
Outer wall	0.30	68	Addition of 50–80 mm wood fibre insulation board	9	7
Basement	0.27	33	Change to ventilated crawl-space basement with 100 mm foam insulation	39	31
Windows	1.2	33	Addition of secondary glazing	6	4
Doors	1.4	7.6	No changes applied	0	0
Linear thermal bridges	0.044	19	Estimated improvements due the proposed refurbishment measures	2	1
	Average: 0.30	Total: 178	All changes of individual elements implemented	60	48

basement, the cumulative effect of all the other measures would reduce the heating energy use by more than 15%. If all proposed refurbishment measures were implemented, the B52 house would fulfill the maximum allowed average thermal transmittance of 0.4 W/m^2 K, specified by the current Swedish building code [4]. This value is indicated as mandatory for new dwellings and it should be strived for if extensive changes to the building envelope are implemented.

Not included in Table 2 are proposals to decrease the ventilation losses, which are likely to be relevant to increase the energy efficiency of this building. Still, the proposed refurbishment measures of the building thermal envelope will contribute to lower the heat losses through air leakages. Nevertheless, if the air leakages in the thermal envelope are to be sealed, such as the air vents in the existing windows, alternative measures are required to ensure sufficient ventilation. A possible solution could be the installation of an air-to-air heat exchanger, taking advantage of the existing chimneys for the positioning of the ducts.

Furthermore, given the overheating detected by the monitoring, a simple and reliable measure to reduce the energy use for space heating would be to provide the existing radiators with electronically controlled thermostatic valves with a set-point of the air temperature of around 21 °C.

The development of these additional proposed measures, together with an accurate analysis assessing the consequences of the increased thermal performances of the envelope on the moisture transfers, could be the next steps for this study.

7 Conclusions

The main conclusions for the paper are:

- It was shown that the measures for the building envelope would allow a considerable reduction of the heating energy use of the Bläckhorn house B52, with a very limited impact on the preservation of its heritage values. A further reduction of energy use could be achieved by improving the ventilation system, which was not specifically addressed in this paper.
- The multidisciplinary method proposed for this study, merging qualitative and quantitative assessments and using a value-based approach in the design of energy retrofitting measures for historic buildings, was proved effective for this purpose. Still, more specific studies on the building physics would be required for a deeper understanding of the long-term consequences of the proposed measures, also considering climate change.
- The results obtained for B52, and the proposed measures, can, with minor modifications, be considered representative for the other variants of Bläckhorn houses. They can thus foster a re-discussion of the objectives of moving the next Bläckhorn houses and turn the current heritage safeguarding operation into an opportunity to increase the energy efficiency of Kiruna's built environment.

The presented study of the B52 house demonstrated that, if correctly assessed and designed, energy retrofit measures are not necessarily a threat for the preservation of cultural heritage buildings. In upgrading their livability, attractiveness and sustainability, these added layers of materials and value constitute a distinctive and meaningful sign of the needs, expectations and aims of our time and should therefore be considered an enrichment of the cultural significance of our built heritage.

Acknowledgements Funding for this study were provided by the Hjalmar Lundbohm Research Centre (HLRC), the Swedish Energy Agency through Spara och Bevara (project Smart energieffektivisering av kulturhistoriska byggnader i kallt klimat) and Interreg Nord programme (project EEBAK - EnergiEffektiva Byggander I Arktiska Kommuner).

References

1. E. Avrami, R. Mason, M. de la Torre (eds.), *Values and Heritage Conservation, Research Report* (The Getty Conservation Institute, Los Angeles, CA, 2000)
2. F. Bedoire, *Gustaf Wickman som sjukhusarkitekt: En studie i det tidiga 1900-talets vårdbyggande* (Fritzes hovbokh, Stockholm, 1974)
3. B. Berggren, M. Wall, Moisture conditions in exterior walls for net zero energy buildings in cold climate considering future climate scenario, in *Proceedings from 7th International Cold Climate HVAC Conference*, Calgary, (2012) p. 8
4. BFS 2011:6 Boverkets byggregler, BFS 2011:6 med ändringar t.o.m. BFS 2017:5, BBR 25. Boverket, Karlskrona
5. C. Björk, L. Nordling, L. Reppen, *Så byggdes villan: svensk villaarkitektur från 1890 till 2010* (Formas, Stockholm, 2009)
6. Boverket, *Teknisk status i den svenska bebyggelsen-resultat från projektet BETSI* (Boverket, Karlskrona, 2010)
7. L. Brunnström, *Kiruna - ett samhällsbygge i sekelskiftets Sverige* (Umeå universitet, Umeå, 1981)
8. EN 16883:2017, in *Conservation of Cultural Heritage—Guidelines for Improving the Energy Performance of Historic Buildings* (European Committee for Standardization, Brussels, 2017)
9. ICOMOS, *The Venice Charter: International Charter for Conservation and Restoration of Monuments and Sites* (International Council on Monuments and Sites (ICOMOS), Paris, 1964)
10. ISO 13370:2007, *Thermal Performance of Buildings—Heat Transfer Via the Ground Calculation Methods* (International Organization for Standardization, Geneva, 2007)
11. ISO 13789:2017, *Thermal Performance of Buildings—Transmission and Ventilation Heat Transfer Coefficients Calculation Methods* (International Organization for Standardization, Geneva, 2017)
12. ISO 6946:2017, *Building Components and Building Elements—Thermal Resistance and Thermal Resistance Calculation Methods* (International Organization for Standardization, Geneva, 2017)
13. T. Johansson, M. Vesterlund, T. Olofsson, J. Dahl, Energy performance certificates and 3-dimensional city models as a means to reach national targets—A case study of the city of Kiruna. Energ Convers. Manage. **116**, 42–57 (2016). https://doi.org/10.1016/j.enconman.2016.02.057
14. P. Joseph, *Bedömning av kulturvärden för byggnader inom Gruvstadsparken* (Kiruna kommun, Kiruna, 2010)

15. Kiruna kommun, *Beslut om bygglov. Bygglov uppförande flerbostadshus byggnad B52, D 2016-000376*. Kiruna kommun, Kiruna (2016a)
16. Kiruna kommun, *Beslut om bygglov. Bygglov uppförande flerbostadshus byggnad B53, D 2016-000378*. Kiruna kommun, Kiruna (2016b)
17. A. Krus, *Kulturarv-funktion-ekonomi: Tre perspektiv på byggnader och deras värden, Licentiatuppsats* (Göteborg, Chalmers tekniska högskola, 2006)
18. M. Legnér, A. Luciani, The historical indoor climate. A long-term approach to conservation environments within heritage buildings, in *Online Proceedings of the Conference BH 201*, ed. by M. Boriani, R. Gabaglio, D. Gulotta (Politecnico di Milano, Milan, 2013), pp. 1321–1328
19. D. Lowenthal, Stewarding the past in a perplexing present, in *Values and Heritage Conservation, Research Report*, ed. by E. Avrami, R. Mason, M. de la Torre (The Getty Conservation Institute, Los Angeles, CA, 2000), pp. 18–25
20. S. Muñoz Viñas, *Contemporary Theory of Conservation* (Routledge, London, New York, 2005)
21. G. Nordström, S. Lidelöw, H. Johnsson, Comparing energy signature analysis to calculated U-values in wooden houses in cold climate, in *Eco-Architecture IV* (WIT Press, Southampton, 2012), pp 1–9
22. V. Pracchi, Historic buildings and energy efficiency. Historic Environ. Policy **5**(2), 210–225 (2014)
23. RAÄ (2008) Det byggda kulturarvet i framtidens Kiruna. Sammanfattning av seminariet i Kiruna, 18–19 September 2008. Riksantikvarieämbetet, Stockholm
24. RAÄ, *Reviderad riksintressebeskrivning för Kiruna-Kirunavaara [BD 33] (Kiruna stad, Jukkasjärvi sn) 2010-05-19, dnr 331-00556-2009* (Riksantikvarieämbetet, Stockholm, 2018)
25. Regeringen, Möjlighet till undantag från de tekniska egenskapskraven på byggnadsverk. Lagrådsremiss (2007)
26. SFS 2010:900 Plan- och bygglag
27. J.-U. Sjögren, S. Andersson, T. Olofsson, Sensitivity of the total heat loss coefficient determined by the energy signature approach to different time periods and gained energy. Energy Build. **41**, 801–808 (2009). https://doi.org/10.1016/j.enbuild.2009.03.001
28. J. Sjöholm, Heritagisation, Re-Heritagisation and De-Heritagisation of Built Environments. The Urban Transformation of Kiruna, Sweden Doctoral Dissertation, Luleå University of Technology, Luleå (2016)
29. K. Yliniemi, A device and a method for measurement of energy for heating tap water separated from the building's heating energy-usage. Patent no. PCT/SE2006/001250, World Intellectual Property Organization, Geneva (2007)
30. K. Yliniemi, J. Delsing, J. van Deventer, Experimental verification of a method for estimating energy for domestic hot water production in a 2-stage district heating substation. Energy Build. **41**(2), 169–174 (2009). https://doi.org/10.1016/j.enbuild.2008.08.008

Renovating the Housing Stock Built Before 1945: Exploring the Relations Between Energy Efficiency, Embodied Energy and Heritage Values

Paula Femenias, Liane Thuvander, Pär Johansson, Paula Wahlgren and Petra Eriksson

Abstract Swedish multi-family buildings constructed before 1945 constitute an important part of the national built heritage. However, the majority does not have a formal heritage protection. Part of this building stock has already been renovated, notably through earlier energy saving programmes where additional exterior insulation, new façades and windows were frequently installed with little consideration for the original architecture. Now, 40 years later, these buildings face new renovations. This provides opportunities to improve the energy efficiency, indoor climate and user comfort. At the same time, the original architectural and historical characters lost in previous renovations could be recreated. In this paper, an inter-disciplinary research team illustrates the challenges met in practice to reach a sustainable renovation based on three cases. The case buildings are so-called "Landshövdingehus", constructed in the 1930s and owned by a public housing company. The relations between building physics, energy efficiency, embodied energy, and the effect on heritage aspects in renovation are studied. The results demonstrate the potential to reach 30% calculated energy efficiency without investing in ventilation systems. When comparing embodied energy to savings in operational energy a short payback time is achieved. However, focusing on the replacement of windows, the cases illustrate difficulties to recreate heritage values at same time as achieving an air-tight and energy efficient construction. In order to improve the results from the heritage point of view, there is a need for quality assurance of the renovation and building permit process.

Keywords Multi-family buildings · Energy efficiency · Embodied energy
Heritage values · Value conflicts

P. Femenias · L. Thuvander · P. Johansson (✉) · P. Wahlgren
Department of Architecture and Civil Engineering, Chalmers
University of Technology, 412 96 Gothenburg, Sweden
e-mail: par.johansson@chalmers.se

P. Femenias
e-mail: paula.femenias@chalmers.se

P. Eriksson
Department of Art History, Uppsala University, 621 67 Visby, Sweden

D. Johansson et al. (eds.), *Cold Climate HVAC 2018*,
Springer Proceedings in Energy, https://doi.org/10.1007/978-3-030-00662-4_25

1 Introduction

Swedish public housing owners must conform to European, national and local directives to reduce the energy use in their existing stocks. Energy savings can be addressed through an overhaul of systems and installations but can also require improvements of the building envelope. To reach cost-efficiency, energy saving measures are often planned in relation to extended maintenance and renovation [1]. Energy saving renovations involve a number of conflicts between different objectives, stakeholders and value areas which need to be considered when choosing a renovation strategy: the conflict between architecture, heritage values and energy savings [2, 3], the landlord-tenant dilemma, i.e. the distribution of value and costs as a result of renovations [4], or the relation between the decreased operational energy and the embodied energy of the added building materials in a renovation project [5].

In this paper, we focus on renovation of Swedish multi-family buildings constructed before 1945. These buildings constitute an important part of the built heritage. However, the majority does not have a formal heritage protection. A share of this stock has already been renovated, notably through energy saving programmes in the 1970s and 1980s. These energy renovations usually involved additional exterior insulation with a new façade and new windows. The alterations were frequently applied without consideration of the original architecture. It also appears that many renovations were carried out to compensate for neglected maintenance and to reduce the need for future maintenance, for example, by replacing wooden façades with corrugated metal sheets. Now, after 40 years, these buildings face new renovations as part of their maintenance cycle. This provides the opportunity not only to improve the energy efficiency but also to recreate or restore original architectural and historical characters lost in previous renovations.

Based on three case buildings, the paper illustrates how renovation measures will affect moisture safety, energy efficiency, embodied energy, and heritage values. The research is carried out by an interdisciplinary team involving architects, building physicists and a building conservator. The research is still on-going and what is presented is work in progress. The overall objective of the project is to provide a basis for guidelines about energy renovation of this part of the housing stock. The end-users' perception is studied in the research project but not presented in this paper. The research aims at enriching the knowledge base about viable and sustainable renovation strategies for historic buildings by extending the discussions beyond a domination of building conservation and energy saving criteria [6].

2 Method

The overall method is a case study with rich data descriptions and what can be regarded as typical cases with potential for replication [7]. Studies have been carried out post-renovation (Case A and B) to compare results before and after renovations.

The cases are part of a systematic development carried out by the public housing company Bostads AB Poseidon in Gothenburg in search for renovation solutions to be used as reference or prototype for housing stocks constructed before World War II. These buildings need measures addressing energy saving and improved indoor climate, which also permit the transformation of existing attics to additional apartments and rentable space. There is a high interest in transforming attics into apartments due to a general housing shortage in Sweden.

We have studied three blocks built in the 1930s, case A–C. Case A recently went through a renovation of the building envelope with new windows and additional exterior insulation of walls and the roof. The exterior insulation method permits the creation of new apartments in the attic. This possibility is not explored in Case A due to the small size of the attics in that building, but can be used in other part of the housing company's stock. In Case B, only an interior insulation of the attic/roof was performed together with installation of new windows. Case C is proposed to be renovated using the same concept as in Case A and is used as a reference for the status of the buildings before renovation. More information is found in Table 1.

Table 1 Overview of Cases A to C

Building	A	B	C
Year of construction/ renovation/ re-renovation	1937/1970/2015–2016	1939/1976/2014	1938/1970
Number of apartments	36	12	30 + 30
Attic	Insulated on the outside with 100 mm phenolic foam insulation, mineral wool at the wall to roof connection	Insulated on the inside, loose fill insulation, careful sealing	Ventilated cold attic, new roofing and a non-insulated outer roof
Windows	U-value new windows 1.1 W/m^{2}K. U-value old windows in the range of 3 W/m^{2}K	U-value new windows 1.1 W/m^{2}K. U-value old windows in the range of 3 W/m^{2}K	U-value existing windows in the range of 3 W/m^{2}K
Façade	Added 70 mm mineral wool, 25 mm ventilated air space. New façade with wooden boards and battens. Plastered ground floor	Façade repainted, wooden panels on all three floors	Asbestos plates on wooden boards on the two upper floors, bottom floor plastered
Heated area (m^2)	2,674	755	4,090
Energy use (kWh/m^2)	Before renovation 154[a] After renovation 93 (calculated)	Before renovation 182[a] After renovation 130[b]	Today 169[a]

[a]From EPCs

[b]Based on the delivered heating energy. 24 kWh/m^2 was added for energy use for hot water production and property electricity to enable comparisons

An inventory of the case buildings was made using different methods. Energy figures before the renovations were collected from energy performance certificates (EPC) and calculated figures for the energy use after the renovation were provided by the housing company. Information about renovation measures with respect to thermal performance, airtightness (thermal comfort), heritage values and aesthetics have been studied through observations, drawings and documents and interviews with involved actors. Data on the materials used for the renovation were provided by the contractor. For the assessment of heritage values, we have focused on visual aspects of the exterior appearances and the historical authenticity with reference to generic elements of heritage valuation [2] and to the heritage preservation programme of the City of Gothenburg [8]. The embodied energy of building materials has been calculated for building A, defined as the total primary energy use for the extraction, processing, manufacturing, delivery of building material to the building site and construction. The input data is generic from databases such as Ökobilanzdaten im Baubereich. Comparisons have been made for a 50-year scenario, including maintenance, indicating the break-even between savings in operational energy and embodied energy of the renovation.

3 The Case Buildings

The case buildings, constructed between 1937 and 1939, are located in Gothenburg and of the type "Landshövdingehus", a local recurrent housing type originally built for working class citizens, with one floor in brick or stone and two floors in wood. In the area of the case buildings, the influence by emerging functionalistic ideals of the 1930s is visible in few decorations and that they are constructed as lamella blocks in a strict parallel order breaching with the earlier typology of Landshövdingehus constructed around closed courtyards.

Although many buildings in the neighbourhood of the case buildings have façades altered by renovation, the area is identified with heritage values in the heritage preservation programme for the built environment [8]. The motivation for the heritage values is that the urban plan is a clear example of the functionalistic ideals and, in Gothenburg, the last large area built with "Landshövdingehus" (Fig. 1). Case A and B were specifically designed for families with many children and thus also has a value from a sociohistorical perspective.

In 1970, the façades of case Case A and C (Fig. 2) were covered in asbestos boards to reduce maintenance costs. No insulation was added. In the 2010s, renovations of Case A and B were initiated due to high energy use. In Case A, exterior insulation was added to the façades and the windows were replaced. Initially the housing company wanted to replace the asbestos façade with a new board material. After negotiations with the city planning office, a wooden façade was chosen instead, with the goal to reconstruct the appearance of the original façade. For the roof, a solution with exterior insulation was chosen. One advantage with this solution is that the tenants do not have to empty their attic storage during the renovation process.

Fig. 1 The housing area today and in the late 1930s. Case B can be seen in the upper right part. Case C has not yet been built. Photo: Eniro (left), and Gothenburg city museum (right)

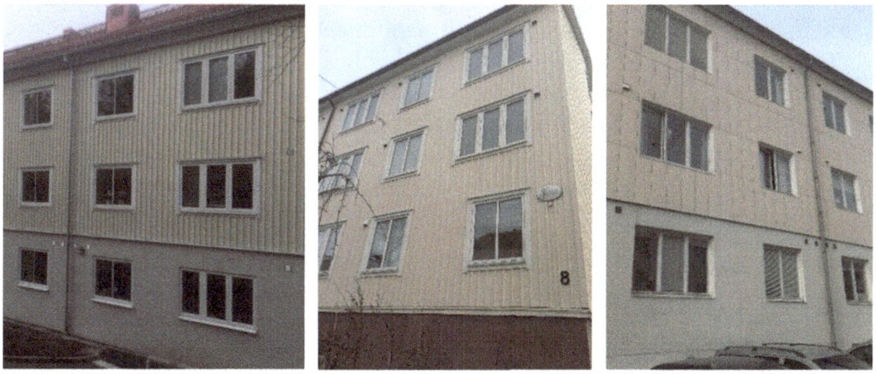

Fig. 2 Façade of Case A after the reconstruction of the wooden façade (left), Case B with unchanged façade (middle), and façade of Case C with the asbestos boards from 1970s (right)

In Case B, an alternative to the renovation method used in Case A was tested applying an internal insulation of the attic and replacing the windows. The status of the case buildings and the measures for the latest renovation are presented in Table 1.

4 Results and Discussion

Energy savings have mainly been achieved by insulating the thermal envelope. The heating and ventilation systems are unchanged, i.e. hot water radiators are connected to district heating and the ventilation is based on natural ventilation (stack effect and wind). After renovation, the indoor air temperature has been monitored in the different apartments (sensors placed on an inner wall in the living room).

In Case A (most recently renovated), the indoor temperatures are substantially higher than in Case B. As an example, 23 of February 2017, the outdoor temperature was 5 °C, and the indoor temperatures were ranging from 20 to 26 °C in Case A compared to 19–22 °C in Case B. However, overheating has decreased (in top floor) after insulating the attic in Case B. After the renovation in Case A, the energy use is expected to decrease from 154 to 93 kWh/m^2 (calculated). Additional savings are expected when the heating system is adjusted to the new indoor conditions.

4.1 Insulation, Windows and Outer Façade Measures

The largest renovation was made in Case A where the attic was insulated from the outside with 100 mm insulation (vapour open phenolic foam, $\lambda = 0.020$ W/(m K), system Clima Comfort, Monier) with an outer ventilated air gap under brick tiles, see Fig. 3. Walls were insulated with 70 mm mineral wool on the exterior.

The new windows are pivot windows with a U-value of 1.1 W/m^2K. The windows were, with respect to the original façade from the 1930s, placed in line with the outer wooden panel except for the ones in the basement (masonry walls). They were left in their original position because of the storage rooms in the basement and that moving the tenants' belongings for this operation was considered too complicated and costly. The new placement of the windows in the wooden walls proved to be difficult as it left an uninsulated air gap around the window causing problems with both conductive and convective heat losses. It is essential for energy, thermal comfort and moisture protection to have a good air barrier when renovating.

Fig. 3 Detail of the roof to wall connection and the wall design (left, building permit, 2015-04-09), Case A with recessed windows in the basement floor (middle) and new doors (right)

In Case A, one part of the building had an air barrier (wind barrier on the exterior) and one part lacked air barrier. In order to investigate how this affected the airtightness, one apartment of each type was tested (airtightness and air leakage search). The apartments had similar measured airtightness rates and penetrations were a large cause of air leakage (for example air inlets that were not properly connected). There were draught and cold surfaces around the windows (in total 8 examined windows). All windows except one had been changed and the original window was the only one without leakages. The fitting of new windows during renovation has often been proved problematic with respect to airtightness. Since there were leakages around several windows, the sealing was redone and extra sealing was applied using expanded polyurethane on the exterior and extra joint sealants on the inside.

The details in the original panel were more in proportion and thought through than the details on the new façades. The original panel is equipped with both window ledge and a decorative transition from panel to ground floor, while Case C only has window ledge. Almost all windows in the three buildings have been replaced and it is no longer possible to find original window settings and boards.

4.2 *Operational and Embodied Energy*

Case A has been used for a comparison of operational and embodied energy. To assess the impact of the embodied energy, the whole renovation process was studied, including raw material extraction and transportation, manufacture, transportation of building materials to the building site and construction work. Generic input data from sources like Ökobilanzdaten im Baubereich were used [9]. The operational energy was estimated using data from the EPC before renovation and calculated energy use data after renovation. The weight of material that was used or removed during the renovation were provided by the contractor.

The quantity of removed and added new materials reflects the main changes in the renovation. In Table 2, the top five materials (in terms of weight) that were removed or added are listed. The climate impact of the waste material only refers to

Table 2 The five most used components/materials in terms of weight during the renovation

Component	Material	Removed material (kg)	New material (kg)	Life span (year)
Roof tiles	Clay	38,600	35,640	30
Plaster	Plaster (gypsum)	22,500	22,500	30
Wood panels	Pine wood		16,360	30
Roofing frame timber	Pine wood		13,280	50
Windows	Glass, pine wood, aluminium	11,100	10,930	30

the transportation of the material (no production), and this is estimated to be less than 6% of the total embodied energy added during the renovation. Furthermore, the amount of embodied energy from manufacturing is strongly dependent on the choice of database [9]. In particular wood data is very different in the databases and, consequently, wood can be estimated to have a very large, or a minor, effect on embodied energy.

In terms of weight, roof tiles and plaster are important, but they have a relatively low impact on the embodied energy from the manufacturing process. On the contrary, insulation materials represent a small amount in terms of weight, but have a large impact in terms of embodied energy. As an example, insulation is 6% of the added material weight and 9% of the embodied energy. Roof tiles is 30% of the added material weight and 7% of the embodied energy. Windows, doors and balconies represent almost 25% of the embodied energy due to manufacturing of new materials.

The payback time (with respect to embodied and operational energy) of the studied renovation is estimated to 1–2 years using four different databases. Results from the Ökobilanzdaten im Baubereich database are shown in Fig. 4 (payback time 2 years).

4.3 Architectural and Heritage Values

In the early discussions for the building permit in Case A, the city planning office emphasised on a reconstruction of the original wooden façade with reference to the heritage protection programme and by indicating a quicker handling process. The wooden façade would be more expensive than the initial idea of using a board

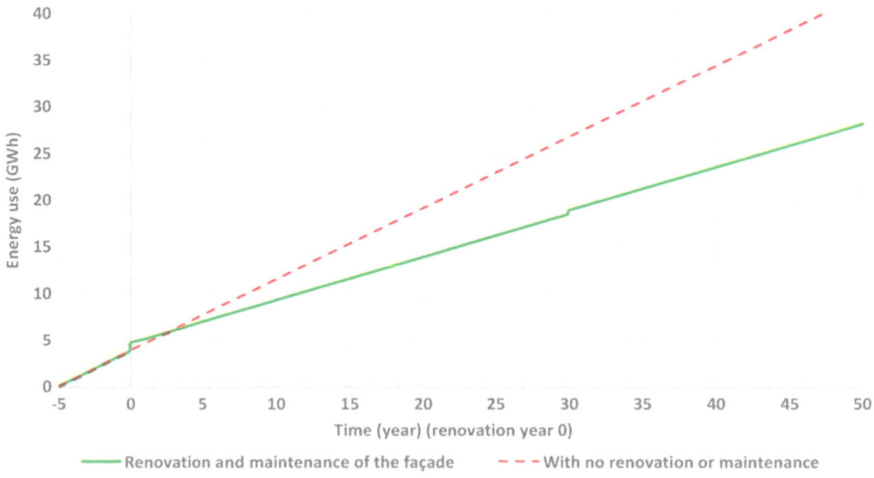

Fig. 4 Primary energy use for Case A over a 50-year scenario estimated with the Ökobilanzdaten im Baubereich database

material, but the property owner managed to get the tenants and the tenants association (responsible for rent negotiations in Sweden) approval for a small rent increase of approx. 25 €/month or just under 5% with reference to several value-additions for the tenants e.g. better indoor comfort due to new windows and new outdoor areas. The renovation process was halted for two years before an agreement with the tenants was reached.

In the design of the new façade no inventory of the original façade under the asbestos boards were carried out. The city museum was not approached and no building conservation officer was consulted. The design was made by an architect who searched inspiration from the surrounding area proposing a yellow wooden façade with laths covering the interstice between the wooden planks (as is found on for example Case B) and a grey coloured basement. The original façade revealed in the demolition showed a pale green colour and without any covering laths. However, according to the contractor, such a smooth surface would anyway have been difficult to recreate with contemporary material and production methods. The heritage preservation programme gives no detailed guidance regarding the façades, focusing instead on the importance of keeping the original urban plan. A heritage inventory carried out by the city museum in 1979, after the façade of Case A and C were covered in asbestos, proposes the original façades to be recreated, a wish that is now carried out.

The architect proposed the existing double casement vertically hinged windows to be recreated. These were also to be aligned with the new outer limit of the façade, not to create deep window sills, which is an important characteristic of local traditional housing. No detailed drawings were provided by the architect as the contractor took over the design as part of a design-build contract. The client, maybe out of concern for cost or future maintenance, instead choose pivot hanged windows with a false mullion to recreate the visual of a double casement window. The same kind of windows had earlier been chosen for the renovation of Case B. The choice of windows largely lowers the overall architectural impression of the buildings and is not true to the original heritage characteristics.

5 Conclusion

The renovation strategies in Case A and B demonstrate an interest in finding renovation solutions for public housing where heritage values are protected or even recreated. At the same time a substantially lower energy use is reached without investing in ventilation systems that usually also lead to necessary interior changes. The renovation strategies are also good choices with respect to the short payback when comparing embodied energy to savings in operational energy. While the strategy used in Case B has the advantage of being smaller and less resource intensive (with reference to material use), the thermal comfort is potentially better in Case A due to the insulated walls, which is also of importance.

However, in practice, there have been problems to reach the goals. In Case A, the calculated energy, has yet to be reached. There have been difficulties to achieve an air-tight construction. Partly, the higher energy use (compared to calculations) might also be explained by unnecessary high indoor temperatures. As for the windows, our studies illustrate problems to reach heritage and energy goals. Since the U-value of the window affect energy use, in Case B it might have been possible to improve the U-value without changing the whole window. Keeping the original windows would have been of interest both from a heritage perspective and from thermal comfort perspective. In Case A, additional insulation is added to the outside of the façade and the window location is adjusted for heritage and aesthetical reasons. There were substantial difficulties to achieve good airtightness around the windows (and prevent draught) in their new position, and to avoid large thermal bridges. The construction company would have preferred the original position of windows. Our results point to a need of adequate detailing in drawings, and instructions for the construction workers.

The housing company initiated Case A and B to test and evaluate renovation strategies for this part of their stock. The company considers the roof renovations having potential for replication in renovation of the remaining stock from the same construction period. An example of their learning process is an improved roof solution for Case C, where a more simple and cheaper roof solution than Case A is discussed.

However, the company still have some improvements to make regarding learning and innovation. The former energy strategist and a property engineer have left the housing company. This is problematic since documentation of the building process is somewhat sparse which makes renovation information dependent on staff knowledge. Therefore, processes for documentation during and after renovation projects need to be improved.

Finally, with respect to protecting and recreating heritage values in renovation, the cases point to the need for quality control in the building permit process. The final choice of windows should have been supervised by the city planning office that, when needed, can bring in building conservators as expert support. Detailed drawings of the windows should have been requested in order to assess the impact from a heritage point of view.

Acknowledgements This research has been funded by the Swedish Energy Agency within the Programme Save & Preserve, project number 40461-1. We thank Bostads AB Poseidon and Calles Bygg for access to data on the properties and the renovation.

References

1. K. Mjörnell, P. Kovacs, L. Hägered Engman, T. Gustavsson, P. Ylmén, in *Monitoring of indoor environment and energy use in the renovated buildings at Brogården in Alingsås*, in Proceedings from The Nordic Passive House Conference, 7–8th of October 2010, Aalborg, Denmark (2010)
2. T. Broström, P. Eriksson, L. Linn, P. Rohdin, F. Ståhl, B. Moshfegh, A method to assess the potential for and consequences of energy retrofits in swedish historic buildings. Hist. Environ. **5**(2), 150–166 (2014). https://doi.org/10.1179/1756750514Z.00000000055

3. H. Norrström, Sustainable and balanced energy efficiency and preservation in our built heritage. Sustainability **5**(6), 2623–2643 (2013). https://doi.org/10.3390/su5062623
4. B. Ástmarsson, P.A. Jensen, E. Maslesa, Sustainable renovation of residential buildings and the landlord/tenant dilemma. Energy Policy **63**, 355–362 (2013). https://doi.org/10.1016/j.enpol.2013.08.046
5. N.W. Brown, S. Olsson, T. Malmqvist, Embodied greenhouse gas emissions from refurbishment of residential building stock to achieve a 50% operational energy reduction. Build. Environ. **79**, 46–56 (2014). https://doi.org/10.1016/j.buildenv.2014.04.018
6. A.L. Webb, Energy retrofits in historic and traditional buildings: a review of problems and methods. Renew. Sustain. Energy Rev. **77** (Supplement C), 748–759 (2017). https://doi.org/10.1016/j.rser.2017.01.145
7. B. Flyvbjerg, Five misunderstandings about case-study research. Qual. Inq. **12**(2), 219–245 (2006). https://doi.org/10.1177/1077800405284363
8. G. Lönnroth, *Kulturhistoriskt värdefull bebyggelse i Göteborg* (Ett program för bevarande. Del I, Göteborg, 1999)
9. A. Jerome, Rapport de stage de recherche: renovation of the old multi-family building stock: energy and carbon impact of two cases (2017)

What Is the Minimum District Heating Supply Temperature in Residential Buildings in Norway?

Maria Justo Alonso and Igor Sartori

Abstract High temperature district heating systems are connected to high losses, therefore, new district heating systems have lower temperatures. In Zero Emission Neighborhoods (ZEN) the heat demands are reduced and 4th generation district heating (4DH) solutions opens as a solution to provide the heating demands with lower transmission losses. However, district heating (DH) has to cover demands in new, renovated and old buildings connected to the supply. Therefore the supply temperature cannot be reduced without further consideration. One strategy to cover demands is to connect older buildings with higher temperature requirements on the supply pipe and the new/renovated buildings on the return pipe. A second strategy is to use local boosters (at building level), e.g. heat pumps or boilers, in a low temperature DH network, to supply the buildings that require high temperature supply (old radiators, leaky envelopes…). Either way, the fundamental constraint is what is the minimum heating supply temperature in different building types? In turn, this holistic approach will map the minimum DH supply temperatures that the heating demands of all the buildings forming the ZEN can be covered. This paper analyzes for the case study of residential buildings in Norway according to the archetypes defined in the Tabula/Episcope projects. The focus of this paper is on single-family houses of the seventies. Each case is subdivided based on energy performance (original, standard renovation and ambitious renovation). The buildings were simulated in IDA Indoor Climate and Energy (IDA ICE). The effect of renovation on the achievement of thermal comfort levels compared to minimum supply temperature is studied.

Keywords Zero emission neighborhoods (ZEN) · Supply temperature
District heating · Single-family house (SFH)

M. J. Alonso (✉)
SINTEF Building and Infrastructure, Høgskoleringen, 7B 7034 Trondheim, Norway
e-mail: maria.justo.alonso@sintef.no

I. Sartori
SINTEF Building and Infrastructure, Forskningsveien, 3B 0373 Oslo, Norway

D. Johansson et al. (eds.), *Cold Climate HVAC 2018*,
Springer Proceedings in Energy, https://doi.org/10.1007/978-3-030-00662-4_26

1 Introduction

Energy use in buildings represents 40% of the energy use in Norway [1] and a big share is drawn from the electricity grid. In addition, oil furnaces and fossil fuels are being phased out so electricity demands may even increase. Addressing the supply of thermal demands through district heating is in practice a grid solution. Flexible and efficient use of peak power and energy is an important national issue [2]. The use of 4th Generation District Heating is seen as the key for making heat supply environmentally friendly. Lowering the supply temperature enables different renewable energy sources to be used. In addition, surplus heat from industrial processes or local overproduction could be sold to the space heating (SH) supplier, district heating (DH) in this study, reducing the needs for larger peak boilers. Thus, reducing needs for burning fuels and therefore bringing us closer to the environmental challenges related to cutting CO_2 emissions.

In Norway, there are some limitations regarding reducing the temperature in the DH network. Firstly, DH providers cannot supply temperatures below 70 °C as this is what is required to avoid *legionella* problems. Secondly, the DH networks in Norway have to supply heating to all the buildings in the concession and the supply temperature depends on the temperature requirements of the worst and furthest building. In order to use lower temperatures in the DH net there are two strategies that could be used to reduce supply temperature: One could connect the inefficient buildings that require higher temperature to the supply pipe and the buildings that require lower energy or supply temperature in the return pipe. Secondly, local boosters to supply the buildings with demands for high temperature can be considered so that the general DH temperature can be reduced yet the demands are covered. The main question is then how much can the DH temperature be reduced and yet cover all the heating demands of every type of building constituting the ZEN given that legionella problems are solved otherwise.

In this paper the effect of the states of the buildings, and their renovation level is studied. A single-family house (SFH) connected to the DH grid is simulated fixing the supply and return temperatures during the coldest day and the maximum mass flow rate through the radiators. The supply temperature would be fixed by the compensation curve of the DH. The return temperature would be calculated based on the heating demands to maintain a fixed comfort temperature, the supplied water flow rate and the heat gains. By doing this study on the non-renovated building and sequentially on the various stages of renovation, one could study the effect of the supply temperature of DH in the maintenance of comfort and on the return temperature to the DH. The temperature request of the buildings connected to the network could also be examined.

2 Simulation Procedure

In order to simulate the needs of the buildings connected to a DH network the heat load profiles are simulated in IDA Indoor Climate and Energy (IDA ICE). A study of the typologies and age classes has been made. The Norwegian building stock is divided in 21 segments [1], consisting of:

- 3 types of buildings: single-family house (SFH), terraced house (TH) and apartment blocks (AB)
- 7 age classes: Before 1955, from 1956–1970, 1971–1980, 1981–1990, 1991–2000, 2001–2010, 2011-afterwards.

A synthetic average building is defined for each segment. Its characteristics are representative of the most common features found in the segment, based on the best knowledge available. Each synthetic average building is described in 3 levels of energy performance (original, standard renovation, ambitious renovation). In this work, the focus is set on single-family houses. According to [2] most of the buildings built before 1950 are already renovated or most probably will not be renovated as they possibly are historical buildings. Most of the buildings built between 1950 and 1980 would have gone through renovations by 2015. Most of the buildings between 1981 and 2000 will be renovated by 2030. Therefore, this article presents a building built in the seventies and the renovations are done to the standard level in 2000 and to the standard defined in the Norwegian regulation TEK10 [3].

2.1 Building Specifications

The simulated buildings are based on the data available from [1] and [4] for a single-family house constructed in the seventies. Each building is simulated in IDA ICE as a multi-zone model where each room (12 in total) represents a zone. IDA ICE is an equation based simulation program for the study of indoor air quality of individual zones within a building. It also calculates energy consumption for the whole building or for a single zone [5].

The initial case (before renovation) is dimensioned based on a design outdoor temperature (DOT) of −20 °C in Oslo. Heating is provided to each building by non-idealized water radiators. Ventilation systems and infiltration rates are dimensioned based on relevant standards or normal practices as summarized in Table 1.

Buildings in the period 1971–1980 used to have exhaust ventilation only. In order to simulate this, extraction is foreseen in the bathrooms but no extraction is anticipated in the other rooms. The normal practice of window opening during night time is neglected for all the presented simulations. A constant extraction of 0.5 ACH is run constantly in the bathrooms and all the other rooms are ventilated by

Table 1 Summary of renovation actions and simulated ventilation and infiltration

Building standard	Actions taken	Ventilation	Infiltration rate
Not renovated		Only exhaust ventilation in the bathroom	Constant 4 ach
New windows	Windows renovated	Only exhaust ventilation in the bathroom	Constant 4 ach
Standard renovation (Std renovation)	Windows and tightness renovation	Balanced ventilation 0.5 ach	Constant 2 ach
TEK 10 renovation	Full renovation	Balanced ventilation 0.5 ach. Heat recovery efficiency 80%	0.6 ach at 50 Pa

means of a leaky building envelope (infiltrations). A constant infiltration rate of 4 ACH is introduced to account for the normal infiltrations based on [6]. This value is reduced after the renovation actions. No heat recovery for ventilation is simulated for the three first cases. Only when the house is renovated to TEK 10 level, a heat wheel with an efficiency of 80% is simulated.

The initial district heating in the original building is simulated with the supply and return temperatures for the coldest day, Tin = 90 °C and Tout = 70 °C, respectively. The outdoor compensation curve is simulated as shown in Fig. 1. DH in Oslo today is outdoor temperature compensated with a supply temperature of 120 °C for the coldest periods and it never goes below 80 °C as legionella protection.

The radiators are dimensioned based on the heating demands for the coldest day in January (−20 °C in Oslo). "Ideal heaters" are simulated to dimension heating needs and sizes. Based on this simulation, sizing of the emitters is done for the initial state and this size is kept for the case of the new windows and the standard renovation. For the year simulation, standard heat loads defined in NS 3031 are considered. For the ambitious renovation to TEK 10, the sizing of the radiators is recalculated as a change on the heating system is normal practice. The simulated control is a proportional control. Thermal bridge values are assumed to be large, and are set according to typical "poor" values in IDA ICE for the three first cases.

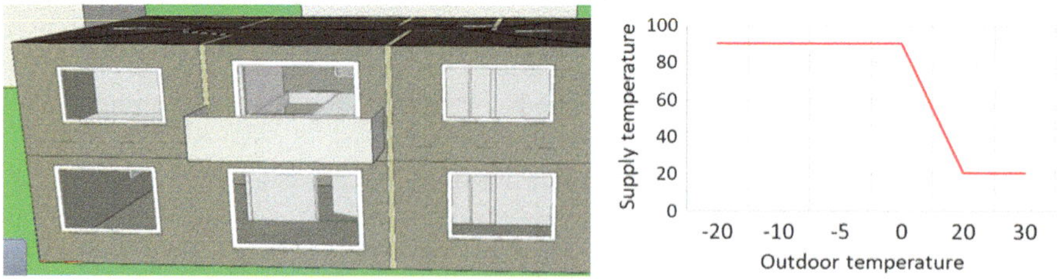

Fig. 1 Drawing of the simulated building (left), SH supply temperature compensation curve (right)

For the TEK 10 level, this value is changed to "good". The building envelope for the initial and renovated stages are simulated as summarized in Table 2.

Simulations are performed based on four different district heating supply temperatures for DOT: 90, 80, 55 and 40 °C. The radiator supply and return temperatures are set to 90/70, 80/60, 55/35 and 40/21 °C, while the set point for indoor air temperature is maintained at 20 °C everywhere in the house. The second step is to repeat the same supply temperatures with stepwise standard renovation. First the windows are renewed, then the insulation level and airtightness are improved and third, both are changed. Thirdly, the same is repeated for the case of renovation to TEK 10 with insulation and window quality as defined in Table 2. For readability purposes only the non-renovated case presents the results for all the temperatures. The process is summarized in Fig. 2. The simulations purpose is to check how low the supply temperature from the district heating may be, while ensuring thermal comfort at 20 °C and considering the constrain that the mass flow cannot be risen over 800 kg/h. This means that is the heating needs are such that if the maximum mass flow limit is reached, the comfort requirement would not be satisfied.

3 Results

3.1 Simulations of the Non-refurbished House with Lower Supply Temperature

The results from Figs. 3 and 4 are used for the discussion. For the coldest periods, the supplied mass flow increases when the outdoor temperature drops. The mass flow rate from the DH to the house is limited to 800 kg/h. Figure 4 shows that for the coldest periods (below −5 °C) when lowering the supply temperature, this is the circulated flow. Figure 3 shows the supplied temperature to the radiators and the return to the district heating. For the coldest days, the return temperature is as fixed. During outdoor temperatures ranging from −10 to −5 °C the return temperature increases due to the high water flow rate and lower heating demands. Over −5 °C the return temperatures drop as a result of lower heating demand and lower mass flow rate. Over +20 °C the temperature oscillates as a result that there is no more heat supplied by the DH network (see Fig. 1, compensation curve) and the indoor temperature will fluctuate with solar radiation and internal heat loads.

During periods with low outdoor temperatures, the comfort temperature in the coldest room (bedroom North, second floor) is not achieved when the DH temperature is reduced as shown in Fig. 4. When the DH supply temperature is reduced to lower temperatures than design (90/70 °C), problems related to indoor thermal comfort arise.

When reducing DH supply temperature, the water flow rate through the radiators needs to be increased to provide the same heat to the rooms. However, due to limitations related to vibrations and maximum velocities through the installed

Table 2 Construction and respective U-values for the different components of the building for the initial and following renovation levels

Component	Specification	U-value initial built (W/m²K)	Specification	U-value standard renovation (W/m²K)	Specification	U-value TEK10 renovation (W/m²K)
Roof	48 × 198 mm beam 200 mm min wool	0.20	50 mm additional min wool in cold attic	0.16	50 mm additional min wool in cold attic	0.16
External wall	Light timber framework, 48 × 98 mm	0.41	50 mm additional min wool on the outside	0.29	150 mm additional min wool on the outside	0.19
Windows	Double-glazed window, regular glass, air	2.60	Double-glazed window, one LE-coating, air	1.90	TEK10-window	1.20
Floor	48 × 148 mm beam 150 mm min wool	0.22	50 mm additional min wool in cold basement	0.20	100 mm additional min wool in cold basement	0.15

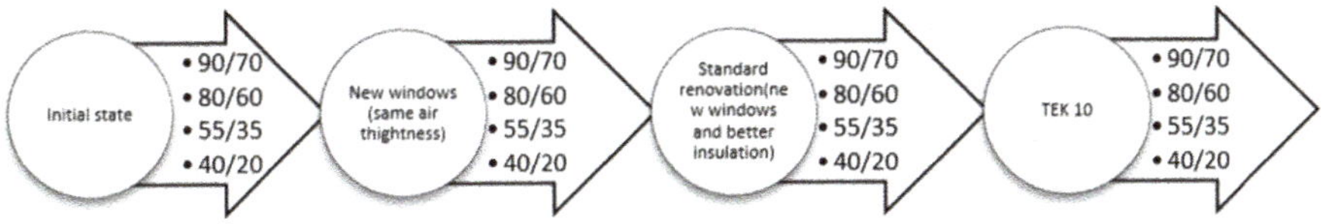

Fig. 2 Simulation process

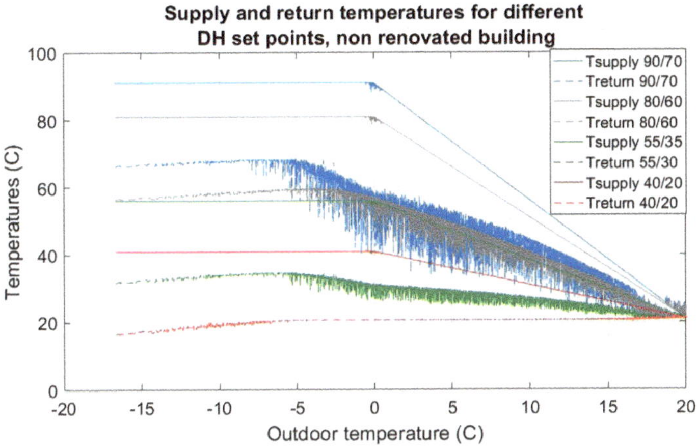

Fig. 3 Supply and return temperatures when the house is not renovated

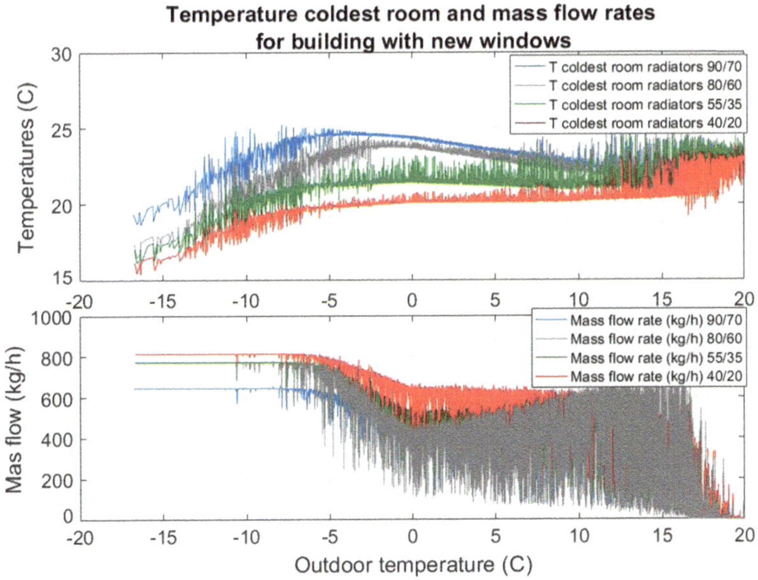

Fig. 4 Temperature in the coldest room of the non-renovated house for decreasing supply temperatures in the DH network below: mass flow supplied to the non-renovated house for decreasing supply temperatures in the DH network

piping this is not feasible. In order to provide the same thermal comfort, the area of the radiators would need to be increased, but as constrain of the project this is not done in the simulations.

Figure 4 shows the temperatures in the coldest room of the non-renovated house. For a DH supply of 40 °C the air temperature in the coldest room is below 20 °C when outdoor temperature drops below −5 °C. This happens 7% of the time. For a room with higher occupancy, as for example the living room, the temperature drops below 20 °C only 3% of the time as higher internal loads are considered. If the heating set point in these rooms would increase to 22 °C with a maintained supply of 40 °C, thermal comfort would not be achieved 72% of the time. These results are in line with those obtained by [7] for the Danish case.

Figure 4 shows the simulated mass flows to try to supply the heating demands. In the simulation, the supply and return temperatures are fixed for the coldest day but the mass flow is floating with the maximum limitation of 800 kg/h. The break point of this non-refurbished SFH is about 18 °C which agrees with [8]. After 0 °C outdoor temperature, the supply temperature starts to decrease. This is compensated by the regulation system by increasing the water flow rate, but as the needs are so low, the system oscillates. A better control should be considered analyzing these oscillations and a PI controller would have been more suitable.

Note that in these simulations the infiltration rates are very high, as we are considering houses constructed in a period where the importance of airtightness was neglected. For this period [6] measurements have shown infiltration rates ranging between 2 and 10 ach. This has a very strong effect in the heating needs.

When reducing the DH supply temperature from 90 to 40 °C, the mass flow rates through the radiator would need to be increased. In this case, even after doubling the water flow rates, the temperature in the coldest room would be below 18 °C during 5% of the time.

3.2 *Simulations of the "New Windows" and "Standard Refurbished" SFH Standard*

Table 2 describes the materials used for the refurbishment. The case called "new windows", only considers that the windows are replaced. However, the airtightness and insulation are kept equal. The "standard renovation" case considers that both windows, building envelope and airtightness are improved according to Tables 1 and 2.

Figure 5 shows that when new windows are installed but the airtightness is not improved, the comfort temperature is not reached if the radiators are not changed and the mass flow restriction is kept. In this case for 4% of the time, an indoor air temperature of 20 °C is not achieved. When a standard renovation is considered: the comfort temperature threshold is reached during 100% of the time.

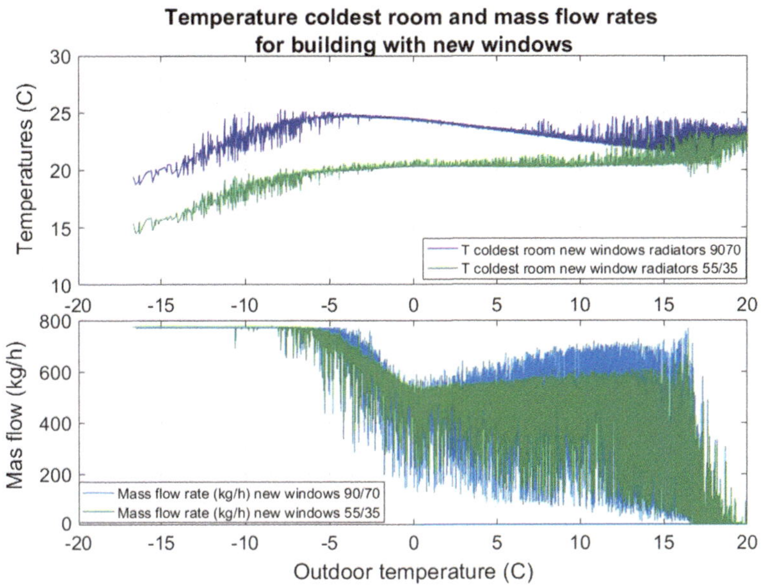

Fig. 5 Temperature in the coldest room (top) and mass flow in supply and return pipes (bottom) when the windows are changed

As for the previous SFH standard, there is a large oscillation of the mass flow during the warmest periods. Over 10 °C there are still heating demands, however, they are low and may be covered by solar heat loads and internal gains. Therefore, based on the existence of these heat loads or not, the water flow rates oscillate too largely, proving that a different controller should be considered for the regulation of water flow rates. In this case the control was a proportional, but using a PI controller would have been better.

Regarding the supply and return temperatures from DH, Fig. 6 shows a similar profile as Fig. 3. As the heat losses related to ventilation and infiltration are not improved for this renovation state, the profile of heating demands is similar to the previous case. The same evaluation can be done for this case.

When the building is further tightened and ventilation and infiltration heat losses are reduced through renovation, the heat demand profiles are changed. Figure 7 shows that during the coldest period the comfort temperature is achieved despite a reduction of the DH supply temperature. This is achieved by almost doubling the water flow rates. From a DH provider point of view, to cover the comfort requirements would mean large investments on larger piping systems for very little economical gains. During these cold periods, the high circulated mass flow effects as well as having a larger return temperature (Fig. 8) is not positive for the DH providers. In this case, a local solution as a heat pump or storage should be considered to reduce DH investment. The results obtained here are in line with those of [7]. For the remaining part of the year, the return temperature as shown in Fig. 8 is low, which is positive.

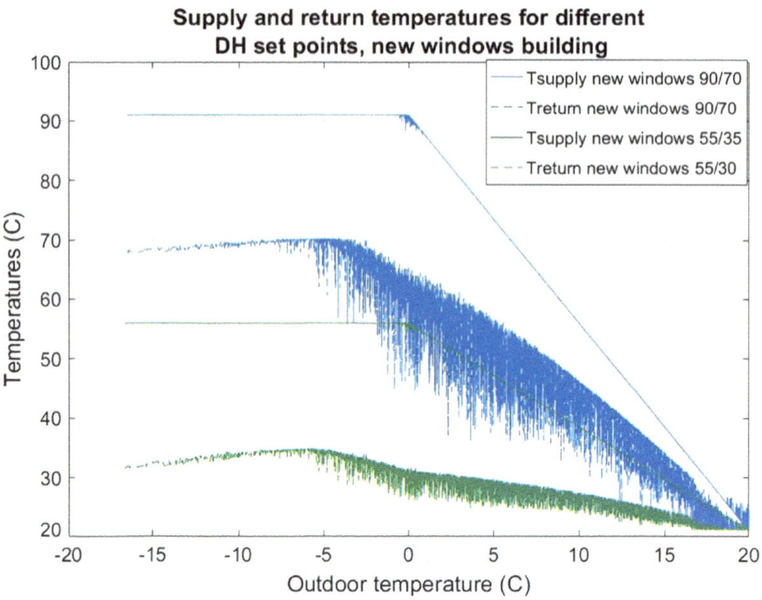

Fig. 6 Supply and return temperatures when the windows are changed

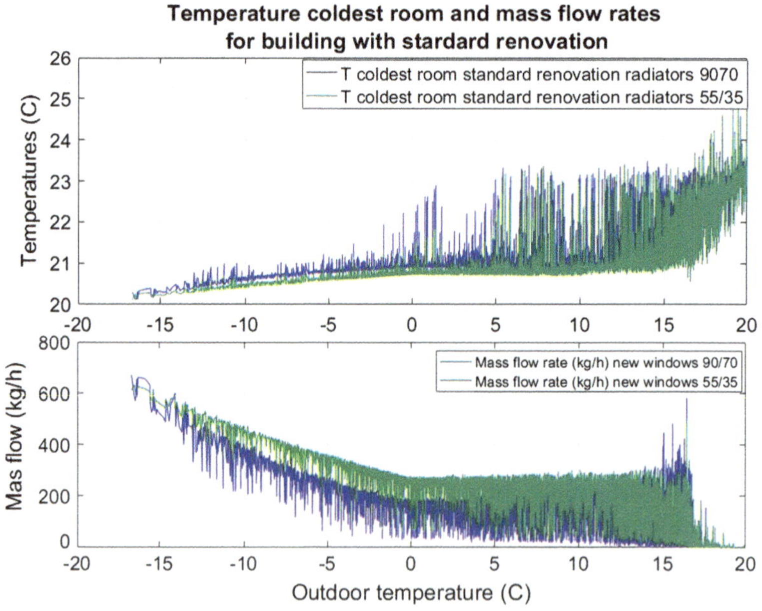

Fig. 7 Mass flows and temperatures in the coldest room (secondary axes) in supply and return after the standard renovation

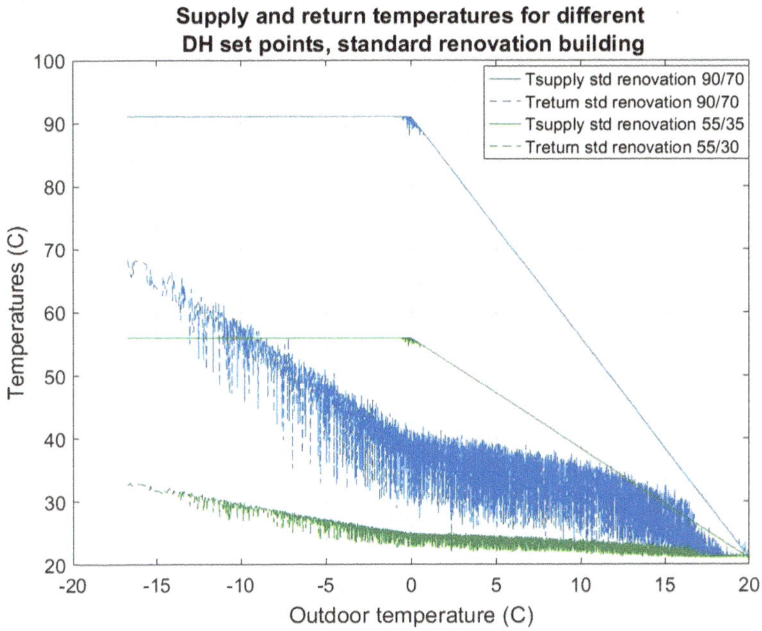

Fig. 8 Temperatures in supply and return after the standard renovation

3.3 *Simulations of the SFH Refurbished to TEK 10 Level*

When renewing to TEK 10, it is no problem to achieve the comfort temperature. However, the same problem as in Fig. 7 is encountered in Fig. 9. In order to see possible improvements in this case, the return temperature has been left floating and the flow limitation has been reduced to 300 kg/h, but the control is still proportional. In this case we assume that radiators are changed, the airtightness is reduced to 0.6 ach and balanced ventilation with 80% heat recovery is installed.

For this case, when the coldest period hits, the flow is again increasing from about 50 kg/h when the supply is at 90 °C to 100 kg/h when the outdoor temperature decreases to −15 °C. Again, for a brief period of supply, and little energy sold, the DH company would have to install larger piping, making the investment larger and resulting in a longer payback time. For this kind of situation, it seems to be a real need to think about new ways of supply. Regarding the warmer period we see the same challenge regarding the regulation and again a PI for the regulation of radiators is recommended.

In addition, Fig. 10 shows that the return temperature from this building to the DH net is low for both cases, enabling to completely use the installed capacity of the DH supply. The coverage of control demands with lower temperature supply

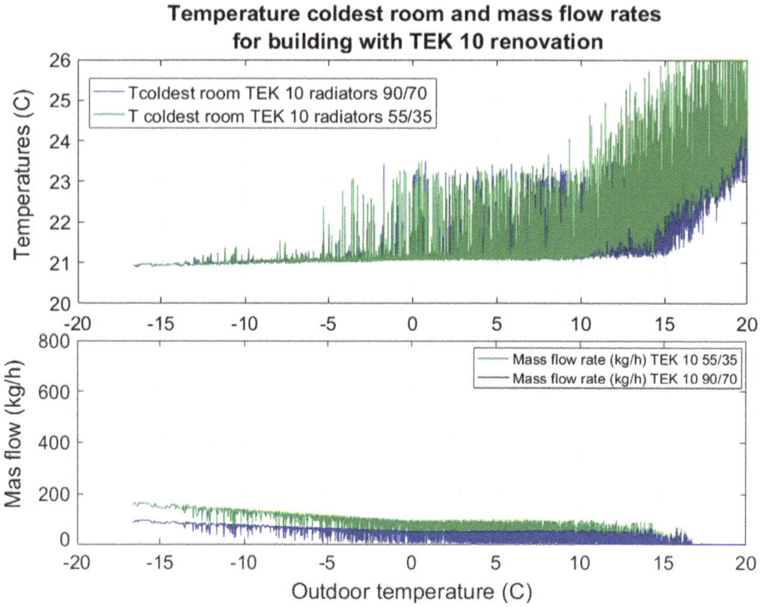

Fig. 9 Mass flow needed and temperature achieved in the coldest room when DH is supplied at 90 and 55 °C

from DH opens for the possibility to connect this type of building to the return piping. This ensures a larger utilization of the supplied energy from the DH provider. However, for these cases, the challenge would lie on the production of domestic hot water. When these highly efficient buildings would be connected to the return, they would require local solutions for the domestic hot water. The health authorities set as a requirement that the domestic hot water is supplied at 70 °C to avoid *legionella*. However, there is an ongoing discussion about the *legionella* requirements. These vary within the Scandinavian countries, where Norway has the most restrictive regulations.

The solution of connecting highly efficient buildings to the return flow is suitable as long as the whole building mass is not fully renovated. The larger the share of renovated buildings connected to the return flow, the lower energy would be available on the return pipes, and new grid solutions need to be implemented (probably reducing the overall temperature in the network and connecting to supply).

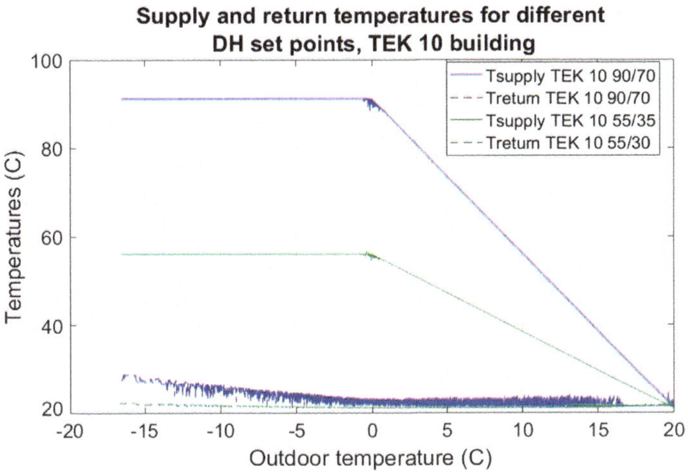

Fig. 10 Supply and return temperature when DH is supplied at 90 and 55 °C

4 Conclusions

From this study, it becomes clear that lowering the grid temperatures cannot be done without reservation. When a non-renovated area connected to high-temperature DH suffers supply temperature reduction, there are some challenges that need to be considered. New business models need to be developed for the development of DH. In detail:

- Lowering the temperature of the network when buildings are not adapted and radiator change is not applied, may yield risk of thermal discomfort. When the space heating emitter area is not increased, the design comfort temperature may not be achieved. A studied solution is to increase the mass flow rates, however, for some cases, even doubling the mass flow rate would not be enough. If in addition, the comfort temperature is raised, these buildings would not achieve thermal comfort during a large share of the year.
- For the non-renovated buildings, when supply temperature is decreased, solutions as extra peak load boilers should be considered for the coldest part of the year.
- For renovated buildings, depending on the extension of the renovation, the comfort may be achieved even for 100% of the time. However, it seems that in the coldest periods, again there would be a need to increase the temperature locally to avoid oversizing the DH network beyond economically interests.
- Finally, in the SFH renovated to the highest standard, the demands for heating would be very low. However, the peak demand would still be high in the coldest days of the year.

- In this paper, the supply of domestic hot water is not considered, however, we can foresee a challenge to supply the 70 °C that the Norwegian authorities demand, when lowering the supply temperatures. In these cases, local solutions need to be studied and optimized.
- Knowledge of how the types of buildings connected to the DH would react given a DH supply temperature reduction is needed for all the building classes before one considers reducing the supply temperatures in the DH network.

Acknowledgments The authors gratefully acknowledge the support from the Research Council of Norway and several partners through the Research Centre on Zero Emission Neighbourhoods in Smart Cities (FME ZEN). Grant number 257660.

References

1. H. Brattebø et al., *Typologier for norske boligbygg. Eksempler på tiltak for energieffektivisering* (2014)
2. N.H. Sandberg, I. Sartori, H. Brattebø, Using a dynamic segmented model to examine future renovation activities in the Norwegian dwelling stock. Energy Build. **82**, 287–295 (2014)
3. DiBK, D.f.b., *Byggteknisk forskrift med veiledning (TEK10)* (2010)
4. C. Mjønes et al., *Potensial- og barrierestudie Energieffektivisering av norske boliger-* (2012)
5. Equa, ed. by *User Manual IDA Indoor Climate and Energy* (2013)
6. P. Blom, S. Uvsløkk, *Bygg tett!* Prosjektrapport 98, SINTEF serie (2012)
7. M. Brand, S. Svendsen, *Renewable-based low-temperature district heating for existing buildings in various stages of refurbishment.* Energy (2013)
8. T.H. Dokka, C. Grini, *Etterprøving av bygningers energibruk.* Enova (2013)

An Assessment of the QUB/e Method for Fast In Situ Measurements of the Thermal Performance of Building Fabrics in Cold Climates

Johann Meulemans

Abstract The QUB/e method is a dynamic measurement method developed to estimate the whole heat loss coefficient and local U-values of a building in a single night without occupancy. The ability of the method to provide reliable results was demonstrated experimentally in a climate chamber with controlled conditions in a previous work. This paper presents the findings from a series of in situ measurements carried out in a circa 1960s multi-family housing located in Stockholm area (Sweden). The U-values estimated with the QUB/e method were in good agreement with the quasi steady-state (ISO 9869-1) values (i.e., the relative differences were within the uncertainty bound of the measurement methods). It was thus demonstrated that the QUB/e method can deliver a good estimation of the thermal performance of building fabrics within just one night, significantly less than the 2–4 week period required for quasi steady-state methods.

Keywords Heat loss coefficient · U-values · QUB/e method · ISO 9869-1

1 Introduction

The performance gap between the design and the as-built thermal performance of buildings is an issue of high importance for the construction industry and reducing this gap has been the subject of significant research (e.g., see [1–3] and references therein). Assessing the design thermal performance is straightforward; however determining the as-built performance of a building presents several technical challenges that often restrict its wider implementation. One of the most common challenges is the time required to perform such testing [4].

The whole Heat Loss Coefficient (HLC, in W/K) of a dwelling is the total power needed to maintain a constant interior/exterior temperature difference at steady-state (i.e., the inverse of a thermal resistance). The whole HLC is thus the aggregate of

J. Meulemans (✉)
Saint-Gobain Recherche, Aubervilliers, France
e-mail: johann.meulemans@saint-gobain.com

D. Johansson et al. (eds.), *Cold Climate HVAC 2018*,
Springer Proceedings in Energy, https://doi.org/10.1007/978-3-030-00662-4_27

transmission heat losses (including thermal bridges) and infiltration heat losses; it characterises the thermal performance of the building fabric. It should be noted that its value may be prone to variations due to the influence of the exterior environment (e.g., wind) on either the transmission losses or the infiltration losses.

The QUB/e method is a dynamic measurement method developed to estimate the whole HLC and local U-values of a building in a single night without occupancy [5, 6]. This makes it suitable for large scale use by industry to test the actual thermal performance of buildings. The ability of the method to provide reliable results was demonstrated experimentally in a climate chamber with controlled conditions for both the whole HLC and local U-values [5, 6], and in a real climate for the whole HLC only [7–9].

In this study, a comprehensive set of in situ measurements were performed in a circa 1960s multi-family housing (MFH) located in Årsta (Stockholm area, Sweden) to (i) evaluate the thermal performance of the building fabric (i.e., whole HLC and local U-values) before a full retrofit programme, and (ii) validate, in a real climate, the QUB/e method by cross-comparison with quasi steady-state measurements.

This paper is organised as follows. The materials and methods used in this study are described in Sect. 2. The results obtained from in situ measurements are presented and discussed in Sect. 3. Concluding remarks can be found in Sect. 4.

2 Materials and Methods

2.1 Description of the Building

The measurements took place in an apartment on the 11th floor of a circa 1960s multi-family housing (MFH) located in Årsta (Stockholm area, Sweden). A view of the building and the layout of the apartment are shown in Fig. 1.

The wall construction is aerated concrete and the windows are double glazing units (DGU) with wood frames. The apartment has South-West orientation, floor area, attached area and net heated area of approximately 90, 224 and 269 m^2, respectively. It should be noted that the window-to-external wall ratio is quite high (i.e., 35%) and the proportion of the net heated area in contact with the exterior is low (i.e., 45 m^2 or 17% of the total net heated area) but typical for an apartment in a MFH.

2.2 Heat Flowmeter (HFM) Method (ISO 9869-1)

The heat flowmeter method (HFM) [10] is a quasi-static approach used to estimate the thermal transmission properties of plane building components, primarily

Fig. 1 (Left) View of the external façade (the red rectangle corresponds to the location of the apartment). (Right) Layout (BR = bedroom, K = kitchen, LR = living room)

consisting of opaque layers perpendicular to the heat flow and having no significant lateral heat flow (i.e., relatively far from thermal bridges). The appropriate location (s) may be investigated by infrared thermography (in accordance with standard EN 13187 [11]). The heat flux and the temperatures are monitored and the thermal transmittance U (in W/m^2-K) can be computed with the following formula (average method):

$$U = \frac{\sum_{j=1}^{n} q_j}{\sum_{j=1}^{n} \left(T_{int,j} - T_{ext,j}\right)} \quad (1)$$

where q_j, $T_{int,j}$ and $T_{ext,j}$ are, respectively, the heat flux density (in W/m^2), the interior environmental (ambient) temperature (in K) and the exterior environmental (ambient) temperature (in K) of the jth individual measurement. The duration of the test should exceed 72 h (normative). Usually, the thermal transmission properties can be determined after a couple of weeks for a building in the field. The total uncertainty of the measurements is expected to be between 14 and 28% if the standard ISO 9869-1 is followed carefully [10].

The HFM (ISO 9869-1) can be performed during a co-heating test [12, 13], the whole heat loss coefficient and local U-values can be thus determined simultaneously (e.g., [14]). A better accuracy of the in situ U-values can be obtained thanks to a stable elevated temperature during heat flux measurement period.

The evolution of air temperatures and heat flux densities passing through a glazing and an external wall during an ISO 9869-1 measurement period in Årsta (Sweden) is illustrated on Fig. 2.

2.3 QUB/e Method

The QUB method [7–9, 15–17] is a dynamic method developed to determine the as-built whole HLC of dwellings within one night without occupancy. The test commences after sunset and finishes before sunrise of the following day (see Fig. 3). The principle of the QUB method is based on a single resistance and capacity model where the building is represented by a global thermal resistance R (the reciprocal of HLC of the building) and a global capacitance, C (the internal heat capacity). Internal and external temperature nodes (T_{int} and T_{ext} respectively) are considered homogeneous and heat exchange between the two nodes occurs through the thermal resistance, R. Therefore, the energy input, P(t), is heat lost through the envelope and stored/released by the thermal mass of the fabric (see Eq. 2).

$$P = \frac{T_{int} - T_{ext}}{R} + C\frac{dT_{int}}{dt} \tag{2}$$

The two unknowns, R and C in Eq. 2, can be determined by using two different constant powers in two different phases (respectively noted 1 and 2 in Fig. 3). The HLC can then be determined by the following formula:

$$HLC = \frac{T_2' P_1 - T_1' P_2}{T_2' \Delta T_1 - T_1' \Delta T_2} \tag{3}$$

where P_i is the power input during phase i, T_i' is the slope of the temperature profile at the 'end' of phase i (i.e., for $t \in [t_{i0} + d_i - \min(d_i/2, \tau), t_{i0} + d_i]$ where t_{i0} is the beginning of phase i, d_i the duration of phase i and $\tau = 4$ h) and ΔT_i is the internal to external temperature difference at the end of phase i. For increased accuracy it is aimed that the test is carried in an empty unoccupied dwelling without additional heat sources and in most cases there is almost no power input in phase 2 ($P_2 \approx 0$ W), i.e. this is the free cooling phase.

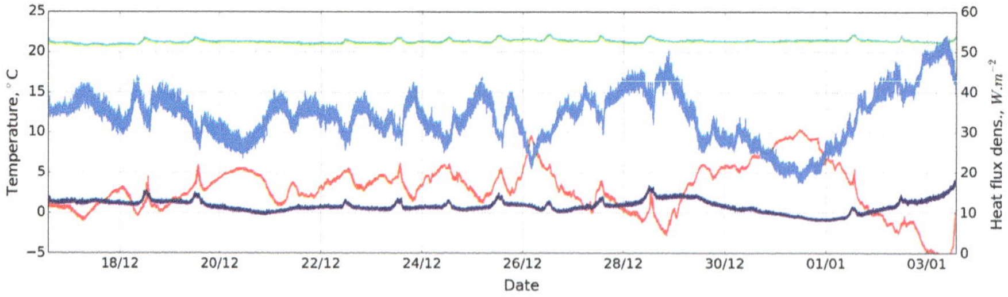

Fig. 2 Evolution of air temperatures (green solid line = internal air temperature, red solid line = external air temperature) and heat flux densities (light blue solid line = glazing, navy blue solid line = external wall) during an ISO 9869-1 measurement period in Årsta (Sweden)

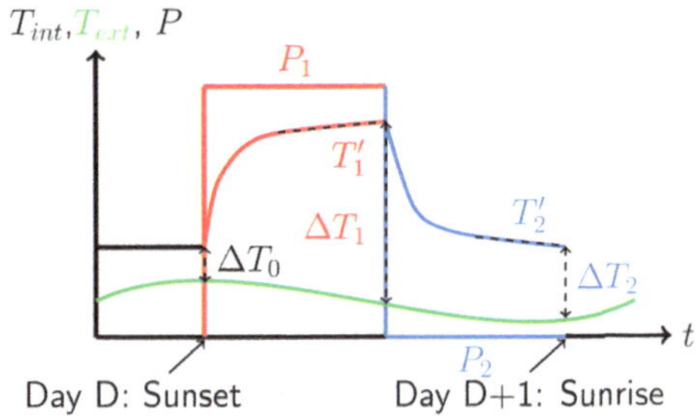

Fig. 3 Evolution of temperature and power during a QUB test

The error on the estimated HLC with the QUB method depends on a dimensionless parameter α [5, 6, 8, 9, 17] defined as follows:

$$\alpha = 1 - \frac{HLC_{ref} \Delta T_0}{P_1} \tag{4}$$

where HLC_{ref}, P_1 and ΔT_0 are a reference heat loss coefficient (theoretical or determined experimentally, in W/K), the heating power (in W) and the initial temperature difference (in K) between the internal and external environment (i.e., at the beginning of a QUB test), respectively. The QUB method is able to provide reasonably consistent results (i.e., with a coefficient of variation of ±10%) provided the dimensionless parameter α lies within the recommended range (i.e., between 0.4 and 0.7) [5, 6, 8, 9, 17].

With the QUB/e method [5, 6], heat flux densities and nearby air temperatures for each building element of interest are monitored during a QUB test. The QUB analysis procedure is then used to derive the U-values of each building element:

$$U = \frac{T_2' q_1 - T_1' q_2}{T_2' \Delta T_1 - T_1' \Delta T_2} \tag{5}$$

where q_i, T'_i and ΔT_i are, respectively, the mean heat flux density, the slope of the inside air temperature and the inside/outside air temperature difference at the 'end' of the phase i (i.e., for $t \in [t_{i0} + d_i - \min(d_i/2, \tau), t_{i0} + d_i]$ where t_{i0} is the beginning of phase i, d_i the duration of phase i and τ = 4 h).

The Taylor series method for uncertainty propagation [18] is used to compute the relative uncertainty associated to the HLC and the U-values.

The evolution of the heating power, the air temperatures and the heat flux passing through a building element during a QUB/e test in Årsta (Sweden) is illustrated on Fig. 4.

The whole HLC and the local U-values of a building can thus be estimated in one night using the QUB/e method. For more details on the QUB/e method and previous work, the interested reader should refer to [5–9, 15–17] and references therein.

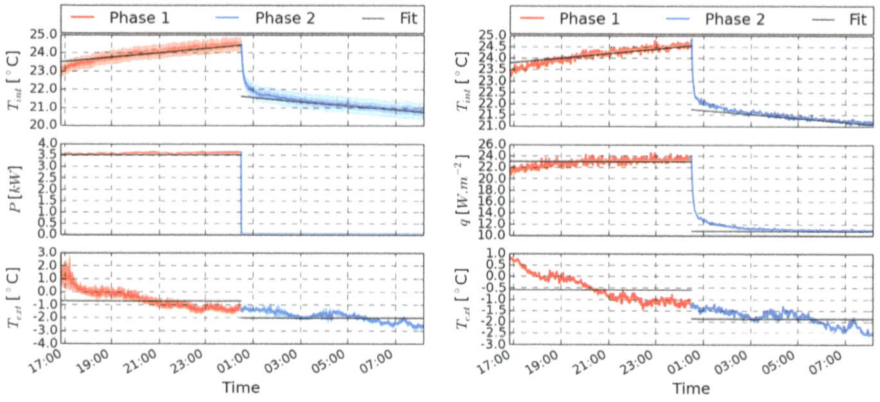

Fig. 4 Evolution of air temperatures, heating power and heat flux density during a QUB/e test (test n°2) in Årsta (Sweden)—Whole house heat loss coefficient (left) and in situ U-value of an external wall of bedroom 1 (right). The red, blue and black solid lines correspond to the heating phase, the free cooling phase and the linear regressions used to derive the quantities used in the QUB/e formula, respectively

2.4 Monitoring Equipment and Testing Protocol

The tests were carried out when the apartment was unoccupied between December 13, 2016 and December 31, 2016. Three consecutive QUB/e tests were first undertaken at night followed by quasi steady-state measurements (ISO 9869-1) during 15 days. The mechanical ventilation was switched off. The heating system of the apartment was switched off only during the QUB/e tests. Electrical heaters (i.e., fan heaters) were placed within the apartment in order to provide a uniform heating source for the QUB/e tests. The total duration of each QUB/e test was 16 h (i.e., between 4 pm and 8 am).

Heat flux plates (Hukseflux HFP01) and type K thermocouples were used to monitor the heat flux densities on building elements and the air temperatures. A silicone paste was used to ensure a good thermal contact between the heat flux plates and the building elements. All sensors were connected to data loggers (Graphtec GL820). Weather conditions (i.e., temperature, relative humidity, wind orientation and speed, solar radiation) were recorded with a Davis Vantage Pro2 weather station installed on the balcony of the apartment. The data acquisition rate was set to one minute.

In situ U-value measurements were also undertaken in accordance with standard ISO 9869-1 [10] (average method) while the set point temperature of the apartment was approximately 21 °C (see Fig. 2). In situ measurements of heat flux density, from which in situ U-values are derived, were taken at 25 locations on the thermal elements (7 on the external walls, 6 on the internal walls, 4 on the floor, 4 on the ceiling and 4 on the glazings) of the exhibition apartment using heat flux plates (HFPs). Only measurements of heat flux density obtained from those locations that were considered not to be significantly influenced by thermal bridging at junctions with neighbouring thermal elements (typically at distances greater than 500 mm from the junction) were used in the calculation of the in situ U-values. HFPs were also placed at the centre pane of the windows, i.e. only the U_g-value of these glazing units could be derived from our measurements. The U-values estimated in situ with the quasi-steady state (ISO 9869-1) and the QUB/e methods were compared.

Unfortunately, the whole HLC could not be estimated during this measurement period because the energy use of the apartment could not be monitored; i.e. only the overall energy use of the building was available.

The indoor air temperature was not kept constant during the QUB/e test so that there might have been a temperature difference across the internal walls, the floor and the ceiling separating the exhibition apartment and the neighbouring internal zones (apartments, corridor). Since the heat fluxes were monitored during the QUB/e tests, these internal heat losses (or gains) could be accounted for. The HLCs were thus corrected in order to report only heat losses (or gains) to the exterior environment and have a sound comparison with theoretical calculations which assume a uniform temperature within a building (i.e., there are no heat losses/gains).

The HLC with respect to the exterior environment was computed with the following equation:

$$HLC_{ext} = HLC_{raw} - \sum_{j} U_{eff,j} A_j \tag{6}$$

where HLC_{ext}, HLC_{raw}, $U_{eff,j}$ and A_j are the HLC w.r.t. the exterior environment only (in W/K), the 'raw' HLC (in W/K) obtained from the standard QUB analysis, the effective U-value of the jth internal element (in W/m^2-K) obtained from the QUB/e method and the area of the jth internal element (in m^2), respectively.

3 Results and Discussion

3.1 Whole Heat Loss Coefficient (HLC)

Figure 5 shows the whole heat loss coefficient (HLC) estimated with the QUB/e method for three consecutive nights. The internal heat losses (or gains) through the ceiling, the floor and the internal walls were accounted for with the estimation of the effective U-values. The experimental tests covered values of the parameter α between 0.4 and 0.7 (the value of the mean HLC was used as the reference HLC). The estimated values of the whole HLC with QUB/e method are thus deemed reliable. The QUB/e method yielded a robust estimation of the whole HLC with an estimated mean value of 47.4 $\pm$ 2.6 W/K. This value can be used to estimate the space heating needs of the apartment (e.g., with degree-days [19]).

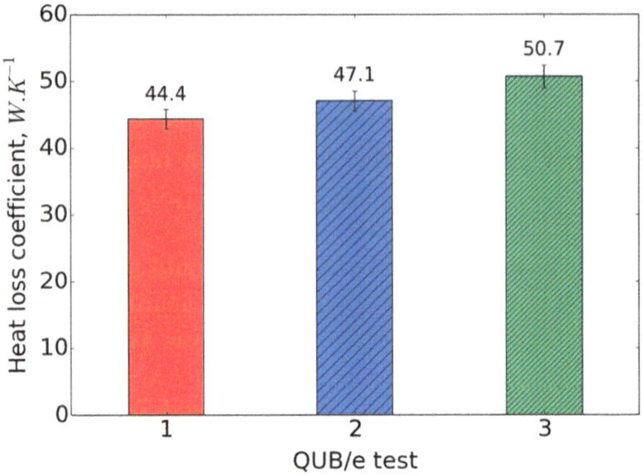

Fig. 5 Whole heat loss coefficient (HLC)

The main source of uncertainty arose from the correction of the internal heat losses (or gains) derived from the estimated effective U-values. The effective U-values of the floor and the internal walls (neighbours, corridor) were almost nil while the effective U-value of the ceiling was quite high (i.e., around 0.27 ± 0.12 W/m^2-K). It was attributed to a much lower internal air temperature set point chosen by the upstairs neighbour(s). Unfortunately, we could not access the apartment to verify this assumption.

3.2 *Local U-Values*

The U-values estimated with the quasi-steady state (ISO 9869-1) and the QUB/e methods are plotted against each other in Fig. 6. Each symbol corresponds to the U-value of each HFP placed on the external walls (7 different locations) and the glazings (4 different locations).

The in situ measurements showed a good agreement between the quasi-steady state (ISO 9869-1) and the QUB/e methods (i.e., the relative differences were within

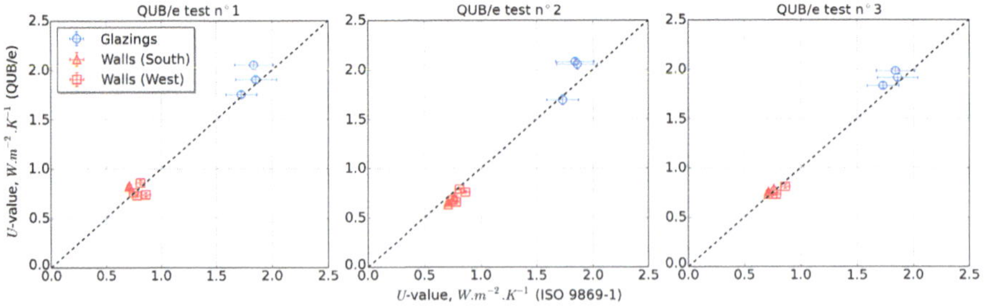

Fig. 6 Local U-values: cross-comparison between ISO 9869-1 and QUB/e methods. For comparison, the y = x reference curve is plotted for each case (dashed black lines)

the uncertainty bound of the measurement methods). It should be noted that the U-values measurements undertaken in accordance with ISO 9869-1 were obtained after a period of 15 days whereas each QUB/e test was performed in a single night. It was thus demonstrated that the QUB/e method can deliver a good estimation of the thermal performance of the building fabric in a single night without occupancy for this type of building.

The external walls seemed fairly homogenous thermally-wise with an estimated mean U-value of 0.77 ± 0.05 W/m^2-K with the QUB/e method. This statement was backed-up with a thermographic survey undertaken in accordance with EN 13187:1999 [11] (not reported here for the sake of brevity). The mean U_g-value of the double glazing units (DGU) was estimated at 1.93 ± 0.12 W/m^2-K with the QUB/e method. The slight discrepancy observed between the estimates based on the QUB/e and ISO 9869-1 tests (i.e., around 6%) should be further investigated.

4 Conclusion

The findings from a series of in situ tests in a circa 1960s multi-family housing (MFH) carried to assess its thermal performance were presented in this work. The objectives of the study were to (i) evaluate the thermal performance of the building fabric (i.e., whole heat loss coefficient and local U-values) before a full retrofit programme, and (ii) validate the QUB/e method by cross-comparison with quasi steady-state measurements.

The in situ measurements showed a good agreement between the quasi-steady state (ISO 9869-1) and the QUB/e methods (i.e., the relative differences were within the uncertainty bound of the measurement methods). It was thus demonstrated that the QUB/e method can deliver a good estimation of the thermal performance of building fabrics within just one night without occupancy, significantly less than the 2–4 week period required for quasi steady-state methods. Although further comparison with quasi steady-state methods in real climates for different types of buildings is needed, the QUB/e method is a promising method to test the actual performance of buildings.

Acknowledgements The author would like to thank Mika Hakosalo, Jan-Ulric Sjögren (Stockholms stad) and Harry Matero (Stockholmshem) for allowing us to perform these measurements in Årsta. The help of colleagues at Saint-Gobain (Tomas Pühringer, Theo Kiriakou, Andrea Mazzucco and Jean-Benoît Racaud) is also greatly appreciated.

References

1. A. Stafford, M. Bell, C. Gorse, *Building Confidence—A working paper* [Report 008]. The Centre for Low Carbon Futures, Birmingham, UK (2012)
2. Zero Carbon Hub, *Closing the Gap Between Design and As-built Performance—Evidence Review Report* (Technical report, London, UK, 2014)

3. D. Johnston, D. Miles-Shenton, D. Farmer, Quantifying the domestic building fabric a 'performance gap'. Build. Serv. Eng. Res. Technol. **36**(5), 614–627 (2015). https://doi.org/10.1177/0143624415570344
4. A. Janssens, *International Energy Agency, EBC Annex 58 – Reliable Building Energy Performance Characterisation Based on Full Scale Dynamic Measurements* [Report of Subtask 1b: Overview of methods to analyse dynamic data]. Technical report, KU Leuven, Leuven, Belgium, ISBN: 9789460189890 (2016)
5. J. Meulemans, F. Alzetto, D. Farmer, C. Gorse, QUB/e—a novel transient experimental method for in situ measurements of the thermal performance of building fabrics, in *International SEEDS Conference 2016: Sustainable Ecological Engineering Design for Society*, ed. by C. Gorse, M. Dastbaz (Leeds, UK, 2016)
6. J. Meulemans, F. Alzetto, D. Farmer, C. Gorse, QUB/e—a novel transient experimental method for in situ measurements of the thermal performance of building fabrics, in *Building Information Modelling, Building Performance, Design and Smart Construction*, ed. by M. Dastbaz, C. Gorse, A. Moncaster (Springer International Publishing, AG, 1st edn., 2017), ISBN: 978-3-319-50345-5
7. G. Pandraud, E. Mangematin, D. Roux, E. Quentin, *QUB: a new rapid building energy diagnosis method*, in Proceedings of CLIMA 2013. The 11th REHVA World Congress and the 8th International Conference on IAQVEC, Prague, Czech Republic (2013)
8. A. Alzetto, G., Pandraud, R., Fitton, The QUB method: a fast and reliable building envelope thermal diagnosis. Preprint submitted to Energy and Buildings (2016)
9. V. Sougkakis, J. Meulemans, F. Alzetto, C. Wood, M. Gillott, T. Cox, An assessment of the QUB method for predicting the whole building thermal performance under actual operating conditions, in *International SEEDS Conference 2017: Sustainable Ecological Engineering Design for Society*, ed. by C. Gorse, M. Dastbaz (Leeds, UK, 2017)
10. International Organization for Standardization, ISO 9869-1:2014. Thermal insulation—Building elements—In-situ measurement of thermal resistance and thermal transmittance—Part 1: Heat flow meter method, Geneva, Switzerland (2014)
11. British Standards Institution, BS EN 13187:1999. Thermal performance of buildings—Qualitative detection of thermal irregularities in building envelopes—Infrared method, London, UK (1999)
12. D. Johnston, D., Miles-Shenton, D. Farmer, J. Wingfield, *Whole House Heat Loss Test Method (Coheating)* [Technical report]. Leeds Metropolitan University, Leeds, UK (2013)
13. G. Bauwens, S. Roels, Co-heating test: a state-of-the-art. Energy Build. **82**, 163–172 (2014). https://doi.org/10.1016/j.enbuild.2014.04.039
14. A. Stafford, D. Johnston, D. Miles-Shenton, D. Farmer, M. Brooke-Peat, C. Gorse, Adding value and meaning to coheating tests. Struct. Surv. **32**(4), 331–342 (2014). https://doi.org/10.1108/SS-01-2014-0007
15. E. Mangematin, G. Pandraud, D. Roux, Quick measurements of energy efficiency of buildings. C R Phys. **13**(4), 383–390 (2012). https://doi.org/10.1016/j.crhy.2012.04.001
16. G. Pandraud, R. Fitton, QUB: *Validation of a rapid energy diagnosis method for buildings*, in International Energy Agency, EBC Annex 58, 4th Expert meeting, Holzkirchen, Germany (2013)
17. G. Pandraud, D. Gossard, F. Alzetto, *Experimental optimization of the QUB method*, in International Energy Agency, EBC Annex 58, 6th Expert meeting, Ghent, Belgium (2014)
18. H.W. Coleman, W.G. Steele, *Experimentation, Validation, and Uncertainty Analysis for Engineers*, 3rd edn. (Wiley Inc, Hoboken, New Jersey, 2009). ISBN 978-0-470-16888-2
19. T. Day, *Degree-days: Theory and Application* [Technical Report TM41: 2006], Chartered Institution of Building Services Engineers (CIBSE), London, UK (2006). ISBN: 1-903287-76-6

Part III
Efficient HVAC Components

Sol-Air Thermometer Measurement of Heat Transfer Coefficient at Building Outdoor Surfaces

K. E. Anders Ohlsson, Ronny Östin and Thomas Olofsson

Abstract There exists a building energy performance gap between theoretical simulations and the actual energy usage as measured. One potential reason for this gap might be a mismatch between predicted and measured values of the heat flux q through the building envelope. There is therefore a need to develop accurate and more cost-efficient methods for measurement of q. The standard ISO 9869-1 states that, at the outdoor surface, $q = h_o(T_s - T_{env})$, where h_o is the overall heat transfer coefficient, including both convective and radiative components, T_{env} is the environmental temperature, and T_s is the temperature of the building surface. It has previously been shown that the sol-air thermometer (SAT) could be used for convenient measurement of T_{env} under dark conditions. In the present work, two SAT units, one heated and the other unheated, were employed for accurate outdoor measurements of h_o in cold winter climate. Validation was performed by comparison of results from the new method against measurements, where previously established methodology was used. With current operating conditions, the measurement uncertainty was estimated to be 3.0 and 4.4%, for h_o equal to 13 and 29 $\mathrm{Wm^{-2}K^{-1}}$, respectively. The new SAT steady-state method is more cost-effective compared to previous methodology, in that the former involves fewer input quantities (surface emissivity and infrared radiation temperature are unnecessary) to be measured, while giving the same h_o results, without any sacrifice in accuracy. SAT methodology thus enables measurement of both T_{env} and h_o, which characterizes the building thermal environment, and supports estimation of q.

Keywords Heat transfer coefficient · Sol-air thermometer · Environmental temperature · Building energy performance gap

K. E. A. Ohlsson (✉) · R. Östin · T. Olofsson
Department of Applied Physics and Electronics, Umeå University,
90187 Umeå, Sweden
e-mail: anders.ohlsson@umu.se

R. Östin
e-mail: ronny.ostin@umu.se

T. Olofsson
e-mail: thomas.olofsson@umu.se

D. Johansson et al. (eds.), *Cold Climate HVAC 2018*,
Springer Proceedings in Energy, https://doi.org/10.1007/978-3-030-00662-4_28

1 Introduction

It is widely recognized, that there is often a gap between predicted and measured energy performance of buildings [1, 2]. Several different causes has been suggested for this energy performance gap. One possible cause is the use of inaccurate or inadequate input data for model predictions. Data which describes the thermal environment of the building is often of importance both at the design stage, for modelling of heat transfer between the building and its surroundings, and at the operational stage, as input to the heating (or cooling) control system. For evaluation of the energy performance gap, it is necessary to acquire relevant and accurate on-site measurement data for comparison against predicted values. In the present work, we developed a new measurement method, which can be used for accurate characterization of the building outdoor thermal environment. We start by showing how the thermal environment of the building could be described by a linear thermal network model, and how its parameters could be estimated, using the sol-air thermometer (SAT).

The heat flux q (Wm^{-2}) between the outer surface of the building envelope and the surrounding outdoor thermal environment could be modelled as:

$$q = h_o(T_s - T_{sa}), \qquad (1)$$

where h_o is the heat transfer coefficient, T_s is the building surface temperature, and T_{sa} is the sol-air temperature [3]. The heat flux q consists of three components: convective and net infrared radiative heat transfer, and incoming absorbed solar irradiation. Figure 1 illustrates how q could be expressed as in Eq. (1), by simplification of a linear thermal network model of the thermal environment, and applying the Thevenins′ theorem of the circuit theory [4]. In this model, T_{sa} is the

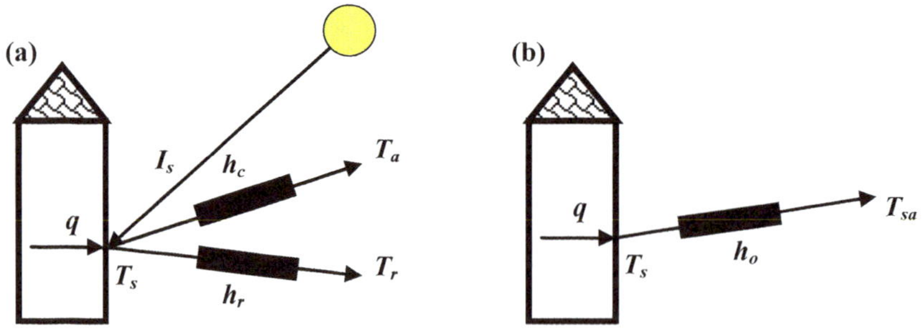

Fig. 1 Thermal network model for building surface and its environment. Thermal environment modelled **a** by solar irradiation, air temperature, and mean (infrared) radiant temperature, or **b** by the sol-air temperature, as driving forces for heat transfer

equivalent temperature of the thermal environment of the building surface, and is expressed as:

$$T_{sa} = T_{env} + \frac{\alpha I_s}{h_o}, \tag{2}$$

where T_{env} is the environmental temperature, α is the solar absorptivity, and I_s is the solar irradiation (Wm^{-2}). T_{env} consists of the weighted sum of the air temperature T_a and the mean radiant temperature T_r (longwave infrared radiation). The weights include the convective heat transfer coefficient h_c and the radiative heat transfer coefficient h_r, respectively. With $h_o = h_c + h_r$, we obtain:

$$T_{env} = \frac{h_c T_a + h_r T_r}{h_o}. \tag{3}$$

Under dark conditions, $I_s = 0$, and $T_{sa} = T_{env}$.

The SAT was originally designed for measurement of T_{sa} [5, 6], and its performance for that purpose has been evaluated [7, 8]. The SAT simply consists of a metal plate, with its back and side surfaces thermally insulated, i.e. there is ideally no conductive heat loss through the insulation material. The SAT is inserted into the building envelope, with the building surface aligned to the SAT front surface. These two surfaces are thereby exposed to the same thermal environment. The temperature of the SAT plate, T_{SAT}, is measured, often by insertion of a temperature probe into the plate. By applying Eq. (1) to the SAT front surface, we obtain:

$$q = h_o(T_{SAT} - T_{sa}) = 0, \tag{4}$$

where $q = 0$ due to the insulation layer. We now see that $T_{SAT} = T_{sa}$, i.e. the SAT measures the sol-air temperature.

At dark conditions, the SAT should enable direct measurement of T_{env}. The standard ISO 9869-1 claims that direct measurement of T_{env} cannot be performed (cf. section A.3.1 in [3]). However, we have recently used the SAT to directly measure T_{env} under dark and stable winter conditions [8].

SAT methodology could also be used for measurement of h_o at steady-state heat transfer. For this purpose, two SAT units (SAT1 and SAT0) were employed, and positioned side by side to expose them to the same thermal environment. These SAT units were heated electrically using resistive heater foils, and the supplied heater powers were p_1 and p_0 (Wm^{-2}), respectively. The energy balances for these units were:

$$p_1 = h_o(T_{SAT1} - T_{sa}) \tag{5}$$

$$p_0 = h_o(T_{SAT0} - T_{sa}). \tag{6}$$

Solving Eqs. (5) and (6) for h_o yields:

$$h_o = \frac{p_1 - p_0}{(T_{SAT1} - T_{SAT0})} \tag{7}$$

Equation (7) is the basis for the new SAT method for measurement of h_o presented in this work (here named "SATss", the steady-state SAT method). The SATss method is a slight modification of the method used by Ito et al. [9] for measurement of the convective heat transfer coefficient h_c, given as:

$$h_c = \frac{p_1 - p_0 - \varepsilon_s\sigma\left(T_{SAT1}^4 - T_{SAT0}^4\right)}{(T_{SAT1} - T_{SAT0})}, \tag{8}$$

where ε_s is the emissivity of the SAT surface, and σ is the Stefan-Boltzmann constant ($\sigma = 5.67\times10^{-8}$ $Wm^{-2}K^{-4}$). We notice that, with the SATss method for measurement of h_o [Eq. (7)], there is no need for an estimate of ε_s, as is necessary for measurement of h_c using Eq. (8). In the present work, we selected p_0 to be zero (for measurement of T_{sa}, or T_{env} under darkness), i.e. SAT0 was left unheated.

In the present work, the objective was to demonstrate the accurate measurement of h_o, using two SAT units, and the SATss method expressed by Eq. (7). For validation of the SATss method, we performed simultaneous measurement of h_o using previously established methodology (here named the "ItoHr" method). In the ItoHr method, h_o was obtained as the sum $h_o = h_c + h_r$, where h_c was measured using the Ito method [Eq. (8)] and h_r was estimated as follows [3]: The infrared radiative heat flux q_r at the SAT front surface is given as:

$$q_r = \varepsilon_s\sigma\left(T_{SAT}^4 - T_r^4\right) = h_r(T_{SAT} - T_r) \tag{9}$$

By linearization of the T^4 terms around the average temperature $T_m = (T_{SAT} + T_r)/2$, we obtain:

$$h_r = 4\varepsilon_s\sigma T_m^3 \tag{10}$$

The measurement of h_r, using Eq. (10), thus requires estimates of both ε_s and T_r. In the present work, T_{SAT0} was used in T_m of Eq. (10).

2 Experimental

A SAT unit, used in the present work, consisted of two circular copper plates, with 80 mm diameter, and with 12 and 1.5 mm thickness for the top and bottom plates respectively (see Fig. 2). A 0.3 mm thick resistive heater foil (polyimide Thermofoil™, Minco, Mineapolis, USA; 50.8 × 50.8 mm^2) was placed in between

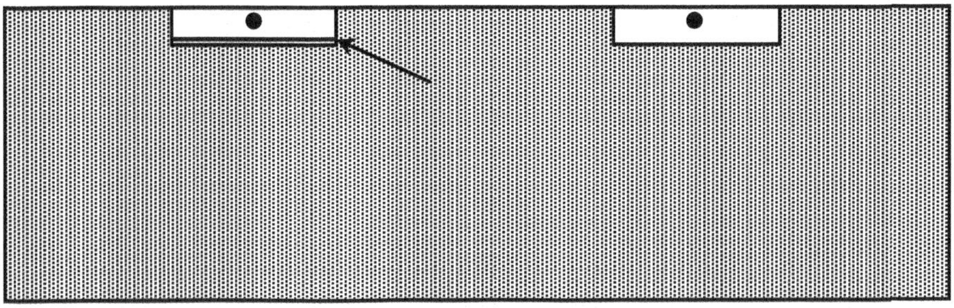

Fig. 2 Cross-sectional schematic view of the SAT pair, embedded in the insulation plate (grey). The arrow points to the heater foil (shown only for the heated SAT, but present in both). The black dots shows the positions for the SAT temperature sensors

the plates, with thermal contact ensured by application of thermal paste (Wacker, paste P12). The heater resistance R_h was nominally 661 Ω. The R_h, and its temperature dependence, was accurately determined in separate experiments.

The SAT unit was inserted into plates of extruded polystyrene insulation material (Sundolit XPS300BE, Sunde Group, Norway). The thermal conductivity of the insulation plate was ca 0.035 $Wm^{-1}K^{-1}$, and its thickness was 100 mm. Since the SAT front surface was aligned with the surface of the insulation plate, there was a 86 mm thick insulation layer at the rear side of the SAT.

The front surface of the SAT was coated with a 0.1 mm thick layer of matt black paint (Nextel-Velvet coating 811-21; Mankiewicz Gebr., Hamburg, Germany). The emissivity of this Nextel coating has previously been determined to equal ε_s = 0.943 [10]. The temperature of the SAT was measured by insertion of a Pt-100 probe (4-wire, class A, 4 mm diameter, Jumo, Fulda, Germany) into a hole in the side of the top SAT plate.

The heat supplied (p_1) to the SAT1 unit was obtained by measuring the electrical DC voltage U supplied to its heater, and applying:

$$p_1 = \frac{U^2}{A_{SAT} R_h}, \tag{11}$$

where A_{SAT} (m^2) is the SAT front surface area. In Eq. (11), R_h was corrected to its value at the T_{SAT1} temperature. T_r was measured using a pyrgeometer (Hukseflux, model IR20-T1, Netherlands), while the absence of daytime levels of solar radiation was confirmed by measuring the solar irradiation I_s using a pyranometer (Hukseflux, model SR11) (see Fig. 3).

All measurements (T_{SAT0}, T_{SAT1}, T_r, I_s, and U) were performed simultaneously using a data acquisition system (CompactDAQ and Labview software; National Instruments, USA), at a 0.02 Hz sampling frequency. Further details of the measurement system is given in [8, 11].

Fig. 3 Experimental setup. Three SAT units shown as black circle areas to the right in the figure, with SAT1 (heated) at the top and SAT0 (unheated) at the middle position). (The third SAT shown was not used in this work.) To the left side, the pyrgeometer (top) and the pyranometer (bottom) are seen

The series of measurements were performed outdoors on a roof terrace at Umeå University campus, Sweden, during the period from December 2016 to January 2017 (the present measurement series is identical to the one used as basis for work in ref. [8]). The insulation plate, with the two SAT units separated by a distance of ca 300 mm, was mounted side by side with the pyranometer and the pyrgeometer. These four measurement devices were all exposed to the same thermal environment (see Fig. 3).

The measured quantities (T_{SAT0}, T_{SAT1}, T_r, and U) were all estimated as the average value over a sampling period of time Δt, where environmental steady-state conditions prevailed. The criteria for selection of Δt also included a constant wind speed (measured at an on-site weather station, assuming that wind speed is constant also at the SAT surfaces), and the absence of daylight and precipitation (snow or rain). We also required that the Δt to always be larger than the time constant τ of the SAT, to allow time for its approach to a steady-state heat balance with its surroundings.

To ensure fulfillment of this requirement, we estimated τ, using a thermal network model. In this model, the SAT thermal capacitance C was estimated as:

$$C = \rho c_p V = \rho c_p L A_{SAT}, \tag{12}$$

where ρ is the SAT plate density (kg m^{-3}), c_p is its specific heat capacity (J kg^{-1}K^{-1}), L its thickness, and V its volume. The network model of SAT, and its thermal environment, then consists of the capacitor C in series with the resistance $R = 1/(A_{SAT}h_o)$, and with the time constant given by:

$$\tau = RC \tag{13}$$

In the present work, the SAT plate was made of copper, with $\rho = 8.93 \times 10^3$ kg m^{-3} and $c_p = 385$ J kg^{-1}K^{-1} (material property data obtained from [12]), and had $V = 6.78 \times 10^{-5}$ m^3. Finally, C was estimated to the value equal to 233 J K^{-1}.

3 Results and Discussion

Several series of measurements were performed outdoors under various winter weather conditions, and under darkness (cf. [8], where also wind speeds and weather observations are given). From these series, run sequences were extracted from time periods where measured temperature and radiation levels were constant, and where there was no precipitation (rain or snow). Table 1 lists the average temperature values obtained from each of 13 such runs, along with the heat supplied in each instance. Based on these data only, h_o was predicted by the SATss method [Eq. (7)], where two SAT units were used.

For purpose of method validation, h_o results from using the SATss method were compared against results obtained by using the ItoHr method (see Sect. 1). In addition to the measurement results shown in Table 1, the ItoHr method requires measured values for T_r, and an estimate of ε_s (see Table 2).

Table 1 Prediction of h_o, using the SAT steady-state (SATss) method [Eq. (7)], with $p_O = 0$

Run	T_{SATO} (K)	T_{SATI} (K)	p_I (Wm^{-2})	$h_{o,ss}$ (Wm^{-2}K^{-1})
1	254.49	262.99	123.85	14.57
2	269.42	284.30	245.11	16.47
3	272.28	281.85	245.19	25.63
4	275.21	286.66	245.03	21.39
5	273.91	282.69	245.17	27.95
6	269.28	282.19	245.19	19.00
7	267.54	279.64	245.29	20.27
8	268.23	278.51	245.31	23.87
9	272.78	283.83	245.16	22.18
10	272.84	282.08	245.22	26.54
11	273.19	282.69	245.20	25.83
12	273.60	282.69	245.12	21.28
13	256.71	274.39	245.54	13.89

Table 2 Prediction of h_o, using the ItoHr method, where $h_o = h_c + h_r$, and data consisting of measured values of T_r, in addition to measured values given in Table 1. $\varepsilon_s = 0.943$

Run	T_r (K)	h_c ($Wm^{-2}K^{-1}$)	h_r ($Wm^{-2}K^{-1}$)	$h_{o,itohr}$ ($Wm^{-2}K^{-1}$)
1	244.82	10.86	3.33	14.19
2	264.99	11.93	4.08	16.01
3	261.14	21.08	4.06	25.14
4	274.74	16.65	4.45	21.10
5	272.40	23.34	4.36	27.70
6	257.57	14.52	3.91	18.43
7	255.49	15.89	3.83	19.71
8	255.10	19.50	3.83	23.33
9	269.08	17.57	4.25	21.83
10	259.86	21.97	4.04	26.01
11	259.69	21.24	4.05	25.29
12	273.06	16.61	4.37	20.98
13	255.74	9.88	3.60	13.48

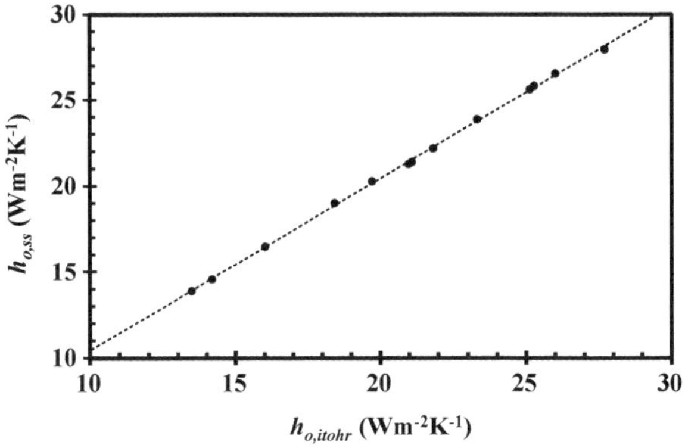

Fig. 4 Comparison of predicted values of h_o, obtained using the new SAT steady-state method ($h_{o,ss}$), against predictions obtained from using previously established methodology ($h_{o,itohr}$; the ItoHr method). An ordinary linear regression line is also given (dashed line)

The method comparison is shown in Fig. 4. An ordinary linear regression yielded the equation $h_{o,ss} = 1.000 \cdot h_{o,itohr} + 0.43$, with standard errors equal to the values 0.008 and 0.16, for slope and intercept, respectively. The conclusion is here that h_o can be accurately measured, using the SATss method, except for a small constant bias, which is barely significant at the 95% confidence level.

The advantages with the SATss method for measurement of h_o in comparison to the established ItoHr method are that, in the former, T_r and ε_s need not to be known. Since the estimation of these two quantities requires use of extra resources, there is cost reduction associated with changing to the SATss method. Note also that here, for both the SATss and ItoHr methods, the power supplied to the heater was measured electrically, and not by using a heat flow meter, as has previously been performed with the Ito method [9]. This probably improve the accuracy of the estimate of the supplied power, for both methods.

We here expressed measurement uncertainty following the GUM standard [13]. The relative measurement uncertainty for h_o, as estimated from the Eqs. (7) and (11), was equal to 3.0 and 4.4% for the lower and upper values, respectively, as shown in Fig. 4. At the lower end, the measurement uncertainty was limited by the uncertainty in estimation of A_{SAT}, while, at the upper end, the uncertainty in the measurement of SAT temperature difference was the limiting factor.

The time constant τ was estimated to equal between 28 and 56 min, for the range of h_o values spanned by the 13 runs. The sampling time period Δt was always $\geq \tau$, with an average value of 3.6 τ.

4 Conclusions

We have designed a new and accurate method (named SATss) for measurement of h_o, using two SAT units. The SATss method requires less number of input data, compared to what was necessary using previously established methodology (ItoHr method), and could therefore be considered as more cost-efficient. Although the SATss method is simpler in its design, there is no sacrifice in accuracy.

SAT methodology now enables accurate measurement of both T_{env} and h_o. These two quantities together give an improved characterization of the building thermal environment, in terms of the simple network model, with the driving temperature (T_{env}) connected to the building surface by one single resistance ($R = 1/(A_{SAT}h_o)$). By providing accurate on-site measurement data, the SAT methodology has potential to reduce the building energy performance gap.

Acknowledgements We gratefully acknowledge the financial support for this project from the Swedish Energy Agency, through IQ Samhällsbyggnad and the E2B2 program (project no. 39699-1), and the Kempe Foundations. We are also indebted to Fredrik Holmgren and Johan Haake for technical support in the construction of the experimental equipment.

References

1. A.C. Menezes, A. Cripps, D. Bouchlaghem, R. Buswell, Predicted vs. actual energy performance of non-domestic buildings: using post-occupancy evaluation data to reduce the performance gap. Appl. Energy **97**, 355–364 (2012)
2. P. De Wilde, The gap between predicted and measured energy performance of buildings: a framework for investigation. Autom. Constr. **41**, 40–49 (2014)
3. ISO 9869-1, in *Thermal insulation—Building elements—In-situ measurement of thermal resistance and thermal transmittance*; Part 1: Heat flow meter method. 1 edn. ISO, Switzerland (2014)
4. M.G. Davies, The use of flux temperatures in thermal design. Build. Serv. Eng. Res. Technol. **2**, 160–164 (1981)
5. C.O. Mackey, L.T. Wright, The sol-air thermometer—a new instrument. Trans. Am. Soc. Heat. Vent. Eng. **52**, 271–282 (1946)

6. R.W. Muncey, T.S. Holden, The calculation of internal temperatures—a demonstration experiment. Build. Sci. **2**, 191–196 (1967)
7. K.P. Rao, E.R. Ballantyne, *Some Investigations on the Sol-Air Temperature Concept* [CSIRO, Division of Building Research Technical Paper No. 27], Australia (1970)
8. T. Olofsson, K.E.A. Ohlsson, R. Östin, Measurement of the environmental temperature using the sol-air thermometer. Energy Procedia **132**, 357–362. (2017)
9. N. Ito, K. Kimura, J. Oka, A field experiment study on the convective heat transfer coefficient on exterior surface of a building. ASHRAE Trans. **78**, 184–191 (1972)
10. J. Lohrengel, R. Todtenhaupt, Wärmeleitfähigkeit, gesamtemissionsgrade und spektrale emissionsgrade der beschichtung Nextel-Velvet-Coating 811-21 (RAL 900 15 tiefschwarz matt). PTB-Mittelungen **106**(4), 259–265 (1996)
11. K.E.A. Ohlsson, R. Östin, S. Grundberg, T. Olofsson, Dynamic model for measurement of convective heat transfer coefficient at external building surfaces. J. Build. Eng. **7**, 239–245 (2016)
12. F.P. Incropera, D.P. DeWitt, *Fundamentals of Heat and Mass Transfer*, 5th edn. (Wiley, London, 2002)
13. JCGM 100, Evaluation of measurement data—Guide to the expression of uncertainty in measurement (GUM 1995 with minor correction). BIPM, ISO, Geneva (2008)

What Should the Minimum Ventilation Rate Be in a Demand-Controlled Ventilation Strategy?

Mads Mysen, Sverre Holøs, Aileen Yang, Kari Thunshelle and Peter Schild

Abstract Demand-Controlled Ventilation is emerging as a dominant ventilation strategy in non-residential buildings in Norway. The ventilation airflow rate is controlled between pre-set minimum (V_{min}) and maximum (V_{max}) values, based on the signal from room-sensors. The choice of V_{max} is based on current knowledge about necessary airflow rate to reach an acceptable IAQ (indoor air quality) with maximum likely personal load and emission load from building materials. The choice of V_{min} has an obvious impact on energy use, but there are few studies about its impact on IAQ. V_{min} varies typically from 0.7 to above 2 (l/s)/m^2 in Norway. In several buildings, V_{min} is set to the upper range of this interval due to technical limitations of the specific equipment used. We have performed blind cross over intervention-studies with an untrained test panel to evaluate PAQ (perceived air quality) when entering 20 *PAQ-rooms*. All the rooms have low-emitting building materials, but extra pollution sources were introduced in some of the rooms for this study. Supplementary, intervention studies were performed in a dedicated *test room* to assess the impact of airflow rate on PAQ, performance and well-being during the first 20 min of occupation. We found that increasing V_{min} has a significant positive impact on PAQ in rooms with extra pollution sources. This effect was not consistently present in the low-emitting rooms. Airflow rates did not noticeably affect PAQ, performance and well-being during the first 20 min of occupation. This indicates that V_{min} above 1 (l/s)/m^2 has limited benefit to IAQ in low emitting rooms.

Keywords Demand-controlled ventilation · Indoor air quality · Performance test Low-emitting materials · Pollution level · Ventilation strategy

---

M. Mysen (✉) · S. Holøs · K. Thunshelle
SINTEF Building and Infrastructure, Trondheim, Norway
e-mail: mads.mysen@gk.no

M. Mysen · A. Yang · P. Schild
OsloMet—Oslo Metropolitan University (Formerly Oslo and Akershus University College of Applied Sciences), Oslo, Norway

D. Johansson et al. (eds.), *Cold Climate HVAC 2018*,
Springer Proceedings in Energy, https://doi.org/10.1007/978-3-030-00662-4_29

1 Introduction

1.1 The Scope of This Paper

The amount of ventilation necessary to maintain good IAQ (indoor air quality) in a room depends on the strength of the pollution from interior surfaces, furniture and occupants. These pollution sources are either stationary or variable. The variable sources are mainly users and user-related activities. The purpose of demand-controlled ventilation (DCV) is to continuously follow these changes in ventilation requirement.

DCV achieves the greatest energy savings in buildings with rooms that are unoccupied for a significant part of the operating hours [1] and when the ventilation rates are significantly reduced in unoccupied rooms [2]. DCV has thus emerged as a dominant ventilation strategy for such non-residential buildings in Norway.

The ventilation airflow rate in modern DCV systems is controlled between pre-set minimum (V_{min}) and maximum (V_{max}) limit values, based on the signal from one or more room sensors (Fig. 1). The V_{min} and V_{max} limit values that can be set to

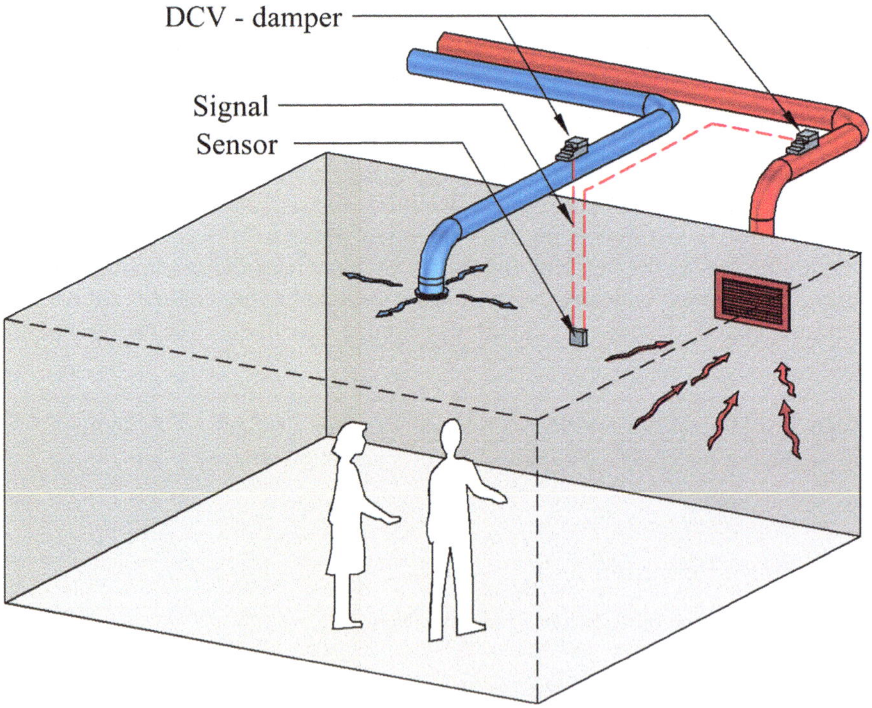

Fig. 1 DCV-controlled room. A sensor picks up the pollution level and signals DCV-dampers about a corresponding ventilation volume to maintain IAQ (Illustration SINTEF)

account for changes in, for example, pollutant load from materials, room size, or the maximum likely number of occupants [3].

The choice of V_{max} is based on current knowledge about necessary airflow rate to reach an acceptable IAQ taking into account maximum likely personal load, cooling load, and the total emission load from materials in the room. Recommendations about necessary airflow rates, e.g. HealthVent review report [4], are mainly based on studies with constant ventilation rates per person. These recommendations are not fitted differences in emission sources or DCV-system dynamics.

The choice of V_{min} has an obvious impact on energy use, but there are few studies on the impact on IAQ. Thus, there are no scientifically-based optimal guidelines for V_{min}. The present rationale for choosing V_{min} is to maintain a constant IAQ in accordance with Fanger's olf-based "addition principle" [5]. This implies that V_{min} is set to achieve an acceptable olfactory concentration accounting for the total pollutant load from materials plus occupants in the room. For unoccupied rooms, this minimum ventilation demand varies typically from 0.7 to over 2 $(\mathrm{I/s})/\mathrm{m}^2_{\mathrm{floor}}$ in Norway. V_{min} is often set to the upper end of this range due to risk off high emitting furniture, or technical limitations in the equipment such as flow-measurement. However, we should depart from this practice, and instead acknowledge that unoccupied rooms do not need to be intensively ventilated primarily for olfactory comfort. The remaining question is whether reducing V_{min} has any negative impact on PAQ (perceived air quality), indoor-air related symptoms, health, or human performance, when an occupant enters the empty room. This has been investigated in the R&D project BEST VENT. In this paper, we report the results from the first experiments carried out in 2016. BEST VENT will continue experimentation for two more years.

2 Methods

2.1 Experimental Design Overview

The experimental design in BEST VENT consists of five major steps:

1. Identify a test school with dedicated test room with controllable ventilation and close to project partners' laboratory facilities
2. Establish a test panel (occupants)
3. Develop sensitive and reliable surveys and performance tests
4. Perform adequate measurements
5. Rational data collection and analysis.

Each step is briefly reported in this paper.

2.2 Selection of Test School and Test Panel

DCV-systems in 5 schools were audited. Fernanda Nissen School in Oslo was chosen as our field laboratory, as it has all the required features, including easily adjustable airflow rates, and temperature and CO_2-control. The interior surfaces in this passivhaus-standard school was completed in 2016, approx. 6–8 months before the experiments. The school has concrete floor slabs covered with linoleum, walls are timber frame, 300 mm mineral wool insulation, and are generally clad with 13 mm plasterboard with acrylic paint. Materials and paint are either M1-certified or implicitly low-emitting. The use of sealants was limited, with no sealants visibly exposed to rooms. All classrooms have balanced supply and exhaust mechanical ventilation. The ventilation system has bag air filters of class F7 in accordance with EN 779:2012.

One 60 m^2 classroom, denoted the *Test room*, was loaned for research purposes year-round. Other rooms were available for research during holidays, including week 40 (autumn holiday). 20 un-occupied rooms were carefully selected for week 40, and are denoted *PAQ-rooms* in this paper. All of the education rooms at Fernanda Nisses school have mixing ventilation with well distributed ceiling-integrated air supply devices (Fig. 2).

30 students from Oslo and Akershus University college of Applied Science, were recruited as an untrained test panel. It was on a voluntary basis, with book gifts for those who attended 6 or more experiments. The number of students each test day varied between 15 and 22. 10 of them participated in all 8 experiments.

The test panel left the Test room for minimum 30 min between experiments (Table 1). Figure 2 shows the test room prior to experiment.

Fig. 2 The test room at Fernanda Nissen School, Oslo. All of the education rooms have mixing ventilation with similar air inlet and outlet as the test room. *Photo* SINTEF

Table 1 Time schedule for testing in the dedicated *test room*

Day	MONDAY	TUESDAY	WEDNESDAY	THURSDAY	FRIDAY
00:00 –07:00					
7:00–8:45					
8:45–9:45		(1) Testing with panel		(3) Testing with panel	(7) Testing with panel
9:45–10:15					
10:15 – 11:15		(2) Testing with panel		(4) Testing with panel	(8) Testing with panel
11:30 – 12:30					
13:00 – 14:00				(5) Testing with panel	
14:00– 14:30					
14:30 – 15:30				(6) Testing with panel	Removing
15:15 – 16:30					
16:30 – 17:30					
17:30–00:00:00					

V_{min} varied between1 (l/s)/m^2 [yellow and orange] and 2 (l/s)/m^2 [blue]. Eight experiments were conducted. White hatched area means ventilation off (Wednesday night and Thursday night)

2.3 *Experimental Design, Surveys, Performance Tests and Measurements in the* Test Room

Intervention-studies with the test panel present, were performed in the *Test room* to assess the impact of airflow rate on PAQ, performance and well-being during the first 20 min of occupancy. These experiments were repeated 8 times with different airflow rates from 3 to 16 (l/s)/person when the room was occupied. V_{min} was set to 1 or 2 (l/s)/m^2 before the test panel entered the room, hence they were blind to the intervention. The physical lower limit of the DCV-damper was 1 (l/s)/m^2.

The eight experiments in the *Test room* were performed without any obvious negative acoustic, actinic or mechanical factors that may confound the results. Indoor temperature varied within limits of ± 1 °C during experiments. Tables 1 and 2 show the time schedule for the test week and for each experiment, respectively. Only results from the first 20 min are reported in this paper. The colour codes show ventilation periods before each experiment. The room was empty between the experiments, expect for necessary time to read or adjust instruments and ventilation settings.

PAQ and building-related symptoms were reported by the test panel and scored using digital tablets. The survey questions are based on the Ørebro questionnaire [6]. The score was on a continuous-scale slider. For PAQ the extremes were "clearly unacceptable" to "clearly acceptable". For SBS-related questions like "Do you have headache", the extremes were "Yes, very" to "Not at all". It was not allowed to score at the midpoint. The scores were stored in the Microsoft cloud solution, prepared for data-analysis performed with statistical program R (Fig. 3).

The *OK Tick-off Test* ("OK-Tekstkryss" in Norwegian) measured sustained human performance. The test is a visual detection task designed to assess the ability

Table 2 Time schedule for testing in the dedicated *test room* after test panel entrance

Minutes after entrance	Activity
0–5	PAQ, IAQ-survey
5–20	Performance test
15–45	Lecture
44–55	Performance test
55–60	PAQ, IAQ-survey

The test panel was in the *test room* for 60 min

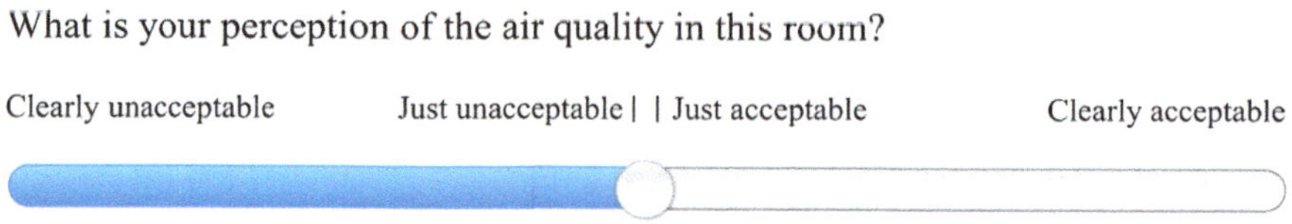

Fig. 3 Scores were reported on digital tablet. The figure shows the PAQ score slider

of individuals to maintain visuo-cognitive alertness for an extended period of time. The test contains meaningless, but readable, words. The task is to tick off as many Os and Ks as possible in 10 min (Fig. 4). The paper-version of the OK Tick-off Test has shown satisfactory reliability [7]. A digital version for the digital tablet was developed and used in the BEST VENT-project.

Measurements: Air temperature, relative humidity, CO_2, Ozone, NO_x (NO_2 + NO), volatile organic compounds (total VOC and specific compounds), formaldehyde and particulate matter (PM) at different size fractions from 3 nm to 32 μm, were measured during the test week with calibrated instruments. Supply and exhaust air rates were measured by a differential pressure sensor in the DCV-damper with sufficient straight duct length upstream to have a distributed airflow stream throughout the duct cross-section. The values were continuously logged by the BMS (building management system). The accuracy of the measurements was controlled prior to the experiments with calibrated funnels. Only measurements result with scope relevance are reported in this paper.

Table 3 shows V_{min} and average ventilation rate per person during the first 20 min. In experiments 1, 2, 3, 7 and 8 the ventilation rate was maintained at V_{min} for the first 20 min.

Test hypotheses: Before the experiments, it was believed that low V_{min} has negative impact on immediate PAQ when entering, and on SBS-symptoms and performance during the first minutes of occupancy. To falsify the test-hypothesis: "*V_{min} of 1 or 2 (l/s)/m^2 has no impact on PAQ, SBS-symptoms or performance*", V_{min} was set to 1 or 2 (l/s)/m^2 before entering the room, and then increased to between 3 and 16 (l/s)/person for the first 20 min of occupancy.

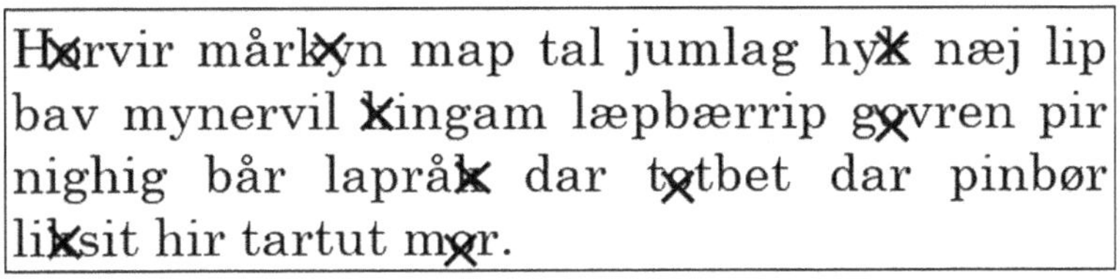

Fig. 4 Example taken from the OK Tick-off Test

Table 3 The 8 *test room* experiments

Expt. no:	V_{min} (l/s)/m^2	$V_{pers20min}$ (l/s)/person	Number of persons present
1	1	3	22
2	2	6	21
3	1	3	21
4	1	13	21
5	2	8	15
6	1	16	15
7	1	3	22
8	2	6	22

$V_{pers20min}$ is average suppled airflow the first 20 min of occupancy divided by the number of room occupants

2.4 *PAQ in 20 Rooms*

The test panel assessed PAQ when entering the 20 un-occupied *PAQ-rooms* with mixing ventilation similar with the test room. Room 13, 16 and 19 had a floor area of 30 m^2. The rest of had a floor area of 60 m^2. All rooms were built with certified low-emitting materials. Extra pollution sources were introduced to some of the rooms—i.e. old shoes were hidden in rooms 5 and 9. Room 7 had initially a peculiar smell from an unidentified source. The rooms were ventilated with V_{min} of 0, 1, 1.5 or 2 (l/s)/m^2 before entering. The last three are typical V_{min}-values. Individuals in the test panel entered the rooms one by one with at least 30 s intervals. Air temperatures were steady at 21.5 ± 1 °C. The test panel were blind to the different ventilation rates and pollution sources. Airflow rates, air temperature, relative humidity and CO_2 were logged during the test week.

Test hypotheses: Before the experiments, it was believed that low V_{min} has a negative impact on immediate PAQ when entering, and this impact is enhanced with additional pollution load in the room. To falsify the test-hypothesis "*V_{min} of 1, 1.5 or 2 (l/s)/m^2 has no impact on PAQ immediately after entering*", V_{min} was set to 1, 1.5 or 2 (l/s)/m^2 before entering. The ventilation rates were crossed between the experiments in randomly chosen rooms. Some rooms were kept unchanged to reliability test the test panel and experimental design.

3 Results

3.1 Test Room *Results*

Before the experiments, it was believed that low V_{min} negatively affects PAQ, SBS-symptoms and performance. Table 3 shows V_{min} and average ventilation rate per person the first 20 min. The test-hypotheses was not falsified. The result analysis revealed no consistent tendency of ventilation impact during this first period of occupancy. The detailed results are not given in this paper since the data material is vast and shows no pattern.

3.2 20 PAQ-Rooms Results

PAQ-results were linearly scored from −1 (clearly unacceptable) to +1 (clearly acceptable). Figure 5 shows the average score for the different air flow rates used in the specific PAQ-rooms. Table 4 highlights significant differences.

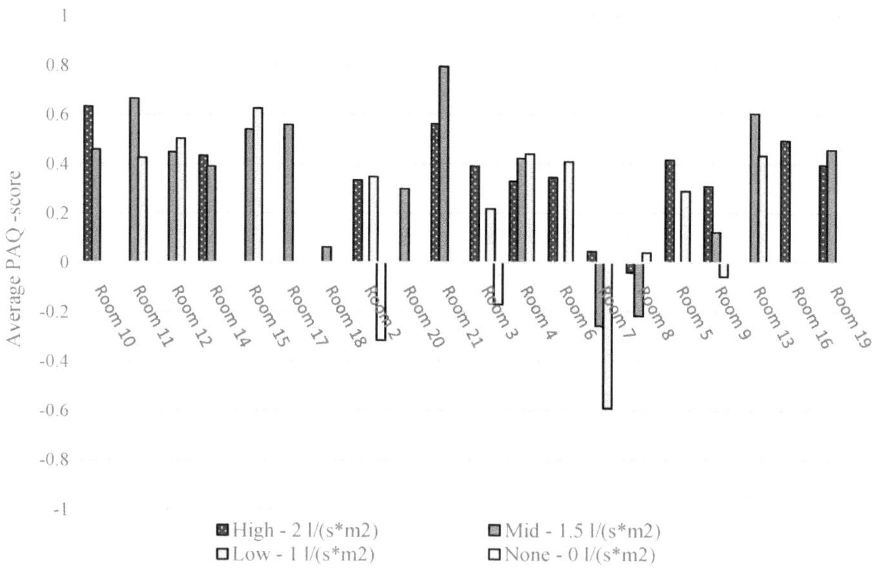

Fig. 5 Average PAQ-score by the test panel for test room nr 3–21. PAQ-results were linearly scored from −1 (clearly unacceptable) to +1 (clearly acceptable)

Table 4 Paired sample *t*-test of the differences in PAQ with different V_{min}-values

	High–mid 2–1.5 (l/s)/m^2	High–low 2–1 (l/s)/m^2	Mid–low 1.5–1 (l/s)/m^2	High–none 2–0 (l/s)/m^2
Rom 2:	–	0.113, (−0.145, 0.561)	–	**0, (0.307, 0.96)**
Rom 3:	–	**0.037, (−0.017, 0.329)**	–	**0.002, (0.189, 0.765)**
Rom 4:	0.854, (−0.405, 0.13)	0.927, (−0.408, 0.067)	0.54, (−0.203, 0.184)	–
Rom 5:	–	**0.035, (−0.015, 0.349)**	–	–
Rom 6:	–	0.798, (−0.289, 0.121)	–	–
Rom 7:	**0.02, (0.016, 0.608)**	**0, (0.277, 0.745)**	0.157, (−0.152, 0.443)	–
Rom 8:	0.078, (−0.061, 0.351)	0.567, (−0.429, 0.365)	0.97, (−0.548, 0.013)	–
Rom 9:	**0.02, (0.013, 0.455)**	**0.027, (−0.005, 0.492)**	0.353, (−0.236, 0.342)	–
Rom 10:	0.473, (−0.19, 0.203)	–	–	–
Rom 11:	–	–	**0.004, (0.078, 0.403)**	–
Rom 12:	–	–	0.496, (−0.264, 0.266)	–
Rom 13:	–	–	0.168, (−0.168, 0.452)	–
Rom 14:	0.365, (−0.236, 0.328)	–	–	–
Rom 15:	–	–	0.91, (−0.273, 0.057)	–
Rom 19:	0.663, (−0.411, 0.274)	–	–	–
Rom 21:	0.96, (−0.436, 0.028)	–	–	–

P-values and confidence intervals are shown. Significant differences are in bold

4 Discussion

4.1 Test Room

PAQ, SBS and performance was assessed by a joint test panel in accordance with the time schedule in Table 2. The statistical analysis reveals no consistent tendency of the impact of V_{min} during the first 20 min of occupancy, with neither significant impact on PAQ, SBS nor performance.

We believe that the *test room* is a low-emitting room. Material emissions are probably diluted below our sensory threshold level by 1 (l/s)/m^2 or more of supply air. This explains why V_{min} above 1 (l/s)/m^2 had no impact on PAQ. Keep also in mind that the temperature was kept approximately constant in these experiments.

Another explanation of the PAQ results is that this room was taken into use by the complete test panel counting from 15 to 22 persons and it took a few minutes before PAQ was scored on the digital tablet. PAQ could have been dominated by human bioeffluents or the influence of a few minutes of higher ventilation rate. However, experiments 1, 2, 3, 7 and 8 do not support this explanation, since the ventilation rate was not increased from V_{min} in these experiments.

SBS-symptoms and performance were not influenced by ventilation rate the first 20 min. This indicates that human beings need more than 20 min exposure to be influenced by the ventilation differences in these experiments. This implies that it is not necessary for DCV-systems to respond quickly to olfactory pollution load changes.

A field laboratory experiment might always be influenced by factors not controlled for. These experiments must be repeated before any final conclusions can be made.

4.2 *PAQ-Rooms*

PAQ was assessed in the PAQ-room by individuals from the test panel one at the time. We found that ventilating with 2,0 (l/s)/m^2 significantly improved immediate PAQ compared to no ventilation (Table 4). This was expected.

We found that increased V_{min} above 1 (l/s)/m^2 significantly improved PAQ in rooms 5, 7 and 9, all of which had additional pollution sources (Table 4).

We did 17 comparisons between different V_{min}'s in 13 rooms with two significant outcomes in room 3 and 11 (Table 4). In five of the rooms (2, 4, 6, 8, 12 and 15) the average PAQ-score was elevated with V_{min} = 1 (l/s)/m^2 compared to V_{min} = 2 (l/s)/m^2 (Fig. 5). This indicates that the positive impact of V_{min} above 1 (l/s)/m^2 was not present in the low-emitting rooms without additional pollution sources. These indications are supported by the results from the *Test-room*.

However, the results are not consistent. Further experimentation on pollution loads from interior and user activity is required.

5 Recommendations for V_{min}

We found that increasing V_{min} had a significant positive impact on PAQ in rooms with extra pollution sources. This effect was not present in the low-emitting rooms with temperatures at 21.5 ± 1 °C. We could not see that different airflow rates had an impact on PAQ, performance or well-being in the dedicated *test room* during the first 20 min of occupation. This indicates that V_{min} above 1 (l/s)/m^2 has limited impact on IAQ in real low-emitting rooms. Our preliminary recommendation is to restrict V_{min} to 1 (l/s)/m^2 in rooms designed to be low-emitting. Additional ventilation due to uncertain pollution load from materials or equipment should be included in the choice of V_{max}.

This preliminary recommendation is based on the prerequisite that a DCV-system has easily adjustable V_{min} and V_{max}. This means that V_{min} can be easily adjusted in the case of rooms not successfully fitted as a low-emitting room, if the pollution load is changed by different use of the room or changes in room size.

Acknowledgements This study is funded by the Research Council of Norway EnergiX program under Grant 255375/E20 together with the industry partners Undervisningsbygg Oslo KF, GK Inneklima AS, DNB Næringseiendom AS, Erichsen & Horgen AS, Hjellnes Consult AS, Multiconsult AS, Interfil AS, Camfil Norge AS, Swegon AS, Belimo Automasjon Norge AS, NEAS AS and the foundation Norsk VVS Energi- og Miljøteknisk Forenings Stiftelse for forskning.

Compliance with Ethical Standards

Formal consent was given by the volunteers who participated in this study. Permission was given by the school where this study was conducted. We did not collect any identifiable or sensitive information that would require ethical approval. The research has been conducted in compliance with the ethical standards at OsloMet—Oslo Metropolitan University (formerly Oslo and Akershus University College of Applied Science) and Norwegian Law.

References

1. M. Mysen, S. Berntsen, P. Nafstad, P.G. Schild, Occupancy density and benefits of demand-controlled ventilation in Norwegian primary schools. Energy and Buildings **37**, 1234–1240 (2005)
2. J. Halvarsson, Occupancy pattern in office buildings—Consequences for HVAC system design and operation. Doctoral Thesis, at NTNU, 2012:37 (2012)
3. M. Mysen, P.G. Schild, A. Cablé, in *Demand-controlled ventilation—requirements and commissioning*. SINTEF Research 24 Guidebook published by SINTEF. ISBN: 978-82-536-1414-4 (63pp) (2014)
4. O. Seppänen, N. Brelih, G. Goeders, A. Litiu, *Existing buildings, building codes, ventilation standards and ventilation in Europe* [HEALTHVENT WP5 final report] (2012)
5. P.O. Fanger, Introduction of the olf and the decipol units to quantify air pollution perceived by humans indoors and outdoors. Energy Build. **12**, 1–6 (1988)
6. K. Anderson, *The MM-Questionaires—A Tool when Solving Indoor Climate Problems* (Örebro Medical Center Hospital, Sweden, 1993)
7. K.I. Fostervold, J. Nersveen, Proportions of direct and indirect indoor lighting—The effect on health, well-being and cognitive performance of office workers. Light. Res. Technol. **40**(3), 175–200 (2008)

Dynamic Thermal Performance and Controllability of Fan Coil Systems

Taha Arghand, Anders Trüschel, Jan-Olof Dalenbäck and Saqib Javed

Abstract In order to characterize and control a system properly, it is inevitably necessary to understand and define the interactions between various design parameters and the controllability of the system. This study experimentally investigates effects of three design parameters, including supply water temperature, fan speed, and room heat load, on the dynamic response of a fan coil system. The experiments have been performed in a mock-up of an office room equipped with a fan coil unit (FCU). A direct ground cooling system has been used to supply the FCU with high-temperature chilled water ranging between 15 and 23 °C. The dynamic response of the system to step changes in the design parameters has been studied using room operative temperature as an indicator. The results of this study indicate that the system behaves as a first-order system without a time delay. The results also suggest that the time constant and the characteristic of the dynamic response of the fan coil system are not affected by the initial room temperature. Among the design parameters, fan speed is observed to have the most significant effect on the dynamic response of the system. Supply water temperature and room heat load are both found to have insignificant effects on the dynamic response characteristics of the system.

Keywords Controllability · Fan coil unit · Dynamic response

1 Introduction

Fan coil units (FCU) are among the most viable options for heating and cooling of residential and commercial buildings. Their advantages include smaller or no plant-room requirements, individual room temperature control, and high cooling capacity due to forced convection, among others.

T. Arghand (✉) · A. Trüschel · J.-O. Dalenbäck · S. Javed
Building Services, Chalmers University of Technology, 412 96 Gothenburg, Sweden
e-mail: arghand@chalmers.se

D. Johansson et al. (eds.), *Cold Climate HVAC 2018*,
Springer Proceedings in Energy, https://doi.org/10.1007/978-3-030-00662-4_30

A conventional FCU system, when operating in cooling mode, uses chilled water between 4 and 8 °C, generally provided by a chiller, to provide both sensible and latent cooling. However, in recent years, FCUs operating with high-temperature chilled water, between 15 and 20 °C, have also gained in popularity [1–4]. These systems, sometimes referred to as dry FCUs, only provide sensible cooling, as the chilled water temperature supplied to the FCU's cooling coil is higher than the room air dew point temperature to provide any latent cooling. These FCUs can be used with direct ground cooling, without operating the chiller, and thus offer high potential for energy efficiency and savings.

The overall thermal and operational performance of a FCU system depends both on steady-state and dynamic behaviors of the system. Research shows that the cooling capacity of a FCU, operating with high-temperature chilled water, is lower than an equivalent conventional system [5]. The change in supply water temperature (T_s) to the FCU not only affects the steady-state parameters, e.g. cooling capacity, but it also alters the dynamic parameters of the cooling system, e.g. response time of the system. This is particularly important from a control point of view as the controllability of the system depends both on the steady-state and the dynamic characteristics of the system. Thus, FCUs operated with high-temperature chilled water require a different control approach than their conventional counterparts.

The primary goal of this study is to study the controllability of a FCU supplied with high-temperature chilled water from a direct ground cooling system. The study experimentally investigates how the controllability of the FCU is affected by design parameters including supply water temperature, fan speed, and room heat load.

2 Method

2.1 Experimental Facility

The experiments for this study were performed in a laboratory-based test room (4.3 m × 3.0 m × 2.4 m), which was built as a mockup of a single-office room. The room had a light weight construction and was made up of polystyrene panel walls with gypsum board finishing, and light weight panels ceiling with external glass wool insulation. The test room was built inside an existing large room, with relatively stable temperature, to minimize the effects of external disturbances. The heat load in the room consisted of a heated dummy, electric lights, and electrical heating foils, see Fig. 1.

A cassette-type FCU installed in the test room was provided with high-temperature chilled water from a direct ground cooling system. The FCU circulated the room air through its cooling coil using a variable speed fan.

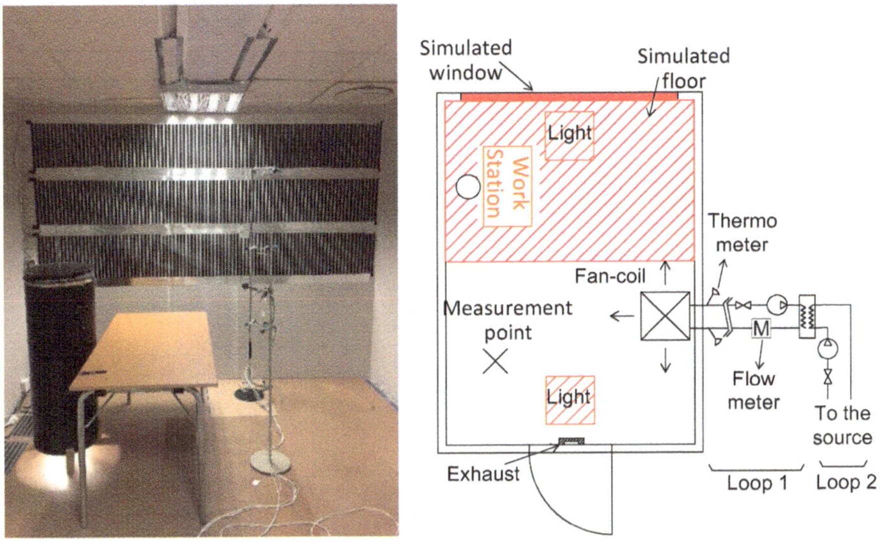

Fig. 1 Test room layout

2.2 *Experimental Conditions*

Three design parameters, supply water temperature, fan speed, and room heat load were studied and their effect on the controllability of the cooling system was analyzed. These parameters were studied using three different sets of tests shown in Table 1. The first set of tests (Test 1–1 to Test 1–4), referred to as Case 1 in Table 1, was performed with varying supply water temperatures, while maintaining the room heat load and the fan speed constant. On the other hand, the second set of tests (Test 2–1 to Test 2–4), referred to as Case 2 in Table 1, was performed with fixed supply water temperature and constant room heat loads. The fan speed though varied for each test of this case. Finally, the third set of tests, referred to as Case 3 in Table 1, was performed by altering the room heat loads at prescribed supply water temperatures. The fan speed, however, remained unchanged for this case.

2.3 *Instrumentation and Measurement Procedure*

As shown in Fig. 1, the water circuit of FCU consisted of two loops. The water flow rate in each loop was controlled independently by either regulating the pump speed or by adjusting the control valve position or a combination of both. The step changes in water temperature supplied to the FCU were accomplished by altering

Table 1 Experimental conditions

Experimental condition	Test number	Supply water temperature [°C] ± SD		Room heat load [W]	Fan speed [%]
		Before change	Final value		
Case 1	Test 1–1	15.5 ± 0.10	17.7 ± 0.00	850	100
	Test 1–2	17.7 ± 0.00	19.7 ± 0.00	850	100
	Test 1–3	19.5 ± 0.00	21.6 ± 0.00	850	100
	Test 1–4	20.9 ± 0.00	23.0 ± 0.06	850	100
Case 2	Test 2–1	19.2 ± 0.07	19.2 ± 0.07	850	100–80
	Test 2–2	19.2 ± 0.07	19.2 ± 0.07	850	80–60
	Test 2–3	19.2 ± 0.07	19.2 ± 0.07	850	60–40
	Test 2–4	19.2 ± 0.07	19.2 ± 0.07	850	40–20
Case 3	Test 3–1	15.3 ± 0.00	17.1 ± 0.04	370	100
	Test 3–2	17.1 ± 0.04	19.2 ± 0.01	370	100
	Test 3–3	19.2 ± 0.01	21.3 ± 0.10	370	100
	Test 3–4	15.2 ± 0.00	17.4 ± 0.05	530	100
	Test 3–5	17.4 ± 0.05	19.4 ± 0.13	530	100
	Test 3–6	19.4 ± 0.13	21.3 ± 0.06	530	100
	Test 3–7	15.3 ± 0.26	17.2 ± 0.05	700	100
	Test 3–8	17.2 ± 0.05	19.1 ± 0.12	700	100
	Test 3–9	19.1 ± 0.12	21.1 ± 0.01	700	100

flow rate of the water in loop 2. However, it took a certain amount of time for the fluid temperature in loop 1 to reach the targeted value after adjusting the flow rate in loop 2. The reasons for this delay included high thermal inertia of loop 2 due to large mass of water in it, and the time delay in controlling supply water temperature from the direct ground cooling system. The flow rate in loop 1 was measured using a Vortex flow meter. It is worth mentioning that for all experiments reported in this paper, the water flow rate in loop 1 was kept constant. Two PT-100 temperature sensors, installed in very close proximity to the FCU, were used to measure supply and return water temperatures to and from the FCU. Measurements of operative temperature (T_{op}) were performed using a PT-100 temperature sensor placed in a gray ping-pong ball of 4.0 cm diameter, and positioned 1.1 m above the floor level.

The FCU was equipped with a speed control motor connected to an inverter unit, which allowed for adjusting the fan speed from 0 to 100%. The heat loads from the human dummy and electric lamps were 70 and 100 watts, respectively. These heat loads were kept unchanged during all experiments. Additional heat gains included those from electrical heating foils installed on the floor and one wall. All changes in room heat loads, as shown in Table 1, were accomplished by regulating the voltage of the electric heating foils.

2.4 Evaluation

All experiments were performed under dynamic condition. However, for the sake of this study, the dynamic condition was assumed to be between the initial state and 95% of the final state [6]. This is because, it is not always possible experimentally to reach the stable steady-state condition in the room.

In order to characterize and control a system properly, it is inevitably necessary to understand and define the interactions between various design parameters and controllability of the system. It is also equally important to identify and select an appropriate parameter to study the dynamic response of the system. In this study, room operative temperature (T_{op}) was adopted to study the dynamic response of the system. It is not only relatively easy to measure, but also indicates, to some extent, the thermal climate perceived by the human body.

The dynamics of the cooling system was analyzed in relation to time constant (τ) and dead time of the system. The time constant characterizes the time response of a step change. For first order systems, it is the time taken to reach 63.2% of the final value after a step change has been applied to the system [7]. The dead time represents the delay in the response of the system to a stimulus in the input. The time constant and dead time of a system suggest how fast the system responds to a change in its conditions.

3 Results

3.1 Case 1—Step Change of Supply Water Temperature

Figure 2a presents the response of room operative temperature to changes in supply water temperature. The room operative temperature increases with each step increase in supply water temperature. Figure 2b shows the net increase in operative temperature in response to a step change in supply water temperature for each test. No specific pattern of increase in room operative temperature with supply water temperature is observed.

Table 2 provides further details of the dynamic behavior of the system. The table shows the experimentally deduced values of dead time, time constant and stabilization time for all tests of Case 1. Stabilization time is equal to three times time constant, and is the time required to reach stable room temperature after applying a step change. As noticed earlier for room operative temperatures, there exists no clear pattern of increase in time constant with changing values of supply water temperature, i.e. Tests 1–1 to 1–4. However, as seen from Fig. 2b and Table 2, the net increase in steady-state room operative temperature and the stabilization time for Tests 1–1 and 1–4 are higher than Tests 1–2 and 1–3. Hence, the time constant for these two tests is also larger than the other two tests.

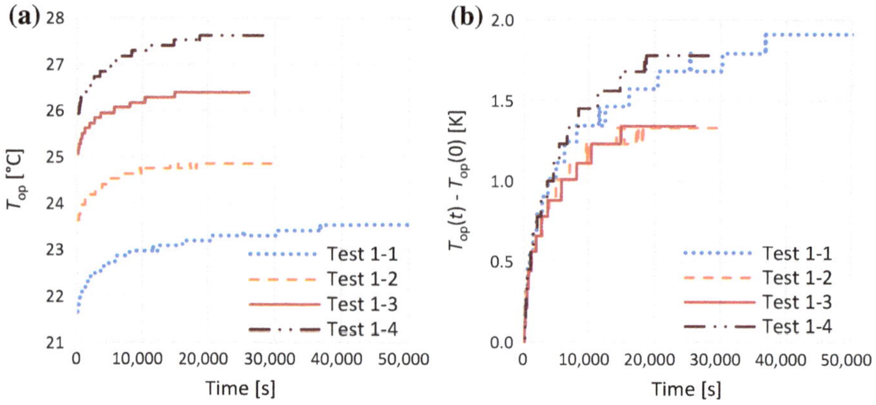

Fig. 2 Effect of step changes in supply water temperature (T_s) on room operative temperature (T_{op}), **a** absolute temperatures and **b** net increase

Table 2 Dynamics of tests of Case 1

Test	T_{op} (final)–T_{op} (initial) [K]	Dead time [s]	τ [s]	2.3 τ [s]	3 τ [s]
Test 1–1	1.9	60	6440	14,812	19,320
Test 1–2	1.3	140	3060	7038	9180
Test 1–3	1.3	140	2720	6256	8160
Test 1–4	1.8	220	4580	10,534	13,710

It can also be observed from Table 2 that the dead time of the FCU is relatively short compared to the time constant and the stabilization time of the system. A slight increase in dead time can be observed with increasing supply water temperature, which is probably due to the increased difference between the supply water temperature and the room operative temperature.

3.2 Case 2—Step Change of Fan Speed

Figure 3a shows the response of room operative temperature to step changes in fan speed. As seen from Table 1, all changes in fan speed were implemented in a descending order, i.e. from higher to lower speeds, thus reducing the air flow rate drawn from the room across the cooling coil. For all tests of Case 2, the difference between the FCU supply temperature (i.e. T_s) and the return temperature (T_r) remains nearly constant, see Table 3. However, the convective cooling from the FCU reduces due to the decreasing fan speed, which in turn reduces the room air flow.

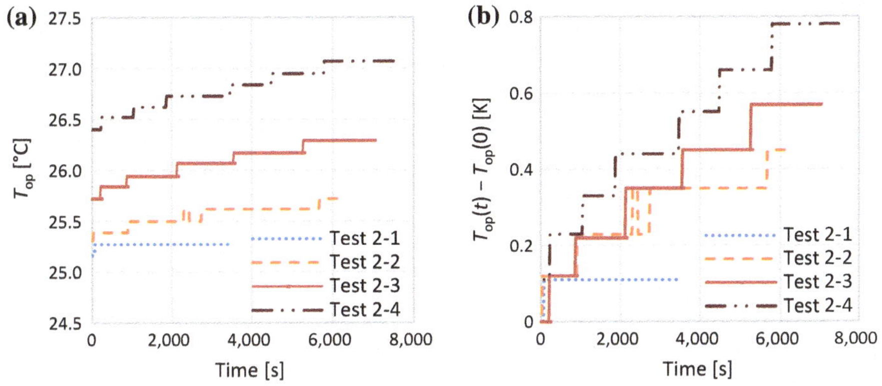

Fig. 3 Effect of step changes in fan speed on room operative temperatures (T_{op}), **a** absolute temperatures and **b** net increase

Table 3 Dynamics of tests of Case 2

Test	T_{op} (final)–T_{op} (initial) [K]	T_s [°C]	T_r [°C]	Dead time [s]	τ [s]	3 τ [s]
Test 2–1	0.0	19.1	21.2	0	0	0
Test 2–2	0.5	19.2	21.1	220	2940	8820
Test 2–3	0.6	19.2	21.1	400	2220	6660
Test 2–4	0.8	19.2	21.0	160	3860	11,580

Figure 3b shows the net increase in the room operative temperature following step changes in the fan speed. The time constant of the cooling system increases with the decreasing fan speed due to net increase in room operative temperatures and the stabilization time.

3.3 Case 3—Step Change in Room Load

Each of Figs. 4a–c show the net increase in the room operative temperature, i.e. T_{op} (t) − T_{op} (0), for three different room heat load conditions each with a prescribed supply water temperature. This case focuses on the changes in the time constant of the system. Hence, for reasons of clarity, the plots of Fig. 4 are shown only for short testing times.

The results of Fig. 4, do not exhibit any clear pattern of the impact of the room heat load on the rate and the magnitude of the increase in the room operative temperature. It can, however, be noticed that for a prescribed supply water temperature, the time constant of the system does not change significantly with the room heat load. In other words, the magnitude of difference between supply water temperature and room operative temperature does not affect the time constant of the system significantly.

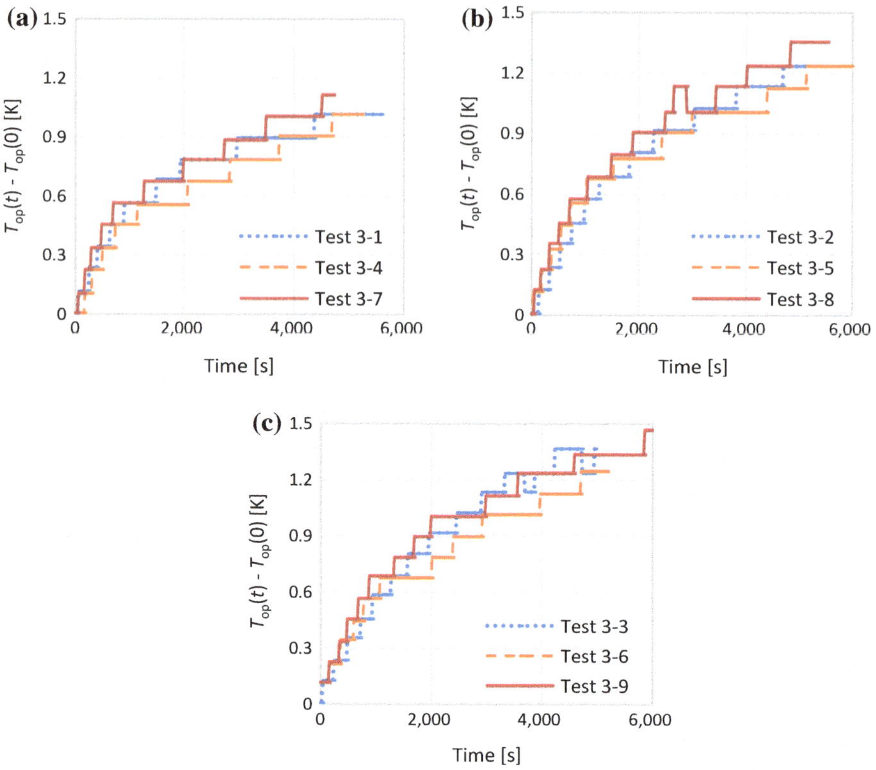

Fig. 4 Effect of step changes in room load on room operative temperatures (T_{op}) for prescribed supply water temperatures, **a** $T_s \approx 15.3$ °C, **b** $T_s \approx 17.2$ °C, and **c** $T_s \approx 19.2$ °C

Table 4 shows experimentally deduced values of dead time and time constant for all tests of Case 3. It can be observed from the table that the relatively short dead time is inherent in the system. The dead time of the system remains between 2.5 and 6.5 min for all 9 tests performed with varied test conditions. This is so, it seems, because of the forced convection effect caused by the FCU. However, no clear effect of the room heat load on the magnitude or order of the dead time can be observed. The same is also true for the time constant.

4 Discussion

As mentioned earlier in the introduction section, this study aims to investigate how and to what extent changes in design parameters, including supply water temperature, fan speed, and room heat load, affect the dynamic behavior of a fan coil system, operated with high-temperature chilled water. In this study, the dynamic behavior of the overall system has been characterized using a systematic series of experiments measuring the response of the system to step changes in the design inputs. Based on the experimental results, the increase in room operative temperature can be modelled using the following equation.

$$\Delta T_{op} = \left(T_{op\ max} - T_{op\ min}\right) \times \left(1 - e^{-\tau \cdot t}\right)$$

where, $T_{\text{op min}}$ and $T_{\text{op max}}$ are, respectively, the minimum and maximum operative temperature before and after imposing a step change, τ is the time constant, and t is the time. The value of time constant is deduced in form of a regression coefficient applying the "least squares" statistical method on the experimental results.

Figure 5a, b, respectively, present increase in the room operative temperature to step changes in supply water temperature and fan speed using two tests from Case 1 and two tests from Case 2. For all tests presented, a very good agreement between the values estimated from the above equation and measured through experiments can be observed. It can also be seen that the room operative temperature response to the step changes is exponential. This suggests, the fan coil system, operated with high-temperature chilled water, behaves approximately as a first-order system without a time delay. The system has a very quick control response, which facilitates the control operation. The same pattern has also been observed for all tests presented in Table 1. This result is of paramount importance for designing robust controllers for FCUs operated with high-temperature chilled water.

Being a first-order system, the dynamic behavior of the fan-coil system can be characterized by its time constant. For all step changes in supply water temperature, the measurements undertaken under different experimental conditions suggest that the supply water temperature or the initial room temperature does not affect the time constant of the system considerably. Instead, these parameters may affect the dead time of the system, and the temperature increase between the steady-state conditions before and after the step change. On the contrary, the time constant of the system is very sensitive to step changes in the fan speed. The results of tests with step changes in fan speed indicate that for optimum control it is equally important to reduce the time constants of both the cooling system and the room.

Table 4 Dynamics of tests of Case 3

Test	Room load [W]	Dead time [s]	τ [s]	3 τ [s]
Test 3–1	370	260	1700	5100
Test 3–2	530	160	1280	3840
Test 3–3	700	260	1480	4440
Test 3–4	370	340	2040	6120
Test 3–5	530	240	2640	7920
Test 3–6	700	260	2100	6300
Test 3–7	370	240	2160	6480
Test 3–8	530	380	3060	9180
Test 3–9	700	200	1900	5700

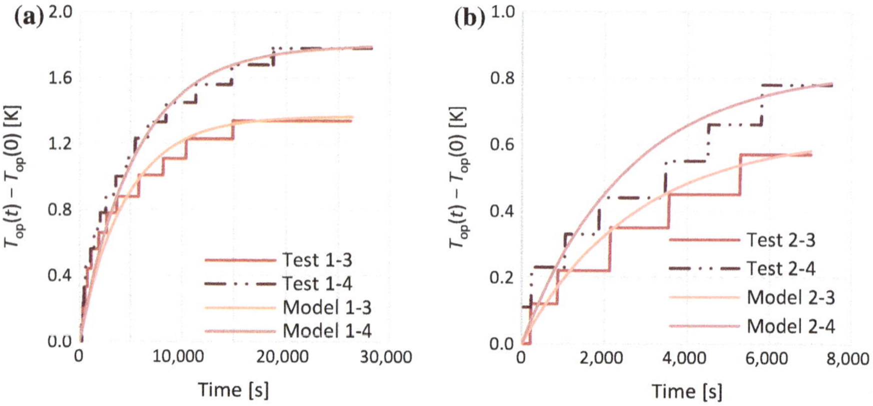

Fig. 5 Measured and model response to a step change in **a** supply water temperature, and **b** fan speed

5 Conclusions

The design parameters, including supply water temperature, fan speed, and room heat load, have been studied to analyze the dynamic behavior of a fan coil system, operated with high-temperature chilled water. Based on the dynamic response of the FCU to step changes made in the design parameters under different experimental conditions, it can be concluded that the system behaves as a first order system with no time delay. This makes the control response quick, which, in turn, also facilitates the control operation.

The results of this study demonstrate that characteristics of dynamic response of the FCU system to changes in room thermal conditions are not greatly influenced by the supply water temperature. The FCUs operated with high-temperature chilled water are clearly capable of adapting to changes in the room thermal conditions. Finally, when designing controllers for FCUs operated with high-temperature chilled water, the dynamics of the cooling system and the thermal characteristics of the building shall both be accounted for.

Acknowledgements The support of all authors was provided by the Swedish Energy Agency (Energimyndigheten) through their national research program EFFSYS Expand. The funding source had no involvement in the study design; collection, analysis and interpretation of data; or in the decision to submit the article for publication.

References

1. Y.M. Xuan, F. Xiao, X.F. Niu, X. Huang, S.W. Wang, Research and applications of evaporative cooling in China: a review (II)—systems and equipment. Renew. Sustain. Energy Rev. **16**, 3523–3534 (2012). https://doi.org/10.1016/j.rser.2012.02.030

2. K. Zhao, X.-H. Liu, Y. Jiang, Application of radiant floor cooling in a large open space building with high-intensity solar radiation. Energy Build. **66**, 246–257 (2013). https://doi.org/10.1016/j.enbuild.2013.07.014
3. K. Zhao, X.-H. Liu, T. Zhang, Y. Jiang, Performance of temperature and humidity independent control air-conditioning system in an office building. Energy Build. **43**, 1895–1903 (2011). https://doi.org/10.1016/j.enbuild.2011.03.041
4. R. Lizarte, M. Izquierdo, J.D. Marcos, E. Palacios, Experimental comparison of two solar-driven air-cooled LiBr/H2O absorption chillers: indirect versus direct air-cooled system. Energy Build. **62**, 323–334 (2013). https://doi.org/10.1016/j.enbuild.2013.03.023
5. X. Liu, Y. Jiang, T. Zhang, *Temperature and Humidity Independent Control (THIC) of Air-Conditioning System* (Springer Science & Business Media, Berlin, 2014)
6. W.K. Roots, *Fundamentals of Temperature Control* (Academic Press, London, 1969)
7. W. Altmann, *Practical Process Control for Engineers and Technicians* (Newnes, London, 2005)

Evaluation Study of the Performance of Dual Core Energy Recovery System for Dwellings in the Arctic

Boualem Ouazia and Ganapathy Gnanamurugan

Abstract In extreme cold climates, conventional single core heat/energy recovery ventilators experience major problems with ice formation on the extracted airside and frequently under-perform or fail. A dual core design could overcome the performance of single core design units and meet the rigorous requirement for operation in the North. The evaluation of the performance of a dual core unit was undertaken using dual environmental chambers capable of reproducing the typical outdoor and indoor conditions with regard to temperature and relative humidity. Experiments were conducted under a range of outdoor cold temperatures, and indoor conditions identified for housing in the Arctic. The dual core ERV was found to be capable of withstanding a temperature of down to −37 °C without deteriorating its thermal performance, and therefore was more frost-tolerant than conventional single core units. In conclusion, the dual core design system could be a feasible option for extreme cold climates.

Keywords Energy recovery ventilator · Dual core · Ventilation
Residential · Arctic

1 Introduction

Effective ventilation is essential to provide acceptable indoor air quality (IAQ) and control moisture in homes. However, ensuring proper ventilation while minimizing energy costs in dwellings can be a challenge in Canada's North and Far North (Arctic) due to several factors including a harsh cold climate and frequent overcrowding. Space heating requires an enormous amount of energy in an extremely cold climate with long winters. To reduce energy costs building techniques have been improved in the Arctic regions making building envelopes more air tight to decrease the heating demand. As homes across Canada's North are built tighter,

B. Ouazia (✉) · G. Gnanamurugan
National Research Council Canada, Ottawa, ON, Canada
e-mail: Boualem.ouazia@nrc-cnrc.gc.ca

D. Johansson et al. (eds.), *Cold Climate HVAC 2018*,
Springer Proceedings in Energy, https://doi.org/10.1007/978-3-030-00662-4_31

there has been an increase in the poor IAQ. Without ventilation air, carbon dioxide, odors, dust, airborne pollutants and excess humidity are kept indoors, potentially causing or aggravating problems to occupants' health and comfort, and encouraging mold growth. An increasingly common method to provide a required ventilation rate, control moisture build-up and reduce energy costs is to install a balanced ventilation system that uses a heat or enthalpy recovery ventilator (HRV or ERV). HRV/ERV ventilation systems (single core design) installed in dwellings in Canada's North and Far North frequently underperform and fail [1, 2]. Ventilation of houses can be very problematic in the North where frosting poses a significant challenge for the heat/energy exchangers. The design winter temperatures in the far North are much colder than the outdoor temperature of −25 °C that is typically used by HRV/ERV manufacturers for their certified products rating. Frost formation in exchangers is common in cold regions where the outdoor temperature is below −10 °C for the majority of the cold season. Conventional problems created by the formation of frost in heat/energy exchangers [3] are:

- Partial or full blockage of air flow passages,
- Increase in pressure drop through the exchanger or decrease in airflow rate,
- Increase in electric power for the fans,
- Decrease in the heat transfer rate between the two air streams and
- Cold draughts in the space due to low supply air temperatures.

In cold climates, conventional HRVs/ERVs require a defrost strategy to remove frost that can form within the air-to-air heat exchanger core. The conventional frost protection mechanisms in single core units can incidentally create an intermittent ventilation system that causes cross contamination of exhaust and supply air streams. These shortcomings can undermine the outdoor air requirements of residential ventilation standard [4], a residential ventilation standard commonly used as a baseline for building codes and programs. This can result in uncomfortable and unhealthy living conditions for occupants. At present, there are no HRVs/ERVs specifically designed and manufactured to meet the rigorous requirements for operation in the Far North. A new design of energy recovery ventilation system with dual core exchangers addresses the frost protection concerns by periodically warming the return air through one of the two heat exchanger cores while the outdoor air gains heat from the other. The new design could overcome the challenges faced by conventional single core heat/energy recovery ventilators in the Arctic. This paper presents the performance evaluation results from laboratory testing of the dual core unit for realistic Northern indoor and outdoor conditions.

2 Methodology

The experimental work consisted of laboratory testing of a dual core energy recovery unit using a combination of two environmental chambers and an HRV/ERV test rig.

2.1 Tested Technology

A dual core energy recovery unit is based on the working principle of cyclic storage and release of heat in the heat exchanger cores of corrugated sheets alternately exposed to exhaust and intake air. The unit includes a supply and an exhaust fan and two cores filled with specially corrugated 0.7 mm thick aluminum plates which act as heat accumulators. In between the cores is a patented damper section which changes over every 60 s to periodically direct warm air through one of the two cores while outside air gains heat in the other core. Before each fan is a filter section to filter the air. Heat recovery is automatically activated when called upon. The schematic of the dual core unit with the two sequences is presented in Fig. 1, with a description of the 2 sequences below.

Sequence 1—Exhaust air (RA) charges *Core B* with heat from return warm air (RA) from indoor and *Core A* discharges heat to supply air (SA).

Sequence 2—Exhaust air (RA) charges *Core A* with heat from return warm air (RA) from indoor and *Core B* discharges heat to supply air (SA).

In addition to the innovative heat recovery strategy described above control logics are implemented to regulate internal damper to ensure that comfortable air delivery temperatures are achieved in all conditions. The damper is controlled by the 2 internal thermostats GT1 in the supply air and GT2 in the exhaust air. GT1 is set to 15 °C; GT2 is set to 20 °C. The sequence of operation is as follows:

1. If exhaust air temperature is lower than 20 °C, unit in energy recovery mode (cycling every 60 s).
2. If exhaust air temperature is higher than 20 °C and supply air temperature higher than 15 °C, unit in free cooling mode (cycling every 3 h). (This mode is not expected to be used often in Arctic conditions. Monitoring in the Arctic planned for future phases should help verify this.)
3. If exhaust air temperature is higher than 20 °C and supply air temperature lower than 15 °C, unit in energy recovery mode until the supply air temperature becomes higher than 15 °C then it will revert to free cooling mode.

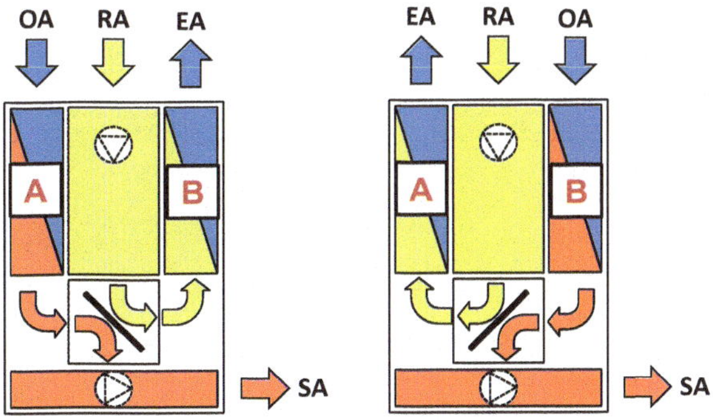

Fig. 1 Principle of function—sequence 1 (left) and sequence 2 (right)

2.2 Experimental Facility

The following experimental facility and associated setup were used to perform a first round of short-term cold-climate performance tests of the dual core heat exchange unit under controlled laboratory conditions at NRC installations in Ottawa. The NRC's Environmental Chambers are capable of simulating interior and exterior climatic conditions over an extended period of time. The exterior climatic conditions can be varied over a range of −40 to +40 ± 1.0 °C with the capability of maintaining a steady state set point. Simulated interior climatic conditions can be varied from 15 °C to 30 ± 1.0 °C with the capability of maintaining a steady state ambient humidity (30 to 60% RH). The HRV/ERV test rig shown in Fig. 2 and installed between the indoor and outdoor climatic chambers consisted of four rigid straight ducts for airflow measurements, connected to the unit under test and to the climatic chambers using flex ducts. The ducts are insulated with a low vapour permeability sleeve to minimize heat transfer and vapour accumulation in these long ducts.

2.3 Instrumentation

Air is drawn from two environmental chambers at desired conditions. In order to determine the efficiency and the conditions leading to frosting in the dual core unit, several properties are measured at different locations in the test facility. Two airflows were measured using Nailor type airflow elements. Airflow elements were installed in the supply (SA) and exhaust (RA) ducts to measure the mass flow rates of dry air in the supply and return airstreams. Pressure taps were placed at the inlet and outlet of unit (SA, RA, EA, OA) to measure the static pressures, connected by PVC tubes to the pressure transducers integrated in a designed pressure transducer box. The air temperature and relative humidity were measured using RH and T

Fig. 2 HRV/ERV test rig

Table 1 Specifications of instrumentation

Type	Model	Range	Accuracy
RH and T probe	Vaisala HMP60	−40 to +40 °C	±0.2 °C
		10–90%	±3%
Pressure transducer	Setra C264	0–125 Pa/−125 to +125 Pa	±1% FS
Airflow element	6″ Nailor 36FMS	30–200 cfm	±5%

probes, calibrated over a temperature range of −40 to +40 °C and over an RH range of 10–90%. The temperature and RH were measured at the inlets and outlets of the unit. The required sensors were calibrated; specifications of the sensors used to measure temperature, relative humidity, pressure and flow, and their accuracy are presented in Table 1.

3 Test Procedure

A series of experiments were conducted to gather data on the thermal/ventilation behavior and performance of the tested unit when subjected to steady state climatic indoor and outdoor conditions. Collected data was used to calculate the apparent sensible/total effectiveness and assess the impact of potential build-up of frost on the thermal and ventilation performance of the dual core technology. The research was limited to the evaluation of innovative technologies for the Arctic, thus the testing was done for the heating mode only. The living conditions in the north shows that the homes can be more crowded and that the level of indoor relative humidity would be higher than in homes located in the south part of Canada. The two main air properties that may affect frosting are temperature at supply inlet (from outdoor) and RH at the air exhaust inlet (from indoor). The indoor air temperature (exhaust inlet) is kept constant in all experiments, and RH at the supply inlet (from outdoor) does not play a significant role in the frosting. The conditions in the indoor chamber were set at realistic indoor conditions in northern homes. The temperatures in the outdoor chamber were varied in a range from +10 to −35 °C to challenge the test unit with extreme cold conditions. The experimental evaluation of the performance of a dual core unit was done with supply and exhaust airflows calculated for the Canadian Centre for Housing Technologies (CCHT) houses [5]. Outdoor air shall be mechanically supplied to each dwelling unit using a ventilation system providing no less than the rate specified by Eq. (1) taking into account people and house air needs [4].

$$Q_{tot} = 0.03A_{floor} + 7.5(N_{br} + 1) \quad (1)$$

where Q_{tot} is the required outdoor ventilation flow rate, A_{floor} is the floor area and N_{br} is the number of bedrooms, not to be less than one.

Table 2 Experimental design

Test	Indoor T [°C]/RH [%]	Outdoor T [°C]	Duration [hr]
1	25 °C/55%	10	7
2		0	7
3		−5	7
4		−10	7
5		−15	8
6		−20	8.5
7		−25	8.5
8		−30	9
9		−35	14

The CCHT twin houses have a floor area of 210 m^2 (2260 ft^2) with 4 bedrooms, which require a total ventilation rate of 47.2 L/s (105 cfm). The dual core unit was tested at balanced supply/exhaust airflows at 47.2 ± 2.4 L/s (105 ± 5 cfm) calculated using Eq. (1), and following the heating mode experimental design presented in Table 2. Nine tests were undertaken between October 17th and 31st, 2016 with identified Northern indoor conditions of 25 °C and 55% RH. After each test the outdoor chamber was reset at a temperature higher than zero to allow the meltdown of any potential frost build up in the two heat exchangers before the next run. This allowed the same initial condition at the start of each test; i.e., consistently no presence of frost in heat exchangers at the start of each test permits comparison between test results.

4 Data Reduction

The performance of the dual-core unit is primarily determined by its effectiveness (sensible and total) as described in ventilation and certification standard [6–8], its pressure drop and mass flow characteristics, and supply air temperature to indoor. The heat exchanger effectiveness includes the following two terms:

Apparent Sensible Effectiveness (ASE) —The measured temperature rise of the supply airstream divided by the difference between the outdoor temperature and entering exhaust system air temperature, then multiplied by the ratio of mass flow rate of the supply airflow divided by the mass flow rate of the lower of the supply or exhaust system airflows.

Apparent Total Effectiveness (ATE) —The measured enthalpy change (sensible plus latent) of the supply airstream divided by the difference between the outdoor enthalpy and entering exhaust system enthalpy, then multiplied by the ratio of the supply airflow divided by the mass flow rate of the lower of the supply or exhaust airflows.

The temperature and relative humidity across the dual core were used to estimate the sensible and total (enthalpy) efficiencies. The supply and exhaust flows were assumed equal (based on balancing the flows at the beginning of the test). The sensible effectiveness and total effectiveness can be calculated using Eq. (2) and are more equivalent to the apparent sensible or total effectiveness because the heat transfer from each fan or through the cases is disregarded.

$$\varepsilon = \frac{m_s(X_{SI} - X_{SO})}{m_{min}(X_{SI} - X_{EI})} \tag{2}$$

where ε is either the sensible, latent, or total heat effectiveness, X is either, the humidity ratio, w, the total enthalpy, h or the dry-bulb temperature, T, respectively, at the supply inlet (SI), supply outlet (SO) and Exhaust inlet (EI). m_s is the mass flow rate of the supply air, m_e is the mass flow rate of the exhaust air and m_{min} is the minimum value of either m_s or m_e.

5 Results

A total of 9 tests were performed to investigate the performance of the dual core unit under heating mode with northern indoor conditions. Each test started with initial parameters that posed no risk of frosting (outdoor chamber at T > 10 °C). Outdoor/supply temperatures are then lowered for each subsequent test to increase the possibility of frosting. In all experiments, the exhaust inlet conditions (return air temperatures from indoors) were maintained at 25 °C and ~55% RH. The outdoor temperature varied between +10 and −37 °C and the supply inlet temperature to the unit varied between +8 and −17 °C. The mass flow rate of dry air was ~47.2 ± 2.4 L/s (~100 ± 5 cfm) for both supply and return airstreams.

5.1 *Airflow*

The supply and exhaust airflows measured during the test done with the coldest outdoor temperature of −35 °C are presented in Fig. 3. The results showed a limited decrease in supply and return airflows over time at low outdoor temperature. The decrease started at outdoor temperatures below −20 °C and for testing period longer than 7 h. The longest test period over 14 h and at an outdoor temperature of −35 °C, the decrease of flows are more pronounced and could be caused by frost occurrence in the heat exchangers. The mass flow rate is reduced as ice or frost forms within the heat exchangers at the exhaust side. The supply and return flows could decrease further for longer test period under extreme cold outdoor conditions.

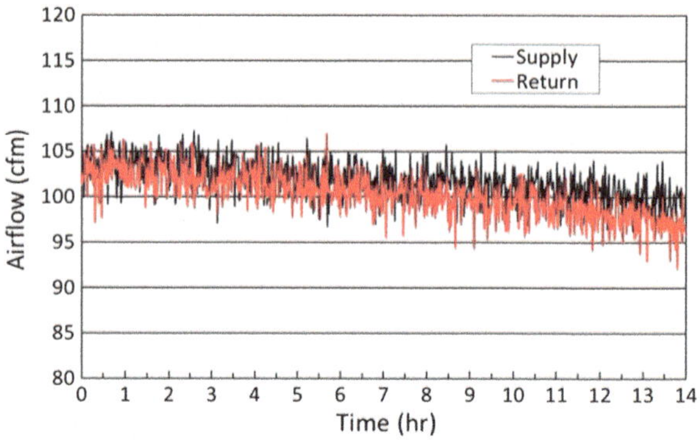

Fig. 3 Measured supply airflow rate at outdoor temperature of −35 °C

The time it would take to choke the flow completely, assuming linearity of the result, could be only confirmed through testing done for more than 14 h or through extended monitoring under extreme cold conditions in the North.

5.2 *Effectiveness*

The calculated apparent sensible efficiencies calculated for the Northern operating indoor conditions are presented in Fig. 4 nine outdoor temperatures varying from −35 to +10 °C. The calculated values ranged from 79 to 93%. The calculated values for different outdoor conditions were in a same order around 85%, much higher than the claimed sensible efficiencies for conventional single core HRVs/ERVs.

The calculated apparent total (sensible + latent) efficiencies for the same Northern operating indoor conditions and outdoor temperatures are presented in Fig. 5 and the calculated values ranged from 59 to 91%. Same trend has been seen for the total effectiveness increasing at lower outdoor temperatures and getting closer to the sensible values at outdoor temperatures below −20 °C. The results showed that the sensible effectiveness is stable (no reduction) with supply air temperature (outdoor temperature) variations. However, the results showed an improvement of the apparent total effectiveness with decreased supply outdoor temperature (decreased outdoor temperature). This could be explained by the fact the total effectiveness depend on the latent heat (moisture in the air) and with the dual core technology, condensation forms on the exhausting heat exchangers (melting the frost build-up by the exhaust warm humid air). When the cycle changes, the outdoor air is passed over the heat exchanger and that moisture is added back to the airstream. The indoor humidity is retained by this technology and reduces the need for added humidity in the conditioned home.

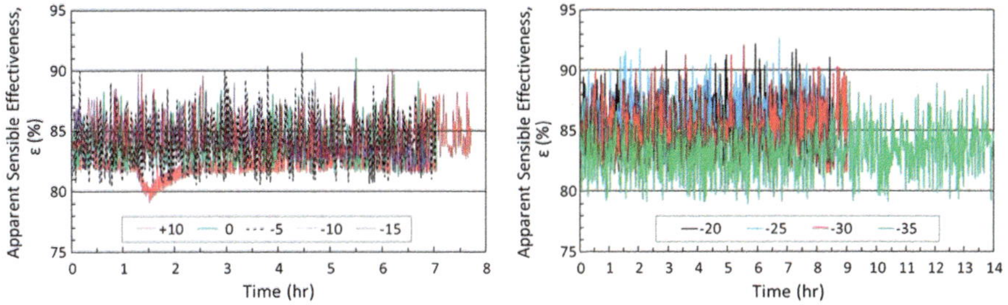

Fig. 4 Apparent sensible effectiveness

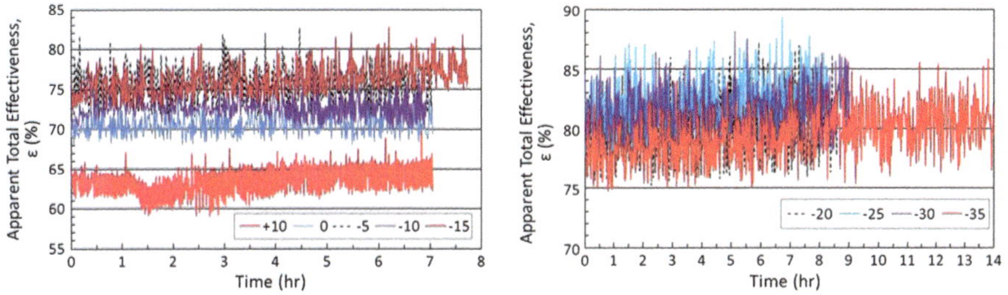

Fig. 5 Apparent total effectiveness

5.3 Supply Air Temperature to Indoor

The efficiency of heat/energy recovery systems should be high enough that tempering of the supply air would not be needed to maintain comfortable interior conditions, even in cold weather. The measured supply air temperature for Northern operating indoor conditions to the indoor are presented in Fig. 6 and the values ranged from 15.3 to 22.8 °C and averaged from 15.5 to 22.6 °C.

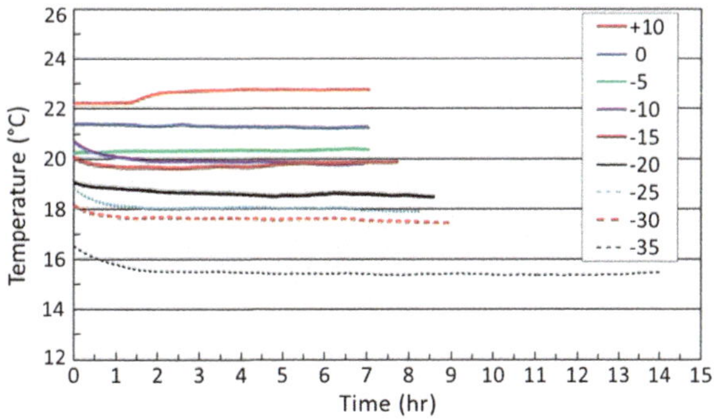

Fig. 6 Supply temperature to indoor with northern indoor conditions

As expected again, the values decreased with decreasing supply temperature (and outdoor temperature), providing a supply air temperature as low as 15.5 °C which is fairly low and will require either a provision for tempering by blending the supply air with room air or preheating before direct delivery to the occupied spaces.

5.4 Frost Occurrence

Depending on exposure time and physical circumstances, frost forming on cold heat exchanger surfaces may enhance or reduce heat fluxes while airflow rates and pressure drop may remain nearly constant or sometimes change drastically. Many parameters may affect frosting, including air inlet conditions, flow rates, exchanger effectiveness and design. In this research, and air flow rates were kept constant. The two main air properties are the supply inlet temperature (T_{OA}) and exhaust inlet relative humidity (RH_{RA}). T_{RA} is kept constant in all experiments, and RH_{OA} does not play a significant role in frosting. A visual observation was conducted to show if there was or not presence of frost in the two heat exchangers. At the end of each test, the frontal door of the dual core unit was opened and photos were taken to visualize potential frost formation. Photos taken during tests are presented in Fig. 7 for the top part (frontal) of the exchangers. The photos show formation of frost on the top front part starting at outdoor temperature condition of −25 °C, and no frost formation on the lower front portion of the exchangers. The length of the frontal portion with frost formation increased at lower outdoor temperature (OA), with frost formation on almost the half top of the front of exchangers at −35 °C.

Although photos taken at temperatures below −25 °C may look like the heat exchangers are fully blocked, the frost build up shown in those photos does not

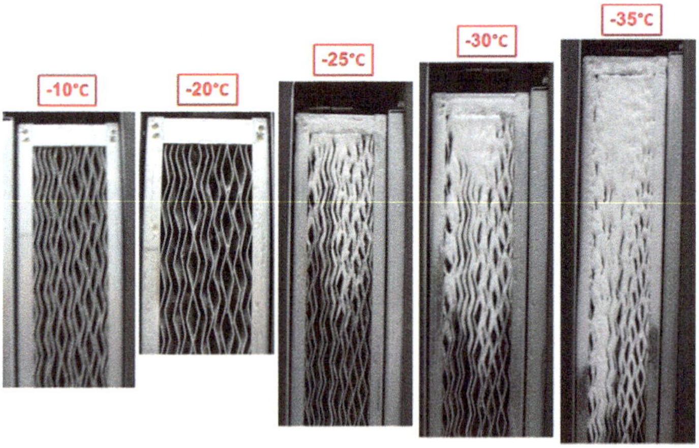

Fig. 7 Photos of the front top of the heat exchangers

extend through the depth of the heat exchangers in the third dimension. The frost formation was largely concentrated on the front and sides of the heat exchangers, with a small amount on the back and only a very small amount at the top of the heat exchangers where most of the flow occurs. The frost formation was limited to front and sides of the heat exchangers and almost none at the inlet/outlet (of the airstreams) and inside (between plates), confirming the limited flow reduction at low outdoor temperatures.

6 Overall Thermal Performance

The average values of the calculated sensible and total efficiencies, and measured supply air temperatures from tests done with Northern indoor conditions (25 °C and 55% RH) are presented in Fig. 8 versus outdoor temperature. The average value of the sensible effectiveness was constant and did not vary with the outdoor temperature. However, the total effectiveness increased with lower outdoor temperature which was likely due to condensation formed on the exhausting heat exchanger. When the cycle changed, the outdoor air was passed over the heat exchanger and that moisture was added back to the airstream. The lowest supply air temperature of 15.5 °C was reached at outdoor temperature of −35 °C which would require provision of tempering the supply air before delivery to the indoor spaces. The majority of HRVs/ERVs are designed with a single core (one heat exchanger) and few new models with dual core that have two heat exchangers in series or two plate heat exchangers in parallel. They are designed for a range of airflows to cover wide residential applications. The claimed apparent sensible efficiencies provided by manufacturers are from certification tests (under very controlled indoor and outdoor conditions) done at the low speed (lowest airflow) and at outdoor temperature of 0 °C (mandatory) and −25 °C if requested by the manufacturers.

The claimed apparent sensible efficiencies for single core HRVs/ERVs varied between 73 and 84% at their lowest flows and at outdoor temperature of 0 °C. These efficiencies decreased at higher flows down to ~70% and at lower outdoor

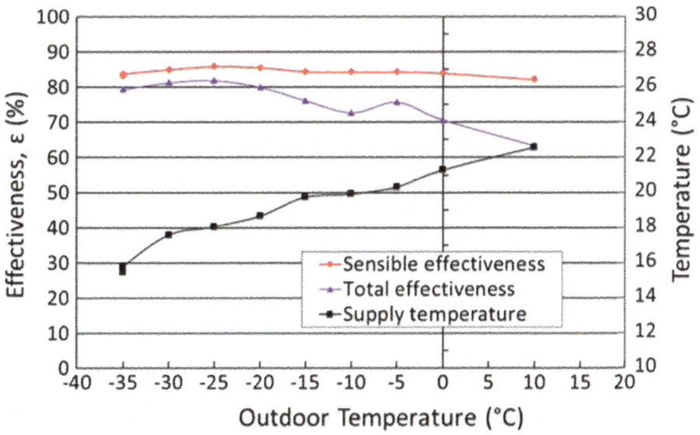

Fig. 8 Average efficiencies and supply temperatures for a range of outdoor temperatures

temperature of −25 °C to 76% (lowest flow) when provided. The calculated apparent sensible efficiencies from experimental data obtained for a dual core unit at a rate of airflow of 100 cfm were between 84 and 86% at 0 °C, 65 and 86% at −20, −25 °C, between 84 and 86% at −30 °C and between 83 and 87% at −35 °C. In general the NRC calculated values were higher than claimed values at 0 °C and much higher (10–15% higher) than the few values claimed for outdoor temperature at −25 °C.

7 Conclusion

Laboratory testing has shown that a dual core energy recovery unit designed with a damper that periodically directs warm air through one of the two heat cores, is capable of continuously addressing the frost protection concerns for identified northern indoor conditions of 25 °C and 55% and outdoor temperature as low as −37 °C.

The calculated apparent sensible and total efficiencies were much higher than values claimed for available single core heat/energy recovery technologies at not only 0 °C and low fan speed, but also at colder outdoor temperature of −25 °C and below.

The dual core unit was found to be capable of operating at a temperature as low as −37 °C without deterioration in its thermal performance, and therefore was more frost-tolerant than conventional single core units. The dual core unit also achieved a supply air temperature above 15 °C at the lowest outdoor temperature of −37 °C. Provision of tempering by blending the supply air with room air or through pre-heating before delivery from ventilation duct might be required for greater indoor comfort in extreme cold conditions. The need for pre-heating will be assessed under extreme climate conditions, over the next phases of this project. The dual core design system could be a feasible option for extreme cold climates. However, extended monitoring of the technology in harsh cold climate is recommended to prove its performance and resilience in Canada's Far North.

References

1. CMHC/SCHL Research Report, *Survey of HRV/ERV Performance Issues in Canada's Near North and Far North* [Final Report] (2016)
2. C. Zaloum, *Technical advice to task force on northern mechanical ventilation equipment design and testing* [CMHC Report] 24pp
3. M.N. Rafati, M. Fauchoux, R.W. Besant, C.J. Simonson, A review of frosting in air-to-air energy exchangers. Renew. Sustain. Energy Rev. **30**, 538–554 (2014)
4. ANSI/ASHRAE 62.2, *Ventilation for acceptable indoor air quality in Low-Rise Residential Buildings. American Society of Heating, Refrigerating and Air-Conditioning Engineers*, Atlanta, GA (2016)

5. M.C. Swinton, H., Moussa, R.G. Marchand, Commissioning twin houses for assessing the performance of energy conserving technologies. In: *VIII Integration of Buildings Envelopes* (Clearwater, Florida 2001), pp. 1–10
6. ANSI/ASHRAE Standard 84, *Method of testing Air-to-Air Heat/Energy exchangers, American Society of Heating, Refrigerating and Air-Conditioning Engineers*, Atlanta, GA (2013)
7. CAN/CSA-C439, *Standard Laboratory Methods of Test for Rating the Performance of Heat/ Energy Recovery Ventilators*. Canadian Standard Association (2010)
8. ANSI/AHRI Standard 1060 (1-P), *Performance rating of Air-to-Air Heat/Energy Exchangers for Energy Recovery Ventilation equipment*. Air-conditioning Heating and Refrigeration Institute (2014)

Experimental Comparison of Performance Between Single and Dual Core Energy Recovery Ventilation Systems

Boualem Ouazia, Ganapathy Gnanamurugan, Chantal Arsenault and Heather Knudsen

Abstract In cold climates, the defrost strategies of the dominant conventional single core heat/energy recovery ventilators (HRVs/ERVs) involve the recirculation of exhausted (return) warm air across the heat exchanger core and back into the supply air for the house. In harsh cold climates, this type of defrost strategy can undermine ventilation performance. A dual core ERV design could overcome this issue by providing continuous ventilation. It has a potential to perform better than a conventional single core unit and meet the rigorous requirements for operation in the Canada's North. This paper provides an experimental comparison of the efficiency between single and dual core ERVs using the twin houses at the Canadian Centre for Housing Technology (CCHT) as a test bed. The side-by-side testing also provided valuable data to measure the impact of the installation of a dual core ERV in a real residential application. Both houses were operated identically and monitored for a range of winter weather conditions. Changes in test house performance due to the innovative dual core design were observed and compared to the performance of the reference house. The parameters that were compared include: efficiency, airflow, indoor temperature and relative humidity, and energy consumption. This paper discusses the effectiveness of the dual core design ERV, energy use of the house and the ventilation rate.

Keywords Energy recovery ventilator · Dual core · Residential
Cold climate

1 Introduction

Mechanical ventilation systems are essential for providing required outdoor air to keep the occupants healthy, remove odors, dilute indoor pollutants, and control indoor relative humidity. There are many ventilation options available, such as

B. Ouazia (✉) · G. Gnanamurugan · C. Arsenault · H. Knudsen
National Research Council Canada, Ottawa, ON, Canada
e-mail: Boualem.ouazia@nrc-cnrc.gc.ca

D. Johansson et al. (eds.), *Cold Climate HVAC 2018*,
Springer Proceedings in Energy, https://doi.org/10.1007/978-3-030-00662-4_32

balanced ventilation system with an air-to-air heat/energy recovery system (HRV/ERV). Air-to-air heat recovery is the process of recovering energy from an airstream at a high temperature to an airstream at a low temperature. This process is important for maintaining acceptable indoor air quality (IAQ) while reducing overall energy consumption. However, ensuring proper ventilation while minimizing energy costs can be a challenge for housing in Canada's North due to several factors including harsh cold climate conditions and frequent overcrowding. Conventional single core units require de-icing in cold climate applications and when installed in a house in Canada's North, they frequently underperform and fail [1–4]. The conventional frost protection mechanisms in an HRV/ERV is to recirculate the exhaust warm air (interrupting the flow of the exhaust air and redirecting the stale warm air back into the house) to defrost the core. The exhaust air heat loss is a considerable part of the total heat loss in cold climates. During the defrost cycle, no outdoor air is delivered to the house by the HRV/ERV. These shortcomings can undermine the outdoor air requirements of residential ventilation standard [5], and can result in deteriorated IAQ and unhealthy living conditions for the occupants. Most single core heat recovery technologies suffer from the same basic drawback. In extreme cold temperatures, frost formed on the exhaust side of the heat exchanger, dramatically reduces the heat recovery effectiveness and drives up true operating costs. A new design of ERV system with dual core exchangers periodically direct warm air (return air) through one core (heat exchanger), while outdoor air gains heat from the other core, could address the frost protection concerns and overcome the challenges faced by conventional single core heat/energy recovery ventilators in Canada's North. This paper presents the performance and operational aspects of a dual core ERV through a comparative side-by-side testing with a conventional single core ERV.

2 Methodology

The side-by-side testing methodology in the Canadian Centre for Housing Technology (CCHT) twin houses enabled a whole house evaluation of the impact of the dual core ERV. Both single core and dual core ERVs were operated for a week to undertake a comparison between the whole house energy performances of the test house with the dual core ERV to the reference house with the single core ERV.

2.1 Experimental Facility—CCHT

The CCHT research facility shown in Fig. 1, consists of two identical research houses the r*eference house* and the *test house*. The facility is located on the National Research Council's campus in Ottawa, the CCHT is co-sponsored by the National

Research Council, Canada Mortgage and Housing Corporation, and two branches within Natural Resources Canada—the CANMET Energy Technology Centre and the Program on Energy Research and Development.

The Centre consists of two Research Houses (Reference House and Test House) and a display-and-demonstration building called the InfoCentre. All three buildings were designed and built by Minto Developments Inc., Canada's largest production builder of R-2000 homes.

The Reference and Test Houses are side-by-side and identical in orientation, size and construction. Both are built to R-2000 specifications [6] and Healthy Housing principles, to establish a "best practices" example of current construction. Their design is a typical 2-storey wood-frame house with standard wall with two-by-six construction, and feature approximately 200 m^2 of liveable area, and a two-car garage that is fully inside the house's footprint. Control rooms have been located inside the garage, to minimize the interference of computer equipment with the house's energy system.

The Reference House serves as a control unit, while the Test House can be modified according to the research requirements. Features such as windows, heating systems, ductwork, and controls can be altered or replaced, allowing for an assessment of their effect on house performance.

Both houses use automated controls to simulate human occupancy and sophisticated computer equipment for data gathering. The twin-house research facility features a "simulated occupancy system". Each house features a standard set of major appliances typically found in North American homes. The houses have an air tightness characteristic of 1.5 ach @ 50 Pa, which is 30% tighter than R-2000 requirement. The simulated occupancy system, based on home automation technology, simulates human activity by operating major appliances (stove, dishwashers, washer and dryer), lights, water valves, fans, and a host of other sources simulating typical heat gains. The schedule is typical of activities that would take place in a home with a family of two adults and two children. Electrical

Fig. 1 CCHT houses

consumption is typical for a family of four and hot water draws are set in accordance with ASHRAE standards for sizing hot water heaters. The heat given off by humans is simulated by two 60 W (2 adults) and two 40 W (2 children) incandescent bulbs at various locations in the house [7–9].

The houses are unoccupied and extensively monitored for energy performance and indoor environmental conditions. Each research house is equipped with a data acquisition system consisting of over 250 (T and RH) and 23 m (gas, water and electrical). A computer in the control room (garage) reads the sensors every five minutes (and provides hourly averages) and record meter data every five minutes. Weather data is available from a nearby weather station.

2.2 Tested Technology

A dual core ERV unit comes with a regenerative cyclic dual core heat exchanger, based on the cyclic storage and release of heat in the corrugated sheets alternately exposed to exhaust and intake air. It includes a supply and an exhaust fans and two cores filled with specially corrugated 0.7 mm thick aluminum plates, which act as heat accumulators. In between the cores is a patented damper section, which changes over every 60 s to periodically direct warm air through one of the two cores, while outside air gains heat from the heated aluminum plates in the other core. Before each fan, there is a filter section to filter the air. Heat recovery is automatically activated when called upon. Figure 2 shows the schematic of the dual core unit with the two sequences.

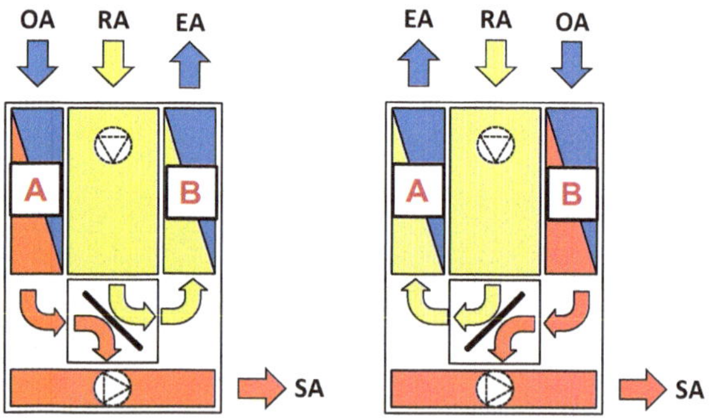

Fig. 2 Principle of function —sequence 1 (left) and sequence 2 (right)

Sequence 1—Exhaust air (EA) charges *Core B* with heat from return warm air (RA) from indoor and *Core A* discharges heat to supply air (SA) from outdoor (OA).

Sequence 2—Exhaust air (RA) charges *Core A* with heat from return warm air (RA) from indoor and *Core B* discharges heat to supply air (SA) from Outdoor (OA).

In addition to the innovative heat recovery strategy described above, an additional control system involving internal damper regulation is also used to ensure that comfortable air delivery temperatures are achieved in all conditions.

The damper is controlled by internal thermostats GT1 (set to 15 °C) in the supply air and internal thermostat GT2 (set to 20 °C) in the exhaust air. When the exhaust air temperature is lower than 20 °C, the unit is in "energy recovery mode" cycling every 60 s. When the exhaust air temperature is higher than 20 °C and supply air temperature is higher than 15 °C, the unit is in "free cooling mode" cycling every three hours. If exhaust air temperature is higher than 20 °C and supply air temperature is lower than 15 °C, the unit is in "energy recovery mode" until the supply air temperature becomes higher than 15 °C then it will return back to "free cooling mode".

3 Test Procedure

The dual core ERV and conventional single core ERV are shown in Fig. 3 and were respectively installed in the Test House and Reference House respectively. Both houses were monitored while they are operated identically. Changes in whole house performance due to the dual core ERV were observed through comparison of test house performance to the reference house performance, for a range of winter

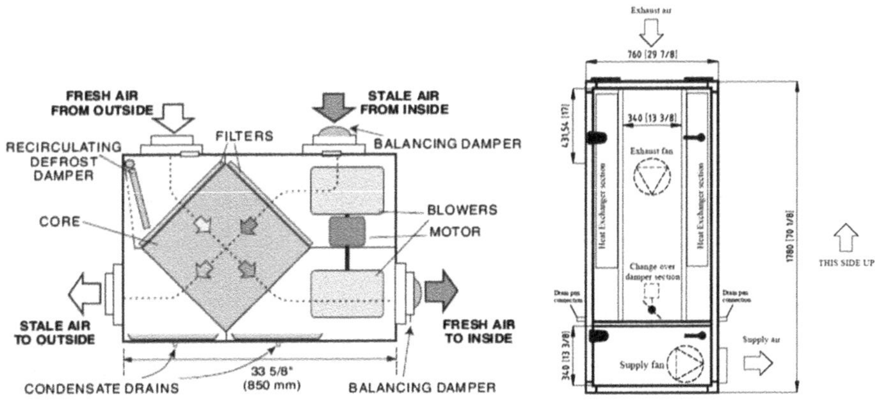

Fig. 3 ERV design—conventional single core (left) and dual core (right)

weather conditions, over a test period of seven days. During the test, both single and dual core ERVs had airflow of 47.2 L/s (105 cfm). The indoor temperature of the houses was regulated by a standard programmable thermostat located in the main floor hallway at mid-wall height, and set to maintain indoor temperature at 22.5 °C and had a narrow band in both houses of ±0.5 °C. These thermostats were used throughout the benchmark and any small inconsistencies between the thermostats in each house form part of the benchmark error and are therefore fully accounted for in the experiments. The following performance indicators were monitored on an hour-by-hour basis in both houses during the experiment: airflow, sensible effectiveness, house energy consumption (including electricity for furnace fans, ERV, and heating energy requirements), house air temperature and humidity.

4 Data Reduction

4.1 Effectiveness

The apparent sensible or total effectiveness of a heat/energy recovery system is a standard measure of performance [10–12]. Apparent sensible effectiveness measures the ability of an HRV/ERV unit to recover available sensible heat. Apparent total effectiveness measures the ability of an HRV/ERV unit to recover available total (sensible + latent) heat. It is calculated by taking the heat/energy recovered (in supply air stream) divided by the total available heat/energy (difference from the interior to the exterior). The effectiveness based on the configuration shown in Fig. 4 is calculated using the Eq. (1).

$$\varepsilon = \frac{m_s(X_1 - X_2)}{m_{min}(X_1 - X_3)} \tag{1}$$

where ε is the latent, total or sensible heat effectiveness, X is the humidity ratio, w, total enthalpy, h, or T is the dry-bulb temperature, respectively, at the supply inlet (1), supply outlet (2) and exhaust inlet (3). m_s is the mass flow rate of the supply air, m_e is the mass flow rate of the exhaust air and m_{min} is the minimum value of m_s or m_e.

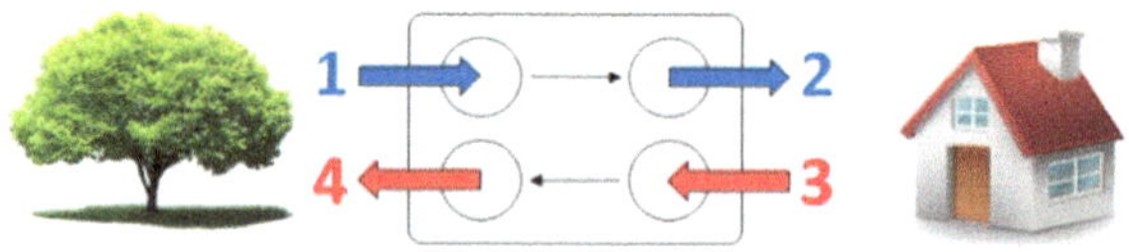

Fig. 4 Air-to-air heat exchanger configuration

4.2 Energy Saving

The real impact of the installation of the dual core H/ERV unit in Canadian houses was evaluated for a range of winter temperatures in Ottawa. The energy analysis compared the total space heating and ventilation energy used in the test house with the benchmark heating equipment and the dual core ERV unit operating against that of the reference house with the benchmark heating equipment and benchmark ERV operating. The energy analysis took into account small difference in energy performance of the twin houses when both are operating in benchmark conditions.

The recorded Reference House and Test House energy consumptions included:

- Heating energy consumption (furnace natural gas consumption) for both houses
- Furnace fan electrical consumption for both houses
- ERV fans electrical consumption for Test House
- Dual core H/ERV fans electrical consumption for Reference House.

The energy analysis (total daily heating and ventilation energy consumption) and savings for the dual core ERV are presented graphically in Fig. 5. Each point on the graph represents a single day of total space heating and ventilation consumption, with the reference house consumption plotted on the x-axis, and the test house consumption on the y-axis.

The expected test house consumption in benchmark configuration (i.e. operating the benchmark ERV equipment) is calculated first from developed characteristic performance line and shown (fitting curve in blue line) in Fig. 5. This produced fitting line was used to calculate the overall energy savings when the dual core ERV system is operating is calculated by subtracting the measured test house experiment consumption from the calculated test house benchmark consumption (vertical distance between the experiment data point and the benchmark trend) for each experiment day.

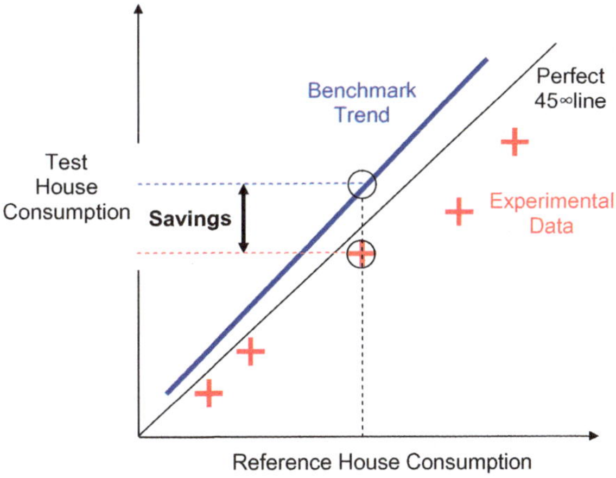

Fig. 5 Graphic representation of the saving method

The reasoning behind this calculation process is as follows. On any given day, some scatter is expected in the results both for the Reference House and the Test House. This scatter appears to be random error. One possible cause is that the houses' heating systems cannot be synchronized. When one house may be at the end of a heating cycle at midnight on one day, and the other house may be at the beginning, resulting in small and opposite errors on both the first and second days when this occurs.

When studying moderate change to the house, on any given day, the error in one house can be the opposite direction of the other house and vice versa. When looking at differences in experimental performance due to a technology change, that relative error can be large compared to difference due to the technology. To reduce the effect of this random scatter, a statistical technique is used to determine a characteristic line of best fit for the benchmark configuration. This reduces the risk of incorporating random errors on any given day of the benchmark, and also benchmarks systematic errors that are not due to the technology being studied. The line of best fit is much more accurate than the individual points, as they are made up of many points, tending to minimize the effect of random errors. So the benchmark data is first used to develop a characteristic performance line. Once the line is produced at a high level of statistical confidence, it is used to calculate the difference in performance, by subtracting the performance of the technology from the benchmark for each experiment day. The uncertainty of the energy saving calculation depends strongly on the length of the test.

5 Results

The results comparing the performance and operation of the two houses are presented in terms airflows, apparent sensible and total efficiencies, delivered air temperature to the indoor, indoor air temperature and relative humidity, and all components of the energy consumption of both houses.

5.1 Ventilation

Effective ventilation is a vital system in a healthy home. Heat or energy recovery units should continuously deliver the minimum outdoor flow rate required for people and house air needs (total ventilation requirement) and set by the building codes and ventilation standards [5, 13]. A balanced ventilation system, for both exhaust and supply, will give the best results for distributed fresh air through the house. Unbalanced system will exacerbate the negative or positive pressure in the

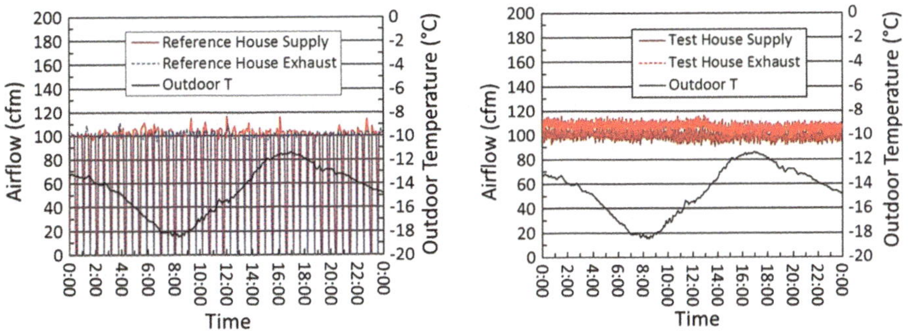

Fig. 6 Daily supply and exhaust airflows

house (pressurization or depressurization of the house), which could have a negative impact on the energy consumption and IAQ. A frequent defrost cycles will also reduce the amount of outdoor air delivered to the house, leading to a situation where the house is under-ventilated and does not meet the ventilation rate requirement. The supply and exhaust airflows through the dual core ERV and the single core ERV were measured during the side-by-side testing period. Daily results for single core ERV (excluding the defrost recirculation of warm stake air) and dual core ERV airflows of a typical day are presented in Fig. 6.

The single core ERV presented fairly balanced supply and exhaust flows and experienced de-icing periods (defrost cycle) to melt ice build-up during the day with outdoor temperature below −10 °C. The dual core ERV, with supply and return airflows cycling between two heat exchangers presented also a balanced supply and exhaust flows and did not stop to defrost cycle in the same day under the same weather conditions—so it continued to provide fresh air throughout, unlike the single core ERV.

The single core ERV unit was often in the defrost mode, recirculating stale exhaust warm air to melt frost build-up in the core, and not continuously supplying outdoor air to the house. The duration that a single core ERV spent in defrost mode ("defrost time") that day was 315 min (or 5.25 h). That day the single core ERV was 22% of the time defrosting and not supplying outdoor air to the house. The de-icing cycle duration is strongly dependent on outdoor temperature. Under extreme cold weather conditions single core ERV/HRV unit could have been in defrosting mode for a long period per day, during which time the house would be seriously under-ventilated and not providing the required ventilation rate set by ASHRAE ventilation standard [5] and the National Building Code of Canada [13].

5.2 *Effectiveness*

The apparent sensible, total efficiencies ε [calculated using Eq. (1)] of a single and dual core ERVs are plotted in Fig. 7.

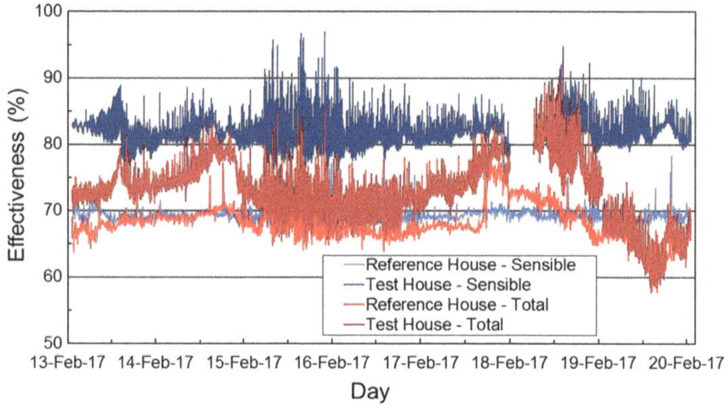

Fig. 7 Apparent sensible and total efficiencies

The calculated sensible effectiveness of the dual core ERV during the side-by-side testing period had an average value of 82%, ranged from 76 and 97%, and with an uncertainty of 3.2%. The single core ERV in the Reference House had an average apparent sensible effectiveness of 69% during the same testing period, ranged from 66 and 78%, and with an uncertainty of 2.8% (difference of at least 10 percentage points). The apparent total effectiveness which takes into account the latent heat of the single core ERV varied between 61 and 78%, with an average value of 68% and uncertainty of 3.9%. The dual core ERV unit had an apparent total effectiveness between 58 and 92%, with an average value of 73% slightly higher than the single core ERV and an uncertainty of 2.5%. The daily apparent sensible and total efficiencies during the side-by-side testing showed clearly that the dual core ERV unit over performed the single core ERV in terms of apparent sensible effectiveness, and consistently performed better in term of apparent total effectiveness.

5.3 *Outdoor Air Supply Temperature to the Indoor*

The supply and exhaust airstream conditions were measured both for the single core ERV unit installed in the Reference House and the dual core unit installed in the Test House. The supply outlet air temperature and RH from single and dual core ERVs to the indoor (house) are presented in Fig. 8.

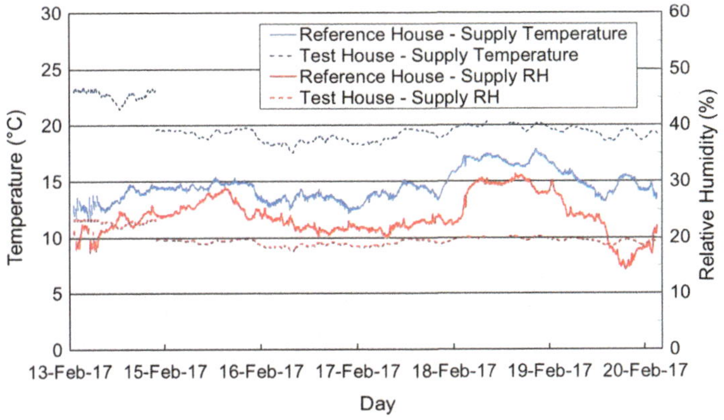

Fig. 8 Outdoor supply temperature and RH to the indoor

The temperature of the supplied air to the house was higher (3–6 °C) from the dual core unit than the single unit as shown in Fig. 8. This was due to the much higher sensible effectiveness of the dual core unit (higher than 80%) from regenerative cyclic dual core heat exchangers. The air supplied to the return furnace in the reference house is cooler than air supplied to the furnace return plenum in the test house. The daily average difference between the supply from single core ERV (Reference House) and the dual core ERV unit (Test House) ranged from 3 to 6 °C. The supply air to the test house would require less tempering by the furnace to meet the thermostat set point of 22 °C, which means that a dual core unit provided more pre-heating than a single core ERV and would lead to higher energy saving. Ventilation air must be introduced into the occupied zone in a way that avoids causing discomfort to the occupants and at acceptable minimum temperatures, 17 °C for floor distribution and 13 °C for ceiling distribution, as specified by the National Building Code of Canada [12]. The supply outlet air temperature from the single core ERV in the Reference House varied between 11.5 and 17.9 °C and daily average values ranged from 13.4 to 16.6 °C. The average value over testing period was 14.6 °C, which is below the acceptable minimum temperature of air that would be introduced to occupied zone; requires tempering of the supplied air. The supply outlet air temperature from the dual core ERV in the Test House varied between 17.5 and 20.3 °C and daily average values ranged from 18.7 and 19.6 °C. The average value over the testing period was 19.2 °C, which is higher than the acceptable minimum temperature of air that would be introduced to occupied zone; requires no or less tempering of the supplied air.

5.4 Space Heating Energy Saving

Benchmarking the houses (both with single core ERV) to show that they were thermally identical occurred over a heating periods in winter of 2017, before and after the dual core ERV assessment. The benchmark periods were done with central thermostat set at 22.5 °C and the energy performance of the test house was shown to be very close to the reference house. This performance comparison formed the benchmark against which dual core ERV can be compared. The energy impact of the installation of the dual core ERV in Canadian houses was evaluated by recording reference house and test house energy consumptions; heating energy consumption (furnace natural gas consumption), furnace fans electrical consumption, single core ERV fans electrical consumption and dual core ERV fans electrical consumption. The total daily heating and ventilation energy consumptions are presented in Fig. 9.

Each point on the graph represents a single day of total space heating and ventilation consumption, with the reference house consumption plotted on the x-axis, and the test house consumption on the y-axis. The average energy savings when operating the dual core ERV compared to the benchmark single core ERV over the period of the study was 6.2%.

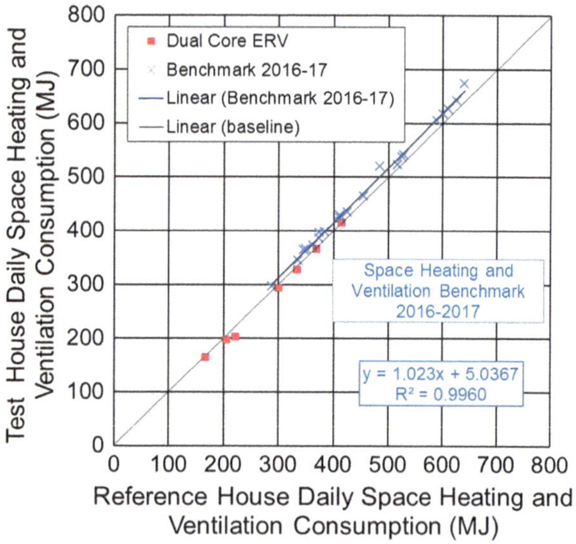

Fig. 9 Space heating and ventilation daily total consumption

6 Conclusion

A dual core ERV (designed with two parallel heat exchangers) was tested during the winter of 2017 in side-by-side testing using the twin houses facility of NRC's Canadian Centre for Housing Technology. In comparison with conventional single core ERV, the dual core ERV had much higher apparent sensible effectiveness (81.2% compared to 69.5%), and slightly higher apparent total effectiveness (72.7% compared to 68.1%). In addition to the better effectiveness, it also showed no sign of frost problems after the seven days of weather testing and successfully continued to provide outdoor air throughout cold days without stopping to defrost, unlike the conventional single core ERV which spent hours defrosting over the coldest test days. Under much colder weather conditions (outdoor temperature below −10 °C), the single core ERV would spend hours in the defrost mode which could have led to a situation where the house would be under-ventilated. This is a common situation for single core units installed in Canada's Far North. The dual core design ERV with much higher heat transfer effectiveness was capable to provide air at the supply outlet at temperature up to 6 °C higher than air temperature supplied by a single core ERV. In comparison with the Reference House with a conventional single core ERV, the Test House with a dual core ERV had similar indoor air temperature (as regulated by the thermostat) and relative humidity and showed saving in heating and ventilation energy consumption, ~6.2% (18.7 MJ/day) decrease on average over the seven day test period. The uncertainty in quantifying potential energy savings depend strongly on the duration of the testing period. As part of this investigation the dual core ERV unit has been deployed in Canada's Arctic (Nunavut) for extended monitoring to provide evidence on the performance and resilience of the technology in the North.

References

1. C. Zaloum, *Technical advice to task force on Northern mechanical ventilation equipment design and testing* (Canada Mortgage and Housing Corporation, Ottawa, 2010)
2. M. Nasr, M. Fauchoux, R. Besant, C. Simonson, A review of frosting in air-to-air energy exchangers. Renew. Sustain. Energy Rev. **30**, 538–554 (2014)
3. E.G. Phillips, R.E. Chant, D.R. Fisher, V.C. Bradley, Comparison of freezing control strategies for residential air-to-air heat recovery ventilators. ASHRAE Transactions **95**(2), 484–490 (1989)
4. E.G. Phillips, D.R. Fisher, R.E. Chant (1992) Freezing-control strategy and air-to-air energy recovery performance. ASHRAE J. 44–49
5. ANSI, ASHRAE 62.2, *Ventilation for acceptable indoor air quality* (American Society of Heating, Refrigerating and Air-Conditioning Engineers, Atlanta, GA, 2016)
6. R-2000 Homes Standard (2014) Natural Resources Canada
7. M.C. Swinton, H. Moussa, R.G. Marchand, Commissioning twin houses for assessing the performance of energy conserving technologies, *VIII Integration of Buildings Envelopes* (Clearwater, Florida, 2001), pp. 1–10

8. B.K. Ouazia, M.C. Swinton, M. Julien, M. Manning, Assessment of the enthalpy performance of houses using energy recovery technology. ASHRAE Trans. **112**(1), 26–33 (2006)
9. Canadian centre for Housing technology (CCHT). www.ccht-cctr.gc.ca
10. ANSI, ASHRAE Standard 84, *Method of testing Air-to-Air Heat/Energy Exchangers* (ASHARE, Atlanta, GA, 2013)
11. CAN/CSA-C439, Standard Laboratory Methods of Test for Rating the Performance of Heat/ Energy Recovery Ventilators. Canadian Standard Association (2010)
12. ANSI/AHRI Standard 1060 (1-P), in *Performance rating of Air-to-Air Heat/Energy Exchangers for Energy Recovery Ventilation equipment.* Air-conditioning Heating and Refrigeration Institute (2014)
13. National Building Code of Canada (NBCC) 2015

Part IV
Heat Pump and Geothermal Systems

Design of Horizontal Ground Heat Exchangers in Sub-arctic Conditions—Sensitivity to Undisturbed Ground Temperatures

Robbin Garber-Slaght and Jeffrey D. Spitler

Abstract Application of ground source heat pumps (GSHPs) in extreme cold climates can be challenging due to the long heating season even with heat pumps designed and marketed for colder climates. One challenge (of several) is design of the ground heat exchanger under conditions where the desired minimum heat pump entering fluid temperature (EFT) is close to the undisturbed ground temperature (UGT). Furthermore, ground heat exchanger (GHE) models used for design purposes and/or energy calculation purposes don't usually incorporate freezing and thawing of the soil. This is the case whether the freezing/thawing is induced by surface conditions or by heat transfer between the ground heat exchanger and the surrounding soil. A recently developed model (Xing and Spitler in Sci Technol Built Environ 23(5), 809–825, 2017, [1]) implemented in a simulation-based ground heat exchanger design tool (Oklahoma State University in GLHEPro 5.0 for Windows—Users' Guide. Stillwater, 2016, [2]) incorporates the effect of surface-condition-induced freezing/thawing when calculating a 2nd-order harmonic approximation for the undisturbed ground temperature. This paper examines the suitability of this harmonic model in GHE design applications by comparing the predicted GHE design to the actual design of a GHE at the Cold Climate Housing Research Center (CCHRC) in Fairbanks, Alaska. Field measurements of UGT and GHE performance are used to examine the limitations of the harmonic model.

Keywords Ground-source heat pumps · Horizontal ground heat exchangers

---

R. Garber-Slaght
Cold Climate Housing Research Center, Fairbanks, AK 99709, USA

J. D. Spitler (✉)
Oklahoma State University, 218 Engineering North, Stillwater,
OK 74078, USA
e-mail: spitler@okstate.edu

D. Johansson et al. (eds.), *Cold Climate HVAC 2018*,
Springer Proceedings in Energy, https://doi.org/10.1007/978-3-030-00662-4_33

1 Introduction

The design of ground heat exchangers, particularly horizontal ground heat exchangers, is sensitive to the undisturbed ground temperatures (UGT). For shallow horizontal ground heat exchangers, the time-varying aspect of the UGT may be of importance. Time-varying UGT have generally been estimated with some variation of the Fourier [3] model. Kelvin [4] used a 2nd order model to represent ground temperatures in Edinburgh. The ASHRAE Handbook series gives a 1st order model, with annual average temperatures and amplitudes given on small hard-to-read maps that cover North America. These maps are adapted from research done in the 1920s [5] and 1950s [6].

An improved method [1, 7, 8] that provides 2nd order harmonic model coefficients for over 4000 locations worldwide relies on a numerical model coupled with typical meteorological year-type weather data to provide a temperature data set for each location that is, in turn, used to tune the coefficients of the 2nd order harmonic model. The numerical model accounts for freezing in the ground, but relies on a heuristic model [1] to estimate the accumulation of snow.

It should be emphasized here that the purpose of the 2nd order harmonic model and accompanying worldwide data set is to provide reasonable estimates of UGT to support design of ground heat exchangers and calculation of heating loads. The procedure for determining the model coefficients for each location necessarily made a number of approximations in addition to the heuristic model for snow accumulation. E.g., fixed values of soil thermal diffusivity were used; the surface evapotranspiration model was limited to short grass and tall grass conditions; and likely most importantly, the weather was assumed to be a repeated typical meteorological type year. The model is intended for use in situations where detailed ground temperature measurements are not available.

Grundmann [9] implemented the improved UGT model in the GHE design software GLHEPRO [2] along with a database of coefficients. The program also includes a model of Slinky-type horizontal GHE, developed by Xiong et al. [10]. This model is based on superposition of ring-sources and is implemented to support sizing of the Slinky-type GHE.

At roughly the same time as these new models were being developed and implemented, the Cold Climate Housing Research Center (CCHRC) installed, instrumented, operated, and monitored a ground-source heat pump system utilizing horizontal Slinky-type GHE in Fairbanks, Alaska.

This gives the opportunity to check the applicability of the UGT and Slinky-type GHE sizing capability against real-world data in a subarctic climate. Therefore, the purpose of this paper is to compare predictions of undisturbed ground temperature and ground heat exchanger design results to actual field measurements and system operation.

2 Facility and Field Measurements

CCHRC's Research and Testing Facility (RTF) is 2044 m^2 of office and lab space. The building has 3 distinct heating areas, which are defined by their heat delivery mechanism. The 464 m^2 office section is heated with the ground source heat pump [11] and a supplementary wood fired masonry stove. Heat in the office section is distributed via an in-floor hydronic system embedded in concrete; it is zoned using 9 thermo-statically-controlled valves. The office section has a 17.6 kW heating design load.

The soils around the RTF have been studied extensively and that information was used to inform the GHE design. The soil underlying the site is moist silt and either degraded or degrading permafrost. Boreholes drilled on the site in 2006 prior to the construction of the RTF found the site underlain with sloping permafrost. The top of the layer started at 3 m on the south side of the building near the undisturbed vegetation and sloped down to 9.1 m on the north side where the native vegetation had been cleared. Data collected under the east end of the RTF since 2006 shows that the permafrost table has further degraded by an additional 0.6 m. A test borehole northwest of the RTF in 2012, prior to installing the GHE, did not find permafrost within 9.1 m of the surface. In addition to changing permafrost, the ground water levels across the site are changing. It is estimated that the ground water has risen 1.5 m since 2006; most of that rise is since 2013. The proximity of permafrost with ground temperatures close to freezing and the changes in ground water are having an impact on the GHE.

The heat pump is a commercially available residential unit with nominal capacity of 21 kW. It was sized entirely based on the heating load. The heat pump heats a 303-l buffer tank of water to a temperature determined by the outdoor set point curve. The minimum temperature set point for the buffer tank is 26.7 °C and the maximum is 42.8 °C. The GHE side of the heat pump is charged with a 20% methanol, 80% water mixture. The building hydronic side of the heat pump is charged with water.

The horizontal GHE was designed based on the technology available in Fairbanks in 2012. Since directional drilling was not an option and deep vertical wells would have run into cold, frozen bedrock, horizontal "slinky" coils were installed. Six 30.5 m long by 1 m wide slinky coils with a 0.5 m pitch were installed 1.8 m apart. Overall, 1463 lineal m of 1.9 cm HDPE was installed at 2.7 m depth to create the in-ground heat exchanger.

The GHE size and depth were determined by knowledge of past installations in the area, in conjunction with ground thermal conductivity test data, and information from a finite element model. The ground heat exchanger was designed by the local GSHP installer using his modified version of IGSHPA methodology [12, 13]. Due to the underlying permafrost, he assumed 0 °C ground temperature and a minimum −3.3 °C entering fluid temperature. The tested thermal conductivity was lower than the typical estimates in the Fairbanks area so the FEM model was used to support increasing coil spacing from the usual 0.9 to 1.8 m.

The heat pump system is monitored in order to evaluate how the heat pump affects the soils around the GHE and also how the changing entering fluid temperatures affect the heat pump efficiency. It is expected that the large imbalance of heat extraction without enough thermal recharge will depress the temperatures in the GHE over time. The CCHRC installation was designed to investigate this and to determine if the GHE temperatures stabilize over time. Temperature data is being collected throughout the GHE, as well as outside the GHE. Five vertical thermistor strings measure temperatures at depths up to 3.7 or 4.1 m. Three vertical thermistor strings measure temperatures inside GHE; two strings are installed outside of the GHE. All of these strings record soil temperatures on an hourly basis. The manifold for this system is located inside the building and the fluid temperature from each slinky coil is tracked on an hourly basis. The energy from the GHE, to the buffer tank, and to the building is also monitored, as is the electrical use of the system. Figure 1 shows the locations for the vertical thermistor strings. The system came online in October 2013 and our data collection system became fully operational a year later.

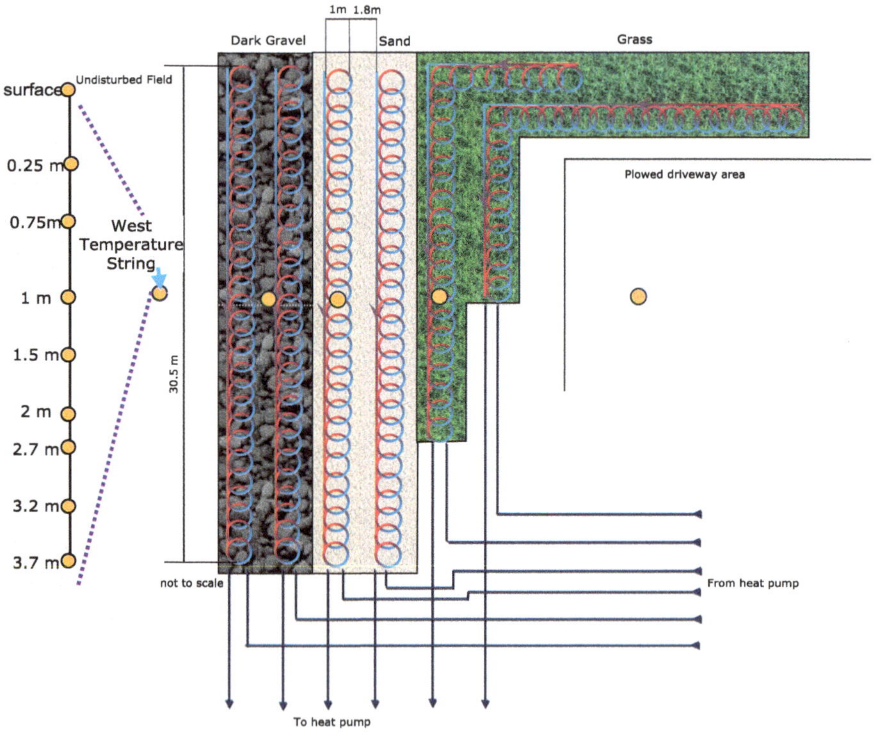

Fig. 1 Ground heat exchanger layout with vertical temperature sensor string locations marked

3 Ground Temperature Measurements and Model Results

As will be shown in the Sect. 4, the design of the ground heat exchanger is very sensitive to the undisturbed ground temperatures. An initial comparison can be made for the depth of the ground heat exchanger, 2.7 m, as shown in Fig. 2. The measured UGT covering the period from 10 October 2014 to 2 June 2017 come from the vertical temperature string 4 m west of the west-most slinky GHE. Two model curves are plotted, representing short grass (Model-SG) and tall grass (Model-TG) conditions. The temperature string is under short grass cover. The first and most obvious difference is that, even with tuning of the coefficients using a numerical model that considered freezing and snow accumulation, the harmonic models don't match the shape very well, with maximum errors of about 3 °C.

This is because, even with the tuning, the harmonic model itself represents a pure heat conduction solution, with 2nd order harmonic boundary conditions. The harmonic model's ability to account for freezing at the surface is very limited.

Furthermore, the model relies on standard boundary conditions (e.g. "short grass") and typical meteorological data, so even the annual average temperature cannot be predicted perfectly at every location. The model has an average temperature of 2.4 °C for every year. The measured average temperature for 2015 (excluding a 40-day period when the temperature sensor failed) was 3.2 °C. The measured average temperature for 2016 (excluding an 8-day period when the temperature sensor failed) was 3.1 °C.

Figure 3 shows for 1 m depth the measured temperatures, the 2nd order harmonic model-predicted temperatures ("Model-SG"), and the temperatures

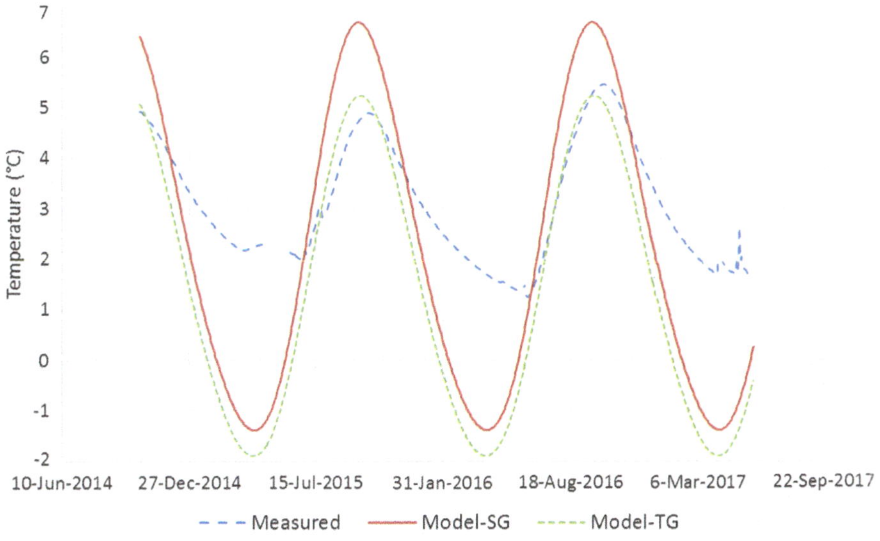

Fig. 2 Undisturbed ground temperatures at 2.7 m depth—measured and modeled

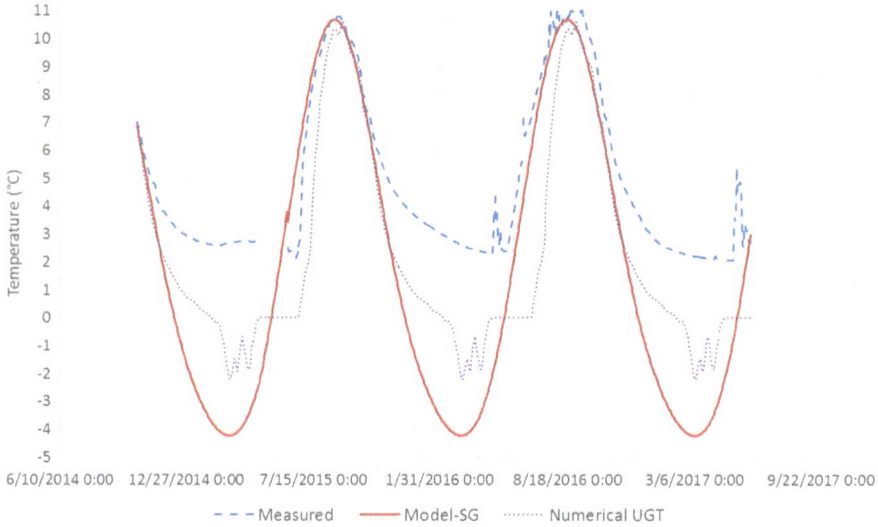

Fig. 3 Undisturbed ground temperatures at 1 m depth—measured and modeled with both harmonic model and numerical model

("Numerical UGT") predicted by the numerical model [1, 7] that were used to fit coefficients for the 2nd order harmonic model. The effect of freezing can be clearly seen in the numerical model, but there is still significant deviation from the measured data. A likely explanation is that the heuristic snowfall model underpredicted the amount of snow cover; both winters were heavy snowfall years with early snowfall that insulated the ground. This is consistent with measurements from a second site, 1 km away, which shows that the ground temperatures during 2014–2016 were somewhat warmer than 2008–2015. From 2008–2013, annual minimum temperatures at 2 m depth were 0.1–0.2 °C. Then from 2014 to 2017, the temperatures rose: 0.34, 0.42, 0.63, 0.84 °C [14].

Keeping in mind that the harmonic model is intended to be globally-applicable and it is intended to be applied in designs where some oversizing of a GHE is acceptable, and that the model has been shown to be more accurate than similar previous models, perhaps it may be considered adequate. Nevertheless, in the authors' opinion, it would be desirable to have a better, but still simplified, design model that better included freezing and snowfall effects.

In order to investigate the advantage of a better model when applied to GHE design, a surrogate location was sought that would better match the measured UGT at 2.7 m depth. This was accomplished by fitting the measured data to the harmonic model, which gave a mean annual surface temperature of 3.1 °C and amplitude of 4.6 °C. Then, all other locations worldwide were sorted by how closely they matched these values. Specifically, a goodness-of-fit parameter was calculated as a weighted sum of the errors squared—the annual temperature was assigned a weight of 2/3 and the first-order amplitude was assigned a weight of 1/3. The selected

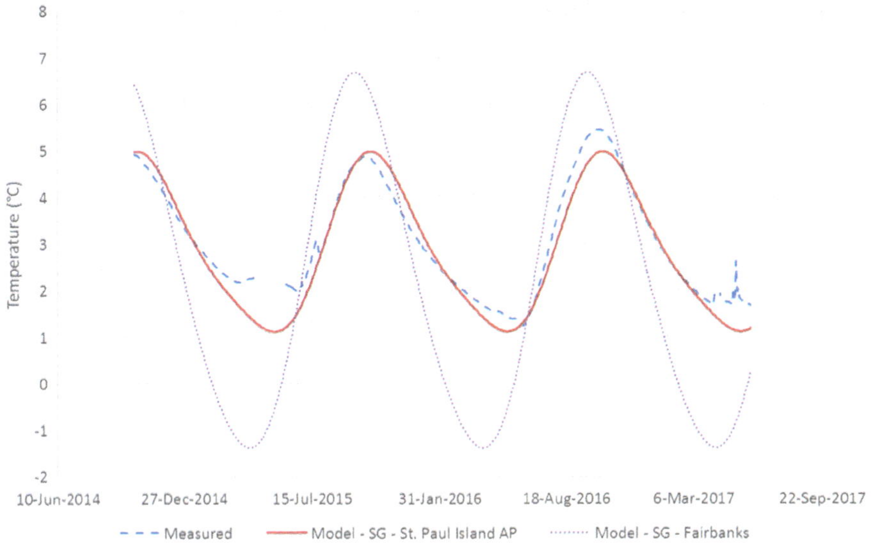

Fig. 4 Undisturbed ground temperatures at 2.7 m depth—measured and harmonic model results for both Fairbanks and St. Paul Island

surrogate location was St. Paul Island, Alaska—one of the Pribilof Islands—and it has a mean annual temperature of 2.9 °C and a first-order amplitude of 4.4 °C. The model-predicted UGTs at 2.7 m depth for both Fairbanks and St. Paul Island are compared to the measured UGTs in Fig. 4. At least at 2.7 m depth, the St. Paul Island UGTs are a much better representation of the measured UGTs in Fairbanks.

4 Ground Heat Exchanger Design

Ground heat exchanger design methodologies may be broadly grouped into two categories: simplified analytical approximations and simulation-based design tools. The first category consists of methods where an analytical approximation to the line-source solution has been algebraically solved to give a design length. The second category consists of methods where the GHE size is treated as an input to a simulation; the size is then automatically adjusted so that the user-specified minimum or maximum heat pump entering fluid temperature is reached, but not exceeded.

GLHEPro, is a simulation-based design tool, and, specifically, a simulation of Slinky-type heat exchangers is made and the size of the ground heat exchanger is then adjusted to meet the minimum heat pump EFT. The simulation can account for most of the features of the installation, but not all. As shown in Fig. 1, two of the trenches take a right turn, so we expect that there will be less thermal interference in the facility than predicted by the simulation. Also, there are three different surface covers; only two of the trenches are covered by grass. Garber-Slaght and Peterson [11] discuss the effect of the surface cover on the returning GHE temperatures. At the beginning of the 2015 heating season, the gravel-covered return temperatures were about 1 °C higher than the grass-covered return temperatures and about 0.5 °C warmer than the sand-covered return temperatures. But these differences were reduced over the winter, so that by the end, all return temperatures were within a 0.3 °C band.

An initial comparison might be made by using the GLHEPro monthly/monthly peak simulation [15] to estimate the average heat pump EFT at month end and the minimum monthly heat pump EFT. These temperatures are plotted along with measured hourly heat pump EFT, using both Fairbanks and St. Paul Island to provide the UGTs for the simulation in Fig. 5. As might be expected, using the St. Paul Island UGTs gives a much better match to the actual measured heat pump EFT. Even using the St. Paul Island UGTs, other deviations between the hourly measurements and the monthly/monthly minimum estimates are caused by a number of factors: heat transfer differing from pure conduction due to freezing around the tube and water movement; limitations of the hybrid time step representation [15]; use of a mixed monthly load profile in GLHEPro; differences in geometry; inhomogeneity in the soil; and, other factors. The degree to which this affects the predicted GHE size will be examined next.

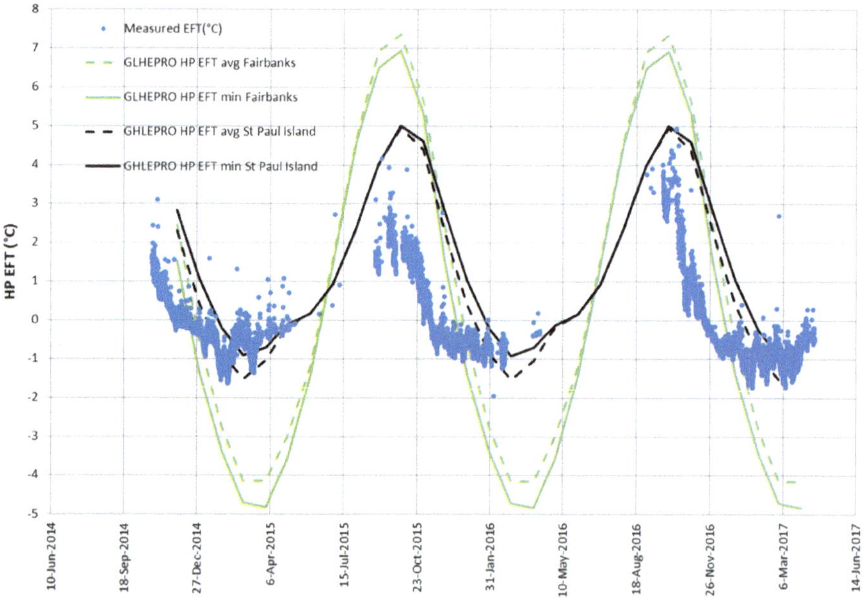

Fig. 5 Measured and simulation-predicted heat pump EFT using UGTs for both Fairbanks and St. Paul Island

In general, the GHE size becomes more sensitive to all parameters as the required minimum heat pump EFT approaches the UGT. This is illustrated in Fig. 6, where the required GHE size (given as the trench length for each of the six trenches) is plotted as a function of the user-specified minimum heat pump EFT. The GHE has been specified as consisting of six Slinky heat exchangers with the pitch, spacing between trenches, depth, flow rate, etc. all set by the user. The trench length and amount of pipe are then adjusted automatically to meet the minimum heat pump EFT.

Four curves are given, representing four different UGT conditions—Fairbanks and St. Paul Island, as already discussed, and two fixed temperatures (3.15 °C, representing the measured annual average UGT at the GHE depth, and 2 °C representing an approximate average value of measured UGT at the GHE depth during the heating season.) In addition, two points are shown. The first represents the conditions during the 2014–2017 period of measurements, where the minimum heat pump EFT is about −1.7 °C and the actual trench length is 30.5 m. The second point represents the contractor's design minimum heat pump EFT of −3.3 °C and the actual trench length.

To return to the importance of the UGTs used in the design, Fig. 4 shows that the 2nd order harmonic model, using parameters for Fairbanks, underpredicts the minimum annual ground temperature at the GHE depth by a little less than 3 °C. Yet, as can be seen in Fig. 6, this has a substantial effect on the required GHE size, and, in practice, the design engineer does not have the luxury of multiple years of measured data on site when the system is being designed. At the minimum heat

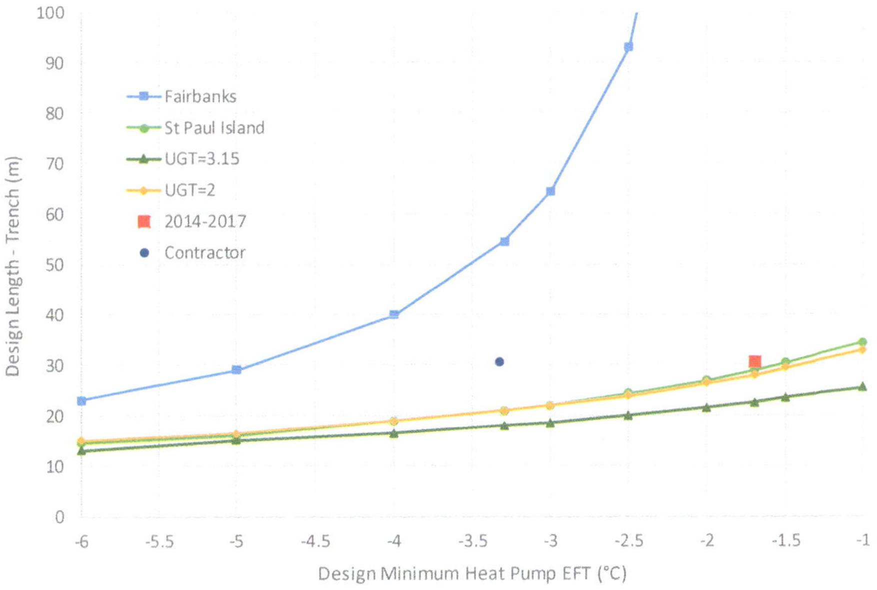

Fig. 6 Sensitivity of GHE design to UGT

pump EFT used by the contractor (−3.3 °C), the GHE sized with the Fairbanks data would be 2.6 times larger than the one sized with the St. Paul Island data. So, yes, the 3 °C difference is very significant under these conditions; specifically, the small difference between the minimum UGT and the design heat pump EFT.

We may also compare the actual performance for 2014–2017. During these years, the minimum heat pump EFT actually reached was −1.7 °C and the actual trench length is 30.5 m. The sizes predicted by the St. Paul Island data and constant 2 °C UGT are slightly smaller than actual (5 and 8% respectively); and the size predicted by the constant 3.15 °C UGT—the annual measured average temperature—is 26% lower. Since using the average annual measured UGT underpredicts the size, it suggests that accounting for the annual variation in UGT is important.

One other observation is related to the fact that the Xiong et al. [10] model used by GLHEPro does not account for freezing around the piping. It seems likely that this should be important, but it is not clearly evident from the results observed here. Looking at Fig. 5, it seems that each heating season, the measured temperatures decrease sooner than GLHEPRO predicts, and then flattens out around heat pump EFTs of −1 °C. It is possible that the measured decrease in temperature is partly due to residual ice around the tubes and that the flattening-out observed is due to further freezing around the tube. But this is somewhat speculative.

5 Conclusions and Recommendations

Though an improvement over what was previously available, the 2nd order harmonic model of undisturbed ground temperatures [1, 7, 8] implemented in GLHEPRO underpredicts the UGT in Fairbanks' sub-arctic climate by around 3 °C, leading to an overprediction of required horizontal ground heat exchanger size. GHE size is shown to be increasingly sensitive as design EFT approaches the UGT.

The form of the 2nd order harmonic model has limited applicability to sub-arctic conditions with a substantial amount of annual freezing and thawing at the ground's surface. Furthermore, examination of the numeric model results used to fit the 2nd order harmonic model constants suggests that the heuristic snow cover model needs improvement. Development of the heuristic snow cover model [1] was hampered by a lack of a precipitation data in the typical meteorological year type weather files used to generate the worldwide database. Therefore, a possible improvement could be along the lines of:

- Investigate alternative sources that provide global precipitation data, with as fine a time resolution as possible. E.g. monthly average precipitation data should be better than annual average data.
- Develop an improved, but probably still heuristic, model of snow cover and implement this in the numerical procedure used to develop the tuning data. For the 2nd order harmonic model, the tuning data are daily temperatures at four depths.

- Develop an alternative to the harmonic model that is still simple enough for widespread design applications, but which provides a better fit to the numerical model.

Finally, the Xiong et al. [10] model gave quite reasonable results—predicted size within 5% of actual size when the temperature inputs were adjusted to better match measurements at the site. However, the impracticality of making long-term field measurements at multiple depths for most design applications reinforces the need for improved UGT models.

References

1. L. Xing, J.D. Spitler, Prediction of undisturbed ground temperature using analytical and numerical modeling. Part II: Methodology for developing a world-wide dataset. Sci. Technol. Built Environ. **23**(5), 809–825 (2017)
2. Oklahoma State University, *GLHEPro 5.0 for Windows—Users' Guide. Stillwater* (2016)
3. J. Fourier, *The Analytical Theory of Heat* (Translation of the 1822 work Théorie analytique de la chaleur.) (University Press, Cambridge, 1878)
4. W. Thomson, On the reduction of observations of underground temperature; with application to Professor Forbes's Edinburgh observations, and the continued Calton Hill Series. Lond. Edinb. Dublin Philos. Mag. J. Sci. XXII Fourth Ser. **23–34**, 121–185 (1861)
5. W.D. Collins, *Temperature of Water Available for Industrial Use in the United States* (Water Supply Paper, USGS, 1925)
6. J.-H. Chang, *Ground Temperature* (Observatory, Harvard University, Bluehill Met, 1958)
7. L. Xing, J.D. Spitler, Prediction of undisturbed ground temperature using analytical and numerical modeling. Part I: model development and experimental validation. Sci. Technol. Built Environ. **23**(5), 787–808 (2017)
8. L. Xing, J.D. Spitler, A. Bandyopadhyay, Prediction of undisturbed ground temperature using analytical and numerical modeling. Part III: experimental validation of a world-wide dataset. Sci. Technol. Built Environ. **23**(5), 826–842 (2017)
9. R.M. Grundmann, *Improved Design Methods for Ground Heat Exchangers*. M.S. Thesis, Oklahoma State University (2016)
10. Z. Xiong, D.E. Fisher, J.D. Spitler, Development and validation of a Slinky™ ground heat exchanger model. Appl. Energy **141**, 57–69 (2015)
11. R. Garber-Slaght, R. Peterson, in *Can ground source heat pumps perform well in Alaska?*, IGSHPA Technical/ Research Conference Proceedings. Denver, pp. 62–70, 14–16 Mar 2017. http://dx.doi.org/10.22488/okstate.17.000525
12. Oklahoma State University, *Ground Source Heat Pump Residential and Light Commercial Design and Installation Guide* (Stillwater, OK, IGSHPA, 2009)
13. Oklahoma State University, *Closed-Loop Geothermal Systems—Slinky Installation Guide* (Stillwater, OK, IGSHPA, 1994)
14. Data provided by V. Romanovsky, U. Alaska. August 7. (2017)
15. J.R. Cullin, J.D. Spitler, A computationally efficient hybrid time step methodology for simulation of ground heat exchangers. Geothermics **40**(2), 144–156 (2011)

Artificial Neural Network Analysis of the Solar-Assisted Heat Pumps Performance for Domestic Hot Water Production

Alireza Zendehboudi, Xianting Li and Siyuan Ran

Abstract The solar-assisted heat pump devices have attracted ample attention as potential alternatives to the conventional technologies for domestic hot water production. Precise COP estimation of the hybrid system is a prerequisite prior to conducting any effort to design and install as well as a necessity for market support and optimal system assurance. Here, to develop a model along that has a high precision and reliability, but less complexity and computational time, the ability of soft computing approaches for predicting the performance of a solar-assisted heat pump system for domestic hot water production is reported. The experimental data gathered from a real project in China and different intelligent models are developed based on the measurements. First, owing to the complexity of our studied system and the high number of influential parameters, a novel and unique multi-objective optimization technique based on NSGA-II optimization method is proposed to determine a set of optimum variables with the highest influence on the desired output. Based on the Pareto frontier, seven input variables out of forty eight are considered for the desired output. The reliability of the developed models is evaluated via statistical and graphical error analyses. It was inferred that integration of the MLP-ANN with the suggested variable selectin algorithm outperformed the other methods by introducing an $R^2 = 0.9951$ and RMSE = 0.0917. The current investigation can aid as a gear in the direction of improving the precision in estimation of performance of solar-assisted heat pump devices.

Keywords Domestic hot water system · Solar-assisted heat pump
Artificial neural network · Estimation

A. Zendehboudi · X. Li (✉) · S. Ran
Department of Building Science, School of Architecture,
Tsinghua University, Beijing 100084, China
e-mail: xtingli@tsinghua.edu.cn

D. Johansson et al. (eds.), *Cold Climate HVAC 2018*,
Springer Proceedings in Energy, https://doi.org/10.1007/978-3-030-00662-4_34

1 Introduction

Nowadays, environmental pollution and energy shortage are two main challenges in the world. Therefore, saving energy in heating, ventilation, and air-conditioning devices plays a momentous role in the worldwide energy usage. In China, the domestic hot water production is responsible for 23.4% of the total residential building energy consumption in urban areas [1]. Currently, boilers and electric water heaters are commonly utilized in residential and commercial buildings to provide the domestic hot water. One of the main concerns of prevalent utilization of these systems is their low primary energy efficiency as well as contributing to air pollution, climate change, and global warming. Consequently, to avoid the fossil fuels waste and the nettlesome problems that they bring about, renewable energy sources like solar energy have attracted much attention from industries and researchers in recent decades as a substantial resource. Solar energy is able to offer a novel alternative solution due to its clean nature and abundance in the world.

Recently, new-generation hot water devices such as the solar water heater have attracted ample attention as potential alternatives to the conventional technologies. However, the relatively low efficiency makes some limitations to exploit the solar water heaters via various technologies, especially for locations with low solar potential. In this context, the solar-assisted heat pumps were developed and presented based on the integration of solar water heaters and air source heat pumps, which are using the benefits of each single device. In an indirectly connected system, water or a solution is utilized to transfer heat from the solar collector to the evaporator of the heat pump. In recent years, there have been several researches on the experimental and analytical studies of solar-assisted heat pumps [2–4].

Owing to variability of solar radiation, high number of influential parameters, and the considerable complexity that is involved in the solar-assisted heat pumps, the reliable information on the hybrid systems performance is a prerequisite prior to conducting any effort to design, install, and evaluate the aforementioned systems. Generally, the superior way of knowing the performance of the systems under different operating conditions is to establish a highly equipped laboratory and conduct sufficient and nettlesome experiments. Nevertheless, due to a series of restrictions like requiring considerable investments and proper instruments the experimental approaches may not assess performance of these types of complex systems. Therefore, it is of great importance to develop the numerical approaches and computer technologies to tackle the limitations and alternatively evaluate performance of such systems. Soft computing techniques are an alternative, which have the high ability to model complex relationships between input-output pairs. The major merits of artificial neural algorithms are their capability to map nonlinear functions, to forecast with high generalization capability, to deal with multivariable problems, and to process large amount of data, which make them practical for modeling applications [5].

The primary objective of this investigation is to present a reliable, accurate, and practical model to simulate and monitor the performance of the solar-assisted heat pumps. This provides us the possibility to select the correct value of each operating variable to maximize the hybrid systems performance. To fulfill this objective properly, experimental data samples from a real project in China are collected and a Multilayer Perceptron-Artificial Neural Network (MLP-ANN) is developed. Firstly, a novel method for determining the most dominated input variables is introduced to elucidate the influence of each input variable on the output. Ultimately, a comparative study is carried out to appraise the reliability of the proposed MLP-ANN model against the other well-known intelligent models, such as Hybrid-Adaptive Neuro Fuzzy Inference System (Hybrid-ANFIS), Genetic Algorithm-Least Square Support Vector Machine (GA-LSSVM), and Group Method of Data Handling (GMDH) methods. The results obtained are presented in further detail in following sections.

2 Methods

2.1 Experimental Set-up

The principle of the studied solar-assisted heat pump is shown in Fig. 1. The system is a practical project of domestic hot water for a hotel, which consists of three water tanks, i.e., solar water tank, heat storage water tank, and supply water tank. The cold water is pumped into the solar water tank and circulated between solar collector and solar water tank, which is installed at the roof of the hotel. After heated by solar energy, the water is then discharged to the heat storage water tank and is heated by air source heat pumps (ASHP1 and ASHP2). The hot water from heat storage water tank enters the supply water tank and is supplied to the hotel. Another air source heat pump (ASHP3) is set to ensure the temperature of the supply water. The heat storage water tank and supply water tank are located in a cabin beside the hotel, while the air source heat pumps are next to the cabin.

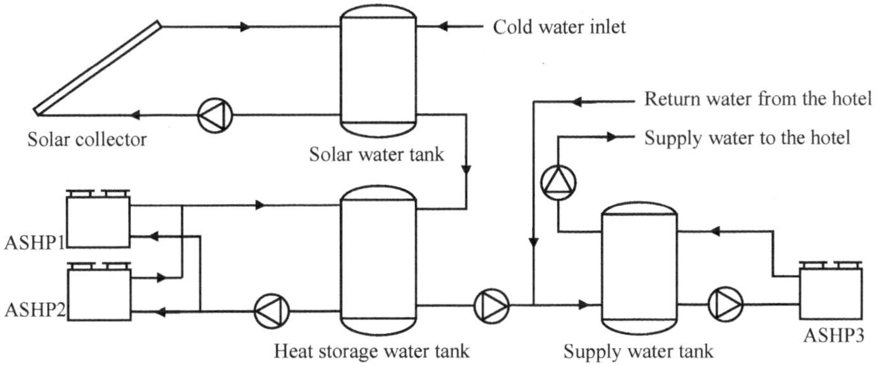

Fig. 1 Schematic diagram of the system

The volume of the solar water tank, heat storage water tank, and supply water tank is 1.5, 12, and 3 m^3, respectively; but the tanks are not always full of water. The solar collector is heat pipe evacuated tube collector with total area of 67.2 m^2. The intercept efficiency and efficiency slope of the collector is 0.760 and 2.284, respectively. The heating capacity of the three air source heat pumps is 28 kW when ambient temperature is 15 °C and electric power consumption is 7 kW. The maximum temperature of solar water tank is 50 °C. When the temperature is above 50 °C, the water is discharged to the heat storage water tank. The temperature range of heat storage water tank is 43–48 °C, which is a bit lower than maximum temperature of solar water tank because of the heat loss of the tank and pipes. The ASHP turns on if the temperature is below 43 °C. The temperature of the supply water tank is 45–50 °C to meet the demand of the hotel.

The data set for the needs of this investigation contains 425 experimental samples derived from a real project in China, which comprises forty eight input variables, as listed in Table 1. These 48 inputs were monitored from 2016/3/1 to 2017/5/31. The output parameter is COP of the system, which defined as the total heating capacity divided by the total electric consumption. The ambient temperature, the temperature of the solar collector, the temperature of water tanks, and COP are shown in Fig. 2. The temperature of solar collector, temperature of the top of solar water tank, ambient temperature of ASHP3, and COP varies with date, while the temperatures of heat storage water tank and supply water tank are nearly constant.

2.2 *Modelling*

Input variable selection. Input variable selection refers to identifying the optimal or a set of optimum variables with the highest influence on the desired output. Owing to the complexity of our system, the high number of influential parameters on the system, and the complex inter-relations among them, proper selection of the most significant input variables for predicting the COP of the system is desperately needed to develop a model along that has a high precision, robustness, and reliability, but less complexity and computational time. It has been proven that more or irrelevant input variables create many drawbacks in the model. Some of the drawbacks are: (1) difficulty in explaining the model; (2) inaccuracies caused by irrelevant parameters; (3) complexity in the developed model due to a high number of required inputs; and (4) the requirement of the time-consuming task of collecting more data. These factors may consequently cause the generalization capacity of the model to deteriorate.

In this research work, a multi-objective optimization method based on non-dominated sorting genetic algorithm, NSGA-II, was developed and utilized to determine the optimum number of input parameters for the desired output. More details about genetic algorithms and their procedures are given elsewhere [6]. The mean square error (MSE), which is the calculated error between the predicted

Table 1 Recorded parameters of the system

Parameters	
Date	Inlet water temperature of ASHP1
Inlet volume of heat storage water tank	Inlet water temperature of ASHP2
Inlet volume of supply water tank	Inlet water temperature of ASHP3
Inlet water volume	On-off ratio of compressor of ASHP1
On-off ratio of the pump between heat storage water tank and supply water tank	On-off ratio of compressor of ASHP2
On-off ratio of the pump between solar water tank and heat storage water tank	On-off ratio of compressor of ASHP3
On-off ratio of the pump for solar collector	On-off ratio of defrosting mode of ASHP1
On-off ratio of water inlet valve of the system	On-off ratio of defrosting mode of ASHP2
Temperature of solar collector	On-off ratio of defrosting mode of ASHP3
Temperature of supply water tank	On-off ratio of fan 1 of ASHP1
Temperature of the bottom of heat supply water tank	On-off ratio of fan 1 of ASHP2
Temperature of the bottom of solar water tank	On-off ratio of fan 1 of ASHP3
Temperature of the electric heater	On-off ratio of fan 2 of ASHP1
Temperature of the heat storage water tank	On-off ratio of fan 2 of ASHP2
Temperature of the supply water tank to the user	On-off ratio of fan 2 of ASHP3
Temperature of the top of solar water tank	On-off ratio of pump between heat pump and heat storage water tank of ASHP1
Temperature of the water from the user	On-off ratio of pump between heat pump and heat storage water tank of ASHP2
Total heating capacity of the system	On-off ratio of pump between heat pump and heat supply water tank of ASHP3
Volume of heat storage water tank	Outlet air temperature of the evaporator of ASHP1
Volume of solar water tank	Outlet air temperature of the evaporator of ASHP2
Volume of supply water tank	Outlet air temperature of the evaporator of ASHP3
Ambient temperature of ASHP1	Outlet water temperature of ASHP1
Ambient temperature of ASHP2	Outlet water temperature of ASHP2
Ambient temperature of ASHP3	Outlet water temperature of ASHP3

values by the developed predictive models and the experimental values considering different sets of input variables, and the number of input variables are considered as the cost functions to be minimized using the evolutionary algorithm. The MSE is described as:

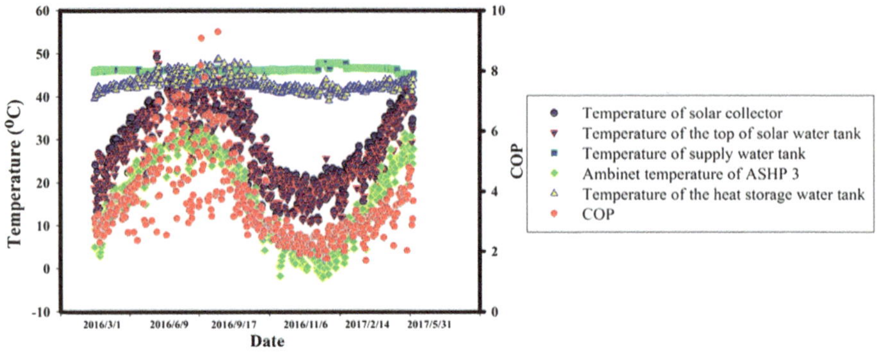

Fig. 2 The input and output variations

$$MSE = \frac{1}{2}\sum_{i \in N}(t_i - p_i)^2 \quad (1)$$

where N represents the training data set, t_i and p_i are the target and predicted vales for the ith training data. In the approach, the final non-dominated solutions are considered as the final responses.

Multilayer perceptron-artificial neural network model. Artificial neural networks (ANNs) represent parallel information processing methods, which are capable of explaining nonlinear and complex logic operations and determining the pattern within data via input-output training pairs from a data set. The adaptive nature of ANNs and their high ability in the learning of system's behavior from the information that introduced to them during training phase make them popular in different research areas to tackle with different large-scale ill-defined problems [7]. The feed-forward neural network is the most common used network architecture. In general, three layers are applied to construct the MLP-ANN model, with every layer performing a specific action. This approach includes simple processing elements called neurons, which are connected to one another in a parallel structure by weights.

The number of neurons in the input and output layers is defined based on the problem of interest, while the optimum numbers and sizes of hidden layers are determined by adding neurons in a systematic form during training process. These numbers are flexible and directly depend on the complexity of problem, the number of training data samples, and the noise in the representative data [8].

Back-Propagation (BP), which is a supervised training algorithm, is widely applied for training MLP-ANNs to update the weights and biases [9]. Suppose an input vector X = $[\chi_1, \chi_2, \chi_3, \ldots, \chi_n]$ and weight vector W = $[w_1, w_2, w_3, \ldots, w_n]$, the signals are fed to the input layer and the input weighted sum $\sum_{i=1}^{n} w_i x_i$ is calculated. This produced sum is transferred to a predefined transfer function, as shown in Eq. (2).

$$y(x_1, x_2, x_3, \ldots, x_n) = f\left(\sum_{i=1}^{n} w_i x_i + b\right) \tag{2}$$

where f (*) indicates the transfer function. At the next step, following the rule minimizing the error between the experimental COP values and the values predicted by the approach, an objective function such as MSE is defined. The calculated error is considered as a criterion for tuning of the network's weights and biases.

To assess and compare the performance of the aforementioned technique, two well-known statistical error metrics like the coefficient of determination (R^2) and root mean square error (RMSE) were employed. These parameters are depicted in Eqs. (3) and (4):

$$R^2 = 1 - \left(\sum_{i=1}^{n}(t_i - p_i)^2 / \sum_{i=1}^{n}(t_i - \bar{t}_i)^2\right) \tag{3}$$

$$RMSE = \left(\frac{1}{n}\sum_{i=1}^{n}(t_i - p_i)^2\right)^{0.5} \tag{4}$$

where n, t_i, p_i, and $\bar{t}_i$ are the total number of data, the *ith* target data value, the *ith* data value predicted by the model, and the average of the target data values, respectively.

3 Results and Discussion

To estimate the COP of the hybrid system, the novel variable selection algorithm was applied and run considering the forty eight recorded variables for selecting the most influential input variables on the output parameter. To highlight the merits and effectiveness of the proposed novel method, an additional predictive model was constructed and developed based on forty eight input variables and the results were compared to appraise the integrity and the potential of the suggested technique. Next, considering the determined input variables, the MLP-ANN predictive method was trained and tested. Ultimately, to achieve the main goal of this research through the current study, a comparative study between the MLP-ANN and the other benchmark models was carried out to ascertain the generality, applicability, and comprehensiveness of the developed MLP-ANN in estimating precise values of the output parameter.

3.1 Implementing the Variable Selection Technique

Figure 3 represents the Pareto frontier solution for the problem of interest associated with the MSE and number of input variables objective functions for the multi-objective optimization.

Obviously observed that at the start the MSE value decreases significantly as the number of input variable increases. It should be mentioned that in multi-objective optimization and the generated Pareto solution, each point has the possibility to be considered as an optimized point. In this context, the best solution for each problem is determined based upon the criteria of the decision maker and might be considered a different point as for the optimum solution according to the desires. Hence, the 0.0201 value for the MSE occurs when seven input variables, i.e., (1) temperature of solar collector, (2) temperature of the electric heater, (3) on-off ratio of fan 1 of ASHP1, (4) on-off ratio of compressor of ASHP2, (5) on-off ratio of fan 2 of ASHP3, (6) on-off ratio of pump between heat pump and heat storage water tank of ASHP3, (7) total heating capacity of the system, were considered for the desired output, while the value decreases slightly as the number of input variable increases. Therefore, this point is considered as the optimal situation by the purpose of constructing and developing a high performance and reliable predictive model.

3.2 MLP-ANN Output Results

The best network topology for forecasting the COP was determined by designing different architectures using the BP algorithm and changing the number of hidden layers, the number of neurons in the layers, and the transfer functions in the hidden and output layers. Table 2 summarizes the functions that used to construct the optimum MLP-ANN scheme.

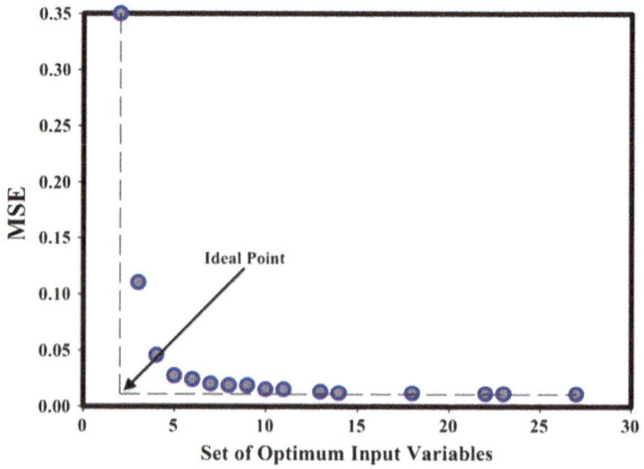

Fig. 3 Pareto frontier: best trade off values for the objective functions

Table 2 User defined functions implemented in the MLP-ANN predictive model by considering seven input variables

Number of layers	Number of neurons			Training function	Transfer function	
3	Input	Hidden	Output	TrainBR	Hidden	Output
	7	7	1		Tansig	Purelin

The regression plot presented in Fig. 4 illustrates the correlation between experimental data samples and estimated by the developed MLP-ANN considering different input vectors. The calculated R^2 values using the MLP-ANN for before and after implementing the variable selection method are 0.9771 and 0.9951 respectively, which ascertain the agreement between actual and estimated values. As is evident, the values of R^2 are quite close to unity, while the computed value after using the variable selection technique and determining the most effective input variables is higher, which shows a favorable and adequate performance of the MLP-ANN model when is combined with the novel algorithm. To support this claim, Fig. 5 presents relative error deviation percentage plot of estimations of both developed models as a function of actual COP values. This figure indicates that the model with seven input variables exhibits lower relative errors compared to another model with forty eight input variables as the relative errors of most data points predicted by the seven input variables model fall in the region bounded by two lines with relative deviations of −11.9 and +7.14%, while another developed model is capable of predicting the COP within −30.2 and +56.4% error band based on the given data. This fact indicates that the model is able to reproduce the whole set of experimental data with the lowest possible error. The computed statistical error indicators are presented graphically in Fig. 6. Once again, it is clear from Fig. 6 that the MLP-ANN model combined with the novel variable selectin technique has the best performance on estimating the COP. It can be observed from Fig. 6 that the selected input variables for developing the MLP-ANN model resulted in a higher R^2 and lower RMSE values.

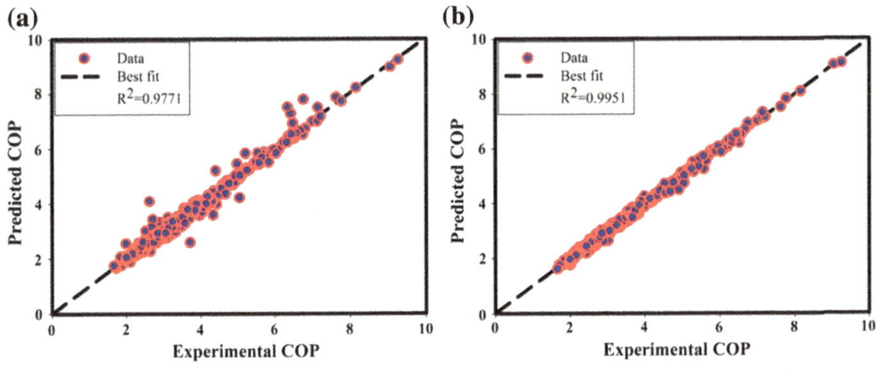

Fig. 4 Estimated COP versus experimental by considering different input variables: **a** 48; **b** 7

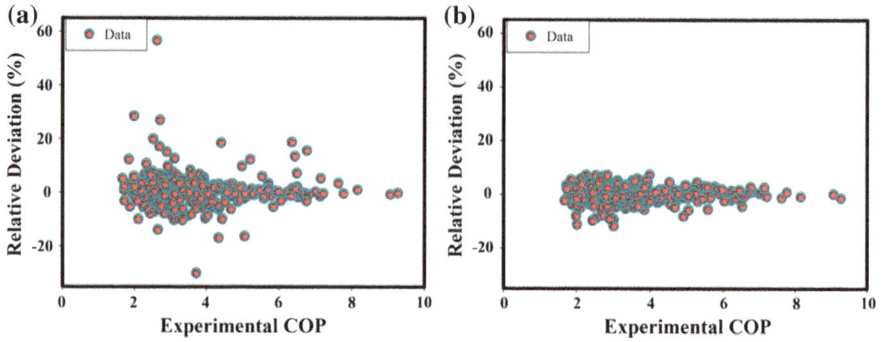

Fig. 5 Relative deviation versus experimental by considering different input variables: **a** 48; **b** 7

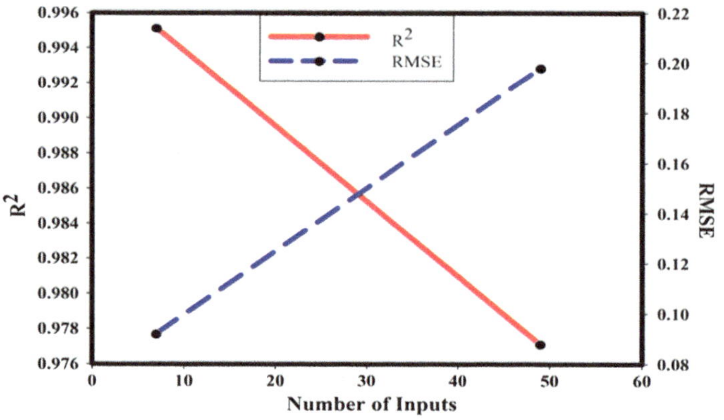

Fig. 6 The variation of statistical criteria for different models by considering different input variables

3.3 A Comparison with the Other Benchmark Models

The well-known statistical error indicators, R^2 and RMSE, are calculated and summarized in Table 3 when the GA-LSSVM, Hybrid-ANFIS, and GMDH techniques are implemented to evaluate the predictive ability of the proposed MLP-ANN model from different aspects. Based on the calculated results reported in Table 3, the MLP-ANN provides the most accurate results and falls in the category of an excellent fit by presenting the highest precision and integrity for estimation of the COP based on the given data, followed by the GA-LSSVM and Hybrid-ANFIS approaches. However, the GMDH technique was not among the best models and it provided lower performance.

Table 3 Accuracy comparison results by considering seven input variables

Parameter	Statistical criteria	Network			
		MLP-ANN	GA-LSSVM	Hybrid-ANFIS	GMDH
COP	R^2	0.9951	0.9929	0.9864	0.9357
	RMSE	0.0917	0.1114	0.1533	0.3335

4 Conclusion

The present contribution computed the performance of a solar-assisted heat pump device, which is based on integration of solar water heaters and air source heat pumps to produce domestic hot water. To determine a subset of the most suitable input variables, a novel method based on NSGA-II optimization method was proposed and implemented on the data set. As the first study of its type, the new technique paves the way for selecting the most predominant input variables among the factors that influence prediction of the desired output. It was indicated that one of the merits of this unique algorithm was decreasing the computational time and costs. Compared to the model including all of forty eight input variables, another positive outcome of this approach was increasing the coefficient of determination from 0.9771 to 0.9951 and decreasing the root mean square error from 0.1977 to 0.0917 when it combined with a MLP-ANN. A comparison of the MLP-ANN technique with the GA-LSSVM, Hybrid-ANFIS, and GMDH based on the same data set and input variables was carried out to assess the estimation robustness and precision. According to the results, the developed MLP-ANN method outperformed the others. Finally, it is safe to assert that the proposed hybrid model is suitable and practical for other complex systems due to its advantages.

Acknowledgements This study was supported by the Innovative Research Groups of the National Natural Science Foundation of China (Grant No. 51521005).

References

1. Tsinghua University Building Energy Saving Research Center, *Annual Report on China Building Energy Efficiency* (China Architecture and Building Press, Beijing, 2011). (in Chinese)
2. G. Panaras, E. Mathioulakis, V. Belessiotis, Investigation of the performance of a combined solar thermal heat pump hot water system. Sol. Energy **93**, 169–182 (2013)
3. S. Poppi, C. Bales, M.Y. Haller, A. Heinz, Influence of boundary conditions and component size on electricity demand in solar thermal and heat pump combisystems. Appl. Energy **162**, 1062–1073 (2016)
4. S. Poppi, C. Bales, A. Heinz, F. Hengel, D. Chèze, I. Mojic, C. Cialani, Analysis of system improvements in solar thermal and air source heat pump combisystems. Appl. Energy **173**, 606–623 (2016)

5. D. Svozil, V. Kvasnicka, J. Pospichal, Introduction to multi-layer feed-forward neural networks. Chemometr. Intell. Lab. Syst. **39**, 43–62 (1997)
6. K. Deb, A. Pratap, S. Agarwal, T. Meyarivan, A fast and elitist multiobjective genetic algorithm: NSGA-II. IEEE Trans. Evol. Comput. **6**, 182–197 (2002)
7. S. Haykin, *Neural Networks. A Comprehensive Foundation* (Prentice Hall Inc, New Jersey, 1999)
8. K.-L. Du, M.N.S. Swamy, *Neural Networks in a Soft Computing Framework* (Springer, London, 2006)
9. A.T.C. Goh, Back-propagation neural networks for modeling complex systems. Artif. Intell. Eng. **9**, 143–151 (1995)

Evaluation of Two Ground Source Heat Pump Systems in Nearly Zero Energy Buildings

Caroline Haglund Stignor, Ola Gustafsson and Jon Persson

Abstract In the future, most buildings will be nearly Zero Energy Buildings (nZEBs). Heat pumps are frequently used as heating system in Swedish single family buildings, but in most cases they are used in buildings with higher heating demand than the nZEBs of tomorrow. In this study, operation parameters such as heating water and brine temperatures were analyzed in real operation in two different nZEBs. The results show that the measured supply temperatures coincide in some cases well with what is described in the standard developed to be harmonized with the ecodesign and energy labelling regulations—EN14825, but in some cases they were higher. However, the brine temperatures were often considerably higher than the test conditions described in EN14825. The results also show how interconnection of a tank affects the operating parameters of a heat pump system. Another finding was that in order to reach the highest overall energy performance the heat pump and the heating system must be optimized together and not separately, which often is the case today.

Keywords Heat pump · nZEB · Energy labelling

1 Introduction

1.1 Background

The updated Energy Performance of Buildings Directive, 2010/31/EU (EPBD2) requires very low energy consumption in all newly or re-constructed buildings from

---

C. Haglund Stignor (✉) · O. Gustafsson · J. Persson
RISE Research Institutes of Sweden, Box 857, 501 15 Borås, Sweden
e-mail: Caroline.haglundstignor@ri.se

O. Gustafsson
e-mail: Ola.gustafsson@ri.se

J. Persson
e-mail: Jon.persson@ri.se

D. Johansson et al. (eds.), *Cold Climate HVAC 2018*,
Springer Proceedings in Energy, https://doi.org/10.1007/978-3-030-00662-4_35

year 2021 and onward. Across Europe there are a number of Nearly Zero Energy Buildings (nZEB) that meet the requirements of the EPBD2, but the concept is still in the pilot or demonstration stage according to Wemhöner and Kluser [1]. Previous research [1–3] has shown that heat pumps are an attractive solution for these buildings seen from energy point of view. Also, in these buidlings heat pumps are often used because of the flexibility they offer—they allow for greater freedom in designing the building envelope, since nZEB definitions are in many cases based on bought energy, and they can provide both room heating/cooling and domestic water heating [2, 3]. Moreover, heat pumps can be effectively linked to various heat sources and sinks and they can provide load balancing in a future smart grid. Wemhöner and Kluser [1] concluded that heat pumps are both an energy-efficient and cost-effective system technology for nZEBs. However, one conclusion from the same study was that there are no commercially available products that are of the right capacity. It was shown by Persson et al. [2] that a liquid/water heat pump was the most efficient heating option from both an energy and cost perspective in a nZEB under Swedish circumstances. Since the total heating demand in the nZEB is small, the cost of the heat pump system can not be too high for the system to have a competitive LCC. A heat pump system in combination with some form of heat storage is also an attractive alternative in future smart energy systems where intermittent renewable energy sources (e.g. wind and solar) are becoming more common. Therefore it is very important to increase the knowledge of how the heat pump's operating parameters (e.g. flow temperature, brine temperature, efficiency etc.) in actual operation in a nZEB are affected by speed control, by the connected storage tank, the choice of heating system etc.

Heat pumps are frequently used as heating system in Swedish single family buildings, but in most cases they are used in buildings with higher heating demand than the nZEBs of tomorrow. In the European Union there are mandatory eco-design and energy label requirements for heat pumps for hydronic heating systems from 2015. These requirements will have a large influence on the design of heat pump systems. Due to these reasons we should learn more about real operation conditions for heat pumps in nZEBs. This is important, first of all to be able to optimize the design of such heat pump system, but also to know how well their operation is reflected by the performance data for the labelling, in order to be able to influence a revision of the regulations if so needed.

1.2 Scope

The scope of this study was to:

- increase the knowledge of how different operation parameters are affected by the type of control (inverter-controlled compared with on-off), the different types of distribution system, to thereby provide data on how well the energy label correspond to reality for a heat pump in a nZEB building

- increase the knowledge of how the interconnection of a tank affects the operating parameters of a heat pump system, in order to obtain data for guiding how heat pump systems can be developed for future smart grids and use of electricity produced on-site (since the latter is important for nZEB definitions in several countries).

2 Method

2.1 Evaluated Objects

This study is based on evaluation of two different heat pump systems in two almost identical nZEBs in Sweden. One of the heat pump systems consists of an on/off controlled heat pump with an extra storage tank (see Fig. 1) and the other nZEB has a heating system with an inverter controlled heat pump. More information about the houses and their heating systems is found in Table 1.

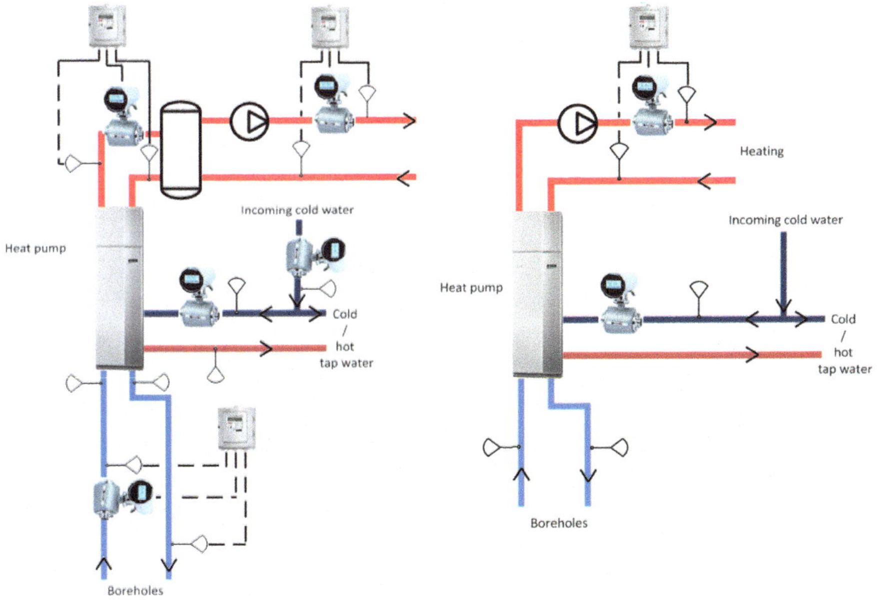

Fig. 1 Schematics of the measurement equipment in the two heating systems. To the left the system with an on/off heat pump and an extra storage tank is shown (Borås). On rhe right the system with the inverter controlled heat pump is shown (Varberg)

Table 1 Technical information about the two different nZEBs and their heating systems evaluated in this study

Place	Borås	Varberg
Size	• 166 m^2, 22 kWh/m^2/year (projected space heating and DHW demand)	• 166 m^2, 20 kWh/m^2/year (projected space heating and DHW demand)
Ventilation	• Balanced ventilation system with heat recovery	• Balanced ventilation system with heat recovery
Heating source	• Ground source heat pump (4.5 kW, on/off controlled) • Storage tank 150 l • Borehole 90 m (81 m active) • Dimensioning temperature: 0 °C	• Ground source heat pump (6 kW, inverter controlled) • Borehole 90 m (71 m active) • Dimensioning temperature: 0 °C
Heating system[a]	• Floor heating on upper and 1st floors • Dimension temperature: 36 °C at dimensioning outdoor winter temperature	• Low temperature radiators, upper floor • Floor heating, 1st floor
Solar	• PV-panels 3000 kWh/year	• PV-panels 3000 kWh/year
Habitants	• Simulated family	• Real family

[a]Controlled by an outdoor sensor and selected heat curves

2.2 *Measurement Plan and Equipment*

Operation parameters such as heating water temperatures, brine temperatures, heating water flow, electric power and outdoor temperature were analyzed in real operation in the two different nZEBs. Sensor type used for each parameter and the estimated expanded measurements uncertainty for the corresponding parameter (including sensor accuracy, sensor mounting, sensor stability, etc.) is listed in Table 2 below, together with expanded uncertainties for calculated parameters.

The results from the two systems were compared to see differences and similarities of the systems with an inverter controlled heat pump and a system with on/off controlled heat pump. The relevant measurement equipment is shown in Fig. **1** including schematic representation of placement of flow meters and temperature sensors. This study presents data evaluated over measurement periods of a year but also presents some examples of specific time events (from hours to days) to show upon cases where the differences of the systems becomes clear.

For the heating system with the on/off heat pump the supply temperature to the tank was also compared to the supply temperature out to the floor heating system to evaluate the efficiency impact of an extra storage tank and the on-off operation.

The evaluation done in this study is based on measurements performed from May 2015 to the April 2016 for the research villa in Borås and from March 2016 to February 2017 for the villa in Varberg. The reason for differing evaluation periods is that the houses were completed at different times. Even though the houses in many ways are identical, there are differing circumstances. First of all, the Varberg villa is placed in a somewhat milder climate. The yearly average climate is 8.0 °C

Table 2 Technical information about the different measurement parameters

Parameter	Sensor type	Expanded measurement uncertainty
Supply temperature, heat pump to tank, $t_{w,hp\text{-}tank}$	Pt100[a]	±0.5 K
Return temperature, tank to heat pump, $t_{w,tank\text{-}hp}$	Pt100[a]	±0.5 K
Supply temperature to heating system, $t_{w,s,hs}$	Pt100[a]	±0.5 K
Return temperature from heating system, $t_{w,r,hs}$	Pt100[a]	±0.5 K
Brine temperature out from heat pump, $t_{b,out}$	Pt100[b]	±0.5 K/±1.0 K
Brine temperature into heat pump, $t_{b,in}$	Pt100[b]	±0.5 K/±1.0 K
Heating water flow rate, heating system, q_{hs}	Armatec AT7500C	±1%
Electric power used by heat pump system, P_{hps}	Velleman EMDIN03 kWh meter	<±4%
Outdoor temperature, $t_{outdoor}$	Pt100	±1.0 K
Specific heat, c_p	Tabulated value	<±0.5%
Density, ρ	Tabulated values	<±0.1%
Cold water temperature $t_{w,DHW,c}$	Pt500	±0.5 K
Hot water temperature $t_{w,DHW,h}$	Pt500	±0.5 K
Domestic hot water flow, q_{DHW}	Armatec AT7080	±1%
Heat losses, water tanks, Q_{losses}	Estimated value	±20%

[a]The Pt100 sensors were placed in thermowells to obtain lowest possible measurement uncertainty
[b] In the research villa in Borås there were both sensors placed in thermowells as well as surface mounted sensors. In the villa in Varberg, there were only surface mounted sensors

compared to 6.6 °C in Borås according to SVEBY [4]. But the largest difference is that there is a real family living in the Varberg villa, while it is a simulated one in the research villa in Borås. During the evaluated periods this has resulted in large differences in ventilation air flow (by choice) and use of domestic hot water. Nevertheless, since the heating systems have many similarities the study stills offers many interesting comparisons.

3 Results and Discussion

In Figs. 2 and 3 the heating demand for space heating and domestic hot water is shown together with the electric power consumption for the two houses. Also the average "SCOP" per month is shown. SCOP (system COP) is calculated according to Eq. 1.

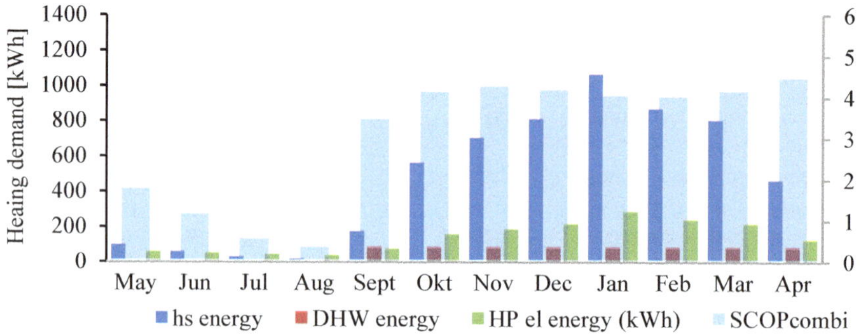

Fig. 2 Measured heating demand and electric power (left axis) and average SCOP per month (right axis) for combined operation in the research villa in Borås. The electricity to the heating system water pump is not included

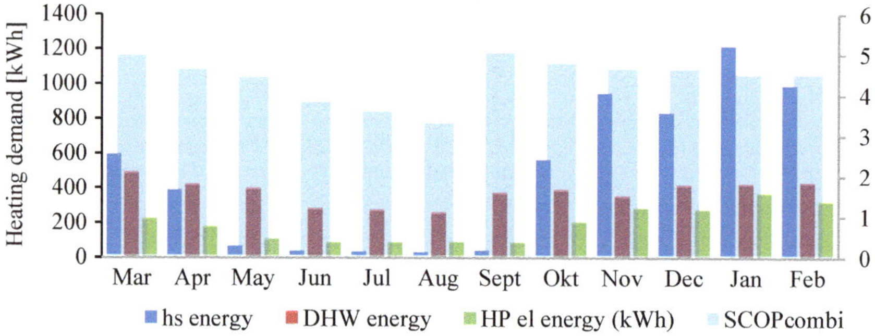

Fig. 3 Measured heating demand and electric power (left axis) and average SCOP per month (right axis) for combined operation in the villa in Varberg

$$SCOP = \frac{\sum\left(\left(t_{w,s,hs} - t_{w,r,hs}\right) \cdot q_{hs} \cdot c_p \cdot \rho + \left(t_{w,DHW,h} - t_{w,DHW,c}\right) \cdot q_{DHW} \cdot c_p \cdot \rho\right) + Q_{losses}}{\sum P_{hps}} \tag{1}$$

As can be seen the SCOP is relatively stable throughout the heating season in both nZEB houses (Figs. 2 and 3) and is only lower during the summer months when the heating demand is very low. The bars for DHW energy include the losses from the DHW tank, Q_{losses} during the months with heating demand and for those months the losses have been included in the SCOP. However, during the months with very low heating demand (May-August), the losses have not been included in the SCOP, since they have been considered as not useful (and for those months they are neither included in the bars for DHW energy). The domestic hot water use varies over the year in the Varberg house, but only moderately. In the research villa in Borås the domestic hot water consumption is very low (almost only losses).

Q_{losses} is not measured, due to difficulties in installing sensors for that, but is instead based on manufacturer data. It constitutes 5–10% of the total, so even if the uncertainty for the value itself is high, it has small effect on the overall uncertainty.

In Fig. 4 the space heating demand is plotted versus outdoor temperature (the measurement interval is 5 min and all values have been backwards averaged for a 24 h period to reduce the scatter in the graph. The same apply for Fig. 5). As can be seen there were no days that are as cold as the coldest hours of the cold climate defined in EN14825, −22 °C during the evaluation periods. The heat demand scatter is relatively large in the Borås villa. However, in the Varberg villa, where there is a real family living in the house, the scatter is much larger. The data points that spread the most are probably a result of adjustments in heating settings made by the family. What also can be seen is that the space heating demand is larger in the Varberg villa compared to the Borås villa, it varies around 55 kWh/day compared to 35 kWh per day at an outdoor temperature of 0 °C, which is due to a higher ventilation air flow (compared to what is stated as minimum constant value in the building regulations) was selected by the family living in the house (observed by monitoring the fan power). The space heating demand approaches zero around an outdoor temperature of 14 °C, which is lower compared to the standard EN14825 [5] which assumes heating demand up to 16 °C, which is the calculation standard that the ecodesign and energy labelling regulations are based on.

In Fig. 5 the supply water temperatures for the two houses are displayed as a function of outdoor temperature. In the left graph there are two sets of values. The upper one is the temperature of the water that is flowing from the heat pump to the tank (see Fig. 1 for schematic drawing) and the lower set of values are the temperature of the water from the tank out to the (floor heating) system. This is partly due to the on-off operation of the heat pump, which forces the heat pump to operate at a higher temperature during its on-periods to compensate for that there is no temperature-lift at all during its off periods (see also Fig. 6). In addition, it is partly due to some extent of mixing in and losses from the tank. In the inverter controlled

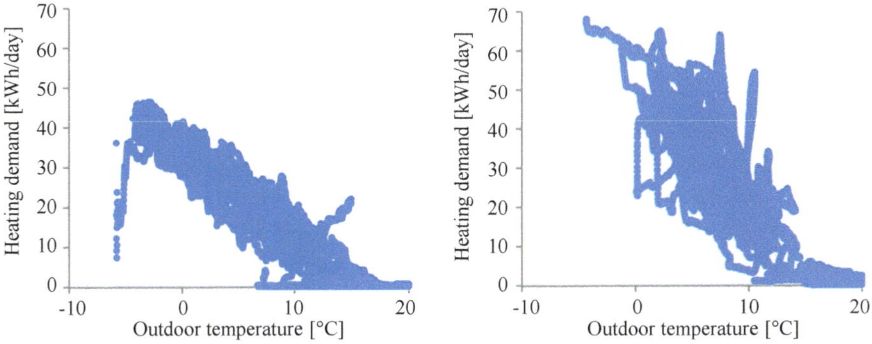

Fig. 4 Space heating demand for the research villa in Borås (left) and in the villa in Varberg (right) as a function of outdoor temperature

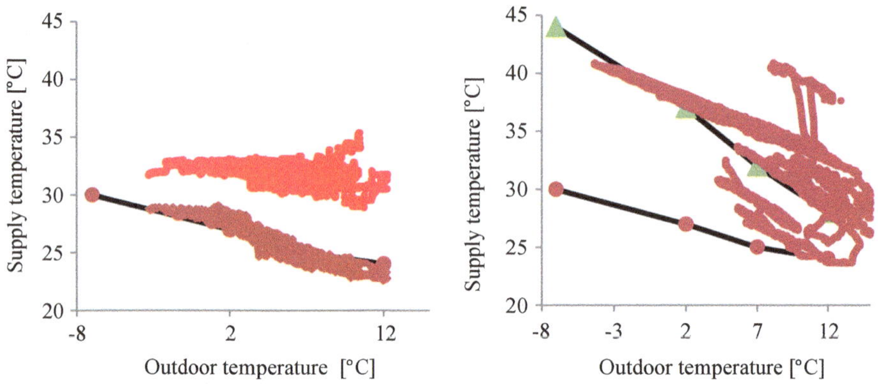

Fig. 5 Heating curves, i.e. supply temperatures (to heating system and to space heating water tank) for the research villa in Borås (left) and the supply temperature to the heating system in the villa in Varberg (right) as a function of outdoor temperature. Also heating curves as described in EN14825 at cold climate and low and medium temperature application is shown

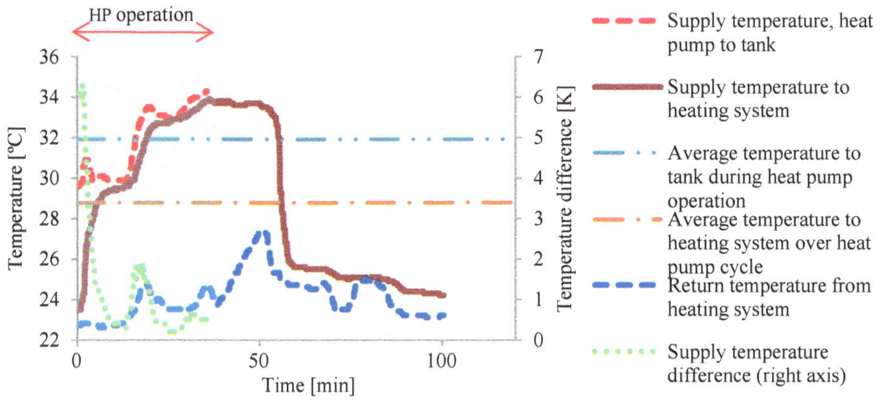

Fig. 6 Supply (and return) temperatures to the tank and to the heating system during an on-off cycle in the research villa in Borås (left) during a period with an outdoor temperature of 2 °C

heat pump system there is no tank and hence only one set of values are shown. The large temperature deviation between the two villas is because different heating systems are used in the houses. In the Borås villa floor heating is used on both floors and in the Varberg villa radiators are used on the upper floor which need a higher supply temperature. The lines in the graphs represent the heating curves in EN14825 [5]. In the Borås house with the on/off controlled heat pump the supply temperature to the heating system coincides well EN14825 heating curve for a cold climate and a low temperature application. In the standard there is an equation correcting for that the heat pump work at a higher supply temperature during the on-periods, so performance data are taken from these higher temperatures when

SCOP is calculated, which seems to be adequate according the measurements. In the Varberg house with the inverter controlled heat pump the heating curve coincides with the EN14825 heating curve for a cold climate and a medium temperature application for the colder part of the measurement period (except for some scatter), but is higher at the higher outdoor temperatures measured values are higher. The reason is probably that the heat pump system has a variable liquid flow operation and lowers the flow rate at lower capacity and the heating curve of the standard assumes constant liquid flow.

Figure 6 shows the on-off operation in detail for one operation cycle. As can be seen the supply temperature from the heat pump to the tank is somewhat higher than the temperature of the water that is leaving the tank out to the heating system during. The difference is shown in detail by the green dotted line and the average difference is 1.2 K. The fluctuations in the difference coincide with fluctuations in the return temperature from the heating system, which probably in turn is caused by closing and opening of the room thermostat valves. During the complete cycle, the average temperature out to the heating system was 28.8 °C while the average temperature from the heat pump to the tank was 31.9 °C, which means that the heat pump has to work at 3.1 K higher temperature compared to what is delivered to the heating system.

Figure 7 below shows two on-off cycles. The measured temperatures out from the heat pump and out from the tank, measured every 5 min, is displayed together with the inlet brine temperature to the heat pump. In addition, the instantaneous measured COP value, the "Carnot" COP and the ratio between those two COP values are shown. As can be seen, the measured COP seems to be instantaneously related to the temperatures of the outgoing heating water and incoming brine. Therefore, it would be beneficial for the efficiency of the heat pump system if the variations of the temperatures of the flows could be dampened. This proves that the heat pump and the heating system should be optimized together and not separately, which often is the case today.

Figure 8 below shows the measured inlet brine temperature to the heat pump in the Varberg house during one year (the measurement interval is 15 min and all values have been backwards averaged for a 24 h period to reduce the scatter in the graph and the same apply for Fig. 9). In EN 14825 heat pumps are tested at an inlet temperature of 0 °C and as can be seen, so low temperature was never measured during the whole year. The consequence of this is that the efficiency displayed on the energy label is underestimated.

In the Borås house, the brine temperature was only measured during the last part of the evaluation period and in Fig. 9 a comparison is made. Since the heating demand of the houses differed the inlet brine temperature is plotted versus electric power input to the heat pump. The on-off system has approximately 1 K lower brine temperature than the other system. Considering that the borehole with the on-off heat pump has been in operation for one more heating season, this difference can be considered as small.

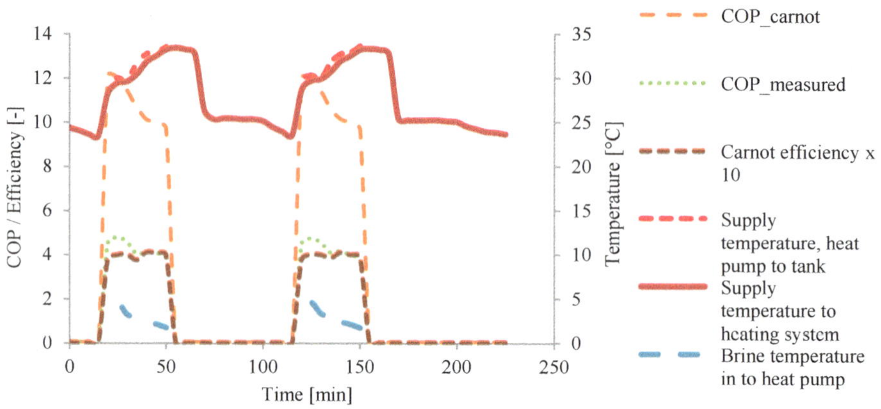

Fig. 7 Supply temperatures to the tank and to the heating system, brine inlet temperature to the heat pump (right axis) and different Coefficients of Performance (COP) and the ratio of these during on-off cycles (left axis) in the research villa in Borås during a period with an outdoor temperature of 2.5 °C

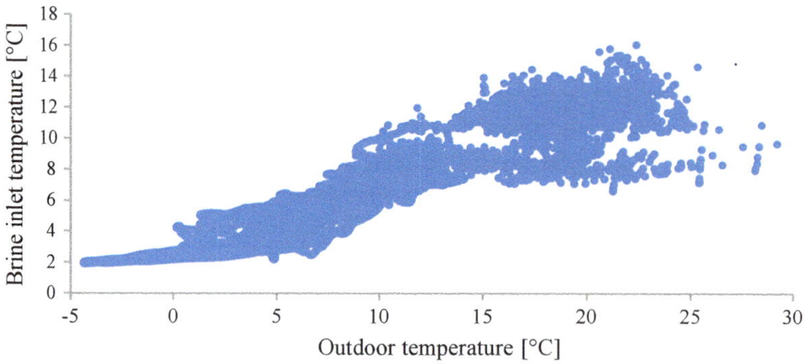

Fig. 8 Brine inlet temperatures during on-periods of the operation cycle in the villa in Varberg as a function of outdoor temperature. Data from March 2016 to February 2017

4 Conclusions

- The brine temperatures were often considerably higher than the test conditions described in EN14825 in the evaluated nZEB-buildings.
- On-off control and a tank in the system results in higher working temperatures for the heat pump compared to variable capacity control which must be accounted for when calculating projected use of energy, especially in "over-sized" heat pumps in houses with low energy demand.

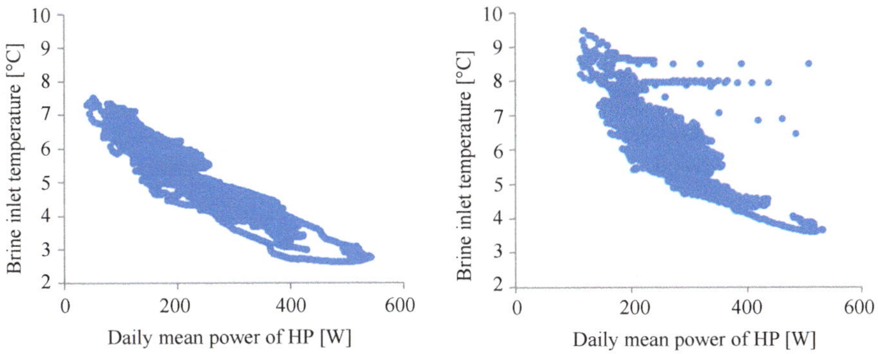

Fig. 9 Brine inlet temperatures during on-periods of the operation cycle in the research villa in Borås (left) and in the villa in Varberg (right) as a function of daily mean power input to the heat pump. Data from 15th of February 2016 to 4th of May 2016

- The heating curves of the standard EN14825 coincide well with the measurements except for variable capacity and flow control in combination with low heating demand.
- The heat pump and the heating system should be optimized together for best overall efficiency.

Acknowledgements The study has been funded by the Swedish Energy Agency (through the research program EffsysExpand), Bosch Thermoteknik, Danfoss Heat Pumps, NIBE, Skanska and TMF companies and they are all kindly acknowledged.

References

1. C. Wemhöner, R. Kluser, in *Heat pump concepts for Nearly Zero Energy Buildings—Interim results of IEA HPP Annex 40*, IEA Heat Pump Conference 2014, Montreal (2014)
2. M. Persson, S. Ruud, J. Benson, C. Haglund Stignor, A. Boss, *Heat pump systems for single family and multi family nZEB*, 11th IEA Heat Pump Conference. Montreal (2014)
3. C. Wemhoener, in *Heat Pump Concepts for Nearly Zero-Energy Buildings: Executive Summary*, IEA Heat Pump Centre, HPP-AN40-SUM (2016)
4. SVEBY Homepage, http://www.sveby.org/. Last accessed 1 Sep 2017
5. EN 14825:2016, Air conditioners, liquid chilling packages and heat pumps, with electrically driven compressors, for space heating and cooling—testing and rating at part load conditions and calculation of seasonal performance, CEN

Cold Climate Heat Pump Using Tandem Vapor-Injection Compressors

Bo Shen, Omar Abdelaziz, Van Baxter and Edward Vineyard

Abstract Conventional air-source heat pumps (ASHPs) experience rather poor performance in cold climate areas. The heating capacity and efficiency of conventional ASHPs decrease significantly as the outdoor temperature decreases. The major R&D challenges are to limit this ASHP heating capacity and efficiency degradation at low and extremely low ambient temperatures. Vapor injection (VI) compressors are able to provide better efficiency and larger capacity at low ambient temperatures. A prototype air-source cold climate heat pump (CCHP), using tandem vapor injection (VI) compressors and inter-stage flash tank, was developed. The CCHP has two identical VI compressors in parallel, which works with a two-stage indoor blower and two-stage thermostat. At moderately low ambient temperatures, only one compressor is called, and at extremely low ambient temperatures, both the compressors are used. The prototype was installed in Fairbanks, Alaska and underwent field testing for six months. The CCHP successfully operated down to −30 °F (−35 °C) and was able to meet the building heating load with good efficiency in a wide range of ambient temperatures. At −30 °F (−35 °C), the CCHP delivered 75% heat pump capacity, relative to the capacity at 47 °F (8.3 °C), and the heat pump COP was 1.8. This paper will introduce the CCHP development and field testing results.

Keywords Vapor injection compressor · Tandem compressors
Cold climate heat pump

This manuscript has been authored by UT-Battelle LLC under Contract No. DE-AC05-00OR22725 with the U.S. Department of Energy. The United States Government retains and the publisher, by accepting the article for publication, acknowledges that the United States Government retains a non-exclusive, paid-up, Irrevocable, worldwide license to publish or reproduce the published form of this manuscript, or allow others to do so, for United States Government purposes. The Department of Energy will provide public access to these results of federally sponsored research in accordance with the DOE Public Access Plan (http://energy.gov/downloads/doe-public-access-plan).

B. Shen (✉) · O. Abdelaziz · V. Baxter · E. Vineyard
Oak Ridge National Laboratory, P.O. Box 2008, Oak Ridge, TN 37831, USA
e-mail: shenb@ornl.gov

D. Johansson et al. (eds.), *Cold Climate HVAC 2018*,
Springer Proceedings in Energy, https://doi.org/10.1007/978-3-030-00662-4_36

1 Introduction

As described by Khowailed et al. [1], in the U.S., the primary target market for cold climate heat pumps (CCHP) is the 2.6 million U.S. homes using electric furnaces and conventional air-source heat pumps (ASHP) in the cold/very cold region, with an annual energy consumption of 0.16 quadrillion (0.17 EJ). A high performance air-source CCHP would result in significant savings over current technologies (greater than 60% compared to electric resistance heating). It can result in an annual primary energy savings of 0.1 quadrillion (0.1055 EJ) when fully deployed, which is equivalent to 5.9 million tons (5.35 million MT) of annual CO_2 emissions reduction. In cold climate areas with limited access to natural gas, conventional electric ASHPs or electric resistance furnaces can be used to provide heating. During very cold periods, the ASHPs tend to use almost as much energy as the electric furnaces due to their severe capacity loss and efficiency degradation. Presently, technical and economic barriers limit market penetration of heat pumps in cold climates. R&D efforts should be employed to overcome these barriers and develop high performance CCHPs that minimize, or even eliminate, the need for backup strip heating.

A typical single-speed ASHP doesn't work well under cold outdoor temperature conditions typical of cold climate locations for three major reasons:

1. Too high discharge temperature: low suction pressures and high pressure ratios at low ambient temperatures cause significantly high compressor discharge temperatures, in excess of the maximum limit for many current compressors on the market. Furthermore, system charge of a heat pump is usually optimized in cooling mode, which leads to overcharge conditions in heating mode, further increasing the discharge temperature.
2. Insufficient heating capacity: heating capacity decreases with ambient temperature. The heating capacity at −13 °F (−25 °C) typically decreases to 20–40% of the rated heating capacity at 47 °F (8.3 °C) (~equivalent to the rated cooling capacity at 95 °F (35 °C)). As such, a single-speed ASHP, sized to match the building cooling load, is not able to provide adequate heating capacity to match the building heating load at low ambient temperatures, and supplemental resistance heat has to be used.
3. Low COP: heating COP degrades significantly at low ambient temperatures, due to the elevated temperature difference between the source side and demand side.

For the CCHP development, cost-effective solutions should be identified to tackle these three issues. US Department of Energy (DOE) has set stringent performance targets for CCHPs as follows: (1) maintain at least 75% of the rated space heating/cooling capacity at −13 °F (−25 °C), relative to the rated heating capacity at 47 °F (8.3 °C), and (2) have a rated heating COP at 47 °F (8.3 °C) greater than 4.0. The 75% capacity criterion would result in a heat pump capacity approximately equal to the building heating load for a well-insulated home at −13 °F (−25 °C) in US climate Region V, for example, Minnesota (assumed to be the DHRmin load condition as defined by AHRI Standard 210/240 [2] for Region V), where the building heating load at −13 °F (−25 °C) is 80% of the building cooling design load at 95 °F (35 °C) ambient temperature.

Researchers have investigated several cycle configurations for CCHP. Wang et al. (2009) [3] studied advanced vapor injection (VI) cycles. A VI cycle uses a vapor injection compressor. In a VI system, liquid from the condenser is expanded to a middle stage between the condensing and evaporating pressures, after phase separation, the vapor is injected to the compressor injection port, and the liquid is further expanded and goes to the evaporator. VI cycles can be classified into two fundamental configurations: (a) Flash tank cycle and (b) Economizing heat exchanger cycle. Figure 1 shows the schematics of a VI cycle for each configuration. In a VI cycle with flash tank, two-phase refrigerant is separated into saturated liquid and vapor by a flash tank after the first expansion. It has the advantage of feeding 100% of saturated vapor to the compressor injection port. The two-stage cycle with economizing heat exchanger allows part of the liquid refrigerant at the condenser outlet to pass through an expansion valve before entering the economizer HX to further subcool the mainstream refrigerant coming from the condenser. The superheated intermediate pressure refrigerant leaving the economizer HX enters the intermediate compressor port. As a result, the separation with economizer HX will never be 100% as compared to the flash tank separation due to the limited surface area involved. The refrigerant flow rate and pressure entering the intermediate compressor port can be easily controlled using thermostatic expansion valves. The VI cycles are able to reduce compressor discharge temperature effectively by directly injecting low enthalpy refrigerant vapor into the compressor compression cylinder. They also increase the evaporating and heating capacities due to the lower enthalpy liquid refrigerant entering the evaporator after the inter-stage phase separation.

Vapor injection (VI) compressors are able to provide better efficiency and larger capacity at low ambient temperatures. The economizing heat exchanger is usually more expensive than the flash tank, but easier to control. The flash tank can lead to

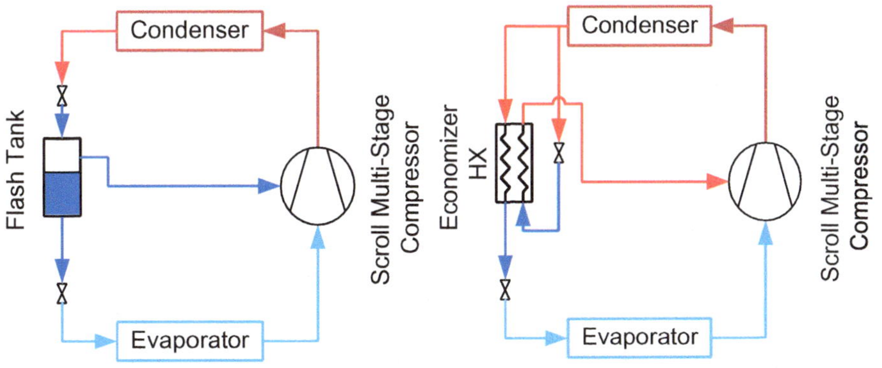

Fig. 1 Vapor injection compressor coupled with a flash tank or economizer

better performance with a lower cost, but it is hard to control in a wide range of ambient temperatures, which may cause vapor underfeeding or overfeeding to the VI compressor. Although VI compressors shows superior performance under low ambient temperatures, a heat pump using a single-speed VI compressor still can't meet the 75% capacity goal at −25 °C, as indicated by Shen et al. [4].

Shen et al. [4] discussed the development of a cost-effective CCHP, using two equal, single-speed compressors (tandem). Note that the system is relatively simple and comparable to conventional ASHPs with the exception of having two compressors in parallel, thus it is considered to be relatively more cost-effective than more complex, variable-speed design approaches. Two options of tandem compressors were studied; the first employs two identical, single-speed compressors, and the second employs two identical, vapor-injection compressors. The investigations were based on system modeling and laboratory evaluation. Both designs have successfully met the performance criteria. Laboratory evaluation showed that the tandem, single-speed compressor ASHP system is able to achieve heating COP = 4.2 at 47 °F (8.3 °C), COP = 2.9 at 17 °F (−8.3 °C), and 76% rated capacity and COP = 1.9 at −13 °F (−25 °C). This yields a HSPF = 11.0 (heating seasonal performance factor, per AHRI 210/240). The tandem, vapor-injection ASHP is able to reach heating COP = 4.4 at 47 °F (8.3 °C), COP = 3.1 at 17 °F (−8.3 °C), and 88% rated capacity and COP = 2.0 at −13 °F (−25 °C). This yields a HSPF = 11.5. This paper introduces a field investigation for the second design, i.e. the tandem, vapor-injection ASHP.

2 Field Study

2.1 System Diagram

For the field testing prototype, ORNL researchers applied an innovative configuration that allows for improved capacity modulation based on equal-size tandem VI scroll compressors. VI compressors are less prone to capacity degradation with decreasing the ambient temperature and the compressor discharge temperature is lower than non-VI compressors, as discussed earlier. The proposed control algorithm facilitates efficient operation over a wide range of ambient conditions.

Figure 2 below illustrates the hardware system configuration for the field investigation.

- The tandem VI compressors are coupled with an inter-stage flash tank to separate phases and feed saturated vapor to the compressors' VI port.
- A suction line accumulator, an injection line accumulator, a liquid receiver and the inter-stage flash tank are used as charge buffers. The liquid receiver is connected to four one-way check valves (3, 4, 5, and 6) to maintain consistent inlet and outlet lines for both heating and cooling modes. The suction line

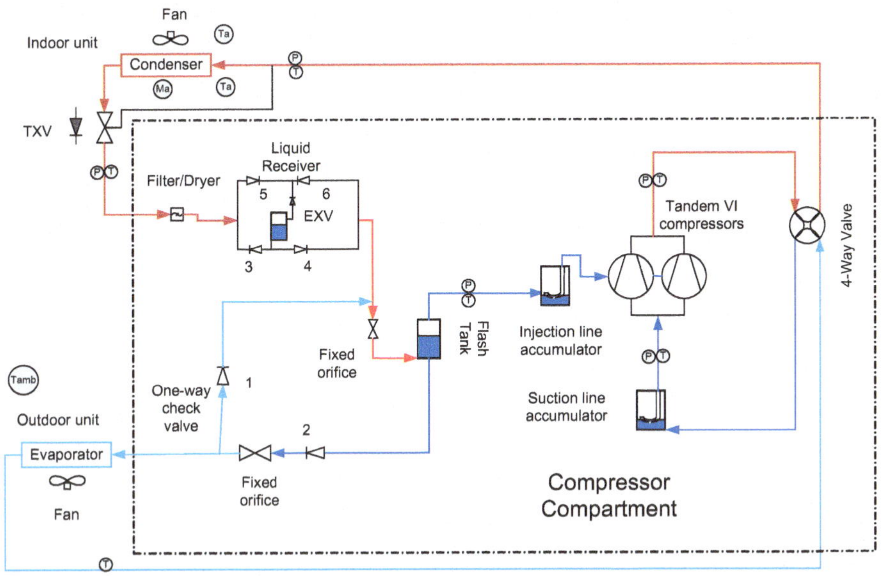

Fig. 2 Refrigeration cycle schematic: air-source CCHP using tandem VI compressors, inter-stage flash tank, and optimum inter-stage pressure control, field testing prototype

accumulator and the injection line accumulator protect the compressors from liquid refrigerant entering.

- Two fixed-orifices are installed upstream and downstream of the flash tank to provide throttling, and control the suction and injection pressures for heating mode.
- An electronic expansion valve (EXV) is installed upstream of the liquid receiver to optimize the injection pressure as a function of the ambient temperature and compressor staging.
- Two check valves (1, 2) are installed to control the flow direction for cooling, heating and defrosting modes.
- A separate TXV is used for cooling mode only.
- A four-way valve is used to change the refrigerant flow direction as a typical heat pump does.
- A two-speed indoor blower is used.

The EXV upstream of the liquid receiver was embedded with a static control function, to adjust the EXV opening as a function of the ambient temperature and compressor staging. That led to optimized head pressure and discharge temperature control as static targets, and prevented underfeeding or overfeeding refrigerant to the VI port. The fixed orifice before the flash tank differentiated the pressures of the liquid receiver and the flask tank, and the liquid level in the receiver can be varied by the EXV than going to the flash tank.

2.2 Field Testing Results

We collaborated with the Cold Climate Housing Research Center (CCHRC) in Fairbanks, Alaska, USA, and used their lab space to host the field investigation. This site doesn't necessarily represent a typical residential or commercial building but was readily available and allowed the CCHP to experience a very harsh winter. Heating season operation started in September 2016. During the field test, the CCHP successfully ran down to −30 °F (−35 °C), with heating capacity >75% of the rated capacity and heating COP > 1.8.

We developed a data acquisition (DAQ) system for the field data monitoring and control implementation. A DAQ controller and measurement devices from National Instruments Inc. were used to monitor the field operations. These were connected to an indoor host computer to process and store the data. The field testing data were sent back through Dropbox via internet connectivity.

Figure 2 describes the system diagram and instrumentation, where P represents refrigerant pressure transducers; T represents refrigerant-side temperature measurements (all inserted probe thermocouples, except the one soldered on the evaporator exit tube surface); Ta means air side temperature measurements; Tamb means an ambient temperature measurement; Ma means the indoor air flow measurement using an air flow monitor having an array of pitot tubes.

The air temperatures in and out of the indoor air handler were measured using T-type thermocouples. Three thermocouples were evenly placed at the entrance of the indoor unit to measure the average return air temperature. At the exit of the indoor blower, three thermocouples were used to monitor the supply air state. Four pressure transducers were used to measure the refrigerant pressures entering and leaving the indoor coil, as well as entering and leaving the compressors. Another pressure transducer was used to measure the pressure entering the injection port. Four watt transducers were used to measure the powers of the outdoor fan, indoor blower and two compressors, individually. The DAQ system scanned all the sensors and recorded the data every half minute.

Figure 3 illustrates the run time fraction of the second compressor to the total heat pump run time, changing with the ambient temperature. The second compressor operated more frequently as ambient temperature fell below 5 °F. At −30 °F (−35 °C), the second compressor was still cycling at a 90% fraction, indicating that the prototype CCHP system still had reserve heating capacity and was able to meet the heating load of the lab space even at this extreme cold condition.

Figure 4 shows the field-measured average delivered heating capacities for each outdoor temperature bin for both single and dual-compressor operation. At −30 °F (−34 °C) with two compressors, the average delivered capacity was 27,000 Btu/h (7.9 kW). This is 75% of the rated "starting point" capacity of the heat pump unit at 47 °F (8.9 °C) or 36,000 Btu/h (10.6 kW). Note that the average heating capacity per temperature bin decreases as the ambient temperature drops to ~5 °F (−15 °C) (with only one compressor operating for majority of the time). As the outdoor temperature decreased from 5 to −30 °F (−30 °C), the second compressor began

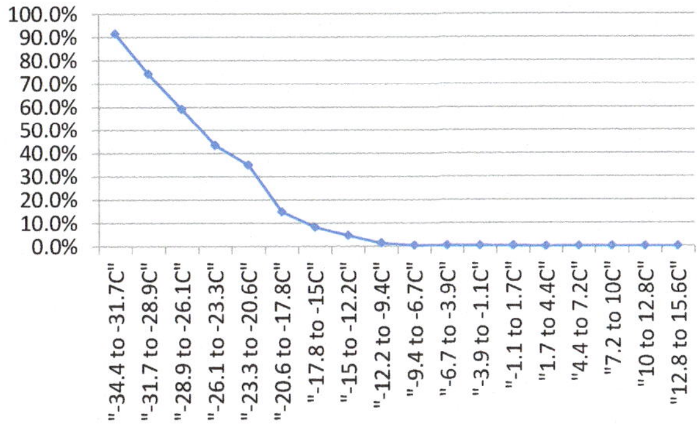

Fig. 3 The second compressor run fraction, relative to the total run time versus ambient temperature

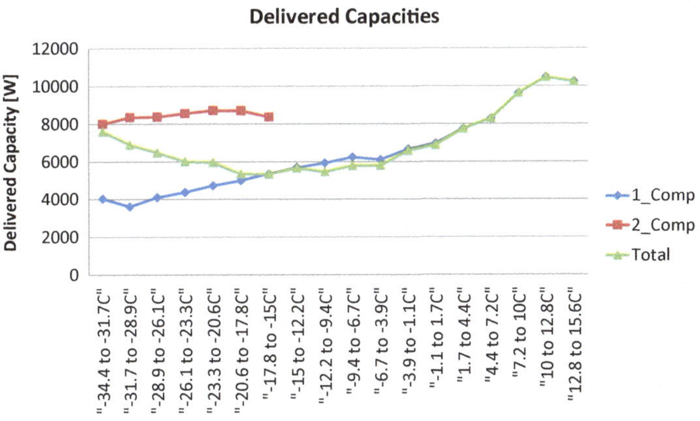

Fig. 4 Delivered heat capacities of running one or two compressors

running for an increasing portion of the time allowing the CCHP system to increase its delivered heating capacity and meet building load down to the lowest temperature bin experienced in the field test.

Figure 5 shows the average measured temperature rise in the indoor coil air stream across the indoor blower unit. The system room thermostat was set at 68 °F (20 °C), which caused the zone temperature to typically change from 66 to 70 °F (18.9 to 21.1 °C) over each heat pump run cycle. The indoor air flow rate was allowed to change with the compressor staging (High/Low). The measured low-stage indoor air flow rate was around 1200 CFM and the high indoor air flow rate was around 1650 CFM. Because the field testing lab space has a lower load profile than a typical building, only a single-speed compressor was normally needed above 5 °F (−15 °C). At moderate ambient temperatures, e.g. above 25 °F (−3.9 °F), the heat pump cycled for most time. All these factors led to an approximately uniform temperature rise from the return to supply air around 15 R (8.3 K).

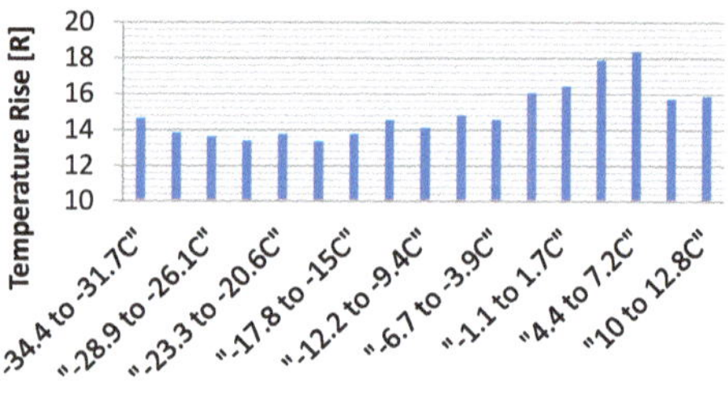

Fig. 5 Average temperature rise from return to supply

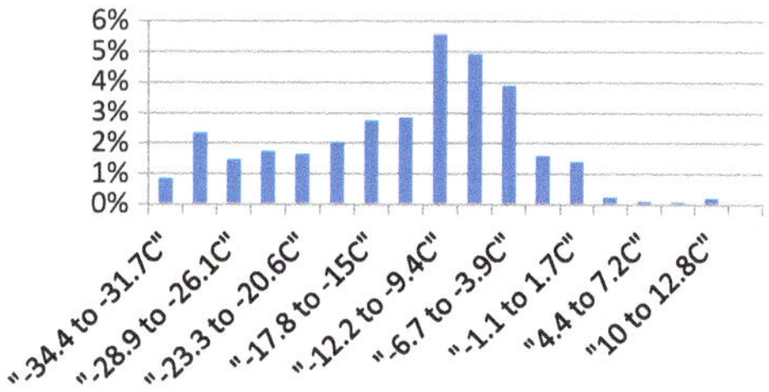

Fig. 6 Defrost time ratio relative to capacity delivered in each bin

Figure 6 presents the defrost cycle run time fraction relative to the total heat pump operation time. It can be seen that the defrosting frequency was minimal for the CCHP for two reasons:

1. During the outdoor temperature range most conducive to outdoor coil frosting, only one compressor was operating most of the time and frost formation was slow. This was due to the outdoor coil being relatively oversized compared to the single compressor capacity leading to higher evaporating temperatures than typical for conventional ASHPs.
2. When two compressors were needed at low ambient temperatures, the outdoor humidity level was very low and hardly any moisture condensed on the outdoor coil.

Because the EXV opening was controlled as a function of the ambient temperature, the defrost frequency appeared to peak in the 10–15°F (−12.2 to −9.4 °C) outdoor temperature bin.

Figure 7 illustrates average field-measured bin COPs for both single- and dual-compressor operation. The total COP (or overall system average bin COP) was calculated by the total energy delivery divided by the total energy consumed, including loss effects due to cycling, frosting/defrosting, and switching between one or two compressor operation. The single- and two-compressor COP curves do not include the defrost loss effect. During the defrosting operation, the heat pump always ran two compressors in the reversed cycle. It can be seen for the 45 to 50 °F (7.2 to 10 °C) bin the total COP is 4.0. This is about 12% lower than that obtained in earlier lab measurements for steady-state operation because of the cyclic loss effects. It is encouraging to see that, at −30 °F (−35 °C), the total COP was about 1.8, i.e. 80% more efficient than resistance heating. The field COP at −13 °F (−25 °C) is around

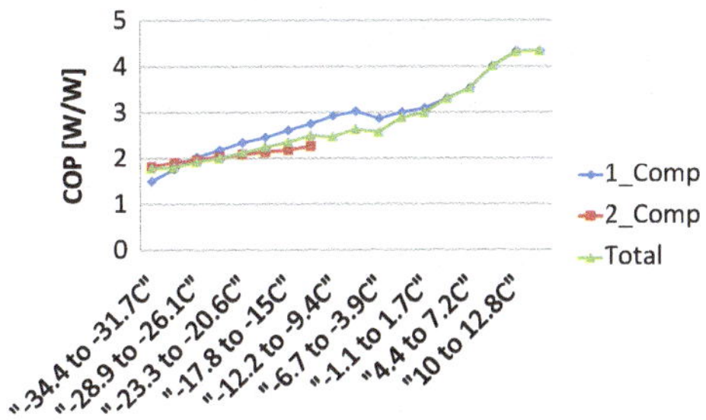

Fig. 7 Field COPs in heating mode

2.0. It is also interesting to note that for ambient temperatures below about −15 °F (−26 °C), single compressor operation appears less efficient than dual compressor operation.

3 Two Lessons Learned

- A low-pressure compressor protection shutoff control prevented system operation below about −30 °F (−34 °C) ambient due to a minimum suction pressure set point >30 psia (207 kPa). However, the compressors could have continued to operate at even lower temperatures. The compressor discharge temperature was around 230 °F at the −30 °F condition, far below the maximum discharge temperature limit of 280 °F (138 °C). A lower suction pressure protection setting would allow the CCHP to operate down to perhaps even −40 °F (−40 °C).
- The field testing site chosen had a low building heating load profile, and the CCHP was sized to match the peak building load around −30 °F (−34 °C). This meant that two-compressor operation typically did not occur until the ambient temperature fell to around 5 °F (−15 °C). The CCHP was able to provide enough heating capacity, but the indoor air flow rates appear too high to deliver comfortable supply air temperatures. A redesign of the indoor blower or reconfiguration of the blower speed settings to reduce the air flow rate would have boosted the supply air temperature and the indoor comfort level. However; this would come at some system efficiency penalty due to higher condensing temperature in heating operation.

4 Conclusions

CCHPs are targeted to the climate zones having a significant portion of heating energy delivered below 17 °F (−8.3 °C) for homes without connection to low cost gas supply. To develop a CCHP working under extremely low ambient temperatures, the first key is to make sure that the compressor(s) can operate without exceeding the compressor discharge temperature and pressure ratio limits. To accomplish this, optimizing the system charge for heating mode, properly sizing the HXs, and controlling discharge temperature are necessary. In addition, the compressor(s) should be able to tolerate high discharge temperature, for example, as high as 250–280 °F (121.1–137.8 °C).

To maintain a good ASHP efficiency at low ambient temperatures, it is most important to provide sufficient heat pump capacity and eliminate most of the resistance backup heating. At low ambient temperatures, heating load increases substantially; however, the heating capacity of a typical single-speed ASHP decreases steadily with lowered ambient temperatures, which degrades to 20–40% of the rated heating capacity. Thus, oversizing is mandatory for a CCHP, i.e. using only part of the compressor capacity to determine the rated heating capacity at 47 ° F (8.3 °C) and meet the building design cooling load. On the other hand, the system should also provide good part-load efficiency in heating mode at moderate ambient temperature, as well as in cooling mode. These requirements point to using tandem compressors or variable-speed compressors with capacity modulation capability, among which the option of tandem compressors is more cost-effective.

We developed lab and field testing prototypes and successfully met the US DOE's performance targets for CCHPs (1) to maintain heating capacity at −13 °F (−25 °C) greater than 75% of the rated heating capacity at 47 °F (8.3 °C), and (2) heating COP at 47 °F (8.3 °C) greater than 4.0.

We developed a lab prototype using tandem vapor injection compressors. The prototype reached 4.4 COP at 47 °F (8.3 °C); 88% heating capacity and 2.0 COP at −13°F (−25 °C), and 3.1 COP at 17 °F (−8.3 °C), having a HSPF of 11.5. Its seasonal performance is uniformly 5% higher than the using just tandem, single-speed compressors.

We fabricated a prototype air-source CCHP system using equal size tandem VI compressors, coupled with an inter-stage flash tank, which was fully instrumented and embedded with a complete control algorithm including defrosting, compressor and fan staging, EXV modulation to control optimum inter-stage compressor as a function of the ambient temperature. The prototype was installed in Fairbanks, Alaska for field testing for the period of 09/2016 to 03/2017. The CCHP successfully operated down to −30 °F (−34 °C) and met the building heating load with good efficiency over a wide range of ambient temperatures. At −30 °F (−34 °C), the CCHP delivered 75% heat pump capacity, relative to the capacity at 47 °F, and the heat pump COP was 1.8.

Acknowledgements The authors thank Mr. Antonio Bouza, Technology Development Manager for HVAC, WH, and Appliances, Emerging Technologies Program, Buildings Technology Office at the U.S. Department of Energy for supporting this research project.

References

1. G. Khowailed, K. Sikes, O.A. Abdelaziz. *Preliminary Market Assessment for Cold Climate Heat Pumps*, ORNL/TM-2011/422, Oak Ridge National Laboratory, August 2011
2. ANSI/AHRI 2012, *Standard 210/240 with Addenda 1 and 2, 2008 Standard for Performance Rating of Unitary Air-Conditioning & Air-Source Heat Pump Equipment, Air Conditioning, Heating*, and Refrigeration Institute, Arlington, VA, USA (2012)
3. X. Wang, Y. Hwang, R. Radermacher, Two-stage heat pump system with vapor-injected scroll compressor using R410A as a refrigerant. Int. J. Refrig. **32**, 1442–1451 (2009)
4. B. Shen, O. Abdelaziz, K. Rice, V. Baxter, H. Pham, in *Cold Climate Heat Pumps Using Tandem Compressor*, Conference Paper in 2016 ASHRAE Winter Conference, Orlando, FL (2016)

Detailed Performance Assessment of Variable Capacity Inverter-Driven Cold Climate Air Source Heat Pumps

Jeremy Sager, Tom Mackintosh, Guillaume St-Onge, Eric McDonald and Martin Kegel

Abstract Enabled by the advancement and incorporation of inverter-driven compressors and controls, cold climate air source heat pumps (CC-ASHP) are being introduced in the Canadian marketplace. Such systems are capable of efficiently meeting space heating loads at much colder ambient temperatures in comparison to single speed compressor air source heat pump (ASHP) technologies and are additionally capable of efficiently modulating to meet heating loads at warmer ambient temperatures without cycling on/off. Coupled with their lower capital costs than ground source heat pump systems, significant interest in these systems has been generated; however their widespread adoption is hindered by the unknown performance and lack of tools to evaluate their energy saving potential. This paper presents the results of detailed performance tests measuring the heating output and power input of two types of CC-ASHPs popular in the Canadian residential marketplace; a centrally ducted 3 ton split ASHP designed for whole house heating and cooling and a 1 ton ductless split ASHP designed for displacing zone heating and cooling requirements. The tests are completed by varying the outdoor temperature and indoor load (or compressor speed/capacity depending on the test approach) in climate controlled test facilities. The test results validate the systems' capability of efficiently heating at low ambient temperatures as well as modulating to meet more moderate part load conditions. However, several factors including the operating mode of the system as well as built-in protection controls for discharge temperature and pressure can serve to limit the maximum available heating capacity. Defrost settings can have a substantial impact on total system performance at cold temperatures. At mild ambient temperatures, relatively small heating load requirements can result in short cycling. The results highlight the importance of properly sizing and commissioning these heat pump systems for their application and indicate the type of data required to better simulate the performance and properly evaluate their energy saving potential.

J. Sager (✉) · T. Mackintosh
CanmetENERGY-Ottawa, Ottawa, ON K1A1M1, Canada
e-mail: jeremy.sager@canada.ca

G. St-Onge · E. McDonald · M. Kegel
CanmetENERGY-Varennes, Varennes, QC J3X 1S6, Canada

D. Johansson et al. (eds.), *Cold Climate HVAC 2018*,
Springer Proceedings in Energy, https://doi.org/10.1007/978-3-030-00662-4_37

Keywords Air source heat pumps · Air to air heat pumps · Cold climate air source heat pumps · Variable capacity · Inverter-driven · Cold climate

1 Introduction

The Canadian residential sector accounts for 17% of Canada's secondary energy end use and 14% of the country's greenhouse gas (GHG) emissions [1]. Space heating represents 63% of this energy end use and space cooling represents only 1% of the residential sectors energy end use although the amount of floor spacc coolcd through air conditioners has tripled since 1990 [1]. The use of heat pumps (HPs) to efficiently meet space heating and space cooling loads can be an attractive solution for reducing residential energy end use, especially if introduced through the space cooling market where HPs have little incremental cost over conventional air conditioners.

Conventional air-source HP (ASHP) systems often have difficulty providing sufficient heating capacity at the low outdoor temperatures common during a Canadian winter. With the advancement of variable capacity technologies, cold climate ASHPs (CC-ASHP) capable of efficiently meeting space heating loads at these low temperatures, have been introduced in the market. Previous studies [2–4] have shown that CC-ASHP systems can be a financially viable option for home-owners to reduce energy, utility costs and ultimately GHG emissions; however their exact performance characteristics and benefits are not widely known, which can hinder increased adoption. Some studies have been performed on field trialing CC-ASHPs; however their exact performance was often estimated from manufacturer performance curves [5] and thus a need was identified to better map the performance of these variable capacity systems.

CC-ASHPs come in two forms in the North American market: ducted and non-ducted systems. Ducted systems have the indoor coils installed in an 'indoor unit', which is connected to the ductwork delivering space heating and cooling throughout the household. Non-ducted systems have indoor coils inside one or several 'indoor unit(s)' mounted on the wall or ceiling to individually heat and cool a space(s). Through the variable capacity technology of CC-ASHPs, space heating loads can be met at low ambient temperatures while still providing the capability to modulate to low speeds (lower capacities) at warmer ambient temperatures. While these systems are claimed as being efficient, their actual performance, particularly at low ambient temperatures, is still largely unknown. Builders and design consultants have identified the need for a better understanding of these systems to increase market adoption.

Natural Resources Canada (NRCan)/CanmetENERGY laboratories in Ottawa, Ontario and Varennes, Quebec have built specialized test facilities to undertake performance tests on ducted and non-ducted CC-ASHP systems respectively, in order to:

1. Compare the manufacturer published performance curves to lab testing.
2. Acquire a wide range of performance data to develop robust simulation models for builders and design consultants.
3. Provide feedback on the proposed Canadian Standards Association Variable Capacity HP test draft standard [6] (CSAEXP07).
4. Identify potential areas of performance improvement.

This paper provides an overview of the test facilities, the measured energy performance and the observed operating limitations of both a ducted and ductless CC-ASHP.

2 Ducted and Ductless CC-ASHP Test Facilities

2.1 Ducted CC-ASHP Test Facility Description

The ducted CC-ASHP testing was undertaken using CanmetENERGY-Ottawa's Climate Controlled Test Facility. The facility consists of a "load and control station" with equipment that provides heating and cooling loads to the "indoor simulation" space, where the indoor side of the HP being tested is installed. The "outdoor simulation" space maintains the simulated outdoor ambient conditions and is where the outdoor side of the HP being tested is installed (see Fig. 1). The "tunnel air enthalpy test method arrangement" was used to measure the HP's capacity as outlined in the ASHRAE—37 standard [7]. The "load and control station" was then used to impose maximum and minimum capacity loads on the HP to acquire the range of performance data.

2.2 Ductless CC-ASHP Test Facility Description

Ductless CC-ASHP performance testing was conducted using an environmental controllable test chamber located at the CanmetENERGY-Varennes laboratory. The test bench consists of two insulated 3.6 m × 4.9 m × 3.6 m sheds equipped with three variable speed exhaust fans and intake louvers in order to induce heating and cooling loads using ambient conditions. The ductless split HP's indoor unit is installed in one shed representing the "demand side" (indoor shed), while the outdoor unit is installed in the second shed representing the simulated outdoor environment (outdoor shed) (Fig. 2). As no chiller was installed to induce a heating load on the demand side shed (indoor shed), performance testing could only be done when ambient temperatures were suitable (below indoor unit set-point).

Performance of the HP is calculated according to the ASHRAE 37 standard [7] measuring the enthalpy difference and the refrigerant mass flow rate through the indoor unit. Enthalpies were estimated based on refrigerant temperatures and

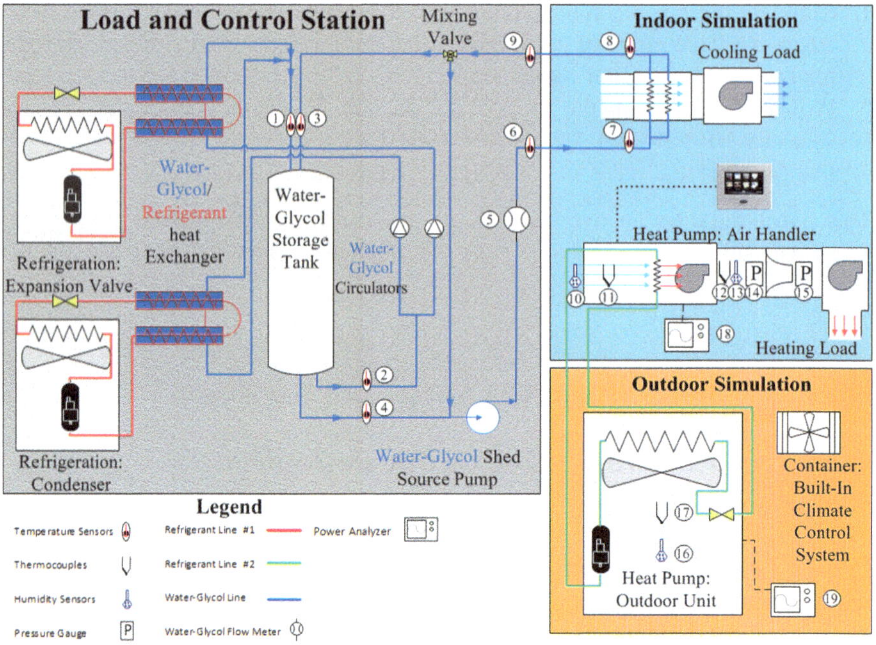

Fig. 1 Climate controlled test facility engineering schematic (heating mode)

pressures measured and calculated using Coolprop [8] (Fig. 3). As the refrigerant temperature could be a liquid-vapour mixture entering the condenser, the heat output of the unit was also validated by measuring the inlet and outlet air temperatures and airflow rate. Power consumption and compressor frequency were measured with watt transducers and current transformers.

3 Ducted CC-ASHP Performance Testing

A market available ducted CC-ASHP system with a rated heating and cooling capacity of 10.3 kW at 8.3 °C and 10.1 kW at 35 °C, respectively, was selected for the ducted system testing. The system was chosen with the objective of meeting the majority of the design heating load of 7.8 kW at −25 °C (estimated from simulation) of the Canadian Centre for Housing Technology (CCHT), a two storey high performance home. Performance tests on the HP were conducted to acquire the maximum and minimum heating/cooling capacity (based on steady state measured airflow and enthalpy difference between supply and return air streams) and COP across a wide range of operating temperatures. Additional longer duration tests were also conducted to facilitate exploration of defrost cycle impacts and different modes of operation.

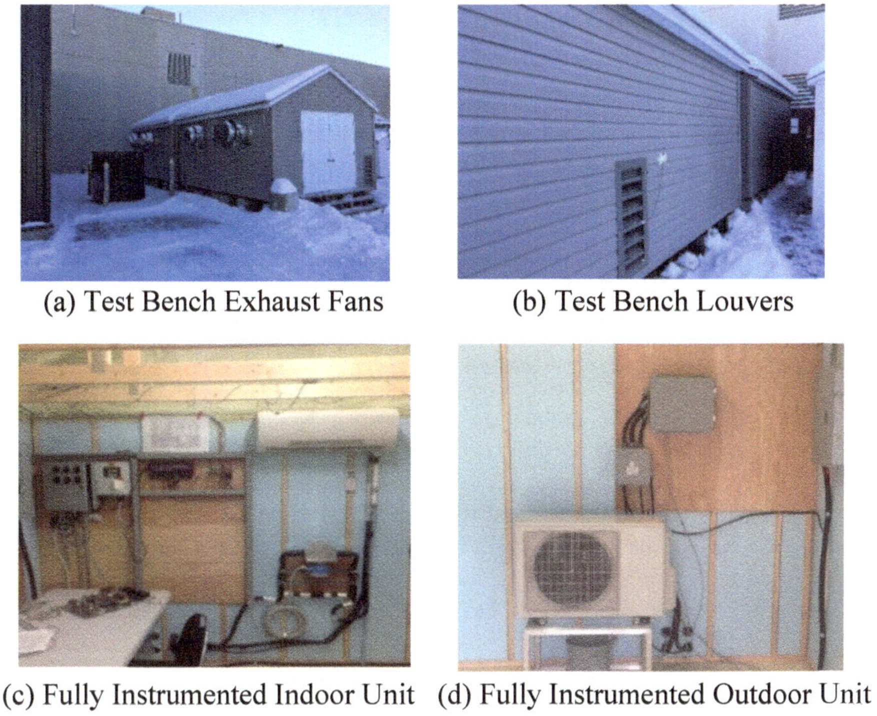

Fig. 2 Ductless CC-ASHP heat pump test bench

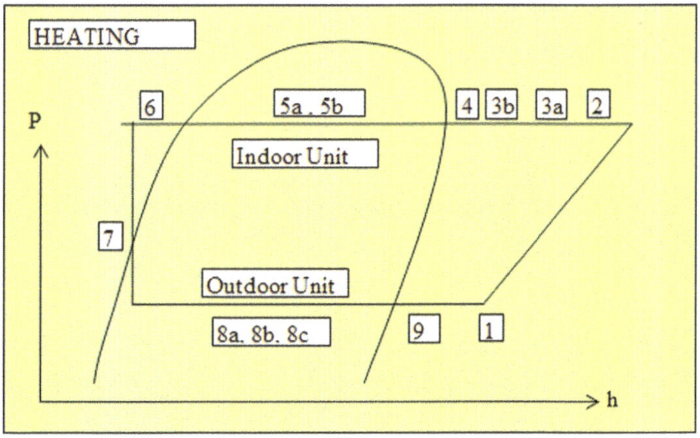

Fig. 3 Ductless CC-ASHP refrigeration cycle and temperature measurements

3.1 Comparison of Capacity with CCHT Heating/Cooling Loads

Figure 4 below compares the CCHT loads to the maximum and minimum steady state capacity using the “as-shipped” default settings of the CC-ASHP tested (as measured during performance testing at the ducted CC-ASHP test facility). The shaded red area of Fig. 4 depicts heating mode while the shaded blue area of Fig. 4 depicts cooling mode. A white shaded area defines the area between the maximum and minimum capacity and includes capacity data for which COP is >1. Any load that occurs above this area would require auxiliary heating or cooling and loads below this area would result in short cycling. Loads inside this area would result in normal operation. The heat capacity degradation from defrost is not taken into account in Fig. 4.

Between 0 and −20 °C ambient temperatures in heating mode, one can see that the total CCHT load is within the maximum and minimum capacities measured. Above 0 °C in outdoor temperature, the total CCHT load as well as individual zone loads are below the minimum capacity, implying the potential for on/off cycling. Below −20 °C, the heat pump would operate at full capacity; however auxiliary heating would be required to meet the entire heating load.

In cooling mode, the total CCHT load is just above the minimum capacity measured at ambient temperatures above 25 °C. However, the individual zone loads are each well below the minimum capacity of the unit, implying that individual zone loads would result in on/off cycling of the CC-ASHP. The degree to which individual zones would call for heating or cooling and thus result in on/off cycling would have to be assessed with detailed simulation.

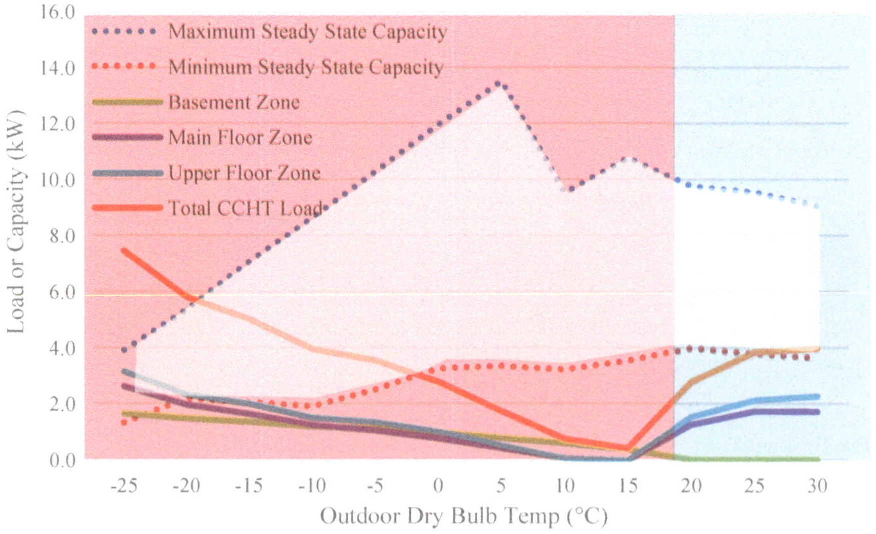

Fig. 4 Comparing zone loads at CCHT to maximum and minimum measured capacity of the CC-ASHP

3.2 *Effects of Defrost Cycles on Maximum Available Capacity*

Figure 5 shows the impacts of defrost cycles on the maximum steady-state capacity when the unit is operated with a field selectable maximum defrost interval setting. Data are based on measured results. Defrost intervals were observed to be between approximately 1 and 1.5 h and would last for a period of 1 to 16 min, depending on outdoor conditions. One can see that in the cold outdoor temperatures (i.e., −25 to −10 °C), capacity is reduced by 30–50% when defrost cycles are integrated into the test data. The degradation in performance is caused by the system injecting cold air into the zone, which needs to be reheated to maintain comfort conditions. Corresponding COPs (shown in green) in the cold outdoor temperatures are reduced by 20–30%. The effects of defrost become less pronounced as the outdoor temperatures increase (less frequent defrost cycles and for shorter durations) with no defrost occurring above 10 °C ambient temperature. Note that return air to the indoor unit during defrost cycles often dropped below the 21 °C typical return temperature, a condition unique to the testing undertaken. This may have served to over-state the impacts of defrost cycles on performance (modifications are being made to the lab to address this limitation).

4 Non-ducted Cold Climate Air Source Heat Pump Assessment

A market available ductless CC-ASHP system with a rated heating and cooling capacity of 4.0 kW (13,600 Btu/hr) at 8.3 °C and 3.5 kW (12,000 Btu/hr) at 35 °C, respectively, was selected for the ducted system testing. Testing was conducted from December 2016 to April 2017 in order to monitor the performance of the

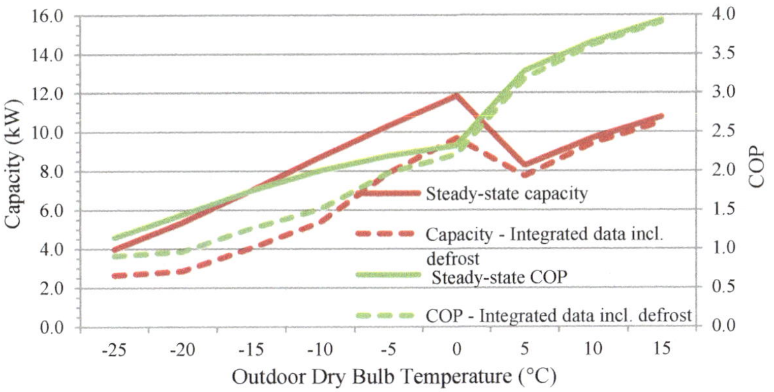

Fig. 5 Capacity and coefficient of performance of the CC-ASHP with and without defrost

system under a wide range of conditions (ambient temperature and induced heating loads) in order to acquire sufficient data to develop a robust simulation model. The tests were also used to observe how the HP would perform if operated well beyond a typical load profile helping demonstrate the importance of properly sizing these systems for their application.

4.1 Monitored Ductless CC-ASHP Performance

Global performance results are given in Fig. 6, showing the heating capacity with respect to outdoor shed dry bulb temperature and compressor frequency. The ductless CC-ASHP was able to operate at very low outdoor temperatures below the manufacturer rated temperature of −25 °C. The inverter-driven compressor also showed ability to modulate its frequency at different outdoor temperatures and heating loads, however, the frequency range seems partially limited. At mild outdoor temperatures, compressor frequencies higher than 75 Hz are rarely obtainable, and moreover, compressor frequencies lower than 20 Hz did not occur under any condition. Due to the limitations in the test bench being dependent on the HP unit to generate the cooling load in the outdoor shed, at test conditions below −20 °C, only the maximum capacity was tested. It was also noted, that if the CC-ASH was not able to attain the desired indoor shed test temperature, the compressor would cycle to 75% of its maximum capacity regardless of the load to prevent the compressor from overheating.

The measured COP was found to be greater than 1.0 during normal test conditions and less than 1.0 when the heat pump would cycle into defrost. Figure 7 plots the COP with respect to outdoor shed dry bulb temperature and compressor frequency. As seen, for equivalent outdoor temperature conditions, the COP of the system increases as the compressor speed reduces. However, it never operates

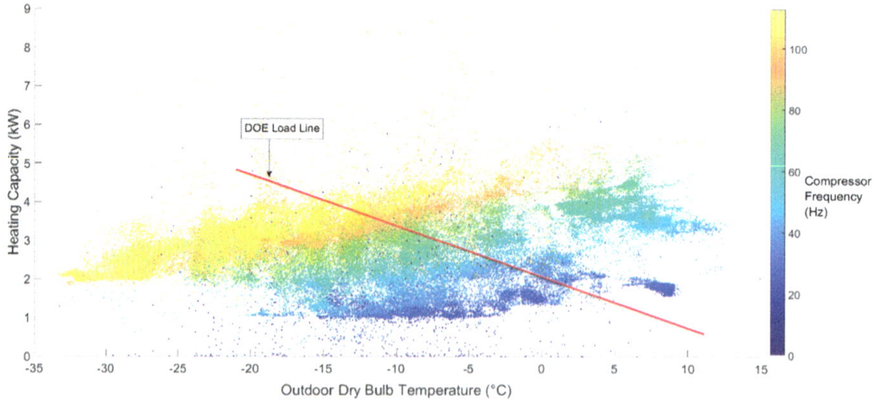

Fig. 6 Measured heat output versus outdoor dry bulb temperature and compressor frequency

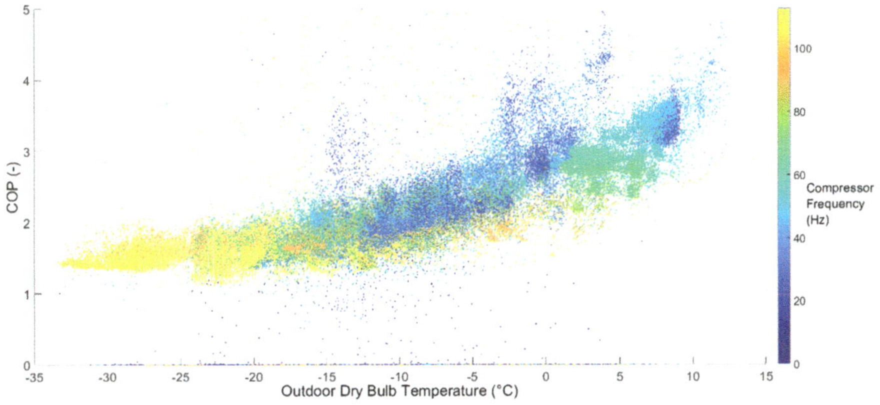

Fig. 7 COP versus outdoor dry bulb temperature and compressor frequency

below 20 Hz (20% of rated maximum compressor speed). This is of particular interest because if the heat pump is sized to meet the maximum heating load, it will begin to cycle at warmer ambient temperatures. Depending on the region, this could significantly impact the seasonal heating efficiency, and affect comfort conditions in cooling mode. Thus, it is critical to select a properly sized HP for the anticipated load to maximize the energy saving potential.

Defrost intervals were observed to be between 2.5 and 2.0 h depending on the ambient conditions and would last for a period of approximately 5 min. As such, the defrost would have minimal impact on the overall steady state capacity. It was noted however that when coming out of a defrost cycle the unit would operate at maximum compressor speed regardless of the ambient temperature condition.

5 Seasonal Coefficient of Performance in Heating (SCOPh)

The second objective of the testing was see how the CC-ASHP systems performed according to CSA EXP07 standard. While the test benches were not designed to meet the tolerances of the test standard, the test chambers could provide feedback on observed trends and challenges when conducting such a performance test. As such, loads were induced on the indoor coils at the respective outdoor air temperature as outlined in the standard. The SCOPh is then calculated based on the selected region and specified operating hours at each temperature bin interval and the annual heating load derived from the load curve prescribed in the draft standard. Table 1 summarizes the SCOPh for various Canadian climatic regions at steady-state (no defrost) and with defrost included. The following additional assumptions were made:

Table 1 Seasonal coefficient of performance for the ductless CC-ASHP tested according to the proposed CSA EXP07 [6]

Climate Zone (Heating season hours)	SCOPh (Ducted)		SCOPh (Ductless)		Canadian region by Province/Territory
	steady-state	with defrost	steady-state	with defrost	
Marine (4630)	3.34	3.21	3.09	3.04	Coastal BC
Cold/Humid (4808)	2.31	1.96	2.23	2.13	Southern ON, QC, NB, NS
Cold/Dry (5017)	2.57	2.23	2.38	2.28	Southern AB
Very cold (6340)	2.06	1.76	2.10	2.02	Northern BC, AB, SK, MB, ON, QC, NB, NS, NL, PEI
Sub Arctic (7758)	1.63	1.42	1.72	1.67	Parts of YK, NWT, NU, Extreme North MB, QC, NL

Note Heating season hours are applied to climate zones according to bin fractions. Refer to [6] for details

1. Backup heat is electric and meets all of the load whenever the unit COPs drop below 1 and/or; the unmet portion of the load if there is insufficient HP capacity.
2. Improved COP at loads below the maximum capacity are interpolated between the COPs at maximum and minimum capacity for the ducted CC-ASHP tests.
3. Performance degradation due to cycling/standby power at loads below minimum capacity are not accounted for.

The SCOPh results demonstrate the energy savings potential for a CC-ASHP in Canadian regions as compared to electric resistance based heating or high performance natural gas heating systems (maximum SCOPh of 1.0 and 0.98, respectively). The detrimental effect of defrost in the cold/humid region is seen, where over a 15% decrease in efficiency occurs due to the length and energy required for the ducted CC-ASHP system. The defrost cycle of the ductless unit did not degrade the performance of the unit as significantly, resulting in approximately a 3% decrease in SCOPh.

6 Conclusion and Future Work

The widespread adoption of CC-ASHPs has been hindered by the unknown performance, lack of tools to evaluate their energy saving potential and lack of guidance as to how to properly select and size systems according to the intended application. NRCan/CanmetENERGY laboratories in Ottawa and Varennes have built specialized test facilities to undertake performance tests on ducted and non-ducted CC-ASHP systems respectively, to acquire performance data, inform on sizing and selection and identify potential areas of improvement.

The maximum and minimum heating and cooling capacities at a wide range of simulated ambient temperature conditions were performed on both ducted and ductless CC-ASHPs. Testing showed that both CC-ASHPs were capable of efficiently operating well above a coefficient of performances of 1.0 confirming their energy saving potential in comparison to electric and natural gas space heating systems. Furthermore, the systems were capable of modulating efficiently to meet low heating load capacities, permitting efficient space heating operation at a wide range of ambient temperatures. However, it was also shown that the CC-ASHPs have a lower limit to their modulation capability before cycling on/off to meet the loads and as such it is important to properly size these units. Therefore, selecting a heat pump to meet the maximum space heating load condition is not always the most suitable design choice.

The effect of taking the defrost cycle into account in the heating capacity and COP was also shown. The higher capacity ducted CC-ASHP experienced a substantial reduction in capacity and COP, particularly at the coldest outdoor temperatures, resulting in up to a 15% decrease to system SCOPh. The smaller capacity ductless CC-ASHP experienced a lower performance degradation of 3% in system SCOPh.

Future work will look to continue detailed assessments of CC-ASHP systems to gain more in-depth knowledge of the performance and control strategies. A design guide and simulation tool are being developed to aid designers and contractors in properly sizing and assessing the energy saving potential of CC-ASHPs for Canadian applications.

Acknowledgements Funding for this work was provided by NRCan through the Energy Innovation Program and Office of Energy Efficiency Equipment Division and SUMARAN Inc.

References

1. Natural Resources Canada, *Energy efficiency trends in Canada 1990–2013* (Natural Resources Canada, Ottawa, 2016)
2. M. Kegel, J. Tamauskas, R. Sunye, D. Giguere, Heat pumps in the Canadian Residential sector, in 11th IEA Heat Pump Conference, paper #O2.4.4. Montreal, Canada (2014)
3. M. Kegel, J. Tamasauskas, R. Sunye, Integration and evaluation of innovative and renewable energy technologies in a Canadian Mid-rise Apartment. In: 9th System Simulation in Buildings Conference, paper #038. Liege, Belgium (2014)
4. J. Sager, M. Armstrong, F. Szadkowski, *Cooling and heating season performance assessment of a cold climate air source heat pump* (Natural Resources Canada, Ottawa, 2013)
5. B. Le Lostec, H.F. Nouanegue, On-site performance of air source heat pumps, in: 11th IEA Heat Pump Conference, paper #O2.4.3. Montreal, Canada (2014)
6. Canadian Standards Association: CSAEXP07—Stakeholder Draft: Energy Performance of split-system and single-package air conditioners and heat pumps, Draft version April 24, 2017. Mississauga, Canada (2017)
7. American Society of Heating and Refrigeration Engineers: ASHRAE 37-2009: Methods of Testing for Rating Electrically Driven Unitary Air-Conditioning and HP Equipment (2009)
8. I.H. Bell, J. Wronski, S. Quoilin, V. Lemort, Pure and pseudo-pure fluid thermophysical property evaluation and the open-source thermophysical property library coolprop. J. Ind. Eng. Chem. Res. **53**(6), 2490–2508 (2014)

Utilisation of Ice Rink Waste Heat by Aid of Heat Pumps

Juris Pomerancevs, Jörgen Rogstam and Agnese Līckrastiņa

Abstract Ice rinks have a simultaneous cooling and heating demand to maintain the ice slab and providing the required heat to different needs in the ice rink facility. This provides a potential for utilising a heat recovery function, which is used in many ice rinks with traditional ammonia (R717) systems, however using carbon dioxide (R744) as a refrigerant in a trans-critical system can provide a fully self-sufficient ice rink in terms of heat. In R744 systems more heat is available at higher temperature levels, whereas in R717 systems the largest heat portion is present at condensing temperature level, which does not fulfill high temperature heating demand. In this study energy saving potential with heat recovery systems in ice rinks is evaluated. A system solution is developed, modelled and evaluated. One objective is to look at using R744 as refrigerant in ice rinks, however, focusing on the heat recovery aspect. In order to check the applicability, a R744 system with heat recovery is compared to traditional R717 refrigeration system with a propane (R290) heat pump. Data from an ice rink in operation is used to map the heating demand during season and to increase credibility. The results show that heat recovery performance in ice rink with R744 direct refrigeration system uses 47 MWh less energy than in indirect R717 systems with R290 heat pump, which constitute to 16%.

Keywords Natural refrigerants · Ice rink · R744 · Heat recovery
Heat pump · Heat export

J. Pomerancevs (✉) · J. Rogstam
Energi and Kylanalys AB, Varuvägen 9, 125 30 Älvsjö, Sweden
e-mail: juris.pomerancevs@ekanalys.se

J. Rogstam
e-mail: jorgen.rogstam@ekanalys.se

A. Līckrastiņa
Riga Technical University, Rīga, Kaļķu iela 1, Riga LV-1658, Latvia
e-mail: agnese.lickrastina@inbox.lv

D. Johansson et al. (eds.), *Cold Climate HVAC 2018*,
Springer Proceedings in Energy, https://doi.org/10.1007/978-3-030-00662-4_38

1 Introduction

1.1 Energy Usage in Ice Rinks

The number of ice rinks in Sweden is about 360 and grows in a rate of 5 to 10 new constructions per year [1]. At the moment, the average annual purchased energy of an ice rink in Sweden is about 1000 MWh/year, where typically about 80% is electricity and 20% is heat [2]. Currently, the total energy usage of indoor ice rinks, in Sweden, reaches a total of 300 GWh/year. Since indoor ice rinks are still increasing in numbers in the country, it is very important to practice a policy of sustainability and search for better energy efficient techniques, given that the amount of energy used to these facilities will continuously to increase.

Results from an investigation in 135 ice rink energy usage in Sweden suggest that the highest energy share is used in the refrigeration system, constituting to around 43%, while the heating system fill almost a third of the whole energy usage [2]. In practice the energy systems are interlinked through different driving forces, without an option of avoiding that completely. However, it is possible to minimize the negative effect by appropriate controls or even benefit from it, which requires a proper design. An obvious possibility is to reclaim waste heat from a refrigeration system to further utilize it for the rink space heating, hot water heating, etc. There are different solutions for partial heating demand coverage by reclaimed heat, while the best practice has proven the ability to cover the whole heating demand.

2 Natural Refrigerant Properties

The most important properties and aspects of the candidate refrigerants will be discussed since that is key to the refrigerant and system selection later on.

The use of natural refrigerants (carbon dioxide—CO_2 (R744), hydrocarbons (HCs), and ammonia—NH_3 (R717)) has increased significantly in the past years. This trend is predicted to continue in the future, as the impact from legislation such as the F-gas regulation in EU, has an increasing influence in the refrigerant choice. Natural refrigerants have no ozone depleting potential (ODP) and neglectable greenhouse warming potential (GWP). The indirect effect, which is the greenhouse gas emission due to power generation, may be reduced due to favorable thermodynamic properties of these fluids.

Ammonia is a widely used refrigerant, suitable mostly in commercial and industrial size applications, i.e., in relatively large refrigeration plants. Thermodynamically the heat of vaporization for ammonia is high and the heat transfer is favorable. On the negative side comes its toxicity and flammability. Well-designed safety measures are necessary and normally ammonia is not allowed to use inside closed spaces with people present such as supermarkets, ice rinks, etc.

Propane is a hydrocarbon that has similar thermodynamic characteristics to the synthetic R22 refrigerant, which is phased out. Because of flammability, charge minimization is a major design challenge when using propane. A study suggests that high evaporation temperature, up to 24 °C, can be achieved in R290 cycle, which makes it suitable as sanitary hot water heating heat pump, for instance [3].

Carbon dioxide is unique among natural refrigerants, because of its good safety characteristics—it is non-flammable, non-explosive, and relatively non-toxic, which makes it a beneficial in applications where relatively large refrigerant quantities are needed. Carbon dioxide is a refrigerant that has a relatively low critical temperature (31 °C) and high working pressure. As the primary refrigerant it can operate in a transcritical mode, rejecting heat without undergoing a phase change and compared to other common refrigerants the heat recovery can be obtained at relatively high temperatures [4].

Historically ammonia has been the most commonly used refrigerant in ice rinks. However, in the past carbon dioxide systems have proved to be an efficient choice, most importantly—due to the heat reclaim function. Both fluids can be compared from a heat recovery perspective using Fig. 1, where the share of used available heat at a certain temperature level is plotted. The comparison is made at an NH_3 condensing temperature of 35 °C and CO_2 head pressure of 80 bar. The relation shows that only 19% of heat for NH_3 is above 35 °C, while for CO_2 the corresponding figure is around 60%. This benefit makes it possible to cover more of, or the whole, heating demand in a typical ice rink by using a CO_2 heat recovery system [5].

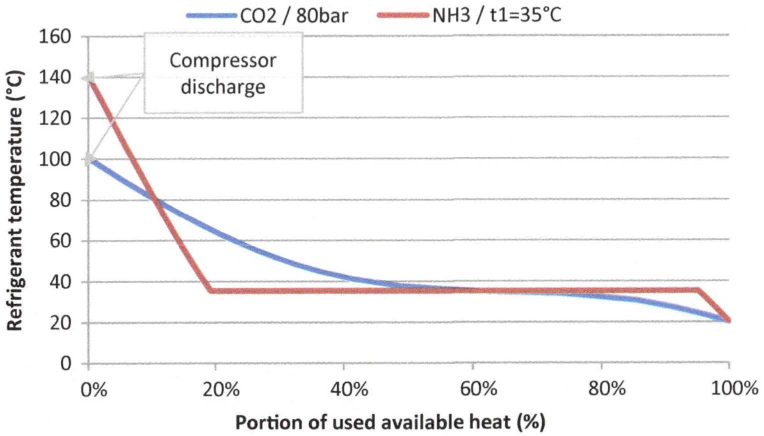

Fig. 1 Comparison between ammonia and carbon dioxide available heat in the temperature range [5]

3 A CO_2 Ice Rink—With a Full Heat Recovery Solution

To generate reference data as far as the heating demand is concerned, an ice rink in Sweden comprising a full direct CO_2 system is used. It fits in a group with around a third of the national stock of ice rinks that are single sheet, mid-size with 500 to 1000 spectator seats, having a heated arena room (5–10 °C). For that reason it can be regarded as typical and further in the text will be referred to as the reference ice rink. Each system/category is measured with "high quality type" (Carlo Gavazzi) energy meters, which are developed and manufactured in full compliance with the most important standard regulations. Measurements are logged in the IWMAC system. It is a permanent measurement/monitoring system, that allows us to analyse the data for the main energy systems as well as the temperature profiles. In this study, data for one season between late July and mid-March, is taken into consideration, to further compare it with the model for an ammonia indirect system, which is discussed more into detail further in this work.

3.1 Refrigeration System

A fully direct CO_2 system has one working fluid circulating in the system. The latter consists of an accumulator tank, where the refrigerant is stored and separated, while a pump is used to circulate the fluid from the accumulator tank into the rink tube system.

The resulting energy usage from Table 1 show that auxiliary refrigeration system equipment has very low relative consumption (CO_2 pump—0.8%; Gas cooler—1.4%), which can be explained with the high volumetric heat of the fluid, low pressure losses due to the high density, as well as the advantages of the direct system solution. The working pressure on the high pressure side is normally above the critical pressure of CO_2, making the system trans-critical.

3.2 Heat Recovery System

The reference ice rink has a heating system that solely utilizes recovered heat to cover the heating demand and the system principle layout is illustrated in Fig. 2.

Table 1 Refrigeration system energy usage in the reference ice rink

Refrigeration energy users (MWh)	Jul.	Aug.	Sep.	Oct.	Nov.	Dec.	Jan.	Feb.	Mar.	Total
Compressors	16.4	41.2	45.5	31.1	31.1	30.8	34.7	33.2	16.6	280.7
CO_2-pump	0.3	0.5	0.5	0.5	0.5	0.5	0.5	0.4	0.2	4.0
Gas cooler	0.6	0.6	0.2	0.2	0.2	0.1	0.2	0.3	0.0	2.4
Total	17.2	42.4	46.2	31.8	31.8	31.4	35.4	34.0	16.8	287.0

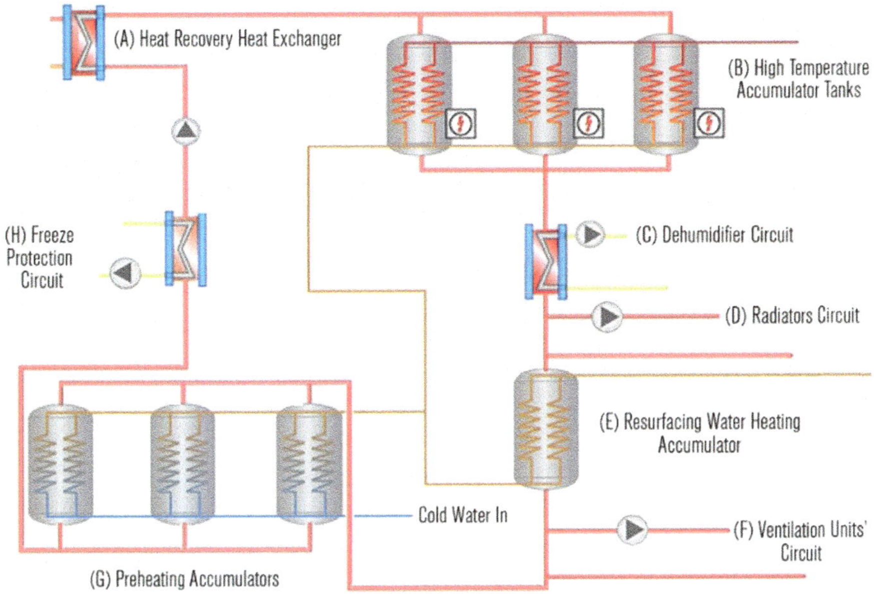

Fig. 2 Reference ice rink heating system principle [5]

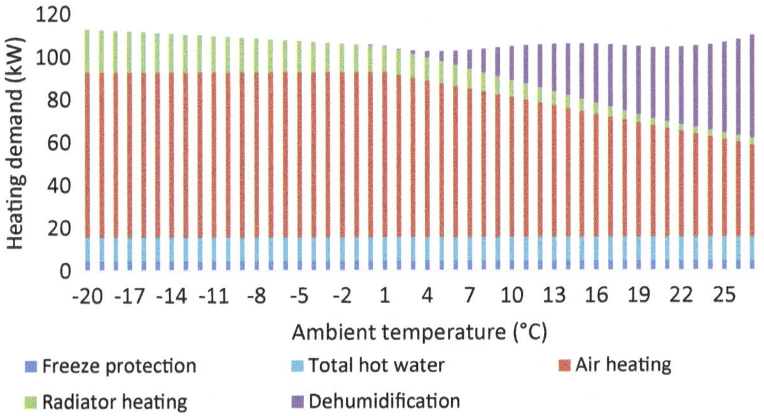

Fig. 3 Heating demand profile in the studied ice rink

In order to illustrate heating demand profile, it is plotted depending on the ambient temperature as shown in Fig. 3, using data from the reference ice rink. The profile is slightly corrected, to disregard fluctuations from the reality. The highest heating need is the one for the arena room heat supply, distributed by the ventilation system and requiring a supply temperature of around 35 °C. When the ambient temperature drops below 1 °C, the air heating demand becomes constant, which can be explained with insufficient air heating system capacity, resulting in decreased indoor temperature. Nevertheless, this is the real profile and the analysis should be based on reality.

The radiator heating demand is linearly depending on the ambient temperature and requires a supply temperature of around 55 °C. The outdoor air, which affects the indoor air, becomes more moist with increasing ambient temperature, which results in higher dehumidification needs. In the studied ice rink a desiccant dehumidifier is used and recovered heat is utilized for the reactivation process. The supply temperature, same as for hot water heating is close to 60 °C. The freeze protection heating demand as well as the water preheating are performed with a supply temperature below 35 °C. Typically the return temperature to the heat recovery is around 25 °C.

4 Evaluation of Heat Recovery Potential in an Ammonia Indirect System

Using heating and cooling profiles from the reference ice rink, a simulation model of ammonia indirect system is created with the aim to evaluate the potential of full heat demand coverage by connecting a heat pump to the refrigeration system's coolant loop.

There are many existing ice rinks with ammonia indirect system in Sweden as well as in the rest of the world. This case is of particular interest since many existing systems could be retrofitted with heat recovery systems. Although this is an environmentally friendly choice considering the direct impact of the substance, the waste heat is either partly or not at all utilized in a majority of the facilities. It is possible to recover heat directly from the condensing heat, however the temperature level is not necessarily high enough, depending on the heating system design.

4.1 Method of Evaluation

A software "Pack Calculation Pro" is used as the simulation tool for the refrigeration system and heat pump analysis. The software credibility was analysed in a previous study where the same CO_2 ice rink, for the same period, has been compared to a CO_2 trans-critical system model results, showing an accuracy within 10% [6]. In this case the system is comparable to the model although a different system solution. The output results are further processed and additional calculations are made in a spreadsheet application based on "Microsoft Excel".

4.2 Refrigeration System

A fully indirect NH_3 system with brine as the secondary fluid is a chosen. As regards to the cooling profile, it is identical to reference ice rink, however, the evaporation temperature profile is deduced from a well monitored indirect NH_3 ice rink system, resulting in lower temperatures than in the reference system for the same capacity, which is mainly due to temperature losses in the heat exchanger.

4.3 Heat Recovery by Aid of a Coolant Heat Pump

As can be seen in the Fig. 4 below, the heat from the refrigeration process is first recovered through the desuperheater. The rest of the heat will pass through the condenser and via the coolant to the accumulator tank. This tank is necessary in practice as a buffer when heat rejection is not simultaneous with heating demand. The remaining heat is rejected to the ambient by means of the dry cooler. The heat pump extracts part of the available heat from the refrigeration system and boosts it to a higher temperature level, and ideally covering the heating demand.

Temperature levels for the heating system are divided into three ranges by aid of accumulator tanks, following the same principle as it is in the reference system. Several studies suggest that a propane cycle efficiency can be increased by implementation of subcooling, which coincides with the heating need of low temperature level and is set to 30 K in the present model.

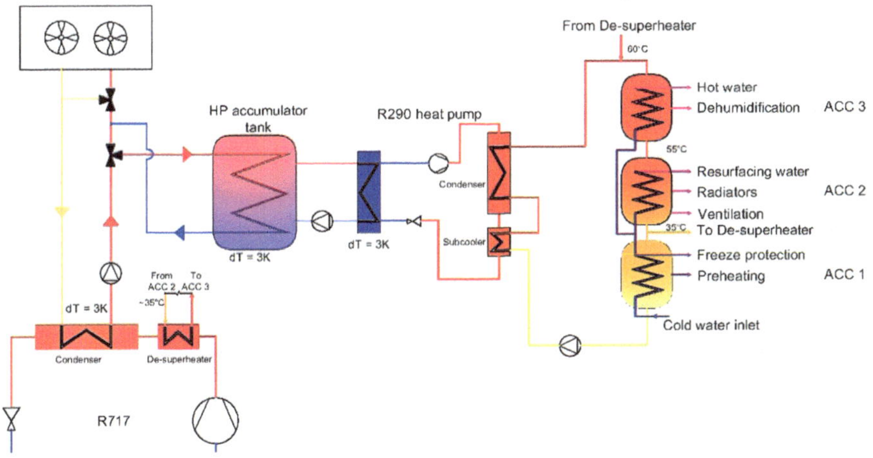

Fig. 4 Schematic for the model case

For the heat pump evaluation heating demand profile, input data is necessary, and it is deduced according to a relation between the ambient temperature and capacity from Fig. 3. The resulting difference in total supplied heating energy between reference system and the model turn out to be only 3.3%.

To understand if the heat pump actually has enough source energy, Fig. 5 is produced, revealing that there is a large portion of excess heat that has to be rejected to the ambient, even after the heat pump has extracted energy.

4.4 Global COP

To compare the CO_2 direct system to an NH_3 indirect system with R290 heat pump in terms of system performance efficiency, a Global COP is calculated.

For a CO_2 direct system, the calculation includes sum of cooling and heat recovery capacities, and this is divided with the refrigeration system compressor and auxiliary equipment power:

$$\mathrm{COP}_{\mathrm{Global\_CO_2}} = \frac{Q_{\mathrm{cooling}} + Q_{\mathrm{HR}}}{E_{\mathrm{comp}} + E_{\mathrm{aux}}} \quad (1)$$

In an NH_3 indirect system—cooling, desuperheater heating and heat pump heating capacities are as the delivered values, while refrigeration and heat pump compressor and auxiliary equipment power is used as necessary input:

$$\mathrm{COP}_{\mathrm{Global\_NH_3}} = \frac{Q_{\mathrm{cooling}} + Q_{\mathrm{HR\_ref\_sys\_desup}} + Q_{\mathrm{heating\_HP}}}{E_{\mathrm{comp\_ref\_sys}} + E_{\mathrm{aux\_ref\_sys}} + E_{\mathrm{comp\_hp}} + E_{\mathrm{aux\_hp}}} \quad (2)$$

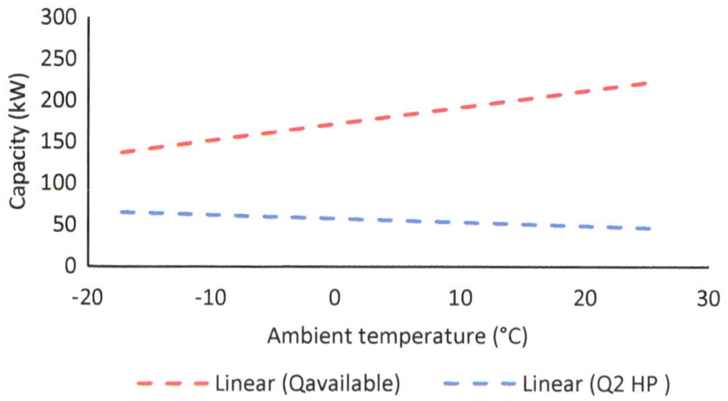

Fig. 5 Available heat from the refrigeration system and heat pump source demand capacity

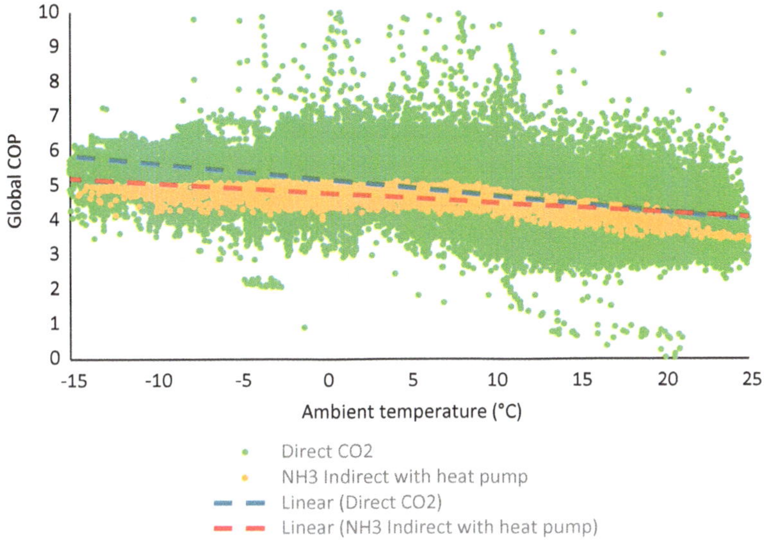

Fig. 6 Global COP depending on ambient temperature

In Fig. 6 the results of the Global COP for both systems are compiled and plotted versus the ambient temperature. Both systems have the same trend—decrease in Global COP with ambient temperature increase. The trendlines in high ambient temperature conditions are in the same level, whereas during low temperatures NH_3 system is showing lower values. The explanation for this is that in the NH_3 system condensing temperature is kept at minimum 19 °C and below 9 °C ambient temperature the COP2 of the refrigeration system becomes stable. The reason why the real data is more fluctuating is that the model assumptions are based on ideal conditions, while in practice the conditions are influenced by many external factors. However, the trendline represents the major data distribution, therefore attention should be concentrated to it.

4.5 *Energy Usage*

The resulting energy usage for the whole season in both systems is shown in Fig. 7, and it suggests that the NH_3 indirect system model predicts a 47 MWh (16%) higher energy consumption than in the existing CO_2 direct system. The reason, why the model results are divided into two parts is that these are separate systems, but providing both cooling and heating functions in the same amount of energy as it is in the reference ice rink using a single system—the CO_2-system.

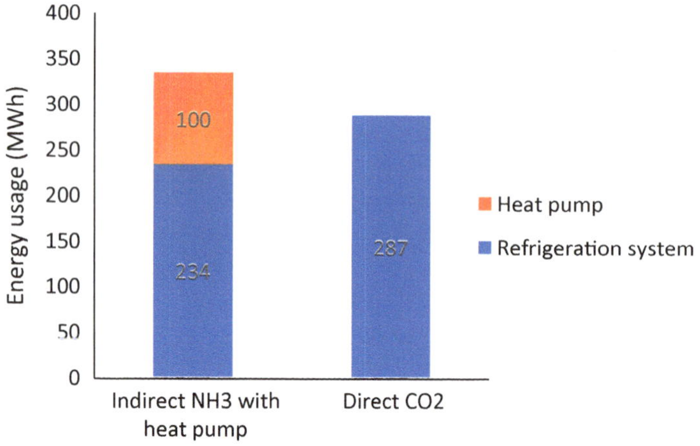

Fig. 7 The total energy consumption providing the same output functions and energy

5 Conclusions

- In this specific study results between simulation model of indirect NH_3 system in junction with propane heat pump and data from existing ice rink with CO_2 direct system suggest that the total energy usage for providing the same function and energy output would lead to 16% lower energy usage in favor to the CO_2 system.
- The general conclusion is that indirect system with a heat pump is also less attractive in terms of more components, more complicated controls and serious safety measures needed due to use of NH_3 and propane. On the contrary CO_2 with one refrigeration system works as a heat pump at the same time and has no hazardous effect on people's health.
- The specific study shows that NH_3 indirect system with a heat pump has a Global COP between 4.1 and 5.3, while the direct CO_2 system shows values in range between 4.0 and 6.0. These results show even better performance than many commercial heat pumps.
- Generally the modelled solution is most likely an attractive solution, when retrofitting an old NH_3 refrigeration system with a heat recovery function.

5.1 Limitations and Future Work

The simulation model in this study is compared to an ice rink in operation with a different system solution. To increase the credibility of the results, a same system arrangement with same refrigerants should be compared to a model.

The auxiliary equipment energy consumption may be calculated using a more sophisticated analysis, which in this study is calculated manually using linear correlations from practice.

The minimum condensation temperature for the refrigeration system in the simulation model was set to 19 °C. Different control strategy, i.e., lowering the heat pump evaporation temperature, thus allowing the refrigeration system to work with lower condensing temperature, might be more energy efficient solution. This possibility should be further evaluated in the future.

It should be also valuable to evaluate a heat pump with CO_2 as refrigerant, as it has proved to be a good solution for sanitary hot water heating in several studies and practical experiences.

References

1. J. Rogstam, S. Bolteau, *Ice Rink of the Future*. Älvsjö (2015)
2. J. Rogstam, C. Beaini, J. Hjert, *Stoppsladd: Energianvändning i svenska ishallar*. Stockholm (2014)
3. M. Pitarch, E. Navarro-Peris, J. Gonzálvez-Maciá, J.M. Corberán, Experimental study of a subcritical heat pump booster for sanitary hot water production using a subcooler in order to enhance the efficiency of the system with a natural refrigerant (R290). Int. J. Refrig. **73**, 226–234 (2017). https://doi.org/10.1016/j.ijrefrig.2016.08.017
4. S. Sawalha, *Carbon Dioxide in Supermarket Refrigeration* (2008)
5. J. Rogstam, S. Bolteau, G. Grönqvist, Cooling and heating ice rinks with CO_2. ASHRAE J. **57**, 48–56 (2017)
6. Y. Kolasniewski, *Evaluation and Modelling of Ice Rink Energy Usage* (2017)

Part V
District and City Energy Systems

Smart EV Charging Systems to Improve Energy Flexibility of Zero Emission Neighbourhoods

A State-of-the-Art for Norway

Åse Lekang Sørensen, Igor Sartori and Inger Andresen

Abstract The increased use of electric vehicles (EVs) calls for new and innovative solutions for charging infrastructure. At the same time, it is desirable to improve the energy flexibility of neighbourhoods. This paper presents state-of-the-art for smart EV charging systems, with focus on Norway. Norway is a leading market for EVs, with more than 110,000 EVs and 2000 charging stations. The paper describes how charging stations can interact with the energy need in buildings and neighbourhoods, local energy production and local electric and thermal energy storage. Examples of commercial smart EV charging systems are described. Smart EV charging systems have the potential to improve energy flexibility in a Zero Emission Neighbourhood (ZEN). Such EV charging systems can also interact with heating loads in neighbourhoods. Piloting of new technologies and solutions can provide more knowledge about smart EV charging systems, and how they can participate in matching energy loads in buildings and infrastructure with local electricity generation and energy storage.

Keywords Zero emission neighbourhoods · Energy flexibility · Flexible heating loads · Smart charging systems · Electric vehicles

1 Introduction

1.1 Background for the State-of-the-Art Study

The increased use of electric vehicles (EVs) calls for new and innovative solutions for charging infrastructure. At the same time, it is desirable to better match the energy needs in buildings and infrastructure, with local energy generation and energy storage.

Å.L. Sørensen (✉) · I. Sartori
SINTEF Building and Infrastructure, Oslo, Norway
e-mail: ase.sorensen@sintef.no

I. Andresen
Norwegian University of Science and Technology (NTNU, Trondheim, Norway

D. Johansson et al. (eds.), *Cold Climate HVAC 2018*,
Springer Proceedings in Energy, https://doi.org/10.1007/978-3-030-00662-4_39

An energy flexible neighbourhood manages the local energy demand, energy production and storage capacity according to local climate conditions, user needs, grid constraints and prices. Flexibility is embedded in both thermal and electric systems and in the interplay between them. Flexibility can further be made available outside the neighbourhood to the grid.

This article describes state-of the-art for smart EV charging systems, with focus on Norway. Norway is a leading market for EVs globally [1], in terms of market share. In 2016, EVs had a 29% market share in Norway [1]. By May 2017, there are more than 110,000 EVs and 2000 charging stations in the country [2, 3].

The study aims to explore if smart EV charging systems can play a role in the pilot areas of the Research Centre on Zero Emission Neighbourhoods in Smart Cities (ZEN Centre) [4]. The article is a first step towards answering the following questions: (1) How does the energy need for EV charging and charging profiles fit with the energy need and energy production in a neighbourhood? (2) How can a smart EV charging system be part of an energy management system which improves energy flexibility of a building and a neighbourhood? (3) Which goals and technologies demonstrated in smart EV charging systems today, can be relevant for the ZEN pilot areas?

The ZEN Centre is a new Centre for Environment-friendly Energy Research from 2017, funded by the Research Council of Norway and 34 partners [4]. The main objective of the ZEN Centre is to develop knowledge, competitive products and solutions that will lead to realization of sustainable neighbourhoods, that have zero emissions of greenhouse gases related to their production, operation and transformation. There are seven pilot areas in the ZEN Centre, where new technologies and systems will be implemented and evaluated. The Centre aims to speed up decarbonisation of existing and new building stock, use more renewable energy sources and create positive synergies among the building stock, energy, ICT, mobility systems and citizens.

1.2 Smart EV Charging Systems

The term "Smart" is used in a number of ways, as exemplified in this chapter and in Chap. 2.4. The added value of the smart-term varies from case to case, and it is therefore difficult to give a general definition of "Smart EV charging systems".

The "Global EV Outlook 2017" [1] describes that as the number of EVs increases, charging could have a sizeable impact on the capacity required by the grid at certain times and locations. At the same time, EVs are well suited to promote synergies with variable renewables. If charging practices strengthen demand-side management opportunities, EVs could allow a greater integration of various renewable energy sources in the power generation mix. Further, the report points out that large-scale electric car charging and demand response will require joint optimisation of the timing and duration of recharging events, the modulation of power delivered by charging outlets and may involve a reliance on bidirectional Vehicle to Grid (V2G) solutions.

The Platform for electro-mobility in the EU states that "Smart charging of electric vehicles should benefit EV owners by reducing their electricity costs in return for the enhanced grid stability and reliability" [5]. Their definition of Smart charging is [6]: *Smart charging consists of adapting EV battery charging patterns in response to market signals, such as time-variable electricity prices or incentive payments, or in response to acceptance of the consumer's bid, alone or through aggregation, to sell demand reduction/increase (grid to vehicle) or energy injection (vehicle to grid) in organised electricity markets or for internal portfolio optimisation.*

Smart charging systems can have several aims, depending on the preferences of the operator. For example, EnergyVille [7] describes three scenarios for the management of a charging process: (1) Peak shaving scenario: Charge when the grid capacity is high (off-peak), or manage the simultaneous charging of several EVs in the same street or car park, by spreading their demand over time; (2) Renewable scenario: Charge when the availability of renewable energy from sun and wind is high; and (3) Balancing scenario: Keep demand/supply balanced. In each scenario, it is guaranteed that the EV will be charged by a certain time, and to the level requested by the owner.

EV flexibility services can be defined as a power adjustment maintained from a particular moment for a certain duration at a specific location [8], characterised by: (1) the direction, (2) the power capacity, (3) the starting time, (4) the duration, and (5) the location. If EV is not V2G capable, the flexibility direction is always the same.

Table 1 describes some control strategies and goals for smart EV charging systems, often found in descriptions of such systems. The possibilities are sorted from low to high "smartness". Methods to achieve intelligent control of charging can vary.

2 Smart EV Charging Systems to Improve Energy Flexibility

2.1 Energy Demand for EV Charging and Charging Profiles

Typical power use during EV charging in a household or commercial building is shown in Table 2. In Norway, the power is typically 2.3 kW when using a household power plug and 3.6 or 7 kW when using a Type 2 connector [9]. For normal charging, Type 2 connectors (EN/IEC 62196) are recommended [10]. However, household sockets are still frequently in use, especially in households. Semi fast chargers are typically 22 kW and fast chargers 50 kW or above [9].

For fast charging, the ten most popular EVs in Norway use CHAdeMO (44–100 kW DC), Combo 2 Charging System (CCS2) (40–50 kW DC) and Tesla supercharger (135 kW DC) [2]. Fast charging stations have fixed external chargers, according to Mode 4 in the standard IEC-61851-1.

Table 1 Examples of common control strategies and goals for smart EV charging systems

Low "smartness"		**High "smartness"**
Uncontrolled EV charging	Active control of charging, by shifting EV charging in time	Building/ neighbourhood energy management incl. energy demand, production and storage
Passive control of charging, by encouraging EV owners	Load management of EV charging	Active use of stationary energy storage (batteries)
	Booking of charging services	Active use of bidirectional V2G solutions

Goals
Charging possibilities also with limited grid capacity
Efficient, practical, cost effective and reliable services for users
Enhanced grid stability and delay grid upgrades
Increasingly powered by local renewable energy sources
Empowering and engaging users
Energy efficient and climate-friendly
New business models and new companies
Secure, e.g. when it comes to fire safety and security of personal data

Table 2 Typical power use during EV charging, based on [11]

Type of charger	Voltage/Current	Power
Power plug for use in households	230 V/10 A/1-phase	2.3 kW
Households/commercial buildings	230 V/16 A/1-phase 230 V/32 A/1-phase 230 V/32 A/3-phase	3.6 kW 7 kW 12 kW
Semi fast chargers	400 V/32 A/3-phase	22 kW
Fast chargers (AC)	400 V/63 A/3-phase	43 kW
Fast chargers (DC)	500 V/>100 A	>50 kW

Examples of gross battery capacity of popular EVs in Norway are 24 or 30 kWh for Nissan Leaf, 24.2 or 35.8 kWh for Volkswagen e-Golf and 75–100 kWh for Tesla [12]. A typical charging profile for a single residential EV charging is shown in Fig. 1, from the project "Low Carbon London" [13]. The illustrated EV has gross battery capacity of 24 kWh. If using 6.6 kWh before recharging, it uses about 2.5 h at 3.7 kW when charging to full capacity.

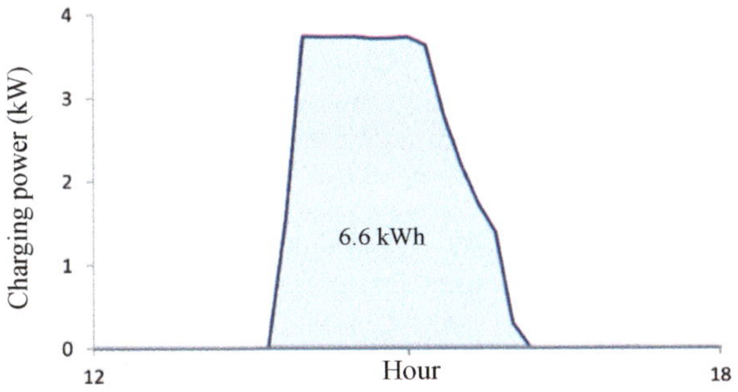

Fig. 1 Demand profile for a typical charging event, where the EV charges at 3.7 kW (i.e. 16 A) for about 2.5 h (charged to full capacity) [13]

Surveys of EV owners in Norway [14, 15] show that owners most frequently charge their vehicles at home or at work, relying on slow chargers. The third most frequent charging choice is publicly available slow chargers. The majority is charging during the night, probably at their home location, and there is also a peak during morning/mid-day, probably at their workplace. Fast charging is used to a little degree, and primarily as planned stops for long distance trips. However, this may be different for professional users of EVs, which was not investigated in detail in the surveys.

When comparing cars with different battery capacity, the charging patterns of owners with Nissan Leaf and Tesla differs. Compared to owners of Nissan Leaf, Tesla owners charge less during the day [15]. This indicates that car owners with larger battery capacity may mainly charge during the night. As the battery capacity of the EVs are increasing, one may therefore expect other EV charging profiles than today. If future EV owner are less anxious about having enough battery capacity for the evenings, even more EV owners may charge their car during the night time. However, this will also depend on other factors, such as convenience and costs.

2.2 Interactions with Neighbourhood Energy System

Beside management of the power demand of the EV charging station itself, it would be desirable if more advanced smart EV charging systems interact with the energy need in the buildings and neighbourhoods, the local energy production facilities and the capacity for local thermal and electrical energy storage.

Energy Need. Households typically have energy peaks in the morning and afternoon/evening, while non-residential buildings have peak loads during office hours [16].

Heating and domestic hot water (DHW) is a large share of the energy load in Norway, especially during the winter. Electricity is the main heating source for households, and accounted for about 73% of household heating in 2012 [17]. Many Norwegian buildings use heat pumps, including close to half of the detached houses [16]. The air-to-air heat pumps are most common. The high penetration of electric heating makes national electricity consumption very temperature dependent and high peak consumption can occur on cold winter days [18].

Some of the energy loads in a building or neighbourhood are flexible [18]. Flexible loads can be shifted in time, or regulated lower or higher. Heating loads and electric water boilers are often suitable for flexibility. Other electricity loads can also be flexible, such as the energy load of white goods, which can be shifted in time.

Energy Production. While there are several options for heating, solar cells (PV) is usually the most relevant energy technology for electricity production in a neighbourhood. PV systems generate electricity during the daytime, and mostly during the spring, summer and autumn. Usually, there are electricity demand in households and non-residential buildings during the day, which can be covered directly by PV. Still, PV systems cannot supply electricity in the evenings, when the demand usually remains high. A prosumer agreement exists in Norway, for locally produced electricity [19]. Smart meters (AMS) measure electricity export and import on an hourly basis. Financially, consumers normally receive less payment for electricity sold to the energy company than what they pay for buying electricity. This makes it beneficial to maximise self-consumption, i.e. minimising export of electricity to the grid.

Energy Storage. Both thermal energy storage and electrical energy storage can be utilized in an energy flexible neighbourhood. A DHW tank is an example of thermal energy storage capacity, which is often already available. If a DHW tank uses electricity as an energy source, this electricity use can be shifted in time or the load can be regulated. The project Linear in Belgium demonstrates how 2.4 kW DHW buffers of 200 litres can interact with EV charging [20]. The project found that the flexibility of DHW tanks remain more stable over time than the studied wet appliances. Further, the flexibility potential of DHW buffers and EVs is significantly higher than the flexibility potential of wet appliances.

In a building or neighbourhood, batteries are usually the most available technology for electricity storage. Beside stationary batteries, batteries in the EVs can be used in a V2G solution. The term V2G covers the solution were the battery in the EV delivers power back to the source. V2G takes advantage of the fact that private vehicles are parked on average around 93–96% of their lifetime [21]. In this configuration, the EVs can provide services like frequency and voltage regulation [22].

2.3 Smart EV Charging Systems to Improve Energy Flexibility

A smart EV charging system can be part of an energy management system, which improves energy flexibility of a building and a neighbourhood. Figure 2 illustrates scenarios for a neighbourhood with and without EV charging and energy management. The figure is developed as an example, with input from hourly electricity consumption and production data from a ZEN pilot area.

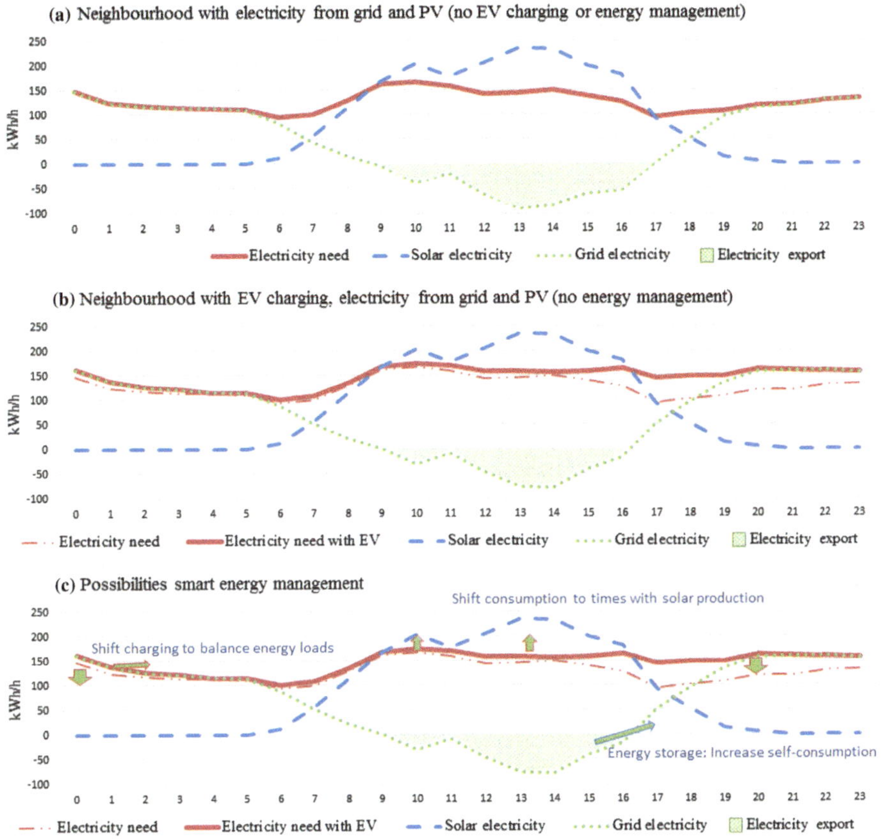

Fig. 2 Daily energy profiles for a neighbourhood without and with smart EV charging, with hourly electricity need, grid electricity, solar electricity, EV charging and battery use

The ZEN pilot area is a university campus with in total around 10,000 m^2 heated area. The electricity consumption data is from a Monday during springtime (May 30, 2016). The electricity production data is from the year after (May 30, 2017), but is up-scaled with a factor of four. After the up-scaling, the illustrated PV plant represents 280 kW_p installed power. The EV charging data is based on [23], with charging measurements from an EV pool in Trondheim (230 V, 16 A). The charging data is up-scaled with a factor of two, illustrating a situation with 30 charging outlets for EVs. In average, each charging outlet provides 15 kWh during the illustrated 24 h.

Three different example scenarios are illustrated. Figure 2a represents a neighbourhood with no EV charging or energy management. Electricity is provided by the grid and PV. Since this is a university campus, the peaks in demand matches quite well with the solar energy production; Better than what could be expected in a neighbourhood with households only. During mid-day, electricity is exported to the grid. Figure 2b represents the same neighbourhood, but with EV charging. The energy need for EVs especially increases the energy consumption in the afternoon. Scenario 2c illustrates possibilities with smart energy management. Energy loads can be shifted from the evening and early night, to reduce the energy peaks. This can be done by delaying EV charging to the night-time, and also by moving other energy loads—such as electricity to the thermal storage tank. More EVs can be charged during daytime, when there is solar electricity available. If also batteries are used in the system, solar electricity can be used in the evenings. With such an energy management system, less electricity is exported and the maximum grid loads are reduced.

2.4 Examples of Commercial Smart EV Charging Systems

Vulkan Garage in Oslo. The charging station Vulkan has the possibility of charging 100 EVs simultaneously. The system is partly financed by the EU project SEEV4-City and is implemented by Aspelin Ramm, Oslo municipality and Fortum. The partners are testing new business models, combining services for residents and professional users. Charging of 3.6 kW power is free during night-time. During daytime, dynamic charging services are available for a fee. Power from 3.6 to 22 kW is available at the charging points, which are managed individually. A battery package of 50 kWh is installed in the garage, as a buffer and to manage peak loads. The garage is prepared for V2G solutions and the next generation fast chargers of 150 kW [24, 25].

Apartment buildings in Norway. The company Zaptec has developed a solution for apartment buildings with limited grid capacity, sharing the power dynamically across several charging stations. Balancing technology makes it possible to charge over 100 EVs in 24 h, on a single 63A circuit. Up to 22 kW of power is available on every charging station. The chargers communicate through a cloud solution.

Frederiksberg Forsyning in Denmark. Ten Nissan eNV200 cars and ten Enel V2G charger stations are located at the headquarters of Frederiksberg Forsyning in Copenhagen [26]. California-based Nuvve provides a platform that controls the power flow to and from the cars. When the EVs are not in use, they can be plugged into the charger stations, receiving energy from and providing energy back to the national grid. The total capacity made available by the ten chargers amounts to about 100 kW.

3 Discussion

With the increased use of EV charging, it is important to investigate if and how smart EV charging systems can improve the energy flexibility of an area. Different situations, e.g. an apartment building, a neighbourhood or a fast charging station, will have different challenges, possibilities and goals, which will require different solutions.

For example, for an apartment building with an old grid, the main goal may be to provide charging possibilities for the residents, even though there is limited grid capacity. Load sharing between the charging stations may solve the issue, or it may be an advantage to also control some of the energy flexible loads in the building or to maximise self-consumption of electricity from PV. For a neighbourhood, the aim might be to avoid grid upgrades in the area. In cooperation with the grid company, energy flexible solutions in the neighbourhood can play an active role in the energy system. A smart EV charging station may play a role as a hub in the neighbourhood energy management system. For a fast charging station, the aim may be to reduce the costs for energy loads, by introducing a battery to the system.

Both in single buildings and in neighbourhoods, there are potential for interaction with heating loads and HVAC facilities. DHW tanks are an example of a flexible load, which can be realized within the comfort requirements of the user [20].

Batteries in the vehicles will most likely play a role in future smart charging systems, though V2G solutions. This is a new technology solution which can provide new opportunities for energy management and participation in the electricity market.

If consumers are energy flexible, this has a value for the grid companies [27]. With the introduction of smart meters (AMS), it is possible to introduce market mechanisms to increase the use of energy flexibility—as an economically attractive alternative to grid investments. As in several other EU countries, the introduction of new market mechanisms for flexibility are investigated in Norway [28]. Such a mechanism, may facilitate for building owners and neighbourhoods to play a more active role in the energy system, together with the grid company.

Piloting of new technologies and solutions will provide more knowledge about smart EV charging systems. Such solutions may therefore be installed in the pilot areas of the ZEN Centre, to better match the energy loads in buildings and infrastructure, with local electricity generation and thermal and electrical energy storage.

4 Conclusion

Increased use of EVs calls for new and innovative solutions for charging infrastructure. At the same time, it is desirable to better match the energy need in buildings and infrastructure, with energy generation and energy storage in a neighbourhood.

This study is a first step towards answering three research questions: (1) How does the energy need for EV charging and charging profiles fit with the energy need and energy production in a neighbourhood? (2) How can a Smart EV Charging System be part of an energy management system which improves energy flexibility of a building and a neighbourhood? (3) Which goals and technologies demonstrated in smart EV charging systems today, can be relevant for the ZEN pilot areas?

If energy management solutions are integrated in smart EV charging systems, such systems can improve the energy flexibility of a neighbourhood. Literature and commercial EV charging systems provide examples of smart EV charging systems, which e.g. provide dynamic load management, offer a variety of charging services and demonstrates V2G solutions. Such smart EV charging systems can also interact with heating loads and HVAC facilities in a neighbourhood.

Smart EV charging systems can play a role in the pilot areas of the ZEN Centre. Piloting of new technologies and solutions will provide more knowledge about how charging systems can play a role in the neighbourhood energy management system.

Acknowledgements This article has been written within the *Research Centre on Zero Emission Neighbourhoods in Smart Cities* (FME ZEN). The authors gratefully acknowledge the support from the ZEN partners and the Research Council of Norway.

References

1. OECD/IEA (2017) Global EV Outlook 2017
2. S. Frydenlund (2017) Historiske tall: De ti mest solgte elbilene i Norge. from http://elbil.no/historiske-tall-de-ti-mest-solgte-elbilene-i-norge/
3. NOBIL (2017) Ladestasjoner i Norge. from http://info.nobil.no/
4. NTNU (2016) The Research Centre on Zero Emission Neighbourhoods in Smart Cities. from www.ntnu.no/zen
5. Platform for Electro-Mobility (2016) Decarbonising transport through electro-mobility - Summary of Recommendations
6. Platform for Electro-Mobility (2016) Accelerating electric recharging infrastructure deployment in Europe—Position paper of the platform for electro-mobility
7. EnergyVille (2017) Smart Charging System for Electrical Vehicles. from http://www.energyville.be/en/sheet/smart-charging-system-electrical-vehicles
8. K. Knezovic (2016) Active integration of electric vehicles in the distribution network—theory, modelling and practice. PhD Thesis, DTU
9. C.H. Skotland, E. Eggum, D. Spilde (2016) Hva betyr elbiler for strømnettet? NVE Rapport nr 74-2016
10. Norwegian Electrotechnical Committee (2015) Lading av elbil – anbefalt kontakt. from https://www.nek.no/lading-av-elbil-anbefalt-kontakt/
11. Ladestasjoner.no (2017) Hvilke elbiler kan lade med hva?, from www.ladestasjoner.no/alt-om-lading-av-elbil/hvilke-elbiler-kan-lade-med-hva
12. elbil.no. (2017). Elbiler i dag. Retrieved 31.07, from elbil.no/elbil-2/elbiler-idag
13. M. Aunedi, M. Woolf, M. Bilton, G. Strbac (2014) Impact & opportunities for wide-scale EV deployment. Report B1 for the "Low Carbon London" LCNF project: Imperial College London
14. E. Figenbaum, M. Kolbenstvedt (2016) Learning from Norwegian Battery electric and Plug-in hybrid vehicle users. TØI rapport 1492/2016
15. H.T. Tveter, *Large scale transition from conventional to electric vehicles and the consequences for the security of electricity supply* (MSc-thesis NHH, Bergen, 2014)
16. T. Ericson, A. Fidje, J.E. Fonneløp, B. Langseth, I.H. Magnussen, W.W. Rode, B. Saugen (2016) Varmepumper i energisystemet - Status og muligheter. NVE Report 60-2016
17. SSB. (2014). Energibruk i husholdningene, 2012. from https://www.ssb.no/energi-og-industri/statistikker/husenergi

18. C. Dromacque, T.N. Mikkelsen, R. Grigoriou (2017) Assessing the potential of home automation in Norway—A report commissioned by NVE. NVE report nr 34-2017
19. NVE (2016) Plusskunder. from https://www.nve.no/elmarkedstilsynet-marked-og-monopol/nettjenester/nettleie/tariffer-for-produksjon/plusskunder/
20. R. D'hulst, W. Labeeuw, B. Beusen, S. Claessens, G. Deconinck, K. Vanthournout, Demand response flexibility and flexibility potential of residential smart appliances: experiences from large pilot test in Belgium. Appl. Energy **155**(2015), 79–90 (2015)
21. H. Turton, F. Moura (2008) Vehicle-to-grid systems for sustainable development: An integrated energy analysis. Technol. Forecast. Soc. Chang. **75** (2008)
22. F. Mwasilu, J.J. Justo, E.-K. Kim, T.D. Do, J.-W. Jung, Electric vehicles and smart grid interaction: A review on vehicle to grid and renewable energy sources integration. Renew. Sustain. Energy Rev. **34**(2014), 501–516 (2014)
23. H. Seljeseth, H. Taxt, T. Solvang (2013) Measurements of network impact from electric vehicles during slow and fast charging, in Conference CIRED 2013
24. Fortum Charge & Drive (2017) Her kommer 100 ladepunkter for elbil i Oslo sentrum." from www.fortum.no/pressemeldinger/kommer-100-ladepunkter-elbil-oslo-sentrum
25. Grønn Byggallianse. (2017) Presentasjoner frokostmøte: 27.april: Elbillading i kontorgarasjer og kjøpesentre. from http://byggalliansen.no/nyside/elbillading-i-kontorgarasjer-og-kjopesentre/
26. NUVVE (2017) Nuvve Operate World's First Fully Commercial Vehicle-to-Grid Hub in Denmark. from http://nuvve.com/portfolio/nissan-enel-and-nuvve-operate-worlds-first-fully-commercial-vehicle-to-grid-hub-in-denmark/
27. European Commission (2013) Incorporing demand side flexibility, in particular demand response, in electricity markets. Commission staff working document SWD 442 final
28. NVE (2015) Ny teknologi og forbrukerfleksibilitet. from www.nve.no/elmarkedstilsynet-marked-og-monopol/sluttbrukermarkedet/ny-teknologi-og-forbrukerfleksibilitet

Hydronic Heating Pavement with Low Temperature: The Effect of Pre-heating and Fluid Temperature on Anti-icing Performance

Raheb Mirzanamadi, Carl-Eric Hagentoft and Pär Johansson

Abstract A renewable method to mitigate the slippery condition on road surfaces is to use Hydronic Heating Pavement (HHP) system. The HHP system starts heating the road when the surface temperature is below both of the dew-point and the water freezing temperatures. Furthermore, in order to improve the anti-icing performance of the HHP system, it is necessary to pre-heat the road surface. The aims of this study are to evaluate the effects of: (i) pre-heating the road surface and (ii) varying the fluid temperature, when the road is pre-heated, on the anti-icing performance of the HHP system. The road surface was pre-heated by adding a temperature threshold (from 0 to 1.6 °C) to the freezing and dew-point temperatures. A two-dimensional numerical simulation model was developed using finite element method in order to calculate the annual required energy and remaining hours of the slippery conditions on the road surface. The numerical solver was validated by an analytical solution associated with an infinite region bounded internally by a pipe with a constant temperature. In order to evaluate the anti-icing performance of the HHP system, the climate data were selected from Östersund, an area in middle of Sweden with cold and long winter period. The results showed that running the HHP system by considering the temperature threshold of 0.1 °C for both freezing and dew-point temperatures led to approximately 110 h shorter slippery conditions on the road surface, compared to the conditions without pre-heating.

Keywords Hydronic heating pavement · Anti-icing · Pre-heating
Fluid temperature · Required energy · Slippery condition

R. Mirzanamadi (✉) · C.-E. Hagentoft · P. Johansson
Division of Building Technology, Department of Architecture and Civil Engineering,
Chalmers University of Technology, 412 96 Gothenburg, Sweden
e-mail: Rahebm@chalmers.se

D. Johansson et al. (eds.), *Cold Climate HVAC 2018*,
Springer Proceedings in Energy, https://doi.org/10.1007/978-3-030-00662-4_40

1 Introduction

In cold regions, winter road maintenance is a necessary service to enhance the level of traffic safety and maintain the accessibility of roads. Traditionally, salting and sanding are main methods to control the friction of winter roads. Salting is mainly used for roads with high traffic volumes when the average annual daily traffic (AADT) ranges from 2000 to 3000 and sanding mainly used for the secondary road networks with less traffic volume [1]. Salting and sanding are not environmentally friendly, i.e. they damage surrounding vegetation, cause corrosion of road infrastructures and salification of fresh water [2]. Moreover, the consumptions of sand and salt, used for winter road maintenance in Scandinavian countries, are over 600,000 and 1,700,000 ton, respectively [3]. Moreover, in the traditional method, before the need for winter road maintenance is reported and authorities start mitigating the slippery conditions, a number of ice/snow induced accidents could occur [4]. Considering the negative impacts of salting and sanding on the environment, their high amount of consumption and delay in response to the presence of the slippery condition, there is a need for alternative solutions. A renewable alternative is to use a Hydronic Heating Pavement (HHP) system. The HHP system consists of embedded pipes in the road. A fluid as thermal energy carrier circulates through the pipes. In summer, the energy is harvested from the road surface and saved in seasonal thermal energy storages. In winter, the energy is pumped back to the embedded pipes in order to mitigate the slippery conditions.

Installing hydronic system in roads is not a new technology. In 1948, the earliest system, which worked based on geothermal hot water, was installed in Oregon in USA and has been on operation for 50 years [2]. Another successful example for the HHP system is the SERSO project in Switzerland, installed on a bridge since 1994 [5]. The surface temperature of the bridge was set just above 0 °C to prevent ice formation and freezing of compacted snow. The SERSO project was annually running approximately 1000 h in summer to harvest energy and another 1000 h in winter to alleviate the slippery conditions [6].

Although there are additional prosperous examples for the HHP system, some questions are still remaining. For example, how the heating control system associated with the HHP system can affect the anti-icing performance of the road. Moreover, the earlier numerical models, used to simulate the HHP system, aimed at either keeping the road surface above 0 °C [6] or heating the road after that the ice is formed on the surface [7].

The aims of this study are to evaluate the effects of: (i) pre-heating the road and (ii) varying the fluid temperature, when the road is pre-heated, on the anti-icing performance of the HHP system. A two dimensional (2-D) numerical model was developed based on the Finite Element Method (FEM) in COMSOL Multiphysics 5.2. The numerical simulation was used to calculate the annual required energy and remaining hours of the slippery conditions on the road surface.

2 Theory

This section presents mass and heat balances of the road surface. The equations are derived from [8].

2.1 Mass Balance

It is assumed that all snowfall is immediately removed from the road surface, rainfall is drained well. Also, the sublimation and deposition are neglected. In this study, mass balance of the road surface is based on condensation and evaporation as Eq. 1:

$$\frac{dm''_{moisture}}{dt} = \dot{m}''_{con} - \dot{m}''_{evp} \tag{1}$$

$m''_{moisture}(\mathrm{kg/m^2})$ is the mass balance of water on the road surface, t (s) is the time, $\dot{m}''_{con}(\mathrm{kg/(m^2\ s)})$ is the condensation rate to the surface and $\dot{m}''_{evp}(\mathrm{kg/(m^2\ s)})$ is the evaporation rate from the surface.

2.2 Heat Balance

It is important to note that the heat flux associated with the latent heat of the snow melting is not considered in the heat balance due to the snow removal from the road surface. However, falling snow, before the snow removal is started, affects the heat balance of the road surface. Hence, the sensible heat flux of snow, q_{snow}, related to the heat capacity of snow is considered in the heat balance. As can be seen in

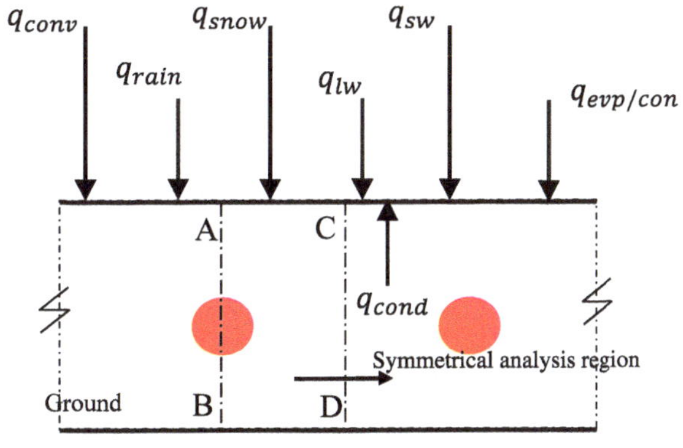

Fig. 1 Heat balance of the road surface

Fig. 1, the heat balance of the road surface consists of seven heat fluxes, including: conductive heat flow from ground and pipes, q_{cond}, convective heat flow from the ambient air, q_{conv}, sensible heat of rain, q_{rain}, sensible heat of snow, q_{snow}, long-wave radiation, q_{lw}, short-wave radiation, q_{sw}, as well as the latent heat of evaporation and condensation, $q_{evp/cond}$. The heat balance is written as:

$$q_{cond} + q_{conv} + q_{rain} + q_{snow} + q_{lw} + q_{sw} + q_{evp/con} = 0 \quad (2)$$

In Fig. 1, all of the heat fluxes directed toward the road surface, however this does not mean that they are always positive. A positive sign (+) indicates that energy is given to the surface from the ambient and a negative sign (−) indicates that energy is taken out from the surface to the ambient.

Table 1 presents the details related to each heat flux. λ (W/(m K)) is the thermal conductivity of the road materials (treated as constant by moisture and temperature variations), T (K) is the temperature, $T_{ambient}$ (K) is the ambient air temperature, $T_{surface}$ (K) is the road surface temperature, T_{sky} (K) is the sky temperature, ε (–) is the emissivity of the surface, α (–) is the solar absorptivity of the surface, $\sigma\left(\mathrm{W}/\left(\mathrm{m}^2\ \mathrm{K}^4\right)\right)$ is the Stefan-Boltzmann constant, I (W/m^2) is the solar irradiation, h_c (W/(m^2 K)) is the convective heat transfer coefficient, $h_e(\mathrm{J/kg})$ is the latent heat of evaporation of water, $\beta(\mathrm{m/s})$ is the moisture transfer coefficient, $\rho_a c_{pa}$ (J/(m^3 K)) is the volumetric heat capacity of the ambient air at atmospheric pressure, $v_{ambient}(\mathrm{kg/m^3})$ is the humidity by the volume of the ambient air, $v_s(\mathrm{kg/m^3})$ is the humidity by the volume of the saturated air at the surface temperature, $\dot{m}_{snow}$ (kg/(m^2 s)) is the snowfall rate per square meter of the surface, $\dot{m}_{rain}$ (kg/(m^2 s)) is the rainfall rate per square meter of the surface, c_{p-snow} $(\mathrm{J/(kg\ K)})$ is the heat capacity of ice crystals in snow and $c_{p-water}(\mathrm{J/(kg\ K)})$ is the heat capacity of water.

In Eq. 4, h_c is calculated as:

Table 1 Different heat fluxes of the road surface

Heat transfer process	Equation
Conductive heat	$q_{cond} = -\lambda \cdot \nabla T$ (3)
Convective heat	$q_{conv} = h_c \cdot \left(T_{ambient} - T_{surface}\right)$ (4)
Long-wave radiation	$q_{lw} = \varepsilon \cdot \sigma \cdot \left(T_{sky}^4 - T_{surface}^4\right)$ (5)
Short-wave radiation	$q_{sw} = \alpha \cdot I$ (6)
Evaporation/Condensation	$q_{evp/con} = h_e \cdot \beta.(v_{ambient} - v_s)$ where $\left(\beta = \frac{h_c}{\rho_a c_{pa}}\right)$ (7)
Sensible heat flux of snow	$q_{snow} = \dot{m}_{snow} \cdot c_{p-snow} \cdot \left(T_{ambinet} - T_{surface}\right)$ (8)
Sensible heat flux of rain	$q_{rain} = \dot{m}_{rain} \cdot c_{p-water} \cdot \left(T_{ambient} - T_{surface}\right)$ (9)

$$h_c = 6 + 4 \cdot v \quad (v \leq 5(\mathrm{m/s}))$$
$$h_c = 7.41 \cdot v^{0.78} \quad (v > 5(\mathrm{m/s})) \tag{10}$$

where $v(\mathrm{m/s})$ is the wind speed. The climate data, used in this study, present the wind speed at the height of 10 m. However, the required value for the wind speed should be closer to the road surface. Therefore, the measured wind speed at the height of 10 m is converted to wind speed at the height of 1 m above the surface. For an open and flat area, the wind speed at different heights is obtained as:

$$v_z = 0.68 \cdot v_m \cdot z^{0.17} \tag{11}$$

where $v_z(\mathrm{m/s})$ is the wind speed at the height of z (m) and v_m (m/s) is the wind speed at the height of 10 m.

3 Validation of the Numerical Solver by Analytical Solution

This section presents the results associated with the heat flow, q (W/m), in an infinite region bounded internally by a pipe with a constant temperature, obtained from an analytical solution and the numerical simulation. The analytical solution is based on Eqs. 12–17 [9].

$$R_p = \frac{1}{2\pi\lambda_{pipe}} \ln\left(\frac{r_{out}}{r_{in}}\right) \tag{12}$$

$$\frac{T_{road}(t) - T_{pipe}}{T_{road}(0) - T_{pipe}} = u\left(\frac{r}{r_{out}}, \tau, \beta\right) \quad \text{if } \frac{r}{r_{out}} > 1 \tag{13}$$

$$u = \frac{2}{\pi} \int_0^{\infty} \frac{Y_0(sr_0)[(s/\beta)J_1(s) + J_0(s)] - J_0(sr_0)[(s/\beta)Y_1(s) + Y_0(s)]}{[(s/\beta)J_1(s) + J_0(s)]^2 + [(s/\beta)Y_1(s) + Y_0(s)]^2} \cdot \frac{e^{-\tau s^2}}{s} ds \tag{14}$$

For $r/r_{Out} = 1$, u (–) will be written as:

$$u_s = \phi(\tau, \beta) = \frac{4}{\beta\pi^2} \int_0^{\infty} \frac{1}{[(s/\beta)J_1(s) + J_0(s)]^2 + [(s/\beta)Y_1(s) + Y_0(s)]^2} \cdot \frac{e^{-\tau s^2}}{s} ds \tag{15}$$

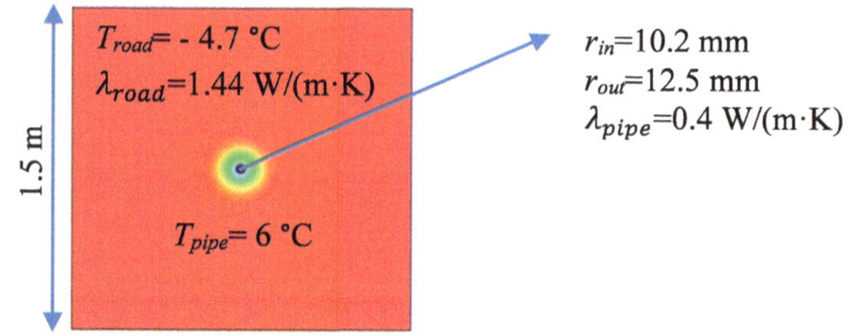

Fig. 2 Scheme of the model used for analytical solution

The heat flow, q (W/m), from the pipe becomes:

$$q(t) = 2\pi\lambda\left(T_f - T_{road}(0)\right) \cdot \beta \cdot \phi(\tau, \beta) \tag{16}$$

The integrated heat loss from the pipe, E (J/m), in time becomes:

$$E(t) = \rho c r_{0ut}^2\left(T_f - T_{road}(0)\right) \cdot 2\pi\beta \cdot \int_0^{\tau} \phi(\tau, \beta) d\tau \tag{17}$$

where J and Y are Bessel functions of the first and second kinds, $a = \frac{\lambda}{\rho c}$, $\tau = \frac{at}{r_{out}^2}$, $\beta = \frac{1}{2\pi\lambda R_p}$, $T_{road}(t)$ (K) is the temperature outside the pipe at the depth D (m) and time t (s), $T_{road}(0)$ (K) is the initial temperature of the road and pipe materials, T_{pipe} (K) is the fluid temperature, r_{out} (m) and r_{in} (m) are the outer and inner radii of the pipe, λ (W/m K) is the thermal conductivity, ρc (J/m^3 K) is the volumetric heat capacity and R_p (m K/W) is the thermal resistance of the pipe. A scheme of the model used for the analytical solution is shown in Fig. 2.

A pipe with the inner radius of 10.2 mm and outer radius of 12.5 mm is embedded in a square area, side length of which is 1.5 m. The size of the area is considered to be large enough in order to resemble the infinite region in the analytical solution. The fluid temperature is 6 °C and the road temperatures is −4.7 °C. The thermal conductivity of the pipe material is 0.4 W/(m K) and that of the road material is 1.44 W/(m K). The results associated with the heat flow, q (W/m), and energy loss, E (kWh/m) are shown in Table 2. As can be seen, the maximum relative errors related to the heat flow and energy loss between the analytical and numerical methods are 4 and 6%, respectively. The relative error for both heat flow and energy loss is less than 4% for when the running time is longer than 20 min.

Table 2 Comparison between analytical and numerical simulation relative error = (Numerical result/Analytical result-1) · 100

	Heat flow, q (W/m)			Energy, E (kWh/m)		
Time (min)	Analytical solution	Numerical solution	Relative error (%)	Analytical solution	Numerical solution	Relative error (%)
10	46.063	47.883	3.95	0.0094	0.0099	5.64
20	40.321	41.400	2.68	0.0165	0.0171	3.58
30	37.442	38.357	2.44	0.023	0.0235	2.28
40	35.593	36.333	2.08	0.0291	0.0296	1.77
50	34.259	34.815	1.62	0.0349	0.0355	1.66
60	33.231	33.668	1.32	0.0405	0.0411	1.60
120	29.769	29.871	0.34	0.0718	0.0727	1.32

4 Numerical Simulation of the HHP System

In this section, the HHP system was numerically simulated. The climate data were obtained from Östersund, an area in middle of Sweden. The area has cold and long winter periods which is interesting for the simulation of anti-icing performance of the HHP system. The climate data included: dry-bulb/air temperature (°C), relative humidity (%), wind speed at the height of 10 m above the road surface (m/s), dew-point temperature (°C), incoming long-wave radiation (W/m^2), short-wave radiation (W/m^2) and precipitation (mm/h). The numerical simulation was modeled using the implicit time-stepping method. The time-step was 1 h in line with the resolution of climate data [10]. To start the investigation, an arbitrary geometry of the HHP system was selected for the simulation. Furthermore, it was assumed that the temperature of the ground at the depth of 5 m from the road surface was equal to the annual mean temperature of the ambient air. The data related to the road and pipes, used for the numerical simulation, are presented in Tables 3 and 4.

Condensation occurs on the road surface if the surface temperature, $T_{surface}$, is lower than the dew-point temperature of ambient air, T_{dew}. If $T_{surface}$ is lower than the freezing temperature of water, $T_{freezing}$, then the moisture on the road will turn to

Table 3 Materials properties of the road layers [11]

Materials	Thickness (mm)	Thermal conductivity (W/(m K))	Density (kg/m^3)	Specific heat capacity (J/(kg K))
Wearing layer	40	2.24	2415	848
Binder layer	60	1.44	2577	822
Base layer	100	1.51	2582	894
Subbase layer	80	0.7	1700	900
Subgrade layer	1000	0.8	1400	900
Ground	3720	0.6	1300	600

Table 4 Information about the simulated HHP system

Parameter	Value	Unit
Thermal conductivity of pipe materials	0.4	W/(m K)
Density of pipe materials	925	kg/m^3
Specific heat capacity of pipe materials	2300	J/(kg K)
Outer diameter of the embedded pipes	25	mm
Pipe thickness	2.3	mm
Distance between the pipes	100	mm
Embedded depth (from center of the pipe to the surface)	87.5	mm
Emissivity of the road surface	0.89	–
Absorptivity of the road surface	0.78	–
Fluid temperature	6	°C

ice. Therefore, if $T_{surface} < 0$ °C, the HHP system is to keep the surface temperature above T_{dew} to avoid condensation. Moreover, to prevent ice formation on the road surface before the heating starts, the HHP system is turned on earlier than $T_{surface}$ is below $T_{freezing}$ and T_{dew}. In this case, even if the surface temperature is lower than the freezing temperature, ice is not formed on the road surface. The mentioned criteria for running the HHP system could be written as:

$$\begin{cases} T_{surface} < T_{dew} + \Delta T_{dew} \\ T_{surface} < T_{freezing} + \Delta T_{freezing} \end{cases} \quad \text{where} \quad \left(\Delta T_{dew} \text{ and } \Delta T_{freezing}\right) \geq 0\ ^{\circ}\mathrm{C} \tag{18}$$

here, $T_{surface}$ is the surface temperature in the middle between the pipes (point C in Fig. 1). ΔT_{dew} and $\Delta T_{freezing}$ are the temperature thresholds for pre-heating (equal to or above 0 °C) related to dew-point and freezing temperatures, respectively. Whenever the heating is started, the temperature of fluid, circulating through the pipes, is set to be 6 °C. Furthermore, whenever the heating is stopped, the boundary condition at the inner surface of the pipe walls is set to be adiabatic. It should be noted that the temperature drop of the fluid along the pipe is assumed to be negligible. By calculating the heat flow from a single pipe, $q_{pipe-heating}$ (W/m), the annual required energy for anti-icing the road surface, E_r (kWh/(m^2 year)), is calculated as:

$$E_r = \frac{1}{c} \cdot \int_{t=0}^{1year} q_{pipe-heating} \cdot dt \tag{19}$$

where c (m) is the distance between the pipes. Furthermore, the number of hours of the slippery condition on the road surface, $t_{slippery}$ (h), are calculated when the temperature of the road surface is lower than both $T_{freezing}$ and T_{dew}.

$$t_{slippery} = \int_0^{1year} f \cdot dt \left(f = 1 \; if \left(\begin{matrix} T_{surface} < T_{dew} \\ T_{surface} < T_{freezing} \end{matrix} \right) \text{else } f = 0 \right) \quad (20)$$

Moreover, in order to investigate how different values of ΔT_{dew} and $\Delta T_{freezing}$ can affect the anti-icing performance of the HHP system, the parameter of *SR* (–) is defined as the relative difference of remaining hours of slippery condition, $\frac{t_{slippery\_inital} - t_{slippery}}{t_{slippery\_inital}}(-)$ to the relative difference of annual required energy, $\frac{E_r - E_{r\_initial}}{E_{r\_initial}}(-)$, see Eq. 21. It should be noted that a higher value of *SR* means that a small increase in the annual required energy results in a large reduction in the remaining slippery hours of the road surface.

$$SR = \frac{\frac{t_{slippery\_inital} - t_{slippery}}{t_{slippery\_inital}}}{\frac{E_r - E_{r\_initial}}{E_{r\_initial}}} \quad (21)$$

5 Results

Table 5 presents the annual required energy for anti-icing the road surface and the mean heat flow from a single pipe for $\Delta T_{dew} = 0$ °C and $\Delta T_{freezing} = 0$ °C. As can be seen, 75.3 kWh/(m^2 year) energy is annually required to mitigate the slippery conditions on the road surface in Östersund. Furthermore, the mean heat flow from a single pipe for anti-icing is 587.8 W/m^2, when the system is running. Using HHP system reduces down the numbers of hours of slippery conditions by 94%, from 2009.1 to 127.6 h.

In order to investigate how pre-heating the road surface can influence the anti-icing performance of the HHP system, six different temperature thresholds for pre-heating the road surface are taken into account. The threshold are: 0, 0.1, 0.2, 0.4, 0.8 and 1.6 °C. The results related to the annual required energy for anti-icing and the remaining slippery hours on the road surface are presented in Fig. 3.

By considering $E_{r\_initial} = 75.3\text{kWh}/(\text{m}^2 \cdot \text{year})$ and $t_{slippery\_initial} = 127.6\text{h}$, the relative difference of the remaining hours of slippery condition to the relative

Table 5 Annual required energy for anti-icing and number of hours of slippery conditions on the road surface with and without heating system (Boundary conditions and geometry of the HHP system are presented in Tables 3 and 4)

Annual required energy for anti-icing the road surface, E_r $(\text{kWh}/(\text{m}^2\ \text{year}))$	Mean heat flow from a single pipe, $q_{pipe-heating}(\text{W}/\text{m}^2)$ when the system is on	Remaining slippery hours without heating the road (h)	Remaining hours of the slippery conditions using the HHP system, $t_{slippery}(\text{h})$
75.3	587.8	2009.1	127.6

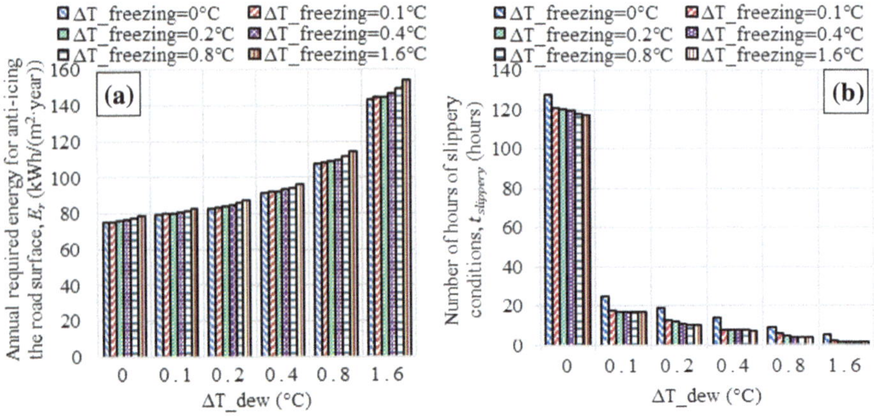

Fig. 3 The effects of pre-heating on the anti-icing performance of the HHP system **a** the annual required energy **b** the remaining slippery hours on the road surface

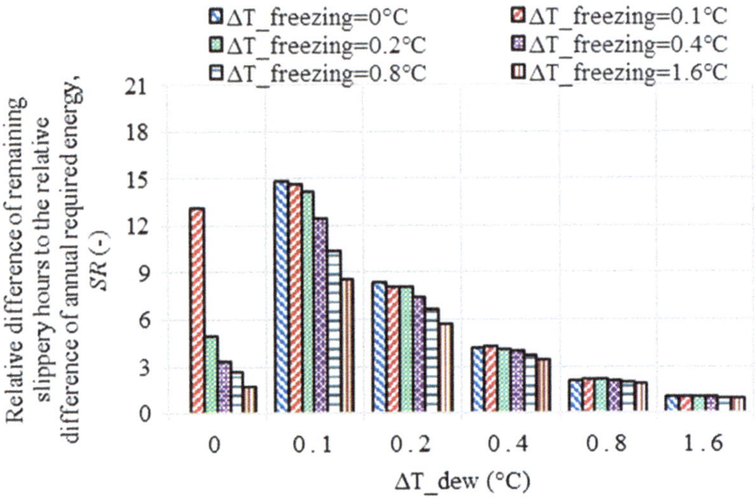

Fig. 4 Relative difference of remaining hours of slippery condition to the relative difference of annual required energy for different cases of pre-heating

difference of the annual required energy, *SR* (–), for each specific combination of ΔT_{dew} and $\Delta T_{freezing}$ will be as Fig. 4.

As can be seen from Fig. 4, the value of *SR* is more sensitive to variation of ΔT_{dew} compared to that of $\Delta T_{freezing}$. For example, keeping $\Delta T_{freezing} = 0.1$ °C constant and varying ΔT_{dew} from 0.1 to 0.2 °C results in a 45% reduction in the *SR*, while keeping $\Delta T_{dew} = 0.1$ °C constant and varying $\Delta T_{freezing}$ from 0.1 to 0.2 °C results in only a 3.5% reduction in the *SR*. It is worth mentioning that the sensitivity

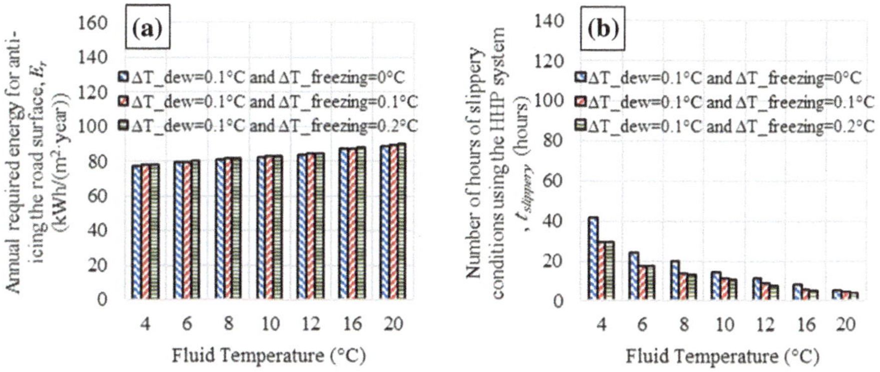

Fig. 5 The effects of varying the fluid temperature, when the road is pre-heated, on the anti-icing performance of the HHP system **a** the annual required energy **b** the remaining slippery conditions on the road surface

of *SR* to variations of ΔT_{dew} and $\Delta T_{freezing}$ reduces down by increasing their values from 0 to 1.6 °C. The minimum value of *SR* is 0.9 which is related to $\Delta T_{dew} = 1.6$ °C and $\Delta T_{freezing} = 1.6$ °C. In addition, the maximum value of *SR* is 14.8 which is related to $\Delta T_{dew} = 0.1$ °C and $\Delta T_{freezing} = 0$ °C.

The recommended fluid temperature for the HHP system is in the range of 25 to 50 °C [2]. This range of temperature is acceptable for snow melting but not for anti-icing due to high consumption of energy. In this study, seven lower fluid temperatures including: 4, 6, 8, 10, 12, 16 and 20 °C were taken into account in order to investigate the effects of fluid temperature on the anti-icing performance of the HHP system, when the road surface was pre-heated. Moreover, three pre-heating conditions which showed the higher impacts on the anti-icing performance of the HHP system were selected, including $\Delta T_{dew} = 0.1$ °C and $\Delta T_{freezing} = 0, 0.1$ and 0.2 °C, see Fig. 4. The results associated with the varying the fluid temperature, when the road is pre-heated, on the annual required energy and the remaining hours of the slippery conditions are shown in Fig. 5.

As can be seen from Fig. 5, increasing the fluid temperature causes an increase in the annual required energy and a decrease in the number of hours of the slippery conditions. In order to evaluate how pre-heating can enhance the performance of the HHP system, it is necessary to calculate the annual required energy, $E_{r\_initial}$, and the remaining slippery hours, $t_{slippery\_initial}$, for each fluid temperature for when there is no pre-heating. Figure 6a, shows the results related to the $E_{r\_initial}$ and $t_{slippery\_initial}$. As can be seen, the $E_{r\_initial}$ values are increasing from 73.4 to 84.2 kWh/(m^2 year) and the $t_{slippery\_initial}$ values are decreasing from 169.6 to 52.5 h, by increasing fluid temperature from 4 to 20 °C. Considering Eq. 21, the *SR* values for the seven fluid temperatures when the road is pre-heated by three different conditions will be as Fig. 6b.

As it is shown in Fig. 6b, the value of *SR* does not follow any special trend when the fluid temperature is varying. Considering all three pre-heating conditions,

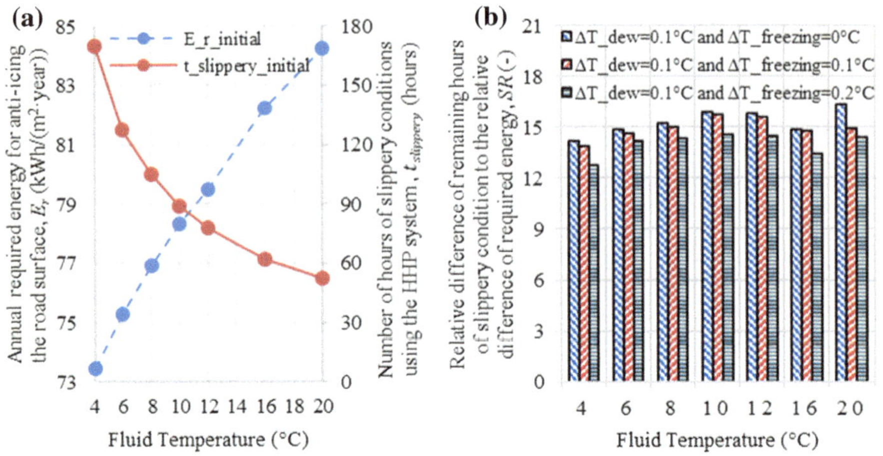

Fig. 6 The effects of fluid temperature on the anti-icing performance of the HHP system **a** the annual required energy and the remaining slippery hours without pre-heating, **b** the relative difference of remaining slippery hours to the relative difference of annual required energy with pre-heating

increasing the fluid temperature from 4 to 10 °C results in a 13% increase in *SR*. However, increasing the temperature from 10 to 16 °C results in a 7% decrease in *SR*. Also, changing the fluid temperature from 16 to 20 °C leads to a 6% increase in *SR*. The minimum value of *SR* is 12.8, the maximum value is 16.4 and the mean value is 14.8. Comparing Figs. 4 and Fig. 6b, the value of *SR* is not very sensitive to the variations of fluid temperatures, compared to the variations of pre-heating conditions.

6 Conclusion

When the road surface temperature is lower than both of the dew-point and the water freezing temperatures, the HHP system is turned on to mitigate the slippery conditions. Furthermore, in order to achieve the highest anti-icing performance of the HHP system, it is of importance to pre-heat the road surface. In this study, the effects of pre-heating the road surface on the anti-icing performance of the HHP system were investigated. The road surface was pre-heated by adding a temperature threshold (from 0 to 1.6 °C) to the freezing and dew-point temperatures; i.e. the heating system was turned on before the surface temperature was lower than both of the dew-point and the water freezing temperatures. A 2-D numerical model was developed, using the FEM, to calculate the annual required energy and the remaining slippery hours of the road surface. The results of the numerical simulation showed that pre-heating the road surface by adding the temperature threshold to the dew-point temperature led to higher reduction in the remaining slippery hours, compared to adding that to the freezing temperature. For example, pre-heating the road by adding 0.1 °C to the dew-point temperature led to 86 h shorter slippery conditions, while pre-heating the road by adding 0.1 °C to the freezing temperature led to only 4 h shorter slippery conditions. Furthermore, the effects of varying the fluid temperature were investigated on the anti-icing performance of the HHP system, when the road was pre-heated. The results showed that pre-heating the road improved the anti-icing performance of the HHP system, regardless of the fluid temperatures. This study was done for a 2-D cross section of road and by assuming that the fluid temperature decline along the pipe is negligible. Further investigations are needed to find out how pre-heating can affect the required energy and the remaining slippery hours on the road surface if there is a temperature decline through the pipe.

Acknowledgements The work was supported by the Norwegian Public Road Administration and Chalmers University of Technology.

References

1. H. Norem, Selection of strategies for winter maintenance of roads based on climatic parameters. J. Cold Reg. Eng. **23**, 113–135 (2009). https://doi.org/10.1061/(ASCE)0887-381X(2009)23:4(113)
2. P. Pan, S.P. Wu, Y. Xiao, G. Liu, A review on hydronic asphalt pavement for energy harvesting and snow melting. Renew. Sustain. Energy Rev. **48**, 624–634 (2015). https://doi.org/10.1016/j.rser.2015.04.029
3. F. Knudsen, K. Natanaelsson, A. Arvidsson, O. Kärki, Á. Jacobsen, G. Guðmundsson, B. Nonstad, K.M. Reitan, Vintertjeneste i de Nordiske land. Statusrapport 2014 (in Norwegian), Norge, 2014
4. T. Liu, N. Wang, H. Yu, J. Basara, Y. (Eric) Hong, S. Bukkapatnam, Black Ice Detection and Road Closure Control System for Oklahoma, (2014) 74. http://www.okladot.state.ok.us/research/FinalRep_2249_FHWA-OK-14-08.pdf
5. W.J. Eugster, Road and bridge heating using geothermal energy. Overview and Examples, in: Proceedings on European Geothermal Congress, Unterhaching, Germany 30 May-1 June, 2007
6. D. Pahud, *Simulation tool for the system design of bridge heatig for ice prevention with solar heat stored in a seasonal ground duct store* (User Manual, Lugano, Switzerland, 2008)
7. M. Abbasi, *Non-skid winter road, investigation of deicing system by considering different road profiles* (Chalmers University of Technology, Gothenburg, Sweden, 2013)
8. C.-E. Hagentoft, *Introduction to building physics, 1:7* (Studentlitteratur, Lund, Sweden, 2001)
9. J.C. Jaeger, T. Chamalaun, Heat flow in an infinite solid bounded internally by a cylinder. Aust. J. Phys. **19**, 475–488 (1966)
10. Meteotest, Meteonorm: meteonorm, global meteorological database. Handbook part II: Theory, version 6.1, Bern, Switzerland, 2010
11. R. Mirzanamadi, *Thesis for the degree of licentiate of engineering ice free roads using hydronic heating pavement with low temperature* (Chalmers University of Technology, Gothenburg, Sweden, 2017)

Life City—A Climate-Conscious Concept for Smart and Sustainable Built Environment

Jussi Rönty, Paula Ala-Kotila and Riikka Holopainen

Abstract Residential construction and land use planning are changing in many countries around the world. There is a clear need for comprehensive, energy-efficient, cost-effective and CO_2-neutral concepts meeting the latest energy and climate policy goals. This opens up new opportunities for the industry. Northern Finland has strong expertise in the field of wooden architecture and wood construction industry. By combining the existing high-tech expertise of enterprises, and strengthening the cooperation between SMEs, new value networks with significant international business potential can be created. This paper presents the Life City-concept, a holistic approach for smart and sustainable built environment, which is being created in the ongoing Life City-project. The concept consists of energy-efficient and modern wooden buildings in CO_2-neutral neighborhoods and includes decentralized energy infrastructure with multiple renewable energy sources and smart networks. Life City-concept is a comprehensive export concept that adapts to the local climate conditions, but also considers the social needs of different communities. Other targets of the Life City-project include enhancing the export business knowledge and technical prerequisites of the collaborating SME companies. In cooperation with the companies, the project aims to develop new design, manufacturing and construction methods that enable improving the cost-effectiveness, environmental impact, energy-economy, microclimate and other living comfort issues of wooden buildings in different climate zones, taking also the climate change and resiliency issues into account. Preliminary project results show that the wood construction industry in Finland is not quite ready for exporting holistic concepts such as the Life City-concept, as more product development, knowledge and resources for export are needed. This finding applies especially to SME companies in the industry.

Keywords Sustainable smart cities · Smart energy solutions · Wooden-town planning · Modern wood construction

J. Rönty (✉) · P. Ala-Kotila · R. Holopainen
VTT Technical Research Centre of Finland Ltd, P.O. Box 1000, FI-02044 VTT Espoo, Finland
e-mail: jussi.ronty@vtt.fi

D. Johansson et al. (eds.), *Cold Climate HVAC 2018*,
Springer Proceedings in Energy, https://doi.org/10.1007/978-3-030-00662-4_41

1 Introduction

Northern Ostrobothnia and the surrounding areas in Northern Finland have strong expertise in the field of wooden architecture and wooden construction industry. However, the development of small-scale industry into international business should be promoted through research and innovations. Residential construction and land use planning are changing in Finland and in many foreign countries. This will open up new opportunities for the industry.

In urban and regional construction areas, there is a clear need for a comprehensive concept, which is energy-efficient and cost effective as well as CO_2-neutral, and which meets the latest energy and climate policy goals. By combining the existing high-tech knowhow of enterprises in Northern Finland, and strengthening the cooperation between SMEs, it is possible to create new value networks that have new business potential. This could also create new enterprises and jobs in the wood product and biomass processing value chain.

1.1 Life City-Project

Life City-project (2015–17) is a research and development project mainly funded by the European Regional Development Fund, coordinated by VTT Technical Research Centre of Finland Ltd, and implemented in cooperation with the University of Oulu, and collaborating companies. The main objective of the project is to develop the Life City-concept, but also to provide knowledge for collaborating companies about sustainable city planning based on modern energy-efficient wooden buildings, and decentralized energy infrastructure with the integration of multiple renewable energy sources and smart networks.

Other objectives include enhancing the export business knowledge and technical prerequisites of the collaborating SME companies mainly in northern Finland. In cooperation with the collaborating companies, the project aims to study and develop new design, manufacturing and construction methods with which it is possible to improve cost-effectiveness, environmental impact, energy-economy, microclimate and other living comfort issues of wooden buildings in different climate zones, taking also the issues of climate change and resiliency into account. These methods include 3D building information modeling integrated more efficiently with manufacturing automation systems (CAD-CAM), and the development of the overall production management process "from forest to cities".

As a result, the project aims to create a CO_2-neutral concept for built environment, consisting of modern wooden buildings in smart and sustainable neighborhoods. Life City is an export concept that adapts to the local climate conditions, but in addition to technical matters also considers the social needs of different communities. It includes both public services (community centers, kindergartens, schools, health centers) and residential buildings (multi-storey apartments - and single-family houses).

Life City as a concept bases on the recent theories on sustainable buildings and communities, and ecologically designed cities with smart functions included. The following chapter briefly describes some of the theories and concepts.

2 Review on the Theories Behind Life City

2.1 Sustainable Communities and EcoCities

The definition of sustainability has been almost the same for the last 200–300 years. However, the subject matter and geographical scope of application have increased during that time. The modern concept of sustainability has three dimensions: economic, social, and ecological. A worldwide implementation is implied repeatedly. The ultimate question is how to reconcile conflicting interests. [1]

An EcoCity can be considered as a community that combines the ecological way of living, resource-wisdom in everyday life, localized economy with self-sufficient systems, and perhaps the utilization of advanced technology. There are several different ways and philosophies in designing an EcoCity, depending on the local conditions such as culture, economic realities, climate and geographical location including natural resources. There have been attempts to create EcoCities, but so far, the level of developments have been mainly just concept creation and the design optimization of different sectors or technologies. Creating an EcoCity is always a comprehensive effort, and technology is just an enabler in this ambitious target [2, 3] (Fig. 1).

An EcoCity is essentially a community built with the principle of living sustainably in a symbiosis with the environment. Most EcoCities strive to eliminate all toxic waste and CO_2-emissions, and to produce only renewable energy for its needs. EcoCities typically endorse local economic growth, localized value chains and services, sustainable agriculture and tourism, ecological traffic solutions such as bicycles and electric vehicles, environmental well-being, reducing poverty, and improving the health of its citizens [2, 4].

VTT's EcoCities approach aims to respond these challenging targets. The main principles of the approach can be described as follows [5]:

- Best combination of technologies and services that form sustainable solutions providing the users and inhabitants a high quality of life and indoor and outdoor comfort.
- Applicable EcoCity solutions depend on local conditions and need to be customized to socioeconomic realities.
- There is no one universal solution that fits all, but a number of possibilities that need to be studied to find the right solutions for each case.

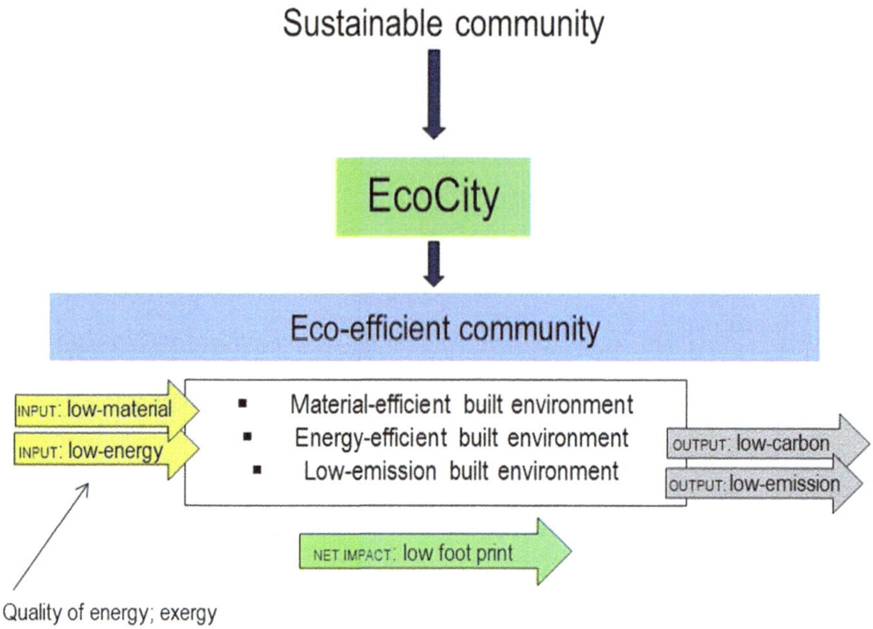

Fig. 1 The key elements of eco-efficiency are material- and energy-efficiency and low environmental emissions [2]

- Requires collaboration with local partners to gain knowledge of local traditions, perceptions, available materials and competent partners.

2.2 *Smart Cities*

There is no single definition for sustainable smart built environment, or *smart cities*. Smart city as a holistic concept is quite controversial and under debate. Something that is smart for some people or stakeholders might be unnecessary or even adverse for others. These are often political questions. However, there are some established definitions on the subject.

A *smart city* invests in human and social capital, and utilize modern information based technologies to gain sustainable economic development and a high quality of life for its citizens [6]. Smart cities use natural resources wisely, endorse citizen participation in decision-making, and strive for healthy and safe living environment with the help of novel technologies and processes.

According e.g. to [6] smart cities can be identified (and ranked) along six main characteristics or dimensions:

- Economy
- Mobility

- Environment
- People
- Living
- Governance

Smart cities combine cutting-edge technologies (such as IoT, sensor networks, data analytics and artificial intelligence) with social enhancements (such as investments in education) in order to reduce the environmental impact, reduce costs, make cities more socially participatory, and offer citizens better lives in many ways. Smart cities aim to bring everyone together for the common cause: city officials, innovative suppliers, national and EU policymakers, academics and civil society [7].

3 Life City Concept

3.1 Basis for the Concept Development

The idea in the beginning of the Life City-concept was to create a Finnish-Japanese network of experts who could design an ecological smart city concept for tsunami-damaged areas in Japan, after the great East Japan earthquake disaster in 2011. Rebuilding in the area was ineffectively organized; solutions were temporary and energy consuming. Part of the area is still today unbuilt as the reconstruction is delayed. There was a serious need to develop the area, which was regressive even before the disaster. Communities need to solve reconstruction problems, authorities and business life need innovative solutions, which can act as a driver for sustainable development.

In the first stage, it was necessary to strengthen the existing relations and co-operation with Japanese and Finnish companies, and local authorities to find possibilities to design a concept for smart and eco-efficient district, and possibly even realize concrete pilot projects. If successful, this could open new markets for Finnish building industry that focuses on ecological solutions.

Life City-concept is an inspiring and healthy way of life. It intends to give an opportunity to live ecological and healthy life easily and without making difficult and complicated choices continuously. In large-scale it also gives the model for the idea that economic growth and eco-efficiency is possible at the same time.

Figure 2 presents the *corner stones* of the Life City-concept. Technical development areas include ecological district planning, low-carbon energy infrastructure, ecological building solutions and human-centric smart community services. With these technical developments, the target is to increase sustainability, smartness, safety, wellbeing and social inspirations in Life City communities.

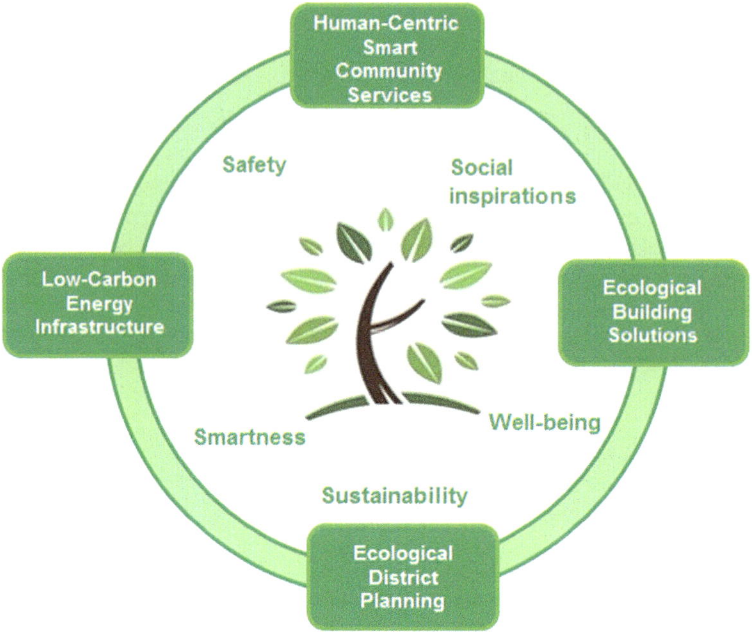

Fig. 2 Life City concept's value propositions and technical development areas

3.2 *Livable City with Ecological Buildings and Smart Solutions*

Buildings are an essential element of city infrastructure, and naturally in Life City-concept as well. In the near future, buildings will be connected seamlessly to their surroundings; other buildings and infrastructure, and especially to the energy networks. This will enable more flexible and sustainable ways to use energy, while reducing overall energy consumption and emissions. IoT-solutions will enable different components and systems in- and outside of the building interact and share information. This will help ensuring the quality of indoor environments, and creating new services and other business opportunities. Buildings in the Life City concept integrate ecological wood-based materials, smart energy production and monitoring technologies and efficient project delivery processes to create greener, safer, more comfortable and productive facilities for the occupants. It is evidently very important to verify the holistic performance of the building for technical performance but also for user experience.

Considering the material and building technology, the concept is based on modular construction units made of massive wood products (i.e. logs or CLT), which are cost-effectively prefabricated in factories. This also guarantees a secure and continuous dry chain. Solar panels, IoT sensors, modern HVAC equipment and other smart solutions can be installed already in the factory, or they can be preinstalled in separate units. Complete modular units are transferred to the construction site, where different sized buildings can be compiled from *plug-and-play* units. The method is speeding up the construction site process significantly, which produces cost saving and secures that the structures stay dry.

The concept bases on sustainable construction methods utilizing wood as much as possible, because massive wood construction has proven to be the best solution for adverse climate conditions, adding resiliency, and can be applied in all climate areas.

Life cycle thinking is another key element in the concept. The design of a *Life City* includes comprehensive life-cycle evaluations, assessments and optimized life-cycle costs, combined with performance monitoring of buildings and infrastructure. VTT has developed tools and methods for extensive monitoring that are ready to be used in Life City area developments.

Life City is an export concept that adapts to the local climate conditions, but also considers the social needs of different communities. It includes both public services (community centers, kindergartens, schools, health centers) and residential buildings (multi-storey apartments - and single-family houses). To ensure the involvement of the people at stake, different working- and design methods have been discussed during the project. These methods give citizens and other stakeholders the opportunity to take part in the development process (co-creation with e.g. charrette method). A summarized description of the overall vision and development process of the concept is presented below in Fig. 3.

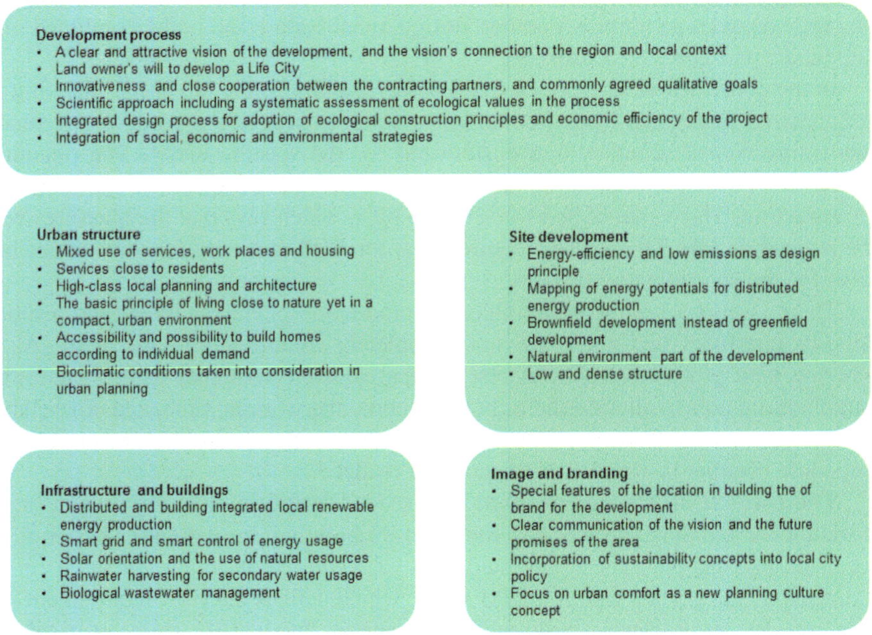

Fig. 3 Overall vision and development process of the Life City concept

3.3 Developing Modern Techniques for Wood Product Manufacturing

Several operators in the field of wood product manufacturing industry have a challenge to find enough skilled designers to use their own products and design tools. In addition, the processes do not allow the full utilization of the products. Late and incomplete plans will cause production errors, downtime and extra waste of time.

3D-model design and Building Information Modeling (BIM) have been found to give invariable cost savings in complex projects. In particular, the savings are realized in the decrease in errors. Benefits of 3D BIM stand out when value added rises (e.g. engineered wood products) and specialty products are used.

The aim is to create—in cooperation with the participating companies—a common concept, which utilizes the strengths of all participants. The conceptualization is developed for an actual construction project, and it combines the strengths of each company, their products and expertise. The aim is to provide a cost-effective and high-quality product package, which primarily concentrates on the production possibilities of a wooden multi-storey building. An important aim is to assure the suitability also for the overseas market.

The process of BIM-based implementation for wooden multi-storey buildings. The objective is to define the needed processes for designing, ordering, manufacturing and executing the wooden multi-storey building and an exemplary schedule. The concept does not only consider the "technical" performance, but aims to create a guideline for how to order, design, manufacture and build, combined into a genuine and controlled package.

In the construction of wooden multi-storey buildings, the advanced planning is emphasized more than in traditional construction, because the products are made with dimensional accuracy and customized for the specific entity. The planning material needs to be ready earlier and it needs to be more detailed than in traditional construction. There cannot be even a single plan, which says that the measures will be revised on site. This causes requirements for BIM, not only for the content but also for the schedule and process.

There is also a new working phase in production planning, where the responsibility interfaces and roles are not completely established in the field. As a pre-specified process, all the interest groups from the customer to designers and suppliers are able to allocate the right production and working time, and are capable of setting the price correctly. Therefore, extra redesigning is avoided and significant schedule benefits in wood construction are realized.

The content of implementation. The needed technical data and project organization for building the wooden multi-storey building are defined. The required decision-making points are presented as a schedule frame connected to designed contents. An important task is a BIM study, which defines the BIM requirements of the wooden multi-storey building in a general level. This study particularly expresses needs about the contents of the adequate and efficient design made in different stages. In addition to design, the role of the collision detection and quality control is viewed as a part of the process. The study does not take a stand between different software programs, but it defines which properties are needed to the implementation in different stages. The study produces different road maps for the overall process, which could be developed to a *ready-to-use* concept for wooden multi-storey buildings in the international level.

In addition to the frame, the goal is also to provide expertise in the design and procurement process, as well as the opportunity to create new business models to subscribe to implementation. The results of the BIM study are:

- Presentation of the design process for a typical project (tools, schedules, created material, responsibilities)
- The definition of the BIM requirements of the wooden multi-storey buildings
- Process description for the entire project, from design, production and construction.

4 Export Knowledge and Skills

One of the main objectives of the Life City-project was to provide market information and knowledge about export business for the companies involved in the project. Life City-concept is designed to be an export product in which the involved companies participate to the process by including their own products and expertise to the concept. The comprehensive export product should be developed together by effective and close co-operation.

Life City-project's one work package concentrated on collecting needed market information by doing market research, creating contacts with several partners in selected countries and making a questionnaire for Finnish SME companies in wood industry sector. This information helps the companies to develop their own export business and products. It also gives the researchers important information about the current situation of The Finnish wood industry and the direction where it should be developed in the future.

The market research was focused on wood product and wood construction markets in Japan, Germany, Sweden and Norway. It mainly examined lumber and massive wood products such as log and CLT structures.

A web-based questionnaire was sent to 150 Finnish SME companies in the field of wood products, manufacturing of massive wood elements, and wooden house construction. The response rate was 14%. Of these respondents, most have had more than 10 years of experience in wood industry and all have less than 250 employees. The export business of the companies is mainly targeted to Europe and Japan and concentrated in building timber products, including log and CLT.

Most of these respondents wanted to increase their export business and those who did not, were lacking the suitable products for that. However, all of the respondents were missing resources for exporting. The biggest need was for skilled sales, marketing and exportation specialist and time. Knowledge of the foreign economic development, political atmosphere, procedures and products of the competitor countries were biggest barriers for exportation. In addition, products and process needed to be developed. Development for the degree of processing, delivery cycle, route network automation, product specialization and variation depending on the export country were seen as important.

Previous projects for developing export in wood construction area were not found very effective as they are now. Some of the respondents required projects to be built around their business and less concentrated on building networks. However, at the same time, more than half of the respondents appreciated the development of co operation and networking between companies. The tight eco nomic situation in Finland has sharpen the competition and complicated co-operating and real networking. Respondents felt that training the export sales experts, cutting down the bureaucracy of the financing the exportation, developing the co-operation networks for small businesses and establishing the professional contact person in target countries create the base of the good export business.

Public financial support for wood product export was mostly highly appreciated; especially export start-up support to small businesses for projects and training.

In general, there is a lot of talk about wood product export and developing export businesses, but the opinion about developing and boosting first the wood construction inside Finland was strongly emphasized in the responds for the questionnaire as well as putting efforts in further processing of wood products.

5 Conclusions

The Life City-concept is not a quick solution for single buildings, but a comprehensive development method for districts, areas, blocks, quarters, neighborhoods or entire cities, according to the needs of the community. The extent, methods and applications are customized as the concept adapts to the local climate, culture, buildings codes, and other conditions in different countries around the world. It contains the pre-planning, design work and alternative suggestions for building structures, materials and energy systems, performance monitoring and even traffic arrangements, adding resiliency to the built environment towards adverse climate events. The solutions aim to increase ecological way of living that is also affordable, comfortable and healthy.

The research results so far show that the wood product and construction industry in Finland is not quite ready for exporting holistic concepts of this magnitude, as it needs more product development and knowledge, and most of all resources for export. This finding applies especially to SME companies in the industry. There is a need in Finland to develop co-operation inside the wood industry and between supporting industries. Export business needs more experience and training, and a clear orientation to meet the customers' needs. The biggest concerns in companies are:

- lack of resources for export
- products are not suitable for exporting
- lack of skilled sales, marketing and exportation specialist and time
- not enough knowledge about the foreign economic development, political atmosphere, procedures and products of the competitor countries
- undeveloped export process: for the degree of processing, delivery cycle, route network automation, product specialization and variation depending on the export country
- tight economic situation, which makes co-operation and networking complex
- bureaucracy of the exportation financing.

One concrete project result is the process description of BIM-based implementation for wooden multi-storey buildings. The development work produces guidelines and knowledge on the needed processes for designing, ordering, manufacturing and constructing wooden multi-storey buildings. The results of the BIM study include at least:

- Presentation of the design process for a typical project (tools, schedules, created materials, responsibilities)
- The definition of the BIM requirements of the wooden multi-storey buildings
- Process description for the entire project, from design, production and construction.

As a summary, the project was challenging from the start. It was realized during the course of the project, that it is quite impossible to reach all the ambitious targets, and this type of comprehensive concept is never completely ready. The concept clearly needs to be developed further in future projects. The next step could be the realization of a concrete pilot project in which the products and expertise of participating Finnish companies could be utilized together in co-operation with local stakeholders.

References

1. S. Vehkamäki, *The concept of sustainability in modern times* (Department of Economics and management, University of Helsinki, Helsinki, 2005)
2. J. Nieminen, P. Lahti, A. Nikkanen, U-M. Mroueh, T. Tukiainen, J. Shemeikka, P. Huovila, S. Pulakka, C. Guangyu, S. Nan, K. Lylykangas, Miaofeng mountain town EcoCity. VTT & Mentougou Science and Technology Office. 259 p. (2010)
3. P. Tuominen, C. Antuña Rozado, Å. Hedman, J. Jantunen, T. Pajula, A. Abdel Monteleb, S. Abd Elhafez, O. Balbaa, W. Dawoud, B. Elboshy, Y. ElMahgary, A. Elkafoury, G. Elshafei, A. ElShazly, M. GamalEldin, A. Kamel, M. Kamel, A. Negm, H. Saeed, A. Samy, B. Salem, M. Shahin, A. El-Nashar, A. Yousry. EcoCity roadmap for Egypt. Actions for eco-efficient urban development. VTT Technology: 215, 2015. VTT, Espoo, 36 p., ISBN 978-951-38-8241-9 (2015)
4. E. Paloheimo, Ecocity? What is an ecocity? Ecocities Emerging, Dec. 2008

5. P. Huovila, C. Antuña Rozado, EcoCities, VTT's concept for sustainable community and neighbourhood regeneration and development. 2nd World Sustainability Forum. 1–30 November 2012
6. A. Caragliu, C. Del Bo, P. Nijkamp, Smart cities in Europe. Serie Research Memoranda 0048, VU University Amsterdam, Faculty of Economics, Business Administration and Econometrics (2009)
7. Smart Cities in EU, http://www.eu-smartcities.eu/faqs#Smart_Cities. Accessed 31 Aug 2017

Energy Pathways for Future Residential Building Areas in Norway

Natasa Nord, Ola Skrautvol, Stian Fossmo Eliassen and Tymofii Tereshchenko

Abstract Due to stricter building energy requirements, future buildings will be characterized with low base loads and occasional high peaks. However, future building areas will still contain existing and historical buildings with high energy use. Additionally, there is a requirement that future building areas have to get energy from renewable energy sources, while existing buildings need to make transition to the renewables. The aim was to analyze different energy systems and technologies that can help to reduce CO_2 emissions in the future building areas in Norway. In this study, different methods were combined: detailed building simulation, energy supply technology simulation, heat demand aggregation, and data post-processing. The results showed that the energy pathways would be very dependent on the CO_2-factors for the energy sources and it is hard to tell which CO_2-factor is correct. An increasing housing stock development would slightly increase the CO_2 emission towards 2050, even though the new buildings used half the energy than the existing buildings and the existing buildings undergone energy efficiency improvements. A constant housing stock would decrease the CO_2 emission by around 22–27% depending on energy supply systems. The results showed that the influence of implementing stricter building codes had a lower impact on the total CO_2 emissions, compared to the influence of the CO_2-factors and energy supply technologies. Regarding the existing buildings, the requirements such as: limited use of direct electric heating, requirements on the service systems, and definition on hot tap water use should be emphasized.

Keywords Energy planning · Building stock · Residential buildings
Energy supply · Building requirements

N. Nord (✉) · T. Tereshchenko
Department of Energy and Process Engineering, Norwegian University of Science and Technology, NO-7491 Trondheim, Norway
e-mail: natasa.nord@ntnu.no

O. Skrautvol
Erichsen & Horgen AS, Oslo, Norway

S. F. Eliassen
Seoul National University, Seoul, South Korea

D. Johansson et al. (eds.), *Cold Climate HVAC 2018*,
Springer Proceedings in Energy, https://doi.org/10.1007/978-3-030-00662-4_42

1 Introduction

Energy planning of the building stock is a highly important topic and highly relevant for energy policy, requirements, and standards development, see Fig. 1. Regardless of its importance, this research topic is still in its infancy. Different tools have been developed, but they have only partially succeeded due to sectional and particular interests when developing the tools [1, 2]. There are several reasons for this such as a fragmented building industry, complex building ownership, divergent interests regarding energy use and supply.

Currently, there are only a few transparent tools internationally with limitations for developing building energy requirements, which links building stock characteristics to aggregate energy demand and the associated greenhouse gas emissions. For example, the discussion on the last Norwegian building technical requirement, TEK17, showed this problem when the requirement on energy flexibility for building energy supply was removed, due to government promise to enable cheaper housing. The requirement on energy flexibility that implied possibility for the water based heating and connection to the district heating was diminished: (1) by decreasing the share of energy that should be delivered by the flexible solutions, and (2) by increasing the building area with the energy supply requirement. All these meant that requirement for the implementation of the renewable sources was weakened. This example from Norway showed that the development of building energy requirements was not performed neither by considering energy analysis of the building stock nor estimation of the effect of the decisions on the future energy demand of the building stock. Current building energy requirements do not guarantee that buildings and building stocks will achieve global environmental requirements, because there are few organizational and institutional structures to support these changes and no sufficiently useful and transparent energy planning methodology [3]. Hence, there is a lack of tools for energy planning on different levels for different decision makers. To achieve global emission requirements, policy developers, energy planners, energy supply companies, and city planners

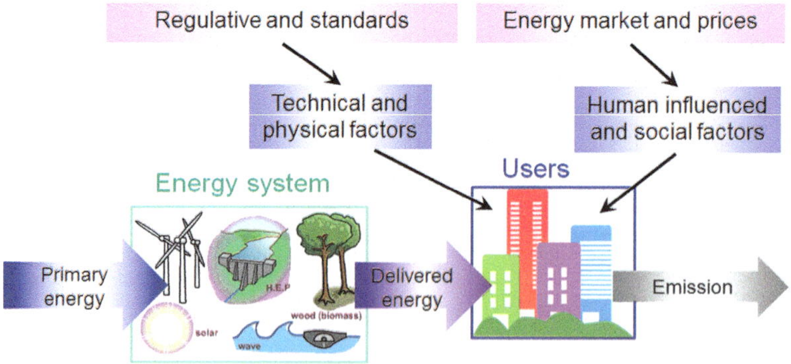

Fig. 1 Big picture of building energy planning

need a transparent tool and methodology to analyze the possibilities, so that their requirements and ideas would provide a renewable society with very low emissions.

Different methods are suggested to model energy use and emissions of the building stocks. The extrapolation method considering the occupant behavior is done by many researchers in Japan. Energy Solar Planning (ESP) tool is a simple tool for municipalities' district planning based on a steady-state monthly energy balance method [1]. Monte Carlo simulation has been used to predict space heating energy use of housing stocks in a bottom-up approach [4], but the model shows to have uncertainties in prediction. At present the only software to deal with this topic is the Sustainable Urban Neighborhood modeling tool (SUNtool) and its recent successor CitySim. Even though these tools are dealing with occupant behavior, they are not treating energy storage associated with buildings or district energy systems. Even though the research field on energy flow modeling of the urban built environment has been growing, it is still in its infancy. Only limited progress has been made with respect to modeling of supply from energy conversion systems [1]. The simulation tool for energy planning called EnergyPLAN treats the supply side, while the heating and electricity load of an urban area are inputs [5]. Therefore, one of the aims of this study was to develop approach for modelling the building area together with the energy supply system.

The aim of the study was to develop energy pathways for residential building areas, taking into account different structure of the building area, implementation of the new requirements for buildings, and energy supply systems. For this purpose, different calculation methods were combined: detailed building simulation, energy supply technology simulation, heat demand aggregation, and data post-processing. Building models covering different building codes were developed in IDA ICE. An imaginary area connected to district heating representing in a well way the Norwegian residential statistic was analyzed.

The paper is organized as follows. Methodology is introduced in brief by presenting the main calculation steps. In the result section, energy use of residential buildings at different standard are firstly introduced. Based on the linear projection of the building development, total energy demand until 2050 was presented. The most important performance data of the energy supply plants were introduced. Finally, the CO_2 emission development of the building area over the years is given. This was a big study. Due to effectiveness of the paper, only the main results are introduced. The sensitivity analysis and criticism on the results is given as comment when relevant.

2 Methodology

This study was combining the methods from building energy use analysis and energy supply system analysis.

2.1 Main Information Flow for the Study

The main information flow for the study is introduced in Fig. 2.

Firstly, the study focused on energy use in buildings and the purpose was to create different simulation models that presented households according to different

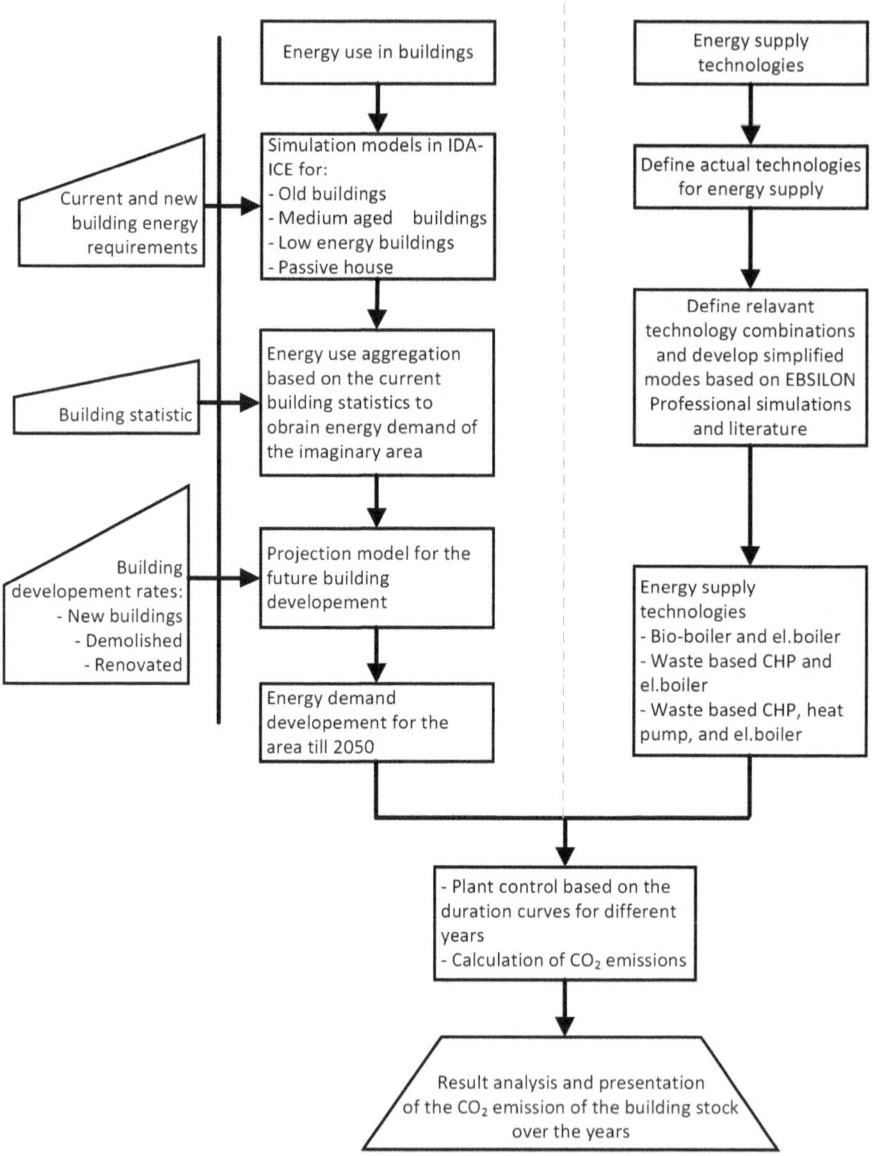

Fig. 2 Main information flow for the methodology combining building energy use and energy supply

standards, thus representing all the building energy classes in Norway. The first part of the method is given at the left side in Fig. 2. The input data are placed in trapezoids. The building models were developed in the IDA ICE simulation tool, and were based on historical and current standards. The models were simulated separately and then aggregated to achieve an imaginary housing area with 72 MW of heat demand and 28 MW of electricity demand.

The second part of the study, the right part in Fig. 2—dealt with the energy supply technologies and thereby the power and energy requirements of the housing stock were used as the input for evaluating different energy supply technologies.

Energy supply technologies were determined through a literature study that deals with current renewable energy sources and associated CO_2 factors [6, 7]. Combinations of energy supply technologies, as well as power distribution between technologies, were based on experience from other facilities. Simplified simulation models based on EBSILON Professional simulations were calibrated with the corresponding real plant data. With the simplified models, it is meant parametric models giving relationship between the heat and power output and fuel input. By combining the parametric models and the load of the building stock, a method has been developed to control input of the energy supply technologies based on the energy requirements. Finally, CO_2 emissions were calculated for today and future years. The technologies were further evaluated against each other to find the most optimal solution.

2.2 Building Energy Demand

In this study, a reference building that is a typical two-storey, three-bedroom. single-family house built before 1980, with an area of 122.2 m^2 was chosen, see Fig. 3. The unit is located at the end of a terraced building to represent the worst-case scenario. A multi-zone model of the building was developed in IDA-ICE [8], and Climate data for Oslo, Norway, were used.

To calibrate the energy use of the reference building in Fig. 3, data from both the Norwegian Energy Efficiency Agency, Enova [9], statistical data, and standards were used. For buildings built around 1969, the average total specific energy use was 178 kWh/m^2. Radiators were chosen as the heating system. Based on NS3031 [10], the space heating system was designed for an desired temperature of 21/19 °C in occupied/non-occupied hours during the coldest period of the year. The internal gains and electrical appliances were modelled based on the standard values [11] and to achieve an average energy use for electrical appliances of 25 kWh/m^2 [12].

After the building simulation model was calibrated based on the national statistics, different measures were introduced to model medium aged buildings, low energy, and passive house buildings. The medium aged buildings presented the existing buildings that have passed some energy efficiency measures. In this study, detail hourly values on the heat and electricity demand were used. Such high quality and high resolution data were difficult to find in the national statistic data.

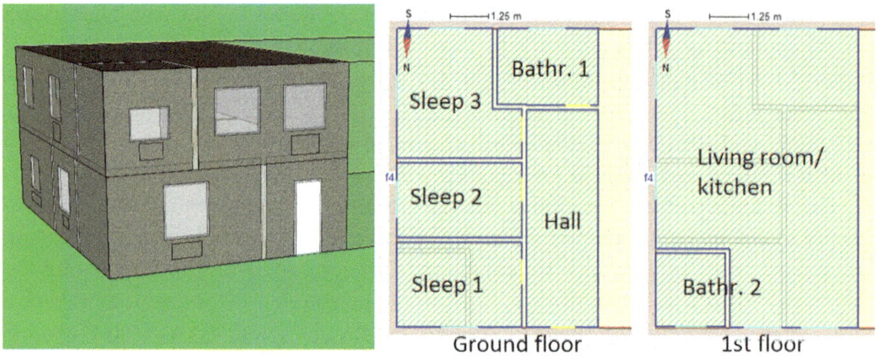

Fig. 3 Residential building model

In general, it was only possible to find specific total building energy use per m^2. A summary of all the relevant input data for modelling of the single residential building are given in Table 1.

From Table 1, it can be noted that there is a large decline in average U-value, from older buildings to passive houses. Further, it can be noted that the average U-value has a declining tendency in recent years, which indicates that the development of dense and compact building bodies has come a long way and that the improvement potential is not as great as before. For the building given in Fig. 3 and the input data given in Table 1, the total specific energy use considering different building standards is given in Fig. 4.

From Fig. 4 it can be noted that energy use was almost halved from the older building for the passive house. The biggest contribution to the reduction came from a significantly lower heating requirement through better insulated building body. The heating requirement for tap water is unchanged, while electricity use was reduced

Table 1 Summary of the input data for the building simulation

Description	Old building	Medium aged building	Low energy building	Passive house
U-value for roof (W/m^{2}K)	0.40	0.20	0.13	0.09
U-value for floor on ground (W/m^{2}K)	0.40	0.299	0.15	0.08
U- value for external wall (W/m^{2}K)	0.50	0.299	0.18	0.12
U-value for windows (W/m^{2}K)	2.889	2.40	1.145	0.78
U-value for doors (W/m^{2}K)	2.00	2.00	1.20	0.80
U-value averaged (W/m^{2}K)	0.548	0.380	0.2723	0.1922
Normalized cold bridges (W/m^{2}K)	–	–	0.03	0.03
Air change at 50 Pa pressure difference (h^{-1})	4.0	4.0	2.50	0.60

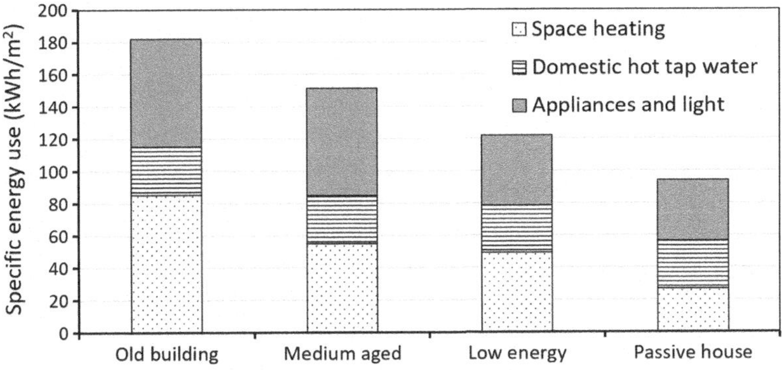

Fig. 4 Total specific single building energy use

due to more energy-efficient equipment and lighting. The results in Fig. 4 were also compared with the real measurements and statistics and were found to fit well with the energy measurements. Hence they were treated as reliable for further analysis.

For the developed building models, influence of occupant behavior on the building stock energy demand was also analyzed by considering differences in light, appliances, and domestic hot tap water use. Different schedules and demand were implemented for the domestic hot tap water. Different operation schedules and installed power for the appliances and light were treated, too.

2.3 Building Stock Model

Energy demand aggregation was based on housing statistics and statistics from the energy labeling system to get accurate weighting of the housing stock. Furthermore, the housing stock was projected with linear growth rates, so that an overview of the power and energy requirements today and in the future was provided. Three different scenarios were made for the projection of housing stock in order to capture the uncertainty associated with housing growth. A summary of the projection rates for the building stock is given in Table 2.

The projection rate in Table 2 were assumed based on a report on future development. Please note that these projection rate are based on a possible economical development, rather than on achieving certain energy demand decrease of

Table 2 Projection rates for building stock

Projection rates (% per year)	Normal	Conservative	Ambitious
New building rate	1.33	1.06 (−20%)	1.66 (+20%)
Rehabilitation rate	1.50	1.20 (−20%)	1.80 (+20%)
Demolition rate	0.60	0.48 (−20%)	0.72 (+20%)

the building stock. Based on the Norwegian building statistics of the building stock and forecasts for the residential building development given in Table 2, an analysis on development of the energy demand until 2050 was made.

2.4 Energy Supply System and CO_2 Emission

To combine hourly heat and electricity demand with the energy supply technologies, parametric models were used as explained in Sect. 2.1. The parametric model included performance data under the part load. Depending on the energy supply scenario, see Table 3, the plants were controlled in the following way. The plants were organized as the base and the peak load plant. The plant with high investment cost and low energy cost were treated as the base load plants, while the plants with the low investment and high energy cost were considered as peak load plants. The sizes of the plants in Table 3 were obtained based on the cost-optimal approach explained in [7, 13].

The control of the plants were organized as: if the heat load at an observed moment was lower than the full base load plant, then that plant was operating under the part load. If the observed load was higher than the full base load plant, the second plant was started. Based on the rest load, the second or the peak load plant was operating on part load. The hourly calculation for each year considering change in the duration curves and thereby energy demand due to building stock change was calculated.

After the fuel and energy input were calculated over different years, the CO_2 emission was estimated. The CO_2 emission was estimated by using the factors given in Table 4 based on [14]. There have been lots of discussion about the values on the CO_2 emission factors. Depending on the institution, the values from 3 gCO_2e/kWh [15], if only the Norwegian hydro power is considered, up to 493 gCO_2e/kWh [16], if the guarantee of origin for the renewable electricity were treated. Regarding the CHP plant, the CO_2 emission related to the heat and electricity production, was allocated by using the energy method [6].

Table 3 Energy supply scenarios

Scenario	Energy production (heat demand)	Required rate (MW)
F1	90% Bio-boiler	33.1
	10% Electrical boiler	–
F2	90% Waste based CHP	33.1
	10% Electrical boiler	–
F3	80% Waste based CHP	25.6
	20% Electrical boiler	–
F4	67.7% Waste based CHP	19.0
	22.3% Heat pump	14.3
	10% Electrical boiler	–

Table 4 Recommended CO_2 emissions factors [14]

Energy soures	Total (gCO_2e/kWh)
Electricity (Nordic production mix)	110
Municipal waste (with sustainable criteria)	11
Municipal waste (without sustainable criteria)	175
Wood chips	18
Pellets and wood powder	19
Heat pump	110

In Table 4, for the municipal waste with sustainable criteria, it was meant waste without fossil waste and this value was used in the study. With the CO_2 emission of the heat pumps, it is meant CO_2 emission of the electricity used by the heat pump. All the values may change over the years, but this was not treated in this study. For analysis purpose, different values for the CO_2 emission were considered and commented.

3 Results

A brief summary of the most relevant results giving the main idea how to decrease the CO_2 emission of the building stock and what are the main issues is given in this Section.

3.1 Energy Demand of the Building Stock and Future Development

Based on the method introduced in Fig. 2 and combining the data on development of building energy use in Fig. 4 and the building stock development in Table 2, the total energy use of the observed building area is obtained as in Fig. 5. Please note that when the structure of the area was defined, the national building stock was considered. Therefore, the results in Fig. 5 may be treated to present national trend in building energy demand. As input on the building stock development, an increased building stock with normal development rate was assumed.

The results in Fig. 5 show that the total energy demand of the building stock will increase in the future in the case of the increased building stock, regardless of a significant decrease in single building energy demand, see Fig. 4. By performing a sensitivity analysis considering occupant behavior, it was found that the results in Fig. 5 may vary for up to 10%. In the case that the building stock was kept constant without new development, the total energy demand would decrease.

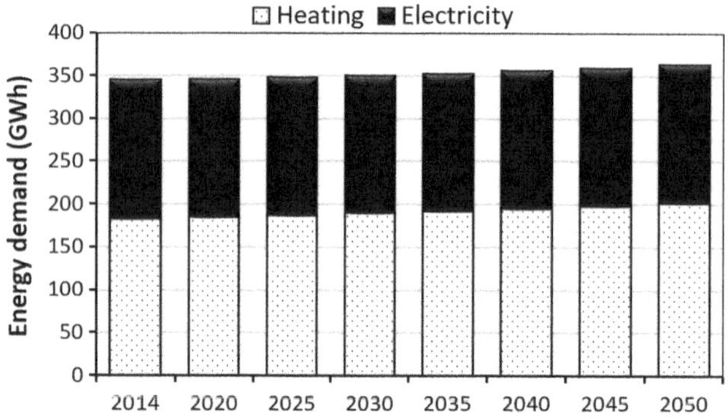

Fig. 5 Total energy demand of the building stock for increased building stock with normal development rate

3.2 CO_2 Emission of the Building Stock

Based on the heat demand and energy supply scenarios defined in Table 3, the CO_2 emission for increased building stock was obtained as in Fig. 6. The shortcuts are used to mark each energy supply scenario. For the reference and easy reading of Fig. 6, please see Table 3. Finally, allocation of the CO_2 emission for the different energy supply scenario is given in Fig. 7.

In Fig. 7, 2014 was consider as a reference year, since the building stock model was developed based on the building stock structure in 2014. The results in Figs. 6 and 7 show an increasing trend in the CO_2 emission. However, depending on the technology choice, the CO_2 emission may be decreased. For example, the CO_2 emission is the lowest in the case of the waste based CHP plant in combination with the electric boiler, scenario F2. In general, it might be noted that the scenarios including CHP provided lower CO_2 emission.

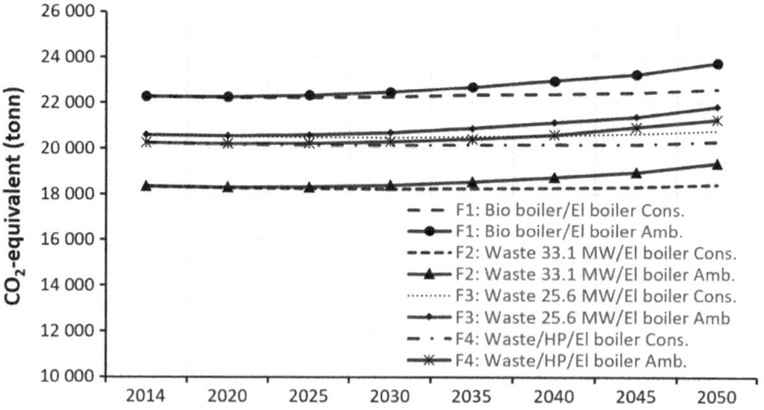

Fig. 6 CO_2 emission of the increasing building stock considering different development rates

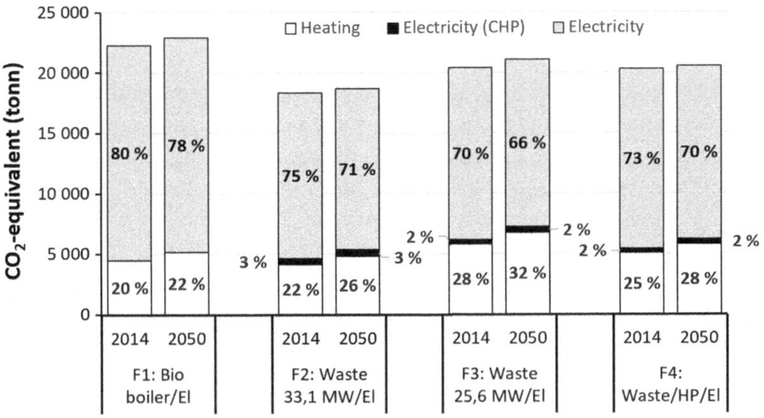

Fig. 7 Allocation of the CO_2 emission for the different energy supply technologies for the increased building stock

Finally, to show what may give the biggest decrease of the CO_2 emission of the residential building stock in Norway, a comparison of the results for the stock without increase in the building number and with the increasing of the buildings is given in Fig. 8. The results show that the constant building stock would show a higher decrease in the CO_2 emission. This means that the implemented energy-efficiency renovation rates and penetration of the new houses would not be enough to give a significant decrease in the CO_2 emission of the building stock.

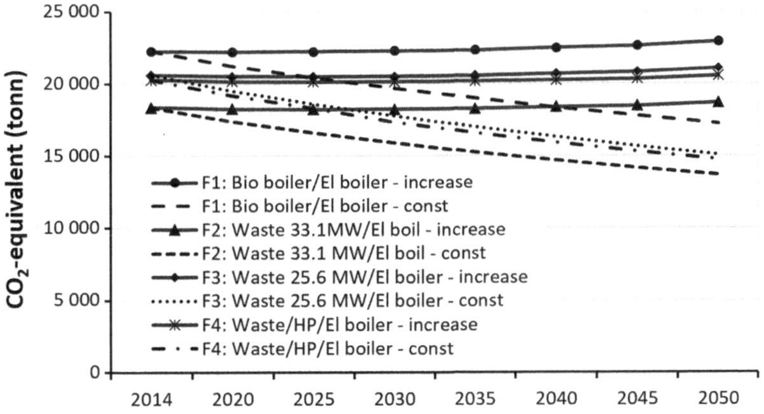

Fig. 8 Comparison of the CO_2 emissions for the increasing and constant building stock

4 Conclusions

The aim of the study was to analyze energy supply systems and technologies for buildings for future building areas. To answer the task, building energy performance models were developed in IDA-ICE with different building standards, which presented different building categories in a housing stock. In addition, four energy supply technologies for the base and the peak load solutions were developed in EBSILON Professional, where the goal was to look at which solutions gave the lowest CO_2 emissions by 2050. Using the CO_2 emission factors recommended in [14], the waste based CHP in combination with the electric boiler gave the lowest CO_2 emission with a margin of 10.3% in 2014. Furthermore, the energy supply system with the waste based CHP in combination with the heat pump and the electric boiler gave good results. The poorest results were obtained for the energy supply system with bio boiler and electric boiler that had CO_2 emission 21.3% higher than the best one. When using the CO_2 factors from other sources, the results would change. If other CO_2 emission factors were used than those in [14], the solution with the waste based CHP in combination with a heat pump and electric boiler that would be the best. This shows that the choice of CO_2 factor is very important for the result, especially the CO_2 factor for electricity. Different companies and organizations often operate with different CO_2 factors to substantiate their own interests, so it can be difficult to assess which CO_2 factors are provided on the most objective basis.

In the scenario of increasing housing building stock, it is apparent from the results that CO_2 emissions increase slightly by 2050 for all energy supply systems, despite the fact that the passive houses that were being built use only half the energy of the existing buildings being demolished. This is a good indication that much stronger activities on the building renovation are highly necessary. The renovation rates used in this study seems to be not enough, even though they are higher than in rest of the world. In a scenario with a constant housing stock, CO_2 emissions were estimated to be reduced by between 22 and 27% depending on which energy supply system were analyzed. This shows that implementation of stricter construction regulations has a positive impact on CO_2 emissions, but that it has less impact on emissions than the choice of energy sources, because the energy sources gave bigger reduction of the CO_2 emissions. The results and conclusions in this study might have some limitations due to all the assumptions made. However, the results might give some recommendations on building energy planning and choosing energy supply sources.

References

1. D. Robinson, Integrated resource flow modelling of the urban built environment, in: *Building Performance Simulating for Design and Operation* (2012), pp. 441–468
2. Y. Chen, T. Hong, M.A. Piette, Automatic generation and simulation of urban building energy models based on city datasets for city-scale building retrofit analysis. Appl. Energy **205**, 323–335 (2017). https://doi.org/10.1016/j.apenergy.2017.07.128

3. D.B. Crawley, Building simulation for policy support, in: *Building Performance Simulating for Design and Operation* (2012), pp. 469–480
4. M. Kavgic, A. Summerfield, D. Mumovic, Z. Stevanovic, Application of a Monte Carlo model to predict space heating energy use of Belgrade's housing stock. J. Build. Perform. Simul. (2014). https://doi.org/10.1080/19401493.2014.961031
5. H. Lund, *Renewable Energy Systems: The Choice and Modeling of 100% Renewable Solutions* (Academic, London, 2010)
6. T. Tereshchenko, N. Nord, Uncertainty of the allocation factors of heat and electricity production of combined cycle power plant. Appl. Therm. Eng. **76**, 410–422 (2015). https://doi.org/10.1016/j.applthermaleng.2014.11.019
7. T. Tereshchenko, N. Nord, Energy planning of district heating for future building stock based on renewable energies and increasing supply flexibility. Energy **112**, 1227–1244 (2016). https://doi.org/10.1016/j.energy.2016.04.114
8. EQUA, IDA-ICE 4.6.2, in, 2015
9. S. Frederiksen, S. Werner, *Fjärrvärme: teori, teknik och funktion* (Studentlitteratur, Stockholm, 1993)
10. NS 3031 - Calculation of energy performance of buildings—Method and data in, Standards Norway, 2014
11. ISO, ISO 13790, Energy performance of buildings, in, 2008
12. SINTEF, ENØK i bygninger, effektiv energibruk, 3. ed., Gyldendal Undervisning, Oslo, 2007
13. S. Frederiksen, S. Werner, *District Heating and Cooling* (Studentlitteratur, Lund, 2013)
14. E. Otterlei, Climate calculation for district heating - available in Norwegian, in, Norsk Energi, Oslo, 2015
15. Varedeklarasjon 2014, Utslipp av CO2 og radioaktivt avfall, in, Norges vassdrags- og energidirektorat, 2015
16. Varedeklarasjon., in, Norges vassdrags- og energidirektorat 2016

Assessing the Potential of Energy Retrofitting and Renewables in the Campus of Lund University

Yurui Huang, Yuchen Yang and Vahid M. Nik

Abstract Several concerns about energy have been discussed during recent decades, such as the shortage of traditional energy resource, increase of energy price and destruction of the living environment. To solve these problems, sustainable development of energy become a preferential task all around the world. Under current circumstances, applying energy saving measures and using renewable energy resource are two of the best choices. The goal of this study is to assess the potential of applying energy saving measures and adding renewable energy resource for the campus of Lund University in Sweden. Energy consumption simulations towards representative buildings were performed for both current and future climatic conditions, meanwhile, investigating the potential of adding solar and wind energy. For energy saving perspective, results show that adding insulation material to old walls, adjusting heating and cooling set points, applying high-efficient heat recovery system, and adding shading devices would have significant effects. Effects of future climatic conditions on heating and cooling energy consumption are considerable. From adding renewable energy resources perspective, results show that the campus of Lund University have enough potential to applying solar and wind energy resources by installing PV systems and small-scale wind turbines. Besides, future climatic condition would not have huge or certain influence on renewable energy applications.

Keywords Energy retrofitting · Renewable energy sources · Climate change

---

Y. Huang
Department of Architecture, Huzhou University, Huzhou 313000, China

Y. Yang · V. M. Nik (✉)
Division of Building Physics, Department of Building and Environmental Technology, Lund University, Lund 22241, Sweden
e-mail: vahid.nik@byggtek.lth.se

V. M. Nik
Division of Building Technology, Department of Architecture and Civil Engineering, Chalmers University of Technology, Gothenburg 41296, Sweden

D. Johansson et al. (eds.), *Cold Climate HVAC 2018*,
Springer Proceedings in Energy, https://doi.org/10.1007/978-3-030-00662-4_43

1 Introduction

The European Union set up the target to reduce the greenhouses gas emissions and innovated the 20-20-20 strategies, which points out that the EU countries must reduce 20% greenhouse gases emissions, 20% building primary energy and increase renewable energy usage by 20% between 1990 and 2020 [11].

With the fast growing of housing demand in Sweden, the energy demand of building consuming increase dramatically. Approximately 40% of total energy use in Sweden were used for building operation and 8% of it were used for construction process [3]. A study by Langer et al. [13] shows that the amount of energy-efficient buildings like passive houses has been increased rapidly to 7.2% of total amount of buildings in Sweden. These indicate that energy demand of buildings has been highly concerned and the energy-efficient building would develop quickly and become popular in Sweden. Applying more renewable energy resource also plays a significant role in having a brighter future. In Sweden, several areas are inside or near arctic circle with extreme cold climate, which demand more energy than other areas to keep a satisfied indoor environment. Thus, it is of immense importance to assessing the renewable energy potential for environment protection and sustainable development.

This work investigates the potential of applying energy saving measures and adding renewable energy resources in the campus of Lund University in Sweden, for current and future climatic conditions.

2 Energy Saving Techniques

The efficiency of retrofitting measures as it would be affected by plenty of parameters, like location, climate and so on. Nik et al. [15] pointed out nine retrofitting measures which are applicable for residential buildings in Sweden for current and future climate. One key point for recognizing the proper building retrofitting techniques is their long-term benefits [15]. Previous study for buildings in the campus of Lund University assessed the long-term performance of various energy retrofitting measures for exterior wall and roof for current and future climate conditions [9]. Results predict a heating energy saving of 20–30% for wall retrofitting and 5% for roof. It shows that the retrofitting measures are economically infeasible but environmentally beneficial.

Except wall retrofitting, window change is also an important measure for energy saving. Poor windows and frames could lead to several problems, such as draught and acoustic. Also, window with poor water tightness would have the risk of moisture problem. Cuce [5] shows that about 33% of heat could be saved by using airtight windows. Many recent researches show that windows and shading devices with optimized design would significantly reduce energy consumption and increase thermal comfort [20]. Due to office buildings are popular to be designed with large glazed windows, most of the related researches were performed for office buildings [1].

3 Renewable Energy Resource Application

Covering the energy demand of buildings, especially in cities and urban areas, through greater application of decentralized energy systems is a developing topic. Perera et al. [17, 18], (2016) have shown the potentials of Energy-Hubs for integrating larger shares of renewable energy sources. Applying solar energy resource is a new rising trend of urban development. The solar photovoltaic (PV) systems have rapidly developed in the 21st century. The domination of the PV module market was modules made by traditional technologies such as crystalline silicon. New raised PV module technologies are also applied worldwide [2]. The method for quantifying the solar potential which presented by Compagnon [4] is setting up threshold values to qualify external envelope of buildings for active or passive heating, photovoltaic electricity production and daylighting. Based on Lund Solar map [12], Kanters and Wall [10] presented a new threshold value for acceptable categories.

Wind energy is also one of the most popular renewable energy resources. Almost 15–30% of total electricity production is generated by wind power in some European countries [6, 8]. Although hydropower and nuclear power take a great proportion in electricity supply in Sweden, an enlargement of renewable energy use has come into sight from the 21st century, in which wind power contribute the most. In 2016, 605 MW of new wind power has been installed in Sweden. The total wind power of 6422 MW was distributed over 3335 projects in Sweden until 2016 (Swedish Energy Agency 2016). Following the development of wind energy application technology, small-scale wind energy generators has made significant and speedy development [19]. The main advantages of investing in small-scale wind turbines are generating electricity for personal need and reduce energy dependence on unrecycled energy source like fossil fuel. Also, it is appreciated to make personal or corporate contribution to the environment of human lives [21].

4 Assessing Energy Retrofitting and Renewables in the Campus

In this study, the main focus of energy simulation was on two representative buildings inside the campus, Maskinteknik huset (M-huset) and Ingvar kamprad designcentrum (IKDC). M-huset is a historical old building and a representative building of all buildings in LTH campus which were built in 1960s. The IKDC was completed in 2002, with a modern construction and large window. The focus of assessing renewable energy potential was on several buildings inside the campus.

Weather data is a necessary element to assess the building energy performance as well as the potential of renewable energy generation. A method for the impact assessment of climate change of buildings was suggested by Nik [14], which is based on synthesizing one typical and two extreme weather data sets out of regional climate models (RCMs): Typical Downscaled Year (TDY), Extreme Cold Year (ECY), and Extreme Warm Year (EWY). More than energy simulations, the application of the method has been also verified for hygrothermal simulations in buildings [16]. Representative weather data were generated for 2009–2038, 2039–2068, and 2069–2098 and used for energy simulation in IDA ICE energy, solar potential simulation in DIVA for Rhino, PV output simulation in System Advisor Models, and wind analysis in WRPLOT View.

4.1 Energy Consumption Simulations

The energy models for IKDC and M-huset (see Fig. 1) were made in IDA ICE and simulated for current and future weather conditions, for more details on modelling buildings, the reader is referred to Huang and Yang [7]. Several retrofitting building technologies (in Table 1) were assessed to reduce the energy consumption based on the simulated model.

4.2 Renewable Energy Potential Assessment

Efficient PV panels has been installed on M-huset in 2014 (MonoX Neon PV panels and inverters from Fronius SYMO). With the total area of 1050 m^2, the estimated peak power is about 191 kWp and annual production is approximately 182,000 kWh. Based on outdoor temperature, solar radiation, and PV output in 2015 recorded by Akademiska Hus, PV panel simulation was run in System Advisor Model (SAM). The losses of electricity generation were adjusted based on the measured PV output data. This model was then used for parametric study of electricity consumption coverage proportion (covering whole consumption in May

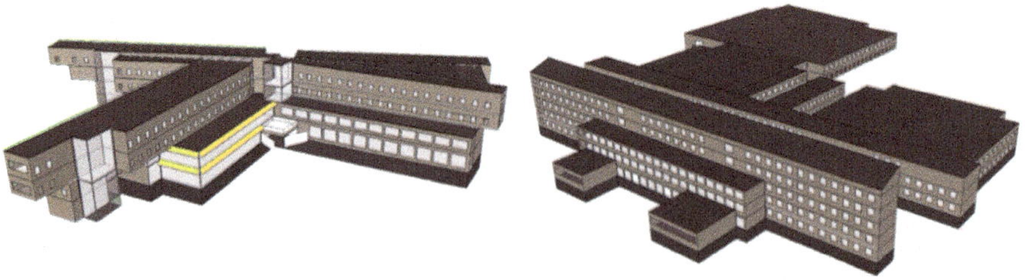

Fig. 1 IDA ICE 3D model: IKDC in the left, M-huset in the right

Table 1 Retrofitting building technologies applied to M-huset

	Type	Original condition	Retrofitting condition
Case 1	Adding insulation material to walls	Thermal transmittance: 2.21 W/(m^{2}K)	Thermal transmittance: 0.103 W/(m^{2}K)
Case 2	Changing windows	Thermal transmittance: 1.37 W/(m^{2}K)	Thermal transmittance: 0.79 W/(m^{2}K)
Case 3	Adjusting temperature set point	15 °C for heating, 27 °C for cooling	14 °C for heating, 28 °C for cooling
Case 4	Improving heat recovery system	Heat recovery rate of 60%	Heat recovery rate of 80%
Case 5	Adding exterior shading	–	Adding "drop arm awing" shading device

when both radiation and consumption are high and in December when radiation is lowest in the year), and future scenarios (2009–2098).

To analyze the wind variation in Lund, the histogram of wind frequency distribution in the reference year was generated in WRPLOT View. As big scale wind turbine is not available to install inside campus for space limitation, small scale wind turbines were taken into comparison of different rated power wind turbines and estimation of wind potential. Three wind turbines with vertical blades and two wind turbines with horizontal blades from Aeolos Company were chosen in this study. For more details on characteristics, the reader is referred to Huang and Yang [7]. Power-Wind speed diagram from product catalogue for different products were used for individualized but representative simulations.

5 Results of Assessment in the Campus

5.1 Energy Consumption Simulations

After comparing measured energy consumption data received from Akademiska Hus and simulated energy consumption data gathered from IDA ICE for M-huset, the simulating model with annual heating demand, annual cooling demand and annual electricity demand of 79.8, 7.37, and 79.6 kWh/m^2 year is referred to the reference model. Results for the IKDC could be found in Huang and Yang [7]. Effects of energy retrofitting measures for M-huset on the annual heating demand and the annual cooling demand is plotted in Fig. 2, including adding insulation material to old walls (case 1), changing windows (case 2), adjusting temperature set point (case 3), Improving heat recovery rate (case 4), and adding shading device (case 5). When applying all five measures together for M-huset, the annual heating consumption reduced by 27% from 2,267,501 to 1,662,628 kWh/year and cooling consumption reduced by 12% from 209,432 to 184,306 kWh/year.

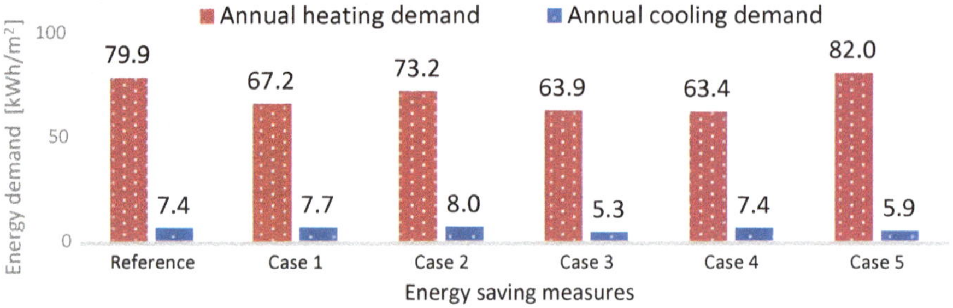

Fig. 2 Energy consumption variations of M-huset when applying energy saving measures

Energy consumption variations for M-huset in future weather condition of typical downscale year (TDY) are shown in Fig. 3, results for the IKDC could be found in Huang and Yang [7].

5.2 Renewable Energy Potential Assessment

The annual solar radiation of M-huset and the IKDC are showed in Fig. 4 (results of other simulated buildings could be found in Huang and Yang [7]).

Although energy generation of the simulated data and the measured data received from Akademiska Hus varies in different months, the accumulated data are

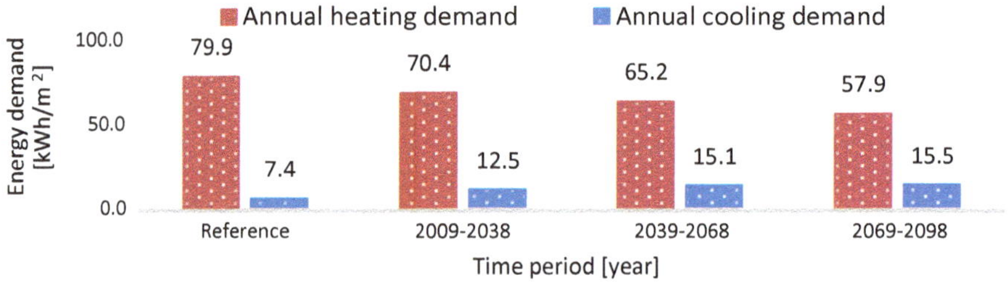

Fig. 3 Energy consumption variations of M-huset in TDY weather file for different period

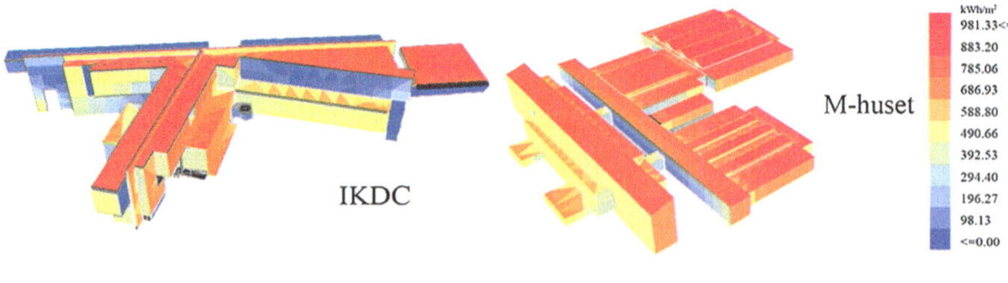

Fig. 4 Annual solar radiation of the IKDC and M-huset

almost overlapping curves. Thus, the simulated case with an annual PV generation of 191,445 kWh is defined as base case for M-huset for further assessment. Parametric simulations for covering different coverage proportion of electricity consumption in M-huset are showed in Fig. 5. Information of the adjusted and simulated PV modules are listed in Table 2. Case a is the scenario when monthly PV generation equal to monthly electricity consumption in M-huset in May. In this case, electricity consumption in M-huset during May to August can be covered by PV panel, which need 7.4 times bigger scale of the existent PV. Case b is the scenario when monthly PV generation equal to monthly electricity consumption in M-huset in December. In this case, electricity consumption in M-huset during the entire year could be covered by PV panel. However, it need 94.8 times bigger scale of the PV already installed.

The PV generation simulation results received from System Advisor Model (SAM) under typical downscale year (TDY), extremely cold year (ECY), and extremely warm year (EWY) are showed in Fig. 6. It indicates that monthly PV generation, annual PV generation, and PV power all are varied from time to time, no certain trend were found.

Annual energy generation by wind turbine in local climatic condition was compared with measured energy generation data at stable wind speed of 6 and 7 m/s condition from Product catalog, showed in Fig. 7.

The wind speed variations gathered from weather file under typical downscale year (TDY), extremely cold year (ECY), and extremely warm year (EWY) were simulated. The series called Triple showed in Fig. 8 is a combination of wind speed

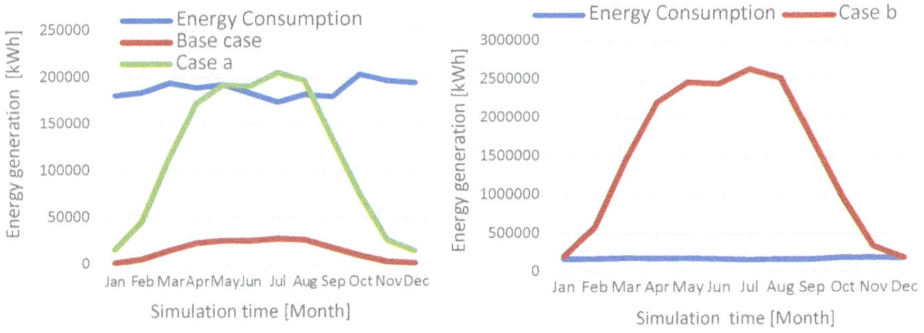

Fig. 5 Comparison of different coverage proportion of energy consumption

Table 2 Information of parametric simulated PV modules

Case	Nameplate capacity (kWdc)	Number of modules	Total module area (m^2)
Reference	192	648	1029
Case a	1420	4800	7622
Case b	18,150	61,440	97,567

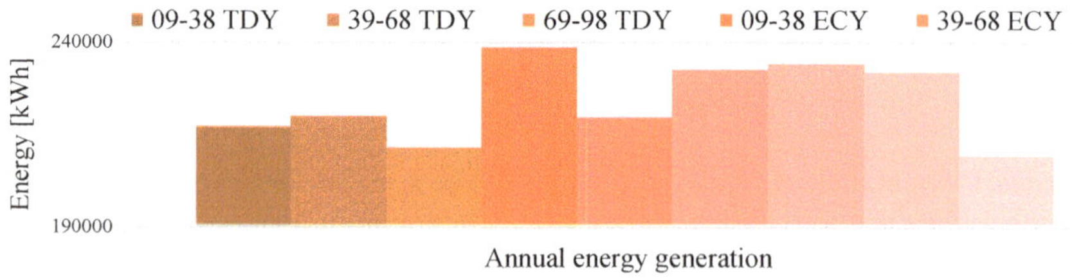

Fig. 6 Annual PV generation in TDY scenario for 2009–2098

Fig. 7 Energy generation of wind turbines

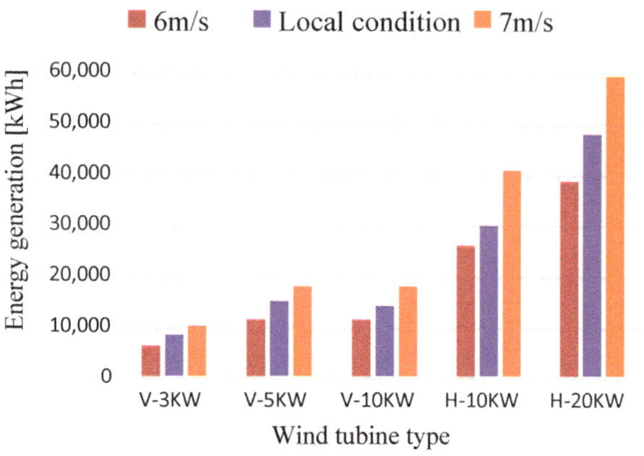

variation result in TDY, ECY and EWY scenarios. It indicates that the interquartile range of wind speed has decreased into a lower interval with the passage of time.

Comparison of wind class frequency distribution in TDY scenario for 2009–2098 and reference model, which generated by WRPLOT View, are showed in Fig. 9. It is obvious that maximum ratio appears in the interval when wind speed is between 5.7 and 8.8 m/s in reference year and period 2009–2038. However, after 2038, the interval when wind speed is between 3.6 and 5.7 m/s become the most likely interval.

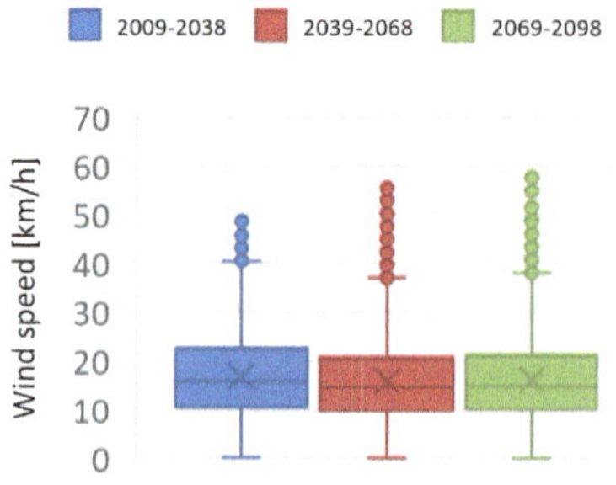

Fig. 8 Wind speed variations in 2009–2098

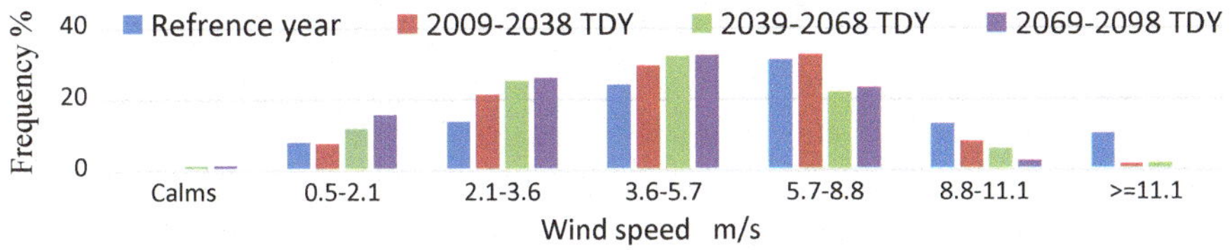

Fig. 9 Comparison of wind class frequency distribution in 2009–2098

6 Discussion

From applying energy saving measures perspective, although several energy saving measures have significant effect on decreasing heating or cooling energy consumption, it is important to consider economic profits together with local climatic condition when choosing retrofitting methods for buildings. Moreover, if building envelope renovating is applied, it need a slightly long time for construction which may affect the daily operation of the building. From assessing the potential of adding renewable energy resources perspective, simulated annual solar radiation is more than 800 kWh/m^2 which reached an available level. Meanwhile, annual wind turbine energy generation are higher than approximate wind turbine output with stable wind speed of 6 m/s from Product catalog. Thus, several simulated buildings inside the campus have enough potential for applying solar and wind energy resources by installing PV systems and small-scale wind turbines. Besides, for solar energy application, to cover the electricity consumption in M-huset only seven times bigger scale of the PV already installed would be needed, which indicates that it is a possible application since M-huset still have plenty of unused roof area with high annual solar radiation. Whereas, the energy consumption in winter is hard to reach due to space limitation. For wind energy, although big scale wind turbines have higher efficiency, shorter investment payback time, and lower environmental burden than small-scale wind turbines, land limitation and noises problem make only the small-scale wind turbines be suitable to install inside the campus. Moreover, although majority of wind speed would slightly lower down and the power generation of wind turbine would decrease in future, no certain variation trend was found for solar energy in this study.

For limitation, some admissible errors would arise when hourly data was needed to do comparison between realistic data and simulation results due to the uncertainties in measuring and recording data. Besides, a lot of input data like heating and cooling set points, supply air temperature etc. were set based on logical assumptions. Although conclusive results were adjusted with the realistic measured data, various uncertain parameters would have positive or negative effect on the conclusive results.

Further research could be focus on the innovative retrofitting techniques with building automation and control or artificial intelligence. As well as other renewable energy resource like biomass, hydro, tidal energy and so on are deserved to do research.

7 Conclusion

From the masses of results, several valuable conclusions were drawn. For applying energy saving measures perspective, results show that adding insulation material to old walls, adjusting heating and cooling set point, applying high-efficient heat recovery system, and adding shading device would have significant effect on decreasing heating or cooling energy consumption. From assessing the potential of adding renewable energy resources perspective, the campus of Lund University has enough potential to applying solar and wind energy resources by installing PV panels, small-scale wind turbines. Moreover, for future climatic condition, heating and cooling energy consumption would reduce and increase respectively. However, future climatic condition would not have huge or certain influence on renewable energy applications.

References

1. A. Atzeri, F. Cappelletti, A. Gasparella, Internal versus external shading devices performance in office buildings. Energy Procedia, 463–472(2014)
2. V. Baharwani, N. Meena, A. Sharma, R. Stephen, P. Mohanty, Comparative performance assessment of different solar PV module technologies. Int. J. Innov. Eng. Technol. **5**(1) (2015)
3. B. Berggren, M. Wall, Thermal bridges in passive houses and nearly zero-energy buildings, in 4th Nordic Passive House Conference (2011)
4. R. Compagnon, Solar and daylight availability in the urban fabric, Energy Buildings, 321–328 (2004)
5. E. Cuce, Role of airtightness in energy loss from windows: experimental results from in-situ tests. Energy Buildings, 449–455(2017, 3 15)
6. P. Higgins, A. Foley, The evolution of offshore wind power in the United Kingdom. Renew. Sustain. Energy Rev. **37**, 599–612 (2014)
7. Y. Huang, Y. Yang, Assessing the potential of applying energy saving measures and renewable energy resources in the campus of Lund University. Master thesis in Energy-efficient and Environmental Buildings, Faculty of Engineering, Lund University, Sweden (2017)
8. IEA. Technology roadmap—wind energy. Paris: IEA (2013)
9. N. Jensen, Hygrothermal Analysis of Retrofitted Buildings in the campus of Lund university. Master thesis in Energy-efficient and Environmental Buildings, Faculty of Engineering, Lund University (2016)
10. J. Kanters, M. Wall, The impact of urban design decisions on net zero energy solar buildings in Sweden. Urban, Plann. Transp. Res. **2** (1), 312–332 (2014)
11. B. Karlsson, M. Gustafssona, M. Rönnelidc, L. Tryggd, CO_2 emission evaluation of energy conserving measures in buildings connected to a district heating system-case study of a multi-dwelling building gin Sweden. Energy, 341–350 (2016, 9)
12. Kraftringen. Kraftringen. Retrieved 2012, from https://www.kraftringen.se/Privat/solceller/solkartan/ (2012)
13. S. Langer, G. Bekö, E. Bloom, A. Widheden, L. Ekberg, Indoor air quality in passive and conventional new houses in Sweden. Build. Environ. 92–100 (2015)

14. V.M. Nik, Making energy simulation easier for future climate—Synthesizing typical and extreme weather data sets out of regional climate models (RCMs). Appl. Energy, **177**, 204–226 (2016, 9 1)
15. V.M. Nik, E. Mata, A. Sasic Kalagasidis, J.-L. Scartezzini, Effective and robust energy retrofitting measures for future climatic conditions—Reduced heating demand of Swedish households. Energy Buildings, 176–187 (2016, 7 1)
16. V.M. Nik, Application of typical and extreme weather data sets in the hygrothermal simulation of building components for future climate—A case study for a wooden frame wall. Energy Build. **154**, 30–45(2017)
17. A.T.D. Perera, V.M. Nik, D. Mauree, J.-L. Scartezzini, Electrical hubs: an effective way to integrate non-dispatchable renewable energy sources with minimum impact to the grid. Appl. Energy **190**, 232–248 (2017)
18. A.T.D. Perera, V.M. Nik, D. Mauree, J.-L. Scartezzini, An integrated approach to design site specific distributed electrical hubs combining optimization, multi-criterion assessment and decision making. Energy **134**, 103–120 (2017)
19. S. Rolland, B. Auzane, The potential of small and medium wind energy in developing countries. Alliance for Rural Electrification position paper 2012 (2012)
20. S. Stevanović, Optimization of passive solar design strategies: a review. Renew. Sustain. Energy, 177–196 (2013)
21. E. Thorstensson, Small-scale Wind Turbines- Introductory market study for Sweden. Master thesis in International Project Management, Chalmers university of technology, Sweden (2009)

Part VI
Buildings in Operation

Inverse Model Identification of the Thermal Dynamics of a Norwegian Zero Emission House

Pierre J. C. Vogler-Finck, John Clauß, Laurent Georges, Igor Sartori and Rafael Wisniewski

Abstract Dynamical model identification is an essential element in the implementation of a model predictive controller. In this work, a control-oriented first order model was identified in a dedicated experiment on a super-insulated single-family house. First, parameters resulting from CTSM and the MATLAB System Identification toolbox were compared. Then, a comparison of model predictions and measurements showed that this simple model captures well the main dynamics of the building-averaged indoor temperature, after one week of training on rich data with sample time below 15 min. It was also observed that this prediction performance was not affected by the configuration of internal doors.

Keywords Zero emission building · Dynamical thermal model identification
Control-oriented building modelling

1 Introduction

1.1 Context

Model predictive control (MPC) is a control technique that has gained rising attention from the HVAC sector over the last years, as a way to increase energy efficiency and reduce environmental impact [1]. MPC is also seen as a promising

P. J. C. Vogler-Finck (✉)
Neogrid Technologies ApS, Aalborg, Denmark
e-mail: pvf@neogrid.dk

R. Wisniewski
Aalborg University, Aalborg, Denmark

J. Clauß · L. Georges
Norwegian University of Science and Technology, Trondheim, Norway

I. Sartori
SINTEF, Oslo, Norway

D. Johansson et al. (eds.), *Cold Climate HVAC 2018*,
Springer Proceedings in Energy, https://doi.org/10.1007/978-3-030-00662-4_44

tool to deploy the flexibility of the building heat demand in order to achieve some greater objectives (e.g. reducing the peak load of a cluster of buildings, reducing greenhouse gas emissions from the power consumed) [2].

A central aspect in MPC implementation is the identification of a robust model, which is among the most demanding tasks in the development of such a control. This identification of a dynamical model relies on a well-established theoretical framework [3], and has been extensively studied [4, 5]. To allow using well-known optimisation methods with low computational cost, linear models are preferred. A stochastic component is often added to explicitly account for their uncertainty.

1.2 State of the Art

Different modelling approaches are possible, depending on the availability of measurement data. If no data is available, a so-called "white-box" model can be built from knowledge of the building structure and materials and physical principles. However, this approach is time consuming and leads to complex models that are often not appropriate for predictive control [4]. On the other hand, when only data is available and no knowledge of physical principles is used, so-called "black-box models" can be used. Examples of such models include ARX, ARMAX, Box-Jenkins models [6], subspace models [7] and neural networks [8]. However, a drawback of this approach is the need for a large amount of data covering the whole range of operating conditions in order to achieve a robust model.

A third approach combining data and physical principles is known as "grey-box modelling" (or "inverse modelling" as it is based upon observations of an actual behaviour). In that case, simplified physical models of the buildings are proposed, and their parameters are identified from experimental data [5]. A method for selecting the best candidate within a set of grey-box models of increasing complexity is presented in [9], where likelihood ratio tests are used as a decision criterion. Furthermore, considerations about the quality of such identified parameters are introduced in [10], where the investigation was made on a simulated building. These works were later extended by a report from IEA EBC Annex 58 providing practical guidance for experimental characterisation of a building's thermal dynamics [11].

1.3 Contribution

A first contribution of this work is in the comparison of two tools for identification of a simple first order (grey-box) model describing the thermal dynamics of a light-weight super-insulated building: CTSM [12] and the MATLAB System Identification toolbox [13]. A second contribution is in assessing the temperature disparities within the building during an identification experiment. A last contribution is found in the assessment of the prediction capability of a single zone first order model for predicting the future thermal behaviour of the building.

2 Building and Experiments

2.1 The LivingLab

The LivingLab is a zero-emission single family house at the Norwegian University of Science and Technology (NTNU). Here 'zero-emission' is used in the sense of compensating the greenhouse gases emissions resulting from operation (including equipment) and most of the materials [14].

The building is made of a highly insulated (rock-wool layer of thickness 35–40 cm) wooden structure, with a significant window area (ca. 20% glazed area). Mechanical ventilation with heat recovery is implemented. Local energy production is installed, and consists of solar thermal and PV panels, and a ground source heat pump (the main heating system of the building was disabled in this experiment). Phase change material (PCM) boards are also installed behind the cladding of the ceiling of the building. More detailed information about the building may be found in [15].

The inside of the building consists of several inhabitable rooms of a total area of ca. 100 m^2 (not counting the attendant technical room), as seen on Fig. 1. Doors can be opened/closed to separate the bathroom and two bedrooms from the main zone.

2.2 Weather and Heating Conditions

Three successive experiments were carried out on the unoccupied building. A single heat emitter composed of electric radiators was used to heat the building (the

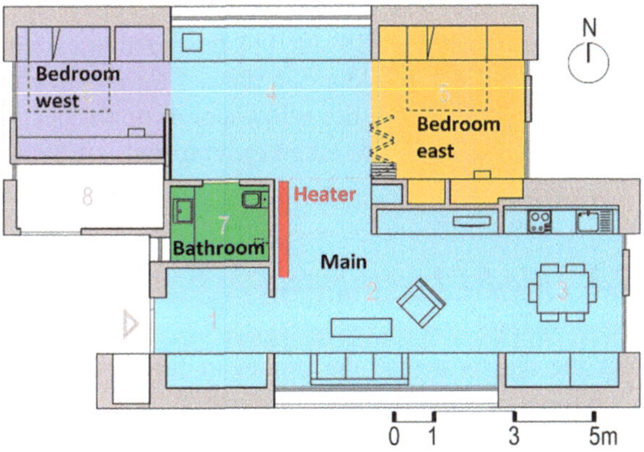

Fig. 1 Layout of the inside of the LivingLab and position of the heat emitter in the experiments (adapted with authorisation from [15] and [16])

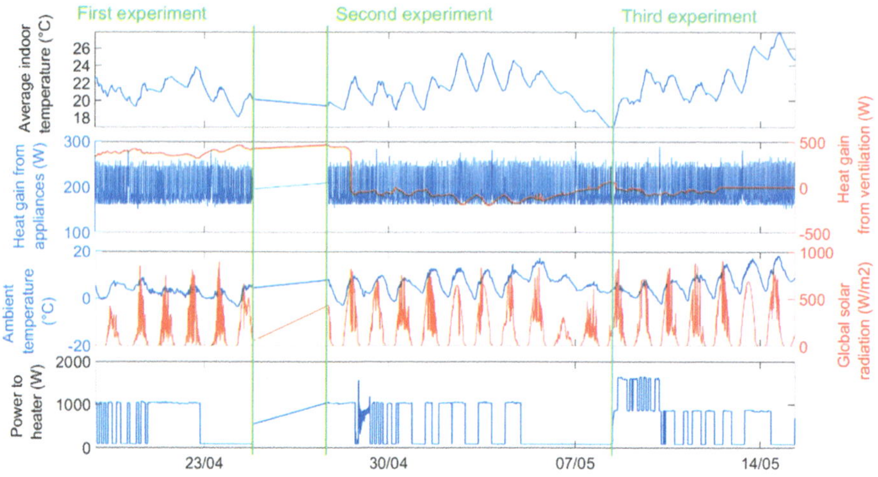

Fig. 2 Experimental data (an interruption happened before the 2nd experiment) [16]

ventilation also contributed in the first experiment, due to a high supply air temperature set-point). This use of a single source is in line with the aim of a simplification of the heat distribution in zero emission buildings (see Chap. 2.8 of [14]).

Data collected included: power to the heater, indoor temperatures, global solar irradiance, ambient temperature, as well as heat gains from appliances/lighting and ventilation. The corresponding data is presented in Fig. 2, and freely available (including details of the measurement instrumentation) on an open platform for further reuse in benchmarking studies [16].

To assess the impact of their configuration, the doors to the bedrooms were opened in the first and the last experiment, and closed in the second. The door to the bathroom was always closed. Ventilation settings were updated during the experiment to increase the variety of conditions. Initially, supply air temperature was regulated to 30 °C. It was then lowered to 18 °C in the beginning of the second experiment. Finally, it was fully stopped in the last days of the third experiment.

The radiator was mostly operated according to successions of pseudo random binary sequences (PRBS) [9, 11], designed to excite the building over a large range of frequencies and yield a rich dataset for model identification. It is however worth noting that this does not allow comfortable occupation of the building—unless specific precautions are taken (which was not the case here).

2.3 Model Investigated

As the aim is to use the model in future implementation of a model predictive controller, a simple linear formulation was adopted. It turned out that a first order model was sufficient to represent the main thermal dynamics of the building, as will be seen in a later section of this article.

This first order model is described by the following differential equation:

$$dT_i = \frac{1}{C_i}(UA_i[T_a(t) - T_i(t)] + A_W\Phi_G(t) + P_{app}(t) + Q_{vent}(t) + P_{rad}(t))dt + d\epsilon_p(t) \quad (1)$$

where T_a is the ambient temperature, T_i the (lumped) indoor temperature, Φ_G the global solar radiation, P_{rad} the power to the radiator, P_{app} the power to appliances and lighting, Q_{vent} the estimated heat gain from ventilation, and $d\epsilon_p$ a stochastic process (typically assumed to be a Wiener process). Parameters of this model are a global (lumped) heat loss coefficient to the ambient UA_{ia}, a (lumped) heat capacity C_i, and an effective window area A_W (reusing the term from [9], although such a gain to the global horizontal radiation is hard to interpret in physical terms). It is worth noting that numerical values of UA_{ia} and A_W are dependent on the value of C_i, since they only appear in ratios in the input/output dependencies of the above equation. Moreover, the choice to focus solely on the global horizontal radiation for sun modelling is because predictions of it are available from weather forecast services (similarly to T_a), therefore allowing its use in predictive control.

The heat gain from ventilation Q_{vent} was estimated using the approximation:

$$Q_{vent}(t) = M_{air}C_{p,air}\rho(t)\left(T_{supply}(t) - T_{extract}(t)\right) \quad (2)$$

where M_{air} is the volume mass of air, $C_{p,air}$ the specific heat of air, ρ the air flow rate, T_{supply} the supply air temperature and $T_{extract}$ the extract air temperature. Infiltration is not included in this, but already integrated in the model (as a part of UA_{ia}).

In the experiments, the building-averaged temperature was used as an observation of this lumped indoor temperature T_i (model output). However, operative temperature measurements could be more physically representative here.

2.4 CTSM and the MATLAB System Identification Toolbox

Several tools are available for dynamical modelling using measurement data. In this study, the choice was made to use and compare the CTSM package for R [12] and the MATLAB System Identification Toolbox [13].

CTSM is a software tool interfaced with the free software statistical modelling environment R. It is meant to identify parameters of continuous time stochastic state space models, using a maximum likelihood approach.

On the other hand, the MATLAB System Identification toolbox is a commercial software. It can identify a variety of model types, including both continuous and discrete time grey-box models, with either a stochastic or deterministic approach.

A *prediction error method* was used for parameter identification (using the *pem* function—but *greyest* yielded similar results). For consistency of the approach with CTSM, a continuous time stochastic approach was used in the modelling with the toolbox.

It is worth knowing that these two tools use a different formal description of stochasticity, which prevents direct comparison of their noise models (however, these two descriptions have a zero mean—so that they result in the same deterministic description).

3 Experimental Results

3.1 Limits of the Single Thermal Zone Approximation

Significant disparities of air temperature were observed within the building, with differences of up to 10 °C between the highest and lowest measured temperature during the experiment. This is due to the superposition of 2 effects: vertical stratification within each of the rooms, and horizontal disparities due to zoning (particularly clear in the case of closed doors, as seen in Fig. 3). In the case of stratification, air temperature measurements taken at several heights showed differences as high as 4 °C in the main zone between floor and ceiling levels.

As reminded by Fig. 3, the indoor air temperatures varied in a large range which was not compatible with comfortable occupation of the building, with the chosen PRBS excitation. In particular, the solar gains in the main (southern) zone led to large excursions of its air temperature.

With these strong inhomogeneities in mind, a single zone approach was adopted in the rest of the work. The building-averaged indoor temperature was computed by averaging the numerous measurements according to volume considerations (to account for combined effects of stratification and zoning). In all 3 experiments, this average included bedroom temperatures and bathroom.

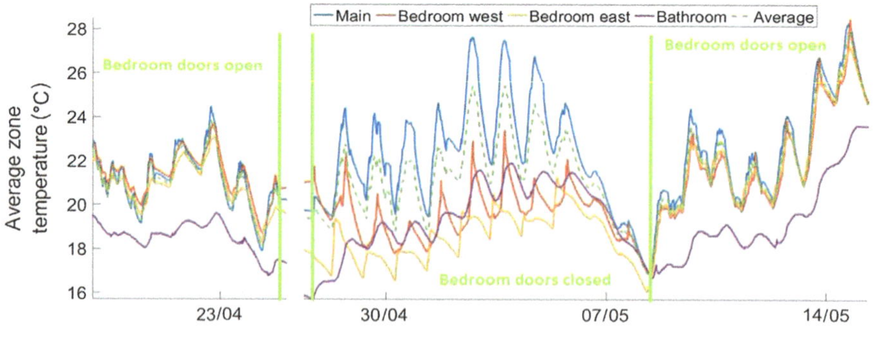

Fig. 3 Zone-averaged air temperatures (an interruption happened before the 2nd experiment)

3.2 Identified Model Parameters

The first order model parameters of the dynamical model described in Eq. (1) were then identified using CTSM and the MATLAB System Identification toolbox. Resulting parameters (and their uncertainties) are plotted in Fig. 4. One may notice the influence of different sample times for the data (values of 5, 15, 30 and 60 min were considered).

In all cases, common bounds and initial values were used for the parameters. UA_{ia} was initially set to 0.1 kW/K, with allowed range 0–5 kW/K. A_W was initially set to 2 m^2 with allowed range 0–30 m^2. C_i was initially set to 4 kWh/K with bounds 0–100 kWh/K (for the third experiment with CTSM, an initial value of 3 kWh/K was used for allowing convergence of the optimisation).

As expected, the uncertainty of the parameters is significantly reduced by considering the whole period instead of a subperiod. Similarly, uncertainty was increased by higher sample times. These results provide an estimation of the long time constant of the building (estimated by the ratio C_i/UA_{ia}) in the range of two days, which is consistent with its well-insulated lightweight wooden structure. As observed, a period of one week (equal to several long time constants) is sufficient to identify a simple first order model of the building's thermal dynamics. Here, it is however important to bear in mind the use of a PRBS type of excitation, and the relatively clear sky days.

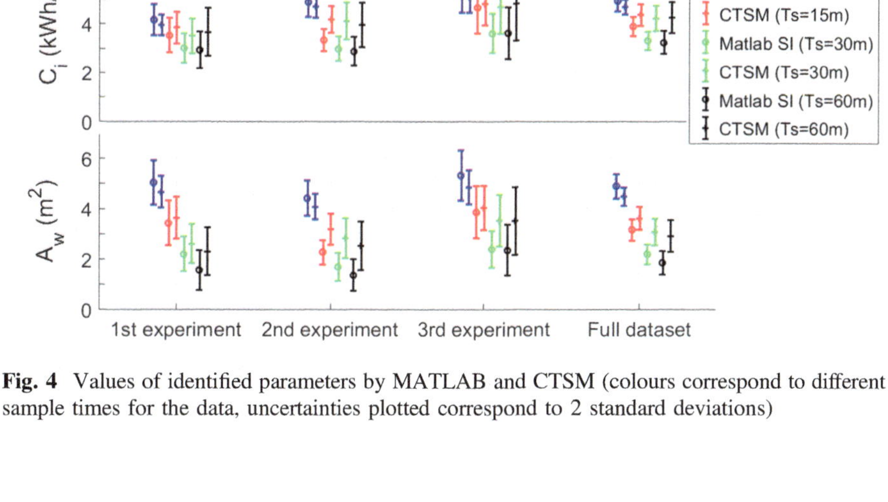

Fig. 4 Values of identified parameters by MATLAB and CTSM (colours correspond to different sample times for the data, uncertainties plotted correspond to 2 standard deviations)

The change in the configuration of internal doors to bedrooms resulted in small reductions of the global heat loss (UA_{ia}) and effective window area (A_W) parameters. The value identified for the global heat loss were in a similar range to the expected value from previous modelling in the simulation software *IDA Indoor Climate and Energy* as part of previous independent work on the LivingLab [17] (0.07 kWh/K—estimated by a product of average U value by envelope area). On the other hand, the heat capacity observed was an order of magnitude above the total indoor air heat capacity (0.12 kWh/K). This confirms usage of the thermal mass of the building that is lumped together with air capacity in the model. Lastly, the identified value of the effective window area was an order of magnitude below the glazed area (36 m^2).

Changes in the sample time of the dataset did affect the value of the identified parameters. A downward trend was observed for the value of the global heat loss (UA_{ia}) and effective window area (A_W) parameters as the sample time increased. This may potentially reflect a tendency to convert some of the solar gains in reduced heat losses, with a similar overall effect on the indoor temperature.

Moreover, it was observed that the System Identification toolbox generally yielded lower values of the parameters for sample times of 15, 30 and 60 min in all three experiments (and their aggregation), compared to CTSM.

3.3 Prediction Capability of the First Order Model

For the sake of conciseness, the analysis of this paragraph is reduced to the models identified by the MATLAB toolbox over the first experiment and the whole dataset.

The prediction capability of the models identified for each sample time were evaluated by predicting the evolution of the indoor temperature over each of the 3 experiments. In every case, prediction was made starting from the initial temperature and assuming perfect knowledge of the disturbances and inputs (namely T_a, Φ_G, P_{app}, P_{rad}, and Q_{vent}). The resulting predictions are presented in Fig. 5.

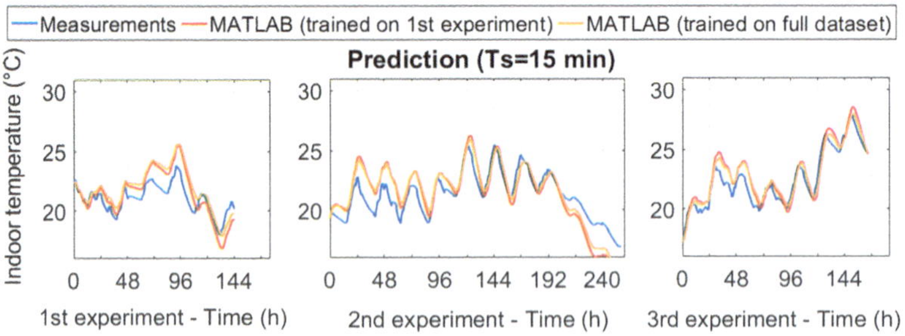

Fig. 5 Prediction of the evolution of the indoor temperature (starting from the initial value) for a model trained on data from the first (red) or all 3 experiments (yellow)

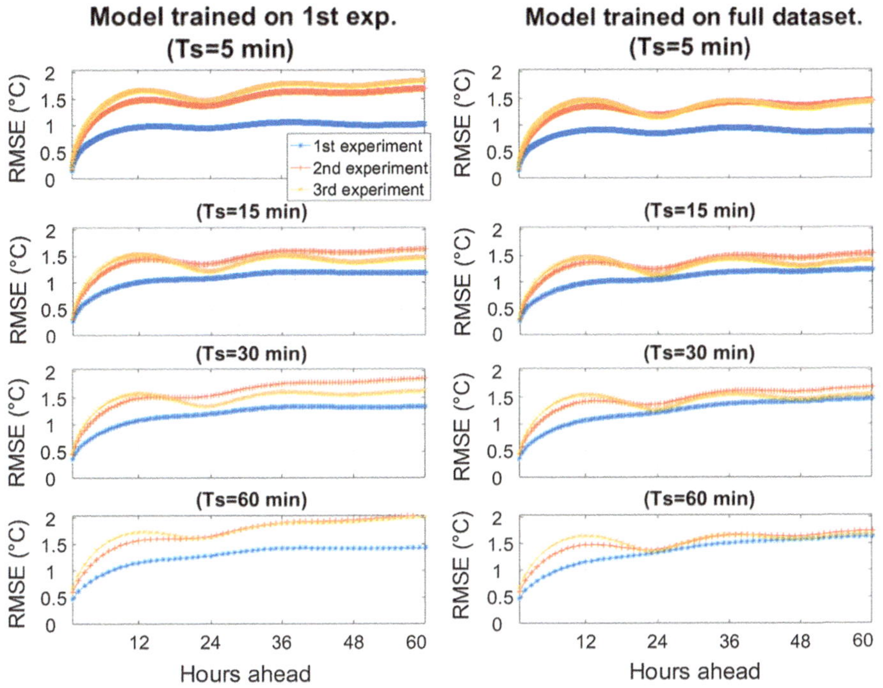

Fig. 6 Evolution of RMSE over prediction horizon for the 3 experiments and different sample times for the model identified on the first experiment (left) and the whole dataset (right). On the left-hand side, blue corresponds to fit to the training data, while red and yellow correspond to validation data, while on the right-hand side all 3 curves correspond to training data

As seen in these results, a simple first order model trained on one week of data was sufficient for predicting the main slow thermal dynamics of this lightweight building over several days. This is supported by Fig. 6 presenting the evolution of root mean square error (RMSE) over the prediction horizon for sample times of 5, 15, 30 and 60 min. It was observed that short sample times of 5 and 15 min appeared to provide better overall prediction capability.

As also seen on Fig. 6, the prediction performance is increased by considering more data than just the first experiment. In other words, one week of data was not sufficient for obtaining the best possible fitting to all experiments, as cross validation did show a gap between training and validation data (see left column, as opposed to right column). Yet the improvement of RMSE resulting from training upon the 2 later experiments was small in the case of 5 and 15 min data.

It is also important to note that the performance of the model identified on the first experiment was similar on validation data with open and closed bedroom doors. This can be explained physically by higher heat transfer between rooms (both through walls and infiltration through doors) than losses to the ambient through the envelope in such a super-insulated building.

4 Conclusion

This work presented experiments to identify the thermal dynamics of an actual super-insulated building. The heating was operated according to a predetermined rich excitation sequence (PRBS) on a single electrical heat emitter. The experiments showed that large disparities of air temperature happened in the building, due to the combination of zoning (using doors) and air stratification.

The model investigated was a first order model, with 3 parameters to be identified: a lumped heat loss to ambient, a heat capacity and a solar gain. Two software packages were compared for achieving the model fitting: CTSM and the MATLAB System Identification toolbox. These two software packages yielded different values (and uncertainties) of the model parameters, despite identical initial values and bounds in the optimisation. Increase of the sample time of the measurements decreased the values of all 3 model parameters (especially for the MATLAB toolbox), while increasing the uncertainty on each of them. In any case, none of the software packages was found to attain a significantly higher performance than the other.

The prediction capability of the first order model identified in MATLAB was analysed, revealing better short-term performances for measurement sample times within 5–15 min, which are therefore recommended in future works.

One week of data under PRBS excitation with an electric radiator allowed identifying such a simple single zone model. This model represents well the main slow thermal dynamics of the building-averaged temperature in this lightweight building case (a different conclusion may be reached on a heavier one). Moreover, this prediction performance was observed to be unaffected by the configuration of internal doors.

Further work should investigate more detailed dynamical models (exploiting the open dataset of the experiment [16] for benchmarking purposes), improvement of the solar radiation modelling (using e.g. a time variating parameter for A_W), and assess the potential of the model for operation of the heating using model predictive control.

Acknowledgements The authors would like to acknowledge valuable input from Eirik Selvnes, Stein Kristian Skånøy, Martin Thalfeldt, Peng Liu, Francesco Goia (NTNU), Maria Justo-Alonso (SINTEF), Glenn Reynders (KU Leuven), Peder Bacher (DTU Compute), and the 2 anonymous reviewers. Access to the LivingLab was provided and funded by the *Research Centre on Zero Emission Buildings project* (and its successor: the *Research Centre on Zero Emission Neighbourhoods in Smart Cities*), while the main author's work was part of the *ADVANTAGE project* funded by the European Union's Seventh Framework Programme for research, technological development and demonstration under grant agreement no. 607774. This work was initiated as part of a collaboration within the IEA EBC Annex 67 "Energy Flexible Buildings".[1]

[1]Project links: http://www.zeb.no/ | http://www.fp7-advantage.eu/ | http://www.annex67.org/.

References

1. A. Afram, F. Janabi-Sharifi, Theory and applications of HVAC control systems—A review of model predictive control. Build. Environ. **72**, 343–355 (2014)
2. Clauß J, et al.. Control strategies for building energy systems to unlock demand side flexibility—A review. Building Simulation Conference 2017, San Francisco: 2017
3. Ljung L. System identification: theory for the user. 2nd ed. (1999) Prentice-Hall
4. X. Li, J. Wen, Review of building energy modeling for control and operation. Renew. Sustain. Energy Rev. **37**, 517–537 (2014)
5. E. Atam, L. Helsen, Control-oriented thermal modeling of multizone buildings: methods and issues. IEEE Control Syst. Mag. **36–3**, 87–111 (2016)
6. Madsen H. Time series analysis. (2008) Chapman & Hall
7. S. Prívara et al., Building modeling as a crucial part for building predictive control. Energy Build. **56**, 8–22 (2013)
8. T. Salque, D. Marchio, P. Riederer, Neural predictive control for single-speed ground source heat pumps connected to a floor heating system for typical French dwelling. Building Serv. Eng. Res. Technol. **35**, 182–197 (2014)
9. P. Bacher, H. Madsen, Identifying suitable models for the heat dynamics of buildings. Energy Build. **43**, 1511–1522 (2011)
10. G. Reynders, J. Diriken, D. Saelens, Quality of grey-box models and identified parameters as function of the accuracy of input and observation signals. Energy Build. **82**, 263–274 (2014)
11. H. Madsen, et al. Thermal performance characterization using time series data; IEA EBC Annex 58 Guidelines 2016
12. CTSM-R Development Team. Continuous Time Stochastic Modelling in R, User's Guide and Reference Manual. 2015, http://www.ctsm.info/
13. Ljung L. MATLAB System Identification Toolbox ™—Getting Started Guide R2016b (Version 9.5). 2016
14. A.G. Hestnes, N.L. Eik-nes (eds.), *Zero emission buildings* (Bergen, Fagbokforlaget, 2017)
15. Goia F, Finocchiaro L, Gustavsen A. The ZEB Living Laboratory at the Norwegian University of Science and Technology : a zero emission house for engineering and social science experiments. 7 . Passivhus Norden | Sustainable Cities and Buildings, Copenhagen: 2015
16. P. Vogler-Finck, J. Clauß, L. Georges, A dataset to support dynamical modelling of the thermal dynamics of a super-insulated building. (2017). https://doi.org/10.5281/zenodo.1034819
17. J. Clauß, P. Vogler-Finck, L. Georges, *Calibration of a high-resolution dynamic model for detailed investigation of the energy flexibility of a zero emission residential building* (Cold Climate HVAC, Kiruna, 2018)

Sustainable Architecture in Northern Subarctic and Artic Climate

Ulf Nordwall and Thomas Olofsson

Abstract Since the antique the Vitruvian virtues three basic qualities, solid, useful and beautiful is central themes when judging architecture. Although the themes have remained, the conceptual idea of these qualities has shifted. In today's modern society, sustainability requirements are an integrated perspective, often expressed as a balanced consideration of ecology, economy and social aspects. There are numerous approaches to measure and evaluate sustainability. Often the perspective of the evaluator's model is predominant. An alternative approach is to ask architects or designers to describe the sustainable idea. This is an approach used to identify alternative perspectives. In this work, we introduce a multiple case study. The scope, to investigate sustainable architecture based on the architects or designers own drawings. The aim is not to define sustainability in architecture but rather to illustrate articulated examples of integration of sustainability issues in architecture. The study is limited to a selection of 22 buildings, proposed as exceptionally sustainable in public media. The northern subarctic and artic regions here are the geographical boundary. The cold climate description used as the background for illustrating a variety of sustainable concepts. We propose that descriptive information, interpreted from the architect or designer perspective, has important contributive factors in the understanding of sustainable housing. We argue that the architect or designers design critique influences the result of conceptual meaning. The attention given to sustainability is the founding architectural factor in these studied buildings, expressed in various forms.

Keywords Sustainable · Architecture · Cold climate

U. Nordwall (✉)
Peab Sverige AB, Umeå, Sweden
e-mail: ulf.nordwall@peab.se

T. Olofsson
Department of Applied Physics and Electronics,
Umeå University, Umeå, Sweden

D. Johansson et al. (eds.), *Cold Climate HVAC 2018*,
Springer Proceedings in Energy, https://doi.org/10.1007/978-3-030-00662-4_45

1 Introduction

When studying examples of sustainable architecture, it is easy to become mesmerized by the differences in the ways it is expressed simply by moving from one place to another. Similarly, the time factor has a clear impact on the depiction of sustainable construction.

An early historical insight into how the significance of the architecture is influenced can be found in the description by the ancient Roman architect Vitruvius in his multi-volume work "Ten books on architecture" [1]. He describes architecture using three basic qualities.

The first one, "Venustas", often translated as beauty. But what is beauty? It is a subject that has caused a lot of concern. The philosopher Wittgenstein expressed it as something immeasurable, the personal experience between the viewer and the object, something that can only be experienced [1]. Some argue that even if beauty is both immeasurable and subjective, there is still something that is more universal, and that is the experience of completeness. We can feel this completeness in situations where we feel that everything is true, when things fit together, when the aesthetic shape is attractive and is independent of time, place and cultural context.

The second quality that Vitrivius emphasizes, "Commoditas", can be translated as convenience or functionality. Good functionality is obviously essential, but if we look at how production has evolved over the last half century, the concept of function has often been regarded primarily as something that can be measured—something that is expected to fulfill important needs. An inherent challenge in the concept is also to find functional solutions that fill the needs of our times and also meet requirements with regard to flexibility so that it is possible to meet future demands for adaptation.

The third quality Vitruvius describes, "Firmitas", can be translated as sustainability. This usually involves technical sustainability. The whole concept of sustainability is usually much broader today. It is based on the qualities of the place of construction, the energy supply systems available and the technical possibilities. Materials are important. What lies beneath, behind the facade and the paint? Is it plaster, porcelain, wood, board or concrete? By allowing the materials to retain their intrinsic and aesthetic values and instead of looking at the material's ageing as a problem, we can look at it as an aesthetic asset.

Vitruvius chose three qualities in order to summarize what he believed architecture should incorporate. The important thing was that all of the parts would be given their importance. Similarly, we may argue today that the concept of sustainability is of importance when it is given a holistic perspective. This may be when the three qualities, good function, sustainability and beauty have a well thought out intention, and especially when great care is paid to design.

Sustainable architecture can vary in design within a geographical area, but there are key elements that unite their resulting sustainability. The scope of this work is to show how building sustainability can be expressed when the architect is allowed to make the description. The study is limited to Northern Sweden (cold climate).

2 Method

2.1 A Subsection Sample

There are numerous approaches to measure and evaluate sustainability. Often the perspective of the evaluator's model is predominating. An alternative approach is to ask architects or designers to describe the sustainable idea with the aimed result of identifying alternative perspectives. In this work, we introduce a multiple case study from a descriptive documentation of a totally 22 sustainable buildings in Northern Sweden. Information of the particular concepts can also be found in [2].

The scope has been to investigate sustainable architecture based on the architects or designers own drawings. The aim is not to find a definition of sustainability in architecture but rather to illustrate articulated examples of integration of sustainability issues in architecture. The analysis of the text is grounded on the three virtues of Vitruvius—function, sustainability and beauty. This is illustrated in Fig. 1 where economy is introduced as a fourth feature.

In this perspective, six different sustainability categories are identified and analysed: Climate, Site, Function, Life cycle, Low energy and Passive house.

From these texts, summarized in [2], a suggestion of integration of the six perspectives in the triangle is suggested based on a simple narrative analysis.

3 Concept

Twenty-two building concepts [2] documents were used in the conducted study. Those building concepts were categorized into the above introduced 6 perspectives. This section is condensed to comprise one selected building concept for each perspective.

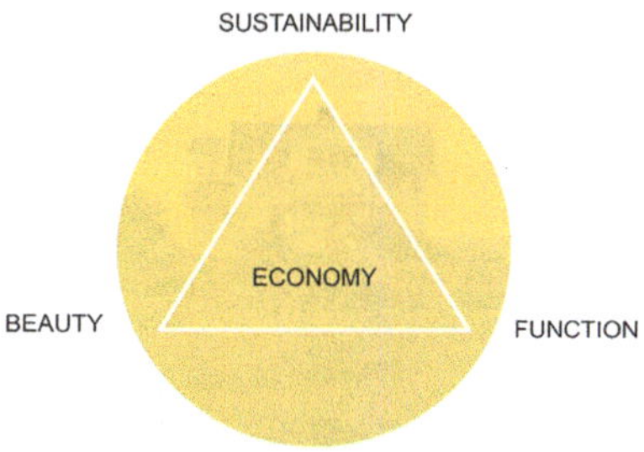

Fig. 1 The Vitruvian triangle with economy as a fourth feature

3.1 Climate

3.1.1 Ice Hotel

The Ice Hotel in Jukkasjärvi, is the first ice hotel and art exhibition of snow and ice. It is today one of Northern Europe's largest tourist attraction. The Ice Hotel is built yearly near the Torne Riverbank, approximately 20 km from the city of Kiruna. Stretching over an area of 5500 m2 the hotel is built of snow and ice collected from the river. Its architecture changes with every year as the foundation must be rebuilt after the warmer seasons. During the month of March, ice is harvested and stored in the production hall with a capacity of over 4000 tons of ice—in preparation for the following season. Production of a strong structural building uses approximately 1000 tons of ice. Snow machines are used for the production of wall elements. Snow is sprayed on to metal structural forms that are left to freeze over a few days. These forms are then removed to reveal a labyrinth of freestanding chambers. Walls are built in to form corridors, rooms and hotel suites. The stored ice blocks are then transferred in to the hotel where artists from all corners of the world participate to produce art and design with ice. The ice hotel is built with natural ice taken directly from Torneälv and is shown in Fig. 2. The ice hotel only exists between December and April and when spring comes, everything is melting away and returning as water to the Torne River. The material is 100% reusable.

Fig. 2 Ice hotel

3.2 *Site*

3.2.1 Fiskevistet

Skagsudde with the fishing villages Skeppsmalen and Skagshamn has long been a popular place for excursions, situated in the outermost archipelago Southeast of Örnsköldsvik. The area has a dramatic landscape and an interesting cultural heritage. Skagsudde is important for the local maritime shipping. The Gävle Fishermen Haxar who arrived by boats founded the beautiful fishing village of Skeppsmalen at the beginning of the 17th century. They fished along the coast of Norrland leasing cabins and land from the farmers in the area.

The area has an abundance of archaeological, as well as contemporary cultural findings. Skagsudde constitutes the northern gate to Höga Kusten, well known for its shingle beaches and unique world heritage remnants. Fishermen huts, fishermen chapels, unique boats and tools are preserved here. Emphasising each other and creating tension between the traditional and the contemporary aspects is the theme of the building that has been erected by Skagsudde. The site and the characteristics of the items of the exhibition, along with the expectations of the building meant that tradition and modernism were forced to consolidate and find balance. Careful attention to craftsmanship and material contributes to the aesthetics of the aged materials without appearing deteriorated, see Fig. 3.

Fig. 3 Fiskevistet

3.3 Function

3.3.1 Architectural School

Both buildings and operational functions within the Artistic campus at Umeå University have long-term sustainability goals and themes. The new buildings are quite different from one another, yet are bound by their common principles, designed for long-term sustainability. An example of this is found in the logistical placement of all meeting points gathered together on the ground level. The new buildings are connected by a common architectural design theme, with facades clad in untreated larch wood panelling, sharing common system solutions and use of local production. The buildings are designed and constructed to last for one century. Further commonality are the smaller outdoor areas, found between various forms create natural space for pause and reflection with spectacular views. There rooms are designed with the same hardwearing and easily cleaned coloured concrete floors found within these new buildings. A large share of the Arkitektskolans components are locally manufactured. The facades Siberian larch wood is imported from Russia while the wood elements are manufactured in Burträsk by the Västerbotten carpenters WBTrä. Window glass is delivered by the local glass business Umeå Glas AB. The design is illustrated in Fig. 4. A curtain wall of glass could not minimise the energy use, here the preferred solution to use renewable resource of wood dominated the project. The building is characterized by its flexible architectural design. The materials have been chosen to be resistant for at least one

Fig. 4 Architecture school

century. The technical systems are designed to provide low energy use and flexibility. The ventilation systems are integrated into the steel piles and steel beams that form the building's structural design.

3.4 Life Cycle

3.4.1 Laggarbergs School

Laggarbergs School in Timrå is world recognised since its rebuild in 1994. During the past twenty-five years it is estimated that more than 500 delegations from near and far have visited and studied all of the advanced innovative ideas found within the school. Examples such as ecocycling, ventilation systems for heating and innovative discussions pertaining to separating toilets. All school children participate in a bathroom course to ensure that the correct discards wind up in the correct holder. The school building gives an impression of healthiness, one that is difficult to wear out. The design choice of building materials and technology is clearly communicated. Wood and excelsior boards are predominant and an example of the sustainable material that withstands time, see Fig. 5.

Here, the most prominent idea is the ventilation system, natural ventilation—a soundless and effective system that aims to replicate qualities of actual anthills. This system creates a perfect indoor climate where the user does not find the need to open windows at any point. This project is an example of the success that may be achieved by rebuilding and transforming an old school to a modern school with a build in eco cycle—rather than demolishing it. Laggarsbergs School is built on an educational, ecological and technical vision developed in consultation with a

Fig. 5 Laggarbergs school

dedicated municipal project group. The school offers, with its sustainable solutions, a physical environment that may be used as case study examples for its own courses.

3.5 *Low Energy*

3.5.1 Geografigränd

After a short and intensive sketching phase during the spring of 2010, a final design solution with the disposition of apartments, plan drawings and exterior design was produced. The house is placed according to the existing block structure for the area with yards opening up to the south. Parallel to the sketching period, a large focus on obtaining economical support from the delegations for sustainable cities, a focus that succeeded in a positive result. In the open area of the property a sauna could be placed accompanied by a winter garden. All 137 apartments have their own balcony or outdoor patio. Each level of the three houses also have glass enclosed balconies. A ground rule for the buildings placement was the need for single sided apartments to be positioned towards a quieter area—the garden. This resulted in the design of narrower houses with corridors in the façade coupled by a characteristic window placement pattern towards the Studentvägen and Gegrafigränd. It is in the connection to the corridors that the common glassed in balconies are found. Each corridor bears its own colour; yellow, green, blue, lilac and red. Colours that can also be found in the balconies front in perforated sheet metal, see Fig. 6. The choice of the appropriate façade material to adapt these buildings to the 'Ålidhemskaraktären' (the character of Ålidhem) was brick. Brick used for the

Fig. 6 Geografigränd

bicycle storage house was reused brick from the original burnt down building. The result is Norrlands largest low energy project. The strategy of sustainable building with energy consensus in cold climate, places the focus on both energy and lowered total energy use. The base idea is to use the reusable district heating systems that exists in Ålidhem and minimise as well as substitute the use of electricity and fossil energy.

3.6 *Passive House*

3.6.1 Tavleliden

From in the beginning of the design phase of the sustainable houses, this project had a project goal of 'Triple zero", which in principle is a zero tolerance for the three categories; non-renewable energy, emissions and waste. No energy should come from non-renewable primary energy sources as well as a general zero tolerance for toxic emissions from such as materials, construction or energy conversion. No dangerous waste means using renewable and plant-based materials that can be easily reused or recycled. All constructions parts and materials should be easily demountable. Composite materials are avoided. The buildings are carefully planned. The execution will be as accurate possible since the project is to be carried out by the students attending Dragonskolans Construction Program. This arrangement also allows students to be trained to build the next Nearly Zero-Energy Buildings, which will be EU requirements starting in 2019. The entire project consists of six single-family villas and is adapted to the students' academic year and curriculum, in two stages. One of the buildings are illustrated in Fig. 7.

Fig. 7 Tavleliden

Process quality has been important in the project. In order for the villas to get the function they intended, a "Sustainability and quality program" was based on the project. The program has been developed with the client and project management, which guides the process, from the idea stage to the completed building and the future use phase. The six villas on Tavleliden is one the world's northernmost certified passive houses 30 miles south of the arctic circle. The project is Sweden's first with a combined certification of both passive house certificate and a more comprehensive sustainability certificate according to SGBC "Environmental Building Gold". Passive housing components have been used, chosen a construction without cold bridges and quality assured built according to PHI's five recommendations.

4 Results and Discussion

Based on the descriptive study of the architects who designed the investigated sustainable projects we find more focus of sustainability in the categories Passive house and Low energy, than in the categories Climate and Function. That category site is found somewhere in the middle. A summary of categorization is illustrated in Fig. 8.

The study takes the starting point in the perspective that depending on the time location and situation, sustainability comes to different expressions.

Developing sustainable architecture and continuing the exploration of new methods within sustainability and its integration is an increasingly valuable issue for society. Sustainability can be seen holistically as Vitruvius proposed: function, sustainability and beauty.

The built environment should be seen as a result of the chosen process form given to each individual project. We have the possibility to introduce new elements and competencies to such existing processes in order to challenge the standard solutions otherwise normally followed.

Current sustainability methods result in higher conceptual meaning when architects or designers with their feedback and theoretical tools are invited to criticize and influence result. This is confirmed in our studies, where the careful attention given to the element of sustainability is seen as a grounded architectural

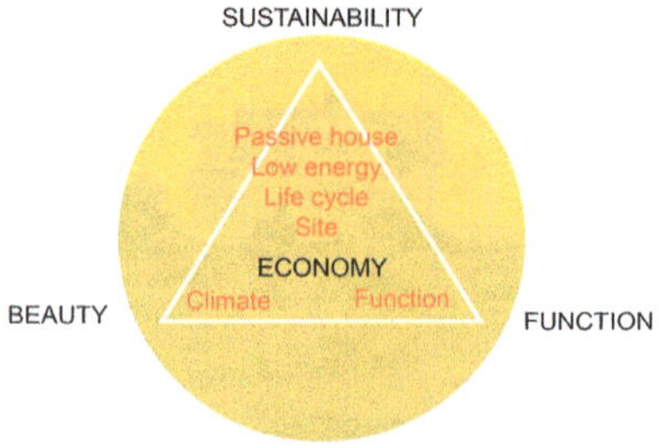

Fig. 8 The Vitruvian triangle with economy and cold climate sustainable perspectives

factor—a factor that has the capacity to be expressed in varying ways and solutions. Whether this be in the physical building elements, technical solutions, chosen materials or its program.

In summary, we propose that descriptive information, interpreted from the architects or designer perspective is an important contribution in the understanding of sustainable housing. The architects and designers role are invaluable with their holistic competencies to be integrated in the traditional process of building and designing sustainable environments.

References

1. Vitruvius, Om arkitektur: 10 böcker, översättare Birgitta Dalgren, Dymlings förlag ISBN ISBN 9789185173129 (2009)
2. Nordwall U. and Olofsson T, Hållbara hus i kallt klimat 22 hållbara hus från Bollnäs till Kiruna 1994–2014, Svensk byggtjänst, Göteborgstryckeriet, Mölndal, Sweden, ISBN 978-91-7333-496-9 (2015)
3. MILJÖBYGGNAD—en svensk certifiering som värnar om människa och miljö. 2011. URL: https://www.sgbc.se/docman/certifieringssystem-1/25-broschyr-miljobyggnad/file
4. SVANENMÄRKT http://www.svanen.se/
5. Om KOLJERN®-tekniken http://www.koljern.se/se/koljern/
6. Boverket, Boverkets byggregler – föreskrifter och allmänna råd, BBR, BFS 2016:13 BBR 24. Boverket, 2016

When Buildings Become Intelligent—A Network Analysis of Building Automation, Operation and Competencies

Marianne Forman and Nils Lykke Sørensen

Abstract The global climate agenda has challenged the construction sector and led to a significant use of technology in buildings. At the same time, new technologies like Internet of Things, Big Data and Artificial Intelligence are beginning to fit into the buildings' technical systems. Knowledge is available about the technical side of the "intelligent building" however, there is insufficient practical experience and knowledge about the competencies that are required in order to interact with the intelligent building. The purpose of this analysis is to elucidate competence requirements in connection with the dissemination of building automation in large buildings. The study is based on qualitative interviews of six industry organizations and four organizations that in different ways relate to the operation of large buildings. They are strategically chosen whereas they represent four different models of building operation. In the study, building automation is related to the construction sector's need for finding optimal solutions and methods that ensure knowledge transfer across building operations and construction projects. The results indicate that (1) building operation can be configured in many ways, (2) the different building operation models work towards optimization of operation using building automation (3) the different building operation models use different strategies for transferring knowledge from operation to building project. The recommendation is that a forward-looking competence strategy can benefit from building on a system understanding of building automation, rather than a sole focus on technical specialization. The analysis also points to the need to explore building automation as a socio-technical system, thereby supplementing the technical research of the past.

Keywords Building automation · FM strategies · Competencies

M. Forman (✉) · N. L. Sørensen
Aalborg University, Copenhagen 2450, Denmark
e-mail: maf@sbi.aau.dk

N. L. Sørensen
e-mail: nls@sbi.aau.dk

D. Johansson et al. (eds.), *Cold Climate HVAC 2018*,
Springer Proceedings in Energy, https://doi.org/10.1007/978-3-030-00662-4_46

1 Introduction

Like other sectors, the operation of large buildings is under pressure to change because of the increasing use of new technologies in buildings. The development is important in Denmark like in other countries, as evidenced by new government research initiatives. In 2016, the Ministry of Higher Education and Science published the report "An OECD Horizon Scan of Megatrends and Technology Trends in the Context of Future Research Policy". The report points to trends and technologies that shape future research needs. Out of ten highlighted areas, there are three areas that are relevant in the context of building automation: Internet of Things, Big Data and Artificial Intelligence. Common for the technologies are that they are all new major focus areas within the construction sector and that the expectation is that they are all part of the future solution due to building performance concerning climate change, energy reduction, increased productivity (economic), increased health etc. However, at the same time, there are already examples of use of the technologies within the operation of major buildings.

In Denmark, the construction project and the operating organization are typically separated. After a period of unfulfilled expectations in connection with deliveries of new buildings and subsequent operations there has been a focus on the interaction between the building operation and new technologies. Consequently transferring of operating organization's experiences to construction projects has become a key issue to ensure solutions that also work in operation. Traditionally the construction project and the operation organization are two different worlds that differ concerning purpose, tasks, organization structures and actors. The construction project aims at designing, executing and delivering a building to the client, usually at a specific price and at a specified date. The organization is a temporary project organization. When the project stops, the organization is dissolved. The operation of a building, also often called facility management (FM), is about operating and maintaining the building in order to meet the requirement of the users and owners. Unlike the construction project, the FM organization is often a stable organization. However, in turn the way the organization is configured can vary depending on the degree of outsourcing of functions to sub-suppliers from few to partly all functions. In general, attention has been paid to FM as a central knowledge base of expertise that can help to meet the challenges during operation, but today it is also pointed out that new solutions must be found in a closer interaction between the construction project and the operating organization during the design process, where solutions should be based on practical experiences from the operation of buildings [1, 3]. Hence the need for a close link between the construction project and the operating organization has increased. The use of technology in FM has also increased the need to control the interaction between the different technologies. Previously it was common to involve the client and the user and their requirements in the construction project, but the FM organization was often absent, a kind of "invisible user" [2]. For that reason large operating organizations has begun to include the FM organization in construction projects and to formulate and systematize operational requirements when ordering new buildings or refurbishments. However, experience shows that it is not necessarily easy and that in practice change needs time. Jensen [3] points out that the transfer of knowledge between construction projects and operating organizations introduces new types of challenges, whereas Forman [2] points to the need for development of new information infrastructures between operating organizations and construction projects based on learning processes. Other strategies are strategic partnerships with suppliers, outsourcing of parts of the entire FM task and the use of commissioning in construction projects.

2 Methods

Building automation is embedded in a technical and social context, why the analysis is based on Science and Technology Studies (STS). An STS approach implies that technological change is studied as processes where technologies and actors interact mutually, creating a socio-technical network. The importance of technology development for the operating supplier system has therefore been investigated, as a basis for understanding the future competence requirements related to building automation.

The survey was based on a single case of embedded analysis units, according to Yin [5]. This approach made it possible to produce context-related and practical knowledge about the importance of the field in question. The case method was chosen as it provides access to real-life situations with a richness of details, which made it possible to study a process in action. Since it is an area, primarily studied from a technical perspective, it was central to gain access to real situations in order to identify the mutual formation of building automation and social contexts in connection with the operation of buildings.

Different data sources were used in the case study. Initially, a feasibility study was carried out for the purpose of uncovering technological trends, strategies and interests at the policy level. Key persons from industry and employee organizations were interviewed about their experience with and expectations of building automation. The chosen organizations all participated actively in a committee on continuing vocational training (VEU), including in the field of building automation.

The organizations/key persons interviewed were:

1. EL-forbundet (Electrician trade union): Secretariat for Education
2. Tekniq: Assistant Director and a Consultant
3. Dansk Industri (Confederation of Danish Industry, Enterprise): Industry Director
4. Dansk Byggeri (The Danish Construction Association): Chief Consultant

5. Arbejdsgiverne (The Danish Employers' Association for Industrial and Construction): Industry Director and Development Manager
6. Dansk Metal (Danish metal trade union): Professional Secretary and a Consultant.

Following the feasibility study, a company-level survey was conducted in order to collect practical and local experience of building automation. The business survey was based on four strategically selected organizations, representing different locations in the operating-delivery system based on a desire for maximum variation in ways of organizing building operation. At the same time, they were chosen because they are exploiting new technological capabilities, thereby ensuring a high level of information about the relationships between new technologies and building operating. This may give a bias towards a more mainstream development, but in this situation, this risk was considered to be offset by the fact that all organizations were well established and reputable, for which reason they were perceived as first movers. Table 1 briefly presents the chosen organizations and key persons that were interviewed:

The case was validated in different ways. Firstly, descriptions of the company practice were prepared based on interviews and documents and subsequently approved by the interviewees. Secondly, the entire analysis was discussed by the VEU committee in Denmark.

The structure of the analysis was as follow. The feasibility study made it possible to develop a network-based model of building automation as well as a concept for the connection between the operating-delivery system, the construction project and building automation. Both models were drawn up as sketches and used to mediate the qualitative interviews at company level. The two models will be elaborated in the next section. Then the results from the company analysis follow. Finally the conclusion suggests the central accommodations.

Table 1 Companies and keypersons

Companies	Interviewees
SIF: Certified electricians, which deliver technical services and operating services within building automation	Development manager
Siemens: Component and system products supplier within building automation, and provider of operating service	Responsible for ESCO activities CTS specialist Senior business development manager
DTU, Campus Service: Operating organization where all operating functions are primarily in house	Head of Building Management System (CAS BMS)
Aalborg University Copenhagen, Campus Service Copenhagen: Operating organization where all operating functions are primarily outsourced	Operating manager, Copenhagen

3 Concept 1: Building Automation—A Network Representation

Typically, the understanding of building automation is based on a representation as a pyramid, where the respective technical function-divided systems are located at the bottom and the management of the overall technology is located at the top, i.e. a hierarchical system. However, technological development means that this representation of building automation no longer reflects the technology development that is currently transforming automation into a network technology. The need for a new representation based on a network's understanding reflects on the one hand the ongoing technology development that moves towards network-based solutions and, on the other hand, the development of knowledge and skills that move towards distributed knowledge systems where knowledge and skills to develop, manage and use the technologies are located at many different actors.

Figures 1 and 2 illustrates the change in perspectives from building automation represented as a hierarchical system to a building automation represented as a network system.

Especially, three technologies have impact on building automation: (1) Internet of Things (IoT), is in a FM context referred to as 'Embedded Technology' and consists of sensors that collect data relevant to the FM system. Data is transferred to a management system for further processing. The use of sensors is growing by the increasing number of embedded measurements. In the report of the Ministry of Higher Education and Science [4], IoT is further linked to energy systems and smart

Fig. 1 Building automation represented as a hierarchical system

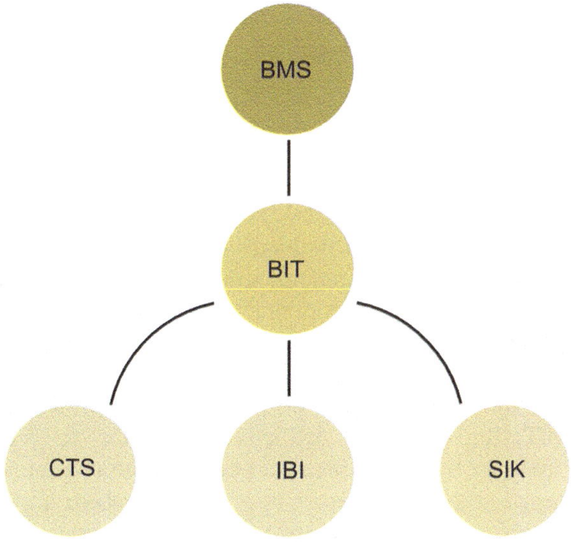

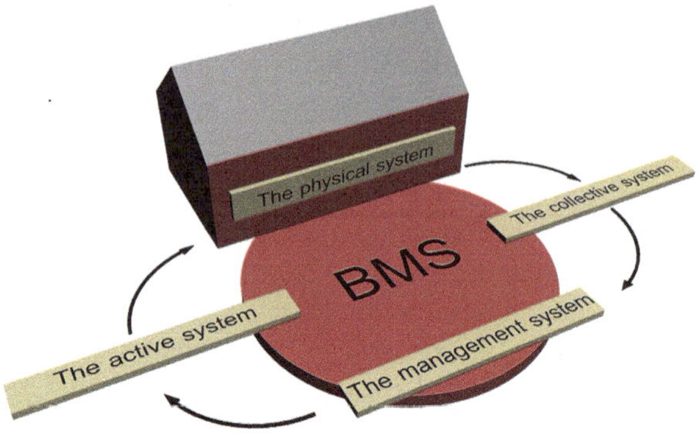

Fig. 2 Building automation represented as a network-based system

cities. The research area will lead to the development of new sensor types, but also address the problems associated with the use of sensors. (2) Big Data is large amounts of data. The increased digitalization and coupling of sensors to the net increases the ability to utilize large amounts of data for optimization of operations in relation to, for example energy optimization, but also for improving decision-making in connection with design, construction and product development. New types of analyzes based on FM data can be used to formulate operational requirements for construction projects and new products in FM. Analyzes of Big data are already used in management systems in supply companies. The research area will undoubtedly lead to the development of more opportunities for working with Big Data analytically, but will also be facing the problems that analyzing Big Data entails such as: Who has a legitimate right to own data and how to ensure privacy and personal protection? (3) Artificial Intelligence deals with systems that process the large amount of data resulting from IoT and Big Data. In an FM organization, this is referred to as decision support systems. Artificial intelligence is already used in several technical systems, such as wastewater plants and production systems. In FM systems, data can be organized, analyzed and reported using various algorithms for both system optimization and planning, and in both cases it supports decision-making. The research area will undoubtedly lead to the development of new forms of decision-making systems, and with increasing intelligence, but will also address problems like: Who should have the right to decide what; the Building, users, operations, management or manufacturers?

In the network representation (see Fig. 2), BMS is perceived as the sum of the four subsystems (the physical system, the collecting system, the management system and the active system) and their mutual impact and exchange effect. In the network representation, the systemic aspect of building automation is weighted and IoT, Big Data and Artificial intelligence are integrated in the representation. Contrary to the hierarchical structure, the respective technical functionalized systems are bound together in a network where also the physical building with its technical installations is included. The four subsystems can be described as follows:

1. The physical system is the physical building and its installations. It can be perceived as a system that is the cause of information delivery and the target of action.
2. The collecting system is all that collects data about the physical system. This may be sensors or meters that are embedded in the physical system, but also other kinds of information like collected user information.
3. The management system is where the collected information is processed and transformed into action, knowledge or perhaps inactivity. It is a decisive system capable of interpreting and analyzing the information received and which can initiate the relevant actions in continuation. The system contains servers, networks, interfaces and algorithms in order to transform the collected information into action.
4. The active system is the executive system that translates the management system's requirements and instructions to actual actions. This can be the person sent to the blasted water pipe or the signal that automatically changes the system's temperature.

However, the network-based representation of building automation can also be seen as a representation of the intelligent building and thereby illustrate how the intelligent building itself contributes to the development and demands new competence needs.

4 Concept 2: Relations Between FM, Construction Project and Building Automation

Figure 3 shows the link between the construction project and the operating system. The figure is inspired by the FM circle and illustrates the interaction/dependencies between the operational function's ability to influence building projects by setting

Fig. 3 Competence needs in the FM delivery system in connection with building automation

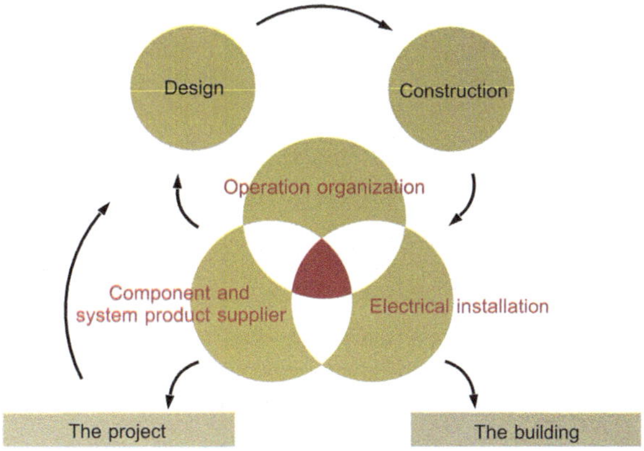

operational requirements and the operational function's ability to make buildings function properly by taking over buildings already optimized.

Operational requirements and knowledge transfer between the construction project and the FM system have long been in focus, but with the technological trends it may appear that new ways of generating operational requirements for the construction project are not only based on practical experience, but also on computer-generated requirements, hence building automation can be related to two challenges in construction. The first challenge is the need to find solutions that ensure optimal operating conditions, see Fig. 3, the arrow to the right, at the bottom. The second challenge is the need to find solutions where building automation supports knowledge transfer from operation to construction project to ensure higher quality in deliveries, see Fig. 3, the arrow to the left at the bottom.

5 Company Analysis and Discussion

All interviewees supported the models (Fig. 2 and Fig. 3) as they were presented to the models during the interview. This indicates that both the technology and the networks that relate to the technology are changing. In the following sections, the results of the company survey will be presented focusing on various aspects of this transformation.

5.1 The Construction Sector's Two Challenges and Building Automation

Building automation in FM can contribute to two key challenges in construction; Optimization of the operation itself and improvement of the construction industry by returning the operational knowledge for upcoming construction projects. The case companies have different options and roles in the operating delivery system, which gives them different possibilities for realizing business strategies in relation to FM. The cases shared a lot of common experience on handling the first challenge, while handling the second challenge has been the cause of different types of strategies.

Relative to the first challenge, the different delivery systems related to operation systematically worked towards continuous optimization of operation using building automation through ongoing troubleshooting and optimization. Relative to the second challenge, several strategies became visible.

DTU CAMPUS Service "FM organization with Competences in House" worked as a learning organization in the field of 'Intelligent Building' closely integrated with DTU's additional organization. The strategy was targeted involvement of the FM organization in construction projects, so that knowledge and experience could be systematically involved. For that reason, the FM organization worked to improve this process through the development of requirements specifications, procedures, guidelines, etc. as a solution to the other challenge.

Siemens Building Technologies "Component, System Products, and Operations supplier" and SIF "Provider of FM Service" both worked toward improving opportunities to enter a dialogue with both customers and construction projects in order to ensure better solutions already in the construction projects. Both companies work with the development of more specialized solutions for different market segments such as development of welfare technology for the healthcare sector. Thus FM experience can be embedded in new products from suppliers, FM consulting services to customers and designers as well as new solution concepts.

AAU Campus Service Copenhagen "FM organization with a high degree of outsourcing" worked with improving contractual relationships, improving partner relationships and improved documentation procedures. Despite outsourcing of tasks, it was stated that the organization needed to systematize FM experience and get access to operational data as a basis for requirements formulations in connection with construction projects. The organization emphasizes their need to make the FM organization both personally and company-independent in order to secure operations. It may seem that suppliers and service engineering should not only turn their attention to increased consulting activity in the construction project, regarding challenge 2, but also increase the interaction of the delivery system in order to qualify the operating delivery system's overall knowledge transfer between FM and construction project.

5.2 *Building Automation as Infrastructure for Different Actors*

Building automation is growing as a new infrastructure with many users. As previously mentioned, building automation is expected to help solving the challenges of the constructions sectors not only in relation to energy and indoor climate, but also in the management of climate changes, sustainability, etc. This development will further link new subjects and new thematic areas to building automation and thereby expand the heterogeneous network of actors that relate in various ways to building automation. At the same time, there is an increasing tendency for building automation not only to be linked to the management of traditional building installations, but also to integrate with the special functions of the buildings, such as when the certified electricians works with welfare technology in the field of care and the DTU Campus Service works with the management and documentation of research conditions at the university. This may affect the ways in which existing professional standards develop, caused by issues such as—"what is healthcare" and "what is research"? This illustrates how new user groups in relation to building automation emerge, and indicates how building automation knowledge is being distributed to new types of actors. Thus automation can create new conditions for the development of practice in different sectors, and without proper reflection, new norms and values can be introduced through the backdoor. In connection with research and development programs, there is a particular focus on security regarding the use of the technologies concerning hacking of building systems, personal safety and rights. Since these are matters that have to be taken into account through choice of solutions, products, organization and service structures etc. these are issues that employees working with building automation should be able to handle reflexively and critically. It is not sufficient that building automation is technically well-functioning; the solutions should also reflect considerations on non-intended applications of systems and personal rights.

Both the certified electricians and the suppliers described how the increased specialization of solutions for different market segments is likely to be embedded in new product services, targeted the use of buildings. This indicates that depending on building function, problems and solutions related to hacking, personal safety and values will require different types of systems and standards, and not least competencies.

In order to counter the new working relationships that arise with the development of building automation, it must be argued that the need of competence cannot be isolated to a well-defined area, but necessarily reflects local competence in the heterogeneous network. One way is to balance the relationship between a necessary understanding of technology and a local competence need, can be to distinguish between 'something common' as that common to all actors, and 'something local' linked to the individual actor's responsibility, e.g. energy, indoor climate, climate measures, sustainability, care or research.

5.3 Significance of Competencies

What is knowledge related to building automation? Everyone can agree that there is some technical knowledge, but apparently there are also a lot of other knowledge and competencies central to the successful expansion of building automation. In the four studied organizations the need for upgrading FM operators, advisors, builders and users was all mentioned; all of whom were perceived as having knowledge for the "right solution" but at the same time lacking in knowledge about building automation so as to be able to ask for the correct solution, use it correctly, design it correctly, etc. Today, a lot of training of these actors is provided by the suppliers and certified electricians in connection with consulting services and delivery. The key to this observation is to understand and analyze building automation in a diverse context. Pursuing the argument mentioned above, 'something common' can be regarded as being the common amount between the three types of organizations in the FM delivery system and "something local" as being the distinctive in the specific organization. Figure 3 illustrates the common competence requirements represented by the red subset. For all three types of organizations, the goal is to optimize the FM operations through the use of building automation. To find the common needs of competence, an analysis has been conducted across organizations. The analysis identified four overall skills of competence actively demanded by all. The four skills outlined in paragraphs 1–4 are connected and cannot be separated.

1. Understanding building automation: An abstract understanding at the system level for use in diagnosing operational problems and analyzing large amounts of data. In this context, it was emphasized that it's all about understanding the individual subject in the system, and keep the professionalism of craftsmanship.
2. Document building automation: Documentation of issues and solutions as a basis for testing, troubleshooting and learning, to ensure person-independent systems.
3. Communicate building automation: Simplification of complex conditions so they can be conveyed to the end users without technical prerequisite. In this context, emphasis was placed on understanding and decoding user's needs, interface that makes sense for the user, and dissemination of technical complexity, enabling the user to engage in complex decision-making processes.
4. Managing development in values, standards and security using building automation: Development and implementation of building automation can create new professional standards such as research standards, care standards, etc. that can change values and work processes. In this regard it is important to visible the choices taken. Digitalization and automation opens for new opportunities for the operation of buildings, but also for new types of crime in the form of hacking and the collection and dissemination of personal data. In this context, emphasis is placed on the fact that new criminal patterns grow along with new solutions and that it is therefore essential to be at the forefront to disarm crime patterns.

Where the four common competence skills describe the system required for all FM employees related to building automation, the local subject-specific competencies skills must reflect the need for specific issues. Then the four common competence requirements can be combined with different problem-oriented competence. Within the various problem areas such as energy, indoor climate, sustainability, climate measures, etc. these skills will varied as they are embedded in different product and production systems and subject to various economic and political priorities.

6 Conclusion

Based on this study, it is suggested that an understanding of the relationships between building automation, building operation and competencies should be based on an understanding of building automation as a network-based technology embedded in a heterogeneous and distributed actor network. This offers the opportunity to identify all the different user practices and work processes that relate to building automation in the various companies and organizations. In addition, it will provide users with a better basis for integrating with the new infrastructure, which is in natural continuation with the idea of the agile organization in response to the challenges and competencies of the intelligent building. This is supported by the fact that much knowledge in connection with building automation is already distributed through networks, which became visible in the survey. Product-specific knowledge is spread through supplier courses, users being upgraded through consultancy and courses provided by suppliers and certified electricians, and not least relevant experiences and knowledge spreading in the delivery system through employees who change jobs between companies and organizations in the operating delivery system.

References

1. F. Bougrain, M. Forman, S.C. Gottlieb, K. Haugbølle (2014) 'Complex performance in construction': Danish Building Research Institute, Aalborg University no. 14
2. M. Forman, Experience with the use of building commissioning advisor—from design to operation. CIB Proceed. **05**(2016), 39–51 (2016)
3. P.A. Jensen, Design integration of facilities management: a challenge of knowledge transfer. Architect. Eng. Des. Manag. **5**(3), 124–135 (2009)
4. OECD (2016) An OECD horizon scan of megatrends and technology trends in the context of future research policy. Danish Agency for Science, Technology and Innovation. ISBN 978-87-93151-92-5
5. R.K. Yin, *Case study research: design and methods*, 4th edn. (Sage Publications, Thousand Oaks, 2009)

Does the Obligatory Ventilation Control Fulfill Its Purpose?

Kristina Fyhr, Caroline Markusson and Svein Ruud

Abstract The regulation "Obligatory Ventilation Control, OVK" has been into practice since 1991 in Sweden. The role of the regulation is to control the ventilation in premises, multi-family houses and single and two-family houses (with certain ventilation systems). The control is done before the ventilation system is taken in operation and after that on regular intervals (except for single and two-family houses). The purpose of the OVK is to secure that the indoor climate is good and that the ventilation system is functioning. This paper will present results on how the OVK works today, both in theory and in practice and, by stakeholders, identified wished development of the OVK. To get a broad picture of how the OVK works, interviews were carried out with persons related to the OVK; building owners, OVK controllers, administrators at the municipality, legislators, organizations etc. Based on this, suggestions were made for how the OVK could develop to better suit its purpose. Results from the interviews show that an approved OVK is not a guarantee that the indoor environment is satisfactory, since the current use of the premises isn't taken into account. Further, the legislation for ventilation is not adapted to new technology such as demand control ventilation, and different authorities' legislation also differs regarding for example reduced/shut off ventilation. The study also shows that the follow up of the OVK from the municipality often are inadequate. Further, energy-saving measures that should be included in the OVK are handled very differently and the level varies considerably.

Keywords Ventilation control · OVK · Indoor environment · Energy efficiency · Ventilation

K. Fyhr (✉) · C. Markusson · S. Ruud
RISE Research Institutes of Sweden, Gothenburg, Sweden
e-mail: kristina.fyhr@ri.se

D. Johansson et al. (eds.), *Cold Climate HVAC 2018*,
Springer Proceedings in Energy, https://doi.org/10.1007/978-3-030-00662-4_47

1 Background

The Obligatory Ventilation Control, in Swedish called "obligatorisk ventilationskontroll—OVK", was introduced in Sweden year 1991. The purpose of the OVK is to "show that the indoor climate is good and that the ventilation systems are functioning". The control must be carried out by an expert performance controller who is certified by an accredited certification body. Every OVK must control that the ventilation system is functioning according to those mandatory provisions that were in force when the system was put into use [1]. In practice, this means that the OVK controller measures the air flows for supply and exhausts air in ducts and diffusers, and compares those measurements with the design values. The OVK control also includes a visual inspection of the ventilation system for pollutants, this in order to prevent pollutants to spread in the building. The OVK also inspects that instructions and maintenance directions for the ventilation system are easily available. The controller must provide suggestions on how to reduce energy consumption for ventilation, the suggested measures are not allowed to cause poorer indoor environment. The OVK shall investigate the ventilation in every premise and multi-family house before taken into operation and after that on regular intervals of three or six years depending on type of building and ventilation system. Single and two-family houses shall only be inspected before the ventilation system is taken into operation, and some types of ventilation systems are completely exempted from the OVK [1].

Since the OVK was introduced, the regulation has been revised several times, and in year 2006 an energy efficiency perspective was added to the OVK. This meant that energy-saving measures should be included and noted in the OVK protocol [1].

The OVK includes many participants/steps, see Fig. 1. The building owner should initiate and order an OVK. A certified OVK controller then carries out the OVK according to the regulation, and writes a protocol containing the results from the OVK [2]. The protocol is then distributed in two copies, one to the building owner and one to the municipality Building Committee. The municipality is responsible for control and supervision, and should ensure that the building owner fulfill their OVK responsibilities.

However, many studies show that the OVK does not work as intended; many premises have complaints about the ventilation and/or lack an approved OVK. Also, the energy-saving part does not work properly.

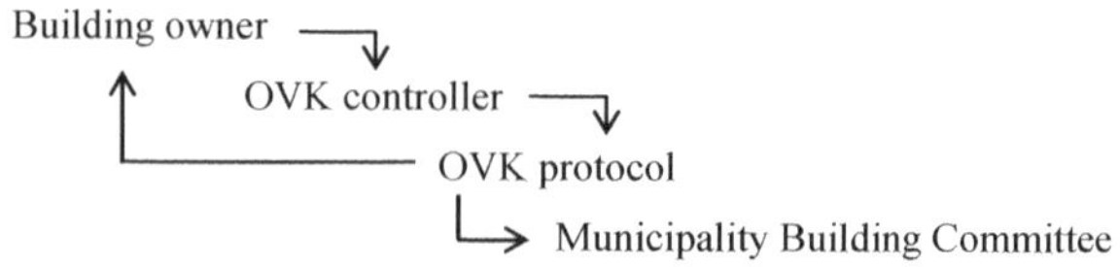

Fig. 1 The participants/steps in the OVK

During the years the OVK has been in practice, several investigations have been made to figure out how the OVK works, and which improvements to be made [3–9]. However, very few of the proposed changes have been adopted in reality.

Further, since the OVK was introduced, much development has been made within control systems, energy management systems as well as regarding ventilation units, sensors and measuring equipment. The possibilities to follow up how the ventilation works in practice and to improve the standard of the indoor environment have increased significantly etc. Therefore there are opportunities to modernize the OVK. A well-functioning OVK can warrant a good indoor environment. A well-functioning OVK is especially important in cold climates, in order to secure both a healthy indoor environment and a low energy use.

2 Purpose

The purpose of the project was to gather information of how the OVK works, functions and is perceived within the whole OVK chain. The result will be used to inform legislators and responsible authorities, to give them more support for implementing needed changes.

3 Method

The project was divided into two main parts. The first part consisted of workshops and interviews in order to identify present situation. Workshops were held with a project group to get their view of how the OVK works in theory and in practice. The project group consisted of building owners, property managers, FUNKIS (an organization representing OVK inspectors and municipal OVK administrators) and Swedish Ventilation (an organization representing the ventilation business in Sweden). Additionally, personal interviews were carried out. In total thirteen interviews were conducted, divided on two OVK controllers, two municipal OVK administrators, one private building owner, one representative from the Swedish Property Federation, one Environmental and health inspector, as well as one person each from the Swedish Work Environment Authority, Asthma and Allergy Association, National Board of Housing, The Public Health Agency of Sweden, County government and SKL—the Swedish Association of Local Authorities and Regions.

The interviews were conducted via telephone or Skype, with exception for one interview that was held in person. During the interview, the whole process from ordering an OVK, execution, reporting, actions were gone through. Questions concerning legislation within the area were highlighted, and if there were different interpretations and ambiguities, these were identified. Proposals of changes in the OVK were also discussed.

After this, the second part of the project was initiated to identify the desired development of the OVK. Based on the outcome from the first part of the project, proposals on how the OVK could be developed were formulated. This work was based on the interviews, and later refined during workshops with the project group.

4 Results

All the interviewed persons agree that it is important with a well-functioning OVK, in order to maintain a well-functioning ventilation system and also to get an objective control. But when it comes to shortcomings with the current OVK, their views are partially different:

- The building owners highlight flaws in the supervision from the municipality Building Committee, and to some extent the OVK controllers who doesn't always match the desired "quality" in execution.
- The OVK controllers state that the weakest link is the supervision from the Building Committee that isn't working, but there also are building owners who doesn't care. It is perceived frustrating when faults and deficiencies are not rectified, and then it's both the building owner and the municipality that fail and do no follow up on the protocols.
- The administrators at the municipality responsible for the supervision, experience a lack of resources (money) and time in many municipalities in order to work actively with OVK, and the control and supervision therefore becomes inadequate. They consider that an active Building Committee is required that sends reminders etc. for it to work properly, at least for smaller, individual building owners.

Further aspects, that are highlighted by the Swedish Work Environment Authority, The Public Health Agency of Sweden and Asthma and Allergy Association, among others, is that there is a discrepancy between what is considered a satisfactory indoor environment/air quality and what gives an approved OVK. This has earlier been highlighted by an investigation from the National Board of Housing [5], where it was concluded that it is not possible to use the OVK to control if the environmental objective "A Good Built Environment" is fulfilled. Sweden has set up in total sixteen environmental objectives and "A Good Built Environment" is one of them. One of the milestone targets in "A Good Built Environment", was that all premises should have a documented effective ventilation in year 2015. This was meant to be achieved by an approved OVK, however not all premises are included in the OVK and an approved OVK does not equate an effective ventilation. The reason for the difference between actual indoor environment and an approved OVK, can be explained (according to several interview persons) that the OVK does not take into account and controls the actual activity and use in the premise, and this becomes especially clear in schools and kindergartens that often have a changing use of the premise. It is neither done an evaluation of the efficiency of the ventilation, for example air exchange in individual class rooms or how the ventilation is perceived by those staying in the room.

Since year 2006 it is included in the periodic inspections of the OVK to examine which measures that can be made in order to improve the energy efficiency of the ventilation system, this without compromising the indoor climate [10]. The extent and level of the energy efficiency measures that are suggested by the OVK controllers varies a lot. Further, it is not something that the administrators at the municipality attach any importance to because it is not clearly part of their mission. One of the administrators didn't even know that is was a demand to include energy saving measures in the OVK protocol.

Thus all flaws, most interviewed think that the OVK is a good system and worth building on. The ambiguities are in several instances of the OVK chain, which rules out one simple solution to correct the flaws, and several changes are necessary. Those activities, or improvement proposals proposed in the following, are based on the project group and the interviews. The development proposals have been discussed within the project group and are considered proposals that could address at least some of the shortcomings found in the current OVK. But in order for them to be implemented, a number of legislative changes are necessary, as well as clarification about the legislation. It would also be desirable with research projects that investigate certain topics, such as reduced and shut-off ventilation and its effect on the indoor environment and the building. As well as which ventilation flows are appropriate for different premises.

The following development proposals are set by the project:

- Enable the municipality Building Committee to charge a fee for the control and supervision of OVK, in order to improve the follow-up.
- Develop a uniform OVK protocol to be used, and gather the protocols in a national register with possibility for follow-up and statistics. Preferable an electronic protocol.
- Introduce the possibility that after-inspections is carried out by the OVK controller instead of the municipality.
- Set minimum requirements for the energy-saving measures to be included in the report.
- Take better account of current activities and current indoor environment in the OVK, so that not only the requirements are compared to the legislation in force when the ventilation system was taken into use.
- Review the legislation in the area and make it consistent, and include how, for example demand control and reduced/shut off ventilation should be handled.

These proposals will be presented in a report later on, and be used as a basis in dialogues with National Board of Housing, Swedish Work Environment Authority, The Public Health Agency of Sweden, SKL—the Swedish Association of Local Authorities and Regions, as well as representatives of parliament and government.

5 Conclusions

Based on the results, it can be stated that there are clear shortcomings in the OVK. In extension, this may implicate, if the OVK is not changed and the ambiguities are not corrected, that the justification of the OVK is reduced. If the OVK shall be relevant, it must be related to the actual indoor environment, and not just control that the ventilation fulfils the demands when taken into use. Further, energy saving measures must be taken into bigger consideration, in order to maintain and improve the energy performance during the buildings life-span.

References

1. Boverket Homepage (National Board of Housing, Building and Planning), http://www.boverket.se/en/start-in-english/building-regulations/building-regulations/national-regulations/obligatory-ventilation-control/. Accessed 23 Aug 2017
2. Boverket (National Board of Housing, Building and Planning).: Regelsamling för funktionskontroll av ventilationssystem, OVK (2012) ISBN 978-91-87131-03-5, http://www.boverket.se/globalassets/publikationer/dokument/2012/regelsamling-for-funktionskontroll-av-ventilationssystem-ovk.pdf. Accessed 23 Aug 2017
3. Arbetsmiljöverket (Swedish Work Environment Authority).: Projektrapport för Arbetsmiljöverkets nationella tillsyn av skolan 2013-2016. Rapport 2017:1. (2017)
4. Boverket (National Board of Housing, Building and Planning).: System för kvalitetssäkring av obligatorisk ventilationskontroll. Redovisning av regeringsuppdraget bättre inomhusmiljö. (2008) ISBN 978-91-86045-12-8, http://www.boverket.se/globalassets/publikationer/dokument/2010/godbebyggdmiljo_radon_ventilation.pdf. Accessed 29 Aug 2017
5. Boverket (National Board of Housing, Building and Planning).: God bebyggd miljö - Utvärdering av delmål för god inomhusmiljö – resultat från projektet BETSI. (2010) ISBN 978-91-86342-55-5, http://www.boverket.se/globalassets/publikationer/dokument/2010/god bebyggdmiljo_radon_ventilation.pdf. Accessed 23 Aug 2017
6. Boverket (National Board of Housing, Building and Planning) and Energimyndigheten (Swedish Energy Agency).: Energianvändning & innemiljö i skolor och förskolor - Förbättrad statistic i lokaler, STIL 2.Rapport ER 2007:11. (2007) ISSN 1403-1892, http://belysningsbranschen.labbplats.se/wp-content/uploads/2012/10/RAPPORT-STIL2_skolor_och_forskolor.pdf. Accessed 29 May 2017
7. Folkhälsomyndigheten (The Public Health Agency of Sweden):Inomhusmiljön i skolan – Ett nationellt tillsynsprojekt 2014-2015. (2015) ISBN 978-91-7603-581-8, https://www.folkhalsomyndigheten.se/pagefiles/22105/inomhusmiljon-skolan-15128-webb.pdf. Accessed 29 May 2017
8. SOU 1997:177: Byggkvalitet för framtiden. Byggkvalitetutredningens slutbetänkande. Stockholm. (1998) http://www.fastighetsagarna.se/aktuellt-och-opinion/remisser/plan-och-byggfragor/byggkvalitetutredningens-slutbetankande-byggkvalitet-for-framtiden. Accessed 29 May 2017
9. SOU 2005:55: Bättre inomhusmiljö. Slutbetänkande av Byggnadsmiljöutredningen. Stockholm. (2005) ISBN 91-38-22375-9, http://data.riksdagen.se/dokument/GTB355, Accessed 29 May 2017
10. SFS 2006:1296: Svensk författningssamling – Förordning om ändring i förordningen (1991:1273) om funktionskontroll av ventilationssystem. (2006)

Vertical Temperature Gradients in Apartments with Hydronic Radiator Heating

Mats Dahlblom, Birgitta Nordquist, Petter Wallentén, Lars-Erik Harderup and Lars Jensen

Abstract A vertical temperature stratification normally exists in rooms during the heating season in cold climates. An expression of the gradient in apartments heated by hydronic radiator heating systems with exhaust ventilation has earlier been developed assuming a dependency of the outdoor temperature. The expression was used by a public real estate owner when re-calculating measured indoor temperature at 2.1 m above floor to 1.2 m above floor representing the occupancy zone and used for individual metering and billing of space heating cost. To validate the suggested expression temperature measurements have been made at four heights in living rooms in apartments built in the 70's. The heights includes 0.0, 0.1, 1.1 and 1.7 m above floor. The theoretical expression has been compared to the full-scale measurements and in general the expression overestimates the vertical temperature gradient. The measured gradients are generally very low. The thermal comfort in the aspect of vertical temperature gradient is good for the studied period.

Keyword Vertical temperature gradient · Indoor temperature · Residential apartments · Hydronic heating system

1 Introduction

A vertical temperature stratification usually arise indoors. The pattern of this gradient depends during the heating season on the type of heating system, and the characteristic of the surrounding surfaces and the type of heat sources inside the room. The type of ventilation system will also influence the temperature distribution

M. Dahlblom (✉) · B. Nordquist · L. Jensen
Building Services, Department of Building and Environmental Technology, LTH, Lund University, Lund, Sweden
e-mail: mats.dahlblom@hvac.lth.se

P. Wallentén · L.-E. Harderup
Building Physics, Department of Building and Environmental Technology, LTH, Lund University, Lund, Sweden

D. Johansson et al. (eds.), *Cold Climate HVAC 2018*,
Springer Proceedings in Energy, https://doi.org/10.1007/978-3-030-00662-4_48

in the room. The vertical gradient is important for thermal comfort aspects, a too high temperature difference may negatively affect the thermal comfort experience.

The gradient is also important in aspects of controlling the heating system. The feed forward control method used for hydronic heating system that has been applied for many years only account for the outdoor temperature. This may result in too high room temperatures and unnecessary energy use during for example sunny days. It may also lead to increased airing resulting in increased energy use. The feed forward control method has drawbacks both in terms of thermal comfort and energy use.

An improved control method; the feed-back method which also accounts for the heat gains and losses inside the rooms has however during the recent years been given more focus. There has been successful implementations of this method in real apartment buildings [1]. The results show a decreased correlation to the outdoor temperature and a more even indoor temperature which is the purpose of the method. If some form of feed-back control is to be applied the indoor temperature must be measured in order to control the supply temperature of the hydronic heating system. As the temperature varies in height, the height of a reference temperature sensor inside a room is also of significance for the outcome of the control and thereby the resulting indoor temperature in the room. This will influence both the thermal climate in the rooms and possibly also the energy use for space heating.

So consequently it is important for both thermal comfort and energy use to have knowledge about the vertical temperature gradient inside rooms.

Overby and Steen-Thøde [2] present a model for calculation of vertical temperature gradient in a room heated by an electrical radiator and without forced air movement. The plume generated by the radiator has in the model been calculated from an equation for a linear heat source also presented in Hansen et al. [3]. The model and the laboratory measurements presented in [2] agrees well for a case with a heat load of 400 W, in which the gradient varies from approximately 0.5 K/m at start to approximately 1.5 K/m after 8 h with this heat load.

Laboratory studies on temperature gradients in rooms can be found, several of them concern rooms with displacement ventilation. Studies with full-scale measurements in real buildings are however limited. One study explores the vertical temperature gradient for a multi-storey building [4]. This investigates how vertically adjacent apartments thermally influence each other. The present study focus on vertical temperature gradient in occupied apartments with radiator heating and mechanical exhaust air ventilation.

In southern Sweden a municipal company has introduced a feedback control method. In the actual buildings a system for individual measuring and billing of space heating costs (IMB) was in place [5]. In this system temperature sensors were placed in every living room and every bedroom. The sensors are mounted just above the doorframe, i.e. 2.1 m above floor on inner walls, not to be blocked by furniture or disturbed by solar heat or internal heat sources. Though, this level is above the occupancy zone, and to get a more fairly value corresponding to the thermal comfort the temperature has been re-calculated to a temperature at 1.2 m height above floor with respect to the vertical temperature gradient with a proposed equation based on linear heat sources. This temperature has then been used as the reference temperature used for billing in the IMB system. Some comparing measurements have been made by Rundberg [6], but the equation used for this recalculation needs to be validated in a more extensive way.

It is also, as mentioned, important to know about the vertical temperature gradient for thermal comfort reasons. Temperature measurements have been performed in apartments owned by the company in a large multi-disciplinary research project called PEIRE [7]. The overall purpose of that project is to examine the indoor climate and energy use before and after renovation as well as resident behaviour related issues. The performed measurements before the renovation allows validation of the suggested equation as well as serving reference values for vertical temperature gradients that could be expected in occupied apartments.

The purpose of the paper is to examine whether the proposed equation for vertical temperature stratification coincides with temperature gradients in apartments and also attain reference values of temperature gradients in apartments heated by a hydronic system with radiators in each room and ventilated by a mechanical exhaust system with air intake terminals through the façade.

2 Methods

2.1 *Proposed Equation of Vertical Gradient in Apartments Heated by Hydronic Heating*

When introducing individual metering and billing of space heating costs, the indoor temperature was for practical reasons measured at a height of 2.1 m. An expression for estimating the vertical temperature gradient in an apartment were developed to be able to estimate the temperature at 1.2 m. This was done to attain more representative values for the thermal comfort in the apartments as 1.2 m is included in the occupancy zone and 2.1 m is not. If the billing were to be based on the measured value at 2.1 m the temperature could be overestimated, knowing that a vertical gradient exists, which could imply that the residents would be paying a too high cost for the space heating.

The indoor temperature was, due to an assumed vertical temperature gradient, recalculated to 1.2 m above floor with the equation.

$$T_i = T_{meas} - 1 + 0.025 \cdot T_e \tag{1}$$

T_e external (outdoor) temperature (°C)
T_i mean indoor temperature (current value) 1.2 m above floor (°C)
T_{meas} measured room temperature at 2.1 m above floor (°C).

At an outdoor temperature of 0 °C the expression gives 1 °C lower temperature at 1.2 m compared to 2.1 m above floor level, i.e. a vertical temperature gradient of 1/0.9 = 1.11 K/m.

The expression is developed based on theories of temperature gradients created by one or more punctiform heat sources and by a linear heat source [3], field measurements in seven residential apartments during the implementation of a measurement system aimed for individual measuring and billing (IMB) of space heating costs and complementary field measurements in two different apartments [6].

The result from [6] was that the vertical temperature gradient varied from approximately 0.5 K/m at no heat load to 1.5 K/m at large heat loads. If the heat load are distributed on more than one source the gradient will be lower. In a room there normally is more than one heat source, except from the radiator there are electrical appliances, persons and insolation.

2.2 *Temperature Measurements in Apartments During Heating Season*

Temperature measurements have been made in ten apartments during the heating season between the 16th of January and 5th of April 2017. The same equipment has been used moving it between the apartments. The measurement period has been daylong, i.e. both days and nights, one week in each apartment, starting in apartment 1 then moving to number 2 on the same day, for a total of ten weeks. Three gradients have been measured in each apartment; one in the master bed room and two in the living room; one closer to the exterior wall and windows and one in the inner part of the room, Table 1. HOBO loggers (HOBO U12-012) with accuracy of ±0.2 °C were used and external temperature sensors (TMC x-HD) mounted inside a black painted spherical globe [8].

The outdoor temperature was measured at a location 2.5 km from the buildings.

The buildings consist of three story rectangular concrete buildings with infill walls of wooden studs with filling of mineral wool. Each building contains 20–25 apartments distributed on three stairwells. The buildings are ventilated by a mechanical exhaust ventilation system. The outdoor air is supplied directly to the bed rooms and living rooms through vents and extracted in kitchens and bathrooms.

Table 1 Sensor type and placements

Height above floor/room	0.0 m	0.1 m	0.6 m	1.1 m	1.7 m
Living room inner	Logger	Globe	–	Globe	Logger
Living room façade	Logger	Globe	Logger	Globe	–
Bed room façade	Logger	Globe	Logger	Globe	–

3 Results

Measurements were performed in ten apartments, from which preliminary results are presented for apartment no. 3, 5, 6, 8, 9 and 10. Short facts about these are presented in Table 2. The outdoor temperature varied during the measurement period, 2017-01-31–2017-04-05, between −3 and +20 °C.

The temperatures are registered every 5th minute and the differences are calculated for hourly mean values. Based on the temperature difference between sensors at two levels have three different vertical temperature gradients been calculated. (A: 0.1–1.7 m, circular marker, dotted regression line. B: 1.1–1.7 m, triangular marker, solid regression line. C: 0.0–1.7 m, squared marker, dashed regression line.)

Figures 1, 2, 3, 4, 5 and 6 present the calculated vertical temperature gradient (K/m) compared to three gradients (A, B and C) as a function of outdoor temperature. As can be seen the conformity between calculated and measured temperature gradient is weak, except for apartment no. 8, where it is remarkably good if calculated between 1.1 and 1.7 m above floor (B).

As the gradients not can be expected to be a linear function of the height there could be a difference depending on which heights are used. Compared to the temperature gradient calculated with the proposed equation the measured gradients are generally lower. The largest measured gradient is B, i.e. between 1.1 and 1.7 m above floor. Note that the registered temperature in all cases are somewhat higher on level 0.0 m (floor level) than on 0.1 m, which means that gradient C is the lowest and in apartment 6, 9 and 10 even negative. Some of the gradients are relatively small, 0.2–0.4 K/m, i.e. in the same order of magnitude as the accuracy of the sensors.

Table 3 presents the results from linear regression for the three temperature gradients. To agree with the equation used for recalculation, k should be $-0.0278\ \mathrm{m}^{-1}$ and l should be 1.11 K/m. Best agreement can be seen for the measured gradient based on temperature difference between 1.1 and 1.7 m above floor. These heights are also those best corresponding to the heights used in the proposed equation (1.2 and 2.1 m). For apartment no. 8 the agreement is very good. Though, the determination coefficient, R^2 is low also in this case. In some cases the slope k is opposite the expected, i.e. the vertical temperature gradient is negative,

Table 2 Short facts about the apartments

Apt no.	Size	Apartment type	Living room orientation
3	4 rooms and kitchen	Double-sided	South
5	3 rooms and kitchen	Double-sided	South
6	3 rooms and kitchen	Double-sided	South
8	2 rooms and kitchen	Single-sided	South
9	4 rooms and kitchen	Double-sided	North
10	2 rooms and kitchen	Double-sided	South

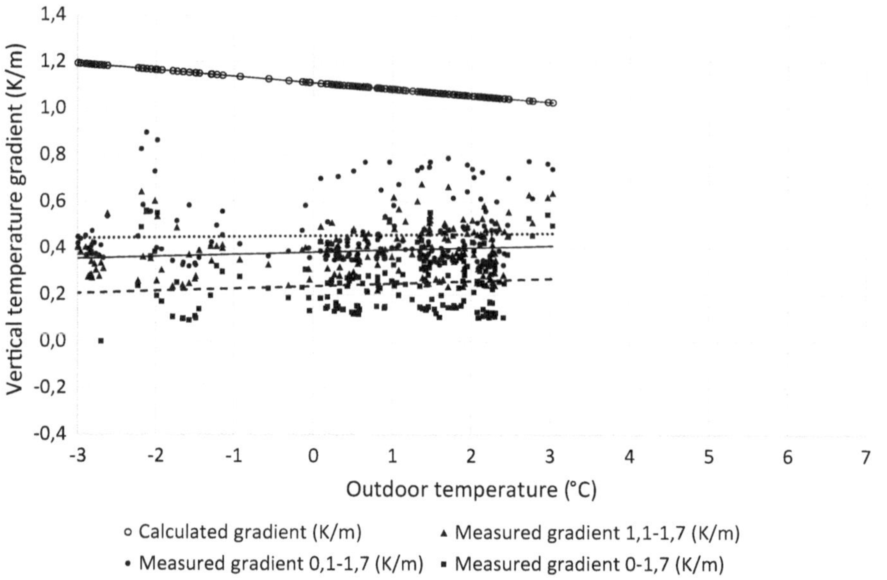

Fig. 1 Vertical temperature gradient as a function of outdoor temperature. Apartment no. 3. 2017-01-31–2017-02-08

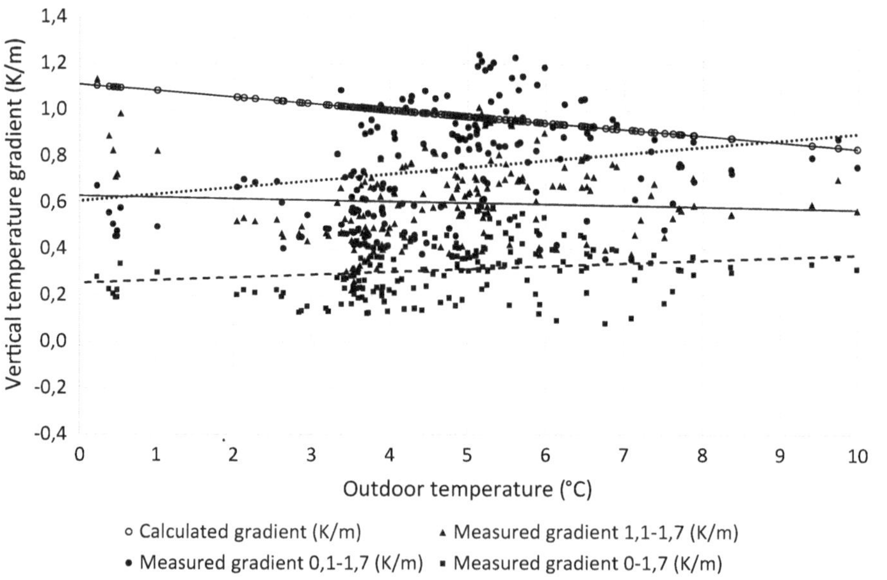

Fig. 2 Vertical temperature gradient as a function of outdoor temperature. Apartment no. 5. 2017-02-16–2017-02-24

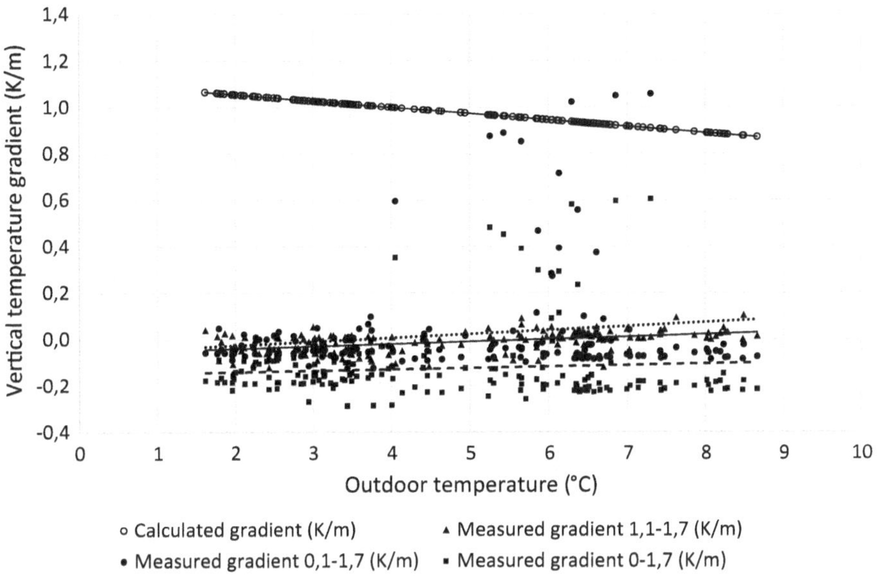

Fig. 3 Vertical temperature gradient as a function of outdoor temperature. Apartment no. 6. 2017-02-25–2017-03-04

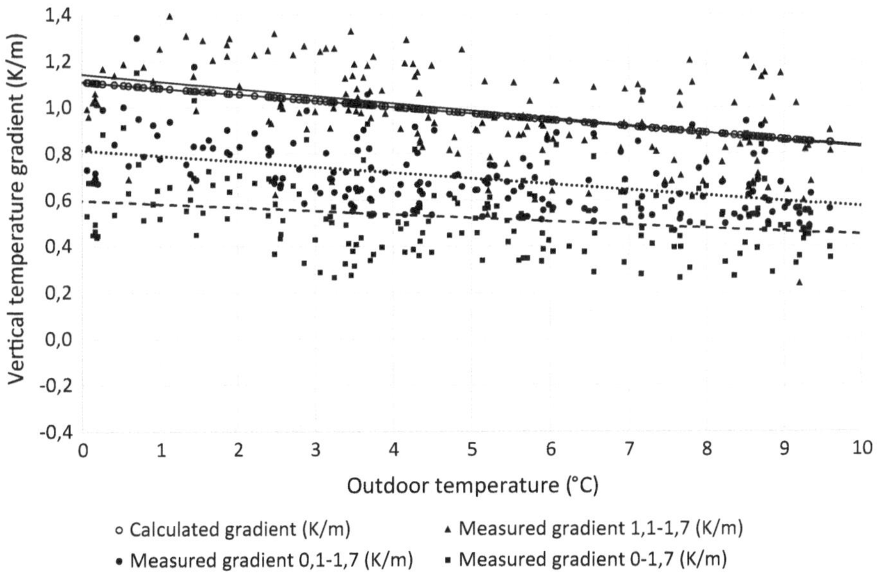

Fig. 4 Vertical temperature gradient as a function of outdoor temperature. Apartment no. 8. 2017-03-12–2017-03-20

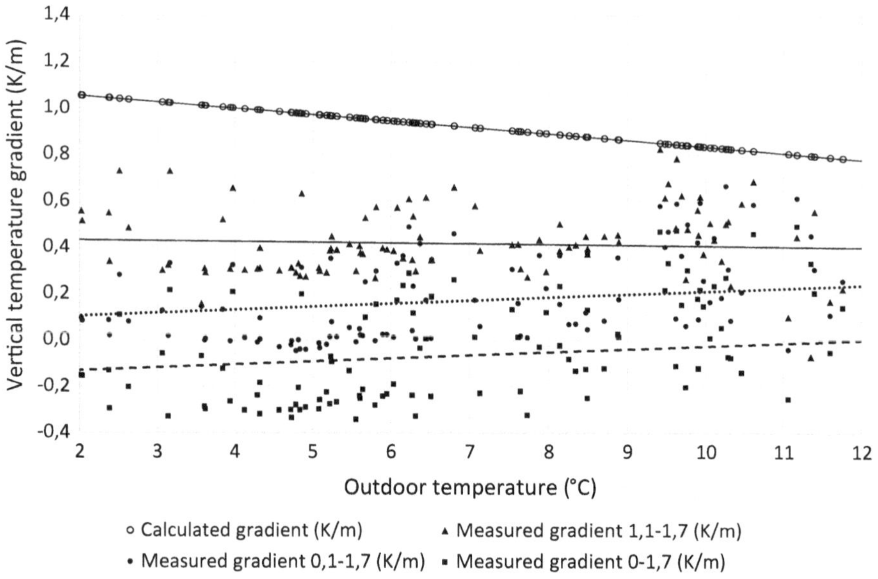

Fig. 5 Vertical temperature gradient as a function of outdoor temperature. Apartment no. 9. 2017-03-23–2017-03-28

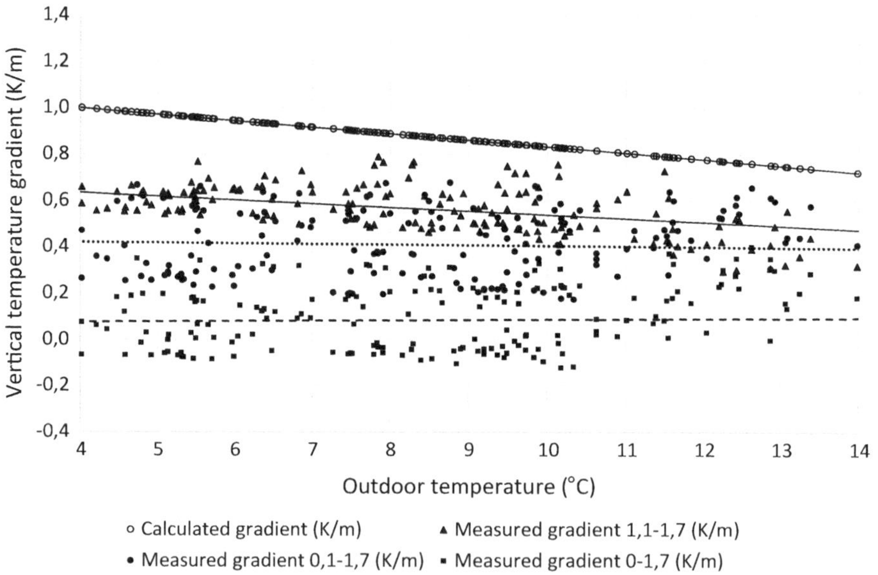

Fig. 6 Vertical temperature gradient as a function of outdoor temperature. Apartment no. 10. 2017-03-28–2017-04-05

Table 3 Results from linear regression

Apt.	0.0–1.7	(C)		0.1–1.7	(A)		1.1–1.7	(B)	
	k	l	R^2	k	l	R^2	k	l	R^2
3	0.0104	0.2375	0.0117	0.0035	0.4543	0.0019	0.0093	0.3829	0.0226
5	0.0122	0.2535	0.0523	0.0291	0.6082	0.0775	−0.0058	0.6302	0.0049
6	0.0059	−0.1524	0.0054	0.0166	−0.0594	0.0215	0.0103	−0.0600	0.2101
8	−0.0150	0.5977	0.0927	−0.0244	0.8146	0.2915	−0.0318	1.1435	0.1873
9	0.0130	−0.1563	0.0632	0.0131	0.0791	0.0869	−0.0032	0.4396	0.0082
10	0.0017	0.0692	0.0025	−0.0026	0.4307	0.0051	−0.0160	0.6989	0.2882

the higher level the lower temperature. At 0 °C outdoor temperature the gradient (l in Table 3) should have been 1.11 K/m, but in nearly all cases it is half or even lower.

4 Discussion

The reference height, 1.2 m, used by the company for calculating the apartments' reference temperature in the used system for IMB, is not the same height as used for the measurements in this study. The heights in this study are selected as these are international reference heights, thus making it possible to compare to international and national standards for thermal comfort. But, as a gradient can be calculated based on the difference in heights this should be sufficient.

The temperature measurements have been performed daylong with tenants living as normal as possible, which of course mean lots of disturbances, such as movements, window airing, different presence for different number of persons, lighting and other electrical appliances, etc. and in daytime possibly solar gain through windows. This is of course also the situation in the apartments where the referred IMB-system is used.

Wallentén [9] did measurements in one room in a full scale laboratory with different heat loads and two different placements of the radiator, below the window and at the opposite inner wall, both with and without supply air. During the measurements there was no disturbances from persons nor from window airing. The pattern for the vertical temperature gradient differs a lot, especially can large differences be seen between night and day. For a night case in January 1998, with a radiator heat load of 180 W, it is approximatively 1.4 K/m, both in the inner part of the room and close to the window. The day after, which was sunny, the heat load was zero and the gradient roughly varies from 0.5 K/m close to the inner wall to 1.0 K/m near the window, in both cases without ventilation.

The measurements both in the apartments in this study and in the apartments with the IMB-system were done in the inner parts of the rooms, why it is most relevant to compare them, well aware that even without occupants, the disturbances are large in the whole room volume. The proposed equation emanates from conditions with a linear heat source. The more distributed the heat sources are and the lower their temperature are, the lower will the gradient be. The radiators should have been sized to give accurate power at design heat load with supply temperature 80 °C and return temperature 60 °C. Very often they, for different reasons, became more or less oversized, which had led to lower maximum temperatures, e.g. between 60 and 70 °C was noticed in the referred buildings with IMB-system. This have probably also influenced the gradients measured in this study.

Overall it can be stated that the measured gradients are low, lower than the proposed equation suggests and much lower than the recommended maximum vertical temperature gradient of 3 K/m (SS-EN ISO 7730). The later indicates that the thermal comfort in this aspect for the inner part of the room is good.

The proposed equation seems to over-estimate the vertical gradient. As this recalculation is used when billing heating costs, it is better that the relation is overestimated otherwise the tenants may have had to pay a too high cost.

In this study both day and night temperatures was used. During days disturbances from insolation might have influenced the temperature gradient, even though the sensors were placed in the inner part of the rooms, thus not hit by direct sunlight. Next step could be to separate day and night measurements to see if there are large differences. Anyhow, to be useful, a compromise between conditions during days and nights would be the best choice for the desired purpose, i.e. to recalculate a temperature from a higher to a lower level, more representative to occupants' perception of the operative temperature in a room.

The measurements were done during a limited and fairly mild period and only in ten apartments. The temperature measurements show that the gradients varies both within and between the different apartments. This is reasonable as so many factors influence the indoor temperature, including occupants' behaviours and habits, e.g. opening and closing the windows and vents. Though the varied result, they indicate that the proposed equation for recalculation from 2.1 to 1.2 m above floor overestimate the gradient. To be able to revise the equation, more extended measurements, also comprising periods with outdoor temperatures below zero, preferable down to minus 10 °C or lower, should be performed.

5 Conclusions

The gradients recalculated with the proposed equation is in general larger than the measured gradients in the apartments, implying that it overestimates the gradients for the studied conditions.

The measured vertical temperature gradients during the heating season in the inner part of the living room, in apartments heated by a hydronic system and ventilated by mechanical exhaust ventilation, is low.

It can be concluded that the thermal comfort in terms of vertical temperature gradients is sufficient in the studied apartments for the studied conditions.

At least, in the referred system for IMB, the risk that a tenant should be billed for a higher indoor temperature than delivered is minimal.

Acknowledgements The projects within PEIRE are financed by the Swedish Research Council for Environment, Agricultural Sciences and Spatial Planning FORMAS (942-2016-79) and the Swedish Energy Agency (43092-1). The municipal property company in Lund who allowed the measurements to be performed in their apartment buildings are also acknowledged as well as the tenants that generously opened their homes for the measurements.

References

1. M. Dahlblom, L. Jensen (2011) Reglering av värmesystem i flerbostadshus med individuell värmemätning. Slutrapport för forskning med stöd från CERBOF [Control of heating systems in apartment buildings with individual heat metering. Final report of research supported by CERBOF]. (Rapport TVIT—11/3006). Lund: Building Services, Lund University. Available at: http://www.hvac.lth.se/fileadmin/hvac/files/TVIT-3000/TVIT-3006webMD_LJ.pdf. Accessed 1 Apr 2014
2. H. Overby, M. Steen-Thøde (1990) Calculation of vertical temperature gradients in heated rooms. Dept. of Building Technology and Structural Engineering, Aalborg. (Indoor Environmental Technology; No. 17, Vol. R9046)
3. H. Hansen, P. Kjerulf-Jensen, O.B. Stampe, Dansk Varme- og Klimateknisk Selskab. (1997). Varme- og klimateknik, grundbog (2. udgave, 1. oplag ed.). Lyngby: Danvak
4. M. Dahlblom, L. Jensen (2014) Vertical temperature increase in multi-storey buildings. Proceedings NSB 2014. 10th Nordic Symposium on Building Physics. Lund, Sweden 15–19 June 2014
5. Lunds Kommuns Fastighets AB (2011) Komfortvärme [Comfort heat]. [Electronic] Available: http://www.lkf.se/ImageVault/Images/id_5605/scope_0/ImageVaultHandler.aspx. Accessed 20 June 2012
6. G. Rundberg (2005) Individuell värmedebitering. Från uppstart i kv Grynmalaren till igångsättning i kv Labben [Individual metering and billing of space heating costs. From start in the property Grynmalaren to implementation in the property Labben]. (Rapport TVIT—05/7005). Building Services, Lund University, Lund
7. E. Pedersen, J. Borell, H. Caltenco, M. Dahlblom, C. Gao, L-E., Harderup, Y. Li, B. Nordquist, K. Stålne, P. Wallentén, A. Wierzbicka (2018) People as part of the energy system in residential buildings—challenges in transdisciplinary research for integrated understanding of technical installations, building performance, and tenants' perception and behavior. Submitted to: Cold Climate HVAC 2018. The 9th International Cold Climate Conference. Sustainable new and renovated buildings in cold climates. Kiruna, Sweden 12–15, March 2018
8. R. Shiltagh (2015) Mätsystem för operativ temperatur – test av hemtillverkade globtermometrar. [Operating temperature measurement system—Test of home-made global thermometers]. (Rapport TVIT—15/5054). Building services, Lund University, Lund
9. P. Wallentén (1998) Heat flow in a full scale room exposed to natural climate. (Report TABK—98/3051). Lund: Building Science, Lund University

Renovation of an Office Building with Prefabricated Wooden Element—Case Hedensbyn

Karin Sandberg, Anders Gustafsson, Anna Pousette, Joakim Norén and Mohsen Soleimani-Mohseni

Abstract There is a major need of cost-effective renovation that leads to lower energy consumption and better environment. This article shows the results from a pilot case of a newly developed prefabricated building system. It is an industrially prefabricated insulated wooden element adapted to renovation and upgrading of building envelopes. The renovated building is a one-story office building located in Skellefteå in the north of Sweden. Energy performance, thermal bridges, risk of moisture problems, LCA, applicability of the renovation method and assembly time were evaluated during the planning and execution of the renovation. Results from this case show that the elements were very light and easy for one person to handle at the building site. There is a great potential to reduce assembly time with improved joints and element sizes adapted to the building as well as improved batch packaging from the factory. With 100 mm insulation, the renovation gives a certain energy savings, and LCA calculations show that the reduction of climate impact due to reduced heating energy used during a service life 50 years corresponds to the climate impact of the renovation measures. The risk of microbial growth can be regarded as small.

Keywords Façade renovation · Building envelope · Prefabricated wood element Energy efficiency · Thermal bridge · Insulation · Climate impact Retrofitting

K. Sandberg (✉) · A. Gustafsson · A. Pousette
RISE Research Institutes of Sweden, Laboratorgränd 2, 93177 Skellefteå, Sweden
e-mail: karin.sandberg@ri.se

J. Norén
RISE Research Institutes of Sweden, Box 5606, 11486 Stockholm, Sweden

M. Soleimani-Mohseni
Umeå University, 90187 Umeå, Sweden

D. Johansson et al. (eds.), *Cold Climate HVAC 2018*,
Springer Proceedings in Energy, https://doi.org/10.1007/978-3-030-00662-4_49

1 Introduction

Worldwide primary energy demand has never stopped increasing during the last decades [1]. The EU Commission has recently stated that one of its highest priority tasks is to address global warming, especially focusing on reducing greenhouse gases. The Commission states that the building sector must decrease its use of energy to reduce CO_2 emissions. In addition, the 2016 Energy Efficiency Directive update established a set of binding measures to help the EU reach its 30% energy efficiency target by 2030 [2].

In this active context, Sweden has decided to take a step forward by completely phasing out net greenhouse gas emissions to the atmosphere by 2045 [3]. In order to meet these targets, many different activities must strive towards the same goal. One major part is the residential and service sectors, which accounts for about 40% of total energy use in Sweden [4]. Reducing the environmental impact and energy consumption of buildings cannot be solely achieved by regulating new buildings; it will also require renovation of existing buildings.

Moisture issues and thermal bridges have to be considered during energy renovation in order to avoid damage after renovation. A recent report [5] highlighted that 751,000 Swedish buildings were exposed to moisture damage to the point of affecting the indoor environment. Because of this, it is important to assess whether there is a risk of moisture in the construction after renovation, and if so, to take preventive actions at an early stage.

The effects of thermal bridges on the overall thermal performance of a well-insulated and energy-efficient building can be significant. A thermal bridge is a part where the thermal conductivity is greater than in other parts of the structure. This can occur, for example, around windows, between floors and walls, in corners, around studs or where there is insufficient insulation. Thermal bridges are a major problem in terms of energy losses as they contribute to increased heat flow out of the building envelope, but they can also create moisture related problems in the wall.

A façade renovation affects the performance of external walls in terms not only of energy performance and life-cycle cost, but also building performance, physical behavior, durability and aesthetic appearance. It is difficult to develop one single system for prefabricated renovation since buildings in need of renovation have been built in different eras with different cultural-historic aspects, with different technologies and materials. Even local climates give rise to varying renovation needs. Economic conditions, ownership of the estates, building regulations and construction methods differ between countries and have been investigated in numerous international projects. Some studies have focused on wooden facades, for example [6, 7].

A common way to renovate façades is to use a contractor on site that removes the outer layers of old façades and builds the new ones on site. The aim in this project was to develop a concept for industrially prefabricated insulated elements for renovation and upgrading of building envelopes. The project with participants from Sweden, Finland and Norway focused on increased prefabrication of building envelopes and solutions based on wood. The prefabricated elements have been described in [8], the energy savings in [9], concept and cost in [10]. This study is about a real case and gives information on the applicability of the new system for renovation and an environmental assessment, and may be used for improvements of the renovation system.

The hypothesis in the project was that a fast installation process and a short period of disturbance to the tenants are advantages of prefabricated solutions. Important for the quality of the job are the indoor production of the wood element in dry conditions and reduced moisture exposure on building due to rationalized work on site. Wood was chosen, as it is a renewable and sustainable material with low climate impact.

1.1 Objective

The aim of this paper is to evaluate renovation with a newly developed renovation concept with prefabricated wall elements. A pilot renovation of an office building in Skellefteå, Sweden, "Case Hedensbyn", was performed to verify the functionality. Evaluation was based on both simulations and measurements during planning and execution stages.

2 Building and Materials

2.1 Case Hedensbyn—The Building

"Case Hedensbyn" is an office building in need of façade refurbishment located in Skellefteå, in the Northeast of Sweden. It was built in 1976 and is owned by Skellefteå Industrihus AB. The tenant is a company with focus on mobile communication.

The building is one story high and connected to a larger machine hall. It was built with a timber frame and had a yellow brick facade (see Fig. 1). Some bricks had started to fall down, and the façade had been complemented with red profiled steel sheet on the north side of the building (see Fig. 2). The original wall was built of 13 mm gypsum board, 0.15 mm polyethylene foil, 140 mm mineral wool of glass fiber and 145 mm wooden wall studs, 12 mm bitumen-impregnated fiberboard, 10 mm ventilated air gap, 65 mm facade bricks or 0.7 mm profiled aluminum sheets above/between windows.

Fig. 1 Photo of the building before renovation, west and south wall

Fig. 2 Photo of the building before renovation, north and west wall

2.2 *Prefabricated Insulated Wood Elements for Renovation*

In “Case Hedensbyn”, a new system of small, modular prefabricated wood elements for renovation of façades was used (patented Termowood As; EP1963593, NO 323561). The prefabricated elements consist of two outer parallel multilayer solid wood panels with insulation and wooden connection rods between the wood panels. The elements are available with 50–250 mm insulation, depending on the requirements to upgrade the insulation in the building. Elements are produced with a width of 200 mm, thickness of 94–330 mm and length up to 6 m (but normally up to 3 m). Thickness of the solid wood panels is 22–40 mm. The elements can be combined by assembling segments together horizontally or vertically using a tongue and groove connection with a polyethylene sealing strip in the groove to ensure airtightness.

2.3 *Condition of the Building Envelope*

Drawings and documents were studied and the building was inspected to find out if it was suitable for renovation with prefabricated wall elements and to check the status of the existing envelope. The attachment of wood elements to the wood studs of the building was studied and there was a concern that the wall would have an irregular surface after the existing brick facade was pulled down. For use of

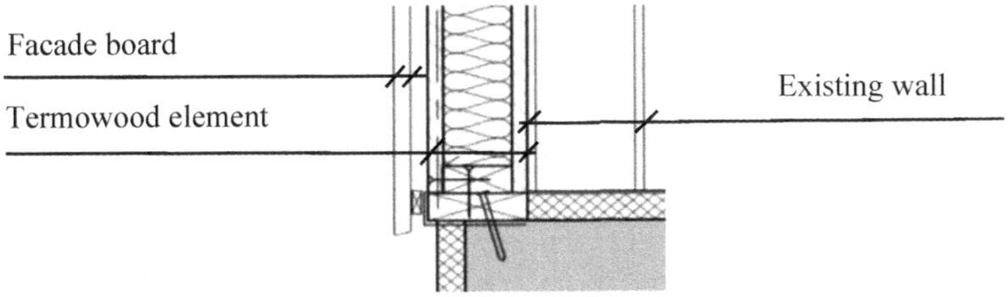

Fig. 3 Wall section of renovation with Termowood elements and facade boards as external cladding

elements on existing walls, the size, attachment and tolerances of the elements are important for the installation of new elements directly to the existing wall. This also requires no moisture damage or mold growth in the existing wall. If the old façade contains a ventilated air gap, the façade cladding must be removed down to the wind barrier and the condition of the existing façade should then be inspected. Damaged materials should be replaced with new materials before attachment of the elements.

The investigation of the building indicated that it was possible to attach the elements to the existing structure, and the work continued with the involvement of an architect and a contractor. The windows were changed some years ago, and it was decided to keep them as they were.

The thickness of the new wall elements had to be adapted to existing roof-overhang and base structure, and therefore the elements had only 100 mm additional insulation. Standard width was 200 mm, but also 98 mm wide elements were used to adapt to windows. The element length was generally 2.05 m, and elements were installed in two layers to get the full wall height. There were also shorter elements to be installed over (1.77 m) and under (0.81 m) windows. The thickness of the solid wood panels in the new wall elements was 22 mm. The elements were fixed to a sill that was screwed to the existing building at the bottom, middle and top of the wall. The new façade cladding was decided to be a mixture of a traditional 22 mm softwood façade and a newly developed wooden façade with milled pattern (patent Sweden nr 1451222-2—Cross-laminated wood facade element).

The facade was built with a ventilated air gap behind cladding, and façade boards were fixed to double battening of horizontal 28 × 70-mm and vertical 12 × 48-mm battens. For a cross-section of the lower bottom part of the wall, see Fig. 3.

3 Procedure of Evaluation

Several actions have to take place in planning an actual renovation to secure the execution at a later stage. For "Case Hedensbyn", the renovation process was divided into three main steps: Planning stage, Execution stage and After completed

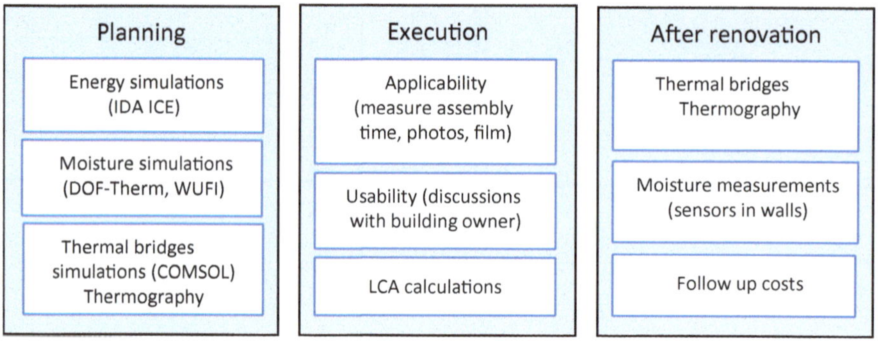

Fig. 4 Steps of evaluation and activities with methods in brackets

renovation. Each step consists of several activities (see Fig. 4). The first two stages are presented in this paper.

Planning stage. At this stage, the renovation was planned and the property owner, the builder and the material supplier were to agree on the terms. They had to evaluate if the building was suitable for prefabricated wall elements, which involved examining the existing building envelope using various methods. Calculations were made of energy savings, energy flow, heat losses due to thermal bridges and moisture transfer. In addition, many decisions had to be made: design, insulation thickness, connections, change of windows, roof connection and design and color of new facade cladding.

Execution stage. The applicability of the renovation methods was analyzed by in terms of the assembly time at the building site and LCA calculations. Documentation was also made with photos and a film. To assess the degree of satisfaction with the renovation concept, there were discussions with the builder and building owner during the renovation and after.

After completed renovation. Thermal bridges were evaluated by thermography images and moisture measurements with sensors in the new wall elements, and follow-up costs were assessed with builder and owner.

3.1 Energy Simulations

The software IDA ICE was used for calculation of energy savings. The height of the building was 4.2 m, and the connected building was considered in the calculations. Two-glass windows with 2.5 W/m K thermal conductivity were assumed and placed in the wall according to the plan drawing. The climate data of the city of Umeå (small difference in climate between Skellefteå and Umeå) were used. The thermal bridges were assumed to give low losses, and indoor temperature was 21 °C.

3.2 Thermal Bridges

Thermal bridges were simulated in 1D, 2D and 3D with COMSOL software. A climate data file for Oulu, Finland was used (small distance to Skellefteå and Umeå).

Thermography with a FLIRT620 camera was done to investigate thermal bridges in the existing building envelope. The thermography was done in May 2017 and carried out at night for the outdoor temperature to be as low as possible. An independent temperature gauge was used to calibrate the temperature of the camera. The temperature was measured on the outside of the building and used in the measurements. The camera was angled to the wall to avoid reflections from the body-heat radiation. Images were taken at windows, in the middle of the wall and at ground level. There was also thermography from the inside of the building, this time at ceiling beams, floor tiles, corners and middle walls.

3.3 Moisture Simulations and Measurements

The DOF-THERM software was used to evaluate moisture transport under stationary conditions. WUFI® software was used to evaluate moisture conditions under nonstationary conditions with coupled heat and moisture transfer in the building components.

HygroTrac sensors S-900-1 from General Electric were installed during the execution stage for measurement of moisture content (MC), relative humidity (RH) and temperature (T) in positions that are believed to have the greatest risk of moisture damage. Measurements are planned to continue during several years after completed renovation.

3.4 LCA

The LCA calculation was made with EPD data [11] and general data. Stages were according to EN 15804 [12]. The refurbishment stage b5 included removal of waste material from demolition of the old façade (brick façade, sheets, etc.), production of new materials and transport of materials. Maintenance of the new façade was also calculated. The functional unit was one office building with an area of 229 m^2 over a lifetime of 50 years. The LCA was used to compare the climate impact of the renovation and the climate impact resulting from a reduction in energy use during the renovated building's service life. The LCA also gives indications of possible improvements that could further reduce the climate impact of the renovation system.

4 Results

4.1 Energy Savings and Thermal Bridges

The energy saving was 9 kWh/m^2 per year when simulating the new façade element in IDA ICE [13]. The existing building had simulated annual energy consumption for heating the building of 258 kWh/m^2, and the renovated building with 100 mm additional insulation got an annual energy consumption of 249 kWh/m^2. The energy saving is only from the extra insulation; existing insulation in the roof structure and windows and doors with high U-values was retained.

The major thermal bridges are usually located at connections. In this case, it was important to study the surface thermal bridges caused by the timber frame and assess how much energy could be saved by reducing these thermal bridges. Thermographic pictures taken before the renovation are shown in Fig. 5.

Simulation results of thermal bridges showed that the temperature difference between points on the inside of the wall at the position of studs and positions between studs (no thermal bridge) was reduced with added insulation (see Fig. 5). In January (coldest winter month), the reduction was from 1.4 to 0.6 °C, and in July (warmest summer month) the reduction was only from 0.2 to 0.1 °C. The average for the year after refurbishment was 0.41 °C. From this, the energy losses were calculated, and the calculated energy saving on average was 3.5 kWh/m^2 of wall surface in a year as a result of reduced thermal bridges at the studs.

4.2 Moisture Simulations and Measurements

The DOF-THERM software calculations showed that for January, with outdoor temperature −12 °C and indoor +20 °C, there is risk of condensation on the outer wooden panels, and therefore a vapor proof barrier in the wall is important.

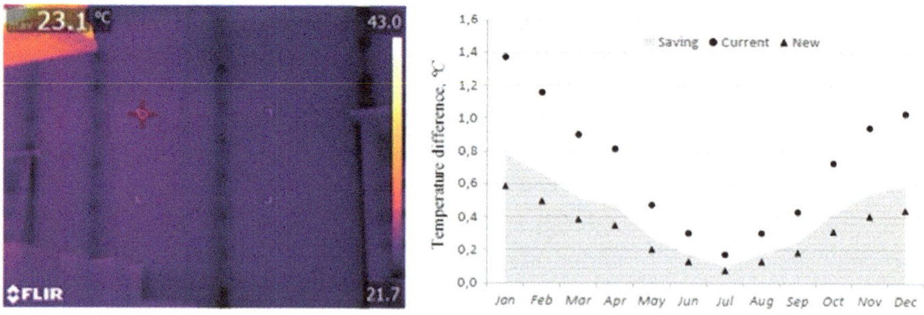

Fig. 5 Thermographic pictures show cold spots along the timber frame (studs). To the right, temperature difference calculated

WUFI simulations of the walls with elements with 22 mm wood panels and 100 mm insulation showed that the outer wood panel in the northern wall will reach RH just over 80% during the winter months and the insulation closest to the outer panel will get RH 80–90% [13].

4.3 LCA

The LCA calculations show the climate impact of the renovation compared with the reduced climate impact of the energy savings according to IDA ICE simulations in Sect. 4.1. Energy for heating of the building comes from a district heating plant (biofuel). The results show that the reduced climate impact of energy savings over the lifetime of 50 years for a wall element with 100 mm insulation corresponds to the climate impact for renovation and maintenance (see Fig. 6).

4.4 Execution Stage—Usability During Renovation

When the old brick façade was removed, the fiberboard behind appeared to be in good condition. The removal was made before delivery of the elements, with all that this implies of exposure to the surrounding climate. Moisture measurements with a hand-held moisture meter (Delmhorst RDM-2S) about 30 mm into the wooden ground sill in several places showed MC between 17.5% and 22.8%, with an average of 19.8% (9 measurements), in the north façade, and between 13.5 and 19.0%, with an average of 15.4% (20 measurements), in the west and south façades.

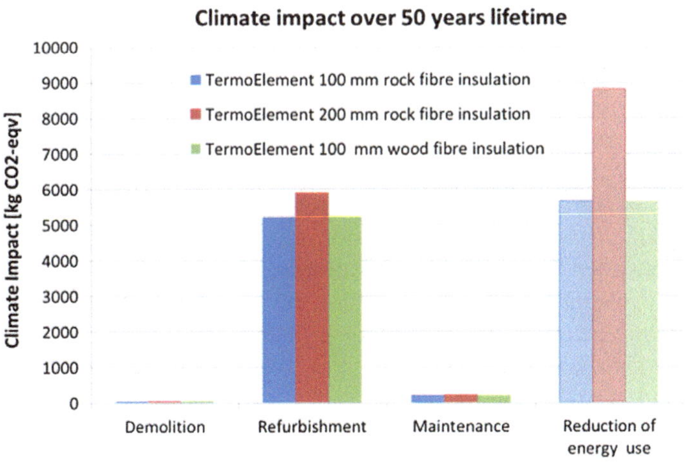

Fig. 6 Climate impact over 50-year lifetime compared to climate impact due to reduced energy use (patterned bar)

Before the installation of the wall elements, a sill was screwed to the concrete foundation. The east wall near the hall had a different construction and was supplemented with studs, insulation and plywood before mounting the elements, and the concrete foundation wall was raised approximately 400 mm to avoid moisture impact from the ground near the garage door.

The elements were prefabricated and stored at the factory before transport, so that all elements were delivered at one time. They were unloaded and stored at the building site, covered with the packing for transport. The renovation started in November, and shortly thereafter, it started to rain and snow.

For old buildings, it is difficult to know precise measurements of windows and doors in advance, and the elements were therefore delivered without exact adaptations. The elements were 2.05 m and had low weight and were easy to handle and lift at the building site. There had to be some adjustments and cutting around windows and doors at the workplace.

There was some trouble with assembling elements because elements were stored and handled in wet and snowy weather on site, resulting in swelling of the wood and making the grooves too narrow and extra effort was needed to assemble the parts. This was labor-intensive banging with a hammer to bring together the element edges. Problems at the site with repacking and storage of elements after finished working days also increased the moisture impact on the elements. The rear was most difficult to fit as elements bent outward from the existing wall when simultaneously pulled together with straps. Average assembly time at the pilot project "Case Hedensbyn" was about 0.5 h/m^2 of façade when everything went well, and the assembly time was about 2–3 times longer where there were problems with the tongue and groove connections of the elements.

4.5 Conclusions and Discussion

The renovation gives a certain energy saving in this case. The overall profit will not be that big, because the extra thermal insulation layer (100 mm) is relatively small due to the building conditions. The LCA calculations show that the saving of climate impact due to reduced heating energy during a service life of 50 years corresponds to the climate impact of the renovation measures. Moreover, the renovated building has, of course, an improved performance and a better indoor comfort level when the thermal bridges decrease. Environmental calculations show that with a thicker insulation, the reduction in climate impact during the use phase of the building increases more than the climate impact of the renovation. It also appears that there is a great potential to reduce climate impact from the wall element by selecting materials produced close to the element factory with a greater share of renewable energy as well as shorter transports.

From the data it can be concluded that, on average, the climatic conditions in the outer wooden layer are adequate in relation to little risk of mold or degradation due to high RH during normal use. Various software solutions have been used to evaluate the risk of condensation in the wall. All calculations show, in principle, similar results. Relative humidity will exceed 75–80% for shorter periods. However, the risk of condensation is very small (inside the outer wooden layer). The risk of microbial growth can be regarded as small. Probably, the wooden panel will absorb this moisture and even up the moisture content of the wooden panel, and no mold growth will occur. This will be followed up with sensor measurements during use.

WUFI® simulations had earlier been made for walls with Termowood elements with 40-mm multilayer wood panels and insulation thickness 250 mm [14]. Simulations over five years with climate data from Oslo, Norway, and indoor temperature 20°–22° did not indicate any risk of condensation on the outer panel. For only shorter periods, the relative humidity (RH) exceeded 80%, and moisture content (MC) in the external panel was less than 14.5%.

Assembly time per façade area was longer than expected, but then it should be taken into account that it was quite a small building, which usually means more time than average buildings, and it was a new type of prefabricated insulated element never tested for renovation before. A small construction project usually has a larger share of common jobs that affect the unit time for each building component. With improvements of tongue and groove joints, improved choice of element lengths and adaptations to fit existing building the installation process would be quicker and easier. A wider choice of element size (width and length) and improved batch packaging would also improve the speed of assembly and logistics at the building site. It appears that there is a great potential to reduce the assembly time with improved joints.

This paper shows the results for a single building, a small office building with quite thin added insulation. This building was suitable for a first test with this building system. Future studies on other types of buildings in other locations and climates are needed to further evaluate the system. Future work is also to improve the tongue-and-groove connection between elements to increase ease of application.

Acknowledgements This paper was written within the project, Nordic Built Concept for renovation and upgrading of residential buildings, supported by Nordic Built, Swedish Energy Agency, The Swedish Research Council Formas and industry partners from Sweden, Norway and Finland. The work with "Case Hedensbyn" was also supported by Nordic Renovation Center, a project financed by an EU Interreg program with participants from Sweden, Norway and Finland to exchange best praxis and information on renovation.

References

1. International Energy Outlook 2017, report number DOE/EIA 0484 September 2017 Issued by International Energy Agency
2. EU Commission, in *Clean Energy For All Europeans—Energy Efficiency Directives*, (11 2016), pp. 4–6
3. http://www.regeringen.se/artiklar/2017/06/det-klimatpolitiska-ramverket/
4. Swedish Energy Agency, Energy in Sweden 2017, ET2017:12, (2017) p. 9

5. The Swedish Research Council Formas, in *SB11 Helsinki—World Sustainable Building Conference*, ed. by G. Bergström, U. Save-Öfverholm, Comprehensive knowledge of building and housing (2011) p. 19
6. TES Energy Façade, www.tesenergyfacade.com/, 19 August 2016
7. Smart TES, www.tesenergyfacade.com/index.php?id=14_smarttes, 19 August 2016
8. K. Sandberg, T. Orskaug, A. Andersson, Prefabricated wood elements for sustainable renovation of residential building façade. Energy Procedia **96**(2016), 756–767 (2016)
9. S. Ruud, L. Östman, P. Orädd, Energy savings for a wood based modular pre-fabricated façade refurbishment system compared to other measures. Energy Procedia **96**(2016), 768–778 (2016)
10. A. Bystedt, M. Knuts, J. Johansson, K. Westerlund, H. Thorsen, Fast and simple—cost efficient façade refurbishment. Energy Procedia **96**(2016), 779–787 (2016)
11. EPD TermoElement, 2016, www.epd-norge.no
12. EN15804:2012 + A1:2013 Sustainability of construction works—environmental product declarations—core rules for the product category of construction products, 2013
13. M. Wikner, Z. Holmstedt, in *Utvärdering av renoveringssystem för ytterväggar* (Evaluation of refurbishment system for external walls) (in Swedish), B.Sc. Thesis—Energy Engineering, Dep. of Applied Physics and Electronics, Umeå University, Sweden
14. P. Blom, in *Termowood elementer*, SINTEF Rapport no SBF2015F0287(in Norwegian), 16 June 2015

How to Extend the Service Life of School Buildings by Improving Their Indoor Climate Conditions?

Antti Knuuti, Jarmo Laamanen, Timo Kauppinen, Ilkka Räinä and Markku Hienonen

Abstract Mold- and moisture problems have damaged and prevented the use of many buildings in Finland; typically, the problems have existed in school buildings, but also in other public buildings like health centers and day care buildings. Even relatively young buildings have had those problems. The reasons are caused roughly by three factors: design errors and deficiencies during construction and use. The learning environment has deteriorated. Both pupils and teachers have had various symptoms, in some cases they have been unable to use the building anymore. The situation has led mostly to a continuous repair cycle—the indoor environment and conditions have not always reached the satisfactory level. In this presentation, firstly, the reasons and solutions for indoor air problems in different cities are introduced, and then is described, how the service life of two school buildings has been extended. The existing problems are described and what measures are needed to carry out that the indoor environment would reach a safe and healthy level. In both cases the recent condition and situation are analyzed, and then reported some realized renovation works based on monitoring and measurements, which are briefly described. After modifications and repairs the buildings have been monitored during 2 years. The aim was to keep both buildings in use during a limited time period. The use of the other school building was discontinued and the other school will change the use. In both cases the community authorities decided to try to extend the service life instead of a new school building. Massive and deep renovation operations were excluded because of the high costs; there should be a way to find how to have more time for planning and financing new schools without

A. Knuuti · J. Laamanen
VTT Technical Research Centre of Finland Ltd, P.O. Box 1000, FI-02044 VTT, Finland

T. Kauppinen (✉)
Tmi Mutsal, PO Box 60, 90651 Oulu, Finland
e-mail: tkk@mutsal.fi

I. Räinä · M. Hienonen
Building Supervision of Oulu, PO Box 38, 90015 Oulu, Finland

D. Johansson et al. (eds.), *Cold Climate HVAC 2018*,
Springer Proceedings in Energy, https://doi.org/10.1007/978-3-030-00662-4_50

using temporary spaces or substitute premises. The other building was overpressurized after repairs, which is not normally used; in other building targeted repairs were done. In this presentation, the research and monitoring program, starting point, renovation measures and monitoring results are presented. Special attention has been paid to ventilation systems and also to the typical reasons for indoor problems—very often one cannot name just one factor, but the performance is depending on more than one factor. The performance of a building depends on how well the building systems are integrated to operate together—envelope, heating, ventilation and automation systems—and also performance depends on the use, users and weather conditions.

Keywords Buildings in operation · Renovation of buildings · Indoor environment and health · Moisture safety and water damage

1 Introduction

1.1 Problems in Existing Building Stock

The Finnish educational building stock is relatively old. The biggest increase was after the 2nd world war and then when the postwar big boomers came to school: 19% of the floor area is built between 1960 and 1969 and 21% in years 1940–1959 [1]. In 2010 the share of educational buildings represented for about 30% of the public building stock [2]. It is estimated that about 25% of school buildings are affected by moisture and mold damages [3]. The significant moisture and mold related damages have been estimated in schools and kindergartens from 12 to 18%, having users of 200,000.

The costs caused by absences, sicknesses, decline of work efficiency, and on the other hand, by repairs, are high. Indoor condition surveys and studies have made in every case using various methods. Generally, only a part of the factors dealing with indoor conditions have been resolved, depending on the tasks and doers. Based on these results they have started to make renovation plans. There are many cases, where the indoor problems or symptoms have continued—a second renovation round has been done, and repairs have been continued. In many cases the school or part of it works in temporary relocation or substitute premises during deep renovation or waiting the new building to be completed.

Building codes and a new healthy housing degree 545/2015 (Ministry of Environment) and the its guidelines for application (Valvira/National Supervisory Authority for Welfare and Health, 8/2016) define acceptable conditions and limit values for indoor climate.

Indoor conditions and the building performance are always a sum of various factors, which depend partly on each other. The performance of buildings must be evaluated on the basis of different factors, and all the reasons cannot always be found (at reasonable costs). The problems and damages have evolved over a long period of time. At least some of the causes, that have led to indoor air problems, can be found through a holistic approach and extensive measurements and surveys. By repairs should set the work environment and conditions into a level, that working can safely be continued in the building. It should be developed a comprehensive commissioning procedure for indoor air problems. The most important thing is to increase the building quality both in new and especially in renovated buildings. "Renovation debt" is a general term (used in all communities describing the investments needed) which shows the gap between the required investment costs and the available funding, when there are no resources to keep the building at an acceptable level. This may lead to demolition [4].

1.2 How the Communities Meet and Have Reacted to Indoor Condition Problems

The indoor air problems in schools (and also in other type of buildings) have led in the worst case to disable from use or demolition of the school building, to the temporary relocation, to closing the whole building or part of it. Teachers and students have got symptoms, and when the symptoms have worsened, they cannot work in certain classes or in the school building. The communities and the building owners react after complaints; cities have launched various programs to identify problems and to prevent them. It is prominent, that these problems have become more pronounced in the last 10–15 years, even those damages have been found also earlier. Publicity has increased; the main reasons for increased discussion are recognized healthy problems which are connected with indoor environment.

Decisions and experiences of a bigger city. In a bigger city Oulu (in Finnish scale, 235,000 inhabitants) the Building Supervision Office (BSO) has developed different quality control concepts for buildings since the beginning of 2000s. One form is quality control guidance- and information cards. BSO has also created a moisture management procedure "Dry-Chain 10" [5].

One new project is a quality card for commissioning of renovated apartment houses. In renovation of residential and other buildings it is important, that the requirements which are set will be met, the building performs as designed and new repairs are not needed.

Even the most buildings perform as designed, there is too much defects and problems, caused by design, implementation and use. The existing building stock is a crucial in improving energy efficiency and indoor environment. In the last years the volume of new building in Finland has been 1–2% from the whole building stock and the buildings which require the most repair is built in 60s–80s.

The building board of the city of Oulu established in June 2017 a group for the review of so called demanding renovation projects [5]. The board deals with part of building renovation permits. The audit team include experts of Building

Supervision Office, experts from private sector, and additionally also other authorities, e.g. health inspectors, if needed. One of the essential results is to form a unified vision of different building sectors dealing with that particular project. At present in Oulu area, there is buildings, also schools, which are not in use, because of indoor air and environment problems.

The measures required by the group, especially in the projects dealing with repair of indoor conditions, will give an important response for the requirement level and what type of subassemblies must be done in the project before the permit can be granted. Too often the parties involved in a building project, deals renovation project only with individual elements. They have not considered the building as a whole by sufficiently broad range. The legislation, which defines the construction, requires anyhow that both the renovation and new building projects must be considered as an entity, not as a sample of single parts in buildings.

The coming regulations require, that ventilation will be designed almost balanced, unless, for example a clear negative pressure drop compared with outdoors in some room space is appropriate. This is concerning both new building and renovated targets. Balancing, means in the first place, supplied and exhaust air flow, but also that there will be no significant pressure differences e.g. between different ventilation service areas (sometimes it is needed to prevent dispersion of contaminants). Typically, in all existing school- and office buildings ventilation is controlled automatically by time of the day and other measured variables. The system is not originally designed to manage that there will be no pressure difference between service areas. A very typical example is pressure conditions between stairway and surrounding spaces.

New air handling units are also more accurate to adjust for managing pressure differences.

Condition and indoor environment assessments, middle-size city 1. A middle-size city (22,000 inhabitants) in eastern Finland was made indoor environment and condition assessments for many public properties.

Results will be used as a base to develop a new strategy for commercial premises and for facility management.

In the strategy for commercial premises a long-term renovation and action program is presented. Based on the assessment results, specific programs were also drawn up for the necessary measures to ensure the availability of buildings up to a larger renovation or replacement investment. The necessary renovation measures to ensure the use of the buildings have partially already been carried out [6].

The biggest reasons for renovation needs have been;

- decreased indoor conditions
- building defects and damages
- especially changes of use (the most general case)

Significant reasons for indoor air problems are;

- concentrations in the air from structures
- these concentrations have increased after e.g. sealing repair

- after repair of building parts, e.g. change of windows
- when the ventilation rate is intensified

Air supply routes have been directed, in addition to existing air supply units, also to the air leakage points, especially if pressure conditions have not been checked. This means that part of supplied air comes through the structures, not from air supply units. The pressure conditions may also vary because of shortcomings in the control system or because of different running times—if there are several different ventilation units in use.

The most important reasons for the presence of indoor air problems in the building stock in that city, according to the studies are;

- over time increased repair debt, because: refurbishments and partly maintenance operations have been moved into the future; there has not been possible to budget enough money for the investment program for renovation/replacement investments
- problems to control and manage make-up air and pressure conditions of buildings; this is seen especially in the targets, where mechanical ventilation is increased or boosted
- supplied air is infiltrating to the room space through wrong places, e.g. through structures which contains impurities
- the selection of a contractor only based on price competition can often lead to the choice of low-quality contractor; too busy timetables in the projects, which leads critically short drying times of materials and too short testing and balancing period before in-use
- deficiencies of weather protection in the building site—building parts are wetted in some stage of building process, e.g. because of improper storage
- defective indoor air related guidance and control in building- and renovation projects—this applies both design-, realization and use stages
- the tightness of ventilation ducts is also important, e.g. in the breakthroughs where ducting goes through different material layers like insulation etc.

The most common reasons are listed above, but depending on the target, also other causes have occurred. They can be summarized, that almost all risk presented in sc. DryChain-10 [4] risk list has been realized in some constructions.

The city authorities have also used overpressurization in some buildings, which have had indoor air problems. By that way the service life of the building has been extended before decommissioning.

Indoor environmental problems, middle-size city 2. In another middle-size city (20,000 inhabitants, in Middle Finland) the users in two schools (high school and secondary school) started to complain about indoor air problems. In these two case-study school buildings, the learning environment could be improved; but unfortunately, the symptoms of some users continued.

Thereinafter is the overview of the indoor problems of the two schools, of the research method and the repairs made [7].

2 School Cases

2.1 *The Aim and Objectives of the Study*

The city authorities (middle-size city 2) decided to find a way to extend the service-life of two schools instead of relocation or demolition for a limited time (2 years), after a new high-school building was planned to build and the secondary school was planned to change the use. The question was how to improve the indoor environment of these schools into an acceptable and safe level to work. The procedure is presented in the next three chapters. In both schools a systematic research program was carried out, and after renovation works based on recommendations of the studies, the indoor environment was monitored during the rest service life. The research program is introduced in the Sects. 2.3 and 2.4.

2.2 *Background*

The first target building 1. *is a high school building (case 1.)* in Middle Finland. The teachers and students had symptoms caused by indoor conditions for years. Some students have not been able to study in the building. The original stone-made building is from 1959. The three-part wood-made extension has completed in years 1964–1968. This extension is on a slope. The property includes a sports facility connected with the older part. The total gross area is 7958 m^2. The building has mechanical ventilation with heat recovery and it is connected with district heating system.

The school has been repaired since the mid-90's, when the ventilation system was renewed. In 90s several roof leaks were found in the new part of the building. Moisture damages have been found in other places, too. In the late 2000s sealing repairs have been made and some changes and new installations in ventilation systems have been made.

Earlier made studies were used to define the need for repair. In performance evaluation has come forward that the building could temporarily be over pressurized. The total performance evaluation of the building was made in terms of indoor environment, including the evaluation of starting point (2014) and a proposal for urgent repairs. The object was also to introduce measures, using which the school could be used during two next years (2014–2016) without to take temporary relocation in use.

The new school building was due to be completed 2016, but it was delayed to 2017. The newer part of the old school building was demolished.

The other school building 2. (case 2.) was also in the same city, a stone-made *secondary school*, built 1953. The users had indoor air-related symptoms. 40% of teaching staff have had symptoms. Part of the teachers and pupils could not work in the building. The building had 4 aboveground floors, an attic and a basement.

The gross area is 690 m^2 and volume 27,200 m^3. The school was renovated in 1975 and repaired during 1993–1996. The ventilation system was renewed during the second repair cycle. Kitchen and dining room were renewed 2012. The building is connected with district heating system and it has mechanical ventilation with heat recovery. The building is locating on very wet ground.

During 2013 a condition survey and condition assessment was made. Microbe damages were found from intermediate floors. The room spaces were usable except two classrooms. However, it was found in order to maintain the usability of the premises, measures should be taken to bring the building to a level that is in line with current indoor air quality standards.

A holistic performance evaluation was made in 2015 in terms of indoor conditions, including the mapping of existing situation and a proposal for urgent repairs. The need for repair was found out, and alternative measures, that the school was possible to keep in use to the end of 2018. After that the building will change the use.

In both cases during the final use stage on-line data collection was used to confirm the proper working conditions [8].

2.3 *Case 1. High School (Target Building 1.)*

To define the starting point and preliminary need f of repair the following program was realized (stage 1, in 2014):

- mapping the recent situation and a proposal for urgent repairs, based on existing material, documents, site visits, measurements and condition assessments in the site
- air tightness measurements by blower door and own ventilation system, air leaks location by IR-camera, main air flow measurements of ventilation system, supplied and exhaust air flow measurements both pressure condition measurements in selected classrooms
- separately defined indoor condition- and concentrations measurements and sampling for microbe surveys
- interviews of users and maintenance staff, conclusions and reporting

During the first stage (2014) new repairs were carried out. Two wings of the newer part removed from the use by a solid partition wall. All materials and furniture in classrooms were cleaned (HEPA-vacuum cleaning, wipeout and fumigation), also sealing of structure seams. In addition, changes have been made to the ventilation system and ventilation was intensified and service areas were isolated from each other at a door. The school building was decided to be overpressurized during the rest of its service life (2 years).

Stage 2 included monitoring of indoor conditions, concentration measurements and analysis of results from 2014 on. Indoor conditions were monitored 2014–2016 in separately selected rooms during 6 periods, the periods being from one week to one month. The monitored factors were:

- pressure conditions between indoors and outdoors, air flow measurements
- indoor conditions: relative humidity (RH), room temperature, carbon dioxide (CO_2)
- microbiological studies by contact plate sampling from surfaces
- thermography and air tightness test before and after repairs.

2.4 *Case 2. Secondary School (Target Building 2.)*

The research plan followed mainly the plan for high school (target building 1.).

The baseline was checked in 2015, including the controls of ventilation systems, the indoor air quality, sealing of structures. The first report was given in autumn 2015 and the building was monitored 2015–2017 like in the case of the high school.

3 Results

3.1 *Case 1. High School (Target Building 1.)*

The air leak number n50 of the newer part was 3 1/h. The air tightness n50 of the older part was 1,5 1/h, without sport facility 1,7 1/h. The values represent typical values of construction time, but 3 1/h is too high considering air leaks, moisture transfer and concentrations the air tightness of these two closed wings in the newer part was worse than in the other part. Based on the measurements, sealing measures for several targets were suggested.

- increased concentrations were measured by contact plates in some classrooms
- ventilation pipe of a sewer channel was too close to air intake of one ventilation unit, possibility of contaminant dispersion to interiors—the position of the vent pipe was moved and the pipe was raised
- in the beginning of the 2. stage, after the changes, one part of the newer school was still slightly under negative pressure, otherwise overpressurized—the room spaces suggested to be overpressurized in the level of +5 Pa, the older part of the school was overpressurized as planned
- CO_2 level of exhaust air in the older part was max 500 ppm, in the newer part 700–800 ppm, in some classes indoor temperatures and CO_2 were relatively high during workdays
- some shortcuts from air supply units to exhaust caused by air distribution method, air exchange efficiency was decreased
- additional changes in ventilation.

Monitoring was launched in autumn 2014 according to the plan. Feedback was sent to the customer (city authorities) after each monitoring period, and changes or control measures were suggested if needed. Indoor conditions had reached a satisfactory level, based on the monitoring results from the view of the learning environment. Although in individual classrooms, at times, they may be close to the limit values. Microbe contents have remained at a low level. Nevertheless, the questionnaire conducted to pupils who entered the school in the autumn of 2015, showed that also the new students had symptoms. According the measurements the ventilation system was performing according to regulations—but in positive pressure difference as adjusted, to prevent that contaminants from the structures would not infiltrate to indoors. According microbe surveys, the level of various microbes was below the limits.

3.2 Case 2. Secondary School (Target Building 2.)

A baseline study was done in the summer 2015. Following measurements were made in selected classrooms:

- air tightness measurements
- indoor thermography to locate air leak points
- air flow measurements, indoor air and concentration measurements
- moisture measurements of structures, structural analyzes and microbe surveys

Air tightness number n50 varied in 2nd–3rd floor between 1,8–2,2 1/h, in 4th floor 2,6–4,6 1/h. The highest value is too high even the other values represent typical values of construction time. The classrooms in the top floor were leakier than the classrooms in the other floors. The higher values in the top floor indicated that there are leaking areas from the roof and the attic. These figures also include internal transfer leaks from surroundings. Part of the measured classrooms were overpressurized. According to the monitoring results and based also on earlier condition surveys, part of the classrooms had been at least part time overpressurized, part time depressurized.

If there are impurities in the structures, changing pressure conditions can drive contaminants to indoor air and, in the other hand, moisture into the structures. The ventilation system must be balanced and there should not be pressure peaks.

Following measures were suggested:

- running time adjustments of ventilation units, performance evaluation of ventilation units (is there fluctuation)
- adjustment and control of air flow rates (must be in balance floor by floor)
- cleaning of the ducts and terminal devices
- recording and reporting of ventilation measurements in accordance with user needs

- regular filter changes and more attention paid to regular maintenance operations (ventilation)
- seaming of the structures

Internal air flows, caused by pressure difference were defined by smoke tests. Industrial mineral fibers were examined from ventilation system, from ventilation units in from the inner surfaces of supply air ducts. In addition, structural analyzes were made.

The inner surfaces of the building were in normal condition, taking into account the age of the building.

The pressure conditions in the building were varying, the reason could be in part changes in air flows (fluctuation). The air flow data was not available afterwards from the building automation system. Based on air flow measurements, in some classrooms the air exchange rate could fall short, possibly because of shortcuts from air supply units to exhaust vents. Microbe levels were within acceptable levels.

4 Conclusions

It is important, that the main air flows and other factors could be monitored in real-time. That's why the data collection and processing capacity should be improved. Also, the size of database should be kept limited, but the data should be available to be processed to useful information according to stakeholder's needs.

The factors which affect the performance and indoor environment of a building are:

- building envelope
- heating, ventilation and automation systems
- use and maintenance
- external and internal loads, weather conditions

The structures and building services (HVAC-systems) must integrated to perform well together. The performance must be at a level, that there will be no renovation debt and it is safe and healthy to use the building. Target 1 (case 1) was built in two parts, the new part was a three-part building, built on a descending slope. Target 2 (case 2) was originally built on a wet ground, and water was pumped continuously out from the basement well. Target 2's (case 2) structures were in a relatively good condition, despite its age. Target 1 (case 1) had roof leaks over time in the lightweight new part. The building was close to a big industrial plant.

The indoor air problems begun (or it was informed) after ventilation systems repairs in both buildings. When pressure conditions changed strongly, the contaminants from the structures are transmitted to the indoor air.

During the monitoring period, the target 1 (case 1) was tried to keep overpressurized, and air flow rates in accordance to the building codes. Some problems occurred in the control of the ventilation system in the target 2 (case 2), pressure conditions varied and the reason for that was some sensor installation problems. The control of pressure conditions is important—it is difficult if the service areas of each units are not isolated from each other.

In school building the use and load of the classrooms is varying; if the control is based on temperature/CO_2 of the total exhaust air flow, the divergent load of one class may remain undetected.

The maintenance of the buildings had properly handled.

After the adjustments, tuning and after targeted light repairs the both schools met well the requirements, but still some symptoms occurred, also among the new pupils who had come to school after the repairs. The repair and research costs were according to the city representatives only approximately 1/10 of the costs compared with relocation.

The sealing of structures in old buildings is difficult. It is almost impossible to attain the current good level of air tightness. The importance of pressure conditions and balancing of ventilation is growing.

A holistic approach is needed, where the structures, HVAC-systems and user's experiences must be examined. The measures should be planned based on them. When selecting research methods, we must often compromise between the extent of measurements and the costs. It would be better to have a covering study in the beginning. Using the recent technology, the building performance could be considerably improved. Facility management is using "management by information"—also in addition to remote control of buildings—which requires to process the measured data to useful information [9]. Data processing and reporting is one part of "management by information". Big amount of data is useless if it not processed (and pretreated) by proper way.

5 Summary

The aim was to extend the service life of two schools, which had indoor air problems. In both schools, some users had symptoms, and some students could not work in the school. Instead to move the activities to a temporary relocation, the city authorities decided to find a way to extend the service-life of both school for 2 years. The new facilities for high school (case 1) was planned to completed 2016 but must postponed to 2017. After that part of this old school was demolished. The secondary school building (case 2) will change the use after 2017. A concept was planned to implement the program—how to keep these schools safe and the learning environment so that the work can continue. The initial situation was mapped in both cases, and after urgent repairs—based on the condition survey—indoor environment was monitored during the rest service life. The repair in both cases were mainly cleaning, sealing (structures) and re-adjustment of ventilation systems. The operation conditions were attempted to change so that the learning environment would be at satisfactory level. The indoor conditions were monitored in aim to ensure and verify the performance of the building. Based on the results and measurements, indoor environments were at the appropriate level during the monitored time. In the high-school building the solution was to keep the building overpressurized, which normally is not recommendable.

The problem was people who already were exposed; even the improved indoor conditions could not recover them. One can say that based on the measures carried out in school buildings the indoor environment was at acceptable level during the rest of the service life—the goal was to keep the schools 2 years in use and reach healthy and safe working conditions.

The reasons for indoor air quality problems were caused by many factors in both cases. The reasons could be divided roughly into two parts: ventilation caused reasons and structures related things. The problems have evolved over a long period of time, and principally are depending on each other. Pressure conditions, which depend on the performance and adjustment of the ventilation system and air leak routes are the most important reasons in these targets—but in each case, there are other factors, too and also reasons due to design and building defects—form the construction stage on. It seems that more attention should be paid to ventilation systems and especially pressure conditions, also to pressure differences inside of the building.

When indoor problems start to occur, it is important that a holistic procedure is in use. Because the performance is the sum of many factors, which affect each other the attention must be paid all the factors—even if one obvious reason is found. There is still lack of concepts and procedures (such as building commissioning).

On-line measurements and monitoring are needed; the problem of many building automation system is that afterwards is difficult to find trends, if these trends are not preset; also, the instrumentation does not necessary serve the needs of facility managers and users. Because of the extent of the problems, and the biggest building stock is built in 60s–80s there will be unfortunately new problem cases expected.

The building performance problems should be prevented rather than to correct them afterwards. By wrong measures the buildings may be unusable, by proper measures the service-life can be extended. Very often we are investigating the exact consequences but we should get also into the causes. The main reasons to close the school or move to relocation facilities are symptoms that are considered to be caused by growths and microbes, but the improvements and solutions demand a sufficiently comprehensive and systematic review.

Acknowledgements This presentation is based on the information and support provided by representatives of cities Äänekoski, Iisalmi and Oulu, Finland.

References

1. K. Reijula et al., in *Rakennusten kosteus- ja homeongelmat. Eduskunnan tarkastusvaliokunnan julkaisu* 1/2012. ISBN 978-951-53-3454-1 (nid.), ISBN 978-951-53-3455-8 (PDF). Kopijyvä, Espoo 2012 (The moisture and mold problems in buildings, The Parliament of Finland, Audit Committee Publications 1/2012)

2. J. Ruokojoki, in *Kuntien omien rakennusten lämmön, sähkön ja veden kulutus v. 2010 (Taulukko-osa). Suomen Kuntaliitto, 2011* (The consumption of heat, electricity and water of public buildings in Finland, Association of Finnish Local and Regional Authorities 2011)
3. U. Haverinen-Shaughnessy et al., Occurrence of dampness, excess moisture and mould in schools in three climatic regions of Europe. Indoor Air **22**(6), 457–466 (2012)
4. see e.g. Avaimet terveelliseen ja turvalliseen rakennukseen (AVATER). Valtioneuvoston selvitys- ja tutkimustoiminnan julkaisusarja 44/2017. The keys to healthy and safe construction (AVATER) Summary report. Publications of the Government's analysis, assessment and research activities 44/2017. Prime Minister's Office, 25.4.2017 and Eero Nippala & Terttu Vainio: The need for renovation in residential buildings in 2006–2035, VTT Technology 274. ISBN 978-951-38-8460-4. http://www.vttresearch.com/impact/publications, Sept. 2016, Finnish, English and Swedish abstract
5. http://www.ouka.fi/oulu/rakennusvalvonta/, http://kuivaketju10.fi/
6. Private communication from the city authorities
7. T. Kauppinen, A. Knuuti, J. Laamanen, in *Ongelmarakennuksen käyttöiän jatkaminen. Rakennusfysiikka 2015*, Uusimmat tutkimustulokset ja hyvät käytännön ratkaisut, ed. by J. Vinha, T. Ruuska. 20–22 Oct. 2015, Tampere. Seminaarijulkaisu 4. TTY, Rakennnustekniikan laitos, Rakennetekniikka, Tampere, 2015, 259–264 (The extension of service life of a problem building) Building Physics Symposium 2015, Tampere University of Technology, Department of Civil Engineering, Structural Engineering & Finnish Association of Civil Engineers RIL, Proceedings, ed. by J. Vinha, T. Ruuska
8. A. Knuuti, J. Laamanen, T. Kauppinen, H. Kääriäinen, in *Sisäilmastoseminaari 2016. Sisäilmaongelmien selvittäminen ja parannusratkaisut*. pp. 219–225. Raportti, Sisäilmayhdistys 2016. (Indoor environment problems and improvement solutions, Indoor Air Seminar 2016, Report, Indoor Air Association of Finland)
9. T. Kauppinen, M. Hienonen, in *Monitoring based commissioning (MBCx) in energy and facility management*, BauSim, Dresden 2016. Proceedings, Central European symposium on building physics. C-04: Energetisches Gebäudemonitoring & Betriebsoptimierung, ed. by J. Grunewald. p. 259–267. ISBN (E-Book): 978-3-8167-9798-2

A Net ZEB Case Study—Experiences from Freezing in Ventilation Heat Exchanger and Measured Energy Performance

Björn Berggren

Abstract Net Zero Energy Buildings constitute one measure to reduce energy use and increase use of energy from renewable sources. Hence, it is important share knowledge and experiences from completed projects. This case study show that it is possible to build Net Zero Energy Buildings with existing techniques. However, a common strategy to prevent or limit the build-up of ice and frost in ventilation heat exchangers, Supply fan shut off, were not suitable this project, since it is air tight buildings. After occurring problems in the first winter, ventilation pre-heater were installed to prevent the build-up of ice and frost. Thanks to placement of temperature sensor after the pre-heater, the increased energy use for pre-heater may be expected to be low, roughly 1 kWh/m^{2}a.

Keywords Net ZEB · Energy use · Freezing · Ventilation

1 Introduction

Buildings account for over 40% of the primary energy use worldwide and 24% of its greenhouse gas emissions [1]. The world's population is growing and also the need for buildings. Hence, reduction of energy use and increased use of energy from renewable sources are important measures for climate change mitigation.

Many studies identify a performance gap between predicted energy use and actual measured energy use once buildings are in user phase [2–10]. Hence, it is important that energy use in user phase is measured and verified to enable dissemination of knowledge. Especially in high performance buildings such as Net Zero Energy buildings (NetZEBs).

B. Berggren (✉)
Skanska Sverige AB, Teknik, Stockholm, Sweden
e-mail: bjorn.berggren@skanska.se

D. Johansson et al. (eds.), *Cold Climate HVAC 2018*,
Springer Proceedings in Energy, https://doi.org/10.1007/978-3-030-00662-4_51

This study presents a Net ZEB neighbourhood in the south of Sweden with verified plus energy performance. The technical solutions used and measured energy performance is presented. Experiences from the user phase is shared, with focus on problems related to freezing in the ventilation heat exchanger.

2 The Case Study

The case study consists of seven one-storey terraced houses (three dwellings in each house), built in the southern part of Sweden, see Fig. 1.

The Net ZEB balance were reached in three steps:

1. Reduction of thermal losses by designing the buildings with an air tight and well insulated building envelope and using balanced mechanical ventilation with high heat recovery, heat recovery ventilation (HRV). The occupants has the possibility to increase the ventilation, manually or set point based. One HRV unit per dwelling.
2. Reduction of need for import of energy by choosing a ground source heat pump (GSHP) to cover space heating, via underfloor heating, and heating of water.

Fig. 1 Left: Location of case study in Sweden. Top right: Layout of terraced house. Bottom left: Facade facing south

During summer, free cooling is taken from the bore holes for the GSHP. Cooling is supplied via the ventilation system. One GSHP per building.
3. Generation of electricity by installing photovoltaic panels (PV-panels), on the roof facing south.

Simulations were conducted with VIP Energy [11], validated with ASHRAE 140 [12]. A summary of a technical description is given in Table 1 and results from simulations are presented in Table 2.

It shall be noted that weighting factors should be used in the Swedish NetZEB balance calculations [13], where 2.5 may be a Primary Energy Factor (PEF) used. However, in this analysis no weighting factors are applied as the building only demands and generates electricity.

Furthermore, electricity for plug loads and lighting are not included in the Swedish NetZEB balance. I.e. the generation from the PV-panels should cover the energy use, excluding plug loads and lighting.

Table 1 Summary of technical description of case study. All values are design values except for air tightness

Type of data/description	Value
Conditioned area	258 m^2
Enclosing area/conditioned area	2.88
Mean U-value for building envelope[a]	0.17 W/m^2K
Air tightness, measured (q_{50} and n_{50})	0.21 l/s, m^2 and 0.84 h^{-1}
HRV (heat recovery and specific fan power)	90% and 1.5 kW/m^3s
Ventilation rate	92 l/s and 0.5 h^{-1}
GSHP, Seasonal coefficient of performance (SCOP)	3.0
Photovoltaic panels (area/power)	66 m^2/10 kWp

[a]Including thermal bridges, windows and doors

Table 2 Results from simulations for the case study

Energy use	kWh/year	kWh/m^2a
Fans	1546	6.0
Pumps (including cooling)	934	3.6
GSHP (space heating and hot water)	5214	13.6
Total energy demand, excluding plug loads and lighting (disregarding PV-panels)	7694	29.8
Plug loads and lighting	7766	30.1
Electricity from PV-panels, direct use	−3832	−14.9
Electricity from PV-panels, exported	−4053	−15.7

3 Failure Description

During the first winter, some residents complained about low indoor temperature when the outdoor temperature dropped below somewhere in-between −5 and −10 ° C. They also complained regarding the supply air temperature, which they said were much too low.

After some investigation the reason for the problem were discovered; the condensing extract air were forming ice and blocking the heat exchanger.

This subject is not new, and HRV manufactures have developed different strategies to prevent or limit the build-up of ice and frost in heat exchangers, which has been highlighted and discussed before [14–16]. Common strategies may be Recirculation, Supply fan shut off and Supply air preheating.

In these ventilation units, the defrost strategy was supply fan shut off. This strategy means that the supply air fan stops, while extract air fan continues to run. I. e. the warm extract air defrosts the heat exchanger.

This strategy assumes, when the supply fan stops, that supply air partly finds its way into the dwelling through imperfections in the building envelope, see Fig. 2. However, in this case, the building envelope were very airtight (see Table 1).

Since the building envelope were air tight, the supply air mainly came via the supply air ducts and inlet, even though the supply air fan were shut off.

This resulted in build-up of frost and ice in the heat exchanger and low supply air temperature. The consequence of the low supply air temperature were initially limited discomfort, due to low supply air temperature. However, in this project, the HVAC design engineer had assumed that the supply air would not drop below +15 °C. When the supply air fell to low temperatures, roughly under +10 °C, the underfloor heating system were not able to compensate for the low supply air temperature, and the temperature in the dwellings dropped, causing high discomfort for the residents.

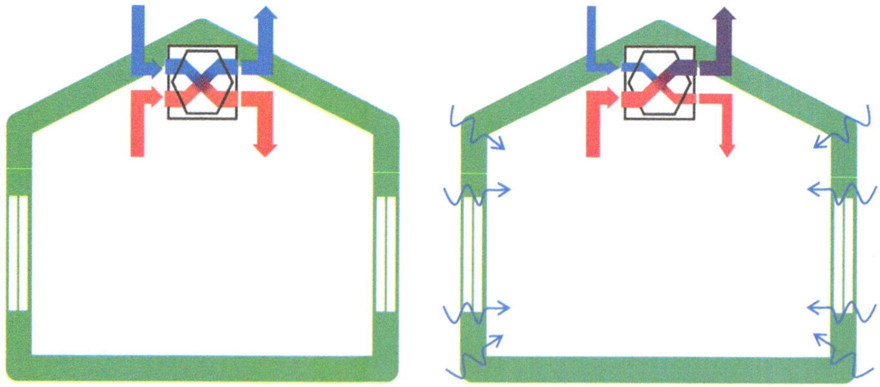

Fig. 2 Schematic description of assumed air flows. Left: Normal use, balanced ventilation. Right: Defrosting mode, Supply fan shut off

4 Action and Evaluation

4.1 Chosen Technical Solution

When the problems occurred. The subcontractor of ventilation and heating (the same subcontractor) were contacted and ask to suggest a technical solution to overcome the problem. Hence, the subcontractor were bound to ensure >+21 °C indoor temperature, at −15 °C outdoor temperature.

The subcontractor contacted the supplier and asked for a solution. Initially the supplier suggested to pre-heat the outdoor/fresh air, with an electric pre-heater (1 kW), to ensure no frosting- and freezing problems. However, in the initial suggested solution, the activation of the preheater would be based on the temperature of the outdoor/fresh air (Left in Fig. 4) and start heating when the temperature dropped under −1 °C. Based on the outdoor temperature a normal year (see Fig. 3) and the suggested installed power. This solution were expected to increase the yearly energy use by 1000–2000 kWh/ventilation unit, and therefore rejected.

After some discussion, the subcontractor found out that the initial given information were wrong/misunderstood. The temperature sensor were actually placed after the pre-heater unit (Right in Fig. 4). This would mean that the preheater would shut off as soon as the temperature after the preheater exceeded −1 °C. This was expected to vastly reduce the energy consumption, and the decision was made to test the solution. The pre-heaters were mounted and measurement and evaluation started in March.

4.2 Evaluation of Increased Energy Use for Pre-heaters

Already before problems occurred, total electricity use were measured in each dwelling. However, including plug loads, lighting and electricity for ventilation units. Evaluation of increased energy use due to installed pre-heaters were decided to be carried out in two different ways:

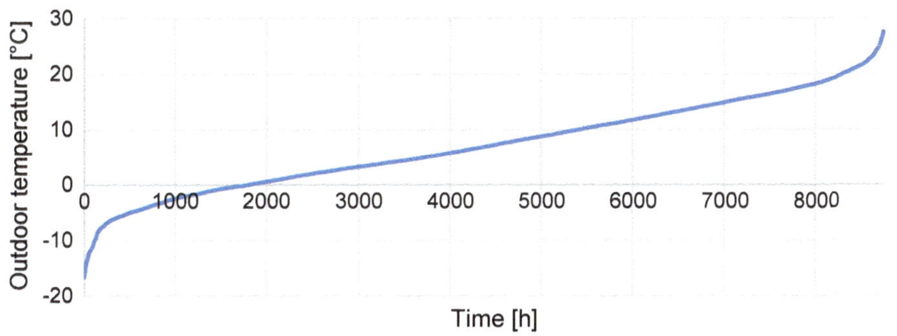

Fig. 3 Duration diagram of outdoor air a typical metrological year (TMY)

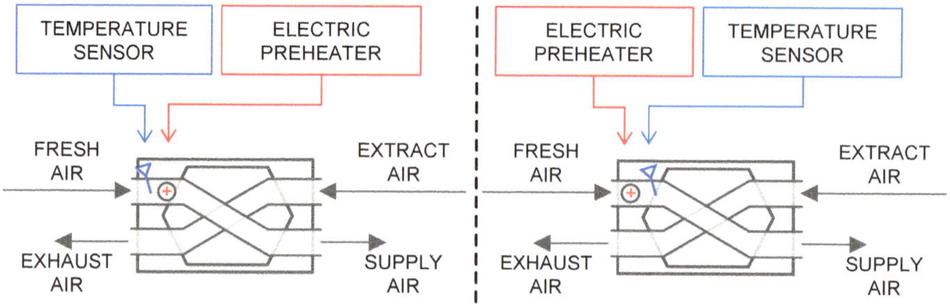

Fig. 4 Schematic description regarding position of temperature sensor and electric preheater. Left: First suggestion given by the manufacturer. Right: The installed and evaluated solution

1. Total electricity use in six of seven houses, between 3 A.M. and 5 A.M. were analysed, as it was assumed that the total electricity use in each dwelling during that time would be rather stabile, except when the pre-heater would be needed.
2. New electricity meters were mounted on two of ventilation units, to get detailed results.

4.3 Evaluation of Energy Use

One of the houses which did not have problems with the ventilation were monitored in detail. Starting in March 2015, energy use and energy generation were measured and is still ongoing.

5 Results from Measurements

5.1 Pre Heating

The results from measurements of total energy use is presented in Fig. 5. Based on the average energy use in all dwellings. It was concluded that the energy use for fans, refrigerators, stand-by for TVs, etc. (I.e. when there were no active use of owens, computers, etc.) were 165 Wh/h, dwelling between 3 A.M. and 5 A.M. This is roughly equal to 1.9 W/m^2, conditioned area.

Based on energy use before the pre-heaters were mounted (left in Fig. 5) it was possible to investigate increased energy use related to outdoor temperature. The average increase of energy use between 3 A.M. and 5 A.M were gathered (right in Fig. 5). Based on the equation for the interpolation (right in Fig. 5) and TMY for the location (Fig. 3), the increased energy use (due to pre-heaters) were calculated to 1.2 kWh/m^{2}a.

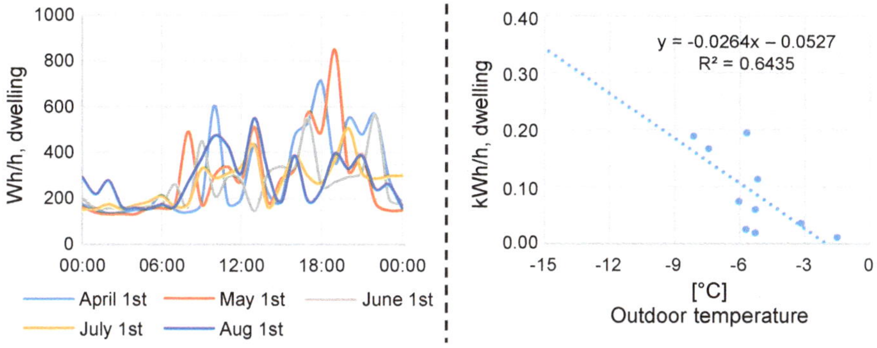

Fig. 5 Left: Average electricity use per dwelling in Solallén, before pre-heaters. Right: Average increased energy use in relation to outdoor temperature, between 3 A.M. and 5 A.M

The standard error (SE) for the equation (right in Fig. 5) is 0.047 kWh/h, dwelling. Using the maximum and minimum values for standard error the uncertainty is calculated to ±0.6 kWh/m^{2}a, or 50%.

The results from the detailed measurements from one of the ventilation unit is presented in Fig. 6. Also here, only data between 3 A.M. and 5 A.M. is included. Hourly data is presented. Energy use at outdoor temperatures below −2 °C is separated from energy use at outdoor temperatures above −2 °C. The mean energy use for the ventilation unit at outdoor temperatures above −2 °C were 0.036 kWh/h. Which corresponds to a specific fan power of 1.2 kW/m^{3}s (This ventilation unit had a ventilation rate of 0.03 m^3/s) or 0.4 W/m^2, conditioned area. Based on the equation for the interpolation (Fig. 6) and TMY for the location (Fig. 3), the increased energy use (due to pre-heaters) were calculated to 0.8 kWh/m^{2}a.

The standard error (SE) for the equation (in Fig. 6) is 0.022 kWh/h, dwelling. Using the maximum and minimum values for standard error the uncertainty is calculated to ±0.2 kWh/m^{2}a, or 25%.

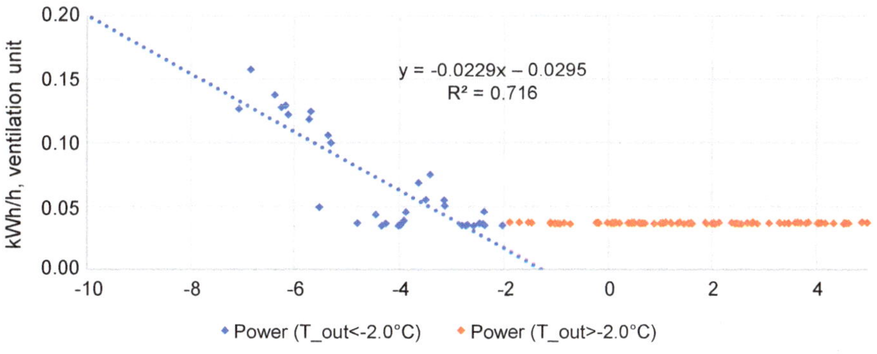

Fig. 6 Energy use for ventilation unit after mounting of pre-heater in relation to outdoor temperature

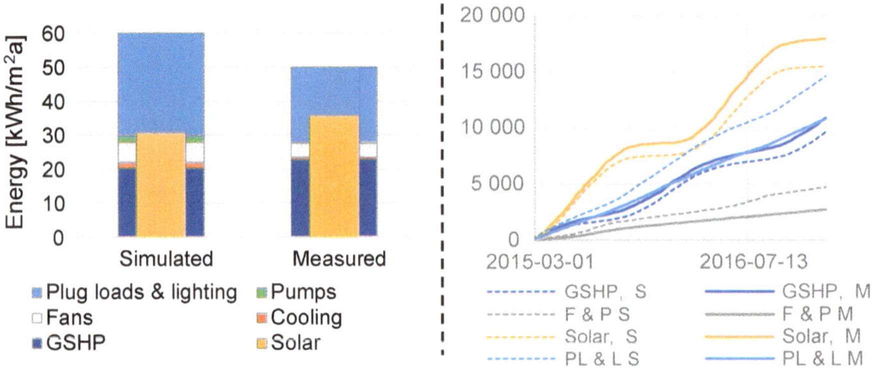

Fig. 7 Left: Comparison of annual energy use and solar energy generation simulated and measured. Right: Comparison of accumulated energy, generation and simulation. GSHP = Ground source heat pump, F&P = Fans and pumps, P&L = Plug loads and lighting

5.2 *Energy Performance*

In Fig. 7, results from simulations and measurements is presented. Energy use for GSHP were 3 kWh/m^{2}a higher compared to simulations. However, the main reason for higher energy use were lower inter heat loads due to plug loads and lighting, which were 8 kWh/m^{2}a lower compared to simulations. Electricity generation from PV-panels were 5 kWh/m^{2}a higher compared to simulations. The main reason for the higher energy generation were higher solar radiation, 10% higher compared to TMY. Energy use for fans were almost 2 kWh/m^{2}a lower compared to simulation. The main reason for the lower energy use were more efficient fans compared to procurement/design requirements.

6 Discussion and Conclusions

The case study clearly shows that it is possible to build Net ZEB with existing technologies. However, it also shows that a previously proven working defrosting strategy, "supply fan shut off", does not work. Hence, it highlights the importance of considering the secondary effects which may occur striving towards Net ZEBs. It is not always suitable to follow "rules of thumb".

Based on measuring of total energy use, the installed pre-heaters may be expected to increase the energy use in this project, by 1.2 ± 0.6 kWh/m^{2}a. Based on detailed measurements from one of the ventilation unit, the installed pre-heaters may be expected to increase the energy use in this project, by 0.8 ± 0.2 kWh/m^{2}a. In relative terms, the deviation/uncertainty is rather high 50 and 25% respectively. However, even in a worst case scenario, the increased energy use is lower than the surplus from energy generation and energy use. I.e. the Net ZEB balance is still reached.

The measurements were conducted in the end of the Swedish winter. I.e. the chosen solution has not been evaluated for outdoor temperatures below −10 °C. However, based on the installed capacity of the pre-heaters (1 kW) and air flow 30 l/s, the chosen technical solution is expected to ensure good indoor comfort when temperature is dropping below −10 °C. The pre-heater should enable a temperature increase of the outdoor air, before it reaches the heat exchanger, of roughly 25 °C, preventing frost down to outdoor temperatures of −25 °C (which is not normal in this part of Sweden).

Secondary effects are hard to predict and investigate. More research is needed and more time is needed in the design phase of building projects, especially in Net ZEB projects.

References

1. International Energy Agency (IEA), SHC Task 40/ECBCS Annex 52 Towards Net-Zero Energy Solar Buildings, 2013 Highlights (2013)
2. A.C. Menezes, A. Cripps, D. Bouchlaghem, R. Buswell, Predicted vs. actual energy performance of non-domestic buildings: using post-occupancy evaluation data to reduce the performance gap. Appl. Energy **97**, 355–364 (2012)
3. C. Demanuele, T. Tweddell, M. Davies, in *Bridging the gap between predicted and actual energy performance in schools*, World Renewable Energy Congress XI, Abu Dhabi (2010)
4. B. Bordass, R. Cohen, J. Field, in *Energy Performance of Non-domestic Buildings—Closing the Credibility Gap*, International conference on improving energy efficiency in commercial buildings, Frankfurt (2004)
5. J. Rekstad, M. Meir, E. Murtnes, A. Dursun, A comparison of the energy consumption in two passive houses, one with a solar heating system and one with an air–water heat pump. Energy Build. **96**, 149–161 (2015)
6. D. Majcen, L.C.M. Itard, H. Visscher, Theoretical vs. actual energy consumption of labelled dwellings in the Netherlands: discrepancies and policy implications. Energy Policy **54**, 125–136 (2013)
7. D. Majcen, L. Itard, H. Visscher, Actual and theoretical gas consumption in Dutch dwellings: what causes the differences? Energy Policy **61**, 460–471 (2013)
8. P. de Wilde, The gap between predicted and measured energy performance of buildings: a framework for investigation. Autom. Constr. **41**, 40–49 (2014)
9. G. Branco, B. Lachal, P. Gallinelli, W. Weber, Predicted versus observed heat consumption of a low energy multifamily complex in Switzerland based on long-term experimental data. Energy Build. **36**(6), 543–555 (2004)
10. N. Kampelis, K. Gobakis, V. Vagias, D. Kolokotsa, L. Standardi, D. Isidori, C. Cris-talli, F. M. Montagnino, F. Paredes, P. Muratore, L. Venezia, M.K. Dracou, A. Montenon, A. Pyrgou, T. Karlessi, M. Santamouris, in *Evaluation of the Performance Gap in Industrial, Residential & Tertiary Near-Zero Energy Buildings*, Energy and Buildings
11. Strusoft, VIP Energy 2.1, www.strusoft.se Visited Sept 10, 2017
12. ASHRAE, Standard Method of Test for the Evaluation of Building Energy Analysis Computer Programs (ANSI Approved) (2014)
13. Sveriges Centrum för Nollenergihus, Kravspecifikation för nollenergihus, passivhus och minienergihus (2012)

14. E.G. Phillips, R.E. Chant, D.R. Fisher, V.C. Bradley, *Comparison of Freezing Control Strategies for Residential Air-to-Air Heat Recovery Ventilators* (ASHRAE Annual Meeting, Vancouver, 1989)
15. J. Kragh, J. Rose, S. Svendsen, in *Mechanical Ventilation with Heat Recovery in Cold Climates*, 7th Symposium on Building Physics in Nordic Countries, Reykjavik (2005)
16. Alonso M. Justo, P. Liu, H.M. Mathisen, G. Ge, C. Simonson, Review of heat/energy recovery exchangers for use in ZEBs in cold climate countries. Build. Environ. **84**, 228–237 (2015)

Part VII
Building Simulation

Assessment of Thermal Bridging Heat Loss by Means of the Infrared Thermography Technique

Małgorzata O'Grady, Agnieszka A. Lechowska, Jacek A. Schnotale and Annette M. Harte

Abstract Thermal bridges influence the thermal standard of a building envelope. They are taken into account in the linear thermal transmittance or Ψ-value, and can be estimated from tabulated values for standard building details or from numerical modelling. However, these methods are not applicable for existing buildings with unknown structure. In these cases, in situ measurement of the Ψ-value is necessary. In this study, a methodology to measure the actual thermal bridging performance is developed. The methodology is based on a infrared thermography technique, which can be used on any existing thermal bridge. To ensure high accuracy, temperature-dependent surface heat transfer coefficients are calculated using the surface temperatures recorded by the infrared camera. The methodology has been validated under steady state conditions in a hot box apparatus with very good agreement.

Keywords Building envelope · Hot box · Quantitative thermography technique Thermal bridging

1 Introduction

Thermal bridging causes additional heat losses that can significantly degrade the building envelope thermal performance and therefore should be minimized. At the design stage, the Ψ-values that describe these heat losses are predicted by numerical simulations, according to the standard EN ISO 10211 [1]. For these simulations,

M. O'Grady · A. M. Harte
Civil Engineering Department, National University of Ireland Galway, University Road, Galway, Ireland

A. A. Lechowska (✉) · J. A. Schnotale
Department of Environmental Engineering, Cracow University of Technology, Cracow, Poland
e-mail: alechowska@quino.wis.pk.edu.pl

D. Johansson et al. (eds.), *Cold Climate HVAC 2018*,
Springer Proceedings in Energy, https://doi.org/10.1007/978-3-030-00662-4_52

construction details must be known which makes this approach not suitable for many existing structures. It these cases an in situ measurement has to be used.

The one of the first researchers to use quantitative infrared thermography technique (ITT) for measurement of the heat loss associated with thermal bridging was Benkő [2]. He defined an energy saving factor as the ratio of the heat losses through a building component with and without a thermal bridge. Likewise, Asdrubali et al. [3] described the heat loss related to thermal bridging as a ratio, the incidence factor I_{tb}, which shows the increase of heat loss caused by a thermal bridge. The I_{tb} obtained by means of the ITT can then be multiplied by the U-value of building component not influenced by thermal bridge U_{1D}, in order to obtain the U-value of a building component influenced by thermal bridging U_{tb}. Asdrubali et al. [3] measured the U_{1D} value using heat flow meter (HFM) whereas Bianchi et al. [4] calculated it.

In many existing building envelopes, the calculation method is not feasible as the construction of the building envelope is not known. On the other hand, the HFM requires special skills to use and is time consuming.

The methodology presented in this paper allows quantifying the heat flow rate through a thermal bridge and the Ψ-value by means of the ITT solely, without using any tabulated values or other measurements methods. The current methodology accounts for variable convective and radiative heat transfer coefficients where a thermal bridge disturbs the temperature distribution. This approach is novel and in contrast with both Benkő [2] and Asdrubali et al. [3], who assumed the surface heat transfer coefficients h as constant at the thermal bridge and outside thermal bridge zone of influence. The methodology was tested in laboratory conditions and validated with hot-box measurements.

2 Methodology

As mentioned previously, measurement is the only approach available for assessing the heat loss caused by thermal bridging in an existing building envelope, where the construction details are not known. Up to now, no such a method has been standardized. This paper presents a methodology prepared to evaluate the actual thermal bridge heat flow rate q_{TB} caused by the thermal bridge. The q_{TB} shows the additional heat loss through the building component due to the presence of the thermal bridge. The q_{TB} value is obtained as a difference between the total heat flow rate q_{tot} and the heat flow rate through the uniform part q_u that would occur if thermal bridge is substituted with a plain component.

The methodology is based on the energy balance, which equates the rate at which energy is transferred to/from the surface in conductive mode to the rate at which it is transferred from/to the surface in convective and radiative modes, under steady state conditions. It is important that the IR image includes the whole range of temperatures disturbed by thermal bridge together with the uniform temperature region typical of a part of the component outside the thermal bridge zone of influence. In Fig. 1, a sample IR image containing a thermal bridge is presented.

This methodology has been developed for indoor conditions with free convection. For methodology modified for the outdoor conditions please refer to [5]. The procedure is illustrated on an example of a vertical thermal bridge but it can be applied, with some alterations, to any linear thermal bridge. For the post processing, five sequential IR images of the same thermal bridge are used. On each of the IR image, an IR line is created (see Fig. 1). In order to generate this line, three rows of pixels at the middle height of the IR image are chosen. Each pixel on the IR line indicates the average surface temperature of the centreline pixel and its eight adjoining pixels. Then, from the five IR lines, a mean IR line is created. The surface temperatures on this line are used for further calculations.

Firstly, the heat flow rate for each pixel q_x needs to be calculated using (1). This equation is based on the assumption that the building internal air temperature T_i is similar to the surrounding temperature T_{sur}. For other cases please refer to [6]:

$$q_x = l_x[(h_{cx} + h_{rx})(T_i - T_{sx})] \quad (1)$$

where

q_x heat flow rate calculated for each pixel, W/m,
l_x pixel length, m,
h_{cx} convective heat transfer coefficient of a pixel, W/(m^{2}K),
h_{rx} radiative heat transfer coefficient of a pixel, W/(m^{2}K),
T_i indoor air temperature,
T_{sx} surface temperature of a pixel, °C.

This methodology includes precise calculation of the convective heat transfer coefficient h_{cx} using (2) from the Nusselt number Nu_x, evaluated from the Churchill-Chu equation [7]. The coefficients are calculated for each pixel on the IR line because the pixels have different surface temperatures:

$$h_{cx} = \frac{Nu_x k_x}{l_{ch}} \quad (2)$$

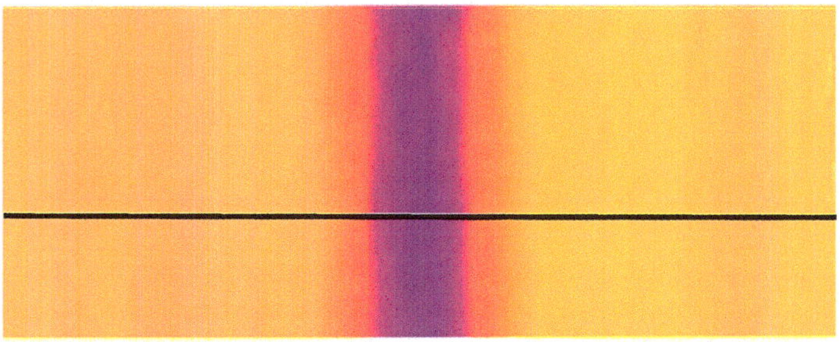

Fig. 1 A part of IR image of component with linear thermal bridge

where

l_{ch} characteristic length in vertical direction over which h_{cx} applied, m,
k air thermal conductivity, W/(mK).

The radiative heat transfer coefficient h_{rx} can be evaluated for each pixel on the IR line with the use of (3). This equation, similarly to (1), is used under the assumption that the building internal air temperature T_i is very similar to the surrounding temperature T_{sur}:

$$h_{rx} = \varepsilon\sigma(T_{sx} + T_i)(T_{sx}^2 + T_i^2) \quad (3)$$

where

ε surface emissivity, measured during the thermographic survey (Sect. 5),
σ Stefan-Boltzmann constant, W/(m^{2}K^4).

Using (4), the thermal bridge heat flow rate for each pixel, q_{xTB}, can be calculated:

$$q_{xTB} = q_x - q_{xu} \quad (4)$$

where

q_{xu} heat flow rate for a uniform component part, a pixel not influenced by thermal bridge, W/m.

By summing up the q_{xTB} for all pixels on the IR line, the value of the thermal bridge heat flow rate q_{TB} can be calculated. Finally, the linear thermal transmittance Ψ-value can be achieved using (5):

$$\Psi = \frac{q_{TB}}{(T_i - T_e)} \quad (5)$$

where T_e is the external air temperature, °C.

A more detailed description of this methodology can be found in [6].

3 Validation of the Methodology

The proposed methodology was validated under laboratory conditions in the hot box apparatus, allowing controlling the ambient conditions. The experimental set up, showed in Fig. 2, enabled testing the same specimen first using the hot box method and then using the ITT, under a steady state. Both the hot box test and the ITT measurements were undertaken under the same conditions, summarised in Table 1.

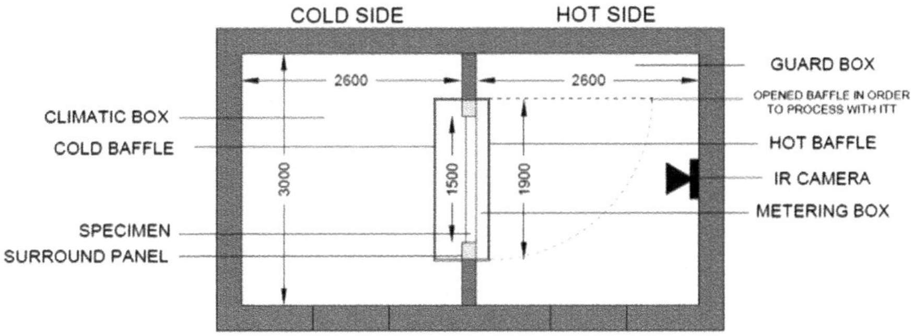

Fig. 2 Experimental set up (dimensions in mm)

Table 1 Hot box measurement at the steady state

Parameter	Unit	Value
T_e	°C	−4.96
T_i	°C	24.82
w_i	m/s	0.1
w_e	m/s	1.55
Φ_{in}	W	22.70
q_{sp}	W/m^2	7.01
T_{ni}	°C	24.50
T_{ne}	°C	−5.01

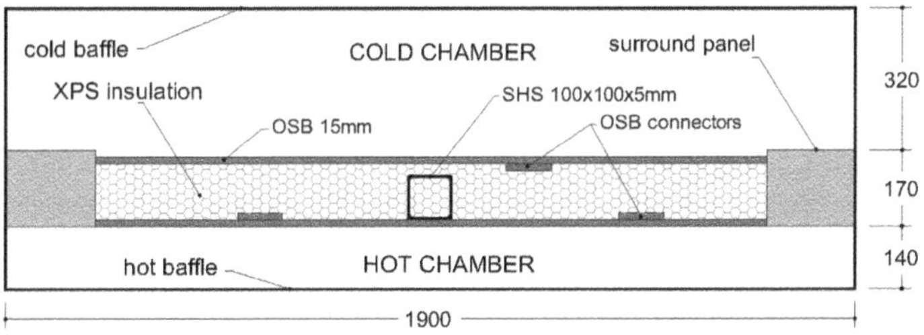

Fig. 3 Cross-section of tested specimen, inserted in hot box (dimensions in mm)

A specimen with the dimensions 1.5 m in length and 1.5 m in height purpose-built for the tests is shown in Fig. 3. It was made of structural insulated panels (SIP) consisting of extruded polystyrene insulation (XPS) 125 mm thick covered with 15 mm thick oriented strandboard (OSB) on both sides. A thermal bridge was created using a steel square hollow section (SHS) post 100 mm × 100 mm × 5 mm.

4 Hot Box Testing

The hot box testing was performed in accordance with EN ISO 8990 [8] at Cracow University of Technology, Department of Environmental Engineering. The hot box is built of a climate box, simulating the outdoor environmental conditions (cold side) and a metering box, simulating the indoor environmental conditions (hot side). The specimen was placed into a surround panel, which is made of insulation to minimize any side heat losses. After the specimen was sealed into the surround panel, the metering box was attached on the hot side of the specimen. The hot air temperature designed for this test was +25 °C. To ensure a uniform air temperature distribution in the metering box, a natural convection with air velocity of 0.1 m/s was created. On the cold side of the specimen, in the climatic box, an isothermal baffle was attached. To simulate outdoor environmental conditions on this side, air velocity of about 1.50 m/s was induced with the air temperature in the region of −5 °C. The measurements were taken after a minimum three hours of steady state conditions. The hot box was fitted with an AMR Ahlborn Wincontrol program that recorded data during the testing including the heat power supplied to the hot box $\Phi_{in,}$ the air temperature, and the wind velocity on the cold side w_e and on the hot side w_i. Based on Φ_{in}, the surface heat flux q_{sp} was calculated. On the hot surface of the specimen, two thermocouples measuring the surface temperature were attached, one in the middle of thermal bridge T_{TB} and another one in a distance of 0.40 m from the middle of the thermal bridge T_u.

T_{ni} and T_{ne} are indoor and outdoor environmental temperatures that, according to EN ISO 8990 [8] and ISO 12567-1 [9], should be used for calculations. They are calculated from the temperatures that were measured in a hot box. Before the testing started, the hot box was calibrated according to EN ISO 8990 and EN ISO 12567-1. The calibrations were undertaken with a calibration panel, with known thermal transmittance, to calculate surface heat transfer coefficients on both sides of the specimen. In order to evaluate the q_{TB} and the Ψ-value, another specimen with the same dimensions but without a thermal bridge was tested. Having the heat flow rate of this plain specimen, which was equal to 14.65 W, the q_{TB} and Ψ can be obtained, using (6) and (7), respectively:

$$q_{TB} = \frac{\dot{Q}_{sp} - \dot{Q}_{spplain}}{H} \tag{6}$$

$$\Psi = \frac{q_{TB}}{T_{ni} - T_{ne}} \tag{7}$$

5 Thermographic Testing

After completing the hot box testing, the thermographic measurements were started under the same wind and temperature conditions as the hot box measurements, summarised in Table 1. The survey began with measuring the reflected ambient temperature and the surface emissivity, as those two factors influence the accuracy of the IR camera reading. The reflected ambient temperature was measured using a direct method, in accordance to the ISO Standard 18434-1 [10] that was previously used by other researchers [3, 11]. The surface emissivity was measured using the contact method, following the ISO Standard 18434-1. Afterward a set of IR images of the hot surface of the specimen was taken. To calculate the q_{TB} and Ψ-value in accordance with the methodology outlined above, a horizontal line (IR line) was established on each of five IR images. The mean IR line was derived from the five IR lines and used for calculations. The specimen was symmetrical, therefore only one half of the line was considered.

6 2D Heat Transfer Simulation

The hot box testing results were also used to validate a numerical 2D steady state heat transfer model, created using Ansys Fluent software. The specimen geometry as presented in Fig. 3 was analysed. A section of mesh in the central part of the specimen can be seen in Fig. 4.

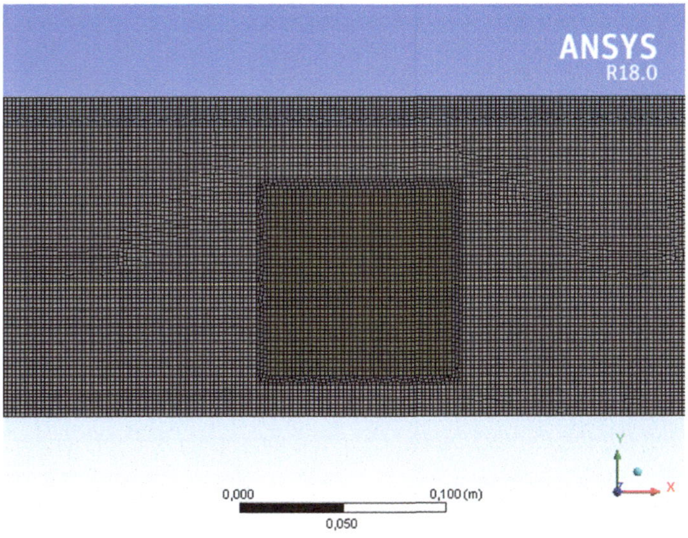

Fig. 4 A section of the meshed specimen

Table 2 Material properties and boundary conditions

Parameter	Unit	Value
XPS—thermal conductivity	W/(mK)	0.033
OSB—thermal conductivity	W/(mK)	0.13
SHS—thermal conductivity	W/(mK)	50.0
Air temperature—cold side	°C	−4.96
Air temperature—warm side	°C	24.82
Surface heat transfer coefficient—cold side	W/(m^{2}K)	25.0
Surface heat transfer coefficient—warm side	W/(m^{2}K)	7.7

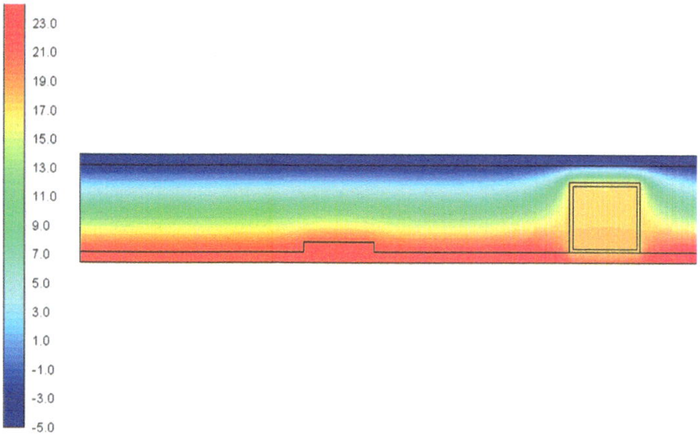

Fig. 5 Contours of temperature in part of the specimen

Material thermal properties as well as the boundary conditions used in the simulations are given in Table 2. In this simulation, the standard boundary conditions, expressed as cold and hot surface heat transfer coefficients, were applied, in accordance with EN ISO 6946 [12].

The calculated temperature field in the part of the specimen impacted by the thermal bridge and of the plain element outside the thermal bridge zone of influence can be seen in Fig. 5.

7 Results

Figure 6 illustrates the temperature distributions obtained by the ITT and from the 2D numerical model together with two temperatures (T_u and T_{TB}) measured by thermocouples during the hot box testing.

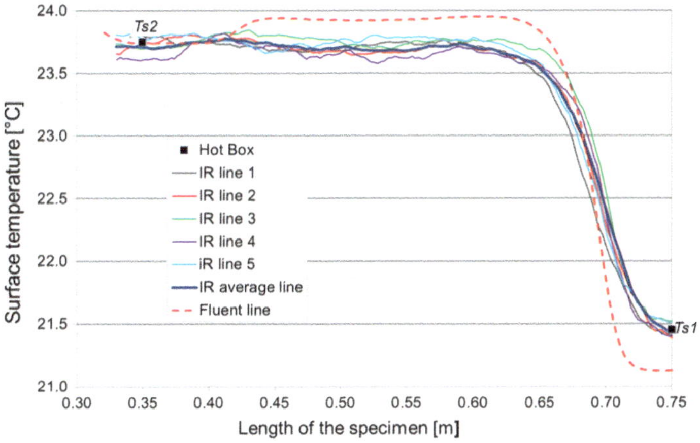

Fig. 6 Surface temperature derived from three methods

Table 3 Thermal bridge heat flow rate q_{TB} and linear thermal transmittance Ψ-value

		Hot box		ITT		2D model	Hot box/ITT	Hot box/ 2D model
		Results	SD (%)	Results	SD (%)	Results	RD (%)	RD (%)
q_{TB}	W/m	2.70	5.21	2.43	13.15	2.54	−10.00	−6.00
Ψ-value	W/mK	0.091	5.21	0.082	12.98	0.085	−9.89	−6.60

The thermal bridge heat flow rate q_{TB} as well as Ψ-value obtained both from ITT and from the 2D model are compared to the hot box results in Table 3.

A relative deviation *RD* in the q_{TB} and the Ψ-value of −10% has been recorded between the ITT and the results of measurements in the hot box. The differences of 0.27 W/m for q_{TB} and 0.009 W/(mK) for the Ψ-value indicate that the results are in a good agreement. It should be underlined that the high level of accuracy of the ITT results has been achieved due to precise calculation of both convective and radiative heat transfer coefficients for each pixel on the IR line. In order to assess how exactly this approach influences the results, the q_{TB} and Ψ-value of the tested specimen was calculated using constant values of h_{cx} and h_{rx} related to the part of the specimen not influenced by the thermal bridge. The q_{TB} and Ψ-value evaluated using these constant values showed greater relative deviation of −18.5% while comparing to the hot box results that illustrates the correctness of the proposed approach of the thorough calculation of h_{cx} and h_{rx}. The values q_{TB} and Ψ-value calculated with the 2D CFD model are also in a good agreement with the results measured in hot box. The relative deviations of q_{TB} and Ψ-values are equal to 6.0 and 6.6%, respectively.

8 Summary and Conclusions

In this paper a methodology for calculating the thermal bridge heat flow rate q_{TB} and the linear thermal transmittance Ψ-value by means of the infrared camera measurements taken from the internal building side has been presented. This methodology is based only on the ITT measurements. It is not necessary to know the building structure to use this method to a thermal bridge located in a building envelope.

This methodology has been validated under laboratory conditions, in a hot box device with good agreement that ensures the correctness of the method and the possibility to apply it in the real building. The current methodology can be practical during building energy efficiency assessment, especially where the building envelope structure is unknown.

The CFD simulations showed that it is possible to calculate the values indicating thermal bridge performance in an accurate way, but only if the specimen data (geometries and materials) are known thus is suitable at the building design stage.

Acknowledgements Financial support acknowledgment to Enterprise Ireland, Innovation Voucher IV-2014–4203 and College of Engineering and Informatics, National University of Ireland, Galway, Postgraduate Scholarship.

The authors would like to thank SIP Energy Ltd., Athenry Co. Galway, and particularly John Moylan, for delivery of the test specimens.

References

1. EN ISO 10211 Thermal bridges in building construction—heat flow and surface temperatures —detailed calculations (2007)
2. I. Benkő, in *Quantitative Analysis of Thermal Bridges of Structures Through Infrared Thermograms*. Proceedings of International Conference on Quantitative Infrared Thermography Qirt 2002, Dubrovnik, Croatia (2002), pp. 203–208
3. F. Asdrubali, G. Baldinelli, F. Bianchi, A quantitative methodology to evaluate thermal bridges in buildings. Appl. Energy **97**, 365–373 (2012)
4. F. Bianchi, A. Pisello, G. Baldinelli, F. Asdrubali, Infrared thermography assessment of thermal bridges in building envelope: experimental validation in a test room setup. Sustainability **6**, 7107–7120 (2014)
5. M. O'Grady, A.A. Lechowska, A.M. Harte, Quantification of heat losses through building envelope thermal bridges influenced by wind velocity using the outdoor infrared thermography technique. Appl. Energy (2017). https://doi.org/10.1016/j.apenergy.2017.09.04
6. M. O'Grady, A.A. Lechowska, A.M. Harte, Infrared thermography technique as an in-situ method of assessing the heat loss through thermal bridging. Energy Build. **135**, 20–32 (2017)
7. M.J. Moran, H.N. Shapiro, B.R. Munson, D.P. Dewitt, *Introduction to Thermal Systems Engineering: Thermodynamics, Fluid Mechanics and Heat Transfer* (Wiley, New York, 2003)
8. EN ISO 8990 Thermal insulation—determination of steady-state thermal transmission properties—calibrated and guarded hot box (1997)
9. EN ISO 12567-1 Thermal performance of windows and doors—determination of thermal transmittance by the hot-box method—Part 1: Complete windows and doors (2010)

10. ISO 18434-1 Condition monitoring and diagnostics of machines—thermography Part 1: General procedure (2008)
11. P.A. Fokaides, S.A. Kalogirou, Application of infrared thermography for the determination of the overall heat transfer coefficient (U-Value) in building envelopes. Appl. Energy **88**, 4358–4365 (2011)
12. EN ISO 6946 Building components and building elements—thermal resistance and thermal transmittance—calculation method (2007)

An Evaluation of the Combined Effect of Window Shading and Thermal Mass to Reduce Overheating

Carlos Jimenez-Bescos

Abstract Thermal mass has the benefit of regulating energy in buildings and generates potential savings in energy and CO_2 emissions. Window and local shading can provide shelter and reduce the severity of overheating during the year and mostly during the summer period. The aim of this study was firstly to evaluate the influence of window shading to reduce overheating and secondly to assess the thermal mass benefits in the presence of shading devices to alleviate the impact of overheating. This study was based on dynamic thermal simulations to analyse the performance of different window and local shading devices to avoid overheating. Twenty building simulation models were performed using the Energyplus plugin in DesignBuilder to evaluate the effect on the thermal mass behaviour to mitigate overheating according to different window shading devices. This study confirmed, as expected, that the use of window shading helps to alleviate the overheating hours in the test room and as such, improving the thermal comfort and reducing the need for cooling. Furthermore, when the window shading devices are coupling with thermal mass and night ventilation, the reduction on overheating hours achieved will reach a reduction of over 50% with respect to not exposing the thermal mass. In conclusion, exposing the thermal mass coupled with a night ventilation strategy provides a reduction on overheating hours, which is increased by using different window shading devices. Exposure of the thermal mass provides a good strategy for reducing the need for cooling and increasing thermal comfort.

Keywords Window shading · Overheating · Dynamic simulation
Thermal mass

C. Jimenez-Bescos (✉)
University of Nottingham, University Park Campus,
Nottingham NG7 2RD, UK
e-mail: carlos.jimenez-bescos@passivcasa.uk

D. Johansson et al. (eds.), *Cold Climate HVAC 2018*,
Springer Proceedings in Energy, https://doi.org/10.1007/978-3-030-00662-4_53

1 Introduction

Overheating issues are becoming more common nowadays in the built environment as a consequence of the effect of climate change, the use of thermal mass presents a good opportunity for non-domestic buildings to regulate the indoor environment, while saving energy and CO_2 emissions. The literature [1, 2] has presented the benefits available by the use of thermal mass and furthermore, when coupled together with night ventilation. According to the Zero Carbon Home [3], combining the effects of thermal mass and night purge ventilation create benefits to reduce overheating. In non-domestic buildings, the thermal mass tends to be hidden behind a false ceiling mainly for acoustic reasons. The suspended ceiling, made of compressed mineral wool tiles, generates a negative effect on the use of the thermal mass, restricting the use of the thermal mass to regulate the internal conditions by loading (day) and unloading (night) itself. By making use of the thermal benefits to regulate the indoor thermal comfort, overheating in summer can be reduced and the need for cooling minimised, allowing a reduction in energy consumption and the respective CO_2 emissions savings [4].

Furthermore to the use of thermal mass coupled with night ventilation, shading should be use to control and reduce solar gains through glazing helping to reduce overheating issues by excluding or minimizing the effect of solar radiation in the indoor environment [5–7]. The importance of adequate shading is even more relevant in well insulated buildings, in which overheating issues are happening during the summer months [8].

Regarding provision of shading, several options are available:

- Window shading by blinds, which could be internal, external or midpane [5].
- Use of electrochromatic glazing for window shading [5].
- Integrating overhangs into the design to provide local shading [3].

The aim of this study was firstly to evaluate the influence of window and local shading to reduce overheating and secondly to assess the benefits of using thermal mass in conjunction with the shading devices to alleviate the impact of overheating.

2 Method

The methodology in this paper is a continuation research building on previous work by the author [4, 9].

An exemplar test room, as shown in Fig. 1, was modeled with dimensions 7.5 m × 7.5 m × 3.5 m. The test room was dynamically simulated using energyplus in DesignBuilder software. U-values for internal floors hidden (with suspended ceiling) were 0.739 W/m^2 K and exposing (without suspended ceiling) the thermal mass were 1.523 W/m^2 K as previously published [4]. The test room was naturally ventilated and a night cooling ventilation strategy was used to cool down

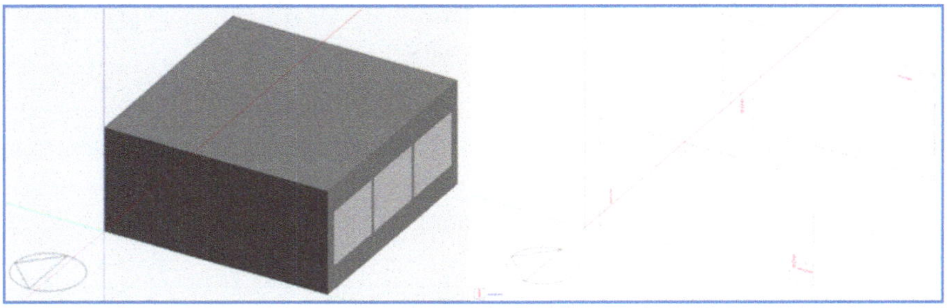

Fig. 1 Test room for dynamic building simulation

the thermal mass and discharge the heat accumulated during the day in accordance to the conditions presented below. No cooling was used in the simulations to be able to isolate and specifically quantify the benefits provided by the thermal mass to reduce overheating on its own. The simulated test room results were audited to confirm corroboration of results with building physics principles.

The dynamic computational simulation in DesignBuilder had the following parameters:

- Simulated location in London (Islington).
- Medium weight construction according to Part L2 2010 (UK).
- All surfaces adiabatic apart from south wall being external with a U-value of 0.26 W/m^2K.
- 50% glazing in south wall with a U-value of 1.978 W/m^2K and g-value of 0.687.
- Office equipment load of 10 W/m^2.
- Lighting load of 0 W/m^2 to focus on the performance of the thermal mass.
- People density of 0.111 people/m^2, following an occupancy schedule from 9:00 to 17:00.
- Constant infiltration of 0.5 air changes per hour (ACH).
- Natural ventilation rate of 1.5 ACH, following a schedule from 8:00 to 19:00.
- Night ventilation rates of 6 ACH, following a schedule from 24:00 to 6:00.
- Heating by gas boiler.
- Heating setpoint temperature of 20 °C and set back temperature of 12 °C.
- No active cooling.
- Simulations run for one year (8760 h).

The following window and local shading devices were simulated with and without suspended ceiling to compare the effect on overheating hours:

- No shading.
- Window shading (implemented according to office occupancy schedule by the UK's National Calculation Method for Non Domestic Buildings).

  - Inside blind with high reflective slats.
  - Outside blind with high reflective slats.
  - Midpane blind with medium reflective slats.
  - Inside shade roll light translucent.
  - Inside shade roll light opaque.
  - Electrochromatic reflective 6 mm glazing.
  - Electrochromatic absortivetive 6 mm glazing.

- Local shading.

  - 0.5 m overhang.
  - 1 m overhang.

The overheating limit was set to 28 °C in accordance with CIBSE definitions [10, 11].

3 Results

Twenty dynamic building simulations were solved using the dynamic Energyplus engine in DesignBuilder to assess the overheating performance, with and without the suspended ceiling, of a range of window and local shading devices. The simulations were modelled without the inclusion of cooling to control overheating, as the main purpose of the simulations were to evaluate the benefits of coupling the thermal mass and purge night ventilation. Two simulations were performed with the test room for each window and local shading device, one simulating the effect suspended ceiling and the other one exposing the thermal mass by non-suspended ceiling.

As previously presented in the literature for London (Islington) [4], the simulations were evaluated following a temperature distribution approach to allocate the number of hours below, at and above every temperature value in the model. The results show the range of temperature distribution with and without suspending ceiling and taking advantage of the thermal mass and purge night ventilation. Table 1 shows the total number of overheating hours above 28 °C, the overheating hours are shown for each window and local shading device and a case of no shading, with suspended ceiling and using the thermal mass and ventilation (non suspended ceiling).

Figure 2. presents a comparison of percentage reduction on overheating hours over 28 °C due to the influence of window shading versus no shading for a suspended ceiling and non-suspended ceiling scenario. The percentage is calculated from the baseline of no shading at all to the reduction in overheating hours above 28 °C due to the use of window and local shading devices as presented in Table 1. Very high overheating reductions, close to 100%, can be achieved by using electrochromatic glazing or outside blinds, while the smallest reductions are achieved by the use of midpane blinds and a 0.5 m overhang. Regardless being the smallest

Table 1 Overheating hours over 28 °C for non-suspended ceilings (thermal mass exposed) and suspended ceilings from simulation results

Shading type	Non-suspended ceiling (h)	Suspended ceiling (h)
No shading	533	730.5
Inside blind	83	229
Outside blind	5	14.5
MidPane blind	161.5	334
Inside shade roll translucent	112	302.5
Inside shade roll opaque	56.5	171.5
Electrochromic reflective	7	30
Electrochromic absorptive	7	39
0.5 m overhang	189	387
1 m overhang	42	56

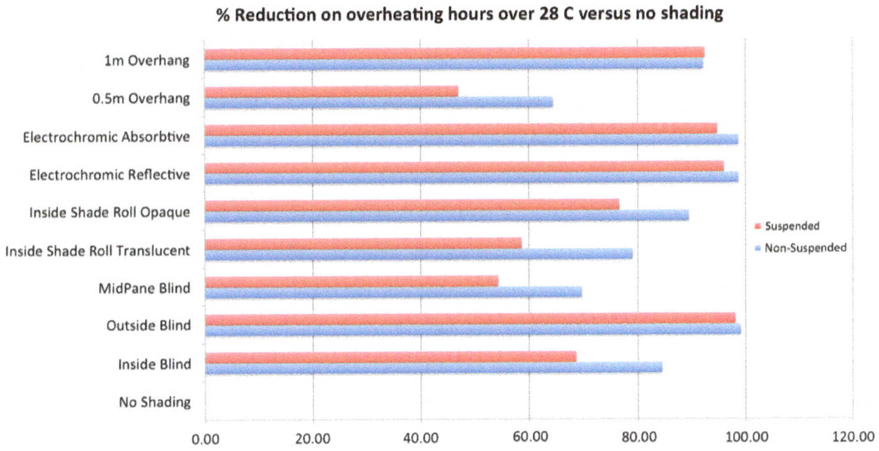

Fig. 2 Comparison of percentage reduction on overheating hours over 28 °C by window and local shading versus no shading for suspended ceiling and non-suspended ceiling scenario

reduction, they still achieve reduction around 50%. It must be noted that in all cases, the use of the thermal mass provided a bigger reduction in overheating, much more pronounced with the use of internal and midpane blinds, as well as shade rolls.

Figure 3 presents the percentage reduction on overheating hours over 28 °C due to exposing the thermal mass in non-suspended ceilings. This percentage reduction is calculate based on the overheating hours above 28 °C when having a suspended ceiling and the overheating hours resulting from exposing the thermal mass (non-suspended ceiling) for each window and local shading devices as presented in Table 1. Overheating reductions above 70% can be achieved by using electrochromatic glazing, while the use of blinds and rolls allow reductions above 60%.

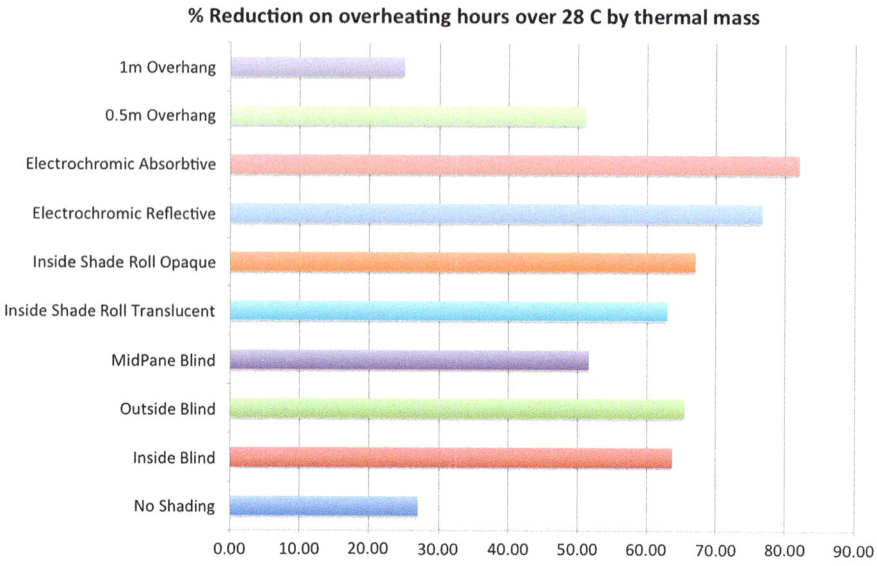

Fig. 3 Percentage reduction on overheating hours over 28 °C by using thermal mass (non-suspended ceiling) versus a suspended ceiling scenario

Little benefit is appreciated by the use of a 1 m overhang but this is due to the overhang already providing a massive reduction of overheating hours by blocking the solar radiation from reaching the glazing, so minimising any radiation effect in the inside room which would be charging the thermal mass and overheating hours being mostly driving by internal gains, form people and equipment. In the case of the smaller contribution of exposing the thermal mass for the case of "no shading", the issue for the thermal mass is completely the opposite. In this case, the thermal mass is fully loaded early in the day, as the number of overheating hours are above 700 h a year. When the thermal mass is fully loaded, it will have no benefit for reduction until the heat contained in the thermal mass is discharged with the night ventilation strategy.

4 Discussion

The use of window or local shading devices to control the solar gain transmission in window can greatly contribute to a reduction of overheating hours over 28 °C as supported by this study and the literature [5–7]. As expected, external or outside blinds provide one of the highest overheating reductions, which agree with the literature [3, 5, 8]. A design limitation on the use of external or outside window shading devices in the United Kingdom is the practice of installing windows, which open outward [3, 12]. On the one hand, electrochromatic glazing provides high overheating reductions, similar to the use of external blind, but avoiding the limitation on opening outwards. On the other hand, electrochromatic glazing has the highest economic cost in terms of new build and retrofit [5], making it the least likely option for window shading regardless of the very good performance.

On top of the benefits achieved by using the window and local shading devices, removing the suspended ceiling will allow the thermal mass to be available in non-domestic buildings to provide further overheating reductions and minimise further the number of overheating hours above 28 °C. This effect is achieved in all window and local shading device simulations for London (Islington), when comparing the overheating hours for the test room with suspended ceiling, blocking the thermal mass, and with non suspended ceiling. The results presented in this paper agree with previous work reporting the advantages of allowing the thermal mass to be used, while coupling with shading devices and purge night ventilation [1, 2, 5–7, 12].

The use of cooling in non-domestic building correlated with the number of overheating hours taking place. Cooling will increase the building energy use and the correspondent carbon emissions to regulate the indoor thermal comfort. An increase of overheating hours will drive upward the cooling demand and as such, a higher energy use for the cooling equipment, which will be reflected on higher carbon emissions for the building.

While this study supports the use of thermal mass and purge ventilation as a mechanism to avoid overheating [1–3] on its own, the benefits of thermal mass can be further extended when window shading devices and local shading is used to limit the solar transmission through glazing [5–7].

These results should be taken into account in the design of new buildings and refurbishment work to avoid overheating and to provide mitigating options to deal with overheating. This study highlights the importance of the thermal mass performance, night ventilation and its effect to reduce overheating also when window shading or local shading is implemented to limit and reduce overheating.

5 Conclusion

This study confirmed, as expected, that the use of window shading helps to alleviate the overheating hours in the test room and as such, improving the thermal comfort and reducing the need for cooling. Furthermore, when the window shading devices are coupling with an optimised used of the thermal mass and night ventilation, the reduction on overheating hours achieved will reach a reduction of over 50% with respect to not exposing the thermal mass. In conclusion, exposing the thermal mass coupled with a night ventilation strategy provides a reduction on overheating hours, which is increased by using different window shading devices. Exposure of the thermal mass provides a good strategy for reducing the need for cooling and increasing thermal comfort.

References

1. Energy Saving Trust. CE129 Reducing overheating—a designer guide (2005)
2. Building Research Establishment (BRE). Thermal mass and overheating (2005)
3. Zero Carbon Home. Understanding overheating—where to start: An introduction for house builders and designers. NHBC Foundation (2012)
4. C. Jimenez-Bescos, in *The Effect of Suspended Ceilings on Thermal Mass to Reduce Overheating*. Proceedings of CISBAT 2015, vol II, ISBN 978-2-9701052-2-0, Lausanne (2015), pp. 821–826
5. Zero Carbon Home. Solutions to overheating in homes (2016)
6. Department of Energy and Climate Change. Guidance on preventing overheating in the home (2015)
7. Adaptation and Resilience in a Changing Climate Coordination Network. Overheating in homes—advice and evidence from the latest research (2013)
8. W. Morten, Strategies for mitigating the risk of overheating in current and future climate scenarios. Encraft Technical Insight (2015)
9. C. Jimenez-Bescos, An evaluation on the effect of night ventilation on thermal mass to reduce overheating in future climate scenarios. Energy Procedia **122**, 1045–1050 (2017)
10. CIBSE. TM52: The limits of thermal comfort: avoiding overheating in European buildings (2013)
11. CIBSE. Guide A: Environmental design (2015)
12. A. Dengel, D. Ormandy, V. Ezratty, in *Guidance document: Overheating in dwellings*, Building Research Establishment (BRE) (2016)

Method for Probabilistic Energy Calculations—Passive House Case Study

Stephen Burke, Johnny Kronvall, Magnus Wiktorsson, Per Sahlin and Anders Ljungberg

Abstract Swedish building regulations require proof that a building fulfills a specific energy use during the design stage. This is done by doing an energy calculation. The result of this calculation is always reported to the nearest integer, for example an energy calculation of a building might predict that it should use 89 kWh/m^2 year when the building regulation limits the actual energy use to maximum 90 kWh/m^2 year. This can lead to conflicts if the measured energy use is greater than the calculated energy use. With the currently available energy calculation tools, if you want to see which risks are associated to the design and material properties, you need to do a parametric study. These types of studies are usually time consuming. Investigations of different buildings show that energy measurements can vary significantly in identical houses. To take into account these differences and avoid costly parametric studies, it is common to add an uncertainty factor to the calculated results. In the project, "Calculation method for probabilistic energy use in buildings" two commercial energy calculation programs developed in Sweden were modified to use Monte Carlo simulations. This method was then tested using a passive house design which was used in the Vallda Heberg passive house development. The calculation results were then compared with the actual measured energy. The results show that a variation of 15 input parameters could explain most of the difference between measured energy use and calculated energy use. The results from the probabilistic energy calculation even showed that the

S. Burke (✉)
NCC Sverige AB, Hyllie Boulevard 10 B, 21532 Malmö, Sweden
e-mail: stephen.burke@ncc.se

J. Kronvall
StruSoft AB, Malmö, Sweden

M. Wiktorsson
Lund University, Lund, Sweden

P. Sahlin
EQUA Simulations AB, Stockholm, Sweden

A. Ljungberg
NCC Sverige AB, Göteborg, Sweden

D. Johansson et al. (eds.), *Cold Climate HVAC 2018*,
Springer Proceedings in Energy, https://doi.org/10.1007/978-3-030-00662-4_54

original energy calculation was on the high-end of the calculated energy distribution. It showed that the probability that the measured energy use would fulfil the Swedish building code was over 95%.

Keywords Energy simulations · IDA-ICE · VIP energy · Statistical input Monte Carlo · Verification

1 Introduction

1.1 Background

This paper is the second article as a part of a project called "Calculation method for probabilistic energy use in buildings". The article Burke et al. [1] provides a background to the project as well as reports the preliminary results obtained before the project was completed. Report Burke et al. [2] is a full report on the project however it is only available in Swedish.

The purpose of this project was to develop and test the applicability of applying Monte Carlo simulations to energy simulations using two dynamic energy simulation programs. The goals of the project were; "to look at which input parameters have the largest influence on the result; to begin defining a realistic spread of the most significant parameters; to study the advantages and disadvantages of probabilistic energy calculations; and to look at the discrepancies between calculated and measured energy use." Burke et al. [1]. Based on the test object study, which is presented in Burke et al. [1], it was determined that the method should be tested with 15 parameters with variable values.

2 Method

This paper shows the results of a blind energy calculation done on a single-family passive house. The calculations used variable parameters and 1000 iterations.

2.1 Mapping Input Parameter Distribution

The goal of the main project was to apply Monte Carlo methods on energy simulations. For this to be realized, a realistic, statistic based variation of input parameters was needed. Fifteen input parameters were chosen and data was collected on the probable variation of the material properties and expected user behavior. Some of the parameters were obtained for the case through a previous

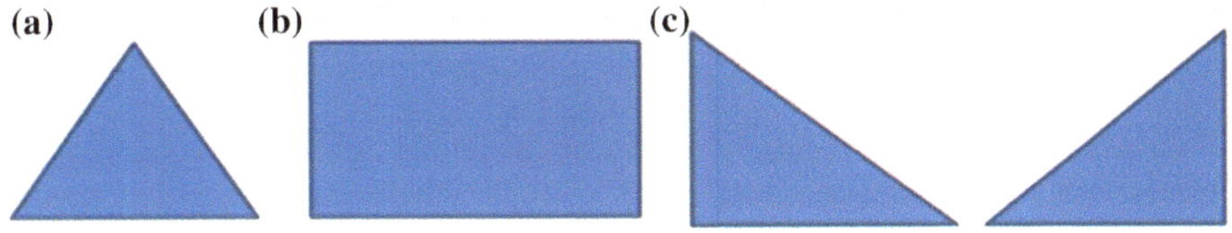

Fig. 1 **a** Triangular distribution, **b** Uniform Distribution, **c** Skewed triangle distribution

study by Fahlén et al. [3]. One limitation is that the distributions could not be clearly defined due to the lack of statistical data. A simplification was made of the distribution based on the available data and experience from the project's working group. This simplification was needed to proceed with the study and test the calculation method. The variations of the distributions are shown in Fig. 1. In the future, more data is required to ensure that the distributions of the real parameters match the theoretical distributions.

2.2 *Energy Calculations*

Energy calculations were done for a passive house single-family home. The design for the passive house was used to build 26 identical houses in the same development. The houses were prefabricated on-site (a mobile indoor construction site was made so the prefabricated parts could be produced indoors then delivered quickly to the worksite and mounted with cranes). This method was used as a means of producing a similar level of quality in all the houses.

Equa and StruSoft, have developed new versions of their energy calculation software (IDA ICE and VIP Energy respectively) for this project. These programs are not available for commercial use. The algorithms used to generate the input-data text file used by the programs were developed by Lund University. The input parameters were defined by energy specialists at NCC based on both the spread of parameters found in the actual houses, and literature. The programs can read the number of calculations to be done from the input file. The programs also load the specific material properties or user behavior values in the current energy calculation and return the result according to the Swedish specific energy use value, as defined by the Swedish National Board of Housing, Building and Planning (Boverket) in the Swedish Building Code (BBR), for each run.

Each run includes random values for each of the input parameters according to the distributions shown in Fig. 1, and parameters presented in Table 1. This results in a distribution for the calculated energy use instead of a single value. This result shows the effects of the uncertainties in the input parameters. The quantiles for the energy calculation can then be calculated. There is an uncertainty remaining in the calculation due to the fact that we use a finite number of random samples for the input parameters. This uncertainty can be taken into account using so called bootstrap techniques.

Table 1 The input parameters for the passive house calculation test case

Parameter	Symbol	Unit	Initial value for calculation	Variation	Distribution
Lambda value of mineral wool insulation[a]	λ	W/mK	0.0438a	±0.00575	Triangular
U-value of windows	U-value	W/m^2K	0.75	±0.2	Triangular
g-value of windows	g	–	0.37	−0.15	Skewed triangular
Thermal bridges	Ψ	% of U*A	25	±5	Uniform
Buildings airtightness[b]	q_{50}	l/sm^2 External surface area @ 50 Pa	0.1	±0.05	Triangular
Indoor temperature	T	°C	21.5	±0.5	Triangular
Supply fan specific fan power (SFP)	SFP_{Sup}	kW/m^3s	0.67	+0.30	Skewed triangular
Extraction fan SFP	SFP_{Ext}	kW/m^3s	0.67	+0.30	Skewed triangular
Heat recovery efficiency of supply and extract ventilation	η	–	0.8	±0.03	Uniform
Supply air flow	q_{Sup}	l/sm^2 Atemp	0.39	±0.13	Triangular
Unbalance (extraction air flow)	q_{Ext}	l/sm^2 Atemp	$q_{Ext} = q_{sup}$ − variation	±0.03	Triangular
Household electricity	Q_{house}	W/m^2 Atemp	4	±1.15	Uniform
Heat generated by people	Q_{pers}	W/m^2 Atemp	1.92	±0.44	Triangular
Domestic hot water use	E_{DHW}	kWh/m^2 Atemp yr	20	±10	Triangular
Kitchen ventilation losses	E_{KV}	kWh/m^2 Atemp yr	3	±1	Triangular

[a]Including wood framing in wall
[b]Based on actual air tightness tests

3 Results and Discussion

The results from the blind energy calculation, which was done on a single-family passive house are shown in Fig. 2. The figure shows the average, median and 95% upper quantile both with and without finite sample correction. The finite sample correction was done using a bootstrap technique. This was done to assess the uncertainty coming from a finite sample size of 1000. For reference, the original energy use (design value) for the passive house was calculated to be 60 kWh/m^2 year.

Figure 3 shows the same information as Fig. 2 but without the average, median or 95% upper quantile. Interestingly both programs had similar results despite small differences in their internal calculation engines.

Figure 4 shows the distribution of the measured energy use for the 26 passive houses. Figure 5 shows a comparison of both the calculated results and measured results relative to the energy use requirement of 60 kWh/m^2 year.

The results seem to show a good correlation between the measured and calculated energy use. This is despite the fact that not all parameters vary. The measured energy use seems to have a few houses which use more energy than predicted, however this could be because of other parameters which were not varied. Other parameters will have to be studied in future work.

Another interesting result is the placement of the energy requirement. The result shows that, already in the calculation, there was a small risk that some of the passive houses would exceed the energy requirement set by the project. This risk

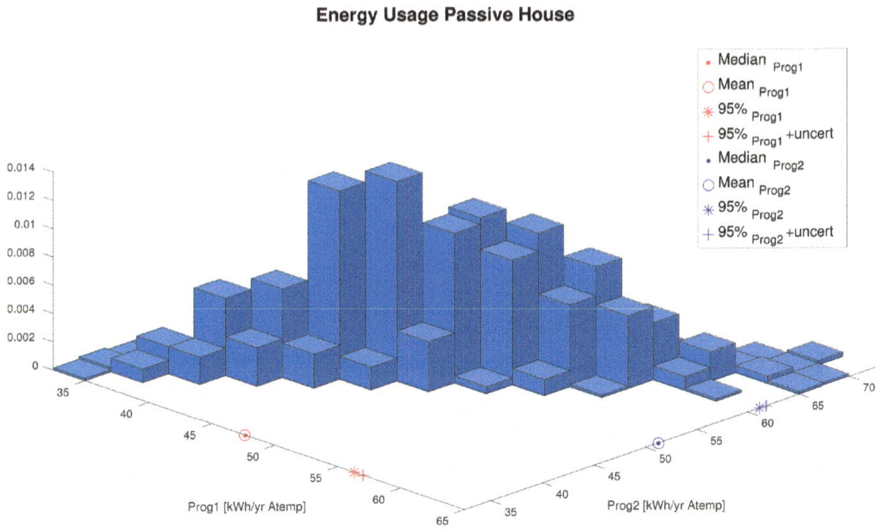

Fig. 2 A three-dimensional histogram of the calculated energy use. The mean, median, and 95% quantiles are marked on the axis. Similar results were obtained with both energy simulation programs

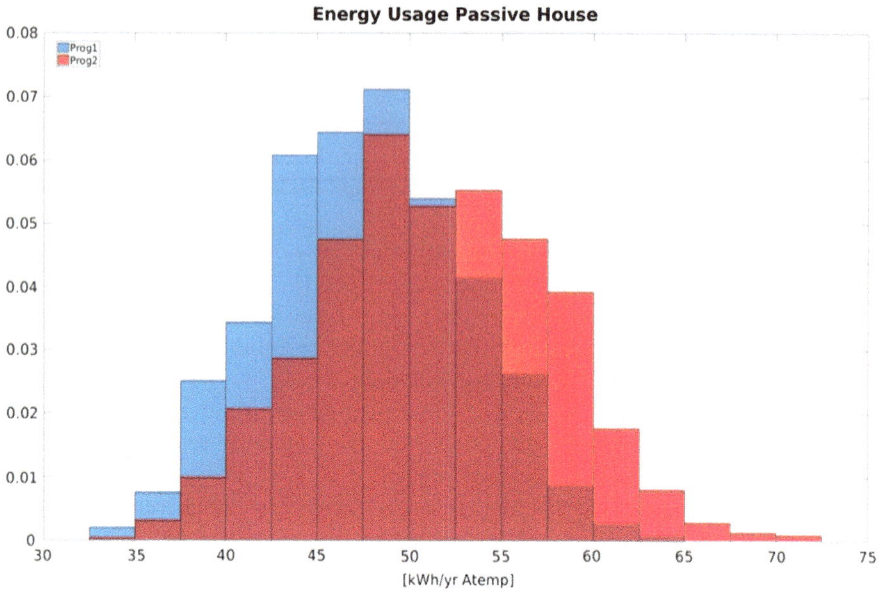

Fig. 3 A two-dimensional diagram showing the results of the energy calculations from the two energy simulation programs

Fig. 4 The measured energy use distribution for the 26 passive houses

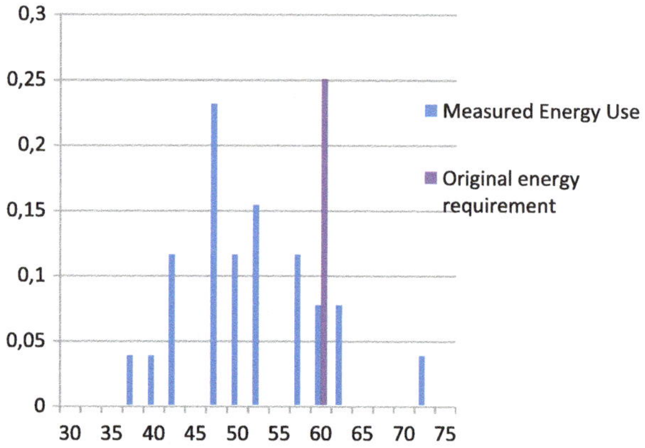

was verified when the calculated energy use was compared to the measured energy use. In the measured houses there were a few more houses over the energy requirement than predicted. It is important to know that an investigation was conducted as a part of the Lågan study [3] which showed why this energy use was higher. It was because of one small detail where the user did not behave as predicted. Each house had a small porch attached to the entrance of their house. This porch is indoors but is designed to be an un-heated space for storing jackets and shoes. Some of the building owners were leaving their interior door open into the porch to warm up this area. This led to a larger energy use in those houses.

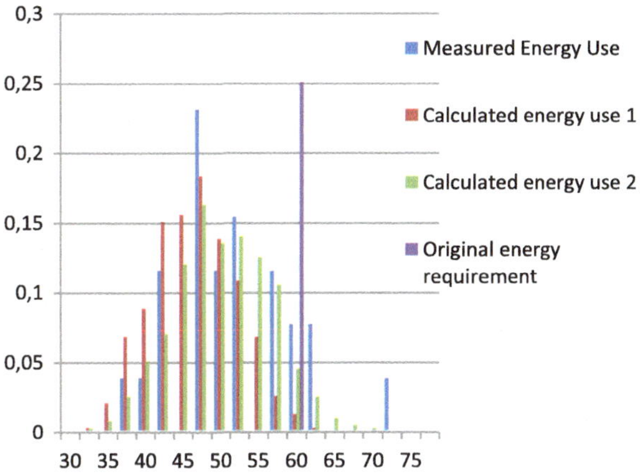

Fig. 5 The calculated energy use from both software compared to the measured energy use and the original energy use requirement of 60 kWh/m^2

In a conventional energy calculation the goal is to predict the energy use as closely as possible without being too low or going over the requirement. This is purely for economic reasons. A house which is much more energy efficient than required probably used more materials than required, increasing its cost. At the same time if the measure energy use is too high, then not enough materials were used and the designer/contractor risks additional costs due to fines or be-building costs. Generally the costs associated with being over the energy use requirement can be higher than when the energy use is far below the required. Safety margins are added to the calculation to try and account for unknown parameters so that the risk of the measured energy use exceeding the calculated energy use is minimized. However, these methods do not provide any indication of how much risk exists.

4 Conclusions

The results of this project showed that it is possible to apply Monte Carlo techniques to dynamic energy calculations; however this process was difficult and complex. The calculated values seemed to agree quite well with the measured values from 26 passive houses.

One interesting conclusion from this project is that with this method it is possible to discuss risks associated with the measured energy use exceeding the calculated energy use without using safety margins. Lowering this risk using random safety margins is potentially more expensive due to the unknown risks associated with the energy calculation. The method presented in this paper shows an alternative way of presenting the calculated energy use and opens up the discussion regarding how much risk is acceptable to the building owner without utilizing safety margins.

Acknowledgements The authors would like to thank SBUF (the Swedish construction industry's organization for research and development), the Swedish Energy Agency under the research programme E2B2, NCC, StruSoft, Equa Simulations AB, Lund University, WSP Sverige AB,

FOJAB Architects, Chalmers Real Estate AB, PEAB, Skanska and The Swedish National Board of Housing, Building and Planning for their time and funding, without which this project would not be possible.

References

1. S. Burke, J. Kronvall, P. Sahlin, in *Method for Probabilistic Energy Calculations—Variable Parameters*. NSB2017 11th Symposium on Building Physics Proceedings, Trondheim, Norway (2017)
2. S. Burke, J. Kronvall, P. Sahlin, A. Ljungberg, in *Beräkningsmetod för sannolik energianvändning i bostadshus* [Method for probabilistic energy calculations in residential housing (In Swedish)], SBUF, At: http://www.sbuf.se/Projektsida/?id=e854c21a-49c2-4eb1-918b-43043a4d6543, Last Accessed 1 Sept. 2017
3. E. Fahlén, H. Olsson, M. Sandberg, P. Löfås, in *Vallda Heberg – Sveriges största passivhusområde med förnybar energi* [Vallda Heberg—Sweden's largest passive house area with renewable energy (original in Swedish)], Lågan, At: http://www.laganbygg.se/UserFiles/Projekt/LAGAN_Vallda_Heberg_slutrapport.pdf, Last Accessed 1 Sept. 2017

Validation of a Zonal Model to Capture the Detailed Indoor Thermal Environment of a Room Heated by a Stove

Martin Thalfeldt, Laurent Georges and Øyvind Skreiberg

Abstract Using wood stoves is a common space-heating strategy in the Nordic countries. Currently, the lowest available nominal power of wood stoves is significantly oversized compared to the design space-heating load of highly-insulated houses. This oversizing might deteriorate the indoor thermal environment by causing overheating and a large vertical temperature stratification. Modelling the indoor thermal environment of rooms heated with a wood stove with acceptable computational time and accuracy, however, is a complex task. The purpose of this study is to analyze the accuracy of a new IDA-ICE zonal model currently under development and to calibrate it against measurements. For this, several experiments were conducted in a test cell, which was heated by an electric stove mimicking a wood stove with a nominal power of 4 kW. Room air temperatures in various positions were measured, while the stove that was placed in the middle of the room was run in cycles with different durations and surface temperature profiles, leading to a thermal stratification of 0.5–2.2 K/m. The zonal model could reproduce the temperatures at the bottom and top layers of the room with good accuracy. However, the model still needs further development and validation to reach good agreement with measurements in the middle layers of the zone. Nevertheless, already at this stage, the model could be used to roughly assess thermal stratification in rooms heated by wood stoves.

Keywords Building simulation · Zonal model · Thermal indoor environment
Wood stoves · Space heating

M. Thalfeldt (✉) · L. Georges
Department of Energy and Process Engineering, Norwegian University of Science and Technology, Kolbjørn Hejes vei 1A, 7465 Trondheim, Norway
e-mail: martin.thalfeldt@ntnu.no

Ø. Skreiberg
Sintef Energy Research, Kolbjørn Hejes vei 1B, 7034 Trondheim, Norway

D. Johansson et al. (eds.), *Cold Climate HVAC 2018*,
Springer Proceedings in Energy, https://doi.org/10.1007/978-3-030-00662-4_55

1 Introduction

Using wood stoves is a common space-heating strategy in the Nordic countries. Currently, the lowest available nominal power of wood stoves is significantly oversized compared to the design space-heating load of highly-insulated houses (as it is difficult to reduce the mean combustion power below 3–4 kW). This oversizing might deteriorate the indoor thermal environment by causing overheating and a large vertical temperature stratification. Modelling the indoor thermal environment of rooms heated with a wood stove is a complex task. Georges and Skreiberg [1] conducted experiments with an electric stove in a passive house and compared the measurements to simulations using TRNSYS. They reached good agreement with operative temperature far from the stove, however this room model was by definition incapable of predicting thermal stratification. For this, a simple simulation method needs to be developed giving acceptable computational time and accuracy. Currently, zonal models that could capture thermal stratification are not implemented in building performance simulation (BPS) softwares, at least not in their publicly available versions. However, a new zonal model is under development in the BPS software IDA-ICE [2] also called a "CFD-free model".

The one-dimensional model splits the room into several horizontal layers with uniform air temperature (i.e. a one-dimensional vertical air temperature field). The mass flow and energy exchanges between layers are computed using so-called flow elements. Flow elements are simplified analytical models for generic flows, such as wall boundary layers, plume or jets. It was originally developed and validated by Togari et al. [3] and has recently been implemented in IDA-ICE in combination with a detailed evaluation of thermal radiation, as described by Eriksson et al. [4]. The shortwave radiation is computed using the radiosity method and detailed view factors computed numerically. This enables to account for non-convex room geometries, such as L-shaped rooms. The model has been implemented also in TRNSYS by De Backer et al. [5, 6]. The implementation of the model in different softwares has shown promise, however it has not been earlier validated against experiments with a strong point heat source such as a stove. The purpose of this study is to analyze the accuracy of the new zonal model and to calibrate it against measurements in a room heated with an electric stove mimicking a wood stove. Several experiments were conducted in a test cell with simple geometry and well-known boundary conditions, which was heated by an electric stove with nominal power of 4 kW.

2 Methodology

2.1 The Test Cell

The experiments in this study were performed in a test cell located in Trondheim, Norway, which is described in Fig. 1. The test cell has one external wall with a window while the other walls are surrounded by a guard zone inside a building. The

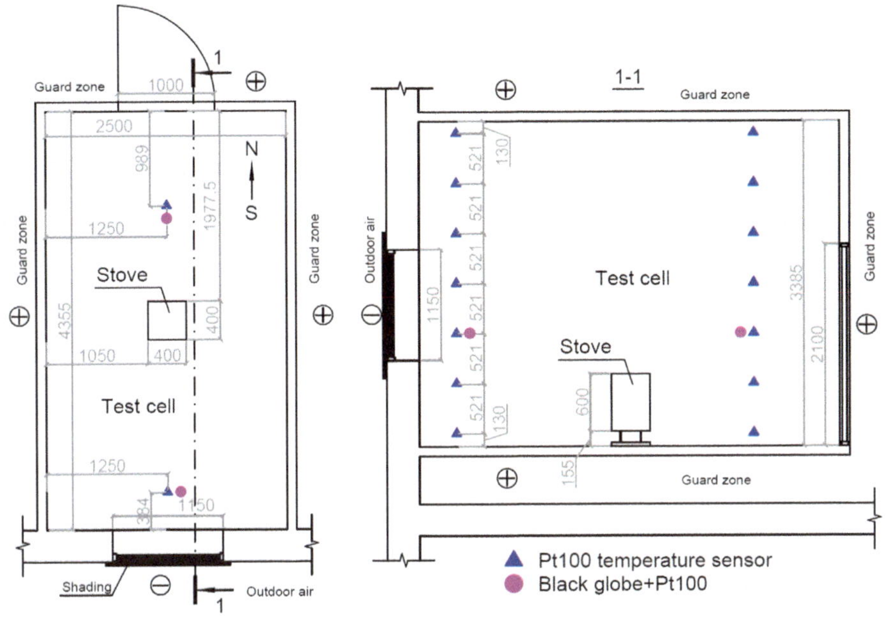

Fig. 1 The ZEB test cell and the location of the stove and Pt100 temperature sensors

air in the guard zone was circulated with fans to assure uniform temperature distribution. The electric stove was installed in the middle of the test cell and two vertical poles with 7 Pt100 temperature sensors and 1 Pt100 inside a black globe were respectively placed between the door and the stove and near the window (see also Fig. 1). The window was covered with a fixed opaque shading to diminish the impact of solar radiation. The space between the shade and window was ventilated with 20 mm vertical air gaps. The construction of the test cell is lightweight as the external wall has a wood-frame construction with 300 mm of glass wool insulation and the internal structures were of sandwich panels with polyurethane foam insulation. A triple pane window has been installed in the external wall. Detailed information about the test cell envelope is given in Tables 1, 2, and 3. The specific heat loss of the combined external wall and window was 2.54 W/K out of which 0.36 W/K were linear thermal bridges. The specific heat loss of internal structures was 13.46 W/K. The emissivity of all internal surfaces was 0.9. T-type thermocouples have been installed on all internal surfaces of the test cell (1–5 sensors per surface).

Additional T-type thermocouples have been installed on the outside surface of the test cell in the guard zone to monitor the ambient conditions of the test cell. The outdoor temperature, relative humidity, wind speed and direction as well as the solar radiation were measured by a weather station on site. Global solar radiation was measured on horizontal and vertical south-facing planes. The parameters of all measurement equipment are given in Table 4.

Table 1 The description of internal structures of the test cell

Material	Thickness (mm)			Thermal conductivity [W/(m K)]	Density (kg/m^3)	Specific heat [J/(kg K)]
	Wall ceiling	Floor	Door			
	Layer order: Test cell to guard zone					
Wood	–	15	–	0.15	1250	1200
Steel	0.6	0.6	0.6	62	7800	500
PUR foam	98.8	98.8	78.8	0.03[a]	35	1600
Steel	0.6	0.6	0.6	62	7800	500

[a]The thermal conductivity of polyurethane (PUR) foam was increased from 0.024 to 0.03 W/(m K) to take into account the thermal bridges in the internal structures

Table 2 Description of test cell window

Total window thermal transmittance [W/(m^2 K)]	Frame		Glazing	
	Thermal transmittance [W/(m^2 K)]	Ratio of window area (%)	Thermal transmittance [W/(m^2 K)]	Solar heat gain coefficient (-)
0.95	2.7	14	0.66/0.59[a]	0.343/0.031[a]

[a]The glazing properties are given with and without shading

Table 3 The description of the external wall of the test cell

Material	Thickness (mm)	Thermal conductivity [W/(m K)]	Density (kg/m^3)	Specific heat [J/(kg K)]
	Layer order: Test cell to outdoor air			
Wood	5	0.15	1250	1200
Glasswool	300	0.035	32	670
Air cavity	0.02	R = 0.18 m^2/(W K)	1.2	1007
Cladding	5	0.5	1250	1200

R of air cavity means thermal resistance of the layer

Table 4 Description of measurement equipment

Type	Position	Physical variable	Accuracy	Frequency
Pt100	Test cell	Air temp	±0.1 K	1 min
TC type K	Stove surface	Surface temp.	±2.2 K	5 s
TC type T	Test cell surface	Air/surface temp.	±0.4 K	5 s/1 min[a]
Pt100	Weather station	Air temp.	±0.15 K	5 s/1 min[a]
Pyranometer	Weather station	Global irradiance	2nd Class	5 s/1 min[a]

TC—Thermocouple

[a]During experiments 5 s and 1 min between experiments

2.2 Electric Stove

The test cell was heated with an electric stove with nominal power of 4 kW and dimensions 400 × 400 × 600 mm (Fig. 2). The stove equipped with electric heating plates on each surface, 50 mm of heatproof insulation and an aluminium structure. The electric heating plates have been covered with 3 mm aluminium plates to make the surface temperature more uniform except for the bottom surface. All stove surfaces were painted to get an emissivity of 0.9, which was verified using a thermal camera. A type K thermocouple was attached in the middle of each stove surface. The thermocouple signal was used to control the respective stove surface temperature using a PID controller.

The heat emission of the stove as a function of temperature difference between stove surfaces and ambient air is illustrated in Fig. 2. The correlation is based on experiments in a large hall, where the stove electric power was measured at different stove temperatures. The electric power of the stove in stead-state was measured using two methods. Firstly, the power was calculated based on measured electric current (Current in Fig. 2). Secondly, the signals of PID controllers were used to calculate the power of each plate based on the maximum power of the heating plates (PID signal in Fig. 2). The measured heat emission showed good agreement with the stove heat emission model described by Georges and Skreiberg [1] (Model in Fig. 2).

2.3 Experiments

A total of 8 experiments with different setpoint profiles of stove surface temperature were conducted during 5 work days to investigate a large range of thermal stratification (Table 5). All used stove temperature setpoint profiles are given in Fig. 3. During experiments, the outdoor temperature ranged between 0.9 and 10.5 °C, with

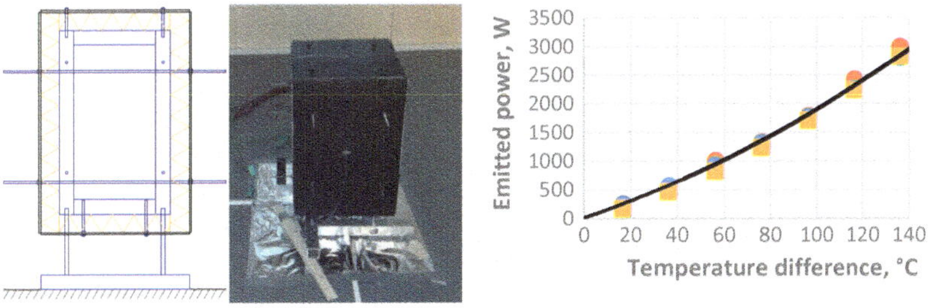

Fig. 2 The electric stove and its heat emission as a function of temperature difference between stove surface and ambient air temperature in a large open space. The prominent rods can be used for attaching radiation shields

Table 5 Overview of conducted experiments and minimum and maximum temperatures during experiments

Exp. no.	Stove temperature setpoint profile	Date	Outdoor temperature (°C)	Guard zone temperature (°C)
1.1	30 min/85 °C	05.04.17	3.4…8.5 (5.7)	19.8…21.7 (20.8)
1.2	30 min/105 °C	05.04.17	3.4…8.5 (5.7)	19.8…21.7 (20.8)
2.1	30 min/125 °C	06.04.17	3.4…5.3 (4.1)	20.3…21.8 (21.3)
2.2	30 min/140 °C	06.04.17	3.4…5.3 (4.1)	20.3…21.8 (21.3)
3.1	20 min/70 °C	09.04.17	9.1…10.5 (9.7)	20.5…22.0 (21.4)
3.2	20 min/85 °C	09.04.17	9.1…10.5 (9.7)	20.5…22.0 (21.4)
4	45 min/140 °C	10.04.17	4.7…6.5 (5.5)	20.6…22.3 (21.6)
5	50 min/150 °C	11.04.17	0.9…4.3 (2.9)	20.5…22.3 (21.6)

Average temperatures are given in brackets

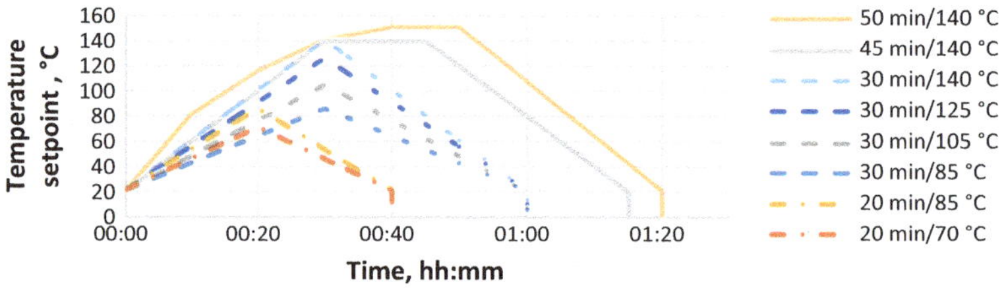

Fig. 3 The setpoint profiles of the stove surface temperature used in experiments

an average value of 5.2 °C. The guard zone temperature was rather stable, ranging between 19.8 and 22.3 °C. Solar radiation did not have significant impact on the experiments due to the shaded window and well-insulated external wall.

2.4 Simulation Model

The test cell was modelled using IDA-ICE version 6.0 Alpha 05, which so far has been distributed to researchers contributing to the development of the software. The test cell zone was surrounded by 5 guard zones as shown in Fig. 4. The temperatures in the guard zones were controlled with ideal heaters and coolers according to the measurements. A steel wall with 50 kg mass and area 10 m^2 corresponding to the objects located in the test cell was added as an internal mass. The convection coefficients of test cell surfaces suggested by Togari et al. [3] were used, which were 2.3, 3.5 and 4.6 W/(m^2 K) for floor, walls and ceiling respectively. The zone was divided into 250 × 250 × 260 mm cells resulting in total 13 horizontal layers. The simulation model was stabilized by simulating 11 days preceding the first experiment based on measured ambient air and surface temperatures.

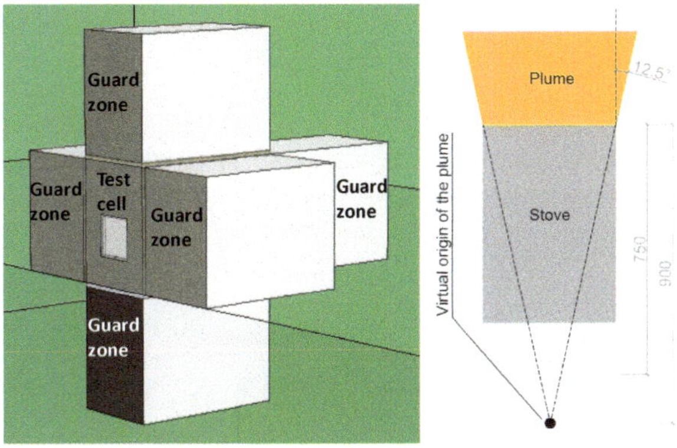

Fig. 4 3D image of the test cell in IDA-ICE (left) and the main parameters of the plume (right)

The stove was inserted in the model in the "Schematic" interface of IDA-ICE. The stove was a closed surface with the same dimensions as the real stove. The imposed boundary condition is the time-varying total emitted power of the stove (i.e. convection and thermal radiation). The starting height of the plume was 0.75 m from the floor and its virtual origin 0.9 m below the plume (see Fig. 4). The convection coefficient of the stove was set at 1.72 W/(m^2 K) (Fig. 5).

3 Results

The experiments resulted in thermal stratification between 0.5 and 2.2 K/m as shown in Table 6 (taken as the maximum value during the stove heating cycle). Thereby, these measurements are a good basis for model validation. The comparison of measured and simulated temperatures at different heights at position between the test cell door and the stove is given in Fig. 6. The match between measured and simulated temperature at the lowest and highest levels was good (blue and dark red lines in the figures). The gap in the case of experiment 1.2 is caused by the initial offset in the beginning of the experiment. However, the simulated temperatures at heights ranging from 651 mm to 2734 mm were significantly higher than the measurements showed. In addition, the simulated temperature at height 2734 mm was slightly higher than temperature at the top layer. In reality, the thermal stratification was rather linear. As an exception, the measured temperatures at heights 651 and 1172 mm were rather similar. Also, there was a time delay in the output files of temperature measurements due to the logging system, but the reason could not be identified. The air temperatures near the window are not presented in this paper, because there were no significant differences between the two locations both in measurements and simulations.

The correlation to calculate stove heat emission shown in Fig. 2 overestimated the heat emission during the experiments, because the correlation was developed based on experiments in a large open space. This mismatch was identified by

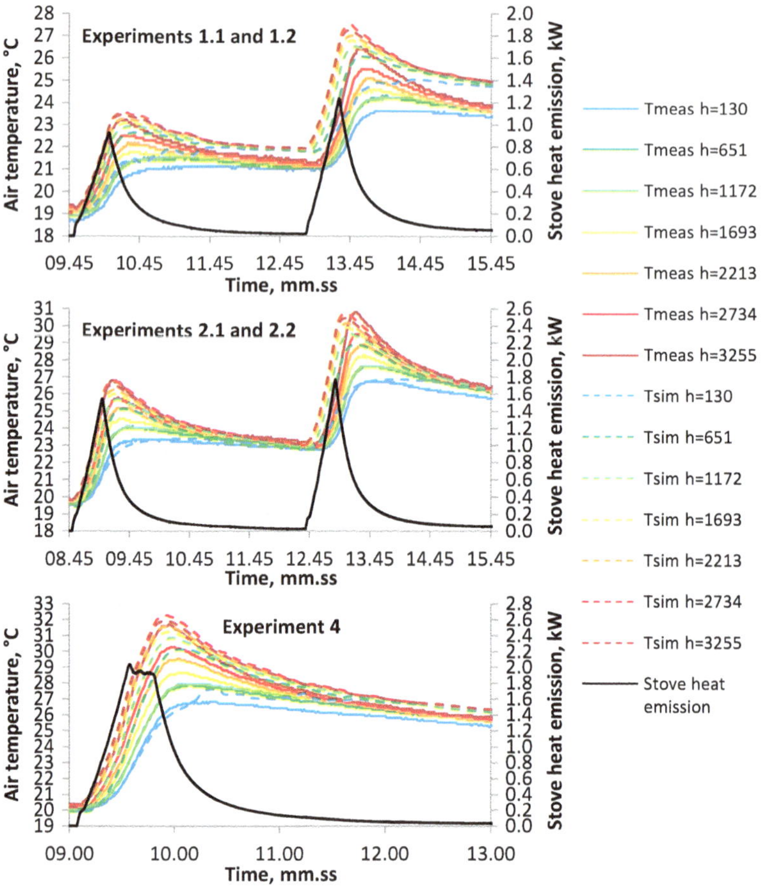

Fig. 5 The stove heat emission, measured (Tmeas) and simulated (Tsim) temperatures between the stove and test cell door at different heights (h) given in millimeters

comparing the calculated total stove heat emission to the test cell and electricity use calculated based on PID controller signals. It might have been caused by the aluminium foil placed under the stove to protect the floor of the test cell and that the surface temperatures of the rest of the room were higher than the temperature measured near the stove. Therefore, the stove power imposed in the model was scaled down so that the heating energy matched the stove electricity use calculated based on PID controller signals.

Due to inaccuracy of temperature simulations in middle layers, the model could not also predict the operative temperature as seen in Fig. 7. Regarding surface temperatures, the simulated wall internal surface temperatures of the test cell were higher than measured ones (Fig. 7). The simulated internal surface temperature of the ceiling was quite accurate during the operation of the stove even though it decreased faster after the stove operating cycle.

Table 6 Measured thermal stratification

Experiment number	Maximum measured thermal stratification[a] (K/m)	
	Between door and stove	Near window
1.1	0.9	0.8
1.2	1.1	1.0
2.1	1.4	1.2
2.2	1.5	1.3
3.1	0.7	0.5
3.2	0.8	0.7
4	1.8	2.0
5	2.1	2.2

[a]Calculated based on temperatures at the lowest and highest layers

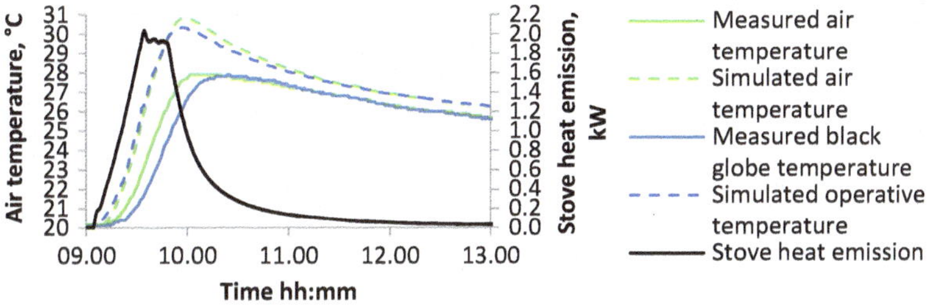

Fig. 6 The measured and simulated air, black globe and operative temperatures, and stove heat emission at height 1172 mm during experiment 4

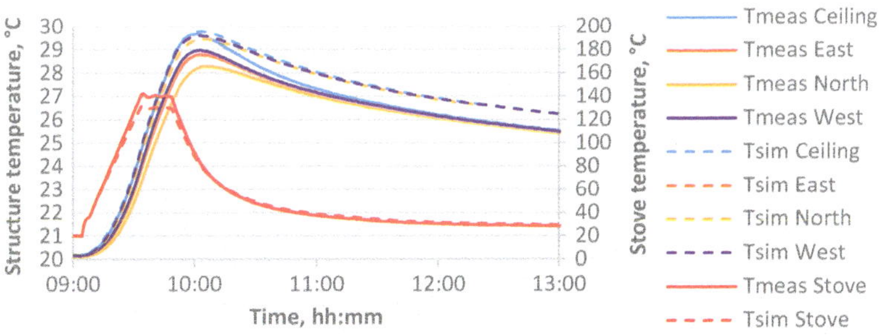

Fig. 7 The measured (Tmeas) and simulated (Tsim) test cell internal surface and stove temperatures during experiment 4

4 Discussion and Conclusions

The purpose of this study was to calibrate a new zonal model of BPS IDA-ICE against measurements in a test cell heated by an electric stove mimicking a wood stove. The experiments resulted in a good set of measurements suitable for validation of the model. The simulation model was able to predict the temperatures at the lowest and highest layers of the zone relatively well. However, the simulated temperatures at the middle layers were significantly higher than measured ones. This is a major limitation as EN ISO 7730 [7] requires comparing vertical temperature difference between heights 0.1 and 1.1 m above the floor to assess thermal comfort. The mismatch might be caused by the fact that only the stove plume is currently considered in the room model, while convection along the surfaces of the stove is not accounted for. Therefore, further research is needed regarding the airflow generated by the stove and how to model it. In the building model, room surfaces are assumed isothermal, a simplification that can also be a source of inaccuracy. In addition, the influence of the coefficient for the thermal diffusion between layers in the Togari model should also be investigated. Results are nevertheless encouraging. Practically, as the overall amplitude of the stratification is well predicted, the vertical temperature difference at the typical user height could be assessed based on temperatures at bottom and top layers of the room and from the vertical temperature distribution reported in studies (see e.g. Georges and Skreiberg [1]).

Based on this study, some model parameters turned to have a large influence on results: the internal thermal mass of the zone, the position and virtual origin of the plume, and the convection coefficients of the room and stove surfaces. In addition, the calibration process requires accurately defined building envelope and internal structures. In future work, the model should be developed further, integrating lessons learned from detailed measurements of stove plumes and should be further validated using field experiments in single-family houses or apartments. Also, attention must be paid to the mismatch in predicted surface temperatures. When validated, the model will be used to establish guidelines for the proper integration of wood stoves in buildings similarly to the work of Georges et al. [8]. This means selecting the right stove power and thermal mass as function of the building thermal properties resulting in a comfortable indoor thermal environment with acceptable operative temperatures and thermal stratification during the entire space-heating season.

Acknowledgements The authors acknowledge the financial support by the Research Council of Norway and a number of industrial partners through the research project WoodCFD ("Clean and efficient wood stoves through improved batch combustion models and CFD modelling approaches", 243752/E20), and good collaboration with EQUA Simulation AB.

References

1. L. Georges, Ø. Skreiberg, Simple modelling procedure for the indoor thermal environment of highly insulated buildings heated by wood stoves. J. Build. Perform. Simul., 1–17 (2016)
2. EQUA AB home page, IDA Indoor Climate and Energy. http://equa.se/en/ida-ice
3. S. Togari, Y. Arai, K. Miura, A simplified model for predicting vertical temperature distribution in a large space. Trans. ASHRAE, 84–99 (1993)
4. L. Eriksson, G. Grozman, P. Grozman, P. Sahlin, M.H. Vorre, L. Ålenius, in *CFD-Free, Efficient, Micro Indoor Climate Prediction in Buildings*. Proceedings of the First Building Simulation and Optimization Conference (Loughborough, UK, 2012), pp. 149–156
5. L. De Backer, J. Laverge, A. Janssens, M. De Paepe, in *The Use of a Zonal Model to Calculate the Stratification in a Large Building*. Proceedings of 35th AIVC Conference (Poznan, Poland, 2014), pp. 522–531
6. L. De Backer, J. Laverge, A. Janssens, M. De Paepe, in *On the Coupling of a Zonal Model with a Multizone Building Energy Simulation Model*. Proceedings of 10th Nordic Symposium on Building Physics (Lund, Sweden, 2014), pp. 71–78
7. EN ISO 7730: 2005. Ergonomics of the thermal environment Analytical determination and interpretation of thermal comfort using calculation of the PMV and PPD indices and local thermal comfort criteria. ISO (2005)
8. L. Georges, Ø. Skreiberg, V. Novakovic, On the proper integration of wood stoves in passive houses under cold climates. Energy Build. **72**, 87–95 (2014)

Validation of TEKNOsim 6 According to CIBSE TM33

Marwan Abugabbara and Saqib Javed

Abstract TEKNOsim is a simple and user-friendly program for analyzing thermal indoor climate and performing annual hourly simulations. The program was originally developed to calculate heating and cooling demands, and indoor climate conditions of a single zone building for peak winter and summer design days. However, TEKNOsim version 6 has been revised to incorporate several new features including annual energy and multi-zone simulations. These features greatly enhance the functionality and the ability of the program. Their implementation in TEKNOsim 6, however, entails a validation procedure to test the accuracy of the new algorithms. This paper presents preliminary results of the ongoing testing and validation of TEKNOsim 6 according to CIBSE TM33 tests for software accreditation and verification. The test results indicate that the new database libraries implemented in TEKNOsim 6 can be easily customized to precisely represent the actual building conditions. The results also show that the program can accurately calculate the total steady-state heat losses for all test cases of CIBSE TM33. The peak cooling loads calculated by TEKNOsim 6 are also in good agreement with CIBSE TM33 for most test cases. Some discrepancies are however observed between TEKNOsim 6 and CIBSE TM33, especially for test cases involving certain building elements. Nevertheless, TEKNOsim 6 is shown to be a fast and reliable program for performing early-design stage simulations.

Keywords TEKNOsim · CIBSE · TM33 · Testing · Validation

M. Abugabbara
Lindab Ventilation AB, 26982 Båstad, Sweden
e-mail: marwan.abugabbara@lindab.com

S. Javed (✉)
Building Services, Lund University, Box 118, 22100 Lund, Sweden
e-mail: saqib.javed@hvac.lth.se

D. Johansson et al. (eds.), *Cold Climate HVAC 2018*,
Springer Proceedings in Energy, https://doi.org/10.1007/978-3-030-00662-4_56

1 Introduction

Building energy simulation programs are nowadays used as a viable tool for the design, performance, and optimization of modern day buildings and building systems. These programs facilitate architects and building services engineers around the world to make decisions which, in turn, improve the energy performance of the buildings. However, today's buildings and building systems are more complex than even before. This entails building energy simulation programs that can make fast and reasonably accurate simulations of thermal loads and energies during the early design phase of a project.

TEKNOsim is a relatively simple yet accurate tool for building simulations developed by Lindab Ventilation AB [1]. It has a large user base among building consultants in Scandinavia. The program was originally developed for indoor climate simulations and calculation of peak heating and cooling demands for winter and summer design days, respectively. However, the most recent version, TEKNOsim 6, has been improved over previous versions in several ways, including a new user interface, updated libraries of weather files, mechanical systems and building components, and the ability to perform annual energy and multi-zone simulations, among others.

A number of studies have been undertaken to compare and validate TEKNOsim against other standards and building energy simulation tools. In the mid-nineties, TEKNOsim 3 was tested and compared [2] against BRIS [3], a then-popular and well-validated Swedish building energy simulation tool. In 2007, TEKNOsim 4 was validated [4] against the German Standard VDI 2078 "VDI Cooling Load Regulation" [5]. TEKNOsim 5 was tested and validated using both standardized and non-standardized methods. Javed et al. [6] compared TEKNOsim 5 with two other building simulation programs using non-standardized test cases. Abugabbara [7] tested and validated TEKNOsim 5 against different standards and building energy simulation tools as a part of his master thesis work at Lund University. All these studies have demonstrated good agreement of TEKNOsim results with the results of other standards and building energy simulation tools.

As mentioned above, TEKNOsim 6 has been extensively updated to incorporate several new features including annual energy and multi-zone simulations. These changes require a comprehensive validation study to test the accuracy of the new algorithms. TEKNOsim 6 is being tested and validated against standard methods of testing building energy simulation programs. This paper presents preliminary results of the ongoing testing and validation of TEKNOsim 6 according to Chartered Institution of Building Services Engineers (CIBSE) TM33 test methods [8]. It also suggests few recommendations and future implementations for improving the accuracy of the program.

2 Methodology

Test cases included in the CIBSE TM33 Standards are divided in three main categories of general purpose, empirical validation and CIBSE-specific tests. The tests considered within the scope of this paper are listed in Fig. 1. The empirical validation test is not included in this study as the asymmetric geometry required by the test cannot be modelled in TEKNOsim 6. In addition, some general purpose and CIBSE-specific tests are not included due to the difficulties in modelling them in TEKNOsim 6 or due to the problems in retrieving the climate data used by these tests. Figure 1 also shows the interdependencies between the tests considered in this paper. As seen, the results of Test G3: Basic thermal calculations are used as inputs to Test G6: Steady state heat loss from rooms. Similarly, results of Test C2: Derived material properties, provide inputs to Test C5: Cooling load, the results of which are in turn used as inputs to Test C6: Summertime temperatures.

2.1 General Purpose Tests

General purpose tests are suitable for internal quality assurance of simulation software. These tests are also a part of the UK's National Calculation Methodology accreditation process [9]. Each test has a reference value to which the results from the tested software are compared for accuracy.

Test G1: Databases. The database tests aim to ensure that a limited subset of available data can be accurately output by the program. The tests are related to material thermal properties (Test G1A), climate data (Test G1B), and loads and schedules (Test G1C).

Test G2: Solar Position. This test demonstrates the ability of the program to determine the solar azimuth and elevation by calculating the position of the sun in the sky with respect to the building's location and the time of the year.

Test G3: Basic Thermal Calculations. This test illustrates the ability of the program to compute steady state (Test G3.1) and transient response (Test G3.2) for different construction types ranging from heavy- to light-weight.

Test G6: Steady-state Heat Loss from Rooms. This test examines the ability of the program to calculate the steady-state heat loss from a room with and without adjacent zones, and with infiltration/ventilation. The boundary conditions of different room surfaces are specified. Figure 2 shows the geometry of the room and the surface numbers.

2.2 CIBSE-Specific Tests

CIBSE-specific tests include additional test cases somewhat similar to general purpose tests category but only meant for validation of programs based on CIBSE methods.

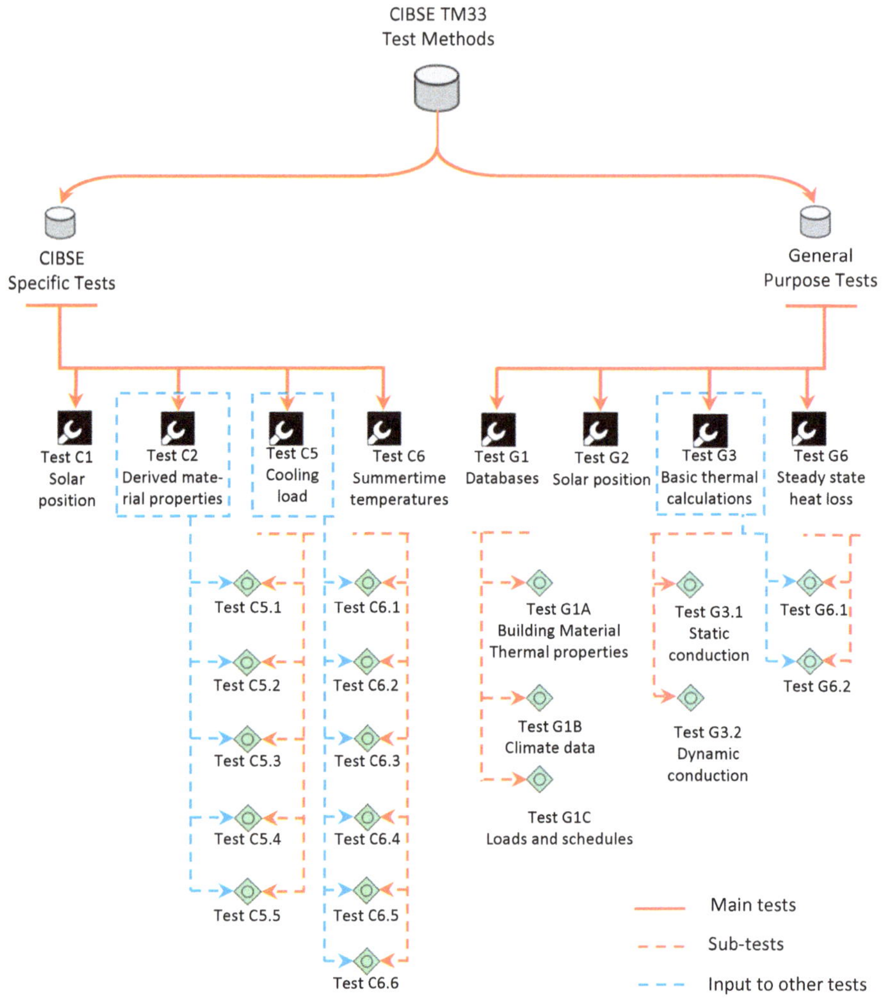

Fig. 1 Test cases from CIBSE TM33 considered in this paper and their interdependencies

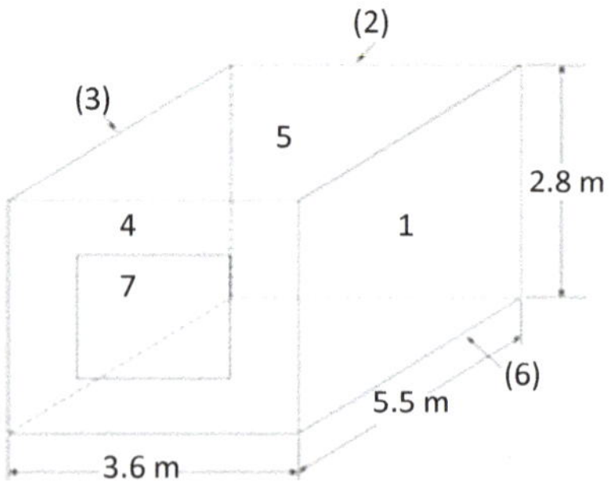

Fig. 2 Isometric view of the room geometry and the surface numbers for Test G6

Test C1: Solar Position. Like Test G2, the objective of this test is to accurately determine the position of the sun in the sky with respect to the building's location and the time of the year. However, the solar azimuth and elevation are calculated for three different locations and four local times.

Test C2: Derived Material Properties. This test demonstrates the ability of the program to derive thermal transmittance and properties of constructions for thermal calculations.

Test C5: Cooling Load. This test examines the ability of the program to calculate design cooling loads. It is based on Standards EN BS 13791 [10] and EN BS 15255 [11] for cooling load calculations. The room geometry is same as shown in Fig. 2.

Test C6: Summertime Temperatures. This test is aimed at calculating peak air and operative temperatures based on Standards EN BS 13791 and EN BS 1379 [12]. The room geometry for this test is same as shown in Fig. 2.

3 Results

3.1 General Purpose Tests

Test G1: Databases. TEKNOsim 6 allows users to provide inputs on material characteristics including density, thermal conductivity, and specific heat capacity, among others. Hence, the building material thermal properties for Test G1A are defined precisely by TEKNOsim 6, with no deviation from the CIBSE TM33 results.

The climate data in TEKNOsim 6 is based on ASHRAE's International Weather for Energy Calculation 2.0 (IWEC2) [13]. In addition, it is also possible to utilize external weather files through the core functionalities of the program. Consequently, for Test G1B, the climate data results from TEKNOsim 6 are in complete agreement with the results from CIBSE TM33. Figure 3 presents the output of climate data results from TEKNOsim 6.

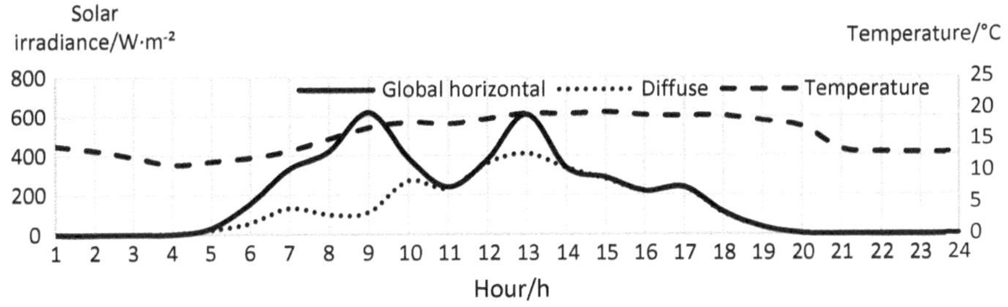

Fig. 3 Solar irradiance and temperature variations during the design day for Test G1B

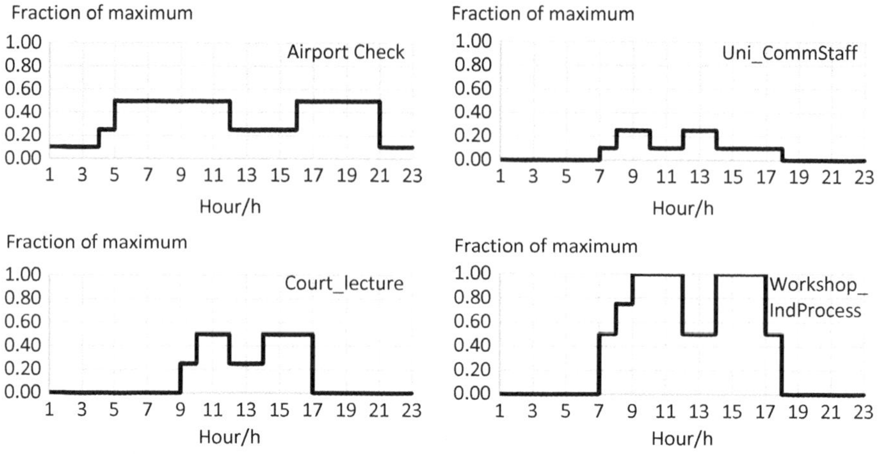

Fig. 4 Daily schedules of space activity and occupancy loads for Test G1C

Figure 4 presents daily schedules for Test G1C for four different space activities. The values in the figures present the fraction of maximum values for stated schedules for different space activities and occupancy loads. A new feature in TEKNOsim 6 enables users to define hourly schedules for space activities and internal heat load generation in a comprehensive manner. Consequently, the daily schedules from TEKNOsim 6 for Test G1C are consistent with the CIBSE TM33 results.

Test G2: Solar Position. The results of this test include the solar azimuth and altitude for a building at three different locations and at four different times of the year. The acceptable tolerance for this test is $\pm 1.5°$ compared to the reference values prescribed in CIBSE TM33. Table 1 shows the sun position calculated by TEKNOsim 6 at the prescribed times for each location. In all cases, the results from TEKNOsim 6 are within the acceptable tolerance limits for both solar azimuth and altitude.

Test G3: Basic Thermal Calculations. The results from this test are divided into two parts. For the static conduction test (Test G3.1), the thermal transmittance values calculated for various construction elements are presented in Table 2. The calculated values should be within 0.01 of the reference values provided by CIBSE TM33. For Test G3.1, most results from TEKNOsim 6 are within the acceptable tolerance range. However, larger differences exist for the cases of Internal wall 2, and Windows 1 and 2. For the case of Internal Wall 2, the difference between TEKNOsim 6 and CIBSE TM33 results is due to the fact that the U-value calculation in TEKNOsim 6 depends on both inner and outer accumulative resistances of the construction layers. When a lightweight construction with no thermal insulation is simulated, the active depth of both inner and outer accumulative layers decreases, resulting in a higher U-value. For the case of Windows 1 and 2, it is not possible to define the pane and the gap properties in the program. Hence, these results are not

Table 1 Results of Test G2: solar position

Time (hh/dd/mm)	London 51.48 N/0.45 W				Manchester 53.25 N/2.27 W				Edinburgh 55.95 N/3.35 W			
	Azimuth		Altitude		Azimuth		Altitude		Azimuth		Altitude	
	Ref.	TS	Ref.	TS	Ref.	TS	Ref.	TS	Ref.	TS	Ref.	TS
1200/22/12	180.0	179.9	15.1	15.1	178.3	178.3	13.3	13.3	177.3	177.3	10.6	10.6
1500/27/02	224.1	223.8	20.4	19.8	221.9	221.6	19.9	19.3	220.2	220.0	18.3	17.7
1200/21/06	178.4	178.4	62.0	61.9	175.2	175.2	60.1	60.1	173.7	173.7	57.4	57.4
1000/20/10	151.1	151.5	24.4	23.0	149.6	150.0	22.3	20.9	149.1	149.4	19.7	18.3

Table 2 Results of Test G3.1: static conduction test

Construction	Transmittance ($W\ m^{-2}\ K^{-1}$)		Construction	Transmittance ($W\ m^{-2}\ K^{-1}$)	
	Ref.	TEKNOsim		Ref.	TEKNOsim
External wall	0.49	0.49	Ceiling	0.23	0.24
Internal wall 1	0.35	0.36	Roof	0.44	0.44
Internal wall 2	1.68	*1.98*	Window 1	2.94	*N/A*
Floor	0.24	0.24	Window 2	1.72	*N/A*

included in Table 2. For real situations, the glass and frame U-values can be retrieved from the window manufacturer, and TEKNOsim can calculate the overall U-value based on the geometry of the window.

For the dynamic conduction test, the room air temperature calculated for four different construction types are presented as Tests G3.2.1 to G3.2.4 in Table 3. These values are calculated for the first five days of the year applying 30 days of pre-conditioning. The calculation results from the program should be within ±0.6 °C of the reference values provided by CIBSE TM33. For these test, over 50% of TEKNOsim6 results are outside acceptable tolerance range. These results are indicated in italics in Table 3. For dynamic conduction tests, the disagreement between TEKNOsim 6 and CIBSE TM33 results is due to the thermal storage in the construction, which in turn is influenced by the arrangement of construction layers. For instance, compared to Test G3.2.4, the difference between TEKNOsim 6 and CIBSE TM33 results for Test G3.2.3 is very large. The two tests have exactly the same construction layers but in opposite order. This implies that in TEKNOsim 6, the heat losses from the construction are influenced by the calculation of the thermal transmittance.

Table 3 Results of Test G3.2: dynamic conduction test

Test	Air temperature after stated time									
	2 h		6 h		12 h		24 h		120 h	
	Ref.	TS	Ref.	TS	Ref.	TS	Ref.	TS	Ref.	TS
G3.2.1	20.04	*21.13*	21.26	*24.61*	23.48	*27.44*	26.37	*29.41*	30.00	29.98
G3.2.2	25.09	*22.61*	29.63	*27.86*	30.00	29.65	30.00	29.97	30.00	29.98
G3.2.3	20.00	*21.58*	20.26	*25.86*	21.67	*28.57*	24.90	*29.82*	29.95	29.98
G3.2.4	20.00	19.67	20.06	19.92	20.25	20.25	20.63	20.89	23.17	*24.69*

Test G6: Steady State Heat Loss from Rooms. The results of this test include the surface and air temperatures and heat losses for four different configurations (Test G6.A1, Test G6.A2, Test G6B1, and Test G6B2). Figure 5 shows the results of surface and air temperatures. The surface numbers refer to the corresponding construction element numbers shown in Fig. 2. The acceptable tolerance range for these results is 0.2 K. The surface temperatures obtained from TEKNOsim 6 are in good agreement with the CIBSE TM33 results. The largest deviation is for surfaces 6 and 7. This deviation can be traced back to the results of Test G3.1, which are used as inputs for this test. In TEKNOsim, the air temperature can only be calculated for transient state simulations. For steady state heat loss calculations, it is assumed to be constant based on user input.

Figure 6 presents the results of fabric, infiltration and total heat losses for Test G6. The acceptable tolerance range is 2.5% of the total heat loss.

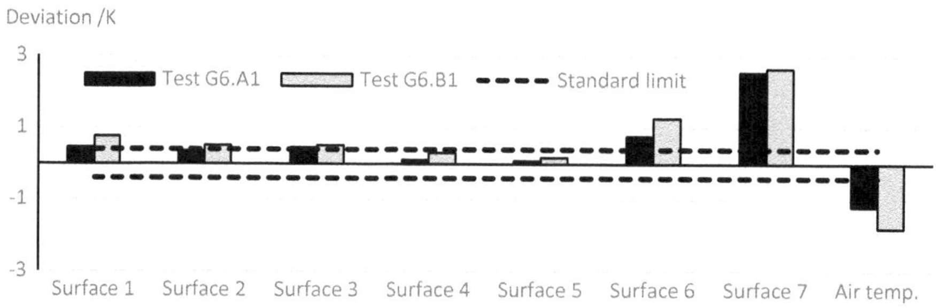

Fig. 5 Results of Test G6: surface and air temperatures

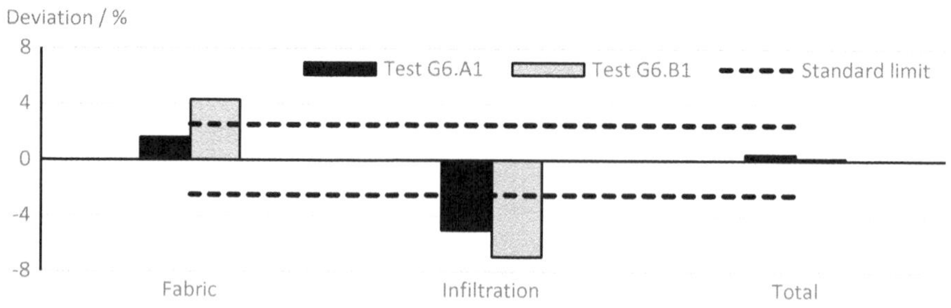

Fig. 6 Results of Test G6: heat losses

Although, TEKNOsim 6 results of fabric and infiltration heat losses have minor deviations from the CIBSE TM33 results, but the total heat loss calculated by the program is in very good agreement with the CIBSE TM33 results.

3.2 CIBSE Specific Tests

Test C1: Solar Position. Like Test G2, the results of this test provide calculations of solar azimuth and altitude. The test is performed for three international locations and for four different times of the year. The acceptable tolerance range for this test is $\pm 1.5°$ compared to the reference values provided in CIBSE TM33. The results of the test are presented in Table 4. TEKNOsim 6 accurately calculates the sun position for all cases prescribed in the test.

Test C2: Derived Material Properties. The results of this test include derived properties of constructions for thermal calculations. This test is similar to Test G3 and applies an acceptable tolerance range of 0.01 of the CIBSE TM33 results. Table 5 presents the results of the test. As for Test G3.1, TEKNOsim 6 results are within the acceptable tolerance range with the exception of Internal wall 2, and Windows 1 and 2.

Test C5: Cooling Load. The results of this test include average and peak cooling loads for five test conditions (Test C5.1–Test C5.5). The acceptable tolerance range for the mean and peak cooling loads is ±5 and ±10%, respectively, of the CIBSE TM33 provided results. The test results are presented in Fig. 7. The peak cooling loads determined by TEKNOsim 6 are in good agreement with the CIBSE TM33 results. The mean cooling loads calculated by TEKNOsim 6 show some discrepancies with the CIBSE TM33 results. The discrepancies can be traced back to the transmission of heat losses through the window systems.

Table 4 Results of Test C1: solar position

Time (hh/dd/mm)	Auckland (GMT+12 h) 37.02 S/174.8 E				Cairo (GMT+2 h) 30.13 N/31.4 E				Reykjavik (GMT-1 h) 64.13 N/21.9 W			
	Azimuth		Altitude		Azimuth		Altitude		Azimuth		Altitude	
	Ref.	TS	Ref.	TS	Ref.	TS	Ref.	TS	Ref.	TS	Ref.	TS
1200/22/12	18.1	18.2	75.8	75.8	182.1	182.0	36.4	36.4	174.1	174.1	2.3	2.3
1500/27/02	301.7	301.1	46.3	46.8	234.1	233.6	33.5	33.0	215.3	215.2	12.9	12.3
1200/21/06	5.8	5.8	29.3	29.3	188.1	188.4	83.2	83.3	169.8	169.9	49.0	49.0
1000/20/10	54.3	55.8	50.9	51.9	145.5	146.4	43.4	42.1	146.8	147.0	11.8	10.4

Table 5 Results of Test C2: derived material properties

Construction	Transmittance ($W m^{-2} K^{-1}$)		Construction	Transmittance ($W m^{-2} K^{-1}$)	
	Ref.	TEKNOsim		Ref.	TEKNOsim
External wall	0.49	0.49	Floor 2	0.74	0.71
Internal wall 1	0.35	0.36	Ceiling 2	0.74	0.73
Internal wall 2	1.74	*1.98*	Roof	0.44	0.44
Floor 1	0.24	0.24	Window 1	2.76	N/A
Ceiling 1	0.23	0.24	Window 2	1.69	N/A

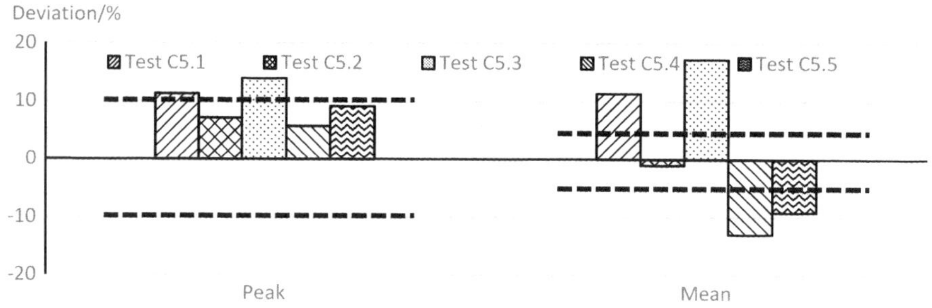

Fig. 7 Results of Test C5: cooling loads

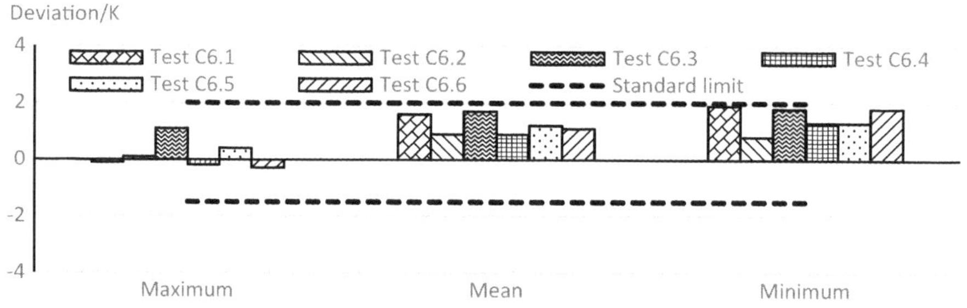

Fig. 8 Results of Test C6: operative temperatures

Test C6: Summertime Temperatures. The results of this test comprise of minimum, maximum, and mean operative temperatures for six test conditions (Test C6.1 −Test C6.6). The acceptable tolerance range for this test is −1.5 K to +2 K of the CIBSE TM33 results. The test results are shown in Fig. 8. For all test cases, the minimum, maximum and mean operative temperatures calculated by TEKNOsim 6 are well within the acceptable tolerance range.

4 Conclusions and Recommendations

This paper has presented preliminary results of the ongoing testing and validation of TEKNOsim 6. The comprehensive validation procedure covers multiple standards and verification methods. This paper has focused on the validation against CIBSE TM33 tests for software accreditation and verification. Some general purpose and CIBSE-specific tests, as well as the empirical validation test, are, however, not covered in this paper due to data or methodological limitations. These tests will be included if and when these limitations are overcome.

Results of Test G1, Test G2, Test G3.1, Test C1, and Test C2 indicate that TEKNOsim 6 allow users to clearly and precisely define inputs related to construction elements, building conditions, and weather parameters, among others. Results of Test G3.2 and Test C6 suggest that TEKNOsim 6 is an accurate tool for predicting the indoor climate conditions in terms of operative and air temperatures. Results of Test G6 and Test C5 imply that TEKNOsim 6 can accurately calculate total heat losses for winter conditions and peak cooling loads for summer conditions. However, as demonstrated through several tests including Test G3 and Test G6, the calculation of transmission losses and solar gains through the window system in TEKNOsim 6 is not sufficiently precise. The algorithms behind the calculations of transmission losses and solar gains through windows should be investigated and revised to provide good comparison with the CIBSE TM33 results.

References

1. Lindab, About Lindab, [Online]. Available: http://www.lindab.com/global/pro/about-lindab/pages/default.aspx. Accessed 28 Aug. 2017
2. Lindab Climate. TEKNOsim Climate Simulation. Lindab, 2001
3. G. Brown, The BRIS simulation program for thermal design of buildings and their servicesm. Energy Build. **14–4**, 385–400 (1990)
4. H. Bley, in *Comparison—Cooling Load Calculation TEKNOsim Versus VDI 2078*. Report: Cologne University of Applied Sciences, Germany (2007)
5. VDI 2078. Cooling Load Calculations of Air-conditioned Rooms, VDI Cooling Load Regulations, 1996
6. S. Javed, R. Lechner, J. Behrens, in *TEKNOsim: A Building Energy Simulation Program.* Proceedings of the 12th REHVA World Congress (Clima 2016), Aalborg, Denmark, 22–25 May 2016
7. M. Abugabbara, in *Validation of TEKNOsim 5: Verification Against Different Standards and Building Energy Simulation Tools*. Master thesis in Energy-efficient and Environmental Buildings, Lund University, Sweden (2017)
8. CIBSE, *Tests for Software Accreditation and Verification—CIBSE TM33:2006* (CIBSE Publications, London, 2006)
9. National Calculation Methodology for the energy performance of buildings: The UK implementation of the requirements of the Energy Performance of Buildings Directive (Office of the Deputy Prime Minster, London, 2006)

10. BSI, Thermal performance of buildings—Calculation of internal temperatures of a room in summer without mechanical cooling—General criteria and validation procedures, BSI Standards Limited, UK, 2012
11. BSI, Energy performance of buildings—Sensible room cooling load calculation—General criteria and validation procedures, BSI Standards Limited, UK, 2007
12. BSI, Thermal performance of buildings. Calculation of internal temperatures of a room in summer without mechanical cooling. Simplified methods, BSI Standards Limited, UK, 2005
13. Y.J. Huang, F. Su, D. Seo, M. Krarti, Development of 3012 IWEC2 weather files for international locations (RP-1477). ASHRAE Trans. **120**, 340–355 (2014)

Real Time Measurement of Dynamic Metabolic Factor (D-MET)

Jakub Wladyslaw Dziedzic, Da Yan and Vojislav Novakovic

Abstract The presented study describes developing a method for observing building occupants' activity. Once their activity is registered, such data can be used to identify typical patterns in their behaviour. The collected information will support development of an occupant-behaviour-energy-related model in residential buildings. Data registration was done with the use of the Microsoft Kinect device as a depth registration camera. This research explores an innovative approach to investigating residents' living and working habits. It supports the already existing thermal comfort models by delivering high resolution information about occupants' activities. The obtained solution and its output will be used in the next stage of developing a dynamic metabolic rate (D-MET) model that will simulate the MET value. With proper data, it will be possible to estimate the real impact of occupants and their behaviour on energy consumption of buildings.

Keywords Occupant behaviour · Metabolic factor · Building performance simulations

1 Introduction

Description of thermal sensation of a human body is a multidimensional problem that has not as yet been thoroughly investigated. Thermal comfort has a different meaning for each person. The main reasons of this absence of consistency are geo-graphical location, culture and many more other factors. The importance of this problem has been previously investigated [1], and it has been proven that overall energy-related behaviour in buildings has an important impact on their energy

J. W. Dziedzic (✉) · V. Novakovic
Department of Energy and Process Engineering, Norwegian University of Science and Technology, 7491 Trondheim, Norway
e-mail: jakub.w.dziedzic@ntnu.no

D. Yan
School of Architecture, Tsinghua University, Beijing 100084, China

D. Johansson et al. (eds.), *Cold Climate HVAC 2018*,
Springer Proceedings in Energy, https://doi.org/10.1007/978-3-030-00662-4_57

performance. The already existing models are able to capture and simulate building energy performance, but their accuracy needs to be improved. In order to do it, it is necessary to capture data that will directly evaluate models' performance. The most common obstacle that makes the description of occupants' behaviour difficult is related to deficiency of real data that researchers could use to verify their findings. Data is scarce because of ethical and privacy issues. It is difficult to obtain reliable data with the resolution allowing interpretation of occupants' behaviour in buildings without crossing certain important boundaries. The most common situation researchers in this field have to deal with is a tradeoff between the quality of measurements and the simplicity of output data and its applicability

A good example illustrating the difficulty of data collection is the comparison be-tween PIR (the abbreviation is not explained, is it obvious for everybody what this means +) detector and video registration camera. The first device gives information about the activity that is deemed sufficiently significant, i.e. one that exceeds the threshold value pre-set in the device and is therefore detected within the monitoring area. This solution allows obtaining brief information about an occupant's activity but it is unable to define the vector of movement. On the other hand, video recordings can be processed to achieve fine, precise data about movement distribution. It can be done with the use of background analyses combined with the Kalman filter. Despite its efficiency, such a solution cannot be used inside residential buildings. The reason is that it would cross important boundaries related to ethical issues, and there-fore the chances of obtaining approval to conduct such a measurement are low due to the potential violation of occupants' privacy.

However, research on this subject still continues, and scientists conduct numerus experiments in various controlled environments, which allows them to develop different models that can be used as approximations for investigating thermal comfort. Most common solutions are proposed by Fanger [2]. ASHRAE standard 55 [3] introduces usage of a calculation formula that enables calculation of thermal condition of occupants. The formula evaluates the energy balance of a human body, but, to do so, it still requires considerable amounts of input information whose source is usually statistical or derivative. Lack of real life recordings makes it difficult to improve the accuracy of the developed models and limits their applicability to standardized scenarios. Such a solution does not guarantee results that will satisfy all of the interested parties. In order to increase the thermal comfort of a building's occupants, it is necessary to understand their preferences, needs and habits. Direct response mechanisms may misunderstand actions performed by occupants and, eventually, they may de-crease the level of satisfaction among them. Such a situation cannot occur. Incorrect functionality of HVAC control systems may lead to a decrease in occupants' confidence in their ability to perform adequately, and this, in turn, may cause a huge in-crease in the operating costs of buildings' HVAC unit. If we want to understand occupants' intentions, it is necessary to monitor their activities. The most common way to do it is to use an additional device—similar to a smartphone or a wrist band—that occupants carry with them all the time. Such appliances can deliver exact information about occupants' positions and it is even able to track the human body core temperature

[4]. The main downside of this solution is that it requires constant carrying of the monitoring device. Although the collected information is highly accurate, once the device is disconnected from the occupant, it may stream faulty information. Additionally, if it is requested from occupants to carry some supplementary device at all times. They may grow excessively self-conscious about it and, in consequence, their "normal" behaviour may be disturbed.

There is, however, a way that allows avoiding all this inconvenience—a tested solution that is able to track occupants' behaviour without interfering directly into their privacy. This kind of measurements can be done with the use of a depth registration camera. A normal camera captures colour picture that are reassembled by a raster picture, and each pixel has a variety of three prime colours intensities. Depth registration camera pixels hold information about the depth of the monitored exposition. In order to obtain this information, the camera is scanning the exposition area via a projector of points in infra-red spectrum and analyses the distance of these points. The range and angle of such scanning depends on the signal amplifier. The principles of operation of these devices is the same as in the case of three-dimensional surface scanners. One of the most popular products that have a depth registration camera built-in as a default setting is the Microsoft Kinect Device [5]. Commonly used for the purpose of entertainment, it also proves to be a useful tool for investigating occupants' behaviour in residential buildings. It seems that this solution delivers quantitative data of sufficient quality, while at the same time avoiding violation ethical boundaries. Observed occupants cannot be directly identified, yet the input data allows differentiating one from another. The obtained data will support further development of HVAC control systems with identification of individual needs. Such a solution can deliver a higher resolution picture about the thermal situation inside a building. Once each occupant's thermal preferences are known, a range of possible decisions will be created to support an indoor air adjustment process that will be more concerned with occupants' needs.

This work is focused on investigating the possibility of approximating an occupant's metabolic rate as a value of generated heat per square meter of human body skin. This estimation will be based on measurements taken with the use of the Microsoft Kinect Device [5]. The obtained data, owing to its quality and amount, can be used for model predictive control (MPC) connected with a control system. It can also be used to build performance profiles that can possibly lead to development of a dynamic metabolic model (D-MET). This research, with its innovative approach, endeavours to contribute to the existing knowledge on the subject. The measurements in question were conducted inside the ZEB Living Lab at NTNU in Trondheim [6], a laboratory that recreates the conditions of a modern residential building. This trial enabled obtaining a 48-h of measurement data, with one occupant living inside the laboratory on normal working days of the week. The participant of the study had been fully informed about the nature and aim of the conducted research and he took part in this research voluntarily.

2 Methods

The greatest advantage of the Kinect device, when compared to other depth registration cameras available on the market, is that it has direct support in the form of the software developer kit (SDK), which allows access to all of the functionalities of the device offering a wide range of options. Other depth registration camera delivers data representation of observed field as a raster picture. Resolution of these pictures depends on device quality, and each pixel holds information about distance from device to observed surface (if the surface is in the registration range of the used de-vice). Kinect device used in this study has an additional build-in software that allows to track people's activity. If it detects human-like shape, it bounds observed person with skeleton model (SM). It is a numerical representation of human skeleton that has a joint points connected to the observed occupant. Some of the most important input that the published SDK has granted, was the access to the SM data. Successful usage of SM was presented while investigating occupants' chosen pathways [7]. A list of SM representative joints have been displayed in Table 1 from Kinect web [5].

When in possession of the whole set of data, it is possible to monitor the activity of all body limbs. Such data has a huge potential in development of a new metabolic rate model that can be used for more accurate prediction of the influence of occupants' behaviour on indoor air quality. Two devices were installed inside the research facility in the discussed measurement trial. The Kinect device is not a stand-alone device able to record and store information, an additional computer is required to capture the recorded data. Due to the data transfer limitation of the connection between the device and the computer unit, it is only possible to connect one device to one unit. The quality of computer (hardware) performance greatly influences the registration procedure, so it is important to use powerful units.

Table 1 Skeleton model joints

Number	Joint name	Number	Joint name
1	Spine base	14	Knee left
2	Spine mid	15	Ankle left
3	Neck	16	Foot left
4	Head	17	Hip right
5	Shoulder left	18	Knee right
6	Elbow left	19	Ankle right
7	Wrist left	20	Foot right
8	Hand left	21	Spine shoulder
9	Shoulder right	22	Hand tip left
10	Elbow right	23	Thumb left
11	Wrist right	24	Hand tip right
12	Hand right	25	Thumb right
13	Hip left		

Precise information about the device that has been used in the trial has been provided in the acknowledgments below. In order to demonstrate the scale of the conducted measurements, Fig. 1 presents the location of the measurement devices and their range. The information about each body limb was stored in chronological order, with a note about the exact sampling time and date. Whenever there is a registered activity of the occupant with a continuous stream, it is called an event. Due to the fact that more than one device was used during the measurement procedure, it is necessary to recalculate some of the recorded data by a transfer vector, to make it more uniform and easier to process.

Kinect bound the observed occupant with an SM, which offered a simple way to monitor the occupant's activity. Once the data is sorted and cleared of artefacts, it is ready for analysis. The first step is to calculate the value of velocity of each body part. While knowing the exact time of each sampling and the exact positions of each joint on the SM, it is possible to calculate travel distances between samplings and the movement velocity. With the view of reducing the amount of data to be processed, all the recorded joint velocities are tested on a correlation matrix. It is a structure composed of n-to-the power of-2 plots, where n it is the number of variables (matrix plot). This analysis visualizes all the correlations between input data. If there is a strong linear signal, it means that they are somehow dependent on each other. If metadata about the joint context (what joints represent) proves that there may be some connection, the data analysis process can be simplified by reducing the included inputs. To make movement correlation more visible only few joints were displayed in from plot matrix format Fig. 2.

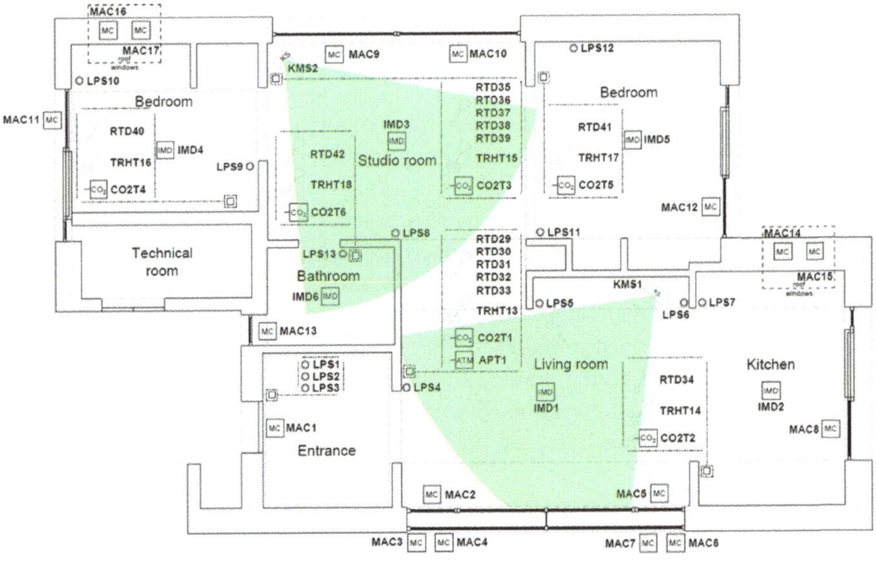

Fig. 1 Living lab layout with localization and range of Kinect devices [8]

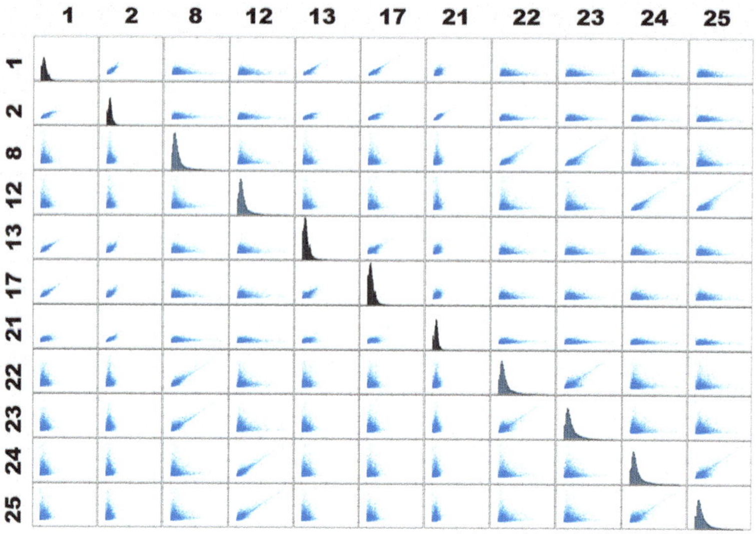

Fig. 2 SM joint movement correlation matrix

The visual analysis enables detection of several data correlations. It is not necessary to cheque and understand each particular plot, but only to find rectangles with clear diagonal lines. The movement correlations are connected via the SM, so, somehow, the bound joints influence each other. For example, records representing hand joints (8 with 22 and 23; 12 with 24 and 25) are similar and they can be reduced to one representative joint. The same can be observed in joints representing SM spine joints (21 with 1, 2, 3) or hips with spine base joints (1 with 13 and 17). Only the scenario about hand movement can be reduced, but the information about thumb orientation supports the approximation of the occupant's observed position. That is why it is possible to reduce the number of analysed joints to twenty-three for the analysis by reducing the data describing the movement of hands tip joints (22 and 24).

With the use of SM combined with a sketch of a typical body shape, it is possible to calculate the mass and surface area of each limb. It can be done by selecting the surface coverage of each body limb. SM provides a map of joint connections, each of the connections being represented by a straight line with joint points at both ends. In the middle of the connection line, a perpendicular symmetrical line has been added, limited by the outline of a typical body shape. Some of the joint connections have to be eliminated, because they provide very small area coverage, with minimum influence on the overall calculations. This is the case of e.g. information about ankles, which has been discarded. Once body parts are separated, with the resulting twenty-one significant surfaces, the percentage of each area value of each limb was calculated. This approach is a simple approximation that does not take into account the non-homogenous structure of the human body or its surface, but it can be treated as a base for further development.

Three series of calibration tests were conducted, first, to answer the question what is a typical movement and, second, to generate the threshold value to filter errors out of the measured data. Each test was aiming to observe a different movement speed (leisurely walk, normal walking, jogging) and how each body part velocity differs. The test was performed in a small hallway to gather a representative number of samples and to get significant data to select the mean value and standard deviation. The calibration process was conducted with one volunteer involved. Participating person was asked to walk back and forth between two selected points on a floor Fig. 3. There were no obstacles on a straight line between selected points. The test participant was advised to stick to the straight line between fixed points. Main purpose of conducted test was to monitor typical movement of each limb during walking. Such data allows to detect measurement errors. If some particular move of the limb was out of the calibration rage, it was discarded from analysis set. Each test took approximately 60 s. The collected calibration data results were displayed with 3 different methods Figs. 3, 4, and 5.

Since this trial measurement included only one participant, there was no need for distinguishing between occupants. If there were more occupants, they would have to be differentiated by their body ratio index (BRI). This value represents simplified information about the body shape, and will vary between different persons. BRI represents the proportion between the person's height and the width of their shoulders. It is highly unlikely that different occupants will have the same BRI value, but if such a situation occurs, the decoding process can be supported by measurements of the occupant's height. Clear information about the BRI value is essential due to the performance of the Microsoft Kinect device. It can register up to six people at the same time. It holds six slots with one SM for each occupant. The choice which SM is bound to which observed occupant is random, and that is why BRI enables detection of which person was connected to which SM at a given time. Finally, BRI delivers a QR–code, which is like a key for personalizing the collected data.

Measurements of occupants' activity can be done whenever they are within the registration range. In that case, there are only two ways to capture activity: by covering the whole potential movement area (apart from private rooms) or by

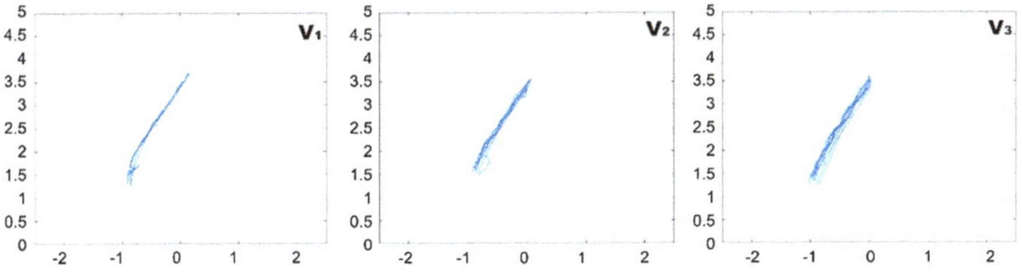

Fig. 3 Movement map of generated pathways

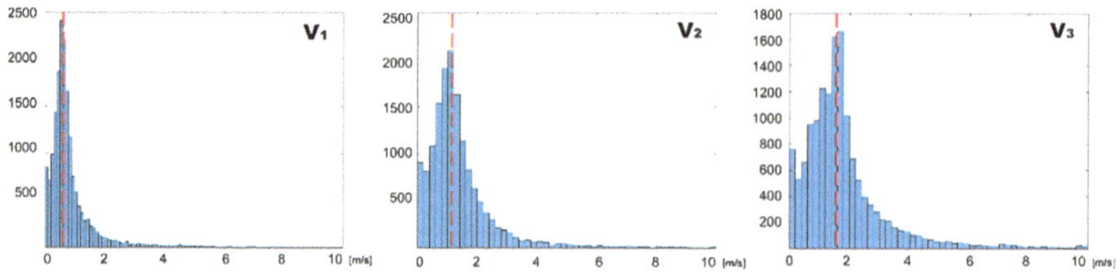

Fig. 4 General histogram of all collected data for each trial

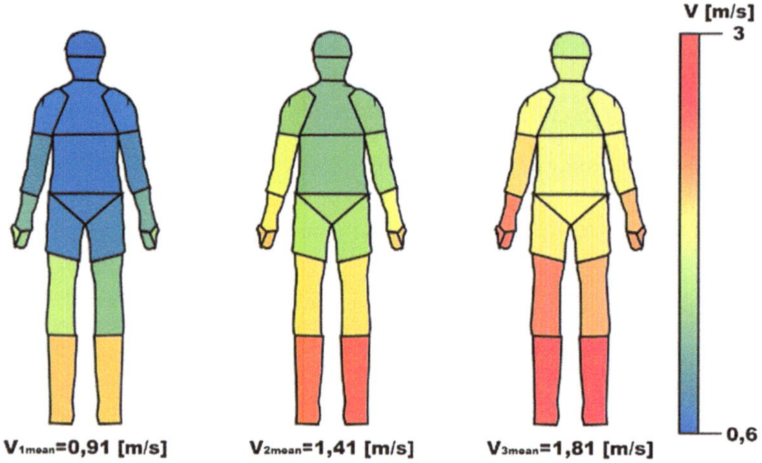

Fig. 5 Mapped colorization of human body by mean velocity, during calibration

approximation based on the assumption that whenever an occupant's activity is registered in one event, his/her behaviour does not change. The first method provides direct data but the second method requires additional knowledge about the spatial distribution of rooms and the location of the device inside the investigated space. Once the distribution of the rooms and location of the occupants is known, their activity can be approximated if they go out of the monitoring zone, by assumption that their locomotion has same pattern and that their activities decrease to a minimum when they reach their destination. This approximation can be done with a simple survival model approximating life span (distance) of the walk, where the main influencing factors are the distance between the device and the entry into the destination zone, the type of zone (living room, kitchen, office, etc.) and the geometrical shape of the zone/room. Description of how this method has been developed goes beyond the scope of this paper.

When the number of joints is reduced to the necessary minimum, data can be recalculated to obtain the acceleration value. This will be the calculation foundation for estimating the metabolic factor. With use of the knowledge about occupants' body general dimensions, it is possible to estimate his/her weight with use of the body mass index (BMI) [9]. For this trial, body weight is estimated with an assumption that involved person had mean value of the BMI factor for Norway, 27 kg/m^2. The data about body weight was taken from The World Health Organization (WHO) [10]. The global body movement vector was used to approximate metabolic rate. While occupant was outside measurement area, the metabolic rate was calculated by a typical value that was connected with a type of zone that was occupied in that time. Functionalities of different zones of the house (like kitchen or bedroom) was set based on a previous research [6]. The values with typical metabolic rates were taken from previously developed tables [11, 12]. To calculate exact value of the heat gains from human body, it was necessary to calculate skin area. It was estimated with most commonly used body surface area (BSA) [13].

3 Results

Results were shown Fig. 6 in various time resolution: native recording (raw data), seconds, minutes and hours. This way allows to display difference between direct measurements and estimated values. Plotted set of graphs shown huge contrast be-tween direct measurements and estimated activates. The main noticeable differences accrued occurred while analyzing the raw data. It stands out from approximated values, even after averaging results to the resolution of minutes. Kinect device is capable of recording with a frequency of 30 frames per second and that is the highest resolution of recording included on this set of graphs. The rest of graphs were developed by averaging of recorded values to the unit of used scale.

4 Discussion

The achieved results prove that it is possible to track an occupant's activity with high accuracy. However, monitoring of the whole body shape can be difficult due to the fact that usually the monitoring device will be placed at a position to cover a maxi-mum area. Additionally, its position will be selected to minimize interaction with the observed occupants and prevent them getting distracted from their normal behaviour. Due to this fact, it may happen during measurement that some parts of the body may not be captured. This can be observed in a movement of joints that should be connected with feet. Where there are no sufficient data (no sight of human feet), SM tries to recreate their position, but sometimes it locates them randomly for a very short time. This is the situation when joints "jump" very quickly to another position, generating noticeable change of joint velocity, which does not match the current movement regime. However, apart from this situation, no other unusual events appeared.

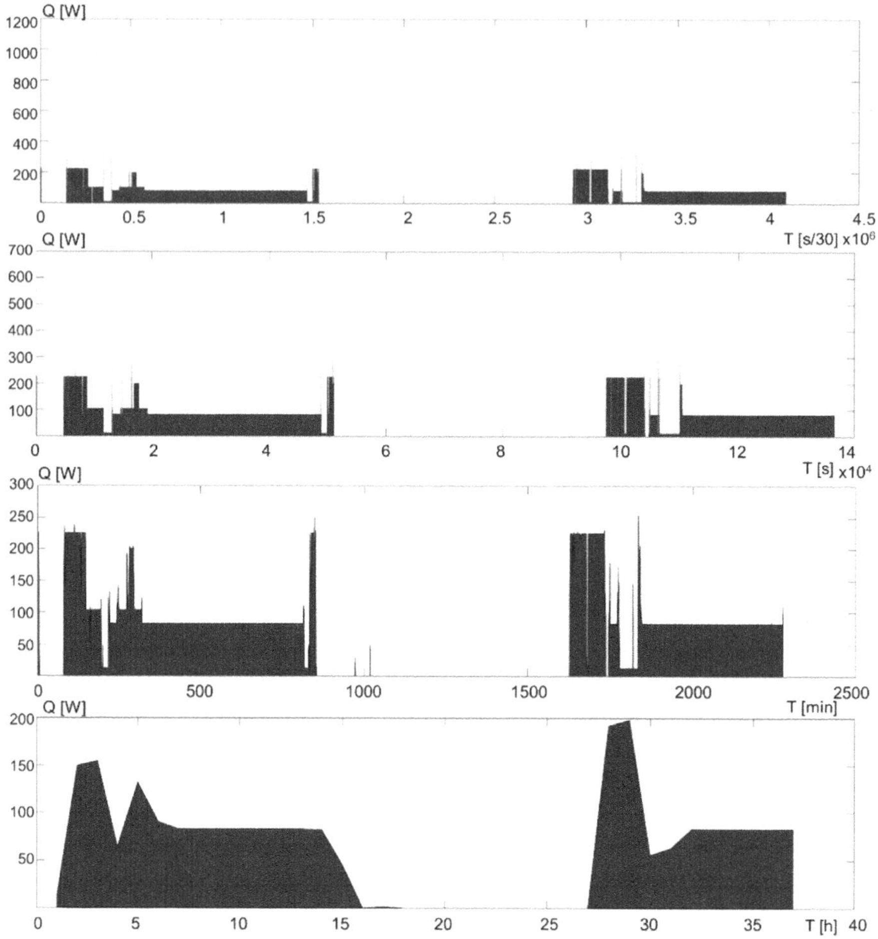

Fig. 6 Heat gains from occupant's activity inside residential building

Access to a better source of data about a human body shape may grant better accuracy of the corresponding description of the observed occupant, without involving that person. With the use of a good special representation of human body metrics, it is possible to approximate the figure of the observed occupants. Different metrics and their distribution about the human body shape may appear due to various factors, such as: geographical location, culture, wealth and many more. Additionally, data that shows typical patterns of an occupant's activity, can be used to support medical research connected with work environment. It can promote developing healthier habits or more ergonomic protocols that support employees (in office type buildings) to prevent development of cardiac diseases and increase their performance. Same profits can be achieved inside residential buildings, where certain specific occupant's activity may be promoted.

5 Conclusions

The obtained results show that dynamic estimation of a metabolic rate and of a met value can be accomplished. With the use of such data, it is possible to develop a metabolic rate reactive model that can be used for building performance simulations. Traditionally, it is assumed what kind of human activity is expected for different building zones. Knowledge about possible dynamics of heat gains from occupants will be granted by developed model. It will put a new light on calculating of energy gains, during building design phase. It has to be checked, how fixed and dynamic values of MET-factor differs in same building zones. Outcomes of such solution are hard to estimate in current state, because it was not tested for a wide range of participants. Besides modeling, a real-time measurement of this factor will benefit building management system, by supporting it, with information of required adjustment of ventilation and air conditioning unit. Such application may increase accuracy of overall simulation, closing the existing gap between energy readings obtained between the design phase and building operation. Additionally, such solution may accelerate developing of new and more sustainable building designs that will fulfil promised energy performance. It seems necessary to find more accurate data about transition of the metabolic energy used for movement to the overall generated energy, and how the amount of generated heat changes during a longer activity. Such data would have tremendous impact on estimation of the metabolic rate. Thus, the situation where an occupant's thermal comfort can be estimated with the use of direct estimation of numerous variables in real time seems to be achievable. It is al-ready possible to describe and establish clothing resistance [14] and certain rules of behaviour related to heat generation can also be captured. Once such estimation is validated with a higher number of participating occupants inside a climate chamber or in another meticulously monitored environment, this measurement method can be safely introduced for use in MPC solutions.

Development of a survival model of potential movement outside the monitoring zone seems to be a step in the right direction in the aspect of further progress. It is understandable that in most cases it will be difficult to get access to a whole house or apartment for registration due to the resistance from occupants, as it will undoubtedly invade their private space. Development of such a model can be done in various ways, but it has to be targeted as a goal for further development.

Acknowledgements Data collection and storing method do not allow to identify participants of the study. That is why this study does not require certification by an ethics board. The authors declare that they have no conflict of interest. For this type of study formal consent is not required and consent of participants was not needed. The authors do not endorse any specific brand or device developer. The study has not been sponsored or influenced in any other manner by private companies. This publication does not seek to promote any specific product or brand.

Computer setting for the measurement procedure: Intel® Core™ i7- 4785T with a CPU of 2.20 GHz: 16 GB DDR3 RAM; Intel HD Graphics 4600. Using a different setting of hardware for the measurement purpose may influence the sampling time.

References

1. D. Yan, W. O'Brien, T. Hong, X. Feng, H. B. Gunay, F. Tahmasebi, A. Mahdavi, Occupant behavior modeling for building performance simulation: Current state and future challenges. Energy Build., 264–278 (2015)
2. P.O. Fanger, *Thermal Comfort* (Danish Technical Press, Copenhagen, 1970)
3. ASHREA Standard 55, *Thermal Environmental Conditions for Human Occupancy* (2013)
4. J. Choi, D. Yeom, Study of data-driven thermal sensation prediction model as a function of local body skin temperatures in a built environment. Build. Environ., 130–147 (2017)
5. Xbox One, Microsoft, http://www.xbox.com/en-US/xbox-one/accessories/Kinect. Last Accessed 24 Aug. 2017
6. L. Finocchiaro, F. Goia, S. Grynning, A. Gustavsen, in *The ZEB Living Lab: A Multipurpose Experimental Facility*. Gent Expert Meeting, Ghent University (2014)
7. J. Dziedzic, D. Yan, V. Novakovic, in *Occupant Migration Monitoring in Residential Buildings with the Use of a Depth Registration Camera*. Proceedings of the 11th International Symposium on Heating, Ventilation and Air Conditioning (ISHVAC 2017) (2017)
8. A. Wagner, W. O'Brien, B. Dong, *Exploring Occupant Behavior in Buildings. Methods and Challenges* (Springer, Berlin, 2018)
9. T.T. Samaras, *Human Body Size and the Law of Scaling* (NOVA Press, Paris, 2007)
10. WHO, Global Health Observatory (GHO) database, Norway. Last Accessed 25 Aug. 2017
11. K. Parsons, *Human Thermal Environment*, 3rd edn. (CRC Press, Boca Raton, 2014)
12. ASHRAE Handbook—Fundamentals (SI Edition), American Society of Heating, Refrigerating, and Air-Conditioning Engineers, Inc (2013)
13. J. Verbraecken, P. Van de Heyning, W. De Backer, L. Van Gaal, Body surface area in normal-weight, overweight, and obese adults. A comparison study. Metab. Clin. Exp., 515–524 (2012)
14. J. Dziedzic, D. Yan, V. Novakovic, in *Measurement of Dynamic Clothing Factor (D-CLO)*. Proceedings of the 4th international Conference on Building Energy & Environment (COBEE 2018) (2018)

Impacts of Common Simulation Assumptions in Sweden on Modelled Energy Balance of a Multi-family Building

Ambrose Dodoo, Uniben Y. A. Tettey and Leif Gustavsson

Abstract Here, we explore key input parameters and common assumptions for energy balance analysis of residential buildings in Sweden and assess their impacts on simulated energy demand of a building. Our analysis is based on dynamic hour-by-hour energy balance modelling of a typical Swedish multi-storey residential building constructed in 1972. The simulation input parameters studied are related to microclimate, building envelope, occupancy behaviour, ventilation, electric and persons heat gains. The results show that assumed indoor temperature set points, internal heat gains and efficiency of ventilation heat recovery systems have significant impact on the simulated energy demand. For microclimate parameters, the outdoor temperature, ground solar reflection and window shading gave significant variations in the simulated space heating and cooling demands. We found that input parameter values and assumptions used for building energy simulation vary significantly in the Swedish context, giving considerably different estimated annual final energy demands for the analysed building. Overall, the estimated annual final space heating demand of the building varied between 50 and 125 kWh/m^2 depending on the simulation dataset used. This study suggests that site-specific parameter values may be appropriate for accurate analysis of a building's energy performance to reduce data input uncertainties, as such factors may have a significant impact on building energy balance and energy savings of retrofit measures.

Keywords Energy simulation · Residential buildings · Input parameter data

A. Dodoo (✉) · U. Y. A. Tettey · L. Gustavsson
Sustainable Built Environment Group, Department of Built Environment and Energy Technology, Linnaeus University, 35195 Växjö, Sweden
e-mail: ambrose.dodoo@lnu.se

D. Johansson et al. (eds.), *Cold Climate HVAC 2018*,
Springer Proceedings in Energy, https://doi.org/10.1007/978-3-030-00662-4_58

1 Introduction

The building sector worldwide is suggested to offer many possibilities for energy savings and climate change mitigation [1]. In the European Union (EU), the building sector accounts for 40% of the total final energy use [2] and is expected to contribute significantly to the goal to reduce total greenhouse gas (GHG) emission by 80–95% by 2050 relative to 1990 levels [3]. The energy performance of buildings directive (EPBD) [2] is a key piece of the legislations to promote improved energy efficiency of European buildings. The EPBD mandates EU member states to report the calculated energy performance of buildings based on a methodology defined at national or regional levels. In Sweden, the government is aiming to reduce total energy use per heated building area by 20% by 2020 and 50% by 2050, using 1995 as the reference [4]. The residential and service sectors in Sweden account for 40% of the national total final energy use [5], with space and tap water heating accounting for 55% of the sectors' final energy use.

Several factors influence the thermal performance of buildings and detailed analysis is important to properly estimate the energy required to operate a building. Increasingly, simulation tools are being used to support the design and construction of high-performance buildings. However, results from simulation tools have been reported to often vary from the actual monitored energy use of buildings [6–9]. Different factors are attributed to this, including inappropriate simulation input data and assumptions for building energy balance modelling. Kalema et al. [10] compared simulation results of a building modelled with different simulation tools by different researchers and noted that discrepancies due to input data were larger than those caused by the use of different tools. Wall [11] noted the simulated energy demand of a terrace house in Sweden to be 33% lower than the monitored energy demand, mainly due to uncertainties related to some input parameter values. Similar discrepancies ranging between 7 and 105% have been reported in different climate contexts, depending on the building type and degree of uncertainties related to various input parameters [6–9].

In this study we explore key input parameters and common assumptions used for energy balance modelling of residential buildings in Sweden and analyse their impacts on simulated energy balance of a building. Our analysis is based on dynamic hour-by-hour energy balance modelling of a typical Swedish multi-storey residential building constructed in 1972.

2 Methods

The general approach consists of the following: (a) modelling the thermal performance of the building in its existing state based on a set of reference input parameters; (b) exploring simulation input parameters and assumptions for the Swedish context and assessing their impacts on the energy balance calculation of the building.

2.1 Analysed Building

The analysed building is located in the Kallinge area of Ronneby municipality in Sweden. It is 3-storey high and made of a concrete-frame structure with façades of brick and wood panels. There are 27 apartments and a basement in the building, with a total heated living floor area of 2000 m^2 and ventilated volume of 5400 m^3. Figure 1 shows a photograph of the building and a more detailed description of the building is given in [12]. Key construction features as well as thermal properties of the various components for the existing building are given in Table 1.

2.2 Energy Balance Simulation

The VIP-Energy simulation program [13] is used to model the thermal performance of the building. The program calculates the final energy demand of a building based on the construction's physical characteristics, internal and solar heat gains, occupancy schedules, heating and ventilation systems, among others. It enables whole building dynamic hourly energy balance calculations with multi-zone and detailed thermal bridges and heat storage capacity as well as one-, two- and three-dimensional modelling features of building envelope components. VIP-Energy is validated by the International Energy Agency's BESTEST, ANSI/ASHRAE Standard 140 and CEN 15265. The reference energy balance of the building is modelled hourly with the 2013 weather data for the city of Ronneby. The 2013 weather data, which reflects the Swedish normal climate, is obtained from the meteonorm database [14].

2.3 Reference and Varied Parameters

The reference and varied input parameter values and assumptions are based on those commonly used for thermal performance analyses of buildings in Sweden and

Fig. 1 The analysed concrete-frame building in Ronneby municipality, Sweden

Table 1 Thermal properties for the existing building

Building element	U-value (W/m² K)	Area (m²)
Attic floor	0.11	688.0
Basement walls	1.33/1.44	57.2/286.9
Doors (clear glass windows in doors)	3.0	84.5
Exterior walls (wood panels and brick facades)	0.311/0.341/0.346	292.0/194.0/565.0
Foundation (slab on ground)	0.26	688.0
Windows (clear glass windows)	2.9	194.5

Table 2 Reference and varied input parameter values and assumptions for energy simulation

Parameter	Reference	Variations	Remarks
Climate	Ronneby 2013	+1 and −1 °C	Varied outdoor temperature
Horizontal angle	20°	10 and 30°	Shading by near-by objects
Wind factor	70%	30 and 100%	Relative to climate file
Solar reflection	0%	25 and 50%	Solar reflection into building
Air pressure outside	1000 hPa	990 and 1020 hPa	Average annual value
Heating set point	21 °C	20, 22, and 23 °C	Living areas, based on [11, 16–18]
Cooling set point	26 °C	27 °C	Living areas, based on [19]
Hot water use	2.85 W/m²	No variation	Standard taps
Electric power use	3.05 W/m²	No variation	Standard equipment
Internal heat gains			
Persons	1.0 W/m²	1.68 and 4.30 W/m²	Based on [11, 20]
Lighting and appliances	3.05 W/m²	2.40 and 4.92 W/m²	Based on [15, 20]
Airtightness	0.8 l/sm²	0.6 and 1.0 l/sm²	Based on [21, 22]
Ventilation, heat exchanger and fans			
Air change rate	0.35 l/sm²	No variation	Based on [23]
VHR efficiency	76%	80 and 85%	Based on [24]
Fan pressure	400 Pa	200 and 600 Pa	Estimated based on [13]
Fan efficiency	50%	33 and 55%	Based on [25, 26]
Airing	Not considered	Considered	Based on suggestions by [13, 15]

are outlined in Table 2. The parameters varied include indoor temperature set points for heating and cooling, outdoor temperature, internal and solar heat gains, micro-climate, building envelope and efficiencies of ventilation heat recovery (VHR) and fan, and fan pressure. The implication of window opening for airing is explored using the suggestion by SVEBY [15] whereby 4 kWh/m^2 is added to the total space heating demand, and suggestion by VIP-Energy [13] whereby a margin of 0.025 l/s m^2 balanced air flow bypassing the heat exchanger is included in the simulation model.

3 Results

3.1 Reference Energy Demand

The reference case simulated energy demands of the building are summarized in Table 3. Space heating accounts for 61% while space cooling and ventilation each account for 2% of the annual total final energy demand of the building.

3.2 Parametric Variations

The effects of variations of various parameter values related to indoor air temperature, envelope airtightness, internal heat gains and microclimate on the simulated space heating and cooling demands of the building are summarized in Table 4. Positive change means the simulated heating demand is greater while negative change means that it is lower, relative to the reference case. The biggest decrease in space heating demand occurs when the reference persons' heat gains of 1.0 W/m^2 is varied to 4.3 W/m^2 while increasing horizontal angle from the reference 20 to 30%, denoting more shading gives the biggest decrease in space cooling demand for the building. The space heating and cooling demands also changed significantly when the reference indoor air temperature of 21 °C for heating and 26 °C for cooling are varied. The electrical heat gains of 2.40 and 4.92 W/m^2 correspond to household electricity demands of 24 and 48 kWh/m^2, (not shown in table) respectively.

Table 5 shows the impact of variations of parameter values related to ventilation fans and the implications of VHR system with different heat recovery efficiencies for the analysed building. Implementation of VHR systems considerably decreased the space heating demand between 28 and 33% and also doubled the ventilation electricity use of the building. The simulated electricity for ventilation is about doubled when the ventilation fan efficiency is decreased to 33% from the reference 50% while an increase of the reference ventilation fan efficiency to 55% has small impact on the simulated ventilation electricity use of the building. Compared to the reference case, the simulated ventilation electricity use is increased by 2–3 kWh/m^2 when the reference fan pressure of 400 Pa is varied to 600 Pa. With a fan pressure of 200 Pa instead of the reference value, the simulated ventilation electricity use is decreased by 1–3 kWh/m^2.

Table 3 Reference simulated annual final operation energy demand (kWh/m^2) of the building

Description	Space heating	Space cooling	Ventilation electricity	Tap water heating	Household electricity	Total
Reference	95	3	3	25	30	156

Table 4 Effects of parameter variations on simulated space heating and cooling demands

Description	Space heating		Space cooling	
	kWh/m^2	Change from ref. (%)	kWh/m^2	Change from ref. (%)
Reference case (ref.)	95	–	3	–
Parameter variations				
Heating set-point				
20 °C	88	−7	3	0
22 °C	104	9	3	0
23 °C	112	18	3	0
Cooling set point				
27 °C	95	0	2	−33
Outdoor temperature				
−1 °C	105	11	2	−33
+1 °C	86	−9	4	33
Horizontal angle				
10°	90	−5	5	67
30°	100	5	1	−67
Ground solar reflection				
25%	92	−3	5	67
50%	90	−5	8	167
Wind load to building				
30%	95	0	3	0
100%	97	2	3	0
Air pressure outside				
990 hPa (−1%)	95	0	3	0
1020 hPa (+2%)	96	1	3	0
Airing				
VIP-energy approach	99	4	3	0
SVEBY approach	99	4	3	0
Airtightness				
0.6 l/sm^2	98	3	3	0
1 l/sm^2	99	4	3	0
Persons heat gains				
1.68 W/m^2	91	−4	3	0
4.30 W/m^2	75	−21	7	133
Electrical heat gains				
2.40 W/m^2	100	5	2	−33
4.92 W/m^2	83	−13	5	67

Table 5 Effects of parameter variations on simulated space heating and ventilation electricity demands

Description	Space heating		Ventilation electricity	
	kWh/m^2	Change from ref. (%)	kWh/m^2	Change from ref. (%)
Reference case	95	–	3	–
Parameter variations				
VHR efficiency				
76%	68	−28	6	100
80%	66	−31	6	100
85%	64	−33	6	100
Fan efficiency				
33%	95	0	5	67
55%	95	0	3	0
Fan pressure				
200 Pa	95	0	2	−33
600 Pa	95	0	5	67

3.3 *Effects of Extremes of Parameter Values*

In Figs. 2, 3, and 4, the combined effects of the extremes of parameter values (from Tables 4 and 5) giving the highest and lowest simulated space heating demand for the building are illustrated. The figures show simulated annual final energy demands, hourly heat load profiles and indoor air temperatures of the building when using the different extreme parameter datasets. To facilitate comparison, the values for the reference case are also shown in the figures. Compared to the reference case, the building's simulated final space heating demand is increased by 32% and decreased by 47% when the extremes of parameter values are used to perform the

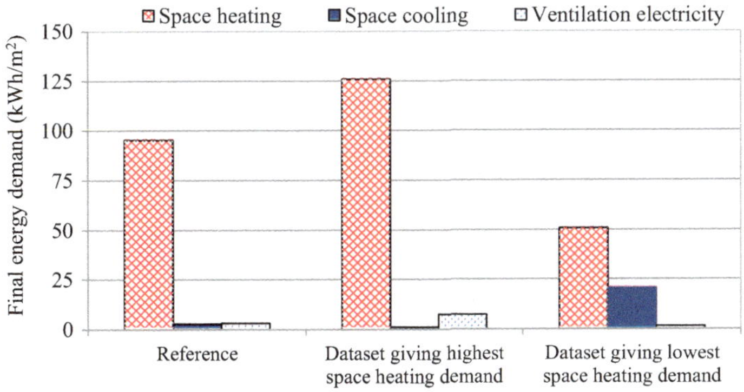

Fig. 2 The reference case energy demands and the combined effects of the extremes of parameter values giving the highest and lowest space heating demand of the building

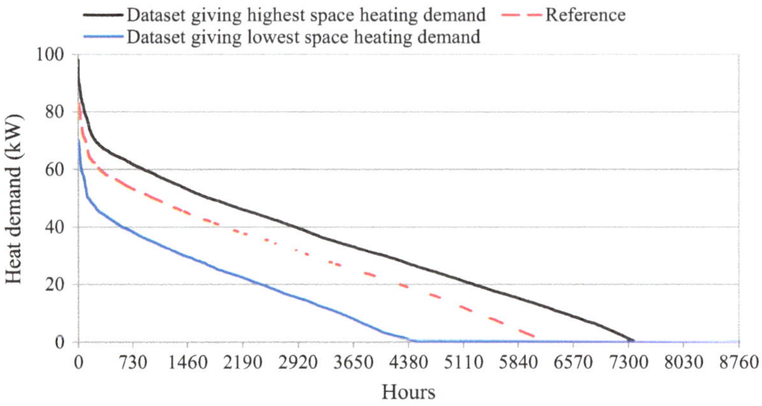

Fig. 3 Simulated annual hourly space heating demand profiles of the building, arranged in descending order

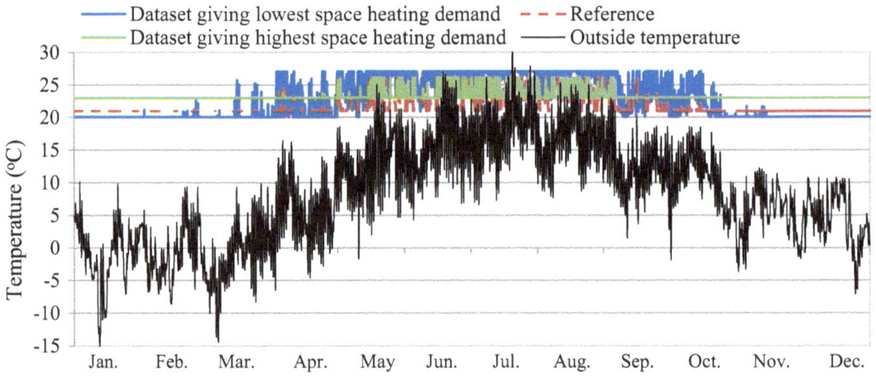

Fig. 4 Simulated annual hourly profiles for indoor air temperature of the building and also outside air temperature, from January to December

calculations. The simulated demand for cooling is significantly greater when the parameters giving the lowest heat demand are used. The extremes of parameter datasets result in about 40% difference in the simulated peak load of the building.

4 Discussion and Conclusions

We conducted this analysis to explore the implications of input parameter values and assumptions commonly used in Sweden for energy balance calculations for residential buildings. The input parameters explored are related to indoor temperature, microclimate, building envelope, ventilation, and internal heat gains from persons and electric appliances. The analysis is based on dynamic hour-by-hour energy balance modelling of a typical Swedish multi-storey residential building from the 1970s. Our findings show that input parameters and assumptions used in energy simulations of residential buildings in Sweden vary significantly and have significant effect on the calculated energy balance of buildings. Our analysis shows that simulated annual final space heating demand of the analysed building varied between 50 and 125 kWh/m^2 depending on the simulation dataset used for the energy balance calculations. Household electricity demands are 48 and 24 kWh/m^2 when the annual final space heating demands are 50 and 125 kWh/m^2, respectively. Hence input parameters used in simulation models need to be characterised, transparent and appropriate, to inform accurate analyses of energy efficiency performance of buildings.

Our calculations show that assumed indoor temperature set points, internal heat gains and efficiency of VHR systems have significant impact on the simulated final energy demand of analysed building. Among the explored microclimate parameters, outdoor temperature, ground solar reflection and window shading gave significant variations in the simulated space heating and cooling demands of the building. In this analysis, variations of air pressure outside and wind load that hit the building had minor impact on the simulated building energy demand.

This analysis highlights the implications of different parameters and strategies for cooling energy demand of Swedish residential buildings. Currently, cooling demands are generally small for Swedish residential buildings as cooling needs are typically managed with strategies as shading, increased ventilation and airing. Notwithstanding, calculations by Persson et al. [18] showed that cooling energy demand for a very low energy building in Sweden can be significant compared to space heating energy demand. Tettey et al. [27] showed that the energy need for cooling of Swedish residential buildings may increase significantly under future climate conditions. Comparison of simulated building energy performance with measured and calibrated data is important, as discussed by Vesterberg et al. [28], and is not within the scope of this paper.

In summary, this study illustrates the implications of different simulation assumptions as well as parameter values used for energy balance analysis in Sweden, and suggests that site-specific parameter values may be appropriate for accurate analysis of a building's energy performance. The findings increase understanding of how various key parameter values, methods and assumptions for energy balance modelling influence simulated energy use of residential buildings in the Swedish context. Accurate analysis of buildings' energy balance is essential to identify the scale, trade-offs and cost-effectiveness of various measures to reduce space heating and cooling demands, and facilitate GHG emissions reductions.

References

1. IPCC, *Climate Change 2014: Mitigation of Climate Change. Contribution of Working Group III to the Fifth Assessment Report of the Intergovernmental Panel on Climate Change* (2014)
2. Directive 2010/31/EU, Directive 2010/31/EU of the European Parliament and of the Council of 19 May2010 on the energy performance of buildings (Recast), Official J. Eur. Union. L 153, **18**(06) (2010)
3. European Commission, 2050 low-carbon economy. Accessed at http://ec.europa.eu/clima/policies/strategies/2050/index_en.htm on 12 Apr 2017 (2016)
4. Swedish Government Bill 2005/06:145, Swedish Government Bill 2005/06:145 National Programme for Energy Efficiency and Energy-smart Construction. Accessed at http://www.government.se/information-material/2006/05/national-programme-for-energy-efficiency-and-energy-smart-construction/ on 12 Apr 2017
5. Swedish Energy Agency, Energy in Sweden 2015. Accessed at www.energimyndigheten.se on 12 Apr 2017, Energimyndigheten (2015)
6. D. Daly, P. Cooper, Z. Ma, Understanding the risks and uncertainties introduced by common assumptions in energy simulations for Australian commercial buildings. Energy Build. **75**, 382–393 (2014)
7. F. Karlsson, P. Rohdin, M.-L. Persson, Measured and predicted energy demand of a low energy building: important aspects when using Building Energy Simulation. Build. Serv. Eng. Res. Technol. **28**(3), 223–235 (2007)
8. A.C. Menezes et al., Predicted vs. actual energy performance of non-domestic buildings: using post-occupancy evaluation data to reduce the performance gap. Appl. Energy **97**, 355–364 (2012)
9. J.F. Karlsson, B. Moshfegh, Energy demand and indoor climate in a low energy building—changed control strategies and boundary conditions. Energy Build. **38**(4), 315–326 (2006)
10. T. Kalema et al., Accuracy of energy analysis of buildings: a comparison of a monthly energy balance method and simulation methods in calculating the energy consumption and the effect of thermal mass. J. Building Phys. **32**(2), 101–130 (2008)
11. M. Wall, Energy-efficient terrace houses in Sweden: simulations and measurements. Energy Build. **38**(6), 627–634 (2006)
12. A. Dodoo, U.Y.A. Tettey, L. Gustavsson, On input parameters, methods and assumptions for energy balance and retrofit analyses for residential buildings. Energy Build. **137**, 76–89 (2017)
13. StruSoft, VIP-Energy, Version 2.1.0 StruSoft AB, Editor. 2012: Sweden. http://www.strusoft.com/products/vip-energy
14. Meteotest, METEONORM 7: Global meteorological database for engineers, planners and education, Meteotest (2015)
15. SVEBY, Brukarindata bostader version 1.1. Accessed at http://www.sveby.org/hur-anvander-jag-sveby/berakning/ 29 June 2015 (2013)
16. Boverkets Byggregler, Boverkets Författningssamling, The national Board of Housing Building and planning, Karlskrona. Accessed at http://www.boverket.se on 10 Apr 2017 (In Swedish) (2015)
17. L. Gustavsson et al., Primary energy implications of end-use energy efficiency measures in district heated buildings. Energy Build. **43**(1), 38–48 (2011)
18. M.-L. Persson, A. Roos, M. Wall, Influence of window size on the energy balance of low energy houses. Energy Build. **38**(3), 181–188 (2006)
19. U.Y.A. Tettey, A. Dodoo, L. Gustavsson, Primary energy implications of different design strategies for an apartment building. Energy **104**, 132–148 (2016)
20. R. Elsland, I. Peksen, M. Wietschel, Are internal heat gains underestimated in thermal performance evaluation of buildings? Energy Procedia **62**, 32–41 (2014)
21. Y. Zou, *Classification of Buildings with Regard to Airtightness, in Department of Civil and Environmental Engineering* (Chalmers University of Technology Götenborg, Sweden, 2010)

22. L. Liu, P. Rohdin, B. Moshfegh, Evaluating indoor environment of a retrofitted multi-family building with improved energy performance in Sweden. Energy Build. **102**, 32–44 (2015)
23. Boverkets Byggregler, Boverkets Författningssamling, The national Board of Housing Building and planning, Karlskrona. Available at http://www.boverket.se (In Swedish) (2015)
24. Swedish Energy Agency, FTX-aggregat hus med 130 m^2 boyta. Available at https://www.energimyndigheten.se/Hushall/Testerresultat/Testresultat/FTX-aggregat-hus-med-130-m-boyta/?tab=1 (2010)
25. Camfil. City pollution. AirMail No. 2. 2014 [cited 3 Sept 2015]; Available from: http://www.camfil.se/FileArchive/Brochures/Airmail/AirMail_2014_2_EN.pdf
26. N. Brelih, How to improve energy efficiency of fans for air handling units. REHVA J., 5–10 (2012)
27. U.Y.A. Tettey, A. Dodoo, L. Gustavsson, Energy use implications of different design strategies for multi-storey residential buildings under future climates. Energy (2017)
28. J. Vesterberg, S. Andersson, T. Olofsson, in *The Influence from Input Data Provided by the User on Calculated Energy Savings*. 10th Nordic Symposium on Building Physics, Lund, Sweden, 15–19 June 2014

A Comparison Between Four Dynamic Energy Modeling Tools for Simulation of Space Heating Demand of Buildings

Amir Vadiee, Ambrose Dodoo and Leif Gustavsson

Abstract Different building energy modelling programs exist and are widely used to calculate energy balance of building in the context of energy renovation of existing buildings or in the design of energy performance of new buildings. The different tools have unique benefits and drawbacks for different conditions. In this study, four different types of building energy system modelling tools including TRNSYS, Energy Plus, IDA-Indoor Climate Energy (IDA-ICE) and VIP-Energy are used to calculate the energy balance of a recently built six-storey apartment building in Växjö, Sweden. The building is designed based on the current Swedish building code. The main outcomes of the software include hourly heating and cooling demands and indoor temperature profiles. We explore the general capabilities of the software and compare the results between them. For the studied building with similar input conditions such as weather climate data file, infiltration and ventilation ratio and internal heat gain, IDA-ICE modeled the highest space heating demand while the TRNSYS the lowest due to the simplification of thermal bridges. The main advance feature of VIP-Energy is the detail thermal bridge analysis while the main drawback is the complexity of using the model. EnergyPlus and TRNSYS can be used for energy supply system integration with the ability to add mathematical sub-modules to the models.

Keywords Building · Energy analysis · Energy simulation tools
VIP energy · Energy plus · IDA · TRNSYS

1 Introduction

Buildings are large user of energy with a global final energy use of about 124 EJ, corresponding to 30% of the total global final energy use. Space heating and cooling dominate the final energy use in buildings and together accounted for 32

A. Vadiee (✉) · A. Dodoo · L. Gustavsson
Department of Built Environment and Energy Technology,
Faculty of Technology, Linnaeus University, Växjö, Sweden
e-mail: amir.vadiee@lnu.se

D. Johansson et al. (eds.), *Cold Climate HVAC 2018*,
Springer Proceedings in Energy, https://doi.org/10.1007/978-3-030-00662-4_59

and 54% of total final energy use of residential and non-residential buildings in 2013, respectively [1]. Furthermore, the energy savings potential is large in the building sector compared to the industrial, transport and power generation sectors.

Analyses of buildings' energy saving potentials are often based on different Building Energy Modeling Programs (BEMPs). The accuracy of the programs is important when estimating potential energy savings of different measures and depends on a program's capacity to reflect the real energy flows in buildings. There are various static and dynamic energy simulation tools which can be used to analyze energy flow of buildings and several tools have been commercially introduced in the last decades such as DOE-2, E-Quest, Energy plus, TRNSYS, IES, Openstudio, Lesosai and Revit. Examples of Swedish commercial building energy simulation tools are Enorm, BV2, VIP-Energy and IDA-ICE. However, various benefits and draw backs are associated with different simulation tools.

1.1 Research Aims and Method

The aim of this paper is to compare the capabilities and outcomes of TRNSYS 16, IDA-ICE, EnergyPlus and VIP-Energy. The tools are used to model the energy flows of a building under cold climate conditions, to investigate how key parameters and energy balance results differ when modeling space heating demand. The analyzed building is located in Växjö, in the county of Kronoberg in southern Sweden. The energy balance of the building is modeled hourly with the same or similar criteria and conditions with all four BEMPs. The monthly and hourly heating demands as well as key parameters used in the building energy balance calculations for the proposed building are assessed. The hourly weather data for Växjö for the year 2013 is used for the analysis and is imported using data generated by the Meteonorm software. In this paper, the selected BEMPs are discussed and used to analyze a building. Input data and assumptions for modelling the energy balance with the BEMPs are described. Finally, the results are compared and the selected BEMPs are assessed regarding their key features.

2 Energy Simulation Tools

2.1 VIP Energy

VIP-Energy is a hourly based time-step and multi-zone dynamic energy simulation program developed by Strusoft [2] and increasingly used by researchers, consultants and construction companies in Nordic countries for analysis of energy balance of buildings. The program has been validated by the International Energy Agency's BESTEST and ANSI/ASHRAE Standard 140 and CEN 15265 as having reliable algorithms and calculation models [3]. VIP-Energy allows detailed analysis of thermal bridges of buildings. It models heat storage of buildings assuming the configurations of the building envelope as a series of thermal resistance and capacitance with finite difference in response to thermal, taking into account various thermophysical properties including thermal conductivity, density and heat capacity of materials. The program has a comprehensive materials and components catalogue and estimates solar radiation available to a building using the Hay-Davies-Klucher-Reindl model [4]. Mathematical descriptions of other key models used in the VIP-Energy program are described by Jóhannesson [5] and Nylund [6].

2.2 IDA-ICE

IDA-ICE is a dynamic whole building indoor climate and energy balance calculation program managed by Equa simulation AB [7]. The program is commonly used in European countries for research and consulting purposes. The program has multi-zone calculation feature and models indoor environment and energy performance of buildings in variable time-steps including hourly and minute time resolutions. It is based on a general system simulation platform with a modular system. The accuracy of the solution in IDA-ICE can be controlled by defining tolerances for calculations. The program is validated with ASHRAE 140-2004 [8] and EN 15255-2007 and 15265-2007 [9], showing that IDA-ICE can give accurate calculations of buildings' energy and indoor climate performances in comparison to other state-of-the art simulation programs. IDA-ICE has a BIM import extension function, allowing importation of 3D CAD objects in open or IFC format.

2.3 Energy Plus

DOE-2 is one of the most known building energy use and cost analysis software with a 25 years history. EnergyPlus is a new generation simulation program built upon the best features of DOE-2 and BLAST although new modeling features are developed in this software beyond the two programs. Sub-hourly time step calculation and dynamic integration of loads and system performance for building energy balance calculations is the most advantages of EnergyPlus over the DOE-2 although it causes much slower processing compared with other DOE-2 based software [2, 3]. The EnergyPlus is an engine for thermal simulation that uses text as an input therefore a "Graphical User Interface" (GUI) such as SKETCH UP or Design Builder can be used together with EnergyPlus in order to have a visual interface for the building models.

2.4 TRNSYS

TRNSYS is a transient system simulation software tool with a modular structure that is designed to develop an energy system with wide range of simple to complex systems [3]. TRNSYS has special module to model the thermal behavior of a building called TRNBuild or Type 56 (subroutine component in the TRNSYS library). The building description is read by this component from a set of external files generated based on user supplied information and generates its own set of monthly and hourly summary output files. The required input data for this subroutine is supplied by the weather data component. The user supplied information to TRNBuild includes e.g. geometric, physical and thermal properties of the building. The simulation in this software can be performed from time interval of 0.1 s [4].

3 Building Energy Models

To assess the features and capabilities of the selected BEMPs, a building is modelled with input data and assumptions described as follows.

3.1 Modelled Building

The modelled building is a 6-storey Swedish multi-family building constructed in 2014 in Växjö, Sweden (lat. 56° 87′ 37″ N; lon. 14° 48′ 33″ E). The concrete-frame building contains 24 apartments and has a total heated floor area of 1686 m^2. The foundation is made up of layers of 200 mm crushed stone, 300 mm Styrofoam insulation and a 100 mm concrete ground floor slab. The external walls consist of 100 and 230 mm concrete on the outside and inside respectively, with a 100 mm layer of cellplast insulation material between them. The roof is made up of 250 mm concrete slab and 500 mm loose fill rock wool insulation with wooden trusses and a roof covering over layers of asphalt-impregnated felt and plywood. Figure 1 shows a ground floor plan and façades of the building respectively and Table 1 gives key envelope characteristics of the building. The window area on each floor is 6.55, 26.87, 12.42 and 6.55 m^2 on south, west, east and north, respectively. The corresponding values for exterior wall area are 205, 329, 353 and 205 m^2 (Fig. 2).

Table 1 Key envelope properties of the modelled building

Description	U-value (W/m^2K)					Air leakage at 50 Pa ($l/s\ m^2$)
	Ground floor	External walls	Windows	Doors	Roof	
Values	0.11	0.32	1.2	1.2	0.08	0.6

Table 2 Building construction material properties

Type of material	Thermal conductivity (kJ/h m K)	Heat capacity (kJ/kg K)	Density (kg/m^3)
Drained Gravel	5.04	1	1800
Cell Plast	0.1296	1.4	25
Concrete normal RH	6.12	0.8	2300
Wood Pine	0.504	2.3	500
Loose Wool	0.1512	0.8	40

Table 3 Annual average, maximum and minimum operative temperatures for living and common areas for the studied models

	VIP Energy	Energy plus	TRNSYS	IDA-ICE	Average of four models	Average for three models[a]
Operative living area temperature						
Average	22.7	22.7	22.1	*23.7*	22.8	22.5
max	27.0	27.4	25.9	*27.6*	27.0	26.8
min	20.4	20.1	20.4	*21.0*	20.5	20.3
Operative common area temperature						
Average	21.5	21.9	21.4	*22.9*	21.9	21.6
max	27.0	27.4	*25.9*	27.6	27.0	27.3
min	17.4	18.4	18.7	*19.9*	18.6	18.2

The value with the highest difference from average for the four models are given in bold
[a]Average of three models excluding the model with largest difference with average of four models
Note The values with the largest differences with the average of four models are highlighted in italics

The main materials which are used for the building construction are wood pine, concrete, cell plast, loose wool and drained gravel. The properties of these materials are given in Table 2.

3.2 Input Data and Assumption

The daily variation and monthly mean values for outdoor air temperature, daily global radiation as well as sunshine hours for the generated and imported 2013 weather file for Växjö are shown in Fig. 3.

The calculations are based on hourly time step in all simulation tools. The ground temperature for all developed models is considered to be 10 °C. The internal heat gains for all models consist of occupancy, lighting system, electrical devices and hot water circulation. The annual average values for all internal heat gains are assumed to be 4.95 and 0.21 W/m^2 for the analyzed building's living and common areas, respectively. The indoor set-point temperatures are 21 and 18 °C in the living and common areas, respectively.

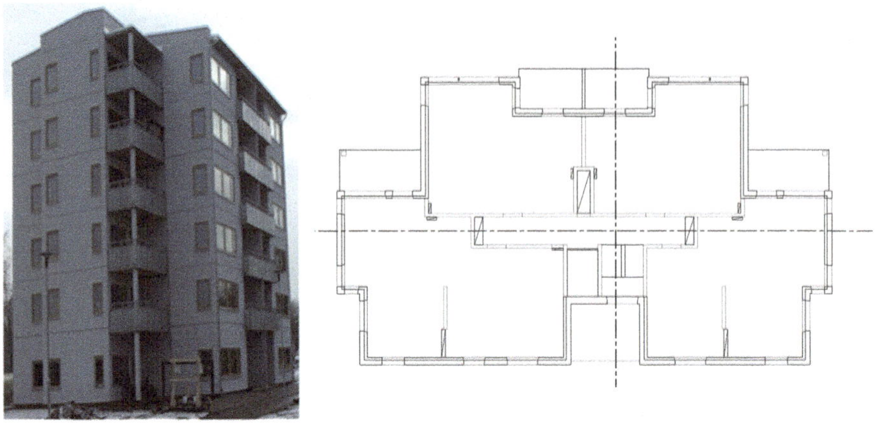

Fig. 1 Photograph (left) and typical floor plan (right) of the modelled building

3.3 *Zone Definition*

The main differences between the models are the zones definitions. VIP-Energy and TRNSYS have two attached zones: Living area and common areas for each floor and all internal walls are considered in the models. In IDA-ICE, the whole building is divided into 25 zones while 4 attached zones represent each single flat for each floor and one single zone for common area for the whole building. EnergyPlus consists of 24 attached zones including 3 living areas and one common area for each floor. The main simplification in EnergyPlus is the balcony, which is assumed to be an exterior part of the building and is not considered in the annual space heating demand analysis but the effect of balcony windows are considered in the simulation. Figure 4 shows the different zones configurations for the building models.

In TRNSYS 16, CAD files or geometry cannot be imported but there is a special module in called TRNBUILD that is used to design a building including almost all details except doors. Therefore, in this study, the effects of doors are just considered in the infiltration section and thermal conductivity of the doors are neglected in the simulation.

4 Result and Discussion

The profiles of the calculated indoor operative temperatures for living and common area zones of the building are shown in Fig. 5.

Table 3 shows average annual operative temperature values for both zones, calculated based on hourly profile for an average day of each simulated month. The annual minimum and maximum operative temperatures for the zones are also given in the table.

Fig. 2 West (**a**) East (**b**) North (**c**) and South (**d**) façades of the modelled building

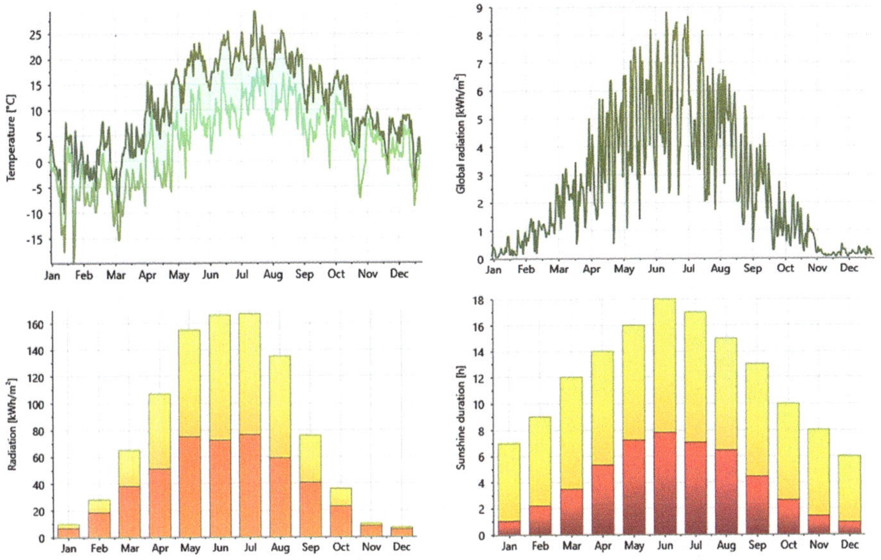

Fig. 3 Climatic key parameters for Växjö based on the meteorological data for 2013 obtained from Meteonorm software

IDA-ICE gives the largest difference in all average, maximum and minimum operative living area temperature compared with the average results of the four models while TRANSYS give the largest difference for maximum operative living area temperature. If the model gives the largest difference compared to average for all four models is not included in the average, the difference is small between the three remaining models and the average of them. The pattern for operative common area temperature is about the same as for living areas. IDA-ICE gives the largest differences in both average and minimum operative common area temperature compared with other models while the TRNSYS gives the largest differences in maximum operative temperature compare to other models. The main reason for these differences is the different definition of common area zone and the boundary conditions. For example, in IDA-ICE the common area is a single zone for the whole building and there is not any inter floor/ceiling through the common area.

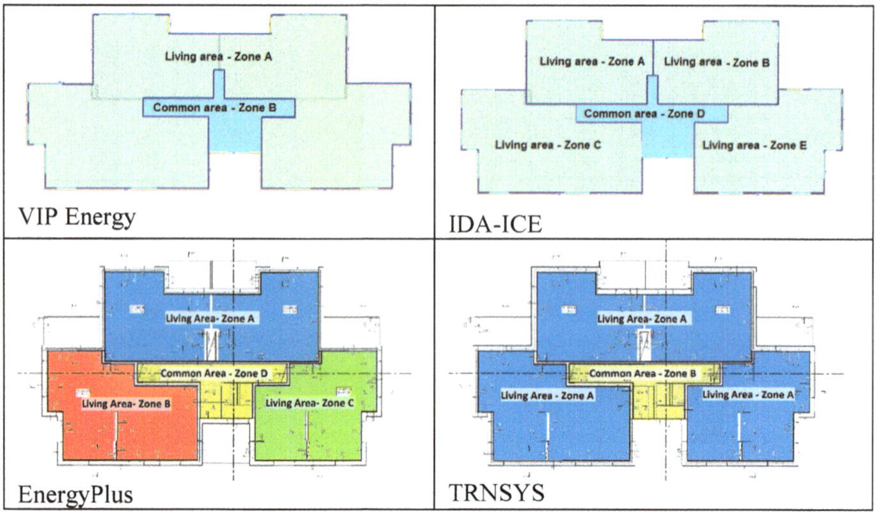

Fig. 4 Configuration of thermal zones at a typical floor level in the models

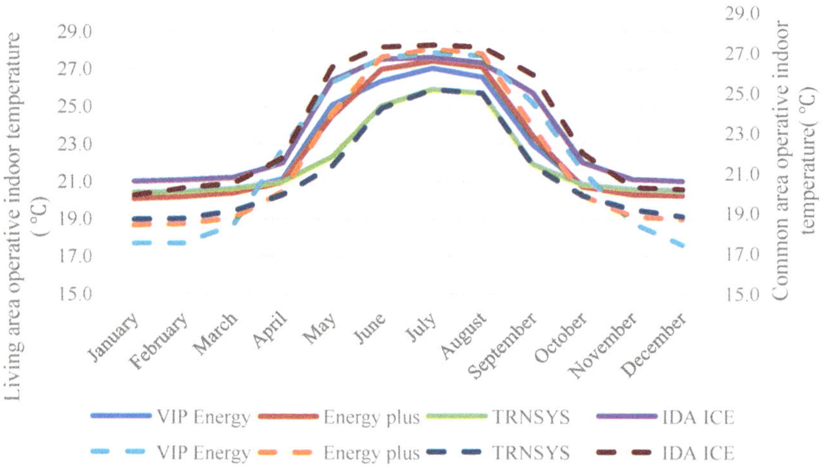

Fig. 5 Operative indoor temperature variation for living area (left axis-solid lines) and common area (right axis-dash lines) for the modeled building

Thus, the effect of internal thermal mass of the inter floors and ceiling in the common area is not considered. The different considered equations for heat transfer convections inside the building and heat losses in each simulation tools may also contribute to these differences.

Figure 6 shows the total monthly space heating demand for the studied building as calculated by the different simulation tools. The annual space heating demand for

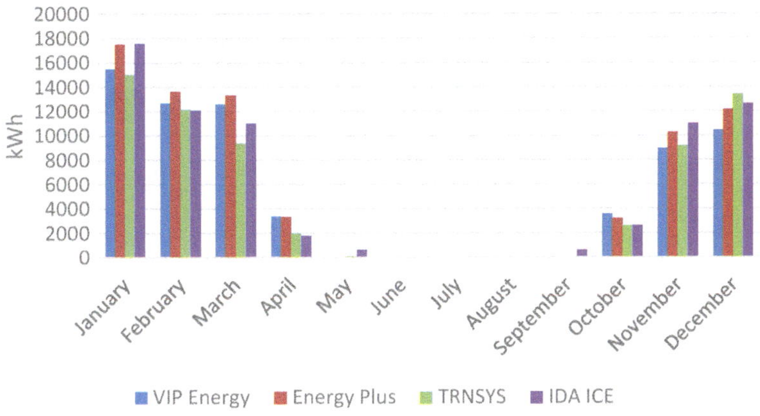

Fig. 6 Space heating demand for the modelled building in the four simulation tools

the studied building is estimated to be 40, 44, 38 and 42 kWh/m^2 by VIP-Energy, EnergyPlus, TRNSYS 16 and IDA-ICE, respectively, giving an average of 41 kWh/m^2. Compared to the average, EnergyPlus gives the largest difference of the models. Excluding the EnergyPlus will give an average of 40 kWh/m^2 as calculated by VIP-Energy.

Eventually, the main features of the four BEMPs are compared and summarized in Table 4 [10–12].

Table 4 Summary of key features of the four BEMPs

Description/feature	VIP-energy	IDA-ICE	EnergyPlus	TRNSYS 16
Multi-zone simulation	✓	✓	✓	✓
Hourly time step	✓	✓	✓	✓
Sub-hourly time-step	✗	✓	✓	✓
BIM or drawing import	✗	✓	✓	✗ (✓ ver.17-18)
3D visualization of model	✗	✓	✗	✗ (✓ ver.17-18)
Material library	✓	✓	✓	✓
Thermal bridge modelling	Detailed	Simplified	Indirect method	Simplified
Visual interface	✓	✓	✗	✓
Incorporate with other tools e.g. (MATLAB, CFD tools, EES,VBA)	✗	✗	✗	✓
Open source code	✗	✗	✗	✓
Ability to add mathematical sub-modules to the model	✗	✗	✗	✓

5 Conclusion

In this study we have compared the calculated space heating demand and operative temperatures of a recently built multi-store residential building when modelled with VIP-Energy, IDA-ICE, EnergyPlus and TRNSYS. The calculations are based on the same or similar specified input data including hourly weather data for the Swedish city of Växjö. The results show that the annual space heating demand calculated by the different tools ranges between 38 and 44 kWh/m^2, indicating about 14–16% differences in the extremes of values calculated by the tools. TRNSYS resulted in the lowest while Energyplus resulted in the highest simulated space heating for the analyzed building. A calculated lower space heating demand by 5% can be caused by simplification of the thermal bridge in these two programs and by different correlations used for estimating the heat transfer coefficient and simulation algorithm for the calculations. For the operative temperature estimation, IDA-ICE has the largest differences of the studied tools with a 4% difference in annual average value for the living area of the building analyzed. The main advance feature of VIP-Energy is the detail thermal bridge analysis while the main drawback of this software is the level of complexity to use the software. EnergyPlus and TRNSYS can be used for energy supply system integration. The main difference between these two programs compared to other two models is the ability to add mathematical sub-modules to the models by adding correlations and equations to modify the inputs or outputs, incorporation with other software's such as MATLAB, component's programming modification due to the open source code and graphic visual interface which motivate the use of TRNSYS.

References

1. IEA, Energy Efficiency Training Week, International Energy Agency, 2016. [Online]. Available: https://www.iea.org/media/training/eetw2016/buildings/A.2_Energy_efficiency_potential.pdf. Accessed 16 Aug 2017
2. StruSoft, VIP+software, Sweden. 2010, Available from http://www.strusoft.com/products/vip-energy
3. VIP-Energy, Validation methods. http://www.vipenergy.net/Validating.htm (2016)
4. J.A. Duffie, W.A. Beckman, *Solar Engineering of Thermal Processes*, vol. 3 (Wiley, New York, 2013)
5. G. Jóhannesson, Active heat capacity: models and parameters for the thermal performance of buildings. Report TVBH, 1981. 1003
6. P. Nylund, Räkna med luftläckningen. Samspel byggnad-ventilation (in Swedish). Swedish Council for Building Research, 1 (1984)
7. EQUA Simulation AB, IDA Indoor Climate and Energy. Available from http://www.equa.se/en/ida-ice/what-is-new/highlights-in-4-7 (2016)
8. Equa Simulation Technical Group, Validation of IDA Indoor Climate and Energy 4.0 build 4 with respect to ANSI/ASHRAE Standard 140-2004, Technical report, Solna, Sweden, 2010
9. Equa Simulation AB, Validation of IDA Indoor Climate and Energy 4.0 with respect to CEN Standard EN 15255-2007 and EN 15265-2007. Equa Simulation AB, Stockholm, 2010

10. D.B. Crawley, J.W. Hand, M. Kummert, B.T. Griffith, Contrasting the capabilities of building energy performance simulation programs. Build. Environ. **43**(4), 661–673 (2008)
11. J. Sousa, in *Energy Simulation Software for Buildings: Review and Comparison*, International Workshop on Information Technology for Energy Applicatons-IT4Energy (Lisabon, 2012)
12. TRNSYS, Using the Simulation Studio, TRNSYS version 16, Thermal Energy System Specialists, Madison

Fast Simulation Platform for Retrofitting Measures in Residential Heating

Philipp Schuetz, Rossano Scoccia, Damian Gwerder, Remo Waser, David Sturzenegger, Peru Elguezabal, Beñat Arregi, Alessandro Sivieri, Marcello Aprile and Jörg Worlitschek

Abstract Energy efficiency aware building owners are facing a massive amount of different retrofitting options. However, a quantitative assessment of the different options requires a high level of technical expertise. In this contribution, a fast and novel simulation platform for the assessment of different residential heating system configurations is presented. This platform enables dynamic simulations of the complete heating system, calculating energy/heat consumption and comfort indicators for different heating systems during a full year in less than 5 s on a recent laptop. Another key feature of the platform is the inclusion of a large variety of different heat sources (oil/gas/biomass/carbon boilers, air/brine-water or sorption heat pumps), sensible thermal heat storages, as well as building models. Shortly, this system will be the core of a platform enabling interested users to calculate the energy consumption of different retrofitting options accurately. To validate the system models, the energy consumption of the three reference buildings (single family houses with an annual heating energy demand of 15, 45 and 100 kWh/m^2) as per the IEA SHC Task 44 is calculated and compared with reference simulations from established simulation frameworks. The energy consumption of these buildings matches the reference values up to 5% for a full year simulation requiring calculations times between 3.3 and 3.7 s on a recent laptop.

P. Schuetz (✉) · D. Gwerder · R. Waser · D. Sturzenegger · J. Worlitschek
School of Engineering and Architecture, Lucerne University of Applied, Sciences and Arts, Technikumstrasse 21, 6048 Horw, Switzerland
e-mail: schutz.p@gmail.com

R. Scoccia · A. Sivieri · M. Aprile
Department of Energy, Politecnico di Milano, Via Lambruschini 4a, 20156 Milan, Italy

P. Elguezabal · B. Arregi
Sustainable Construction Division, Tecnalia, Parque Tecnológico de Bizkaia, Edificio 700, 48160 Derio, Spain

D. Johansson et al. (eds.), *Cold Climate HVAC 2018*,
Springer Proceedings in Energy, https://doi.org/10.1007/978-3-030-00662-4_60

Keywords Assessment of retrofitting measures in residential heating
Fast simulation platform · Economic and ecological assessment tool

1 Introduction

In 2013, around 22% of the overall energy consumption in EU28 was employed for space heating of residential buildings [1]. A reduction of this energy demand often requires a retrofitting of the considered buildings. Despite partially high incentives for renovation, the annual renovation rate in the European Union is currently rather low with around 3.6% estimated [2]. One burden to motivate building owners for retrofitting is the low dissemination of reliable information about different retrofitting options as well as a lack of a platform for comparing different alternatives. However, establishing such a platform poses two competing challenges. On the one hand, the comparison tool should be fast in estimating the energy consumption of the owner's specific building. On the other hand, the estimated consumption values must be reliable/accurate to enable a fair comparison of the different options, and the user must enter many parameters. To estimate the energy consumption over one year, simulations attracted much attention [3–5]. In order to enable simulations of the complete heating system over a long period, often detailed models of individual devices are simplified by analytical models [3, 6, 7] and combined to models of the complete heating system. The major difference between the different approaches is the number of components considered, the complexity of the building and its model and the options of control (three position valve control [3] to complex model predictive control [8]) and the state of validation. The reference frameworks employed for benchmarking include TRNSYS, Matlab/Simulink (Carnot Blockset) [9] and IDA Indoor Climate and Energy framework [10].

In this contribution, a novel framework for the fast assessment of the energy consumption for residential heating systems is presented. Two major applications are envisioned: First, integrated in an online tool, interested users can assess the effect of different retrofitting measures for their situation (location, climate, energy pricing schemes, heating/cooling system). Second, expert users will employ this tool to investigate the effect of different system configurations, size components and quantify the effect of different control options. This contribution is organised as follows: In Sect. 2, the model platform and the models of the individual components are described. In Sect. 3, the validation results and performance evaluations of the simulation platform are presented. In the concluding Sect. 4, a conclusion of the obtained results is drawn.

2 Methods

2.1 Modelling Overview

The dynamics of the residential heating system is described by combining numerical/analytical models of the individual components. For each component, a set of observable functions (e.g. the average temperature in a pipe) is selected. Their evolution is either described by a differential equation derived from an energy balance equation (pipe, building, sensible thermal storage) or by a lookup table (heat pump) or an analytical function (oil/gas/biomass boiler, emitter system, valves).

2.2 Models

Pipes
As observable function the average temperature T_p in the pipe is chosen. From the energy conservation equation, the following governing equation for the temperature evolution is derived:

$$m_p c_m \frac{\partial T_p}{\partial t} = \dot{Q}_{in} + \dot{m}_{in} c_m T_{in} - \dot{m}_{out} c_m T_{out} - \dot{Q}_{loss}$$

with m_p the mass of the medium in the pipe, c_m the specific heat capacity, T_{in}, T_{out} and $\dot{m}_{in}, \dot{m}_{out}$ the temperature and mass flow rate of the medium at the inlet and outlet, respectively. With $\dot{Q}_{in}$ the energy generated by heat sources connected to the pipe inlet is considered. The losses $\dot{Q}_{loss}$ are neglected as short, well insulated pipes are assumed.

Building
The complex dynamics of the building is modelled by a lumped energy balance equation of the building following the work of Burmeister et al. [11]. The change of the room temperature T_{room} is described

$$C \frac{\partial T_{room}}{\partial t} = -H(T_{room} - T_{ambient}) + gI + \dot{Q}_{ES} + \dot{Q}_{internal} \tag{1}$$

by the action of internal losses (parametrised by the loss factor H and the difference between room temperature T_{room} and the ambient temperature $T_{ambient}$), the contribution of solar radiation (g the solar factor considering window area and orientation and I the instantaneous diffuse and direct radiation incident on the windows) as well as the heat transmitted via the emitter system $\dot{Q}_{ES}$ and internal gains $\dot{Q}_{internal}$ from electric/thermal appliances and inhabitants.

The contribution of the diffuse and direct radiation to the transmitted radiation I is calculated based on the Perez sky model [12] and the implementation is adapted from the Carnot framework for Matlab [9]. The shading of the windows is modelled by adjusting the solar factor g and the effect of air exchange is modelled by additional loss terms in the loss factor H.

Oil/Gas/Biomass boiler system

The boiler system model has been implemented according to the European Standard CEN EN 15316-4-1 [13] following the "boiler cycling method" to incorporate the losses during cycling explicitly. This method calculates for each time step the instantaneous generated heat, the consumed energy carrier (oil, gas or biomass depending on the type) based on the current return temperature, and the operation state (burner on/off, loading factor and control strategy). The generated heat is then injected into the heating system analogous to the heat pump model presented above.

Heat pump

The heat pump is modelled based on a performance map approach. For each integration step, the current generated heat and coefficient of performance (COP) of the considered heat pump is determined based on supply and ambient-/brine-temperature by linear interpolation for on/off-controlled heat pumps. For capacity-controlled heat pumps, in addition, the current load status is considered as a third dimension in the performance and COP maps. In the studies reported below, the performance map of an air/water-heat pump are taken from [14].

Emitter system

To model both radiator and underfloor-emitter systems, the heat $\dot{Q}_{room}$ exchanged with the room is calculated with the radiator equation:

$$\dot{Q}_{room} = \dot{Q}_{design} \cdot \left(\frac{\frac{T_{in} + T_{ES}}{2} - T_{room}}{\Delta T_{design}} \right)^{n_r}$$

with $\dot{Q}_{design}$ the design power of the emitter system at the design temperature difference ΔT_{design} between inlet temperature T_{in}, room temperature T_{room} and the outlet temperature in the emitter system T_{ES}. The value of the radiator exponent n_r enables to model a radiator based system ($n_r = 1.3$) and an underfloor based system ($n_r = 1.1$). To capture the inertia of the emitter system, the dynamics of the emitter system is modelled analogous to a pipe with extracted heat $\dot{Q}_{room}$ by the following equation:

$$m_{ES}c\frac{\partial T_{ES}}{\partial t} = -\dot{Q}_{room} + \dot{m}_{in}c_mT_{in} - \dot{m}_{in}c_mT_{ES}$$

In this equation, the mass flow $\dot{m}_{in}$ into and out of the emitter system of the medium with capacity c_m is assumed to be equal.

Valves

The effect of three-point valves is modelled by introducing the mass conservation as a constraint in the mass flow equations. To model a temperature control function, the splitting of the mass flows to maintain the target temperature is added as additional constraint equation.

Sensible thermal storage

The sensible thermal storage is modelled as a vertical standing vessel with constant cross-section A and height h. The height of the system is split into N equal slices of height $\Delta h = h/N$. The mass flows from inlet and outlet pipe to the storage cause internal mass flows $\dot{m}_{up}(i)$ (upwards) and $\dot{m}_{down}(i)$ (downwards) in the slices i between inlet and outlet channels depending on the flow direction. For instance, an inlet mass flow of 1 l/s in a storage vessel with a height of $h = 1m$ discretized in $N = 11$ slices, where the inlet enters at a height of 10 cm and exits at a height of 50 cm, causes a mass flow upwards of 1 l/s for the nodes at 10, 20, 30, 40 and 50 cm. The dynamics within the storage vessel is captured via the evolution of the average temperature T_i in the slice i given by the following energy conservation equation:

$$\begin{aligned} m_ic_m\frac{\partial T_i}{\partial t} &= U \cdot A_{loss} \cdot \left(T_{ambient,storage} - T(i)\right) \\ &+ \frac{\pi d^2}{4}\lambda_F\frac{T(i+1) + T(i-1) - 2T(i)}{\Delta h^2} + \dot{Q}_{in}(i) \\ &+ c_p\left(\dot{m}_{up}(i)(T(i-1) - T(i)) + \dot{m}_{down}(i)(T(i+1) - T(i))\right) \end{aligned}$$

with A_{loss} the surface of the slice in contact with the environment at temperature $T_{ambient,storage}$, λ_F the heat conduction coefficient between adjacent slices, $\dot{Q}_{in}(i)$ the heat injected in slice i by external sources such as heat exchanges and $\dot{m}_{up/down}(i)$ the mass flows in slice i in upward and downward direction, respectively. For the nodes at the bottom and at the top of the storage vessel also the losses through the lids are incorporated with a term analogous to the term on the first line on the right. Effects of free convection within the storage vessel are not directly modelled but may be taken into consideration by the adaption of the thermal conductivity between adjacent slices.

2.3 Calculation of Key Performance Indicators

Energy consumption
The energy consumption of the residential heating system is calculated by integrating the power injected by the heat source into the heating system. As described in detail in Sect. 2.4, the integration is performed stepwise, i.e., every 180 s, the instantaneous power of the heat source is multiplied with the width of the integration time window (unless the remaining simulation time is shorter, an integration interval of 180 s is chosen) and summed over the whole simulation period.

Consumption of energy carrier
The consumption of the energy carrier for the heat source is determined analogously to the energy consumption by stepwise integration of the instantaneous consumption. For the model of the oil/gas/biomass boiler, the instantaneous consumption of the energy carrier is calculated by the model according to EN norm 15316-4-1. In the heat pump model, the instantaneous, average coefficient of performance is determined by an interpolation of the provided COP map. By a division of the heat generated instantaneously and the current COP, the current electrical power consumption can be determined.

Comfort indicators
As comfort indicator, the integrated deviation of the room temperature from the set temperature is considered. To incorporate the human insensitivity to deviations of 0.5 K, the difference is only considered if the room temperature is 0.5 K below the actual set temperature.

2.4 Simulation Framework

The simulation framework is implemented in C++ with Visual Studio Community 2017 from Microsoft. The models of the individual components are implemented in individual classes.

The simulation is performed in three stages:

1. The hydraulic configuration of the system is defined, all component models are loaded with the component parameters and required input and output values of the individual models are connected among the different component models by pointer algebra.
2. The dynamic simulation is performed: The whole simulation time (typically a full year) is split into small windows of 180 s. During this window the ambient temperature, the mass flow through the individual components and the control signals for the components are assumed to be constant. The dynamics of the heating system during this period is simulated by solving the system of coupled differential equations with an implementation of the controlled Runge-Kutta algorithm from the odeint package from the Boost 1.6.4 library (www.boost.org).

3. After each 180 second-window, the mass flows through the individual components are recalculated to account for changes in the operation conditions (e.g. changes in the mixing ratio of a temperature controlled mixing valve). The mass flows are determined by solving the system of mass conservation equations in each point of the hydraulic schema with the Eigen library (eigen.tuxfamily.org). In addition to the mass flow update, the climatic information is updated, changes in the control signal for the individual components are determined, and the changes on the key performance indicators (cf. Sect. 2.3) are calculated.
4. In the last stage, the key performance indicators for the simulation are calculated, and the simulation program is terminated.

2.5 Reference Simulations for the Building Models

For the benchmark of the building model, the reference buildings single family house SFH15, 45 and 100 from the IEA SHC Task 44 are considered [15, 16]. These three reference buildings are models for typical buildings with an annual space heating demand of 15, 45 and 100 kWh/m^2 and represent a modern energy efficient building (SFH15), a conventional, modern house (SFH45) and an existing, non-renovated building (SFH100) located in Strasbourg. To determine the annual energy demand for the models, a simple residential heating system is implemented in the framework: A heat source with constant power output (if running) is connected to the emitter system, which provides the energy to the building for space heating via the term $\dot{Q}_{ES}$. The usage profiles and contributions of the appliances $\dot{Q}_{internal}$ are taken from the reference publication [15, 16].

The simulations are performed for the climatic test reference year (TRY) of Strasbourg as extracted from Meteonorm. To measure the energy demand of the building alone, the heat source is exchanged with a constant power heat source injecting the same power into the heating system whenever activated. The shading control and air exchange are implemented as described in [15, 16].

In the building validation simulation, the heat source is controlled by a two position controller based on the current room temperature (set temperature 20 °C with a death band of 1 K).

To obtain time resolved reference energy consumption profiles of the building during one year, the energy consumption is as well simulated with the two reference software packages Matlab/Simulink (Carnot Blockset) [9] and IDA Indoor Climate and Energy framework (EQUA Simulation AB, Solna, Sweden). These reference simulations have been described elsewhere in detail [14].

3 Results

3.1 Parametrisation of Reference Buildings and Validation Results

Based on the detailed description of the buildings in [15, 16], the values of the lumped losses H, the summative building capacity C and the solar gain factor g have been calculated according to the procedure in [11]. The resulting values are reproduced in Table 1 together with the parameters of the emitter system and heat source in the simple benchmarking heating system.

The calculated cumulative weekly heat demand of the building for one year is shown in Fig. 1 and compared to the reference models. Comparing the three profiles indicates that all three simulations agree. The deviations of the annual space heat demand range between 1.2% (SFH15), 6.2% (SFH45) and 8.5% (SFH100). Although this deviation is non-negligible, it is compatible with the variation among the reference codes (e.g. 11.7% for SFH100). Noteworthily, all simulations of the current model have been performed with a focus on precise temperature control in the building: The comfort indicator (the integrated deviation of the room temperature—when larger than 0.5 K) ranged only between 5 and 12 Kh over a full year (8760 h).

3.2 Running Time Considerations

The running times for the annual simulations are reported in Table 2 and range between 3.5 and 3.6 s. Due to the architecture tailored on speed, these running times are substantially lower than for the reference simulations, where annual simulations require execution times longer than 10 min.

Table 1 Parameters of building models according to Eq. (1), emitter system and heat source for the validation of the reference buildings SFH 15, SFH 45 and SFH 100 considered here

Parameter	Building		
	SFH 15	SFH 45	SFH 100
Loss factor H (W/K)	94.2	160.7	299.8
Capacity C (kJ/K)	169,640	167,123	163,349
Solar gain factor g (m^2)	11.5	12.2	14.8
Radiator exponent n_r (-)	1.1	1.1	1.3
Design power $\dot{Q}_{design}$ (W) (Strasbourg)	1792	4072	7337
Design temperature difference ΔT_{design} (K)	15	15	35
Maximum power of heating system (W)	2330	5293	9538
Ventilation (h^{-1})	0.4	0.4	0.4
Heat exchanger effectiveness	0.6	0	0

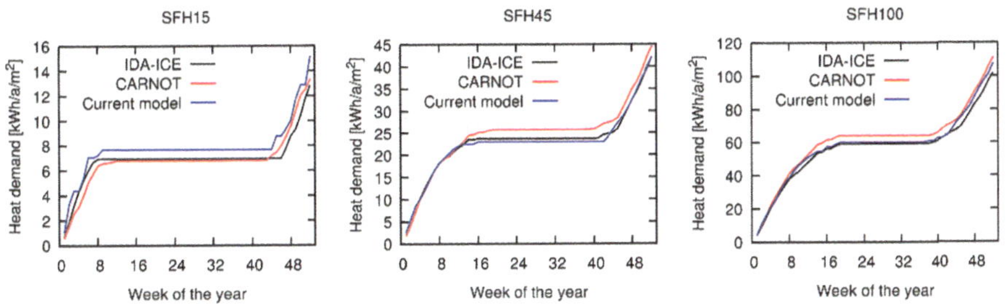

Fig. 1 Comparison of space heating energy demand for the three-considered reference building SFH15 (top, left), SFH45 (top, right) and SFH100 (bottom, left) for all considered simulation frameworks

Table 2 Results of benchmarking study to evaluate annual space heating energy demand

Result	Building		
	SFH 15	SFH 45	SFH 100
Annual energy demand (kWh)	2125	5909	15,194
Specific annual energy demand (kWh/m^2)	15.18	42.2	108.5
Comfort indicator (Kh)	12	5	8.7
Execution time for annual simulation (s)	3.54	3.66	3.61

The calculated specific annual energy demands match the expected values of 15, 45 and 100 kWh/m^2 accurately

3.3 Application on Residential Heating System with Different Heating Systems

As a final application study, the building SFH100 is located in Helsinki (Meteonorm data set FI-Helsinki-Kaisani-29980). To cover the increased heat demand due to lower ambient temperatures, the power of the heat source has been adjusted appropriately (cf. Table 3 for reference). The (specific) energy consumption for the individual heat sources and the comfort indicators are shown in Table 3 as well. The constant power boiler and the condensing boiler are operated with a two level control based on the room temperature (set temperature 20 °C) and a dead band of 1 K. The heat pump operation is controlled based on a set temperature on return temperature to the heat pump according to the heating curve modified by the six-fold deviation of room and set temperature.

Non-surprisingly, the heat demand for space heating almost doubled due to the change of location, which agrees with other simulation studies. The different control algorithms have only a minor impact on the total energy consumption. However, the heat pump control algorithm keeps the temperature in a narrower range yielding an improved value for the comfort indicator. For the considered heat pump, no performance data for ambient temperatures below −10 °C are available. Therefore, the heat capacity and COP values for ambient temperatures below −10 °C have

Table 3 Annual specific energy demand and comfort indicators for an SFH 100 building in Helsinki

Property	Fixed capacity boiler	Heat pump	Condensing boiler
Power heat source at −20 °C at design point (W)	10,931	10,911	10,886
Heat generated per year (kWh)	29,306	29,387	29,147
Specific annual energy consumption (kWh/m^2)	209.3	209.9	208.19
Annual final energy consumption (kWh)	29,305	13,284	31,957
Comfort indicator (Kh)	368.8	46.3	436
Execution time for annual simulation (s)	4.00	4.27	4.10

been derived from a linear extrapolation of the data in the range 0 and −10 °C. This may also be a reason for the rather high values of the seasonal performance factor of 2.19 calculated with the values above.

4 Conclusions

In this paper, a novel simulation platform for residential heating systems has been presented. The key features of this platform are its broad range of covered system components (different types of buildings, oil/gas/biomass boilers, heat pumps, different types of emitter systems as thermal storages) as well as its short execution time. These short execution times are vital for a fast assessment of different system configurations. To validate the building model, the simulation is compared with reference simulations performed with Matlab/Simulink (Carnot Blockset) and the IDA-Indoor Climate and Energy frameworks. The results of the framework agree well with the reference simulations for the SFH 15, 45 and 100 models as introduced by the IEA SHC Task 44. As an application, the energy consumption of the above three buildings has been compared for the different heat sources and different control algorithms with Helsinki and Strasbourg (reference location) as a point of installation.

Acknowledgements The authors would like to thank the European commission for funding of the H2020-project “Heat4Cool” (project ID 723925). The work has also been supported by the Swiss State Secretariat for Education, Research and Innovation (SERI) under Contract No. 16.0082.

References

1. K. Pollier, L. Gynther, B Lapillonne, Energy Efficiency Trends and Policies in the Household and Tertiary Sectors (2015)
2. B. von Manteuffel, C. Petersdorff, K. Bettgenhäuser, T. Boermans, EU pathways to a decarbonised building sector (2016)

3. P. Byrne, J. Miriel, Y. Lénat, Modelling and simulation of a heat pump for simultaneous heating and cooling. Build. Simul. **5**, 219–232 (2012)
4. M. Elci, S. Narmsara, F. Kagerer, S. Herkel, in *Simulation of Energy Conservation Measures and Its Implications on a Combined Heat and Power District Heating System: A Case Study*. 13th Conference of Building Performance Simulation Association (Chambéry, 2013), pp. 104–111
5. E. Georges, G. Masy, C. Verhelst et al., Smart grid energy flexible buildings through the use of heat pumps in the Belgian context. Sci. Technol. Built. Environ. **21**, 800–811 (2015)
6. W. Chung, Review of building energy-use performance benchmarking methodologies. Appl. Energy **88**, 1470–1479 (2010)
7. M. Muratori, M.C. Roberts, R. Sioshansi et al., A highly resolved modeling technique to simulate residential power demand. Appl. Energy **107**, 465–473 (2013)
8. F. Oldewurtel, A. Parisio, C.N. Jones et al., Use of model predictive control and weather forecasts for energy efficient building climate control. Energy Build. **45**, 15–27 (2012)
9. C. Wemhöner, B. Hafner, K. Schwarzer, in *Simulation of Solar Thermal Systems With Carnot Blockset*. Proceedings Eurosun 2000 Conference, ISES (Copenhagen, Denmark, 2000). pp 1–6
10. EQUA IDA Indoor Climate and Energy
11. H. Burmeister, B. Keller, Climate surfaces: a quantitative building-specific representation of climates. Energy Build. **28**, 167–177 (1998)
12. R. Perez, P. Ineichen, R. Seals et al., Modeling daylight availability and irradiance components from direct and global irradiance. Solar Energy **44**, 271–289 (1990)
13. 15316-4-1 CE Heating systems in buildings: Method for calculation of system energy requirements and system efficiency: Part 4-1: Space heating generation systems, combustion systems (boilers)
14. D. Gwerder, P. Schuetz, L. Gasser et al., in *Entwicklung einer optimalen Einheit aus Wärmepumpe und thermischem Energiespeicher*. 21. Wärmepumpentagung BFE Forschungsprogramm. Burgdorf (2015)
15. R. Dott, J. Ruschenburg, F. Ochs et al., The Reference Framework for System Simulation of the IEA SHC Task 44/HPP Annex 38—Part B: Buildings and Space Heat Load. Tech Rep subtask C IEA SHC Task 44 (2013)
16. M. Haller, J. Ruschenburg, F. Ochs et al., The Reference Framework of System Simulations of the IEA SHC Task 44/HPP Annex 38—Part A: General Simulation Boundary Conditions. Tech Rep subtask C IEA SHC Task 44 (2013)

Calibration of a High-Resolution Dynamic Model for Detailed Investigation of the Energy Flexibility of a Zero Emission Residential Building

John Clauß, Pierre Vogler-Finck and Laurent Georges

Abstract A detailed multi-zone building model of an existing zero emission residential building (ZEB) has been created using the software IDA Indoor Climate and Energy (IDA ICE). The model will later be used for investigating control strategies for the heating system to activate the building energy flexibility. The main purpose of this paper is to show how reliable the model reproduces the short-term thermal dynamics and the temperature zoning of the building. This is of particular interest for the control of heating, ventilation and air conditioning (HVAC) systems in order to provide meaningful insights of active demand response (ADR) measures. The model has been validated using data sets from seven experiments. Two dimensionless indicators, the normalized mean bias error (NMBE) and the coefficient of variation of the root mean square error (CVRMSE) were applied in order to evaluate the trend of the average indoor temperatures. The first approach considered standard operating conditions, where the measured indoor air temperature was used as input for the control of the electrical radiator and the total electricity use of the radiator as an output. Excitation sequences have been used in the second approach, where the electric power of the radiator has been imposed and the operative temperature taken as the output. The model shows good agreement between the temperature profiles from the measurements and the simulations based on the NMBE and CVRMSE remaining below 5% for most cases.

Keywords Detailed building model · Model validation · Zero emission building
Energy flexibility · Active demand response

J. Clauß (✉) · L. Georges
Norwegian University of Science and Technology, K. Hejes vei 1b,
7034 Trondheim, Norway
e-mail: john.clauss@ntnu.no

P. Vogler-Finck
Neogrid Technologies ApS, Niels Jernes Vej 10, 9220 Aalborg Ø, Denmark

D. Johansson et al. (eds.), *Cold Climate HVAC 2018*,
Springer Proceedings in Energy, https://doi.org/10.1007/978-3-030-00662-4_61

1 Introduction

1.1 Context of the Work

Calibration of a building model is an essential step to ensure simulation accuracy and thus to increase confidence in simulation results. Calibration can be achieved using monitoring data from the respective building [1]. This is of particular interest for ZEBs which contain advanced technologies that can be challenging to simulate. Model calibration often has, among others, two limitations: model complexity (use of single-zone models) and the use of hourly aggregated data of the energy consumption for space heating [1]. The ASHRAE 14 guideline defines acceptance criteria based on the energy consumption, whereas no standards exist that determine acceptance criteria for model calibration based on indoor temperatures [2, 3]. Nevertheless, several studies use a threshold for the CVRMSE of 5% when calibrating a building model with respect to hourly indoor temperatures [2, 4–6].

A detailed multi-zone numerical model of an existing residential ZEB, the Living Laboratory in Trondheim (Norway) [7], has been created using the software IDA ICE. The calibrated model will later be used to investigate the influence of different control strategies for space heating to activate the energy flexibility using structural energy storage. Knowing the short-term thermal dynamics and thermal zoning of the building envelope as a response to time-varying space heating set-point temperatures is of importance for active demand response measures as well as for the development of resulting control strategies for heat pumps. During structural energy storage, the indoor temperatures should fluctuate within the thermal comfort levels. Therefore, this study aims at investigating the indoor air temperature and operative temperature in different zones of the building.

1.2 Contribution

A multi-zone approach is used to evaluate the differences of the thermal environment between the bedrooms and the living spaces. Monitoring of the indoor thermal environment has been carried out during two periods, in February/March 2017 and April/May 2017. During the experiments in April and May 2017 the dynamics of the indoor thermal environment have been measured as a response to an excitation using a specified pre-defined heating sequence (with sub-hourly resolution). A first qualitative validation of the building model is carried out by a direct visual comparison of the measured building thermal behavior with the predictions from simulations in IDA ICE using identical boundary conditions. In a second step, a quantitative validation against experimental data is conveyed. Two dimensionless indicators, the NMBE and CVRMSE are used to evaluate the accuracy of the calibration based on the average indoor air temperature and operative temperature.

2 The Living Laboratory—A Residential ZEB in Norway

2.1 Short Description of the Building

The case study building (Fig. 2a) is a Norwegian residential zero emission building [7] which is located at the Gløshaugen Campus of the Norwegian University of Science and Technology (NTNU) in Trondheim, Norway.

The on-site electricity generation from the photovoltaic panels is designed to compensate for the building CO_{2eq} emissions from the operational phase as well as for embodied emissions over the lifetime of the building. The floor area is approximately 105 m^2 [8] and the specifications for the building envelope follow the requirements from the Norwegian standard for residential passive houses NS3700. The building has a highly-insulated envelope with a lightweight wooden construction [7] as well as energy efficient windows with low emissivity. Furthermore, it contains 90 m^2 of phase change material (PCM) in the roof construction. The PCM is active between 18 and 26 °C [9] and thus mitigates indoor temperature fluctuations in the building. The Living Laboratory consists of five heated zones (see Fig. 1): two bedrooms, one bathroom, one working room and one large room combining a kitchen, a living room and an entrance.

A single heat emitter is placed between the two living rooms, since such a passive house can theoretically be heated using a single heat emitter [10]. The other rooms are without active heating, except for the bathroom equipped with floor heating. Temperature differences between the rooms are thus expected.

2.2 Building Model in IDA ICE

The building simulation is done in IDA ICE 4.7.1 which is a dynamic multi-zone simulation software. The software applies equation-based modelling [11] and enables the user to evaluate the energy use and the indoor thermal climate of a building. A sketch of the 3-D virtual model is presented in Fig. 2b. The roof

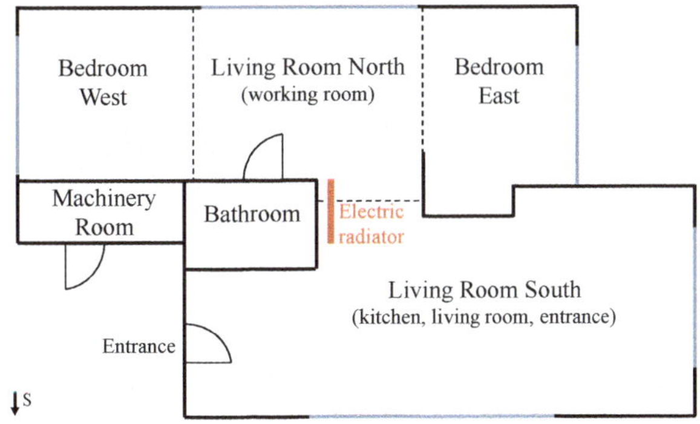

Fig. 1 Floor plan of the ZEB living laboratory (dashed lines show borderlines of the zones; moving doors between bedrooms and Living Room North can be opened)

overhang at the entrance of the real building was not considered in the building model, as this was originally done to increase the roof area for PV installation. It does not affect the heating needs nor the indoor temperatures of the building because this part of the building is not heated up.

3 Experiments

The building model is validated on the basis of two sets of experiments where the first four experiments were conducted between the 16th of February 2017 and the 24th of March 2017, whereas three other experiments were carried out from the 18th of April 2017 to the 15th of May 2017. During the second set of experiments, the indoor air temperatures and operative temperatures (at 0.7 m from the floor) were measured every 5 min in all heated zones, whereas during the first set of experiments the indoor temperatures were measured every minute. The air temperature was measured by Pt100 sensors, whereas the operative temperature was measured using Pt100 sensors enclosed in black globes. Vertical stratification of the air temperature was measured in the two living rooms. The bathroom door was always closed and the building was not occupied, although dummies with incandescent lamps were placed to mimic internal gains according to NS3031 using predefined on/off schedules for the first set of experiments.

3.1 Experiments 1 to 4 from February and March 2017

The aim of these measurements was the investigation of the temperature zoning by closing and opening the bedroom doors and to analyze the thermal dynamics of the Living Lab. An overview of the settings for these four experiments is given in Table 1.

The electricity use of the 1.6 kW electric radiator (with thermostatic on/off control) was logged every 30 s. The constant air volume (CAV) ventilation system continuously supplied air with a temperature of ca. 19 °C. All windows were

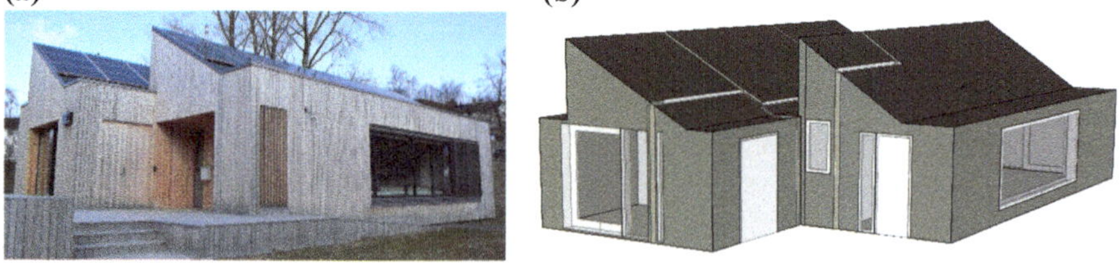

Fig. 2 **a** Photo of the living laboratory at the Gløshaugen campus at NTNU and **b** sketch of the modelled building

Table 1 Overview of the test settings for experiments 1 to 7 (E1–E7) where "Night setback" means that no electric heating is allowed between 23:00–07:00

Experiment	Time period	Bedroom doors	Night setback	Indoor Temperature variation (°C)	Sky condition
E1	17.2. 00:00–20.2. 00:00	Closed	No	15.7–19.9	Overcast
E2	20.2. 09:00–25.2. 00:00	Open	No	17.6–20.1	Clear on 21.2.
E3	16.3. 00:00–20.3. 12:00	Open	Yes	16.9–20.7	Clear on 19.3.
E4	20.3. 12:00–24.3. 12:00	Closed	Yes	14.5–21.6	Clear on 21.3.
E5	18.4. 20:00–24.4. 20:00	Open	No	16.8–24.5	Mostly clear
E6	29.4. 00:00–08.5. 09:00	Closed	No	15.9–26.8	Mostly clear
E7	09.5. 00:00–15.5. 06:00	Open	No	19.0–28.0	Mostly clear

always fully blinded to limit the influence of solar radiation. Weather data was taken from [12], which uses data from major weather stations close to a respective location as well as satellite data to construct the weather conditions with a grid size of 11 × 11 km [13].

3.2 Experiments 5 to 7 from April and May 2017

Electrical radiators with on/off control and a capacity of 0.8 kW (Experiment 7) and 1 kW (Experiment 5 and 6) were operated according to a pseudo random binary sequence (PRBS) [14, 15] in order to investigate the thermal dynamics of the building over a wide range of frequencies. This excitation sequence is typically applied for inverse modelling. A PRBS does not necessarily ensure comfortable thermal conditions inside a building. A more detailed description of these experiments can be found in [16], whereas an overview of the test settings is provided in Table 1.

The ventilation supply air temperature was set to 30 °C for the first experiment (E5) and changed to 18 °C in the beginning of the second experiment (E6). Ventilation heating was deactivated for the last days of Experiment 7. Windows were not blinded. Weather data was measured using the building embedded weather station. The Skartveit-Olsen method has been used to split the global horizontal radiation data (measured on-site) into direct horizontal and diffuse radiation [17].

4 Validation Results

The calibration of the building model was done based on the data set of Experiment 6 and with respect to the operative temperature aiming for a high correlation between the modelled and the measured temperatures. The model parameters (Table 2) were adjusted manually in an iterative manner, starting from their design values. The calibrated model has been validated using the data sets of six other experiments.

Two approaches have been applied to carry out the validation. For Experiments 1–4, the space-average of the measured indoor air temperature of the Living Room North and the Living Room South has been used as a set point for the electric radiator and the time-averaged measured electric power of the radiator was compared to the simulation results ("closed-loop" approach). In Experiments 5–7, the operative temperature of the rooms has been measured as a response to a pre-defined heating schedule ("open-loop" approach).

4.1 Dimensionless Indicators for Model Calibration

Two indicators for evaluating the model accuracy are the NMBE and the CVRMSE. The NMBE gives an indication of the total difference (percent error) between the measured and the predicted value from the simulations [18] and is calculated by

$$MBE = \frac{\sum_{i=1}^{n}(y_i - \hat{y}_i)}{n} \tag{1}$$

$$NMBE = \frac{MBE}{\bar{y}} \cdot 100 \tag{2}$$

where y_i and $\hat{y}_i$ are the measured and simulated value at instance i, $\bar{y}$ is the average of the measured data and n is the number of instances used in the calibration.

Table 2 Overview of tuned model parameters during calibration

Parameter	Starting value	Final tuned value	Range
U-value of external walls [W/(K m^2)]	0.1	0.1591	0.1–0.1591
g-value of windows (–)	0.2	0.3	0.1–0.5
Solar transmittance (–)	0.17	0.24	0.17–0.47
Internal emissivity of windows (–)	0.837	0.6	0.5–0.9
External emissivity of windows (–)	0.837	0.05	0.03–0.837
Thermal bridges [W/(K•m^{2}floor area)]	0.03	0.045	0.025–0.06
Infiltration rates (ACH)	0.3	0.7	0.3, 0.7
C_d flow coefficient for internal openings (–)	0.65	0.80	0.6–0.8

The CVRMSE is a measure for the goodness-of-fit of a model showing the variability between simulated and monitored data [1]. It is calculated by

$$RMSE = \sqrt{\frac{\sum_{i=1}^{n}(y_i - \hat{y}_i)^2}{n}} \tag{3}$$

$$CVRMSE = \frac{RMSE}{\bar{y}} \cdot 100 \tag{4}$$

The model is considered calibrated with NMBE <10% and CVRMSE <30%, if the model is calibrated with respect to hourly energy use for space heating [19]. Since there is no approved guideline that determines acceptance criteria for model calibration with respect to indoor temperatures, the same thresholds were applied in this study, even though the calibration was aiming for a NMBE and CVRMSE <5%. A threshold of 5% has also been applied in other studies [2, 4–6] and is thus used as a benchmark.

4.2 Calibration Data Set: Experiment 6 (Closed Bedroom Doors)

During that experiment, the ambient temperature varied between –2.9 and 17.2 °C, whereas the global solar radiation varied between 0 and 834 W/m^2. An on/off-controlled electric radiator was operated according to a specified PRBS signal (Fig. 3). The bedroom doors were closed during this experiment.

Figure 4 shows the trend of the operative temperature for Living Room South and Bedroom West at the end of the calibration. It can be seen that the model is reliable for predicting the temperature trend for cases with intermittent heating. If the electric radiator is turned off for a longer time (such as in the end of this experiment), the model predicts a faster temperature drop.

The overall UA-value of the building model is 83 W/K and is thus in the range of the values (70–100 W/K) identified during experiments conducted by Vogler-Finck et al. using inverse modelling [20]. If the UA-value of the model was

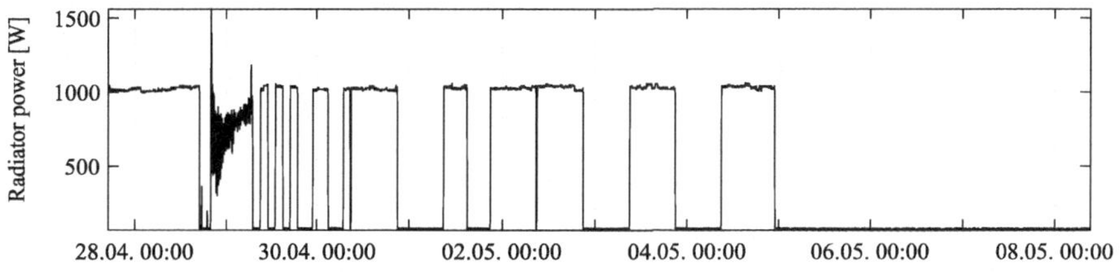

Fig. 3 Power supplied to the radiator during Experiment 6

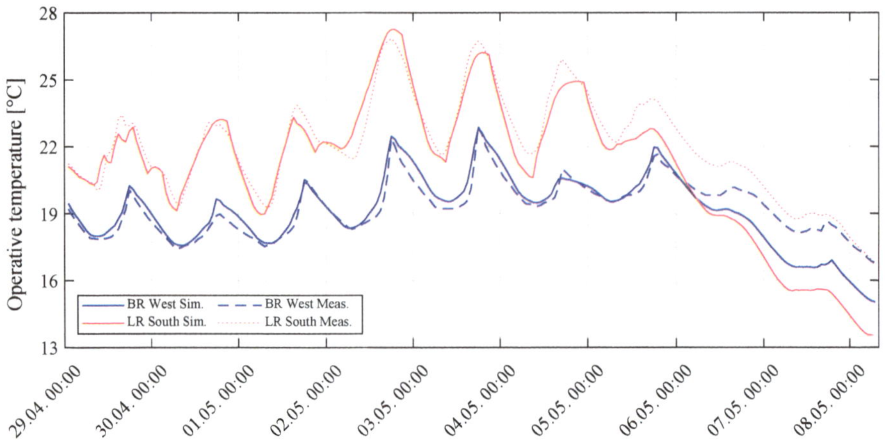

Fig. 4 Operative temperature trend in the Living Room South and Bedroom West at the end of the calibration

increased, the predicted temperature fluctuations would be even larger. The temperature fit for periods without any heating can be improved by considering the thermal mass of technical equipment (white goods) and by adjusting the properties of the PCM. Both will be investigated in further studies. The final NMBE and CVRMSE of Experiment 6 are shown in Table 3.

4.3 Example of a Validation Case: Experiment 7 (Open Bedroom Doors)

The model is also reliable for predicting the thermal dynamics of the building, if bedroom doors are opened. Figure 5 shows the trend of the operative temperatures in Living Room South and Bedroom West for the validation of Experiment 7. The temperature trend is similar for both Living Rooms as well as for both Bedrooms.

The model predicts the temperature fluctuations well for both rooms up to a temperature of 25 °C. It seems to slightly overpredict the maximum room temperature. This can be improved by tuning the thermal properties of the PCM. A summary of the results of all validation cases is given in Table 3.

The calibrated model predicts the temperature trend for all studied cases reliably, based on the NMBE and CVRMSE. Both indicators are below 8% for all studied cases, for most of the cases even below 5%. These results are in good agreement with other validation studies on residential buildings, where the building model was

Table 3 Summary of the NMBE and CVRMSE (for indoor air temperatures and operative temperatures) for all validation cases

Experiment	Room	NMBE (%)	MBE (°C)	CVRMSE (%)	RMSE (°C)	Energy use (kWh)	
						Measurement	Simulation
E1	BR East	5.57	0.94	6.05	1.02	51	42
	BR West	5.95	1.04	6.34	1.11		
	LR North	0.00	0.00	0.18	0.04		
	LR South	−0.56	−0.11	0.62	0.12		
E2	BR East	3.49	0.65	3.67	0.68	112	124
	BR West	2.56	0.47	2.72	0.50		
	LR North	−0.39	−0.07	0.75	0.14		
	LR South	0.74	0.14	1.26	0.24		
E3	BR East	3.47	0.64	3.92	0.72	65	73
	BR West	4.87	0.91	5.16	0.97		
	LR North	1.67	0.31	2.57	0.48		
	LR South	1.93	0.36	3.32	0.62		
E4	BR East	3.10	0.51	3.71	0.61	51	51
	BR West	6.76	1.20	7.00	1.24		
	LR North	1.11	0.21	2.67	0.51		
	LR South	1.60	0.30	2.99	0.57		
E5	BR East	−1.06	−0.22	3.61	0.74	PRBS signal given as model input and thus identical radiator power	
	BR West	0.14	0.03	3.23	0.67		
	LR North	−2.31	−0.47	4.59	0.93		
	LR South	−0.74	−0.15	5.23	1.07		
E6 (used for calibration)	BR East	−0.20	−0.04	3.82	0.71	PRBS signal given as model input and thus identical radiator power	
	BR West	0.80	0.15	4.17	0.80		
	LR North	3.54	0.77	6.46	1.40		
	LR South	4.56	1.01	7.61	1.68		
E7	BR East	−0.85	−0.19	3.71	0.83	PRBS signal given as model input and thus identical radiator power	
	BR West	−0.96	−0.21	3.81	0.84		
	LR North	−3.33	−0.74	4.49	0.99		
	LR South	−1.66	−0.38	3.47	0.79		

calibrated with respect to indoor temperatures [5, 6]. For experiments E2 and E3, the model predicts a slightly higher electricity use for keeping the indoor air temperature.

Only for Experiment 1, the model uses less electricity to keep the temperature set-point in the two living rooms. This discrepancy will be investigated in further studies.

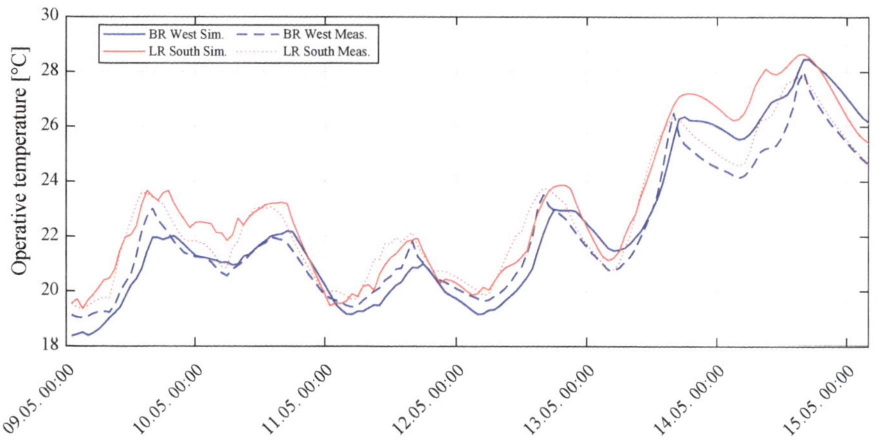

Fig. 5 Operative temperature trend in the Living Room South and the Bedroom West during experiment 7

5 Conclusion

The aim of this study was the calibration of a dynamic multi-zone building model of a super-insulated residential building located in Trondheim, Norway. Seven experiments have been used for the validation of the building envelope model (one for calibration and six for validation). The building model was calibrated on measurement results of a 9-day long experiment with respect to the operative temperature trend in different rooms. An on/off-controlled electric radiator with a capacity of 1 kW heated the building according to a pre-determined excitation sequence.

The calibrated model was validated against measurement data from six other experiments, where the validation was based on the NMBE and CVRMSE, revealing that both indicators are below 8% for all seven cases. The temperature trend of the calibrated model has been studied for cases with opened as well as closed internal doors. The model predicts the temperature trend for the different rooms reliably. The NMBE ranges between −3.33% (MBE is −0.74 °C) and 6.76% (MBE is 1.20 °C), whereas the CVRMSE is in a range of 0.18% (RMSE is 0.04 °C) to 7.61% (RMSE is 1.68 °C). Both indicators are 5% or lower for cases with open bedroom doors (E2, E3, E5, E7), which is in good agreement with other studies on calibrations with respect to indoor temperatures [2, 5, 6].

Compared to the measurements, the model predicts a faster temperature decrease, if the building is not heated (Fig. 4) even though the overall UA-value of the building model (83 W/K) is within the range of 70–100 W/K, which has been determined by Vogler-Finck et al. [20].

Therefore, further work will investigate the impact of additional thermal mass inside the building as well as the influence of the PCM on the indoor temperature trend. Furthermore, the impact of solar radiation on the indoor temperatures will be studied. A similar temperature trend for model predictions and measurements as response to intermittent heating will be important for the evaluation of ADR measures in future studies. The calibrated model will be used for testing different heating control strategies with regards to the energy flexibility potential that residential buildings can provide to the electricity grid.

Acknowledgements The authors would like to acknowledge IEA EBC Annex 67 "Energy Flexible Buildings" which was the platform of this collaboration. Access to the Living Laboratory was provided and funded by the *Research Centre on Zero Emission Buildings* project and its follow-up project *Research Centre on Zero Emission Neighborhoods in Smart Cities*. The PhD position of Pierre Vogler-Finck within the ADVANTAGE project is funded by the European Community´s 7th Framework Programme (FP7-PEOPLE-2013-ITN) under grant agreement no 607774.

References

1. E. Fabrizio, V. Monetti, Methodologies and advancements in the calibration of building energy models. Energies **8**, 2548–2574 (2015)
2. N. Zibin, R. Zmeureanu, J.A. Love, in *A Bottom-up Method to Calibrate Building Energy Models Using Building Automation System (BAS) Trend Data*. Aalborg Universitet CLIMA 2016—Proceedings of the 12th REHVA World Congress (2016)
3. F. Roberti, U.F. Oberegger, A. Gasparella, Calibrating historic building energy models to hourly indoor air and surface temperatures: methodology and case study. Energy Build. **108**, 236–243 (2015)
4. M. Taheri, F. Tahmasebi, A. Mahdavi, B. Ecology, in *Two Case Studies in Optimization-Based Thermal Building Performance Model Calibration*. Central European Symposium on Building Physics (2013)
5. P. Paliouras, N. Matzaflaras, R.H. Peuhkuri, J. Kolarik, Using measured indoor environment parameters for calibration of building simulation model—a passive house case study. Energy Procedia **78**, 1227–1232 (2015)
6. R. Simson, J. Kurnitski, K. Kuusk, Experimental validation of simulation and measurement-based overheating assessment approaches for residential buildings. Archit. Sci. Rev. **60**, 192–204 (2017)
7. F. Goia, L. Finocchiaro, A. Gustavsen, 7. Passivhus Norden | Sustainable Cities and Buildings The ZEB Living Laboratory at the Norwegian University of Science and Technology : a zero emission house for engineering and social science experiments (2015)
8. J. Clauß, I. Sartori, M.J. Alonso, M. Thalfeldt, L. Georges, in *Investigations of Different Control Strategies for Heat Pump Systems in a Residential nZEB in the Nordic Climate*. 12th IEA Heat Pump Conference 2017 (2017)
9. F. Kuznik, J. Virgone, Experimental investigation of wallboard containing phase change material: data for validation of numerical modeling. Energy Build. **41**, 561–570 (2009)
10. L. Georges et al., *Evaluation of Simplified Space-Heating Hydronic Distribution for Norwegian Passive Houses* (Trondheim, 2017)
11. EQUA, EQUA Simulation AB, 2015 [Online]. Available: http://www.equa.se/en/ida-ice
12. OpenStreetMap. Shiny weather data. [Online]. Available: https://rokka.shinyapps.io/shinyweatherdata/. Accessed on 20 May 2017
13. L. Lundström, in *Mesoscale Climate Datasets for Building Modelling and Simulation*. CLIMA 2016—Proceedings of the 12th REHVA World Congress (2016)

14. P. Bacher, H. Madsen, Identifying suitable models for the heat dynamics of buildings. Energy Build. **43**(7), 1511–1522 (2011)
15. H. Madsen et al., Thermal performance characterization using time series data—IEA EBC Annex 58 Guidelines (2015)
16. P. Vogler-Finck, J. Clauß, L. Georges, A dataset to support dynamical modelling of the thermal dynamics of a super-insulated building. Publication in progress (2017)
17. P. Schild, Personal communication (2017)
18. J. Granderson, S. Touzani, D. Jump, Assessment of Automated Measurement and Verification (M & V) Methods (2015)
19. ASHRAE, Measurement of energy and demand savings. *ASHARE Guidel. 14-2002* **8400**, 1–165 (2002)
20. P. Vogler-Finck, J. Clauß, L. Georges, I. Sartori, R. Wisniewski, Inverse model identification of the thermal dynamics of a Norwegian zero emission house (2017)

Simulation of Ventilation Rates and Heat Losses during Airing in Large Single Zone Buildings in Cold Climates

Abolfazl Hayati, Jan Akander and Magnus Mattsson

Abstract Airing can be a solution to introduce extra ventilation in large single zone buildings, especially where there are large aggregations of people such as churches or atriums. In naturally ventilated domestic and ancient buildings, opening of a window or door can introduce extra fresh air and remove particles and other contaminants emitted from people and other sources such as lit candles in churches. However, the energy use might be an issue in cold climates, where airing might lead to waste of heated air, at the same time as indoor air temperatures can be uncomfortably low. In the present study, the energy loss and ventilation rate due to airing in a large single zone (church) building is investigated via IDA-ICE simulation on annual basis in cold weather conditions. The results can be used in order to prepare airing guidelines for large single zone buildings such as atriums, churches, industry halls and large sport halls. According to the results, one-hour of airing in the studied church building resulted in 40–50% of exchanged room air and, if practiced once a week, an increase of around 1% in heating energy.

Keywords Airing (single-sided ventilation) · IDA-ICE simulation
Large single zones

1 Introduction

Natural ventilation is the outdoor flow penetrating into a building through purpose provided openings on the building envelope. There are different types of natural ventilation including single-sided, cross and stack ventilation [1]. Single-sided or cross ventilation are also called airing which is intentional air exchange through large openings such as windows and doors [2]. That is, when the opening(s) are located on the same façade, i.e. single-sided ventilation, or on different facades, i.e.

A. Hayati (✉) · J. Akander · M. Mattsson
Faculty of Engineering and Sustainable Development, Department of Building, Energy and Environmental Engineering, University of Gävle, Gävle, Sweden
e-mail: abolfazl.hayati@hig.se

D. Johansson et al. (eds.), *Cold Climate HVAC 2018*,
Springer Proceedings in Energy, https://doi.org/10.1007/978-3-030-00662-4_62

cross flow. Airing can be used for refreshing the interior air and extracting the pollutants especially after aggregations or occasions when there are many people and—in case of churches—lit candles. Airing can be used alone in naturally ventilated buildings or even as a complement for mechanical ventilation at schools, for example see [3], especially in occasions when there are many people present at the same time and the high amount of CO_2 should be diluted.

Airing is a complicated phenomenon. There are different parameters that affect the airing flow rates, such as the terrain surrounding a building, the position of the opening in a building envelope as well as opening size (height, width and depth), building air tightness, distribution of the leakage on the envelope, weather conditions including wind speed and direction and its turbulence. There are other factors, which might limit the applicability of the airing occasions such as ingress of noise and pollutants.

Driving forces for airing are buoyancy and wind effect in a naturally ventilated building. The temperature difference between inside and outside induces a pressure difference across the opening. However, wind effect is more complicated; and not only the average wind velocity but also turbulences in the wind affect and induce airflows [2]. Wind direction is a deciding parameter. Previously performed tracer gas measurements showed that the single-sided windward flows can be up to more than two times larger than the leeward flow, i.e. when the porch is located in the leeward side [4].

Hayati [2] also investigated the airing flow rate via model studies in a wind tunnel and found higher air flows when the porch is located in the windward side in comparison to the cases when the porch is located in the leeward side. Different models have been developed for combining these effects and make a prediction of the total air flow through openings, see for example [5–7]. A summary of the airing models for single-sided ventilation through large vertical openings is presented by Hayati et al. [4].

Airing and the models for airflow through large vertical openings is validated by Hayati et al. [8, 9], especially the reliability of the models used in IDA Indoor Climate and Energy (IDA-ICE) for predicting the single-sided flows in large single zone buildings. IDA-ICE is software for simulation of energy use and indoor climate for individual zones as well as the whole building. IDA-ICE involves dynamic simulation of a single or multi-zone building with a system of boilers, chillers and air handling unit(s). Room units (heaters/coolers) and local air handling units are also available. Neutral Model Format (NMF) is used as the coding language for implementing the mathematical models in the program. The building model can be supplied with synthetic or actual weather file. An actual weather file in IDA-ICE consists of time (hour), air temperature (°C), relative humidity (%), wind speed (m/s) and direction, direct normal beam radiation (W/m^2) and diffuse radiation on horizontal surface (W/m^2). The models simulate air containing both CO_2 and humidity. In IDA-ICE, airing flow is calculated based on the Bernoulli equation (the so-called orifice flow equation) by assessing the wind pressure difference at the top and bottom of the opening. The program takes into account both buoyancy and wind (caused by averaged wind speed) effects [9]. Air leakage is modelled by means of power law equations [10].

Ancient buildings like churches are naturally ventilated via leakage, i.e. air infiltration or via openings such as windows or doors. Normally there are also limited possibilities to add some mechanical ventilation or tightening the envelope in such buildings because of esthetical and preservation aspects. Other concerns include air humidity and particle deposition, which might deteriorate different pieces of art inside. Napp and Kalamees [11] studied the possibilities of implementing different climate control systems in churches by performing IDA-ICE simulations, addressing above mentioned restrictions.

The energy loss due to airing in cold climate is investigated in different studies with the aim of introducing a national standard value to be used in residential building energy simulations [12]. Different suggestions in order to compensate the airing effect on energy use include the addition of 4 kWh/(m^2·year) to the yearly specific energy use (for residential buildings) or adding the equivalent airing flow to the air infiltration rate or mechanical ventilation rates [13–16]. Among the suggested compensating methods, addition of energy is recommended because airing is a temporary measure in order to increase the flow rate and refresh the interior occasionally when it is needed. In office buildings, there is no default value suggested since airing is considered to be negligible [17]. However, air flow through entrance or garage doors through which many persons are passing may be significant but due to the lack of previous studies, no standard value can be put forth.

The aim of this study is to investigate the effect of airing on energy use in naturally ventilated single-zone church which is located in cold climates. By means of simulations with IDA-ICE, a validated model of the church is used to predict increase in specific energy use due to one hour of airing after Sunday mass (every week per year). This also involves assessment of air exchange in the zone during the airing occasions.

2 Method

The original model was from a church located in Hamrånge, mid Sweden [9]. The same model was simulated, using weather data from Kiruna, in the north of Sweden and Gothenburg (Göteborg), located in the south-west of Sweden. The church constitutes a great hall with thick stonewalls, has rendering on both inside and outside, and with double outer doors to enter the large hall. It is equipped with gable roofs and inner ceilings that are plastered on the inside and well insulated on the outside with wind barrier- coated mineral wool towards a naturally ventilated attic. Windows are double-glazed and weather-stripped. The church has a crawl space underneath a wooden floor, consisting of double boards with a ~15 cm layer of lime sand in between. The church is naturally ventilated through leakages in the building envelope. Size characteristics are summarized in Table 1. The interior zone of the church is not perfectly cuboid since ceilings are vaulted and resemble more or less semi-cylindrical or semi-spherical shapes. Thus, the ceiling height is not a fixed value. However, the simulated model includes simplifications of the actual church regarding the interior shape.

Table 1 Size characteristics of church

Volume (m^3)	Floor area (m^2)	Ceiling area (m^2)	Wall area (m^2)	Max ceiling height (m)	Average ceiling height (m)
7620	695	862	1188	13.7	11.0

The church model was validated against measurements in Hamrånge regarding both the indoor temperatures and the airing flow rates [9]. The weather data for Hamrånge, used in this study, was obtained from Swedish Meteorological and Hydrological Institute (SMHI) available at [18]. The weather data used in the current study for Kiruna and Gothenburg were obtained from the ASHRAE IWEC 2 database, which contains "typical" weather files for 3012 locations available for direct downloading via the IDA-ICE program. The files are also derived from Integrated Surface Hourly (ISH) weather data originally archived at the National Climatic Data Center [10].

Air flow through large vertical openings like doors has been investigated in previous studies [8, 9] and the simulation results were compared with the airing rates, measured by tracer gas decay method in the interior, the main hall of the church. The studies [8, 9] showed that the simulated single-sided flows were in the same order of magnitude of the measured ones, although slight over-prediction was observed when the porch was in the windward as well as slight under-prediction when the porch was on the leeward side. However, larger over-prediction was observed in case of cross flows. But overall, it seems that IDA-ICE simulation regarding single-sided airing flow is adequate enough to be used in this study for other climates, i.e. the cold climate of Kiruna and windy Gothenburg, in order to investigate the energy loss and the ventilation flow rates during the airing period.

The church was modeled and simulated in IDA-ICE simulation program version 4.7.1. There, the building was divided into six different zones, including main hall, crawl space, main entrance and the tower, sacristy, attic room and the storage room as shown in Fig. 1. The thick external walls consist of 0.85 m stone with an assumed U-value of 0.4 W/(m^2·K) and default values were used for roof, floor and inner walls in IDA-ICE. Each window is 2.6 m times 4.7 m large and has a U-value of 1.9 W/(m^2·K), solar heat gain value of 0.68 and solar transmittance value of 0.6. The side porches are 1.9 m tall and 2.9 m wide. The building is quite wind exposed. The default values of pressure coefficient (varying with wind direction) for wind exposed building in the program were used in this study; the data are available directly in the program and are originally obtained from handbook data set of the Air Infiltration and Ventilation Centre [19, 20].

The occupancy schedule for the main hall was set as 20 people present on Sundays from 10:00 to 15:00 and 2 people present during the rest of the days from 8:00 to 16:00 with half an hour lunch break. The side porch of the church (located on the western façade) was opened one hour directly after each aggregation on Sundays (from 15:00 to 16:00) throughout the whole year. The activity level was assumed as 1 MET for all occupants. No occupancy was assumed for the rest of the zones. The only room units were in the main hall, consisting of 20 electrical bench

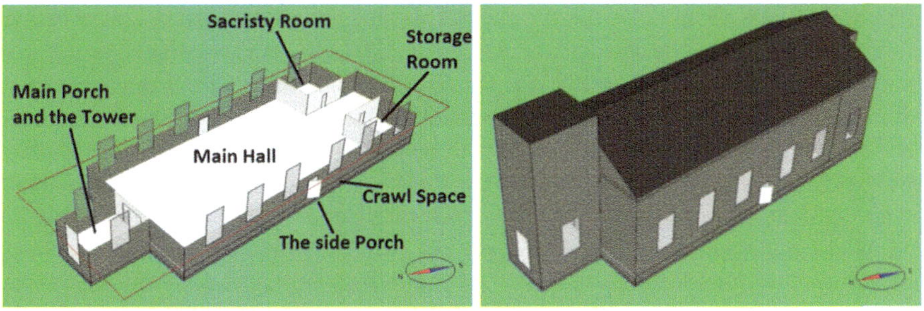

Fig. 1 Hamrånge church model in IDA-ICE

heaters with 120 kW total power. The heaters were coupled with thermostats working between 15 and 16 °C. Total lighting of 2 kW and 1 kW equipment power was set for the main hall with the same schedule for the occupants. For the storage room and sacristy, 0.5 kW electric heaters as well as 0.2 kW lighting was assumed.

From blower door tests, the permeability of the building envelope was obtained as 3.64 l/(s·m^2) at 50 Pa pressure difference (corresponding to 3.42 ACH) and the flow exponent was obtained as 0.86. In IDA-ICE, there is the possibility of entering permeability in form of evenly distributed air leakages. However, the choice was to model air leakage as unevenly distributed effective leakage areas in the facades based on field observations including IR-thermography and analytical model studies so that almost 50% of the leakage occurs through the floor, 25% through the surrounding walls of the main hall and 25% through the ceiling [21]. Thus the total effective leakage area, approximately 0.32 m^2 at 4 Pa (according to [22]) based on blower door test results, was distributed to the different parts at these percentages, see [9].

3 Results and discussion

The results from IDA-ICE simulation are presented and discussed here. As the heating system inside the church is controlled by thermostats and ideal heaters are used in the simulation, the inside temperatures are between 15 and 16 °C, as expected; and this is in line with measured indoor temperatures during airing periods [9]. The ideal heaters in IDA-ICE deliver the given heating capacity to the zone independent of the type of the actual room unit (radiator, convector or etc.); physically, an ideal heater can be considered as a standalone electric heater. Therefore, the focus of the results is on the airflow rates and the percentage of the exchanged room air during the airing period as well as the energy loss due the same airing periods.

The annually averaged air inflow and outflow from the main hall are depicted in Figs. 2 and 3. The diagrams include airflow both through the leakage in the rest of the building envelope and through the open porch. The air in- and exfiltration depicted in Figs. 2 and 3 is the air leakage during the airing periods, i.e. when the porch was opened. The airing model used in IDA-ICE was validated in previous studies [8, 9] which shows that the simulated single-sided air flow rates are of similar magnitude to the measurement data, however the effect of wind direction is less clear in the simulations, maybe due to ignorance of wind turbulence. However, the simulations are run throughout the year and include both windward and leeward cases, and at least regarding the total flow rate and energy loss, the simulated single-sided airing flow rates are judged to be fairly reliable. The difference in the air in- and outflow in Figs. 2 and 3 is due to the difference in the air density between the fresh colder air entering the main hall and the warmer interior air exiting the hall. Noticing Fig. 3, the reason for the high exfiltration in Gothenburg can be due to the dominating western winds (blowing from the sea on the west side of city) that increase the airing through the westward porch, which in turn pushes the air via the leakages in the rest of the building, i.e. air exfiltration.

The climate data (including the wind speed and the outdoor temperature) as well as the exchanged room air during porch airing are also depicted for each location (averaged over each month), see Figs. 4 and 5. The exchanged room air shown in Fig. 5 is the total inflow through both the porch and the leakage in the building envelope.

Noticing Figs. 2 and 3, higher flow rates are observed in Kiruna; the colder climate of Kiruna (see Fig. 4) which increases buoyancy forces, is the likely reason for larger airing values. The outside temperature is slightly lower for the Hamrånge case in comparison to the Gothenburg climate, and despite of having higher wind speeds in the coastal city of Gothenburg, both cities have almost the same airing flows, see Figs. 2 and 3. This can be due to that the wind driven flows are connected with the wind pressure coefficients in IDA-ICE, which in turn depend on the wind direction. As only wind speed is illustrated in Fig. 4, further studies are

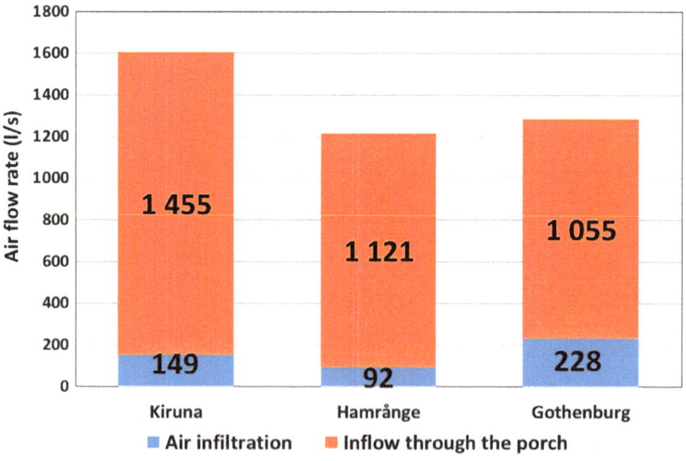

Fig. 2 Air inflow to the main hall, averaged over the whole year due to 1-h airing on each Sunday

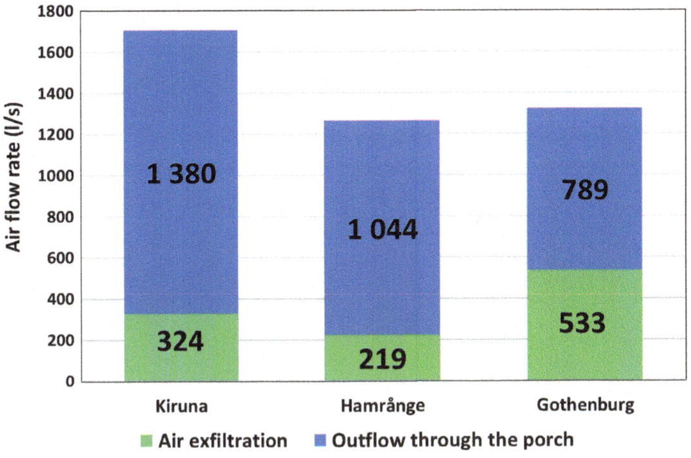

Fig. 3 Air outflow from the main hall, averaged over the whole year due to 1-h airing on each Sunday

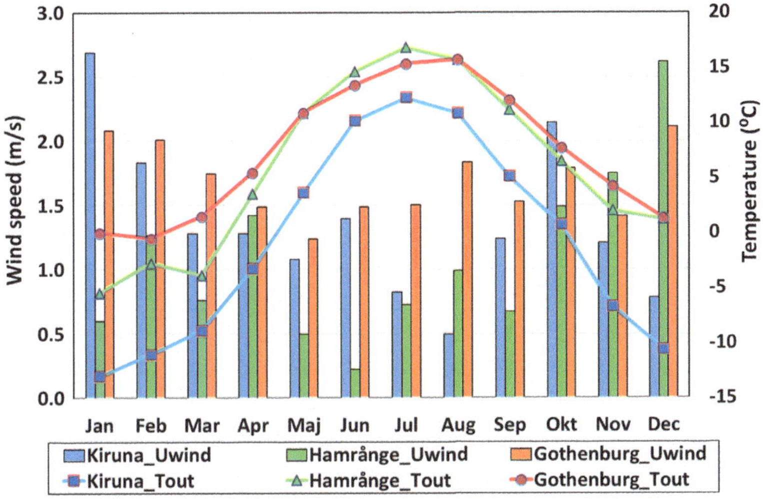

Fig. 4 Monthly averaged weather data including wind speed (Uwind) and outdoor temperature (Tout) for Kiruna, Hamrånge and Gothenburg

recommended in order to include the wind directions as well other cities/climates in order to analyze the energy loss and the amount of airing flows. Moreover, surveying the airing habits in churches is also recommended in order to map the energy loss due to airing. What is evident from this study is that buoyancy forces seem to influence air exchange to a larger degree than what wind driven forces do.

According to Fig. 5, it appears that one hour airing will cause an exchange corresponding to 30–60% of the volume (monthly averaged values). Averaged over

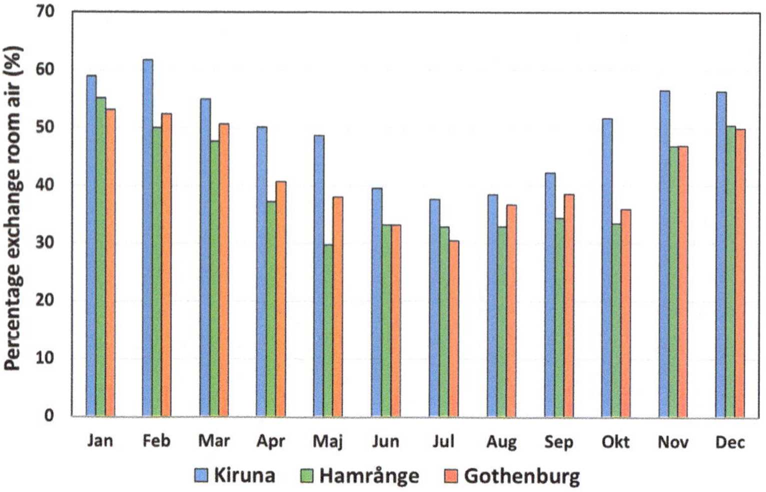

Fig. 5 Percentage exchanged room air after 1 h airing and infiltration into the main hall, averaged over the Sundays of every month

the year for each location, the results show that one hour of airing during the weekend causes between approximately 40–50% exchanged room air in average, which is recommended after each ceremony in the church where there might have been many people and lit candles. Also, bearing in mind that special attention should also be taken to the relative humidity of the outside weather. If the indoor surface temperatures are lower than the indoor air dew-point in that occasion, there is condensation risk which might deteriorate paintings, furniture and other artifacts and therefore airing should be avoided. However, a study by Hayati [2] has shown that there is minor risk of condensation during shorter airing periods in Sweden, even in summer seasons with high relative humidity.

The annual heating of the main hall in Hamrånge church, without considering airing, is simulated as 185, 88 and 83 kWh per m^2 floor area if the church is located in Kiruna (north of Sweden), Hamrånge (mid-Sweden) and Gothenburg (south-west of Sweden), respectively. For comparison, the simulations indicate that one hour porch-airing every week (on Sundays after aggregations) in the church causes an increase in specific energy use of 2.1, 1.1 and 0.8 kWh/(m^2·year) if the church is located in Kiruna, Hamrånge and Gothenburg, respectively. There are no statistics or questionnaires concerning airing habits in churches or similar buildings; however, airing is recommended after each aggregation in order to remove the particles and moisture from people and lit candles and also because the airing effect on energy loss is negligible as increase in the specific energy due to airing is only around 1%, as appears from the data above. The suitable duration of airing depends on many factors, such as the weather conditions outside as well the outside conditions regarding the pollutants, moisture and noise. One-hour duration is used in the current simulations in order to compare different climate regions in a cold climate such as Sweden. The simulated values are lower but of the same magnitude as the standard specific energy use additional value of 4 kWh/(m^2·year) for airing in residential buildings. However, if the number of aggregations is increased to occur on a daily basis, the specific energy use will increase by seven times and consequently the energy loss due to airing will increase to approximately 7%.

4 Conclusion

A church model was built in the IDA Indoor Climate and Energy (IDA-ICE) simulation program with the aim of investigating the airing flow and the related energy loss. One-hour porch airing is simulated for a large single zone church building located in Kiruna (north of Sweden), Hamrånge (mid-Sweden) and Gothenburg (south-west of Sweden). The results indicate that one hour porch-airing every week (on Sundays after aggregations) during a year causes an increase in specific energy use of 2.1, 1.1 and 0.8 kWh/(m^2·year) (corresponding to only 1% of the total heating demand) if the church is located in Kiruna, Hamrånge, and Gothenburg, respectively. That is, the heat loss is roughly two times higher in the cold climate of Kiruna. The magnitude of that heat loss is however quite small compared to the total yearly heat loss.

According to the simulations, one-hour airing results in 40–50% (on average ranging from 30–60%) of exchanged room air. Thus, airing seems to be a workable ventilation method in churches and similar kinds of naturally ventilated buildings, in order to introduce fresh air and evacuate contaminants emitted from e.g. people and lit candles, especially after aggregations.

Further investigations and improvements of the models used for airing are recommended as well as studying the airing habits in churches and similar naturally ventilated buildings.

Acknowledgments The authors would like to thank for the financial support from the Swedish Energy Agency (Dnr 2011-002440) and the University of Gävle. Thanks also to technical engineers Svante Lundström and Elisabet Linden for all the assistance during the field measurements.

References

1. H. Awbi, *Ventilation Systems: Design and Performance* (Taylor & Francis, London, 2008)
2. A. Hayati, Natural Ventilation and Air Infiltration in Large Single Zone Buildings : Measurements and Modelling with Reference to Historical Churches, Gävle University Press, 2017. http://hig.diva-portal.org/smash/get/diva2:1117979/FULLTEXT01.pdf.
3. B. Nordquist, Ventilation and window opening in schools, Experiments and Analysis, Lund University (2002)
4. A. Hayati, M. Mattsson, M. Sandberg, Single-sided ventilation through external doors: measurements and model evaluation in five historical churches. Energy Build. **141**, 114–124 (2017). https://doi.org/10.1016/j.enbuild.2017.02.034

5. W. De Gids, H. Phaff, Ventilation rates and energy consumption due to open windows: a brief overview of research in the Netherlands. Air Infiltration Rev. **4**, 4–5 (1982)
6. T.S. Larsen, P. Heiselberg, Single-sided natural ventilation driven by wind pressure and temperature difference. Energy Build. **40**, 1031–1040 (2008). https://doi.org/10.1016/j.enbuild.2006.07.012
7. P. Stabat, M. Caciolo, D. Marchio, Progress on single-sided ventilation techniques for buildings. Adv. Build. Energy Res. **6**, 212–241 (2012)
8. A. Hayati, M. Mattsson, M. Sandberg, in *A Study on Airing Through the Porches of a Historical Church–Measurements and IDA-ICE Modelling*. ASHRAE AIVC IAQ 2016—Defining Indoor Air Qual. Policy, Stand. Best Pract (Alexandria, Virginia, USA, ASHRAE, 2016) pp. 216–223, 12–14 Sept 2016
9. A. Hayati, Measurements and modeling of airing through porches of a historical church. Sci. Technol. Built Environ. (2017). https://doi.org/10.1080/23744731.2017.1388132
10. EQUA AB, IDA Indoor Climate and Energy (2017). http://www.equa.se/en/ida-ice. Accessed on 1 Aug 2017.
11. M. Napp, T. Kalamees, Energy use and indoor climate of conservation heating, dehumidification and adaptive ventilation for the climate control of a mediaeval church in a cold climate. Energy Build. **108**, 61–71 (2015)
12. SVEBY, Brukarindata bostäder Version 1.0. Branschstandard för energi i byggnader (Svebyprogrammet, Stockholm, 2012). http://www.sveby.org/wp-content/uploads/2012/10/Sveby_Brukarindata_bostader_version_1.0.pdf.
13. J. Eriksson, Å. Wahlström, Reglerstrategier och beteendets inverkan på energianvändningen i flerbostadshus, Rapport från Effektiv 2001:04, SP (2001)
14. Stockholms stads LIP-kansli, MEBY, Bilaga 2. Kommentarer och underlag till kravspecifikationen (2002)
15. Stockholms stads LIP-kansli, Teknikupphandling av energiberäkningsmodell för energy-effektiva sunda flerbostadshus (MEBY). Anbudsunderlag (2002)
16. E. Sandberg, K. Engvall, MEBY-projektet, delrapport 3, Beprövad enkät – hjälpmedel för energiuppföljning (2002)
17. SVEBY, Brukarindata bostäder Version 1.1. Branschstandard för energi i byggnader. Stockholm: Svebyprogrammet (2013). http://www.sveby.org/wp-content/uploads/2013/06/Brukarindata-kontor-version-1.1.pdf
18. L. Lundström, Shiny Weather Data (2016). https://rokka.shinyapps.io/shinyweatherdata/. Accessed 25 Aug 2017
19. AIVC, Wind pressure workshop proceedings, in: AIVC Tech. Note 13.1, Air Infiltration and Ventilation Centre (AIVC), Brussels, Belgium (1984)
20. M. Orme, M.W. Liddament, A. Wilson, *Numerical Data for Air Infiltration & Natural Ventilation Calculation* (Coventry, UK, 1998)
21. A. Hayati, M. Mattsson, M. Sandberg, Evaluation of the LBL and AIM-2 air infiltration models on large single zones: three historical churches. Build. Environ. **81**, 365–379 (2014). https://doi.org/10.1016/j.buildenv.2014.07.013
22. ASHRAE, Chapter 16: Ventilation and infiltration, in: ASHRAE Handb. Fundam. 2013, American Society of Heating, Refrigerating, and Air-Conditioning Engineers, Atlanta, GA (2013)

User Related Input Data for Energy Usage Calculations the Case of Low Energy Schools in Sweden

Branko Simanic, Dennis Johansson, Birgitta Nordquist and Hans Bagge

Abstract In order to calculate building energy usage, apart from the technical characteristics, user related factors needs to be determined. Unless the user related factors are determined by specific project, the idea is to apply a standardized list of input data for a normal operation during a normal year, so the calculated energy value does not depend on variation of these factors. Such list was issued by Boverket (The Swedish National Board of Housing, Building and Planning) in a document named BEN1 in 2016, and updated in BEN2 in 2017. A disadvantage of this list is that, the part about schools is based on references that are rather older and sparse and needs to be updated. This paper investigates the user related input data in 10 newly built low energy schools in Sweden and compares those to BEN2. It also compares the schools' calculated energy performance to the BBR25 requirement, the latest national building codes and recommendations. The schools are investigated in this research as there is a demand for about 1000 elementary schools to be built in Sweden in the coming 10 years. The paper shows significant user related effect to energy usage and importance for the standardized user related input data for energy calculations. Future research aims to verify these calculations and user related input data with measured data for the chosen schools.

Keywords Low energy schools · User related input data · Calculated energy

---

B. Simanic (✉) · D. Johansson · B. Nordquist
Building Services, Faculty of Engineering LTH,
Lund University, Lund, Sweden
e-mail: branko.simanic@skanska.se

H. Bagge
Building Physics, Faculty of Engineering LTH,
Lund University, Lund, Sweden

B. Simanic
Skanska Sverige AB, Svedala, Sweden

D. Johansson et al. (eds.), *Cold Climate HVAC 2018*,
Springer Proceedings in Energy, https://doi.org/10.1007/978-3-030-00662-4_63

1 Introduction

The total energy needed for a building is influenced, not only by technical characteristics, but also by user related factors such as tenant electricity and domestic hot water (DHW) usage. When focusing on the technical design stage, it is therefore important to have the same basic conditions for user related factors, so that the calculated energy value does not depend on variation of these factors. Unless there are specific reference values for the building project, the ideas is to apply the same reference values for all objects. For this reason Boverket (The Swedish National Board of Housing, Building and Planning) has issued user related input data for the calculation of energy usage during normal operation and a normal year for a new building construction or retrofit in a document named, BEN2 [1]. A disadvantage of this list is that the part for elementary and secondary schools is based on older references which needs to be reviewed [2].

The lack of standardized user related input data has been identified as a big challenge for data collection [3]. One line of research has focused on particular parameter such as occupancy [4–6]; DHW usage in multifamily dwellings [7, 8] and artificial light control and another line has focused on probabilistic building simulation models in order to predict occupant behavior influence on energy usage [9, 10]. However, there are few studies considering user related input data for schools [4, 5].

This paper investigates user related input data in several recently built low energy schools and ongoing projects in Sweden, and compares these to BEN2. It also compares the schools' theoretically calculated energy performance to BBR25 (Swedish national building codes and recommendations) [11] requirement, the latest national building codes and recommendations, issued by Boverket, which is heading towards near-zero energy building (N-ZEB) codes. We assume that these low energy schools will meet the N-ZEB requirements which are to be implemented by 2021, according to EU/2010/31 directive, article 2 [12]. Low energy schools are determined as their calculated energy usage is 75% of the applied BBR requirement. Only elementary schools are considered and are presented by their code names as they want to stay anonymous.

This paper considers the elementary schools only as there is a demand for about 1000 elementary schools to be built in Sweden in the coming 10 years which is a consequence of population growth [13]. By 2024 population will be increased from 10 to 11 million [14]. 20 large ongoing school projects were listed in Byggvärlden, a building construction magazine, in Sweden 2016 [15]. 30 new schools will start to be built in Gothenburg, the second largest city in Sweden, in 2019 [16] (Table 1).

Table 1 Nomenclature used in this paper

Atemp	Space heated area above 10 °C in square meters
BBR	Swedish national building codes and recommendations
BEN	Swedish national codes and recommendation in order to determine building energy usage during normal operation and normal year
DH	District heating
DHW	Domestic hot water
E_{BP}	Energy for building property (electricity), not tenant electricity
E_{cool}	Energy for space cooling
E_{DHW}	Energy for DHW production
E_{SH}	Energy for space heating
EP_{pet}	Energy performance requirement
F_{geo}	Geographical correction factor
GSHP	Ground source heat pump
HDH	Heating degree hours (°C h/year)
IAQ	Indoor air quality
$q_{average}$	Average fresh air flow (l/(s m^2))
PE	Primary energy factor for energy source
PV	Photo voltaic
N-ZEB	Near-zero energy buildings
SH	Space heating
VAV	Variable air volume ventilation

2 Swedish National Building Codes and Recommendations for Energy Use, BBR and BEN

In BBR25 in Chap. 9 (dedicated to energy performance), Boverket has introduced a primary energy as the energy usage requirement. Primary energy factors (PE) are 1.6 for electricity and 1 for other energy sources (district heating/cooling, biofuel, oil, and gas). By 2021 these factors will probably turn into 2.5 for electricity and 1 for the other energy sources [17]. In BBR25 EP_{pet} requirement for school building is $80\,\mathrm{kWh}/(\mathrm{m}^2_{\mathrm{Atemp}}\ \mathrm{year})$ plus supplement for air flows, see Eq. 1. EP_{pet} includs energy for space heating/cooling, domestic hot water DHW and building property electricity E_{BP} to drive HVAC system, BMS, elevators, roof defrosting etc., see Eq. 2. EP_{pet} does not include tenant electricity. To EP_{pet} is added a factor that covers geographical/climate differences for space heating F_{geo}, where the lowest value is 0.8 and the highest is 1.9. Overall U value requirement in BBR25 is 0.6 W/(m^2 K).

In order to obtain building construction permit from the local authority a project must submit theoretically calculated EP_{pet} which must comply with actual BBR requirement. To calculate EP_{pet}, a project specific users related input data can be used or a list of input data for e.g. school buildings for normal operation during a normal year. Such list is issued by Boverket in BEN1 [18] and the latest updated in BEN2 [1]. Prior to BEN the Sveby (Swedish building industry standards for energy and buildings), introduced a list of user related input data for educational buildings [2] in May 2016.

There are several types of buildings listed in BEN2: dwellings, residential buildings, offices, children day cares, elementary school and gymnasiums, high schools and universities [1].

Primary energy number EP_{pet} requirement for commercial buildings (which addresses school buildings) is:

$$EP_{pet} = 80 + 70\left(q_{average} \quad 0.35\right) \tag{1}$$

where $q_{average}$ is average fresh air flow liter/(s m^2_{Atemp}) during heating season, where maximum can be 1. EP_{pet} (kWh/(m^2_{Atemp} year)).

Primary energy number EP_{pet} is calculated as:

$$EP_{pet} = \sum_{i=1}^{6} \left(\left(E_{SH,i}/F_{geo} \right) + E_{cool,i} + E_{DHW,i} + E_{BP,i} \right) PE_i / A_{temp} \tag{2}$$

where "i" is type of energy source. EP_{pet}, E_{SH}, E_{cool}, E_{DHW}, E_{BP} in (kWh/(m^2_{Atemp} year)), A_{temp} in (m^2_{Atemp}), F_{geo} and PE are constants.

3 Comparison Among Several Low Energy School Projects

The energy performance of 10 Swedish elementary schools compares to BBR25 energy requirement and their user related input data to BEN2 in Tables 2 and 4. These schools were built or have being built under the older versions of BBR. Neither BEN1 nor BEN2 were used in these school projects. The school "A", has used Sveby's input data list [2] in the later stage of its' construction. Energy engineers in these school projects estimated user related influence usually based on their experience and/or project specific data.

The chosen schools applied earlier versions of BBR. In these versions the energy performance requirement were based on purchased energy and there were different requirements if electricity is used for the space heating or not.

The schools are chosen by several criteria. The most important is to be a low energy school and recently built, and their calculated energy usage must be 75% of actual BBR requirement, as their energy performance can comply with year 2021 energy requirements. Another criteria is an established measurement data export from the building management system, as the schools are going to be monitored

Table 2 Energy performance of the listed schools

School name (code name)	Energy performance according to applied BBR (purchased energy)		EP_{pet} according to BBR25 (primary energy)	
	Requirement/ $(kWh/(m^2_{Atemp}\ year))$	Calculated/ $(kWh/(m^2_{Atemp}\ year))$	Requirement/ $(kWh/(m^2_{Atemp}\ year))$	Calculated/ $(kWh/(m^2_{Atemp}\ year))$
S	55	38	80	70
N	91	62	91	75
K	80	33	80	44
B	75	38	80	60
Va	80	51	84	68
Vi	80	51	80	60
Ve	88	40	88	52
L	60	34	88	56
A	65	26	110	50
G	65	54	80	61
Average	74	43	86	60

during one year period. Each school projects have calculated their energy demand based on the project specific data and the projects estimations.

Among the projects the energy performance requirement varies, as there have been several BBR issued during the last several years, where the requirements for electrically heated and non-electricity heated buildings were separated, see first column in Table 2.

Mainly IDA ICE [19] a dynamic computer simulation tool was used to calculate energy usage. Another simulation tool used in one school was VIP Energy [20].

Table 2 compares calculated energy usage according to applied BBR for each school project and to BBR25 requirement which has been calculated in this paper. The schools' energy calculations were made for normal usage and normal year. Estimated electricity production by solar photovoltaic do not account in calculated energy usage in this paper. Table 2 shows that the calculated energy is about 75% of the applied requirement, so the schools meet BBR25 EP_{pet} requirement which makes them relevant in this study.

Table 3 shows data such as building year, size, climate, shape, type of ventilation, thermal resistance and source of space heating and DHW production.

The shape of the school buildings are pretty similar, usually made of several rectangular blocks joint together by corridors and smaller rooms. They are mainly two floors buildings, a few are with one floor only. Almost all of them have sport hall and fully equipped kitchen facility with dining hall, apart from "Vi" and "L". "Ve" and "G" do not have sport hall. Each block has usually its own centralized ventilation system. The schools do not have separate comfort cooling systems.

Table 3 Building year, size and some other features of the listed schools

	In operation since	Floor area/ (m^2)	HDH/(°C h/y)	Envelop area/ volume ratio/(m^2/ m^3)	Ventilation system	Overall U value/ windows U value/(W/ (m^2K))	Heating energy source/ renewable energy
S	Aut. 2016	4764	102600	0.38	VAV	0.23/0.9	GSHP
N	Spr. 2017	8125	92800	0.44	VAV	0.45/0.85	DH
K	Aut. 2016	11222	102600	Unknown	VAV	0.18/0.8	DH/Solar PV
B	Aut. 2014	8051	111500	0.39	VAV	0.2/0.9	GSHP
Va	Aut. 2017	9000	111500	0.31	VAV	0.37/0.9	DH
Vi	Aut. 2016	1725	118000	Unknown	VAV	0.21/0.9	DH
Ve	Aut. 2015	3233	92800	Unknown	VAV	0.21/0.8	DH/GSHP/ Solar H
L	Aut. 2016	898	99600	Unknown	VAV	0.26/1	GSHP/Solar PV
A	2018	9073	92800	0.41	VAV	0.28/1	GSHP/Solar PV
G	2018	4690	92800	0.37	VAV	0.3/0.9	DH/Solar PV

Average annual outdoor temperature varies from 5.5 to 7 °C. Heating degree hours (HDH) numbers are taken from Swedish standard HVAC tables [21]. Some schools are at the very south of Sweden, two of them are as north as Stockholm, and others are in between.

Table 4 shows user related input data from the chosen schools and BEN2. The air flow column indicates operational time for the ventilation system and its air flow. There are basic (minimum) and forced (maximum) air flows. In the simulation/practice, the air flows are somewhere between basic and forced air volumes as the VAV system is implemented in the schools. The ventilation system in the kitchen and the dining hall can have a shorter and the sport hall longer operational hours. Almost each classroom has its own airflow's control system that can depend on: presence detectors, room air temperature and/or CO_2 level. One or the most two parameters to control the VAV system are usually used.

The sun shading column includes both fixed and variable sun shading. The variable solar shading includes interior and exterior solar shading devices, which can be influenced by the users. Two different numbers per school, in Table 3, means that there are different shading system depending on façade orientation.

The DHW includes thermal losses for the DHW recirculation circuits. BEN2 excludes these losses in the DHW number. The kitchen's energy is included in the tenant electricity and its' DHW usage is a process energy. These two are not included in BBR's energy requirement.

The airing column presents energy losses due to opening of windows/doors by users. The value $4\,\mathrm{kWh/(m^2_{Atemp}\,year)}$ is based on earlier reference [22].

The tenant electricity (lighting and equipment) is not included in EP_{pet} calculation but it emits free heat and contributes to space heating. In some schools lighting and equipment are estimated to be on during 52 weeks/year, in one is 39 weeks/year, the others are in between. The tenant electricity can be as low as $8\,\mathrm{kWh/(m^2_{Atemp}\,year)}$ and as high as $35\,\mathrm{kWh/(m^2_{Atemp}\,year)}$. It shows the complexity to predict user related electricity usage and how large variations can be even though the schools in practice are having the same energy efficient type of lighting and equipment and its amount.

The average values in Table 4 are based on the values of the chosen schools.

3.1 Calculated Values for Space Heating, DHW, Building Property Energy and EP_{pet}

Figure 1 shows calculated values of energy for: space heating E_{SH}, domestic hot water E_{DHW}, building property energy E_{BP}, and EP_{pet} in relation to the average value. The values are calculated according to BBR25. The EP_{pet} are very concentrated and near the average value, $60\,\mathrm{kWh/(m^2_{Atemp}\,year)}$.

The E_{DHW} values are very scattered as the DHW usage are depending on each project estimation. 2 of 10 schools have very high DHW value which increase the

Table 4 User related input data list for elementary/secondary school building and its' implementation in the listed schools

	Indoor air temper.	Air flow	Sun shading	DHW	Tenant electricity	Personnel	Airing
	Minimum air Temperature/(°C)	Basic/forced (l/sm^2)/(l/sm^2) Time (h/d/w)/ (h/d/w)[a]	User related control	Not including losses for DHW recirculation/ (kWh/(m^2 year))	Annual value (kWh/m^2 year) to calculate PET Lighting/(W/m^2) Time/(h/d/w)[a] Equipment/(W/m^2) Time (h/d/w)[a]	Person density/ (persons/m^2) Time/(h/d/w)[a] Person heat effect/(W/person)	Annual value to calculate EP_{pet}/(kWh/ (m^2 year))
BEN 2	22	3/–[f] (10/5/44)/–[f]	0.65	$2/\eta_{DHW}$[b]	22 5 (10/5/44) 5 (10/5/44)	0.067 (6/5/44) 80	$4/\eta_{SH}$[c]
S	21	0.5/2.2 (7/7/52)/(7/7/52)	0.29–0.5	4.6[e]	7.8 2(11/5/52) 2 (7/5/52)	0.06-0.3 (7/7/52) 108	1.86
N	21	0.5/2.7 (6/5/52)/(6/5/52)	0.5	6[e]	19 4(11/5/52) 3.5 (8,5/5/52)	0.06 (8.5/5/52) 120	2
K	21	1.3/1.9 (5/5/40)/(5/5/40)	0.75	4[e]	8 3(8/5/40) 2 (8/5/40)	0.058 (8/5/40) Unknown	Unknown
B	21	1.4/2.8 –/(13/5/52)	0.24–0.51	3.2[e]	35 12(12/5/52) 2 (6/5/46)	0.13 (5/5/46) 108	1.86
Va	21	0.35/2 –/(10/5/52)	0.5	15[e]	30 6(10/5/52) 5.5(10/5/52)	0.032 (5/5/52) 108	4
Vi	22	–/2.6 –/(9/5/39)	0.43	5.3[e]	17 1.5(9/5/39) 8 (9/5/39)	0.11 (9/5/39) 70	Unknown

(continued)

Table 4 (continued)

	Indoor air temper.	Air flow	Sun shading	DHW	Tenant electricity	Personnel	Airing
Ve	21	Unknown	0.1–south	5.2[e]	Unknown 10(11/5/52) –	Unknown (11/5/52) 80	Unknown
L	21	–/1.8 –/(15/5/52)	0.35	10[e]	15 – 5.8 (10/5/52)	0.1 (7.5/5/52) 80	4
A	20	0.94/3 –/(12.5/5/52)	0.33–0.5	6,1[e]	14.4 3.5(9/5/44) 3.5(9/5/44)	0.067 (7.5/5/39) 80	4
G	21	0.8/2.6 –/(10/5/44)	0.2–0.45	20[e]	20.2 3.8(10/5/44)[d] 2,5(10/5/44)[d]	0.067 (8/5/44) 80	4
Average	21	0.9/2.4 –/(10/5/49)	0.41	8.5[e]	18.4 5(10/5/46) 4(8.5/5/46)	0.08 (7.5/5/46) 94	3.1

[a]Operational time: hours per day, days per week and weeks per year
[b]η_{DWH} annual efficiency for DHW production in building
[c]η_{SH} Annual efficiency for space heating production
[d]Part load during holiday time
[e]Including hot water recirculation thermal losses
[f]The value from BEN1 as it was excluded from BEN2

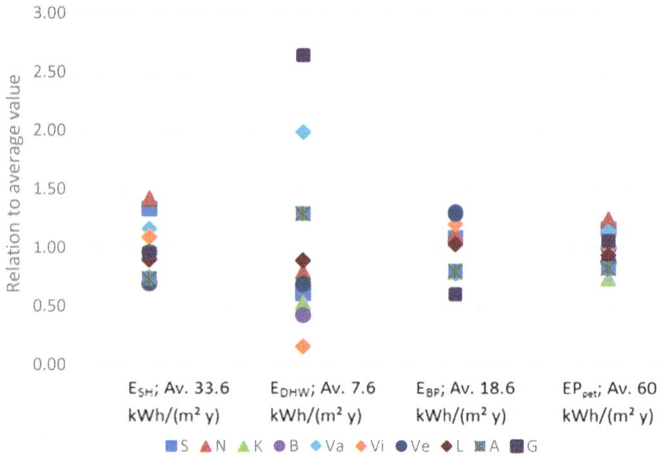

Fig. 1 Calculated values in relation to average value for each school

average value. The difference in the estimated values shows how important is to have the standardized input data.

On the other hand, the E_{SH} values are very much concentrated as they are calculated by the energy simulation programs and based on specific input data from each project which are possible to calculate or obtain from vendors. Climate variations cannot be seen in E_{SH} as the geographical factors F_{geo} are included in the calculations.

The ventilation systems in some schools are in operation 52 weeks and at the others about 39 weeks per year. These variation can be seen in E_{BP} results. It varies between 11 and 24 $\text{kWh}/(\text{m}^2_{\text{Atemp}}\ \text{year})$, the average value is 18.6 $\text{kWh}/(\text{m}^2_{\text{Atemp}}\ \text{year})$. As the specific fan performance has been required in all latest BBR and is 1.5 $\text{kW/m}^3\text{/s}$ or better, the variation mainly depends on operational hours and average air flow. The operational hours is the user related parameter. The air flows are also depending on the users as the VAV system is applied in the schools.

The average values of user related parts: E_{DHW} and E_{BP} contribute for 45% of the total energy EP_{pet}. This give us a picture how strong is the user related impact on the overall energy performance.

4 Discussion and Conclusion

The schools used in this study have been built between 2013 and 2017. They are all very well insulated and have similar: shape, ventilation system, and windows and therefore have similar technical conditions regarding energy performance. The average calculated energy performance for the listed schools is 60 $\text{kWh}/(\text{m}^2_{\text{Atemp}}\ \text{year})$, where 55% is energy for space heating, 15% for DHW, and

30% for building energy property. The user related parameters have tendency to vary a lot, as they are estimated by each energy consultant. The results show that the DHW varies between 3.2 and 20 $kWh/(m^2_{Atemp}\ year)$, the tenant electricity between 8 and 35 $kWh/(m^2_{Atemp}\ year)$, or the ventilation operation varies between 39 and 52 weeks per year. However, the user related part contributes to almost 45% of the total energy, so the estimations on users' related parameters have strong impact on the building energy performance.

The calculated values for the space heating or building property energy are more concentrated. Estimations on technical details (such as: size of cold bridges, fan efficiency etc.) can be calculated or obtained from vendors, and therefor they do not cause large variations on the energy calculations. On the other hand, the estimations on soft, user related, details can cause large variations between projects. They are difficult to obtain, calculate and therefore to estimate. These variations illustrate how important is the standardized user related input data list, such as BEN2, to calculate the energy usage during normal operation and a normal year. However, such list should be based on solid reference values. The reference values in BEN2 for Swedish schools are rather old and sparse and there is a need for newer and more reference values for these type of buildings which is a main goal of this study.

This paper shows significant effect of building users' on energy usage and a need for standardize user related input data for the energy calculations. Another line of researchers investigate probabilistic simulations models of occupants' behavior which also need estimated ranges of impact/input data.

The user related input data has been difficult to summarize for the schools, as pupils and staff work flows are moving around the building between classroom, common areas, labs, music rooms, dining room, sport hall etc. Such flows of pupils and staff can be simulated in different ways in building energy simulation programs. Some of the schools' energy reports are comprehensive and have sufficiently enough information on input data. Some reports are very short and content very few information on input data, and the others are in between. There is not defined standard on energy report format by BBR. Thermal comfort and indoor air quality cannot be read from the energy reports.

Another aspect that can affect the energy performance is pupils/personnel interaction with building management system, such as an interaction with temperature/moving sensors, radiator thermostats, ventilation system with a positive or negative intentions. However, those aspects has not been attended to study in this paper.

The energy calculations will be verified and input data list reviewed by monitoring seven out of listed ten schools during one year period. At the same time thermal comfort and IAQ will be evaluated.

Acknowledgements The authors would like to thank the enterprise Skanska Sverige AB and SBUF (Swedish building industry organization for research and development) for funding this research project. The authors would also like to thank the local municipalities that have provided access and information about the studied schools and last but not least the academic support by divisions Building Services and Building Physics at Faculty of Engineering LTH at Lund University.

References

1. Boverket, BFS 2017: 6 BEN 2. 1–16 (2017)
2. Svebyprogrammet, Brukarindata undervisningsbyggnader (2016)
3. T. Hong, S.C. Taylor-Lange, S. D'Oca, D. Yan, S.P. Corgnati, Advances in research and applications of energy-related occupant behavior in buildings. Energy Build. **116**, 694–702 (2016). https://doi.org/10.1016/j.enbuild.2015.11.052
4. T. Sekki, M. Andelin, M. Airaksinen, A. Saari, Consideration of energy consumption, energy costs, and space occupancy in Finnish daycare centres and school buildings. Energy Build. **129**, 199–206 (2016). https://doi.org/10.1016/j.enbuild.2016.08.015
5. D. Johanssson, in *Measured Occupancy Levels in Twelve Swedish Class Rooms*. Proceedings of Clima, Antalaya. Presented at the (2010)
6. H. Bagge, L. Lindstrii, D. Johanssson, Brukarrelaterad energianvändning:Resultat från mätningar i 1300 lägenheter. LTH, Lunds University (2012)
7. H. Bagge, D. Johansson, L. Lindstrii, BRUKARRELATERAD ENERGIANVÄNDNING; Mätning och analys av hushållsel och tappvarmvatten. LTH, Lunds University (2015)
8. M. Neusser, A. Lederer, C. Harreither, T. Bednar, Identification of the user behavior related influence on the estimated energy performance. Energy Procedia. **78**, 597–602 (2015). https://doi.org/10.1016/j.egypro.2015.11.030
9. T. Buso, V. Fabi, R.K. Andersen, S.P. Corgnati, Occupant behaviour and robustness of building design. Build. Environ. **94** (2015). https://doi.org/10.1016/j.buildenv.2015.11.003
10. K. Tijani, Q.D. Ngo, S. Ploix, B. Haas, J. Dugdale, Towards a general framework for an observation and knowledge based model of occupant behaviour in office buildings. Energy Procedia. **78**, 609–614 (2015). https://doi.org/10.1016/j.egypro.2015.11.035
11. Boverket: BFS 2017:5 BBR 25. 1–16 (2017)
12. DIRECTIVE 2010/31/EU of The European Parliament and of the Council, 13–35 (2010)
13. Skanska Sverige AB: 1 000 skolor på 10 år. Skanska Sverige AB (2016)
14. Statistics Sweden: The future population of Sweden 2016–2060 article no BE18SM1601 (in Swedish) (2016). https://doi.org/URN:NBN:SE:SCB-2016-BE18SM1601_pdf
15. A. Sjöström, Skolbyggandet ökar i hela Sverige (2016)
16. S. Rissvik, Nya skolor behöver byggas för att klara befolkningsökningen. http://sverigesradio.se/sida/artikel.aspx?programid=83&artikel=6753090
17. Boverekt: BFS 2017:xx BBR (B). 2015–2018 (2017)
18. Boverket: BFS 2016:12 BEN 1. 1–16 (2016)
19. IDA Indoor Climate and Energy. http://www.equa.se/en/ida-ice
20. VIP-Energy. http://www.strusoft.com/products/vip-energy
21. J. Andersson, L.-O. Matsson, S. Gehlin, *VVS 2000 Tabeller och Diagram, Meterologi och Klimatologi* (Förlags AB VVS, 2002)
22. B. Nordquist, Ventilation and Window Opening in Schools—Experiments and Analysis (2002)

Part VIII
Trans Disciplinary Connections and Social Aspects

Business Model Analysis of Geo-TABS Buildings with Predictive Control Systems

Qian Wang and Suleyman Dag

Abstract This paper investigates the conceptual framework and impacts of business models in model predictive control (MPC)-based geothermal *Thermally Active Building System* (Geo-TABS). The analysis is done by compiling technical, political, economic, social and environmental analytical frameworks of MPC Geo-TABS. The elements of the business model Canvas are identified and analyzed in this application. Theoretical bases of business model generation are verified by substantiating arguments and potential profit analysis for stakeholders via four demonstration buildings. The focused building types/cases involve office building, schools, elder-care houses and multi-family house. Methods to verify the proposed value propositions in the business model are given special interests. The results show that correctly sizing and combining the four major components: MPC, geothermal, TABS and suitable building types, are the core in both technical and business development perspectives. Complete design guidelines are crucial for promoting MPC Geo-TABS business in its service chains. Transforming the conventional economy-oriented business development method to holistic sustainability-oriented profit matrix can further strength the value propositions of MPC Geo-TABS. The findings aim at supporting decision-makers and further improving engineering guidelines in implementing MPC based Geo-TABS in a larger scale in Europe.

Keywords Business model · Geo-TABS · MPC · Sustainability
EU buildings

Q. Wang (✉)
Division of Fluid and Climate Technology, Department of Civil and Architectural Engineering, KTH Royal Institute of Technology, Brinellvägen 23, 100 44 Stockholm, Sweden
e-mail: qian.wang@uponor.com

Q. Wang · S. Dag
Uponor AB, Hackstavägen 1, 721 32 Västerås, Sweden

D. Johansson et al. (eds.), *Cold Climate HVAC 2018*,
Springer Proceedings in Energy, https://doi.org/10.1007/978-3-030-00662-4_64

Nomenclatures

AHU	Air-handing units
B2B	Business to business
COP	Coefficient of performance
DHW	Domestic hot water
Geo	Geothermal, including both active and passive ground-source systems
GSHP	Ground-source heat pumps
HVAC	Heating, ventilation and air conditioning
LT/MT/HT	Low temperature/Medium temperature/High temperature
IPR	Intellectual property rights
MPC	Model predictive control
PMV-PPD	Percentage of mean vote based predicted percentage dissatisfied
TABS	Thermal active building systems
VP	Value proposition
θ	Temperature, °C

1 Introduction

1.1 The Concept of Geo-TABS

Building sector is one of the largest energy users, which accounts for approximately 32% of global energy usage [1]. In European Union (EU), energy usage in buildings has increased from 400 Mtoe to 450 Mtoe in the past 20 years, and this increase is bound to continue if adequate energy saving measures are not carried out [2]. Development and promotion of new methodologies and technologies to contribute to the future regulations and technical solutions have been addressed by both the loaded EU and International Energy Agency directives [3]. The involved policies require member states to implement energy-efficient measures in both existing and new buildings, aiming to accelerate the transformation of EU buildings towards Net-Zero energy/emission buildings. Additionally, energy performance/ sustainability certificates are to be included in all advertisements for the sale or rental of buildings [4].

In order to meet these EU energy and climate landscapes, Thermal Active Building System (TABS) has emerged as an innovative solution to improve building energy performance and indoor climate. TABS combining cooling and heating system in the structural concrete slabs/walls of a multi-storey building, which is able to operate hydraulic temperature close to ambient temperature, such as 22–28 °C for heating and 16–20 °C for cooling [5]. TABS are primarily for sensible cooling and secondarily for base heating. The whole system works with radiant heating and cooling, which is not any air-conditioning or radiators, and does not commonly substitute any ventilation system. Furthermore, TABS stores heat via building structures themselves and can commonly provide upgraded global thermal comfort than conventional convective heating/cooling methods [5]. Due to the reduced draught, noise levels and improved mean radiant temperature through less fluctuated surface temperature, local thermal comfort is commonly high in TABS buildings [6]. All the above advantages have promoted TABS as a competitive heating/cooling emission system in the current EU building markets.

From a sustainability perspective, TABS-served low-temperature heating (LTH) and high-temperature cooling (HTC) provide wide opportunities for the integrations and applications of renewable energy, such as geothermal energy or ground-source heat pumps (hereafter refer as Geo-TABS), in which the coefficient of performance (COP) can be largely improved. This further reduces the operational costs for heating and cooling. However, due to the nature of large thermal mass in TABS buildings, this system responses very slow comparing to conventional convective heating/cooling system, such as radiators and air conditioning. This leads to a high requirement to the operating conditions and controlling of geothermal systems in certain conditions: such as peak loads during heating/cooling seasons. As introduced above, Geo-TABS buildings need more advanced self-tuning control strategies in order to fully utilize the potentials. Hence, further improve the responding time and energy performance of such system comes into question. In recent years, a model predictive control (MPC)-based Geo-TABS system has been technically developed and implemented in several European buildings [7–9]. However, most of the implementation and analysis of existing MPC Geo-TABS practice were carried out on a case-to-case base [10, 11]. The business guidelines are highly case-driven and lacking of systematic indications. How to characterize the value proposition (VP) of combining MPC with Geo-Tabs, and how to qualitatively evaluate such system in a large scale are still not sufficiently reported. From corporate level, developing a high-level business model that can lead to approaches and conceptual frameworks for engineers, managers and stakeholders to further explore the market values of such systems are not attained. More importantly, few studies have explored the usage of business model in promoting MPC Geo-TABS, and examining barriers to evaluate such system in market, social, regulatory and financial aspects in a systematic approach. A clear business model and analysis scheme for the future supply, design, commissioning, operating and maintenance (O&M), and monitoring are in need.

1.2 Objective

The objectives of this paper can be listed us:

- The system scheme of MPC Geo-TABS components and working principles for the selected building types.
- The proposal of business model for MPC Geo-TABS building and potential evaluation methods.

2 Applications of MPC: State of the Art and System Frameworks

2.1 The Principle of Using MPC in Geo-TABS

As introduced in Sect. 1.1, TABS works with hydraulic temperature that is close to the ambient. For heating seasons, the $\theta_{supply}/\theta_{return}$ are 28 °C/26 °C, given the room temperature of 20 °C. The maximum heating capacity of standard TABS are 25 W/m^2. High performance TABS (embedded pipes close to concrete surface) can provide maximum heating capacity around 40-50 W/m^2 [5]. For cooling seasons, the $\theta_{supply}/\theta_{return}$ are 16 °C/19 °C, given the room temperature of 26 °C. The maximum cooling capacity of standard TABS is 40 W/m^2. High performance TABS can provide maximum cooling capacity around 60–80 W/m^2 [5]. Based on the local climate and heating/cooling loads, the low-temperature heating (LTH) and high-temperature cooling (HTC) working principle of TABS provides broad opportunities to improve the COP of GSHP, which further saves electricity (El) usage of compressors. Figure 1 shows the system scheme of Geo-TABS components and working principles. It can be seen that TABS works with a "zone-to-zone" basis and these "zones" are defined in the early building design stage. When TABS is connected with geothermal, modern GSHP commonly is able to work with diverse working modes based on the sizing and energy demand of TABS "zone" [12]. These modes can be listed as:

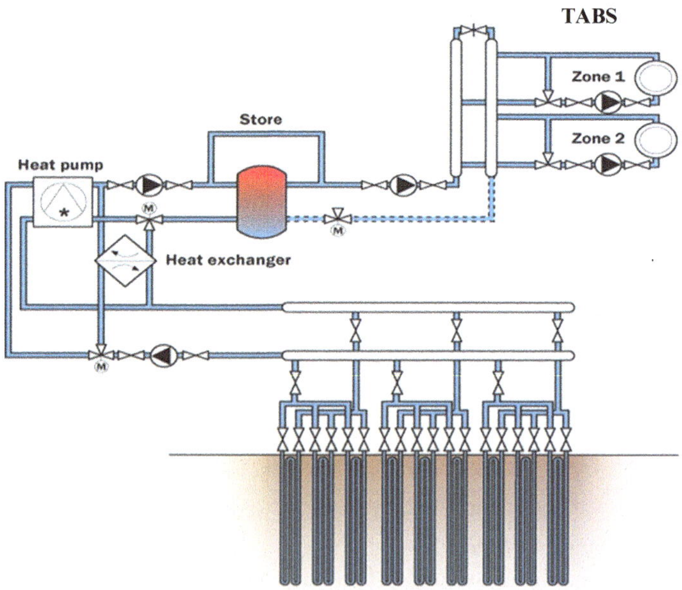

Fig. 1 A system scheme of Geo-TABS components and working principles [5]

- Base heating control mode: Geo-TABS provides base heating during non-peak load time in heating seasons.
- Dual mode: when heating and cooling loads are existing year-round, such as for office buildings in some selected climates.
- Active cooling control mode: Geo-TABS provide cooling during cooling seasons.
- Passive cooling control mode: At sufficient temperature level of the ground, GSHP operates in free cooling mode which only circulation pumps are involved.

Additionally, the controlling strategy may differ in Geo-TABS based on how advanced control system the building needs. For example, three conventional controlling strategies are used jointly with the above working modes [13]:

- 24 h operation: this controlling strategy allows pumps to work through continuous operation in 24 h with a constant θ_{supply}. This strategy largely depends on the behaviors of TABS by its self-regulating effect; it is not demand-optimized and leads to continuous El usage for compressors.
- Day-night operation: this strategy loads the building thermal mass during the night and let the heat be delivered to the room during the day. It highly depends on the storage capacity of the mass, and GSHP works according to the simple principle of "On/Off", which is only suitable for buildings with well-scheduled occupancies and high thermal mass, such as concrete core-activated offices.
- Cyclic operation: this strategy allows TABS to be operated in selected time intervals, which is also considered as "pulse width modulation" [13]. The core of this strategy lies in learning the ambient temperature compensated θ_{supply} with pulse width modulation (via constructing heating/cooling curve with respect to θ_{set}).

As introduced above, MPC was introduced as an innovative application in order to solve the dynamics of building energy demand instead of constructing simple heating/cooling curve by long learning process, as the conventional controls. MPC selects and identifies the building energy demand model, and further optimize the working strategy of Geo-TABS well ahead of time, in order to achieve the targets of: controlling GSHP to supply correct mass flow with sufficient θ_{supply}, under approximate working modes of heat pump groups. The predictions allows the optimization to select the best control parameters that can minimize the energy usage, while at the same time without compensating thermal comfort (or other objective functions). The general MPC optimization principle can be outlined as [14]:

$$Minimize_{u_t,\ldots,u_{t+T-1}} \sum_{k=t}^{N-1} f(x_k, u_k) \quad (1)$$

Subject to:

$$x_{k+1} = g(x_k, u_k, d_k) \quad \text{where } k = t, \ldots, t + T - 1 \quad (2)$$

$$x_k \in X_k, u_k \in U_k, \quad (3)$$

In which the predicted results (at time t) are generated from the prediction model Eq. (2). The predicted input parameters and conditions are constrained to the settings of Eq. (3). The initial condition of $x_{t=0}$, optimizer Eq. (1), generate the results $u_0^{'}, \ldots, u_{N-1}^{'}$ by control data inputs. And these results are optimal with respect to Eq. (1). In this way, a particular choice x_k and u_k is assigned by the function $f(x_k, u_k)$. However, this standard MPC concept may not be applicable in all cases of Geo-TABS in reality. The reason is that disturbances d_k exists. d_k can be subject to outdoor temperature, building types, internal heat gains (from both solar and occupants), as well as building usage schedules, etc. [14]. And some parameters of these d_k can be random, such as outdoor temperatures. Therefore, a good understanding of how MPC should be coupled with Geo-TABS, more importantly, what HVAC components are available and linked in the selected archetypes are important. Existing studies have reported that building control system should be developed in the pipeline of early-stage building design. A good understanding of the implemented building types provides flexibility and low-risk experimentation [9]. A high-level system framework can further assist the development of deterministic formulations of d_k, by either bounded constraint (robust MPC), or probabilistic model (stochastic MPC).

2.2 *High-Level Framework of Applying MPC Geo-TABS in the Building Types*

Based on the MPC concept introduced in Sect. 2.1, it is believed that MPC is mostly suitable as a top-level controller to optimize set-points of lower level controllers [11]. Therefore, a high level framework of applying MPC Geo-TABS in four selected building types are developed, shown in Fig. 2. The included building types cover office, school, care house and residential buildings (multi-family). Different modules provide information and interactions of HVAC components. The framework classifies two types of flows in all Geo-TABS systems: energy flow (black solid line), and communication flow (black dashed line). Constrain settings in Eq. (3) and d_k are generated by back arrows to MPC. Prediction signals based on prediction model Eq. (2) are generated from MPC along the out-going arrows to different modules in Geo-TABS. A benchmark system is further developed in order to verify and compare the advantages of combining MPC with Geo-TABS, based on the selected building types and current HVAC system in EU building market [15], shown in Table 1.

3 Business Model Analysis on MPC Geo-TABS

The application of business model in MPC Geo-TABS are based on the theoretical bases of business model generation Canvas [16]. In principle, business models are designed and executed in specific environments that adapted to this case can be

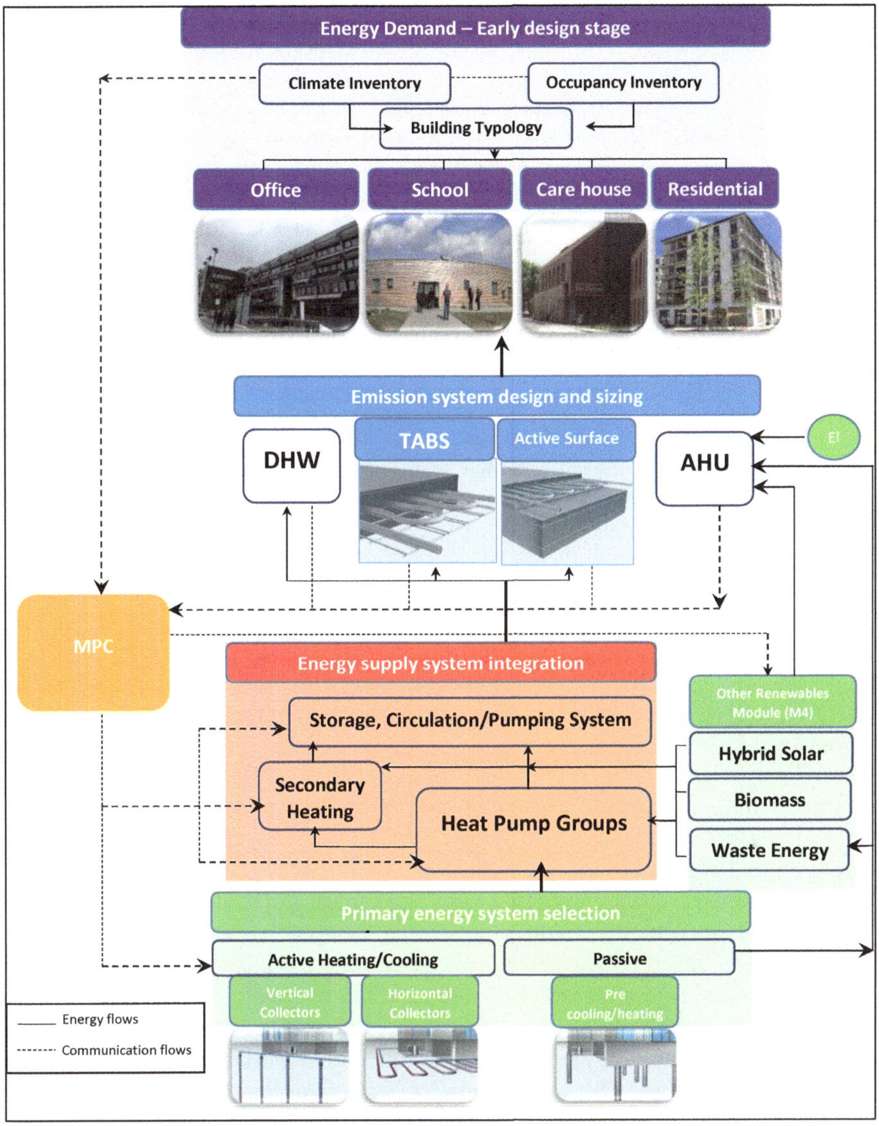

Fig. 2 A system scheme of MPC Geo-TABS components and working principles

depicted mainly on three aspects—market forces, industry forces and key trends. Utilizing a business model approach for the assessment from both source and end-use will ensure a more commercial driven analysis aimed at identifying a hygienic and sustainable water supply system. Information extracted from stakeholder insights analyses, understanding the "jobs they are doing" and the "pains" and "gains" they have, will form a good basis to derive relevant value propositions for realizing the concept. The structure of business model canvas is commonly

Table 1 The benchmark systems for MPC Geo-TABS

Building types	Benchmark	MPC Geo-TABS
Office buildings	District heating/cooling + boilers +Fan coil units/Chilled beams + Mechanical ventilation with heat recovery + Conventional feedback control	GSHP + TABS heating/ cooling + Passive cooling + Boiler as secondary energy for DHW + Mechanical ventilation only for required indoor air quality + other renewable alternatives + MPC
Residential buildings	District heating/ boilers + radiators + Mechanical ventilation with heat recovery + Conventional feedback control	GSHP + Thermal active surface (floor) heating + Exhaust ventilation with heat recovery + other renewable alternative + MPC
Care house	District heating/ boilers + radiators + Fan coil units + Mechanical ventilation with heat recovery + Conventional feedback control	GSHP + TABS for heating/ cooling + Secondary system + other renewable alternatives + Mechanical ventilation with heat recovery + MPC
School	District heating/ cooling + Radiators + Fan coil units + Mechanical ventilation with heat recovery + Conventional feedback control	GSHP + TABS for heating/ cooling + Boiler for DHW + Mechanical ventilation only for required indoor air quality + MPC

composited by 9 key elements, centered by value proposition. The development of these 9 key elements are commonly conducted by brainstorms among the key players in the whole life of the project: such as consortium, stakeholders, developers, investors, customers and marketing agents as well as commissioning partners. Figure 3 shows the business model for MPC Geo-TABS linking these 9 basic elements. These results are expected to provide an early-design stage indication that how MPC Geo-TABS should be promoted from a business perspective. The business model in Fig. 3 was developed based on the principle of sustainability on an interactive and iterative process around value propositions involving multiple cooperation partners.

Three key dimensions, namely, value proposition, distribution channel and cost structure, are believed to be key importance, with respect to building types, current Geo-TABS markets and technical innovations. The core of this business model is the proposed value proposition, presented in the center of Fig. 3, which are formed by six key perspectives. Distribution channel of MPC Geo-TABS is addressed based on two market categories: established market and emerging market, via three representative building types: office building, residential building and public buildings. Three cost structure are addressed by formulating the major and unique costs of MPC Geo-TABS: by cost related to warranties, prefabrication and pre-engineering process, fixed cost related to continuously ensuring compatibility of new modules. Based on this business model, a detailed analysis focusing on how to verify the proposed value proposition is further presented, as shown in Fig. 4.

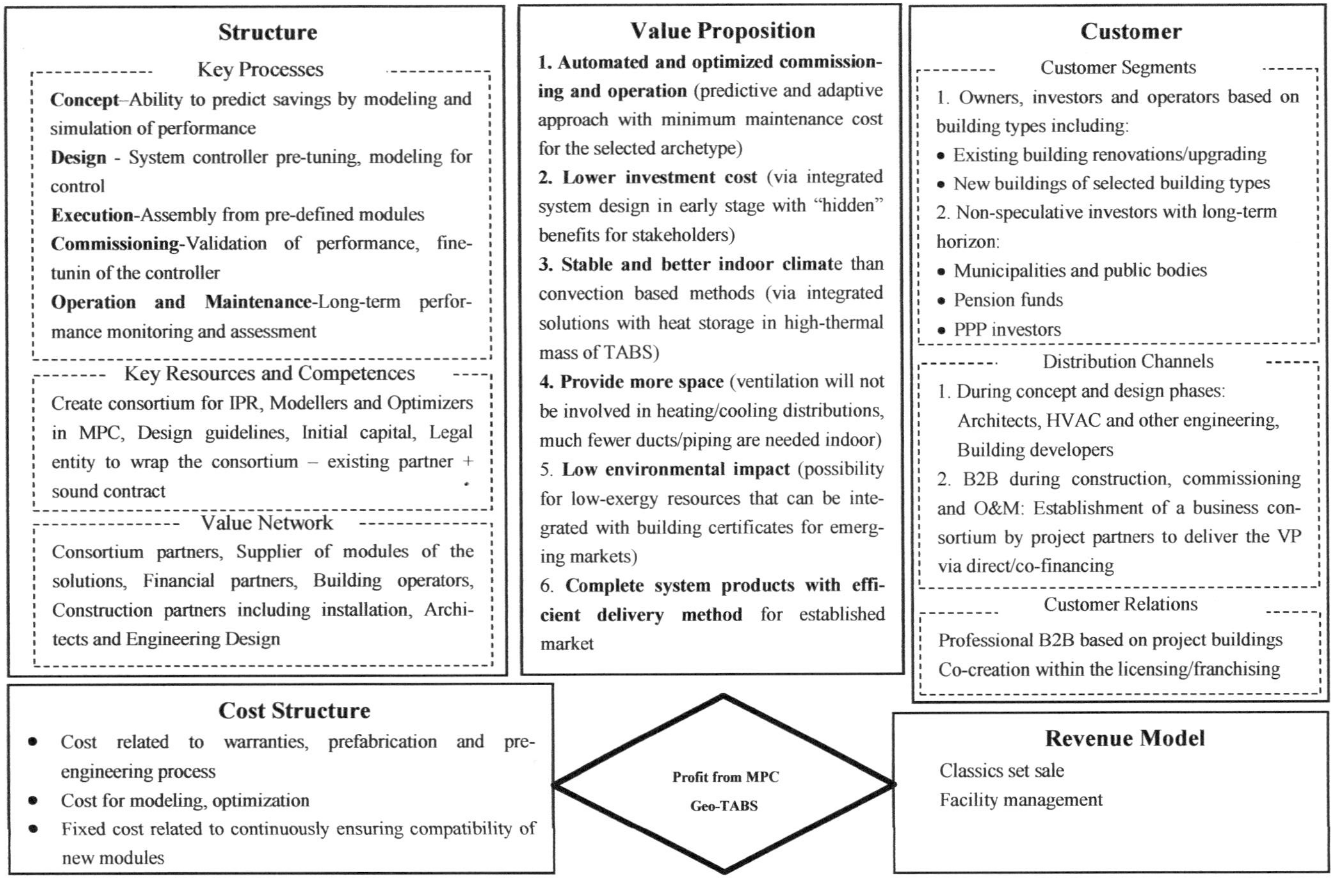

Fig. 3 Business model development for MPC Geo-TABS

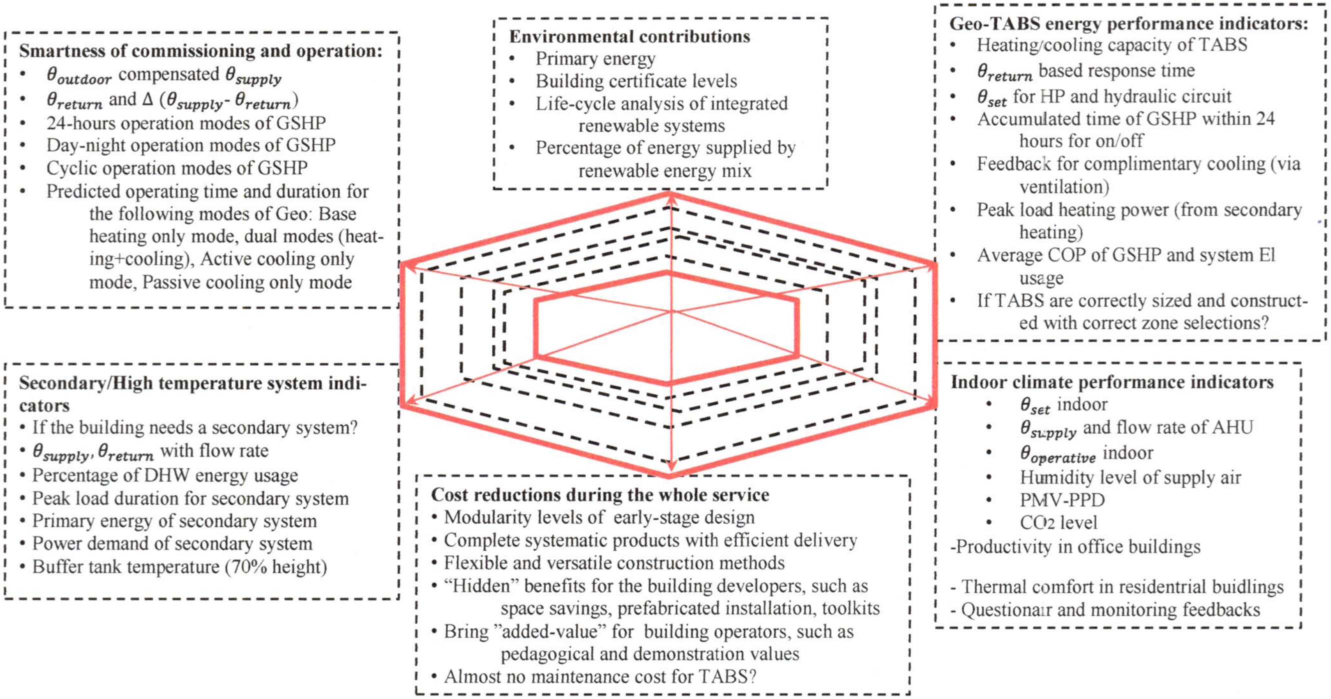

Fig. 4 Analysis and evaluation guidelines for the VP of MPC Geo-TABS based on indicator scoring scheme

The results of Fig. 4 is expected to construct a set of quantitative criterion system that are measureable for stakeholders to further validate the business model, and explore the limitations of current business model for MPC Geo-TABS in the market. This criterion system is structured based on the six key value proposition proposed in Fig. 3, but further adapted to measurable parameters during the whole service life of MPC Geo-TABS. Each of the six criteria is constructed by a set of indicator/parameters, which can be quantified by modeling, on-site output measurements, monitoring, questionnaires and control system metering.

The indicator sets and evaluating guideline, shown in each criterion in Fig. 4, is expected to be then constructed by a holistic rating scheme (shown as different layers in the center of Fig. 4). The out-layer of the circle represents the highest "score" of the correspondent criteria index, which is weighted by the "sub-scores" of each quantitative indicator. In this way, the stakeholder networks are taken into account as a whole, whereas value propositions that occurred during the whole MPC Geo-TABS service life chain is verified and quantitatively evaluated to the entire range of stakeholders, occupants, building owners/developers, engineers and architects, beyond just customers and shareholders. Given the fact that TABS itself is already economically competitive, the sustainability criterion involved in MPC Geo-TABS is more detailed classified in Fig. 4. This approach is expected to drive the transformation of common business model to "sustainability – oriented" business model for future implementing MPC Geo-TABS projects in EU building types. How to supply and verify the studied value proposition for identifying customer segments of MPC Geo-TABS will be further studied in details during the whole process of the project disseminations.

4 Conclusion

It is concluded that business development of MPC Geo-TABS should be done by conducting a continuous dialogue with the main stakeholders of the building, via, e.g. validating anticipated performance, monitoring feedback, interview, occupant questionnaires and public network constructions. A systematic approach to visualize and score the gained experiences/knowledge should be further quantified and identified as references based on the local climate and social values. To have a successful business model for MPC Geo-TABS, correctly sizing and combining the four major components: MPC, geothermal, TABS and suitable building types, are the cores. Complete design guidelines are crucial for supplying MPC Geo-TABS business in its early design stage. Transforming the business development method to sustainability-oriented profit matrix can further identifying the customer segments and verify the value propositions of MPC Geo-TABS for the future large-scale implementation in EU building market.

Acknowledgements The authors are grateful to European Union for providing financial support in Horizon 2020 project "*Model Predictive Control and Innovative System Integration of GEOTABS in Hybrid Low Grade Thermal Energy Systems-Hybrid MPC GEOTABS*". We also gratefully appreciate the support of DTU Technical University of Denmark, Feramat Cybernetics Ltd. for providing valuable information in the project.

References

1. European Commission, Energy roadmap 2050. European Union (2012)
2. National Science and Technology Council, C. o. Federal R&D. Federal R&D Agenda for Net Zero Energy, High-Performance Green Buildings [Internet]. USA; 2008 Sep [cited 2013 Jun 1]. Available from: http://www.whitehouse.gov/files/documents/ostp/NSTC%20Reports/Federal%20RD%20Agenda%20for%20Net%20Zero%20Energy%20High%20Performance%20Green%20Buildings%20Oct2008.pdf
3. Cost optimality calcuation methods within the EPBD framework—ECEEE [Internet]. [cited 2014 Jul 28]. Available from: http://www.eceee.org/policy-areas/buildings/cost_optimality
4. G. Jóhannesson, in *Building Energy–A Practical Design Tool Meeting the Requirements of the EPBD*. Proceedings of the 7th Symposium on Building Physics in the Nordic Countries (Reykjavik, Iceland, 2005)
5. Uponor, Uponor Contect compendium—Thermally active building systems (2014)
6. M. Achermann, Z. Gerhard, RADTEST–Radiant heating and cooling test cases. Subtask C Rep IEA Task 22 (2003)
7. A. Afram, F. Janabi-Sharifi, Theory and applications of HVAC control systems—A review of model predictive control (MPC). Build Environ. **72,** 343–355 (2014)
8. J. Romaní, A. de Gracia, L.F. Cabeza, Simulation and control of thermally activated building systems (TABS). Energy Build. **127**, 22–42 (2016)
9. A. Preglej, J. Rehrl, D. Schwingshackl, I. Steiner, M. Horn, I. Škrjanc, Energy-efficient fuzzy model-based multivariable predictive control of a HVAC system. Energy Build. **82**, 520–533 (2014)
10. M. Schmelas, T. Feldmann, E. Bollin, Savings through the use of adaptive predictive control of thermo-active building systems (TABS): a case study. Appl. Energy **199**, 294–309 (2017)
11. J. Cigler, D. Gyalistras, J. Široky, V. Tiet, L. Ferkl, in *Beyond Theory: The Challenge of Implementing Model Predictive Control in Buildings*. Proceedings of 11th Rehva World Congress, Clima [Internet] (2013) Available from: http://www.opticontrol.ethz.ch/Lit/Cigl_13_Proc-Clima2013.pdf
12. M. Schmelas, T. Feldmann, E. Bollin, Savings through the use of adaptive predictive control of thermo-active building systems (TABS): a case study. Appl. Energy **199**, 294–309 (2017)
13. M. Schmelas, T. Feldmann, P. Wellnitz, E. Bollin, Adaptive predictive control of thermo-active building systems (TABS) based on a multiple regression algorithm: first practical test. Energy Build. **129**, 367–377 (2016)
14. J. Drgoňa, Model Predictive Control with Applications in Building Thermal Comfort Control (2015). Available from: https://www.kirp.chtf.stuba.sk/~drgona/files/presentation/minim_2015.pdf
15. M. Economidou, B. Atanasiu, C. Despret, J. Maio, I. Nolte, O. Rapf, Europe's Buildings under the Microscope. A Country-by-Country Review of the Energy Performance of Buildings (Buildings Performance Institute Europe (BPIE), 2011)
16. A. Osterwalder, Y. Pigneur, Business model generation: a handbook for visionaries, game changers, and challengers [Internet] (Wiley, 2010). Available from: https://books.google.com/books?hl=zh-CN&lr=&id=UzuTAwAAQBAJ&oi=fnd&pg=PA7&dq=business+model+generation+&ots=yXBPAcC51u&sig=elYXHmbNL3KrvtCOME55A9qTSi0

Atrium in Residential Buildings—A Design to Enhance Social Interaction in Urban Areas in Nordic Climates

Itai Danielski, Malin Krook and Kerstin Veimer

Abstract The design concept of conditioned atria receive growing popularity in both commercial and service buildings all over the world, but still not common in the residential sector. This study used a psychological framework to examine if building design with enclosed heated atria in apartment buildings can enhance sense of community and social interactions in Nordic climates. A qualitative study was conducted to understand the perception of residents living in apartment buildings with heated atrium. One of the few examples in Sweden. This was compared to the experience of residents in a "traditional" apartment building without an atrium. The questionnaire was comprised of six parts: (i) socio-demographic aspects; (ii) information about the apartment; (iii) social activities within the building; (iv) social interaction with neighbours; (v) information about principles in life; and (vi) sense of community linked to their homes. The results showed significant social differences between the residents of the atrium and "traditional" buildings, which could not be explained solely by differences in preferences and principles in life. A large proportion of the social differences between the buildings could be explained by the building design, as the common and semi-private areas within the atrium building provide opportunities to establish social interactions. The residents in the atrium building was found to have greater sense of community and higher frequency of interactions, which are both parts of social sustainability.

Keywords Atrium · Residential building · Social interactions

1 Introduction

Two-thirds of the world population is expected to live in cities by 2050. In Sweden, this level of urbanization was already reached during the 60s. Since then, urbanization has increased steadily; currently more than 85% of Swedes live in cities.

I. Danielski (✉) · M. Krook · K. Veimer
Mid-Sweden University, 83631 Östersund, Sweden
e-mail: itai.danielski@miun.se

D. Johansson et al. (eds.), *Cold Climate HVAC 2018*,
Springer Proceedings in Energy, https://doi.org/10.1007/978-3-030-00662-4_65

Such rapid urban growth induces both challenges and opportunities. High population density in cities may benefit, e.g. from interconnection of many social circles forming vast information networks.

A social network refers to a set of individuals, the relationships among them (social ties), which facilitates by social interactions. In his theory on the spread of information in social networks known as "The Strength of Weak Ties", Granovetter [1] discussed the effect of social ties on flows of ideas, influences and information between individuals. He distinguished between strong and weak social ties. He claimed that weak ties are more likely to connect different social circles and to be the source of non-redundant information, whereas strong ties provide redundant information. Strong ties are often characterized as ties among close friends, whereas weak social ties are occasional, e.g. between casual friendship and neighbours.

Neighbourhoods can offer different type of localities for social interactions: public areas, semi-private areas and private areas. Public areas were reported to effect social interaction regarding access to pedestrian [2] and main streets [3], just to name a few. Semi-private spaces like the front yards of terrace house's and front balconies were reported to encourage sense of community and social life in residential neighbourhoods [4, 5]. In private areas like residential buildings, factors such as proximity of apartments in multi-storey buildings, their position and orientation towards other apartments were found to affect the social interactions among residents [6]. However, indoor common space within multi-story apartment buildings are not usually designed to becomes an integral part of the residents' daily activities.

In this context, a courtyard or atrium design within residential buildings may benefit from all of the above three localities: a common space in the "middle" of the residential building (courtyard or atrium) used by all residents. Apartments as private spaces orientated towards the atrium and connected by indoor corridors

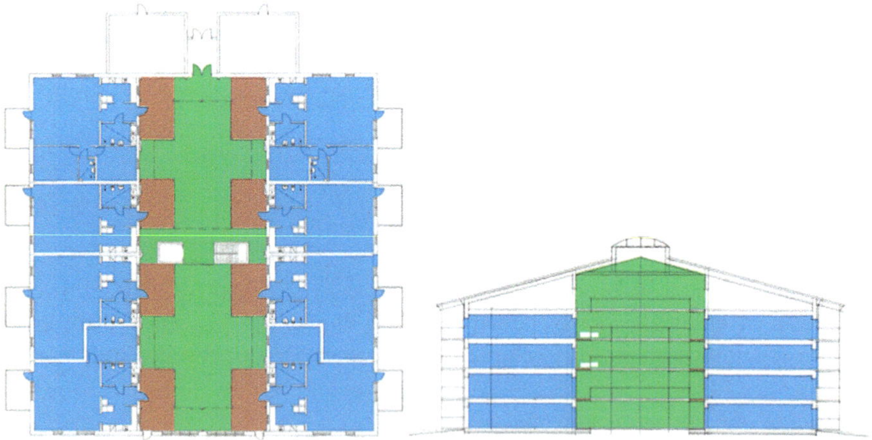

Fig. 1 The atrium building with top view (left figure) and side view (right figure), and the three different areas: atrium (green), apartment (blue), and indoor balconies (brown)

within the atrium space. Indoor balconies (integrated in the indoor corridors) facing the courtyard acting as a semi-private spaces. They are part of the apartment area but one cannot avoid its neighbours (see example in Fig. 1).

An open courtyard within residential buildings may not entail large benefits as a place for social interaction in Nordic climates. This is due to the short daytime and the low outdoor thermal comfort during the heating season. A design with heated enclosed courtyard, so called atrium, may be utilized to a greater extent throughout the year. With a proper design, it may also have lower energy demand for space heating due to lower ratio of thermal envelope to floor area [7]. However, such design in residential buildings is still uncommon in the Nordic regions.

2 Aim

This study used a psychological framework to examine if building design with heated atria in apartment buildings can enhance sense of community and social interactions in Nordic climates. A qualitative analysis using a survey was conducted to understand the perception and experience of residents living in apartment buildings with heated atrium. This is one of the few examples of residential buildings design with atrium in Sweden (Fig. 1). The results were compared to the perception and experience of residents living in apartment building with a "traditional" design, i.e. without an atrium.

3 Methodology

3.1 Case Studies

The atrium-design building was constructed in the northern part of Sweden during 2006. It comprises of two identical five-storey apartment buildings under one roof, joined by a linear atrium in-between. The two buildings have a total floor area of 3830 m^2 accommodating 32 apartments with two, three, and four rooms. At the entrance of each apartment there is an indoor balcony facing the atrium. The balconies in each floor are connected by suspended corridors. Although each indoor balcony can be regarded as a property of the apartment owner, neighbours can still pass through, for example, on the way to their own apartments. In the middle of the atrium there is a staircase and an elevator, which serve both buildings by connecting all corridors. The atrium space is heated during the cold season and can be used by the residents for different activities throughout the year, regardless of the outdoor weather conditions.

The design of the atrium building includes indoor areas with different social characteristic. The apartment areas act as private spaces orientated towards each other. The indoor balconies and indoor corridors acting as semi-private spaces facing the courtyard, i.e. they are part of the private area (apartments) but one cannot avoid its neighbours. The atrium courtyard acts as a common space located in the "middle" of the residential building for the use of all residents, as illustrated in Fig. 1.

The atrium-design building was compared to a building with a "traditional" design with similar ownership type (condominiums), since the type of ownership could affect sense of community [8]. The "traditional" design building is a single apartment building located in the same city district area as the atrium-design building. It was constructed during 2011 and consists of 30 apartments divided between four staircases with two or three apartments on each floor. Each apartment has access to private outdoor balcony. The staircases are heated during the cold season and each includes an elevator.

3.2 Questionnaire Survey of Apartment Owners

A questionnaire was delivered to all apartments in both the atrium and "traditional" design buildings. (62 households in total) during two days in February 2015. The questionnaire were delivered personally to each apartment owner (one per apartment) or alternatively by the mailbox, along with a prepaid return envelope, if the residents were not at home at the time. The choice of which of the individuals in the household providing the answers was left to the respondents. The response rate was 81% (26 of 32) and 87% (26 of 32) for the atrium and "traditional" design buildings, respectively.

The questionnaire (Appendix 1) was comprised of six parts: (i) socio-demographic aspects; (ii) information about the apartment; (iii) involvement in social activities within the building (frequency and type); (iv) social interaction with neighbours (frequency and type); (v) information about principles in life; and (vi) sense of community linked to their homes. At the end of the questionnaire form, the respondents could add optional comments and supplementary information.

The measure of the sense of community used in this Study was a shorter version of McMillan and Chavis [9] scale that was created and validated by Peterson, Speer and McMillan [10]. The scale consists of eight questions that aim to assess dimensions in the experience of the neighbourhood based on the four subdivisions: fulfilment of needs, group membership, influence, and emotional ties.

Questions about guiding principles in life measure if sense of community could be affected by social-altruistic values. Respondents were asked to grade 16 values representing four guiding principles in Life from Schwartz's Value Inventory Scale [11]. These include self-transcendence values like universalism (e.g. social justice, peace on earth, consciousness and equality) and benevolence (e.g. loyalty, responsibility, helpful and forgiving), and opposing values like power (e.g. social power, authority, care for status and wealth) and achievement (e.g. success, skills, influence and ambition). The 16 values were previously used in Swedish context [12].

Questions about social interaction with neighbours include three aspects: (i) questions about services, like babysitting, lending out tools or food, help in emergency, etc. The services are examples of neighbouring behaviour in a study of

Perkins and Long [8]; (ii) question about frequency and locations of interaction, e.g. staircase, corridors, balconies, courtyard, apartments; (iii) questions about how well are the correspondents acquainted with their neighbours.

The measure of involvement in social activities within the buildings consisted of five questions designed to capture what was considered as relevant behaviour for engagement and activities in tenant-owner rights. Involvement in society has been related to the sense of community [10], but because this study was interested in the community within the building, new questions were created to be more relevant in this context. The questions asked respondents to indicate their frequency of participation in various activities and events in their own residential buildings like meetings, community tasks or other social activities.

4 Results

4.1 Socio-demographic Parameters

Table 1 list the socio-demographic parameters of the atrium and "traditional" design buildings. There are three main socio-demographic differences between the buildings. First, more than 50% of the households in the atrium-design building are single-family households, in comparison to 11% in the "traditional" design building. Second, the average age of the respondents in the atrium-design building is also significant higher. Last, the occupancy time of the respondents in the atrium-design building is higher. These three above-mentioned socio-demographic differences were used as control factors for the analysis of variables related to sense of community (see Fig. 2), social factors (see Fig. 3), frequency of meetings (see Fig. 4) and principle of life (see Fig. 5).

Table 1 Description of demographic factors of the atrium and "traditional" design buildings and calculated differences with χ^2 test and t-tests

	"Traditional" design n = 26	Atrium design n = 26	χ^2 (df = 1)
Gender: man/woman	10/16	11/12	0.44
Previous connection with the area	11	13	3.1
Single family household	3	14	11.34**
Post-secondary education	21	24	2.88
Household with children	8	1	6.28*
Working/Studying	20	14	0.47
	Mean (SD)	Mean (SD)	t (df = 49)
Age	45 (17.61)	64 (11.19)	−4.63**
Number of years in the house	2.5 (1.46)	6.8 (2.87)	−5.66**
Size of the apartment	3.3 (0.8)	2.9 (0.57)	2.19*

*p <0 .05; **p <0.001

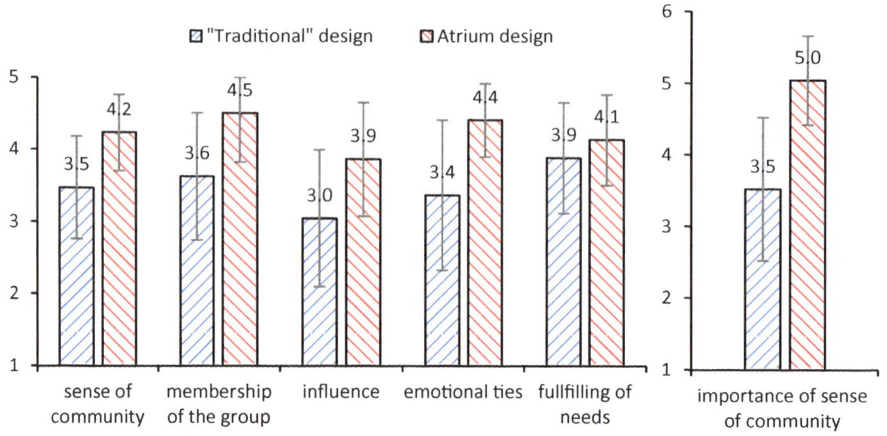

Fig. 2 Differences in variables related to sense of community. The 'importance of sense of community' have six-scale question. The columns represent mean values and the bars represent standard deviation

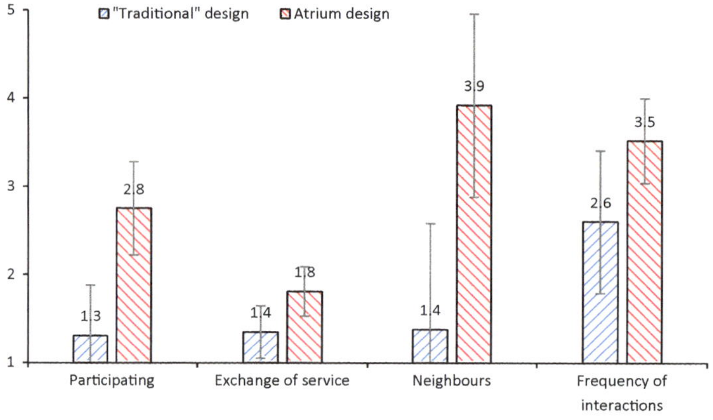

Fig. 3 Differences in social variables between the atrium and the "traditional" design buildings presented with mean value and standard deviation. The columns represent mean values and the bars represent standard deviation

4.2 *Sense of Community*

Figure 2 illustrates the differences in variables related to sense of community. For all the variables, values reported by the respondents of the atrium-design building were higher in comparison to the "traditional" design building. A variance analysis shows a significant difference ($p < 0.001$) between the two buildings regarding the overall 'sense of community', 'membership of the group', 'influence', 'emotional

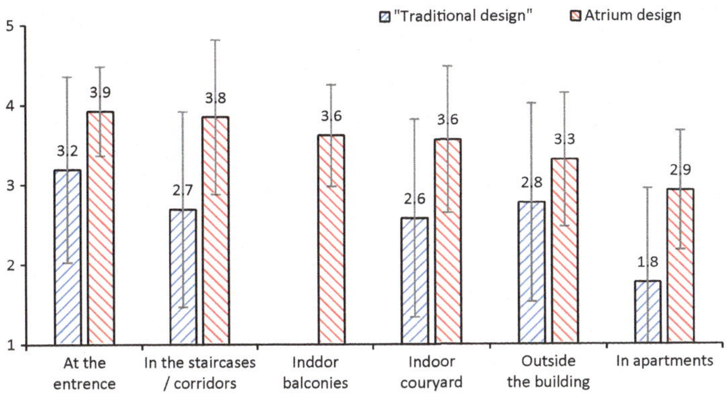

Fig. 4 Places the residents indicated they meet other neighbours and the frequency of meetings. Indoor balconies are only in the atrium-designed building. The columns represent mean values and the bars represent standard deviation

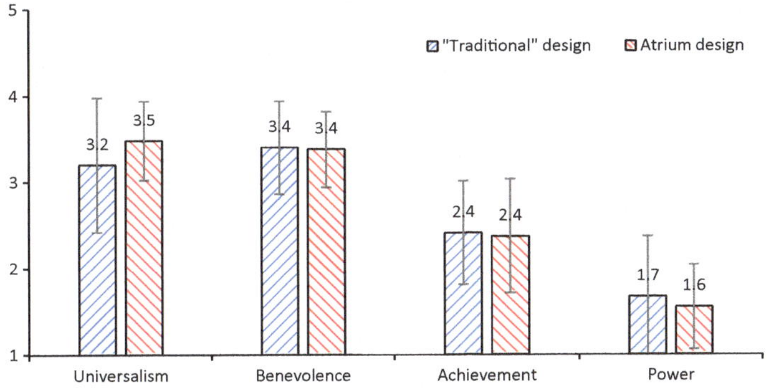

Fig. 5 Differences in guiding principles of life between the atrium and the "traditional" design buildings. The columns represent mean values and the bars represent standard deviation

ties', and the 'importance of sense of community'. No significance difference was found between the two buildings regarding 'fulfilling the needs'.

Significant differences between the two groups' regarding 'sense of community' remained after adjustment for the control variables (variance analysis): age, number of occupancy years in the building, and single households. The variable 'influence' found to correlate to the demographic factor 'Number of years in the house' in both buildings, i.e. respondents with longer occupancy time in the building reported higher influence in the community in comparison to respondents with shorter occupancy time.

4.3 Social Variables

Figure 3 illustrates differences in social variables between the atrium and the "traditional" design buildings. The respondents of the atrium-design building reported higher grade of participation in common activities within the building, higher grade of familiarity with other neighbours, and higher grade of frequency of meetings with their neighbours. Slightly higher exchange of services with other neighbours was also reported in the atrium-designed building in comparison to the "traditional"-designed building. A variance analysis shows a significant difference ($p < 0.001$) between the two buildings regarding all the above-mentioned social factors. The significance also remained after adjustments for the demographic factors.

4.4 Frequency of Meetings

Figure 4 compare the frequency, in which the respondents meet other neighbours in different locations within both the atrium and "traditional" design buildings. The results shows that most of the meeting took place in the common and semi-private areas in both buildings, i.e. the entrance, staircases/corridors and in the case of the atrium-design building also in the indoor balconies and atrium space.

The results also shows that the atrium-design building on average has a higher frequency of meetings then in comparison to the "traditional" design building. A variance analysis shows a significant difference between the two buildings regarding frequency of meetings at the entrance (t (36) = 2.88, $p < .01$), at the indoor corridors (t (50) = 3.77, $p < .001$), in someone's apartment (t (42) = 4.23, $p < .001$) and in the common atrium (T (50) = 3.20, $p < .01$). High meeting frequency was reported also in the indoor balconies of the atrium design building. No significant difference, in meeting frequency, was found between the groups in the open-space around the buildings.

4.5 Guiding Principles of Life

Figure 5 compare the four guiding principles of life used in this study: universalism, benevolence, achievement and power. Residents in both the atrium and "traditional" design buildings tend to have slightly higher self-transcendence values (universalism and Benevolence), but a variance analysis shows no significant difference between the two buildings regarding all the above-mentioned guiding principles of life.

5 Discussion

A previous study, examining the same atrium-design building, showed that atrium designed in apartment buildings in Nordic climates have a potential to reduce the annual energy use for space heating [7]. In this study, a psychological framework was used to examine if building design with heated atria in apartment buildings can enhance also sense of community and social interactions in Nordic climates. The study compare between the perception of residents of two apartment buildings with different designs located in the same city district in northern Sweden; one building with an heated atrium, and the other without.

A significant differences in social factors was found between the residents of the atrium and the "traditional" design buildings. Residents in the atrium-design building reported higher frequency of interactions. Most of the interactions among the residents found to occur in the common and semi-private areas within and around the buildings. According to Gehl [5] common and semi-private areas are the places in which most of the spontaneous interactions occur. Spontaneous interactions facilitates weak ties that may develop to strong ties. Weak ties contribute to the development of social networks by flow of information among different circles or social groups [1]. Such flow of information may even contribute to higher creativity.

The residents living in the atrium-design building found also to know their neighbours better, and are more active in the common activities organized within the building in comparison to the residents in the "traditional" design building. This in turn can facilitates a sense of community within the building. A sense of community is a catalyst for behaviours such as organized participation and emotional ties to the neighbourhood [13], and is an important part of social sustainability.

A sense of community, in this study, was measured by four categories; membership of the group, influence (i.e., ability to express themselves and affect the community within the building), emotional linkage (i.e., extent and quality of interaction between members of the building), and fulfilled needs (i.e., whether needs are perceived as satisfying in their current accommodation). The residents of the atrium-design building reported higher values in all first three categories.

An explanation may be that the needs of the residents are fulfilled in both groups, but those living in the "traditional"-design building are not familiar with the social benefits of the atrium. This can be supported by the 'lower importance of sense of community' (Fig. 2) reported by the residents of the "traditional" design building, i.e. how important it was to be part of the neighbours community.

Significant differences between the groups' sense of community remained after checking for the variables age, number of years of accommodation and single occupancy. Respondents with longer occupancy time in the building reported higher influence in the community in comparison to respondents with shorter occupancy time.

No differences were found concerning principle of life between the residents of the two buildings. Therefore, it is possible to conclude that large part of the differences in social aspects between the two buildings could be explained by their design, i.e. the atrium as a facilitator for social interactions.

6 Conclusions

A heated atrium in apartment buildings in Nordic climates, with appropriate design that includes common, semi-private and private areas, seems to have a potential to facilitate social interactions, engage residents in common activities and provide them with a sense of community, which is an important part of social sustainability.

Residents in the atrium-design building reported higher frequency of interactions with their neighbours. They know their neighbours better, and are more active in the common activities organized within the building. They have stronger connection with their neighbours, and feels higher ability to affect the community within the building, in comparison to the residents in the examined "traditional" design building.

Appendix 1: Questionnaire

Mittuniversitetet
MID SWEDEN UNIVERSITY

Frågeformulär

Sociala aspekter av den fysiska miljön i flerbostadshus

I olika sammanhang framhålls den fysiska miljöns bidrag till social interaktion som en möjlighet till sociala aktiviteter och mänskliga möten. Huvudsyftet med denna enkät är att förstå dina upplevelser och erfarenheter av ditt grannskap kopplat till ditt bostadshus. Denna enkät skickas till alla boende i huset och kan besvaras av hushållets samtliga vuxna (över 18 år) som är boende i lägenheten, antingen i pappersformat eller via länken:

Jhoagugyoughsgcasca.com

Ditt svar kan inte ersättas med någon annans. Inga svar är fel och alla svar är lika värdefulla. Varje enkät har ett kodnummer för varje lägenhet vilket tas bort så fort så fort enkäten skickats in. Ingen annan personinformation samlas in. I redovisning av resultaten framgår aldrig vad enskilda personer har svarat och svaren går inte att spåra till enskilda individer. Namnet på din bostad kommer inte heller att nämnas i resultaten. Enkäten är frivillig och undersökningen görs som grund till ett examensarbete. Svaren i undersökningen blir mer tillförlitliga ju fler som svarar på enkäten.

Oavsett om du väljer att fylla i enkäten i pappersformat eller via länken uppskattar vi om du skickar in ditt svar senast **28 februari 2015.** Portot på kuvertet är betalt i förväg. Fyller du i enkäten via länken använder du det nr som står i vänstra hörnet på det tilldelade formuläret.

Har du frågor, funderingar eller kommentarer kring denna enkät så är du välkommen att kontakta mig, **Malin Krook**, Student, Mittuniversitetet, **e-post:** krookmalin@gmail.com

Handledare på Mittuniversitetet, 831 25 Östersund
Ingrid Zakrisson, Docent, Avdelningen för Psykologi: Ingrid.Zakrisson@miun.se
Kerstin Weimer, Doktorand, Avdelningen för Psykologi: Kerstin.Weimer@miun.se
Itai Danielski, Doktorand, Avdelningen för Ekoteknik- och hållbart byggande: Itai.Danielski@miun.se

A. Bakgrundsinformation

1. **Vilket år är du född?**	År: 1 9 _ _
2. **Jag är …**	☐ Man ☐ Kvinna
3. **Vilken är din högsta utbildning?**	☐ Grundskola, folkskola, realskola eller liknande ☐ Gymnasieutbildning ☐ Universitets – eller högskoleutbildning
4. **Vilken är din nuvarande arbetssituation?**	☐ Arbetande ☐ Arbetssökande ☐ Pensionär ☐ Studerande ☐ Annat
5. **Är du gift/sambo?**	☐ Ja ☐ Nej
6. **Hur många personer bor i hushållet inräknat dig själv?**	……………
7. **Om det bor barn i hushållet, hur många barn mellan 0 och 10 år bor där regelbundet?**	……………
8. **Vad har du för nationalitet?**	☐ Svensk medborgare ☐ Medborgare från annat nordiskt land ☐ Medborgare från annat EU-land ☐ Medborgare av annan nationalitet

B. Frågor om din bostad

1. Vilket år flyttade du till din nuvarande bostad?	År: ☐				
2. Vilken typ av boende har du?	❑ Hyresrätt ❑ Bostadsrätt				
3. Hur många rum förutom köksutrymmet har du?	❑ 1 rum ❑ 2 rum			❑ 3 rum ❑ 4 rum eller mer	
4. Varför valde du just den här lägenheten?	Stämmer inte alls				Stämmer mycket bra
Området har tilltalande omgivningar	❑	❑	❑	❑	❑
Närheten till affärer, butiker, andra bekvämligheter	❑	❑	❑	❑	❑
Kostnaden var lägre än de alternativ jag tittade på	❑	❑	❑	❑	❑
Jag tilltalades av arkitekturen och utformningen	❑	❑	❑	❑	❑
Området har en bra skola, förskola etc.	❑	❑	❑	❑	❑
Den ligger nära min eller min partners arbetsplats	❑	❑	❑	❑	❑
Annat:	❑	❑	❑	❑	❑

5. Vad hade du för relation till området innan du flyttade hit?	❑ Jag hade bekanta i området när jag flyttade hit ❑ Ingen koppling till området sedan tidigare ❑ Jag är uppväxt i området ❑ Jag har bott i området tidigare
6. Har du planerat att flytta från lägenheten inom det närmaste året?	❑ Ja ❑ Nej ❑ Vet ej

7. Om du planerar att flytta från lägenheten inom det närmaste året, vad är orsaken till detta?

____

C. Delaktighet

Vilka aktiviteter förekommer i ditt grannskap? Har det någonsin hänt att du deltagit i följande exempel på aktiviteter?	Ej aktuellt	Aldrig	Någon gång om året	Någon gång i månaden	Någon gång i veckan	Dagligen
Deltagit i möten som ordnas av ditt grannskap?	❑	❑	❑	❑	❑	❑
Engagerat dig i styrelsearbetet i ditt bostadshus?	❑	❑	❑	❑	❑	❑
Deltagit i en aktivitet som anordnats av boende i ditt bostadshus, bortsett från medlemsmöten?	❑	❑	❑	❑	❑	❑
Utfört något arbete för ditt grannskap bortsett från möten?	❑	❑	❑	❑	❑	❑
Träffat någon granne för att umgås över en fika eller någon annan social aktivitet?	❑	❑	❑	❑	❑	❑
Finns det andra aktiviteter i ditt bostadshus som du varit delaktig i? Vänligen specificera: ………………………………………………………………	❑	❑	❑	❑	❑	❑

D. Kontakt med grannar

Vänligen ange hur ofta under det senaste året, du har blivit ombedd att göra var och en av sakerna för en granne i ditt bostadshus	Aldrig	Någon gång om året	Någon gång i månaden	Någon gång i veckan	Dagligen
Passa någons barn	❑	❑	❑	❑	❑
Handla matvaror åt en granne	❑	❑	❑	❑	❑
Låna ut verktyg eller mat	❑	❑	❑	❑	❑
Titta till en grannes hem när denne varit bortrest	❑	❑	❑	❑	❑
Hjälpa en granne i en nödsituation	❑	❑	❑	❑	❑
Erbjuda en granne råd om ett personligt problem	❑	❑	❑	❑	❑
Diskutera ett problem som berör bostadshuset med en granne	❑	❑	❑	❑	❑

E. Vägledande principer i ditt liv

Ange nedan hur väl var och en av följande principer stämmer med din uppfattning.	Stämmer Inte alls	Stämmer dåligt	Stämmer delvis	Stämmer något	Stämmer helt
1. Vara vidsynt (visa tolerans för annorlunda idéer och trosuppfattningar)	❑	❑	❑	❑	❑
2. Vara hjälpsam (arbeta för andras välfärd)	❑	❑	❑	❑	❑
3. Ha social makt (kontroll över andra, dominans)	❑	❑	❑	❑	❑
4. Vara framgångsrik (uppnå mål, lyckas)	❑	❑	❑	❑	❑
5. Sträva efter social rättvisa (förhindra orättvisa, bry sig om de svaga)	❑	❑	❑	❑	❑
6. Vara förlåtande (villig att förlåta andra)	❑	❑	❑	❑	❑
7. Ha auktoritet (rätt att leda och befalla)	❑	❑	❑	❑	❑
8. Vara duglig (kompetent, skicklig, effektiv)	❑	❑	❑	❑	❑
9. Sträva efter jämlikhet (ge alla samma möjligheter)	❑	❑	❑	❑	❑
10. Vara lojal (trogen mina vänner och den egna gruppen)	❑	❑	❑	❑	❑
11. Vara rik (materiellt ägande och pengar)	❑	❑	❑	❑	❑
12. Vara ambitiös (hårt arbetande, målmedveten)	❑	❑	❑	❑	❑
13. Sträva efter fred i världen (undvika krig och konflikter)	❑	❑	❑	❑	❑
14. Vara ansvarstagande (pålitlig, tillförlitlig)	❑	❑	❑	❑	❑
15. Vara mån om mitt rykte (skydda mitt anseende)	❑	❑	❑	❑	❑
16. Vara inflytelserik (påverka människor och skeenden)	❑	❑	❑	❑	❑

F. Känsla för grannskapet

Följande frågor berör din upplevelse av grannskapet och gäller för det bostadshus du bor i.

Hur viktigt är det för dig att uppleva en känsla av gemenskap med andra i ditt grannskap?

1	2	3	4	5	6
Föredrar att inte vara en del av detta grannskap	Inte alls viktigt	Inte särskilt viktigt	Någorlunda viktigt	Viktigt	Mycket viktigt

Hur väl representerar vart och ett av följande påståenden hur du känner för ditt grannskap?	Tar helt avstånd	Tar delvis avstånd	Tveksam	Instämmer delvis	Instämmer helt
1. Jag kan få vad jag behöver i detta grannskap.	❑	❑	❑	❑	❑
2. Detta grannskap hjälper mig att uppfylla mina behov.	❑	❑	❑	❑	❑
3. Jag känner mig som en del av detta grannskap.	❑	❑	❑	❑	❑
4. Jag hör hemma i detta grannskap.	❑	❑	❑	❑	❑
5. Jag har inflytande över vad som händer i mitt grannskap.	❑	❑	❑	❑	❑
6. Människor i det här grannskapet är bra på att influera varandra	❑	❑	❑	❑	❑
7. Jag känner samhörighet med detta grannskap.	❑	❑	❑	❑	❑

8. Jag har en god sammanhållning med andra i detta grannskap	❑	❑	❑	❑	❑

Du är välkommen att göra egna tillägg eller kommentarer till enkäten här:

Hjärtligt tack för din medverkan!

References

1. M.S. Granovetter, The strength of weak ties. Am. J. Sociol. 1360–1380 (1973)
2. L. Wood, L.D. Frank, B. Giles-Corti, Sense of community and its relationship with walking and neighborhood design. Soc. Sci. Med. **70**(9), 1381–1390 (2010)
3. V. Mehta, Look closely and you will see, listen carefully and you will hear: urban design and social interaction on streets. J. Urban Des. **14**(1), 29–64 (2009)
4. B.B. Brown, J.R. Burton, A.L. Sweaney, Neighbors, households, and Front Porches: new urbanist community tool or mere nostalgia? Environ. Behav. **30**(5), 579–600 (1998)
5. J. Gehl, 'Soft edges' in residential streets. Scand. Hous. Plann. Res. **3**(2), 89–102 (1986)
6. C.C. Marcus, W. Sarkissian, *Housing as if People Mattered. Site Design for Medium Density Family Housing* (University of California Press, 1992) 280 pages
7. I. Danielski et al., Heated atrium in multi-storey apartment buildings, a design with potential to enhance energy efficiency and to facilitate social interactions. Build. Environ. **106**, 352–364 (2016)
8. D.D. Perkins, D.A. Long, in *Neighborhood sense of community and social capital.* Psychological Sense of Community (Springer, 2002) pp. 291–318
9. D.W. McMillan, D.M. Chavis, Sense of community: a definition and theory. J. Commun. Psychol. **14**(1), 6–23 (1986)
10. N.A. Peterson, P.W. Speer, D.W. McMillan, Validation of a brief sense of community scale: confirmation of the principal theory of sense of community. J. Commun. Psychol. **36**(1), 61–73 (2008)
11. S.H. Schwartz, Universals in the content and structure of values: Theoretical advances and empirical tests in 20 countries. Adv. Exp. Soc. Psychol. **25**, 1–65 (1992)

12. A. Hansla et al., The relationships between awareness of consequences, environmental concern, and value orientations. J. Environ. Psychol. **28**(1), 1–9 (2008)
13. D.M. Chavis, A. Wandersman, Sense of community in the urban environment: a catalyst for participation and community development. Am. J. Commun. Psychol. **18**(1), 55–81 (1990)

Comparative Evaluation of City Dwellers' Perspectives on Household Energy Use Based on Housing Tenure: Survey Results from Northern Sweden

Gireesh Nair, Thomas Olofsson, Annika Nordlund and Christine Hudson

Abstract The successful implementation of energy efficiency measures in the residential sector will depend to a large extent on the attitudes and perceptions of the end-users since they are the final decision maker. The tenure of the housing could influence the building occupants' perspectives on energy issues. In this study we conducted a comparative evaluation of perspectives on energy use of three categories of households: those living in single family houses, tenants and owners' of apartment. The analysis is based on responses to a mail-in questionnaire by approximately 650 residents in Umeå, Sweden. Majority of the respondents believed that their annual household energy use is less. Residents in single-family houses, as compared to the other two types of tenure of the housing, were more likely to believe their heat energy use as high and likely to take actions to reduce the energy use. Financial incentives such as subsidy or lower interest rate were preferred by most of single-family homeowners (45%) to motivate them to take actions to reduce energy use. While personalized information to reduce energy use and lower interest rate and reduced rent are preferred by more residents in the other two categories. The implications for promoting energy efficient measures based on housing tenure is discussed.

Keywords Perception · Occupants · Energy efficiency

G. Nair (✉) · T. Olofsson
Department of Applied Physics and Electronics, Umeå University, Umeå, Sweden
e-mail: gireesh.nair@umu.se

A. Nordlund
Department of Psychology, Umeå University, Umeå, Sweden

C. Hudson
Department of Political Science, Umeå University, Umeå, Sweden

D. Johansson et al. (eds.), *Cold Climate HVAC 2018*,
Springer Proceedings in Energy, https://doi.org/10.1007/978-3-030-00662-4_66

1 Introduction

Energy efficiency improvements in buildings could significantly reduce the greenhouse gas emissions. Sweden has approximately 4.5 million dwellings, of which single-family houses constitute approximately 45% of the dwellings while the rest are apartments in multi-family buildings [1]. There is a large potential to improve the energy efficiency of residential buildings in Sweden [2]. The energy efficiency investments in existing single-family houses will depend largely on the attitudes and perceptions of the house owners since they are the final decision maker pertaining to their house. The multi-family buildings in Sweden are owned by municipality housing companies, private housing companies or by co-operative housing associations. The apartments in municipality and private owned housing companies are rented out, while the apartments in co-operative housing association is similar to a condominium. The new buildings, which are steered by stringent energy efficiency requirements, are added to the existing building stock very slowly. Hence to reduce the energy use from the building sector it is important to improve the energy efficiency of the existing buildings.

The building occupants could play an important role to reducing the energy use in existing residential buildings. The tenure of the housing could influence the perspectives of the occupants on energy issues. Residents in different home tenure's could have different priorities and limitations related to household energy use. A tenant usually live in a house for relatively shorter duration, for example an American study showed that the renters move every two years [3]. Hence the tenants may not be interested to invest in energy efficiency improvements. A tenant may like to reside in apartment with lower rent, especially when all other important factors that influence their housing choice are similar. If the tenants could reduce the rent by energy efficient behaviour then they may have a positive attitude to reduce energy use. However, if the energy cost is included in the rent which makes the energy cost "fixed" then the tenants may be less inclined to make efforts to reduce their household energy use. In Sweden, more than 80% of multi-family buildings are heated by district heating [2], and usually the apartments do not have a separate heat energy meter and the heating cost is included in the monthly fee/rent. The electricity cost is usually paid separately by the residents. Nevertheless in some co-operative housing associations (and also in some rented apartments) the electricity cost is included in a monthly fee (towards maintenance, common services, loan repayment, if any) the apartment owners pay to the housing association.

The apartment owners also face certain limitations in-terms of their household energy use. For example, an apartment owner as an individual, especially in Sweden, cannot choose the heating energy supplier. Apartment owners' cannot make decision related to building envelope improvement as such decisions are usually made by the executive board of the co-operative housing associations. Moreover, for majority of the apartments in Sweden the heating energy cost is included in the monthly fee the apartment owners pay to the association. Hence, in housing associations unless several households reduce their heating energy and thereby reducing the monthly fee, the apartment owners may not get any financial benefit from their actions to reduce the heating energy use.

Among the different categories of home tenures, owners of single-family houses are better positioned to influence their energy use as they are the direct beneficiary of their actions and also have the ability to choose. Nevertheless, a survey in Sweden suggest that a significant percentage of single-family households did not feel a need to improve the energy efficiency of their building envelope [4]. A prerequisite for the residents to adopt energy efficiency measures is that they should feel a need for it [5]. The need to improve energy efficiency could be trigged by various factors such as physical condition of the building component(s), high annual energy cost. If the residents are satisfied with the existing energy use in their houses then they are less likely to implement/adopt measures to reduce the energy use.

This case study aims to provide pointers on possibilities and barriers in reducing the household energy use among residents in a subsection of a city. The analysis is based on the perception of household energy use by three categories of residents living near to a University in northern Sweden.

2 Methodology

The analysis is based on a mail-in questionnaire survey to 3000 residents living near to the Umeå University, Sweden. The survey, based on a random sampling method, was conducted to study the residents perspectives on various aspects related to the sustainable development of the city. The questionnaire had different sections and include questions on travel/mobility aspects, perception and attitude on household energy use, views on energy use feedback, perspectives on various measures (separate questions on mobility and energy use in buildings) related to sustainable development of the city, safety aspects on the city they live, values important for them, and a few question related to socio-economic aspects. Majority of the responses were evaluated on a 7-point Likert scale. For example, to understand respondents' perception on household heating energy use we asked the question "*What do you think about your household annual heating energy use?*". Respondents answered the question using a 7-point Likert scale (1 = very low, 7 = very high). For simplifying the analysis, we reclassified the responses in a 7 point Likert scale into three (for example, *low, neither nor, high*) by grouping options 5–7 as *high* and options 1–3 as *low*, while option 4 is the neutral option which is *neither nor*.

The addresses were collected from Umeå municipality. The survey was carried out during February–April 2016, and 657 filled-in questionnaire were received. Since the survey was conducted among people living near to a University, some characteristics of the respondents' could be different from that of Swedish population. For example, approximately 60% of the survey respondents had University/college education ≥ 3 years (Table 1), while about 24% of the Swedes in the age group of 20–74 years has a University/college education ≥ 3 years [6]. Hence the survey sample, though may be typical to a University neighbourhood, is not representative for the country. Accordingly, generalization of the study results to the entire population is limited.

3 Results

As the target groups were people living near to the University, a large percentage of respondents were living in apartments, either as tenants or as apartment owners. 23, 24 and 53% of the respondents were single-family homeowners, apartment owners and tenants, respectively. Approximately 70% of single-family homeowners live in a house larger than 125 m^2, whereas 63% apartment owners and 78% tenants live in an apartment size less than or equal to 75 m^2. The demographic composition of the respondents are provided in Table 1.

62% of the apartment owners and 76% of the tenants responded to the survey stated that the monthly fee/rent they pay to the housing association/company included the heating cost. Similarly, 25% of the apartment owners and 31% of the tenants stated that the cost of electricity is included in the monthly fee/rent.

The respondents' perception on annual heating energy and electricity use in their homes is presented in Table 2. Majority of the respondents in all the three categories of home tenures did not consider that their heating and electricity use was high. Among the respondents, residents of single-family houses were more likely to consider that their household energy use was high. Similarly, it was found that majority of the respondents consider that their heating and electricity cost was low (Table 3). Among the three categories of residents, single-family homeowners were more likely to consider that their annual heating and electricity cost was high.

More number of single-family homeowners believed that their heating use and cost were high as compared to their electricity use and its cost. However, among apartment owners and tenants there was not much difference on their perception towards heating energy and electricity use and cost. Though majority of respondents consider their annual household energy use and its cost was not high, a large percentage of them consider it important to reduce their household energy use (Table 4). Single-family homeowners were more likely to have a positive attitude towards reducing the heating energy use. However, no statistically significant relationship was found between respondents' home tenure and their attitude to reduce their electricity use.

A large percentage of respondents were of the opinion that they have done whatever possible to reduce the use of heating energy and electricity in their household (Table 5). Among the respondents, single-family homeowners were more likely to agree that they have done whatever possible to reduce the heating energy use in their home. However, no difference was found between residents' home tenure and their belief that they took whatever actions possible to reduce their electricity use.

Table 1 Demographic composition of the respondents

	% of respondents[a]			
	Single-family homeowners (N = 149)	Apartment owners (N = 154)	Tenants (N = 343)	Total survey respondents (N = 646)
Gender				
Male	48	44	46.5	46
Female	52	56	52	53
Other	–		1.5	1
Age				
<=25 years	4	13	36	23
26–35 years	7	23	36	26
36–45 years	20	11	9	12
46–55 years	24	2	6	11
56–65 years	23	24	7	13
>65 years	22	27	6	15
Marital status				
Married/Living together	88	64	43	58
Single	12	36	57	42
Annual household income				
<200,000 kr	5	15	45	29
200,001–300,000 kr	4	14	14	12
300,001–400,000 kr	7	19	16	15
400,001–500,000 kr	10	10	8	9
500,001–600,000kr	12	16	6	9
>600,000 kr	62	24	11	26
Education				
Up to high school	20	27	33	29
University/college (<3 years)	9	11	12	11
University/college ($\geq$3 years)	71	62	55	60
Employment				
Full time job	57	41	32	40
Part time job	11	10	8	9.5
Student	5	14	45	28.5
Pensioners	24	29	8	16.5
Others	3	6	7	5.5

[a]11 respondents didn't mention the category of their house and their responses were excluded from the analysis

Table 2 Respondents' perception on annual energy use in their home

	% of respondents		
	High	Neither nor	Low
Heating energy use[a]			
Single-family homeowners	25	35	40
Apartment owners	7	38	55
Tenants	10	35	55
Household electricity use[b]			
Single-family homeowners	16	37	47
Apartment owners	6	36	58
Tenants	10	35	55

[a]Statistically significant relationship with chi-square test $p \leq 0.001$

[b]Statistically significant relationship with chi-square test $p \leq 0.1$

Table 3 Respondents' perception on annual household energy cost

	% of respondents*		
	High	Neither nor	Low
Heating energy cost[a]			
Single-family homeowners	30	40	30
Apartment owners	8	33	59
Tenants	6	36	58
Household electricity cost[b]			
Single-family homeowners	20	42	38
Apartment owners	8	30	62
Tenants	9	35	56

[a,b]Statistically significant relationship with chi-square test $p \leq 0.001$

*16% of the single-family households and 11% of respondents living in apartments reported that their dwellings are heated partly/fully by electricity. It may be difficult for these households to differentiate between heating and household electricity cost

Table 4 Respondents' attitude to reduce household energy use

	% of respondents		
	High	Neither nor	Low
Heating energy use[a]			
Single-family homeowners	50	17	33
Apartment owners	34	17	49
Tenants	31	19	50
Household electricity use			
Single-family homeowners	45	17	38
Apartment owners	39	17	44
Tenants	40	18	42

[a]Statistically significant relationship with chi-square test $p \leq 0.01$

Table 5 Respondents view on whether they have done what is possible to reduce energy use in their household

	% of respondents		
	Agree	Neither nor	Disagree
Heating energy use[a]			
Single-family homeowners	54	19	27
Apartment owners	48	16	36
Tenants	39	21	40
Household electricity use			
Single-family homeowners	45	25	30
Apartment owners	48	20	32
Tenants	44	21	35

[a]Statistically significant relationship with chi-square test $p \leq 0.05$

Table 6 Measures residents undertook in the last three years (2014–2016) to reduce household energy use

Options to reduce household energy use	% of respondents		
	Single-family households	Apartment owners	Tenants
Switch off lights when I leave room	69	87	85
Lowering indoor temperature	42	32	23
Purchased energy saving equipment (refrigerators, freezers)	56	46	12
Replace electrical equipment such as TVs	37	40	18
Replaced bulbs with LED lamps	84	79	73
Installed water saving equipment	33	36	8
Use less hot water	22	25	30
Invested in home (for example, added insulation)	40	4	1

The non-investment behavioural measures (for example, switching off lights) was the most common energy efficiency measure adopted among apartment owners and tenants (Table 6). 87% of apartment owners and 85% tenants reported that they switch off the lights when leaving the room. Low-investment measure (replacement of bulbs with LEDs) was also a popular energy saving measure among the residents. As expected, tenants were less likely to implement investment oriented measures to reduce electricity. Hot water reduction and water conservation measures were preferred by less number of households.

As compared to residents of single-family houses more apartment owners and tenants were satisfied with their daily energy use (Table 7). Approximately 25% respondents among the three categories of residents agreed that they would consider reducing their household energy use, however, they were not sure on how and when they could do it. More number of single-family homeowners (27%) were aware of the energy aspects in their house and were reflecting on it. In general, single-family

Table 7 Respondents' plan to reduce daily energy use in their household

	% of respondents		
	Single-family households	Apartment owners	Tenants
I am satisfied with the level of energy use in my household and sees no need to reduce it	22	39	34
I would like to reduce the household energy use but at the moment it is impossible for me to do it	6	5	12
I will consider to reduce my household energy use. At the moment I am not sure on how and when I could do it	27	23	23
I have a target to reduce energy use. I know actions I can take but not gone from words to actions yet	17	9	6
I am aware of energy aspects in my house and I am reflecting on it	27	20	17
I cannot influence the energy use in my house and so reducing it is not an option for me	1	4	8

Table 8 Perspectives on various stimuli that could encourage the respondents to reduce household energy use

	% of respondents agreed		
	Single-family households	Apartment owners	Tenants
Reduce the rent if the indoor temperature is reduced (by 1–2 °C)	5	40	42
Financial support/subsidy to install energy saving measures	46	19	23
Possibility to borrow/rent tools	4	7	14
To have reliable information on costs and benefits of energy efficiency measures	29	33	33
Personalized information on ways to reduce energy use	37	33	42
Possibility to have a lower interest rate on home loans when implementing energy efficiency measures	46	43	28

homeowners seems to be having more favourable plan to reduce their daily energy use. Only a few percentage of respondents thought that they could not influence the energy use in their house.

The respondents' view on various stimuli that may encourage them to reduce household energy use is provided in Table 8. 46% of residents of single-family houses were of the opinion that financial support and possibility to have a lower interest rate on home loans when implementing energy efficiency measures may motivate them to take actions to reduce household energy use (Table 8). 40% of apartment owners and 42% of tenants thought that financial incentives, in the form of reduced rent, may motivate them to reduce the indoor temperature. A large number of respondents in all categories of housing tenure believed that access to reliable and customized information on energy efficiency measures may motivate them to take actions to reduce household energy use.

4 Discussion and Conclusions

Majority of the survey respondents from three categories of home tenures didn't consider their annual household energy cost as high. The perceived lower energy cost may be a barrier among majority of respondents, especially those living in apartments, to take actions to reduce their household energy use. For majority of respondents who are living in apartments, the heating energy cost is included in the rent or monthly payment to the association/housing company. As the heating energy cost is "hidden" in the monthly payment/rent, the residents of apartment may not be aware of their actual heating energy use and cost. This situation may be partly responsible for the large percentage of (approximately 50% among apartment owners and tenants) respondents non-favourable attitude to reduce the heating energy use. Among the three groups of respondents, the residents of single-family houses were more likely to consider that their heating energy and electricity cost is high. Further, relatively large number of residents in single-family houses were likely to take actions to reduce their daily household energy use. The perceived higher energy cost may be a motivation for their energy efficiency improvement plans.

Majority of respondents didn't consider their household energy use as high. Nevertheless, a large percentage of residents consider it as important to reduce their household energy use (both heating energy and electricity). This suggest a positive attitude among a large number of city dwellers towards environment.

Majority of the respondents undertook switching-off lights as an energy saving measure, which supports earlier findings that the homeowners are more likely to undertake non-investment measures as against investment measures [7]. Individuals' preference towards switching-off lights as an energy reduction measure may be partly attributed to the "visibility" aspects associated with that action. Though switching off lights when not needed is a positive action towards sustainability, the energy saving potential for such a measure may be limited. This may have implications on residents' further energy efficiency behaviour especially when they assume that such measures could lead to significant energy cost reduction. For example, as noted by Kempton et al. [8] if the measures undertaken in anticipation of specific cost savings could not results in the desired savings, such as switching off lights, then people may get disenchanted with energy efficiency measures.

Reducing the indoor temperature, which is another non-investment measure is preferred by relatively less number of respondents. Tenants were less likely to reduce their indoor temperature as an energy saving measure. However, a large number of tenants also mentioned that they might do so if they receive financial incentive (lower rent) for their action to reduce the indoor temperature.

A number of residents are interested to reduce their household energy use, however, they are not sure on how and when they could carry out the measures.

Moreover, due to lack of accurate information many residents may consider that their efforts to reduce energy use is sufficient [9] and hence may miss measures that could significantly reduce their energy use. Providing relevant information on energy efficiency measures that are perceived reliable may facilitate these group to take (start) energy reduction initiatives. Approximately, one third of the respondents consider that access to reliable information on costs and benefits of energy efficiency measures could motivate them to implement measures to reduce their household energy use. Moreover, a large number of respondents also believed that customized information could motivate them to take actions to reduce the energy use. A large percentage of respondents, especially single-family homeowners believed that they have done whatever possible to reduce their household heating energy. Accordingly, due to "selective exposure", those residents may ignore information on energy efficiency measures as such information may be inconsistent with their beliefs and needs to reduce energy use. This perception of the end-users may create challenges in communicating to these residents about further energy efficiency improvements, if any. The results from this study reiterates the importance of provision of effective and reliable information, especially customized information as a policy tool to promote energy efficiency in the residential sector.

Sweden uses a mix of regulatory, fiscal and informational policy instruments to promote energy efficiency measures in the residential sector. However, the target group for several such policy instruments are often single-family households. For example, the target group for municipality energy advisers which provide impartial advice on energy issues are mainly single-family households [10]. Major subsides, especially those related to thermal energy, for energy efficient equipment are also usually aimed for single-family houses. Hence apartment residents, which constitute more than 50% of the total dwellings may not be reached sufficiently through policy instruments. It is important to include this segment's concerns and needs while introducing policy tools to facilitate the energy reduction targets in the residential sector.

Acknowledgements The authors gratefully acknowledge the financial support from the Swedish Energy Agency.

Note: The survey was conducted in 2016, in accordance with the APA (American Psychology Association) general principles. The survey included information about the study, about participation being voluntary, that data would be handled maintaining confidentiality, and that data only would be analysed on a group level, in the first section of the survey. Participation in the survey was then taken as informed consent. No formal ethical vetting was conducted since the survey did not contain questions of a sensitive nature.

References

1. SCB (Statistics Sweden), Yearbook of Housing and Building Statistics 2012. [Bostads- och byggnadsstatistisk årsbok 2012] ISSN 1654-0921 (2012)
2. J. Nässén, J. Holmberg, Energy efficiency—a forgotten goal in the Swedish building sector? Energy Policy **33**, 1037–1051 (2005)

3. U.S. Department of Housing and Urban Development: American Housing Survey for the United States: 2007 (2008)
4. G. Nair, L. Gustavsson, K. Mahapatra, Owners perception on the adoption of building envelope energy efficiency measures in Swedish detached houses. Appl. Energy **87**, 2411–2419 (2010)
5. E.M. Rogers, *Diffusion of innovations*, 5th edn. (Free Press, New York, 2003)
6. Statistics Sweden: Population by level of education, sex, age and born in Sweden/foreign born 2016, https://www.scb.se/en/finding-statistics/statistics-by-subject-area/education-and-research/education-of-the-population/educational-attainment-of-the-population/. Last accessed on 13 Oct 2017
7. G. Nair, L. Gustavsson, K. Mahapatra, Factors influencing energy efficiency investments in existing Swedish residential buildings. Energy Policy **38**, 2956–2963 (2010)
8. W. Kempton, J.M. Darley, J. Weihl, Do consumers know "what works in energy conservation? Marriage and Fam. Rev. **9**, 115–133 (1985)
9. S.Z. Attari, M.L. DeKay, C.I. Davidson, W.B. de Bruin, Public perceptions of energy consumption and savings. PNAS **107**, 16054–16059 (2010)
10. K. Mahapatra, G. Nair, L. Gustavsson, Swedish energy advisers' perceptions regarding and suggestions for fulfilling homeowner expectations. Energy Policy **39**, 4264–4273 (2011)

Challenges in Transdisciplinary Research—Example from a Study on People as Part of Energy and Ventilation Systems in Residential Buildings (PEIRE)

Eja Pedersen, Jonas Borell, Héctor Caltenco, Mats Dahlblom, Chuansi Gao, Lars-Erik Harderup, Yujing Li, Birgitta Nordquist, Kristian Stålne, Petter Wallentén and Aneta Wierzbicka

Abstract Energy efficiency measures in residential buildings typically include changes in ventilation and heating systems, and increased thermal insulation of the building envelope. The expected energy efficiency is not always reached, despite large knowledge and professional implementation of each separate measure. There is a lack in understanding of how technical systems interact, and how the occupants are influenced by and in turn influence the systems by their behaviour. A holistic view and a transdisciplinary research approach are needed to understand relevant interactions and propose integrated energy efficiency measures. The aim of this paper is to reveal challenges in transdisciplinary research projects that include real

E. Pedersen (✉)
Environmental Psychology, Department of Architecture and Built Environment, LTH, Lund University, Lund, Sweden
e-mail: eja.pedersen@arkitektur.lth.se

J. Borell · C. Gao · A. Wierzbicka
Ergonomics and Aerosol Technology, Department of Design Sciences, LTH, Lund University, Lund, Sweden

H. Caltenco
CERTEC, Department of Design Sciences, LTH, Lund University, Lund, Sweden

M. Dahlblom · B. Nordquist
Building Services, Department of Building and Environmental Technology, LTH, Lund University, Lund, Sweden

L.-E. Harderup · P. Wallentén
Building Physics, Department of Building and Environmental Technology, LTH, Lund University, Lund, Sweden

Y. Li · K. Stålne
Department of Materials Science and Applied Mathematics, Malmö University, Malmö, Sweden

D. Johansson et al. (eds.), *Cold Climate HVAC 2018*,
Springer Proceedings in Energy, https://doi.org/10.1007/978-3-030-00662-4_67

world studies on both humans and technical systems with measurements before and after renovation of multifamily housing. It is based on experiences from the PEIRE-project (People, Environment, Indoor, Renovation, Energy) carried out by a research team with expertise on environmental psychology, human behaviour, interaction design, universal design, building physics, building services, thermal comfort, aerosol technology, exposure assessment, acoustics, daylight, and complex thinking. Differences in theoretical bases and methodology needed to be dealt with. Metatheory building could help with the transition from a multi- to a transdisciplinary understanding.

Keywords Renovation · Energy efficiency · Indoor climate · Multifamily housing Tenants · Transdisciplinary research · Metatheory

1 Introduction

1.1 Human Understanding and Behaviour as Part of HVAC Research

Energy efficiency measures in residential buildings typically include changes in ventilation and heating systems, and increased thermal insulation of the building envelope. The expected energy efficiency is not always reached, despite large knowledge and professional implementation of each separate measure. There is a lack in understanding of how technical systems interact, and how the occupants are influenced by and in turn influence the systems by their understanding and behaviour. For example, a more air-tight façade could require increased ventilation; increased ventilation could generate higher noise levels and higher energy use, especially if installed without heat recovery measures. Individuals perceiving the noise as disturbing might reduce or even block the ventilation, and compensate this by opening a window; resulting in an increase in energy use instead of a decrease, when in cold-climate. In order to map the energy use and flow in such situations, knowledge on both the technical aspects and human behaviour are required. A holistic view and a multi- or transdisciplinary research approach are needed to understand relevant interactions and propose integrated energy efficiency measures.

1.2 Multi-disciplinary and Trans-disciplinary

The holistic approach will entail cooperation over the boundaries of disciplines, which calls for clarifications on some concepts that relate to different strategies on how to pursue this. First, with a scientific discipline we mean a coherent body of knowledge, methods and theories engaged by a community of researchers pursuing what they regard as meaningful research questions. The discipline is associated with a set of assumptions, criteria and norms on how research should be carried out and regarding good quality research. Multidisciplinary research involves researchers from more than one discipline that share a research question, but operating from within their respective disciplines [1].

In transdisciplinary research activities the research question is shared, but here methods and approaches may be transferred between the different disciplines. This entails higher ambitions of integration between the researchers and their disciplines, and new disciplines or theories might emerge from this. It might also lead to disagreement or even conflicts when treading into other researchers' territories or on differing assumptions on how to perform good research. Some of these problems can be attributed to communication difficulties due to different scientific backgrounds with different terminology that arise despite a fundamentally shared understanding of the studied object or process.

1.3 The PEIRE-project

The PEIRE-project (People, Environment, Indoor, Renovation, Energy) aims at improved understanding of how the building, the technical systems for ventilation and energy provision, and the individual occupying the building interact [2]. A multidisciplinary research team was set up with expertise on environmental psychology, human behaviour, interaction design, universal design, building physics, building services, thermal comfort, aerosol technology, exposure assessment, acoustics, daylight, and complex thinking.

Thorough technical measurements of building performance and the physical indoor environment, and both quantitative (sensors registering occupancy, activity at the stove, and window opening; questionnaires; journals) and qualitative studies (interviews) of the tenants' understanding of the systems and behaviour, were carried out in 10 typical flats built in the 1970's in southern Sweden. The studies were carried out during the winter and spring periods prior to the buildings' renovation, which took place during the summer. The housing company's objectives for the renovation were multiple, i.e. to perform major maintenance on the buildings, to lower the energy used for heating, tap water, and technology, and to improve the indoor environment in terms of thermal comfort and air quality. The renovation is from the research project seen as an intervention that alters building performance as well as qualities of the indoor environment for the occupants. The measurements will be repeated next winter to capture the changes due to the renovation. Data will be co-analyzed with a meta-theoretical perspective with the aim to present a transdisciplinary model of the interaction between building, technical systems, and tenants (Fig. 1).

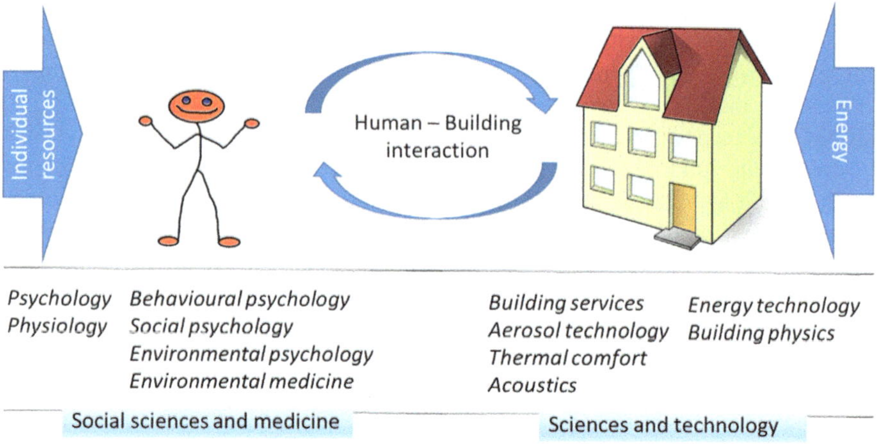

Fig. 1 Individuals are influenced by the indoor environment in the building and, in turn, influence the building performance and therewith the energy needed. An array of traditional research disciplines are needed in studies aiming at a holistic understanding of the interaction, of which some are listed here. Other scientific approaches, such as universal design that deal with the direct interaction between people and technology, could help bridging the gap

The PEIRE-researchers met within the multidisciplinary program Healthy Indoor Environments at the Pufendorf Institute at Lund University (now Centre of Healthy Indoor Environments, CHIE) [3]. There we were given the opportunity to discuss indoor environments and human–environment interactions one day a week for eight months, without any demands for counter deliveries [4]. It was a good opportunity to try to understand the starting points that researchers in other disciplines have, to have collisions of different opinions, to find the common ground and connections of the thoughts, to create new ideas, and to write joint proposals for research projects. Despite this good start, new challenges related to the transdisciplinary approach had to be met once the actual research projects started. Some of these will be discussed here.

2 Aim

The aim of this paper is to reveal challenges in transdisciplinary research projects that include real world studies on both technical systems and humans. The paper is based on our experiences from the PEIRE-project and the workshop that took place after the first round of measurements. We hope to inspire more research that extends beyond traditional disciplinary boundaries and therewith meets the societal demand to provide healthy indoor environments with good comfort, and at the same time low energy use.

3 Experiences from a Transdisciplinary Research Project

3.1 Visualising Theoretical Assumptions and Prejudices

Working in a transdisciplinary research group with the ambition to present a joint model demands willingness from all researchers to understand the contexts that each researcher are acting within, e.g. theoretical base, measurement standards, and terminology. From our experiences from the Pufendorf program, it takes a lot of time to transfer such experiences and knowledge. For one thing, within social sciences there is a firm tradition of always declaring the theoretical point of departure for choice of study methods or how the results are reported. This declaration of theoretical perspective is not as common within science and technology research in which fundamental science laws are taken for granted. The perspectives that the researchers in the program contributed with to the group differed accordingly. From the social sciences, theories of how people understand and act in their environment were presented as the main issue, with empirical data supporting the theories' correctness. Presentations from the scientific and technical side were, on the other hand, often derived directly from empirical measurements, with conclusions based on aggregated observations. This distinction needed to be bridged already when the proposals for financing were formulated.

The large differences between social sciences and natural sciences are easy to observe, but the variances between research fields that for a scholar from 'the other side' appears to be alike are more difficult to detect. For example, within psychology, different theoretical frameworks and methodological traditions rule the research on human behaviour [5]. The differences are reflected in which journal the results are published; there are seldom any cross-over from one branch of psychology to another. The same diversity is found within science and technology, though possibly more based on the study object than differences between theoretical frameworks. Even though the differences between research fields are better bridged in socio-technical sciences, which mix areas in behavioural psychology and technology to better understand technology from the human-socio perspective, the knowledge in the field is far from holistic. One challenge is therefore to clearly understand which disciplines are needed to answer the particular research question in focus. This includes understanding which questions cannot be addressed by a defined set of disciplines. Assumptions that for example all engineers understand all technical systems in the building or that psychologists can study sociological phenomena need to be addressed.

3.2 Adjusting to Disciplinary Customs

Each discipline has its traditions when it comes to measurement standards. One distinction is if the results of the measurement will be used as the main outcome results, for adjusting other results, or as input in different types of modelling. How data should be used also controls whether instantaneous values, logged values over time, or aggregated values are required. For the technical measurements, the goal is to capture the physical environment as well as possible, though always with compromises due to for example the limitations of the measuring equipment, the physical context, and the time available.

In the first data collection within our project, it was clear that we had not foreseen the diversity of measurement modes that the researchers in the group expected to perform, despite that they already had substantially scaled down their sub-studies with respect to other researchers' needs. Temperature is an example of that. Temperature was measured in various places in the same apartment using different instruments and loggers, with different time resolution and different length of measuring periods. It was derived as comfort temperature in the living room during two hours (main outcome), logged temperature in the bedroom (ventilation and humidity study), and logged temperature in the living room, balcony and kitchen (behavioural study). Data give good insight in how the temperature interacts with for example indoor air quality or tenants' window opening. Before the next round of measurement, the possibility to limit the amount of measurements even further, without losing the quality, by identifying data that could satisfy the need of more than one discipline will be discussed.

Another type of measurements that had different purposes for different disciplines were questionnaires that the tenants were asked to fill in. For experts on energy modelling, the questions were seen as a way of obtaining input data for the models. In other cases, they were used to understand technically collected data from, for example, aerosols. In a third case, the questionnaires were the main instrument for behavioural attitudes to study the tenants' perceptions and understanding of their indoor environment. The different approaches created frustration in the design of the questionnaires, especially as all researchers realized that it was not possible to burden the tenants with too many questions.

3.3 Working in Real Environments

Places where researchers perform their studies depends on research questions and in part on traditions within the discipline, but also on type of studies the researcher has developed his or hers skills. The PEIRE study is carried out in real environments [6], i.e. in apartments with the tenants present most times. For researchers who were mainly active in laboratories, it posed new challenges as the equipment had to be adapted to the limited space and if left over time, had to be accepted by the tenants as part of the interior. There was also no time for trying out the measurements on site, and for example getting an additional cord if needed, or redoing measurements that did not work out as expected.

Another challenge was to not interfere with each other measurements within the research group. Eleven senior researchers with doctoral and master students and research assistants were measuring or setting up log measurements, at only three occasions in each apartment to minimize the amount of visits. An examples of a logistic considerations was that information about tenants' understanding of the technical systems that rules their indoor environment needed to be obtained before any of the technical measurements started to avoid influence on their perception. Also, temperature and relative humidity in the apartment is influenced by opening the front door and the number of people in the apartment, which must be considered when scheduling the measurements. Measurements of sound pressure levels are sensitive to ambient noise and could not be carried out when the equipment for airborne particles was operating. It was a demanding task to plan the multi-parallel measurements that had to be carried out simultaneously or overlapping with limited amount of equipment, resource and during a limited period of time, since the measurements needed to be finished before the end of the cold season.

3.4 Ethical Considerations in Human Subjects Research

As a consequence of what was revealed during the Nuremberg trials after the Second World War, ethical rules for human subject research were agreed on internationally in the Declaration of Helsinki 1948 [7]. The declaration has been updated and followed by other international agreements and implementations in national legislation. The main principles are that everyone involved in research volunteers, that the impact on the participants is as small as possible and not harmful, and that the research contributes to the improvement of the society. This means that the participants should be informed about what it means to participate in the research study and give their approval (written informed consent), that they have the right to terminate participation at any time without explaining why, and that the number of participants should be large enough to secure that the results could be generalized, but not larger than needed.

In Sweden, the Central Ethical Review Board monitors compliance with the rules. Researchers apply for ethical vetting to one of their six regional boards. This procedure is well known among medical and psychology researchers, but from our observations, routine mainly among technology researchers collaborating with medicine and psychology. However, applying for ethical vetting concerns all human subject research. Within the PEIRE-project, only few of the researchers were familiar with how to handle the ethical requirements such as how to formulate invitation letters, pointing out that participation was voluntarily without risking to loose participants, or coding of participants identity though still being able to reach them. As the interest in research involving people increases within the sciences and engineering disciplines, the requirement for education in how ethical issues are to be addressed also increases. The issue should be discussed already during the PhD-studies.

3.5 Managing Multiple Data Sets

In Sweden, the University owns all data collected in research projects that belongs to the institute. Universities in Sweden are federal authorities and the publicity principles implies that anyone can request the data after publication. Data therefore need to be systematically stored so that they can be linked to the current project and can be found without delay. In single-group research, it is often clear who is responsible for the collected data and this is therefore not a problem; in practise it is the project leader who takes responsibility for how the data is handled and stored, including deciding who has access to it. Senior researchers who run their own projects might be used to storing their data in personal computers or, when on paper, in their offices. In the case of human subject research, they also take the responsibility to keep data in such a form that they do not reveal who participated in the study.

Empirical studies like the PEIRE-project with the aim to capture many aspects of the indoor environment in multi-family housing, including indoor air quality, human understanding and behaviour, and energy use, generate loads of data. The facts that the researchers involved are active at different departments and that the project has the ambition to analyse data as a whole have complicated data management. A list of managers for each sub-set of data was established, and a routine regarding sharing the data is under development. The closed and secure university web-based platforms are used for sharing files, but it is difficult to determine when a data set is in a state to be up-loaded, i.e. when it can be meaningfully used by other researchers within the group. An additional question is how data can be used by the individual researchers for niched publications within the individual research fields. A pragmatic view, allowing large freedom for publication as long as it does not precede the joint analyses and publications has been applied so far, though as the amount of data increases, a more stringent policy might need to be agreed on.

3.6 Advice for Future Transdisciplinary Collaborations

There is a clear need for transdisciplinary cooperation with the aim of increasing the understanding of the interaction between individuals and the physical environment. Our experience is that it is both fun and fruitful to work in such groups. However, we want to point out some challenges.

- Allow time before the study to learn to know each other's research fields, especially the underlying theoretical assumptions and the terminology. The effort and amount of time needed for this should not be underestimated.
- Explain the purpose of each measurement to each other and how the results should be used.
- Adjust all measuring devices to real environments by limiting the size and creating a design that fits with the environment.

- Make a pre-trail with all measurements in real scenario, with established time-scales. Make sure your planned measurements are feasible and that they do not interfere with each other. Include researchers in the group with logistic skills.
- Be sure that all researchers are aware of ethical issues in human subject research and the implications for treatment of the participants and handing of data.
- Make an agreement already in the beginning of the project on how collected data should be handled, who is responsible for each sub-set, how data could be used for publications, and how it should be shared between the researchers.
- If possible, carry out a pilot study that comprises all the steps from data collection to analysis. This will for example clarify if certain data can be used by multiple disciplines and reveal if additional variables need to be measured.

4 From Multi to Trans: Metatheory Building

In this article, we employ both multidisciplinary and transdisciplinary approaches to address our research question and we argue that the former is an appropriate starting point but the latter may follow as a result of researchers getting more acquainted with the complexity of the research question as well as with each other and the respective research disciplines. A conceptual tool that we have found helpful in this transition is metatheory building [8]. If theories are ways of organising empirical data and coordinate different variables and factors, metatheories coordinate theories and see them as lenses or perspectives on the issue or phenomenon at hand. A research discipline may be considered as representing a certain perspective, and examples of perspectives on indoor environments could be medical, social, psychological, physical or technical. Organising the different perspectives in an appropriate metatheory can clarify and articulate different assumptions that are associated with the respective discipline. Thus, metatheory building can facilitate the transition from a multidisciplinary to a transdisciplinary approach.

Acknowledgements The projects within PEIRE are financed by the Swedish Research Council for Environment, Agricultural Sciences and Spatial Planning FORMAS (942-2016-79) and the Swedish Energy Agency (43092-1).

References

1. J. Fröberg, L. Geschwind, C. Sundström, Att utvärdera tvärvetenskap : reflektioner utifrån Högskoleverkets utvärderingar 2001–2005 (Stockholm, 2007). Retrieved from http://urn.kb.se/resolve?urn=urn:nbn:se:kth:diva-162224
2. PEIRE Homepage, http://www.peire.lth.se/. Last accessed 8 Aug 2017
3. Centre for Healthy Indoor Environments Homepage, http://www.eat.lth.se/indoor-environment. Last accessed 8 Aug 2017

4. Pufendorf Institute Homepage, http:// www.pi.lu.se/. Last accessed 14 Aug 2017
5. M. Johansson, L. Neij, Addressing human behaviour in assessments of energy efficiency in buildings. ECEE, France, June (2017)
6. C. Robson, *Real World Research: A Resource for Users of Social Research Methods in Applied Settings*, 3rd edn. (Wiley, Chichester, West Sussex, 2011)
7. Declaration of Helsinki. Bull. World Health Organ., **79**(4), 373–374 (2001)
8. M. Edwards, Misunderstanding metatheorizing. Syst. Res. Behav. Sci. **31**(6), 720–744 (2014)

Part IX
Indoor Environment and Health

Assessment of Indoor Air Quality and Hygrothermal Conditions of Boarders During Autumn, Winter and Spring in Two of Estonian Straw-Bale Houses

Jane Raamets, Aime Ruus and Mari Ivask

Abstract Indoor air quality affects human health. These effects can be either positive of negative. Straw bale building has been claimed as a sustainable way of building. The aim of this study was to evaluate indoor air quality in two of Estonian straw bale houses and provide solution of monitoring for indoor air quality and hygrothermal conditions of boarders as complex. Samples were collected between October 2014 til March 2015. Sampling media and procedure was designed according to ISO standard 16000-18: *Detection and enumeration of moulds—Sampling by impaction*. Data loggers for collecting the data about CO_2, temperature and humidity were also used. Two of them (recorded temperature and humidity) were installed inside the wall (depth ~20 cm), the third logger was used as a desktop logger. We also collected some straw samples inside the wall to see which kind of microorganisms are living on them. Samples were plated, total colony forming units were counted and identified from the isolated colonies. The results from air samples (CFU) were in one house higher than in the other one. Temperature, humidity and CO_2 levels were also higher in one house. This is probably affected by the different building characteristics (one of the houses is modular wall straw bale house, the other one is timber frame straw bale house). Species, which we found, were similar in both houses. The most fungal genes isolated from samples were *Aspergillus*, *Penicillium*, *Alternaria* and *Cladosporium*.

Keywords Airborne fungi · Indoor air quality · Straw bale houses
Indoor air · Green building

J. Raamets (✉) · A. Ruus · M. Ivask
Tartu College, School of Engineering, Tallinn University of Technology,
Puiestee 78, 51008 Tartu, Estonia
e-mail: jane.raamets@ttu.ee

D. Johansson et al. (eds.), *Cold Climate HVAC 2018*,
Springer Proceedings in Energy, https://doi.org/10.1007/978-3-030-00662-4_68

1 Intoduction

Brasche and Bischof [1] reported in their article that the overall mean time spent at home is around 15.7 h. This time plays a critical role for persons exposure to environmental pollutants [2], so it is important that indoor environment will be healthy for living organisms.

Straw bales are used for building more than 200 years, first in Nebraska [3]. Straw is in most countries easily accessible, it is cheap and mostly usable (if there are no weeds, which can cause rotting). According to Paist et al. [4] every year in Estonia approximately 125 tons of straw will be produced as a byproduct.

Straw has excellent thermal—and sound insulation properties as well it is a low energy process compared to other building materials [5]. Bales are fire resistance (rating F90) and normally the material is inflammable [6].

Three common methods for building with straw bales are used. The traditional load-bearing method, frame and infill system and hybrid technology. There is also a method to build with modular wall panels [7].

Indoor Air Quality is an important issue in the home environment. An inadequately ventilated home environment or a poorly designed ventilation system can lead to the buildup of a variety of problems with indoor air [8].

The actual indoor temperature is the most important data to assess thermal comfort and indoor climate. Relative humidity (RH) and absolute humidity will also play an important role to indoor climate and hygrothermal design conditions [9].

Mold exposures and dampness in buildings are common in buildings [10–12]. A major result caused by moisture damages is health effects on the occupants. The evidence of a causal association between dampness and health effects is strong, unfortunately the mechanisms are unknown [13].

Insufficient ventilation, inadequate insulation or water damage (leakage, pipe bursts) can lead to excess moisture in buildings and to a condensation of water on surfaces such as floors or walls [14].

The aim of this study was to evaluate indoor air quality in two of Estonian straw bale houses via collected data (air and material samples, data about CO_2, humidity and temperature levels) provide solution of monitoring for indoor air quality and hygrothermal conditions of boarders as complex

2 Materials and Methods

The assessment of indoor air quality and hygrothermal conditions of boarders in two of straw bale houses was carried out between October 2014 til March 2015. Occupants were asked not to ventilate the houses 6 h prior to measurements. Sampling media and procedure was designed according to ISO standard 16000-18: *Detection and enumeration of moulds—Sampling by impaction* [15].

Microbiological air samples (collected with Microbio MB2 air samplers) were collected 3 times (during autumn, winter and spring time). Used air samplers were passing 100 l of air per sample. Sampling time was 1 min. A set of four air samples captured in each area was used to obtain average contamination values and representative results. The corresponding outdoor air was measured as reference value.

Sampling was carried out inside 1 m above floor level and outside 1.5 m above the ground. The sample plates were incubated at 32 °C for 72 h and the colony forming units (CFU) were counted. Dichloran glycerol agar (DG18) and malt extract agar (MEA) plates were used as culture media.

Colonies from plates were recultivated periodically. The isolated microorganisms were plated with a spatula on the agar plates. The cultures were grown on the Petri dishes under 32 °C.

Data loggers for collecting the data about CO_2 (about 1.2 m from floor level, measurement was taken after every 30 min), temperature and humidity were also used. Loggers were installed indoors. Two of them (recorded temperature and humidity) were installed inside the wall (depth ~20 cm, height 1.5 m and 20 cm, measurement was taken after every 10 min). To determine microbial growth inside the wall from each house two samples inside the wall were taken. Samples were plated, total colony forming units were counted and identified from the isolated colonies.

Air temperature and relative humidity in indoor air and inside the boarder was studied for evaluating the environmental conditions. It is known from mould growth index model [16] that some conditions are more suitable for fungi than others.

Monitoring of CO_2 enables to describe also the indoor climate but even more important is that the information about production and indoor-outdoor values enables to get an idea about ventilation presence and activity.

All chemicals and reagents (Fluka) used for this experiment were purchased from HNK Analüüsitehnika.

The identification procedure was performed via sample staining and microscopy. Bergey's manuals and online databases were used for the identification.

3 Results and Discussion

The characteristics of two buildings, which we investigated, are shown in Table 1. Both of buildings have been built as post and beam constructions with wooden frame. In first building straw panels are used as insulation material, second building is a classical construction, where straw bales are placed into wooden frame as insulation.

The wall depth in both houses is the same—60 cm, which consists of 50 cm of straw bale and 5 + 5 cm of clay plaster (both houses were plastered from inside and outside as well).

Table 1 Characteristics of buildings

Characteristic	Building 1	Building 2
Indoor area (m^2)	200	250
Occupants	1	6
Bedrooms	1	3
Floors	1	1
Type of construction	Load bearing with straw panels	Load bearing with straw bales

Indoor air quality is one of the main factors affecting productivity, health and wellbeing of people [17]. The results from air samples (CFU) were in one house higher than in the other one. Temperature, humidity and CO_2 levels were also higher in one house.

3.1 Indoor and Outdoor Air Microbiology

As seen from the Table 2, indoor MEA CFU counts are in both buildings higher as in outdoor air. Moisture content inside the bales is important for the microbial activity [18]. Due of the higher moisture content microorganisms can activate from spores and will start growing again. In these two buildings growing conditions for microbiological life are good during almost all the time. Autumn indoor readings from the first house were probably higher because of the mold spots found from bedroom outer wall.

DG18 plates had a little growth from outside air (less than 20 CFU per plate). Indoor DG18 air sample plates had only few colonies (max 5 colonies per plate). All the colonies from outside were identified and indoor identified species corresponded to outdoor ones.

In normal and healthy building the majority of fungi come from outdoor sources [19]. Total number of air fungi is always lower than outdoors [20]. In both of our buildings the total number of air fungi indoors was greater than outdoors. We found a fungal growth from a north side corner of a bedroom. Growth was identified as a member of *Cladosporium* family. Homeowner used a fungicide which we suggested (Biotol© spray, which is effective against mould growth [21]) multiple times

Table 2 Seasonal variation in culturable airborne fungi on media MEA (malt extract agar) indoors and outdoors shown as mean and in parenthesis range of CFU m^{-3} air

Sampling site and media	Autumn	Winter	Spring
Building 1 indoor (MEA)	503 (480–530)	460 (380–610)	153 (60–240)
Building 1 outdoor (MEA)	345 (300–380)	233 (200–280)	105 (80–160)
Building 2 indoor (MEA)	160 (100–220)	158 (100–250)	140 (100–160)
Building 2 outdoor (MEA)	278 (220–320)	113 (100–130)	70 (60–80)

and added a coat of fresh clay plaster. This problem had not reported again until now.

In second building we did not found any visible mould spots, but the owners claimed there was a small pipewater leakage for few hours in early summer 2014—before when we started our monitoring process. Combained with temperature (2014 was a hotter summer than normal in Estonia [22]) it might be the case why there was more fungi present in inside. We also took some new air samples in autumn 2016 and in indoors CFU was lower than outdoors.

Viitanen [16] has been reported that for mould development is the ideal environment with temperatures from 20 to 28 °C and relative humidity more than 55%. In our case we have temperatures and RH% lower than reported before, but still we can see some indications of microbial activity.

Species, which we found, are quite common in outdoor air [19, 23, 24] and were similar in both houses. During autumn was the most abundant fungi from genus *Cladosporium*, followed by *Alternaria*, *Penicillum*, other fungi and *Aspergillus*. Most common genus in indoor air during winter was *Penicillium*, followed by *Aspergillus* and other fungi. During the spring the main genus was *Cladosporium*, followed by fungi from genus *Penicillium*, *Aspergillus*, others and *Alternaria*. Some of the species can be harmful to human health by the ability to produce mycotoxins [25].

3.2 CO2, Relative Humidity, Temperature Indoors

CO_2, relative humidity and temperature levels of buildings are shown on Table 3. In first building the mean CO_2 level was 569.7 ± 164.1 ppm. Mean temperature was 18.8 ± 1.2 and mean RH was 31.4 ± 5.1%. In second building the mean value of CO_2 was 681.1 ± 130.3 ppm. Mean temperature was 19.1 ± 1.2 and mean level of relative humidity was 36.6 ± 3.1%. Indoor vapor content in first building was on average 5.1 g/m^3, in second building 5.9 g/m^3. These levels are suitable for good thermal comfort in bedrooms [26].

Carbon dioxide concentrations in both buildings over a same 24 h period has shown on Fig. 1. All the other measured days were similar. Carbon dioxide levels are similar an hour before midnight, but differ a lot during night and day time. On building 2 this measured bedroom was also a playroom for children during daytime,

Table 3 CO_2, relative humidity and temperature levels of buildings with standard deviatation

Sampling site	CO_2 (ppm)	RH (%)	Temperature (°C)
Building 1	569.7 ± 164.1	31.4 ± 5.1	18.8 ± 1.2
Building 2	681.1 ± 130.3	36.6 ± 3.1	19.1 ± 1.2

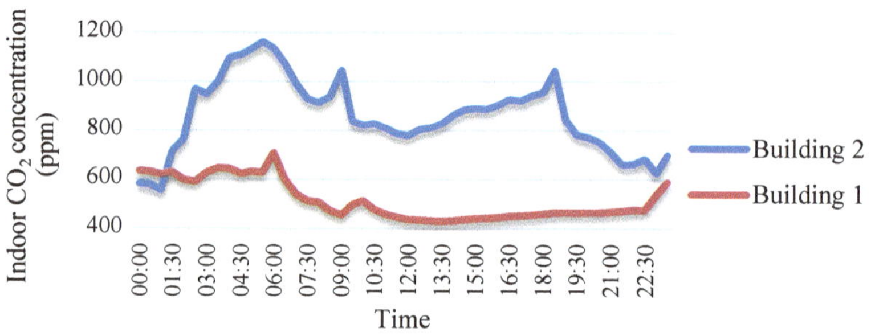

Fig. 1 Carbon dioxide concentration (ppm) dynamics over a 24 h period in buildings

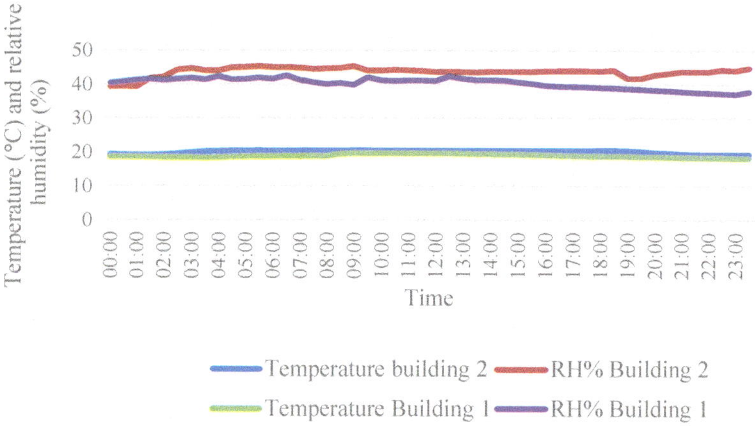

Fig. 2 Temperature (°C) and relative humidity (%) dynamics over a 24 h period in buildings

thus greater CO_2 readings. The owner from building 1 was away from bedroom during daytime and therefore are also lower the CO_2 readings.

On Fig. 2 are shown temperature and relative humidity dynamics over a same 24 h period than on Fig. 1. Relative humidity is higher in second building, temperatures in both buildings differ only a little.

Several factors can affect bedroom air quality and sleep comfort. A major group of these are related to thermal comfort [27]. A common indicator of indoor air quality is CO_2 concentration. Presence of CO_2 are connected with human metabolism and respiration. It has been found earlier that elevated indoor carbon dioxide levels (1000–4000 ppm) increases sick building syndrome (SBS) symptoms, but no direct link had not been found [28–30]. On the figures are shown carbon dioxide concentration dynamics during a 24 h period, measurements were made twice in an hour. CO_2 concentration in first building is higher during nighttime. From Fig. 2 we can clearly see the room airing before sleep which will lower the CO_2 concentrations during night-time (Fig. 2).

Table 4 Relative humidity (%) and temperature (°C) inside the wall with standard deviatation

Sampling site	RH (%)	Temperature (°C)
Building 1 upper logger	44.8 ± 2.3	15.5 ± 2.4
Building 1 lower logger	49.5 ± 4.3	10.1 ± 3.5
Building 2 upper logger	35.9 ± 3.1	19.4 ± 1.5
Building 2 lower logger	34.4 ± 4.6	15.9 ± 2.1

To calculated the ventilation air flow in the room. We used a calculation as follows [31]:

$$Q = \frac{G}{C_{In} - C_{out}}, \tag{1}$$

where

G —emission of CO_2 in the room mg/s,
C_{in} —indoor CO_2 concentration mg/m^3,
C_{out} —outdoor CO_2 concentration mg/m^3.

For building 1 bedroom the airflow was 122.4 m^3/h (34 l/s) and for second building bedroom it was 72 m^3/h (20 l/s). Both of these levels are much higher than required [31] and both of our buildings are belonging to the highest class of indoor climate [32].

3.3 Relative Humidity and Temperature Inside the Wall

We used two loggers to get data from inside the wall. Results of our measurements are shown in Table 4.

Relative humidity level inside the wall were on average. In second building, where was the water leakage, levels of relative humidity are lower than in first building. We assume it is so due higher temperature and room airing.

4 Conclusions

In literature is a limited database for straw bale buildings in northern hemisphere, but they are highly requested, because the interest for building with renewable materials is increasing in time.

An experiment to evaluate indoor air quality in two of Estonian straw bale houses and provide solution of monitoring for indoor air quality and hygrothermal conditions of boarders as complex was carried out from autumn 2014 til spring 2015 in Northern part of Estonia. Two straw bale buildings were monitored.

Mean values of CFU (higher during autumn and spring, lower during winter) and species we found were similar in both cases. There was a seasonal difference between airborne fungal species. Most of found species can present a health risk for humans, especially are endangered elderly people and children, also persons with weakened immune system, respiratory diseases and allergy.

CO_2 concentration levels were in both cases on lower end of the scale, temperature and humidity levels were on comfortable range.

References

1. S. Brasche, W. Bischof, Daily time spent indoor in German homes—baseline data for the assessment of indoor exposure of German occupants. Int. J. Hyg. Environ. Health **208**, 247–253 (2005)
2. J.K. Eom, K.S. Lee, D. Moon, D. Park, K.Y. Yang, Investigating activity patterns and time spent for exposure assessment of college buildings in Korea. Procedia Comput. Sci. **32**, 756–761 (2014)
3. A. Chaussinand, J.L. Scartezzin, V. Nik, Straw bale: a waste from agriculture, a new construction material for sustainable buildings. Energy Procedia **78**, 297–302 (2015)
4. A. Paist, Ü. Kask, L. Kask, A. Vrager, P. Muiste, A. Padari, L. Pärn, Potential of biomass fuels to substitute for oil shale in energy balance in Estonian energy sector. Oil Shale **22**, 369–379 (2005)
5. A. Aciu, N. Cobirzan, Use of agricultural products and waste in the building materials industry. J. ProEnviron. **6**, 472–478 (2013)
6. G. Minke, F. Mahlke, *Building with straw: design and technology of a sustainable architecture* (Birkhäuser, Basel, 2005)
7. E. Milutiene, K. Jürmann, L. Keller, Straw bale building—reaching energy efficiency and sustainability in northern latitudes, www.earthzine.org/2010/04/26/straw-bale-building-%E2%80%93-reaching-energyefficiency-and-sustainability-in-northern-latitudes/. Last accessed 30 Aug 2017
8. S. Magzamen, A.P. Mayer, S. Barr, L. Bohren, B. Dunbar, D. Manning, S.J. Reynolds, J.W. Schaeffer, J. Suter, J.E. Cross, A multidisciplinary research framework on green schools: infrastructure, social environment, occupant health, and performance. J. Sch. Health **87**, 376–387 (2017)
9. T. Kalamees, J. Vinha, J. Kurnitski, in *Indoor Climate Conditions and Hygrothermal Loads in Finnish and Estonian Dwellings*. Closing Seminar Nordic Building Physics Conference, Copenhagen, Denmark, 19 June 2008. Annex 41 Whole Building Heat, Air, Moisture Response, 81–92 (2008)
10. M.I. Gunnbjornsdottir, K.A. Franklin, D. Norback, E. Bjornsson, D. Gislason, E. Lindberg et al., Prevalence and incidence of respiratory symptoms in relationship to indoor dampness: the RHINE study. Thorax **61**, 221–225 (2006)
11. D. Mudarri, W.J. Fisk, Public health and economic impact of dampness and mold. Indoor Air **17**, 226–235 (2017)
12. M.J. Mendell, A.G. Mirer, K. Cheung, M. Tong, J. Douwes, Respiratory and allergic health effects of dampness, mold, and dampness-related agents: a review of the epidemiologic evidence. Environ. Health Perspect. **119**, 748–756 (2011)
13. C.G. Bornehag, G. Blomquist, F. Gyntelberg, B. Järvholm, P. Malmberg, L. Nordvall, A. Nielsen, G. Pershagen, J. Sundell, Dampness in buildings and health. Indoor Air **11**, 72–86 (2001)

14. National Research Council, *Damp Indoor Spaces and Health* (The National Academies Press, Washington, DC, 2004)
15. ISO, *ISO 16000-18: Detection and Enumeration of Moulds—Sampling by Impaction* (European committee for Standardisation, Bruxelles, 2011)
16. H. Viitanen, A. Hanhijärvi, A. Hukka, K. Koskela, Modelling mould growth and decay damages healthy buildings. Espoo, 6–10 Aug 2000. FISIAQ 3, 341–346 (2000)
17. Y. Al horr, M. Arif, M. Karafygiotou, A. Mazroei, A. Kaushik, E. Elsarrag, Impact of indoor environmental quality on occupant well-being and comfort: a review of the literature. Int. J. Sustain. Built Environ., **5**, 1–11 (2016)
18. T. Ashour, H. Georg, W. Wu, Performance of straw bale wall: a case study. J. Energy Build. **43**, 1960–1967 (2011)
19. W.E. Horner, A.G. Worthan, P.R. Morey, Air - and dustborne mycoflora in houses free of water damage and fungal growth. Appl. Environ. Microbiol. **70**, 6394–6400 (2004)
20. J.P.S. Cabral, Can we use indoor fungi as bioindicators of indoor air quality? Historical perspectives and open questions. Sci. Total Environ. **408**, 4285–4295 (2010)
21. J. Raamets, S. Kutti, A. Vettik, K. Ilustrumm, T. Rist, M. Ivask, in *The Antimicrobial Effect of Three Different Chemicals for the Treatment of Straw Bales Used in Housing Projects*. Proceedings of the International Conference on Sustainable Housing Planning, Management and Usability: Sustainable Housing (2016), 537−545
22. Estonian Weather Service, https://www.ilmateenistus.ee/kliima/weather-events/2014-weather/?lang=en. Last accessed 30 Aug 2017
23. B.G. Shelton, K.H. Kirkland, W.D. Flanders, G.K. Morris, Profiles of airborne fungi in buildings and outdoor environments in the United States. Appl. Environ. Microbiol. **68**, 1743–1753 (2002)
24. E. Mędrela-Kuder, Seasonal variations in the occurrence of culturable air borne fungi in outdoor and indoor air in Craców. Int. Biodeterior. Biodegradation **52**, 203–205 (2003)
25. J.W. Bennet, M. Klich, Mycotoxins. Clin. Microbiol. Rev. **16**, 497–516 (2003)
26. C. Song, Y. Loi, X. Zhou, J. Liu, Investigation of human thermal comfort in sleeping environments based on the effects of bed climate. Procedia Eng., **121**, 1126–1132 (2015)
27. P. Batog, M. Badura, Dynamic of changes in carbon dioxide concentration in bedrooms. Procedia Eng. **57**, 175–182 (2013)
28. M.G. Apte, W.J. Fisk, J.M. Daisey, Associations between indoor CO2 concentrations and sick building syndrome symptoms in US office buildings: an analysis of the 1994–1996 BASE study data. Indoor Air **10**, 246–257 (2000)
29. C.A. Erdmann, K.C. Steiner, M.G. Apte, in *Indoor Carbon Dioxide Concentrations and Sick Building Syndrome Symptoms in the Base Study Revisited: Analyses of the 100 Building Dataset*. Indoor Air 2002—9th International Conference on Indoor Air Quality and Climate, Monterey, California, USA, 30 June–5 July 2002
30. A.N. Myhrvold, E. Olsen, O. Lauridsen, in *Indoor Environment in Schools—Pupils Health and Performance in Regard to CO2 Concentrations*. Proceedings of Indoor Air '96: The 7th International Conference on Indoor Air Quality and Climate, Nagoya, Japan, 4, 369–374, July 1996
31. EN15251, Indoor environmental input parameters for design and assessment of energy performance of buildings addressing indoor air quality, thermal environment, lighting and acoustics (European committee for Standardisation, Bruxelles, 2007)
32. T. Kalamees, Eesti eluasemefondi puitkorterelamute ehitustehniline seisukord ning prognoositav eluiga: uuringu lõpparuanne. Puitmajade raport (Technical condition and service life of Estonian wooden apartment buildings). Tallinn Technical University Press (2011)

Investigating Recommended Temperature Zones and Clothing Assumptions in the Assessment of Classrooms' Thermal Environment

Despoina Teli and Jan-Olof Dalenbäck

Abstract There has been a lot of research over recent years on children's thermal comfort, which highlighted the different needs of young children compared to adults. These findings pose a challenge to designers on how to best meet these needs. This paper focuses on recommended temperature zones and assumptions used in standards through a case study in a grade school in Gothenburg, Sweden. Six classrooms were investigated in three buildings of the same school. The indoor temperature was measured using small-scale data loggers programmed to log at 5-minute intervals for a period of 5 months (mid-December to early-June). Thermal comfort questionnaires were also distributed to children throughout the monitoring period. A total of 45,000 temperature readings corresponding to assumed occupied hours and approximately 2000 thermal sensation votes and clothing insulation values are used in the analysis. Results indicate that assumed occupancy schedules may differ to real use, leading to overestimation of time when indoor environmental parameters are outside recommended ranges. Children's clothing insulation was found to be lower than assumed in standards in both winter and summer. Omitting to account for such differences may lead to misinterpretation of indoor environment assessments and design solutions.

Keywords Thermal comfort · Indoor environment quality
School buildings · Children · Clothing insulation

D. Teli (✉) · J.-O. Dalenbäck
Division of Building Services, Department of Architecture and Civil Engineering, Chalmers University of Technology, 412 96 Göteborg, Sweden
e-mail: teli@chalmers.se

D. Teli
Division of Energy and Climate Change, Sustainable Energy Research Group, Faculty of Engineering and the Environment, University of Southampton, SO17 1BJ Southampton, UK

D. Johansson et al. (eds.), *Cold Climate HVAC 2018*,
Springer Proceedings in Energy, https://doi.org/10.1007/978-3-030-00662-4_69

1 Introduction

Over the last years a number of studies in schools found differences in thermal sensation and comfort between children and adults [1–6] and highlighted that methods and criteria currently used in the design and evaluation of school environments may not match the children's needs or the conditions experienced in classrooms. Part of the issue may lie in assumptions used when assessing children's thermal comfort, which is the focus of this paper.

For the prediction of thermal comfort six parameters are used; four environmental: air temperature, radiant temperature, air speed and relative humidity, and two personal: metabolic rate and clothing insulation [7]. Considering possible differences in these parameters in spaces occupied by adults or children, the personal parameters would be the most influential. A sensitivity analysis has also shown that the majority of current indices used for predicting thermal comfort are most influenced by personal variables [8]. Guidance on calculating metabolic rate and clothing insulation can be found in standards ISO 8996 [9] and ISO 9920 [10] respectively. However, in most applications direct measurements of these parameters are not possible and therefore tables with standard values are used [7, 11]. In relation to school environments, there is lack of specific data for children [12], hence the same tables are used under the assumption that children's clothing, metabolic rate and adaptive behavior are comparable to adults' in offices.

This study investigates assumptions used in the assessment of the thermal environment in schools through a case study in six classrooms in Sweden, focusing on the recommended temperature zones in current standards and the role of clothing insulation.

1.1 Clothing Insulation in Thermal Comfort Assessment

The thermal insulation provided by clothes has a considerable impact on thermal comfort and clothing adjustment is probably the most powerful behavioral action to restore comfort. Clothing insulation is the "equivalent uniform thermal resistance" on a human body [10] expressed in clo (1 clo = 0.155 $m^2 \cdot K \cdot W^{-1}$). The unit was introduced for the representation of the insulation required to keep a seated person comfortable at 21 °C and corresponds to the insulation of a suit with normal underclothes [13], a work outfit that was common in office buildings in the 1970s.

In the standards, recommended temperature ranges are typically calculated for clothing insulation equal to 0.5clo and 1clo for summer and winter respectively [7, 11, 14]. If there is no information on clothing, thermal and energy evaluations are performed with these two values, which affect the design of buildings and HVAC systems, the energy use and the operation of buildings [15].

European Standard EN15251 on 'Indoor environmental input parameters for design and assessment of energy performance of buildings' recommends

Table 1 Example of the relationship between activity, clothes and appropriate indoor air temperature for comfort, taken from [17], translated to English

Activity	Clothing combination			
	Sports clothing	Summer clothes (shirt, shorts)	Winter light indoor clothing (sweater, trousers)	Winter warm indoor clothing (costume)
Schoolwork, sitting	28.0 °C	24.0 °C	23.0 °C	21.0 °C
Schoolwork, standing	26.5 °C	23.0 °C	20.0 °C	17.0 °C

temperature ranges for classrooms based on the above standard assumptions. For "clothing ~1.0clo" the temperature range for heating in a category II building (normal level of expectation) is 20.0–24.0 °C, whilst the range for cooling with "clothing ~0.5clo" is 23.0–26.0 °C [14]. The Public Health Agency of Sweden provides temperature ranges which also apply to classrooms (20.0–23.0 °C) [16], without any reference to corresponding clothing levels. In reality, these ranges should be adjusted based on actual clothing levels and activities found in schools. Such approach is included in a Swedish document with guidelines for school environments developed in the 1990s [17]. An example can be seen in Table 1. The proposed relationships however are still based on Fanger's PMV model, which was developed with adult subjects and has been found to be inappropriate for young children [1–6]. Being developed in the 90s, it may also be rather outdated regarding typical school clothing.

2 Methods

The study presented here includes long-term measurements and thermal comfort surveys in a grade school in Gothenburg, Sweden. The school is housed in 9 buildings, seven of which were built in the turn of the 18th to the 19th century and two in the end of the 20th century, which have all been refurbished. The surveys took place in 6 classrooms located in 3 of the 9 buildings.

The study follows in most parts the methodology previously used in UK schools [2, 18]. However, this time an extended version of the same thermal comfort questionnaire was used, translated into Swedish. The study was approved by the ethics committee of the University of Southampton, which is partner organisation in the project (submission 17626). No identifying or sensitive information was included in the questionnaire. Consent by the children's parents was not required. Instead, informed consent was obtained from the school headteacher, as the parents' representative. The questionnaire consists of 9 questions covering thermal perception, comfort and adaptive behaviour. Approximately 16 surveys were conducted in each classroom during class and at least 20 min after breaks or other non-sedentary activities. This ensures as much as possible consistent metabolic rates throughout the study. At the time of the surveys measurements of the environmental parameters affecting thermal comfort were also taken (air temperature, globe temperature, air speed, relative humidity). Ventilation rates were not measured as the focus of the study is on thermal comfort and the parameters directly affecting it. Details on the instruments and measuring procedures can be found in an earlier publication [19]. In this paper answers in two of the questions are used: (1) thermal sensation vote on a 7-point scale and (2) clothing items worn, using a clothing item checklist.

For the long-term measurements of air temperature and relative humidity in the classrooms small Madgetech dataloggers were used. They were placed on pinboards in the classrooms at a height of approximately 1.5 m and away from heat sources and direct solar radiation. The loggers were programmed to take and store measurements every 5 min.

3 Results

The long-term measurements were processed so that only temperatures during school days are used. All the holidays, weekends, out-of-term dates and out-of-school hours were removed. The remaining data reflect the hours that classrooms are expected to be occupied, between 8:30 and 15:00. However, breaks or other activities outside the classrooms within school hours could not be subtracted from the dataset as every classroom had breaks and classes at different times.

It should be noted that for the assessment of the thermal environment operative temperature data should be used so that both convective and radiative effects are taken into account. However, in this study only the air temperature was monitored continuously in the classrooms. In a previous investigation in three school buildings with different construction and thermal mass, the difference between operative and air temperature was found to be very low [20]. The same investigation was conducted here, using 90 sets of simultaneous measurements of air (T_a) and globe (T_{op}) temperatures for various outdoor weather conditions. The average difference between T_{op} and T_a was 0.1 ($\sigma = 0.1$), which is very small and lower than the manufacturer-stated accuracy of the air temperature sensor. Furthermore, the measured air speeds during surveys were always below 0.1 m/s. Therefore, the monitored air temperatures are used without any correction.

3.1 Classrooms' Thermal Environment

The average air temperature of the classrooms in the heating period (Dec–Apr) was 21.5 °C ($\sigma = 1.1$) and in the non-heating period (May–Jun) 24.5 °C ($\sigma = 2.1$). Figure 1 shows the classrooms' average daily temperatures in relation to the

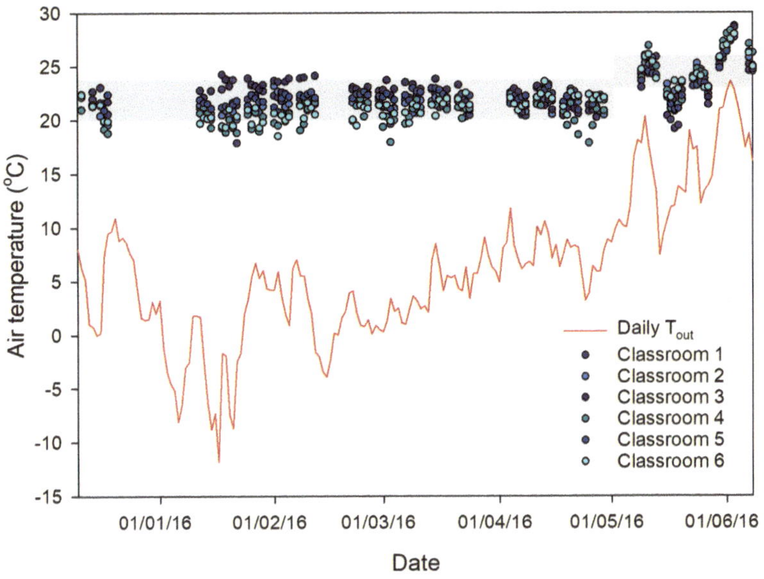

Fig. 1 Average daily ambient temperature and classrooms' average daily indoor temperature for the duration of the study (Dec 2015–Jun 2016). Recommended temperature ranges are based on EN15251 [14]

outdoor daily temperature and the recommended ranges based on EN15251 [14]. The indoor temperatures on average in most classrooms and on most days were within recommended ranges. There are however a number of datapoints outside the ranges, mainly in classrooms 4 and 6. It appears that these classrooms' average daily temperature follows the outdoor temperature changes more closely than the others, which is most probably due to building characteristics. E.g. classroom 4 is naturally ventilated (through window opening) at the top floor of the building with three sides exposed to outside conditions.

Figures 2 and 3 show the distributions of the 5-minute measurements of air temperature and relative humidity per classroom during expected occupied hours for the two seasons investigated (heating/non-heating). In winter there are several measurements below 20 °C, e.g. 33% in classroom 4 and 23% in classroom 6 (Fig. 2). In the non-heating season there is exceedance of the recommended upper limit in all classrooms, ranging from 18% of the measurements in classroom 5–35% in classroom 4. Relative humidity is overall relatively low in winter with an average of 31.5%, while in the summer it increases to 38.5%. However, when the air temperature is within the comfort range, as in the majority of the time here, the effect of humidity on thermal sensation is modest [21].

Overall, the measurements shown in Figs. 2 and 3 point to a wide range of conditions in the classrooms during the assumed occupancy period. However, the actual schedule of occupancy may vary significantly based on educational practices. It is therefore likely that more time than assumed is spent outside the classroom, which

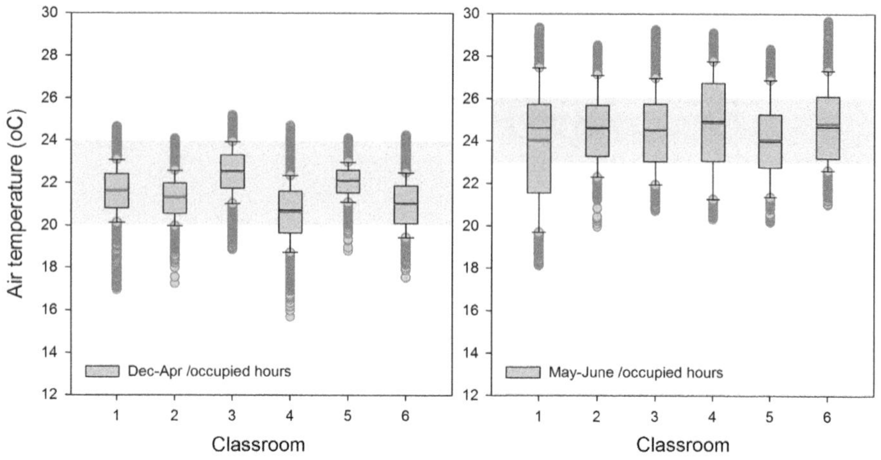

Fig. 2 Distribution of the measured air temperature per classroom (1–6) during occupied hours in the winter and summer months. Box: the 50% of the values; whiskers: the 10th and 90th percentile; dots: outliers; black line: median, red line: mean. Recommended temperature ranges are based on EN15251 [14]

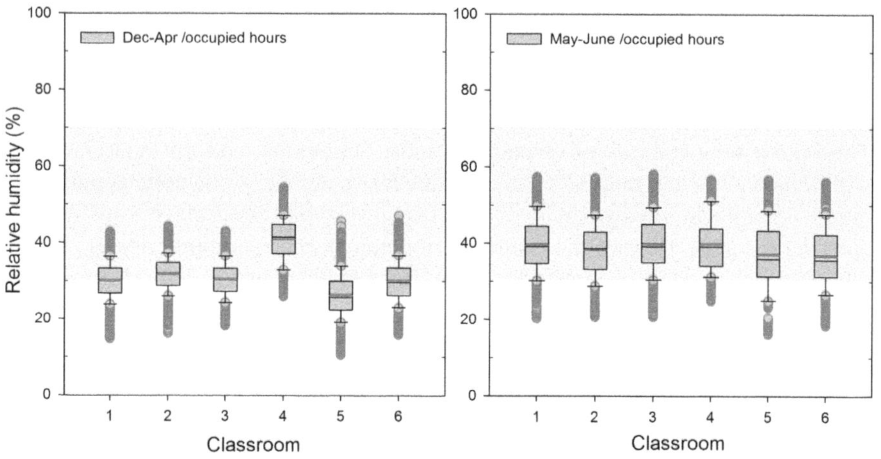

Fig. 3 Distribution of the measured relative humidity per classroom (1–6) during occupied hours in the winter and summer months. Box: the 50% of the values; whiskers: the 10th and 90th percentile; dots: outliers; black line: median, red line: mean

agrees with anecdotal evidence from the teachers and visual inspection of the data. Furthermore, the recommended ranges used above assume typical metabolic rates of around 70 W/m^2 (based on adults) and clothing insulation values of 1 and 0.5clo for winter and summer respectively. The effect of thermal adaptation, which leads to higher tolerance to temperature typically experienced [13], is also not taken into account. These results therefore do not provide a clear indication of children's comfort.

3.2 Thermal Sensation, Temperature and Clothing Insulation

For the comparison of the thermal environment with children's thermal sensation and clothing insulation in both investigated seasons the data from the surveys are used. The operative temperature during the 90 surveys conducted was always above 20 °C, which supports the hypothesis that the lower temperatures seen in Fig. 2 are from breaks or other activities outside the classroom, when teachers would open windows for ventilation.

As can be seen in Tables 2 and 3, the average clothing insulation is consistent between classrooms in the same season, with averages of 0.7clo and 0.38clo in winter and spring/summer respectively. These values are lower than the assumed values of 1clo for winter and 0.5clo for summer used in standards [7, 11, 14] and that found in Dutch school classrooms 10 years ago [12]. However, this study's average clo for winter is close to the median winter clothing insulation value of 0.69clo found in office buildings [15], suggesting a possible trend towards lower clothing insulation values indoors. At the same time, the average operative temperatures during surveys were between 22.0–23.4 °C, which lies in the warm end of the recommended zone but is reasonable, considering the relatively low clothing insulation. There appears to be a trend towards lower clothing insulations and higher indoor temperatures. The causal effect however is unknown, i.e. whether clothing choices led to the temperature increase or, conversely, whether children adapted their clothing to their classroom's thermal environment.

In the non-heating season the relationship is clearer, with the children adapting their clothing to the warmer temperatures (Table 3). The classrooms' average operative temperatures during surveys were between 24.3 and 26.4 °C, again in the warm side of the spectrum but clothing insulations are again lower than assumed. The average thermal sensation votes of the children are between 0.4–0.9 scale

Table 2 Mean and standard deviation (SD) of the children's thermal sensation vote (TSV), the operative temperature (T_{op}) and clothing insulation (clo) during surveys in the heating season (winter) by classroom

Classroom	TSV		T_{op}		Clo	
	Mean	SD	Mean	SD	Mean	SD
1	0.5	1.3	23.0	1.1	0.65	0.15
2	0.3	1.2	22.0	0.7	0.67	0.15
3	0.7	1.2	23.4	0.6	0.70	0.15
4	0.4	1.1	22.8	0.6	0.71	0.15
5	0.0	1.2	22.3	0.3	0.68	0.14
6	0.0	1.1	22.7	0.9	0.71	0.14

Table 3 Mean and standard deviation (SD) of the children's thermal sensation vote (TSV), the operative temperature (T_{op}) and clothing insulation (clo) during surveys in spring/summer by classroom

Classroom	TSV		T_{op}		Clo	
	Mean	SD	Mean	SD	Mean	SD
1	0.7	1.1	25.3	0.9	0.35	0.13
2	0.7	1.1	25.3	1.0	0.36	0.13
3	0.9	1.0	26.1	0.7	0.35	0.14
4	0.6	1.1	26.4	1.0	0.40	0.14
5	0.4	1.2	24.4	0.8	0.39	0.15
6	0.4	1.2	24.3	0.2	0.45	0.12

points, which lie between 'neutral' and 'slightly warm'. Considering the findings that children feel warmer than adults at the same temperatures [2], clothing adaptation succeeded in restoring comfort to a large extent.

4 Conclusions

Standards and guidelines for the indoor environment often provide recommended temperature ranges based on assumed clothing insulation values, metabolic rates and basic adaptive behavior. Although designers and practitioners should adjust these recommendations based on the specific applications encountered, in practice these ranges are mostly used as fixed thresholds. This could lead to misinterpretation of indoor climate and energy assessments.

This paper explored this through a case study in a Swedish primary school. Assessment of the classrooms' thermal environment during assumed occupied hours and using the recommended ranges highlighted issues of temperatures outside the recommended comfort zones both for winter and summer. However, further investigation indicated that part of these values were outside occupancy hours or were due to windows being intermittently opened for ventilation. It is likely that changes in educational practices towards more flexible curricula may have also contributed to this difference. Such developments may be overlooked, especially when it comes to analysing large amounts of data where behaviours and occupancy schedules are difficult to identify.

It was observed that children's clothing insulation was on average lower than the assumed values both in winter and spring/summer (0.7 and 0.4 respectively). Even though this helps to achieve comfort in warm summer conditions without mechanical cooling, in winter it is coupled with high indoor temperatures, which lead to higher heating demand. In either case, these values do not reflect those assumed in standards. It is clear that personal parameters and occupant behavior (teachers' and children's) should be more adequately addressed in guidelines for schools' thermal environment, to improve decision making in the design stage.

Acknowledgements The authors would like to thank the teachers and children who participated in this study. This work has been performed with support from VINNOVA (Swedish Governmental Agency for Innovation Systems), the Profile 'Energy in Urban Development' within the Area of Advance 'Energy' at Chalmers University of Technology and the Sustainable Energy Research Group (www.energy.soton.ac.uk) at the University of Southampton.

References

1. Mors St, J.L.M. Hensen, M.G.L.C. Loomans, A.C. Boerstra, Adaptive thermal comfort in primary school classrooms: creating and validating PMV-based comfort charts. Build. Environ. **46**, 2454–2461 (2011)
2. D. Teli, M.F. Jentsch, P.A.B. James, Naturally ventilated classrooms: an assessment of existing comfort models for predicting the thermal sensation and preference of primary school children. Energy Build. **53**, 166–182 (2012)
3. M. Trebilock, R. Figueroa, *Thermal comfort in primary schools: a field study in Chile*, in Counting the Cost of Comfort in a changing world, 10–13 April 2014. Cumberland Lodge, Windsor, UK (2014)
4. R. de Dear, J. Kim, C. Candido, M. Deuble, Adaptive thermal comfort in Australian school classrooms. Build. Res. Inf. **43**, 383–398 (2015)
5. S. Haddad, P. Osmond, S. King, Revisiting thermal comfort models in Iranian classrooms during the warm season. Build. Res. Inf., 1–17 (2016)
6. A. Montazami, M. Gaterell, F. Nicol, M. Lumley, C. Thoua, Developing an algorithm to illustrate the likelihood of the dissatisfaction rate with relation to the indoor temperature in naturally ventilated classrooms. Build. Environ. **111**, 61–71 (2017)
7. ISO, EN ISO 7730:2005 Ergonomics of the thermal environment- Analytical determination and interpretation of thermal comfort using calculation of the PMV and PPD indices and local thermal comfort criteria. Geneva: International Standardisation Organisation (2005)
8. S. Gauthier, in *The Role of Environmental and Personal Variables in Influencing Thermal Comfort Indices Used in Building Simulation*. Conference Proceedings: 13th Conference of International Building Performance Simulation Association (BS2013) (2013), 2320–2325
9. ISO, EN ISO 8996:2005 Ergonomics of the thermal environment-determination of metabolic rate. Geneva: International Standardisation Organisation (2004)
10. ISO, EN ISO 9920:2009 Ergonomics of the thermal environment. Estimation of thermal insulation and water vapour resistance of a clothing ensemble (ISO 9920:2007, Corrected version 2008-11-01). Geneva: International Standardisation Organisation (2009)
11. ASHRAE, ANSI/ASHRAE Standard 55- Thermal Environmental Conditions for Human Occupancy. Atlanda: American Society of Heating, Refrigerating and Air-Conditioning Engineers (2013)
12. G. Havenith, Metabolic rate and clothing insulation data of children and adolescents during various school activities. Ergonomics **50**, 1689–1701 (2007)
13. F. Nicol, M. Humphreys, S. Roaf, *Adaptive Thermal Comfort: Principles and Practice* (Routledge, London, 2012)
14. CEN, EN 15251:2007 Indoor environmental input parameters for design and assessment of energy performance of buildings addressing indoor air quality, thermal environment, lighting and acoustics. Brussels: CEN (European Committee for Standardization) (2007)
15. S. Schiavon, K.H. Lee, Dynamic predictive clothing insulation models based on outdoor air and indoor operative temperatures. Build. Environ. **59**, 250–260 (2013)
16. Folkhälsomyndigheten. FoHMFS 2014:17. Folkhälsomyndighetens allmänna råd om temperatur inomhus (In Swedish). (Folkhälsomyndigheten, Stockholm, Sweden, 2014

17. Boverket och Arbetarskyddsstyrelsen. Att se, höra och andas i skolan, Handbok H255 (In Swedish) (1996)
18. D. Teli, P.A.B. James, M.F. Jentsch, Thermal comfort in naturally ventilated primary school classrooms. Build. Res. Inf. **41**, 301–316 (2013)
19. D. Teli, J-O. Dalenbäck, L. Ekberg, in *Winter Thermal Comfort and Indoor Air Quality in Swedish Grade School Classrooms, as Assessed by the Children*. Indoor Air 2016: 14th International Conference of Indoor Air Quality and Climate 2016
20. D. Teli, L. Bourikas, P.A.B. James, A.S. Bahaj, Thermal performance evaluation of school buildings using a children-based adaptive comfort model. Procedia Environ. Sci. **38**, 844–851 (2017)
21. D.A. McIntyre, *Indoor climate* (Applied Science Publishers, London, 1980)

Indoor Temperature Variations in Swedish Households: Implications for Thermal Comfort

Despoina Teli, Sarka Langer, Lars Ekberg and Jan-Olof Dalenbäck

Abstract Everyday thermal environments affect people's comfort and wellbeing, with extreme conditions affecting human health. A strong focus on avoiding the extremes along with the introduction of tight thermal comfort criteria over the years has led to design strategies and behaviors that promote thermally stable indoor environments. However, recent research has shown that indoor temperature variation has significant health benefits, e.g. it could help tackle diabetes and obesity. These findings suggest that it is important to investigate not just the average temperature levels in households but also their distribution and variation over different periods. In Sweden, indoor temperatures are considered to be on average high and constant due to a combination of the heating provision mechanism and the high building standards compared to other countries. This paper investigates the temperature distributions in Swedish households using detailed 15-minute indoor air temperature measurements from the 2008 BETSI-survey, provided by the Swedish National Board of Housing, Building and Planning (Boverket). Approximately two million measurements from 1306 households taken during two-week periods in winter 2007/08 are used in this investigation. Indoor temperature variation is investigated in two levels: (i) over the 2-week monitoring period and (ii) within-day. Results showed a considerable range in average dwelling temperatures of 9 K, highlighting a substantial variability between homes in heating temperature and most likely in thermal comfort preferences. Regardless the different

D. Teli (✉) · S. Langer · J.-O. Dalenbäck
Division of Building Services Engineering, Department of Architecture and Civil Engineering, Chalmers University of Technology, 412 96 Göteborg, Sweden
e-mail: teli@chalmers.se

D. Teli
Division of Energy and Climate Change, Sustainable Energy Research Group, Faculty of Engineering and the Environment, University of Southampton, SO17 1BJ Southampton, UK

S. Langer
IVL Svenska Miljöinstitutet AB, 53021, 40014 Göteborg, Sweden

L. Ekberg
CIT Energy Management AB, 412 88 Göteborg, Sweden

D. Johansson et al. (eds.), *Cold Climate HVAC 2018*,
Springer Proceedings in Energy, https://doi.org/10.1007/978-3-030-00662-4_70

temperature levels, the majority of dwellings maintain stable thermal conditions, as demonstrated from the very low temperature variations found. Differences in daily temperature patterns were also observed.

Keywords Thermal comfort · Indoor temperature · Housing · Indoor environment quality · Heating

1 Introduction

People spend considerable amounts of time at home and the indoor climate they experience affects both the buildings' energy consumption and their wellbeing and health. After many years of one-sided focus on the energy performance of buildings it is becoming clear that more emphasis should be placed on the quality of buildings' indoor environment [1].

Occupants' thermal comfort is influenced by four environmental parameters: air and radiant temperature, relative humidity and air velocity. From these, temperature is considered as the most important factor for comfort [2]. Over recent years there has been an increase in data collection of indoor temperatures in households, which can help to better understand trends, preferences and patterns in populations. For example, a historic upward trend in winter indoor temperatures has been identified especially in bedrooms, with an increase in mean dwelling indoor temperature in the UK of 1.3 °C per decade between 1978 and 1996 [3]. In Sweden, average indoor temperature were estimated at 21.2 °C in single-family dwellings and 22.3 °C in multi-family dwellings using data from 2007/08 [4, 5], whilst in 1984 the estimated averages were 20.4 and 21.8 °C respectively [6]. Whilst average values are helpful to understand general levels and historic trends, they may hide or smooth information, especially when derived from very large datasets. This work aims to investigate detailed indoor temperature measurements in Swedish homes and compare them with thermal comfort guidelines and recent findings on healthy indoor temperature variations.

1.1 Indoor Climate Requirements in Dwellings

Standards and guides provide design criteria for the thermal environment in living spaces, including homes [7–9]. Some of them provide recommended indoor temperatures in the form of either a threshold (minimum for winter and maximum for summer) or a comfort range (Table 1). In terms of thresholds, the UK's 'Cold Weather Plan' recommends a minimum indoor temperature at home of 18 °C for a sedentary person, wearing suitable clothing [10]. However, it is highlighted that temperatures up to 21 °C may be beneficial for health. The World Health Organization suggests lower limits of 21 °C for living rooms and 18 °C for

Table 1 Design winter indoor temperatures for residential buildings

Study	Design minimum (°C)	Range (°C)
WHO (World Health Organization)	21.0 (living room)/18.0 (other)	–
EN 15251/1.0 clo//Category I[a]	21.0	21.0–25.0
EN 15251/1.0 clo//Category II	20.0	20.0–25.0
EN 15251/1.0 clo//Category III	18.0	18.0–25.0
The Public Health Agency of Sweden	18.0/21.0 (sensitive persons)	20.0–23.0

[a]EN 15251 categories represent different levels of expectation

bedrooms [11, 12] while The Public Health Agency of Sweden a minimum of 18 °C (21 °C for sensitive persons) [13].

The above recommended thresholds and ranges are mainly based on experimental research from the 1970s [14] and therefore may not reflect the variability of indoor conditions experienced in real everyday environments [15], especially in different locations around the world. The recommended ranges for heated spaces are typically rather narrow, within 3–4 °C, as it is assumed that this reflects the thermal preferences and needs of occupants, considered to be approximately the same. For example energy and comfort modelling are using a set-point value for the indoor temperature, i.e. 21 °C, with standard schedules of use. However, research in the UK has shown that these assumptions are not necessarily representative, with temperature profiles varying significantly [16], while increased exposure to thermo-neutral conditions might be a contributing factor to weight gain [3].

Similarly, it has been assumed that stable indoor temperatures are the ideal situation for most people, while recent research has shown that exposure to mildly cold or warm environments has significant health benefits, supporting dynamic and drifting temperatures for healthy indoor environments [17]. In addition to the health benefits, asymmetrical and transient thermal environments can lead to more pleasurable thermal experiences than achieved by the isotheral and static conditions, due to the effect of 'thermal alliesthesia' [18]. In a dynamic situation, a daytime variation of 8 K (e.g. 17–25 °C) has been found acceptable by both young adults and elderly [19]. There is scope for further investigation of the types of indoor climates experienced in domestic buildings and their relation to comfort and health. This paper is focusing on temperature variability within the Swedish building stock using the BETSI database, as described below.

2 Methods

For the analysis presented here data from the BETSI dataset are used, which were collected in the heating season 2007/08. The BETSI program involved inspection of 1800 buildings, from which 1400 were residential [4]. The buildings were selected as representative of the Swedish building stock and include both single-family dwellings and apartments in multi-family buildings.

The data include air temperature and relative humidity measurements at 15 min intervals over a period of two weeks. The corresponding outdoor dry bulb temperature and relative humidity were also included in the dataset, taken from the SMHI station (Swedish Meteorological and Hydrological Institute) closest to the dwelling. The dataset was cleaned from cases with missing or incorrect data. Values of average indoor temperature below the lowest acceptable value of 18 °C were also excluded, as maintaining temperatures that low for prolonged periods suggests that the homes may have been unoccupied for a big part of the monitoring period. A total of 1306 dwellings remained for analysis. The statistical analysis of the dataset was conducted in SPSS Statistics 22.

3 Results

The analysis was undertaken in two steps: the first focuses on temperature level and variation differences between dwellings over the 2-week monitoring as a representation of a short heating period. It should be noted that the two-weeks are only a snapshot of a full heating period. The second step investigates the daily temperature variations and examples of daily patterns of air temperature in the sample.

3.1 *Summary Statistics and 'Temperature Cloud'*

Table 2 shows summary statistics of the dwellings' data. The average dwelling temperature ranges between 18.2 and 26.9 °C, with a mean of 22.0 °C (excluding values below 18 °C). The mean standard deviation is 0.7 K, with a minimum of 0.1 (negligible indoor temperature variation during the two weeks of monitoring) and a maximum of 6.5 (high indoor temperature variation).

The range of almost 9 K in average temperature between dwellings in heating mode is considerable. The scatter can be seen in Fig. 1, which shows the relationship between mean indoor and mean outdoor temperature. This result agrees with the observations of Nicol [15] of a surprisingly wide range of indoor

Table 2 Summary statistics of dwellings' temperature data

	Mean	Minimum	Maximum
Dwellings' 2-week average temperature (°C)	22.0	18.2	26.9
Standard Deviation (SD)	0.7	0.1	6.5
2-week average outdoor temperature (°C)	3.8	−8.8	13.7
Standard Deviation (SD)	3.3	0.0	6.6

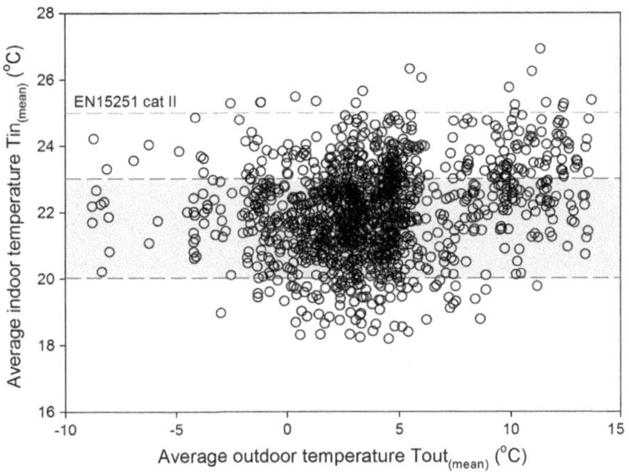

Fig. 1 Relationship between dwellings' average indoor temperature and the average outdoor temperature during the corresponding two-week monitoring period. Included is the recommended range by The Public Health Agency of Sweden (grey zone) and the EN15251 higher upper recommended value

temperatures in heated or cooled buildings. He even found that buildings with a mechanical conditioning system had a wider range of temperatures than free running buildings (non-heated/non-cooled), unlike common assumptions in standards for the opposite.

3.2 Within-'Two Weeks' Temperature Variations (Monitoring Period)

The dwellings' mean air temperatures and standard deviations were binned at 1 °C and are presented in the histograms of Fig. 2. Although the average air temperatures are nearly normally distributed, the standard deviations are clearly skewed towards very low values below 1 K, highlighting that in the vast majority of the dwellings (1100 out of 1306) there was very little variation in indoor temperature over the two-week monitoring period. It can also be seen that around 30% of the values are outside the temperature range of 20–23 °C recommended for comfort by The public health agency of Sweden, with the majority (25%) above 23 °C.

In order to determine whether temperature variations are associated with specific levels of indoor temperature (e.g. if people maintaining on average high indoor temperatures experience high variation or vice versa), the standard deviations were plotted against the corresponding average air temperatures (Fig. 3). It can be seen that there is no correlation between the two variables, with the high standard deviations being spread across the temperature range. This suggests that temperature variation at home is not necessarily linked to a particular temperature preference and very low variations are experienced over the entire range of average dwelling temperatures (18–27 °C).

Table 3 shows studies in dwellings analyzed by Nicol [15]. Whilst direct comparison cannot be made due to different measurement protocols and study

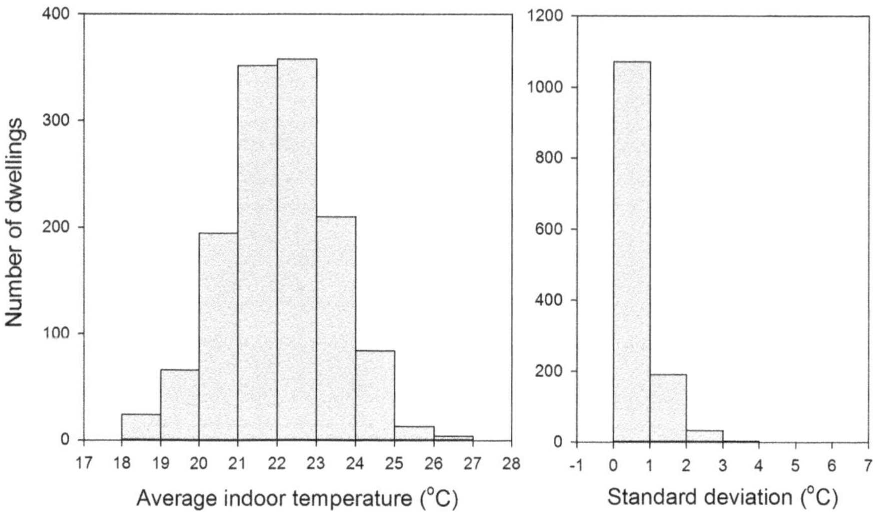

Fig. 2 Histograms of average dwelling temperature (left) and standard deviation (right) during the monitoring period

Fig. 3 Dwellings' standard deviations of measured air temperatures against the corresponding average temperatures

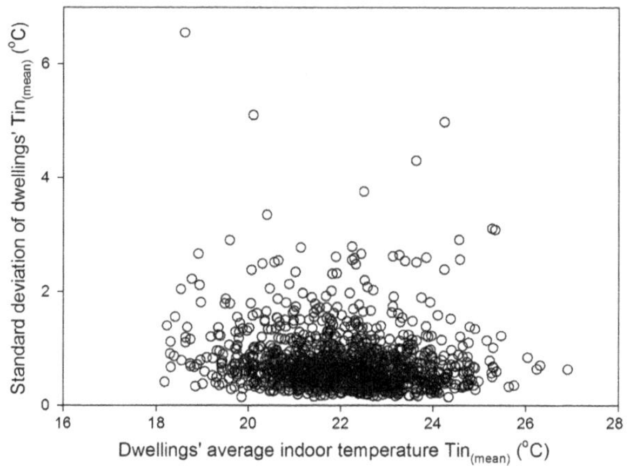

durations, it is interesting to see the difference in both levels and variation in indoor temperature. Overall, the indoor temperatures in Swedish dwellings are higher and considerably more stable than in the other locations. Based on Mata et al. [20], the reasons for the high and constant temperatures in Sweden are: that "the share of centrally heated buildings is much higher than in other countries; that the outdoor temperature in winter is rather stable due to low solar radiation; and that the buildings have good insulation and air-tightness (compared to other regions)". In addition to the above, it is likely that occupants have adjusted their behavior as they

Table 3 Dwelling temperatures and SD in other studies in heating season (based on [15])

Source	Location/year	T_{in} mean	SD
[15]	Tokyo, Japan	19.6 °C	2.8 K
[16]	UK	19.0 °C	2.5 K
[22]	Harbin, China/2000–01	20.1 °C	2.4 K
[23]	Beijing and Shanghai/2012–13	21.4 °C	2.7 K
This study	Sweden/2007–08	22.0 °C	0.7 K

adapt to these indoor conditions, such as their clothing (lighter) and use of controls (less frequent or no use) [21].

3.3 Within-Day Temperature Variations

The data were then analyzed by day in order to see whether there are temperature drifts in the dwellings during a day. The standard deviation is used as way of expressing how much the temperature varies around the daily mean. As shown in the boxplots of Fig. 4, there is the same tendency of constant temperatures, with the standard deviation in 90% of the days being below 1 K. Based on a visual inspection of the data, the higher values correspond to a number of situations: (a) dwellings with individual heating where it is possible to set it lower/higher, program or switch off, (b) days in the transition from heating to no-heating or vice

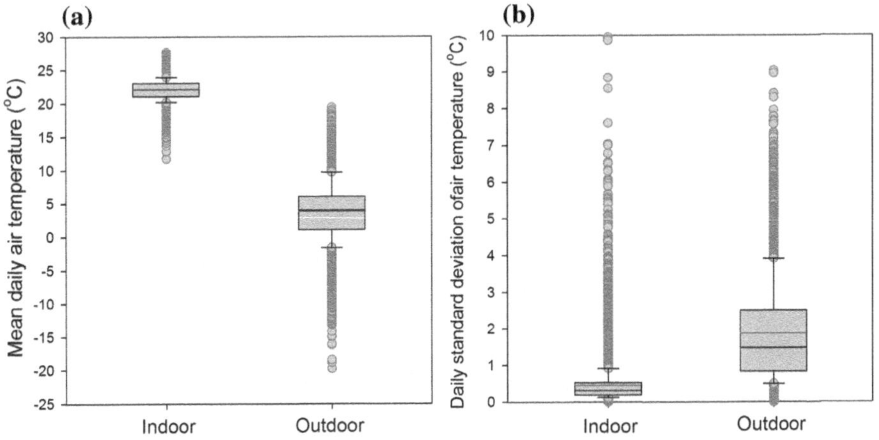

Fig. 4 **a** Daily air temperature boxplots indoors (dwelling) and outdoors (SMHI) and **b** standard deviation boxplots of daily air temperatures. Box: the 50% of the values; whiskers: the 10th and 90th percentiles; dots: outliers; black line: median; red line: mean

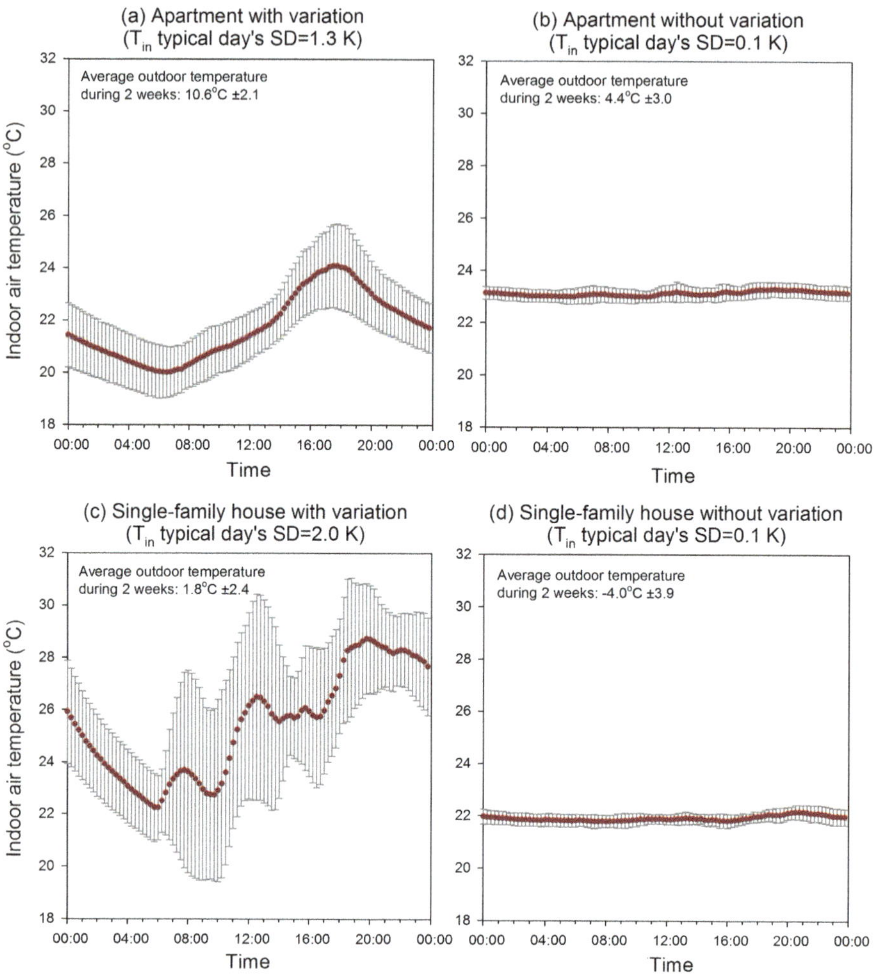

Fig. 5 Typical indoor temperature variations for **a** apartment with variation (typical day SD = 1.3 K), **b** apartment without variation (typical day SD = 0.1 K), **c** single-family house with variation (typical day SD = 2.0 K) and **d** single-family house with variation (typical day SD = 0.1 K). Standard deviation for each time-step is also included

versa, (c) false readings or readings on particularly atypical day patterns (e.g. SD = 10.7 K after a day with SD = 0.7 K) or d) particular weather conditions, e.g. solar radiation, wind etc.

For an indication of what these standard deviations mean in terms of temperature experience within a day, four examples were selected from the dataset: two for each of the main dwelling types: two apartments in multi-family buildings [(a) and (b)] and two single-family dwellings [(c) and (d)]. Typical daily variations for weekdays were created by averaging the measurements for each 15-minute time-step. Figure 5 shows the four temperature profiles created, with the standard deviation for each time-step of the day also plotted on the graphs. As can be seen, from SD values of indoor temperature for a typical day of 0.1–2.0, the environments experienced are very different. The sources of variation in figures (a) and (c) are unknown but patterns can be identified, with similarities to the temperature profiles found in Huebner et al. [16]: flat temperature profile [(b) and (d)], occupancy pattern with peaks (c) and 'steady rise' (a). It is clear that there are differences between dwellings both in temperature levels and shapes that need to be explored further.

4 Conclusions

This exploratory paper presents results from a two-step analysis of high-resolution measurements of air temperatures in Swedish homes. A wide range of temperatures was found among dwellings, with 30% being outside the range recommended by The public health agency of Sweden. The majority of average temperatures were between 20–24 °C, which lies in the high end of recommended values in standards (Table 1) and higher than those found in other studies (Table 3). From the 1306 households investigated, 90% of them had standard deviations of air temperature during the monitoring period below 1 K, suggesting that the majority experienced very small variations. Although such thermal stability is typically seen as positive, there may be comfort and health implications, as suggested by recent research. Future work will focus on identifying causes of temperature variation, i.e. its relationship with building type, location, climate, heating system and occupant behaviour. Furthermore, associations with perception, thermal preference, health and wellbeing will be explored, using occupants' responses to surveys from the BETSI program. Future research should also look into how healthy variations can be incorporated in indoor climate control without causing discomfort, especially to vulnerable people (e.g. elderly).

There are certain limitations in the analysis presented here: the data is almost 10 years old, the measurements correspond to a complete 24 h day without distinction between occupied and non-occupied hours and the location of monitoring is not consistent in all dwellings. However, the large size of the dataset has enabled the derivation of interesting observations regarding the level and variation of indoor temperatures in Swedish dwellings.

Acknowledgements The data analysed in this paper is drawn from the BETSI database, which was made available by Björn Mattsson at Boverket. D.Teli is a VINNMER Fellow supported by VINNOVA (Swedish Innovation Agency), Marie Curie Actions and the Profile 'Energy in Urban Development' within the Area of Advance 'Energy' at Chalmers University of Technology.

References

1. European Commission, Promoting healthy and energy efficient buildings in the European Union: National implementation of related requirements of the Energy Performance Buildings Directive (2010/31/EU). Luxembourg: Publications Office of the European Union (2016)
2. F. Nicol, M. Humphreys, S. Roaf, *Adaptive Thermal Comfort: Principles and Practice* (Routledge, London, 2012)
3. A. Mavrogianni, F. Johnson, M. Ucci, A. Marmot, J. Wardle, T. Oreszczyn et al., Historic variations in winter indoor domestic temperatures and potential implications for body weight gain. Indoor and Built Environ. **22**, 360–375 (2013)
4. Boverket. Så mår Våra Hus. Redovisning av Regeringsuppdrag Beträffande Byggnaders Tekniska Utformning m.m. (in Swedish). Karlskrona, Sweden (2009)
5. S. Langer, G. Bekö, Indoor air quality in the Swedish housing stock and its dependence on building characteristics. Build. Environ. **69**, 44–54 (2013)
6. M. Holgersson, U. Norlén, Domestic indoor temperatures in Sweden. Build. Environ. **19**, 121–131 (1984)
7. ASHRAE, ANSI/ASHRAE Standard 55- Thermal Environmental Conditions for Human Occupancy. Atlanda: American Society of Heating, Refrigerating and Air-Conditioning Engineers (2013)
8. ISO, EN ISO 7730:2005 Ergonomics of the thermal environment- Analytical determination and interpretation of thermal comfort using calculation of the PMV and PPD indices and local thermal comfort criteria. Geneva: International Standardisation Organisation (2005)
9. CEN, EN 15251:2007 Indoor environmental input parameters for design and assessment of energy performance of buildings addressing indoor air quality, thermal environment, lighting and acoustics. Brussels: CEN (European Committee for Standardization) (2007)
10. Public Health England, The Cold Weather Plan for England 2014. Protecting health and reducing harm from cold weather. In: Department of Health, editor (2014)
11. World Health Organization (WHO), Housing, Energy and Thermal Comfort. A review of 10 countries within the WHO European Region. Copenhagen (2007)
12. Marmot Review Team, *The Health Impacts of Cold Homes and Fuel Poverty* (Friends of the Earth & the Marmot Review Team, London, 2011)
13. Folkhälsomyndigheten. FoHMFS 2014:17. Folkhälsomyndighetens allmänna råd om temperatur inomhus (In Swedish). (Folkhälsomyndigheten, Stockholm, Sweden, 2014)
14. P.O. Fanger, *Thermal Comfort: Analysis and Applications in Environmental Engineering* (McGraw-Hill, New York, 1970)
15. F. Nicol, Temperature and adaptive comfort in heated, cooled and free-running dwellings. Build. Res. Inf., 1–15 (2017)
16. G.M. Huebner, M. McMichael, D. Shipworth, M. Shipworth, M. Durand-Daubin, A. J. Summerfield, The shape of warmth: temperature profiles in living rooms. Build. Res. Inf. **43**, 185–196 (2015)
17. W. van Marken Lichtenbelt, M. Hanssen, H. Pallubinsky, B. Kingma, L. Schellen, Healthy excursions outside the thermal comfort zone. Build. Res. Inf., 1–9 (2017)
18. R.J. de Dear, Revisiting an old hypothesis of human thermal perception: alliesthesia. Build. Res. Inf. **39**, 108–117 (2011)
19. L. Schellen, W.D. Van Marken Lichtenbelt, M.G.L.C. Loomans, J. Toftum, M.H. De Wit, Differences between young adults and elderly in thermal comfort, productivity, and thermal physiology in response to a moderate temperature drift and a steady-state condition. Indoor Air **20**, 273–283 (2010)
20. É. Mata, A. Sasic Kalagasidis, F. Johnsson, Energy usage and technical potential for energy saving measures in the Swedish residential building stock. Energy Policy **55**, 404–414 (2013)

21. D. Teli, S. Gauthier, V. Aragon, L. Bourikas, P. James, A. Bahaj, in *Thermal Adaptation to High Indoor Temperatures During Winter in Two Uk Social Housing Tower Blocks*. 9th Windsor Conference: Making Comfort Relevant 07–10 Apr 2016 (Cumberland Lodge, Windsor, UK, 2016)
22. Z. Wang, A field study of the thermal comfort in residential buildings in Harbin. Build. Environ. **41**, 1034–1039 (2006)
23. M. Luo, B. Cao, X. Zhou, M. Li, J. Zhang, Q. Ouyang et al., Can personal control influence human thermal comfort? A field study in residential buildings in China in winter. Energy Build. **72**, 411–418 (2014)

Wood as an Exposed Building Material for Indoor Climate Adaptation

Kristine Nore, Dimitrios Kraniotis and May-Linn Sortland

Abstract Use of massive wood has increased during the last decade. The concept of massive wood, mainly as cross laminated timber elements (CLT), has become a popular building method for new constructions, both in public and private sector. Massive wood elements take advantage of wood as building material, also as an indoor climate buffer. Moholt 50|50 is a new student-housing project in Trondheim, Norway, which consists of five mass timber towers. Each of them with eight stories built in CLT on top of a concrete storey. Apart from the student homes, the buildings host facilities, such as activity center, kindergarten, commercial areas and a library, also built in CLT. This makes Moholt 50|50 a significant wooden living lab in Trondheim. The building technique follows the development from the first Norwegian CLT student housing built in Ås in 2012, and reproduced later on in similar patterns in other Norwegian cities, as Tromsø, Haugesund, Drammen, Fredrikstad, Halden, Hønefoss, Porsgrunn and Trondheim. Research on comfort and operation cost coupled to indoor surfaces are included in project Moholt 50|50. The towers are built according to Norwegian low energy standards. All surfaces are treated with water solvent varnish, apart from two stories in one of the Moholt timber towers. Four stories are instrumented to document the difference in the behavior of untreated and treated wooden surfaces. Measurements show a different indoor climate of the stories with untreated surfaces. The measurements presented give preliminary results of a measurement period which, when finished, will include one year of inhabited studios from the date of moving in.

Keywords Wood · Moisture · Ventilation · Control · Indoor climate

---

K. Nore (✉) · M.-L. Sortland
Norwegian Institute of Wood Technology, PO box113 Blindern, 0314 Oslo, Norway
e-mail: kristine.nore@treteknisk.no

D. Kraniotis
Oslo Metropolitan University, PO box 4 St. Olavs Plass, 0130 Oslo, Norway

D. Johansson et al. (eds.), *Cold Climate HVAC 2018*,
Springer Proceedings in Energy, https://doi.org/10.1007/978-3-030-00662-4_71

1 Introduction

Solid timber constructions has increased during the last decade, passing a production volume of 500,000 m^3 per year in Europe [1]. The development of cross laminated timber (CLT) realize robust engineered building systems in wood. Many projects, both public and private, built in CLT, have been completed around the world, also tall wood buildings [2]. Norway is not an exception: with a breakthrough of almost 2500 student studios in ten Norwegian cities. Nowadays tall wood buildings are built with complete residential standards.

Hygroscopic materials have the ability to accumulate and release moisture due to change in the surrounding humidity [3–5]. The moisture buffer capacity is regarded as this ability to moderate, or buffer, the indoor humidity variations. The indoor relative humidity (RH) is closely related to indoor comfort, more specifically to thermal and respiratory comfort and perceived air quality. Both persistently low RH (<20%) and high RH (>75%) can cause health threats for humans such as respiratory infections or the growth of mould [6]. Wood is naturally hygroscopic, which enables it to act passively and to stabilize the indoor humidity and thus, temperature variations, since thermal and hygric phenomena are coupled [7]. Woloszyn et al. [8] confirmed that the use of gypsum-based moisture-buffering materials, combined with a relative humidity sensitive (RHS) ventilation system, could reduce the mean ventilation rate by 30–40% and generate 12–17% energy savings while during the heating period. The combined effect of ventilation and wood, as buffering material, make it possible to maintain a stable indoor RH between 43 and 59%.

Moisture buffering is closely related to latent heat. The latter refers to the heat of sorption due to the phase change from vapour to bound water in the material and vice versa. Osanyintola and Simonson [9] pointed the potential for direct energy savings-through latent heat- when a hygroscopic material, as wood, is combined with a well-controlled heating, ventilation and air conditioning (HVAC) system: around 2–3% of the total heating energy, but significant for cooling, i.e. 5–30% of the total cooling energy. The potential indirect savings from adjusting the ventilation rate and indoor temperature, while maintaining adequate indoor air quality and comfort, are in the order of 5% for heating while they range from 5 to 20% for cooling. In particular, in rooms with high moisture generation and consequently significant potential of heat of sorption, as bathrooms, hygroscopic surfaces can save up to 320 kWh/year from the energy demand of such spaces compared to a bathroom with non-permeable surfaces, by adjusting the heating system 3° lower [10].

Mendes et al. [11] used a heat and mass transfer model to show the effects of moisture on sensible and latent conduction loads. The results showed that excluding moisture transport in the building envelope from the whole-building energy simulation models potentially results in overestimation of conduction peak loads, but also in underestimation of the yearly integrated heat fluxes. The latter may lead to over dimensioning the HVAC systems, especially in dry climates, as well as to underestimate energy consumption, primarily in humid climates.

Previous experimental studies revealed the potential of untreated wooden surfaces to increase their surface temperature during moisture uptake [12, 13]. A mathematical elaboration of these findings showed that there a potential for reducing heat losses by transmission through opaque elements under certain conditions and limitations [14].

CLT elements exposed to indoor environment have an additional advantage: not only the hygroscopic properties of exposed wooden surfaces can be exploited but in addition the significant thermal mass of CLT elements. In this context, untreated CLT can have beneficiary results for indoor climate by both buffering the variations of RH [15] and the variations of air temperature indoors.

The potential for energy savings by reducing the demand for heating and ventilation without compromising indoor comfort can be great. Moholt 50|50 provides the opportunity to study such phenomena in situ and therefore give a better understanding and more deliberate implementation of these properties. This paper provides the initial experimental findings which present the initial measurements of the moisture buffering and latent sorption heat capacities in exposed cross-laminated timber (CLT). Moholt 50|50 has wooden surfaces at inner walls and the ceiling. On the 2nd and 3rd storey the all interior wooden surfaces are treated, while on 4th and 5th storey are untreated. The scientific question is how do these surfaces influence indoor climate and energy consumption with and without surface treatment.

2 Method

The unique in situ measurements of indoor climate in wooden surroundings are set up **at Europe's greatest CLT** neighborhood, at Moholt in Trondheim see Fig. 1a. The measurement project was early planned to be set up just prior to inhabitation. The research tower, as shown in Fig. 1b, unfortunately got a delivery stop on the tailored furniture order. A six-month period with the tower is ready for inhabitation, just lacking interiors. Thus, a unique period to align measurement equipment with behavior of each story, and perform a zero analysis.

2.1 Moholt Student Village

Moholt 50|50 student village in Trondheim includes five 9-storey buildings-towers, solely built in CLT. The total built area is 21,800 m^2, including the commercial areas in the first storey in concrete. The CLT is used in exterior walls, interior walls, slabs, elevator shafts and stairwells. Apart from the 632 studios, the buildings host other facilities, such as an activity center, a kindergarten, commercial areas and a library, forming thus a 'community'. The construction started in March 2015,

(a) (b)

Fig. 1 **a** Moholt student village with five timber towers and several lower and older brick-clad blocks; **b** the research tower of Moholt 50|50

and the first building was erected in August 2015. The entire project was completed in 2016, only lacking the activity house which is planned in the middle of the five timber towers.

The mass timber towers have a passive house standard, with a 200 mm thick mineral wool cover, clad with ventilated Kebony boards. The U-value of the outer walls are calculated to 0.10 W/m^{2}K as an insulation layer. The inner side of the CLT consists of a 50 mm mineral wool cover and a gypsum sheeting.

Each storey is 484 m^2 and holds 15 studios from 13 m^2. Each studio has a separate bathroom. About 250 m^2 common rooms with hallways, kitchen, stairs and toilet. The outer walls are covered, but the inner wooden walls are exposed on one side, while the other side has a sound dampening wall construction with gypsum sheeting. The shared rooms also have exposed wooden ceilings.

Ventilation is based on temperature and CO_2. The research block is just inhabited, which leaves the ventilation rate to a minimum. The ventilation rate surpasses $n = 0.5\ h^{-1}$. Heat is provided by electricity ovens close to the windows, but temperature is con-trolled by ventilation air. The ventilation system is, as normal, not set to include the potential energy buffering from latent heat exchange in the exposed wood. However, with the low ventilation effect, a normal adaption to the diurnal cycle is that the rooms may naturally adjust to absorbing solar radiation by drying out wooden surfaces and regain moisture and thus deliver sorption heat during the night. This heat pump effect is sought documented in the measurements in the research tower of Moholt 50|50.

2.2 Measurement Set up

Figure 2a show interiors of the kitchen and common living area. The shared room area has about 400 m^2 exposed wood area in total. Figure 2b shows the setup of the measurement equipment in the ceiling of the kitchen. The measurement sensors are integrated in a custom made logger. Table 1 display the sensor, including measurement specter and tolerance.

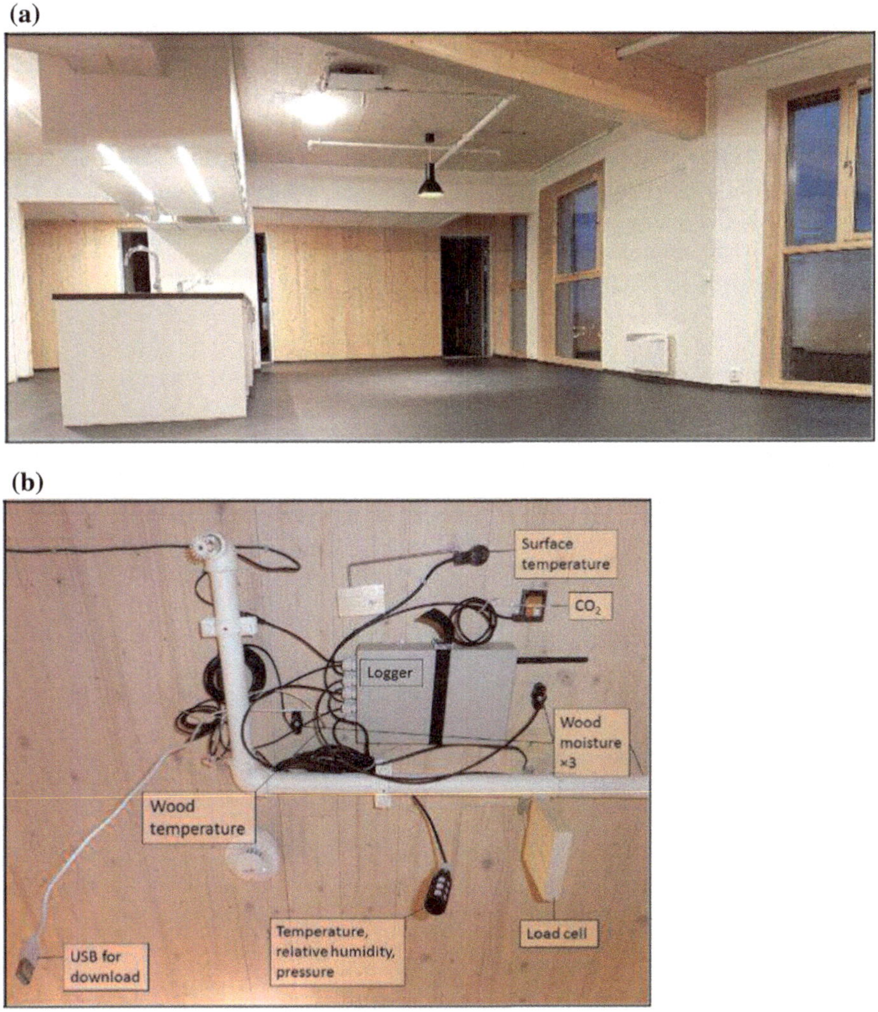

Fig. 2 **a** Interiors at the research stories at Moholt student village, measurement equipment in ceiling in the middle of the picture, covered by a white plate; **b** the measurement equipment with the sensors

Table 1 Sensors included in the measurement set up at Moholt with the respective measurement area and tolerances

Sensor	Unit	Measurement area	Tolerances
Temperature air	°C	−40−80	±0.5
Temperature wood surface (IR)	°C	−40−120	±0.5
Temperature wood (pt100)	°C	−30−100	±0.4
Load cell	gram	0–1000	±0.05%
CO_2	ppm	70–2000	±50
Relative humidity	%	0–99	±3
Wood moisture	MΩ	200–10,000	±2%
Pressure	hPa	300–100	±50

The measurement sensors are plugged in a proprietary logger. The set up includes normal indoor climate parameters like T_{air}, RH and $CO_{2,}$ and in addition wood specific measurements. T_{wood}, $T_{surface}$, wood moisture content and a load cell measure wood dynamic according to the indoor climate fluctuations. The load cell detects moisture up-take in dry wood when the electrical resistance measurements measuring wood moisture are difficult to detect, also used in [13].

3 Results

The preliminary results from Moholt 50|50 corresponds to two measurement weeks in 2017. The first week was uninhabited, from June 28th 2016 to July 5th, whereas the second week was inhabited, from Aug. 16th to Aug 23rd. The climate of the two weeks are shown in Fig. 3. Both weeks were warm and humid due to precipitation (Table 2).

The average values are given for all parameters except for the weight cell, which are only meant to study weight uptake and reduction, that is the water balance of the wood. The results clearly show an increased variation in the untreated floors.

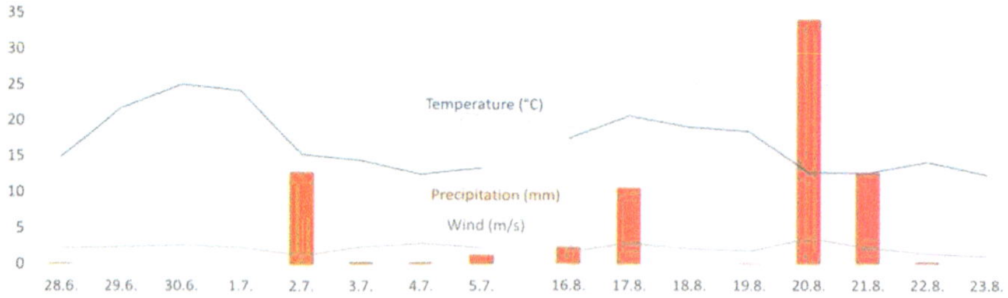

Fig. 3 Climate with temperature, wind velocity and precipitation in the first, uninhabited, and second, inhabited week

Table 2 Average and standard deviation of the $T_{operative} = (T_{air} + T_{surface})/2$, weight cell and Rel-ative Humidity of the four floors during the uninhabited and inhabited week

		Temperature operative (°C)		Weight cell (g)		Relative humidity (%)	
		Uninhab.	Inhab.	Uninhab.	Inhab.	Uninhab.	Inhab.
2. floor	Avg.	22.3	24.7			33.3	38.3
Treated	St. dev.	0.29	0.33	0.11	0.12	2.82	3.32
3. floor	Avg.	21.3	24.1			33.9	37.7
Treated	St. dev.	0.49	0.31	0.17	0.20	2.74	3.27
4. floor	Avg.	21.6	25.0			33.8	37.9
Untreated	St. dev.	0.29	0.41	0.44	0.51	3.36	3.63
5. floor	Avg.	22.9	26.4			31.6	37.0
Untreated	St. dev.	0.26	0.53	0.63	1.01	2.92	3.99

Initially, a reduced variation was expected whit the abundant moisture buffer capacity.

The combined temperature, $T_{operative}$, which is the average of indoor air temperature and the wood surface temperature, the relative humidity and the weight cell measurements provide insight in the provisional differences.

Figure 4 and Table 2 shows the variation of the $T_{operative}$, which has similar variation and averages for the uninhibited week. However, for the inhibited week the temperatures are higher and have higher variation for the untreated floors. The temperature increase in the storey without surface treatment can be justified by the latent heat of sorption that is generated because of the moisture absorption in CLT elements.

Figure 5 present the buffering capacity which results in moisture uptake in the wooden samples. The untreated floors show larger variation. The difference in the moisture up-take is approximately in average 0.2–1.2 g per day in the load cell

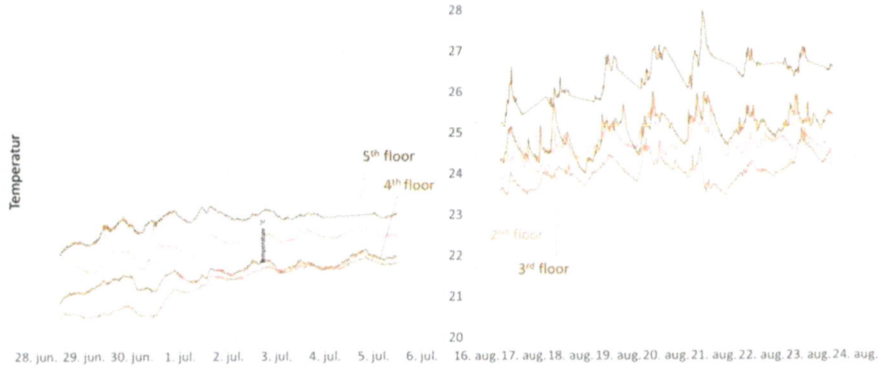

Fig. 4 The combined temperature $T_{operative}$ show increased temperature and slightly increased variation for the untreated floors when the studios are inhibited

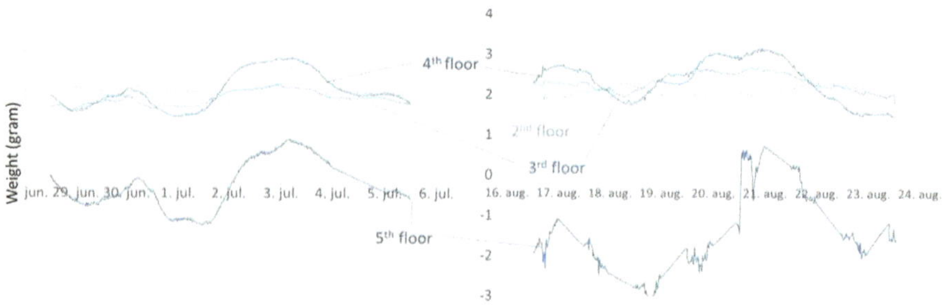

Fig. 5 The weight change of the treated and untreated floors which clearly differ with greater variation for the untreated floors

wood piece in 4th and 5th floor, the untreated walls and ceiling. The wooden dummy in the weight cell corresponds to an absorption surface of 250 cm^2. If this magnitude gets extrapolated to the total surface area of CLT elements in a storey without surface treatment, i.e. 400 m^2, then the total amount of moisture uptake in the walls and the ceiling can be estimated as about 15 kg per day. The absorbed amount of vapors is responsible for the latent heat of sorption generated in the mass of CLT elements without surface treatment. Based on the study of Hameury [14], the total energy (E) per day that is passively stored in massive wood in each of the floors without surface treatment is approximately E = 37 000 kJ $\approx$ 10 kWh or 100 J/m^2 $\approx$ 25 Wh/ m^2 floor area.

The relative humidity, presented in Fig. 6, does not show influence of the buffer capacity found in the temperature and moisture uptake in the weight cell. The ventilation system is, like every other ventilation system nowadays, not tuned into provide humidity on demand. The measured and provide potentially cooling for heating effect of the moisture sorption could have been included in the energy balance if the control system were instrumented to provide such input data.

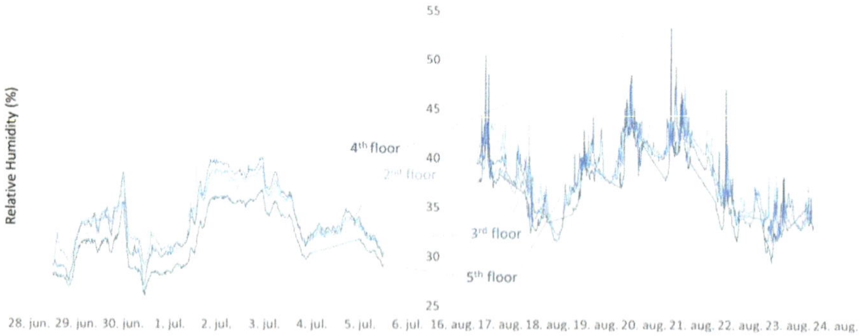

Fig. 6 Relative humidity in the floors. A clear influence of inhabitance in the second week, but no clear difference between the treated and untreated

4 Preliminary Conclusions

The studios, of the research tower of the Moholt student village, are just recently inhabited in. However, some preliminary findings are reported and discussed.

The results indicate a difference in the indoor climate in the stories with exposed, untreated surfaces and stories with surface treatment. The most significant effect is shown in the weight cell with the wood piece, where the response differs in the untreated stories.

A different response on indoor climate is expected when moisture production is in balance in the building. Moisture, e.g. from people, showers and cooking, give different exposure to this wooden living lab and the influence on ventilation systems, comfort and wooden behavior may provide more clear indications of the influence of wood on indoor climate.

The untreated stories have at this point a diurnal moisture fluctuation of about 15 litres each. This equals an energy buffer of around 10 kWh per storey per day. Note that this effect is 100% efficient due to the proximity of the energy storage.

It is of importance to note the influence of the ventilation system, which up till now does not link with the material potential in indoor climate adaptation.

Acknowledgements The authors greatly acknowledge the student housing organization in Trondheim (SiT) for always supporting and helping the research in progress and the Norwegian Research council for supporting the contractor through Skattefunn to finance the research activity.

References

1. G. Schickhofer, R. Brandner, H. Bauer, in *Introduction to CLT, Product Properties, Strength Classes*. Conference: Cross Laminated Timber—a competitive wood product for visionary and fire safe buildings: Joint Conference of COST Actions FP1402 and FP1404, At KTH, Stockholm, Sweden. March 2016
2. Survey Report: Survey of International Tall Wood Buildings. Forest Innovation Investment (FII) and Binational Softwood Lumber Council (BSLC). RethinkWood. Canada (2014)
3. C. Rode, R. Peuhkuri, L.H. Mortensen, K.K. Hansen, B. Time, A. Gustavsen, T. Ojanen, J. Ahonen, K. Svennberg, L.E. Harderup, J. Arfvidson, Moisture buffering of building materials. 264 Project No.: 04023. Nordic Innovation Centre (2005)
4. C. Rode, K. Grau, Moisture buffering and its consequence in whole building hygrothermal modeling. J. Build. Phy. **31**, 333–360 (2008)
5. C.J. Simonson, M. Salonvaara, T. Ojanen, Improving Indoor Climate and Comfort with Wooden Structures. Technical Research Centre of Finland, VTT Publications 431; Espoo, Finland (2001)
6. ASHRAE. Standard 55—Thermal Environmental Conditions for Human Occupancy (2013)
7. A. Karagiozis, H. Künzel, A. Holm, in *WUFI-ORNL/IBP—A North American Hygrothermal Model*. In Proceedings: Conference on Performance of Exterior Envelopes of Whole Buildings VIII, ASHRAE, Atlanta, USA (2001)

8. M. Woloszyn, T. Kalamees, M.O. Abadie, M. Steeman, A.S. Kalagasidis, The effect of combining a relative-humidity-sensitive ventilation system with the moisture-buffering capacity of materials on indoor climate and energy efficiency of buildings. Build. Environ. **44**, 515–524 (2009)
9. O.F. Osanyintola, C.J. Simonson, Moisture buffering capacity of hygroscopic building materials: experimental facilities and energy impact. Energy Build. **38**, 1270–1282 (2006)
10. K. Nore, A. Nyrud, D. Kraniotis, K.R. Skulberg, F. Englund, T. Aurlien, *Moisture buffering, energy potential and VOC emissions of wood exposed to indoor environments. Science and Technology for the Built Environment* (in press)
11. N. Mendes, F.C. Winkelmann, R. Lamberts, P.C. Philippi, Moisture effects on 269 conduction lads. Energy Build. **35**, 631–644 (2003)
12. K. Nore, D. Kraniotis, C. Brückner, The principles of sauna physics. Energy Procedia **78**, 1907–1912 (2015)
13. D. Kraniotis, K. Nore, C. Brückner, A.Q. Nyrud, Thermography measurements and latent heat documentation of Norwegian spruce (Picea abies) exposed to dynamic indoor climate. J. Wood Sci. **62**, 203–209 (2016)
14. D. Kraniotis, K. Nore, Latent heat phenomena in buildings and potential integration into energy balance. Procedia Environ. Sci. **2017**(38), 364–371 (2017)
15. S. Hameury, Moisture buffering capacity of heavy timber structures directly exposed to an indoor climate. A numerical study. Build. Environ. **40**, 1400–1412 (2005)

Black Carbon Concentrations Inside and Outside Occupied Residences

Aneta Wierzbicka, Hamza Licina, Yuliya Omelekhina, Patrik T. Nilsson and Anders Gudmundsson

Abstract Black carbon (BC, soot, elemental carbon) is a component of airborne particulate matter, which has been linked with negative effects on respiratory and cardiovascular systems. BC is considered an indicator of combustion related component of particulate matter, which can be emitted from both outdoor and indoor sources. Pollution control measures focus mainly on outdoor concentrations, whereas control of indoor levels seem to be neglected despite the fact that we spend on average 90% of our time in indoor environments. The aim of this study was to assess the differences in BC concentrations inside and outside occupied residences during weeklong real time measurements in ten residences. BC concentrations were measured simultaneously indoors and outside of ten occupied residences using two microAeth® AE51 (AethLabs, USA) instruments. Continues measurements inside and outside lasted at least seven consecutive days in each residence. Comparisons of BC concentrations were conducted for the times when occupants were present at home and when there was no one in the residence. Average concentrations of BC during occupancy time were comparable between indoors and outdoors. However, a significant contribution of indoor sources to measured BC was observed. High concentrations of BC indoors were due to cooking and candle burning. Concentrations of BC during non-occupancy time were higher outdoors than indoors as expected, as there were no indoor activities which may have contributed to observed levels indoors. Obtained results indicate that in order to minimize exposure of occupants to BC concentrations, efficient control measures of indoor emissions might be as important as prevention of outdoor pollution infiltration to indoor environments.

Keywords EC · Indoor air quality · Outdoor pollution

A. Wierzbicka (✉) · H. Licina · Y. Omelekhina · P. T. Nilsson · A. Gudmundsson
Ergonomics and Aerosol Technology, Lund University, 221 00 Lund, Sweden
e-mail: aneta.wierzbicka@design.lth.se

D. Johansson et al. (eds.), *Cold Climate HVAC 2018*,
Springer Proceedings in Energy, https://doi.org/10.1007/978-3-030-00662-4_72

1 Introduction

Epidemiological studies confirm association of negative health effects with exposure to black carbon [1]. BC is also of concern as a carrier of chemicals to human lungs due to large surface area and adsorption proprieties. BC is emitted from variety of combustion sources that originate from both outdoors (e.g. diesel exhaust) and indoors (e.g. candles burning, cooking) [2]. Pollution control measures focus mainly on outdoor concentrations whereas control of indoor levels seem to be neglected despite the fact that we spend on average 90% of our time in indoor environments. The aim of this study was to assess the differences in BC concentrations inside and outside occupied residences during weeklong real time measurements in ten residences.

2 Methodology

BC concentrations were measured simultaneously indoors and outside of ten occupied residences using two microAeth® AE51 (AethLabs, USA) instruments. Continues measurements inside and outside lasted at least seven consecutive days in each residence. MicroAeth® AE51 enables real-time measurements of BC with 1 min resolution. MicroAeth® measures the transmission of infrared (IR) light (880 nm) through the aerosol sample collected on a filter. The accumulation of particles on the filter over time increases the absorbance, which is calculated relative to a reference cell. The attenuation IR is then transferred to mass concentration of black carbon [3]. The Optimized Noise reduction Averaging (ONA) algorithm has been used to post-process the data [4].

Preliminary results from ten residences are presented here. Residences comprised three detached single family houses with natural ventilation, one apartment with natural ventilation and six mechanically ventilated apartments. All residences were situated in urban areas in southern Sweden. Air exchange rates were assessed and varied between 0.4–0.6 h^{-1} in naturally ventilated residences and 0.8–1.6 h^{-1} in residences with mechanical ventilation.

Occupants kept logbooks recording their presence in the residences. Data analysis was performed for occupancy time, when at least one person was present in the residence and non-occupancy when no one was present in the residence. Occupants were asked to maintain their typical indoor activities including food preparation and candle burning. Occupants kept also logbook with performed activities that are prone increase the level of BC indoors. Specific instructions were given as well as logbooks with listed activities of interest.

3 Results

Indoor to outdoor (I/O) ratios were used to compare differences in concentrations of BC in occupied residences. I/O ratios were calculated on the basis of average concentrations for occupancy and non-occupancy time. The occupancy time concentrations are the most relevant for the exposure assessment, while the non-occupancy concentrations indicate infiltration of outdoor BC to indoors. The average occupancy time for the weeklong measurements was 86%, ranging from 72–98%. The average occupancy time I/O ratios of BC concentrations for studied residences ranged from 0.7 to 2.4 (average 1.1). I/O ratios greater than 1 prove that the contribution of indoor sources to BC concentrations indoors was greater than infiltration from outdoors. Non-occupancy time I/O ration ranged from 0.3 to 0.9 (average 0.6). Lower I/O ratios during non-occupancy were expected, since there are no occupants indoors and no indoor activities that may generate BC indoors

The highest maximum concentrations of BC were reached indoors accounting to 775 μg/m^3, while maximum concentrations measured outdoors reached 6 μg/m^3. High concentrations of BC indoors were identified to originate from cooking and candle burning.

In Fig. 1 an example of measured concentrations inside and outside of a residence A is presented. In residence A, on average BC concentrations inside were higher than outside, I/O ratio was 2.4. Distinct peaks that were identified on the basis of logged occupants' activities illustrate a strong influence of indoor sources such as cooking, frying, using the microwave and candle burning. During operation of these indoor sources concentration of BC inside were higher than outdoors.

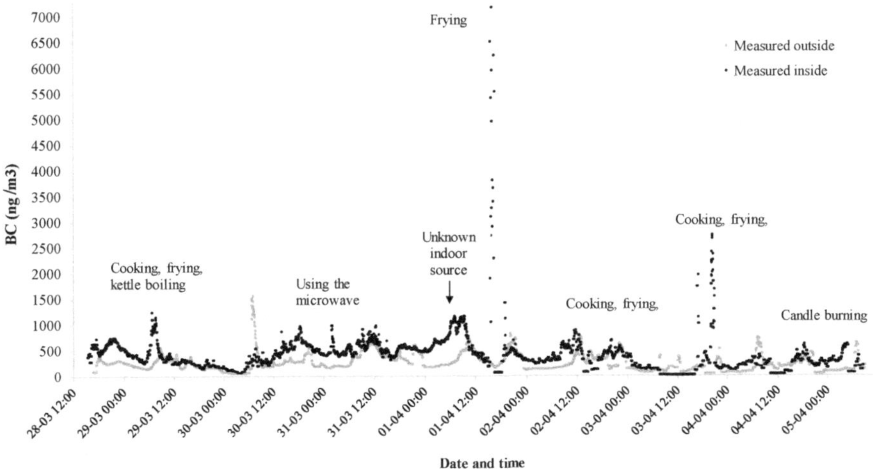

Fig. 1 Black carbon concentrations measured inside (black colour) and outside (grey colour) an occupied residence A. Concentrations inside higher than outside, strong influence of indoor sources is visible. I/O ratio 2.4

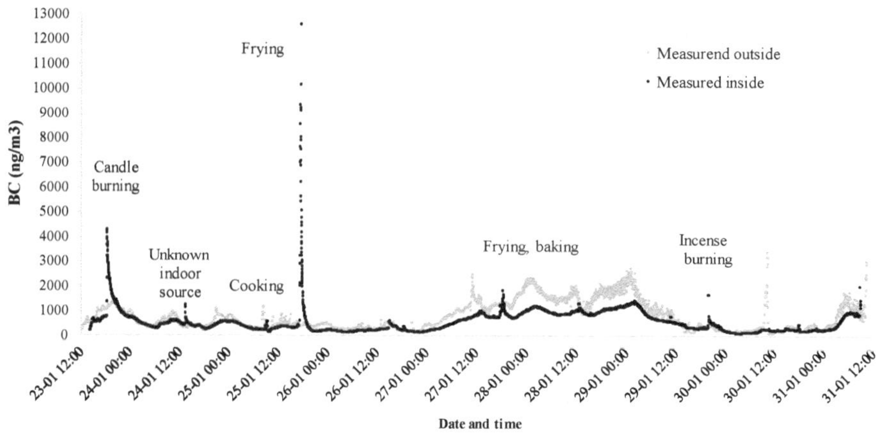

Fig. 2 Black carbon concentrations measured inside (black colour) and outside (grey colour) of an occupied residence B. On average concentrations outside were higher than inside, high peaks inside (higher than outdoor concentrations) due to candle burning and frying. I/O ratio 0.7

In Fig. 2 an example of measurements of BC inside and outside of a residence B, where I/O ratio was 0.7, is given. On average the concentrations outside were higher than indoors, thus the influence of indoor sources was not as pronounced as in residence A. However still two very high peaks due to candle burning and frying could be observed. On 27th and 28th of January, it can be seen that the increase in outdoor BC concentrations is followed by increase in indoor concentrations due to infiltration from outdoors. Concentrations indoors on these days are lower than outside due to losses upon particle penetration through building envelope and ventilation system.

4 Conclusions

In general, average concentrations of BC during occupancy time were comparable between indoors and outdoors, about 0.5 μg/m^3. However, observed BC concentrations indoors during occupancy time do not originate only from outdoors. During occupancy time both higher I/O ratios and very high peaks during cooking and candle burning (reaching 775 μg/m^3) prove a significant contribution of indoor sources to measured BC indoors. The influence of indoor sources to BC concentrations indoors have been neglected in the past and the control measures (if any) are focused on prevention of infiltration of outdoor pollution to indoor environments. However, the results of this study indicate that in order to minimize exposure of occupants to BC concentrations, efficient control measures of indoor emissions might be as important as prevention of outdoor pollution infiltration to indoor environments.

Acknowledgements This work was financed by the Swedish Research Council FORMAS (Project Dnr 942-2015-1029 and Dnr 2016-20079) and the Swedish Energy Agency (Dnr 43092-1).

Compliance with Ethical Standards

Authors declare no potential conflicts of interest. Results presented here were compiled from three projects. In projects 2016-200079 and 43092-1, which involved studying behavior of the occupants (not described here), the ethical approval has been obtained from the regional ethical review board at Lund University. In project 942-2015-1029 ethical approval was not required, it comprised of technical measurements only. In 942-2015-1029 collected particles on the filters (not described here) were used in toxicological study in mice. All animal procedures were reviewed and approved by the local Animal Welfare Body at NRCWE in Denmark as well as by the Animal Experiments Inspectorate under the Danish Ministry of Justice. The studies were performed in accordance with the Declaration of Helsinki. Informed consent was obtained from all participants in the studies.

References

1. United Nations, Effects of Air Pollution on Health. Report from the fifteenth meeting of the Joint Task Force on the Health Aspects of Air Pollution. Economic and Social Council—Economic Commission for Europe, Geneve (2012)
2. C. Isaxon, A. Gudmundsson, E. Nordin, L. Lönnblad, A. Dahl, G. Wieslander, M. Bohgard, A. Wierzbicka, Contribution of indoor-generated particles to residential exposure. Atmos. Environ. **106**, 458–466 (2015)
3. Y.H. Cheng, M.H. Lin, Real-time performance of the microAeth® AE51 and the effects of aerosol loading on its measurement results at a traffic site. Aerosol Air Qual. Res. **13**(6), 1853–1863 (2013)
4. G.S. Hagler, T.L. Yelverton, R. Vedantham, A.D. Hansen, J.R. Turner, Post-processing method to reduce noise while preserving high time resolution in aethalometer real-time black carbon data. Aerosol and Air Qual. Res. **11**(5), 539–546 (2011)

Should We Differentiate Ventilation Requirements for Different User Groups?

Nora Holand, Aileen Yang, Sverre Holøs, Kari Thunshelle and Mads Mysen

Abstract The aim of our study is to investigate whether it is necessary to adjust the ventilation requirements according to different user groups. This study is focusing especially on teenagers, who might have a higher odour load than children due to increased hormone and sweat production during puberty. The odour intensity (OI) and the perceived air quality (PAQ) were evaluated in four classrooms in Oslo, Norway. Two control classrooms of 9–11 years olds (children) were compared with two case classrooms of 12–15 years olds (teenagers). A sensory panel of 18 untrained people visited the four classrooms three times during a three-hour period and were asked to evaluate PAQ and OI upon entering the classrooms. The classrooms were supplied with a constant ventilation rate of 7 l/s per person, with no additional ventilation for building materials. We found that the classroom with children had a significant better PAQ-score than both classrooms with teenagers. Furthermore, although the ventilation rate per person was reduced, the percentage of panellists dissatisfied with OI and PAQ was lower (<20%) than expected. Our results indicate that children and teenagers have different sensory pollution loads, and therefore might need differentiated ventilation rates if the ventilation rates were to be optimised. However, more research is needed.

Keywords Perceived air quality · Odour intensity · Percentage of dissatisfied Ventilation rate · IAQ · Sensory pollution load · Bioeffluents · Untrained panel · School

N. Holand · A. Yang · M. Mysen
OsloMet—Oslo Metropolitan University (formerly Oslo and Akershus University College of Applied Sciences), P.O. Box 4 St. Olavs plass, 0130 Oslo, Norway

N. Holand
Multiconsult Norway AS, Nedre Skøyen vei 2, 0276 Oslo, Norway

S. Holøs · K. Thunshelle · M. Mysen (✉)
SINTEF Building and Infrastructure, Forskningsveien 3B, 0373 Oslo, Norway
e-mail: mads.mysen@gk.no

D. Johansson et al. (eds.), *Cold Climate HVAC 2018*,
Springer Proceedings in Energy, https://doi.org/10.1007/978-3-030-00662-4_73

1 Introduction

The main purpose of ventilation is to reduce and dilute the indoor pollution sources, such as building material emissions and body odour intensity. Especially for children at school, it is important to have good indoor air quality (IAQ) in the classrooms as it can improve their learning and performances. Considering the complex structure of indoor air, with thousands of chemical compounds in the air, using chemical analysis has so far not answered the question of what makes a good indoor climate. Sensory pollution analysis, introduced by Fanger [1], has during the last 40 years been used as the basis for ventilation standard and guidelines [2, 3]. The "olf" unit defined as the emission rate of air pollutants (bioeffluents) from a standard person has been used to quantify air pollution sources. Any other pollution source is then expressed by the equivalent source strength, defined as the number of standard persons (olfs) required to cause the same dissatisfaction as the actual pollution source. Table 1 summarizes the pollutant load from different sources used in the European Standard CEN 1752 [2].

Currently, the ventilation standard distinguishes between the emission rates of different building materials, but not for the ventilation rates per person of different user groups [3]. According to ASHRAE, the ventilation rate required for acceptable IAQ can be defined with <20% of people dissatisfied (PD) [3]. For a standard person (1 olf), this is equivalent to a ventilation rate of 7 l/s per person. However, as shown in Table 1, the sensory pollution load for children differs from the standard person, but the recommended ventilation rates does not take this factor into consideration. Seemingly, the resulting sensory pollution load, based on different user groups, hasn't implicated any practical meaning considering the ventilation requirements. For example, using Fanger's [1] equation to calculate the required ventilation rate to remove the sensory pollution from children (1.3 olf) and to achieve maximum 20% of people dissatisfied, the ventilation rate would be approximately 9.1 l/s per person.

During puberty, due to increased hormone and sweat production, teenagers might have an even higher sensory pollution load than children. Based on this, the purpose of this paper was to assess whether it is necessary to adjust the ventilation requirements according to different user groups, focusing on teenagers and children.

Table 1 Sensory pollution load from humans and buildings [2]

Source	Sensory pollution load (olf)
Sedentary adult (1–1.5 met)	1
Person exercising, low activity (3 met)	4
Person exercising, medium activity (6 met)	10
Kindergarten, 3–6 years (2.7 met)	1.2
Schoolchildren, 14–16 years (1–1.2 met)	1.3
Building, low pollution	0.1 per m^2
Building, polluting	0.2 per m^2

2 Methods

2.1 Study Design

The study was carried out in four classrooms at a school in Oslo, Norway. Two control classrooms with pupils in the 5th and 6th grade (children aged 9–11) were compared with two case classrooms with pupils in 8th and 10th grade (aged 12–15). The characteristics of the classrooms are summarized in Table 2. The airflow rates in the four classrooms were kept constant during the experiments, even though the school has demanded-controlled ventilation. The classrooms were supplied with a constant ventilation rate of 7 l/s per person with no additional ventilation for building materials. The four classrooms had an average floor area of 60 m^2, height of 2.8 m and similar furnishings.

An untrained sensory panel of 18 people evaluated the odour intensity (OI) and perceived air quality (PAQ) in the four classrooms during a school day. The school day consisted of a morning teaching period (8:30–11:50), lunch break (11:50–12:30) and an afternoon teaching period (12:30–15:55). The experiments were carried out on the 13th March 2017 in the morning teaching period; during the first, second and third teaching hour. To ensure that the experiments were performed close to steady state conditions, each visit was done at minimum 30 min into the teaching hour. The four classrooms were visited three times in a random order. The untrained sensory panel consisted of six females and 12 males aged 24–41 (mean ± standard deviation: 28.3 ± 4.3). Each panellist received assessment forms for each round and the evaluations were done according to the ASHRAE Standard 62 [3]. The panellists entered the classrooms at the same time and gave their scores for PAQ and OI within maximum 30 s to counteract sensory adaptation. PAQ was evaluated using a continuous acceptability scale divided in two parts [4]. Odour intensity was measured using a 6-point scale [5]. Figure 1 shows the scales used for subjective assessment of PAQ and OI.

Data on temperature, CO_2 concentrations and outdoor airflow rate were collected for each classroom by the Building Management system (BMS). Relative humidity was measured using Tinytag Plus 2 (Gemini Data Loggers, UK).

Table 2 Characteristics of the classrooms, with estimated number of people (N)

Classroom	N	Floor area (m^2)	Airflow rate (m^3/h)	Air change rate (h^{-1})
5. grade	28	60.1	706	4.20
6. grade	24	60.0	605	3.60
8. grade	17	60.9	428	2.54
10. grade	25	60.5	630	3.75

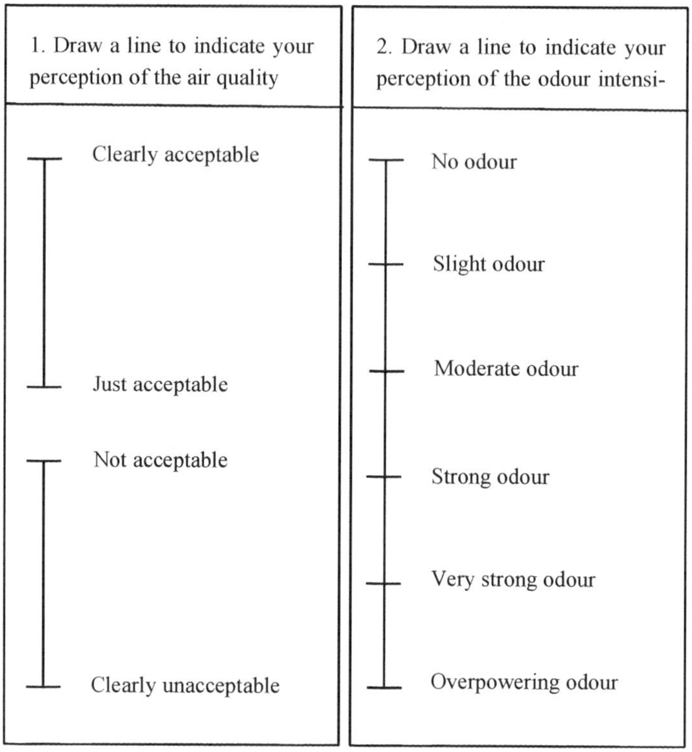

Fig. 1 Scale used for subjective assessment of PAQ (left) and odour intensity (right). The panellists were asked to draw a line. Note that the scale for PAQ is divided into two parts

2.2 *Data Analysis*

For data analysis, both the PAQ acceptability scale and OI-scale was converted into numbers. The PAQ acceptability scale was divided in two parts and coded as following: 1 = "Clearly acceptable", 0.01 = "Just acceptable", −0.01 = "Not acceptable", −1 = "Clearly unacceptable". The OI-scale was coded as following: 1 = "No odour" and −1 = "Overpowering odour". The scores of odour intensity and perceived air quality were used to calculate the percentage dissatisfied (PD) with air quality [1].

The statistical differences between case and control classrooms were analysed using the non-parametric tests Friedman's ANOVA by ranks or Wilcox Signed-Rank test. Statistical analysis was performed with SPSS version 24 (SPSS Inc, Chicago, USA).

3 Results

Table 3 summarizes the measured indoor climate parameters obtained from the BMS loggers. We also calculated the personal ventilation rate and enthalpy. The actual number of occupants in classrooms of 6th and 10th graders deviated the most, resulting in a higher ventilation rate per person. Generally, there were not many variations in temperature, relative humidity and CO_2-level between the visits across the four classrooms. As seen in Table 3, the difference in temperature is most likely due to the reduced ventilation rates as the enthalpy in each classroom is similar. The first visit, which was made approximately 30 min after the school started, resulted in the lowest CO_2-level and highest RH in all four classrooms.

3.1 *Odour Intensity*

Figure 2 shows the variation of the scores for odour intensity in the four classrooms. The OI-score of 0.1629, corresponding to "*moderate odour*", was set to indicate acceptable odour intensity. The median OI-scores for the classrooms with teenagers were mostly lower than the median OI-scores for the classrooms with 6th graders. Surprisingly, the classroom of 5th graders received the lowest overall OI-score (median = 0.48), which corresponds to moderate to slight odour. The percentage

Table 3 Overview of the actual number of occupants/pupils (N), outdoor air supply rate ($\dot{V}_{supply}$), calculated ventilation rate per person ($\dot{V}_{pers}$), room temperature (T), CO_2, relative humidity (RH) and calculated enthalpy during the three visits in the classrooms

	N	$\dot{V}_{supply}$ (m^3/h)	$\dot{V}_{pers}$ (l/s)	T (°C)	CO_2 (ppm)	RH (%)	Enthalpy (kJ/kg)
5th grade							
1.	26	677	7.2	22.5	609	30.4	38.5
2.	26	642	6.9	23.2	1006	27.5	35.8
3.	26	736	7.9	23.7	847	24.9	35.5
6th grade							
1.	18	591	9.1	23.3	588	32.3	38.2
2.	15	547	10.1	23.5	863	26.7	36.0
3.	20	572	7.9	23.3	809	24.8	34.7
8th grade							
1.	16	432	7.5	24.0	531	26.5	36.8
2.	18	397	6.1	24.8	607	24.4	37.1
3.	17	397	6.5	25.4	599	22.7	37.3
10th grade							
1.	25	672	7.5	23.7	593	29.9	37.8
2.	20	582	8.1	24.2	827	25.4	36.6
3.	26	588	6.3	24.2	814	26.1	37.3

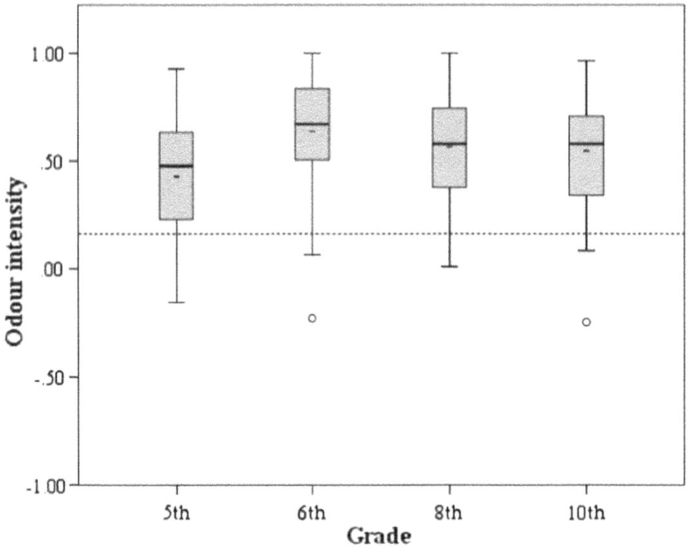

Fig. 2 Boxplot of the scores for odour intensity by grade. The dotted line indicates moderate odour (score = 0.1629). The dark line in the middle of the boxes is the median, the short line is the mean. The top and bottom of the box are the 75th and 25th percentiles. Whiskers indicate the 10th and 90th percentiles and individual outliers are shown as points

dissatisfied was also highest in this classroom (PD = 16.7–22.2%). We were later made aware that people had complained about unpleasant odours occurring in this classroom. Consequently, the classroom with 5th graders was excluded from further analysis since it was not considered a representative classroom.

Generally, we found no significant differences in OI-score between the classroom with children and the two classrooms with teenagers. The classroom with 6th graders, with the highest personal ventilation rate, received the highest overall OI-score (median = 0.67) and the lowest percentages of dissatisfied with the IAQ. During the three visits, the panellists were largely more dissatisfied with the odour intensity in the classrooms with teenagers than the classroom with 6th graders (see Table 4).

3.2 *Perceived Air Quality*

An overview of descriptive statistics for each visits is shown in Table 4. Figure 3 shows the variation of PAQ-scores across the four classrooms, with a PAQ-score of 0.01 set to "Just acceptable". Generally, the classrooms with teenagers received lower PAQ-scores than the classrooms with children. The classroom of 6th graders received the highest overall PAQ-score (median = 0.73), and the classroom of the 8th graders the lowest (median = 0.39). We also found that the PAQ-score for the

Table 4 Descriptive statistics of the odour intensity and perceived air quality (PAQ) during the three visits in the four classrooms

	Odour intensity (OI)				PAQ			
	Mean ± SD	Median	Min, max	% PD	Mean ± SD	Median	Min, max	% PD
5th grade								
1.	0.41 ± 0.29	0.43	−0.14, 0.87	22.2	0.55 ± 0.26	0.58	0.10, 0.96	0
2.	0.39 ± 0.31	0.45	−0.12, 0.91	22.2	0.50 ± 0.27	0.51	0.06, 0.98	0
3.	0.49 ± 0.34	0.55	−0.16, 0.93	16.7	0.46 ± 0.28	0.52	0.03, 0.87	0
All	0.43 ± 0.31	0.48			0.50 ± 0.27	0.53		
6th grade								
1.	0.60 ± 0.31	0.64	−0.23, 1.0	5.6	0.51 ± 0.33	0.61	−0.06, 0.98	5.6
2.	0.65 ± 0.28	0.72	0.08, 1.0	5.6	0.70 ± 0.24	0.79	0.21, 1.0	0
3.	0.66 ± 0.29	0.70	0.06, 1.0	11.1	0.68 ± 0.26	0.75	0.12, 1.0	0
All	0.64 ± 0.29	0.67			0.63 ± 0.29	0.73		
8th grade								
1. visit	0.54 ± 0.28	0.55	0.10, 0.98	11.1	0.33 ± 0.34	0.34	−0.17, 0.93	11.1
2.	0.61 ± 0.29	0.64	0.01, 0.98	16.7	0.44 ± 0.36	0.35	−0.06, 1.0	5.6
3.	0.54 ± 0.23	0.55	0.10, 1.0	11.1	0.48 ± 0.31	0.51	0.10, 1.0	0
All	0.57 ± 0.26	0.58			0.41 ± 0.34	0.39		
10th grade								
1. visit	0.51 ± 0.22	0.57	0.10, 0.83	11.1	0.52 ± 0.25	0.54	0.01, 0.93	0
2.	0.53 ± 0.36	0.61	−0.25, 0.96	16.7	0.38 ± 0.39	0.31	−0.51, 0.96	11.1
3.	0.60 ± 0.24	0.59	0.10, 0.94	5.6	0.49 ± 0.30	0.56	−0.08, 0.91	11.1
All	0.55 ± 0.28	0.58			0.46 ± 0.32	0.51		

classroom of 6th graders differed significantly from the two classrooms with teenagers (Friedman's ANOVA, $p < 0.01$). We did not find any significant difference in PAQ-score between the classroom of 5th graders and the classrooms with teenagers.

As seen in Table 4, during the three visits, the panellists were generally more dissatisfied with the air quality in the classrooms with teenagers than the classrooms with children.

4 Discussion

The use of sensory measurements to calculate the necessary ventilation to achieve a certain PAQ is generally accepted [1, 5]. The main research question asked in this paper is whether differences in user groups should be considered when estimating the required ventilation rates for optimal indoor air quality. According to the technical report CEN CR 1752 [2], the sensory pollution load from kindergarten children and school children differs from the standard person, this due to children

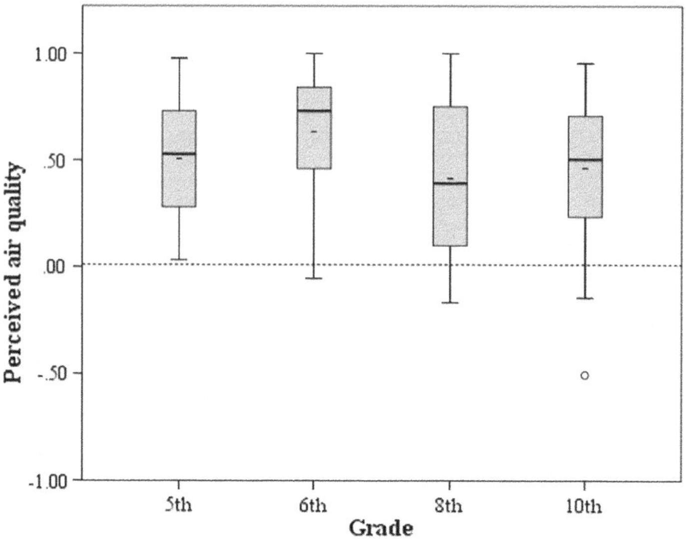

Fig. 3 Boxplot of PAQ-scores by grade. The dotted line indicates acceptable PAQ (score = 0.01). The dark line in the middle of the boxes is the median, the short line is the mean. The top and bottom of the box are the 75th and 25th percentiles. Whiskers indicate the 10th and 90th percentiles and individual outliers are shown as points

having a higher activity level and possibly poorer hygiene. Subsequently, it is reasonable to assume that this difference influences the indoor air quality, and thus ventilation requirements should be differentiated between different user groups. Our findings do not provide any indications of whether the pollution load from children and teenagers are lower or higher than the standard person. It would be of interest to compare the pollution load of the standard person with children and/or teenagers.

Nevertheless, we did find that the OI-scores was overall lower for the classrooms with teenagers compared to those with children, although this difference in OI-scores was not significant. Furthermore, the PAQ-scores for the classrooms with teenagers were also significantly lower than the classroom with children. This indicates that the panellists were generally more satisfied with the air quality in the classrooms with children. Yaglou et al. [5] recommended higher ventilation rates per person for children to achieve similar acceptable odour intensity as for adults. Similarly, we found that higher personal ventilation rates for children in the 6th grade resulted in the lowest percentage dissatisfied and highest OI-score.

Compared with other studies, we found a lower percentage dissatisfied with OI and PAQ in our study. This could be due to that majority of the studies on sensory measurements and ventilation rates also included pollution load from tobacco smoke and building materials in addition to occupancy [1, 6–8]. Fanger [1] also postulated that the sensory unit olf for individual pollution sources can be added if they occur at the same time, but also recommended further research about this. In our study, low-emitting building materials have been used in all classrooms, thus it is reasonable to assume that the pollutant load from the materials is rather low. The focus of our study was mainly on the pollution load from occupants, assessing pollutant loads from building materials is beyond the scope of this study.

During field experiments, it is hard to control every parameter that can influence the results. The sensory panel visited the classrooms during a normal school day. As shown in Table 3, during the three visits, the number of occupants in the classrooms varied. This obviously affected the amount of supplied air per person ($\dot{V}_{pers}$). The school has demand-controlled ventilation with optimizers. Even though the four classrooms were programmed to deliver a constant air volume, the supplied airflow rate might vary due to pressure changes in the main ventilation duct when the ventilation in the classrooms nearby changed during the school day. This factor, and the 15-min long break before the third visit, might have caused the different CO_2-levels during the day. Nevertheless, none of the pupils left the classrooms during the 30 s the untrained panel evaluated the PAQ and the OI.

As seen in Table 3, with a ventilation rate of 6.3 l/s per person and a temperature of 24.4 °C, we only achieved a PD of 5.6% during the third visit in the classroom with 10th graders. With the exception of the classroom with 5th graders, which was excluded due to known unpleasant odours, the percentage of panellists dissatisfied with PAQ (<12% dissatisfied) and OI (<17% dissatisfied) was considerably lower than expected. With the supplied ventilation rates based on the occupants only, the PD was expected to be higher than 20% if the pollution loads in olf are summarised. Based on this, even with lower ventilation rates, it should be possible to accomplish an expected level of 20% dissatisfied in three of the four classrooms we visited. Even though only four classrooms were assessed, our results indicate that the olf values shown in Table 1 might be outdated. However, further research is needed.

5 Conclusion

Our results indicate that there might be a need to differentiate between user groups in regards to ventilation rates.

Acknowledgements This paper is based on the master thesis by Nora Holand, and was a part of the BEST VENT project. BEST VENT is funded by the Research Council of Norway EnergiX program under Grant 255375/E20 together with the industry partners Undervisningsbygg Oslo KF, GK Inneklima AS, DNB Næringseiendom AS, Erichsen & Horgen AS, Hjellnes Consult AS, Multiconsult AS, Interfil AS, Camfil Norge AS, Swegon AS, Belimo Automasjon Norge AS, NEAS AS, Norsk VVS Energi- og Miljøteknisk Forenings Stiftelse for forskning.

Compliance with ethical standards

This study involves no more than minimal risk and involves no procedures for which consent is required from parents/LAR. Formal consent was given by the school and the teachers to visit the classrooms during teaching hours. We did not collect any identifiable or sensitive information that would require ethical approval. The research has been conducted in compliance with the ethical standards at OsloMet—Oslo Metropolitan University (formerly Oslo and Akershus University College of Applied Science) and Norwegian Law.

References

1. P.O. Fanger, Introduction of the olf and the decipol units to quantify air pollution perceived by humans indoors and outdoors. Energy Build. **12**, 1–6 (1988). https://doi.org/10.1016/0378-7788(88)90051-5
2. CEN. Technical Report CR 1752, *Ventilation for Buildings. Design Criteria for Indoor Environments* (European Committee for Standardization, Brussels, 1998)
3. ASHRAE. ANSI/ASHRAE Standard 62, *Ventilation for Acceptable Indoor Air Quality* (American Society of Heating, Refrigeration and Air-Conditioning Engineers, Atlanta, USA, 2001)
4. L. Gunnarsen, P.O. Fanger, Adaptation to indoor air pollution. Environ. Int. **18**, 43–54 (1992). https://doi.org/10.1016/0160-4120(92)90209-M
5. C.P. Yaglou, E.C. Riley, D.I. Coggins, Ventilation requirements. ASHVE Trans. **42**, 133–162 (1936)
6. P. Wargocki, Sensory pollution sources in buildings. Indoor Air **14**, 82–91 (2004). https://doi.org/10.1111/j.1600-0668.2004.00277.x
7. W.S. Cain, B.P. Leaderer, R. Isseroff, L.G. Berglund, R.J. Huey, E.D. Lipsitt, et al., Ventilation requirements in buildings—I. Control of occupancy odor and tobacco smoke odor. Atmos. Environ. 1983 **17**, 1183–1197 (1967). https://doi.org/10.1016/0004-6981(83)90341-4
8. P.O. Fanger, J. Lauridsen, P. Bluyssen, G. Clausen, Air pollution sources in offices and assembly halls, quantified by the olf unit. Energy Build. **12**, 7–19 (1988). https://doi.org/10.1016/0378-7788(88)90052-7

Influence of the Thermal Environment of a Bathroom After Renovation on Blood Pressure of Residents

Chiemi Iba, Shuichi Hokoi, Nobuyuki Masugi, Daisuke Ogura, Satoru Takada and Kenji Iida

Abstract Many Japanese people are dissatisfied with the coldness of dressing rooms and bathrooms. It is well known that accidents due to changes in blood pressure during bathing may lead to casualties, particularly in elderly people. However, measurements of blood pressure changes along with evaluation of the thermal environment in residential buildings are very limited. This study aimed to survey the influence that the thermal renovation of the dressing rooms and bathrooms has on the health of elderly residents during winter in residential buildings in Hokkaido, a cold northern region in Japan. Three detached residences built before the 1990s, whose residents were over 60 years of age, were renovated. Bath units were replaced, or additional thermal insulation was added to the openings and floor of the dressing rooms and bathrooms. Before and after the renovation, the thermal environment around the bathroom was evaluated and the blood pressure of residents during bathing was measured. From undressing in the dressing room until dressing after bathing, blood pressure was measured sequentially by the residents themselves using a handy-type hemadynamometer. After renovation, the increase in the highest systolic blood pressure during undressing time and the range of the blood pressure change during the whole bathing process were lower than those before renovation. Systolic blood pressure decreased with the increase in dressing room temperature. Correlation analysis showed that systolic blood pressure decreased up to 20 mmHg when the dressing room temperature increased by 10 °C.

C. Iba · D. Ogura
Kyoto University, Kyoto 615-8540, Japan

S. Hokoi (✉)
Southeast University, Nanjing, China
e-mail: hokoi@maia.eonet.ne.jp

N. Masugi
Central Japan Railway Co, Nagoya 450-6101, Japan

S. Takada
Kobe University, Kobe 657-8501, Japan

K. Iida
LIXIL Corporation, Noda 270-0237, Japan

D. Johansson et al. (eds.), *Cold Climate HVAC 2018*,
Springer Proceedings in Energy, https://doi.org/10.1007/978-3-030-00662-4_74

Keywords Bathing · Renovation · Thermal environment · Blood pressure Elderly

1 Introduction

Bathing is very effective in relieving fatigue and is thus one of the most important daily actions for Japanese people. However, the number of people, especially elderly people, whose deaths result from an accident in the bathroom is large. Most of them die due to heart and blood vessel discascs cxacerbated by a sudden change in blood pressure due to changes in the thermal environment around the human body during bathing [1] and dizziness caused by postural changes [2]. Several experiments examining body temperature, blood pressure, and pulse during bathing showed that elderly people experience larger changes in blood pressure and a slower cardiovascular response than young people, and the temperature of hot water in the bath and dressing rooms have a significant influence on the change in body temperature and blood pressure during bathing [3–5]. Kajii [6] carried out experiments to examine the relationship between bathing posture and blood flow and changes in body temperature. These results indicated that an improvement of the bathing environment may be effective to achieve safer and more comfortable bathing and that the prediction of physiological response during bathing is essential to design a suitable environment in bathrooms and dressing rooms.

However, research on actual situations that include blood pressure data has been very limited, probably because continuous and detailed measurement of blood pressure is difficult, particularly for elderly people. This study investigated the influence that an improvement of the indoor thermal environment due to thermal renovation of dressing rooms and bathrooms has on the safety of bathing in Taikicho, Hokkaido, where winter is very severe. An inquiry survey on basic information regarding heating and living environment was conducted and then blood pressure during bathing was measured along with evaluation of the thermal environment in dressing rooms and bathrooms before and after the renovation.

2 Outline of Survey

2.1 Surveyed Residences and Subjects

Three detached houses in Taikicho, a town in northern Japan, built before 1990, were surveyed. The residents of these houses are couples (a husband and a wife) over 60 years of age. Table 1 lists some attributes of the residents such as age and treated disease.

Table 1 Attributes of the residents

Residence	Resident	Age	Occupation	Bedtime	Wake-up	Treated disease	Renovation
House A	Husband	62	–	22:00	6:00	Diabetes	From 13 to 21
	Wife	62		22:00	6:00	High blood pressure	Jan. 2015
House B	Husband	85	–	22:00	6:00	High blood pressure	From 18 to 28
	Wife	83		23:00	7:00	High blood pressure	Jan. 2015
House C	Husband	64	Office work	22:00	6:00	High blood pressure Nasal catarrh	From 8 to 15
	Wife	60	–	22:00	6:00	Sensitivity to cold	Jan. 2015

The survey was conducted from April 2014 to March 2015. In the first half of the period, a preliminary survey was conducted and detailed designs of the renovations were produced. The renovation was done in January 2015. Before and after the renovation, the thermal environment was evaluated and blood pressure measured.

2.2 *Renovation of Residences*

(1) **House A**. The old bath unit with poor thermal insulation and an inefficient ventilator for dehumidification was replaced by a more efficient one (Fig. 1). The wall and floor of the bathroom were also insulated (Fig. 2). Because the door of the dressing room leading to the outside was poorly insulated, causing the dressing room to be cold and condensation to accumulate on the door, it was insulated.

(2) **House B**. The tiled bathtub was replaced with a bath module with thermal insulation to prevent a drop in temperature of the hot water (Figs. 3 and 4). The tiled bathroom was also insulated to increase the bathroom temperature (Figs. 3 and 4). The service door leading to the corridor adjacent to the dressing room was also insulated.

(3) **House C**. The undersurface of the floor not only of the dressing room and bathroom but also of the kitchen and entrance areas was insulated to slightly warm the whole house (Figs. 5 and 6). In addition to the insulation of the floor, the external wall of the space connecting the car garage and the entrance hall was insulated and made airtight.

Fig. 1 New bath unit

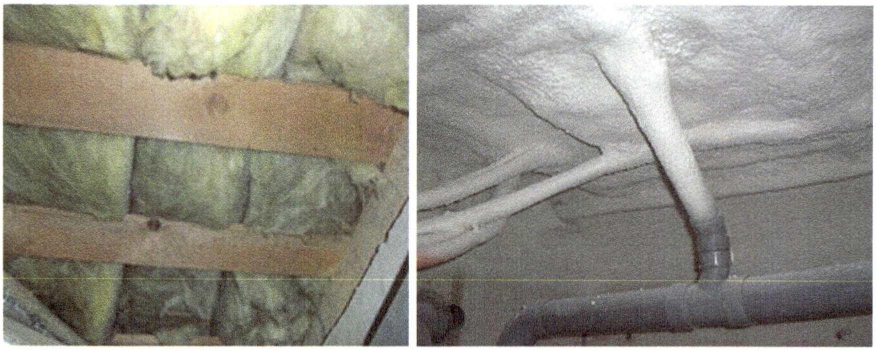

Fig. 2 Wall insulation (left: before renovation) and undersurface of the floor (right: after renovation)

Fig. 3 Removal of inner finishing tile

Fig. 4 Setting of new bathtub

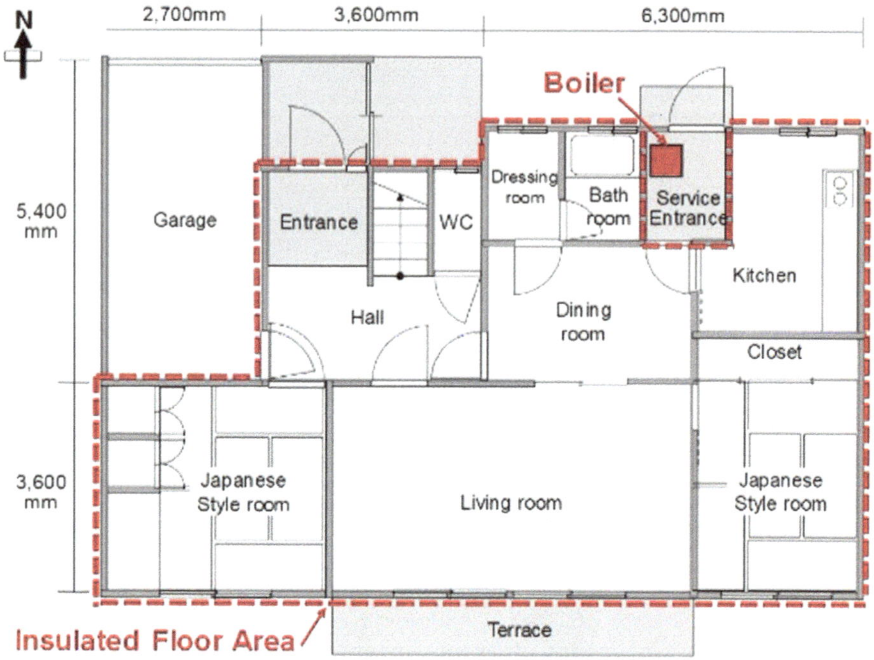

Fig. 5 Plan of first floor and insulated area

Fig. 6 Insulation of undersurface of floor

2.3 Measurement of Blood Pressure

Outline of measurement. Blood pressure, along with the temperature and humidity of the dressing rooms and bathrooms, was measured sequentially in a clothed condition in a dressing room before bathing and in a clothed condition again after bathing to delineate the blood pressure changes under actual bathing conditions and to clarify the relationship between the indoor thermal environment and blood pressure. The results were expected to clarify the influence, via the change in thermal environment, of the insulation renovations of the dressing room, bathroom, and bathtub on the health (blood pressure) of the residents. Because the residents of the surveyed houses were aged from 60 to 80 years, the influence of the thermal environment on their blood pressure during bathing was expected to be not insignificant. The participants agreed to cooperate in their blood pressure measurements and the written consent for this was not needed. There was no need for ethical approval.

(1) **Survey and measurement methods**. First, the residents were asked about their usual actions when bathing. Subsequently, the timing of blood pressure measurement appropriate for each resident was determined and a questionnaire sheet was designed. In addition to entries of the measured values of blood pressure, entries of other factors such as heating of the bathroom, reheating of hot water, alcohol drinking before bathing, and set point temperature of hot water were included in the questionnaire sheet. The subjects were also asked to register their posture (standing or sitting) and whether clothed by him/herself, when measuring blood pressure. The measurements were made at least twice (2 days), before and after the renovation by the residents (subjects) themselves.

(2) **Measurement procedures**. The measurements were carried out in the following order: ① in the living room, ② in the dressing room in a clothed condition, ③ in the dressing room in an unclothed condition, ④ just after moving to the bathroom, ⑤ before getting into the bathtub after washing their body or hair, ⑥ just after getting into the bathtub, ⑦ just before getting out of the bathtub, ⑧ repeat processes ⑤–⑦ if getting into the bathtub more than once, ⑨ just after getting out of the bathroom, ⑩ repeat processes ③, ②, ① after moving to the dressing room.

Blood pressure was measured sequentially using a handy-type hemadynamometer (OMRON HEM-6310F) by the residents themselves. Each blood pressure measurement took approximately 40 s, and the data were stored in the hemadynamometer. After bathing, the residents were asked to register the measuring time, systolic and diastolic blood pressure, and pulse rate into the questionnaire sheet.

3 Survey and Measurement Results

3.1 House A

Figures 7 and 8 show the systolic and diastolic blood pressure and pulse rate of the husband and wife in House A, along with the room temperature of the living room, dressing room, and bathroom. The orange shaded area indicates the period during which the resident was in a bathtub.

Husband (Fig. 7). Before the renovation (January 10, 2015), systolic blood pressure of the husband in House A increased from 165 to 175 mmHg in the dressing room, and decreased from 165 to 125 mmHg in the bathtub. It increased again to 160 mmHg in the dressing room after bathing. The dressing room temperature was almost constant at 20 °C through the whole bathing process. After the renovation (January 30, 2015), the systolic pressure increased only up to 145 mmHg in the dressing room, and decreased to 80 mmHg in the bathtub. Through the entire bathing process, systolic blood pressure decreased from 50–45 mmHg. The dressing room temperature was between 21.5 and 22 °C, higher than it was on January 10.

Wife (Fig. 8). Before the renovation, the systolic blood pressure of the wife in House A increased up to 140 mmHg in the dressing room, and the total change in the systolic pressure during the whole bathing process was 60 mmHg. After the renovation, systolic pressure increased only up to 120 mmHg, and the total change was less than 30 mmHg. The dressing room temperature increased from approximately 21 °C before the renovation to 22.5 °C after the renovation.

3.2 House B

Figures 9 and 10 show the systolic and diastolic blood pressure of the husband and wife in House B during bathing, respectively, along with the temperatures of the living room, dressing room, and bathroom. The set point of the hot water

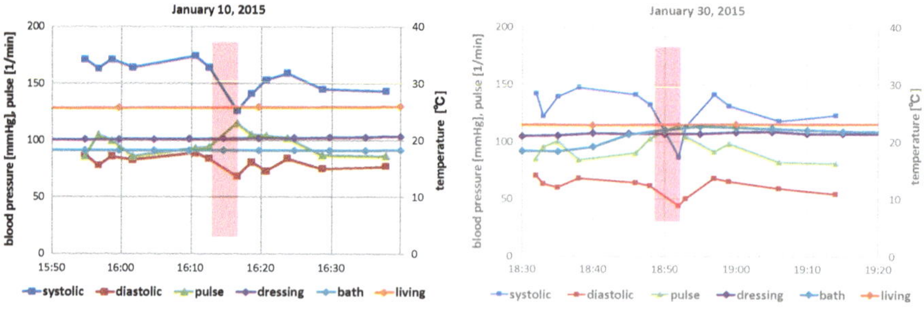

Fig. 7 Systolic and diastolic blood pressure of House A husband (left: before renovation; right: after renovation)

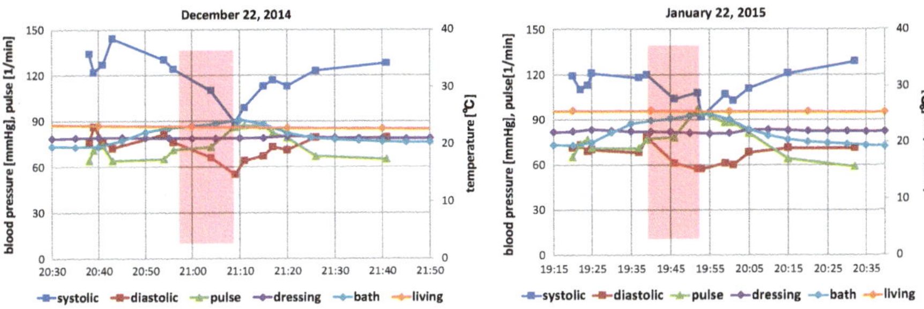

Fig. 8 Systolic and diastolic blood pressure of House A wife (left: before renovation; right: after renovation)

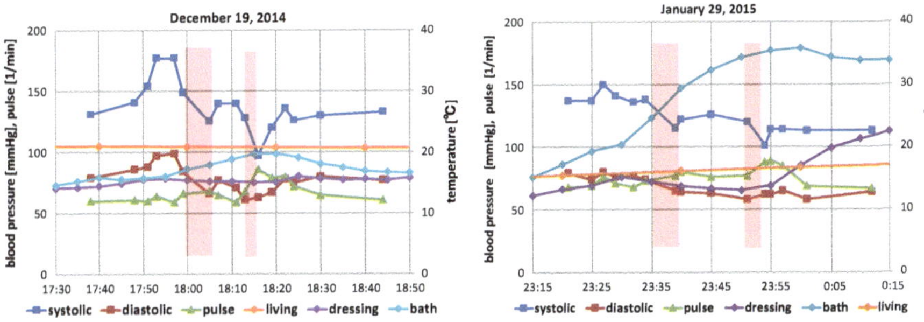

Fig. 9 Systolic and diastolic blood pressure of House B husband (left: before renovation; right: after renovation)

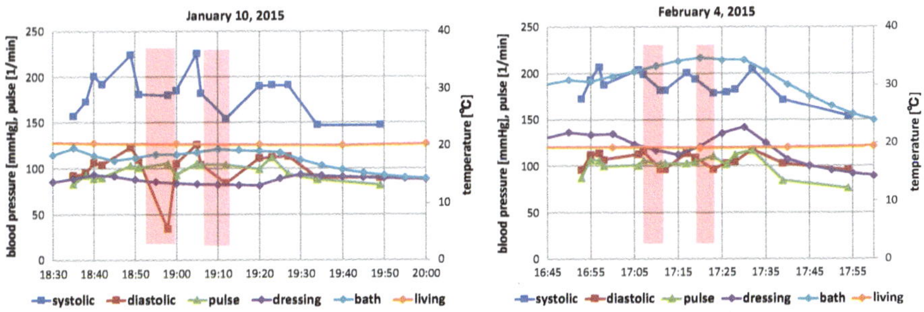

Fig. 10 Systolic and diastolic blood pressure of House B wife (left: before renovation; right: after renovation)

temperature was between 40 and 42 °C, and neither the husband nor the wife reheated the hot water.

Husband (Fig. 9). Before renovation (December 19, 2014), the systolic blood pressure of the husband increased from 130 to 180 mmHg in the dressing room and decreased to 100 mmHg in the bathtub. But after the renovation (January 29, 2015), it increased only up to 150 mmHg in the dressing room. Through the whole bathing process, systolic blood pressure decreased from 80 to 50 mmHg. Before the renovation, the dressing room temperature before bathing was approximately 15 °C whereas it was 16 °C after bathing, and these values increased to 16 °C and approximately 18 °C, respectively, after the renovation.

Wife (Fig. 10). The systolic blood pressure of the wife normally fluctuated between 150 and 200 mmHg even aside from bathing, indicating that she had high blood pressure as an illness. Before the renovation, her systolic blood pressure reached up to 230 mmHg in the dressing room (January 10, 2015), and the total change in the systolic pressure during the bathing process was 80 mmHg. However, the systolic pressure increased only up to 200 mmHg, and the total change was less than 25 mmHg after the renovation (February 4, 2015). The temperature in the bathroom was between 16 and 20 °C before renovation, but it increased to between 20 and 36 °C after the renovation due to the use of the heating system of the unit-bath.

3.3 *House C*

Husband (Fig. 11**).** Before the renovation, the systolic blood pressure of the husband in House C increased from 125 to 140 mmHg in the dressing room, and after the renovation, it increased from 110 to 140 mmHg, and decreased to 90 mmHg in the bathtub, much lower than before the renovation. Through the whole bathing process, systolic blood pressure increased from 30 to 50 mmHg. The dressing room temperature changed from approximately 12.5 °C before the renovation to 13.5 °C after the renovation, slightly higher. Therefore, in this case, both the systolic blood pressure and the total pressure change increased after the renovation, although the dressing room temperature increased.

Wife (Fig. 12). Before the renovation (January 6, 2015), the systolic blood pressure of the wife in House C was approximately 130 mmHg in the dressing room and decreased to 100 mmHg in the bathtub. The total change in the systolic pressure

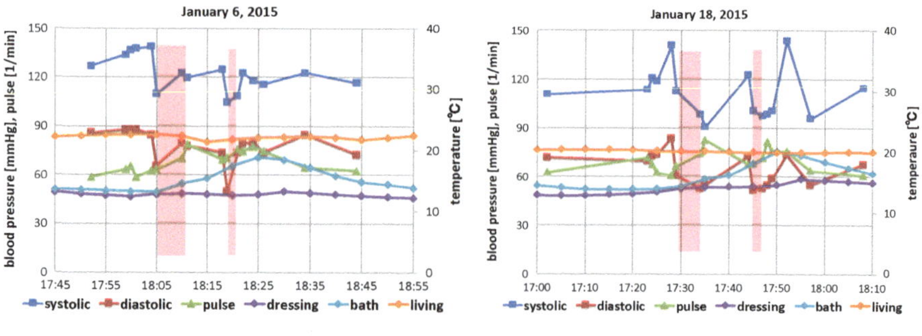

Fig. 11 Systolic and diastolic blood pressure of House C husband (left: before renovation; right: after renovation)

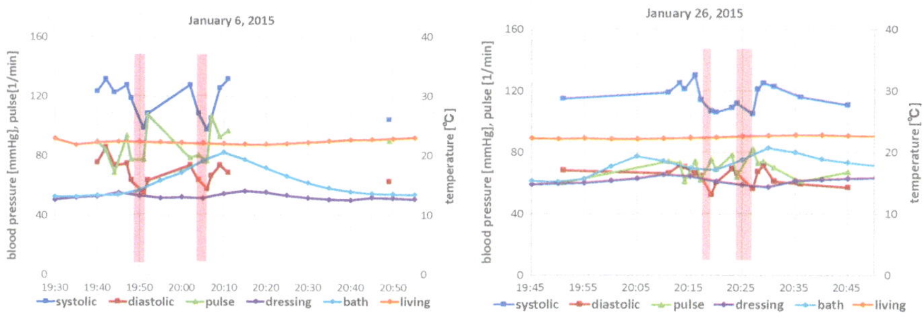

Fig. 12 Systolic and diastolic blood pressure of House C wife (left: before renovation; right: after renovation)

was 30 mmHg. After the renovation (January 26, 2015), the total change in the systolic pressure slightly decreased to 25 mmHg, although the dressing room temperature became higher from 13 to 15 °C after the renovation.

Because the temperatures of the dressing room and bathroom depend on the outdoor temperature and other factors, these factors should be taken into consideration when evaluating the effect of the renovation.

3.4 Relationship Between Dressing Room Temperature and Systolic Blood Pressure

Although most of the obtained results so far showed a tendency for systolic blood pressure to become lower after the renovation, and the total change throughout the whole bathing process decreased, the temperature increase in the dressing room and bathroom due to the renovation differed between houses and on each day probably due to the outdoor weather conditions. Therefore, the relationship between dressing room temperature and systolic blood pressure was examined. The dressing room temperatures both before and after bathing were examined.

Figure 13 shows the relationship between the dressing room temperature and the systolic blood pressure of each resident. The dressing room temperatures before and after bathing are distinguished by different symbols. The straight lines in the figure are the correlation curves.

In every case, systolic pressure is negatively correlated with dressing room temperature. Furthermore, the systolic blood pressure after bathing was lower than that before bathing, and the (negative) gradient of the correlation line is steeper after bathing in many residents. This indicates that the systolic pressure decease with the increase in dressing room temperature is larger after bathing than before bathing. This might be related to the dilatation of the skin blood vessel caused by heat exchange with the hot water in the bathtub.

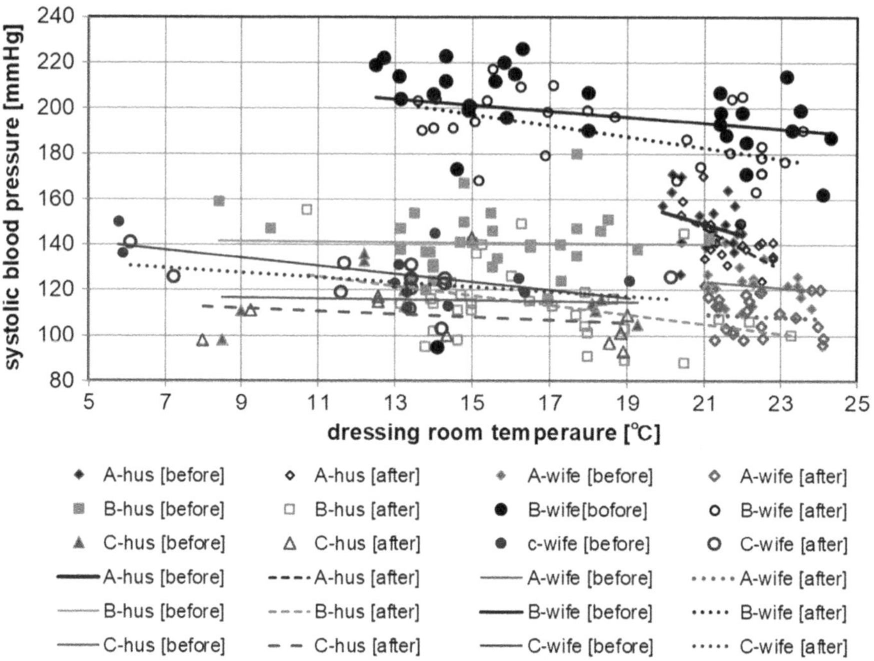

Fig. 13 Relationship between dressing room temperature and systolic blood pressure before and after bathing of all residents

The gradient of the correlation line of each resident is relatively similar, showing that systolic blood pressure decreases by 0–20 mmHg with each 10 °C increase in the dressing room temperature.

4 Conclusion

In this study, the influence that the thermal renovation of dressing rooms and bathrooms has on the health of elderly residents during winter was surveyed in residential buildings in Hokkaido, a cold northern region in Japan.

Bath units were replaced or additional thermal insulation was added to the openings and floor of the dressing room and bathrooms. Before and after the renovation, the thermal environment around the bathroom and blood pressure of residents during bathing were measured, along with an inquiry survey on the heating system of the residence and the lifestyle of the participants.

After the renovation, it was found that there was a tendency for the increase in the highest systolic blood pressure during undressing time and the range of the blood pressure change during the whole bathing process to be lower than those before the renovation. It was clarified that systolic blood pressure decreased with increased dressing room temperature. The correlation analysis showed that systolic blood pressure decreased up to 20 mmHg when the dressing room temperature increased by 10 °C.

The number of the subjects is only six in this preliminary study, and more research is needed to draw firm conclusions.

References

1. R. Nishikawa, E. Osaka, A measurement survey on the environment of bathing in Tohoku Area in winter, Summary of AIJ (Architectural Institute of Japan) Tohoku Chapter Research Meeting, Environmental Engineering, (76), 19–26, (in Japanese) (2013)
2. T. Hongo, T. Hiroshige, Standard physiology, Igaku-syoin, (in Japanese), 545–555, (2000)
3. Y. Asakawa, F. Takahashi, F. Endo, Cardiovascular response of elderly people while bathing, Rigaku Ryoho Kagaku, **21**, 433–436, (in Japanese) (4), (2006)
4. Y. Tochihara, N. Hashiguchi, I. Yadoguchi, Y. Kaji, S. Shoyama, Effects of room temperature on physiological and subjective responses to bathing in the elderly. J. Hum. Environ. Syst. **1**(15), 13–19 (2012)
5. H. Koshimizu, Y. Tochihara, T. Ohara, Effects of water temperatures and room air temperatures on thermal responses at bathing, Summary of AIJ Annual Meeting, D-2, 367–368, (in Japanese), (1996)
6. K. Kajii, The study of blood flow, body temperature and posture in the bath, Summary of AIJ Annual Meeting, D-2, 381–382, (in Japanese), (2001)

Assessment of the Effects of Using Wood Stoves on Indoor Air Quality in Two Types of Norwegian Houses

Mathieu Hamon, Guangyu Cao, Øyvind Skreiberg, Laurent Georges, Morten Seljeskog, Roger Khalil, Alexis Sevault and Hans Martin Mathisen

Abstract This study aims to assess the effects of using wood stoves on indoor air quality (IAQ) regarding fine PM, ultrafine PM, CO2 and relative humidity in Norwegian residential houses. Measurements were performed in an old natural ventilated house and a new mechanical ventilated house. Three locations for PM measurements were selected: close to the stove opening, in the middle of the room and at the supply air inlet, with original installed stoves typical for the buildings' time of construction. Each measurement lasted 3 h, which includes monitoring of the background concentration, the light up process, the burning and the refill processes. The results show peaks of fine and ultrafine PM emissions during the light up and refill phases, connected to opening of the wood stove door. The ultrafine PM peaks were higher and occurred more frequently than the fine PM ones, indicating that not only the opening of the wood stove door caused these PM peaks. Significant differences were found between the two houses regarding the relative distribution between fine PM and ultrafine PM. Peak concentrations of ultrafine PM took longer time to fall back towards background levels compared to the fine PM concentrations. No clear correlations were found between the load of the stove and PM emissions, and further research is required to assess why. Yet the situation was not alarming as the 24 h mean PM2.5 concentration in both houses was below the WHO guideline. CO2 emissions in both houses were on average always at a healthy level.

Keywords Indoor air quality · Wood stove · Ultrafine and fine particulate matter pollutants · Ventilation · CO_2

M. Hamon (✉)
National Institute of Applied Sciences of Lyon–INSA Lyon, Villeurbanne, France
e-mail: mathieu.hamon@insa-lyon.fr

M. Hamon · G. Cao · L. Georges · H. M. Mathisen
Norwegian University of Science and Technology, Trondheim, Norway

Ø. Skreiberg · M. Seljeskog · R. Khalil · A. Sevault
SINTEF Energy Research, Trondheim, Norway

D. Johansson et al. (eds.), *Cold Climate HVAC 2018*,
Springer Proceedings in Energy, https://doi.org/10.1007/978-3-030-00662-4_75

1 Introduction

The airtightness of new buildings is the outcome of the search for energy savings and emission reductions. However if the airtightness is not compensated by a proper ventilation system, deficiencies in IAQ (Indoor Air Quality) may occur. Products of human activities, as PM (Particle Matters) in high concentration can influence health: SBS (Sick Building Syndrome), respiratory system (asthma, etc.) [1] and cardiovascular diseases [2]. It has already been concluded that ventilation rate and IAQ are related [3, 4]. A higher ventilation rate leads to a higher air exchange rate in the room and thus a better dilution and faster removal of the pollutants. In Europe, all building codes have requirements regarding the ventilation rate in order to ensure a reasonable IAQ. The actual ventilation rate requirements in Nordic countries [5–8] are gathered in Table 1.

Those requirements are the minimum to ensure a good IAQ in the dwelling and prevent health issues. It corresponds usually to an air exchange rate [4] of 0.5 h^{-1}. Nordic countries have stricter requirements: 0.5 h^{-1} compared to 0.3 h^{-1} in other European countries. Those requirements are relevant if no special human activities disturb the indoor environment. Indeed, main sources of indoor particle emissions are from inside the house: the bathroom (shower, bath), the kitchen (cooking) or from heating systems like wood stoves in living rooms. It is therefore hard to manage all activities with the ventilation system to ensure a good IAQ because the concentration of PM varies a lot. In Nordic countries, the use of wood stoves for heating is quite common, both in old houses and new low energy houses. In 2012, about 12% of the population used wood as a main source of heating [9]. The popularity of this heating device is due to the low price of the wood compared to oil and to its environmental-friendliness [10]. Yet wood stoves could influence occupants' health, as wood incomplete combustion produces several pollutants such as particulate matter (PM10, PM2.5), ultrafine particles, carbon monoxide (CO),

Table 1 Nordic ventilation rate requirements

	Air flows						
	General (supply) $m^3/h.m^2$		Kitchen hood m^3/h (extract)	Bathroom m^3/h (extract)	Toilet m^3/h (extract)	Bedroom (extract)	Laundry room m^3/h (extract)
	In use	Not in use	/	/	/	/	/
Norway	1.2	0.7	108	54	36	26 m^3/h per person	36
Sweden	1.26	0.36	/	/	/	/	/
Finland	1.8	/	90	36	36	1.8 $m^3/(h.m^2)$	3.6$m^3/(h.m^2)$
Denmark	1.08	/	72	51	51	/	36

nitrogen dioxide (NO_2), several aldehydes (HCHO), polycyclic aromatic hydrocarbons and other free radicals [11, 12]. In the wood smoke, emissions of fine PM are up to 9.5 g/kg of wood burned, emissions of CO around 130 g/kg of wood burned, and emissions of aldehydes up to 4.45 g/kg [13]. If several studies have shown the relationship between IAQ and wood stoves, the impact of each PM characteristic (size, composition, concentration, etc.) on occupants' health is still not precisely quantified [13]. The spread of those pollutants in the indoor environment is influenced by the air infiltration rate, the balance of the ventilation and the tightness of the stove envelope [12]. Also experience in lighting the stove does not necessarily significantly impede the formation of particles, but rather contribute to improved particle burnout.

A previous study [12] has already been conducted, investigating the effect of wood stove operation on indoor air quality regarding fine PM emissions. It was performed in an old natural ventilated house and a new mechanical ventilated one. The wood smoke particle emissions present a peak in the size distribution between 0.15 and 0.4 μm and contain a large number of particles less than 100 nm [13]. Thus to broaden the investigation to the ultrafine domain (less than 0.3 μm) two Pegasor ultrafine particles counters were used. This domain needs to be studied more as most of the particulate pollutants in the air are ultrafine [14] and can influence health as their sizes allow them to be translocated into the lungs cells [15].

The present study is a continuation work and aims to continue the earlier investigation on the effect of using wood stoves on IAQ also regarding ultrafine PM, CO_2 and relative humidity in residential houses in addition to fine PM.

2 Materials/Method

2.1 *Buildings for Case Studies*

The two selected houses are the same ones as in the previous study. The one with only natural ventilation (named "old house" along the article) was built in 1953, equipped with a Jøtul 606, an old wood stove model. The other one (named "new house" along the article) was built in 2013 and was equipped with a Jøtul F373, a modern wood stove, and mechanical ventilation according to the Norwegian TEK10 building code. More detailed information about the buildings can be found in the previous article [12].

2.2 *Measurement Setup*

Three locations were selected to measure the peak concentration of particulate matter pollutants: close to the stove opening, the middle of the room and at the supply air inlet. The supply air inlet corresponded to the window in the old house, and the air inlet from the mechanical ventilation system in the new house. Three Aerotrak® particle counters were placed in three locations whereas two Pegasor ultrafine particle counters were placed at the supply air inlet and in the middle of the room, only.

2.3 Measurement Conditions

The article considers one 3 h measurement for each house, including monitoring the background concentration, the light up process, the burning processes, and the refill processes (two refills). Measurements were done under similar outdoor conditions. The stove run at part load operation i.e. reduced power output, when the air inlet opening of the stove was set below 50%. Figure 1 shows the wood stove operation. The wood type used was pine.

2.4 Measurement Instruments

In this study, three TSI AeroTrak® Handheld Particle Counters Model 9306-v2 were used. The particle counters measure particles in the range of 0.3–10 μm with a flow rate of 0.1 CFM (2.83 L/min). The counting efficiency of this instrument is 50% at 0.3 μm and 100% for particles >0.45 μm (per ISO 21501-4 and JIS). The three AeroTrak counters have been calibrated by the manufacturer [12].

Two Pegasor AQTM Indoor ultrafine particles counters that can measure particles in the range of 0.01–2.5 μm with a flow rate of 3 L/min were used. The resolution is 0.001 mg/m^3 (with 1 min integration time). Those devices also measure the concentration of CO_2 (ppm), the temperature (°C) and the relative humidity (%) simultaneously.

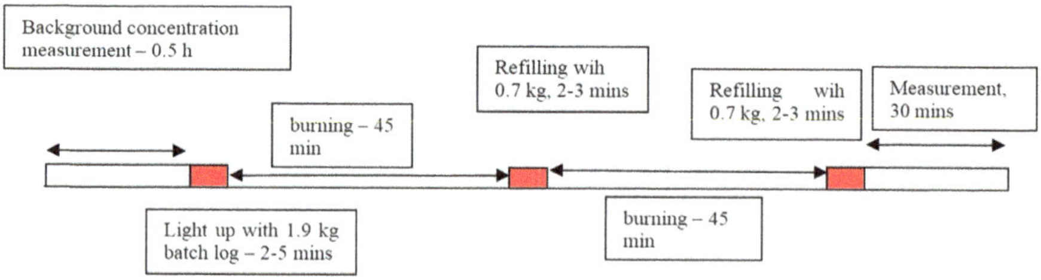

Fig. 1 The wood stove operation with reduced power output

3 Results and Discussion

3.1 Temperature and Relative Humidity

For the new house, the temperature rose regularly during the operations from 20 to 27 °C. For the old house, it started at 16 °C and reached a maximum of 18.5 °C.

The relative humidity was measured at the supply air inlet and in the middle of the room. At the supply air inlet, in the old house, the humidity decreased during the measurement almost linearly from 26 to 23% whereas in the new house the humidity decreased from 31 to 16% with a drop of 5% in the middle of the operation. In the middle of the room, a similar drop is observed for the new house whereas a considerably smaller drop of 1.5% happened in the old house. The load of the stove seems to increase the dryness of the room (considering relative humidity).

3.2 Fine Particles

To investigate the influence of the wood stove on the fine PM concentrations in the room, ratios of fine PM concentrations have been calculated with background measurements taken as the reference:

$$Particlesratio(PR) = \frac{PM\ concentration(at\ time\ t)}{PM\ concentration(background)} \quad (1)$$

The background concentration was measured for a period of 30 min before starting the measurement with the stove.

Table 2 shows the average background PM concentration for each house, sorted by particle size. The old house contains more than five times the concentration of fine particles than the new house.

The ratios used in the graphs in Fig. 2 are based on the average background concentration for the corresponding house and particle size. Each part figure represents a measurement and is composed of three plots: one measurement close to the stove, one in the middle of the room and one at the air supply inlet of the room (i.e. the mechanical ventilation supply for the new house, the window for the old house). The different events that occurred during the measurements are shown on the plots. A particular case happened in the new house though. Besides the programmed openings, the stove door had to be opened several times after the light up in order to keep the flame alive.

Table 2 Average background PM concentration in the old house (30 min average)

0.3–0.5 μm (#/m^3)		0.5–1 μm (#/m^3)		1–2.5 μm (#/m^3)	
Old house	New house	Old house	New house	Old house	New house
22.7E+06	3.50E+06	17.2E+05	4.19E+05	1.19E+05	1.85E+05

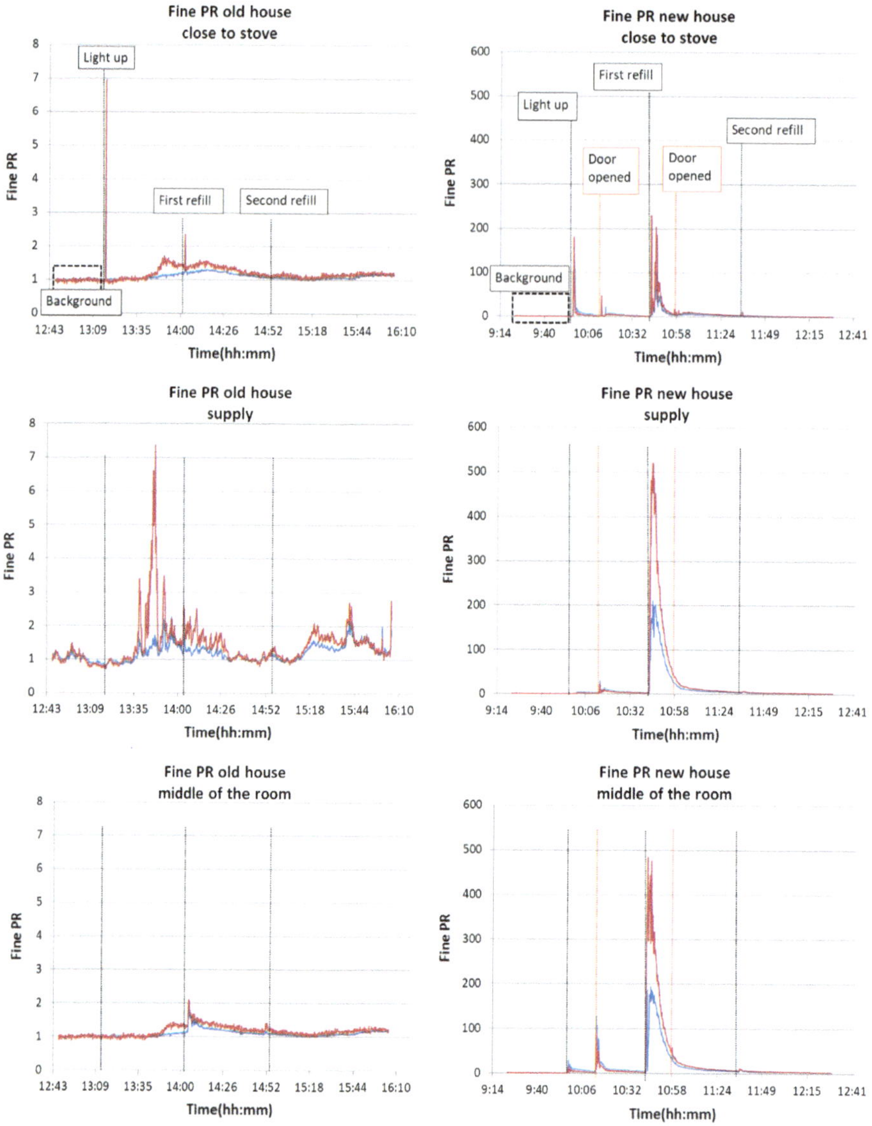

Fig. 2 Fine particles ratio (PR) in the old house (left) and new house (right)

In both houses PR close to the stove is above 1, which means that fine particulates have been emitted from the stove. This rise in PM concentration has been observed in different studies [10, 12, 13]. The other ratios are also above 1, which means that the fine particulates emitted have been transported through the room, affecting the middle of the room and even the supply air inlet, which was alimented in fresh air. The concentration peaks occurred usually during user actions on the stove: lighting, refills, other door openings, etc. In the new house, stove door openings led to a small emission increase. Once the stove door was closed, the concentration in the whole room tended to return to the background level. However, the intensity of each peak varies a lot based on the combustion conditions, the time the door is open, etc. and could not be predicted. In the new house, the first refill seemed to have a greater impact than the other events, especially at the supply air inlet and in the middle of the room. The ratios are much higher in the new house than in the old house (30 to 200 times higher). Even though the background concentration was higher in the old house it does not explain such a high amount of particulates in the new house. Thus, the fine PM emissions in the new house were significantly higher during the measurements. One explanation of this difference may be a draft issue of the stove/chimney reported by the experimenter. Indeed, modern buildings may cause poorer draft for the flue gas, especially when the stove is cold, releasing much more particles when the stove door is open.

The WHO 24-h mean guideline for PM2.5 concentration is 25 $\mu g/m^3$. The 24-h mean have been found seven times lower for both houses, with the old house concentration higher than the new house one. The concentration peaks are however from 7 to 34 times higher than the advised 24-h average values during light up and first refilling for the new house while they keep under the limits for the old house. The wood stove/chimney draft issue may be one of the causes of this phenomenon.

3.3 *Ultrafine Particles*

The same method with ratios is used for the ultrafine particles measurements done in the middle of the room and at the supply air inlet. The background concentration for the ultrafine PM is much higher than for the fine PM. The background concentration of ultrafine PM was more or less the same for both houses (Table 3).

The ratios presented in Fig. 3 are based on the average background concentration for the corresponding house and size. Each measurement is composed of two plots: one measurement in the middle of the room and one at the supply air inlet. The events occurring during the measurements are the same as presented in the fine PM PR plots.

For each measurement, ultrafine PR ratios in the middle of the room and at the supply air inlet are above 1, which means that ultrafine particles were produced and transported across the room during the use of the wood stove. The concentration peaks are related to user actions with the stove: light up, refills, other door openings, etc. As for the fine PM, the intensity of the peaks seems to be unpredictable with the

Table 3 Average ultrafine PM background concentration for the old house and the new house

	Air supply (30 min) (#/m^3)	Middle of the room (30 min) (#/m^3)
Old house	3.17E+08	3.05E+08
New house	2.94E+08	2.94E+08

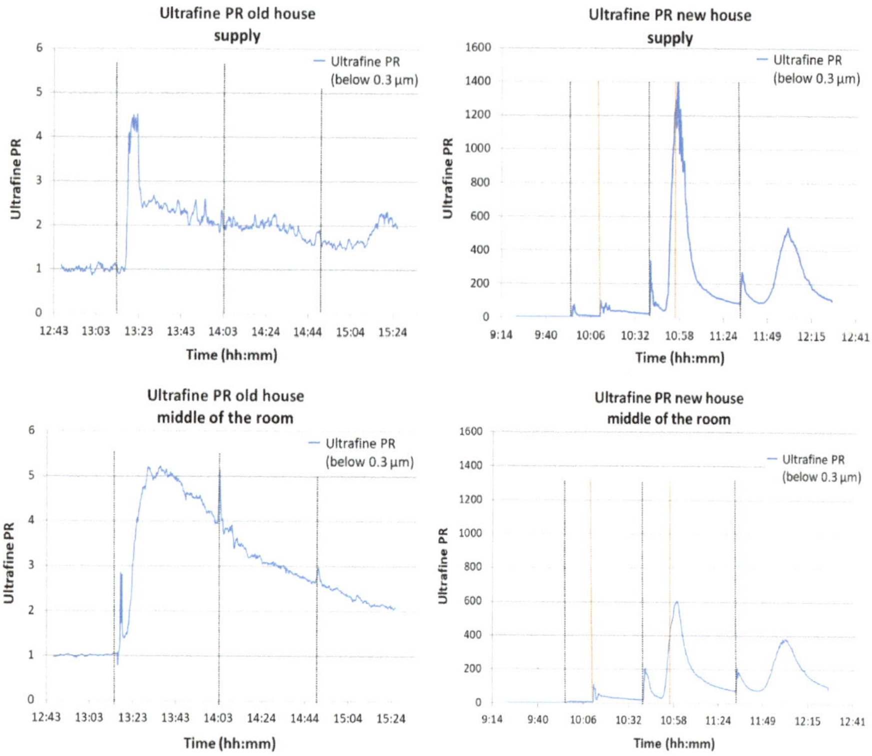

Fig. 3 Ultrafine particles ratio (PR) in the old house (left) and new house (right)

information available. The ratios are much higher in the new house than in the old house (up to 300 times in this study). The background concentration was about the same in both houses. Thus, it means that the ultrafine PM emissions in the new house were significantly higher during the measurements. The draft issue of the stove/chimney may be one of the causes. Also, for both houses, the ultrafine PM concentration is much higher than the fine PM concentration (up to hundreds of billions particles per m^3).

However, differences can be observed between ultrafine particles peaks and fine particles peaks. For example, in the middle of the room in the old house, a peak corresponding to the light up (at 13:15) appears for the ultrafine particles (ratio of 5) where no peak is observed for the fine particulates. Thus, an increase of ultrafine particles concentration may not imply a rise in fine particles concentration and vice-versa. The peaks seem steeper for the fine particles than for the ultrafine particles. The concentration of ultrafine particulates seems to take more time to get back to the background level than that of fine particles. Indeed, ultrafine particles are not easily removed by gravitational settling and therefore could stay airborne and be transported over long distances [14]. This observation is clearer in the old house than in the new house. This may be due to the better, constant ventilation provided in the new house. The ventilation was two times 25 m^3/h supply air and 36 m^3/h exhaust, which complies with the national requirements (based on 1.2 m^3/(h.m^2)) and the size of 34.6 m^2 of the living room, giving a requirement of 41.5 m^3/h). Then, appropriate constant ventilation may prevent the airborne effect of the ultrafine particles. Ultrafine particles ratios are either in the same range (supply air inlet) or two times higher (middle of the room) than the fine particles in the old house. In the new house, ultrafine particles ratios are more than twice higher at the supply air inlet and a bit higher in the middle of the room.

3.4 CO_2 Concentration

The CO_2 concentration was measured at the supply air inlet and in the middle of the room (Fig. 4). Table 4 gathers the average concentration and maximum peak.

In the old house the CO_2 concentration does not seem to be much affected by the stove. The average concentration is around 440 ppm for both locations and the peak never exceeded 500 ppm. In the new house, the concentration seems to increase due to the light up process and the first door opening to keep the flame alive. The

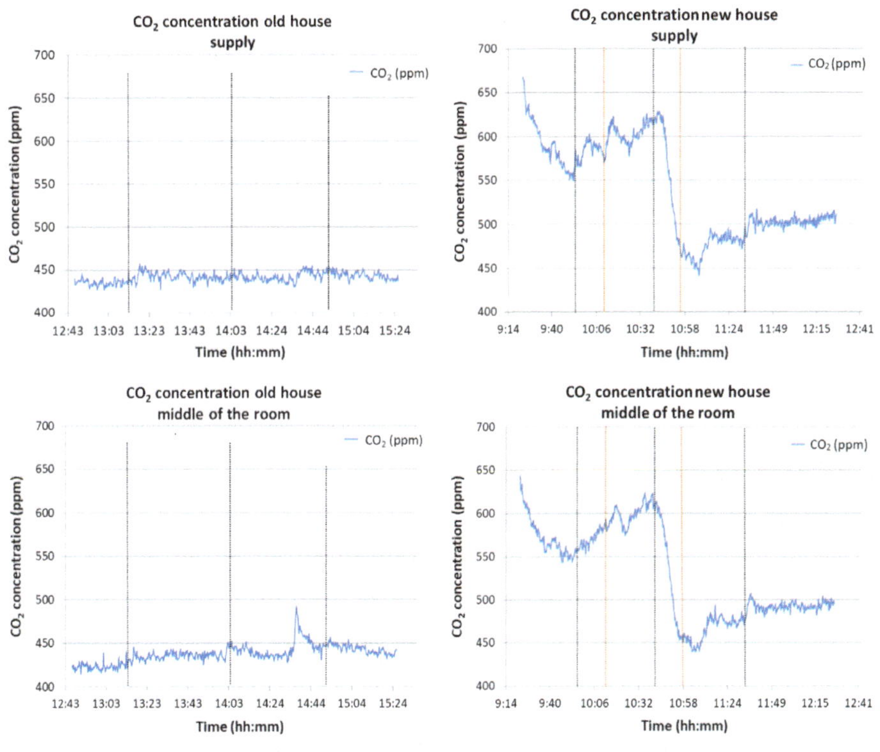

Fig. 4 CO_2 concentrations in the old house and new house

Table 4 CO_2 concentrations in the old and new house

CO_2	Air supply (ppm)		Middle of the room (ppm)	
	Average	Max peak	Average	Max peak
Old house (old wood stove)	440	457	437	492
New house (modern wood stove)	540	668	537	813

draft issue may have its consequence also here. A drop of 150 ppm can be observed just after the first refill. Afterwards the concentration stays more or less around 500 ppm. In general, the CO_2 concentration in the new house was higher than in the old house.

4 Conclusion

The present study constitutes a preliminary work in the continuation of Cao's [12] one. The influence of wood stove on the IAQ has been confirmed. Peaks of fine PM and ultrafine PM emissions were observed during the light up phase, refills phases and other openings of the door. The ultrafine PM concentrations showed to be much higher than the fine PM concentrations, and seemed to take longer time to get back to the background level compared to the fine PM concentrations. Ultrafine PM may stay airborne longer due to the difficulty of removing them by gravitational settling. This effect may be counterbalanced by the mechanical ventilation. The stove type was also an influent factor, with notably the draft issue in the new house, which may have caused high particles release during door openings. Yet, in both houses the 24-h mean PM2.5 concentration stayed below the WHO guideline.

CO_2 concentrations did not seem to be much influenced in the old house whereas peaks and higher average concentration were observed in the new house. The relative humidity dropped constantly. It led to a dry environment in both houses.

These observations on the influence of wood stoves on the IAQ revealed important physical phenomena that should be investigated further at better controlled operating conditions. To do so, the two houses' experiments should be more standardized: same wood stove, same wood type, same operational procedure, etc. It would then help to reveal the influence of the ventilation system itself.

References

1. R. Ruckerl, A. Schneider, S. Breitner, J. Cyrys, A. Peters, Health effects of particulate air pollution: a review of epidemiological evidence. Inhalation Toxicol. **23**(10), 555–592 (2011). https://doi.org/10.3109/08958378.2011.593587
2. K.A. Miller et al., Long-term exposure to air pollution and incidence of cardiovascular events in women. N. Engl. J. Med. **356**, 447–458 (2007). https://doi.org/10.1056/nejmoa054409

3. P. Wargocki et al., Ventilation and health in non-industrial indoor environments: report from a european multidisciplinary scientific consensus meeting (EUROVEN). Indoor Air **12**(2), 113–128 (2002). https://doi.org/10.1034/j.1600-0668.2002.01145.x
4. J. Sundell, H. Levin, W.W. Nazaroff, W.S. Cain, W.J. Fisk, D.T. Grimsrud, F. Gyntelberg et al., Ventilation rates and health: multidisciplinary review of the scientific literature. Indoor Air **21**(3), 191–204 (2011). https://doi.org/10.1111/j.1600-0668.2010.00703.x
5. Direktoratet for byggkvalitet TEK10, 13.2 Ventilasjon i boenhet. https://dibk.no/byggeregler/tek/3/13/i/13-2/
6. Sweden building code, Boverket, BBR19 BFS 2011:26. http://www.boverket.se/en/start-in-english/publications/2012/building-regulations-bbr/
7. D2 Suomen rakentamismääräyskokoelma Ympäristöministeriö, Rakennetun ympäristön osasto. http://www.finlex.fi/data/normit/37187/D2-2012_Suomi.pdf
8. Danish Building Regulations BR15. http://bygningsreglementet.dk/file/591081/br15_english.pdf
9. A.B. Wyss et al., Particulate matter 2.5 exposure and self-reported use of wood stoves and other indoor combustion sources in urban nonsmoking homes in Norway. PLOS One **11**(11), e0166440 (2016). https://doi.org/10.1371/journal.pone.0166440
10. R.W. Allen et al., The impact of wood stove technology upgrades on indoor residential air quality. Atmos. Environ. **43**(37), 5908–5915 (2009). https://doi.org/10.1016/j.atmosenv.2009.08.016
11. B. Lévesque et al., Wood-burning appliances and indoor air quality. Sci. Total Environ. **281** (1–3), 47–62 (2001). https://doi.org/10.1016/s0048-9697(01)00834-8
12. G. Cao et al., Experimental study of indoor air quality in two types of Norwegians houses heated by wood stove, NTNU, SINTEF (2015)
13. L.P. Naeher, M. Brauer, M. Lipsett et al., Woodsmoke health effects: a review. Inhalation Toxicol. **19**, 67–106 (2007). https://doi.org/10.1080/08958370600985875
14. B.F. Yu et al., Review of research on air-conditioning systems and indoor air quality control for human health. Int. J. Refrig. **32**(1), 3–20 (2009). https://doi.org/10.1016/j.ijrefrig.2008.05.004
15. J. Froines, Ultrafine Particle Health Effects. (2006). http://www.aqmd.gov/docs/default-source/technology-research/ultrafine-particles-conference/pre-conference_2_froines.pdf?sfvrsn=2

How Does Low Relative Humidity Affect Perceived Air Quality, Thermal Comfort and Symptoms in Modern Office Buildings in Cold Climates?

Merethe Lind, Sverre Holøs, Kari Thunshelle, Aileen Yang and Mads Mysen

Abstract To assess how people are influenced by relative humidity (RH) in cold climates, a study was conducted in an open office landscape in Oslo, Norway. The study took place during three cold days in February 2017. Fourteen subjects were blindly exposed to different levels of RH in the order of low (14 ± 1%), high (38 ± 3%), and medium (24 ± 4%). The subjects received emails twice a day (at 12:00 and at 14:30) with a link to a webpage where they were asked to: (1) assess perceived air quality (PAQ), (2) respond to a questionnaire about indoor environment quality and symptoms. The subjects performed normal office activity in between the two sessions. We found no significant impact of the level of RH on PAQ. Nevertheless, there were significantly more complaints about dry air at low RH than at medium and high RH. Furthermore, the air was perceived to be significantly more stuffy and heavier at high RH than at medium RH. There were no significant differences in thermal comfort at different RH, yet more people complained that it was cold on the day with low RH and warm on the day with high RH. Generally, there were few complaints related to symptoms at different RH. There were however significantly more complaints of itching and burning in the eyes at low RH than at medium and high RH.

Keywords Relative humidity · Perceived air quality · Thermal comfort
Dry air · Symptoms

M. Lind · A. Yang · M. Mysen
OsloMet—Oslo Metropolitan University (Formerly Oslo and Akershus University College of Applied Sciences), P.O Box 4 Olavs Plass, 0130 Oslo, Norway

S. Holøs (✉) · K. Thunshelle · M. Mysen
SINTEF Building and Infrastructure, Forskningsveien 3B, 0373 Oslo, Norway
e-mail: Sverreb.holos@sintef.no

D. Johansson et al. (eds.), *Cold Climate HVAC 2018*,
Springer Proceedings in Energy, https://doi.org/10.1007/978-3-030-00662-4_76

1 Introduction

Relative humidity (RH) can be as low as 10% during cold and dry winters in Nordic climates, and complaints about dry air happen frequently, especially in office buildings. Studies have shown associations between low humidity and discomfort in the eye, skin and nose of occupants [1]. Furthermore, low humidity could have an impact on respiratory health effects, and thus the Norwegian Institute of Public Health recommends a relative humidity above 20% [2]. However, humidification of air is generally discouraged due to (I) risk of condensation in the building envelop or on windows, (II) risk of *Legionella* and other microbial growth in humidification systems and (III) high energy and maintenance costs for humidification system [1]. With the current trends toward better insulating windows, low-emitting materials and more air-tight building envelops as well as more advanced demand-controlled ventilation systems, it is conceivable that strategies for avoiding very low humidity without significant trade-offs can be designed. Even without dedicated humidification, reduced ventilation in the coldest period of the year could increase the minimum RH. The goal of this study was to investigate the relation of low relative humidity with performance, perceived air quality (PAQ), thermal comfort and symptoms in a group of office workers in a newly built office building with passive house standard and demand-controlled ventilation.

2 Methods

2.1 Study Design

To assess how people are influenced by relative humidity (RH), an intervention study was conducted in an open-plan office in Oslo, Norway. The office is situated in a BREEAM "Very good" certified office building from 2012, with a heated floor area of 14 300 m^2, and measured airtightness of 0.23 air change rate at 50 Pa. The office has balanced demand-controlled ventilation with infrared and temperature sensors. Furthermore, it is equipped with steam humidifiers (Condair plc, UK), which inject steam to the local supply air to achieve different levels of humidity. Each steam humidifier incorporates a touch screen system to control the humidity level. The study took place during three consecutive cold days in February 2017 with a stable average outdoor temperature of −5 °C. Fourteen subjects (seven females), who were the employees of the companies in the office building, participated in the study. We aimed to blindly expose the subjects to three levels of relative air humidity in the order of low (15 ± 5%), high (35 ± 5%), and medium (25 ± 5%). Indoor temperature, relative humidity and CO_2 concentration were measured and collected using a Rotronic CP 11 (Rotronic AG, Bassersdorf, Switzerland) with a declared accuracy of ±2.5%RH, and a Q-trak 7565-X (TSI

Incorporated, Minneapolis, USA). The Norwegian Meteorological Institute provided data on outdoor temperature.

The subjects received emails twice a day (at 12:00 and at 14:30) with a link to a webpage where they were asked to: (1) assess PAQ, (2) respond to a questionnaire about indoor environment quality and symptoms. The schedule was made to maximize attendance, with the test at noon representing a situation short after entry to the room, and 14:30 representing the situation after 2.5 h of exposure. The subjects performed normal office work in between the two sessions.

2.2 *Perceived Air Quality (PAQ)*

PAQ was evaluated using a continuous acceptability scale divided in two parts [3]. The PAQ-acceptability scale was coded as following: 1 = "Clearly unacceptable", 5 ="Just unacceptable/Just acceptable" and 10 = "Clearly acceptable". It was not possible to score at the midpoint (see Fig. 1).

2.3 *Questionnaire*

The online questionnaire is based on the MM-questionnaire developed at the Department of Occupational and Environmental Medicine in Örebro, Sweden [4], modified to record current intensity on a continuous scale rather than frequency in a recall period.

The questionnaire consists of 25 questions related to subjective assessment of general perceptions of the indoor environment, thermal comfort and sick building syndrome (SBS) symptoms. The questionnaire also included questions on gender, allergy/asthma and location of the work space. A continuous scale slider was used to record the responses to the questions, where the response "No, not at all" was converted to a score of 0 and "Yes, very" to a score of 10. It was not possible to score at the midpoint. Figure 2 shows an excerpt of the online questionnaire.

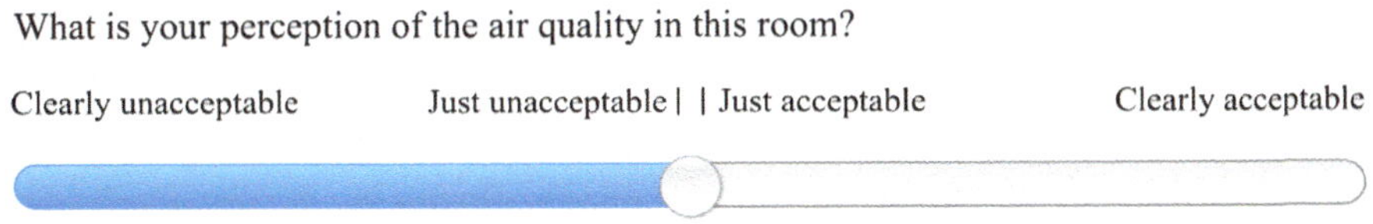

Fig. 1 Assessment of perceived air quality (PAQ). The acceptability score 0 corresponds to "Clearly unacceptable", 5 = "Just unacceptable/Just acceptable" and 10 = "Clearly acceptable"

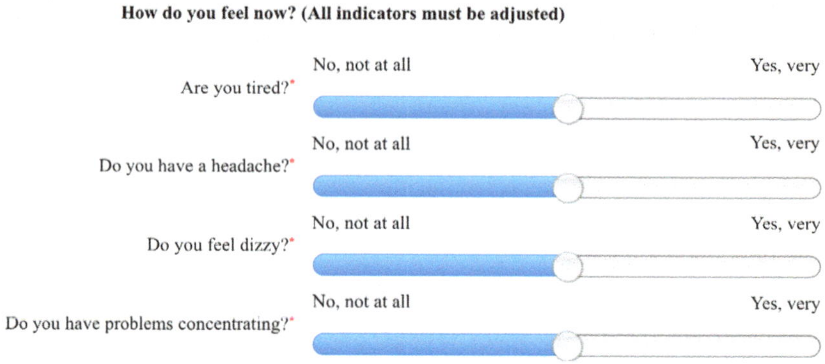

Fig. 2 Excerpt of the online questionnaire. The score 0 corresponds to "No, not at all" and 10 corresponds to "Yes, very"

2.4 Operation Span Task (OSPAN)

The test Operation span task (OSPAN) was developed by Turner and Engle [5] and is used to determine people's work memory or span of attention. The task consists of deciding whether a mathematical equation is true/false, followed by a four-letter word that needs to be memorized. The test consisted of 12 rounds in random order, with three to seven words per round to memorize. The number of correctly recalled words in the correct order for all 12 rounds, which gives a total of 54 words, was measured.

2.5 Statistical Analysis

The responses of the questionnaire were automatically converted into scores where the value 0 corresponds to "No, not at all" and 10 corresponds to "Yes, very". The PAQ-score was coded as following: 10 = "Clearly acceptable" and 0 = "Clearly unacceptable".

The non-parametric test Friedman's ANOVA by ranks was used to check for differences in the responses to the questionnaires (PAQ and indoor climate factors) between the three RH-levels. Whenever significant differences were found, paired comparisons were done with Sign test. OSPAN was analysed using repeated-measures ANOVA. Statistical analysis were performed with SPSS version 24 (SPSS Inc, Chicago, USA).

3 Results

As seen in Table 1, the measured RH levels were well within the intended levels for the three experimental days. The indoor temperature was stable during all three days. The CO_2 concentrations were somewhat higher on the day with low RH, whereas the outdoor airflow rate was higher on the day with medium RH.

3.1 PAQ

Figure 3 shows the variations of the PAQ-scores at different RH-levels. A decrease in the score as well as a broader range in individual scores as the relative humidity increases is apparent. However, this decrease in PAQ-score was not statistically significant. Nevertheless, this could indicate that the subjects perceived the air quality to be less acceptable as the relative humidity increased.

3.2 Questionnaire

Indoor climate factors

The responses to the questions related to indoor climate factors are summarized in Table 2. The median score was 0 for majority of the questions, indicating that the majority of the subjects found the indoor climate generally to be comfortable at all three experimental conditions. The level of RH had a statistically significant impact on sensation of dry air and stuffy air (Friedman's ANOVA, $p < 0.05$). The score for dry air was significantly higher at low RH than at medium and high RH (Sign Test, $p < 0.01$) at 12:00, but not at 14:30.

The score for the question related to stuffy air increased with increasing RH. Interestingly, the score for stuffy air was only significantly higher at high RH compared with medium RH (Sign Test, $p < 0.05$).

Table 1 Average (standard deviation) and [min, max] values of the actual measurement data of relative humidity (RH), indoor and outdoor temperature, outdoor airflow rate ($\dot{V}_{supply}$), and CO_2 concentrations during the experimental days at low, medium and high RH

RH	Actual RH (%)	T_{indoor} (°C)	T_{out} (°C)	CO_2 (ppm)	$\dot{V}_{supply}$ (m^3/h)
Low	14 (0.2) [13, 14]	22.7 (0.2) [22.2, 22.9]	−6.6 (0.2) [−6.8, −6.4]	696 (33) [588, 729]	363 (31) [301, 424]
Medium	24 (2.5) [20, 28]	22.6 (0.1) [22.4, 22.8]	−4.3 (0.2) [−4.6, −4.0]	627 (28) [552, 662]	444 (66) [355, 612]
High	38 (1.5) [35, 41]	22.7 (0.1) [22.6, 22.9]	−5.7 (0.2) [−5.9, −5.4]	614 (30) [528, 653]	377 (25) [342, 444]

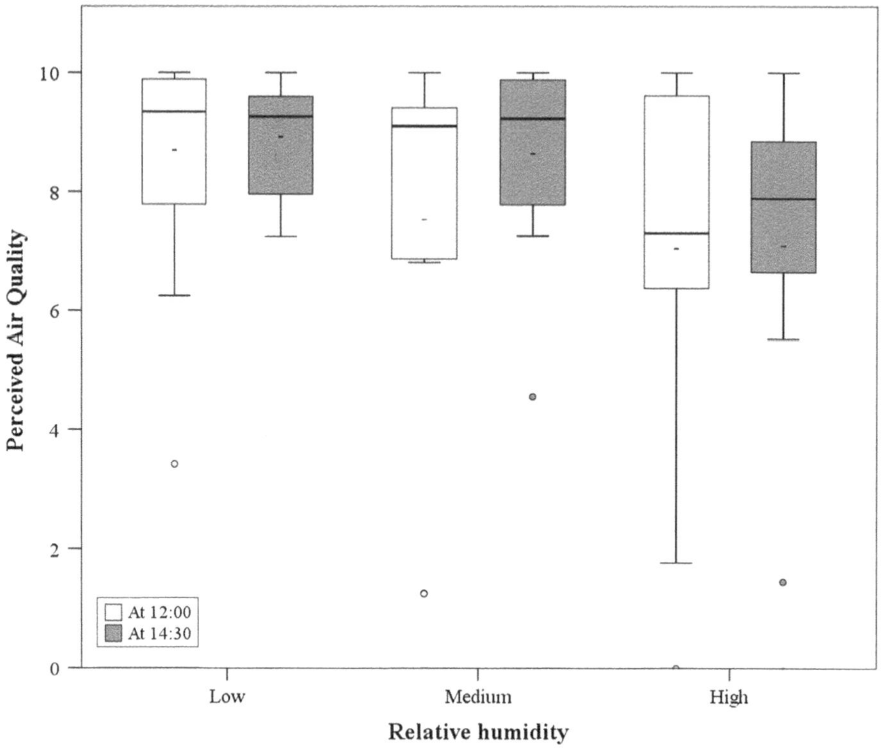

Fig. 3 Boxplot of the PAQ acceptability scores (0 = "Clearly unacceptable", 10 = "Clearly acceptable") by RH-level and session. The dark line in the middle of the boxes is the median, the short line is mean. The top and bottom of the box are the 75th and 25th percentiles. Whiskers indicate the 10th and 90th percentiles and individual outliers are shown as points

Although no impact of RH was found on factors related to thermal comfort, there were indications of more subjective complaints that it was cold on the day with low RH (mean score of 2.53) and warm on the day with high RH (mean score of 3.32).

SBS symptoms

The responses to the questions related to SBS symptoms are summarized in Table 3. Overall, there were few complaints related to symptoms at different RH-levels as the median score for majority of these questions were 0. The average symptom intensity was highest at low RH for all symptoms except difficulties concentrating and nausea at 14:30. The score for itching and burning of the eyes was significantly higher at low RH compared with medium (Sign test, $p < 0.01$) and high RH (Sign test, $p < 0.05$) in the afternoon. The score for fatigue was significantly higher (Sign test, $p < 0.05$) at low RH than at medium RH, both at noon and at 14:30.

Table 2 Mean and median values of the responses to the questionnaire related to indoor climate factors at low, medium and high RH at 12:00 and at 14:30

	At 12:00 (Mean/median)			At 14:30 (Mean/median)		
	Low N = 14	Medium N = 12	High N = 12	Low N = 14	Medium N = 12	High N = 12
Dry air*	1.67/0.23	0.09/0	0.06/0	1.95/0.27	1.71/0	0.02/0
Stuffy air*	0.71/0	0.96/0	3.48/1.55	0.43/0	1.33/0	3.22/0.79
Unpleasant odor	0.07/0	1.40/0	1.43/0	0/0	1.72/0	2.34/0
Too cold	2.53/0	1.24/0	1.4/0	2.65/0.73	1.55/0	1.29/0
Too warm	0.34/0	0.61/0	3.32/2.12	0.56/0.08	0.94/0	1.47/0.03
Draught	1.54/0	0.64/0	0.07/0	1.86/0	1.44/0	0.17/0
Varying temperature	1.77/0	1.44/0	2.08/0	1.33/0	1.28/0	1.83/0
Heat from sun	0.18/0	0/0	0/0	0.11/0	0/0	0/0

*$p < 0.05$, Friedman's ANOVA

Table 3 Mean and median values of the responses to the questionnaire related to SBS symptoms at low, medium and high RH at 12:00 and at 14:30

Symptoms	At 12:00 (Mean/median)			At 14:30 (Mean/median)		
	Low N = 14	Medium N = 12	High N = 12	Low N = 14	Medium N = 12	High N = 12
Fatigue	2.84/ 2.30	1.21/0.63	2.73/ 1.26	3.72/ 3.77	2.00/0.84	2.77/1.7
Heavy-headed	2.20/ 1.05	0.46/0	1.60/ 0.24	2.55/ 2.42	1.62/0.3	2.52/ 1.46
Headache	0.63/0	0.03/0	0.10/0	0.11/0	0.83/0	0.87/0
Dizziness	1.18/0	0.64/0	0.25/0	1.26/0	0.89/0	1.05/0
Difficulties concentrating	2.93/ 1.60	1.96/0.27	2.66/ 1.32	2.16/ 1.17	3.26/1.52	3.42/ 2.45
Itching, burning eye*	1.63/0	0.01/0	0.19/0	1.75/ 0.32	0.11/0	0.35/0
Hoarse, dry throat	0.87/0	0.02/0	0.11/0	1.18/0	0.12/0	0.63/0
Itching hands/face	1.14/0	0.10/0	0.07/0	0.62/0	0.35/0	0.07/0
Nausea, unwellness	0.63/0	0.46/0	0.55/0	0.64/0	1.39/0	1.14/0

*$p < 0.05$ Friedman's ANOVA

3.3 OSPAN

The results of the OSPAN test varied widely between individuals as illustrated in Fig. 4. Only nine subjects completed the test for all three days. The total number of accurate words in correct order is 54 per test round. We did not find any significant effect of RH-level on the cognitive performance of the subjects.

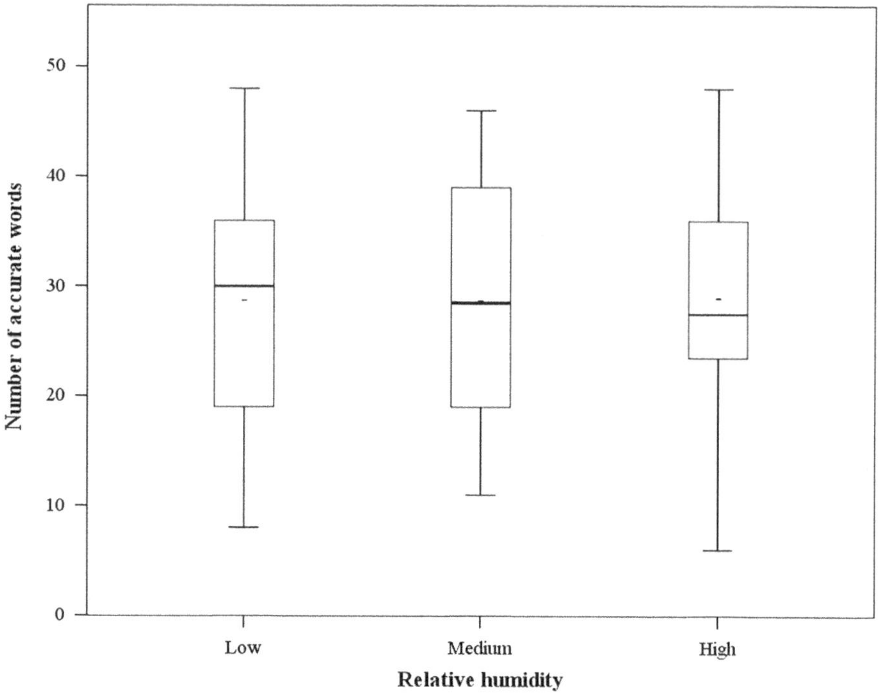

Fig. 4 Number of accurate words in correct order (total N = 54 words) by RH-level. The dark line in the middle of the box is the median, the short line is the mean. The top and bottom of the box are the 75th and 25th percentiles. Whiskers indicate the 10th and 90th percentiles and individual outliers are shown as points

4 Discussion

The aim of this study is to assess the impact of RH level on PAQ, indoor climate factors related to human comfort and cognitive performance, in a humidity range relevant for office buildings in Nordic winter situations. The lowest level of 14% corresponded to normal operation of the study building at an outdoor temperature of −5 °C. In such a weather situation, 38% RH was considered representative for the highest relevant humidity for an office building.

A recent review by Derby et al. [1] identified several studies on the effect of low humidity on comfort, health and indoor environmental quality (IEQ). They noted that perceived quality or acceptability of air generally decreases with increasing temperature and RH. Oftentimes, also on the odor intensity, which is related to the concentrations of volatile organic compounds (VOCs). In order to disentangle the effect of temperature from the effect of RH, we kept the temperature constant during the three experimental days.

In our study, we found no statistically significant impact of RH on the PAQ-score as assessed according to the method by Gunnarsen and Fanger [6], although PAQ-score did decrease with increasing humidity. Also, the scores for stuffy air and unpleasant odor showed an increasing tendency with increasing RH. This is in line with previous observations in several studies that increasing enthalpy decreases acceptability of the air and individual air quality indicators with RH in the range 20–70% [3, 7–9]. Possible causes for such a relationship include increased olfactory sensitivity or increased emissions from pollution sources at higher enthalpy, or perhaps a satisfying experience resulting from the cooling of the airways at lower enthalpy. The humidification process could also lead to increased pollution of the air if the water or equipment is not free from pollutants. In our experimental setup, we have no way of assessing the relative contribution of different causes.

Wyon et al. [10] observed no effects of humidity on PAQ when thirty subjects were exposed for 5 h to clean air at 5, 15, 25 and 35% RH at 22 °C. Unlike our study, the exposure to RH were done in experimental chambers [10]. It would be of interest to repeat the study by Wyon et al. [10] in an actual work situation with more subjects, in order to examine if the results are representative for newer office buildings.

Previous studies have reported the perception of dry air to be poorly correlated with humidity [1, 7]. The literature is inconclusive on whether participating subjects are able to perceive low humidity, albeit studies have shown that people prefer rather dry and cool air and decreased humidity (down to 20% RH) has a beneficial effect on PAQ [7, 11]. The sensation of dry air increases with increased temperature and air velocity. The most probable cause of sensation of dry air could be elevated levels of pollutants such as particulate matter and dust [11]. In contrast, our results indicate that the subject are able to perceive dry air as they found the air to be too dry at 14% RH compared to an RH level above 24%. We measured similar outdoor airflow rates during the three experimental days, and the indoor temperatures were very similar. Even if particulate pollutants were not measured, it is reasonable to assume a constant and low level of particulate concentrations in the supply air for all three days.

Derby et al. [1] also reported an increase in skin dryness, eye irritation as the humidity decreased in a review of several studies. Generally, these studies indicate a breakpoint between 20 and 30% RH for discomfort to skin, eyes and membrane irritation [1, 12]. Our findings are in line with these previous results, as fewer subjects complained of itchy and burning eyes at 24 and 38% RH.

Theoretically, the thermal balance of a person is affected by the enthalpy of the air, meaning that more heat is lost to dry than moist air at the same temperature due to evaporative heat loss. The subjects reported being too cold at low RF, and too warm at high RF. However, thermal sensation is highly affected by clothing and activity level, and the study design did not control for this.

Overall, our results indicate that RH-levels at 14–24% may reduce comfort in a well-ventilated building with low-emitting materials, while it is less obvious that increasing RH to 24–38% increases comfort. Given the inherent weakness in study design where interventions are examined sequentially on different days with a limited number of subjects, these indications obviously need confirmation with a larger study population before they are used in designing strategies for building operation. However, our results support previous findings that even relatively small increases in RH from the very low levels commonly observed in well-ventilated office buildings at below-zero outdoor temperatures may have beneficial effects. Thus, further examination on possible measures for increasing indoor RH without negative effects is called for.

5 Conclusion

We found no significant impact of level of relative humidity on PAQ and cognitive performance. However, we found more complaints of dry air at 14% RH compared with an RH level above 24%. Furthermore, more subjects complained about itchy and burning eyes at 14% RH. Our study indicates that increasing relative humidity can reduce complaints and symptoms due to dry air. However, increased complaints of stuffy air were observed at 38% RH.

Acknowledgements We would like to thank the participants from GK A/S for their assistance and participation in this project. This paper is based on the master thesis by Merethe Lind, and was a part of the BEST VENT project. BEST VENT is funded by the Research Council of Norway EnergiX program under Grant 255375/E20 together with the industry partners: Undervisningsbygg Oslo KF, GK Inneklima AS, DNB Næringseiendom AS, Erichsen & Horgen AS, Hjellnes Consult AS, Multiconsult AS, Interfil AS, Camfil Norge AS, Swegon AS, Belimo Automasjon Norge AS, NEAS AS, and Norsk VVS Energi- og Miljøteknisk Forenings Stiftelse for forskning.

Compliance with Ethical Standards

Formal consent was given by the volunteers who participated in this study. We did not collect any identifiable or sensitive information that would require ethical approval. The research has been conducted in compliance with the ethical standards at OsloMet—Oslo Metropolitan University (formerly Oslo and Akershus University College of Applied Science) and Norwegian Law.

References

1. M.M. Derby, M. Hamehkasi, S. Eckels, G.M. Hwang, B. Jones, R. Maghirang et al., Update of the scientific evidence for specifying lower limit relative humidity levels for comfort, health, and indoor environmental quality in occupied spaces (RP-1630). Sci. Technol. Built Environ. **23**, 30–45 (2017). https://doi.org/10.1080/23744731.2016.1206430
2. Folkehelseinstitutt. Anbefalte faglige normer for inneklima. Revisjon av kunnskapsgrunnlag og normer—2015. Oslo, Norway: Folkehelseinstituttet (2015)
3. L. Fang, G. Clausen, P.O. Fanger, Impact of temperature and humidity on the perception of indoor air quality. Indoor Air **8**, 80–90 (1998). https://doi.org/10.1111/j.1600-0668.1998.t01-2-00003.x
4. K. Andersson, epidemiological approach to indoor air problems*. Indoor Air **8**, 32–39 (1998). https://doi.org/10.1111/j.1600-0668.1998.tb00005.x
5. M.L. Turner, R.W. Engle, Is working memory capacity task dependent? J. Mem. Lang. **28**, 127–154 (1989). https://doi.org/10.1016/0749-596X(89)90040-5

6. L. Gunnarsen, P.O. Fanger, Adaptation to indoor air pollution. Environ. Int. **18**, 43–54 (1992). https://doi.org/10.1016/0160-4120(92)90209-M
7. L. Fang, D.P. Wyon, G. Clausen, P.O. Fanger, Impact of indoor air temperature and humidity in an office on perceived air quality, SBS symptoms and performance. Indoor Air **14**, 74–81 (2004). https://doi.org/10.1111/j.1600-0668.2004.00276.x
8. L. Fang, G. Clausen, P.O. Fanger, Impact of temperature and humidity on chemical and sensory emissions from building materials. Indoor Air **9**, 193–201 (1999). https://doi.org/10.1111/j.1600-0668.1999.t01-1-00006.x
9. L.M. Reinikainen, L. Aunela-Tapola, J.J. Jaakkola, Humidification and perceived indoor air quality in the office environment. Occup. Environ. Med. **54**, 322–327 (1997)
10. D.P. Wyon, L. Fang, L. Lagercrantz, P.O. Fanger, Experimental determination of the limiting criteria for human exposure to low winter humidity indoors (RP-1160). HVACR Res. **12**, 201–213 (2006). https://doi.org/10.1080/10789669.2006.10391175
11. Fanger P. Ole, What is IAQ? Indoor Air **16**, 328–334 (2006). https://doi.org/10.1111/j.1600-0668.2006.00437.x
12. P. Wolkoff, J.K. Nøjgaard, C. Franck, P. Skov, The modern office environment desiccates the eyes? Indoor Air **16**, 258–265 (2006). https://doi.org/10.1111/j.1600-0668.2006.00429.x

Effect of Filter Type in Ventilation Systems on NO_2 Concentrations in Classrooms

Aileen Yang, Kristian Fredrik Nikolaisen, Sverre Holøs, Kari Thunshelle, Franck René Dauge and Mads Mysen

Abstract This study was conducted to assess how different filter types in the ventilation system affect the indoor NO_2 concentrations. Measurements were carried out in two classrooms and air intakes in a primary school located in Oslo, Norway. A regular F7 particle filter and an F7 combination filter with activated charcoal lining were compared. NO_2 concentrations were measured for five weeks during winter 2017 in a cross-over study design to compare: (1) NO_2-levels in classrooms with regular filter (RF) versus combination filter (CF); (2) indoor/outdoor ratio with regular filter versus combination filter. One-hour average concentrations are reported during operating time of the ventilation system (6:00–23:00) and during hours with high (>40 $\mu g/m^3$) outdoor NO_2 concentrations. The measured average NO_2 concentrations in both classrooms with an RF were significantly higher than with a CF. The median CF/RF ratios for the two classrooms were 0.50 and 0.81 during hours with high NO_2 concentrations, and 0.48 and 1.00 during the period the ventilation system was operational. During hours with high NO_2 concentrations, the median indoor/outdoor ratios for the two classrooms with an RF were above 1.00, while the corresponding ratios with a CF were 0.78 and 0.75. Our results demonstrate that a combination filter is more efficient than a regular filter in reducing NO_2 concentrations in classrooms during hours with high outdoor concentrations.

Keywords NO_2 · Ventilation · Filtration · School · Filter efficiency Indoor · Outdoor

---

A. Yang (✉) · K. F. Nikolaisen · M. Mysen
OsloMet—Oslo Metropolitan University (Former Oslo and Akershus University College of Applied Sciences), P.O. Box 4 St. Olavs Plass, 0130 Oslo, Norway
e-mail: aileen.yang@oslomet.no

K. F. Nikolaisen
Erichsen & Horgen AS, Nydalsveien 36, 0484 Oslo, Norway

S. Holøs · K. Thunshelle · M. Mysen
SINTEF Building and Infrastructure, Forskningsveien 3B, 0373 Oslo, Norway

F. R. Dauge
NILU—Norwegian Institute for Air Research, Instituttveien 18, 2007 Kjeller, Norway

D. Johansson et al. (eds.), *Cold Climate HVAC 2018*,
Springer Proceedings in Energy, https://doi.org/10.1007/978-3-030-00662-4_77

1 Introduction

Studies have shown associations between high levels of traffic-related pollutants and respiratory health effects, especially among children as they are a sensitive subgroup [1]. Road traffic is one of the major sources of nitrogen dioxide (NO_2). In Nordic countries, exceedances of NO_2 concentration are recurrent during winter and specially associated with inversion periods. Children spend majority of their daytime indoors in schools and very limited of their time outdoors. Especially for schools situated near roadways with high traffic intensity, it is important to minimize the levels of air pollutants in the classrooms. The ventilation system is one of the factors influencing the concentration of NO_2 in classrooms in Norway. Conventional bag air filters are installed in most ventilation systems in Norwegian schools and studies have shown that ventilation systems with air filtration are efficient in reducing particle concentrations in classrooms [2–4]. However, these filters do not account for gaseous components such as NO_2. Furthermore, majority of the existing studies on indoor air quality in schools provide limited information on ventilation systems [5, 6].

Given that buildings, such as schools, are more frequently built near roads with high traffic intensity, the aim of our study was to investigate how different filter types in the ventilation system can influence the NO_2 concentrations in classrooms in a school closely situated to a highway.

2 Methods

2.1 Study Design

The study was carried out at a primary school situated 500 m southwest of a highway (Ring 3, see Fig. 1). The school was completed in August 2016 after passive-house standards and has a capacity of 840 students. It has concrete floor slabs covered with linoleum, walls are timber frame insulated with 300 mm mineral wool, and are generally clad with 13 mm plasterboard with acrylic paint. Materials and paint are either M1-certified or implicitly low-emitting. The school has demand-controlled ventilation (DCV) with CO_2- and temperature control in each classroom. All classrooms have balanced supply and exhaust mechanical ventilation with heat exchange. The ventilation system has bag air filters of class F7 type Standard-Flo XLT 7. An overview of the characteristics of the school and the classrooms is provided in Table 1.

Inversion periods can occur during winter, which can result in elevated concentrations of outdoor air pollutants. Subsequently, continuous measurements were taken between 24 January and 1 March 2017, as high outdoor NO_2 concentrations can be expected. To compare the effectiveness of two different types of filters, measurements were taken inside two classrooms and in the corresponding air

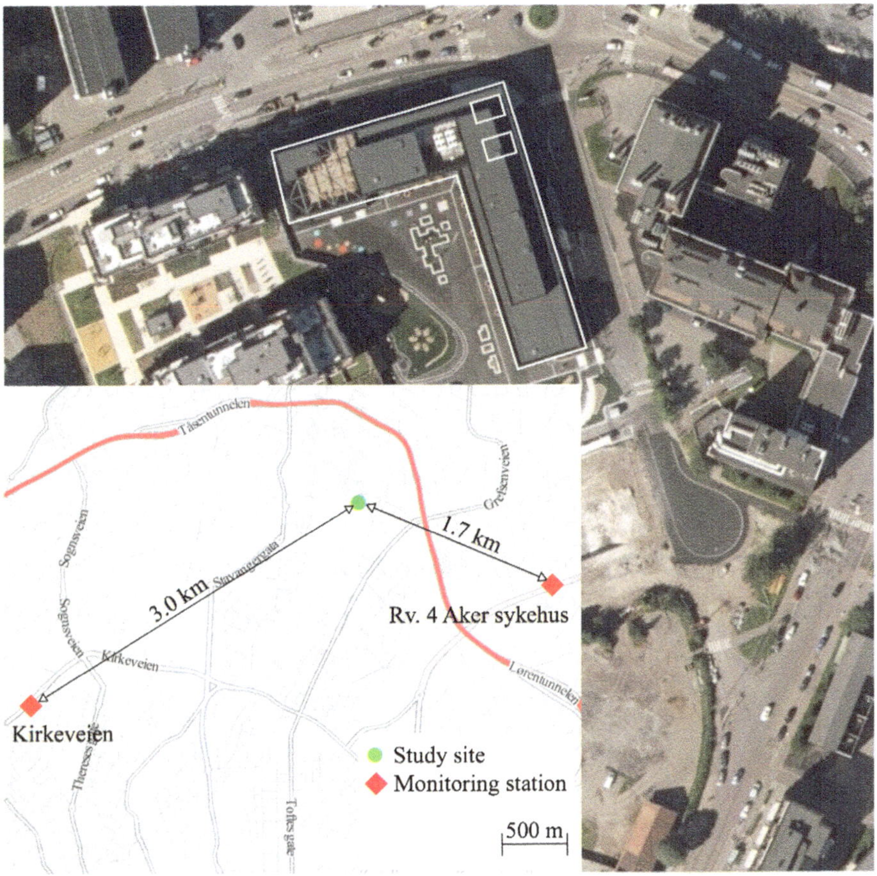

Fig. 1 Location of the school (white L-shaped, green circle) in Oslo, Norway, and the two classrooms (white square). The monitoring sites (diamond shaped) are situated 3.0 km southwest (Kirkeveien) and 1.7 km southeast (Rv. 4 Aker sykehus). The Ring 3 highway is marked in red. *Image source* norgeibilder.no ©2018 Kartverket (Screenshot by author)

Table 1 Characteristics of the study site

Characteristics	
Traffic intensity	57,526 vehicles/day
Distance to highway	500 m
Floor area of study class rooms	61 m^2
Height of study classrooms	2.8 m
Volume of study classrooms	170.8 m^3
Air change rate	7.3 h^{-1}
Airflow rate	20.5 m^3/h per m^2

intakes. The two classrooms are similar in size and situated on the third floor, separated by a room in between. One of the classrooms was occupied by elementary pupils, whereas the other one was used sporadically. The ventilation system was operational between 6:00 and 23:00. To ensure similar air change rates in both classrooms during the sampling campaign, the ventilation was set to a maximum airflow rate of 20.5 m^3/h per m^2.

New bag air filters were installed in the ventilation system at the study site. The filter types used in this study, provided by Camfil AS, consisted of a regular F7 particle filter (RF; Hi-Flo II XLT 7) and an F7 combination filter seeded with activated charcoal lining (CF; City-Flo XL7).

Two Teledyne Advanced Pollution Instrumentation (API), T200 and 200A models, were used to measure NO_2 at 1-min intervals. For the *indoor* measurements, the instruments were placed in a small room adjacent to the classrooms to prevent noise nuisance. The air inlet tubes of the sampling instruments were placed between the supply air vents to ensure that the sampled air is representative for the classroom. To keep the air inlet tubes out of reach of the pupils in the classrooms, the measurements were done at a height of approximately 2 m. For the *outdoor* measurements, the instruments were placed outside the air intake chambers in the maintenance room at the 4th floor to protect them from rain and humidity. The air inlet tubes of the sampling instruments were placed inside the air intake chamber.

To make the most use of the two measurement instruments, the sampling plan, as shown in Fig. 2, was divided into two parts:

- Compare effect of filter on indoor concentrations (sampling period 1–2)
- Compare effect of filter on indoor-outdoor concentrations (sampling period 3–6).

2.2 *Data Collection and Analysis*

The raw NO and NO_x data were extracted and calculated into 1-h average NO_2 concentrations and used for statistical analysis. Separate analyses were done for the period the ventilation system was operational (6:00–23:00) and for hours with high (>40 $\mu g/m^3$) outdoor NO_2 concentrations. The annual mean NO_2 limit value of 40 $\mu g/m^3$ was used to indicate high outdoor concentrations.

	Jan			Feb						Mar
	23	24-27	28.1-3.2	04-10	11-14	15-17	18-20	21-25	26.2-3.1	2
Sampling period			1	2	3	4		5	6	
Classroom 1			RF	CF	CF	RF				
Classroom 2			CF	RF				CF	RF	
Air intake room 1										
Air intake room 2										

Fig. 2 Measurement plan. CF = combination filter, RF = regular filter

To assess the effect of filter type on the indoor concentrations, we calculated the median ratios of (1) NO_2-levels in classrooms with combination and regular filter (CF/RF), and (2) the indoor/outdoor (I/O) with combination and regular filter. The median is less influenced by outliers. To see how representative the measurements taken at the air intakes at the study site are with the outdoor concentrations, we also collected NO_2 measurement data from two monitoring stations situated in a distance of 1.6 and 3.0 km (see Fig. 1).

The statistical differences between mean concentrations were calculated by t-test. Statistical analyses were performed with R (http://www.R-project.org).

2.3 Data Quality and Control

Measurements and quality control procedures were done in accordance with the EU's air quality directive 2008/50/EC. The instruments were calibrated and scaled with zero-gas and span-gas between each sampling period. Parallel measurements were done at the beginning and the end of the study in the air intake of one classroom to compare the two instruments. Co-location of the two instruments indicated an average measured NO_2 concentration difference of 7%. A very high correlation was found between the measurements obtained from the two instruments ($R^2 = 0.99$).

3 Results

3.1 NO_2 Concentrations

During the entire measurement period, the average NO_2 concentration measured in the air intakes of the two classrooms was 26.51 $\mu g/m^3$ (range: 0.32–74.41 $\mu g/m^3$). In comparison, the average concentrations measured at the monitoring stations were higher, 44.16 $\mu g/m^3$ at the Kirkeveien monitoring station (3.0 km southwest) and 42.15 $\mu g/m^3$ at the Aker sykehus monitoring station (1.4 km southeast). NO_2 measurements taken at the study site were moderately to highly correlated with those taken at monitoring stations (Kirkenveien: $R^2 = 0.61$; Aker: $R^2 = 0.52$).

3.2 Impact of Filter Type on Concentrations in Classrooms

Figure 3 shows the measured NO_2 concentrations inside the two classrooms during sampling period 1 upper, combination filter in classroom 2 and regular filter in classroom (1) and period 2 lower, combination filter in classroom 1 and regular filter in classroom (2). For both sampling periods, the average NO_2 concentration in the classrooms were significantly lower with the combination filter (14.6 and 14.8 μg/m^3) than with the regular filter (29.6 and 15.4 μg/m^3) in the ventilation system, especially during sampling period 1 (see Table 2). The concentrations in the two classrooms with different filters were similar during sampling period 2, where also few hours with high outdoor NO_2 concentrations were observed (see Table 2, N = 4).

As seen in Table 2, the median CF/RF ratios were lower for sampling period 1 than for sampling period 2, indicating a 52% reduction in indoor NO_2 concentrations with CF compared to RF during the period the ventilation system was operational. Furthermore, the reduction in NO_2 concentrations in the classrooms with CF was generally higher during hours with high outdoor NO_2 concentrations than during the period the ventilation system was operational.

3.3 *Comparison of Indoor/Outdoor* NO_2 *Concentrations*

The measured indoor and outdoor concentrations and the median I/O ratios are shown in Table 2. The average outdoor concentrations were lowest (27.7 μg/m^3) during sampling period 5 and highest (42.1 μg/m^3) during sampling period 4.

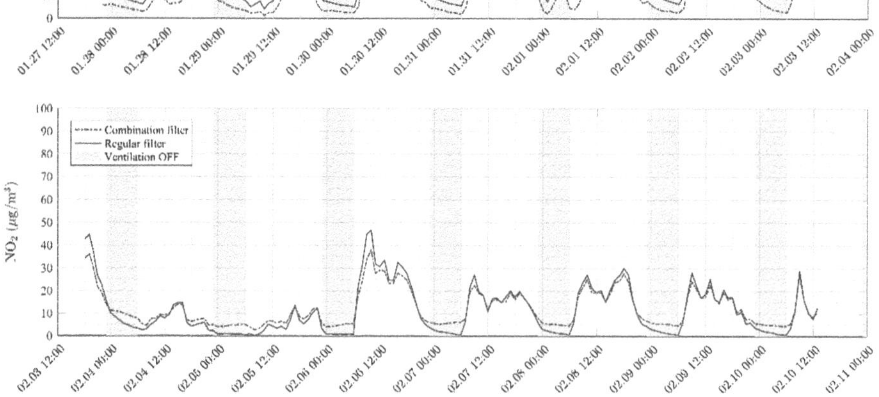

Fig. 3 NO_2 concentrations measured simultaneously inside two classrooms during sampling period 1 and 2. The filters in the ventilation systems of the two classrooms were switched between the measurement periods. Values are 1-h averages

Table 2 Mean (standard deviation) NO_2 concentrations measured indoor (with CF and RF) and outdoor during the period the ventilation system was operational and during hours with high (>40 μg/m^3) outdoor concentrations, and the number of 1-h measurements (N)

Sampling period	N	Indoor		Outdoor	Median CF/RF ratios	Median I/O ratios
		Combination filter (CF)	Regular filter (RF)			
Ventilation operating hours (6:00–23:00)						
1	111	14.6 (7.9)**	29.6 (14.8)		0.48	
2	115	14.8 (8.1)*	15.4 (10.6)		1.00	
3	72	24.8 (14.9)**		29.6 (21.6)		0.90
4	39		45.6 (9.2)**	42.1 (8.1)		1.07
5	86	27.2 (12.4)		27.7 (15.4)		0.97
6	46		32.2 (12.9)**	28.1 (12.1)		1.14
Hours with high outdoor NO_2 concentrations						
1	25	25.8 (4.1)**	50.4 (8.8)		0.81	
2	4	35.5 (1.9)**	44.7 (1.7)		0.50	
3	25	41.0 (6.2)**		53.6 (10.6)		0.78
4	28		50.0 (3.2)**	46.1 (3.1)		1.06
5	16	42.1 (8.8)**		55.3 (9.1)		0.75
6	9		51.2 (3.9)*	48.0 (3.7)		1.07

$^*p < 0.05$, $^{**}p < 0.001$

During the period the ventilation system was operational, the median I/O ratios were 0.90 and 0.97 with combination filter in the ventilation system. These median I/O ratios decreased with 13 and 23%, respectively, during hours with high outdoor NO_2 concentrations. These results indicate that a higher filter efficiency is achieved during hours with outdoor NO_2 concentrations. In comparison, the median I/O ratios were above 1.00 with the regular filter in the ventilation system, resulting in higher NO_2 concentrations indoors than outdoor. During sampling period 6, where few hours of high outdoor concentrations were observed, the median I/O ratio was 1.14 during the operational hours of the ventilation system. In comparison, the median I/O ratio for sampling period 5 was 1.07.

To evaluate how the CF reduces the NO_2 concentrations in the two classrooms compared with RF, median I/O ratios during sampling periods with CF (3 and 5) were compared to those during the sampling periods with RF (4 and 6). This resulted in a reduction of 15 and 16% in median I/O ratio by the CF during the period the ventilation system was operational, and a reduction of 26 and 30% during the hours with high outdoor NO_2 concentrations.

Scatterplots of indoor/outdoor NO_2 concentrations are shown in Fig. 4. The slope of the regression equations represents the fraction of NO_2 penetrating into the classrooms. The high R^2 (0.56–0.91) demonstrates that the outdoor concentrations explained a large part of the total variation of the NO_2 concentrations in the classrooms.

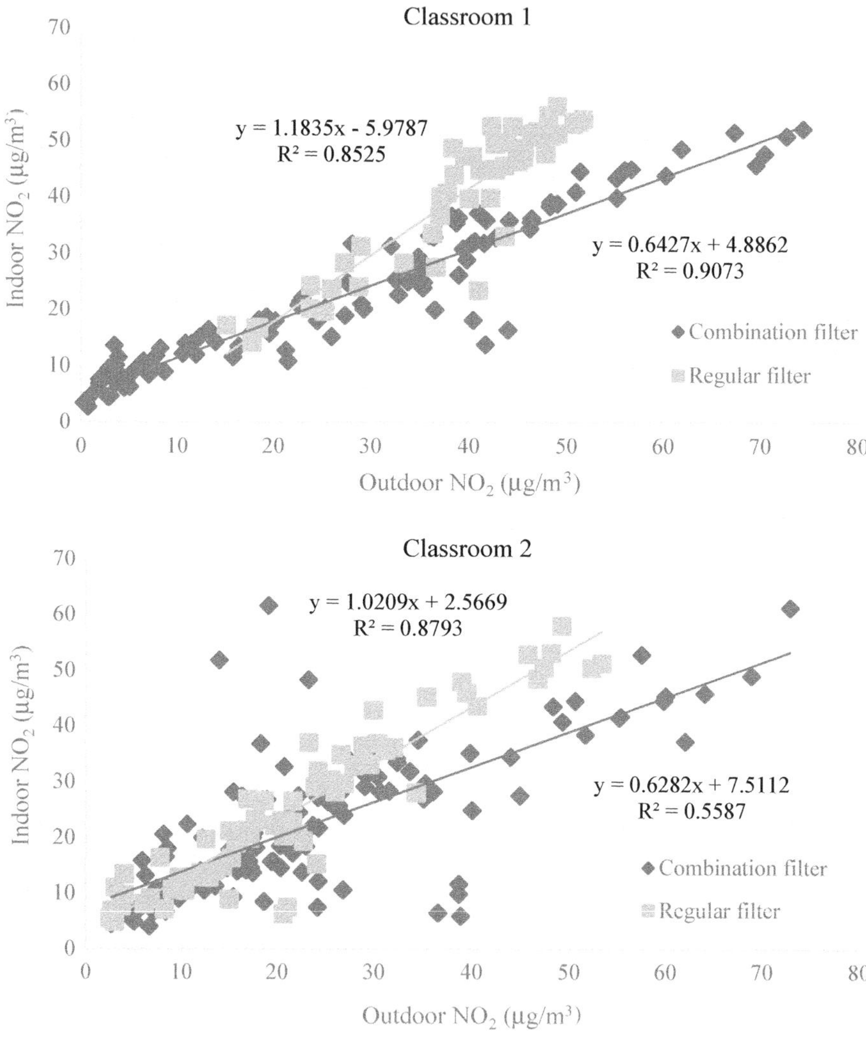

Fig. 4 Scatterplots of 1-h average (00–24 h) NO_2 concentrations measured outdoor and indoor during sampling period 3–4 (upper) and 5–6 (lower)

4 Discussion

4.1 Outdoor Versus Monitoring Sites

The outdoor NO_2 concentrations measured in the air intakes were on average 34 and 37% lower compared to the monitoring stations. This is probably due to the two monitoring stations being situated next to streets with higher traffic intensity, as traffic is a major source of NO_2. Nevertheless, the measurements at the monitoring sites correlated highly with the measurements taken at the air intakes of the classrooms. Similarly, a study in Norway found a higher correlation ($R^2 = 0.76$) between NO_2 measurements taken at the location of the municipality instrument and at the office building [7]. The lower correlations found in our study could be due to the longer distances between our study site and the monitoring stations (1.7 and 3.0 km). In contrast, the distance between the office building and the monitoring station in the study by Reyes-Lingjerde [7] was only 300 m.

4.2 Effect of Filter Type

Our study has demonstrated that using a combination filter in the ventilation system results in lower NO_2 concentrations in the classrooms compared with a regular filter. Furthermore, the combination filter is able to remove a large percentage of the outdoor NO_2 penetrating indoor.

The regular filter resulted in increased NO_2 concentrations in the classrooms, with median I/O ratios of 1.07–1.14 during the hours the ventilation system is operational. This is in line with the findings by Reyes-Lingjerde [7] who also observed that the indoor NO_2 concentrations with a regular filter was 8% higher than outdoor. As there were no known NO_2 sources indoors, the observed higher indoor concentrations can be attributed to the chemical reactions between NO and background ozone. A study which assessed the I/O NO_2 concentrations in 16 schools in Sweden reported an average I/O ratio of 0.96 for schools and 1.07 for pre-schools [6]. These schools had mechanical ventilation. As there were no indoor NO_2 sources present, inefficient filter in the ventilation system was provided as a possible explanation to the high I/O ratios. Blondeau et al. [5] reported average I/O ratios of NO_2 in the range of 0.88–1.17 for six schools in France. With the exception of one school that had mechanical ventilation, all others had natural ventilation. Our findings and these studies suggest that ventilation systems without a proper filter are ineffective against infiltration of outdoor NO_2. Majority of studies on I/O ratios for NO_2 lack information on the ventilation system, especially for commercial buildings and schools. The type of ventilation system and air change rates are important factors that can influence the indoor NO_2 concentration.

Compared with a few studies that have assessed the effect of different filters in the ventilation system on indoor NO_2 concentrations, we found a somewhat lower filter efficiency with the combination filter. Reyes-Lingjerde [7] found the average indoor NO_2 concentrations in an office building to be 68% lower when a combination filter was used compared to a regular filter. They also compared the indoor-outdoor concentration difference and found the indoor concentrations to be on average 72% lower with a CF compared with an RF. Partti-Pellinen et al. [8] assessed the effect of ventilation and air filtration systems on indoor NO_2 concentrations in a children's daycare center in Finland. They found that the combination of mechanical filters (type EU1, EU5 and EU7) and a chemical filter with added gas filtration unit was most efficient in removing outdoor NO_2, with a 48% reduction during weekdays and 66% during periods with high (>50 μg/m^3) outdoor NO_2 concentrations. In comparison, the filtration efficiency of a mechanical filter (EU1) was 22% during weekdays and 50% during periods with high (>50 μg/m^3) outdoor NO_2 concentrations. The lower reduction observed in our study (26–30% during hours with high outdoor NO_2 concentrations) could be explained by that filter efficiency increases with increasing outdoor concentrations. As seen during sampling period 2, where there were only 4 h with high outdoor concentrations, we found little impact of filter type on the NO_2 concentrations in the classrooms compared to sampling period 1. The outdoor concentrations measured in our study were in the range of 2.6–72.9 μg/m^3, whereas the outdoor concentration in the study by Reyes-Lingjerde [7] were in the range of 2.5–154.0 μg/m^3.

Compared to previous years, the measured outdoor concentrations at the monitoring stations during the study period were moderate, with the highest observed concentration of 118 μg/m^3, well below the 1-h mean NO_2 limit value of 200 μg/m^3 set by WHO. Nevertheless, during cold winter periods in Nordic countries, it would be beneficial to install combination filters in the ventilation systems, especially in buildings near areas with high traffic. However, the cost-benefits of converting from regular to combination filter could be a deciding factor. Finally, it is also important to take into consideration the lifetime (6–12 months for the filters used in our study) and maintenance of the filters, as any type of filter is only effective if frequently replaced.

4.3 Strengths and Limitations of the Study

The sampling plan was set up systematically to make the best use of the two sampling instruments. The study rooms were similar in layout, size and ventilation conditions. Although there were differences in occupancy and usage of the two classroom, we found no obvious contributions of indoor sources to the NO_2 concentrations, which could have influenced our results. Furthermore, the outdoor concentrations explained a large part of the total variation of the NO_2 concentrations in the classrooms. During the study period, the outdoor temperature ranged from −9.5 to 5.4 °C, with an average temperature of −1.2 °C. Although temperature influences the level of outdoor air pollutants, the outdoor NO_2 concentrations in the sampling period 3–6 did not vary much.

5 Conclusions

Our study demonstrates that installing a combination filter in the ventilation system could reduce the NO_2 concentration in classrooms. During hours with high outdoor concentrations, the combination filter can reduce the NO_2 concentrations in the classrooms with 26–30%.

Acknowledgements We would like to thank Camfil AS who provided the filters and guidance in how to install them in the ventilation system. This paper is based on the master thesis by Kristian Fredrik Nikolaisen, and was a part of the BEST VENT project. BEST VENT is funded by the Research Council of Norway EnergiX program under Grant 255375/E20 together with the following industry partners: Undervisningsbygg Oslo KF, GK Inneklima AS, DNB Næringseiendom AS, Erichsen & Horgen AS, Hjellnes Consult AS, Multi-consult AS, Interfil AS, Camfil Norge AS, Swegon AS, Belimo Automasjon Norge AS, NEAS AS, and Norsk VVS Energi- og Miljøteknisk Forenings Stiftelse for forskning.

References

1. WHO, Review of evidence on health aspects of air pollution—REVIHAAP project. Bonn, Germany (2013)
2. A. Polidori, P.M. Fine, V. White, P.S. Kwon, Pilot study of high-performance air filtration for classroom applications. Indoor Air **23**, 185–195 (2013)
3. S.C. van der Zee, M. Strak, M.B.A. Dijkema, B. Brunekreef, N.A.H. Janssen, The impact of particle filtration on indoor air quality in a classroom near a highway. Indoor Air **27**, 291–302 (2017)
4. M.C. McCarthy, J.F. Ludwig, S.G. Brown, D.L. Vaughn, P.T. Roberts, Filtration effectiveness of HVAC systems at near-roadway schools. Indoor Air **23**, 196–207 (2013). https://doi.org/10.1111/ina.12015
5. P. Blondeau, V. Iordache, O. Poupard, D. Genin, F. Allard, Relationship between outdoor and indoor air quality in eight French schools. Indoor Air **15**, 2–12 (2005). https://doi.org/10.1111/j.1600-0668.2004.00263.x
6. J. Wichmann, T. Lind, M.A.-M. Nilsson, T. Bellander, PM2.5, soot and NO_2 indoor–outdoor relationships at homes, pre-schools and schools in Stockholm, Sweden. Atmos. Environ. **44**, 4536–4544 (2010). https://doi.org/10.1016/j.atmosenv.2010.08.023
7. A. Reyes-Lingjerde, Reduction in NO_2-concentration across ventilation filters in an office building located close to heavy traffic, Master thesis. University of Bergen, 2016
8. K. Partti-Pellinen, O. Marttila, A. Ahonen, O. Suominen, T. Haahtela, Penetration of nitrogen oxides and particles from outdoor into indoor air and removal of the pollutants through filtration of incoming air. Indoor Air **10**, 126–132 (2000). https://doi.org/10.1034/j.1600-0668.2000.010002126.x

Part X
Moisture Safety and Hygrothermal Aspects

Determination of Maximum Moisture Zone on Enclosing Structures

Vladimir Gagarin, Vladimir Kozlov and Kirill Zubarev

Abstract Engineering methods of determining the position of the maximum moisture zone in enclosing structures have been theoretically justified. Calculation formulas have been determined with the use of moisture potential. This method has been used for studying positions of the maximum moisture zones in a enclosing structure wall made of aerated concrete blocks, with facade composite thermal insulation, external plaster with various layer thicknesses. It has been calculated that with the thermal insulation thickness over 37 cm the maximum moisture zone is located in the internal layer of the aerated concrete blockwork. The phenomenon has been called "the over-insulation effect".

Keywords Enclosing structures · Moisture potential · Maximum moisture zones

1 Introduction

In the course of Building Rules 50.13330.2012 updating [1] the modifications were made to the "Protection Against Enclosing Structures Over-Moistening" calculation procedure. This comprises introduction of the maximum moisture zone position in enclosing structures into the calculation procedure. This method was first proposed by Kozlov in 2004 [2]. It was further confirmed in the regulatory document in 2005. Building Rules [1] has demonstrated the applicability of this method for enclosing structure with facade composite thermal insulation with external plastering. This article reviews the theoretical justification of this method, and analyses the influ-

V. Gagarin (✉) · V. Kozlov
Research Institute of Building Physics (NIISF RAACS),
Locomotivny pr, 21, Moscow, Russian Federation
e-mail: gagarinvg@yandex.ru

V. Gagarin · K. Zubarev
HVAC Department, Moscow State University of Civil Engineering (MSUCE),
Yaroslavskoe shosse 26, 129337 Moscow, Russian Federation

D. Johansson et al. (eds.), *Cold Climate HVAC 2018*,
Springer Proceedings in Energy, https://doi.org/10.1007/978-3-030-00662-4_78

ence of thermal insulation layer thickness on the maximum moisture zone location in energy-saving enclosing structure.

2 Maximum Moisture Zone Location Calculation

The calculation procedure as per [1, 2] consists in the following. The value of $f_i(t_{m.m.})$ array shall be calculated for each layer of a multi-layer enclosing structure, which value shall be representative of the temperature in the maximum moisture zone.

$$f_i(t_{m.m.}) = 5330 \cdot \frac{r_{t,v}\left(T_{in} - T_{ext,neg}\right)}{R_t\left(e_{in} - e_{ext,neg}\right)} \cdot \frac{\mu_i}{\lambda_i} \quad (1)$$

where

- R_t enclosing structure heat transfer total resistance, (m^2 · °C)/W;
- $f_i(t_{m.m.})$ function that corresponding to the temperature of the layer i in the maximum moisture zone;
- $t_{m.m.i}$ maximum moisture zone temperature of the layer i, °C;
- $r_{t,v}$ enclosing structure water vapor permeability total resistance, m^2 · s · Pa/g;
- T_{in} inside air average temperature, °C;
- $T_{ex,neg}$ outdoor air average temperature in the period of monthly average temperature below zero, °C;
- e_{in} partial pressure of inside air water vapor, Pa;
- $e_{ext,neg}$ partial pressure of outdoor air water vapor in the period of monthly average temperature below zero, Pa;
- μ_i vapor permeability coefficient of i-th layer material, g/(m · s · Pa);
- λ_i thermal conductivity coefficient of i-th layer material, W/(m^2 °C);
- 5330 coefficient of the semi-empirical equation for the dependence of the quantities E on T (see Eq. 10).

Based on the determined values of $f_i(t_{m.m.})$ array with the help of Table 11 in Building Rules [1] the maximum moisture zone temperature $t_{m.m.i}$, shall be defined for each layer of the multi-layer structure. The structure has only one maximum moisture zone—the maximum moisture zone in one of the structure layers. Item 8.5 of Building Rules [1] describes the procedure of determining this zone.

This method has been proven by the moisture potential F theory [2, 3]. The theory of this method is explained in Chaps. 3 and 4

3 Applied Moisture Potential of Structure Materials

The method of determining the maximum moisture zone in enclosing structures is based on the use of moisture potential. In this case, the moisture potential F developed by [2, 3] has been used as the functional relation F between material moisture content and temperature determined with the formula:

$$F(w,T) = \frac{1}{\mu}\int_{0}^{w} \beta(\varsigma)d\varsigma + E(T)\varphi(w) \quad (2)$$

where

F(w,T)	moisture potential depending on the moisture content and temperature, Pa;
w	material moisture content, kg/kg;
T	temperature, K;
μ	material water vapor permeability, g/(m · s · Pa);
β	material water content conductivity coefficient, g/(m · s · kg/kg);
ς	argument in the integral;
E	saturated water vapor pressure, Pa;
φ	relative air humidity or relative pressure of water vapor in material pores, unit fractions.

The potential is comfortable to use because it allows to consider the liquid moisture carryover and water-vapor diffusion in enclosing structure materials simultaneously. Input data needed to calculate the potential F shall be collected as per the standard techniques for measuring coefficients of vapor permeability [4], water content conductivity [5] and thermal conductivity [6]. Therefore, when working with the potential F the data on material characteristics collected in the previous years shall be used.

The practical application of the potential F in the calculations is similar to the application of water vapor partial pressure, however it allows to make the calculations both in the sorption zone of structure materials moisturizing and in the supersorption zone of moisturizing simultaneously. This potential, unlike other moisture potentials such as [7, 8 etc.], considers the various energies of moisture bond to the material structure in the above mentioned moistening zones. This was the purpose, for which Professor V.N. Bogoslovsky created and developed the theory of moisture potential in early 1950th as described in [7].

Moisture potential simplifies considerably the calculation of moisture steady-state distribution. In the steady state (without air leakage considered) the equation of moisture carryover in the structures shall be formulated with the potential F:

$$\frac{\partial^2 F}{\partial x^2} = 0 \quad (3)$$

Therefore, the solution to this equation is the linear function.

$$F = Ax + B \quad (4)$$

where A and B are the constants determined on the basis of boundary conditions and conditions at material joints respectively, Pa/m and Pa.

In Eq. (4) the constant A is equal to the gradient of the potential F. Consequently $\mu \cdot A$ is the specific moisture flow with dimensionality of mg/(m^2 · h). In the multilayer enclosing structures Eq. (3) solution is piecewise.

The Building Rules [1] Chapter "Protection Against Enclosing Structures Over-Moistening" calculation procedure is based on the calculation of the moisture carryover under steady conditions that consider only vapor permeability of construction materials. It is necessary to consider the supersorption moisture transfer in materials as well. One of the interesting aspects of this approach used in the regulatory documents was the search for the maximum moisture zone position, with respect to which the enclosing structures moisture balance was calculated and for which the verification of non-availability of over-moistening was made. The potential F application makes it possible to find the exact position of the maximum moisture plane in the used model.

4 Method of Calculating the Maximum Moisture Zone Location

To find the maximum moisture zone it is necessary to determinate the points in the drawing of the section of the enclosing structure, where the coordinate derivative of moisture function is equal to zero, as well as the points where the moisture function has breaks. The use of potential F allows to find such points avoiding the determination of moisture distribution as per the civil structure thickness, which makes the calculation much simpler.

The x coordinate derivative of potential F.

$$\frac{\partial F}{\partial x} = \frac{\partial F}{\partial w}\frac{\partial w}{\partial x} + \frac{\partial F}{\partial T}\frac{\partial T}{\partial x} \tag{5}$$

Let us find the moisture coordinate derivative.

$$\frac{\partial w}{\partial x} = \frac{\frac{\partial F}{\partial x} - \frac{\partial F}{\partial T}\frac{\partial T}{\partial x}}{\frac{\partial F}{\partial w}} \tag{6}$$

To find the point coordinates, where $\frac{\partial w}{\partial x} = 0$, it is necessary to solve the following equation:

$$\frac{\partial F}{\partial x} - \frac{\partial F}{\partial T}\frac{\partial T}{\partial x} = 0 \tag{7}$$

Potential F dependence on the coordinate is defined in Eq. (4), and the same on the temperature in Eq. (2). Therefore (7) can be rearranged with the use of (2) and (4):

$$A - \varphi(w)\frac{\partial E}{\partial T}\frac{\partial T}{\partial x} = 0 \tag{8}$$

Temperature distribution along the enclosing structure thickness in the one-dimensional steady state case is defined by the formula:

$$T = T_{ext} + \frac{T_{in} - T_{ext}}{R_t \lambda(x)} x \tag{9}$$

where

R_t	enclosing structure heat transfer total resistance, $(\mathrm{m}^2 \cdot \mathrm{K})/\mathrm{W}$;
T_{in} and T_{ext}	temperature of inside and outdoor air respectively, K;
$\lambda(x)$	thermal conductivity coefficient of the cladding layer material (accepted as constant within the cladding homogeneous layer), W/(m · K).

To determine $\frac{\partial E}{\partial T}$ it is undesirable to use various formulas involving the approximation of *E(T)* dependence in the form of polynomials with various powers, as the *E* function derivative is being approximated. That is why the semi-empirical expression [4] determined on the basis of the Clapeyron-Clausius equation [4] is used:

$$E = 1.84 \times 10^{11} \exp\left(-\frac{5330}{T}\right) \tag{10}$$

After inserting the formulas (9) and (10) Eq. (8) will look as follows:

$$A - \varphi \frac{5330}{T^2} E \frac{T_{in} - T_{ext}}{R_t \lambda(x)} = 0 \tag{11}$$

The first component of the formula (11) is the moisture potential gradient in the structure material layer, and the second component is the moisture potential gradient at constant moisture content (caused by the temperature drop). The difference is the isothermal gradient of the moisture potential allowing to find out the moisture distribution in the structure layers.

To determine the location of the maximum moisture zone, Eq. (11) components are grouped in such a way that all values depending on the structure and boundary conditions are on the right,

$$\frac{T^2}{E} = \frac{(T_{in} - T_{ext})}{R_t A} \cdot \frac{5330}{\lambda} \tag{12}$$

On the left of Eq. (12) there remains the physical value of water vapor in the atmosphere with the certain pressure. In our case of the standard atmospheric pressure this value depends on temperature only. For the easy use of the formula (12) the T^2/E dependence on temperature is shown in Table 1.

Table 1 Dependence of $\frac{T^2}{E}$ value on temperature

t (°C)	$\frac{T^2}{E}$ (K^2/Pa)	t (°C)	$\frac{T^2}{E}$ (K^2/Pa)	t (°C)	$\frac{T^2}{E}$ (K^2/Pa)
−30	1554	−12	313.9	6	83.25
−27	1187	−9	245.4	9	69.27
−24	898.6	−6	193.2	12	57.89
−21	682.8	−3	153.15	15	48.65
−18	520.2	0	121.98	18	41.03
−15	403.4	3	100.36	21	34.74

SP 50.13330.2012 [1] contains the similar table of different format, with different interval and other limits needed for engineering calculations.

In case of steady conditions, it is a lot easier to find the moisture flow being transferred through the enclosing structure than the moisture distribution throughout the enclosing structure. That is why it is assumed that during initial consideration stage the moisture flow shall be found, and the coordinate of the maximum moisture shall be determined with formula (12) and Table 1.

In a general case there are two boundary condition variants, which lead to two different formulas describing moisture flow thought a structure. However checking a structure over-moistening as per [1] assumes a structure to be dry, besides, the entire method of further calculation as per [1] allows only to consider the water vapor diffusion. That is why one of the variants, which corresponds to the over-sorption moistening of the external edge of the structure and high level of the structure moistening in general, can be disregarded. Therefore, the formula for moisture flow through a structure (11) shall be considerably simpler, and it is necessary to proceed to determining the coordinate of the moisture maximum in the structure.

$$\frac{T^2}{E} = 5330\frac{r_{t,v}(T_{in} - T_{ext})}{R_t(F_{in} - F_{ext})} \cdot \frac{\mu}{\lambda} \tag{13}$$

Under the conditions, for which the structure is being checked for over-moistening as per [1] the potential F is equal to the relevant partial pressure of outdoor and inside air water vapor. No additional concepts are to be introduced for the use of the proposed method. In the final revision, the right side of Eq. (13) looks as in the formula (1).

Item 8 of [1] stipulates the algorithm of determining the maximum moisture zone that foresees the sequential calculation on all structure layers. It is interesting that this calculation can be made substantially simpler as the enclosing structure layers moistening is changed systematically, and the method helps to determine the direction of the moistening change in the structure. Three variants are possible, when we analyze any of the structure layers. The first variant is that where the temperature of the maximum moisture zone $t_{\mathrm{m.m.}i}$, is achieved in the layer being studied. This means that the moisture maximum has been found. The second variant is that where the temperature of the maximum moisture zone of the layer being studied is higher than its temperature calculated with the formula (9). Therefore the maximum moisture zone is located to the warm side of the layer being studied. The third variant is that where the temperature of the maximum moisture zone of the layer being studied is lower than its temperature calculated with the formula (9). So in this case the maximum moisture zone is located to the cold side of the layer being studied. In the most of the up-to-date Enclosing structures, the maximum moisture zone is usually located in the thermal insulation layer or in the layer neighboring to it.

5 Practical Application of Moisture Conditions Calculation Method in Various Enclosing Structures with Facade Systems

Several calculations of the maximum moisture zone position have been made for energy-saving structures. Below you can find the results of calculations for the wall structure of a residential house in Moscow; the wall was made of aerated concrete blocks, with facade composite thermal insulation with external plaster layer. The internal side of the wall was coated with cement and sand mortar. The thermal insulating material was mineral wool with thickness varying in the course of study.

The thermal insulation layer was varying from 0.37 to 0.65 m. As a result, in case of small insulation thickness values the maximum moisture zone was located between the external plaster layer and the mineral wool layer. It is interesting that when thermal insulation layer thickness achieved 37 cm, the maximum moisture zone shifted to aerated concrete and remained there with further increase of thickness (Table 2). It is proposed to name this effect as "the effect of over-insulation" of enclosing structure.

"The over-insulation effect" is substantiated by the fact that in case of increasing the mineral wool layer thickness the gradient of moisture potential F in the thermal insulation decreases, and its value becomes insufficient for moisture carryover trough the mineral wool layer. That is why moisture remains in the layer of aerated concrete.

Table 2 The calculated values of the maximum moisture zone coordinates in enclosing structure with the increase of thermal insulation layer thickness from 0.37 to 0.65 m—"the effect of over-insulation" of enclosing structure

Item No.	Thermal insulation layer thickness (m)	Distance from the mineral wool and aerated concrete contact layer to the maximum moisture zone in aerated concrete (m)
1	0.37	0.027
2	0.45	0.073
3	0.5	0.123
4	0.55	0.176
5	0.6	0.231
6	0.65	0.289

Substantial thicknesses of enclosing structure thermal insulation layer reaching 0.40–0.50 m are applied in "passive houses". Along with some energy saving effect, the high thermal insulation layer thickness results in negative consequences, such as increased materials moistening. Hagersedt and Harderup [9] describe this phenomenon in wooden passive houses in Sweden. The method described in the Russian Building Rules [1] allows making calculation of such phenomenon as "the over-insulation effect".

6 Conclusion

The substantiation has been given for the procedure of calculating the protection against over-moistening as per SP 50.13330.2012. This procedure yields results without sophisticated calculations with specialized computer software, unlike the known methods used in Russia and abroad. The results of determining the maximum moisture zone for facades with composite thermal insulation with thin plaster layers have been given as an example of the SP procedure use. "The over-insulation effect" having great importance for "passive houses" designing has been described. This effect has been substantiated within the framework of the present theory. The further development of the theory based on the moisture potential F will make it possible to quantify materials moisture in the maximum moisture zone.

References

1. Building rules 50.13330.2012 "Thermal performance of the buildings" (rus)
2. V.V. Kozlov, Engineering assessment of moisture condition of modern wall structures with increased heat-insulating level accounting for vapor permeability, moisture conductivity and air filtration. Ph.D. Abstract. Moscow, 2004. p. 24 (rus)
3. V.G. Gagarin, The theory of state and moisture transfer in building materials and thermal performance of enclosing structures of buildings. The thesis for the degree of Doctor of Technical Sciences. Moscow (2000) (rus)
4. Standard GOST 25898-2012, Building materials and products. Methods of vapor permeability and resistance to water vapor permeability determination (rus)
5. Standard GOST R 56504-2015, Building materials. Methods of water content conductivity coefficient determination (rus)
6. Standard GOST 7076-99, Building materials and products. Method of determining thermal conductivity and heat transfer resistance under steady temperature conditions (rus)
7. V.N. Bogoslovsky, *Fundamentals of Material Potential Moisture Theory Used in Wall Structures* (Moscow State University of Civil Engineering, 2013), p. 112 (rus)
8. H.M. Künzel, Verfahren zur ein- und zweidimensionalen Berechnung des gekoppelten Wärme- und Feuchtetransports in Bauteilen mit einfachen Kennwerten. Dissertation des Doktor-Ingenieurs. Stuttgart. 1994. 68 S
9. S.O. Hagersedt, L.-E. Harderup. Control of moisture safety design by comparison between calculations and measurement in passive house walls made of wood, in XII DBMC, in *International Conference on Durability of Building Materials and Components*, Porto, Potugal, April 12th–15th, 2011

Moisture Safety of Tall Timber Facades—LCA and LCC Calculations of Damage Scenarios

Joakim Norén, Anna Pousette and Karin Sandberg

Abstract The use of timber structures in tall buildings increases the demand of moisture safety in facades. Moisture in the facade could result in unwanted consequences such as mold, decay and distortions in wood materials. This might have impact on the indoor climate and the building quality. The aim of this study was to evaluate the moisture safety regarding the composition of the façade and connection details such as windows and balconies. Scenarios with possible damages were evaluated with LCC (Life Cycle Cost) and LCA (Life Cycle Assessment). The scenarios were developed based on experiences from manufacturers and insurance companies and on research investigations of damages. LCC and LCA includes replacement of damaged building materials, transports of new materials and damaged materials for recycling or energy recovery and the use of energy for drying of moisture in the structure. Both light frame structures and CLT (Cross Laminated Timber) structures were included. Improvements of detail connections to increase moisture safety were also evaluated regarding risk of damage, costs and environmental impact. The results show that even small and inexpensive improvements will increase the moisture safety and significantly reduce the risk of damage.

Keywords Timber facades · Moisture safety · LCA · LCC

J. Norén (✉)
RISE Research Institutes of Sweden, Stockholm, Sweden
e-mail: joakim.noren@ri.se

A. Pousette · K. Sandberg
RISE Research Institutes of Sweden, Skellefteå, Sweden
e-mail: anna.pousette@ri.se

K. Sandberg
e-mail: karin.sandberg@ri.se

D. Johansson et al. (eds.), *Cold Climate HVAC 2018*,
Springer Proceedings in Energy, https://doi.org/10.1007/978-3-030-00662-4_79

1 Introduction

A significant number of multi-storey office and housing projects have been and are currently developed all over Europe. In order to preserve and develop the chances on the market, wood construction must be reliable, durable, flexible and strong in off-site production/prefabrication. To fulfil most of these requirements a prolonged moisture-safety is necessary, which was in the focus of the project “Tall timber facades”. The project was funded under the European WoodWisdomNet research program with the objective to facilitate safe design of sustainable and hence cost-effective design solutions for the building envelope, tall timber facades, by the combination of existing best-practice with a risk-based concept [1].

With an increasing height of timber buildings, the challenge is growing to provide moisture-safe conditions for the expected lifetime of building envelopes. Tall buildings are particularly exposed to high wind pressures combined with driving rain. Compared to fire safety and static demands, the risk of moisture damages today is often underestimated in planning, building processes and quality management.

1.1 Goal and Scope

This study is part of the project Tall timber facades, with the goal to show the environmental impact and cost of moisture damages in façades and thereby enable comparison of improvement measures for increased moisture safety with consequences of moisture damage. Scenarios with possible damages were evaluated with LCC (Life Cycle Cost) and LCA (Life Cycle Assessment).

2 Method and material

2.1 Consequences of Moisture

The consequence of moisture load on the construction and parameters were analysed as direct and indirect consequences of moisture.

Mold is caused by excessive humidity in wooden constructions, and can result in financial loss and unfavorable social problems such as discomfort and health risks. Mold is a very complex biological phenomenon, which is highly dependent of the interrelation between humidity, temperature and time [2].

Decay is caused by moisture content above saturation point and decay fungi that grow through the wood cells and release enzymes that break down the wood components. This results in loss or significant reduction of many of the wood properties. Most common are brown rot, white rot and soft rot degradation [3, 4]. All decay fungi attacks on wooden elements will influence the mechanical resistance.

2.2 *Experiences of damages in timber facades*

Several damage investigations of building facades have been reported [5]. However, most of the results describe the amount of damages but not the exact cause and spread [6]. A review of errors and damages to facades was carried out by an insurance company that provides assurances for errors/damages that occur during the first ten years after completion of the building. The review included 185 damage cases that occurred during the period 2009–2016. Most houses were 1- or 2 storey timber frame houses with wood or plaster cladding. The main causes of damage were shortcomings in window flashings and other metal sheets and their connections to the facade, mainly around windows and doors. These damages are also found in studies of timber walls with plaster façade [7].

A questionnaire was sent to the timber house manufacturers in Sweden to collect their experiences of moisture damages in buildings.

2.3 *Wall Structures*

Two typical walls were studied; one was a CLT (Cross Laminated Timber) structure and the other a timber frame structure, see Fig. 1. These walls represent best practice constructions that fulfil all other requirements on facades, such as strength, acoustics, fire and insulation. The environmental impact and the cost with respect to moisture damages in these walls have been evaluated.

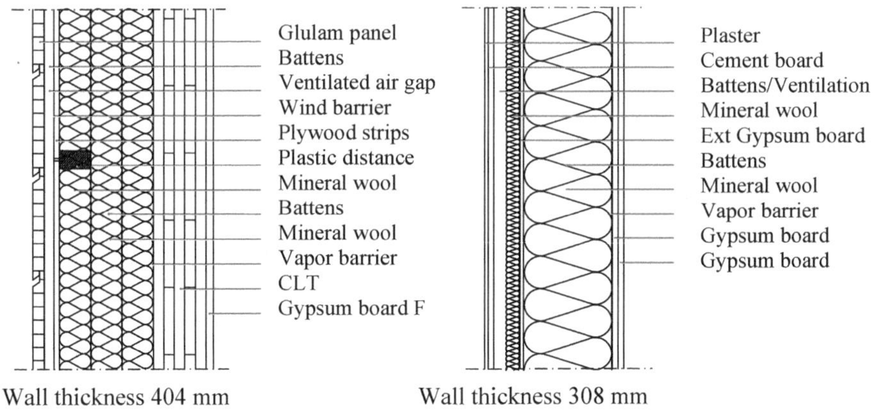

Fig. 1 Analysed walls. To the left is CLT structure and to the right structural timber frame

2.4 LCA and LCC

LCA/LCC calculations were used to evaluate the environmental impact and costs of moisture-related damages in the wooden facades over the building life cycle.

LCA Calculations

The calculation follows the modular set up of the life cycle stages in accordance with EN 15804 [8]. The inventory includes the following stages of the building life cycle:

- Product stage, A1-3
- Construction process, A4-5
- Use stage, B3 Repair and B4 replacement

Due to missing data and other information about the processes, the inventory is not comprehensive for all stages. However, for the product stage module A1-3 and the repair module B3, most of the relevant data of the processes that have the biggest environmental impact are included in the inventory. These processes are:

- Production of new materials/component and ancillary materials
- Use of energy (machines for drying and heat during repair B3)
- Production and transport of wastage of materials during repair or replacement
- Transportation of new materials/components and ancillary materials

The functional unit is the base for calculation of the material flows and environmental impacts over the products life cycle stages. For calculation of the indirect consequences of moisture damages in the facade, the functional unit was “a wall element with length 3 m and height 2.4 m including one window 1.2 m × 1.4 m or one balcony door 1.0 m × 2.1 m and with a life span of 50 years”.

LCC Calculations

Life-Cycle Costing (LCC) is used for assessing the cost performance of constructed assets and is described in the standard ISO 15686-5[9]. Both generic data from Sektionsfakta®—ROT [10] and specific data assigned to a specific company was used. Future costs were converted to a net present value. The expected real discount rate per annum was chosen 4%. The probability of when moisture damage will occur is difficult to know. LCC was calculated with a time of 2 years to discover a small damage, 5 years to medium damage and 9 years to extensive damage. Maintenance of façades was included once during the life of the frame structure with plaster and twice during the life of the CLT façade with wooden cladding. Scenarios with improvements include sealing tape and steel profile for the window connection and sealing tape for the balcony door. Life cycle stages and functional unit was the same as for LCA calculations.

3 Results

3.1 Consequences of Moisture

The input and output parameters for calculations of consequences, grouped into direct and indirect consequences, are shown in Fig. 2. The input parameters to simulate moisture and failure modes in a construction are climate, wind, rain, temperature, initial moisture, building moisture and all the material layers of the facade. Direct consequences of moisture can be moisture creep, decay (fungi), mold and façade deterioration. This will indirectly affect the insulation, risk of failure and deformation, the aesthetics and air quality. This may in turn lead to increased environmental impact and costs from replacement, repair, etc.

3.2 Experiences of damages

The results of the questionnaire to house manufacturers showed that most companies have experience of damages in facades due to water penetration from outside. It was specific details in the structure that caused the damages due to outside penetration. The main cause was horizontal interruptions (balcony, loggia, etc.), some were caused by openings (windows, doors, etc.), and there were also a few damages due to penetrations (pipes, cables, etc.). No damages were reported in buildings with CLT structures since this is a rather new type of building system. Based on the experience of the house manufactures the main risk areas in a building façade were identified, see Fig. 3.

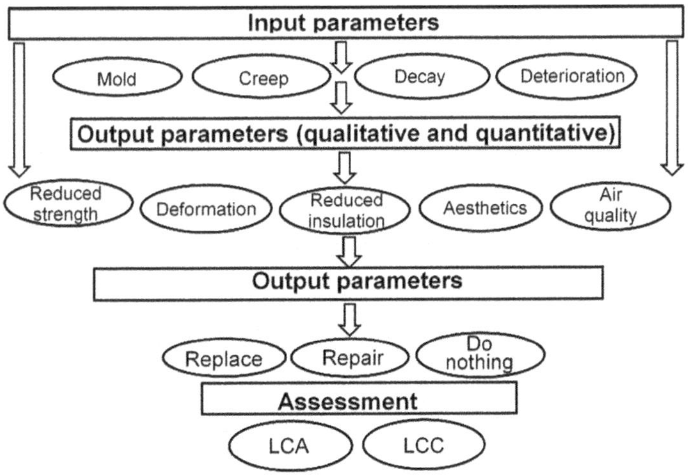

Fig. 2 Input and output parameters for calculations of direct and indirect consequences of moisture

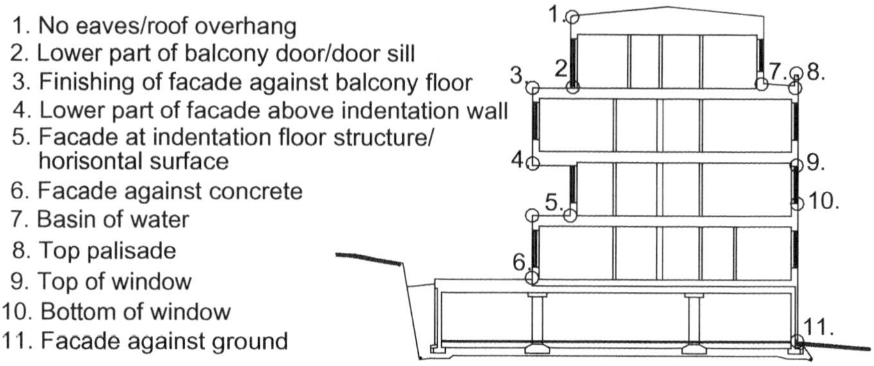

Fig. 3 Risk areas in the building façade

3.3 Damage Scenarios

Different damage scenarios were developed to enable LCA and LCC calculations. Scenarios for three different extents of moisture damage, small damage, medium damage and extensive damage were based on the experience from experts, house manufactures and insurance company. The different extent of damages refers to the time of damage detection. The scenarios are illustrated and defined for window and balcony door connection for both CLT structure and timber frame structure. However, since there is little information of damages in CLT structures these scenarios are estimated based on the construction and the materials included. In Fig. 4 examples of scenarios for moisture damage in window connection in timber frame structure are shown.

The consequence of a leakage will be to remove all moist materials and moisture damaged materials. After that the moist structure must be dried using dehumidifier and the wall is then rebuilt and restored. If there is decay in structural parts they are normally kept and supplemented if it is difficult to unload and replace, for example bearing studs and CLT elements.

3.4 Improvement of Details

A standard connection usually works well. On more exposed facades though, an improved connection with a steel profile behind reveal can provide increased safety against leakage, see Fig. 5. An improved connection at the balcony can be obtained with a sealing layer, as a second protection layer on top of the sill and on the side of the wall opening. If the technical improvement results in lower environmental impact and lower costs over the building life cycle the improvement is favourable to include in the connection detail.

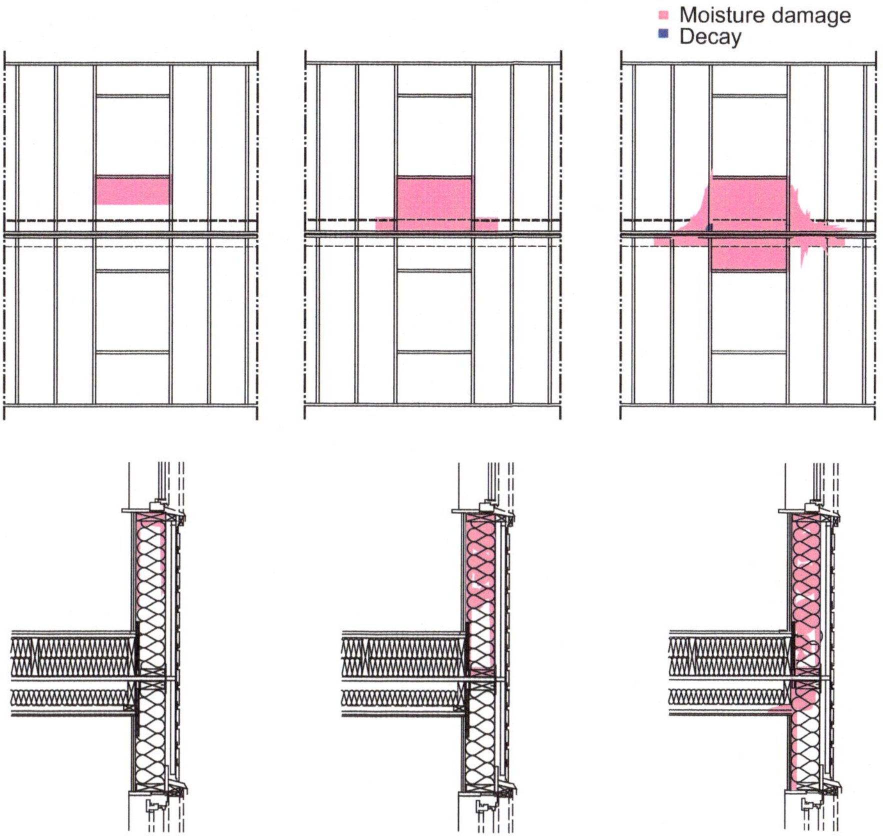

Fig. 4 Scenarios for three different extent of moisture damage in window connection in wall with timber frame structure. Upper: Front view of water spread in wall. Lower: Section of water spread in wall

3.5 LCA

The analyses are based on the damage scenarios with different extent of moisture damage. Climate impact was used to describe the environmental impact of the production of wall, maintenance and repair during service life. In Figs. 6 and 7 the environmental impact of the production of exterior wall elements and regular maintenance are presented. The figures also include the environmental impact of moisture damages in window and balcony connections in the wall.

Based on the damage scenarios the climate impact of damages in the connections is much lower than the production of the wall. This is mainly because the spread of water is limited in the wall and the damage area much smaller than the plain wall element. The balcony connection in the wall with structural frame results in a

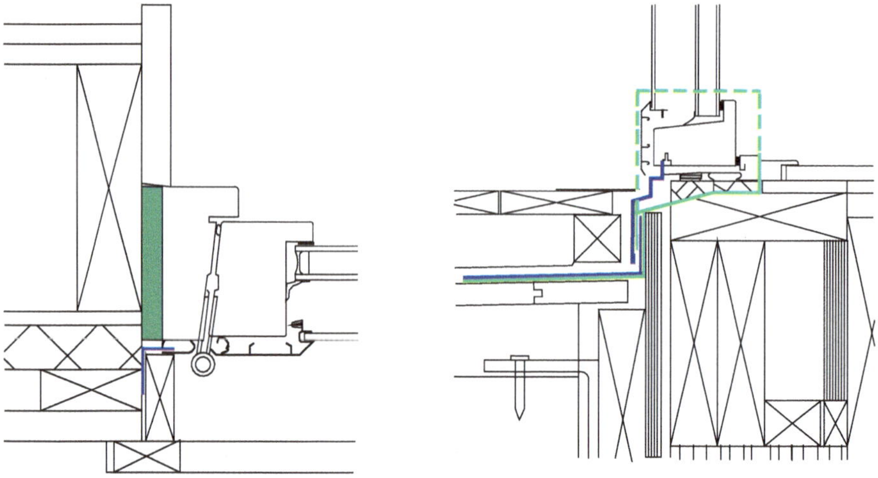

Fig. 5 Left: Improved connection with steel profile behind reveal (blue line). Right: Improved connection with a sealing layer (green line), on top of the sill and on the side of the wall opening (dotted green line)

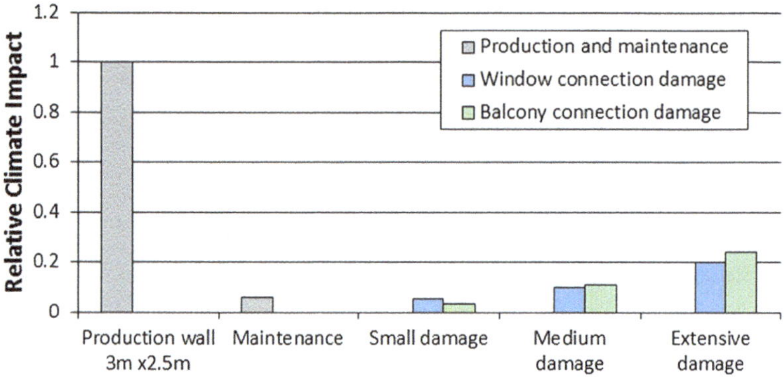

Fig. 6 Climate impact of damages in window and balcony connection in wall with CLT structure compared to the impact of wall production and regular maintenance

slightly higher impact due to more materials involved in the repair, see Fig. 6. In this wall the repair needs to be done from both sides of the wall.

In Fig. 8 climate impact of modification of window connection according to Fig. 5 is compared to impact of damages. The figure shows that the impact of the modification is much lower than the impact of the damages.

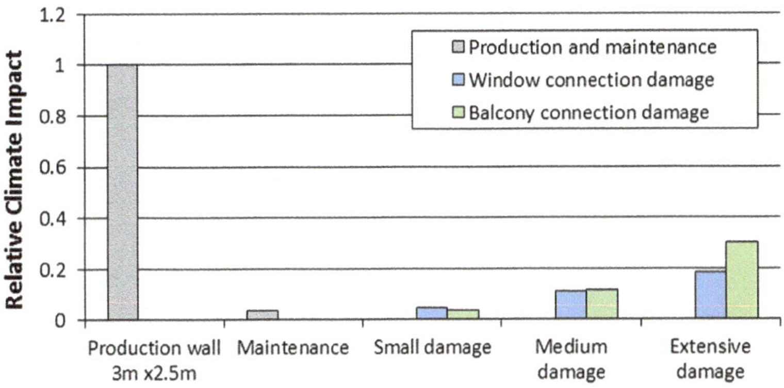

Fig. 7 Climate impact of damages in window and balcony connection in structural frame wall compared to the impact of wall production and regular maintenance

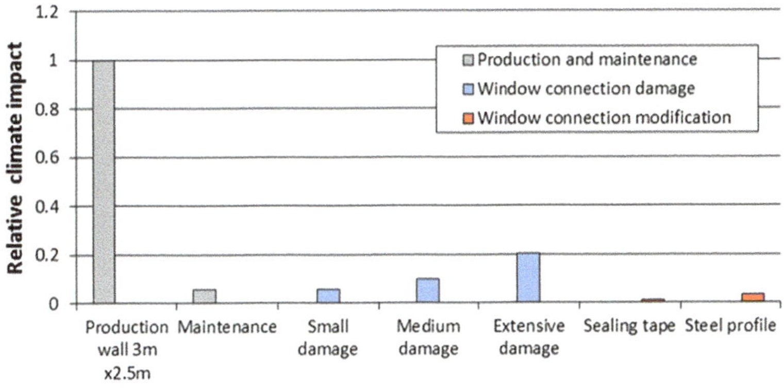

Fig. 8 Climate impact of damages in window connection compared to modification of connection

3.6 LCC

In Figs. 9 and 10 the life cycle costs of the production of exterior wall element and regular maintenance are presented. The figures also include the costs of moisture damages in window and balcony connections in the wall.

Fig. 9 indicates that repair cost of damage in a window connection will be higher than the costs of the damages in a balcony connection based on the used scenarios. This is because the repair must be done from the outside which requires scaffolding or lifts. The repair costs of the timber frame wall are higher than of the CLT structure. This is due to more materials involved in the repair, see Fig. 10.

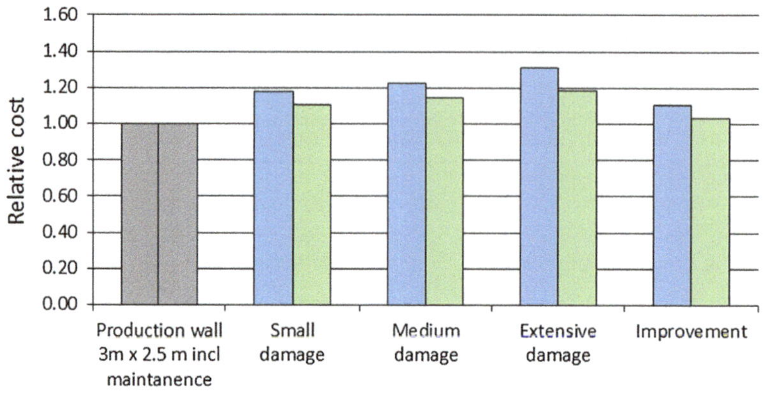

Fig. 9 LCC of CLT structure. Wall element (grey), window connection (blue) and balcony connection (green)

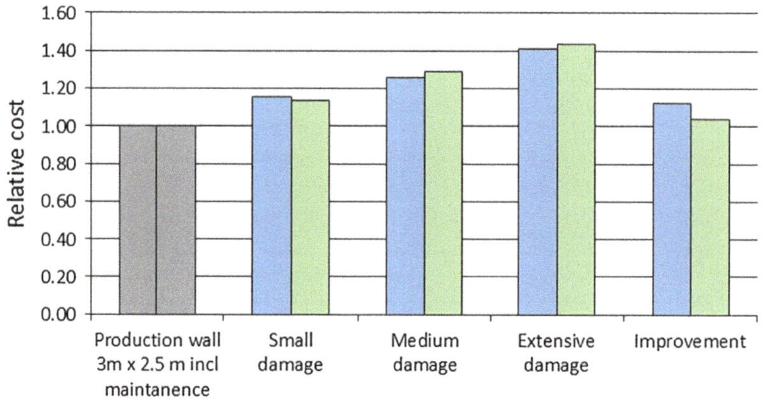

Fig. 10 LCC of timber frame structure. Wall element (grey), window connection (blue) and balcony connection (green)

4 Discussion and Conclusions

The scenario method used in this study was based on experience from builders, timber house producers, insurance companies and experts in timber building. The scenario model contains some uncertainties due to design, execution, and exposure time and if the structure has the ability to dry between exposures. However, the use of scenarios makes it possible to evaluate the consequence of a damage caused by failure in a connection detail.

Different improvements of detail connections to increase moisture safety were evaluated regarding risk of damage, costs and environmental impact. The results showed that even small and inexpensive improvements that will increase the moisture safety only have a small impact on the environment compared to damages over the building life time.

The monetarization of consequences demonstrated the relevance of moisture safety measures in order to avoid high costs for timber building industry and house owners. The findings are relevant due to the high monetary impact of possible moisture damages on envelopes of tall timber buildings.

Acknowledgement This study was supported by WoodWisdomNet+ research program; project Tall Timber Facades (TallFacades). We would like to thank Martinsons Byggsystem, Moelven Byggmodul and Vinnova for supporting the work.

References

1. S. Ott, et al., Final Project Report. Tall timber facades—Identification of Cost-effective and Resilient Envelopes for Wood Constructions (TallFacades) 31.07.2017
2. K. Gradeci, N. Labonnote, B. Time, J. Köhler, Mold models applicable to wood-based materials—A generic framework, (Accepted manuscript) Energy Procedia, 2017
3. C. Brischke, L. Meyer-Veltrup, Modelling timber decay caused by brown rot fungi, Materials and Structures, 2015
4. C. Brischke, E. Frühwald Hansson, Modeling biodegradation of timber—Dose-response models for above-ground decay and its climate-dependent variability, International Conference on Structural Health Assessment of Timber Structures, Lisbon, 16–17 June, 2011
5. Institut für Bauforschung e.V., Analyse der Entwicklung der Bauschäden und der Bauschadenkosten, Gemeinschaftsprojekt vom Bauherren-Schutzbund e.V., der AIA AG und dem Institut für Bauforschung e.V, Forschungsbericht: 19.02.2015 IFB – 14553
6. L. Olsson, Laboratoriestudie av träfasaders täthet mot slagregn, SP Rapport 2012:45. ISBN 978-91-87017–63-6. Borås 2012 (In Swedish)
7. A. Jansson, Putsade enstegstätade regelväggar. Erfarenheter från undersökningar som SP har utfört. SP Rapport 2015:01. ISBN 978-91-88001–27-6. Borås 2015 (In Swedish)
8. EN 15804, Sustainability of construction works—Environmental product declarations—Core rules for the product category of construction products, 2012
9. ISO 15686-5, Buildings and constructed assets—Service life planning—Part 5: Life-cycle costing, 2008
10. Sektionsfakta®–ROT, Teknisk-ekonomisk sammanställning av rot-byggdelar, 15/16, Wikells Byggberäkningar AB, 2015 (In Swedish)

Determination of Clay-Sand Plaster Hygrothermal Performance: Influence of Different Types of Clays on Sorption and Water Vapour Permeability

Erik Altmäe, Aime Ruus, Jane Raamets and Ernst Tungel

Abstract Eight different clay-sand plaster mixtures were studied. Mineral content and particle size distribution were estimated for all specimens. Hygroscopic sorption properties were determined (in climate chamber) at temperature of 23 ± 0.5 °C. The specimens were weighed at 1, 2, 3, 6, 12, 24 h until stabilisation at RH level of 30, 50 and 80% Moisture uptake (kg/m^2) and moisture uptake rate kg/(m^{2}h); moisture content and; points at sorption curve were monitored. There were large differences in sorption properties depending on clay type and plaster recipes. Total uptake of moisture at 30, 50 and 80% of RH for 2.5 cm plaster was 9.4–301.1, 17.5–465.9 and 41.6–744.9 g/m^2 accordingly. Standard (EN 1015-19) procedure was followed. Water vapour diffusion equivalent air layer thickness S_d = 0.08–0.12 m was declared. Strong positive correlation was found between the amount of calcite and sorption properties of plasters.

Keywords Clay plaster · Hygroscopic sorption · Water vapour permeability Moisture buffering

1 Introduction

The necessary ventilation rate depends on several aspects - heat production, the presence of CO_2 or other gases, body odor, moisture, smoke, airborne particles, etc. Ventilation rates calculated on the basis of CO_2 balance can be different from the rate attained on the basis of moisture balance. Several household activities have influence on moisture production and it can result in too low or high relative humidity.

The buffering ability of indoor finishing materials enables to smooth the peaks of indoor humidity. The Moisture Buffer Value of materials (MBV) was studied by

E. Altmäe · A. Ruus (✉) · J. Raamets · E. Tungel
Tallinn University of Technology, School of Technology, Tartu College, 78, Puiestee, 51008 Tartu, Estonia
e-mail: aime.ruus@ttu.ee

D. Johansson et al. (eds.), *Cold Climate HVAC 2018*, Springer Proceedings in Energy, https://doi.org/10.1007/978-3-030-00662-4_80

several authors [1, 2], and divided into three descriptive levels: material, system and room level. Fluctuating variation of moisture buffering can be diurnal or seasonal, also differences between workdays and weekends can be significant. In real life attention is mostly paid to diurnal variations of moisture buffering like proposed by NORDTEST method [1].

Janssen [3] introduced the formula for $MBV_{practical}$ [g/(m^2 %RH)] calculations including moisture mass (max-min) in sample and range of RH levels. **Thermal effusivity** [3] of material is calculated as the square root of density, thermal conductivity and specific heat capacity, When introducing **moisture effusivity b_m** [kg/(m^{2}Pa · s½)] in a comparable way, it describes the ability of a material to absorb or release moisture. A daily hygroscopic inertia index, $I_{h,d}$, was introduced by Ramos [5, 6] as a function of MBV which takes ventilation and time into account. All the parameters described above need quite exact numbers for calculation.

McCregor [7] found out that soil selection (minerals and particle size distribution) is more important for the moisture buffering of clay product than changes can be made to a particular soil (density etc.). The clay mineral kaolinite is known as a mineral which absorbs most of water within the first minute and at temperature under 100° C, and it hardly gives up any water [8]. Montmorillonite (Smectite) is known as a mineral easily absorbing water and also its expansion is remarkable. But for example clays in Estonia do not seem to be rich in smectite [9], only illite-smectite and kaolinite can be found.

Schneider and Goss [10] introduced a revised pedotransfer function (PTF) enabling to predict (based on the clay content of the soil) sorption isotherms for dry soils. PTF seems to be suitable to soils containing more than 7% clay and with a clay fraction ratio of 2:1. Soils containing mostly kaolinite as the dominant clay mineral absorb less water than predicted and cannot be described by the log linear function proposed by Campbell and Shiozawa [11]. Arthur [12] studied the possibilities of prediction of clay content from water vapour sorption isotherms considering hysteresis and soil organic matter content. He presented regression relationships for determining clay content from water contents by using the relationship between clay and soil water content (3–93% RH).

The study presented focuses on the possible differences within the "same material" i.e. different products known by the common name "clay plaster". The main components of clay plaster are sand and clay—both of which are mineral resources with very different properties because of their chemical composition or particle size distribution.

2 Materials and Methods

The study focuses on the hygrothermal properties of clay plasters. Eight different clay-sand plaster (6 specimen each) mixtures were used (diameter 100 mm, thickness 25 mm). Mineral content and particle size distribution were estimated for all specimens. All the plaster mixtures tested were produced by manufacturers and the exact content of recipes was not available for study. Only water was added in the laboratory. Short description of mixtures is presented in Table 1. All materials are relevant for indoor use and usually have some surface coating and the influence of coatings must be studied, but separately. The study focuses on plasters but some preliminary data about coatings is gathered by authors already [13].

The consistence of fresh mortar was estimated by standard EVS-EN 1015-3:2004 + A2:2007 [14] and flow table Cooper TCM-0060/E was used. Hygroscopic sorption properties were studied at levels 30, 50 and 80% of RH following standard [15] procedure (in principle). Studied levels were found from literature [16, 17]. RH hardly exceeds value of 80% in living room, but often in bathroom and sometimes in kitchens too.

To describe the dynamics, the specimens were weighed at 1, 2, 3, 6, 12, 24 h during the first day and with a 24-h interval until stabilization [17] at every RH level. The standard [18] procedure was followed when determining plasters properties for water vapour permeability.

The same specimens (embedded into a plastic cylinder and tightened with plastic film and silicon hermetic) were used for both tests. Climate chamber method (RUMED 4101 RH = 20−95%, accuracy ±2 to 3%, temperature 0 to +60 °C, accuracy ±0.5 °C) was used.

Mineralogical composition was estimated at the Institute of Ecology and Earth Sciences of Tartu University by a professor of the Department of Geology Kalle Kirsimäe. X-ray diffraction method with XRD spectrometer Bruker D8 Advance was used. Particle size distribution was estimated by wet sieving analyses by standard EVS-EN1015-1:2004 [19]. The measurement of the 0.063 mm sieve enables only to separate the mixture of silt and clay as a total from sand particles. The calculation formula of fineness modulus for sand was used to describe plasters with one distinguishable number and make results easily comparable.

Table 1 Description and properties of clay plaster specimens

Clay plastr mixture	Specimen group	Fresh mortar specimen on flow table (cm)	Hardened mortar specimen		
			Dry weight (g)	Volume (cm^3)	Dry density (kg/m^3)
Red 0–2 mm cattail	I	19.0	374	200	1864
Brown 0–2 mm cattail	II	18.0	383	203	1892
Dark red 0–2 mm cattail	III	17.5	347	190	1830
Blue 0–2 mm cattail	IV	17.5	363	199	1826
Grey 0–2 mm cattail	V	17.8	374	206	1812
Dark red 0–2 mm	VI	16.5	319	176	1808
White 0–2 mm	VII	16.8	353	202	1748
Dark red 0-2 mm hemp shives	VIII	–	258	190	1357

3 Results

Moulds used are uniform by size. However—dry weight, volume and dry density of specimens have variance (Table 1) because of shrinkage (volume of groups III and VI) and/or additives (dry density of group VIII) (Fig. 1). All mixtures were prepared with a minimum water content indicated in the recipes as first choice. Mixture I received diameter of 19 mm on flow table with minimum water content. Mixture with hemp shives had not flowability. For other recipes some water was added. Shrinkage during the drying was very different and can be seen in photos (Fig. 1) being about 1 mm for white (VII) up to 6 mm for dark red (III, VI).

Sorption. The specimens with 0% of moisture content were placed to the climate chamber with air temperature 23 °C with RH of 30%. A problem occurred—after drying and during cooling there was some water uptake (0.17–0.35 g per specimen).

This has some influence on the results and desiccants must be used more effectively in the future. Moisture uptake rate kg/(m^{2}h) describes the speed behavior within the first hours and total moisture uptake (kg/m^2) describes water vapour gathered at RH = 0–30, RH = 30–50 and RH = 50-80%.

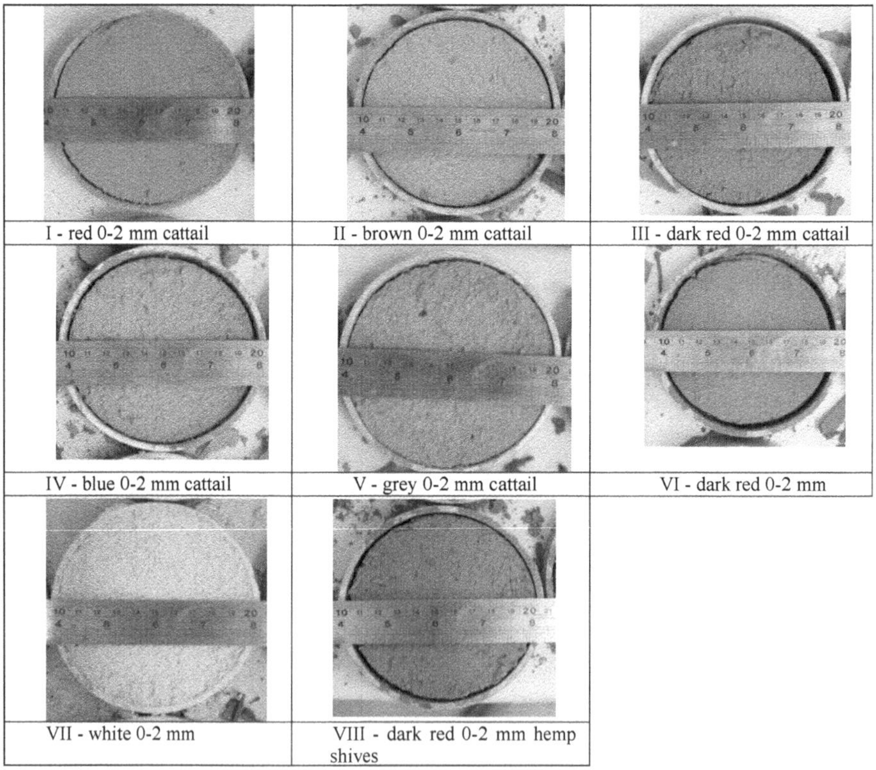

Fig. 1 Examples of specimen groups

Sorption range RH = 0–30%. The moisture uptake rate (Table 2) was highest during the first hour (6.0–14.8 g/(m^2h). Higher was rate for plaster groups I, III, VI, VIII [14.2–14.8 g/(m^2h)] being 5–20% of total moisture uptake.

For group I stabilization time was 48 h and for III, VI and VIII 120 h. The lowest rate was presented by group VII [6.0 g/(m^2h)] being 64% of total moisture uptake (9.4 g/m^2) and stabilizing at 24 h. It can be seen that plasters behave very differently.

The range of **RH = 30–50%** is presenting ventilated indoor living environment most exactly. (Table 3). The range of **RH = 50–80%** but could occur in rooms with high occupation and unsatisfying ventilation rate (Table 4). For both ranges it can be seen that during the first hour the most active is plaster group I and the least active group is VII. There are some negative values of adsorptions during the second and third hour. That is because environments in laboratory (RH = 22–27%) and climate chamber (RH = 50% or RH = 80%) are so different from each other and during weighing time some moisture is already released. On the other hand, it indicates the sensitivity of plasters. The problem must be solved in the future and test must be repeated in a more reliable way. Reliable data used in calculations are total moisture uptake within the given ranges. For desorption results are more adequate. In groups III, VI and VIII the total moisture uptake within the range (RH = 30–50%) was 147.8–164.8 and 250.8–298.9 g/m^2 (RH = 50–80%). In group VII the total moisture uptake within the range (RH = 30–50%) was 8.1 and 24.1 g/m^2 (RH = 50–80%). Results are similar at desorption (Tables 3 and 4). In real life probably 24 h cyclic fluctuation could be more adequate than total moisture uptake or release within 5–6 days.

Anyway, the tendency described above is seen already within the first hours (especially at desorption) and there are large differences between the plasters—up to 20 times (8.1 vs. 164.8 g/m^2 at RH = 30–50%). In Table 5 water uptake per specimens (g) and moisture content of plasters (%)—points on sorption curve are presented.

Diffusion. Difference from standard was thickness of air layer used in calculations—approx. 5 cm instead of 1 ± 0.5. That aspect was taken into account in calculations but as the material is highly water permeable it has some influence on results. The

Table 2 Sorption at RH = 0–30%.

	Sorption								
	1 h	2 h	3 h	120 h		1 h	2 h	3 h	120 h
	g/(m^2h)	g/(m^2h)/g/m^2	g/(m^2h)/g/m^2	g/m^2		g/(m^2h)	g/(m^2h)/g/m^2	g/(m^2h)/g/m^2	g/m^2
I	14.4	7.5/21.7	4.3/25.9	71.9	V	11.8	3.0/14.8	3.6/18.4	89.7
II	8.6	5.0/13.4	4.1/17.5	114.2	VI	14.7	8.8/23.5	6.8/30.3	301.1
III	14.8	6.6/21.4	6.2/27.6	283.7	VII	6.0	2.2/8.2	−1.7/6.5	9.4
IV	7.5	3.6/11.1	4.2/15.3	82.1	VIII	14.2	7.9/22.1	6.9/29.0	269.6

Moisture uptake rate g/(m^2h) and total moisture uptake g/m^2

Table 3 Sorption and desorption at RH = 30–50%. Moisture uptake/release rate g/(m^{2}h) and total moisture uptake/release g/m^2

	Sorption				Desorption			
	1 h	2 h	3 h	120 h	1 h	2 h	3 h	120 h
	g/(m^{2}h)	g/(m^{2}h)/ g/m^2	g/(m^{2}h)/ g/m^2	g/m^2	g/(m^{2}h)	g/(m^{2}h)/ g/m^2	g/(m^{2}h)/ g/m^2	g/m^2
I	9.4	−1.2/8.2	4.1/12.3	45.3	−8.3	−6.3/−14.5	−6.3/−20.8	−53.3
II	4.7	0.5/5.2	1.6/6.8	63.1	−6.5	−4.4/−10.9	−4.0/−14.9	−71.3
III	7.8	3.3/11.1	1.2/12.3	147.8	−11.9	−7.0/−18.9	−4.3/−23.2	−148.2
IV	5.5	0.8/6.3	0.9/7.2	69.5	−6.6	−4.3/−10.9	−2.8/−13.7	−78.0
V	3.9	−0.3/3.6	1.9/5.5	55.7	−6.7	−4.8/−11.5	−2.8/−14.3	−65.1
VI	6.5	7.4/13.9	−0.8/13.1	164.8	−15.3	−6.8/−22.1	−5.9/−28.0	−165.1
VII	3.4	−1.2/2.2	0/2.2	8.1	−3.5	−2.3/−5.8	−1.9/−7.7	−13.7
VIII	7.4	4.3/11.7	0.8/12.5	164.1	−12.7	−6.3/−19.0	−5.1/−24.1	−161.6

Table 4 Sorption at RH = 50–80%. Moisture uptake/release rate g/(m^{2}h) and total moisture uptake/release g/m^2

	Sorption				Desorption			
	1 h	2 h	3 h	120 h	1 h	2 h	3 h	120 h
	g/(m^{2}h)	g/(m^{2}h)/ g/m^2	g/(m^{2}h)/ g/m^2	g/m^2	g/(m^{2}h)	g/(m^{2}h)/ g/m^2	g/(m^{2}h)/ g/m^2	g/m^2
I	9.6	−14.8/−5.2	8.1/2.9	110.7	−23.3	−25.8/−49.1	−4.7/−53.8	−116.2
II	4.7	−3.4/1.3	−0.6/0.7	146.7	−14.5	−11.6/−26.1	−7.3/−33.4	−128.4
III	4.9	0.1/5.0	−6.4/−1.4	250.8	−30.6	−12.1/−43.7	−14.7/−58.4	−209.6
IV	5.0	−1.8/3.2	2.9/1.1	187.3	−26.3	−10.0/−36.3	−9/−45.3	−154.0
V	4.6	−5.7/−1.1	−0.7/−1.8	145.6	−15.6	−13.5/−29.1	−7.9/−37.0	−135.8
VI	5.5	6.9/12.4	−12.8/−0.4	279.0	−30	−11.6/−41.6	−22.3/−63.9	−228.4
VII	2.9	−11.4/−8.5	5.8/0.4	24.1	−8.8	−7.3/−16.1	−1.6/−17.7	−24.2
VIII	5.3	3.7/9.0	8.6/0.4	298.9	−29.9	−11.0/−40.9	17.4/−58.3	−213.8

water vapour permeance of 2.5 cm layer varied between 1680 and 3076 $\times$ 10^{-9} kg/(m^{2}sPa) and water vapour permeability was from 44 to 76 $\times$ 10^{-12} kg/(msPa). Water vapour resistance factor μ was found to be μ = 2.7–4.6. Water vapour diffusion equivalent air layer thickness S_d is presented in Table 7. *T*-test analysis was performed to study water vapour resistance. It proved that group III (r = 0.33 10^9 m^2 s Pa/kg) is significantly different from all other groups. For group VI p equalled 0.016. All other values were smaller than that. Groups VI and VIII were also statistically different from other groups (p = 0.014).

Mineralogical composition including sand and additives (e.g. cattail) in clay plaster mixture was estimated The results are presented in Table 6 where “tr” means that the content of the mineral is less than accuracy.

Table 5 Water uptake per specimen (g) and moisture content of plasters (%)

	RH = 0–30% (120 h)			RH = 30–50% (120 h)			RH = 50–80% (120 h)		
	Water uptake		Moisture content RH = 30%	Water uptake		Moisture content RH = 50%	Water uptake		Moisture content RH = 80%
	Range	Aver.		Range	Aver.		Range	Aver.	
	g	g	%	g	g	%	g	g	%
I	0.51–0.68	0.58	0.16	0.35–0.38	0.36	0.25	0.82–0.95	0.89	0.49
II	0.85–0.95	0.92	0.24	0.47–0.54	0.51	0.37	1.11–1.21	1.18	0.68
III	1.96–2.32	2.19	0.63	1.07–1.23	1.14	0.96	1.85–2.05	1.93	1.52
IV	0.61–0.67	0.65	0.18	0.54–0.57	0.55	0.33	1.44–1.54	1.48	0.74
V	0.54–0.91	0.72	0.19	0.42–0.48	0.45	0.31	1.12–1.21	1.17	0.63
VI	2.17–2.32	2.24	0.70	1.19–1.31	1.23	1.09	2.02–2.14	2.08	1.74
VII	0.02–0.11	0.08	0.02	−0.02–0.1	0.07	0.04	0.13–0.23	0.20	0.10
VIII	2.07–2.17	2.13	0.83	1.22–1.38	1.30	1.33	2.32–2.42	2.36	2.24

Table 6 Mineralogical composition of plaster mixture, % of mass

Component	I	II	III	IV	V	VI	VII	VIII
Quartz	54.7	58.7	49.1	52.8	54.7	45.6	55.8	49.8
K-feldspar	9.8	8.4	9.6	7.3	9.8	6.6	10.5	16.2
Plagioclase	9.5	7.5	7.8	8.8	9.5	7.9	5.6	9.1
Chlorite	0.6	1.0	1.0	0.8	Tr	1.5		0.9
Ill./Ill-smect.	13.0	12.0	14.0	16.4	13.0	20.9	8.0	8.0
Kaolinite	5.4	2.8	4.2	3.8	5.4	4.1	15.7	3.6
Calcite	4.2	5.2	8.6	5.7	4.2	8.5		7.2
Dolomite	2.2	3.6	4.2	4.2	2.2	4.0		3.5
Hematite			Tr			0.5		0.5
Amphibole	0.6	0.8	0.9		0.6	0.5	0.9	1.2
Cristobalite							1.9	
Anatase							tr	

Clay mixture VII which had the highest kaolinite content absorbed less water than others—according to data known from literature (Fig. 2a). On the other hand, it can be seen that other mixtures which have similar kaolinite content do not absorb water the same way and no strong correlation can be found $R^2 = 0.42$. From literature data it is known that smectite has a good ability to absorb water.

Table 7 Summarized data collected

Specimen	Specimen moisture content g				S_d m	Kaolinite (%)	Calcite (%)	FM
	0–30%	30–50%	50–80%	0–80%				
I	0.58	0.36	0.89	1.83	0.10	5.47	4.2	1.5
II	0.92	0.51	1.18	2.61	0.11	2.8	5.2	1.6
III	2.19	1.14	1.93	5.26	0.07	4.2	8.6	1.7
IV	0.65	0.55	1.48	2.68	0.12	3.8	5.7	1.7
V	0.72	0.45	1.17	2.34	0.12	5.4	4.2	1.6
VI	2.24	1.23	2.08	5.55	0.08	4.1	8.5	1.2
VII	0.08	0.07	0.20	0.35	0.12	15.7	0	2.9
VIII	2.13	1.29	2.36	5.78	0.08	3.6	7.2	1.2

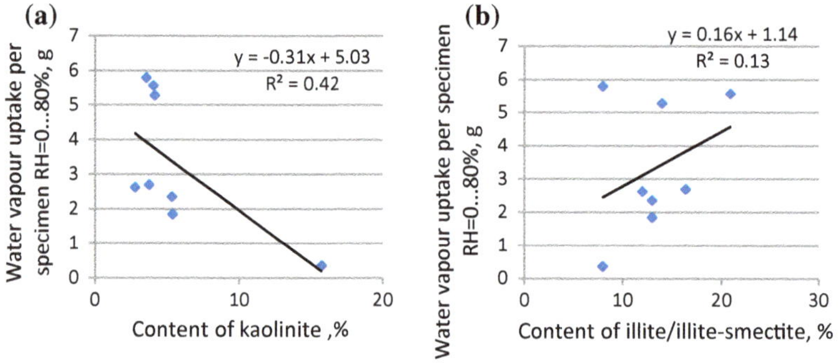

Fig. 2 Water vapour uptake g per specimen depending on the content of kaolinite (**a**) and content of illite/illite-smectite (**b**)

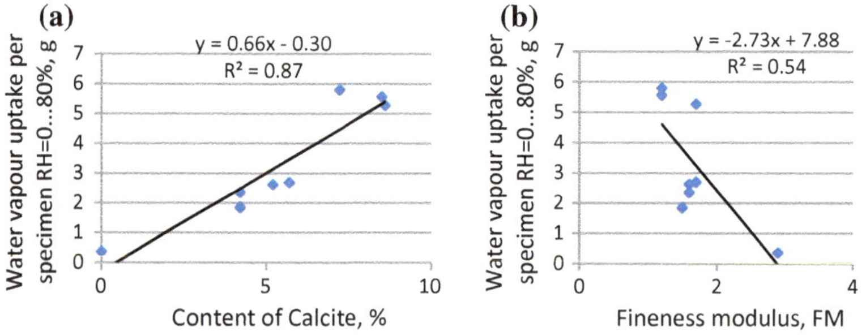

Fig. 3 Water vapour uptake g per specimen depending on content of calcite (**a**) and FM (**b**)

In mixture under investigation there was no smectite separately present. No correlation $R^2 = 0.12$ (Fig. 2b) can be found between illite/illite-smectite content and water vapour uptake. Strong correlation $R^2 = 0.87$ (Fig. 3a) was found—between content of calcite in plaster and water vapour uptake ability.

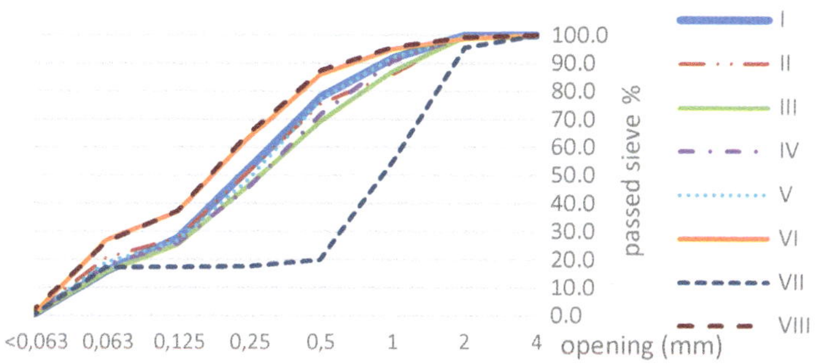

Fig. 4 Particle size distribution curves

Particle size distribution was estimated by wet sieving analyses. Curves can be seen in Fig. 4 where Group VII (FM = 2.9) again has differences with others (FM = 1.2–1.7). In group VII bout ¾ of the material has particle size 0.5–2 mm. In other groups about ½ of material has particle size 0.125–0.5 mm. There was no strong correlation between fineness modulus and water vapour uptake—$R^2 = 0.54$ (Fig. 3b), but it is clearly seen that the mixture (VII) which has bigger particle size and therefore less specific surface area has the smallest water adsorption ability.

4 Discussion and Conclusions

All the results are summarised in Table 7. Group VII (white clay) differentiates from others in most aspects except water vapour permeability. There kaolinite and FM = 2.9 (e.g. coarse sand) content is the largest. There is no calcite in the clay and almost no ability to absorb water vapour compared to other clay plaster mixtures (0.35 g vs. 1.8–5.8 g per specimen). All other groups having FM = 1.2–1.7 could be compared to fine sand. According to literature smectite is well known because of its good moisture buffering ability. There was no pure smectite in the plasters studied.

By sorption properties clay mixtures can be divided into three groups. The largest amount of water vapour was absorbed by specimen groups III, VI and VIII (>5 g per specimen) all basing on the same dark red clay. There calcite content is 7.2 to 8.6%—clearly bigger than in others and S_d = 0.07–0.08 m vs. S_d = 0.1–0.12 m.

No other positive or negative correlation was found. Similar (medium) sorption properties (1.83–2.68 g per specimen) presented specimen groups I, II, IV and V basing on clays from different sources (brown, grey, blue, red). There calcite content is 4.2–5.7%—clearly smaller than in specimen groups III, VI and VIII.

Moisture uptake rate properties show that group I is very active (Tables 2, 3 and 4) and has almost largest initial moisture uptake rate (1 h). Stabilization time is about 48 h. Group VII has the lowest initial moisture uptake rate and the smallest stabilisation time (6–24 h). Groups III, VI and VIII have the longest stabilization period. Activity within the first hour depends on the range of RH.

Moisture buffering ability of material is strongly related with sorption properties. By using formula introduced by Janssen [3] it can be calculated from data presented that MBV_{8h} at RH = 30–50% range is 0.3 (VII) to 1.9 (VI, VIII) and at RH = 50–80% range is 0.6 (VII) to 2.3 (VI, VIII) g/(m^2 %RH)@8/16 h). According to $MBV_{prcatical}$ classification introduced by Rode [4] different clay plasters could be estimated as "limited" ($MBV_{practical}$ = 0.2–0.5 g/(m^2 %RH)@8/16 h), "moderate" (0.5–1), "good" (1–2) or even "excellent" (2–3.5 g/(m^2 %RH)@8/16 h)).

By water vapour permeability plasters can be divided into two groups, where differences are not as clear as at sorption, but tendency is the same. Groups III, VI and VIII have higher water vapour uptake/release and water vapour permeability.

The study presented proved that there are differences within the "same material" i.e. different products known by the common name "clay plaster". The main components of clay plaster are sand and clay—both of which are mineral resources with very different properties because of their chemical composition or particle size distribution.

The clearest correlation was found between the calcite content of plaster mixture and water vapour uptake. The properties of products must be monitored carefully. If a plaster is included into hygrothermal calculations of boarders or ventilation rate, the properties must be declared and the user has to be aware of them. Also, the user who just wants to use plaster with good moisture buffering ability must be supplied with relevant information if the product does not have commonly expected properties.

Acknowledgements This study was supported by Tartu College of TUT. Special thanks to Marko Kikas from Saviukumaja and Prof. Kalle Kirsimäe from Tartu University.

References

1. C. Rode, Moisture buffering of building materials, BYG DTU-126 Report, Department of Civil Engineering, Technical University of Denmark, 2005, http://www.byg.dtu.dk/upload/institutter/byg/publications/rapporter/byg-r126.pdf
2. K. Svennberg, Moisture Buffering in the Indoor Environment. Thesis Building Physics LTH Lund University, 2006
3. H. Janssen, A. Roels, Qualitative and quantitative assessment of interior moisture buffering by enclosures. Energy Build. **41**(4), 382–394 (2009)
4. C. Rode, R. Peuhkuri, B. Time, K. Svennberg, T. Ojanen, Moisture buffer value of building materials. J. ASTM Int. **4**(5), 1–12 (2007)
5. H.M.M. Ramos, J.M.P.Q. Delgado, V.P. de Freitas, Influence of finishing coatings on hygroscopic moisture buffering in building elements. Constr. Build. Mater. **24**, 2590–2597 (2010)
6. N.M.M. Ramos, V.P. Freitas, Laboratory testing for daily hygroscopic inertia assessment, in *Proceedings of the 8th Symposium on Building Physics in the Nordic countries*, Copenhagen, Denmark (2008), pp. 809–816

7. F. McGregor, A. Heath, A. Shea, M. Lawrence, The moisture capacity of unfired clay masonry. Build. Environ. **82**, 599–607 (2014)
8. W.A. White, E. Pichler, *Water Sorption Characteristics of Clay Minerals*, USA (1959)
9. M. Palolill, Looduslike savide ja savipinnaste geotehniliste omaduste sõltuvus saviminer-aalsest koostisest: ekvivalentse basaalse distantsi kontseptsioon. Master Thesis, Tartu University Tartu, 2007, (In Estonian) http://dspace.ut.ee/bitstream/handle/10062/2752/palolill_margus.pdf?sequence=1
10. M. Schneider, K.-U. Goss, Prediction of the water sorption isotherm in air dry soils. Geoderma **170**, 64–69 (2012)
11. G.S. Campbell, S. Shiozawa, Prediction of hydraulic properties of soils using particle size distribution and bulk density data, in *International Workshop on Indirect Methods for Estimating the Hydraulic Properties of Unsaturated Soils*, University of California Press, Berkely (1992) referred by Scheider
12. E. Arthur, M. Tuller, C. Moldrup, D.K. Jens, L.W. De Jong, Prediction of clay content from water vapour sorption isotherms considering hysteresis and soil organic matter Content. Eur. J. Soil Sci. **66**, 206–217 (2015)
13. O. Vares, A. Ruus, J. Raamets, E. Tungel, Determination of hygrothermal performance of clay-sand plaster: influence of covering on sorption and water vapour permeability. Energy Procedia **132**, 267–272 (2017)
14. EVS-EN 1015-3:2004 + A2:2007, *Methods of Test Mortar for Masonry Part 3: Determination of Consistence of Fresh mortar by Flow Table*
15. EVS-EN ISO 12571, *Hygrothermal Performance of Building Materials and Products—Determination of Hygroscopic Sorption Properties*
16. G. Minke, *Building with Earth—Design and Technology of a Sustainable Architecture* (Birkhäuser, Basel, 2012)
17. M. Maddison, T. Mauring, K. Kirsimäe, M. Mander, The humidity buffer capacity of clay-sand plaster filled with phythomass from treatment wetland. Build. Environ. **44**, 1864–1868 (2010)
18. EVS-EN 1015-19:2005, *Methods of Test for Mortar for Masonry—Part 19: Determination of Water Vapour Permeability of Hardened Rendering and Plastering Mortars*
19. EVS-EN1015-1:2004, *Methods of Test for Mortar for Masonry. Part 1: Determination of Particle Size Distribution (by sieve analyses)*

Hygrothermal Performance of Timber External Walls Insulated with Natural and Industrial Materials

Martti-Jaan Miljan and Jaan Miljan

Abstract The test house with special timber structure was built by the department of rural building of Estonian University of Life Sciences. The hygrothermal properties of sawdust and cellulose, used as insulation materials in building envelope, were investigated in long term. The test was carried out during four winters. The wind and moisture-vapor barriers were not applied in the first testing period. In the second testing period the wall was covered inside with the moisture barrier and outside with the wind barrier and the layer of OSB plates. In wintertime the test house was heated with electric batteries to keep the indoors temperature stable. The aim of this research was to measure and determine the thermal transmittance of differently insulated wall sections. To fulfil the task the devices to measure relative moisture and temperature were placed in five different positions (in three depths inside the insulation) and indoors and outdoors. The heat flux plates were applied on the inner surface of the wall. Data were recorded with 10 min interval. The analysis of these test periods was carried out and presented in this paper.

Keywords Thermal insulation · Thermal transmittance · Hygrothermal performance · Sawdust · Cellulose insulation

1 Introduction

The theme became actual because a lot of envelopes (walls and ceilings) of old residential houses built before the II WW and even in 1950–60-ies were insulated with saw dust and the hypothesis was posed either to change the old saw dust with new saw dust or it will be better to change old insulation against the new material come to the markets—loose-placed cellulose. First of all we wanted to investigate and compare the hygro-thermal behavior of those two materials during longer

M.-J. Miljan (✉) · J. Miljan
Estonian University of Life Sciences, Kreutzwaldi 5, 51014 Tartu, Estonia
e-mail: martti-jaan.miljan@emu.ee

D. Johansson et al. (eds.), *Cold Climate HVAC 2018*,
Springer Proceedings in Energy, https://doi.org/10.1007/978-3-030-00662-4_81

exploitation. Saw dust is waste from sawn timber production and loose cellulose is done from recycled old newspapers, so both materials are renewable and sustainable ones.

About sawdust [1] it was said that the material has big settlements during time and due to cold bridges which will form into the upper part of walls the thermal transmittance will increase. Our experience in opening old walls built 50–60 years ago [2] and insulated with sawdust show that if the work has done correctly there were no gaps in the upper parts of walls. The sawdust was standing in the wall as if it was a monolith. So the idea of using it was worth of testing.

The second tested material was new in Estonian market—dry blown cellulose [1], nowadays quite a widespread and offered by several producers in Estonia.

1.1 The Structure of Test House and Used Materials

In 2008 the test house (Fig. 1) was built by the department of Rural Building in Estonian University of Life Sciences. The plan dimensions of the test house were 4.0 × 6.0 m. The walls of the test house were composed from timber sections where inner and outer wallboards (board thickness 32 mm) were connected with veneer boards and between them mounted the insulation material.

Get sections were 600 mm wide and the thickness of insulation layer inside was 227 mm. The floor was insulated (187 mm) and covered from inside with 28 mm boards and below some kind of timber board plates like OSB3 (8 mm) or RKL (30 mm) were applied. The insulation layer on ceiling (300 mm) was covered from attic side with paper and in room it was covered with OSB3 boards (8 mm). As you see from Fig. 2 half of the wall was insulated with saw dust and another half with cellulose, so the floor and ceiling were insulated accordingly. Applied materials

Fig. 1 Timber test house to investigate different insulation materials

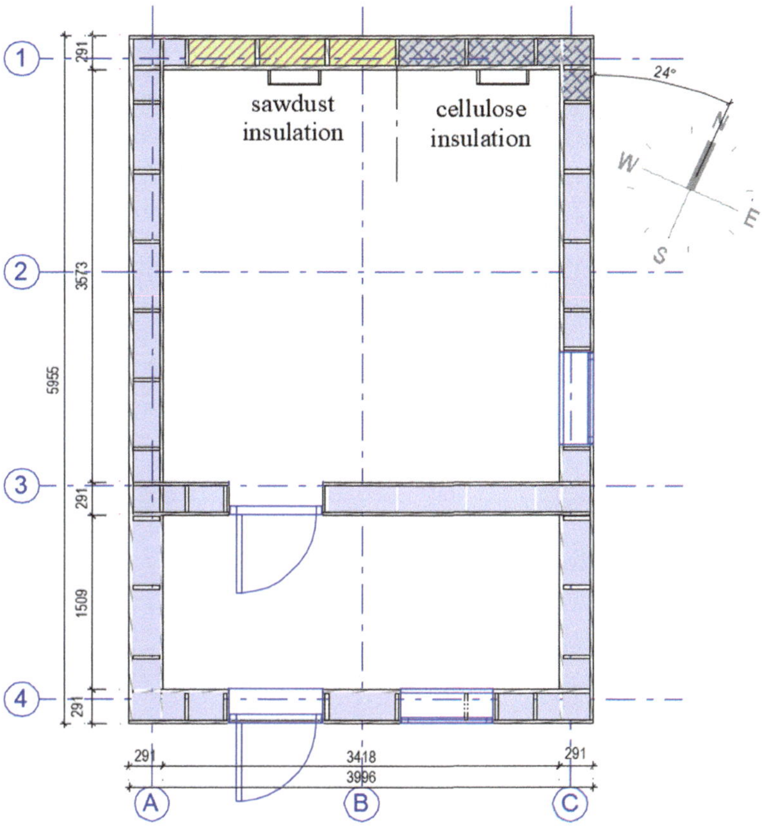

Fig. 2 The plan of the test house and the placement of elements insulated with sawdust and cellulose

were also tested beforehand in climate chamber test to get general imagination about their thermal conductivity. Our measurements showed that the thermal conductivity of sawdust was $\lambda = 0.044$ W/(mK) and of cellulose was $\lambda = 0.062$ W/(mK) [3]. These two wall sections for longtime testing were situated on the northward of the house. During winter the test house was heated by 2 kW electric battery and in very cold weather another 2 kW battery was added. After 1.5 years of exploitation (in autumn 2010) the water vapor barrier ($S_d = 7.5$ m) was applied inside and wind barrier ($S_d = 0.05$ m) outside of the walls to protect the structure from precipitation. Both wall structures are presented on Fig. 3.

Sawdust was bought from sawmill beside town. Cellulose was (Fig. 2) applied using dry blown method.

Windows were from 3-glassed packets and roof was covered with steel.

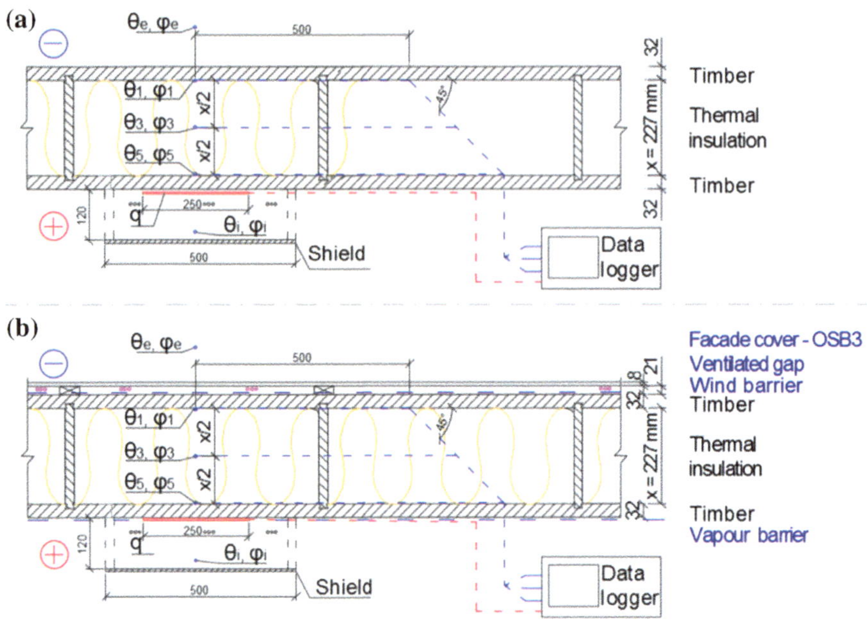

Fig. 3 **a** The wall structure in the first period of testing; **b** The wall structure in the second period of testing

2 The Aim and Description of the Test

Long-time measurements began on October 2008 and finished in February 2012. The thermal transmittance, temperature and relative humidity of the wall panels were measured.

The measurements were done in non-steady state, i.e. the temperatures and RH changed continuously in time and hence the heat flow through the wall varied. Temperatures were measured parallel to the heat flow in five places. The cross section of the tested walls are presented on Fig. 3 and all readings were recorded with a 10-min interval in Almemo data logger and Hobo data loggers.

The following characteristics were measured to determine the thermal transmittance of the wall.

q—heat flow rate through the wall (heat flow plate Ahlborn FQA019C) [W/m^2];
θ_e, φ_e—external air temperature and external air relative humidity (Temp & RH data logger Hobo U12-011) [°C, RH%];
θ_1, φ_1—the temperature and the relative humidity of the outer surface of the insulation layer (Temp & RH sensors Ahlborn FHA646R) [°C, RH%];
θ_3, φ_3—the temperature and the relative humidity in the middle of the insulation layer (Temp & RH sensors Ahlborn FHA646R) [°C, RH%];

θ_5, φ_5—the temperature and the relative humidity of the inner surface of the insulation layer (Temp & RH sensors Ahlborn FHA646R) [°C, RH%];
θ_i, φ_i—internal air temperature and relative humidity (Temp & RH data logger Hobo U12-011) [°C, RH%].

In the first period of testing the walls of the test house were without wind barrier outdoors and vapor barrier indoors. In summer of 2010 these protective layers were applied with wind gap 20 mm and 8 mm OSB plate to protect the wall's surface from outside. All measurement devices stayed in former places, but devices to measure internal and external temperature and relative humidity were removed.

3 Calculations

Thermal transmittance. Thermal transmittance calculations are done by Formula 1 given in standard ISO 9869-1 [4]. All data is taken from measurement devices.

$$U = \frac{\sum_{j=1}^{n} q_j}{\sum_{j=1}^{n} \left(\theta_{i.j} - \theta_{e.j}\right)}, \quad (1)$$

where

U thermal transmittance of the wall [W/(m^2 K)];
q heat flow rate through the wall [W/m^2];
θ_i internal air temperature [°K];
θ_e external air temperature [°K]

Hygrothermal behavior of tested wall. Calculations of hygrothermal performance have been done using simplified method presented in standard EN ISO 13788 [5, 6]. External and internal temperature and relative humidity are really measured values (θ_e, θ_i, φ_e, φ_i) see Figs. 4 and 5. Values presented in Table 1 are taken from different literature sources. Index d1 thermal conductivity and water vapor diffusion resistance factor values is taken from EN-ISO 10456 [7] standard or from our researches [3]. d2 values is from the research work of prof J. Vinha [8] Graphs describing hygrothermal performance of walls are presented on Figs. 8 and 9.

4 Results

Data recorded with 10 min interval were recalculated to mean values of a day to follow the graphs better.

External and internal measurements. Temperatures measured in- and outdoors of the test house are presented on Fig. 4.

From Fig. 4 we can see that the winter 2008/2009 was the warmest and 2010 summer was considerably warmer than other summers. Data of measuring relative humidity at that time is presented on Fig. 5.

From Fig. 5 we can see that relative humidity during winter months is low because additional supply of water was missing. The aim was also to watch does the insulation dries out.

Measured relative humidity on the inner and outer surfaces of the thermal insulation layers. Changes of relative humidity during 3 years of the walls insulated with saw dust are presented on Fig. 6. Measurements were done on the inner and outer surfaces of insulation layer.

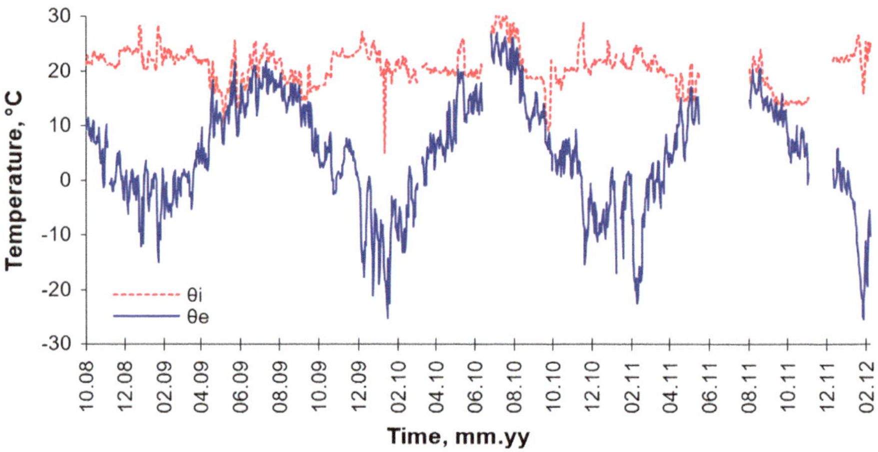

Fig. 4 Temperature (°C) indoors (θ_i) and outdoors (θ_e) during longtime test (daily means)

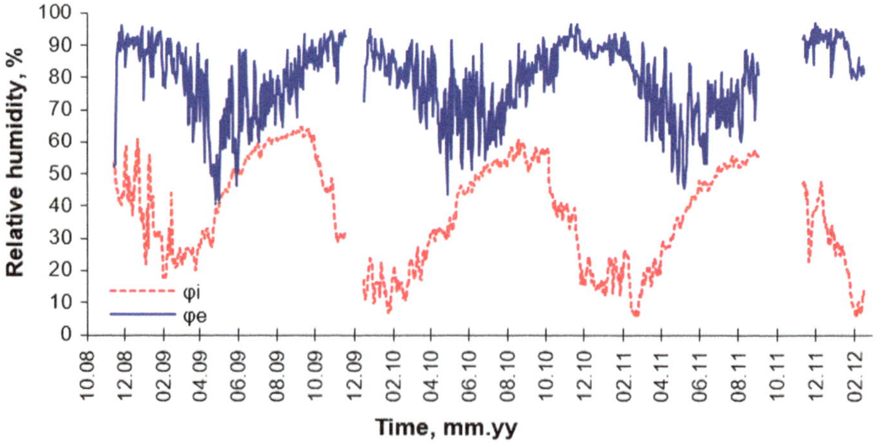

Fig. 5 Relative humidity (%) indoors (φ_i) and outdoors (φ_e) during longtime test (daily means)

Table 1 Physical properties of materials by different sources [3, 7, 8]

Material	Calculation index	Density	Thermal conductivity	Water vapor diffusion resistance
		ρ (kg/m³)	λ [W/(K*m²)]	factor μ
Timber (softwood)	.d1	450	0.12 [7]	−50 [7]
	.d2	530	0.12 [8]	−49 [8]
Sawdust	.d1	197	0.062 [3]	−2[a]
	.d2	168	0.08 [8]	−2.1 [8]
Cellulose (loose)	.d1	50	0.045 [3]	−2 [7]
	.d2	60	0.041 [8]	−1.3 [8]

[a]In calculating water vapor diffusion resistance factor of sawdust and cellulose are taken as equal

From Fig. 6 we see that condensation period between outer board and sawdust insulation layer lasted until the middle of July 2009. Dry out of inner surface of the insulation layer began later, on the end of September. In the middle of 2010 additional protective barriers were applied on the wall and dry out process improved and RH of the wall stabilized. There was no condensation threat and RH decreased on the outer surface of saw-dust layer from 90 to 80%. Humidity content on the inner surface of sawdust layer decreases also due to dry out process. During later demolition we didn't notice any traces of mould on the inner surface of the wall boards. The temperature on the outer surface of the insulation layer was low. Looking the picture describing ambient temperature changes, we see that only in summer for a very short time the ambient temperature is high. Data of insulation layer's outer surface showed that the temperature there rose over 20 °C for some days only, too short time for mould formation.

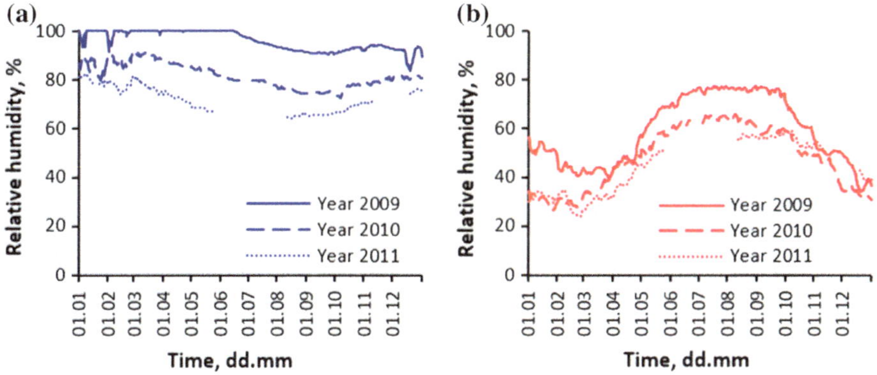

Fig. 6 Relative humidity on the outer surface (**a**) and on the inner surface (**b**) of sawdust insulation layer at points (φ_1 and φ_5 Fig. 3.) in 2009, 2010 and 2011

Changes of relative moisture during 3 years of the wall insulated with cellulose are presented on Fig. 7. Measurements were done on the inner and outer surfaces of the insulation layer.

On Fig. 8 we see that outer surface of cellulose insulation dried out during all three years. After checking and replacing the measuring device on March 2011, 28% change was observed at point φ_1 (look line E1). After applying additional protective layers on the wall in the summer 2010 the relative humidity stabilized.

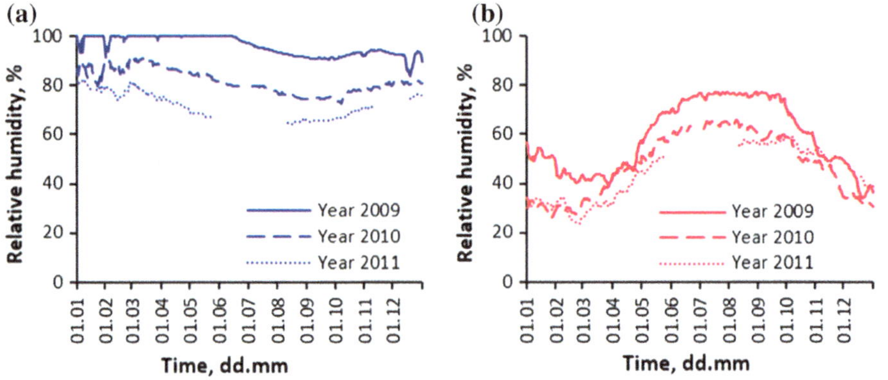

Fig. 7 Relative humidity on the outer surface (**a**) and the inner surface (**b**) of the cellulose insulation layer at points (φ1 and φ_5 Fig. 3) in 2009, 2010 and 2011

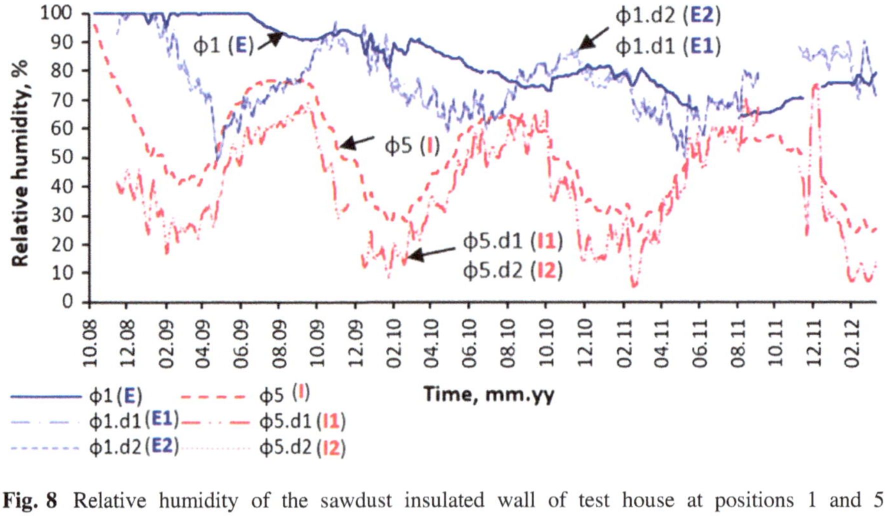

Fig. 8 Relative humidity of the sawdust insulated wall of test house at positions 1 and 5 (placement of measure devices on Fig. 3). Graphs φ_1 (E) and φ_5 (I) are measured, graphs $\varphi_{1.d1}$ (E1), $\varphi_{1.d2}$ (E2) and $\varphi_{5.d1}$ (I1), $\varphi_{5.d2}$ (I2) are calculated results

Comparison of calculated and measured relative humidity on the inner and outer surfaces of the thermal insulation layer. Explanation for graphs on Figs. 8 and 10: lines E and I present changes of the measured temperature and relative humidity accordingly on the outer and inner surface of the thermal insulation layer. Lines E1 and E2 as well as I1 and I2 present calculated results.

Calculations of hygrothermal performance have been done using simplified method presented in standard EN ISO 13788 [5, 6]. External and internal temperature and relative humidity content are really measured values (θ_e, θ_i, φ_e, φ_i) see to Figs. 4 and 5. Values presented in Table 1 are taken from different literature sources and thicknesses of separate layers taken from Fig. 3.

On Fig. 8 we see that measured relative humidity on the outer surface of sawdust thermal insulation differed from calculated results considerably. Looking changes of relative humidity on inner surface of saw dust layer we see that calculated and measured graphs are at least parallel. After placement of additional protective layer in summer 2010 those graphs are a bit more close to each other.

On Fig. 9 we see that measured and calculated results of relative humidity on the outer surface of cellulose thermal insulation are also considerably different. Looking changes of RH on inner surface of the cellulose layer we see that calculated and measured graphs are almost coinciding. The reason of great differences between calculated and measured results notable on outer surface of both insulation material layers might be caused because EN ISO 13788 does not consider the moisture absorption capacity of insulation materials and the moisture dry out is delayed. Although this calculation method is correct in steady state conditions.

Thermal transmittance of the walls. Thermal transmittance of tested walls is presented on Fig. 10.

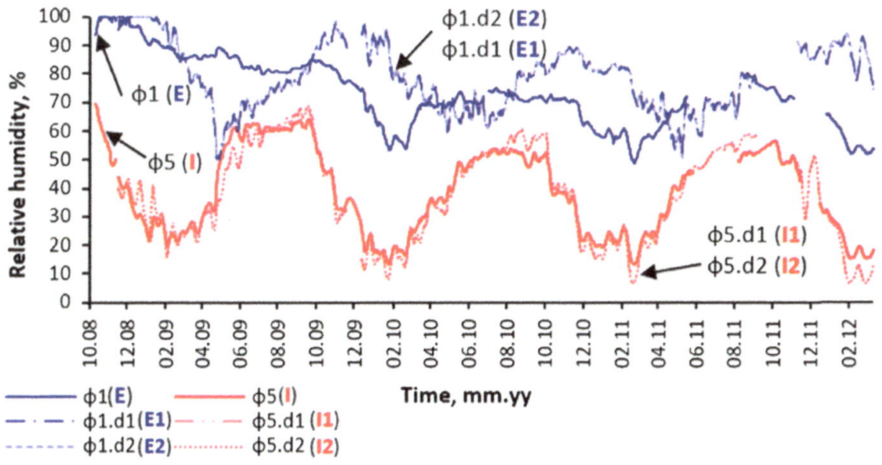

Fig. 9 Relative humidity in the cellulose insulated wall at positions 1 and 5 (placement of measure devices on Fig. 3). Graphs φ_1 (E) and φ_5 (I) are measured, graphs $\varphi_{1.d1}$ (E1), $\varphi_{1.d2}$ (E2), and $\varphi_{5.d1}$ (I1), $\varphi_{5.d2}$ (I2) are calculated results

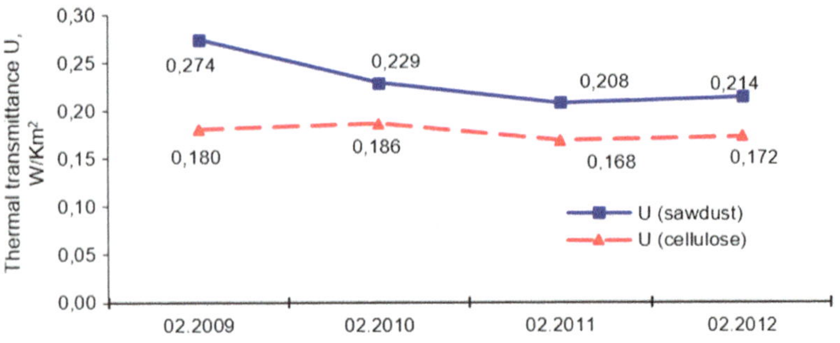

Fig. 10 Thermal transmittance of the walls on February during four years testing

In the first winter of exploitation, on February, the thermal transmittance of sawdust insulated wall was influenced by the water content in the insulation. Dry out process of the material continued during next years and thermal transmittance of this wall element improved about 10% after the mounting of protective layers. Thermal transmittance of cellulose didn´t change much. As the wall was during the first exploitation year opened to precipitations thermal transmittance of cellulose was worst—0.186 W/(m^2 K) on February of 2010. After placing additional protective layers on the wall the situation improved. Comparing calculations of thermal transmittance of sawdust and cellulose using measured heat flow rate on the object (see results on Fig. 10) and calculations [9] using values from Table 1, we see that thermal transmittance of sawdust is 0.229 W/(m^2 K) using.d1 and 0.282 W/(m^2 K) using.d2 and thermal transmittance of cellulose is 0.174 W/(m^2 K) using .d1 and 0.160 W/(m^2 K) using .d2. Calculations done using d1 values from Table 1 matches better with the results which were measured on the test house using heat flow plate.

5 Summary

Comparing walls insulated with sawdust and cellulose we see that calculated and measured results on the inner surface of the insulation material are comparable. On outer surface of the insulation material measured and calculated results are partly reversed.

One reason is that simplified calculation method [5] will not give adequate results. Especially with sawdust. The water content (kg/m^3) of the sawdust is 3–5 times greater than of the cellulose on the same level of the relative humidity [8]. Determining the thermal transmittance of walls we noticed that thermal conductance of cellulose didn't change notably as the thermal transmittance of the wall didn't change much during test. Thermal conductivity of sawdust evolved in the wall during test, as the thermal transmittance of this wall element improved from 0.274 W/(m^2 K) to 0.208 W/(m^2 K). We need to mention that indoors relative moisture depended from the ambient moisture content. Required thermal transmittance of the outer walls in Estonia is 0.12–0.22 W/(m^2 K) [10], so both tested walls answered the energy performance requirements set in Estonia. As a result we should admit that using sawdust is still problematic, because the material is not homogeneous and so the hygrothermal properties of it vary a lot.

Acknowledgements The building of the test house was supported by Ministry of Education and Research as a part of the research "Assessment, analysis and modelling of environmental sustainability of buildings life cycle" (2008–2012).

This article was supported by the Estonian Center of Excellence in Zero Energy and Resource Efficient Smart Buildings and Districts, ZEBE, (2014–2020) funded by the European Regional Development Fund.

References

1. T. Kalamees, Alternatiivseid soojustusmaterjale puitkarkasshoonetele. Ehitaja **7–8**, 71–72 (2002)
2. B. Berge, *The ecology of building materials* (Architectural Press, Oxford, 2000)
3. M.-J. Miljan, J. Miljan, Soojustusmaterjalide omaduste määramine maaehituse osakonna ehitusfüüsika laboris (Determination of thermal properties of insulation materials in the laboratory of building structures), in *Kohalikud looduslikud ehitusmaterjalid ja nende kasutamine (Local Natural Building Materials and Their Use)*, ed. by J. Miljan (Compendium of scientific studies of Estonian University of Life Sciences, EULS, Tartu, (in Estonian), 2012), pp. 27–34
4. ISO 9869-1:2014, *Thermal Insulation—Building Elements—In-situ Measurement of Thermal Resistance and Thermal Transmittance—Part 1: Heat Flow Meter Method* (2014)
5. EN ISO 13788:2012, *Hygrothermal Performance of Building Components and Building Elements—Internal Surface Temperature to Avoid Critical Surface Humidity and Interstitial Condensation—Calculation Methods* (ISO13788:2012), Brussels (2012)
6. EVS 908-1:2016, *Guidance for Calculation of Thermal Transmittance of Building Envelope Part 1: Opaque Building Envelope in Contact with Outdoor-Air* (2016)
7. EN ISO 10456:2008, *Building Materials and Products—Hygrothermal Properties—Tabulated Design Values and Procedures for Determining Declared and Design Thermal Values* (2008)
8. J. Vinha, *Hygrothermal Performance of Timber-Framed External Walls in Finnish Climatic Conditions: A Method for Determining the Sufficient Water Vapour Resistance of Interior Lining of a Wall Assembly*, Tampere (2007). p. 338, appendix 3
9. EN ISO 6946:2017, *Building Components and Building Elements—Thermal Resistance and Thermal Transmittance—Calculation Methods* (ISO 6946:2017) (2017)
10. Minimum requirements for energy performance, The State Gazette I, 28.02.2017, 2, §12 (4) the Government of the Republic of Estonia (2017)

Part XI
Codes, Regulations, Standards and Policies

Needs of Support for Swedish Property Owners to Implement More Energy-Efficiency Improvements During Renovations

Åsa Wahlström

Abstract It is a well-known fact that the built environment in Europe is responsible for approximately 40% of the overall energy end usage and that this has to be reduced drastically for the building stock to be sustainable in the long term. A prerequisite to meet the EU 2020 and 2030 targets is significantly increased ambition among property owners regarding energy-efficient renovation. However, it is not easy to regulate that large-scale energy-efficiency measures should be included in the ordinary renovation of buildings. In order to design the right policy instruments to influence property owners towards energy-efficiency-oriented renovation, deeper knowledge is needed on how property owners act and argue today. This study reviews several recently performed studies with the aim of obtaining an overview of how the decision-making process is conducted, and it makes comparisons between different categories of buildings. The aim is to allow conclusions to be drawn on whether a financial incentive is important to bringing about energy-efficiency renovation and, if so, how such an incentive should be formulated. The study shows that direct and simple subsidies are needed to bring about major energy-efficiency improvements in housing in connection with renovations, especially when it comes to additional insulation for facades. For non-residential property, there is little need for financial support, while other forms of support are needed. Current energy-efficiency improvements in connection with renovations are modest and a long way off the technical potential.

Keywords Energy-efficiency · Buildings · Incentives · Control means
Subsidy

Å. Wahlström (✉)
Building Services, Lund University, 221 00 Lund, Sweden
e-mail: asa.wahlstrom@hvac.lth.se

Å. Wahlström
CIT Energy Management, 412 88 Göteborg, Sweden

D. Johansson et al. (eds.), *Cold Climate HVAC 2018*,
Springer Proceedings in Energy, https://doi.org/10.1007/978-3-030-00662-4_82

1 Introduction

In Europe, the built environment is responsible for approximately 40% [1] of energy usage, and this has to be reduced drastically for the housing stock to be sustainable in the long term. In the recast directive on energy performance of buildings [2], there is a demand for 'nearly zero-energy buildings' to be a requirement for the construction of new public buildings from 1 January 2019 and for all new buildings from 1 January 2021. However, a prerequisite to meet the energy-efficiency targets is significantly increased ambition among property owners regarding energy-efficient renovation. The recast directive also states that member states must undertake the measures needed to ensure that when buildings undergo a major renovation, the energy performance of the building, or the renovated part of the building, is improved so that it meets the minimum requirements regarding energy performance to the extent that this is technically, functionally and economically feasible. The requirement is to be applied to the renovated building, or the renovated unit, as a whole.

Many of the modernist, multi-dwelling buildings built in the boom years between the end of the Second World War and the start of the oil crisis are in need of, and awaiting, substantial renovation work [3]. The next opportunity, after this profitable rebuilding, may not be available for another 40 years.

Even though there appears to be considerable potential for energy improvements, the national statistics show only about 10–15% energy improvement in the residential and service sector in the last ten years in Sweden [4]. To motivate complete renovation concepts in place of separate component measures, new renovation incentives are needed that raise the building owners' ambitions to carry out the renovation down to the 'nearly zero-energy buildings' level.

To determine if there is a need for a financial incentive and, if so, how it should be formulated, the government has set up an enquiry. In an initial report, it is confirmed that there is a need for financial incentive but that a continued enquiry is needed [5]. The present work is part of the continued enquiry.

2 Objective

This study aims to specify if Swedish building owners need financial incentives in order to make more extensive energy-efficiency improvements in connection with a renovation. If so, also specify how an incentive can be formulated, if requirements are needed for energy measures that otherwise will not be performed and how achieved efficiency improvements could be verified.

3 Method

The work is performed by a review of recent studies on how property owners in Sweden act with regard to energy-efficiency improvements of buildings. The reviewed studies were selected based on that they should have been considered implementation of energy efficient measures in renovation of Swedish buildings or they should have studied how Swedish property owners makes decisions regarding energy efficient buildings. The studies should not be older than five years. The methods in the reviewed studies are mainly interviews and questionnaires with property owners and users.

Focus of the review was to get an understanding of the process in which decisions are made and what the technical and financial options are. The review was focused on the following questions:

How are decisions taken on energy-efficiency levels for renovation? What supporting data are the decisions based on? Which measures are cost-effective and which measures are implemented? Is there a need for a financial stimulus such as loans, credit guarantees or subsidies?

Based on the review results the author discuss and makes conclusions on the need of financial incentives, how a stimulus for energy-efficiency improvements in renovations could be formulated and how achieved efficiency improvements could be verified.

4 Results

4.1 How Are Decisions Taken on Energy-Efficiency Levels for Renovation?

Many property companies prioritise new production ahead of renovation as it is seen as more profitable, and it is common for buildings in need of renovation to be sold. Public property owners try to use existing buildings first, before considering new production, and tend to want to renovate to a greater extent than the private sector [6]. Energy-efficiency improvements on their own rarely lead to renovation.

Of the factors that lead to renovation of multi-dwelling buildings, the maintenance need dominates, with the components having reached the end of their life and the maintenance costs being high [7, 8]. Sometimes a renovation can be a result of an urgent need for maintenance measures, but the property companies try to avoid this as far as possible through good planning [6, 9]. This is followed by high operating costs and high energy use. Raising the standard is also an important reason for renovation. However, reasons such as changing function or flat partitioning, improved indoor environment, complaints from tenants and improved availability are among the factors that are least likely to initiate a renovation [7].

In the non-residential sector, adaptations for new or old tenants (business needs) or indoor environment problems are the main reasons for renovation [6, 10, 11].

This is followed by maintenance needs for climate shells and installations mainly associated with the indoor environment. Environmental certification can sometimes be a contributing reason [10, 11].

In companies with multi-dwelling buildings, renovation decisions are sometimes taken by the companies themselves (the MD or management group) and sometimes by the company board [9, 12]. In many companies, the board has set strategic decisions on required returns on investments and production or renovation volumes, while decisions on energy and environment issues are taken directly by the company board. The management department can highlight renovation needs but usually only decide on minor maintenance measures [9].

Some non-residential property companies have to comply with strategic decisions when carrying out renovations, but these are usually general and the level of ambition varies. It tends to be more common to have overall targets for all the building stock. Most non-residential property companies have voluntarily specified measurable energy targets, though they are expressed in slightly different ways. Targets are 10–20% reduction in energy use, heating or CO_2 emissions by 2020. Most have met or expect to meet their targets. One trend to meet the targets is to set tighter requirements for energy use for new production than the building regulations [6].

During project planning, before decisions are taken, the budget can be increased with extra investments if they meet the profitability requirements. If profitable solutions are identified after the start of a project, they rarely lead to any change in the plan [12].

4.2 *What Supporting Data Are the Decisions Based on?*

The companies do not have any rules on or practices for the supporting data that must be available before a decision can be taken on energy-efficiency measures for new production or renovation other than that financial data are always required [8, 12, 13]. Even if the supporting data vary, they will include some form of investment calculations and value estimates. The calculation methods used for renovations are life cycle cost, cash flow, return, alternative cost and repayment period, and sometimes no method is used at all [6].

The financial calculations used by the property companies include operating costs, but they often use a standard value. This means that the return on the investment on a small net operating income cannot be seen and therefore no consideration is given to an increase in property value due to reduced energy costs or a reduced risk of future increases in the energy price [12, 13]. The motivation is that the renovation often involves complex systems that increase the staff operating costs [12].

Consideration is rarely given to an increase in property value from an improved indoor environment. Public property owners also do not usually consider the effect on property value as they are not going to sell their buildings.

In most big companies, it is the divisions for property development or management that produces the supporting decision data [6, 13]. In small companies, cooperative apartments and one-family houses, it is usually a consultant (sometimes a certified expert) who produces the decision data. Regardless of company, the MD, management or board rarely has detailed knowledge about the energy issue, and the supporting data are therefore crucial to the decision, and the level of knowledge of those who produce the data is of utmost importance. As well as the property owners themselves and their support organisations, other actors such as energy consultants and contractors need to have sufficient knowledge to implement and present an investment calculation in which the value of energy-efficiency measures can be shown as quantifiable comparisons [6, 9, 12, 13].

In almost all cases, there is a building envelope and a technical installation status assessment of the building [6, 7]. When it comes to energy-efficiency measures, the supporting data vary [6, 13]. Most non-residential properties have carried out some kind of mapping of their renovation needs, but there may not be a renovation or energy-efficiency plan at all [6].

A status assessment of maintenance needs is the basis for renovation decisions, and in addition, data are needed for energy-efficiency measures. The energy certificate and its proposed cost-effective measures can serve as a starting point for compiling decision data but is rarely sufficient, as every energy-efficiency measure is assessed individually. For larger renovations, the data for the energy-efficiency measures need to be assessed together in a package. This is because different measures affect each other practically and in terms of energy, and there can be further measures that would be profitable in a package or in connection with a renovation and its base costs [14, 15].

Other aspects that are important to renovation decisions are the possibility of evacuation or if the renovation can be carried out with the tenants in the building, technical experiences of different technical solutions, access to competent staff and labour and the ability to get loans even if the renovation is in itself profitable [6–8].

4.3 Which Measures Are Cost-Effective and Which Measures Are Implemented?

Transmission losses from walls and windows as well as ventilation losses are usually the biggest energy losses from residential properties. That is where the energy-efficiency potential is greatest, but they are also measures that require a high investment cost. Other energy-efficiency measures such as loft insulation, energy-efficient hot water fittings, energy-efficient lighting and adjustments are usually more profitable, though they do not generate such great energy savings individually. With regard to windows, there are as-good-as maintenance-free ones with good U-values at reasonable cost, so even if the measure in itself is not profitable, the previous maintenance costs are reduced. Adjustments to different systems, changing the lighting, insulating the loft and replacing windows are the measures that are carried out most [6, 7, 13, 14, 16, 17].

Facade insulation is a measure that often has high investment costs and is frequently set aside despite its big energy-saving potential. The high investment costs together with the fact that the measure often requires building permission and may have historic aspects that need to be considered means that property owners avoid it. More cost-intensive, non-standard-raising measures and partly technically and architectonically challenging energy-efficiency measures such as facade insulation are measures that are shown as rarely implemented in a number of surveys [6, 7, 13, 17].

Heat recovery can be installed using primarily two different technologies. The installation of an exhaust air pump is relatively easy and often profitable [14, 16]. It provides good energy efficiency but does not improve the character of the building itself to a low energy need. The other option is a heat exchanger that recovers energy from the exhaust air and returns it to the building in the supply air. Such an installation requires supply air channels, which, practically, is relatively complex and costly but, on the other hand, significantly reduces the risk of indoor environment problems compared with supply air inlets in the facade. When installing the technology, the supply air inlets in the facade are sealed, which improves sound problems from the outside, reduces the risk of draughts and the building is provided with warm filtered supplied air. It provides conditions to maintain a regulated air flow all the year round without risk of thermal discomfort [18]. Preheated air can also be combined with an exhaust air heat pump.

In most cases, it is finances, and to some extent the customers' ability to pay, i.e. chance to increase the rent, that control the choice of renovation measures. One study shows that if the households in multi-dwelling buildings choose from different renovation options with regard to cost and technical performance, they will choose a lower level of renovation despite a higher level being more cost-effective [19].

In non-residential properties, there are usually a number of measures that are relatively profitable and can be carried out together in a package in which, for example, profitable ventilation measures often help to support less profitable ones. Many different types of non-residential properties that have used the Total Concept method show that it is often possible to find relatively big energy savings with reasonable profitability [20]. However, in some types of non-residential properties, as well as multi-dwelling buildings, building envelope measures (for example facade insulation) often have great energy-saving potential but are a relatively expensive investment.

An interview study with property owners who together represented 25% of the multi-dwelling buildings, 15% of the offices and 18% of the schools in the total area of Sweden confirmed that a very low level of the potential for energy-efficiency improvements in connection with renovation is currently applied. The energy-efficiency measure that are carried out are a long way off the technical potential [6].

4.4 Need for Support

When it comes to support, a few private companies stated that they do not need direct financial support and would rather see other control instruments such as housing allowances or support in the form of direct discounts on contract costs for renovation, which are much easier to administer. Most private and public residential property owners think there is a need for subsidies or grants [6]. A few property owners say that support is needed, especially if the legal energy-efficiency requirements increase. Support are primarily needed for building envelope improvements [6]. One study among non-residential property companies shows that the need for tools and support for decision-makers for renovations is seen as more important than tangible forms of financial support. They believe it is better that it is controlled by the market but that support may be needed to introduce new technology and for innovations to be tested by property owners [6].

In less attractive locations, it is more difficult to make the calculations add up and it can also be hard to get loans due to the uncertain market value of the building [6, 12]. There is therefore currently energy-efficiency support for rental housing in areas with socio-economic challenges [21].

Many public as well as private property companies mention one of the big obstacles as a lack of competent labour on the market. They generally have difficulties getting responses to tenders and good-quality tenders, and the tender prices are often very high due to the low level of competition. They also have difficulties recruiting competent staff, leading to a staff shortage to run renovation projects [6].

To raise the competence within the branch the Swedish Energy Agency supports several competences raising education programs, for example Energilyftet [22], Energibyggare [23] and Beställarkompetens [24] and the National Board of Housing Building and Planning has recently started an information centre [25].

5 Discussion and Conclusions

Based on the review of previous studies, a few preliminary conclusions have been drawn that should not be seen as a final proposal for support or control instruments but a proposed direction for continued investigation. In further investigations and formulation of the subsidy rules it is important to assure that the energy renovation will be performed while the indoor climate quality and the function of the building must remain the same or be improved.

5.1 Financial Support Must Be Easy to Administrate, Have Clear Requirements and Be Guaranteed

Financial support, whether in the form of loans, credit guarantees or grants, must be easy to administrate and apply for. A property owner who plans a renovation often has many aspects to consider in a limited period of time. There is no time for long drawn-out application procedures.

For a financial incentive to influence a decision and get the property owner to implement energy-efficiency measures in connection with a renovation, it must be included in the decision data from the start. It is unlikely that a financial incentive that is not considered until after the renovation plan has been decided will lead to a change in the project plan even if it leads to an energy-efficiency measure becoming profitable.

Financial support must also have clear requirements to be granted and be guaranteed. There has previously been support for solar installations that involved the property owner joining a queue at the county administrative board to maybe get support. Experience shows that this does not contribute to decisions on investment. In the best case, the financial incentive contributed to the energy measure being included as a possible measure in the decision data, but if there are no guarantees that it will be paid it will be considered a bonus and do not influence the decision on whether the renovation measure should be implemented. In the event that a decision is taken on the renovation budget and scope, the financial incentive must be part of the calculation basis, and it must be guaranteed that if the set requirements are met, the support will be paid.

5.2 Packages of Measures Need to Be Considered

Direct support in the form of direct discounts on contract costs for renovation is recommended by several property owners. However, this assumes that the requirements to be implemented should be verified by a contractor, which may be difficult with regard to both competence and supporting data. Here, it may be possible to use the energy certificate directly for small houses, but this would hardly be sufficient for multi-dwelling buildings or non-residential property. For example, a subsidy for facade insulation would be expressed as a number of centimetres of additional insulation so that a contractor can easily verify that the requirement has been met. This would not be a sufficient contribution for the energy-efficiency targets, however, as every renovation should be preceded by a comprehensive investigation in which facade insulation is part of a package of other measures that together provide financial benefits. Practical aspects should also be considered, for example whether windows also need to be replaced or renovated when scaffolding is put up anyway. The energy certificate alone is considered too simple as a basis for granting subsidies. There is a need for more comprehensive supporting data with the building undergoing a thorough basic inspection in which the energy and cost savings of different packages of measures can be assessed in relation to the investment and maintenance requirements.

5.3 Subsidy Is Needed for Residential Buildings

Subsidies are needed to get residential property owners to implement more extensive energy-efficiency measures in connection with a renovation. Better loan terms are not sufficient since it, today, is quite easy to get good loan terms. Furthermore, energy-efficiency measures that are important from an energy-saving point of view but that are rarely carried out should be included in the package of measures to be granted support.

These could be, for example, additional insulation of the facade of residential property and some types of non-residential property, and installation of supply air channels in multi-dwelling buildings. Insulation of the facade is an important energy-efficiency measure that is cost-effective, non-standard-raising, partly technically and architectonically challenging and requires a financial incentive to be carried out. The installation of supply air inlets and sealing of old air inlets provides conditions for good energy economy with an improved indoor climate and conditions for good air quality.

5.4 *Requirements for a Financial Subsidy*

A subsidy should have the following requirements to be really useful and contribute to meeting the energy-efficiency targets:

- the building should have relatively high energy use before the renovation
- a certain share of the energy use should be reduced
- energy-efficiency measures that are important from an energy-saving point of view but that are rarely carried out should be included
- a maximum amount per square metre, possibly dependent on a reduction in energy use.

If the financial subsidy is formulated with different maximum amounts depending on the reduction in energy use, it could satisfy the requirements of different property owners. A small maximum amount and a low requirement for the total energy-efficiency improvement could be beneficial for small houses and cooperative apartments that rarely renovate in big packages but prefer to implement separate measures in stages, while a high maximum amount and requirements for total energy efficiency could be beneficial for private and public multi-dwelling buildings that make extensive renovations.

5.5 *Generally Non-residential Buildings Do Not Need Financial Support*

Energy-efficiency improvements of non-residential property are generally deemed sufficiently profitable for financial support not to be needed. The implementation of more energy-efficiency improvements in connection with renovations requires good decision data and competence of the actors involved. Here the sector may need support to produce decision data, for the industry to develop better tools and for actors in the sector to be trained rather than direct subsidies.

A financial incentive, similar to that for residential property, could be effective for additional insulation for facades of buildings with similar conditions, i.e. with a significant heating requirement and limited internal loads from the operation, for example schools, retirement homes and prisons. For schools, such support could be effective as it represents 31% of the area of non-residential property in Sweden.

5.6 Supporting Data for Granting Subsidies and Reporting Achieved Results

For small houses, supporting data that show that the property owner is entitled to the subsidy may consist of an energy certificate with the building's energy performance and relevant proposed measures. For multi-dwelling buildings and non-residential property that may be entitled to subsidies, the energy certificate must be complemented with a basic report that assesses the cost and energy savings of different packages of measures together with the investment costs of the measures. In recent years, different methods have been developed to produce and compile the required data. There are a number of different variants, but most are similar and include inspection, energy and cost calculations and analysis as well as planning for verification [26–28]. The result can then be presented in slightly different ways. With the help of such data, the profitability of different packages of measures can be quantified and valued.

A new energy certificate with measured energy use may be sufficient for reporting on the achieved results in small houses. For multi-dwelling buildings and any non-residential properties entitled to subsidies, an energy-efficiency improvement report based on the supporting data report and the new energy certificate is needed. The report should show that all the measures included in the original application have been implemented correctly. The energy savings shall be reported and, if they do not meet the set requirements, a report should account for why the energy use deviates. There should not be requirements to repay the subsidy because the expected energy saving is not achieved, if all planned measures are implemented correctly, as that would act as a deterrent to applying for the subsidy. The energy saving made in connection with the renovation is not always directly measurable as the building may have had, for example, poor air flows before the renovation. When new ventilation systems are installed, the ventilation flow are set to requirements according to the regulation all the year round, which may lead to no reduction in energy use or a smaller reduction than in the energy declaration. Here, a comparison needs to be made with a reference case in which the building has required air flows before the measures. The energy-efficiency report is an important element for experience feedback for future renovations.

5.7 *Other Limiting Factors*

To meet the energy-efficiency targets that have been set, the ambitions and the speed of every renovation must increase. Financial incentives can help bring about more extensive energy-efficiency improvements in connection with a renovation. Financial incentives are, however, unlikely to speed up the pace of renovation, which is limited by the lack of competent labour on the market.

Acknowledgements Bengt Nyman and Rolf Westerlund, secretaries of the governmental assignment "Investigation about energy-saving loan (N 2016:02)" are acknowledged.

References

1. European Commission, Energy-Efficient Buildings PPP, Multi-Annual Roadmap and Longer Term Strategy; European Commission: Brussels, Belgium (2010)
2. Directive 2010/31/EU of the European Parliament and of the Council, of 19 May 2010, on the energy performance of buildings (recast). Official Journal, L 153 (2010)
3. The National Bord of Housing Building and Planning, Teknisk status i den svenska bebyggelsen – resultat från projektet BETSI, Boverket (2010)
4. Swedish Energy Agency, Energiläget i siffror (2017)
5. B. Nyman, R. Westerlund, Government public investigation: Energisparlån – ett incitament för energieffektivisering? Rapport från Utredningen om energisparlån, N 2016:02 (2016)
6. Å. Wahlström, A. Persson, K. Glader, K. Westerbjörk, A. Göransson, Energieffektivisering vid renovering av flerbostadshus, skolor och kontor - En intervjustudie och analys i HEFTIG, Rapport till Energimyndigheten (2017)
7. L. Thuvander, P. Femenías, M. Xygkogianni, B. Brunklaus, Renoveringsbarometern: Omfattning och karaktär av renoveringar i bostadshus, Bygg och Teknik nr 2 (2016)
8. L. Thuvander, P. Femenías, K. Mjörnel, P. Meiling, Unveiling the process of sustainable renovation. Sustainability **4**, 1188–1213 (2012)
9. P. Femenías, L. Thuvander, K. Mjörnell, A.-L. Lane, Koll på hållbar renovering, Bygg & teknik Nr 2 (2015)
10. Energy Renovations of Non-residential Buildings in Northern European Countries—National non-technical barriers and methods to overcome them, Report of IEE project Total Concept, www.totalconcept.info (2014)
11. M.-L. Maripuu, P. Wennerhag, Å. Wahlström, P. Pasanen, T. Bruce, P. Krawczyk, A. Afshari, Totalmetodikens marknadspotential och affärsmöjligheter, projektrapport www.nordictotal-concept.info (2016)
12. Å. Wahlström, B. Berggren, J. Florell, R. Nygren, T. Sundén, Decision making process for constructing low-energy buildings in the public housing sector in Sweden. Sustainability **8** (10), 1072 (2016). https://doi.org/10.3390/su8101072
13. S. Hastig, S. Tapper Jansson, Val av energieffektiviserande åtgärder - En studie av fastighetsägarnas investeringsbedömning vid renovering, examensarbete Lunds Tekniska Högskola, ISSN 1651-0100 (2014)
14. Å. Wahlström, P. Filipsson, C. Heincke, in *Cost Optimal Energy Efficiency in Multifamily Houses*, Proceeding of the Sixth Nordic conference on Passive Houses, Göteborg, Sweden, Passivhus Norden, 15–17 Oct 2013
15. Å. Wahlström, E. Abel, M.-L. Maripuu, in *Total Concept-for Better Decision-Making About Energy Efficiency Investments in Non-residential Buildings*, Proceeding of ECEEE Summer Study 2015, Toulon, France, paper 6-103-15, p. 1239, 1–6 June 2015

16. K. Högdal, Halvera Mera Slutrapport, BeBo-rapport, www.bebostad.se (2013)
17. F. Bonakdar, A. Sasic Kalagasidis, in *An Optimum Renovation Strategy for Swedish Single-Family House Envelopes: The Implications of Climate Zones and the Age of the Houses*, Proceeding of ECEEE Summer Study, Toulon, France, paper 5-165, 29 May–3 June 2017
18. Å. Wahlström, T. Berggren, T. Rydstedt, in *Technical Procurement of Heat Recovery Systems in Existing Apartment Blocks in Sweden*, Proceeding of ECEEE Summer Study, Toulon, France, paper 5-104, p. 1165–1172, 3–8 June 2013
19. H. Lind, K. Annadotter, P. Björk, P. Högberg, T. Af Klintberg, Sustainable renovation strategy in the Swedish million homes programme: a case study. Sustainability **8**, 388 (2016)
20. Å. Wahlström, M.-L. Maripuu, A. Afshari, in *Market Potential and Business Opportunities for Mayor Renovation of Non-residential Buildings with the Use of the Belok Total Concept Method in Nordic Countries*, Proceedings of the 12th REHVA World Congress, Clima 2016, paper 460, Aalborg Denmark, 22–26 May 2016
21. National Board of Housing Building and Planning, REN 1, Boverkets föreskrifter om stöd för renovering och energieffektivisering i vissa bostadsområden; BFS 2016:8 (2016)
22. http://energilyftet.learnways.com/
23. http://energibyggare.se/
24. http://www.byggherre.se/utbildning/bestaellarkompetens/
25. https://www.boverket.se/sv/byggande/uppdrag/avslutade-uppdrag/genomfora-upphandling-for-att-uppratta-och-driva-informationscentrum-for-hallbart-byggande/
26. Notice 2012/C 115/01, Guidelines for a comparative methodology framework for calculating cost-optimal levels of minimum energy performance requirements for buildings and building elements, accompanying Commission Delegated Regulation (EU) No 244 (2012)
27. Totalmetodiken konsultrapportmall Etapp 1, www.belok.se (2016)
28. Rapportmall Halvera mera 3, www.bebostad.se (2016)

Criteria for Sustainable Buildings in Sweden

Åsa Wahlström and Catarina Warfvinge

Abstract To continue to be an effective guidance tool for the property and construction sectors and to meet the environmental quality objectives, the Swedish certification system Miljöbyggnad needed to be developed. Over the years, new research results have been published and political decisions made that affect sustainable building and construction. What was rated as 'best available technology' ten years ago has now become standard. In this project, industry and academia have collaborated to implement research findings into practice. Miljöbyggnad considers requirements of energy, indoor environment and material use. The criteria give high scores for low heating power need, low heating loads from the sun, energy efficiency, high share of renewable energy, good sound levels, low radon exposure, good ventilation performance, moisture safety, indoor comfort in winter and summer, good daylight, low risk of legionella, documentation on material used, avoidance of hazardous substances and evaluation of the framework's life cycle effects on climate change. Compliance with the criteria in Miljöbyggnad improves the potential for sustainable building design. The criteria give special consideration to fitting in with the outdoor climate throughout Sweden as well as Swedish building regulations and practice in the property and construction sectors. The work to evaluate and improve the criteria in Miljöbyggnad has taken two years and involved over 250 people from research and industry. The criteria are based on scientific values and can be verified, and they all support meeting different environmental objectives.

Keywords Certification scheme · Energy · Indoor environment
Material use

Å. Wahlström (✉)
Building Services, Lund University, 221 00 Lund, Sweden
e-mail: Asa.wahlstrom@hvac.lth.se

Å. Wahlström
CIT Energy Management, 412 88 Göteborg, Sweden

C. Warfvinge
Sweden Green Building Council, 117 33 Stockholm, Sweden
e-mail: Catarina.warfwinge@sgbc.se

D. Johansson et al. (eds.), *Cold Climate HVAC 2018*,
Springer Proceedings in Energy, https://doi.org/10.1007/978-3-030-00662-4_83

1 Introduction

The housing and service sector is responsible for 40% of Sweden's total final energy use, and meeting the long-term environmental quality objectives requires an overall reduction in energy use as well as an energy supply with low environmental impact [1].

At the same time, 1.2 million adults have health problems that they relate to one or more indoor environmental factors in the home, school or workplace. As many report that they live in homes with visible damp damage, mould or a mouldy smell, and thousands of people suffer respiratory problems as a result [2]. The ventilation in Swedish homes is often poor [3], which can lead to health problems. Examples of serious health problems, and in the worst case death, include lung cancer caused by radon exposure and Legionnaire's disease caused by legionella bacteria. Less serious problems such as poor thermal climate, noise disturbance and lack of daylight are also important.

Health problems can also arise from chemical exposure to materials indoors [2], and chemical substances in the indoor environment can pose a health risk of respiratory illnesses and allergies in children [4]. The main substances pointed out are formaldehyde-emitting materials, softened plastics (phthalates) and newly painted surfaces. There is a national environmental quality objective for 'an environment free of poisons' [5]. Still relatively little has improved in the last decade regarding the environmental problems connected to the buildings.

To improve this situation and support meeting the environmental quality objectives set by the Bygga-bo dialogue (a partnership between the property and construction sector, authorities and academia [6]), a national environmental classification system was produced in 2009. The aim was to provide a strong incentive for building contractors, property owners, property managers and users of the buildings to speed up development towards a sustainable property and construction sector through a practically applicable and well-established method for environmental classification of buildings. In 2011, the method was developed into a certification system called Miljöbyggnad (Environmental Building). The certification involves a third-party inspection of documents to prove the environmental rating that is applied for.

Miljöbyggnad is a voluntary system of environmental certification for buildings considering energy, indoor environment and material issues. It is used for homes and offices, for new building, renovation and existing buildings. Miljöbyggnad has been operational for certification since 2011 and has become the dominant environmental certification in Sweden, with more than 1000 certified buildings [7]. During the years that Miljöbyggnad has been in operation, new research and development results have been published and political decisions made that affect building and management. What was seen as 'best available technology' in 2011 have in many respects become standard solutions. Miljöbyggnad with its areas, indicators, methods and certification needed to be assessed in substance and revised in order to continue to drive the development. This has been done with the drawing up of the new Miljöbyggnad version 3.0.

2 Method

The work started with the collection of concrete experiences and views including of the use of the method by those who had used previous versions of Miljöbyggnad. This was done in six work seminars in May 2015 in six different locations around Sweden. A total of about 160 public and private property owners, architects, energy companies, contractors, consultants, material manufacturers, politicians, etc. took part for one working day. The results of the work seminars were used in the planning of the development work of Miljöbyggnad 3.0. Different subjects that needed to be analysed for the continued work were identified as well as a number of basic principles for Miljöbyggnad, intended to facilitate and guide the continued development work [8].

In autumn 2015, ten expert groups were formed to work on the different subject areas in Miljöbyggnad and analyse, evaluate and propose different indicators to assess the environmental performance of a building. The expert groups consisted of one highly qualified group leader and about ten experts from the industry and academia. A further two expert groups were formed to develop the method used for verification and consolidation of the different subject areas. Their task was to balance the different subject areas into a whole in the Miljöbyggnad system. A total of 110 experts took an active part in this development. Subject-specific consultation groups with a further 120 persons were also linked to the expert groups. For more in-depth analyses, individual experts in academia and the industry were consulted. Coordination was done between the group leaders and the two main project leaders.

The expert groups' proposals were handled in parallel in a reference group consisted of 20 property owners to obtain practical views, primarily with regard to suitability and usefulness in relation to verification costs for different indicators. The work started in November 2015, and in August 2016 a comprehensive and considered proposal for a completely new system was presented at seven hearings around Sweden. A total of 36 organisations submitted comments before the consultation deadline of 30 August 2016.

All the views from the seminars were then considered by subject experts, group leaders and different consultative bodies and, at the same time, a number of specific analyses were conducted. In February–March 2007, a proposal for manual texts was presented. Ninety-six organisations responded with 1214 comments that have been considered in the final manual. These were split as follows: 23% in the energy area, 39% in the indoor environment area, 22% in the material area and 26% in methods. Approximately 25% of the responses were of an editorial nature.

3 Basic Criteria for Development Work

The six work seminars conducted in the beginning identified basic criteria to which the development work for a new Miljöbyggnad should relate.

Building and renovation projects place huge demands on builders, property owners and building contractors as they are costly, take a long time and are often subject to great time pressures during the planning as well as the building phase. The criteria therefore need to be clear, easy to use and cost-effective. It is important that the criteria being controlled are meaningful, provide real use and clearly contribute to meeting the different environmental objectives. The criteria should comply with existing laws, rules and regulations in order to avoid duplicating work that would result from a separate set of rules and regulations. There would also be a risk of two sets of rules and regulations conflicting. It is also important that property owners can trust that the authorities' requirements are met if they comply with Miljöbyggnad. Deviating from laws, regulations and provisions as a platform for Miljöbyggnad is not reasonable as it would increase the costs of the projects and probably lead to property owners and builders lowering their environmental ambitions.

All the criteria in Miljöbyggnad therefore have three different score levels: Bronze, Silver and Gold. Bronze corresponds to the authorities' requirements or building practice. It can for example be the Swedish building regulations set by the National Board of Housing, Building and Planning, the lowest acceptable limits set by the Swedish Work Environment Authority for work places and by the Public Health Authority of Sweden for residential areas, or be an interpretation of environmental quality objectives. Silver provides significantly higher environmental performance than Bronze. Gold is the best function that can be obtained with available and commercial technology or through goal-oriented partnership between the building contractor, planners and contractors. Gold should be challenging but achievable, as the main aim of Miljöbyggnad is to get many property owners to raise their ambitions significantly. By influencing many in the right direction, a bigger contribution is made to meeting the environmental quality objectives than a few individual cutting-edge projects, which can be rewarded in other ways.

All the criteria that are set must be objective and based on science. This is to ensure that parties' pleas aimed at a specific product or solution are not successful. The criteria must therefore be expressed as functional requirements. It should also be possible for the property owner to influence a score, and criteria that the property owner cannot control should be avoided.

As it is meaningless to set requirements that cannot or will not be verified, it must be possible for all criteria to be verified at a reasonable cost. Quality assurance and verification are an important factor for the property owner to comply with Miljöbyggnad as it clearly shows that the building that has been supplied meets the requirements set when it was ordered. To make it easy and to minimise extra costs, the documentation required to verify the requirements should, as far as possible, be the same as that normally used for the building and management process.

Miljöbyggnad has three main areas (energy, indoor environment and material), each with a number of different indicators. The scores for all the indicators in an area are aggregated into an area score, which in turn is aggregated into a final score for the building. As the criteria set should be meaningful and contribute to meeting the different environmental objectives, all the criteria are considered important, and it should therefore not be possible to skip any indicators or compensate for a poor score with a good score for another indicator. All the indicators in Miljöbyggnad must therefore meet the Bronze score, and the scores from all the indicators are aggregated in such a way that the lowest score has a bearing on the score for the whole building. A building with a total Gold score can therefore not have any indicator with a Bronze score, and all three main areas must have reached the score for Gold.

Miljöbyggnad is essentially the industry's own tool to drive aims and knowledge development of environmental issues in the industry and the projects. In principle, it should therefore be possible to use Miljöbyggnad for all buildings whatever the type of owner, contract, category and stage (new production, existing and renovation). Miljöbyggnad should be usable by everyone, whatever the size and experience of the actor or property owner. This assumes that the whole system has limited scope. One of the most important principles of the continued development work was therefore not to let the system grow but to prioritise the most important issues.

4 New Areas for Analysis in the Development Work

The participants in the six work seminars initially felt that Miljöbyggnad had development potential and that each of the indicators and their score criteria needed to be reviewed. New subject areas that were proposed for continued analysis were *Land and infrastructure*, *Building stage* and *Management*.

4.1 Land and Infrastructure

The expert group for land and infrastructure has proposed two indicators called Land pollution and Green area factor or multifunctional outdoor environment. The latter is intended to assess the size, quality and function of the outdoor environment and its green areas by thinking in terms of multifunctionality to deal with climate change (precipitation, rising water levels, heat waves), provide social values and contribute to biological diversity. The area involves deviating from the current principle for Miljöbyggnad, i.e. that only the building's status and that which the property owner directly controls and can influence should be assessed. Influencing this score requires collaboration with the municipality and city planners in the early stages of the building process. Miljöbyggnad also has a principle that all indicators should be assessed, which makes it difficult for buildings without land to meet the criteria. All in all, the area was considered too far from Miljöbyggnad and was recommended for consideration in other environmental certification systems.

4.2 Building Stage

In the expert group for the building stage, about twenty indicators were discussed and a final three proposed for which requirements can be set and that can be verified: waste management and minimising waste, energy use on building sites and fuel use for vehicles. The construction sector generates more than 10 million tons of waste annually [9] and there are not always enough sorting fractions on the building site. Much of the energy-related environmental impact of building also occurs during the building stage compared with the operational stage [10]. There is also huge potential for energy savings and to use a bigger share of renewable energy in site offices [11]. Vehicles on the building site account for 3% of the climate emissions from the building process [10] and many of them could be replaced with vehicles that use more environmentally friendly fuels such as RME, HVO or electricity.

This area requires good planning very early on for the requirements of the indicators to be met. If the area were to be included as mandatory in Miljöbyggnad it would mean that it may not be possible to certify new production if the aim to achieve Miljöbyggnad is adopted at a later stage. The area also involves some deviations from the current principle for Miljöbyggnad, i.e. that only what the property owner is in direct control of and can influence should be assessed, as some collaboration with the contractor is required. It was established that there is a great need to somehow work on environmental issues also at the building stage but that the building stage needs a special environmental building system to be most effective, and it should thereby not form part of the standard Miljöbyggnad.

4.3 Management

The expert group for management primarily worked on two indicators: waste and self-monitoring of management. Setting requirements to reduce the amount of waste or increase sorting of waste is considered to go against the fundamental principles of Miljöbyggnad that only that which is under the direct control of and can be influenced by the property owner should be assessed, as the tenants can also influence this. However, the property owner could be responsible for preparing facilities for sorting at source. After investigating the fundamental issues, it was established that according to Swedish law it is the packaging industry that is responsible for residual material being recycled and for the provision of suitable recycling facilities together with the municipalities. The introduction of an indicator for waste sorting was thereby ruled out as it would move the responsibility of cost from the packaging industry to the property owners and, in the long term, the tenants.

Good management is not only important to maintain the performance of Miljöbyggnad but also a legal requirement. According to the Environmental Code, an operator shall continually plan and control the activity to counteract and prevent nuisance to people's health or harm to the environment [12]. A review of the Environmental Code shows that there are requirements for self-monitoring checks by the property owner for most of the indicators in Miljöbyggnad, and it was therefore deemed more suitable to include these routines with the ordinary indicators than to produce new indicators for management. The relevant indicators have thereby been expanded with requirements for management routines. Additionally, for further quality assurance, Miljöbyggnad 3.0 introduces requirements for the property owner to report results showing that the building maintains its performance according to the score requirements every five years to retain the certification.

5 Development of the Energy Area

Even if a building uses renewable energy with little environmental impact, high energy use is always negative as the same amount of energy could be sufficient for more buildings, and there is also a risk that the renewable energy will not be available for the lifetime of the building. As many aspects need to be considered for a sustainable system design when erecting a building, Miljöbyggnad follows the priorities in the so-called energy triangle:

1. Minimise energy losses

Miljöbyggnad's first indicator, heating power need, rewards a well-insulated and tight climate shell with few thermal bridges and an efficient system for heat recovery from the ventilation air that allows the energy losses to be minimised, i.e. lowers the energy requirements of the building. The second indicator, heating loads from the sun, rewards passive measures such as sun protection inside and outside, choice of windows with regard to the sun factor and adapting the size and orientation of windows to reduce the need for comfort cooling or opening windows due to high indoor temperatures.

2. Meet the energy requirements efficiently

The third indicator of Miljöbyggnad, energy use, rewards the building if it is fitted with highly energy-efficient installations so the building's energy requirements can be met efficiently with low energy use.

3. Use renewable energy sources

The fourth indicator of Miljöbyggnad, share of renewable energy, rewards the building if a large share of the energy it will need can be supplied by renewable energy.

The expert groups have analysed whether Miljöbyggnad should exclude any of these indicators with the motivation that the others are so important that they should always be governing when planning a building. This has been rejected, however, as it could in itself have negative consequences for the indicators that are excluded. Miljöbyggnad therefore continues to base energy optimisation on these four indicators, which together determine the score for the energy area. For a clearer balance, the score requirement for the indicator for heating power need is tightened.

The indicator for energy use is directly related to the Swedish building regulations, which were changed in 2017 from setting requirements for bought energy and actual occupants use to setting requirements for primary energy and normal occupants behaviour. The score boundaries have changed slightly to correspond to this new way of setting the requirements. For verification, however, Miljöbyggnad requires continued accounting of bought energy for actual occupants' use.

The expert group has analysed whether resource efficiency, primary energy or carbon dioxide emissions could be used for the fourth indicator but have established that the concept 'share of renewable energy' is the best and an easy way to reflect resource withdrawal and the effect on climate change. Miljöbyggnad is divided into three categories. *Renewable flowing energy* (sun, wind and water) causes no or minimal resource withdrawal from nature. It also includes waste heat, i.e. heat that cannot be avoided and if unused would be lost and that cannot be used by the process or product itself, as this is considered important to use. *Renewable fund energy* (biofuels and waste with organic origins) causes a withdrawal of a finite resource and use of land that could be used for other purposes. Biofuel burning also leads to some resource withdrawal in connection with the transport of the fuel and handling of the ash. *Non-renewable energy* covers all other energy such as natural gas, oil, peat, coal, nuclear (uranium), waste with fossil origins and energy of unknown origins.

When it comes to mixed energy types in an energy carrier such as electricity or district heating, the Nordic electricity system considered electricity residual and district heating as the average value for the network. Miljöbyggnad permits purchases of allocated or origin-labelled electricity and district heating. This is motivated by the fact that if many buy renewable energy, in the long term it will be a driving factor for the energy companies to convert their energy production to include a greater share of renewable energy. To also be a driving factor for the building owners to invest in new locally produced renewable energy, such as solar energy, a new requirement has been added for a Gold rating. Five per cent of the energy used should be renewable flowing energy newly locally generated. This includes solar energy from solar panels and solar cells, wind and water energy and new use of waste heat nearby, for example, in the housing area or city district to which the building belongs.

6 Development of the Indoor Environment Area

Miljöbyggnad rewards indoor conditions in a building that maintain a certain level of operation, as this is of most interest to the buildings' users. The expert group's analysis showed that the previous indicators in Miljöbyggnad continue to be relevant but that several needed to be changed slightly and clarified. These refer to sound level, moisture safety, thermal climate in summer, daylight and legionella requirements, for which the requirements continue to be high. The exception is the indicator for nitrogen dioxide, which has been excluded as the building owner cannot usually affect the location of the building in relation to, for example, roads. It continues to be important to place air intakes to minimise nitrogen dioxide levels indoors, but this is no longer included in the score setting as analysis shows that the indicator means that a number of reports are required without promoting a more sustainable design of the building.

The radon indicator has been complemented with a requirement for gamma radiation in the residential zone as this is a growing problem with imported building materials. The requirement can be met by setting requirements for the level of gamma radiation of supplied building material and verified indirectly via radon measurements.

The ventilation indicator has been rewritten to focus on functional requirements for air quality instead of specific technical choices, and therefore fully complies with the basic principles of Miljöbyggnad. It requires control of estimated or measured carbon dioxide levels depending on the score, to show that the air quality also depends on the ventilation flow pattern and not only the flow size. The indicator for thermal climate in winter has previously been based on requirements in the Swedish building regulations but has now been made more stringent to correspond to lowest accepted limits set by the Swedish Work Environment Authority and the Public Health Authority of Sweden.

Verification by questionnaire of how the users experience the indoor environment was previously a requirement for the Gold rating for most of the indicators but has shown to have had an inhibiting effect on development. Many property owners worry that a questionnaire does not correctly reflect the actual conditions of the indoor climate but that other conditions that affect it also shine through. Many have therefore chosen the Silver rating to avoid the questionnaire, which has thereby impeded their ambitions to reach Gold also for other indicators. The questionnaire has therefore been removed from the indicator for moisture and daylight. For sound, ventilation and thermal climate it can now be replaced with control measurements.

7 Development of the Material Area

In the material area there is an indicator aimed at documenting which building products and materials are built into the building. This is done in a so-called logbook that should facilitate the identification of materials that are currently considered harmless but, with greater knowledge in the future, may prove to be problematic. Previous versions of Miljöbyggnad have had requirements to document building products such as construction components and other building envelope parts. This is now being expanded. The higher scores now also require heating, ventilation and sanitation products such as equipment with piping or canal systems and insulation of installations to be documented. The expert group deems that there is now good availability of environmental product declarations for these products. For existing buildings, the documentation is not needed. Instead an indicator for slum-clearance is required.

The material area also has an indicator that rewards buildings that are planned, built and managed with a minimum of hazardous materials and building products. This indicator is now being expanded from avoidance of substances that are being phased out to include also avoidance of candidate substances for the Bronze rating, endocrine disruptors for the Silver rating, and priority risk reduction substances for the Gold rating. Emissions of volatile organic compounds to the indoor environment are also considered for Gold. Much clearer deviation documentation is now also required.

The expert group has confirmed that for new production, the buildings are beginning to have such good energy management that the environmental impact of the actual operation of the building must be set against the amount of energy used for extraction, production and transport of building materials. A new indicator is therefore being introduced with a limited life cycle analysis of climate impact from the material used in the frame and basic construction. The introduction of the indicator is partly to increase knowledge and experience of life cycle analyses and raise supply and demand for environmental product declarations (EPDs) for different materials and partly to reward measures that reduce the climate impact of the frame and foundations. The indicator will in this stage mainly be informative, but the aim is to tighten the indicator within a few years to become governing.

8 Discussion and Conclusions

The certification system Miljöbyggnad has undergone a fundamental overhaul, and a version called Miljöbyggnad 3.0 has been produced in collaboration with more than 250 actors from academia and the property and construction sector. It can be confirmed that Miljöbyggnad continues to be the industry's own system to ensure that the quality of a building meets legal and authority requirements and industry practice or the property owner's higher ambitions. The new version has continued to be restricted in size in order to set requirements and verify the most important environmental aspects for a building in a cost-effective way. All the requirements are objective, based on functional requirements and have criteria over which the property owner or building contractor has control. This leads to many using the system and can thereby contribute to meeting the aim for which the system was originally produced, namely to clearly contribute to meeting national and international environmental objectives.

In Miljöbyggnad 3.0, all the indicators have been changed or clarified to correspond to the new rules and current building practice. Many of the criteria have been tightened so that Miljöbyggnad can drive sustainability further. One of the most important ways in which the criteria have been tightened is through increased quality assurance, partly through requirements for management routines for follow-up and partly through the property owner reporting the building's environmental status every five years after verification. In the energy area, Miljöbyggnad 3.0 has been restructured to clarify the links between the performance of the building envelope, the installations and the share of renewable energy. The requirements for investment in renewable energy have been tightened to achieve the Gold rating. In the area of indoor environment, the indicator for nitrogen dioxide has been removed and the requirements for thermal comfort in the winter have been tightened. The indicator for ventilation is now based on functional requirements for air quality instead of specific technical choices. All the indicators have clearer requirement descriptions. An option to replace the questionnaire with measurements has been introduced. In the material area, several requirements have been tightened. More building materials (also building services installations) must be declared and more hazardous substances should be phased out, such as endocrine disruptors, risk reduction substances and emissions from materials. A new indicator is also introduced that assesses the climate impact of the frame and foundations: the house should not only have little environmental impact when it is operational, the choice of material that is built into it should also be considered.

Miljöbyggnad has traditionally consisted of three areas (energy, indoor environment and material) and the work confirms that these three areas should not be expanded with more areas. The comments indicate that there is similar interest in all the areas. The work shows that there is also a need for an environmental system for the building stage but that it is too big and complex to fit within the scope of the regular Miljöbyggnad.

Acknowledgements All experts who contributed to the work to rewrite Miljöbyggnad, the Swedish Energy Agency and its research fund E2B2 and SBUF (the construction industry's organization for research and development) are acknowledged.

References

1. Swedish Environmental Protection Agency, www.naturvardsverket.se. Last access 30 June 2017
2. Public Health Agency of Sweden, Miljöhälsorapporten (2009)
3. National Board of Housing Building and Planning, Teknisk status i den svenska bebyggelsen – resultat från projektet BETSI (2010)
4. M. Mendell, Indoor residential chemical emissions as risk factors for respiratory and allergic effects in children: a review. Indoor Air **17**, 259–277 (2007)
5. Swedish Environmental Protection Agency, www.naturvardsverket.se. Last access 30 June 2017

6. National Board of Housing Building and Planning, Bygga-bo-dailogens årsredovsining (2009)
7. Sweden Green Building Council, https://www.sgbc.se/blogglayout/1393-over-1000-byggnader-certifierade-i-miljobyggnad. Last access 20 July 2017
8. C. Andersson, E. Rönnbacke, En studie om att reducera mängden blandat avfall inom byggsektorn, Examensarbete 2014:46, Chalmers Tekniska Högskola (2014)
9. Å. Wahlström, C. Warfvinge, Miljöbyggnad PM om användarnas synpunkter 150914, Sweden Green Building Council (2015)
10. Royal Swedish Academy of Engineering Science and Swedish Construction Federation, Klimatpåverkan från byggprocessen (2014)
11. C. Heincke, Förstudie - Energianvändning under byggtiden, BELOK-rapport (2014)
12. Environmental- and energy ministry, Miljöbalken (1998:808) 26 kap. 19§

Normalisation of Measured Energy Use in Buildings—Need for a Review of the Swedish Regulations

Björn Berggren, Maria Wall, Henrik Davidsson and Niko Gentile

Abstract Normalisation of measured energy use in buildings is important in order to verify their performance in user phase. Two methods for normalisation have been presented in Sweden, static and dynamic normalisation. The static normalisation considers deviating hot water use, indoor temperature, internal loads and external climate. The dynamic normalisation is based on repeated simulation, meaning that the initial simulation, carried out during the design phase, is repeated with updated conditions regarding actual use of the building and exterior climate. The ratio between the first and second simulation is used as a factor for normalisation. A pre-study has been initiated in Sweden to enable further development of the two methods. This paper present the two methods, the initiated pre-study, and some early findings. The early findings show that there is need for further development of the methods presented.

Keywords Normalisation · Energy use · Swedish regulations

1 Introduction

While pushing boundaries of energy efficiency in buildings, it is of growing importance that predicted energy performance is actually achieved during user phase.

Performance gaps have been identified in earlier studies [1–15], showing that predicted energy use is often not achieved during user phase. Some of the studies show a very large performance gap [3–5, 11], some show a lower performance gap [6, 8].

B. Berggren (✉) · M. Wall · H. Davidsson · N. Gentile
Department of Architecture and Built Environment, Division of Energy and Building Design, Lund University, Lund, Sweden
e-mail: bjorn.berggren@ebd.lth.se

B. Berggren
Skanska Sverige AB, Teknik, Stockholm, Sweden

D. Johansson et al. (eds.), *Cold Climate HVAC 2018*,
Springer Proceedings in Energy, https://doi.org/10.1007/978-3-030-00662-4_84

One way to overcome and to identify actual performance gaps is to normalise the measured energy use. Indeed, in the cited works, a smaller performance gap is generally found when measured energy use is normalised.

Some studies normalise the measured energy use due to either internal or external deviating boundary conditions [6, 8], the latter being investigated and discussed in other studies [1, 2, 9, 14], which however do not attempt to normalise the measured energy use. A Swedish study investigated the uncertainty of different methods for normalizing energy use for deviating external boundary conditions and found that different methods may have a major impact. Furthermore, they concluded that the tested methods needs to be further developed, especially in order to be suitable for low-energy buildings [16].

However, none of the studies [1–16] attempts to normalise measured energy use for both internal and external deviating boundary conditions.

Normalisation of energy use allows comparison and verification of energy use in buildings, clarifying if a deviation is generated by different conditions of use or by an actual performance failure.

The Swedish Board of Housing, Building and Planning (Boverket) recently published regulations regarding verification of energy performance of buildings [17]. These regulations introduce two different methods for normalisation, where it is possible to choose one of these.

The first method is a static approach where the normalisation is carried out in four steps. The second method is a dynamic approach using a simulation tool. These methods have not been evaluated and may both have strengths and weaknesses.

To increase the knowledge on normalisation methods for the measured energy use in buildings a pre-study has been initiated, founded by the Swedish construction industry's organisation for research and development, SBUF [18].

It should be noted that the pre-study is still ongoing. The main purpose of this paper is to present the methods introduced by Boverket, the initiated pre-study, and some early findings.

Boverket has presented two methods to standardise normalisation of measured energy use. However, more work may be needed to improve the methods. The initiated pre-study may be an important first step.

2 Methods for Normalisation from Boverket

2.1 Static Normalisation

The static normalisation is carried out in four steps, including effect of hot water use, deviating indoor temperature, deviating internal loads and deviating external climate. The static normalisation is graphically summarised in Fig. 1 and it follows Eq. 1.

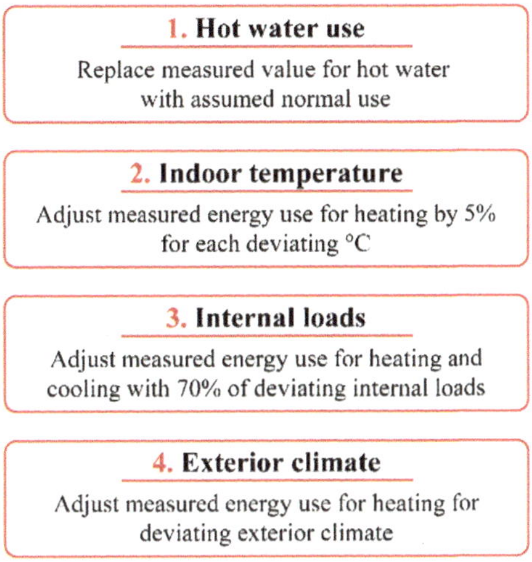

Fig. 1 Summary of static normalisation according to the Swedish national board of planning and housing (Boverket)

$$E_{norm} = E_{meas,DHW} - E_{corr,DHW} + \frac{E_{meas,SH} \cdot TAF + E_{meas,C} - E_{corr,IL}}{OCD} + E_{aux} \quad (1)$$

where E_{norm} is normalised energy performance based on static normalisation, $E_{meas,DHW}$ is the measured energy use for domestic hot water (excluding energy losses for hot water circulation), $E_{corr,DHW}$ is used to normalise energy use for domestic hot water (Eq. 2), $E_{meas,SH}$ is measured energy use for space heating, TAF is used to normalise energy use due to deviating indoor temperature (Eq. 4), $E_{meas,C}$ is the measured energy use for cooling, $E_{corr,IL}$ is used to normalise energy use due to deviating internal loads from plug loads and lighting (Eq. 5), OCD is used to normalise energy use due to deviating outdoor climate (Eq. 6), and E_{aux} is auxiliary energy used, e.g. fans, pumps, elevators [19].

Hot water use

The first step of static normalisation is related to hot water use, see Eq. 2.

$$E_{corr,DHW} = E_{\alpha,DHW} + E_{meas,DHW} \quad (2)$$

where $E_{\alpha,DHW}$ is the normal energy use for domestic hot water and $E_{meas,DHW}$ is the measured energy use for domestic hot water.

If $E_{meas,DHW}$ is measured including energy losses for hot water circulation, Boverket requires that 25% of the energy use for domestic hot water heating should be assumed to be energy losses due to hot water circulation. These energy losses are expected to heat the building and should therefore be included in space heating energy.

If domestic hot water is measured by volume; $E_{meas,DHW}$ may be calculated according to Eq. 3.

$$E_{meas,DHW} = \frac{(V_{DHW} \times 55)}{SCOP_{DHW}} \tag{3}$$

where V_{DHW} is the measured annual volume of domestic hot water (m^3) and $SCOP_{DHW}$ is the seasonal coefficient of performance (SCOP) for the heat source. The equation is based on an assumption that incoming cold water from the municipality on average needs to be heated 47 °C, from 8 to 55 °C.

Indoor temperature (Temperature Adjustment Factor)
The second step of static normalisation is related to indoor temperature, see Eq. 4.

$$TAF = 1 + (T_{\alpha} - T_{meas}) \times 0.05 \tag{4}$$

where T_{α} is the normal indoor temperature during heating season and T_{meas} is the measured indoor temperature during heating season.

Internal loads
The third step of static normalisation is related to internal loads, see Eq. 5.

$$E_{corr,IL} = \frac{(E_{\alpha,IL} - E_{meas,IL}) \times I_h}{SCOP_{heating/cooling}} \tag{5}$$

where $E_{\alpha,IL}$ is the normal energy demand for plug loads and lighting, $E_{meas,IL}$ is the measured energy use for plug loads and lighting, I_h is the share of internal loads assumed to affect the heating or cooling and $SCOP_{heating/cooling}$ is the SCOP for space heating or cooling. According to Boverket, $E_{corr,IL}$ is applied/used if energy for plug loads and lighting deviates more than 3 kWh/m^2a. Furthermore, they recommend that I_h may be assumed to be 70% when adjusting energy use for heating. No recommendation is given for adjustment of cooling.

Outdoor climate (Outdoor Climate Divisor)
The last and fourth step relates to deviating exterior climate. Boverket recommends normalisation by using the energy index [20] from SMHI [21]. The energy index, OCD_{EI} gives a weighted adjustment divisor based on outdoor temperature, solar radiation and wind.

$$OCD_{EI} = \frac{EI_{meas}}{EI_{\alpha}} \tag{6}$$

where EI_{meas} is the measured heating degree days adjusted for solar radiation and wind and EI_{α} is the normal heating degree days adjusted for solar radiation and wind.

2.2 *Dynamic Normalisation*

It is also allowed to normalise the measured energy use based on repeated simulation. This means that the initial simulation, carried out during the design phase, is repeated with updated conditions regarding actual use of the building and exterior climate. The ratio between the first and second simulation is used as a factor for normalisation. Boverket states that the initial simulation and the repeated simulation has to be carried out in the same way. Furthermore, they clarify that technical parameters, such as quantities of insulation etc., must not be changed and this method of normalisation is only allowed when actual use (plug loads, lighting etc.) is verified.

3 The Pre-study

The purpose of the pre-study is to create a knowledge basis for further work. This is done by examining different methods for normalisation and highlighting areas which could benefit from further development. The work is carried out in three phases, see Fig. 2.

3.1 *Literature Review*

The literature review will examine previous studies focusing on identification of important boundary conditions and parameters which may affect buildings' energy use during user phase and how deviating conditions may be accounted for by normalisation.

If possible; the identified conditions/parameters will be ranked based on their impact on energy use.

Fig. 2 Pre-study phases

3.2 Stakeholders' Engagement

Public seminars will be carried out with consultants, practitioners and experts within the field. The purpose of the seminars is to gather input regarding important parameters which should be considered for normalisation of measured energy use.

3.3 Dissemination

The results from the literature review and seminars will be gathered in a report to highlight important areas for further work. The results will also be presented in a Swedish technical journal.

4 Early Findings, Review of Methods for Normalisation

4.1 Static Normalisation

In Table 1, early findings regarding different energy use and aspects which are included/excluded in the static normalisation from Boverket are summarised. As can be seen, there is a large number of aspects influencing the energy use that are not included.

Based on Table 1, the static normalisation method by Boverket has the following limitations with respect to different use of energy:

- Heating; aspects such as deviating hot water use, increased/decreased ventilation, occupancy, and system losses are excluded.
- Cooling; aspects such as exterior climate, indoor temperature, hot water use, increased/decreased ventilation, occupancy presence and system losses are not included.
- Hot water; aspects such as system losses, indoor temperature and set points are not included in the normalisation.
- Ventilation, lighting, plug loads, auxiliary energy and renewable energy production; no aspects are included, there is no method for normalisation.

There are also examples where the factors used in the static normalisation lacks scientific basis. One example is the factor for deviating indoor temperature (5% per deviating °C). Previous studies have shown that deviating indoor temperature has a greater effect than the stipulated 5% per °C [5, 8, 14].

Table 1 Summary of early findings regarding energy use and aspects of normalisation which are included/excluded in the Boverket static method for normalisation

Energy use	Aspects included in Swedish normalisation	Aspects excluded in Swedish normalisation
Heating	Exterior climate Set points/Indoor temperature Plug loads Lighting	Hot water Ventilation Auxiliary Occupancy System losses
Cooling	Plug loads Lighting	Exterior climate Set points/Indoor temperature Hot water Ventilation Auxiliary Occupancy System losses
Hot water	Hot water use	Set points/Indoor temperature System losses
Ventilation		Exterior climate Set points/Indoor temperature Plug loads Lighting Occupancy
Lighting		Exterior climate Occupancy
Plug loads		Occupancy
Auxiliary energy		Occupancy
Renewable energy		Exterior climate

4.2 *Dynamic Normalisation*

Regarding dynamic normalisation there are no instructions regarding parameters which may be included or excluded when the initial "design simulation" should be repeated for the actual conditions regarding use of the building. E.g. is there a need to take into account relative humidity in outdoor air?—If so, it would also mean that it needs to be measured.

5 Discussion and Conclusions

The static normalisation from Boverket tries, and succeeds to some extent, to include both deviating internal and external boundary conditions. The method is simple and straight forward but most likely at the expense of accuracy.

Many important aspects, such as occupancy, are not included in the normalisation. Furthermore, the terms and factors used need to be further developed and clarified. One example may be that the share of internal load that affects the heating or cooling most likely varies in relation to the energy-efficiency of the building. A second example is the normalisation due to deviating external climate; the energy index from SMHI may be applied using one divisor for a whole year, month by month or in a higher resolution, and Boverket does not stipulate which resolution should be used.

Regarding dynamic normalisation, there is much work needed to clarify this method. If the method is allowed to be vague, there is a big risk that different stakeholders will apply and use the method differently.

References

1. A.C. Menezes, A. Cripps, D. Bouchlaghem, R. Buswell, Predicted vs. actual energy performance of non-domestic buildings: using post-occupancy evaluation data to reduce the performance gap. Appl. Energy **97**, 355–364 (2012)
2. C. Demanuele, T. Tweddell, M. Davies, in *Bridging the Gap Between Predicted and Actual Energy Performance in Schools*. World Renewable Energy Congress XI (Abu Dhabi, 2010)
3. B. Bordass, R. Cohen, J. Field, in *Energy Performance of Non-domestic Buildings—Closing the Credibility Gap*. International Conference on Improving Energy Efficiency in Commercial Buildings (Frankfurt, 2004)
4. Carbon Trust, Closing the gap—Lessons learned on realising the potential of low carbon building design, https://www.carbontrust.com/media/81361/ctg047-closing-the-gap-low-carbon-building-design.pdf. Last accessed 31 Aug 2017
5. J. Rekstad, M. Meir, E. Murtnes, A. Dursun, A comparison of the energy consumption in two passive houses, one with a solar heating system and one with an air–water heat pump. Energy Build. **96**, 149–161 (2015)
6. K. Mahapatra, S. Olsson, Energy performance of two multi-story wood-frame passive houses in Sweden. Buildings **5**(4), 1207 (2015)
7. M. Norbäck, Å. Wahlström, Sammanställning av lågenergibyggnader i Sverige (In Swedish) (2016)
8. U. Janson, Passive houses in Sweden, From design to evaluation of four demonstration projects, Doctoral thesis, Lund University (2010)
9. D. Majcen, L. Itard, H. Visscher, Actual and theoretical gas consumption in Dutch dwellings: what causes the differences? Energy Policy **61**, 460–471 (2013)
10. D. Majcen, L.C.M. Itard, H. Visscher, Theoretical vs. actual energy consumption of labelled dwellings in the Netherlands: discrepancies and policy implications. Energy Policy **54**, 125–136 (2013)
11. G. Branco, B. Lachal, P. Gallinelli, W. Weber, Predicted versus observed heat consumption of a low energy multifamily complex in Switzerland based on long-term experimental data. Energy Build. **36**(6), 543–555 (2004)
12. O. Guerra-Santin, L. Itard, The effect of energy performance regulations on energy consumption. Energy Effi. **5**(3), 269–282 (2012)
13. P. de Wilde, The gap between predicted and measured energy performance of buildings: a framework for investigation. Autom. Constr. **41**, 40–49 (2014)
14. M. Wall, Energy-efficient terrace houses in Sweden: simulations and measurements. Energy Build. **38**(6), 627–634 (2006)

15. N. Kampelis, K. Gobakis, V. Vagias, D. Kolokotsa, L. Standardi, D. Isidori, C. Cristalli, F.M. Montagnino, F. Paredes, P. Muratore, L. Venezia, M.K. Dracou, A. Montenon, A. Pyrgou, T. Karlessi, M. Santamouris, Evaluation of the Performance Gap in Industrial, Residential & Tertiary Near-Zero Energy Buildings, Energy and Buildings
16. P. Isaksson, P. Carling, Normalisering av byggnadens energianvändning (In Swedish), Svebyprogrammet, Projektrapport 2012-01-30 (2012)
17. Boverket, Boverkets föreskrifter om ändring av verkets föreskrifter och allmänna råd (2016:12) om fastställande av byggnadens energianvändning vid normalt brukande och ett normalår BFS 2017:6 BEN 2, Boverket, Karlskrona (2017)
18. SBUF, Svenska Byggbranschens Utvecklingsfond, https://www.sbuf.se. Last accessed 31 Aug 2017
19. Boverket, Boverket's building regulations—mandatory provisions and general recommendations, BBR 23 BFS 2016:6, Boverket, Karlskrona (2016)
20. SMHI, SMHI Energi Index, https://www.smhi.se/professionella-tjanster/professionella-tjanster/fastighet/smhi-energi-index-normalarskorrigering-1.3494. Last accessed 31 Aug 2017
21. SMHI, Sveriges meteorologiska och hydrologiska institut, https://www.smhi.se/en. Last accessed 31 Aug 2017

Environmental Sustainability Building Criteria for an Open Classification System

Nicolas Francart, Eje Sandberg and Martin Erlandsson

Abstract Existing classification systems linked to the environmental performance of buildings provide limited added value for practitioners. A survey among Swedish construction entrepreneurs showed that there is a real demand for better formulated criteria and clearer guidance. At the same time, critical investigation of requirements based on fixed average values for primary energy factors (such as in the EU Environmental Performance of Buildings Directive) shows that they are insufficient to provide guidance towards environmental sustainability building practices. They fail to take into account a number of methodological issues, including seasonal and hourly variability of energy supply and demand, and the future evolution of energy mixes. This is illustrated in the case of Sweden. The outline of an Open Classification System, currently under development, is then presented. This system focuses on methodological transparency and validity, as well as ease of use for practitioners. It addresses specifically issues where other existing systems were found to be lacking, and its methodology will be assessed to ensure that it provides optimal guidance towards environmentally sustainable practices. The system is based on three criteria: the energy resource index and global warming potential, calculated with attributional and consequential life cycle approaches, and a heat loss factor to assess the building's energy performance independently from the supply side.

Keywords Sustainable construction · Building certification · Heat losses
Primary energy · Global warming potential

N. Francart (✉) · M. Erlandsson
KTH Royal Institute of Technology, Stockholm, Sweden
e-mail: francart@kth.se

M. Erlandsson
e-mail: martin.erlandsson@ivl.se

E. Sandberg
ATON Teknikkonsult AB, Stockholm, Sweden
e-mail: eje.sandberg@aton.se

M. Erlandsson
IVL Swedish Environmental Institute, Stockholm, Sweden

D. Johansson et al. (eds.), *Cold Climate HVAC 2018*,
Springer Proceedings in Energy, https://doi.org/10.1007/978-3-030-00662-4_85

1 Context and Objective

The newly-started project "Open classification system" aims at developing easy-to-use criteria for environmentally sustainable construction, refurbishment and energy use over a building's life cycle. These criteria could be used in certification schemes such as FEBY [1] or Miljöbyggnad [2], or as an inspiration for further development of regulated energy requirements from the National Board of Housing, Building and Planning (Boverket). They could also serve as a basis for much needed guidelines for Green Public Procurement (GPP) related to construction, as Sweden has been formally notified by the European Commission for failing to implement EU rules on Public Procurement [3]. Project results will be delivered in spring 2018.

Assessment criteria in several certification systems for sustainable buildings are based on a Life Cycle Assessment (LCA) methodology, like in BREEAM [4], LEED [5] and Miljöbyggnad [2]. European standards on Construction Product Regulations (CPR) and Environmental Product Declarations (EPD) include guidance on assessing environmental impacts from construction and operation in a LCA perspective [6, 7]. They are based on attributional LCA (ALCA), meaning that the inputs and outputs of all processes of a system are inventoried as they occur and add up to the global result. An alternative approach is consequential LCA (CLCA), describing all processes happening in the background system in consequence of decisions made in the foreground system, e.g. "how does the energy supply change if demand increases by 1% ?" [8]. Requirements in the EU Energy Performance of Buildings Directive (EPBD) [9] and the Swedish building code [10] are based on ALCA using fixed average values for primary energy factors (PEF). Our working hypothesis is that this method provides unclear guidance and does not steer towards environmentally sustainable practices.

The challenge is therefore to develop scientifically based criteria to describe and quantify desirable building properties that guide practitioners towards environmentally sustainable solutions. The method must be valid, verifiable, and relate to common practices by developers and constructors. These properties should be expressed in a classification system giving clear guidance and creating a communicative platform where technical requirements are translated into market values and used to benchmark environmental performance. To ensure these criteria are developed in a way suitable for the target audience, a survey among practitioners was carried out (see Sect. 3.1).

This paper will first highlight the current regulation and methodological issues in LCA of buildings. The outline of a classification system to answer these issues is then presented, as well as how it will be assessed to ensure it reached its objectives.

2 Background

Performing an understandable and scientifically valid LCA of a building raises several methodological issues. Before assessing operational energy use, at least three background parameters must be set: the geographical scope of the energy supply, the composition of the energy mix, and impact factors per kWh for fuels and other energy carriers. At each step, assumptions must be made and the uncertainty increases.

The assessment of operational energy demand could be based on a Swedish, Nordic or European electricity mix. Transmission capacity is high on the Nordic electricity market, and is likely to increase further in the future, both between Nordic countries and with the rest of Europe, as a result of European policies [11, 12]. It is already common to use a Nordic mix [13], but an European mix might be suitable in the long term. On the other hand, a local or regional scale can be relevant for district heating.

Once the scope of the grid has been decided, the composition of the energy mix can be set. In an ALCA, the energy mix includes all production within the scope of the grid. In a CLCA, the so-called marginal mix corresponds to the generation displaced by a small change in demand. It can be set by selecting the most expensive plants in operation at any given time, but there are disagreements on how "wide" the margin should be (e.g. should it be the 10% most expensive plants or 5%?) [14, 15]. Historical production data for all plants can also be used [16], but assessing future energy use requires predicting future production with a high time resolution.

Once the composition of the energy mix is set, the impact of energy use may be calculated using impact factors per kWh for all energy carriers. The impact of energy use on energy resources can be assessed using primary energy factors (PEFs), in kWh primary energy used per kWh final energy used. The Energy Performance of Buildings Directive [9] and the Swedish building code [10] use fixed average values for PEFs and a yearly time resolution. However, the method they use is problematic for several reasons:

- PEFs are set currently in a manner that is opaque and out of reach for practitioners.
- PEFs do not take into account the variability of electricity and heat production depending on season and time of day. The energy mix varies greatly between peak times and periods of low demand, and the necessity of using hourly data to increase the validity of assessments of energy use is well documented [17–19].
- PEFs do not consider the future evolution of the energy supply: a building will use energy for several decades, but the way this energy will be supplied in the future might be very different from what it is today.

For all these reasons, fixed PEFs as per the EPBD might encourage unsustainable practices. For example, bioenergy is considered to have a PEF of 0 kWh/kWh in the EPBD[1]. Most multi-family dwellings in Sweden are heated by district heating, where the share of bioenergy is already large and will keep increasing. Large- and middle-size producers also produce electricity from biomass in combined heat and power plants (CHP). If an energy requirement based on a PEF of zero for bioenergy is to be used, the calculated impact of energy use would be so low that it would allow unconstrained energy wasting. Poorly insulated buildings shouldn't be prioritized just because a simplified assessment drives down the PEF of the energy mix.

One way to provide a more suitable assessment would be to consider variable impact factors instead. The assessment could be performed with hourly values instead of yearly ones, to consider that peak hour electricity and heat production in winter is both more expensive and more environmentally damaging. The assessment should also consider the future evolution of the energy supply. However, integrating future scenarios with intermittent electricity production from renewable power plants and interconnected European grids is a complex task, yielding an uncertain result.

Another solution is using an additional quality factor at the building level, independent from the supply side, which remains robust regardless on assumptions about the energy grid. One such factor, used in the Swedish FEBY passive house criteria for about ten years, is the Heat Loss Factor (HLF) [1]. Reducing heat losses is a strategic priority, as it decreases the burden on the supply side, especially during peak hours.

The aforementioned solutions do not directly address environmental issues. Primary energy use doesn't consider which and how resources are used. More direct indicators of environmental impact seem needed to address environmental issues. An energy resource index (ERI) has been previously developed that moves beyond primary energy use to also consider the characteristics of energy resources in terms of scarcity and carrying capacity [20]. Global Warming Potential (GWP) is also a common indicator in LCA. Recent developments in LCA literature suggest that GWP from construction can be of the same magnitude as GWP in the operational phase [21–23].

A combination of three criteria based on heat losses, energy resources use and GWP would comply with European directives, provide a robust indication of a building's inherent thermal properties, and inform about its environmental impact.

[1]It is considered to have a PEF of 1 kWh/kWh in the current Swedish building code.

3 Preliminary Results

3.1 Survey Among Practitioners

To help set an initial direction for the design of environmental sustainability criteria, a survey was carried out amongst developers and entrepreneurs with previous experience in energy-efficient buildings and certification schemes. 26 answers were gathered. Relevant results are presented below:

- *Why work with classification systems?* Practitioners want to show that they actively work towards sustainability targets. Criteria also provide a structured procedure.
- *What important areas should be covered by such a system?* Buildings heat demand/losses, electricity use, and climate impact from operation rank the highest. Costs related to energy use, cooling demand, distribution losses and climate impact from construction materials are also considered important. Hot water and energy use for construction processes are considered the least important.
- *What are important areas related to indoor environment?* Almost all answers were rated high. This could mean that construction legislations are too lax, or that clients have high requirements that should be considered in classification systems.
- *How should a classification system be structured?* Practitioners tend to prefer systems where several subsystems can be chosen, and where better results in one subsystem can compensate for lacks in other subsystems.
- *How should certification results be communicated?* Criteria are more impactful when expressed as "grades" (gold, silver or bronze) than when expressed as concepts (passive house, low-energy house or standard house).
- *How should requirements be followed-up?* Third-party control is suggested for air tightness, energy signature and energy use. Self-monitoring is suggested for other aspects such as thermal transmittance or documentation of moisture.
- *Is it economically profitable to use certification schemes?* Lower costs appear to be desirable, and so do better formulations of the requirements and a better guidance to help apply criteria and measurements.

3.2 Outline of the Classification System

Sustainable construction can be defined as addressing environmental, social and economic sustainability aspects. However, the present project focuses on developing a tool to assess environmental sustainability only. This encompasses both a building's properties and external conditions that impact its performance (energy supply, local transport infrastructure, etc.). It is however necessary to set boundaries

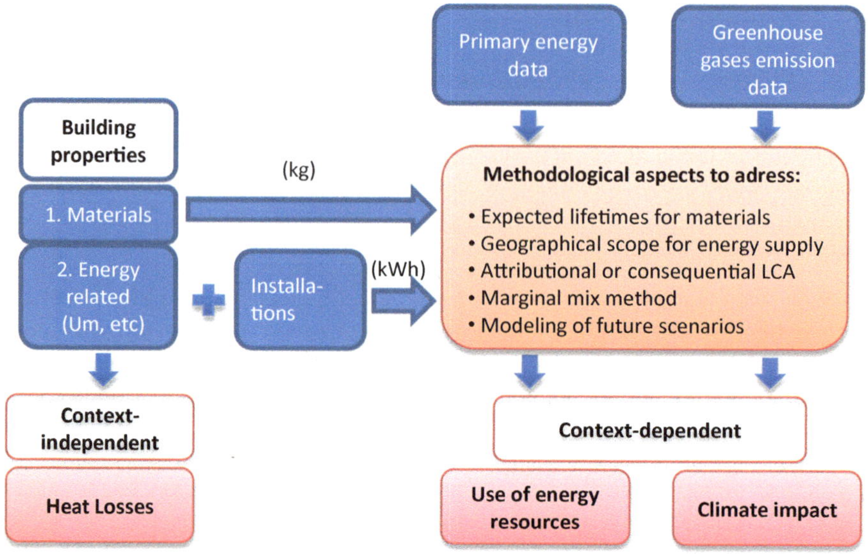

Fig. 1 Methodology and subsystems in the open classification system

to distinguish between different domains of responsibility, as the developer often has little control over the surrounding infrastructure. The classification method developed in this project addresses three subsystems, as illustrated in Fig. 1:

1. Building properties, independent from the supply side
2. Use of energy resources from construction works and operational energy demand
3. Global warming potential from construction works and operational energy demand.

Properties of buildings and materials can be used as a starting point to develop criteria related to each of the 3 subsystems. The Heat Loss Factor (HLF) represents the inherent quality of a building's climate shell. Climate impact is expressed as a Global Warming Potential (GWP) in kgCO2e. To indicate whether energy-intensive systems lead to a sustainable use of energy resources, primary energy is linked to indicators of sustainability, scarcity and carrying capacity in an Energy Resource Index (ERI).

Criteria can be formulated in two different ways:

- As mandatory minimum levels of performance, required to apply for a construction permit or in architectural competitions.
- As a voluntary complementary indicator to be communicated to the consumer (such as a color scale to represent environmental performance).

The latter is more suitable when the methods or data assessment are less robust. Figure 2 illustrates an example: a building could have to reach a minimum thermal

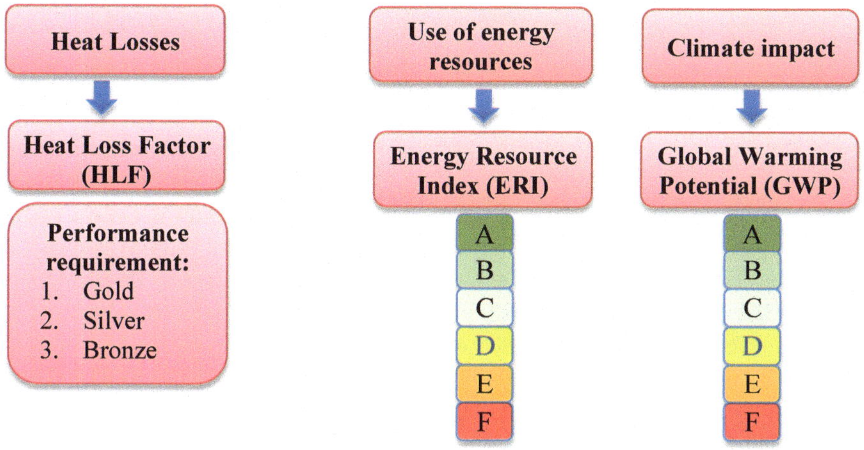

Fig. 2 Criteria used and expression of results in the classification system

performance (maximum value of HLF) to be granted a construction permit, while the ERI and GWP are rated on a A–F scale to be communicated to the consumer.

The Open Classification System will include both ALCA and CLCA perspectives, as the two approaches answer different questions, and two different time horizons:

- ALCA as per EN 15804 [7], using current average data for the energy supply
- ALCA with a high time resolution inventory linking the building's energy performance and the total energy supply, with inventory data based on future scenarios
- CLCA with a high time resolution inventory that links the building's energy performance and the marginal energy supply, including future scenarios.

3.3 *Building Level: The Heat Loss Factor to Assess Energy-Related Properties*

Definition and Measure of the HLF. The HLF is the sum of heat losses by transmission, ventilation and infiltration per square meter (in W/m^2) with an outside temperature equal to the design winter outside temperature (DWOT):

$$HLF_{DWOT} = H_T \cdot (21 - DWOT)/A_{temp}(W/m2) \quad (1)$$

where H_T is the building's heat loss coefficient (W/K) and DWOT is the design winter outside temperature. For buildings with significant heat losses through the ground, it can be necessary to use a separate heat loss coefficient for these losses:

$$\mathrm{HLF_{DWOT}} = \mathrm{H_{T'}} \cdot (21-\mathrm{DWOT})/\mathrm{A_{temp}} + \mathrm{H_{T''}} \cdot \left(21-\mathrm{T_{ground}}\right)/\mathrm{A_{temp}} \quad (2)$$

where *HT'* is the heat loss coefficient without losses to the ground and *HT"* the heat loss coefficient to the ground. A HLF criterion can use hourly, weekly or monthly data, and compensate for small buildings and temperature differences in cold regions.

Relevance of the Heat Loss Factor. Heat losses are directly related to winter peak demand, when delivering energy is the most expensive and environmentally damaging. The HLF is therefore strategically and environmentally relevant.

The HLF is easily calculated and only relates to aspects the developer has influence over (the building's climate shell, air tightness, and ventilation). European standards exist for HLF calculation, and no additional data beyond what is used in yearly energy calculations is required. Calculations can be performed in a spreadsheet with an easy-to-review template. The HLF is therefore easy for practitioners to adopt.

Net heat, the final energy used for heating, is another alternative that could be considered, but it entails a more complicated calculation of energy balance, based on more uncertain behavior-related data. The Swedish building code also includes a complementary requirement on the average energy transfer coefficient (Um), which is linked to heat losses. In practice, this requirement only impacts buildings with a low form factor (relation between enclosing area and heated area). It has virtually no effect on smaller buildings with a high form factor, and doesn't limit losses from ventilation and infiltration, which makes it too rough to provide good guidance.

3.4 Assessment of Climate Impact and Resource Use

A well-established methodology already exists to calculate the HLF. The process is more complicated for the ERI and GWP indicators. Various methodological choices (scope, future scenarios, etc.) affect the results, and the sensitivity of criteria to these choices will be assessed and will be in itself a result of the project.

ALCA based on the current energy mix will use a Nordic electricity mix and hourly data. When including future scenarios, differences between a Swedish, Nordic or European electricity mix and how they impact the assessment results will be investigated. Only the most relevant mix will be used in the final classification system.

Hourly data is required, and such a high time resolution can't be found in reports for future energy scenarios. Two solutions are currently being explored. Ideally, authors of the three scenarios could provide either more detailed results, or a list of parameters they used so that we can run the model ourselves with a higher time resolution. Otherwise, hourly and seasonal variations will be extrapolated from current production data and applied to future yearly mixes found in the reports (assuming similar variations in 2050 and today).

Several approaches are possible to select the marginal mix in a CLCA. It can for instance be considered that the most expensive operated plants at a given time constitute the operational margin. The model for future energy mixes can also be run twice, with and without the additional demand from the building. The influence of this choice on the results will be assessed.

4 Evaluation of the Method and Use of the Results

4.1 Evaluation of Suggested Methods for the Classification System

Methodological choices that impact the classification results must be evaluated. Results must be accurate, robust and give unequivocal guidance, but the method must also be straightforward and cheap to carry out. Simplifications are needed but must be justified. This will be done by running the classification system with different methods on a range of test measures and comparing the results along the following criteria:

- Will the Open Classification System guide practitioners towards investing in measures that reduce impact on primary energy use and GWP?
- Will the method lead to suboptimal investments like overproduction of renewable energy at the building site?
- Will the method encourage measures that store energy from periods of overproduction from renewable power plants to use during demand peaks (winter)?
- How will peak hour demand be affected?
- How will the method address flexibility issues (ease of adapting, switching to new energy carriers, opening/closing plants, etc.). Energy systems have shorter life times than buildings, so they will need to change during the building's lifetime.
- Are the methods transparent and easy to control (even by non-experts)? This is especially important if they are to be integrated in a national building code. Workshops with building contractors will provide feedback on these subjects.

4.2 Using the Classification System to Assess Construction Measures

Changing building and installation properties affects the HLF, ERI and GWP indicators. This will allow the constructor to easily assess the impact of any measure, and choose the optimal solution. Four types of measures linked to energy use can be distinguished, depending on whether they affect:

1. Heat losses (insulation, heat exchanger, etc.) and the HLF
2. Electricity demand, and possibly heat load
3. Heat load from electrical installations that aren't coupled to outdoor temperature (fans, elevator), which changes the balance temperature when heating season starts
4. Hot water demand and heat recovery from hot water use.

Other types of measures concern conversion or production:

5. Electricity production (for internal use or exported to the grid), usually PV panels
6. Heat pump solutions for heat and/or hot water
7. Solar heat recuperation and storage.

These measures will change the energy demand of the building, which is reflected in the indicators. The benefits provided are then easily compared to each measure's cost.

5 Conclusion

The foremost objective of a classification system for environmental performance should be to guide practitioners towards adopting sustainable practices. As such, the system should be user-friendly and provide added value. Furthermore, its criteria should be scientifically valid, robust, and steer only towards sustainable practices. Existing criteria have been found to be somewhat opaque for practitioners and invalid due to oversimplifications (using a single indicator ignoring seasonal and hourly variations and future evolutions of the energy supply). The Open Classification System is currently being developed specifically to address these issues. Using a heat loss factor provides a robust indicator to assess energy performance that can be used to set minimum performance levels. The calculated energy resource index and global warming potential provide useful additional information directly related to resource use and environmental impact, which could be used to inform consumers. A number of methodological issues remain, but choices will be assessed using test measures to ensure the system provides correct guidance. Consultations with practitioners will ensure ease of use and transparency. Further, it should be noted that additional aspects of social and economic sustainability could be part of such a classification but have been excluded here. Seinre [24] points out that indoor environmental quality (IEQ) affects productivity, which in turn has a significant effect on economic sustainability, while Brown [25] showed that environmentally certified buildings are perceived as having a better IEQ but no significant impact on productivity.

Acknowledgements Björn Berggren, from Skanska AB, was responsible for carrying out the initial survey. His help and comments on this article are gratefully acknowledged.

References

1. Swedish Center for Zero-Energy Houses (2012) Kravspecifikation för nollenergihus, passivhus och minienergihus - Bostäder
2. Swedish Green Building Center (2017) Miljöbyggnad - Bedömningskriterier för nyproduktion 3.0
3. European Commission (2016) May infringements' package : key decisions. http://europa.eu/rapid/press-release_MEMO-16-1823_en.htm. Accessed 17 Aug 2017
4. BRE Global Limited (2017) BREEAM International New Construction 2016
5. United States Green Building Council (2014) LEED v4 for Building Design and Construction (Updated April 14, 2017)
6. European Standards (2011) EN 15978:2011 Sustainability of construction works. Assessment of environmental performance of buildings. Calculation method. https://doi.org/10.3403/30204399
7. European Standards (2012) EN 15804:2012 Sustainability of construction works, Environmental product declarations, Core rules for the product category of construction products
8. European Commission – Joint Research Centre – Institute for Environment and Sustainability (2010) International Reference Life Cycle Data System (ILCD) Handbook – General guide for Life Cycle Assessment – Detailed guidance. Constraints. https://doi.org/10.2788/38479
9. European Parliament, European Council (2010) Directive 2010/31/EU of the European Parliament and of the Council of 19 May 2010 on the energy performance of buildings (recast). Off J Eur Union. doi:https://doi.org/10.3000/17252555.l_2010.153.eng
10. Boverket, *Regelsamling för byggande* (BBR, Boverket, Stockholm, 2015)
11. D. Stoerring, S. Horl, European Parliament fact sheet: internal energy market (2017). http://www.europarl.europa.eu/atyourservice/en/displayFtu.html?ftuId=FTU_5.7.2.html. Accessed 18 Aug 2017
12. International Energy Agency (2016) Nordic energy technology perspectives 2016. https://doi.org/10.1787/9789264257665-en
13. J. Gode, F. Martinsson, L. Hagberg et al., *Miljöfaktaboken 2011 Uppskattade emissionsfaktorer för bränslen, el, värme och transporter (in Swedish)* (Värmeforsk Service AB, Stockholm, 2011)
14. World Resources Institute, World Business Council for Sustainable Development (2007) Guidelines for Quantifying GHG Reductions from Grid-Connected Electricity Projects
15. B. Peuportier, C. Roux , Moving towards CLCA to model electricity consumption in buildnings (2015)
16. A.D. Hawkes, Estimating marginal CO_2 emissions rates for national electricity systems. Energy Policy **38**, 5977–5987 (2010). https://doi.org/10.1016/j.enpol.2010.05.053
17. C. Roux, B. Peuportier, Considering temporal variation in the life cycle assessment of buildings, application to electricity consumption and production. World suistainable Build **2014**, 1–9 (2014)
18. J. Gode, A. Lätt, T. Ekvall, et al., Miljövärdering av energilösningar i byggnader - metod för konsekvensanalys. Stockholm (2015)
19. C. Roux, P. Schalbart, B. Peuportier, Accounting for temporal variation of electricity production and consumption in the LCA of an energy-efficient house. J. Clean. Prod. **113**, 532–540 (2016). https://doi.org/10.1016/j.jclepro.2015.11.052
20. M. Erlandsson, Hållbar användning av naturresurser - andelen nedströms klimatpåverkan för byggnader (2014)

21. C.K. Anand, B. Amor, Recent developments, future challenges and new research directions in LCA of buildings: a critical review. Renew. Sustain. Energy Rev. **67**, 408–416 (2017). https://doi.org/10.1016/j.rser.2016.09.058
22. T. Ibn-Mohammed, R. Greenough, S. Taylor et al., Operational vs. embodied emissions in buildings—A review of current trends. Energy Build. **66**, 232–245 (2013). https://doi.org/10.1016/j.enbuild.2013.07.026
23. C. Liljenström, T. Malmqvist, M. Erlandsson, et al., Byggandets klimatpåverkan - Livscykelberäkning av klimatpåverkan och energianvändning för ett nyproducerat energieffektivt flerbostadshus i betong. Stockholm (2015)
24. E. Seinre, J. Kurnitski, H. Voll, Quantification of environmental and economic impacts for main categories of building labeling schemes. Energy Build. **70**, 145–158 (2014). https://doi.org/10.1016/j.enbuild.2013.11.048
25. N. Brown, T. Malmqvist, H. Wintzell, Value creation for tenants in environmentally certified buildings. Build Res Inf 1–16 (2016). https://doi.org/10.1080/09613218.2016.1207137

Part XII
Other Aspects of Buildings in Cold Climates

Sensitivity Analysis of Melting and Freezing of Snow on Roofs

Anker Nielsen

Abstract The paper describes a statistical analysis of a mathematical model for calculation of the melting and freezing of snow on roofs. Parameters are roof length, overhang length, heat resistance of roof and overhang, outdoor and indoor temperature, snow thickness and thermal conductivity. If the snow thickness is above a limit value, then part of the snow will melt. This gives water flow to the overhang. Part of the water will freeze on the overhang and a part will drip from the roof. If the water flow is small, then all the water will freeze on the overhang otherwise there will be dripping and icicles. The paper uses sensitivity analysis with the Morris method to find parameters that are negligible, linear or non-linear. The Sobol sensitivity indices are also calculated. By means of sensitivity analysis, it is possible to determine, which parameters are the most important e.g. the thickness of the snow or indoor and outdoor temperatures and roof thermal resistance. In practice, some parameters are difficult to change, but the analysis shows where the effect is most efficient.

Keywords Snow · Statistical analysis · Roofs

1 Introduction

In winter snow will accumulate on the roof of buildings. This will add insulation to the roof if the snow does not melt. In most periods part of the snow will melt and the water flow can generate icicles at the eaves. This occurs if there is solar radiation on the roof, as it will melt part of the snow. But this is not the only important factor. Investigations in United States by Tobiasson [1] showed that the most important factor for the generation of icicles is melting caused by a heated building. The problem is smaller for a thermally well-insulated building. An unheated

A. Nielsen (✉)
Department of Building and Process, Danish Building Research Institute, Aalborg University, A. C. Meyers Vænge 15, 2450 Copenhagen SV, Denmark
e-mail: ani@sbi.aau.dk

D. Johansson et al. (eds.), *Cold Climate HVAC 2018*,
Springer Proceedings in Energy, https://doi.org/10.1007/978-3-030-00662-4_86

building presents few problems. A Canadian report by Straube [2] discusses formation of ice dams on the roof as the melting water can freeze on the roof overhang. Melting snow on glass roofs has been discussed in Dreier [3] and Nielsen [4]. It is therefore interesting to analyze melting of the snow on the roof, freezing on the overhang and dripping from the eave.

For analyzing the problem with snow melting and freezing the most important factors are climate and the insulation level of the roof. A HVAC system in the building can worsen the problems. An example from Norway [5] is a low building where large icicle were formed. In this case the ventilation system was placed in the attic above the room. The thermal insulation was on the attic floor. As there where little or no thermal insulation of the ventilation system the heat from the air ducts and heat exchangers gave off so much heat that it increased the attic temperature. The result was that the snow on the roof melted in cold periods and large icicles were created all the way from the gutter to the ground. The solution for the problem was thermal insulation of the technical systems.

2 Energy Balance—Calculation Method

Figure 1 is a sketch of the energy and water balance of a typical roof with snow in the winter. We assume that the attic is not ventilated. The effect of a ventilated attic is discussed later. The heat flows from the interior of the house to the surface of the roof through the roof construction. The heat will flow through the snow to the outdoor air and possibly melt some of the snow. With small snow thicknesses, there

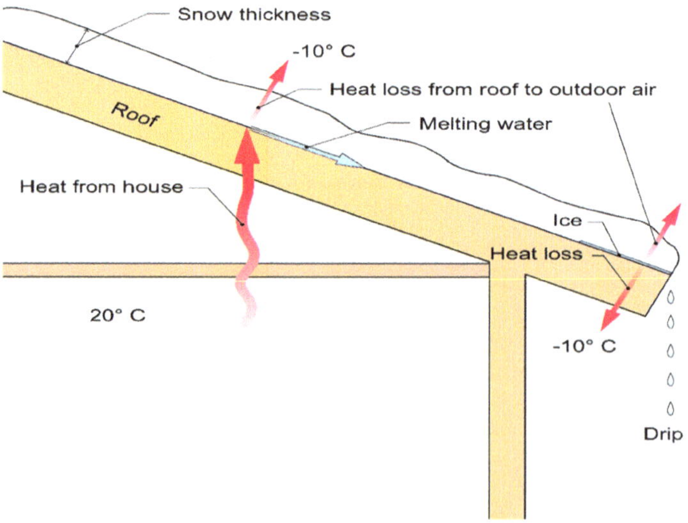

Fig. 1 Heat balance and water flow on roof

will be no melting of the snow as the surface temperature on the roof is below 0 °C. With large snow thicknesses the bottom of the snow layer will melt and result in melt water that follows the slope of the roof. It is therefore interesting to calculate the snow thickness that corresponds to the limit between melting and no melting. This is the case if the roof surface temperature is 0 °C. This is called the melting limit. With known indoor and outdoor temperatures as well as the U-value of the roof, the snow melting thickness can be found. Part of the melting water will freeze on the overhang and the rest will drip from the roof. A special case is if all the melting water is freezing on the overhang so there is no dripping. This is called the dripping limit. It is possible to calculate the snow thickness for this case. This thickness is always larger than the melting limit.

The calculation of the energy balance and the water flows is performed as follows; The heat flows Q is pr. unit of width of the roof. The thermal conductivity of the snow is calculated from the snow density. The thermal resistance of the snow layer (R_s) is then:

$$R_s = \text{snow thickness/snow thermal conductivity} \quad (1)$$

The heat flow (Q_r) from the building to the upper surface of the roof and the melting snow (with a temperature of 0 °C) is:

$$Q_r = l_r/\ R_r * (T_i - 0) \quad (2)$$

The length of the roof is (l_r). The indoor temperature is (T_i). The thermal resistance between the room and upper surface of the roof is (R_r). The heat flow (Q_s) from the underside of the snow to the outside air (T_u) is:

$$Q_s = l_r/\ R_s * (0 - T_u) \quad (3)$$

The thermal resistance of the snow is (R_s). The heat flow (Q_m) for melting is:

$$Q_m = Q_r - Q_s \quad (4)$$

The value (Q_m) must be positive or zero (no melting). An interesting temperature (T_{melt}) is the theoretical surface temperate of the roof if we ignore the melting:

$$T_{melt} = T_i - (R_r/\ (R_r + R_s)) * (T_i - T_u) \quad (5)$$

If $T_{melt} > 0$ then use formula (6) to calculate the flow of melting water m_{melt} else $m_{melt} = 0$:

$$m_{melt} = l_r/\ MH * (T_i/\ R_r + T_u/\ R_s) \quad (6)$$

MH is the latent melting heat from ice to water. The amount of water freezing m_{freeze} on the overhang can be calculated from formula 7:

$$m_{freeze} = -l_o / MH * (T_u / R_s + T_u / R_o) \tag{7}$$

The length of the overhang is (l_o). The resistance from the surface of the roof to the outside air under the overhang is (R_o). The amount of water dripping m_{drip} is calculated with formula 8.

$$m_{drip} = m_{melt} - m_{freeze} \tag{8}$$

The amount of water dripping can either drip or freeze as icicles. In case we have a gutter, then a part will freeze in the gutter and the rest will go into the drainpipe.

In the model is the outdoor heat resistance set to zero instead of using the normal 0.04 m^2K/W. That is acceptable as the heat resistance of the snow in most cases is 20–100 times larger. An alternative is to make a small change in either the snow thickness or the outdoor temperature to account for the heat resistance between the roof and outdoor. Using a model with convection and radiation heat transfer complicate the model without changing the result of the statistical analysis.

2.1 *Melting and Dripping Limits*

The melting limit and dripping limit can be calculated from these equations. In Table 1 is an example for 20 °C indoor and –10 °C outdoor temperature and a roof length of 7 m an overhang of 0.4 m.

It is seen that for an old house the dripping limit is nearly the same as the snow melting limit. For a new insulated house the dripping limit is much higher than the melting limit, as a larger part of the melt water will freeze on the overhang. The result is that a well-insulated house has a lower risk of icicles and an overhang is important to reduce the risk of icicles. More information is found in [6].

2.2 *Non-stationary Cases*

The calculations shown are based on stationary conditions. In reality, the amount of melting water will be reduced when the snow melts since the thickness is then

Table 1 Examples of limits

Parameter	U-value roof (W/m^2K)	Melting limit (cm)	Dripping limit (cm)
Old house	1	3	3.3
Insulated house	0.3	10	12
New insulated house	0.15	20	31
Future house	0.1	30	63

reduced over time. A mathematical model for dynamic conditions can be found in Claesson and Nielsen [7]. This model has also been expanded to include a case with a roof window in the construction. This will typically give more melting water as the U-value of the window is higher than the roof U-value and will result in higher risk of icicles.

3 Methods for Sensitivity Analysis

Sensitivity analysis can be performed using different methods. In this case we use the Morris and Sobol methods as described in Saltelli [8]. The following gives a short description of the methods. For a more detailed description of the methods see the more specialized statistical literature mentioned in Saltelli.

The statistical simulation is done with the Sensitivity Analysis Library SALib [9]. SALib is an open source library written in Python for performing sensitivity analysis. SALib provides a workflow where the mathematical model (in this case the energy balance of the roof) is in a separate module, which in this case is also written in Python [10]. SALib is responsible for generating the model inputs, using one of the sample functions, and computing the sensitivity indices from the model outputs, using one of the analyze functions. A sensitivity analysis using SALib has four steps:

1. Determine the model inputs (parameters) and their sample range.
2. Run the sample function to generate the model inputs.
3. Evaluate the model using the generated inputs, saving the model outputs.
4. Run the analyze function on the outputs to compute the sensitivity indices.

The second part generates for instance 1000 cases, which are run through the model in the third part and generate a similar number of output results. There are special sample and analyze functions for each method as Morris or Sobol.

3.1 Morris Method

The Morris method is a screening method to find parameters that are important or negligible. The method varies one parameter at a time. Each parameter is divided into a discreet number of values that are chosen within the range of variation. The method calculates two sensitivity measures for each parameter. The measure for the overall effect, Morris μ of the parameter on the output, can be called the mean value. The other measure, Morris σ estimates the second and higher order effects in which the parameter is involved and can be called the standard deviation for the parameter.

The method calculates Morris μ and σ for each parameter. A high μ indicates a parameter with an important overall influence on the output. A high σ indicates either interaction with other parameters or a parameter that is non-linear. If both μ and σ are low then this parameter is negligible. The method tends to be qualitative as for ranking the input parameters in order of importance.

The calculation from the statistical module also gives a μ^*, that is the absolute mean as a supplement to the normal μ value. The μ^* is best for comparing the effect of the different parameters.

3.2 Sobol Method

The Sobol method is a quantitative method that gives the percentage of total output variance that each parameter accounts for. The method is a variance-based method to quantify the impact of uncertainties in random variables on the uncertainty of the model output. This method is more computationally expensive than the Morris method. The Sobol method for variance-based estimation is based on decomposition of the variance of a response to its variation sources. The Sobol method makes estimates of first-order sensitivity indices, higher–order indices and total indices. The first-order term represents the partial variance in the response due to the individual effects of a random variable. The higher-order terms show the interaction between two and more variables. The total effect relates to all direct and indirect variance from other variables.

3.3 Parameter Variations for the Model

The input parameters (Table 2) were selected as typical variation. The heat resistance for the roof varies from an old house with low level of thermal insulation to a highly insulated house. The length of the overhang varies from 10 to 90 cm as realistic values. The thermal resistance of the overhang is low as thermal insulation here is not common. The snow density varies from new snow as 80 kg/m^3 to older more compacted snow with 200 kg/m^3. The snow thickness ranges from 1 to 30 cm. The indoor temperature ranges from 5 °C in a room or attic with little heating to a heated room of 22 °C. The outdoor temperature ranges from –2 to –16 °C as typical in periods with snow on the roof and melting.

The analysis is based on at least 10,000 cases as the calculation is fast. Using other limits for the parameters will change the calculated indices, but in most cases this will not change the order for the important parameters.

Table 2 Uniform variation for the input parameters in the analysis

Parameter	Minimum	Maximum	Unit
Roof length	7	15	m
Roof heat resistance	1	5	m^2K/W
Overhang length	0.1	0.9	m
Overhang heat resistance	0.5	1	m^2K/W
Snow density	80	200	kg/m^3
Snow thickness	0.01	0.30	m
Temperature indoor	5	22	°C
Temperature outdoor	−16	−2	°C

3.4 Analysis of Snow Melting on Roof

The first resulting parameter is the snow melting on the roof. Table 3 gives the Morris μ^* and σ values for melting. The μ^* values give a ranking for the influence of each parameter. Two parameters related to the overhang have a value of zero as they have no influence on the melting on the roof. In order the most important is the roof heat resistance, the Indoor temperature and finally the Snow thickness. The μ value for roof heat resistance and Snow density is negative. This indicates that an increase in these parameters will reduce the amount of melting water.

Table 4 gives the Sobol indices for each parameter where S1 are the first order indices. The highest value is the roof heat resistance 0.56 and then the indoor temperature and finally the snow thickness. This is the same order as in the Morris

Table 3 Morris indices for snow melting

Parameter	μ^*	μ	σ
Roof length	0.432	0.432	0.781
Roof heat resistance	1.795	−1.795	2.526
Overhang length	0	0	0
Overhang heat resistance	0	0	0
Snow density	0.083	−0.083	0.119
Snow thickness	0.656	0.656	1.270
Temperature indoor	0.968	0.968	1.371
Temperature outdoor	0.361	0.361	0.518

Table 4 Sobol indices for snow melting

Parameter	S1	ST
Roof length	0.0190	0.0596
Roof heat resistance	0.5558	0.7624
Snow density	0.0021	0.0038
Snow thickness	0.0628	0.1479
Temperature indoor	0.1144	0.2568
Temperature outdoor	0.0247	0.0563

analysis. The total Sobol indices ST gives the total effect of the interaction from two and more other variables. This will always be larger than the S1 indices. The most important is as before the roof heat resistance that increase from 0.55 to 0.76. The parameter with the lowest value is the Snow density, so an average value could have been used without much effect on the results. The Roof length and the Outdoor temperature have a low S1 value but if we include the higher order effects these clearly have more influence.

3.5 Analysis of Water Freezing on the Overhang

The second resulting parameter is the water freezing on the overhang. Table 5 gives the Morris μ^* and σ values for freezing. The much lower values compared to melting is caused by the absolute value of the freezing being much lower than the melting. The most important factor is the outdoor temperature and after that, the snow thickness and finally the overhang length. But the next two parameters, i.e. the roof heat resistance and the indoor temperature, cannot be ignored. Here both the roof length and the snow density could be put in as average values.

Table 6 gives the Sobol indices for each parameter where the S1 is the first order indices. The highest value is the roof heat resistance 0.14 and second the overhang length and third the snow thickness. For the total Sobol indices ST most important

Table 5 Morris indices for freezing

Parameter	μ^*	μ	σ
Roof length	0.003	0.003	0.015
Roof heat resistance	0.053	−0.053	0.102
Overhang length	0.060	0.060	0.080
Overhang heat resistance	0.019	−0.019	0.033
Snow density	0.012	−0.006	0.037
Snow thickness	0.064	0.045	0.101
Temperature indoor	0.034	0.034	0.081
Temperature outdoor	0.070	−0.031	0.105

Table 6 Sobol indices for freezing

Parameter	S1	ST
Roof length	−0.0003	0.0053
Roof heat resistance	0.1414	0.4785
Overhang length	0.1259	0.2801
Overhang heat resistance	0.0198	0.0395
Snow density	0.0035	0.0048
Snow thickness	0.1238	0.4471
Temperature indoor	0.0742	0.3212
Temperature outdoor	0.0300	0.3539

is as before the roof heat resistance that increases from 0.14 to 0.48. In the case of freezing many parameters interact as for instance outdoor temperature that has a S1 of 0.03 and an ST 0.35. The total values of ST are very important for a realistic evaluation. The parameters with the lowest values are the snow density and the roof length, so an average value could have been used without much effect on the results.

3.6 Analysis of Water Dripping from the Roof

The third resulting parameter analysis gives information on the water dripping from the roof. It is from this water that icicles can be formed—depending on the outdoor temperature, wind speed and the amount of water. Too much water could melt the icicles. Table 7 gives the Morris indices. The most important is the roof heat resistance and then the Indoor temperature and then the Snow thickness. The influence from the freezing on the overhang has nearly no effect, so the results are similar to those for melting.

Table 8 gives the Sobol indices for each parameter where the S1 is the first order indices. The highest value is the roof heat resistance 0.55 and then the indoor temperature and then the snow thickness. This is the same order as found in the

Table 7 Morris indices for dripping/icicles

Parameter	μ*	μ	σ
Roof length	0.514	0.514	0.918
Roof heat resistance	1.739	−1.739	2.332
Overhang length	0.061	−0.061	0.081
Overhang heat resistance	0.019	0.019	0.033
Snow density	0.078	−0.078	0.158
Snow thickness	0.731	0.731	1.412
Temperature indoor	1.058	1.058	1.574
Temperature outdoor	0.407	0.407	0.687

Table 8 Sobol indices for dripping/icicles

Parameter	S1	ST
Roof length	0.0198	0.0627
Roof heat resistance	0.5475	0.7728
Overhang length	0.0006	0.0018
Overhang heat resistance	0.0001	0.0003
Snow density	0.0014	0.0048
Snow thickness	0.0531	0.1429
Temperature indoor	0.1064	0.2605
Temperature outdoor	0.0307	0.0684

Morris analysis. The most important ST is as before the roof heat resistance that increase from 0.55 to 0.77. The roof length and the outdoor temperature have low S1 values but if we include the higher order effects these will have a more important influence as seen in ST. The parameter with the lowest value is the snow density, so an average value could have been used without much effect on the results. The influence of the overhang is very low and can be ignored.

4 Example of Use

The statistical analysis can be used for practical examples. In case of an old house, what is the best solution for reducing the risk of icicles? We cannot change the climate or the roof and overhang length. By using Table 8 we can see that the largest total Sobol indices (ST) 0.77 is for the roof heat resistance. A better thermal insulation is the best solution. If this is not possible then reduce the indoor temperature, but the effect is much less as ST is 0.26. An alternative is to ventilate the attic with outdoor air if that is possible. In that case we use the attic temperature instead of the indoor temperature and the heat resistance R_r from the attic to the top of the roof ms. A change in the overhang (which is typically difficult to do) as length and heat resistance has in practice no effect.

For a new house we could use a high thermal insulation in the roof to reduce the risk and if possible a longer overhang as melting water can freeze on the overhang, see Sect. 2.1. But the most important is the thermal insulation and the indoor temperature.

More information on when icicles are growing and when icicles have the highest risk for falling down can be found in Nielsen [6, 11] and [12].

5 Conclusions

The statistical analysis methods are a very useful method to find the most important factors in a model. For the melting of snow on roofs the most important factor is the heat resistance for the roof (ST = 0.76) and second most important factor is the indoor temperature (ST = 0.25) and third the snow thickness (ST = 0.15). For freezing is it interesting to note that overhang parameters are not the most important. For dripping is the important parameters the same as for melting.

To avoid dripping and icicles is it important to have a good thermal insulation and low indoor temperatures in an attic. In a house the indoor temperature will in most cases be decided by the occupants, so it is the thermal insulation that has to be good. As seen in Sect. 3.1 the overhang is more important in a good insulated house.

References

1. W. Tobiasson, S. Buska, A.R. Greatorex, Attic Ventilation Guidelines to Minimize icings at eaves. J. Roof Consult. Inst. **XVI**(1), 17–24 (1998)
2. J. Straube, Ice Dams, Building Science Digest 135 (2006)
3. C. Dreier, T. Gjelsvik, J.R. Herje, T. Isaksen, A. Nielsen, Glasstak. Konstruksjoner. Klimapåvirkninger og løsninger for nordiske forhold. (Glass roofs—constructions. Climate conditions and solutions for Nordic Conditions), Norwegian Building Research Institute, Oslo, Norway, NBI Håndbok 36 (in Norwegian) 2nd printing (1985)
4. A. Nielsen, in *1988: Snow-melting and Snow Loads on Glass Roofs*. First International Conference on Snow Engineering, July 1988, Santa Barbara, USA. Printed as a special report 89-6 from Cold regions research & engineering laboratory (USA, 1989), pp. 168–177
5. H. Juul, T. Böhlerengen, Ising på tak. En studie av et skadetilfelle (Icing on a roof—case study) Norwegian Building Research Institute, Oslo, Norway Prosjekt rapport 68 (1990)
6. A. Nielsen, *Snow, Ice and Icicles on Roofs—Physics and Risks*. Sixth Nordic Conference on Building Physics in the Nordic Countries (Reykjavik, 2005), pp. 562–569
7. A. Nielsen, J. Claesson, *Melting of snow on a roof: Mathematical report*. Chalmers tekniska högskola, Göteborg, 37p (2011)
8. A. Saltelli et al., *Sensitivity Analysis in Practice. A Guide to Assessing Scientific Models* (Wiley, England, 2004), ISBN 978-0470-87093-8
9. SALib—Sensitivity Analysis Library in Python, Homepage http://salib.readthedocs.io/en/latest/version1.0.3
10. Python—homepage https://www.python.org/version3.6.1
11. A. Nielsen, in *Snow, Ice and Icicles on Roofs—Physics and Risks*, Sixth Nordic Conference on Building Physics in the Nordic Countries, Reykjavik (2005)
12. A. Nielsen, J. Claesson, *Snow and Freezing Water on Roofs, Cold Climate HVAC 2009 Conference*, Sisimiut, Greenland (2009)

The Method of Determining Climatic Loads on the Enclosing Structures Taking into Account Global Climate Change

D. Y. Zheldakov and V. G. Gagarin

Abstract In 1997, in accordance with the UN framework Convention on climate change (UNFCCC) the Kyoto Protocol was adopted. In the Committee on adaptation (2012) States the need for all countries participating in the UNFCCC to develop national plans and programs for adaptation to conduct a technical study of the process of adaptation in different spheres and primarily in energy-intensive industries such as construction and operation of residential and administrative buildings. Based on the numerical solution of the differential equation that determining one-dimensional heat transfer under nonsteady conditions with constant coefficients, the method of calculation of the temperature distribution over the cross section of the enclosing structure was developed. On the basis of the developed method of determining the number of cycles of freezing and thawing of moisture in the cross sections of the outer wall of the building are calculated. The developed method was tested in the experiment on the exterior walls of operated buildings. The results showed good convergence of the real and calculated temperature values. The calculation of the number of cycles of freezing and thawing on the cross section of the outer wall of the building according to the developed methodology and the experiment showed the same results. The method of numerical assessment of the impact of global climate change on the enclosing structures was developed. The concept of temperature intensity of the year was introduce. The method uses meteorological data of outdoor air temperature for the previous period and the results of calculation of temperature regime of the enclosing structures. The use of this method allowed to calculate the number of cycles of freezing and thawing in cross sections of the outer wall at any time interval, and, therefore, more accurately predict the durability of the enclosing structures.

Keywords Climate changes · Durability · The building envelope
Method of calculation

D. Y. Zheldakov (✉) · V. G. Gagarin
Research Institute of Building Physics of RAACS, Stuttgart, Germany
e-mail: djeld@mail.ru

D. Johansson et al. (eds.), *Cold Climate HVAC 2018*,
Springer Proceedings in Energy, https://doi.org/10.1007/978-3-030-00662-4_87

1 Problem Statement

When developing the method for calculating the maximum durability (life cycle) of various enclosing structures, i.e. determining the period, during which an enclosing structure performs its functions, it is necessary to assess correctly the rate of the enclosing structure material destruction. In the high latitude, the main criterion of material resistance to destruction is its freeze-thaw resistance. The process of polythermal crystallization leading to the destruction of material of the external building wall is the counting function of the number of cycles of structure material temperature transitions through zero.

The process of freezing and thawing goes not only on the surface of the material external wall but in the depth of it, which results in the strength decrease and even material destruction in the external wall cross-sections at a certain distance from the surface. The decrease of structure material strength reduces the strength of enclosing structure in general. That is why knowing the exact number of cycles of temperature transitions through zero in every cross-section of the external enclosing structure is necessary to determine the maximum durability (life cycle) of the external building wall.

To estimate the number of cycles of temperature transitions through zero in various cross-sections parallel to enclosing structure external surface, it is necessary to resolve the matter of temperature waves passing the wall at outdoor air temperature fluctuations, i.e. to calculate the temperature distribution over the wall mass under nonsteady conditions.

However, the task of developing the method for determining the number of freezing and thawing cycles should be stated and resolved at a broader level. In the calculations, it is necessary to consider the global climate change and, therefore, the climatic component should be introduced into the calculation method along with the thermotechnical component,

The Kyoto Protocol adopted in 2005 in accordance with the UN Framework Convention on Climate Change (UNFCCC) and materials of the committee on adaptation suggest that all the UNFCCC countries are to develop the national plans and programs for adaptation, including development of areas and promotion of programs on assessment of impact, vulnerability, and adaptation to the climate change, first of all, to the growth of outdoor air temperature, and to conduct a technical study of the process of adaptation in different spheres in 2016–2020 [1–3].

One of the main areas of programs for adaptation to global warming is studying the impact of varying environmental parameters on structural materials durability. This conditions the necessity of developing new methods for calculation of climatic impact on strength and durability of enclosing structures taking into account both the temperature change, and the number of cycles of temperature transitions through zero, as the part of the program on adaptation to the climate change or the buildings and structures durability.

On Fig. 1 the graphs are shown that characterize the warming both at the global level of the planet climate change and at the territory of Russia and, more

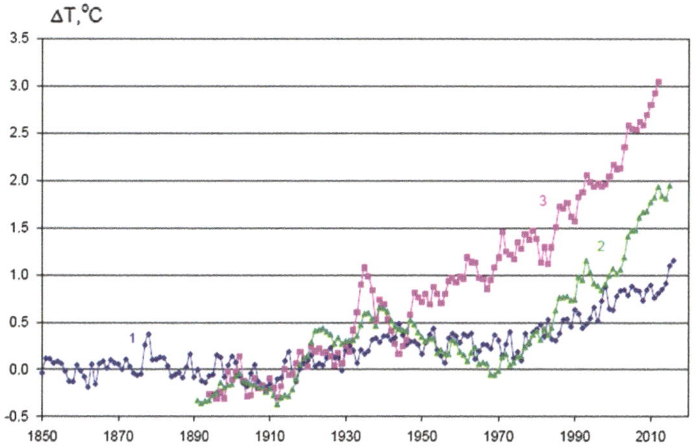

Fig. 1 Changes in average annual temperature of the surface layer of air ΔT in 1850 to 2015, the average for the planet (1) according to [4], on the territory of Russia (2) and in Moscow (3) according to [5]

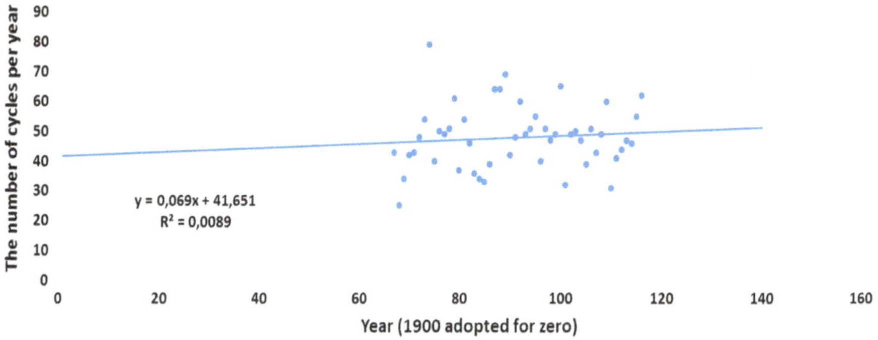

Fig. 2 The Number of cycles of the zero crossing of outdoor air temperature in Moscow (according to data from 1966 to 2016)

specifically, at the territory of Moscow as a metropolis. Starting from the middle of XX century, the annual average temperature increase in Moscow has been more substantial than that on the average throughout the planet. Along with the average annual temperature increase, the number of cycles of outdoor temperature transitions through zero grows. On Fig. 2 the calculations are given regarding cycles of temperature transitions through zero as per the results of observations in Moscow since 1966 till 2016 made by the authors. It should be noted that the graph slightly tends towards the increase of the number of cycles of temperature transitions through zero, however the joint consideration of the graphs shown on Figs. 1 and 2 sets the trends of outdoor air characteristics.

The climate parameters [6, 7] considering standards that are currently in force in Russia describe the temperature fluctuations in winter and summer season with such parameters as mean monthly values of air temperature in January and July as well as the coldest (hottest) day temperature departures from average monthly values. These parameters do not allow reflecting the climatic changes as they operate with average calculated values. The use of "typical year" parameters in the calculation does not make it possible to consider the trends of outdoor air characteristics change as well, because leveled annual data are used for forming the "typical year" characteristics. Similar standards are in force in other countries of the world as well.

Receiving the reliable results on the number of freezing and thawing cycles over the cross section of external enclosing structure is important for the analysis of the structure durability and for the possibility to calculate its life cycle. Thus, the problem shall be stated in the following way: it is necessary to develop the method of calculating the number of cycles of freezing and thawing of enclosing structure that would allow to determine the number of cycles of freezing and thawing in any cross-section of the brickwork considering moisture and solar irradiation. At that, the developed method should allow considering the global climate change in any region.

2 Calculation Method

Method of calculating the number of cycles of freezing and thawing of enclosing structure in any brickwork cross-section considering the global climate change consist of several stages. At the first stage, the problem is solved on temperature distribution over the brickwork section under nonsteady heat transfer conditions for one-dimensional heat flow. At the second stage, the corrections are introduced into calculated data for considering the structure moisture and, therefore, the heat consumption for ice formation and melting in the structure, the solar irradiation influence on calculated outdoor temperature, the temperature of freezing and thawing commencement is adjusted in accordance with the chemical composition of electrolyte inside the brick fragment. At the third stage, based on the statistical analysis of calculated data on the number of freezing and thawing cycles in various sections of enclosing structure, the schedules of change in the number of freezing and thawing cycles in each cross section are defined, which makes it possible to identify the number of freezing and thawing cycles during any time interval considering the global climate change in any region and for any structure. This article described the method itself in brief, as the main focus is on the analysis of results and possibility of estimating the design solutions of enclosing structures from their durability viewpoint.

2.1 Method for Calculating Temperature Distribution Over Brickwork Cross-Section

The problem of temperature distribution over the brickwork cross section is resolved on the basis of the mathematical solution for the one-dimensional problem of nonsteady process, when heat is transferred via the flat wall with indefinite length proposed by Vlasov [8] and Fokin [9]. To resolve the set problem of the temperature distribution over the cross section of external enclosing structure it is necessary to solve the Fourier's differential heat equation that defines the one-dimensional heat transfer under nonsteady conditions

$$\frac{\partial t}{\partial z} = \alpha \frac{\partial^2 t}{\partial x^2}, \tag{1}$$

where a—thermal diffusivity coefficient, m^2/s;

$$\alpha = \frac{\lambda}{c\gamma}, \tag{2}$$

where

- λ thermal conductivity of material $W/(m^2\ {}^\circ C)$;
- c specific thermal capacity of material, kJ/(kg K);
- γ material density, kg/m^3;
- t temperature, °C;
- z time, s;
- x coordinate along heat transfer axis, m.

The boundary conditions of the third kind for the inner surface of the wall is the law of heat transfer between the wall and the internal environment with t = 20 °C.

The general formula for defining the temperature in any plane after the time interval Δz per the temperatures in the same plane and in two proximate planes in the previous moment z shall be as follows:

$$t_{n,z+1} = t_{n,z} + a\frac{\Delta z}{\Delta x^2}\left(t_{n+1,z} + t_{n-1,z} - 2t_{n,z}\right) \tag{3}$$

This scheme is today recognized as Crank-Nicolson scheme [10–16].

The initial temperature distribution over the wall cross section is calculated based on the assumption that as of the beginning of calculation the steady conditions have been reached in the wall.

For reliable sampling, the graphs were plotted regarding temperature distribution in the given cross sections since October 1966 till April 2011. The reference year was taken since the month, when the outdoor air temperature was below zero degrees, till the month, when no temperature below zero was recorded. Thus, in the period under consideration about 600,000 reference points were obtained, based on which 2450 graphs of temperature distribution over brickwork cross section were plotted.

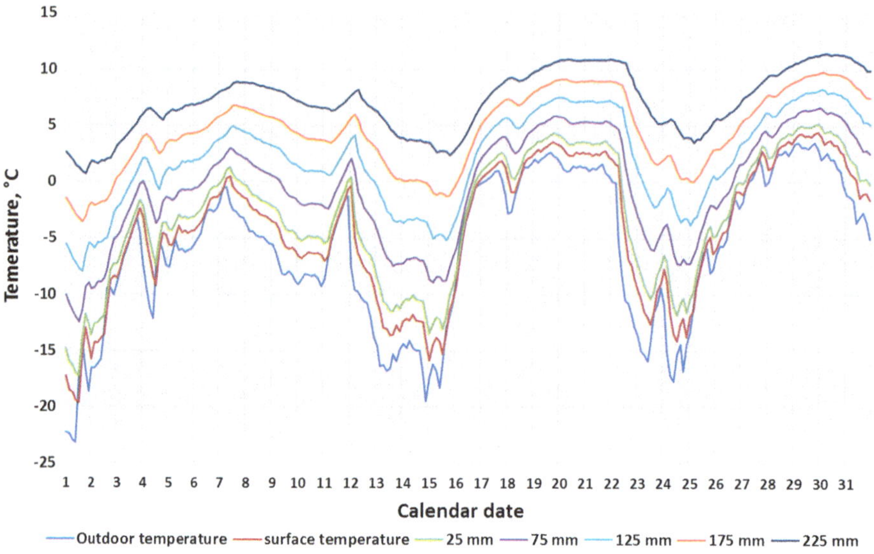

Fig. 3 The temperature distribution over the cross section of the external wall (brick 510 mm) in December, 1998

On Fig. 3 the results of temperature distribution over the cross-section of the external building wall are shown. The graphs of outdoor temperature, external surface temperature and temperature in the relevant cross-sections of the brickwork are shown on the Figure. The number of cycles of temperature transitions through zero in any cross section of brickwork shall be determined by the double crossing of the horizontal line corresponding to zero degrees. Calculation results for the entire basic reference period since October 1966 till April 2011 are summarized in Table 1.

The developed method allows to calculate the temperature distribution over the brickwork cross-section under nonsteady conditions not only in case of outdoor temperature change but also considering the change of the indoor temperature.

2.2 *Experimental Verification of Developed Calculation Method*

Field experiment was conducted to verify the accuracy of the developed heat distribution calculation method at a cross-section of the external building wall. For the experiment, the building constructed in 1905 was chosen, with 950 mm thick external walls made of solid loan brick. The wall, on which the experiment was conducted, was oriented to the south orientation and faced a closed yard. The nearest building wall is located 4.7 m away [17, 18].

To measure temperature and heat flow density, the device was used for measuring and logging the heat flow rate going through heat exchange surfaces of

Table 1 Number of transitions through zero over the cross-section of the external wall of the building (510 mm solid bricks) per months (Number/percent of total transitions in this cross-section)

Designed cross-section of external wall	Month										Total transitions in is cross-section
	VIII	IX	X	XI	XII	I	II	III	IV	V	
Outdoor temperature	2	38	283	301	249	214	256	639	395	21	2398
Surface temperature (mm)	1	10	137	223	242	196	218	490	156	4	1677
25	0	6	77	175	222	184	206	424	71	1	1366
		0.44	5.64	12.81	16.25	13.47	15.08	31.04	5.20	0.07	
75	0	0	22	103	185	159	203	259	8	0	939
			2.34	11.14	19.70	16.93	21.62	27.59	0.85		
125	0	0	1	51	137	153	227	141	0	0	710
			0.14	7.18	19.30	21.55	31.97	19.86			
175	0	0	0	16	65	111	134	12	0	0	338
				4.73	19.23	32.84	39.65	3.55			
225	0	0	0	1	11	31	22	0	0	0	65
				1.54	16.92	47.69	33.85				

thermal power facilities as well as temperature of those surfaces and their surrounding gaseous and granular medium. Measurement range for heat flow density is 10–999 W/m^2, for temperature −30 to 100 °C.

Device operating principle involves the measurements of thermo-electromotive force of temperature sensitive heat flow sensors and temperature sensors resistance.

Temperature sensitive sensors consist of galvanic copper–constantan thermobattery made of several hundred thermocouples connected in series, bifilarly folded down in spiral and filled with epoxy compound with various additive agents. The sensor has two outputs (one on each side of the sensitive element). Sensor function is based on "additional wall" principle. The sensor is fixed on heat transfer surface of the subject forming an additional wall. Heat flow going through the sensor forming a temperature gradient and a corresponding thermoelectric signal inside it.

Platinum thermoresistance sensor locked in a leak-tight disk-shaped case are used as remote temperature sensors in the device; they provide for measuring surface temperature of solid bodies by way of fixing (sticking) them to the studied surfaces, and for measuring air and granular media temperature by way of immersion.

Temperature sensors were installed for measuring outdoor (t_1) and indoor (t_7) temperature at the distance of 300 mm away from the wall surface, external (t_2) and internal (t_6) wall surfaces temperature, and three sensors were installed inside the external wall at the distance of 85(t_3), 185(t_4) and 280(t_5) mm from external wall surface (Fig. 4).

During the installation process of (t_3)–(t_5) sensors several holes of 22 mm were drilled on the internal wall surface. Sensors were installed on the end surface of the drilled hole using thermal compound and fixed with a special device. Open end of

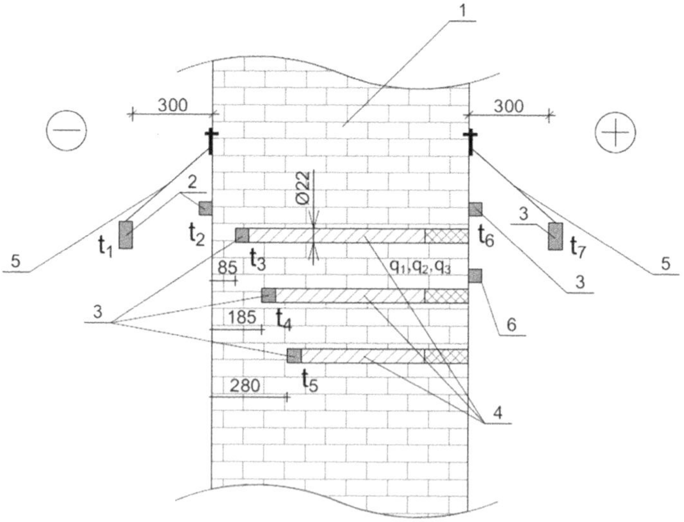

Fig. 4 The arrangement of sensors for experimental confirmation of the calculation of the temperature distribution over the cross section of the outer wall

the holes was insulated at 150 mm depth using an insolent. Three heat flow sensors (q_1, q_2, q_3) were installed on the inner surface 250 mm apart from one another. The arithmetic average of the readings was taken as the heat flow value.

The experiment was performed for over 3.5 months, since early January till mid-April. The interesting feature of this period is that during it the lowest temperatures of outdoor air and the period of active temperature transition through zero were recorded.

During the field experiment in automatic mode, every twenty minutes outdoor (t_1) and indoor (t_7) temperature, temperature of external (t_2) and internal (t_6) wall surfaces was measured, as well as heat flow values at three points (q_1, q_2, q_3). The arithmetic average of three readings was taken as the heat flow value.

Real values of material thermal conductivity coefficient of external wall as well as heat transfer coefficient of the internal and external surfaces were calculated on the basis of those values. Besides, heat conduction coefficient was calculated based on the temperature values t_3 and t_5 in cross-sections at 85 and 280 mm depth from the external surface. The calculation results demonstrated the good convergence: the average values of two calculations were 0.85 and 0.88 W/(m K). Further, the calculations of heat transfer coefficient of external and internal walls were made. Due to the massive amount of experiment data, Fig. 5 demonstrates the results of experiment and interim calculated values for 16 days, since 4th till 20th of January 2017.

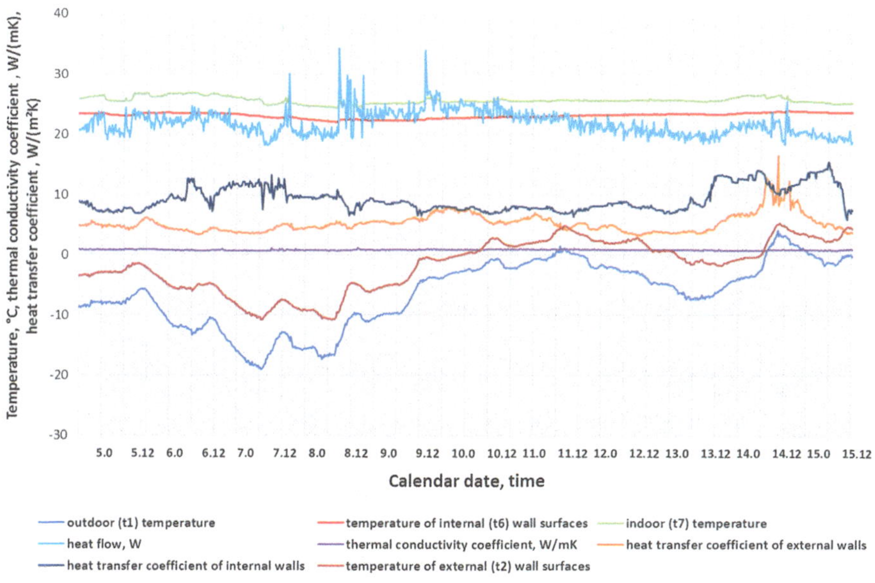

Fig. 5 Experimental data of the temperatures of external and internal air, temperatures of internal and external wall surface and the calculated values of heat transfer coefficients and thermal conductivity coefficient

The thermal conductivity was calculated as per the formula (4)

$$\lambda = \delta q/(t_6 - t_3) \tag{4}$$

where

- δ wall thickness, m;
- q average heat flow value, W/m^2;
- t_6, t_3 temperature of internal surface of the wall and at the depth of 85 mm from external surface, °C.

To verify the solution, the heat conduction coefficient was calculated based on the temperature values t_3 and t_5 in cross sections at 85 and 280 mm depth from the external surface. The calculation results demonstrated the good convergence, the graphs λ_1 m λ_2 were almost identical. The average values of two calculations were 0.83 and 0.88 W/(m °C).

Further, the calculations of heat transfer coefficient of external and internal walls were made. Due to the fact that thermal resistivity calculated as per the formula (4) is known and equal to $R = \delta/\lambda$ [m^{2}K/W], the heat transfer coefficients were determined as per the formulas

$$\propto_{\mathrm{B}} = \frac{\left(\frac{t_{\mathrm{B}} - \tau_{\mathrm{H}}}{t_{\mathrm{B}} - \tau_{\mathrm{B}}} - 1\right)}{R} = \frac{\left(\frac{t_7 - t_2}{t_7 - t_6} - 1\right)}{R}, \tag{5}$$

$$\propto_{\mathrm{H}} = \frac{\left(\frac{\tau_{\mathrm{B}} - t_{\mathrm{H}}}{\tau_{\mathrm{H}} - t_{\mathrm{H}}} - 1\right)}{R} = \frac{\left(\frac{t_6 - t_1}{t_2 - t_1} - 1\right)}{R} \tag{6}$$

where

- $\propto_{\mathrm{B}}$, $\propto_{\mathrm{H}}$ heat transfer coefficients of internal and external surfaces respectively, W/(m^2 K;)
- τ_{B}, τ_{H} temperatures of enclosing structure internal and external surfaces respectively, °C;
- t_{B}, t_{H} temperatures of indoor and outdoor air respectively, °C;
- R thermal resistivity, m^2 K/W.

It should be noted that when calculating the heat transfer coefficient, not the average R values were used in the formulas (5) and (6) but the current values estimated directly on the basis of temperatures values.

It should be noted that temperature wave passing the wall has the phase of temperature and heat flow oscillations damping, and the experiment data should be taken for calculations considering the phase of temperature wave lagging when passing the enclosing structure mass.

A.M. Shklover presented the damping of temperature and heat flow oscillations as the decrease of harmonic oscillations of temperature and heat flow in the course of wave passing the wall [19]. He also noted zero phase offset between oscillations of temperature and heat flow in the course of wave passing the wall. Temperature oscillations damping from the plane with x coordinate till the wall border β for the particular case, when both planes (starting and finishing along the wave movement) are located at the section of regular oscillations, with $Rs \geq 1$, shall be written as follows:

$$\beta = e^{Rs\sqrt{i}} \tag{7}$$

where s—periodic penetration depth of the material introduced by O.E. Vlasov, calculated as per the formula

$$s = \sqrt{\frac{2\pi\lambda c\gamma}{z}} \tag{8}$$

where

- λ thermal conductivity of material, W/(m^2 K);
- c specific thermal capacity of material, kJ/(kg K);
- γ material density, kg/m^3;
- z oscillation period, s.

It should be noted that temperature dumping presented as complex number is very convenient as the number module shows how many times the oscillations amplitude has decreased on the distance under consideration, and the argument shows how much initial phase angles have reduced. Then, the value of temperature oscillations decrease at the regular oscillations section shall be calculated as per the following formula

$$\beta_t = e^{Rs/\sqrt{2}} \quad [\text{times}], \tag{9}$$

and the lagging phase shall be calculated as per the following formula

$$\beta_\varphi = \frac{Rs}{\sqrt{2}} \quad [\text{rad}], \tag{10}$$

To deduce the above mentioned lagging in hours, it is necessary to use the formula

$$\beta'_\varphi = \frac{\beta_\varphi Z}{2\pi} \quad [\text{hour}] \tag{11}$$

Calculations as per the formulas (10) and (11) allowed to determine the temperature wave and heat flow lagging time: lagging time when passing throughout the wall was equal to 25.4 h, from t_3 temperature measurement point to the internal wall surface—to 22.6 h, between t_3 и t_5 temperature measurement points—to 5.1 h, q value was determined with the offset of 22.6 h at that.

The calculated values of heat transfer coefficients of external and internal surface are given in the form of graphs shown at Fig. 5. The average coefficient values as

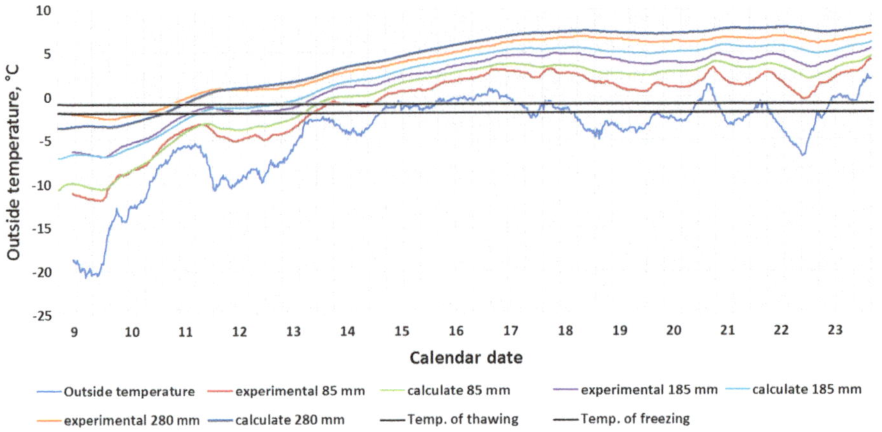

Fig. 6 Comparison of the calculated and experimental values of temperature distribution in various sections of the exterior walls of the building

per 1000 reference points were assumed: for internal surface 9.1 W/(m^2 K), for external surface 5.1 W/(m^2 K). It is possible that the surface heat transfer coefficient on inside of the wall is larger than on the outside as the studied wall was in terminal, air flow was absent.

Considering the interim calculations shown above, the values of temperature in cross sections of the external building wall at the depth of 85, 185 and 280 mm from external surface were calculated with the use of the developed method based on finite differences and compared with actual temperatures in these cross sections measured during the experiment. Figure 6 shows calculated and experimental graphs at the minor time interval of 12 days. During this interval the external temperature grew abruptly after reaching the minimum, after which it was changing more smoothly. The calculated values with the good convergence repeated the experimental curves in all sections at that.

3 Method for Calculating Year Temperature Intensity Parameter for Walls with Various Structures and Results Analysis

Analysis of results on the number of freezing and thawing cycles in various external wall cross-sections in nonsteady process given in Table 1 allows to proceed to thermal forces analysis on the studied structure.

Important parameter N_i, which characterizes the thermal impact on enclosing structure of the building taking into account global temperature changes, defines number of freezing and thawing cycles in all external wall sections during ith year. This parameter is called "year temperature intensity" for external wall structure and it characterizes the impact on this structure caused by outdoor temperature changes. "Year temperature intensity" parameter links outdoor air characteristics: temperature changes caused by climate changes, and changes in the number of cycles of temperature transition through with technical specifications of the enclosing structure: thickness and thermophysical properties of the structure material.

Year temperature intensity graph for external walls of solid common bricks with thickness 510 mm N_i, was constructed based on the analysis of temperature distribution graphs similar to the one in Fig. 3, and it is shown in Fig. 7. For further calculations convenience, the year 1900 was taken as the zero point on X-axis.

The graph in Fig. 7 demonstrates that distribution of value N_i—year temperature intensity—follows simple negative linear regression.

The correctness of our applying the linear regression for describing the year temperature intensity can be analyzed with the help of residual analysis statistical method —the graphic method allowing to evaluate the accuracy of the regression model. It is possible to reveal the potential noncompliances with the regression analysis conditions. Residue, or error of estimate, is the difference between the observed and forecasted values of dependent variable under the given value of variable *x*.

To evaluate the applicability of regression empirical model, the residues shall be singled out in the vertical axis and *x* values—on the horizontal axis. In case the empirical model is applicable, the graph should not display the explicit regular pattern. But in case the empirical model is not applicable, the figure will show the dependence between *x* values and residues. It is evident from the graphic residual analysis made by us that the distribution of residuals is chaotic and does not display the regular pattern (Fig. 8). Residuals often have both positive and negative values. This allows to conclude that the linear regression model is applicable for solving the problem.

Graphs on the figures result from the calculation of the number of cycles of freezing and thawing, and the number of cycles of freezing and thawing in per cent of the total number of cycles per year in the cross section at 25 mm from the external wall surface. The same graphs were constructed for the calculated cross sections of the external building wall structure.

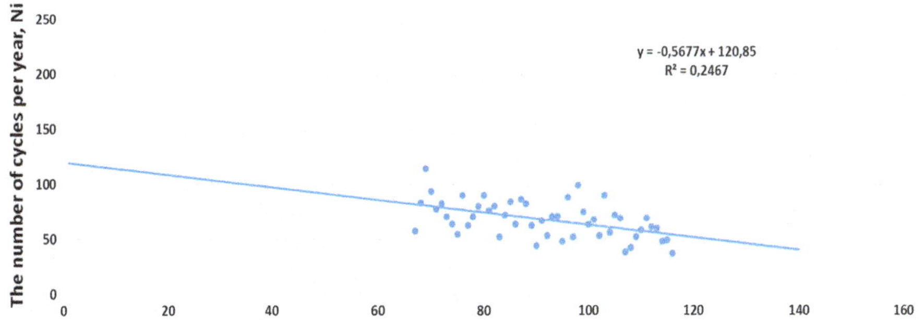

Fig. 7 The number of cycles of the transition temperature of zero in all the estimated cross sections of the outer walls of solid red brick with a thickness of 510 mm N_i (temperature the intensity of years for this material)

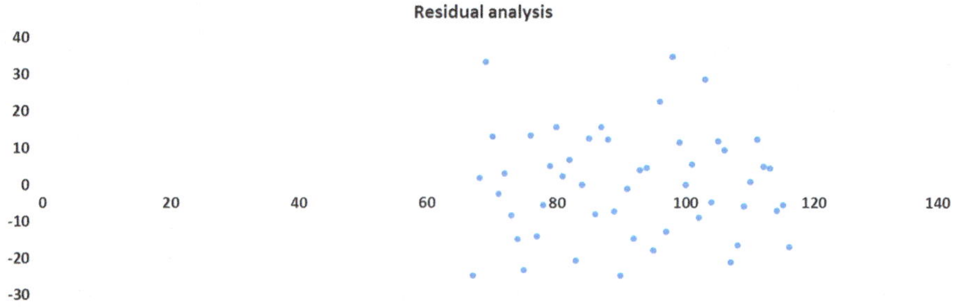

Fig. 8 The residual analysis of the graphic in Fig. 7

4 Analysis of Temperature Impact on Enclosing Structures

At the moment, the evaluation of possibility to use a particular design of external building wall from the viewpoint of its durability is made on the basis of various parameters, such as thermal conductivity, structure humidifying, freeze-thaw resistance. The conclusion on the structure suitability is made on the basis of these parameters in their totality. However until now there was no numeric parameter used to provide the integral evaluation of the temperature impact on enclosing structure.

Let us consider the efficiency of the developed method for evaluation of temperature impact on external walls of the following structure: bearing external wall of the building is made of red loan brick and has the thickness of 510 mm and 700 mm. External wall is reinforced with mineral isolation plates with thickness of 100 mm and thermal conductivity of 0.05 W/(m K) on the internal side. This structure is applicable, for example, for reconstruction of old buildings with lofts arrangement.

For each structure, the graphs similar to that on Fig. 3 are plotted, analysis results are summarized in tables similar to Table 1, based on which graphs for each combination are plotted similarly to those shown on Figs. 9 and 10.

As it was mentioned before, to analyze the temperature impact on the enclosing structure considering the global climate change, we have introduced the year temperature intensity parameter. On Fig. 11 the year temperature intensity graphs are shown for the 510 mm external wall structure with thermal insulation on the internal side and without it. It is evident than in case of the wall with no thermal insulation the polythermal load on the wall will reduce in time, and for the structure with inner thermal insulation it will grow. In case of the external wall structure with no thermal insulation the reduction of freezing and thawing cycles is explained with the fact that the global warming influence (Fig. 1) impacts the external wall more than the certain increase of cycles of outdoor temperature transitions through zero shown on Fig. 2.

This conclusion is confirmed by freezing and thawing graphs as per the structure cross sections plotted similarly to those shown on Figs. 9 and 10. For convenience of analyzing, the characteristics of the linear regression graphs are shown in Table 2.

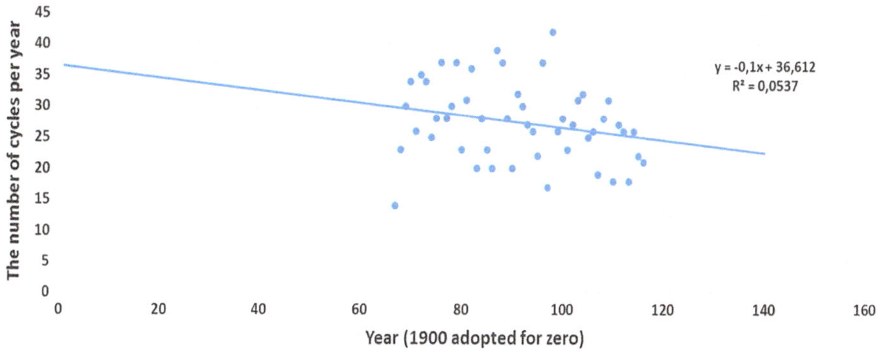

Fig. 9 Changing the number of cycles of freezing and thawing per year in the cross-section 25 mm from the outer surface of the wall

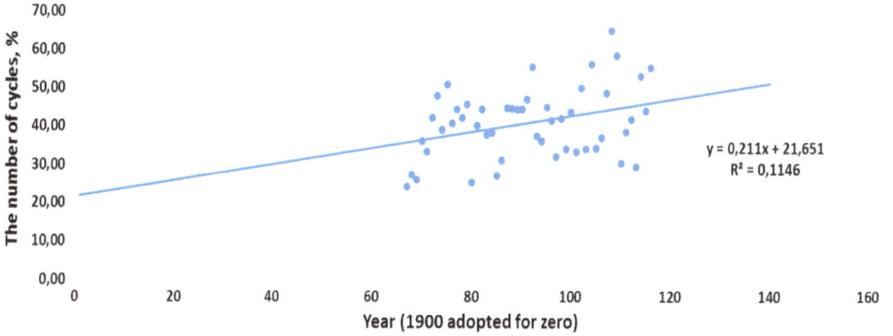

Fig. 10 Changing the number of cycles of freezing and thawing, in percent of the total number of cycles per year, in the cross-section 25 mm from the outer surface of the wall

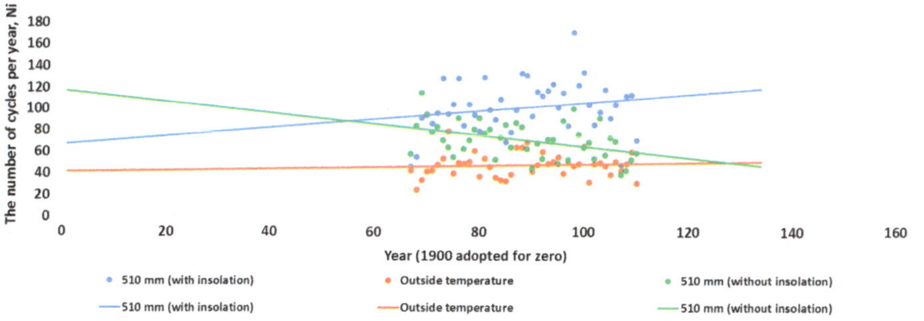

Fig. 11 Graphs of intensity of temperature of the year (N_i) for exterior brick walls 510 mm with and without insulation

Table 2 Characteristics of linear regression graphs for wall with brickwork thickness of 510 mm

Coordinate of external wall cross-section	Graph type	Coefficients of linear regression graphs $y = ax + b$ (*with insulation*)		Coefficients of linear regression graphs $y = ax + b$ (*without insulation*)	
		a	*b*	*a*	*b*
25	Number of cycles	0.10	13.17	−0.10	36.61
	% of the total number of cycles	0.21	21.65	0.21	21.65
75	Number of cycles	0.12	4.84	−0.19	36.13
	% of the total number of cycles	0.21	21.65	−0.08	34.47
125	Number of cycles	0.12	1.98	−0.19	32.02
	% of the total number of cycles	0.08	4.67	−0.12	30.86
175	Number of cycles	0.06	4.69	−0.05	11.58
	% of the total number of cycles	0.03	6.99	0.01	8.98
225	Number of cycles	0.02	5.86		
	% of the total number of cycles	−0.0005	7.85	−0.02	3.88
275	Number of cycles	0.007	6.16		
	% of the total number of cycles	−0.02	8.56		
325	Number of cycles	0.005	5.30		
	% of the total number of cycles	−0.02	7.64		
375	Number of cycles	−0.003	5.48		
	% of the total number of cycles	−0.03	8.27		
425	Number of cycles	−0.004	5.47		
	% of the total number of cycles	−0.04	8.42		
475	Number of cycles	−0.01	6.58		
	% of the total number of cycles	−0.05	9.58		
510	Number of cycles	−0.03	7.80		
	% of the total number of cycles	−0.06	10.08		

It is evident that in the structure with internal thermal insulation the coefficient *a* is much more than zero for sections up to 175 mm from the external wall surface, which indicates the increase of the number of freezing and thawing cycles in these sections as time passes. For sections of 225 mm to 425 mm the coefficient *a* is close to zero, i.e. the number of cycles in these sections is almost constant; for sections of 475 mm and 510 mm *a* is less than zero, the number of cycles decreases, which confirms the conclusion on the decrease of brickwork freezing depth due to the global warming.

For the brick wall without thermal insulation, coefficient *a* is negative for all cross sections, which determines the decrease of number of freezing and thawing cycles throughout the external wall.

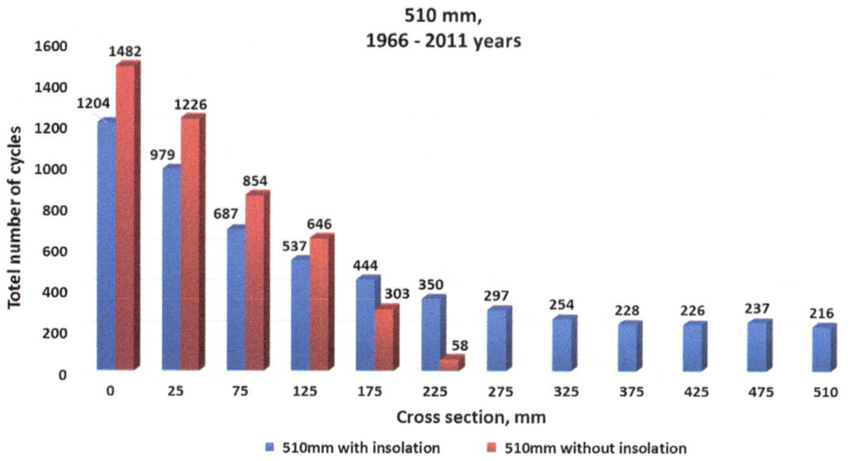

Fig. 12 The number of cycles of freezing and thawing in the cross section of the exterior brick walls with thickness of 510 mm 1966–2011 years

The bar graph in Fig. 12 makes it possible to compare the distribution of the number of freezing and thawing cycles in various cross sections of the brickwork with the thickness of 510 mm with thermal insulation on the internal side and without it. It is important to note that in the brickwork cross sections located closer to the external surface, the number of freezing and thawing cycles is much lower for the structure with thermal insulation, than for the structure without thermal insulation on the internal side of the wall.

It is important to note that the zone of the maximum brickwork humidifying is considered to be the zone in the first third from the external surface of the enclosing structure, i.e. for sections up to 125 mm. In these sections, for the structure with thermal insulation 3407 freezing and thawing cycles will take place in the reference period, and for the structure with no thermal insulation—4208 cycles, i.e. 23.5% more. In the brickwork sections located deeper in the structure with thermal insulation, the number of freezing and thawing cycles will be minor and rather constant, i.e. the certain redistribution of the cycles number throughout the brickwork cross-section will take place with the decrease of the load in external sections to internal ones. It can be defined that 61% of all brickwork freezing and thawing cycles in the reference period will take place in the sections located in the first third of the enclosing structure section in case the internal surface is insulated. For the structure with no thermal insulation this parameter will be equal to 92%.

Figure 13 shows the bar graph for comparing the number of brickwork freezing and thawing cycles for the structure with thermal insulation and without it with breakdown per months. In general, this bar graph confirms the conclusions made earlier, however it gives the data on temperature impact in various months of the year. This is important to know, because the maximum structure humidifying due to the diffusive moisture accumulation occurs in March–April [20–26]. It is evident

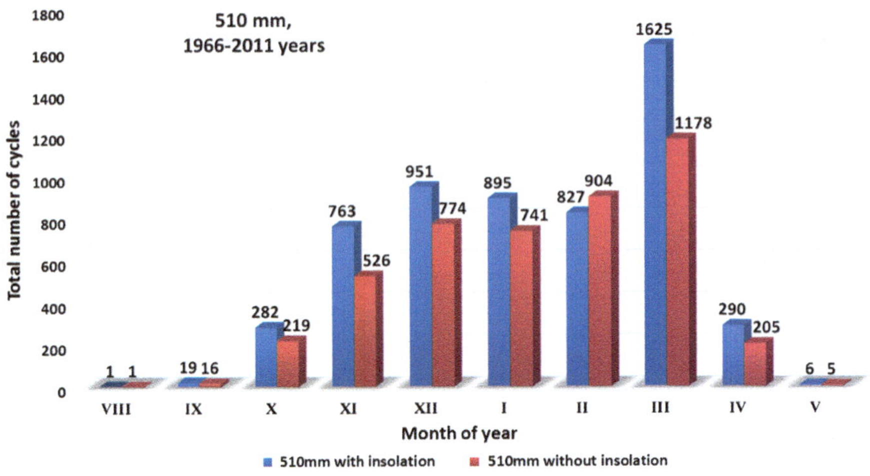

Fig. 13 The number of cycles of freezing and thawing of the exterior brick walls with thickness of 510 mm in 1966–2011 years by month

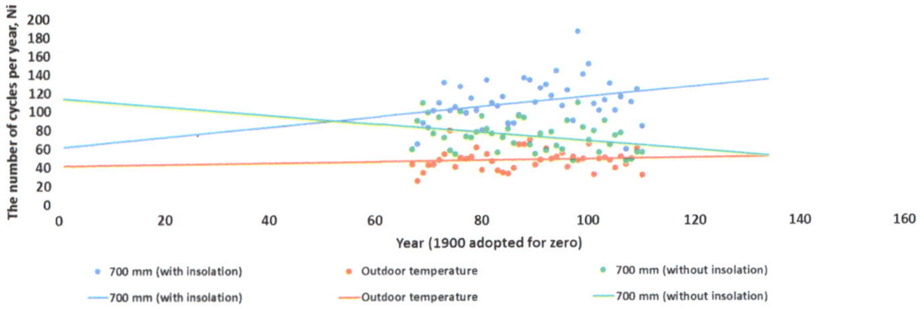

Fig. 14 Graphs of intensity of temperature of the year (N_i) for exterior brick walls 700 mm with and without insulation

that in March the number of freezing and thawing cycles for the structure with thermal insulation is 1625 instead of 1178 cycles for the structure without thermal insulation, which is 38% more.

The graphs of the year temperature intensity parameter change for the structure of external wall with 700 mm thick brickwork and without thermal insulation are shown in Fig. 14. In terms of pattern, these graphs repeat the graphs for 510 mm brick wall structure, which confirms the correctness of conclusions made concerning the previous structure.

Characteristics of linear regression graphs are shown in Table 3.

It should be noted that the number of cycles of freezing and thawing for structure with thermal insulation increases up to 425 mm cross section as time passes, coefficient a is almost equal to zero in the cross section 475, i.e. the number of

Table 3 Characteristics of linear regression graphs for structure with brick wall thickness of 700 mm

Coordinate of external wall cross-section	Graph type	Coefficients of linear regression graphs $y = ax + b$ (*with insulation*)		Coefficients of linear regression graphs $y = ax + b$ (*without insulation*)	
		a	*b*	*a*	*b*
25	Number of cycles	0.13	10.91	0.02	20.65
	% of the total number of cycles	0.21	21.65	0.21	21.65
75	Number of cycles	0.13	3.66	−0.07	21.58
	% of the total number of cycles	0.21	21.65	0.21	21.65
125	Number of cycles	0.14	−0.16	−0.14	25.36
	% of the total number of cycles	0.08	4.01	−0.08	25.05
175	Number of cycles	0.08	2.45	−0.12	20.35
	% of the total number of cycles	0.04	4.98	−0.09	20.79
225	Number of cycles	0.06	3.20	−0.03	9.37
	% of the total number of cycles	0.02	5.57	0.01	7.87
275	Number of cycles	0.008	5.92	−0.04	7.02
	% of the total number of cycles	−0.02	7.71	−0.01	6.23
325	Number of cycles	0.03	2.57	−0.06	7.00
	% of the total number of cycles	0.005	4.14	−0.07	7.97
375	Number of cycles	0.04	1.04	−0.02	2.10
	% of the total number of cycles	0.02	2.61	−0.02	2.43
425	Number of cycles	0.02	2.31		
	% of the total number of cycles	−0.004	4.01		
475	Number of cycles	0.005	3.47		
	% of the total number of cycles	−0.02	5.46		
525	Number of cycles	−0.01	4.70		
	% of the total number of cycles	−0.03	6.54		
575	Number of cycles	−0.01	4.75		
	% of the total number of cycles	−0.04	6.59		
625	Number of cycles	−0.02	5.37		
	% of the total number of cycles	−0.05	7.25		
675	Number of cycles	−0.03	5.94		
	% of the total number of cycles	−0.05	7.38		
700	Number of cycles	−0.02	5.00		
	% of the total number of cycles	−0.04	6.28		

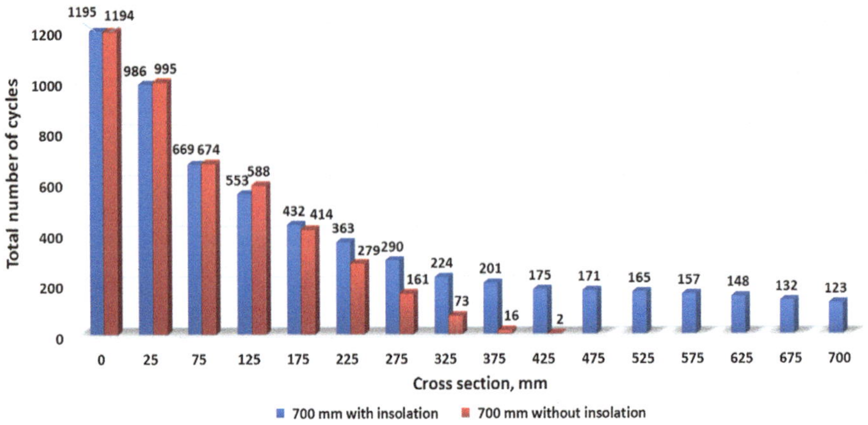

Fig. 15 Distribution of the number of cycles of freezing and thawing on the cross section of the exterior brick walls 700 mm with and without insulation in 1966–2011

cycles in this section is constant during a year, and starting from the cross section at 525 mm the number of cycles of freezing and thawing decreases. For the structure with 700 mm brick wall without thermal insulation, the number of cycles of freezing and thawing for the cross section at 25 mm will increase as well; starting from the section at 75 mm from the external wall surface the number of cycles will decrease as time passes.

When considering the bar graph showing the number of cycles of freezing and thawing in various sections of 700 mm thick brickwork with thermal insulation and without it (Fig. 15), it is necessary to note a certain difference for structures with brick wall thickness of 510 mm and 700 mm. If the number of cycles of freezing and thawing for the structure with 510 mm brick wall was 23.5% less for the structure with thermal insulation, for the structure with 700 mm brick wall the number of cycles in the first triens is almost the same for the structure with thermal insulation and without it. In the deeper layers of the structure, the number of freezing and thawing cycles is also constant and minor, same as for the structure with 510 mm brickwork.

The developed method allows to analyze the dynamics of temperature impact on enclosing structure in the course of time. Figure 16 shows the calculated data on the number of freezing and thawing cycles as per brickwork cross sections for the enclosing structure with 510 mm brickwork with thermal insulation on the internal side and without it, in the period since 2011 till 2055. It is evident that the impact on enclosing structure, especially in the first triens as per the brickwork cross section, changes abruptly versus the basic reference period of temperature impact calculation since 1966 till 2011 (Fig. 12). During the basic reference period the number of freezing and thawing cycles for the structure with thermal insulation in sections within the first triens of the enclosing structure cross section was lower than that for the structure without thermal insulation, however in the period since

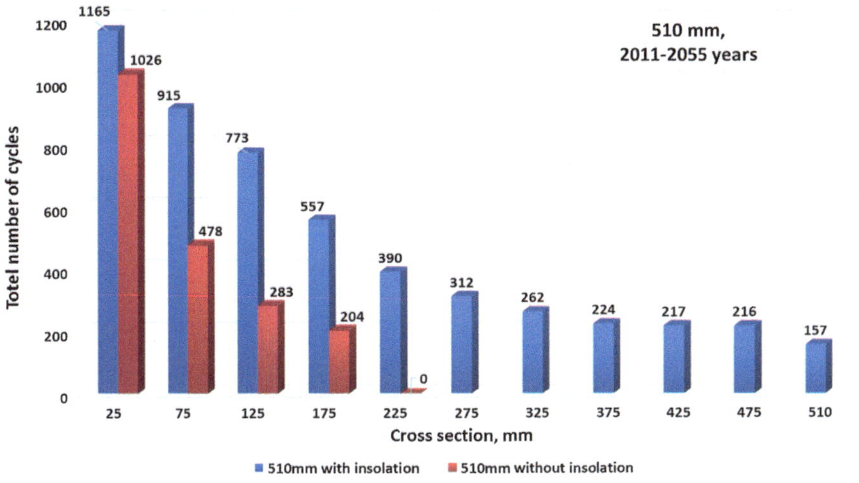

Fig. 16 Distribution of the number of cycles of freezing and thawing on the cross section of the exterior brick walls 700 mm with and without insulation in 2011–2055

2011 till 2055 the number of freezing and thawing cycles in the first triens of cross section of the enclosing structure with thermal insulation is 2853 freezing and thawing cycles, and for the structure without thermal insulation—1787, i.e. 59.7% less. In 2011–2055, 55.0% of all brickwork freezing and thawing cycles in the reference period will take place in the sections located in the first third of the enclosing structure section in case the internal surface is insulated. For the structure with no thermal insulation this parameter will be equal to 89.8%.

5 Conclusions

The method has been developed for calculating the temperature distribution over the cross section of enclosing structure under nonsteady heat transfer conditions. The method allows determining the number of cycles of freezing and thawing in each cross section of the external building wall with a high degree of confidence.

Field study on the method in combination with developed technique for processing results obtained in the field conditions have demonstrated good convergence of calculated temperature with actual temperature measured in various sections of the external building wall.

The developed method for analyzing the results of heat distribution calculation at a cross-section of the external building wall allowed to get the numerical characteristics of the change in temperature impact on various enclosing structures and to define he dynamics in the change of temperature impact per the structure cross section in the next 50–100 years interval. The authors' introducing the "Year

Temperature Intensity" parameter allowed to link the outdoor air characteristics: temperature changes caused by climate changes, and changes in the number of cycles of outdoor temperature transition through with technical specifications of the enclosing structure: thickness and thermophysical properties of the structure material.

The method ensures the ample opportunities for analyzing the durability of various enclosing structures, for example concrete walls, taking into account the adaptation to the global climate change.

Yes, the study can be use for any types of envelopes and now we use it for concrete walls, for example.

References

1. The Second Assessment Report on Climate Change and Their Consequences on the Territory of the Russian Federation/VM Katsov, SM Semenov (Roshydromet, Moscow, 2014), 58p
2. IPCC Fifth Assessment Report, WG II Report "Climate Change 2014: Impacts, Adaptation and Vulnerability" (IPCC 2014)
3. LEG, National Adaptation Plans. Technical Guidelines for the National Adaptation Plan Process (2012)
4. O.N. Bulygina, V.N. Razuvaev, L.T. Trofimenko, et al., *An Array of Average Monthly Air Temperature Data at Russian Stations* (VNIIGMI-WDC, Obninsk). Access: http://www.meteo.ru/climate/sp_clim.php (2016)
5. Russian hydrometeorological portal. Hydrometeorological data of the Russian State Data Foundation on the state of the natural environment http://meteo.ru/
6. SP 20.13330.2011 "Loads and effects (updated version of SNiP 2.01.07–85*)"
7. SP 131.13330.2012. "Set of rules. Building climatology (Updated version of SNiP 23–01-99*)"
8. O.E. Vlasov, Flat heat waves—Izvestiya teplotehnicheskogo institut **3**(26) (1927)
9. K.F. Fokin, *Building Thermophysics of Enclosing Parts of Buildings* (M.: ABOK-press, 2006), 256p
10. L.G. Miller, Calculating vapor and heat transfer through walls. Heat Ventil. **35**(11), 56–58 (1938)
11. H.S. Carslow, J.C. Jaeger *Conduction of Heat in Solids* (Oxford, 1959)
12. J. Crank, *The Mathematics of Diffusion* (Oxford, 1975), 414p
13. W. Buxuan, F. Zhaohong, A theoretical study on the heat and mass transfer in wet porous building materials. J. Eng. Thermophys. **6**(1), 60–62 (1985)
14. Daniel A. De Vries, The theory of heat and moisture transfer in porous media revisited. Int. J. Heat Mass Transf. **30**(7), 1343–1350 (1987)
15. A. Kerestecioglu et al., *Theoretical and Computational Investigation of Algorithms for Simultaneous Heat and Moisture Transport in Buildings* (Florida, 1989)
16. J.-T. Hu, X.-H. Ren, F.-Y. Zhao, D. Liu, H.-Q. Wang, Natural convective heat and moisture transfer in an inclined building enclosure with one slender wall of finite thickness: analytical investigation and non-unique steady flow solutions. Int. J. Heat Mass Transf. **104**, 1160–1176 (2017)
17. D.Y. Zheldakov, A.A. Frolov, S.Y. Ivanov, Investigation of strength of masonry walls in the building Kadashevsky baths. Build. Mater. **6**, 55–57 (2016)
18. D.Y. Zheldakov, A.A. Frolov, Segment method for calculating the temperature distribution along the section of the enclosing building structure. Hous. Constr. **6**, 36–39 (2017)
19. A.M. Shklover, *Heat Transfer with Periodic Thermal Effects* (M.: Gosenergoizdat, 1952), 98p

20. V.G. Gagarin, P.P. Pastushkov, Determination of the calculated humidity of building materials. Ind. Civil. Constr. **8**, 28–33 (2015)
21. V.G. Gagarin, V.V. Kozlov, K.P. Zubarev, Analysis of the location of maximum moisture in building envelopes with different thickness of the insulating layer. Hous. Constr. **6**, 8–12 (2016)
22. M. Bomberg, *Moisture Flow Through Porous Building Materials* (Lund Institute of Technology, 1974), Report No. 52, 188p
23. A.C. Andersson, *Verification of Calculation Methods for Moisture Transport in Porous Building Materials* (Lund, 1985)
24. D.H. Everett, The thermodynamics of frost damage to porous solids. Trans. Faraday Soc. **57** (9), 1541–1551 (1961)
25. W.G. Gray, A derivation of the equations for multi-phase transport. Chem. Eng. Sci. **30**, 229–233 (1975)
26. J. Berger, N. Mendes, D. Dutykh, On the optimal experiment design for heat and moisture parameter estimation. Exp. Thermal Fluid Sci. **81**, 109–122 (2017)

Review of Current Practice of Building Foundations in the Canadian North

Christine Harries

Abstract The Canadian North is experiencing several building engineering problems which must be addressed now to avoid worsening problems in the future. The permafrost layer is retreating downward in the Canadian North. This is resulting in many unstable homes in the 14 villages of Quebec's Nunavik. Current tripod adjustable footings, while the most popular, are not the way of the future. Several **other** methods such as concrete footings on bedrock, or on deep permafrost, exist. When bedrock is very deep, steel piles represent a good way to stabilize buildings. Some cases require add freeze piles when bedrock is too deep to reach economically. Lattice truss systems on tripod footings have also been used. This report explores the most prominent solutions of the future and reinforces which methods are believed to best perform going forward.

Keywords Foundations · Structural engineering · Global warming

1 Introduction

The foundation is arguably the most important element for consideration in the construction of a new building. Excessive repair jobs and architectural reports are showing that current design practices may not be adequate for the future in the Canadian North. Climate change is causing the retreat of the permafrost across the world, leaving previously stable buildings sitting on soft unstable ground. Structural building engineers must reasses the current design of foundations and provide a solution that will ensure stable buildings now and into the future.

C. Harries (✉)
SNC-Lavalin, Montreal, QC, Canada
e-mail: christine.harries@snclavalin.com

D. Johansson et al. (eds.), *Cold Climate HVAC 2018*,
Springer Proceedings in Energy, https://doi.org/10.1007/978-3-030-00662-4_88

2 Heat Transfer to Permafrost

The active layer is a narrow top layer of soil that undergoes freeze thaw action seasonally. At a certain depth however under the surface, and going down to bedrock, the composition is completely frozen year round. In Canada, this active layer normally measures between 0.6 and 1.5 m, meaning the permafrost starts at that depth and descends to bedrock. Permafrost is characterized as essentially frozen water with small amounts of soil particles [1]. It is not a frozen mass of earth. This poses a greater risk, because if permafrost is allowed to thaw, it has almost zero compressive capacity, since it is essentially water. Globally, scientists believe with global warming, the active layer will increase over time, meaning the permafrost will retreat downwards [2]. For Canada, here lies the important design condition, since all 14 villages in the Quebec north, are located on permafrost. Nunavik is an area located above the 55th north Parallel. When a building is built, it is heated a large part of the year. A major problem consists of the heat travelling from the building downwards and thawing the permafrost on which the building sits. Several different foundation types are currently being used, and each has their advantages and disadvantages, as will be studied. The challenge however to prevent heat transfer to the soil under the building is among the greatest design considerations for ensuring an appropriate life cycle cost of the building, and preventing major overhauls over the life of the building. Since land is so readily available there is no incentive to build high buildings, rather buildings are generally limited to a ground floor and a second story. Some buildings have a basement; however this is rather rare in the north.

3 Problems with Foundations: Instability of Tripod Footings

The most commonly used method of supporting buildings in the Canadian north is via adjustable steel tripods sitting on wooden plank bases. The main reason for this is that it is initially the least costly alternative, and due to the fact that all buildings in the north are low rise, the loads to be supported are relatively small. The tripods raise the building 2 ft above the ground, preventing the heat from the building to descend into the ground. The 2 ft also allows the wind to disperse the snow, allowing it to pass under the building and prevent accumulation. The wind passing under the home also assists to keep the space cool, thus prevent the ground from warming up. Major problems result when a minimum of 2 ft is not left under the building. Traditionally, the design consists of 12 tripods under a duplex home, which is the most common. It has been seen that should 2 ft not be left, snow accumulates triangularly up to the roof of the home the residents cannot enter without unreasonable shoveling. A disadvantage of this widely used system that is seen in almost every home, is differential movement of the top soil and permafrost variations, resulting in cracks in the walls and opening of joints which allows snow and debris to enter into the envelope and causes major renovations to be needed a few years after the building is handed over to the residents [3, 4]. Over the past 3 years, 20 homes were completely renovated following a 27 year life span in Nunavik. The home was reduced down to its wood studs. Every home had nails being pulled apart, requiring realignment. Every tripod footing had to be readjusted. The tripods are adjustable and the attempt to balance the buildings each spring is conducted; however the shortage of trained technicians who do a competent job is a huge problem. Often the adjustments are done by people who do not know how to perform this task and the problem becomes worse. A different option is needed to provide a better solution for the northern community. The current strategy necessitates we send trained technicians to the communities ever year to lead the adjustments of the tripods, repair architectural damages and other damages that have resulted from the primary damages. For example, when the joint in the building envelope opens up because of excessive stress due to the building shifting, snow can enter and rot the wood. Mold forms and wears away the structure. Moisture gets on steel from the ventilation system and can cause rust. These are all secondary issues that arise and must be treated yearly resulting in a very high lifecycle cost and large follow up efforts, coordination of professionals and yearly work [5] (Figs. 1 and 2).

Fig. 1 Steel tripod footing

Fig. 2 Steel footing under wood beam

4 Alternative: Concrete Footings on Bedrock

Other alternatives exist, such as digging down to the bedrock and sitting concrete footings on solid bedrock and having piers and foundation walls that come up to the surface. Often the concrete footing is also anchored into the bedrock. This design can be with or without a basement. A slab on grade can sit on a drainage permitted backfill, and a superstructure be built atop the foundation walls which go down to bedrock, or a structural slab can sit on the foundation walls, with an open basement below. This foundation type is an excellent option as the building will not shift. The problem arises when the bedrock is too deep not making it economically feasible to excavate to such a depth. This option should therefore be considered when the depth of bedrock is not very deep (within 2 m of the surface).

A second option that employs concrete footings, in cases where the bedrock is too deep, is to sit the concrete footings on permafrost. This method is employed when the permafrost is structurally sound, at a depth of just over 2 m. Sand is placed under the footing with rigid insulation above and below the footing. This method is less used due to its reliance of the long term performance of permafrost which determines the building stability. Even during excavation, contractors must be careful to not permit the permafrost to thaw as the warmer air blows over the

Fig. 3 Building basement in the North

permafrost. Global warming also leads designers to believe despite the grounds current stable nature, the permafrost may retreat and cause instability for buildings over the next 30 years. This method is therefore only recommended if the depth of the footing is sufficient to know the permafrost will not retreat below this support level (Fig. 3).

5 Alternative: Bored Piles

Another very viable option is to use bored piles. This option results in no settlement, and can be used when the bedrock is very deep. It accompanies pile caps on which a steel structure is connected. This method is the one most commonly used in Europe, such as Norway. In Norway, deeper piling is being used, and buildings are sitting on the piles, leaving a space between the underside of the building and the ground. This allows the air to circulate underneath, and keep the ground frozen. This conserves the permafrost, even if the building stability is not dependant on it, because it is resting on piles [2]. When the bedrock is very deep, it prevents having to excavate such a large area to attain solid rock. The bored piles generally descend and are anchored into the bedrock. The cost of the piles thus depends on the depth of the bedrock. This encourages performing a geotechnical reports with sample boreholes. This is a challenge since it is difficult to get qualified geologists to perform accurate tests. The samples indicate the depth of the initial soil layer, the permafrost depth and the depth at which the bedrock is found. The tests are critical in determining what type of foundation system to use. Should the bedrock be very deep making it uninteresting to bore piles to such a depth, the piles can be anchored into the permafrost in a special type of modified pile called add freeze piles [1]. This type of system consists of boring a hole slightly larger in diameter than the pile diameter to be placed. Once the hole is made, the pile is inserted in the center of the hole and the surrounding area is filled with water. The water freezes and adheres to the permafrost. The friction built between the pile and to the surrounding permafrost acts to hold the pile in place. This design however requires a minimum 2 ft space between the pile cap and the building, to ensure no thawing of the area around the pile. This would be a major problem as the pile is withstanding the compression and tension forces via the ice bond. This system is therefore not suitable where basements are required. The depth of the freeze pile is calculated taking into consideration the frictional resistance of the permafrost and the pile. The retreating of the permafrost, thus shortening of the frictional length over time, must be taken into consideration.

6 Alternative: Lattice System of Trusses

The final system that is a viable alternative in the northern community is a lattice system of trusses resulting in many supports that attempts to be able to absorb variable settlement. This system however is difficult to obtain due to limited suppliers capable of providing this concept, and the coinciding high cost. The design is custom and is performed by a manufacturer. It cannot easily be mass produced and with each manufacturer having their distinct details, the purchasing is difficult to manage. The system also has been found to be congested and the system is dependent on the gravel base underneath. Should the base not be well compacted or should there be a problem with the structural integrity of the gravel, the system will not perform well (Fig. 4).

Fig. 4 Lattice structure under northern home

7 Alternative: Use of Monitoring Sensors

These issues were not big problems in the past, but are becoming more and more alarming to designers moving forward. Within the past 5 years, the permafrost has retreated significantly in several northern villages, especially Salluit which is known to have very unstable ground. The method of tripod footings may have worked in the past; however it is becoming clearer that it will not work for the future, and another method to stabilize buildings will be required [5].

It is a known fact that the geography and soil behavior is changing each year in the Canadian north. Some communities are suffering more than others, such as Salluit. Another possible solution, is to continue using tripod footings however to install monitoring sensors near each sector of footing to follow the temperature of the ground. Should the temperature rise, as it will due to global warming, and permafrost retreat, adjustments can be made through inserting shims and using the adjustors. This will allow for a more accurate portrait of the building movement at prevent damage, through faster detection. Many providers and manufacturers exist and the ground temperature sensors can and should be equipped with wireless information transmission. With large quantity purchases, the price would inevitably be lower, however even currently, unit sales are very affordable. This would provide continuous data being sent back to a main control center, possibly located in main cities with qualified staff able to analyze the data. Should a trend be determined requiring the addition of shims or other work, it can quickly be tented to. It would be important to also follow best practice for the compaction of gravel under the tripod footings. This requires that the contractor wait 1 full year after compacting the gravel before building on top, or placing any load. This allows the gravel to have its initial and most important settlement. The continuous monitoring via ground temperature sensors could provide data for scientists to better understand the future behavior of the land in the North.

8 Alternative: Use of Thermosyphon Foundations

With growing concern of climate change and the warming of the planet, researchers are seeking ways to ensure the ground under buildings in the north remains stable and safe. This brings the option of Thermosyphons. Originating in Alaska in the 1960's, and coming to Canada in the 1970's, their application is vast, however not common. The main way to ensure stable foundations is to keep the ground frozen. One such technique is via thermosyphons and thermopiles. Currently there is only one company producing such a system in Canada. The closed system is a passive one that works only in winter when the ambient air temperature is colder than the subsoil temperature. The system is a closed and pressurized one, ranging between 2100 and 4800 kPa [1]. Thermopiles incorporate the thermosyphon technology, in a vertical set up. The radiators, to disperse the heat, are in an air space between the ground and the building structure. In thermopiles, the condensate circulation occurs by gravity. The load of the structure rests on the head of the thermopile therefore they must be designed knowing the soil characteristics and the load of the structure [1].

While other companies have tried gases such as nitrogen, the only company to successfully market their product in Canada, use carbon dioxide. During the winter, the colder ambient air causes the carbon dioxide to condense on the inside of the closed system. This is because cold air can hold less moisture, and therefore condenses. As the liquid carbon dioxide drips down the closed system, it turns back to gas and absorbs heat from the ground and rises back to the top of the closed pipe. This process continuously extracts the heat from the ground keeping it cold, thus frozen. The heat in the warm gas is dissipated via evaporator fins. Insulation is placed above the closed system to prevent heat from the warm ambient air and building, from descending into the cold ground.

The planning and execution of a good geotechnical study is most important for the thermosyphon type of foundation, as compared to all other types of foundations [1]. The Canadian Standards Association identifies several factors of utmost importance to consider when deciding if the system is appropriate, the first being the depth and variation in the active soil layer. The second consideration is the temperature of the ground and the composition of the ground, either ice or water. Ground temperature instruments should be placed and the temperature monitored to have an accurate temperature profile. The third factor to consider is to identify if there is deep seasonal thawing, or potential presence of a soil composition of talik, which is a layer of soil that is permanently unfrozen, located between the active layer and the permafrost. This would not allow a thermosyphon system to work. Finally, it is important to know if there exists groundwater flowing within the active layer. All this is determined via the geotechnical report, confirming its utmost importance. Very commonly, project managers attempt to save money by limiting the number of boreholes in a geotechnical report, assuming that the ground composition is the same over a large area. This can be very misleading, and costly, as the reality can show varying soil structures over a small area. Some project managers will use what are called test pits. This strategy only excavates to the top of the permafrost and looks only at the active layer. This is highly discouraged and considered poor engineering practice. It does not give an accurate profile of the permafrost and the soil composition and properties [1].

The use of thermosyphons allows buildings to be built on grade yet not worry about the heat being transferred from the building to the permafrost which would otherwise cause instability. It has been proven that insulation under the slab alone, is not enough to prevent the heat from transpiring and melting the permafrost. Insulation merely retards the passage of heat, it does not prevent it. Thermosyphons are not ideal for freezing the ground rather they can help maintain the ground frozen. They are not recommended in areas with large underground water currents [1]. While it is rare to find a site that would remain stable should the permafrost melt, this is possible. Such a case would arise if the bedrock is very close to the surface, or if the permafrost layer was gravel or frozen sand. Should the building be located on ground that is not susceptible to becoming unstable in the case of a thaw, thermosyphons would not be required. It is essential that the building be on stable permafrost, with fully functioning thermosyphons before construction begins. The challenge with this approach is that most often construction materials such as the thermosyphons, gravel, and insulation would arrive via sea shipment in the later summer. By the time everything is installed, it would be fall, at which point the ground would be partly thawed. Contractors would have to wait until later in the winter to begin construction, or spring, at which point the ground would have re-frozen. In order for this system to be feasible with construction delays, improved coordination would be required to have the equipment supplied on the ship in the first sea delivery of early spring in order to have the system installed by late spring and begin construction while the ground is still frozen. When using thermosyphons, construction is recommended to be above ground. Very often crawl spaces below homes end up becoming filled with water, causing flooding.

While thermosyphons are available for use in the great north, their limited freeze capacity limits the time frame in which construction can occur. They require specialists to properly install them, which are limited in such environments, and should there be a mishap in the working condition, the integrity of the building could be jeopardized without much warning. For these reasons, while possible uses in the future after technological advances might be very advantageous, currently, thermosyphons do not represent the best option. Their usage is limited, however they still do exist, and are seen in dams located on a permafrost foundation, where they prevent the permafrost from thawing, and prevent water channels from forming under the foundation through the permafrost, resulting in seepage. They are also seen in road embankment construction. Thermosyphons are also seen in Alaska for the trans-Alaska pipeline project among others [2]. Should a road or railway embankment be thawing, thermosyphons have successfully been used to prevent further thaw. The lifecycle of a thermosyphons is 20–30 years [1, 6].

9 Challenge of Permafrost Thaw and Water Pooling During Summer Excavation

A serious problem that must be anticipated and controlled during excavation, for concrete footings, is that of water pooling during summer excavation due to permafrost thawing. Excavation is required for big footings and in commercial buildings such as school, pools and water facilities for example. The ground composition consists of a thin top soil layer, followed by a thick permafrost layer. This permafrost layer, as studied, consists of mostly water, with little soil particles. When excavating for a new building, often one with a basement, the excavation can descend several meters. Even if the excavation does not descend into the permafrost layer, the exposed soil at a depth of several meters allows the heat from the air to easily access the permafrost. These conditions lead to the melting of the permafrost, thus the transition from frozen state to liquid state. This means that the capacity at the bottom of the excavation drastically decreases. Significant water flooding can occur. Great effort must be made to prevent the permafrost from thawing, and should some thawing be inevitable, a sophisticated pump system must be in place to remove water to keep a dry excavation to allow the workers acceptable working conditions. Concrete cannot be poured in a puddle of water either, therefore if excavation down to bedrock is required, through permafrost, construction managers must anticipate the water to be pumped. Pumps must be in place to remove the water preventing further damage, and preventing compromising the concrete quality which cannot be poured in water.

10 Conclusion

Based on the preceding information, comparing the pros and cons of each method, it can be concluded that it is best design practice to build concrete footings on bedrock or use anchored piles when bedrock is too deep to excavate economically. This strategy will ensure the stability of the building now and in the future. Care must be taken to prevent permafrost thaw, with adequate pumping planned for during excavation. The foundation of a building is arguably the most important element of the building construction, hence why its consideration must be of utmost importance.

Acknowledgements With thanks to Professor Hua Ge, for her guidance and support.

References

1. C. Group, *Thermosyphon Foundations for Buildings in Permafrost Regions* (Standards Council of Canada, Montreal, 2015)
2. C.C. Post, *Permafrost Norway*. Norway (12 Oct 2017)
3. C. Group, *Managing Changing Snow Load Risks for Buildings in Canada`s North* (Standards Council of Canada, Montreal, 2015)
4. S. Tannoury, *Lead Architect with FGMA—Nordic Housing* (C. Harries, Interviewer) (Apr 2017)
5. S. Duguay, *Directeur des Projets Nordiques* (C. Harries, Interviewer) (Mar 2017)
6. TACATC, *Guildlines for Development of Transportation Infrastructure in Permafrost Regions: Drainage and Erosion Control*. TACATC

Creating State of the Art? A Passive House University Hospital North of the Polar Circle

Christian Koch and Martine Buser

Abstract The recent Norwegian passive house legislation has raised concerns as to whether the building industry was able to build cost-efficient buildings, without overspending tax payers' money and having negative consequences for peripheral areas in particular in the very north part of the country. This paper aims at exploring and analyzing how these challenges created by the new legislation has been met during building of a new hospital block in Tromsø, the A-wing. Building on sustainable transition theory which identifies several recombinant dynamics, both public and private, we define the building of passive houses as a societal development encompassing dynamics like company development, personnel competences, as well as architectural, engineering and production methods. The empirical material draws on interviews, analysis of documents relative to the project and public media material. The case study revealed a mixture of recurrent, and specific cold climate challenges: some are directly related to passive house technologies, such as issues with the façade, others indirectly, such as Tromsø being a remote market for material and labour. The project encountered delays, shift in contracts and cost augmentations. Competences had to be developed and combined to achieve the standards of passive house building and the local workforce was complemented by adding workers from other regions and markets; the south and middle Norwegian, Nordic, Baltic and East European countries. Accordingly, the project was not isolated in the northern part of Norway and its challenges appeared to be rather organizational and managerial than technical.

Keywords Passive house · Hospital · Sustainable transition

C. Koch (✉) · M. Buser
Construction Management, Chalmers University of Technology, 42196 Gothenburg, Sweden
e-mail: kochch@chalmers.se

M. Buser
e-mail: buser@chalmers.se

D. Johansson et al. (eds.), *Cold Climate HVAC 2018*,
Springer Proceedings in Energy, https://doi.org/10.1007/978-3-030-00662-4_89

1 Introduction

In the Norwegian process towards passive house legislation for new built, it has been claimed that the building industry was not technically ready for building according to such norms, and that passive houses standards would impose unduly high level of costs on society, that in turn would imply segregation and poorer living conditions for low resource inhabitants [1]. Most of this debate has also implicitly assumed that Norway building conditions were similar for the whole country disregarding challenges for specific geographic situations. However, climate differences 1000 km north of Oslo, do create special demands for sustainable buildings. Besides, the Norwegian state expects that public initiatives are to lead the transition towards climate change mitigation by first building show case examples, second by generally tighten the demands for public buildings. Similarly, there are political ambitions about a fair distribution of public infrastructure leading to the design and construction of hospitals north of the polar circle. In this specific context, the present paper aims at exploring how one of these hospital projects was realized, what were the actual challenges it faced during the design and production phases, and how were they overcome creating a state of the art university hospital north of the polar circle. To do so the contribution of the various players to meeting the passive house requirement is analyzed. The paper draws on transition and innovation theories as framework of understanding of sustainable building. The paper presents a case study of the A-wing (A-fløya) project at the university hospital north Norway (UNN) placed in Tromsø. It is located north of the polar circle, yet following the ASHRAE classification, in zone 7, "very cold", and not arctic [2].

2 Theoretical Framework

The research questions demand a frame of understanding that can study challenges when creating a sustainable building, in this case a passive house building. And conceptualize the role of various players in this process. Sustainable transition theory provides a broad and sociological frame for understanding such innovation. The sustainable transition theory literature developed as a response to the societal challenges of climate change [3]. It encompasses looking at the drivers, emerging actor constellations, technologies and barriers in play. The arguments of "classical" transition theory, the Multi-Level Perspective [4, 5] and the Technological Innovation System [6] is in brief the following: The multi-level perspective views transition as a dynamic of an upcoming niche containing an innovation. The niche is challenging an incumbent regime and technological innovation system. The technological innovation system is defined as "a dynamic network of agents interacting in a specific economic/industrial area under a specific institutional infrastructure and involved in the generation, diffusion, and utilization of technology" [7, p. 93]. Further the multilevel perspective looks upon innovation in a sector as a socio-technical phenomenon and identify three levels of socio-technical interaction within which sectorial innovation can be explained [8]. The micro level is niches where innovations and new designs emerge protected from market mechanisms. The socio-technical regime forms the meso-level, which accounts for the dominating stabilized socio-technical regime which is reproduced by institutionalized learning processes. The macro-level is formed by the socio-technical landscape, an exogenous environment beyond direct influence of niche and regime actors. Niches are important as driver of innovation and can develop new socio-technical regimes towards a sustainable building sector and a sustainable society [4, 8]. The "classical" transition theories feature some important limits that need to be considered: (1) the use of levels to analyze the different dynamics. These levels risk producing illusions on separate worlds with different dynamics. (2) Conceptualization of many, if not too many, dynamics, which makes its explanatory power weaker (3) An unclear role of (human) agency implies that organized and managed change appear less appreciated among the dynamics. Here the two theories are adopted to provide a broad frame of reference around the building of sustainable buildings.

3 Method

The method answers to the main research questions by doing interviews, literature studies, document analysis, and presence at joint meetings. The literature search gathered material on Norwegian passive house development. The written documents used here encompassing the project's own documentation and project plans, scientific [9–11] and governmental [12], publications as well as public information. The latter sources are not referenced here. The hospital case was chosen for quite pragmatic reasons as the case was first developed for another study and therefore known by the authors. We followed two years of the development from October 2015, the early building phase to February 2017 where the construction was half way. We carried two rounds of interviews one in March 2016 and the second in February 2017. The design phase commencing in 2014 was documented by a desk study and the first round of interviews. In total, 13 semi-structured interviews were carried out mainly face to face but also as telephone interviews Interviewees were the client (the hospital) project manager, the architects, consulting engineers, the design built contractor, the technical installations contractor and representatives of the future facilities management. The players are identified in the text through their functional role as the interest here is in the building process of a passive house hospital and not their personal profile in line with the theoretical framework. The case material includes a case study published elsewhere (Koch 2017). It is a limitation that the research project has limited resources compared to the long high resource efforts of the A-wing project located in five main organisations in four quite spread geographical locations. It has been necessary to limit the data collection to a few occasions, relying a lot of "ex post" information, information that is built on how actors interpret something that happened in the past.

4 Case: Passive House University Hospital in Tromsø

The A-Wing project is not the first passive house building to be erected in Tromsø. The first passive house was a small single-family house, I-BOX 120, finalized in 2005. This started a series of experiences of construction of passive houses in cold climate. The pioneer architect Steinsvik, who designed the first I-Box 120 house, commented on the national development of passive house in 2008: "An irresponsible low competence level of the technical consultants in Norway. That is the main problem of building passive houses in Norway" [13]. Nevertheless, passive competences and interest grew, also supported by national incentives and standards [14–16]. It was in this context that in 2008–9 a conceptual sketch for the A wing was contracted by UNN and developed by four companies, two Scandinavian players and two Trondheim based. Here the companies noted that the demands of energy performance were to create a passive house realizing high energy performance, a max net energy consumption of 160 kWh/m^2 and energy label "A". The issue of energy performance was integral, yet peripheral for the UNN strategy considerations in 2009. Location and costs dominated the strategy considerations. UNN decided to go for development of the A-wing and the Patient hotel on the existing site in the town of Tromsø. Upon the sketch project and the strategic decision, UNN initiated a concept project for the A-wing, that was conditioned again to an energy performance at label "A" and 160 kWh/m^2, yet the demand for passive house level was not included. This was for a hospital building at contemporary Norwegian standard. The concept project was developed accordingly. In parallel to this a sketch project for the patient hotel was contracted in 2010 and continued with a design build contract in 2011. The main engineering consultants, a large national player, stated: "The patient hotel should satisfy demands for passive house according to NS3701, something that put strict demands on energy design. [Our company] has contributed with consulting and simulation to reduce the need of cooling…. We … also controlled the heat loss of the building through the building envelope (inclusive thermal bridges) is compliant with passive house demands and through collaboration with the architect contributed to making critical details moisture proof" (project presentation sheet). There thus appears to have been a slight difference in the energy performance demands for the two new buildings; the A-wing and the Patient hotel.

In 2012 a competition for the design of the A-wing was won by two Scandinavian companies an engineering company and an architectural company. The task was to do a functional specification project of the A-Wing. This involved specification of rooms. The building envelope stay largely as it was designed in the pre-project. In parallel the construction of the patient hotel commenced giving the design built contractor further experience building large scale passive houses. In 2012 the Norwegian government announced its future energy plan where passive house should become the legal demand for new built by 2016. In the same period, southern and mid Norway reached a considerable amount of realized passive house, SINTEF estimated a 1000, and a solid body of scientific documentation for selected projects [9, 10].

In 2014 the architectural and engineering design of the A-wing commenced. The latter was organized according to contracts for each technical specialism, preparing for a similar contractor structure. The client UNN shifted project manager and contract strategy however, to design build contracting, mainly to mitigate perceived risks in the project. At this time, the consulting engineer interpreted that the design was about 90% ready. At later stages tensions occurred on the degree of finalization. The client carried out a prequalification and tender and appointed of the later design build contractor and HVAC contractor that teamed up for offering the bid. Detailed design was ongoing in parallel at Architect and Engineering consultant. In 2015 Enova supported the A-wing as passive house at 6 mill kr (Enova is a body owned by the Norwegian Ministry of Petroleum and Energy and administer a series of incentives that contributes to reduced greenhouse gas emissions, development of energy and climate technology and a strengthened security of supply [17]). This assists in keeping the hospital at its goal of doing a passive house. In the summer, a design-built contract was signed at 683 million kroner, changing the concept to a double length and another facade. The costing estimations in this unusual contract, mostly reflect a risk management concern whereas passive house technologies were considered as known. In total 22,000 square meter, and a budget at 1.6 bill. Kroner, constituting a cost increase at around 400 million NOK compared to half a year earlier. The design-built contractor received a very detailed material from the architects and engineers and took over the risk on that basis, but also had to calculate material costs, that elevated the price. This was a general issue and not directly related to passive house elements. Yet the new approach to the building shell leads to roughly 9 months' design of the steel structure and a new prefabricated façade concept also involving the supplier. Part of this design was done by a sub supplier specialized in facades, which came up with a simpler and modular façade design than the one proposed by architects and engineering solving the insulation challenges. The building site activities commence in the autumn with demolition of the existing A-wing. While the construction continues in 2016, also auditing of work-drawings of HVAC involves negotiations between architects, engineers and the HVAC contractor. By August the contractors deliver a "tight building". Subsequently installation works and others worked downwards in the building commencing with the 10th floor. The complex technical installation of the surgery theatres was carried out in parallel. An external HVAC expert was hired to audit the HVAC design. Many issues were found and corrected. In 2017 construction continue and the two main contractors and their sub-contractors hire construction workers from middle and southern Norway, Scandinavia, the Baltics and Eastern Europe. The lean based takt planning means that carpenters, plumbers, electricians, tile setter and painters follow each other. And the project was on schedule until October 2017. By May a gradual handover for operations began, planned to commence by January 2018. The performance at passive house standards is yet to be proved when electricity consuming equipment [18], employees and patient enters the building in 2018.

5 Discussion

The challenges of creating a state of the art university hospital at passive house standard north of the polar circle proved to be multiple and both of indirect and direct character. The encountered challenges that can be viewed as classical for public infrastructure projects. A prolonged time period, shift in contract form and cost augmentations. The shift in contract form meant that part of the knowledge with two Scandinavian operating consultancy companies was substituted with the design -built contractors competences and network. In terms of the specific creation of passive house performance, a series of actors' competences had to be developed and combined. The early architect competences of passive house technology and design approaches was transferred through documents to larger players, that focused on other aspects along with the client. But a foreign façade supplier also played a role through simplifying the building envelope design and offering a modularized solution. This supplier had competences on passive house techniques and Norwegian legislation from previous contracts in Norway. The HVAC and design built contractors craftsmen and building worker where a combination of mostly local, but also some migrant workers. There were clear signs of an impact of a local market of buildings, in particular a core of craftsmen only availed at two or three companies. The hospital building required specific competences related to for example mounting of lead enforced gypsum walls and wall mounted toilets. Also, coordination issues occurred when basic electrical and ventilation installations were to be coordinated with major foreign equipment suppliers (such as surgery theatre suppliers). The competences mobilized were therefore quite a mixture of local, national (further south based), Scandinavian and Baltic origin. The A-wing project commenced and was conceptualized, while passive house technology was early days in Norway. Notably, the first passive house in Tromsø was made 2005–2007. This is contemporary with the very first in Sweden [19] and earlier than the first in Denmark [20]. Moreover, even if Tromsø hosted an early mover in creating small passive house, most of the Norwegian trajectory preparing for the 2016 law enforcement have been carried out in southern regions of Norway [9, 10]. Clearly a demographic explanation would appreciate that most Norwegian people indeed live there. 35% thus lives in the largest six town in the south: Oslo, Bergen, Stavanger, Trondheim, Fredrikstad, and Drammen. In an international perspective, such a concentration is not unusual. And it also leads to concentration of passive house competences.

Enova, the national administrator of the incentive program for energy innovations supported the A-wing project and its passive house aspect. This support was instrumental for keeping the ambition. Not only industry experience, but also much of the documented research come from Trondheim and further south [9]. Even if single family houses were constructed early (2007), the transition from this niche to larger buildings took time. In multilevel perspective sustainable transition theory terms, the early single-family houses are the early niche experiments and subsequently the design become "dominant" and can move to larger buildings. The Patient hotel and the Alta fjord panorama were finished in 2013. The A-wing immediately after consolidated the innovation. Especially the patient hotel, at the hospital campus in central Tromsø developed competences by the actors. The possibilities of making sustainable innovation in public hospital projects is constrained by the long time span from conceptualization to finalized building. In this case the option of innovating to passive standard was adopted around 2009. Once construction began in late 2015, sustainable innovation of buildings in Norway focused on energy plus innovations [11] and the design built contractor found passive house technology a "walk in the park". The process emerged from very little knowledge in 2005 to the public energy plan of 2012 prescribing passive house as obligatory for new build by 2016. A relatively stiff push forward of passive house regulation thus risk hampering further steps of sustainable innovation, such as energy plus. There is a continual need for innovative steps with/ without public support, going beyond present legislation. The Norwegian state have indeed initiated further R&D, but also companies should advance their strategies.

6 Conclusion

This paper set out to scrutinize what special challenges has been met when creating a state of the art university hospital north of the polar circle? and how were they overcome? Also considering the contribution of the different players. The theoretical framework adopted understands sustainable transitions of building as occurring as result of a number of recombinant dynamics, two mains being important. First a niche development of sustainable buildings that come to challenge the incumbent regime of traditional building. Second a pressure of European and national law and regulation challenging the incumbent regime. There were many challenges of creating a state of the art university hospital at passive house standard in a cold climate. The hospital project encountered prolongation in time, shift in contract form and cost augmentations. The prolongation decoupled the design from further energy innovations. The shift in contract form created barriers for knowledge transfer between players, but at a time opened for another type of risk sharing, that was more expensive but also potentially safer. A series of actors' competences had to be developed and combined to create a building at passive house standard. For some passive house technology was well known, for others it was their first time using it. The early architect competences of passive house technology and design approaches was transferred through documents to larger players, that focused on other aspects along with the client. A foreign façade supplier, that was hired at a late stage change the building envelope design an assured the compliance with passive house standards. The HVAC and design built contractors craftsmen and building worker were a combination of mostly local, but also some migrant workers predominantly with little passive house skills. There were clear signs of local market impact, in particular of craftsmen. The competences mobilized, were thus quite a mixture of local, national (further south based), Scandinavian, Baltic and Eastern European origin. The development of the A-wing hospital project did not occur in a northern vacuum. As the sustainable transition theory posits, building passive houses is seen here to be a societal development. The project is embedded in societal dynamics like company development, personnel competence development, architectural, engineering and production methods and this is where the most challenges emerge from, rather than adopting and using passive house techniques in a narrow sense. And paradoxically, once finished, the hospital will be modern it is level of sustainability yet not state of art anymore as buildings in 2017 should contribute to energy production, be sustainable over a life cycle and be economically, environmentally and socially sustainable.

References

1. Stortinget: Innstilling til Stortinget fra energi- og miljøkomiteen klimapolitikk. Innstilling 390 S. Stortinget. Oslo (2011)
2. L. Georges, M. Berner, H.M. Mathisen, Air heating of passive houses in cold climates: investigation using detailed dynamic simulations. Build. Environ. **74**, 1–12 (2014)
3. J. Markard, R. Raven, B. Truffer, Sustainability transitions: an emerging field of research and its prospects. Res. Policy **41**(6), 955–967 (2012)
4. F.W. Geels, Processes and patterns in transitions and system innovations: refining the coevolutionary multi-level perspective. Technol. Forecast. Soc. Chang. **72**, 681–696 (2005)
5. F.W. Geels, The multi-level perspective on sustainability transitions: responses to seven criticisms. Environ. Innov. Societal Transitions **1**(1), 24–40 (2011)
6. S. Jacobsson, A. Bergek, Innovation system analyses and sustainability transitions. Environ. Innov. Societal Transitions **1**(1), 41–57 (2011)
7. B. Carlsson, R. Stankiewicz, On the nature, function and composition of technological systems. J. Evol. Econ. **1**(2), 93–118 (1991)
8. J. Schot, F.W. Geels, Strategic Niche management and sustainable innovation journeys: theory, findings, research agenda, and policy. Technol. Anal. Strateg. Manage. **20**(5), 537–554 (2008)
9. M. Klinski, T.F. Berg, M. Maltha, S. Mellegård, T. Kristjansdottir, M. Berge, S.Holøs, T.H. Dokka, Systematisering av erfaringer med passivhus – oppfølging. Sintef. Oslo (2012)
10. L. Müller, T. Berker, Passive house at the crossroads: the past and the present of a voluntary standard. Energy Policy **60**, 586–593 (2013)
11. H. Nykamp, A transition to green buildings in Norway. Environ. Innov. Societal Transitions **24**, 83–93 (2017)

12. C. Koch, The design, build and operation of the new A-wing of University Hospital North Norway Tromsø. Case report in the BISI project (Building Information Standards and Innovation). Chalmers University of Technology. Gothenburg (2017)
13. O. Iglebæk, Utfordringer ved passivhus. https://www.arkitektnytt.no/ (2008)
14. T. Dokka, I. Andresen, in *Passive Houses in Cold Norwegian Climate*, Paper at the 10th International Passive House Conference, Hannover (2006)
15. I. Dokka, T. Andresen, in *New Norwegian Standard for Low Energy and Passive Houses*. Proceedings 1st Nordic Conference on Passive Houses. Trondheim (2008), pp. 240–247
16. B.J. Wachenfeldt, I. Sartori, in *Midt-Norge som pilotregion for passivhus satsing: Potensialstudie*. Passivhouse Norden Conference Proceedings. Trondheim (2008), pp. 82–92
17. Enova: Enovas homepage; https://www.enova.no/ (2017)
18. N. Christiansen, M. Kaltschmitt, F. Dzukowski, F. Isensee, Electricity consumption of medical plug loads in hospital laboratories: identification, evaluation, prediction and verification. Energy Build. **107**, 392–406 (2015)
19. U. Janson, *Passive houses in Sweden. Experiences from design and construction phase* (Lund University, Lund, 2008)
20. M. Buser, C. Koch, Multiple institutional dynamics of sustainable housing concepts in Denmark—on the role of passive houses. Sociol. Study **2**(10), 725–741 (2012)

Mixing Hot and Cold: Fiery Soul Architects Creating Sustainable Buildings in the Arctic

Christian Koch and Thongchai Lapthananchaiwong

Abstract A small Swedish architectural firm performs a strategy to internationalize in Sweden, Denmark and Norway, even into the most northern parts, Lofoten and Svalbard. These countries and areas exhibit similarities and differences, providing opportunity and barriers for architectural businesses to expand. Three strands of literature are used, business of architectural firm, internationalization, and strategy. The empirical material from the Swedish architectural firm, is gathered through semi-structured interviews and desktop research. The findings indicate that, when the firm executes its business outside local boundaries, three dimensions of strategy must be performed: (1) firm's services are required by international clients. (2) Opportunity has to be provided and firm must response with flexibility. (3) The service, finally, is executed through an international project where synergy and customization are performed. The strategizing should inseparably relate to the multifaceted situation.

Keywords Architect · Strategy · SME · Business model · Green house concept

1 Introduction

Energy consumption in buildings and housing in the arctic, especially in post second world war construction, has been extensive. Strategies of transforming arctic buildings in a sustainable direction are therefore needed. Such steps would often be viewed as an issue of the national institutional set up, which is expected to set up the right incentives and obligations. Yet the arctic transition can also be contributed to by entrepreneurial initiatives of smaller players [1]. This paper aims to map and

C. Koch (✉) · T. Lapthananchaiwong
Construction Management, Chalmers University of Technology, 42196 Gothenburg, Sweden
e-mail: kochch@chalmers.se

D. Johansson et al. (eds.), *Cold Climate HVAC 2018*,
Springer Proceedings in Energy, https://doi.org/10.1007/978-3-030-00662-4_90

analyze a strategizing process of an entrepreneurial architect firm contributing to arctic sustainable transition. The framework is based on concepts of strategizing and internationalization of professional services, building companies and architectural firms, as well as entrepreneurship. The case is a Swedish architectural firm realizing green buildings in Norway (labelled "Green" here). The process is mapped and analyzed. The findings and implications are further elaborated.

2 Method

The research is based on a qualitative research method. The empirical material is collected through semi-structured interview and desktop research, then, scrutinized with a theoretical framework. Furthermore, the interpretive technique is based on an abductive research approach [2]. A Swedish firm "Green arkitektur" was chosen according to the research aim, an entrepreneurial architectural firm. A central selection criterion was that the company should have projects or other business activity in Norway and Denmark. For the analysis, the material was transcribed and digested, additional theory was added. A qualitative analysis and systematic discussion led to research findings as well as answers to the research question. The paper built on a master thesis [3].

3 Framework of Understanding

The framework is composed by three conceptual areas, which are architectural business, internationalization, and strategy.

3.1 Architectural Companies as Professional Business

Architectural companies are a knowledge-intensive professional service firm [4–8]. Cramer and Simpson [9] conceptualize the "design enterprise" with four key success factors. They are operations, professional services, marketing, and finance. The operation is how the projects are executed. The efficient operation within right situation can lead to optimization of results [9]. In terms of professional services, the important tangible parts are products such as specifications, drawings, and models but the less tangible value of services relate to idea generation and problem-solving. Marketing is a process which connects firm and client; it builds on reputation [10] and in addition, it leads to projects [9] For finance, it is about collecting and managing money, income and expenses [9].

3.2 Internationalization

In general, firm has to internationalize to provide international services to cross-border clients [11]. However, more reasons will be identified below. Winch et al [12] identify and analyze different modes of entry to foreign market, mode of association or organization of architectural exports, and the difference of performance. Four different modes of entry into foreign markets are prevalent [12]:

(1) follow the domestic client with investment abroad,
(2) participate and win an architectural design competition,
(3) develop bids in formal or informal selective tenders, and
(4) use a network member as reference.

Mode of association is the next problem firms have to pass through. The new way of operation is required for dealing with cross-border services. Winch et al. [12] identify five forms of association in a foreign context. Those are (1) create a temporary association with a local practice, (2) open a temporary project office, (3) join a network of architectural companies, (4) create a joint venture with a local company, and (5) develop an owned subsidiary. There is also a link between associations forms and strategy.

3.3 Strategy

Strategy is here defined as "a course of action for accomplishing an organization's purpose" [13]. Wit and Meyer [13] employs a problem-driven approach to strategy which elaborate strategy in three dimensions; content, process and context.

In terms of strategy content, the focus is on business level strategy. Wit and Meyer [13] claim that strategy content determines "what" strategy should be for firm. The "fit" between organization and its environment is required in order to accomplish the business purpose. Wit and Meyer [13] claim that aligning resources, activities and service offerings contributes to the business model which companies employ for creation of value and competitive advantage through creating "superior value" to clients. Moreover, it come to operate in a tension between the resource base and value chain in one direction and market demand in opposite direction. For the process of strategy, strategy formation is emphasized. The perspective of strategy as "course of action" can be seen in two complementary directions: "intended" and "realized" actions. The intended strategy refers to a "pattern of decision", prior to action, made by individuals or organizations. The intended strategy consists of strategy formulation process and is followed by a strategy implementation process. These two together constitutes "strategy formation" [13]. Strategy formation is therefore undertaken in a tension of deliberateness and emergence. The deliberateness is the part of realized strategy which is intended while the emergence covers strategy that develop as processes unfold.

The last dimension is strategy context. The strategy context cover environments or occasions where strategy is formed [13]. Managers must take these contexts into account, since they are limitations as well as opportunities, a tension between "shaping" and "adapting to" the context. The firm has to face a tension between globalization and localization that obliges the firm to configure its activities [13].

4 Case: Green Arkitektur and the Glass-House Concept

Green, a Swedish architectural firm based in Gothenburg, was established in 2013. It employs two architects in many projects in Sweden and some in Norway, the arctic area. The projects are in different phases. Green has introduced "glass-house" as core concept based on sustainable architecture and greenhouse building. It is initiated by a partner in the network who had designed and built his own residence, the Eco-house, by employing a greenhouse concept. It gained attention from many people included the Green's owner. Thus, it led to the connection and collaboration. Then, the network was formed and the concept has been developed. The house also gained attention from Sweden-A project's owner which, later on, Green designed her facility as a tourist attraction in central Sweden. This project was considered one of Greens most successful projects. Following the success, it was published in professional media and it reached international clients. Two contacts directed the firm to projects abroad. Both these cases in the northern part of Norway.

Regarding the core concept, it is an integration between a residential and a greenhouse building. However, when transformed into a realized building, the actual integration is based on more solutions, leading to different designs. The residential building area can serve as indoor living space with multiple functions such as living, conference, or work space. The semi-outdoor space, the glasshouse, is for plantation. The outdoor, or semi-outdoor activities, can be done within the glasshouse. Inside the glasshouse, the environment is controlled automatically. The temperature in the glasshouse is slightly higher than the outside temperature. The control is done by adjustable ventilation lids installed in the glasshouse roof. The warmer state is longer than in normal seasoning and is an appropriate circumstance for plantation also of warmer climate plants. The duration for planting is expanded around 4–5 months compared to normal cold-climate condition in Sweden.

The "internal" residential building can be built with materials available locally, such as timber. The greenhouse is built of glass with a steel structure. The building envelope consists of both roof and wall, as in glasshouse buildings (Fig. 1).

The glass enclosure and steel structure is pre-fabricated by a Danish manufacturer, and transported to site for installation. The building is also supplied with an ecological wastewater treatment system which produces recycled water for plantation and other activities.

The Norway-A project is placed in Lofoten, northern Norway. The owner intended to create a tourist attraction point. After finding and studying the Sweden-A project on the internet, he made a first contact. He explained his

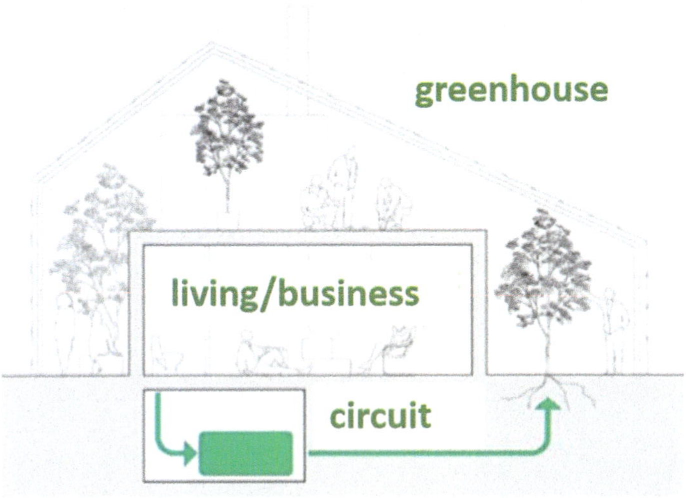

Fig. 1 Conceptual drawing of the greenhouse [14]

intention to renovate an old barn and to build a big glasshouse around it as a new facility. The Green provided professional services in the brief design phase. The project is awaiting financial approval. Another project is Norway-B, in Svalbard, the most northernly part of Norway. The client is a local artist with an intention to build an art gallery. In addition, the glass-house will be utilized as plantation area. It will be planted with nutrition plants or vegetables which might not be able to grow normally despite harsh outdoor conditions. Before making a contact, the client found the firm on internet. He has been attracted and convinced by the Sweden-A and Norway-A project.

5 Analysis

5.1 Business

In terms of marketing, there are two different approaches which are "network marketing" and "process marketing". They are exemplified by the Sweden-A and Norway-A projects. Sweden-A was acquired through Green's network after the client was attracted by the Eco-house whereas Norway-A was by walk-in contact. This type of marketing can be understood as "process marketing" [9]. It can be viewed as a traditional marketing since the client's buying decision is linked with firm's offering. For operation, the operational activities are analyzed in two significant aspects; organization and network. A two-person staff can be considered as almost the smallest possible organizational unit. It is also a typical size of

Fig. 2 Mock up of the greenhouse [14]

architectural firms. More than 90% of Swedish architectural firms have less than ten staffs [15, 16]. The "Sustainable building" is a network, key resources or strategic alliance with different knowledge and specializations. This represents the characteristic of knowledge-intensive firm. The professional services are in the form of architectural design. The glass-house concept is employed. Outcomes and value are delivered in the end of the process. The services also represent the intangibility, heterogeneity and inseparability as specific characteristics of service business. Furthermore, in many projects, such value can be transmitted to be a significant part of client's business (Fig. 2).

Finance occurs as revenue, expense as well as profit or deficit of the company. The major expense is employee's wage. At a time, the employees are also main resource and capability.

5.2 *Internationalization*

Green's internationalization occurs to meet international client's needs. The two projects in Norway develop when clients contact the firm, positing that their intentions cannot be accommodated by local Norwegian architects. Once started, the internationalization consists of several processes transforming the firm's operation from domestic to international including mode of entry and form of association. The mode of entry to the foreign market is by utilizing marketing as a key component based on their specialization. The firm is then invited through informal selective tenders. In the Swedish market, the firm gets projects based on personal relationship and network. At the international market, in this case, the cross-border clients occur as walk-in clients. The firm appears to obtain contact and contracts upon a specialization acknowledged by the clients. This occurs in domestic and international markets. In terms of association and project execution, the firm forecasts to search for and ally with a local partner, possibly an engineer, to carry out the work that relates to the local Norwegian context. And by doing this, to overcome the differences in building regulation and collaborative culture between Sweden and Norway. This represents the extension of the area of (local) knowledge [12]. In addition, it also reflects a characteristic knowledge-intensive organization.

5.3 Strategy

In terms of strategy content, per Wit and Meyer [13], the firm's competitive advantage is mainly contributed by glass-house concept which is the mixing between a residential and a greenhouse. Such concept provides specific characteristics, for instance, controlled environment or warmer condition. In addition, the concept of sustainable architecture is delivered through greenhouse building. In another perspective, these service offerings based on firm's resource and activities lead to "superior value proposition" as well as differentiation. Both in terms of energy performance and esthetical features. The Greenhouse enable growing plant in a northern climate and can be perceived as an attractive and creative frame around customer oriented commercially oriented activities. For Green, this can be obtained by making variants of the firm's core concept. Further aspects of competitive advantage are customer intimacy obtained through a design customization process in interaction with the client. The value also includes service activities after the design process.

The strategy (process) is gradually realized in the domestic market. Nevertheless, some situations emerge that provides unplanned opportunities to enter foreign market as a result of the marketing effort which reach international clients and actually contribute with a lot of unintended opportunities. The strategy is deliberately processually strategized, additionally, direct in connection with single project execution. The strategy is thus also formed during activities the firm undertakes in each project. Therefore, the strategy formation consists of both deliberateness and emergence at the same time. Adding to the intended strategy emergent situations occurred, followed by the international projects. To seize that, the strategist has to consider various factors; resource, capability as well as differences and similarities between two countries. Ultimately, the strategizing process also leads to foreseeing a need for a Norwegian partner.

The Scandinavian market, as the context, determines scope and level of internationalization for Green since the firm considers international projects in Scandinavia as an emergent opportunity to perform its operation. For Green, the synergy exists, since the core concept is sustainable architecture which is extensively accepted in Scandinavia. Once the firm commence to internationalize, it experiences differences between countries. These contribute to tensions during the internationalizing process. Scandinavian regulation in 2017 shares many similarities, but some regulations are still national and different between countries. The design process works as a localization process and assures the value adding for the client. Thus, the design process provides local responsiveness as well as global synergy.

6 Conclusion

The aim of the paper is to map and analyze a strategizing process of an entrepreneurial, fiery soul, architect firm contributing to arctic sustainable transition. The firm performs a strategy to internationalize into arctic parts of Norway, Lofoten and Svalbard. This area exhibit similarities and differences to the rest of Scandinavia, providing opportunity and barriers for architectural businesses to expand. The framework of understanding put together three types of literature, first on the business of architectural firms, second on internationalization and third on strategy. The empirical material from the firm, was gathered through semi-structured interviews and desktop research. The findings show that, when a Swedish architect company internationalizes to the outskirts of the Scandinavian market, the company must strategize in three dimensions: First in a business dimension; the company should have a product or service that is demanded by clients in new context abroad. Second "opportunity" should be viewed as a process, more than an occasion. And flexibility and adaptation is needed. Third the firm should perform a crossnational project and create synergy with existing competences and customizing the service to cross-border clients. The strategizing should be specific, fit, flexible and align with a multidimensional situation.

References

1. J. Schot, F.W. Geels, Strategic niche management and sustainable innovation journeys: theory, findings, research agenda, and policy. Technol. Anal. Strateg. Manag. **20**(5), 537–554 (2008)
2. M. Alvesson, K. Sköldberg, *Reflexive methodology: new vistas for qualitative research* (SAGE, Thousand Oaks, 2009)
3. T. Lapthananchaiwong, in *Strategy at Scandinavian Architectural Firm*. Master Thesis. Chalmers University of Technology, Gothenburg (2017)
4. G. Winch, Strategic business and network positioning for internationalisation. Serv. Ind. J. **34**, 715–728 (2014)
5. G. Winch, E. Schneider, The strategic management of architectural practice. Constr. Manage. Econ. **11**, 467–473 (1993)
6. B.R. Løwendahl, *Strategic Management of Professional Service Firms* (Copenhagen Business School Press, Frederiksberg, 2005)

7. A. Nordenflycht, What is a professional service firm? Toward a theory and taxonomy of knowledge-intensive firms. Acad. Manage. Rev. **35**, 155–174 (2010)
8. R.L. Shell, *Management of professionals* (Marcel Dekker, New York, 2003)
9. J.P. Cramer, S. Simpson, How firms succeed. Principal's Report, 02, 4. (2002)
10. Y. Aharoni, L. Nachum, *Globalization of Services: Some Implications for Theory and Practice* (Routledge, London, 2002)
11. G.M. Winch, Internationalization strategies in business-to-business services: the case of architectural practice. Serv. Ind. J. **28**, 1–13 (2008)
12. G.M. Winch, D. Grezes, B. Carr, Exporting architectural services: the English and French experiences. J. Architectural Plann. Res. **19**, 165–175 (2002)
13. B.D. Wit, R. Meyer, *Strategy: an international perspective* (Cengage Learning, Andover, 2014)
14. T. Lapthananchaiwong, in *Strategy at Scandinavian Architectural Firm*. Power Point Presentation of Master Thesis. Chalmers University of Technology, Gothenburg (2017)
15. SCB. Statistics Sweden. Available: http://www.scb.se/. Accessed 8 June 2017 (2016)
16. STD. Sector Review [Online]. Available: https://www.std.se/in-english/sector-review. Accessed Dec 2016

Communicating the Acoustic Performance of Innovative HVAC Solutions

Soheila Bahrami, Juan Negreira, Stefan Olander and Anne Landin

Abstract Recent years have seen considerable advancements in Demand Controlled Ventilation (DCV) systems aimed to improved energy efficiency and indoor environmental quality in buildings. A significant aspect of DCV systems is their impact on the acoustic comfort in buildings. This study is part of the Urban Tranquility project with a focus on innovative DCV systems. The objective is to add more understanding about communicating the acoustic performance of innovative HVAC solutions during the diffusion phase of the innovation. The research method is a case study on an innovative DCV system that shows how the acoustic performance of the system is communicated with the stakeholders and in what ways the applied methods can be improved. Data collection has been performed through reviewing relevant technical documents and software as well as semi-structured interviews with different stakeholders. The data has been analyzed with reference to three types of knowledge about an innovation. The results indicate that the acoustic performance of this new DCV system has not been effectively communicated due to inconsistent methods of expressing the information. This has revealed the need for developing a systematic method of communicating the acoustic information on DCV systems with the key stakeholders.

Keywords Innovation · Knowledge · Acoustic information · Demand controlled ventilation (DCV) · Indoor environmental quality

---

S. Bahrami (✉) · S. Olander · A. Landin
Division of Construction Management, Lund University, 22100 Lund, Sweden
e-mail: soheila.bahrami@construction.lth.se

J. Negreira
Division of Engineering Acoustics, Lund University, 22100 Lund, Sweden

D. Johansson et al. (eds.), *Cold Climate HVAC 2018*,
Springer Proceedings in Energy, https://doi.org/10.1007/978-3-030-00662-4_91

1 Introduction

According to the World Health Organization (WHO), exposure to high noise levels harms human health causing sleep disturbance, cardiovascular, and psychophysiological problems [1]. In 2014, environmental noise was reported as the primary cause for hospitalization of 43,000 Europeans annually [2]. Such problems can be more severe in dense urban areas without appropriate plans. According to the Vision Sweden 2025 [3], urban densification has been adopted as a development strategy in Sweden. Noise and vibration (N&V) issues are emphasized as the environmental aspects which must be considered by different stakeholders during the planning and implementation processes of urban densification. Sound insulation in building envelopes can be used as a barrier against outdoor noise in dense urban areas [4]. However, this solution cannot be efficient without sufficient levels of indoor air quality. It means, stale indoor air may prompt users to open the windows for natural ventilation which would expose them to the outdoor noise, rendering the sound barrier useless. Therefore, reliable HVAC systems are needed to provide acceptable levels of indoor air quality and avoid the need for additional ventilation. Moreover, an HVAC system itself can be a source of N&V problems in the building. Thus, innovative HVAC solutions with holistic approach to indoor environmental quality are needed to deal with these issues.

During recent decades, Demand Controlled Ventilation (DCV) systems have been used to condition the indoor air and reduce the energy consumption in different types of buildings. In a DCV system, supply air flow rate and the associated energy consumption is controlled by the actual demand in the space [4]. Several HVAC manufacturers have adopted different approaches to improve their DCV systems [4, 5] while a few of them have marketed innovative solutions using wireless communication [6, 7]. As any innovation involves some levels of uncertainty, the related knowledge and information communicated with customers influence their decisions on adopting an innovative solution [8, 9]. Rogers [8] classified the knowledge about an innovation into three types. The first type, *awareness-knowledge*, is the information about the existence of an innovation; this may motivate an individual to seek other two types of knowledge; *how-to knowledge* and *principles-knowledge*. How-to knowledge is the information required to use an innovation. Inadequacy of this type of knowledge may result in rejecting an innovation; particularly for more complex innovations. Principles-knowledge is the information on fundamental and theoretical principles which underpins the functioning of an innovation. It is possible to adopt an innovation without principles-knowledge; however, this runs the risk of incorrect use and termination [8]. For example, a customer might stop purchasing a manufacturer's DCV system due to noise problems caused by improper design of a building HVAC system.

The process of decision making depends on the available knowledge and the receiver's perception [10]. Perception of an innovation is the process through which an individual receives and interprets related information. This process is directly affected by selecting and presenting the information as well as the credibility of the source of information [11]. Moreover, sustainable competitive advantage can be achieved by managing the information about how different stakeholders use, or respond to innovations [12]. It is also necessary to differentiate between the stakeholders [13] and distinguish the perception and interests of the key stakeholders [11].

1.1 Research Objective

The research objective is to add more understanding about communicating the acoustic performance of innovative HVAC solutions during the diffusion phase of the innovation. This study is part of the Urban Tranquility research project which aims to explore new approaches to support innovative solutions to N&V problems in dense urban areas [14]. The focus of this paper is to investigate how the acoustic performance of an innovative DCV system is communicated with different stakeholders and how the applied methods can be improved.

2 Research Methods

A new DCV system produced by a company called Company A in this paper has been selected as a case to study. Single case study on innovation is a useful approach to reproduce the existing knowledge beyond its original context and provide supplementary knowledge for creating improvements [15]. The data has been collected through reviewing relevant literature, technical documents, and software as the primary source of information for different stakeholders. In order to study the collected information from a wider perspective, similar information available on two other main manufacturers' websites has been reviewed. These manufacturers are called Company B and Company C in this paper. The collected information has been complemented by conducting semi-structured interviews with the system development director, research and development, product, and marketing managers, sales, laboratory, and control systems engineers, HVAC designers and consultants. The information was collected from April to July 2017. A qualitative analysis has been performed with reference to the aforementioned classification of the information about an innovation into three types of knowledge. The acoustic information has been considered at different levels of offerings including solution and products with regard to the indoor and outdoor noise. The scope of this study is limited to the acoustic information on DCV components in a room including comfort modules, diffusers, and dampers. In order to identify the gaps in the knowledge provided by manufacturers in communicating the innovative products, the analysis has been performed on the information which is freely available to the stakeholders.

3 Three Generations of DCV Systems

As one of the main producers of ventilation systems in Sweden, Company A offers a wide range of HVAC products and solutions. On the company's website, "using innovation as a mindset" has been mentioned as one of the assets that enables the company to meet the customer requirements and to be "The Indoor Climate Company" [16]. The company has four R&D departments in Sweden. To date, three generations of DCV systems have been developed and manufactured in one of these departments.

The company's first DCV system was launched in 2000. Noise reduction was highlighted as an advantage of the product. It was achieved by lowering the air velocities and pressure drops in the system [17]. In 2008, the next-generation of the company's DCV system came on the market. The aim was to simplify the installation and commissioning processes as well as the connection to the air handling unit. Configuring the components in the factory and the cabling in the control system increased the risk of human errors during installation processes and further required corrective actions that were costly and time consuming. Customer complaints arising from such complications prompted Company A to solve the problem. Considering the increasing demand for DCV systems, the company needed an effective action in order to fulfil the customers' expectations and sustain its competitive advantage. Consequently, a radical change in the control system was decided for the third-generation DCV system. A control system based on radio-frequency identification (RFID) was developed through collaborations with other companies [18].

The new system was first introduced at Nordbygg, the construction industry event, in April 2016 and sales processes started in spring 2017. It is named the Demand Controlled Indoor Climate (DCIC) system at the highest level of customer needs, which includes ventilation as well as air and hydronic heating and cooling. The lower levels are DCV, Variable Air Volume (VAV)/Constant Air Volume (CAV), and product [7] (see Fig. 1). The acoustic information about the new system products is provided in the catalogs (in Swedish) available on the company's website [19]. The parameters used to communicate this information are briefly described in the following section.

3.1 HVAC Acoustic Parameters

In HVAC systems, rotating equipment and movement of air and fluids generate vibration and noise. Vibration can radiate noise defined as undesirable sound; noise generated by the air flow is known as airborne noise while structure-borne noise is generated by sources in direct contact with a building structure [20]. Sound is defined as the fluctuations in the atmospheric pressure which can be sensed by our eardrums. It is described by three dimensions as follows: time (s), frequency (Hz),

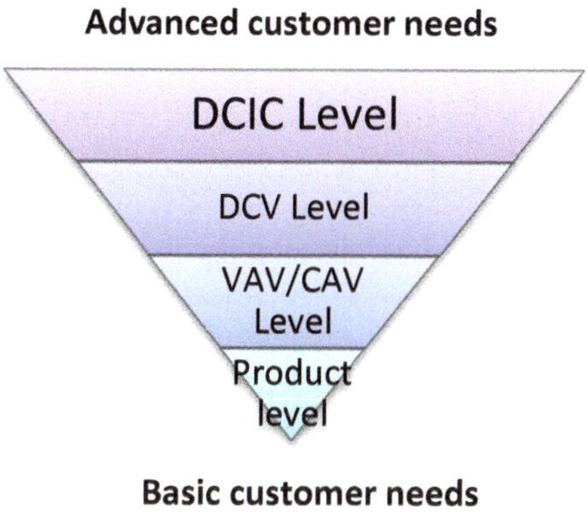

Fig. 1 Different levels of offerings by Company A based on customer needs

and level (dB) [21]. The latter is used as the unit of both sound power level and sound pressure level, although they are different parameters. Sound power is the sound energy released by a source per unit of time. It is expressed as $L_w = 10 \log (W/W_{ref})$, where $W_{ref} = 10^{-12}$ Watts [22]. The sound power level of the HVAC equipment is determined by laboratory measurements according to the relevant standards. Sound pressure, in turn, is expressed as $L_p = 20 \log(P/P_{ref})$, where $P_{ref} = 0.00002$ Pa [22]. Sound pressure level can be either measured in a space using a sound level meter or estimated if the space conditions and the sound power level of the equipment are known. Sound pressure level (L_p) can be expressed as a function of the sound power level (L_w) of the equipment and the distance (d) from the equipment to the receiver as:

$$L_P = L_W + 10 \log[4(1-\alpha)/A + Q/(4\pi d^2)] \quad (1)$$

where Q is the directivity of the source, α is the mean absorption coefficient, and A is the effective absorption area in square meters [21]. A simple expression of the Eq. 1 is $L_P = L_W$ + Room Effect or Attenuation + End Reflection Loss [23].

In Sweden, the acceptable sound pressure level of the HVAC equipment in buildings is legislated by the National Board of Housing, Building and Planning (BBR), which also refers to the sound classes of buildings stated in SS-EN ISO 25267:2015 and SS-EN ISO 25268:2007 standards. Sound levels are specified as frequency-weighted sound pressure levels. Since the sensitivity of the human hearing changes is dependent on the frequency, the sound measurements are weighted as a function of frequency to account for human perception [24]. The maximum A-weighted sound pressure level and equivalent C-weighted sound pressure level are determined for noise from HVAC systems in institutional and commercial buildings [25]. A-weighted sound pressure, in dB(A), is widely used as a single-number measure of the relative loudness of sound. The C-weighted curve, dB(C), is more sensitive to low-frequency sound [21].

4 Results

As inferred from the catalogs [7, 26] and confirmed by the interviewees, the new DCIC system has not been considered as a potential solution to the outdoor noise problems. Therefore, the acoustic information is not highlighted as awareness-knowledge in introducing the system, but rather as how-to knowledge and principles-knowledge at the product level. This means that the innovative solution has been undervalued due to incomplete awareness-knowledge. Hence, this study focuses on the acoustic information of this DCV system's components i.e., products.

At the product level, Company A has different customers amongst whom knowledge of acoustics varies accordingly. According to the interviewees, customer complaints about noise are mainly caused by incorrect applications. Therefore, it is crucial to supply complete principles-knowledge to designers and consultants as well as appropriate how-to knowledge to the users.

Customers can access the acoustic information about the products through the product data sheets, commercial software, and customized simulations. For special applications, mock-up room tests can be performed upon the customer request. According to the laboratory engineer, the acoustic tests are performed according to the Swedish versions of ISO standards such as SS-EN ISO 3740 series and SS-EN ISO 16032. Nonetheless, such information is not included in the technical documents and software that limits the users' information about the applied measurement methods. Consequently, the presented sound data are difficult to analyze. Moreover, terms, notations and units used in presenting the sound data are not consistent.

Knowing the sound power level of the HVAC equipment, a direct comparison can be made between equipment by different manufacturers; however, the sound pressure levels cannot be determined until the space and the HVAC installations are designed. The sound pressure levels can be used to compare the acoustic quality of two pieces of equipment if the same test conditions are applied [22]. Comparing the acoustic performance of the products is challenging due to inconsistent methods for performing measurements and presenting the results adopted by different manufacturers. Three components of the new DCV system (by Company A) used in a room are the comfort module, the air diffuser and the damper. The challenges of communicating the acoustic performance of those products are discussed in the following.

5 Discussion

The Comfort Module is an induction terminal unit which means it operates without a fan or blower. In the catalog of this integrated heating, cooling and ventilation unit, the acoustic data is tabulated for each size of the unit [27]. The equivalent sound pressure levels, in dB(A), are given for different airflow rates based on the nozzle pressure and settings. These values are calculated for a room attenuation of

4 dB (an equivalent sound absorption area of 10 m^2). According to the catalog, calculations for different nozzle settings can be done in two computer programs which will be discussed later.

The term "sound level" is used to express the sound pressure level which can be confused with the sound power level. For each model/size of the product, natural attenuation including end reflection, ΔL (dB), is given in octave bands[1] at different nozzle settings. The sound pressure levels can be converted to sound power levels by using the simplified form of the Eq. 1; however, this information is not given in the catalog. Moreover, the information on the applied measurement methods and standards is not provided. Another table shows the weighted sound reduction index, R_w (dB). The values show the airborne sound insulation performance of different types of partition walls and suspended ceilings in a space without the terminal unit and with the terminal unit installed.

A quick selection table on the cover page of the air diffuser catalogue [28] shows the sound pressure levels in dB(A) as a function of the airflow for different sizes of the diffuser. In the catalog, the sound pressure levels, in dB(A), are plotted against the airflow, pressure drop, and throw length. The factor, K_{ok}, and sound attenuations, ΔL, are tabulated for different sizes of the diffuser. The information for calculating sound power levels is confusing as the equation is given twice with both adding and subtracting the factor K_{ok} without any reference to ΔL.

Improvement in the acoustic performance of the new DCV system is not expected by some of our interviewees reasoning that the components are the same as in the previous generation. However, as inferred from the catalogs and confirmed by the R&D manager and the laboratory engineer at the factory, the acoustic performance of the air diffuser has improved by nearly 20%. This improvement is worth communicating with internal and external stakeholders.

For the circular damper, the A-weighted sound power level, L_{WA}, is plotted as a function of the air flow rate and pressure drop across the damper at positions from 20% to fully open [29]. The A-weighted sound power level obtained from the diagram, can be converted to the sound power level by adding a correction factor, K_{ok}(dB), tabulated in octave bands. In the same way, the correction factor, K_{trans}, is used to calculate the sound power level transmitted through uninsulated casing. For the rectangular damper, L_{WA} is plotted as a function of air velocity while it is mentioned as the air flow in the text. Furthermore, different notations are used to illustrate the A-weighted sound power levels. The L_{WA} found from the diagram is converted into the sound power level by adding the constant K_{ok}. The correction factor, K_k, is also added to include the effect of the cross section area of the damper.

Reviewing similar data from two other Swedish manufacturers' websites, shows that Company B presents the acoustic data of its circular damper in a similar way for the damper positions limited to fully open and closed. Company C provides a similar diagram without any information about the damper positions and for some

[1]The sound spectrum is divided into eight equal octave bands, each defined by its mid-frequency. The first band is centered on 63 Hz and the eighth band on 8000 Hz.

models of its dampers, only tabulated values of L_{WA} are given without further information. For the rectangular damper, Company B gives a simplified plot of air velocity against A-weighted sound power level while the tabulated correction factors are given in limited scales. The only information given by Company C is tables of A-weighted sound power levels. It should be noted that A-weighted sound power levels as single numbers are meaningful when applied to sound pressure level values. The single numbers cannot be verified by measurements in the field. Overall, a common method is not adopted by the three reviewed manufacturers to illustrate the acoustic information of their DCV components [4, 30–32].

The software introduced by Company A for selecting the new DCV system products has not been available within the time frame of this study. According to the system guide [26], the sound data can be calculated based on the room type, form, and size. After selecting products in the software, capacity and energy consumption can be calculated in another type of software in order to find the optimum design for the best achievable indoor climate and energy efficiency. In fact, the software features cannot be reviewed until its market release.

According to the terminal unit catalog, calculations for different nozzle settings can be done in the forthcoming software and the one currently available on the company's website. It should be noted that letter "a" has been added as an indicator to the product name in the list of documents on the website [19]; however, it is not mentioned on the catalog cover page [27]. None of these titles can be found in the drop-down menu of products in the current software. Therefore, the user might not be able to perform the calculations due to inaccuracy of the basic information such as the name of the product. This is an example of limited access to the principles-knowledge caused by inaccurate awareness-knowledge at the product level.

6 Conclusions

This paper has investigated the methods used to communicate the acoustic performance of an innovative DCV system with the key stakeholders. The collected information has been analyzed with reference to three types of knowledge about an innovation i.e. awareness-knowledge, how-to knowledge, and principles-knowledge. The results indicate that although the acoustic performance of Company A's new DCV system is a valuable aspect that has been improved throughout the three generations of the system, it has not been emphasized as such in introducing the system by the company. Moreover, the role of the system in benefiting from a sound insulated building envelope against the outdoor noise is neglected. Therefore, the system can be undervalued due to incomplete awareness-knowledge supplied by the company. The inadequacy of this type of information is further reflected in the product selection software, where access to how-to knowledge and principles-knowledge has been limited by inaccurate awareness-knowledge.

Company A's catalogs include acoustic information in the form of diagrams, tables, equations, and short descriptions, which is more informative compared to the acoustic information on DCV systems provided by the other reviewed manufacturers. Nevertheless, the methods used to express the acoustic information are inconsistent and, in some cases, difficult to understand. An implication is that improved methods of expressing the data would enable the company to communicate its innovative solution more effectively. Reviewing the acoustic information provided by the other manufacturers shows that they have not followed a uniform method to illustrate the information. In addition, they have occasionally used the A-weighted sound power levels as single numbers, which is not sufficient to evaluate the acoustic performance of a product. Therefore, the available information is inadequate for comparing the acoustic quality of the new DCV products with the other reviewed products.

This study has revealed areas for improvement in communicating the reviewed acoustic information that might apply to the HVAC industry in a wider scope. Firstly, there is a considerable need for a systematic approach focusing on the key stakeholders as the receivers of different types of knowledge. Secondly, standardized terms, notations, and units can be defined based on the terminology used in relevant standards. For example, sound power level appears to be the most applicable parameter for all stakeholders to compare the acoustic performance of the DCV products. More importantly, HVAC designers and consultants need this parameter to estimate the acoustic performance of the equipment in a designed space. Thirdly, citing the applied standards in the technical data sheets can give acoustics professionals a clear understanding of procedures and influence customers' perception of the innovation by showing credibility of the source of information. In addition, it is necessary to check the documents for errors before publishing and revise them regularly considering the changes in products, standards, and regulations. The findings suggest that a standardized method can be developed by the Swedish Standards Institute for providing the acoustic information by HVAC manufacturers. Further studies are needed to evaluate the quality of acoustic data calculated by the forthcoming software and the impact of RFID wireless communication on the acoustic performance of the new DCV system by performing measurements during the operation phase of the system.

References

1. WHO, *Noise* http://www.euro.who.int/en/health-topics/environment-and-health/noise/noise. Last accessed 7 July 2017
2. European Environmental Agency: Quiet areas in Europe, Luxembourg (2016)
3. Swedish National Board of Housing, B. and P, *Urban Density Done Right* (2017)
4. Lindinvent, *DCV-RC—Room Climate Control Unit*, https://www.lindinvent.com/media/191193/dcv-rc_c12_pb28_hpt472_eng.pdf. Last accessed 10 July 2017

5. Systemair, *Demand Controlled Ventilation—AIAS* | www.systemair.com, https://www.systemair.com/Global/Solutions/Demand-Controlled-Ventilation—AIAS/. Last accessed 10 July 2017
6. Trane, *Air-Fi^{TM} Wireless*, http://www.trane.com/commercial/north-america/us/en/controls/building-Management/Air-Fi.html. Last accessed 10 July 2017
7. *Swegon*, WISE—The New Generation, http://www.swegon.com/Products/Demand-Controlled-Ventilation1/WISE/. Last accessed 10 July 2017
8. E.M. Rogers, *Diffusion of Innovations* (Free Press, New York, 2003)
9. R. Caiazza, T. Volpe, Innovation and its diffusion: process, actors and actions. Technol. Anal. Strateg. Manag. **29**, 181–189 (2017). https://doi.org/10.1080/09537325.2016.1211262
10. J.H. Rankin, R. Luther, The innovation process: adoption of information and communication technology for the construction industry. Can. J. Civ. Eng. **33**, 1538–1546 (2006). https://doi.org/10.1139/l05-128
11. U. Weisenfeld, Engagement in innovation management: perceptions and interests in the GM debate1. Creat. Innov. Manag. **12**, 211–220 (2003). https://doi.org/10.1111/j.0963-1690.2003.00284.x
12. J.S. Harrison, D.A. Bosse, R.A. Phillips, Managing for stakeholders, stakeholder utility functions, and competitive advantage. Strateg. Manag. J. **31**, 58–74 (2010). https://doi.org/10.1002/smj.801
13. K. Widén, S. Olander, B. Atkin, Links between successful innovation diffusion and stakeholder engagement. J. Manag. Eng. **30**, 4014018 (2014). https://doi.org/10.1061/(ASCE)ME.1943-5479.0000214
14. European Regional Development Fund: Urban Tranquility, http://interreg-oks.eu/projektbank/projekt/urbantranquility.5.2089ae6e15244fe02f31a1c5.html. Last accessed 4 July 2017
15. Ø. Pålshaugen, How to generate knowledge from single case research on innovation? Int. J. Action Res. **5**, 231–254 (2009)
16. Swegon, *Our Values*, http://www.swegon.com/en/About-us1/Our-Values/. Accessed on 09 July 2017
17. Swegon: e.r.i.c. Technical Section, http://www.swegon.com/Global/PDFsArchive/FlowControl/e.r.i.c/General/_en/e.r.i.c._system_technology.pdf. Last accessed 9 July 2017
18. Lumen Radio: MIRA, https://www.lumenradio.com/mira/. Last accessed 13 July 2017
19. Swegon: Homepage, http://www.swegon.com/sv/Support/Hitta-PDF/?selectedfolder=&sort=type&desc=0&p=1&doctype=catalogue&lang=swedish&q=wise. Last accessed 31 July 2017
20. H. Kuttruff, *Acoustics [Electronic Resource] : An Introduction* (Taylor & Francis, New York, 2006)
21. M. Asselineau, *Building Acoustics* (CRC Press, Boca Raton, Florida, 2015)
22. Daikin, *HVAC Acoustic Fundamentals*, http://salesportal.daikinapplied.com/bizlit/DocumentStorage/AirCooledChiller/ApplicationGuides/AG31-010_HVAC_Acoustic_Fundamentals.pdf. Last accessed 26 July 2017
23. R. Zimmerman, Sound Advice [Acoustical Design]. ASHRAE J. **55**, 16–18 (2013)
24. A. Bhatia, *Overview of noise control and HVAC acoustics in buildings* (PDH Center, Fairfax, VA, 2012)
25. L. Åkerlöf, *BULLERSKYDD: Byggvägledning 11, En handbook i anslutning till Boverkets byggregler* (AB Svensk Byggtjänst, Stockholm, 2015)
26. Swegon, *System Guide: Swegon WISE*, http://www.swegon.com/Global/PDFs/Flowcontrol/WISEgen.2/_en/WISE_Systemguide_gen2.pdf. Last accessed 31 July 2017
27. Swegon, *WISE Parasol*, http://www.swegon.com/Global/PDFs/Flowcontrol/WISEgen.2/_sv/WISE_Parasol_a.pdf. Last accessed 25 July 2017
28. Swegon, *WISE Colibri*, http://www.swegon.com/Global/PDFs/Flowcontrol/WISEgen.2/_sv/WISE_Colibri_a.pdf. Last accessed 25 July 2017
29. Swegon, *WISE Damper*, http://www.swegon.com/Global/PDFs/Flowcontrol/WISEgen.2/_sv/WISE_Damper_a.pdf. Last accessed 25 July 2017

30. Systemair, *OPTIMA-R Single or Double Skin Circular VAV Units*, http://catalogue2.systemair.com/FileHandler.axd?hash=i5xyl6WZaYQIiVrWY9pkkA!!. Last accessed 31 July 2017
31. Lindinvent, *DCV-BL—Airflow control*, https://www.lindinvent.com/media/191201/dcv-bl_b03_da_pb22_eng.pdf. Accessed 31 July 2017
32. Systemair, *Constant & Variable Air Volume Controllers*, file:///C:/Users/so1783ba/Downloads/116853-E8250_VAV-Catalogue_GB.pdf. Last accessed 31 July 2017

Printed by Printforce, the Netherlands